System of Two Linear Equations

The linear system $\begin{cases} A_1x + B_1y + C_1 = 0 \\ A_2x + B_2y + C_2 = 0 \end{cases}$ can have

One Solution

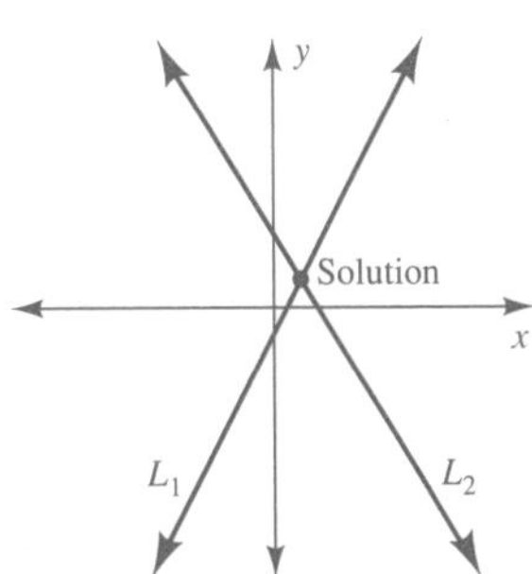

Consistent system of independent equations

No Solution

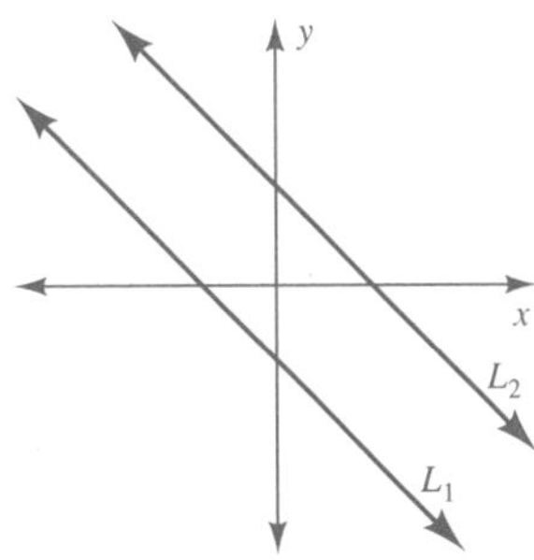

Inconsistent system of equations

An Infinite Number of Solutions

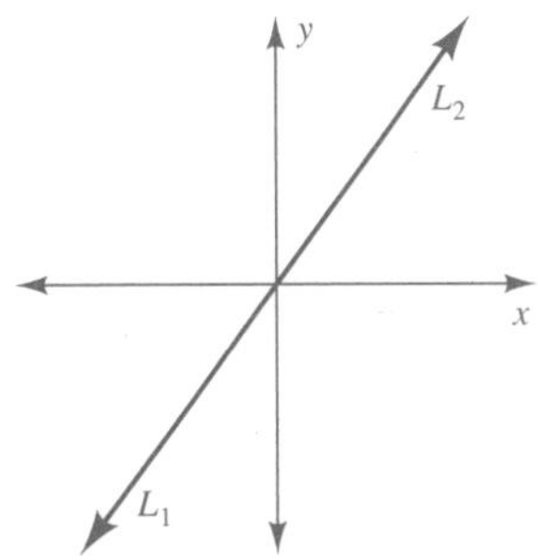

Consistent system of dependent equations

Graphs of Basic Functions

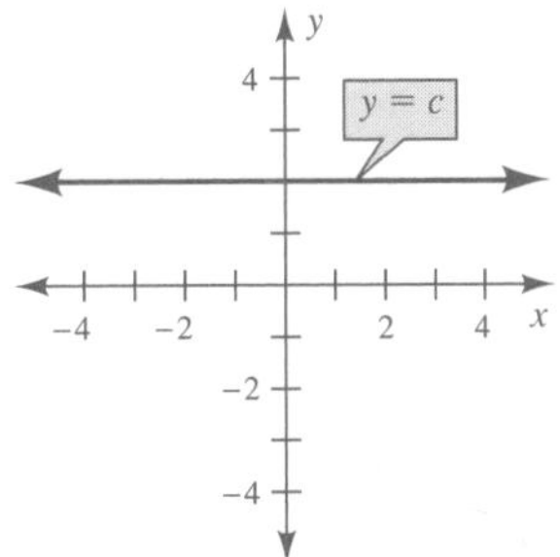

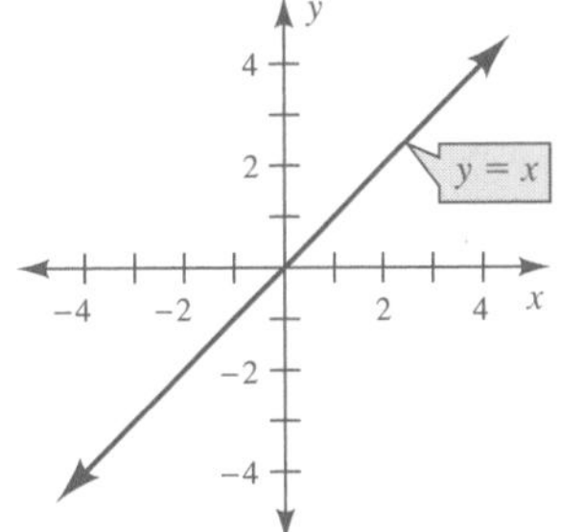

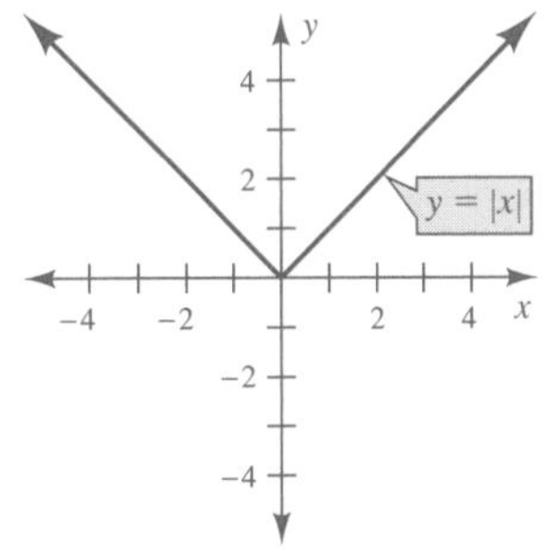

$y = \sqrt{x}$

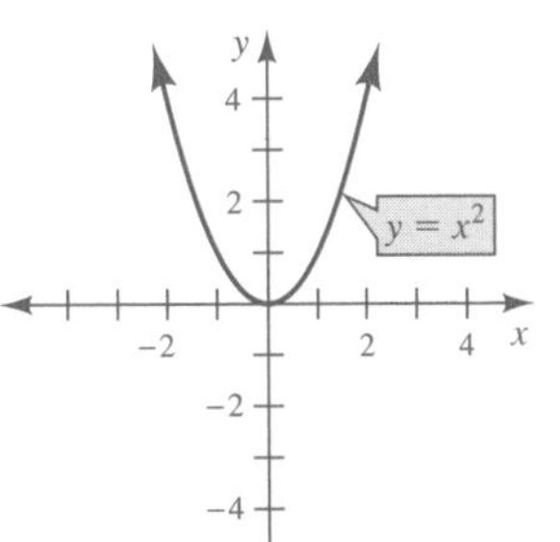

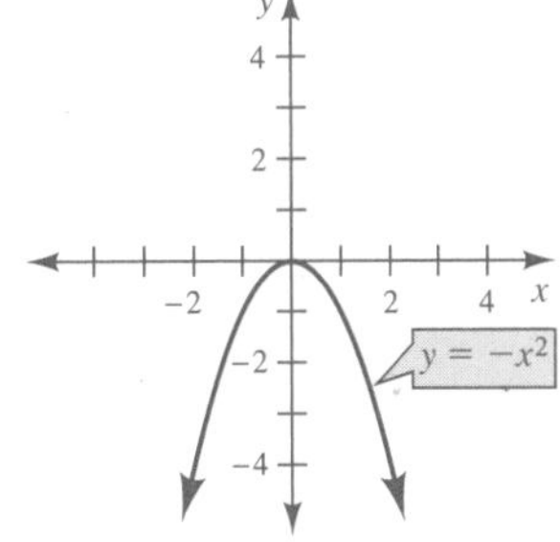

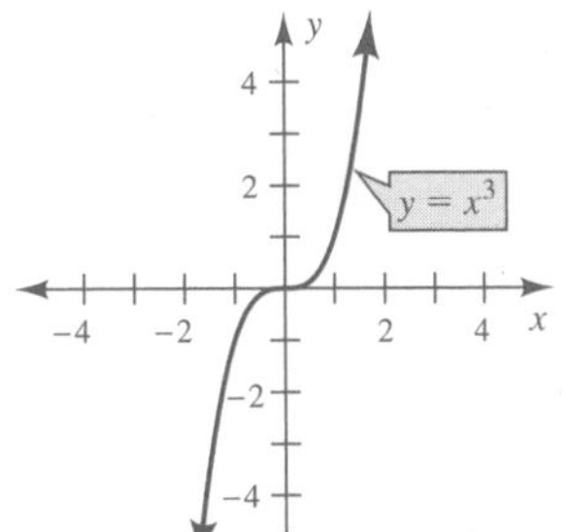

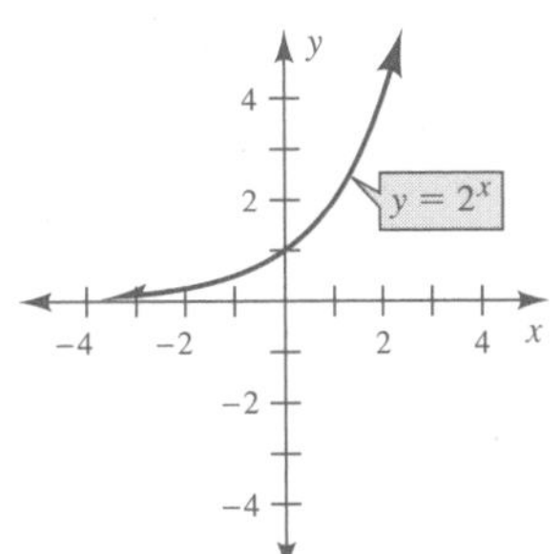

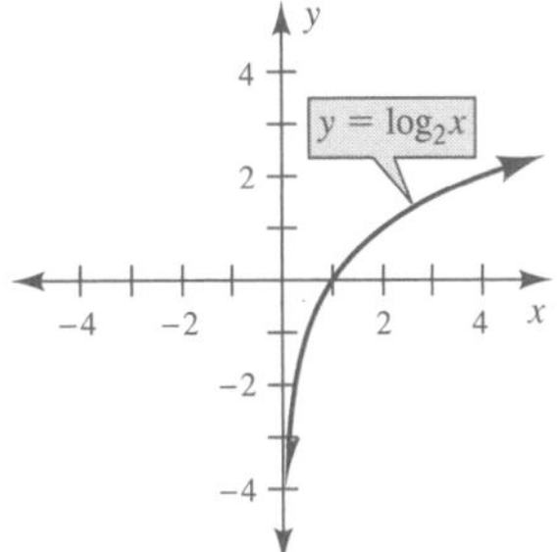

INTERMEDIATE ALGEBRA WITH EARLY FUNCTIONS

INTERMEDIATE ALGEBRA WITH EARLY FUNCTIONS

SECOND EDITION

James W. Hall
Parkland College

PWS PUBLISHING COMPANY

I(T)P An International Thomson Publishing Company

Boston • Albany • Bonn • Cincinnati • Detroit • London • Madrid Melbourne • Mexico City • New York • Paris • San Francisco Singapore • Tokyo • Toronto • Washington

PWS PUBLISHING COMPANY
20 Park Plaza, Boston, MA 02116-4324

International Thomson Publishing
The trademark ITP is used under license.

For more information, contact:

PWS Publishing Co.
20 Park Plaza
Boston, MA 02116

International Thomson Publishing Europe
Berkshire House I68-I73
High Holborn
London WCIV 7AA
England

Thomas Nelson Australia
102 Dodds Street
South Melbourne, 3205
Victoria, Australia

Nelson Canada
1120 Birchmount Road
Scarborough, Ontario
Canada MIK 5G4

International Thomson Editores
Campos Eliseos 385, Piso 7
Col. Polanco
11560 Mexico D.F., Mexico

International Thomson Publishing GmbH
Königswinterer Strasse 418
53227 Bonn, Germany

International Thomson Publishing Asia
221 Henderson Road
#05-10 Henderson Building
Singapore 0315

International Thomson Publishing Japan
Hirakawacho Kyowa Building, 31
2-2-1 Hirakawacho
Chiyoda-ku, Tokyo 102
Japan

Photo Credits: p. 2, © Hideki Fiyii/The Image Bank; p. 8, (top) Texas Instruments, (bottom) The Terry Wild Studio; p. 24, The Terry Wild Studio; p. 57, Courtesy of General Motors Corporation; p. 82, © Will Crocker/The Image Bank; p. 85, James W. Hall; p. 152, © Laurie Rubin/The Image Bank; p. 196, © Barrie Rokeach/The Image Bank; p. 211, The Terry Wild Studio; p. 254, © Andy Caulfield/The Image Bank; p. 304, © David W. Hamilton/The Image Bank; p. 356, © Dag Sundberg/The Image Bank; p. 385, © Chris Hackett/The Image Bank; p. 396, James W. Hall; p. 410, © Eric Schweikardt/The Image Bank; p. 435, Courtesy of Rhone-Poulenc, Inc., Princeton, NJ; p. 440, AP/Wide World Photos; p. 441, Peter Menzel/Stock Boston; p. 454, © Yellow Dog Productions/The Image Bank; p. 481, James W. Hall; p. 496, © Bill Varie/The Image Bank; p. 527, © Murray Alcosser/The Image Bank; p. 546, AP/Wide World Photos; p. 551, Courtesy of NASA; p. 552, © Kaz Mori/The Image Bank; p. 567, Bibliographisches Institut GMBH Leipzig; p. 606, Texas Instruments.

Sponsoring Editor: David Dietz
Editorial Assistant: Julia Chen
Developmental Editors: Maureen Brooks/Barbara Lovenvirth
Production Coordinators: Susan Caffey/Patricia Adams
Production: Lifland et al., Bookmakers
Interior/Cover Designer: Julie Gecha
Interior Illustrators: Scientific Illustrators/Tech-Graphics
Cover Photo/Art: © 1994 Ken Morigami/Panoramic Images, Chicago
Marketing Manager: Marianne C. P. Rutter
Manufacturing Coordinator: Lisa Flanagan
Compositor: Clarinda Company
Cover Printer: John P. Pow Company, Inc.
Text Printer and Binder: Quebecor/Hawkins

Library of Congress Cataloging-in-Publication Data

Hall, James W.
Intermediate algebra with early functions / James W. Hall.—2nd ed.
p. cm.
Includes index.
ISBN 0-534-94026-9
1. Algebra. I. Title.
QA152.2.H364 1995
512.9—dc20 94-31430
CIP

Printed and bound in the United States of America.
94 95 96 97 98 99 — 10 9 8 7 6 5 4 3 2 1

Contents

Functions and graphing coverage indicated in blue

*Optional section

Preface

Intermediate Algebra with Early Functions, Second Edition, is intended to prepare students for a course in college algebra, finite mathematics, or statistics or for other further study in mathematics. It will also equip students for employment in a world where the importance of quantitative reasoning skills is continually increasing. The text features the introduction of graphs and linear functions in Chapter 2; this early introduction of linear functions ensures that fresh material is presented to the student. Coverage of functions and graphing continues throughout the text. See the Table of Contents for an overview of how the coverage is thoroughly integrated.

The text's organization allows for the appropriate use of graphics calculators throughout. The main emphasis is on making mathematical concepts the message, with the graphics calculator used as the messenger wherever it is relevant. This early and graduated use of graphics calculators will provide important background for other courses. This book is designed to work both in classes that fully incorporate the graphics calculator and in those where the instructor uses it occasionally (if only for demonstration purposes). Where appropriate in examples that utilize the graphics calculator, the text illustrates keystrokes for the TI-82 graphics calculator as a model for the student. However, any graphics calculator can be substituted.

Presentation of new ideas in early chapters, appropriate use of modern technology, and inclusion of more real-life applications are all features of this edition designed to produce improvement both in student attitudes and in student performance.

The fundamental organization of this edition builds on the successful presentation of the first edition. New material is introduced clearly and concisely. Examples give the student both a clear mathematical model of the given problem and detailed side-bar explanations of this model. Student aids are abundant and are integrated with the concepts being taught. The exercises contain adequate drill on the basics, as well as material to help students grow mathematically and construct their own bridges between skills and concepts.

Careful pedagogy is employed to ensure that the sections flow and are appropriate for a single lecture. For instance, the general case for the distance formula is developed side by side with a special case. Throughout the text, concepts are related to ideas the students have seen before; then these concepts are developed to a level that will enable students to apply them easily.

FEATURES

New

- *Early Introduction to Functions and Graphs* Early introduction of functions and graphing allows the appropriate use of graphics calculators throughout the book. It also serves to present fresh material to the student who has already completed a Beginning Algebra course. Early in the text many graphs are approximated by broken-line graphs. Thus students can approximate the graphs of a variety of functions without doing anything more complicated than connecting two points with a line segment. This introductory approach opens the door to graphing functions without confronting students with a difficult discussion.
- *Appropriate Use of Technology* Use of scientific and graphics calculators is illustrated wherever appropriate. The emphasis is on the mathematical concepts, with the technology included when it serves a purpose.
- *Emphasis on Estimation Skills* Skills in estimating and checking for the reasonableness of answers are emphasized through exercises designed specifically for this purpose. The appropriate use of technology should be accompanied by a corresponding concern for what is reasonable. The number of exercises on estimation skills has been increased in this edition.
- *Focus on Problem Solving* Consistent with NCTM recommendations, the text focuses on problem solving. There is an early introduction to translating word statements into algebraic statements, and this approach is reinforced throughout the book. Word problems are worked by first forming a **word equation,** which is then translated into an algebraic equation. The use of general principles and tables is emphasized to help the student see the connections that exist between different problems.

New

- *Chapter-Opening Applications* Each chapter opens with an application carefully developed to present the chapter topics in interesting and motivational situations using real-life data provided by industry experts. Students generally work harder when they perceive the material as relevant and important. These chapter openers illustrate the relevance of the mathematics in the chapter and also show possible career opportunities in mathematics.

New

- *Active Applications* These collaborative learning projects, which follow up on selected chapter-opening applications, provide opportunities for students to practice working cooperatively.

New

- *Mathematical Notes* Students should understand that the mathematics they are studying was developed by many civilizations over a long period of time. The Mathematical Notes show where much of the notation we use today came from. They also place mathematics in a more interesting context.

New

- *Geometric Viewpoints* It is appropriate for students who are first viewing new concepts to meet these concepts geometrically, since many of the algebraic methods that we use today were first discovered and used in geometric form.
- *Multiple Representations* Students really understand mathematical principles when they have internalized them and can interpret them from multiple representations. Thus many exercises in the text ask about the same concept from a variety of perspectives. Students will experience the rule-of-three approach, as they see examples and exercises that present real-life functions in numerical, graphical, and algebraic forms.

New

- *Discussion Questions* Students should be able to organize their mathematical ideas and to express them clearly to others. The practice they will receive when responding to the Discussion Questions will not only improve their understanding of mathematics but also contribute to their total education.

New

- *Challenge Questions* These exercises are designed to increase the interest of some students and to foster a variety of problem-solving skills.
- *Self-Check Exercises* Self-check exercises are interspersed throughout for students to check their progress.

New

- *Section Skill Reinforcers* Each section after Section 1-1 contains five exercises that review key skills from earlier in the text.

New

- *Chapter Reviews, Mastery Tests, and Cumulative Reviews* Each chapter has a set of Review Exercises reviewing the entire chapter and a Mastery Test keyed to the chapter objectives. A new Cumulative Review is located after every third chapter.

New

- *Titling of Examples and Exercises* The examples and application exercises have been titled to facilitate the referencing of material by both students and instructors.

New

- *Full Color* The use of full color enlivens the book for the student and facilitates the emphasis of key features. The palette of colors was carefully selected to enhance the pedagogical features of the text.

AN APPLICATION

Time Out for Calculating?

Many professional athletes and coaches now use sophisticated mathematics, implemented on computers, to gain a competitive edge over their opponents. Baseball managers use computers with statistics programs to determine which pitches batters are poorest at hitting. Marathon runners use computer-based motion analysis to fine tune their strides for maximum efficiency. And in mid-race yacht captains use laptop computers to compute optimal sail configurations, based on changing weather conditions.

But not all mathematical applications in sports require the use of computers. The motion of a ball in flight, such as a basketball shot toward a hoop or a football kicked toward a goalpost, can be described fairly well using a quadratic equation. The same sort of equation can describe the motion of a javelin or shot put. Analyzing such an equation can yield useful information. Suppose, for example, that you are a track-and-field coach looking at films of one of your shot-putters at practice; you are watching for the angle at which the shot put leaves the thrower's hand. Using techniques from calculus, it is possible to derive the angle that gives maximum distance. Intuitively, what do you think that angle would be?

The actual path of a thrown or kicked object can be illustrated by the graph of its quadratic function. The graph provides a means of analyzing certain aspects of the flight of the object, including the hang time of a punted football, given its initial upward velocity.

Applications present chapter topics in interesting situations.

Active Application 493

ACTIVE APPLICATION

We will now analyze the motion of a baseball pitch—the familiar curve ball. Place the pitcher's mound at the origin of a coordinate system, and center the front edge of home plate 60 feet away on the positive x-axis. A curve ball thrown by a right-handed pitcher will have a counterclockwise spin, as viewed by the batter, and the pitch will curve toward the first quadrant. A curve ball's motion is actually quite complicated, but the curving path has several reasonable approximations.

1 We can treat the ball's path as a part of a parabola. Suppose the ball leaves the pitcher's hand at $(0, -1)$, initially traveling parallel to the x-axis and reaching home plate over the exact center of its front edge. If the ball's departure point is the vertex of the parabola, what is an equation for the ball's path?

2 If, as many baseball people say, a curve ball hooks more sharply toward the end of its flight than at the beginning, then an ellipse with major axis parallel to the x-axis may be a still better way to describe a curve ball's flight. Suppose the ball travels along

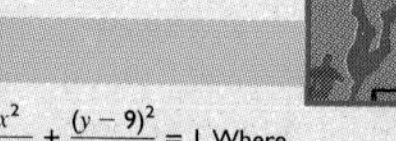

the path $\frac{x^2}{100^2} + \frac{(y-9)^2}{10^2} = 1$. Where will it be when it passes the front edge of home plate? If the ball has to pass the front edge of home plate within $\frac{3}{4}$ ft of its center for a strike, will the pitch be a strike or a ball?

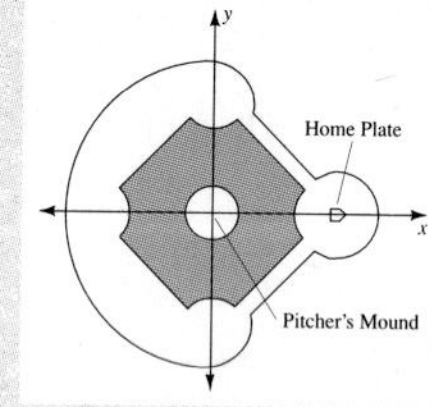

Active Applications, keyed to chapter-opening applications, follow selected chapters.

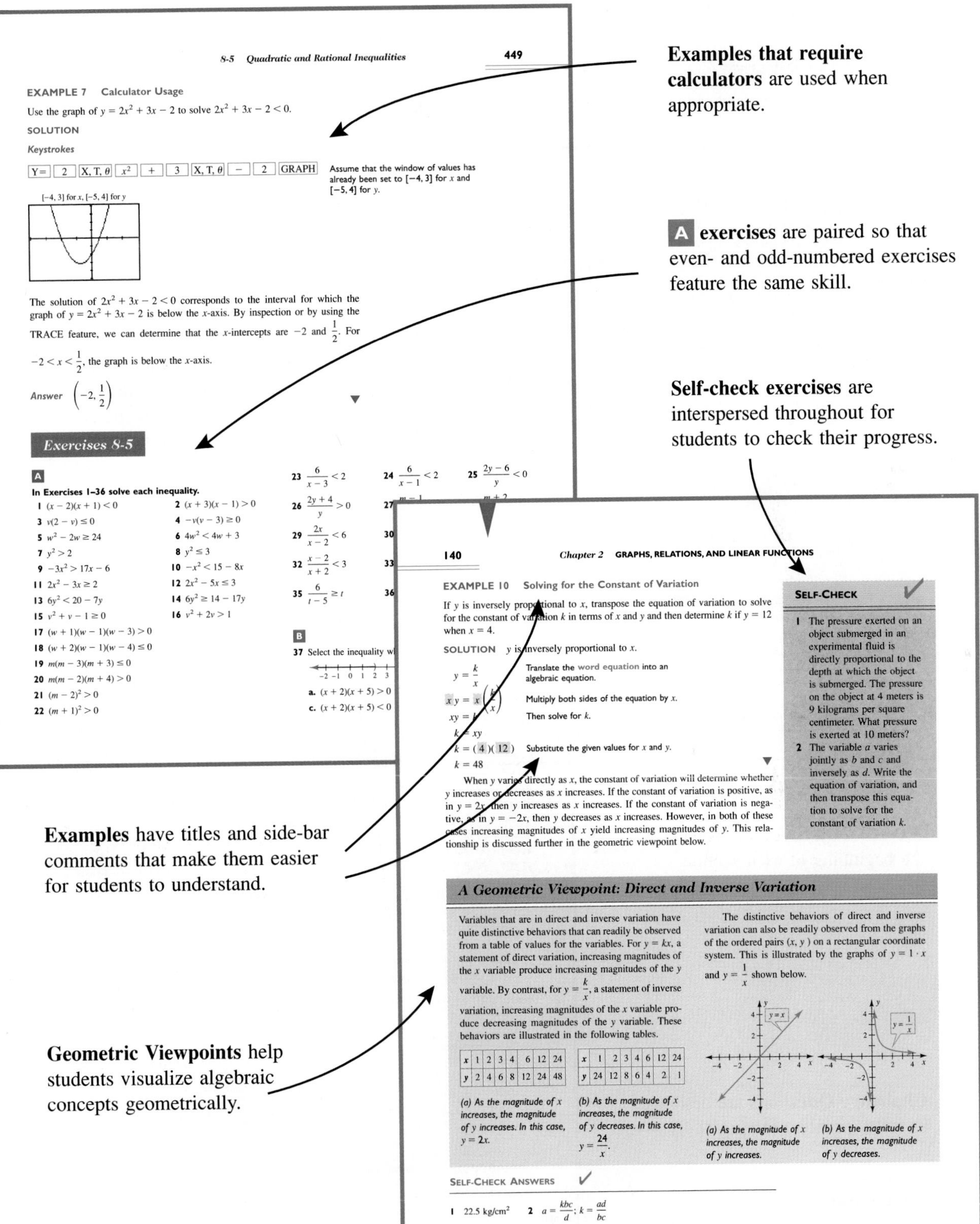

Examples that require calculators are used when appropriate.
A exercises are paired so that even- and odd-numbered exercises feature the same skill.
Self-check exercises are interspersed throughout for students to check their progress.
Examples have titles and side-bar comments that make them easier for students to understand.
Geometric Viewpoints help students visualize algebraic concepts geometrically.
8-5 Quadratic and Rational Inequalities 449
EXAMPLE 7 Calculator Usage
Use the graph of $y = 2x^2 + 3x - 2$ to solve $2x^2 + 3x - 2 < 0$.
SOLUTION
Keystrokes
Y= 2 X, T, θ x² + 3 X, T, θ − 2 GRAPH
Assume that the window of values has already been set to [−4, 3] for x and [−5, 4] for y.
[−4, 3] for x, [−5, 4] for y
The solution of $2x^2 + 3x - 2 < 0$ corresponds to the interval for which the graph of $y = 2x^2 + 3x - 2$ is below the x-axis. By inspection or by using the TRACE feature, we can determine that the x-intercepts are −2 and $\frac{1}{2}$. For $-2 < x < \frac{1}{2}$, the graph is below the x-axis.
Answer $\left(-2, \frac{1}{2}\right)$
Exercises 8-5
A
In Exercises 1–36 solve each inequality.
1 $(x-2)(x+1) < 0$
2 $(x+3)(x-1) > 0$
3 $v(2-v) \le 0$
4 $-v(v-3) \ge 0$
5 $w^2 - 2w \ge 24$
6 $4w^2 < 4w + 3$
7 $y^2 > 2$
8 $y^2 \le 3$
9 $-3x^2 > 17x - 6$
10 $-x^2 < 15 - 8x$
11 $2x^2 - 3x \ge 2$
12 $2x^2 - 5x \le 3$
13 $6y^2 < 20 - 7y$
14 $6y^2 \ge 14 - 17y$
15 $v^2 + v - 1 \ge 0$
16 $v^2 + 2v > 1$
17 $(w+1)(w-1)(w-3) > 0$
18 $(w+2)(w-1)(w-4) \le 0$
19 $m(m-3)(m+3) \le 0$
20 $m(m-2)(m+4) > 0$
21 $(m-2)^2 > 0$
22 $(m+1)^2 > 0$
23 $\frac{6}{x-3} < 2$
24 $\frac{6}{x-1} < 2$
25 $\frac{2y-6}{y} < 0$
26 $\frac{2y+4}{y} > 0$
29 $\frac{2x}{x-2} < 6$
32 $\frac{x-2}{x+2} < 3$
35 $\frac{6}{t-5} \ge t$
B
37 Select the inequality w
a. $(x+2)(x+5) > 0$
c. $(x+2)(x+5) < 0$
140 Chapter 2 GRAPHS, RELATIONS, AND LINEAR FUNCTIONS
EXAMPLE 10 Solving for the Constant of Variation
If y is inversely proportional to x, transpose the equation of variation to solve for the constant of variation k in terms of x and y and then determine k if y = 12 when x = 4.
SOLUTION y is inversely proportional to x.
$y = \frac{k}{x}$ Translate the word equation into an algebraic equation.
$xy = x\left(\frac{k}{x}\right)$ Multiply both sides of the equation by x.
$xy = k$ Then solve for k.
$k = xy$
$k = (4)(12)$ Substitute the given values for x and y.
$k = 48$
When y varies directly as x, the constant of variation will determine whether y increases or decreases as x increases. If the constant of variation is positive, as in y = 2x, then y increases as x increases. If the constant of variation is negative, as in y = −2x, then y decreases as x increases. However, in both of these cases increasing magnitudes of x yield increasing magnitudes of y. This relationship is discussed further in the geometric viewpoint below.
SELF-CHECK
1 The pressure exerted on an object submerged in an experimental fluid is directly proportional to the depth at which the object is submerged. The pressure on the object at 4 meters is 9 kilograms per square centimeter. What pressure is exerted at 10 meters?
2 The variable a varies jointly as b and c and inversely as d. Write the equation of variation, and then transpose this equation to solve for the constant of variation k.
A Geometric Viewpoint: Direct and Inverse Variation
Variables that are in direct and inverse variation have quite distinctive behaviors that can readily be observed from a table of values for the variables. For y = kx, a statement of direct variation, increasing magnitudes of the x variable produce increasing magnitudes of the y variable. By contrast, for $y = \frac{k}{x}$, a statement of inverse variation, increasing magnitudes of the x variable produce decreasing magnitudes of the y variable. These behaviors are illustrated in the following tables.
The distinctive behaviors of direct and inverse variation can also be readily observed from the graphs of the ordered pairs (x, y) on a rectangular coordinate system. This is illustrated by the graphs of $y = 1 \cdot x$ and $y = \frac{1}{x}$ shown below.
x 1 2 3 4 6 12 24
y 2 4 6 8 12 24 48
x 1 2 3 4 6 12 24
y 24 12 8 6 4 2 1
(a) As the magnitude of x increases, the magnitude of y increases. In this case, y = 2x.
(b) As the magnitude of x increases, the magnitude of y decreases. In this case, $y = \frac{24}{x}$.
y = x
$y = \frac{1}{x}$
(a) As the magnitude of x increases, the magnitude of y increases.
(b) As the magnitude of x increases, the magnitude of y decreases.
SELF-CHECK ANSWERS
1 22.5 kg/cm² 2 $a = \frac{kbc}{d}$; $k = \frac{ad}{bc}$

Chapter Objectives, keyed to sections, begin each chapter.

CHAPTER 9 OBJECTIVES

1 **Calculate the distance between two points (Section 9-1).**
2 **Determine the midpoint between two points (Section 9-1).**
3 **Graph a parabola (Section 9-2).**
4 **Determine the maximum or the minimum of a quadratic function (Section 9-2).**
5 **Graph a circle and an ellipse (Section 9-3).**
6 **Graph a hyperbola (Section 9-4).**
7 **Identify the type of conic section from its equation (Section 9-4).**

René Descartes developed analytic geometry in 1619 (see Section 2-1). Prior to Descartes, algebra was concerned with numbers and calculations, and geometry was concerned with figures and shapes. By associating points in the plane with ordered pairs of numbers, Descartes merged the power of these areas into analytic geometry. In this chapter we will use algebra to calculate both the distance between two points and the midpoint between two points. The distance and midpoint formulas are important algebraic tools for analyzing geometric shapes.

Some of the geometric shapes that we will examine in this chapter are parabolas, circles, ellipses, and hyperbolas. These shapes are referred to collectively as *conic sections*, because each of these figures can be obtained by intersecting a plane and a cone.

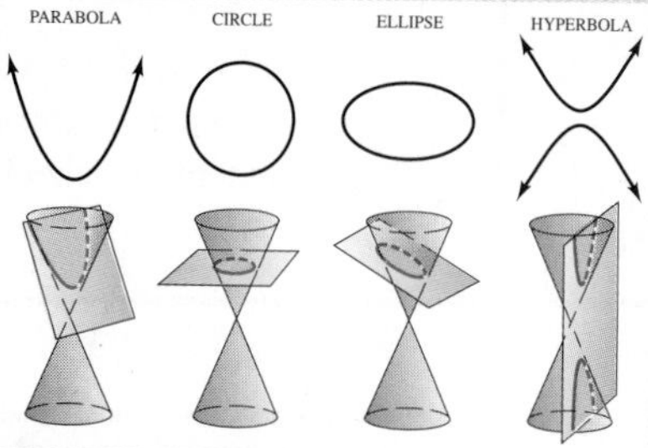

9-1 The Distance and Midpoint Formulas

SECTION OBJECTIVES

1 **Calculate the distance between two points.**
2 **Determine the midpoint between two points.**

An important measurement along a straight line is the distance between two points. To develop the formula for the distance between two points, we will first consider the horizontal and vertical changes between these points and then use

Exercises in Calculator Usage and Estimation Skills are featured in most sections, as are **Discussion Questions.**

Section Objectives are listed at the beginning of each section.

41 $3w^{5/11}(2w^{17/11} - 5w^{6/11} - 9)$
42 $4w^{-5/3}(3w^{11/3} - 7w^{8/3} + 2w^{7/3})$
43 $(a^{1/2} + 3)(a^{1/2} - 3)$
44 $(2a^{1/2} - 3b^{1/2})(2a^{1/2} + 3b^{1/2})$
45 $(b^{3/5} - c^{5/3})(b^{3/5} + c^{5/3})$
46 $(b^{3/5} + c^{5/3})^2$
47 $(b^{3/5} - c^{5/3})(b^{3/5} - c^{5/3})$
48 $(x^{1/2} - x^{-1/2})^2$
49 $(x^{-1/2} + x^{1/2})^2$
50 $(x^{2/3} + x)(x^{-2/3} - x)$
51 $(y^{1/3} + 2)(y^{2/3} - 2y^{1/3} + 4)$
52 $(3y^{1/3} - 5)(9y^{2/3} + 15y^{1/3} + 25)$
53 $[(3^{1/3} + 5^{1/3})(3^{2/3} - 15^{1/3} + 5^{2/3})]^{5/3}$
54 $[(14^{1/3} - 5^{1/3})(14^{2/3} + 70^{1/3} + \ldots)]^{3/2}$

ESTIMATION SKILLS (55–58)

55 The best mental estimate of $145^{1/2}$ is ____.
a. 73 b. 72 c. 14 d. 13 e. 12

56 The best mental estimate of $170^{1/2}$ is ____.
a. 96 b. 97 c. 14 d. 13 e. 12

57 The best mental estimate of $1003^{1/3}$ is ____.
a. 333 b. 300 c. 133 d. 100 e. 10

58 The best mental estimate of $124^{1/3}$ is ____.
a. 42 b. 43 c. 41 d. 5 e. 10

7-1 *Rational Exponents* 365

C

In Exercises 59–61 simplify each expression. Assume that x and y represent positive real numbers and that m is a natural number.

59 $x^{m/3}x^{m/2}$ 60 $\frac{16x^{m/2}}{2x^{m/3}}$ 61 $(x^{m/3}y^{m/2})^6$

62 Factor $x^{5/2} + 3x^{3/2} - 4x^{1/2}$ by first factoring out $x^{1/2}$.

63 Factor $x^{3/2} - 25x^{-1/2}$ by first factoring out $x^{-1/2}$.

64 Complete the following table for $y = x^{1/2}$. Then use these points to draw a line graph of $y = x^{1/2}$.

x	0	0.25	0.81	1.00	1.44	2.25	4	9
y								

65 Complete the following table for $y = x^{1/3}$. Then use these points to draw a line graph of $y = x^{1/3}$.

x	0	0.027	0.125	1	8
y					

CALCULATOR USAGE (66–70)

In Exercises 66–70 use a calculator with a power key labeled y^x or ^ to approximate each number to five significant digits.

66 $37^{1/3}$ 67 $73^{1/4}$ 68 $12^{2/3}$ 69 $128^{2/7}$

70 The third power of the fifth root of 86

71 Give an example of a real number x for which $x^{1/2}$ is defined but $x^{-1/2}$ is undefined.

DISCUSSION QUESTIONS (72 AND 73)

72 Explain why $x^{1/2}$ is defined to be $\sqrt{x}$.

73 Give an example of a real number x for which $(x^2)^{1/2} \neq x$. Discuss why this does not violate the power rule which states $(x^a)^b = x^{ab}$.

CHALLENGE QUESTION

a. Which is larger: $(2^2)^2$ or $2^{(2^2)}$?
b. Which is larger: $(3^3)^3$ or $3^{(3^3)}$?
c. Which is larger: $2^{(3^4)}$ or $4^{(3^3)}$?

SECTION SKILL REINFORCERS

1 Which is a rational number: $\sqrt{9}$ or $\sqrt{10}$? 2 Which is an irrational number: $\sqrt{63}$ or $\sqrt{64}$? 3 List one rational number between $\sqrt{5}$ and $\sqrt{6}$. 4 Simplify $|-7| + |7|$. 5 Simplify $|23| - |-23|$.

Challenge Questions are designed to foster problem-solving skills.

Section Skill Reinforcers review key skills from previous sections.

Key Concepts are summarized at the end of each chapter.

A Mastery Test, keyed to chapter objectives, ends each chapter.

Review Exercises for Chapter 8 **451**

Key Concepts for Chapter 8

1 The standard form of a quadratic equation in x is $ax^2 + bx + c = 0$, where a, b, and c represent real constants and $a \neq 0$.

2 A quadratic equation has either two distinct solutions or a solution of multiplicity two. These solutions may be either real or imaginary.

3 Methods of solving quadratic equations:

a. Factoring

b. Extraction of roots

c. Completing the square

d. The quadratic formula

$$x = \frac{-b \pm \sqrt{b^2 - 4ac}}{2a}$$

4 Nature of the solutions of a quadratic equation if a, b, and c are rational constants and $a \neq 0$:

Discriminant, $b^2 - 4ac$	***Nature of the Solutions***
Positive and a perfect square	Two distinct rational solutions
Positive and not a perfect square	Two distinct irrational solutions
Zero	A rational solution of multiplicity two
Negative	Two imaginary solutions that are complex conjugates

Review Exercises for Chapter 8

In Exercises 1–22 solve each equation by the most appropriate method.

1 $y^2 = 144$

2 $w^2 = 9w - 20$

3 $x^2 - 4x + 2 = 0$

4 $(v - 3)^2 = 36$

5 $10m^2 = 21m + 10$

6 $2y^2 = 6y - 9$

7 $-9v^2 + 24v = 16$

8 $(6x - 3)(x + 5) = 26x$

9 $y^4 = 81$

10 $z^4 = 14z^2 + 32$

11 $(m^2 - 3m)^2 - 44(m^2 - 3m) + 160 = 0$

12 $(5x^2 + 1) - 22\sqrt{5x^2 + 1} = -21$

13 $\left(\frac{2w - 1}{w - 2}\right)^2 + 2\left(\frac{2w - 1}{w - 2}\right) = 3$

14 $15m^{-2} = 17m^{-1} + 4$

15 $(n + 5)(2n - 3)(3n - 10) = 0$

16 $y^3 - 8 = 0$

17 $\sqrt{x + 12} - x = 0$

18 $\sqrt{2x - 1} = x - 2$

19 $m - 3 = \frac{4}{m}$

20 $\frac{x}{2x - 9} = \frac{x + 2}{x - 2}$

21 $v^{2/3} - 26v^{1/3} = 27$

22 $\frac{5y - 1}{3y + 5} - \frac{4y + 1}{2y - 3} = \frac{18}{6y^2 + y - 15}$

Review Exercises for each chapter review current and previous concepts.

Mastery Test for Chapter 8 **453**

58 Construct a quadratic equation in x that has the given solutions.

a. $-\frac{3}{4}$ and 5

b. $\frac{1 + \sqrt{3}}{2}$ and $\frac{1 - \sqrt{3}}{2}$

c. $2 - 5i$ and $2 + 5i$

59 Construct a cubic (third-degree) equation in x that has solutions 0, 4, and 7.

60 **Error Analysis** Examine the student's work shown below, and correct all of the errors that you find.

$$(x - 2)(2x + 3) = 2x - 6$$
$$2x^2 - x - 6 = 2x - 6$$
$$2x^2 - x = 2x$$
$$2x - 1 = 2$$
$$2x = 3$$
$$x = \frac{3}{2}$$

Mastery Test for Chapter 8

Exercise numbers correspond to Section Objective numbers.

[8-1] 1 Solve each of these quadratic equations by extraction of roots.

a. $v^2 = 19$ …

c. $(2x - 3$ …

[8-1] 2 Solve each … the square.

a. $m^2 + 4$ …

c. $-4v^2$ …

[8-2] 3 Solve each … dratic form…

a. $6x^2 - 1$ …

c. $3v^2 + 1$ …

[8-2] 4 Use the dis… lutions of … the solution… tinct irratio… plicity two…

a. $5x^2 + 5$ …

c. $-3w^2$ …

[8-3] 5 Solve each …

a. $x^4 - 34$ …

c. $\left(\frac{v}{v - 2}\right.$ …

[8-3] 6 Solve each of these equations.

a. $\sqrt{x + 42} + x = 0$

b. $\sqrt{x + 4} = x + 1$

Cumulative Review of Chapters 7–9

The limited purpose of this review is to help you gauge your mastery of Chapters 7, 8, and 9. It is not meant to examine each detail from these chapters, nor is it meant to focus on specific sections that may be emphasized at any one particular school.

In Exercises 1–20 perform the indicated operations, and express the answer in simplest form.

1 $\left(\frac{8}{125}\right)^{2/3}$

2 $(36)^{-3/2}$

3 $\sqrt[3]{1{,}000{,}000} - \sqrt[6]{1{,}000{,}000}$

4 $(\sqrt{5} + \sqrt{2})^2$

5 $\frac{14}{\sqrt{7} + \sqrt{5}}$

6 $3\sqrt{8} - 5\sqrt{18}$

7 $(3x^{2/3}y^{1/4})(5x^{1/3}y^{-5/4})$

8 $\frac{12x^{5/3}y^{3/5}}{6x^{2/3}y^{-2/5}}$

9 $(16x^{8/3}y^{4/3})^{3/4}$

10 $\frac{(x^3y^3z^3)^{1/4}}{(xyz)^{1/2}}$

11 $\sqrt[3]{x^3y^5}$

12 $\sqrt{x^3y^5}$

13 $5(3 - 7i) + 2(4 - 2i)$

14 $(3 - 7i)(4 + 2i)$

15 $\frac{\sqrt{3x} + \sqrt{5y}}{\sqrt{3x} - \sqrt{5y}}$

16 $\frac{3 + 5i}{3 - 5i}$

17 $\sqrt{-36} - \sqrt{16}$

18 $\frac{130}{4 + 7i}$

19 $\frac{x^3 - y^3}{6x^2 - 6y^2} \div \frac{2x^2 + 2xy + 2y^2}{9xy}$

20 $\frac{2z + 11}{z^2 + z - 6} + \frac{2}{z + 3} - \frac{3}{z - 2}$

In Exercises 21–30 solve each equation.

21 $(2x - 3)(x + 4) = 0$

22 $6v^2 + 11v - 10 = 0$

23 $(2x - 3)(x + 4) = -12$

24 $2w^3 - 8w = 0$

25 $\sqrt{4a - 6} = 8$

26 $\sqrt[3]{4m + 12} = 2$

27 $\sqrt{5x + 6} = x$

28 $\frac{12}{x + 2} = \frac{6}{x - 1}$

29 $\frac{4}{m^2 + 2m - 3} - \frac{3}{m + 2} = \frac{1}{m - 1}$

30 $x^4 = 29x^2 - 100$

31 Solve $x^2 + 3xy - 10y^2 = 0$ for x.

32 Construct a quadratic equation in x with solutions -3 and $\frac{4}{5}$.

33 **Length of a Rafter** Find the length in feet of the rafter r shown in the figure. The lengths of the two braces are $r - 2$ and $r - 9$.

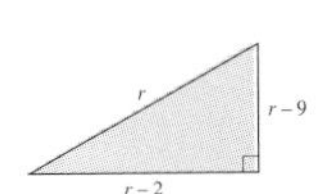

Figure for Exercise 33

34 **Average Speed** An automobile completed a 480-kilometer trip. If the average rate had been increased by 10 kilometers per hour, the trip would have taken 2 hours less than if the average rate had been decreased by 10 kilometers per hour. What was the average rate for this trip?

35 Solve $x^2 - 6x \geq -8$.

36 Solve $\frac{x^2 - x - 20}{x - 2} < 0$.

37 Calculate the distance between $(-2, -2)$ and $(3, 10)$.

38 Determine the midpoint between $(-2, 5)$ and $(6, -7)$.

39 Calculate the slope of the line through $(-1, -8)$ and $(5, 4)$.

40 Write the equation of a line through $(3, 7)$ and perpendicular to $y = -\frac{1}{5}x + 3$.

41 Write the equation of the perpendicular bisector of the line segment connecting $(1, -2)$ and $(5, 4)$.

42 Write the value of the discriminant of $2x^2 + 4x + 3 = 0$, and determine the nature of the solutions of this equation.

43 **Average Cost** The average cost of producing x units of a radio is given by $C(x) = x^2 - 85x + 1850$. Determine the minimum average cost.

44 **Dog Food Mixture** A mixture of dog food is created by mixing two sources. Source A contains 12% protein and 40% carbohydrates, and source B contains 25% protein and 30% carbohydrates. How many grams of each source should be mixed to obtain a supplement containing 272 grams of protein and 480 grams of carbohydrates?

45 **Speed of an Airplane** Two planes with the same air speed depart 1 hour apart and travel in opposite directions. The plane that departs first flies directly into a 50-mile-per-hour wind, and the second plane flies with this wind. After a period of time radar indicates that the first plane has traveled 375 miles and the second plane has traveled 450 miles. Determine the air speed of each plane.

46 Determine the vertex of the parabola $y = 2(x - 3)^2 + 4$.

47 Determine the center of the circle $x^2 + y^2 + 8x + 10y + 40 = 0$.

48 Determine the x- and y-intercepts of the ellipse $\frac{x^2}{100} + \frac{y^2}{49} = 1$.

A Cumulative Review is placed at the end of every third chapter.

Supplements to Accompany the Text

For Instructors:

- *Instructor's Edition* contains answers to all exercises in a section at the end of the book.
- *Instructor's Manual* has answers to all exercises and suggested procedures, where appropriate, for Challenge Questions.
- *Test Bank with Transparency Masters* contains all questions and answers from the computerized test bank, three ready-to-use tests (two open-ended and one multiple choice) for each chapter, two final examinations, and a set of transparency masters.
- *Computer Testing Software* is available both for the IBM and compatibles and for the Macintosh. These computerized test banks contain multiple-choice and open-ended questions and allow the user to edit, rearrange, and add to the question bank. *EXPTest* runs on Windows or DOS, and *Exam-Builder* runs on the Macintosh.

For Students:

- *Student Solutions Manual* contains a complete worked-out solution for every other odd problem.
- *MathQuest Tutorial Software* is an interactive, text-specific tutorial which runs on both Windows and Macintosh platforms. The program poses fill-in and multiple-choice questions. If a student answers a question incorrectly, the program will first respond with hints; if the student answers incorrectly a second time, the program will supply a step-by-step solution. Scorekeeping capabilities enable students to monitor their progress. *DOS Tutorial Software* with similar capabilities is also available.
- *Videotape Series* features professional math instructors and reviews key topics in the text. These tutorial tapes, produced by Educational Video Resources, are free to adopters for use in math labs.
- *Video Guidebook* provides exercises that link the content of the text to the videos. For each video that accompanies the text, a three-part lesson ("Pretest," "While You Watch," and "Posttest") engages the student in active learning. All answers are provided, making this worktext especially helpful for independent learners.

ACKNOWLEDGMENTS

I wish to express my appreciation to all the reviewers for their thoughtful criticisms and suggestions. Their careful attention to detail has ensured that this edition has a sense of continuity and flow that will benefit the students who use it. Special thanks go to users of the text and reviewers who evaluated the manuscript for this second edition, as well as those who contributed to the first edition:

Linda Crabtree
Longview Community College

Susan Marie Dimick
Spokane Community College

Arthur P. Dull
Diablo Valley College

James M. Edmondson
Santa Barbara Community College

Thomas Vanden Eynden
Thomas More College

Henry C. Graves
Trident Technical College

William L. Grimes
Central Missouri State University

Glen E. Johnston
Morehead State University

Jamie King
Orange Coast College

Laura Coffin Koch
University of Minnesota

Mary Ann Koehler
Cuyahoga Community College

Mike Mears
Manatee Community College

Gael Mericle
Mankato State University

James E. Moran
Diablo Valley College

Linda J. Murphy
Northern Essex Community College

Nancy K. Nickerson
Northern Essex Community College

Carol O'Loughlin
Northern Essex Community College

Kenneth Schoen
Worcester State College

Donna Szott
Allegheny Community College

Hubert C. Voltz
Slippery Rock University

I am especially appreciative of the many comments and suggestions made by the students and staff of Parkland College. Other important contributors to this edition were my wife, Peggy, who performed many of the typing and proofing chores, and Diana Pennock, who double-checked the answers to all the exercises.

My special thanks go to editors Barbara Lovenvirth and David Dietz, production coordinators Susan Caffey and Patty Adams, and production editor Sally Lifland. I also appreciate the work of Scientific Illustrators and Tech-Graphics in creating the figures that accompany this text. Thanks also to Laurel Technical Services, Redwood City, California, for consulting with industry experts to provide real-life data for chapter-opening applications.

James W. Hall
Parkland College

Chapter 1

Review of Beginning Algebra

An Application

Basic Math in Speech Therapy

Much of this century's progress in medicine and psychology has been achieved by scientists' studying large numbers of patients and making inferences based on these studies. This area of mathematics is called *statistics.* A special field of statistics, known as *nonparametric statistics,* can be used to analyze data obtained from a limited sample, such as a small group of people. Nonparametric statistics is used in speech pathology, the study and treatment of speech disorders. Although highly developed, the mathematics used in nonparametric statistics often requires only basic skills in adding and comparing numbers—skills that you will be reviewing in this chapter in preparation for more challenging work.

One technique commonly used in nonparametric statistics is to convert data points into rankings before doing further analysis. For example, a speech therapist studying a group of eight 3-year-olds and eight 4-year-olds with delayed speech development might compile the following data:

Speech Error	3-year-olds	4-year-olds
Stopping	10	4
Gliding	15	9
Backing	3	5
Fronting	12	7

Each entry represents the total number of errors in each category made by children of the specified age during a timed conversation. On a separate sheet of paper redraw the table and replace the entries with rankings: Use 1 for the lowest number, 2 for the next lowest—up to 8 for the highest. Now add the rankings vertically to get a rank total for the 3-year-olds and a rank total for the 4-year-olds. Which total is larger? What might you conclude?

If you are interested in pursuing this topic further, a suggested reading is Chapter 11 of *Beginning Statistics, A to Z,* by William Mendenhall, Wadsworth Publishing Company, 1993.

CHAPTER 1 OBJECTIVES

1 **Identify numbers from important subsets of the real numbers (Section 1-1).**

2 **Use absolute-value notation (Section 1-1).**

3 **Identify and use the basic properties of the real number system (Section 1-1).**

4 **Add, subtract, multiply, and divide real numbers (Section 1-2).**

5 **Use natural number exponents (Section 1-2).**

6 **Use the correct order of operations (Section 1-2).**

7 **Evaluate an algebraic expression for given values of the variables (Section 1-3).**

8 **Simplify algebraic expressions by combining like terms (Section 1-3).**

9 **Solve linear equations in one variable (Section 1-4).**

10 **Transpose an equation to solve for a specified variable (Section 1-4).**

11 **Solve word problems using the mixture and rate principles (Section 1-5).**

12 **Use interval notation (Section 1-6).**

13 **Solve linear inequalities in one variable (Section 1-6).**

14 **Solve absolute-value equations and inequalities (Section 1-7).**

The word *algebra* comes from the book *Hisab al-jabr w'almugabalah,* written by an Arab mathematician in A.D. 830; translations of this text on solving equations became widely known in Europe as *al-jabr.* Algebra, however, is concerned with more than solving equations; it is a generalization of arithmetic.

One of the primary uses of algebra is to manipulate mathematical expressions in order to take a given form and rewrite it in a more desirable form. Three reasons that we rewrite an algebraic expression follow.

1 To put the expression in simplest form.

2 To simplify an equation in order to solve the equation.

3 To place a function in a form that reveals more information about the graph of the function.

In this chapter we will begin our emphasis on rewriting expressions. We will also focus on the relationship between algebra and geometry. Geometry gives us the ability to visualize what is being algebraically manipulated. Fundamental to the development of problem-solving skills is the mastery of a few basic principles. We will introduce these principles early, use them to solve problems, and then build on problem-solving skills in incremental stages throughout the book.

1-1 Properties of the Real Number System

Section Objectives

1 Identify numbers from important subsets of the real numbers.

2 Use absolute-value notation.

3 Identify and use the basic properties of the real number system.

Arithmetic uses only constants, such as -13, $-1\frac{1}{2}$, 0, $\frac{2}{3}$, 1.14, $\sqrt{2}$, 3, and π. **Constants** have a fixed, or constant, value. Algebra uses not only constants but also variables, such as a, b, P, V, x, x_1, x_2, and x'. The variables x_1 and x_2 are read as "x sub 1" and "x sub two," respectively; x' is read as "x prime." **Variables** can be used to represent any element in a set of two or more elements. In arithmetic the basic operations are used to form **numerical expressions,** such as $17 - 5$, 11^2, and $\frac{13+5}{7}$. In algebra we can use operations with both constants and variables to form **algebraic expressions,** such as $2x - 5$, $x + y$, $x^2 - 3y$, and $\sqrt{x+1}$.

The set of numbers that may replace a variable is called the **domain** of the variable. Often this domain is not specified in a given problem but is simply understood from the context of the problem. The set of numbers understood in such problems is usually the set of real numbers or one of its important subsets, which we will examine now.

A Mathematical Note

Francois Viete (1540–1603) is considered by many to be the father of algebra as we know it today. Prior to Viete, algebra was expressed rhetorically. Viete promoted the use of variables to represent unknowns, which led to the algebraic notation we now use.

The **number line** is an infinite straight line with a fixed point of reference called the **origin.** Once a convenient unit of measure has been chosen, equally spaced points can be labeled, as illustrated in Figure 1-1. There is a one-to-one correspondence between all the points on the line and the set of real numbers. The number associated with a point on the number line is called the **coordinate** of the point. The coordinates of points to the right of the origin are **positive** numbers, and the coordinates of points to the left of the origin are **negative** numbers (see Figure 1-2).

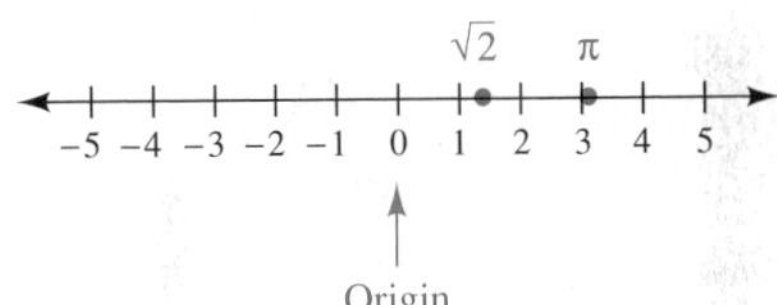

Figure 1-1
Number Line

Each real number is either rational or irrational (see Figure 1-3). The first five letters of the word *rational* spell ratio—this is exactly what a rational number is: the ratio of two integers. (The symbol $\mathbb{Q}$, used to denote the set of rational numbers, stands for quotient.) Every real number can be written in decimal form. This decimal form can then be used to characterize the number as either rational or irrational.

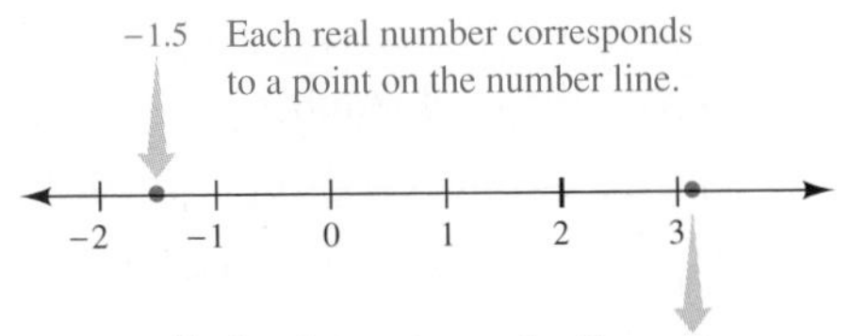

Figure 1-2

Rational and Irrational Numbers

Rational: The decimal form of a rational number is either

a. a terminating decimal or

b. an infinite repeating decimal.

Irrational: The decimal form of an irrational number is an infinite nonrepeating decimal.

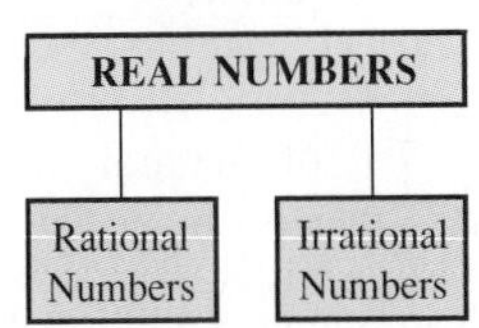

Figure 1-3

Two types of notation are commonly used to represent infinite repeating decimals: Both 0.333 . . . and $0.\overline{3}$ are equal to the rational number $\frac{1}{3}$.

Irrational numbers can be characterized either as real numbers that cannot be written as fractions or as numbers whose decimal form is an infinite nonrepeating decimal. Thus irrational numbers are often represented by some special notation. The square root of any natural number that is not a perfect square is an irrational number. For example, $\sqrt{2}$, $\sqrt{3}$, and $\sqrt{5}$ are well-known examples of irrational numbers. Another well-known irrational number is π.

A Mathematical Note

π is defined as the ratio of the circumference of a circle to its diameter: $\pi \approx 3.14159265$ ($\approx$ means approximately equal to). However, the decimal form of π does not terminate or repeat. In 1897 House Bill #246 was introduced in the Indiana legislature to make 3.2 the value of π in this state.

EXAMPLE 1 Identifying Rational and Irrational Numbers

Classify each of these real numbers as either (a) rational or (b) irrational:

$$-7, -\pi, 0, 0.873, \sqrt{2}, 4.04004000400004\ldots, \sqrt{25}, 5\frac{1}{6}, 7.\overline{3}$$

SOLUTIONS

(a) Rational: $-7, 0, 0.873, \sqrt{25}, 5\frac{1}{6}, 7.\overline{3}$.

The terminating decimals -7, 0, 0.873, and $\sqrt{25}$ can be written, respectively, as $-\frac{7}{1}, \frac{0}{1}, \frac{873}{1000}$, and $\frac{5}{1}$.

Since $5^2 = 25$, $\sqrt{25}$ is the same number as the terminating decimal 5.

The infinite repeating decimals $5\frac{1}{6}$ and $7.\overline{3}$ can be written, respectively, as $\frac{31}{6}$ and $\frac{22}{3}$.

$5\frac{1}{6}$ in decimal form is the infinite repeating decimal 5.1666 . . . , or $5.1\overline{6}$. (Convert to decimal form by dividing 31 by 6.) $7.\overline{3}$ represents the infinite repeating decimal 7.333 . . . .

(b) Irrational: $-\pi, \sqrt{2}, 4.04004000400004\ldots$
All have an infinite nonrepeating decimal form.
They cannot be written as the ratio of two integers.

$-\pi$ is approximately -3.14159265. However, the decimal form of this number does not terminate or repeat. $\sqrt{2} \approx 1.4142136$, but the actual value does not terminate or repeat. Note that 4.04004 . . . does not repeat, since the number of 0s between successive 4s changes.

IMPORTANT SUBSETS OF THE REAL NUMBERS

$\mathbb{N}$ = Natural numbers = {1, 2, 3, 4, 5, 6, 7, . . .}
$\mathbb{W}$ = Whole numbers = {0, 1, 2, 3, 4, 5, 6, . . .}
$\mathbb{I}$ = Integers = {. . . , −3, −2, −1, 0, 1, 2, 3, . . .}
$\mathbb{Q}$ = Rational numbers = $\left\{\frac{a}{b}: a, b \text{ are integers and } b \neq 0\right\}$
$\tilde{\mathbb{Q}}$ = Irrational numbers = The set of real numbers that are not rational
$\mathbb{R}$ = Real numbers = The set of coordinates of all the points on the number line

The set of natural numbers {1, 2, 3, 4, 5, . . .} is given in **roster notation**—that is, by a listing of the elements. The set of rational numbers $\left\{\frac{a}{b}: a, b \text{ are integers and } b \neq 0\right\}$ is given in **set-builder notation**—that is, by specifying a condition that the elements in the set must satisfy. This set-builder notation is read as "The set of all fractions $\frac{a}{b}$ such that a and b are integers and $b \neq 0$." This notation is given below for the prime numbers and the composite numbers.

Prime numbers
P = $\{n$: n is a natural number with exactly two distinct natural number factors, 1 and the number itself$\}$

Composite numbers
C = $\{n$: n is a natural number with more than two distinct natural number factors$\}$

The relationship of important subsets of the real numbers is summarized in Figure 1-4.

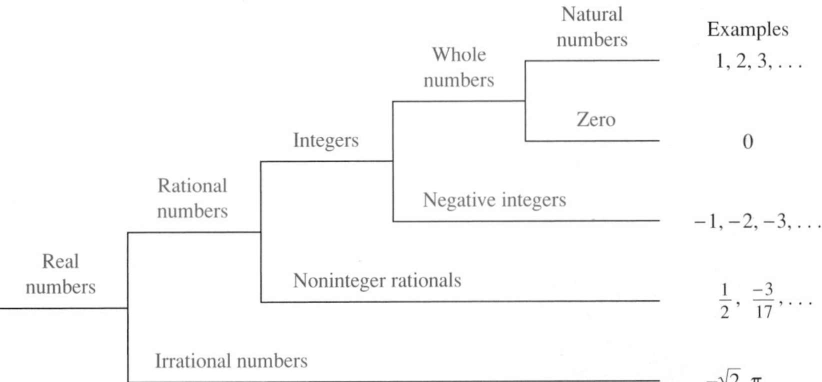

Figure 1-4 *Tree Diagram*

SELF-CHECK 

1 What prime number is between 50 and 58?
2 Are all natural numbers also whole numbers?
3 Are all whole numbers also natural numbers?

Two real numbers that are the same distance from the origin but on opposite sides of the origin on the number line are called **opposites,** or **additive inverses,** of each other. As Figure 1-5 illustrates, 4 and -4 are opposites of each other. The opposite of a positive number is always negative, and the opposite of a negative number is always positive. In general, the opposite of the number a is denoted by $-a$. Thus the opposite of -4 can be denoted by $-(-4)$, which is read as "the opposite of negative four." As noted above, the opposite of -4 is 4; that is, $-(-4) = 4$.

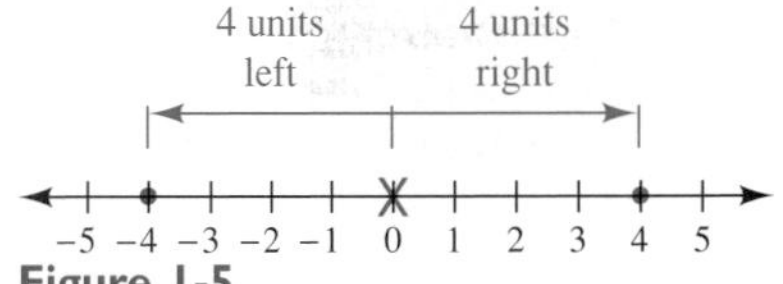

Figure 1-5

OPPOSITES

For any real number a,

1 the additive inverse of a is denoted by $-a$.

2 $-(-a) = a$.

3 the opposite of 0 is 0.

SELF-CHECK ANSWERS 

1 53 **2** Yes **3** No, 0 is not a natural number.

EXAMPLE 2 Writing the Opposite of an Algebraic Expression

Write an algebraic expression for each of these numbers. Then write each expression in simplest form.

SOLUTIONS

	Algebraic Expression
(a) The opposite of six	-6
(b) The opposite of negative seven	$-(-7) = 7$
(c) The opposite of x	$-x$
(d) The opposite of $-x$	$-(-x) = x$

Caution: Read $-a$ as "the opposite of a," not as "negative a," since $-a$ can be either positive, negative, or zero. Also, do not read $-a$ as "minus a," because the word *minus* is used to indicate that one number is to be subtracted from another. On some calculators the (−) key is used to indicate the opposite of a number and the − key is used to perform subtraction (see Figure 1-6). On other calculators the +/− key is used to form the opposite of a number (see Figure 1-7).

Figure 1-6

Figure 1-7

On the real number line the distance from the point whose coordinate is a to the origin is called the **absolute value** of a and is denoted by $|a|$. Because distance is never negative, $|a|$ is always greater than or equal to 0. For example, both -4 and $+4$ are 4 units from 0, so $|-4| = 4$ and $|4| = 4$ (see Figure 1-8).

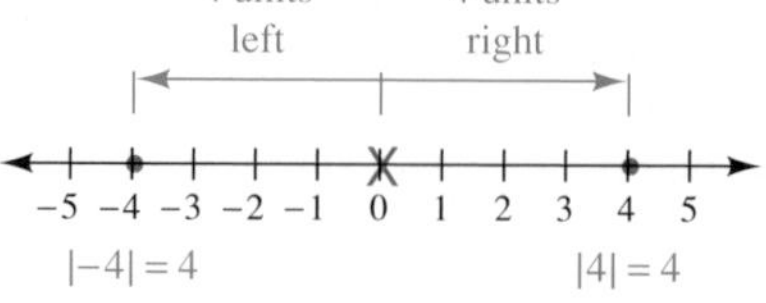

Figure 1-8

ABSOLUTE VALUE

The absolute value of a real number a is the distance between the origin and the point with coordinate a on the number line.

$|a| = a$ if $a \geq 0$

$|a| = -a$ if $a < 0$

$|a|$ units left $|a|$ units right

0

EXAMPLE 3 Evaluating Absolute-Value Expressions

Use the definition of absolute value to simplify each of the expressions below.

	SOLUTIONS	
(a) $\|6\|$	$\|6\| = 6$	$\|a\| = a$ if $a > 0$.
(b) $\|0\|$	$\|0\| = 0$	$\|a\| = a$ if $a = 0$.
(c) $\|-6\|$	$\|-6\| = 6$	$\|a\| = -a$ if $a < 0$.
(d) $-\|6\|$	$-\|6\| = -6$	The absolute value of 6 is 6, and the opposite of this number is -6.
(e) $\|x\|$	$\|x\|$	This expression cannot be simplified unless we know whether $x \geq 0$ or $x < 0$.

▼

Caution: In general, we cannot say that $|x|$ is x; neither can we say that $|-x|$ is x. In each case the simplification will depend on whether $x \geq 0$ or $x < 0$.

The properties of our real number system are the laws that users of algebra must obey. Failure to obey these laws can result in incorrect results. We will start with some of the properties of equality and inequality. It may help you to remember the names of these properties if you compare them to the common words *reflect, symmetrical,* and so on. Note how the names describe the properties.

PROPERTIES OF EQUALITY AND INEQUALITY

For all real numbers a, b, and c:

Reflexive Property: $a = a$

Symmetric Property: If $a = b$, then $b = a$.

Transitive Property of Equality: If $a = b$ and $b = c$, then $a = c$.

Substitution Property: If $a = b$, then a may be substituted for b in any expression without changing the value of that expression.

Transitive Property of Inequality: If $a < b$ and $b < c$, then $a < c$. If $a > b$ and $b > c$, then $a > c$.

Trichotomy Property: Exactly one of these three statements is true: $a < b$, $a = b$, or $a > b$.

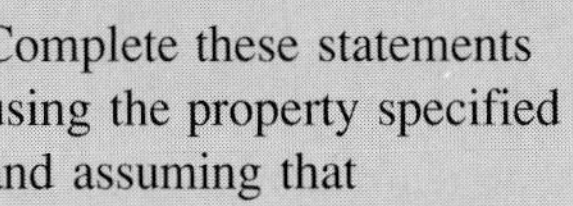

SELF-CHECK

Complete these statements using the property specified and assuming that $5x - 23 = 2x + 75$ and $2x + 75 = 9x - 33$.

1 $5x - 23 = ?$ Reflexive property

2 $5x - 23 = ?$ Transitive property

3 $9x - 33 = ?$ Symmetric property

The relationships $\leq$ (less than or equal to) and $\geq$ (greater than or equal to) are also transitive.

SELF-CHECK ANSWERS 

1 $5x - 23$ **2** $9x - 33$ **3** $2x + 75$

PROPERTIES OF THE REAL NUMBERS

For all real numbers a, b, and c:

Closure Property:	$a + b$ is a real number. $a - b$ is a real number. ab is a real number. For $b \neq 0$, $a \div b$ is a real number.	The real numbers are closed with respect to addition, subtraction, and multiplication, but not with respect to division.
Commutative Property:	$a + b = b + a$	Commutative property of addition
	$ab = ba$	Commutative property of multiplication The commutative properties allow a and b to be moved.
Associative Property:	$a + (b + c) = (a + b) + c$	Associative property of addition
	$a(bc) = (ab)c$	Associative property of multiplication The associative properties allow a, b, and c to be regrouped.
Distributive Property:	$a(b + c) = ab + ac$ $(b + c)a = ba + ca$	Multiplication distributes over addition. The distributive property involves two operations.

A Mathematical Note

The French mathematician Servois (c. 1814) introduced the terms *commutative* and *distributive*. The term *associative* is attributed to the Irish mathematician William R. Hamilton (1805–1865).

Note that the distributive property is the only property given above that involves two operations. It provides a choice of two different ways to write certain expressions that involve both multiplication and addition.

A Geometric Viewpoint: The Distributive Property

Adding the rectangle of ab square units to the rectangle of ac square units produces a rectangle of $a(b + c)$ square units.

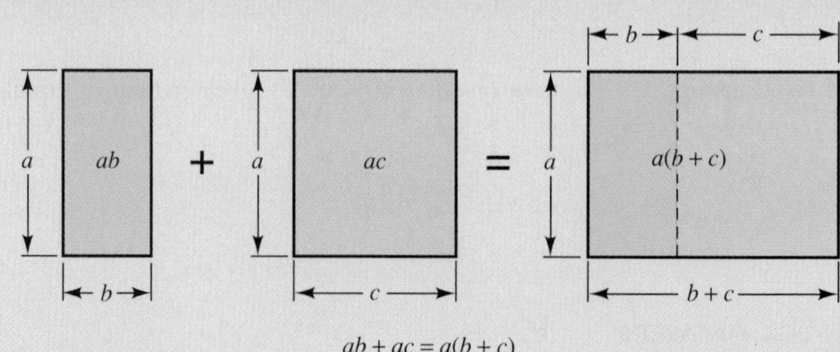

EXAMPLE 4 Identifying Properties of the Real Numbers

Identify the property that justifies each statement.

	SOLUTIONS
(a) $x(y + z) = x(z + y)$	Commutative property of addition
(b) $x(y + z) = xy + xz$	Distributive property
(c) $x(y + z) = (y + z)x$	Commutative property of multiplication
(d) $(2x + 3y) + 4z = 2x + (3y + 4z)$	Associative property of addition
(e) $5xy + 15xz = 5x(y + 3z)$	Distributive property

Use $ab + ac = a(b + c)$ with $a = 5x$, $b = y$, and $c = 3z$.

▼

One reason that the set of real numbers is the set of numbers most frequently used in algebra is that the sum, the difference, and the product of two real numbers are always real numbers too. We say that *a set is closed with respect to an operation* if performing that operation on its elements always yields a unique answer in that set. Thus the set of real numbers is closed with respect to addition, subtraction, and multiplication. The real numbers are not closed with respect to division, since division by zero is undefined.

If $\frac{5}{0}$ is defined to be some number x, then it must be true that $5 = 0 \cdot x$. However, it is impossible for $0 \cdot x$ to equal 5 for any value of x, so $\frac{5}{0}$ is undefined. The expression $\frac{0}{0}$ is called indeterminate, since any real number x will make $0 = 0 \cdot x$ true. Thus there is no reason to select or determine one value of x as preferable to any of the other values of x. The indeterminate expression $\frac{0}{0}$ is also undefined, since there is not a unique value to assign to $\frac{0}{0}$. Therefore we say, in general, that division by zero is undefined.

SELF-CHECK 

Complete these statements using the property specified.

1 $2x(3y + 4z) = 2x(?)$ Commutative property of addition
2 $2x(3y + 4z) = (?)(2x)$ Commutative property of multiplication
3 $2x(3y + 4z) = (2x)(3y) + ?$ Distributive property
4 $2x + (3y + 4z) = (?) + 4z$ Associative property of addition

DIVISION BY ZERO IS UNDEFINED

$\frac{x}{0}$ is undefined for all real numbers x.

$\frac{0}{x} = 0$ for $x \neq 0$.

SELF-CHECK ANSWERS 

1 $4z + 3y$ 2 $3y + 4z$ 3 $(2x)(4z)$ 4 $2x + 3y$

IDENTITIES AND INVERSES

Additive Identity: 0 — 0 is the only real number for which $a + 0 = a$ for every real number a.

Multiplicative Identity: 1 — 1 is the only real number for which $1 \cdot a = a$ for every real number a.

Additive Inverse: $-a$ — For each real number a there is an **opposite,** $-a$, such that $a + (-a) = 0$.

Multiplicative Inverse: $\frac{1}{a}$ — For each real number $a \neq 0$ there is a **reciprocal,** $\frac{1}{a}$, such that $a\left(\frac{1}{a}\right) = 1$.

EXAMPLE 5 Identifying Additive and Multiplicative Inverses

Give the additive inverse and the multiplicative inverse of each of these numbers. State any restriction on x.

SOLUTIONS

	Additive Inverse	*Multiplicative Inverse*
(a) 3	-3	$\frac{1}{3}$
(b) -3	$+3$	$-\frac{1}{3}$
(c) $-\frac{\pi}{7}$	$+\frac{\pi}{7}$	$-\frac{7}{\pi}$
(d) 1	-1	1
(e) $x - 3$	$-(x - 3)$ or $-x + 3$	$\frac{1}{x-3}$ for $x \neq 3$

The additive inverse of a nonzero number is formed by changing the sign of the number. The multiplicative inverse of a nonzero number is formed by taking the reciprocal of the number.

The opposite of $-a$ is a. In symbols $-(-a) = a$.

The following result, which stems directly from the meaning of a multiplicative inverse, is used so frequently that we have placed it in a box for special emphasis.

THE MULTIPLICATIVE IDENTITY

If A is an algebraic expression and $A \neq 0$, then $\frac{A}{A}$ is the multiplicative identity 1.

Caution: Remember that division by zero is undefined.

Each of the following expressions equals 1:

$$\frac{7}{7}, \quad \frac{-4}{-4}, \quad \frac{\pi}{\pi}, \quad \text{and} \quad \frac{x-7}{x-7} \text{ for } x \neq 7$$

The properties we have listed so far can be used to establish many other results. It is important to prove mathematical statements that are used frequently, so that we can use them with confidence, knowing that our answers will be correct. Important statements that can be proven are called **theorems.** We shall show one of these proofs to illustrate how they are done; however, the main emphasis of this text is on using well-established results—not on proofs or theory.

ZERO-FACTOR THEOREM

If a is any real number, then $a \cdot 0 = 0$.

Proof

$0 = a + (-a)$	Additive inverses
$= a \cdot 1 + (-a)$	Multiplicative identity, 1
$= a \cdot (0 + 1) + (-a)$	Additive identity, 0
$= (a \cdot 0 + a \cdot 1) + (-a)$	Distributive property
$= (a \cdot 0 + a) + (-a)$	Multiplicative identity, 1
$= a \cdot 0 + [a + (-a)]$	Associative property of addition
$= a \cdot 0 + 0$	Additive inverses
$= a \cdot 0$	Additive identity, 0
$0 = a \cdot 0$	Transitive property of equality

The following theorems are frequently used as tools in solving equations.

ADDITION THEOREM OF EQUALITY

If a, b, and c are real numbers and $a = b$, then $a + c = b + c$.

MULTIPLICATION THEOREM OF EQUALITY

If a, b, and c are real numbers and $a = b$, then $ac = bc$.

Exercises 1-1

A

1 List the elements of $S = \left\{-7, -\sqrt{7}, 0, \frac{3}{7}, \sqrt{7}, 4.\overline{7}, 7\right\}$ that are elements of the following sets.

a. Natural numbers **b.** Whole numbers
c. Integers **d.** Rational numbers
e. Irrational numbers **f.** Real numbers

2 List the elements of $S = \left\{-5, -\frac{4}{5}, 0, 0.\overline{5}, \sqrt{5}, 5, 5\pi\right\}$ that are elements of the following sets.

a. Natural numbers **b.** Whole numbers
c. Integers **d.** Rational numbers
e. Irrational numbers **f.** Real numbers

In Exercises 3–6 list the numbers that satisfy the given conditions. If necessary, use a calculator to approximate the square roots.

3 a. The even integers between -5 and 9
b. The odd integers between -8 and 10
c. The prime numbers between 30 and 40
d. The rational number halfway between 1.54 and 1.55

4 a. The even natural numbers between -5 and 9
b. The odd natural numbers between -8 and 10
c. The prime numbers between 20 and 30
d. The rational number halfway between $\frac{2}{7}$ and $\frac{4}{7}$

5 a. A rational number between $\sqrt{5}$ and $\sqrt{6}$

b. All integers n for which $|n| < 3$

c. All whole numbers m for which $|m| \leq 6$

d. All composite numbers between 11 and 19

6 a. A rational number between $\sqrt{10}$ and $\sqrt{11}$

b. All integers n for which $|n| = 3$

c. All natural numbers m for which $|m| < 5$

d. All composite numbers between 23 and 31

In Exercises 7 and 8 give the additive inverse and the multiplicative inverse of each number. State any restriction on x.

7 a. 9 **b.** -9 **c.** $\frac{-3}{7}$ **d.** $x + 4$

8 a. 5 **b.** -5 **c.** $\frac{4}{9}$ **d.** $x - 2$

In Exercises 9–12 simplify each expression as completely as possible.

9 a. $|-8|$ **b.** $|8|$

c. $|8 - 8|$ **d.** $|8| + |-8|$

10 a. $|5|$ **b.** $|-5|$

c. $|5 - 5|$ **d.** $|5| + |-5|$

11 a. $-|2|$ **b.** $|-2|$

c. $-|-2|$ **d.** $|2| + |-2|$

12 a. $-|7|$ **b.** $|-7|$

c. $-|-7|$ **d.** $|-7| + |7|$

In Exercises 13–22 complete the statement of the given property by replacing the question mark with the correct value. Assume that all variables used represent real numbers.

13 a. $3(x + y) = (x + y) \cdot ?$
Commutative property of multiplication

b. $3(x + y) = 3(y + ?)$
Commutative property of addition

c. $3(x + y) = ?$
Reflexive property

14 a. $r + (s + t) = ?$
Associative property of addition

b. $r + (s + t) = (s + t) + ?$
Commutative property of addition

c. $r + (s + t) = r + ?$
Commutative property of addition

15 a. $[r + (s + t)] + 0 = ?$
Additive identity

b. $[r + (s + t)] = [r + (s + t)][?]$
Multiplicative identity

c. $-x + x = ?$
Additive inverse

16 a. $mp + mn = mp + ?$
Commutative property of multiplication

b. $mp + mn = mn + ?$
Commutative property of addition

c. $mp + mn = ?$
Distributive property

17 a. $(tv)w = (?)w$
Commutative property of multiplication

b. $(tv)w = t(?)$
Associative property of multiplication

c. For $y \neq 0$, $y\left(\frac{1}{y}\right) = ?$
Multiplicative inverse

18 If $a > x$ and $x > z$, then ?
Transitive property of inequality

19 $n(r + s) = ? + ns$
Distributive property

20 $18xy + 9xw = 9x(2y + ?)$
Distributive property

21 $7ab + 7ac = (?)(b + c)$
Distributive property

22 $x < \pi$, $x = \pi$, or ?
Trichotomy property

In Exercises 23–46 identify the property that justifies each statement.

23 $1 + \sqrt{2} = \sqrt{2} + 1$

24 $-\sqrt{2} + \sqrt{2} = 0$

25 $\sqrt{2} \cdot \frac{1}{\sqrt{2}} = 1$

26 $1 \cdot \pi = \pi$

27 $\pi + 0 = \pi$

28 $e\pi = \pi e$

29 $\pi(x + y) = \pi(y + x)$

30 $(m + n)\pi = \pi(m + n)$

31 $\pi(x + y) = \pi x + \pi y$

32 $(m + n)\pi = m\pi + n\pi$

33 $(x + 2y) + z = x + (2y + z)$

34 $(x + 2y) + z = (2y + x) + z$

35 $3(\sqrt{2}x) = (3\sqrt{2})x$

36 $(x + y) + z = x + (y + z)$

37 $3(\sqrt{2}x) = (\sqrt{2}x)3$

38 $(x + y) + z = z + (x + y)$

39 $(x + y)(z + w) = (z + w)(x + y)$

40 $(x + y)(z + w) = (y + x)(z + w)$

41 $(x + y)(z + w) = (x + y)z + (x + y)w$

42 $(x + y)(z + w) = x(z + w) + y(z + w)$

43 If $(x + y) = 0$, then $(x + y)(z + w) = 0$.

44 If $x = 5$ and $xy - x = 7$, then $5y - 5 = 7$.

45 If $x = y$, then $x + 5 = y + 5$.

46 If $x = 2$, then $3x = 6$.

B

47 What whole number is not a natural number?

48 Which of the following integers is a natural number: $-18, -6, 0, 7$?

49 Which of the following integers is not a whole number: $-7, 0, 7, 17, 71$?

50 Which of the following integers is not a natural number: $-5, 5, 51, 517$?

51 Which of the following numbers is a natural number: $\sqrt{2}, \sqrt{7}, \sqrt{9}, \sqrt{11}$?

52 Which of the following numbers is irrational: $\sqrt{4}, \sqrt{5}, \sqrt{9}, \sqrt{16}$?

53 What real number is neither negative nor positive?

54 What prime number is even?

55 What natural number is neither prime nor composite?

56 What real number has an absolute value that is not positive?

57 Which of the following numbers is composite: 37, 39, 41, 43?

58 Which of the following numbers is prime: 21, 27, 31, 35?

59 Which of the following is a repeating decimal: $\frac{3}{4}, 3.4, 4.33, 3.\overline{4}$?

60 Which of the following is a terminating decimal: $7.6, 6.\overline{7}, 5.6\overline{78}, 7.666\ldots$?

61 What is the one operation that is not defined for all real numbers? Specifically, what is not defined?

62 What property guarantees that there is always some number that can be added to a number to produce a sum of 0?

63 What number equals its additive inverse?

64 What number equals its multiplicative inverse?

65 Does every real number have an additive inverse?

66 Does every real number have a multiplicative inverse?

67 What property says that any real number equals itself?

68 What property says that you can add two real numbers in either order?

69 What property says that you can multiply two real numbers in either order?

70 What property says that if one number equals a second number, then the second number equals the first number?

71 What property says that if you multiply any two real numbers, you are guaranteed to obtain an answer that is a real number?

72 What property says that you can always replace a mathematical expression by the number it equals?

73 What property says that if you see a more convenient way to group numbers before adding them, then it is okay to do so?

74 What property says that if you see a more convenient way to group numbers before multiplying them, then it is okay to do so?

C

In Exercises 75–80 write an algebraic expression for each statement.

75 v is less than or equal to w.

76 The opposite of y is less than negative two.

77 m is not equal to thirty-three.

78 Five is greater than or equal to the opposite of x.

79 The absolute value of seventeen is equal to the absolute value of negative seventeen.

80 The opposite of negative eleven equals eleven.

81 List all real numbers that equal their opposite.

82 Give a number x for which $-x$ is not a negative number.

83 Explain why the statement "The absolute value of every real number is positive" is false.

84 Explain why 1.212112111211112 . . . is an irrational number.

85 A prime number has exactly ________ distinct natural numbers as factors. The number 1 is not prime because it has only ________ factor(s).

86 Fill in the inequality symbol that completes this statement: $|-8|$ ______ $|5|$.

87 **a.** Give a number for which $|x - 3| = x - 3$.

b. Give a number for which $|x - 3| \neq x - 3$.

88 Reorder $(57 + 76 + 32 + 14 + 28 + 23)$ as $(57 + 23) + (76 + 14) + (32 + 28)$, and perform this computation mentally.

89 Use the fact that $(1 + 2 + 3 + 4 + \cdots + 97 + 98 + 99 + 100)$ equals $[(1 + 100) + (2 + 99) + (3 + 98) + \cdots + (50 + 51)]$ to compute this sum mentally.

90 Is the set of odd integers closed with respect to multiplication? If not, give an example showing that it is not closed.

91 $(1 + \sqrt{2}) + (1 - \sqrt{2}) = 2$ is an example showing that the set of ________ is not closed with respect to addition.

92 What other elements would have to be included with the whole numbers for the resulting set to be closed with respect to subtraction?

93 Give the property that justifies each step of the proof that $(a + b) + (-b) = a$.

Proof $a = a$ ________________

$a + 0 = a$ ________________

$a + [b + (-b)] = a$ ________________

$(a + b) + (-b) = a$ ________________

DISCUSSION QUESTIONS (94–97)

94 A shopper places several items in a shopping cart. Will the total price be affected in any way by the order in which these items are entered into the cash register at the checkout counter? Use the appropriate properties of the real numbers to explain your answer.

95 Both the [−] key and the [+/−] key occur on many calculators. Describe the different purposes of these keys.

96 Write a paragraph explaining why $\frac{0}{5} = 0$, but $\frac{5}{0}$ is undefined.

97 Explain why a calculator *cannot* represent exactly an irrational number. What does this mean to you in terms of the accuracy of your calculator computations?

CHALLENGE QUESTION

a. Determine the largest number your calculator can represent without using scientific notation. (Scientific notation will be discussed in Section 4-2.) What happens when you add 1 to this number?

b. Can your calculator represent all rational numbers exactly? Explain why or why not.

c. Is it possible that your calculator can represent exactly both of the real numbers a and b but not be able to represent $a + b$?

d. Is the set of numbers that your calculator can represent closed with respect to addition? Explain why or why not.

1-2 Operations on Real Numbers

SECTION OBJECTIVES

4 **Add, subtract, multiply, and divide real numbers.**

5 **Use natural number exponents.**

6 **Use the correct order of operations.**

This section reviews the addition, subtraction, multiplication, and division of real numbers. We will also review the use of natural number exponents and the order of operations.

The result of adding the real numbers a and b is called their **sum.** In the sum $a + b$ both a and b are called **addends,** or **terms,** of the sum. Before you examine the following rules, note that each nonzero real number has two attributes: its sign (positive or negative) and its absolute value (distance from the origin). Both of these attributes are used when real numbers are added.

Like Signs: Add the absolute values of the numbers, and use their common sign for the sum.

Unlike Signs: Subtract the smaller absolute value from the larger absolute value, and use the sign of the number having the larger absolute value for the sum.

Since the commutative and associative properties permit us to regroup terms and add them in any order, it is sometimes more convenient to regroup terms with like signs together than to add from left to right. This regrouping can be done mentally (you may find it helpful to underline terms with like signs), or the problem can be rewritten as illustrated in part (c) of Example 1.

EXAMPLE 1 Adding Integers

Calculate each sum.

SOLUTIONS

(a) $(-7) + (-9)$

$(-7) + (-9) = -(7 + 9)$
$= -16$

The terms have like signs, so we add the absolute values. The sum is also negative.

(b) $(+7) + (-9)$

$(+7) + (-9) = -(9 - 7)$
$= -2$

The terms have unlike signs, so we subtract the smaller absolute value from the larger absolute value. The sum is negative because the sign of the number with the larger absolute value is negative.

(c) $3 + (-4) + 5 + (-6) + 7$

$3 + (-4) + 5 + (-6) + 7$
$= (3 + 5 + 7) + [(-4) + (-6)]$ Regroup terms with like signs together.
$= 15 + (-10)$ The sign of the number with the larger absolute value is positive, so the sum is positive.
$= 5$

The result of subtracting the real number b from the real number a is called the **difference** of a and b. The subtraction $a - b$ can be defined as $a + (-b)$ with terms a and $-b$. That is, to subtract b from a, we add the opposite of b to a. Thus we can subtract real numbers by transforming a subtraction problem into its corresponding addition problem.

SELF-CHECK 

Calculate each sum.
1 $19 + 17$
2 $19 + (-17)$
3 $-19 + (-17)$
4 $-19 + 17$

RULE FOR SUBTRACTING REAL NUMBERS

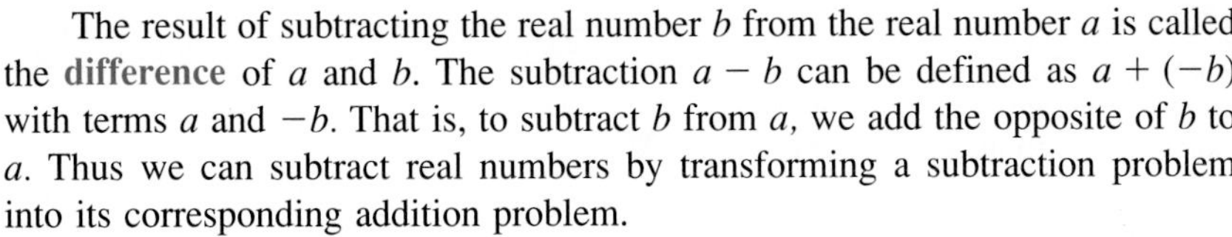
$a - b = a + (-b)$ To subtract b from a, add the opposite of b to a.

The process of rewriting $a - b$ as $a + (-b)$ is shown in Example 2. When working your problems, you should do this step mentally.

EXAMPLE 2 Subtracting Integers

Calculate each difference.

SOLUTIONS

(a) $(+9) - (+7)$

$(+9) - (+7) = 9 + (-7)$
$= 2$

SELF-CHECK ANSWERS 

1 36 **2** 2 **3** −36 **4** −2

(b) $(-9) - (-7)$ $\quad$ $(-9) - (-7) = -9 + (+7)$

$$= -2$$

(c) $(-9) - (+7)$ $\quad$ $(-9) - (+7) = -9 + (-7)$

$$= -16$$

(d) $(+9) - (-7)$ $\quad$ $(+9) - (-7) = 9 + (+7)$

$$= 16$$

EXAMPLE 3 An Application of Subtraction

The following table gives Celsius temperatures at 8 A.M. and 8 P.M. for four consecutive days in February, as recorded in Champaign, Illinois. Calculate the change from the 8 A.M. temperature to the 8 P.M. temperature.

	Monday	Tuesday	Wednesday	Thursday
8 P.M.	$+5°$	$+5°$	$-5°$	$-5°$
8 A.M.	$+2°$	$-2°$	$-2°$	$+2°$

Subtract the 8 A.M. temperatures from the 8 P.M. temperatures.

SOLUTION

	Monday	Tuesday	Wednesday	Thursday
Difference	$+3°$	$+7°$	$-3°$	$-7°$

A positive difference indicates a rise in temperature, and a negative difference indicates a drop in temperature.

The symbol $-$ is used for three distinct but related purposes. Noting these distinctions and using the symbol correctly will improve your understanding of the algebra topics throughout this book.

THREE USES OF THE $-$ SYMBOL

Negative: $-$ is used to indicate that a number to the left of the origin on the number line is negative.

Opposite: $-$ is used to indicate the opposite of a real number—that is, a number the same distance from the origin on the number line but on the opposite side of the origin.

Minus: $-$ is used to indicate the operation of subtraction.

The expression below shows all three uses of the $-$ symbol:

8 minus the opposite of negative 5

$$8 - [\quad - \quad (-5)]$$

Be extremely careful when using the $-$ symbol in conjunction with absolute-value symbols.

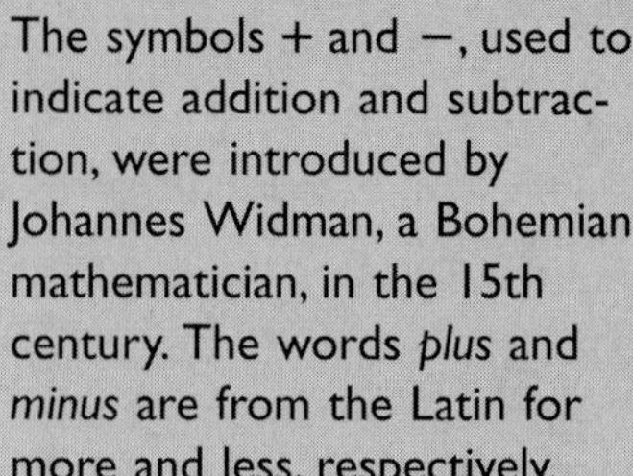

A Mathematical Note

The symbols + and $-$, used to indicate addition and subtraction, were introduced by Johannes Widman, a Bohemian mathematician, in the 15th century. The words *plus* and *minus* are from the Latin for more and less, respectively.

The symbol × for multiplication was introduced by William Oughtred (1574–1660), an English mathematician.

The symbol ÷ for division is an imitation of fractional division; the dots indicate, respectively, the numerator and the denominator of a fraction. This symbol was invented by an English mathematician named John Pell (1611–1685).

EXAMPLE 4 **Subtraction of Absolute-Value Expressions**

Calculate the value of $|-9| - |-7|$.

SOLUTION $|-9| - |-7| = 9 - 7$ First evaluate each absolute value.

$= 2$ Then perform the subtraction. ▼

Remember that in order to add or subtract fractions, you should express them in terms of their least common denominator (LCD). The result should be reduced to lowest terms. If the least common denominator of two fractions is not obvious from inspection, it can be determined by factoring the denominators of both fractions. This process is illustrated in part (b) of Example 5.

SELF-CHECK 

Calculate each difference.

1 $29 - 13$
2 $29 - (-13)$
3 $-29 - 13$
4 $-29 - (-13)$

ADDITION AND SUBTRACTION OF FRACTIONS

If a, b, and c are real numbers and $c \neq 0$, then

$$\frac{a}{c} + \frac{b}{c} = \frac{a+b}{c} \quad \text{and} \quad \frac{a}{c} - \frac{b}{c} = \frac{a-b}{c}$$

EXAMPLE 5 **Adding and Subtracting Fractions**

Calculate the value of each expression.

SOLUTIONS

(a) $\frac{5}{24} - \frac{7}{24}$

$\frac{5}{24} - \frac{7}{24} = \frac{5-7}{24}$ Subtract the numerators.

$= \frac{-2}{24}$

$= \frac{-1 \cdot \overset{1}{\cancel{2}}}{12 \cdot \underset{1}{\cancel{2}}}$ Then reduce the fraction to lowest terms.

$= -\frac{1}{12}$

(b) $\frac{17}{30} + \frac{5}{12}$

$\frac{17}{30} + \frac{5}{12} = \frac{17 \cdot 2}{30 \cdot 2} + \frac{5 \cdot 5}{12 \cdot 5}$ $30 = 2 \cdot 3 \cdot 5$; $12 = 2 \cdot 2 \cdot 3$; $\text{LCD} = 2 \cdot 2 \cdot 3 \cdot 5 = 60$

$= \frac{34}{60} + \frac{25}{60}$ Express each fraction in terms of the LCD, 60.

$= \frac{34 + 25}{60}$ Then add the numerators.

$= \frac{59}{60}$ ▼

SELF-CHECK ANSWERS 

1 16 **2** 42 **3** −42 **4** −16

Note that the answer to Example 5(a) can be written as either $\frac{-1}{12}$ or $-\frac{1}{12}$. The relationship of the sign of the numerator, the sign of the denominator, and the sign of the fraction is shown in the following box.

THREE SIGNS OF A FRACTION

$$-\frac{a}{b} = \frac{-a}{b} = \frac{a}{-b} = -\frac{-a}{-b} \quad \text{and} \quad \frac{a}{b} = \frac{-a}{-b} = -\frac{-a}{b} = -\frac{a}{-b}$$

Remember: Each fraction has three signs associated with it, and any two of these signs can be changed without changing the value of the fraction.

Fractional answers generally are written so that they do not contain more than one negative sign and so that the denominator is positive. For example, $-\frac{8}{-11}$ is written as $\frac{8}{11}$, and $-\frac{-8}{-11}$ is written as $-\frac{8}{11}$.

The result of multiplying the real numbers a and b is called the **product** of a and b. In the product ab both a and b are called **factors** of the product. The rule for multiplication treats the cases for like signs and unlike signs separately.

MULTIPLYING TWO REAL NUMBERS

Like Signs: Multiply the absolute values of the numbers, and use a positive sign for the product.

Unlike Signs: Multiply the absolute values of the numbers, and use a negative sign for the product.

Zero Factor: The product of 0 and any other factor is 0.

EXAMPLE 6 Multiplying Integers

Calculate each product.

	SOLUTIONS	
(a) $(+9)(+8)$	$(+9)(+8) = 72$	Like signs, so the product is positive.
(b) $(-9)(-8)$	$(-9)(-8) = 72$	Like signs, so the product is positive.
(c) $(+9)(-8)$	$(+9)(-8) = -72$	Unlike signs, so the product is negative.
(d) $(-9)(+8)$	$(-9)(+8) = -72$	Unlike signs, so the product is negative.

The product of several nonzero factors is either positive or negative, depending on whether the number of negative factors is even or odd. Thus it may be most efficient to determine the sign of the product first and then multiply the absolute values of the factors.

PRODUCT OF NEGATIVE FACTORS

The product is positive if the number of negative factors is even.
The product is negative if the number of negative factors is odd.

EXAMPLE 7 Multiplying Negative Factors

Calculate each product.

	SOLUTIONS	
(a) $(-1)(-2)(+3)(-4)(-5)$	$(-1)(-2)(+3)(-4)(-5) = +120$	The product is positive, since the number of negative factors is even (four).
(b) $(-7)(+8)(0)(-9)(-5)$	$(-7)(+8)(0)(-9)(-5) = 0$	Because one factor is 0, the product is 0 by the Zero-Factor Theorem.
(c) $(-1)(+2)(-3)(-4)(-5)(-6)$	$(-1)(+2)(-3)(-4)(-5)(-6) = -720$	The product is negative, since the number of negative factors is odd (five).

SELF-CHECK

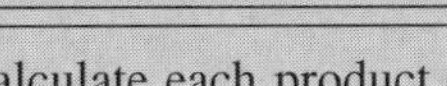

Calculate each product.

1 $19 \cdot 3$ **2** $-19 \cdot 3$
3 $19(-3)$ **4** $-19(-3)$

The result of dividing the real number a by the real number b, $b \neq 0$, is called the **quotient** of a and b. In the division $a \div b$ the **divisor** is b and the **dividend** is a. The division $a \div b$, $b \neq 0$, can be defined as $a \cdot \left(\frac{1}{b}\right)$, or $\frac{a}{b}$. That is, to divide a by b, $b \neq 0$, we can multiply a by the reciprocal of b. Thus the rule for the sign of the quotient in a division problem is the same as the rule for the sign of the product in a multiplication problem.

DIVIDING TWO REAL NUMBERS

Like Signs: Divide the absolute values of the numbers, and use a positive sign for the quotient.

Unlike Signs: Divide the absolute values of the numbers, and use a negative sign for the quotient.

Zero Dividend: $\frac{0}{x} = 0$ for $x \neq 0$.

Zero Divisor: $\frac{x}{0}$ is undefined for every real number x.

EXAMPLE 8 Dividing Integers

Calculate each quotient.

	SOLUTIONS	
(a) $(+36) \div (+9)$	$(+36) \div (+9) = 4$	Like signs, so the quotient is positive.
(b) $(-36) \div (-9)$	$(-36) \div (-9) = 4$	Like signs, so the quotient is positive.
(c) $(+36) \div (-9)$	$(+36) \div (-9) = -4$	Unlike signs, so the quotient is negative.
(d) $(-36) \div (+9)$	$(-36) \div (+9) = -4$	Unlike signs, so the quotient is negative.

SELF-CHECK ANSWERS 

1 57 **2** -57 **3** -57 **4** 57

The rules for multiplying and dividing fractions are given below. Before actually multiplying the factors in the numerator and those in the denominator, remember to examine the numerator and the denominator for a common factor. Divide the numerator and the denominator by any common factor to reduce the result, since $\frac{f}{f} = 1$ for any common factor $f \neq 0$.

SELF-CHECK ✓

Calculate each quotient.

1 $72 \div 12$ **2** $-72 \div 12$
3 $72 \div (-12)$
4 $-72 \div (-12)$

MULTIPLICATION AND DIVISION OF FRACTIONS

If a, b, c, and d are real numbers, $b \neq 0$ and $d \neq 0$, then

$$\frac{a}{b} \cdot \frac{c}{d} = \frac{a \cdot c}{b \cdot d}$$

and if $c \neq 0$,

$$\frac{a}{b} \div \frac{c}{d} = \frac{a}{b} \cdot \frac{d}{c} = \frac{a \cdot d}{b \cdot c}$$

EXAMPLE 9 **Multiplying and Dividing Fractions**

Calculate the value of each expression.

SOLUTIONS

(a) $\left(-\frac{2}{3}\right)\left(\frac{-15}{22}\right)$

$$\left(-\frac{2}{3}\right)\left(\frac{-15}{22}\right) = +\left(\frac{\not{2} \cdot \not{3} \cdot 5}{\not{3} \cdot \not{2} \cdot 11}\right)$$
$$= \frac{5}{11}$$

The product is positive because the number of negative factors is even (two). Reduce by dividing out the common factors.

(b) $-\left(\frac{-6}{7}\right) \div \left(\frac{-9}{14}\right)$

$$-\left(\frac{-6}{7}\right) \div \left(\frac{-9}{14}\right) = -\left(\frac{6}{7}\right)\left(\frac{14}{9}\right)$$

The product is negative because the number of negative factors is odd (three).

$$= -\frac{2 \cdot \not{3} \cdot 2 \cdot \not{7}}{\not{7} \cdot \not{3} \cdot 3}$$

Reduce by dividing out the common factors.

$$= -\frac{4}{3}$$

▼

EXAMPLE 10 **Estimation Skills**

Mentally estimate the value of each of the following expressions.

SOLUTIONS

	Estimated Answer		*Actual Answer*
(a) 12.75(3.1)	$12.75 \approx 13$ $3.1 \approx 3$	$13 \times 3 = 39$	12.75×3.1: 1275 3825 39.525

SELF-CHECK ANSWERS 

1 6 **2** −6 **3** −6 **4** 6

(b) $463.5 \div 1.03$	$463.5 \approx 460$ $1.03 \approx 1$	$460 \div 1 = 460$	$1.03_\wedge \overline{)463.50_\wedge}$ = 450. 412 515 515 00
(c) $843.17 + 108$	$843.17 \approx 840$ $108 \approx 110$	$840 + 110 = 950$	$843.17 + 108.00 = 951.17$
(d) $489.23 - 57.069$	$489.23 \approx 490$ $57.069 \approx 60$	$490 - 60 = 430$	$489.230 - 57.069 = 432.161$

Repeated multiplication by the same factor is usually denoted by exponential notation. For example, in

$$b^5 = b \cdot b \cdot b \cdot b \cdot b$$

the exponent is 5 and the base is b. In this chapter we will examine only exponents that are natural numbers.

Exponential Notation

For any real number b and natural number n,

$$b^n = \underbrace{b \cdot b \cdot \cdots \cdot b}_{n \text{ factors}} \text{ with base } b \text{ and exponent } n$$

b^n is read as "b to the nth power."

EXAMPLE 11 Evaluating Exponential Expressions

Calculate the value of each of these exponential expressions.

SOLUTIONS

(a) $(-3)^4$

$$(-3)^4 = (-3)(-3)(-3)(-3) = 81$$

The product is positive, since the number of negative factors is even (four).

(b) $\left(\frac{-2}{5}\right)^3$

$$\left(\frac{-2}{5}\right)^3 = \left(\frac{-2}{5}\right)\left(\frac{-2}{5}\right)\left(\frac{-2}{5}\right) = -\frac{8}{125}$$

The product is negative, since the number of negative factors is odd (three).

(c) -5^2

$$-5^2 = -(5 \cdot 5) = -25$$

Caution: The base is 5, not −5. To denote a base of −5, use parentheses, as in part (d).

(d) $(-5)^2$

$$(-5)^2 = (-5)(-5) = 25$$

The **area** of a rectangle is determined by multiplying its length times its width ($A = l \cdot w$). The area of the square with side s is $A = s \cdot s = s^2$. The volume of a rectangular box is $V = l \cdot w \cdot h$. Thus the volume of the cube with side s is $V = s^3$.

A Geometric Viewpoint: Terminology for Exponents

x^2, which can be read as "x to the second power," is usually read as "x squared" because this is the area of a square with sides of length x.

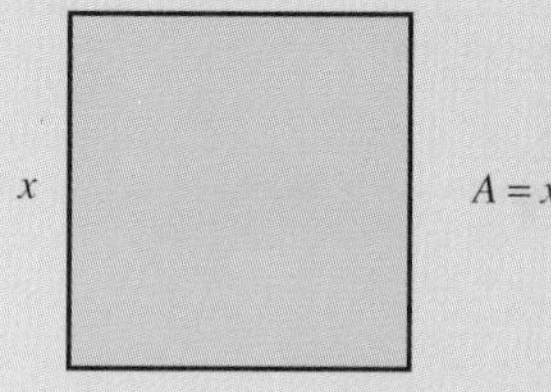

x^3, which can be read as "x to the third power," is usually read as "x cubed" because this is the volume of a cube with each side of length x.

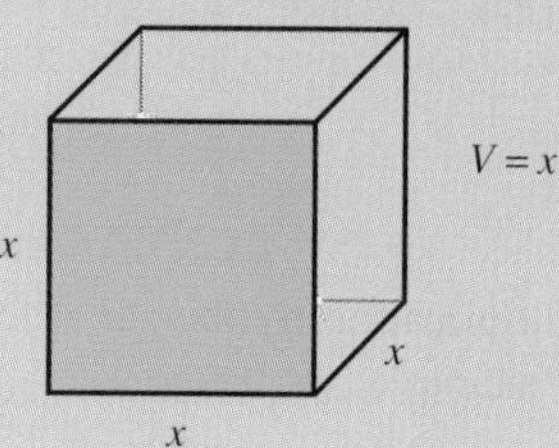

EXAMPLE 12 Determining the Volume of a Cube

Determine the volume of a cube that has a length, width, and height of 5 centimeters.

SOLUTION

Volume = length · width · height

$V = (5 \text{ cm})(5 \text{ cm})(5 \text{ cm})$

$V = (5 \text{ cm})^3$

$V = 125 \text{ cm}^3$ The notation cm^3 denotes a cubic centimeter.

The volume is 125 cm^3. ▼

Many common formulas and algebraic expressions require more than one operation. Consider the expression $2 + 6 \cdot 3$ (see Figure 1-9). Since $(2 + 6) \cdot 3$ yields 24, whereas $2 + (6 \cdot 3)$ yields 20, it is crucial that we all agree on the same interpretation. Otherwise, we might obtain different values for the same expression. In mathematics the order in which operations are performed is universally agreed to be as follows.

Figure 1-9

ORDER OF OPERATIONS

Step 1 Start with the expression within the innermost grouping symbols.

Step 2 Perform all exponentiations.

Step 3 Perform all multiplications and divisions, working from left to right.

Step 4 Perform all additions and subtractions, working from left to right.

This order of operations has also been adopted by the manufacturers of most calculators and has been incorporated into many computer languages. You should verify this hierarchy on your calculator before undertaking any calculations.

The most common grouping symbols are parentheses ( ), brackets [ ], braces { }, and the fraction bar —. If an expression has a fraction bar, first simplify the numerator and the denominator separately (including all additions and subtractions), and then reduce the fraction to lowest terms.

EXAMPLE 13 Evaluating Expressions by Following the Correct Order of Operations

Calculate the value of each expression below.

SOLUTIONS

(a) $15 - 4(3)$

$15 - 4(3) = 15 - 12 = 3$

Multiplication takes priority over subtraction.

(b) $(3 - 5)^2 - (3^2 - 5^2)$

$$\begin{aligned}(3 - 5)^2 - (3^2 - 5^2) &= (-2)^2 - (9 - 25)\\ &= 4 - (-16)\\ &= 4 + 16\\ &= 20\end{aligned}$$

Simplify the expressions within the parentheses.

(c) $100 - 3 \cdot 5^2$

$$\begin{aligned}100 - 3 \cdot 5^2 &= 100 - 3(25)\\ &= 100 - 75\\ &= 25\end{aligned}$$

First square the 5. Then multiply before subtracting.

(d) $\dfrac{5 - 4 \cdot 3}{(5 + 4)(3)}$

$$\begin{aligned}\frac{5 - 4 \cdot 3}{(5 + 4)(3)} &= \frac{5 - 12}{9 \cdot 3}\\ &= \frac{-7}{27}\end{aligned}$$

Simplify the numerator and the denominator separately.

In mathematical applications a variety of words and phrases are used to indicate operations. Some of these words and phrases are listed on the next page in Table 1-1.

The list of phrases in Table 1-1 is certainly not complete. Some applications may include a particular vocabulary item that implies a specific operation. For example, the term **perimeter** is frequently used in geometry to indicate the distance around a polygon. The perimeter of a polygon can be calculated by finding the sum of the lengths of the sides.

SELF-CHECK 

Perform each of the indicated operations.

1 $7 - 5^2$ **2** $7^2 - 5^2$

3 $(7 - 5)^2$

4 $8 - 3(-5 + 2 \cdot 7)$

SELF-CHECK ANSWERS 

1 -18 **2** 24 **3** 4 **4** -19

TABLE 1-1 *Phrases Used to Indicate the Basic Operations*

Phrase	Algebraic Expression	Phrase	Algebraic Expression
Addition		**Division**	
The sum of x and y	$x + y$	x divided by y	$x \div y$, or $\frac{x}{y}$
x plus y	$x + y$	x divided into y	$y \div x$, or $\frac{y}{x}$
The total of x and y	$x + y$	The quotient of x and 2	$x \div 2$, or $\frac{x}{2}$
Five more than x	$x + 5$	The quotient of 2 and x	$2 \div x$, or $\frac{2}{x}$
x added to five	$5 + x$	The ratio of x to y	$x:y$, or $\frac{x}{y}$
Five added to x	$x + 5$	**Exponentiation**	
x increased by five	$x + 5$	x squared	x^2
Subtraction		x cubed	x^3
x minus seven	$x - 7$	x to the nth power	x^n
x take away seven	$x - 7$		
x decreased by seven	$x - 7$		
Seven less than x	$x - 7$		
Seven less x	$7 - x$		
Seven diminished by x	$7 - x$		
Seven subtracted from x	$x - 7$		
The difference between x and seven	$x - 7$		
Multiplication			
x times y	$x \cdot y$		
The product of x and y	$x \cdot y$		
Multiply x by 5	$5x$		
Double x	$2x$		
Twice x	$2x$		
Triple x	$3x$		
One-half of x	$\frac{1}{2}x$		

EXAMPLE 14 Perimeter of a Triangle

Determine the perimeter of the triangle shown.

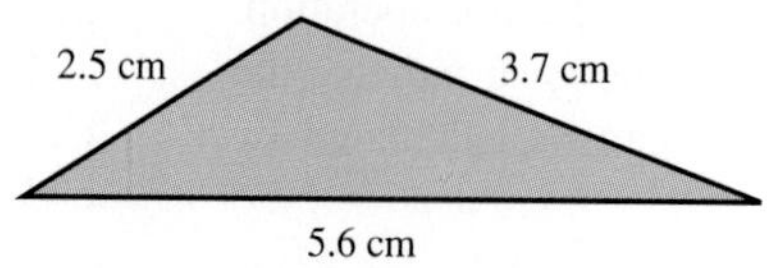

SOLUTION Perimeter = Sum of the lengths of the sides

$$P = 5.6 \text{ cm} + 2.5 \text{ cm} + 3.7 \text{ cm}$$
$$P = 11.8 \text{ cm}$$

It is often wise to condense problems to a **word equation** and then substitute algebraic symbols for these words.

The perimeter of the triangle is 11.8 cm.

Exercises 1-2

A

In Exercises 1–50 perform all the computations without using a calculator.

1 a. $8 + (+5)$ **b.** $-8 + (+5)$ **c.** $8 + (-5)$ **d.** $-8 + (-5)$

2 a. $-31 + (-45)$ **b.** $31 + (+45)$ **c.** $31 + (-45)$ **d.** $-31 + (+45)$

3 a. $-78 - (+173)$ **b.** $-78 + (+173)$ **c.** $78 - (-173)$ **d.** $-78 - (-173)$

4 a. $93 - (-81)$ **b.** $-93 - (-81)$ **c.** $-93 - (+81)$ **d.** $93 - (+81)$

5 a. $2(-3)$ **b.** $-2(3)$ **c.** $-2(-3)$ **d.** $(-3)^2$

6 a. $3(-4)$ **b.** $-3(4)$ **c.** $-3(-4)$ **d.** $(-4)^3$

7 a. $-12 \div 4$ **b.** $12 \div (-4)$ **c.** $-12 \div (-4)$ **d.** $-4 \div 12$

8 a. $15 \div (-3)$ **b.** $-15 \div (-3)$ **c.** $-15 \div 3$ **d.** $3 \div (-15)$

9 a. $-17 - (-19)$ **b.** $16 + 0 + (-12)$ **c.** $-73 + 0 - (+14)$ **d.** $-25 - (-25)$

10 a. $-1(-2)(-3)$ **b.** $-1(2)(-3)$ **c.** $-1(0)(-5)$ **d.** $-3(0)(7)$

11 a. $-2(3)(-5)$ **b.** $-2(-3)(-5)$ **c.** $2(0)(-4)$ **d.** $-2(0)(-4)$

12 a. -4^2 **b.** $(-4)^2$ **c.** $2^2 + 3^2$ **d.** $(2 + 3)^2$

13 a. -6^2 **b.** $(-6)^2$ **c.** $3^2 + 4^2$ **d.** $(3 + 4)^2$

14 a. $5^2 - 2^2$ **b.** $(5 - 2)^2$ **c.** 2^5 **d.** 3^4

15 a. $7^2 - 4^2$ **b.** $(7 - 4)^2$ **c.** 2^4 **d.** 4^3

16 a. 1^{15} **b.** $(-1)^{16}$ **c.** -1^{16} **d.** $(-1)^{17}$

17 a. 1^{20} **b.** $(-1)^{20}$ **c.** -1^{20} **d.** $(-1)^{21}$

18 a. -1^4 **b.** $(-1)^4$ **c.** -2^4 **d.** $(-2)^4$

19 a. -1^6 **b.** $(-1)^6$ **c.** -2^6 **d.** $(-2)^6$

20 a. $2 - 5(6)$ **b.** $(2 - 5)(6)$ **c.** $4 + 2(3)$ **d.** 0^5

21 a. $6 - 2(4)$ **b.** $(6 - 2)(4)$ **c.** $7 + 3(5)$ **d.** 0^6

22 a. $3(4)^2$ **b.** $(3 \cdot 4)^2$ **c.** $3(4)(2)$

23 a. $4(5)^2$ **b.** $(4 \cdot 5)^2$ **c.** $4(5)(2)$

24 a. $48 \div 8 \div 2$ **b.** $48 \div (8 \div 2)$ **c.** $48 \div 8 \cdot 2$

25 a. $24 \div 6 \div 2$ **b.** $24 \div (6 \div 2)$ **c.** $24 \div 6 \cdot 2$

26 a. $(4 + 12) \div (3 - 7)$ **b.** $4 + 12 \div 3 - 7$

27 a. $(8 - 12) \div 4 - 16$ **b.** $8 - 12 \div 4 - 16$

28 a. $10 + 5[3 - 2(3 - 5)]$ **b.** $(10 + 5)[3 - 2(3 - 5)]$

29 a. $-8 + 3[4 + 5(9 - 6)]$ **b.** $(-8 + 3)[4 + 5(9 - 6)]$

30 $-21 + 11 + (-9) - (+4) - (-14)$

31 $37 - (+73) - (+45) + (-15) + 0 - (-7)$

32 $-21 - (-18) - (+17) + (-12) + 32$

33 $-19 + (18 - 23) - (14 - 9 - 13)$

34 $24 - (6 - 8) + (9 - 11)$

35 $-21 - [-11 - (13 - 14) - 3]$

36 $-34 - [-7 + (8 - 11) - 15]$

37 $-6 + |-7| - |-8|$

38 $-15 - |-15| + |-4|$

39 $-|-19 - (-12)| - (|-14| - |-9|)$

40 $|-21| - |-11| - (|-24 - (-17)|)$

41 $-5 + 3(7 - 2) + 12 \div 3 \cdot 4 - 1$

42 $14 - 8(12 - 7) + 10 \div 5[3 - 2(3 - 5)]$

43 $11 - 5^2 \cdot 3 + 16 \div (-5 + 3)^3$

44 $90 - 5(7^2 - 51)^3 - 4(7 - 11)$

45 $6 - 5 + 8 \cdot 3 \div 2^2$

46 $12 - 36 \div 3^2 - 2^3 + 11$

47 $-4^2 + 3[(6 - 11) - 2(4 - 2 \cdot 3)]$

48 $25 - 2(9 - 4^2 \div 2 + 3 \cdot 7)$

49 $(-9)(-91)(0)(43)(-21)$

50 $(-1)^{115} + 115^1 - (1)^{114} - 114^1$

B

In Exercises 51–60 rewrite each phrase as either $a + b$, $a - b$, or $b - a$.

51 a plus b

52 a minus b

53 a units less b units

54 The sum of a and b

55 a units less than b

56 a increased by b

57 b units more than a

58 a decreased by b

59 Take a away from b

60 The total of a and b

In Exercises 61–70 write an algebraic expression for each verbal expression. Do not simplify these expressions.

61 Negative three

62 The opposite of x

63 Three minus x

64 Negative three minus the opposite of x

65 Negative four minus the opposite of y

66 The square of x

67 Opposite the square of x

68 The square of the opposite of x

69 x cubed

70 x minus the quantity y minus z

In Exercises 71 and 72 write each of the expressions in words. In each case express the − symbol as either "negative," "opposite," or "minus."

71 a. -8 **b.** $-x$ **c.** $x - 8$ **d.** $-(x - 8)$

72 a. -7 **b.** $-m$ **c.** $m - 7$ **d.** $-(m - 7)$

Estimation Skills (73–81)

In Exercises 73–76 mentally estimate the value of each expression, and select the answer that is closest to your estimate.

73 $17.4893 + 22.50134$

a. 38.5 **b.** 39.0 **c.** 39.5 **d.** 40.0

74 $175.841 - 34.1027$

a. 141.7 **b.** 140.7 **c.** 140.5 **d.** 141.5

75 $\dfrac{36.0143}{-2.9987}$

a. -27 **b.** -12 **c.** 12 **d.** 27

76 $(36.0143)(-2.9987)$

a. -97 **b.** -98 **c.** -108 **d.** 108

In Exercises 77–79 mentally estimate the perimeter of each figure. Then select the answer that is closest to your mental estimate.

77 a. 30 cm **b.** 3 cm **c.** 300 cm **d.** 28 cm

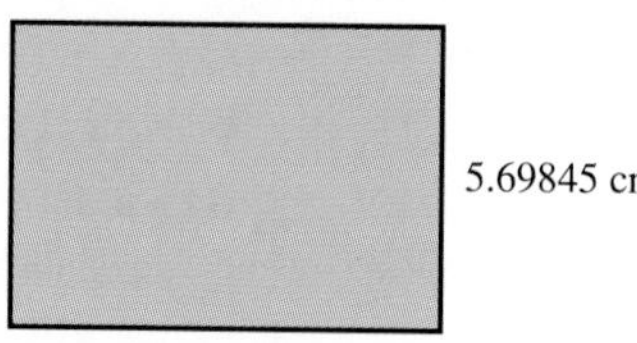

Figure for Exercise 77

78 a. 38 m **b.** 39 m **c.** 40 m **d.** 41 m

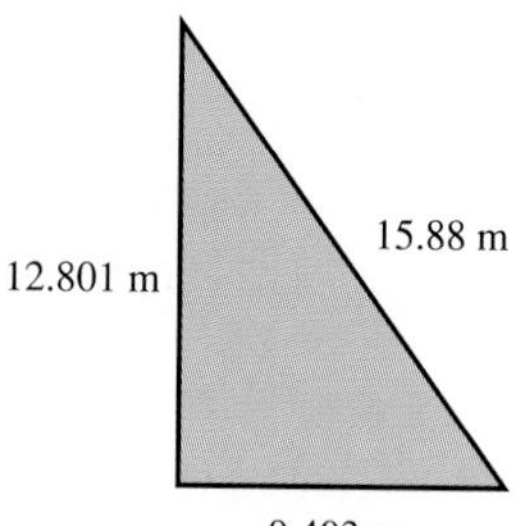

Figure for Exercise 78

79 a. 210 m **b.** 220 m **c.** 420 m **d.** 440 m

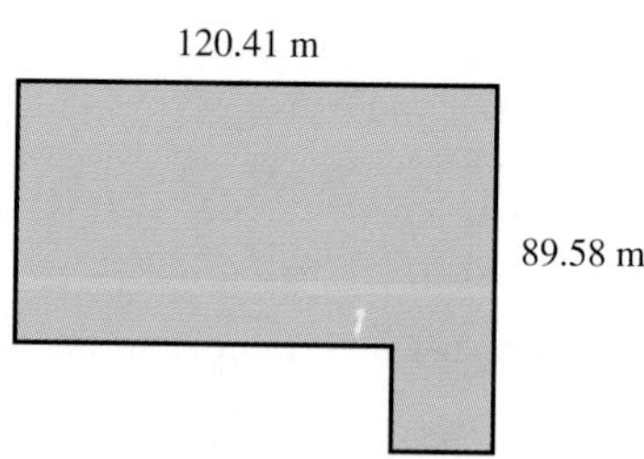

Figure for Exercise 79

80 Mentally estimate the area of the square shown, and then select the answer closest to your mental estimate.

a. 6 cm^2 **b.** 8 cm^2 **c.** 9 cm^2 **d.** 10 cm^2

Figure for Exercise 80

81 Mentally estimate the volume of the cube shown, and then select the answer closest to your mental estimate.

a. 6 cm^3 **b.** 8 cm^3 **c.** 9 cm^3 **d.** 10 cm^3

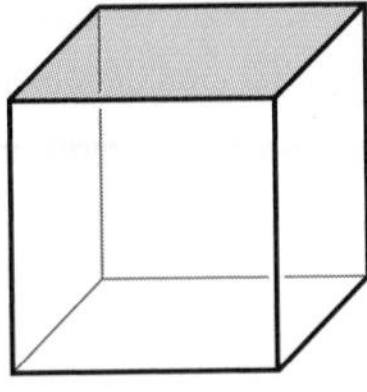

Figure for Exercise 81

C

82 Profit for Two Shifts Fill in the last row of this table with values representing the total daily profits for a business that runs two shifts. The results of each shift are given in dollars per day for Monday through Thursday, with each negative number indicating a loss and each positive number indicating a gain.

	Mon.	Tue.	Wed.	Thur.
Day	545	545	−545	−545
Night	315	−315	315	−315
Total				

83 Comparison of Shift Profits Using the table in Exercise 82, fill in the last row with the difference between the day-shift profits and the night-shift profits. A positive result indicates that the day shift did better.

84 Altitude Death Valley, the lowest spot in the continental United States, is 282 feet below sea level. Mt. Whitney has an altitude of 14,495 feet above sea level. How high above Death Valley is the top of Mt. Whitney?

85 Measuring Ingredients A recipe calls for $\frac{3}{4}$ cup of sugar. If you have only $\frac{2}{3}$ cup, how much more sugar do you need?

86 Meter Reading At the end of February a utility meter read 378.49 units. By the end of March it read 405.34 units. How many units were used in March?

87 Federal Taxes A husband and wife filing a joint federal income tax return had a combined income of $43,000 last year. Of this income, $2000 is tax-exempt; the rest is called the *adjusted gross income*. A $6300 *deduction* is subtracted from the adjusted gross income for the husband, the wife, and each of their three children. The remaining figure is called the couple's *taxable income*, on which they must pay an income tax of 16%. A total of $1490 has already been paid by deductions from their salaries throughout the year. Do they owe more tax, or do they get a refund? What is the amount of the additional tax or refund?

In Exercises 88 and 89 determine the value of the missing entry in the pie chart.

88 **89**

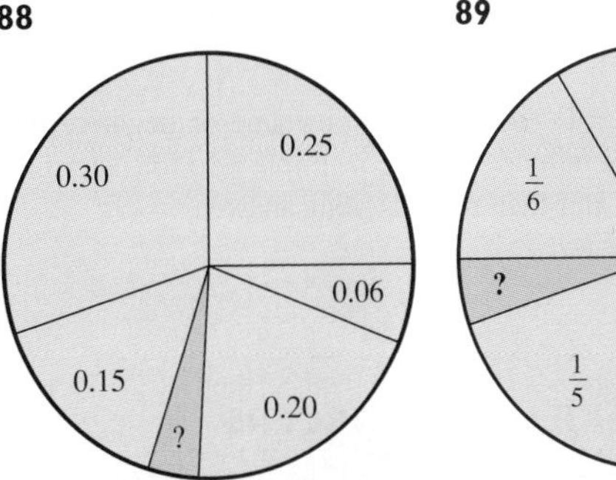

In Exercises 90–95 perform the indicated operations, and simplify the results.

90 a. Add $\frac{5}{33}$ to $\frac{2}{3}$. **b.** Find the sum of $\frac{3}{10}$ and $\frac{2}{15}$.

c. Subtract $\frac{1}{2}$ from $\frac{1}{3}$.

91 a. Subtract $\frac{5}{6}$ from $\frac{4}{15}$.

b. Subtract the sum of $\frac{5}{7}$ and $\frac{7}{8}$ from $\frac{15}{8}$.

c. Add $\frac{2}{5}$ to the difference of $\frac{3}{4} - \frac{5}{12}$.

92 a. Multiply $\frac{-4}{15}$ times $\frac{-25}{24}$.

b. Square the fraction $\frac{-5}{7}$.

c. Triple the fraction $\frac{-11}{39}$.

93 a. Double the fraction $\frac{-7}{10}$.

b. Cube the fraction $-\frac{4}{5}$.

c. Raise $-\frac{2}{3}$ to the fourth power.

94 a. Divide $\frac{-15}{28}$ by $\frac{-9}{49}$. **b.** Divide $\frac{-15}{28}$ into $\frac{-9}{49}$.

c. Find two-thirds of $\frac{15}{22}$.

95 a. $\frac{-5^2 - (-5)^2}{1 - 2 \cdot 3}$ **b.** $\frac{-9}{55} \div \left(\frac{21}{-8} \cdot \frac{33}{40}\right)$

c. $\frac{5}{9} - \frac{21}{-8} \div \frac{7}{4} + \frac{7}{12}$

Discussion Questions (96 and 97)

96 Describe the rule for adding signed numbers. Represent positive and negative numbers by using the terms *profit* and *loss*.

97 The value of $(-1)^n$ is either positive one or negative one for each natural number n. Give a simple rule for determining this value, and then discuss your answer.

Challenge Question

Calculate the value of
$(80 - 100)(80 - 99)(80 - 98) \cdot \cdots \cdot (80 - 3)(80 - 2)(80 - 1)$.

Section Skill Reinforcers

Identify the property that justifies each statement.

1 $a(b + c) = a(c + b)$ **2** $a(b + c) = (b + c)a$ **3** $a(b + c) = ab + ac$ **4** $a + (b + c) = (a + b) + c$
5 $a + (b + c) = (b + c) + a$

1-3 Evaluating and Simplifying Algebraic Expressions

Section Objectives

7 Evaluate an algebraic expression for given values of the variables.

8 Simplify algebraic expressions by combining like terms.

To **evaluate** an algebraic expression for given values of the variables means to replace each variable by the specific value given for that variable. A variable is replaced by the same value every time it occurs in the expression. When you are substituting a constant for a variable, it is often wise to use parentheses to avoid making an error in sign or doing operations in an incorrect order. The ability to evaluate expressions is especially important when formulas are used to calculate results for a specific problem.

EXAMPLE 1 Evaluating an Algebraic Expression with One Variable

Evaluate $3x^2 - 5x + 7$ for $x = -2$.

SOLUTION

$$\begin{aligned} 3x^2 - 5x + 7 &= 3(-2)^2 - 5(-2) + 7 \\ &= 3(4) + 10 + 7 \\ &= 12 + 10 + 7 \\ &= 29 \end{aligned}$$

Note the use of parentheses when -2 is substituted for x each time x occurs. Simplify the expression, following the correct order of operations.

EXAMPLE 2 Evaluating an Algebraic Expression with Three Variables

Evaluate $\dfrac{x - y}{-x + y + z}$ for $x = -2$, $y = 12$, and $z = 7$.

SOLUTION $\dfrac{x - y}{-x + y + z} = \dfrac{-2 - 12}{-(-2) + 12 + 7}$ Again, note the use of parentheses to prevent an error in sign.

$= \dfrac{-14}{2 + 12 + 7}$ Simplify the numerator and the denominator separately, and then reduce the fraction to lowest terms.

$= -\dfrac{14}{21}$

$= -\dfrac{2}{3}$ ▼

EXAMPLE 3 Evaluating a Formula for Given Values

Use the formula $A = \frac{1}{2}bh$ to determine the area of the triangle shown in the figure.

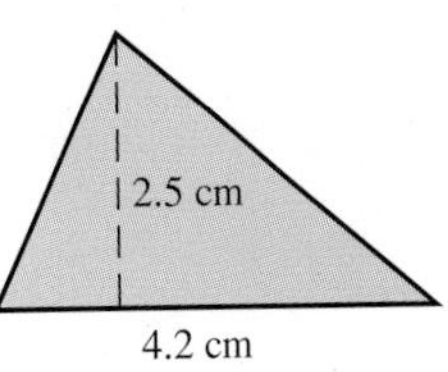

SOLUTION $A = \frac{1}{2}bh$ First write the formula for the area of a triangle.

$A = \frac{1}{2}(4.2 \text{ cm})(2.5 \text{ cm})$ Substitute 4.2 cm for the base, b, and 2.5 cm for the height, h.

$A = 5.25 \text{ cm}^2$ Then simplify.

The area of this triangle is 5.25 cm^2. ▼

SELF-CHECK 

Evaluate these expressions for $x = -1$ and $y = -2$.

1 $-3x^2 - 8x + 11$
2 $x^2 - 5xy - y^2$

The notation $f(x)$ can be used to denote a formula, as in the equation $f(x) = 3x^2 - 5x + 7$. To calculate $f(-2)$, read as "f of -2," we substitute the value of -2 into the formula for x, as in Example 1. The formula in the next example gives the cost per unit of production for a factory making x units of custom boots per day.

EXAMPLE 4 Cost Formula

Given the formula $C(x) = x^2 - 55x + 775$ for the dollar cost per unit of custom boots, determine the cost of making 10 units and 30 units of custom boots.

SOLUTION $C(10) = (10)^2 - 55(10) + 775$ Substitute the given values into the cost formula.

$= 100 - 550 + 775$

$= 325$

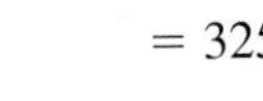

$C(30) = (30)^2 - 55(30) + 775$

$= 900 - 1650 + 775$

$= 25$

Answer At 10 units of production the cost is \$325 per unit, and at 30 units of production the cost is \$25 per unit. ▼

SELF-CHECK ANSWERS ✓

1 16 **2** −13

Calculators are often used to evaluate formulas. The next example illustrates one way to evaluate formulas on a TI-82 graphics calculator. Although the keystrokes shown are representative, keystrokes will vary some, depending on the model of calculator you have.

EXAMPLE 5 Calculator Usage

Use a calculator to evaluate $f(-2.45)$ for $f(x) = 10x^2 - 8x + 7$.

SOLUTION

Keystrokes

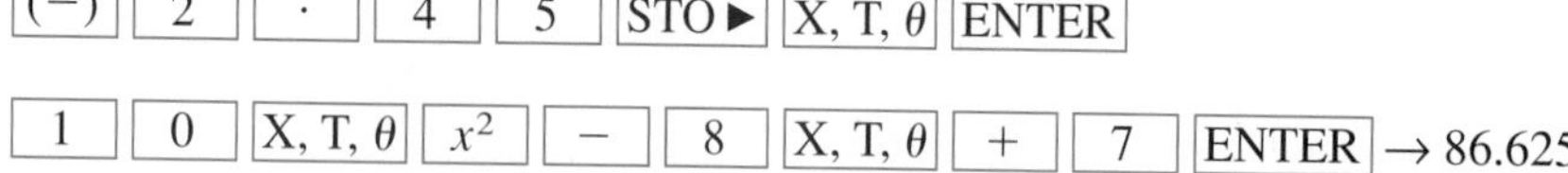

First store the value of -2.45 for x.

Then enter the formula into the calculator.

Answer $f(-2.45) = 86.625$

Recall that numbers that are added or subtracted are called **terms** and numbers that are multiplied are called **factors.** A single term can consist of more than one factor.

TERM

> A **term** of an algebraic expression can be a single constant, a variable, or a product of constants and variables.

The following expressions illustrate this definition:

-13 is a single term.

x is a single term.

$8xy$ is a single term with three factors: 8, x, and y.

$x + 8$ has two terms: x and 8.

$x^2 - 4x + 3$ means $x^2 + (-4x) + 3$ with three terms: x^2, $-4x$, and 3.

The constant factor of a term is called the **numerical coefficient,** or just the **coefficient.** For example, the numerical coefficient of $-3xy$ is -3. The term $\frac{x}{7}$ equals $\frac{1}{7}(x)$ with coefficient $\frac{1}{7}$. Since $1 \cdot x = x$ and $-1 \cdot x = -x$, the coefficient of x is understood to be 1 and the coefficient of $-x$ is understood to be -1.

Terms with exactly the same variable factors are called **like terms,** or **similar terms.** For example, $5x$ and $-3x$ are like terms. We can add like terms by using the distributive property to add their coefficients, as illustrated in Example 6. In practice some of the steps shown here for clarity may be done mentally.

EXAMPLE 6 **Combining Like Terms**

Simplify these expressions by combining like terms.

SOLUTIONS

(a) $15x + 7x$

$$15x + 7x = (15 + 7)x$$
$$= 22x$$

Distributive property

(b) $6m - 2m + m$

$$6m - 2m + m = (6 - 2 + 1)m$$
$$= 5m$$

Distributive property. Note that the coefficient of m is understood to be 1.

(c) $3x - 4y + 8x + 7y$

$$3x - 4y + 8x + 7y = 3x + 8x + 7y - 4y$$

Distributive property

$$= (3 + 8)x + (7 - 4)y$$

Only like terms can be combined.

$$= 11x + 3y$$

▼

Sometimes the distributive property must be used to remove parentheses before like terms can be combined.

EXAMPLE 7 **Combining Like Terms after Removing Parentheses**

Simplify $-2(3x - 4y) - 5(2x + 3y)$ by combining like terms.

SOLUTION

$$-2(3x - 4y) - 5(2x + 3y) = -2(3x) + (-2)(-4y) + (-5)(2x) + (-5)(3y)$$
$$= -6x + 8y - 10x - 15y$$
$$= -6x - 10x + 8y - 15y$$
$$= -16x - 7y$$

First use the distributive property to remove parentheses. Then add like terms.

▼

SELF-CHECK  

Simplify

1. $5a - 3b + 2a$
2. $3(4x - 5y) - 2(3x - y)$

A value of a variable that will make an equation a true statement is called a **solution** of the equation.

EXAMPLE 8 **Checking Possible Solutions of an Equation**

Test -3 and 3 to determine whether either of these values is a solution of $4(2x - 1) = 9x - 1$.

SOLUTION

$4(2x - 1) = 9x - 1$	$4(2x - 1) = 9x - 1$
$4[2(-3) - 1] \stackrel{?}{=} 9(-3) - 1$	$4[2(3) - 1] \stackrel{?}{=} 9(3) - 1$
$4[-6 - 1] \stackrel{?}{=} -27 - 1$	$4[6 - 1] \stackrel{?}{=} 27 - 1$
$4(-7) \stackrel{?}{=} -28$	$4(5) \stackrel{?}{=} 26$
$-28 = -28$ checks.	$20 = 26$ does not check.

Substitute the test value into the equation for x.

Then evaluate each side of the equation.

Answer -3 is a solution; 3 is not a solution.

▼

SELF-CHECK ANSWERS 

1. $7a - 3b$ 2. $6x - 13y$

Exercises 1-3

A

In Exercises 1–10 evaluate each expression for $x = -3$ and $y = 5$.

1 a. $4x - 2$ **b.** $x^2 - 2x + 3$ **c.** $\frac{x + 7}{1 - x}$

2 a. $2x - 4$ **b.** $x^2 + 3x - 2$ **c.** $\frac{-4x}{3 - x}$

3 a. $-2x + 3y$ **b.** $-7(x - 2y)$ **c.** $x^2 - xy - y^2$

4 a. $4x - 5y$ **b.** $-5(2x - y)$ **c.** $-x^2 + 2xy + y^2$

5 a. $(x + y)^2$ **b.** $x^2 + y^2$ **c.** $2x^2 - (2x)^2$

6 a. $(x - y)^2$ **b.** $x^2 - y^2$ **c.** $2y^2 - (2y)^2$

7 a. $\frac{2x + y}{x - y}$ **b.** $\frac{2(x + y)}{x - y}$ **c.** $\frac{2x}{x} - \frac{y}{y}$

8 a. $\frac{x + 2y}{2x - y}$ **b.** $\frac{x + 2y}{2(x - y)}$ **c.** $\frac{x}{2x} - \frac{2y}{y}$

9 a. $|3x - 4y|$ **b.** $3|x - 4y|$ **c.** $|3x| - |4y|$

10 a. $|4x + 2y|$ **b.** $4|x + 2y|$ **c.** $|4x| + |2y|$

In Exercises 11 and 12 write the numerical coefficient of each term.

11 a. $7m$ **b.** $-5m$ **c.** m

12 a. $-4n$ **b.** $9n$ **c.** $-n$

In Exercises 13–30, simplify each expression by combining like terms.

13 $4m - 11m$ **14** $-5m + 3m$

15 $-a + 8a$ **16** $b - 7b$

17 $-2y - (-9y)$ **18** $3w - (-10w)$

19 $7x - 11x + 8x$ **20** $4y + 12y - 3y$

21 $5x - 7y + 9x - 8y$ **22** $3x - 4y - x + y$

23 $3m + 2n - m - n$ **24** $-9m + 8n + 7m - 4n$

25 $4(3x - 2) - 3(2x + 1)$ **26** $5(2x + 1) - 4(3x - 2)$

27 $2(x + 3y) - 4(2x - y)$ **28** $3(2x - 5y) + 2(x - 7y)$

29 $-(5m - n) + (3m - 7n)$ **30** $-(m - 3n) + (9m - n)$

In Exercises 31–36 use the formula $f(x) = -x^2 - 3x - 6$ to evaluate each of these expressions.

31 $f(0)$ **32** $f(2)$ **33** $f(-2)$

34 $-f(2)$ **35** $f(10)$ **36** $f(-10)$

In Exercises 37–42 determine whether the given value is a solution of the given equation.

37 $7x - 3 = 2x + 2$; 1 **38** $5x + 2 = 3x - 2$; -2

39 $2(x - 1) = 3(2x + 7)$; 3 **40** $7(x - 3) = 2(x + 2)$; 1

41 $x^2 + 5x + 6 = 0$; -2 **42** $x^2 + 5x + 6 = 0$; -3

In Exercises 43–48 use the given formulas to calculate the quantities requested.

43 Area of a Rectangle Use the formula $A = l \cdot w$ to calculate the area of the rectangle shown.

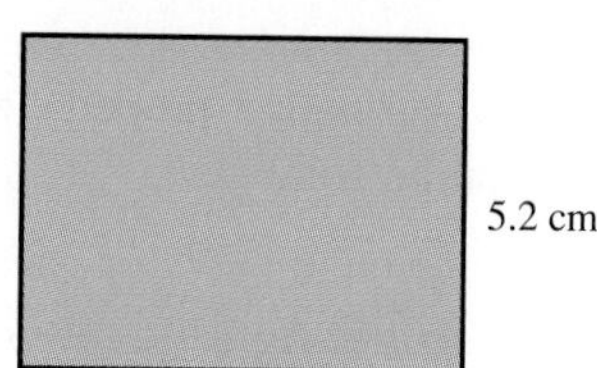

Figure for Exercise 43

44 Circumference of a Circle Use the formula $C = 2\pi r$ to calculate the circumference of the circle shown. Then use a calculator to approximate this circumference to the nearest tenth of a centimeter.

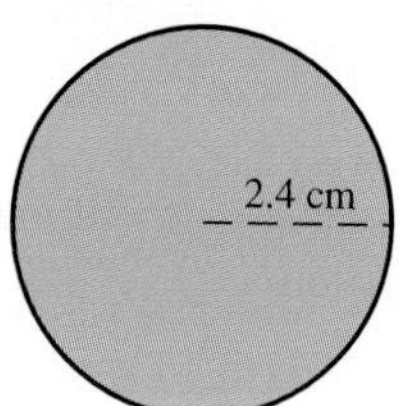

Figure for Exercise 44

45 Simple Interest Use the formula $I = PRT$ to compute the simple interest on a principal of \$1000 at an interest rate of 8.5% for one year.

46 Area of a Square Use the formula $A = s^2$ to calculate the area of the square shown.

Figure for Exercise 46

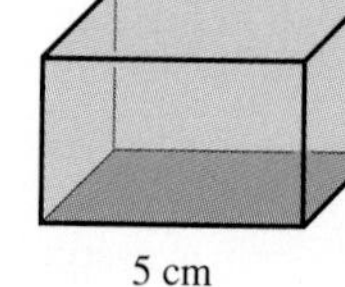

Figure for Exercise 47

47 Volume of a Box Use the formula $V = l \cdot w \cdot h$ to calculate the volume of the rectangular box shown.

48 Body Temperature Use the formula $C = \frac{5}{9}(F - 32)$ to convert a body temperature of 98.6° Fahrenheit to degrees Celsius.

B

ESTIMATION SKILLS (49–54)

In Exercises 49–54 mentally estimate the value of each expression, and select the answer that is closest to your estimate.

49 Given $f(x) = x^2 + 2x$, estimate $f(9.9)$.

a. 29 **b.** 39 **c.** 99 **d.** 120 **e.** 210

50 Given $f(x) = x^2 + 2x$, estimate $f(-9.9)$.

a. 100 **b.** 90 **c.** 80 **d.** −90 **e.** −120

51 Given $f(x) = x^2 + x + 3$, estimate $f(5.01)$.

a. 29 **b.** 33 **c.** 35 **d.** 39 **e.** 93

52 Given $f(x) = x^2 - 2x - 1$, estimate $f(4.96)$.

a. 14 **b.** 12 **c.** 10 **d.** 9 **e.** −2

53 Given $f(x) = -x^2$, estimate $f(99.9)$.

a. −20,000 **b.** −10,000 **c.** −200 **d.** 200 **e.** 10,000

54 Given $f(x) = -x^2$, estimate $f(-10.05)$.

a. −100 **b.** −90 **c.** −50 **d.** 100 **e.** 200

In Exercises 55–62 simplify each expression by combining like terms.

55 $13x - 18y - 19(x - 2y)$

56 $\frac{x}{2} - \frac{y}{3} + \frac{x}{3} - \frac{y}{4}$

57 $\frac{x}{5} + \frac{y}{7} - \frac{2x}{3} + \frac{y}{2}$

58 $2.8v - 9.3v + 7.2v - 1.4v$

59 $-2(3a - 4b + 2c) - (2a - 5b + 6c)$

60 $-3[a - 2(a + 3b)] + 4(3a - 2b)$

61 $-7(v - 13) - (v - 3) + 4(7 - 2v)$

62 $-5(2w - 1) - (w - 17) - 3(4 - 2w)$

In Exercises 63–70 evaluate each expression for $a = -2$, $b = -5$, and $c = 6$.

63 $a - b + c$ **64** $a - (b + c)$ **65** $\frac{ab - c}{ac}$

66 $\frac{a}{3} - \frac{b}{4} - \frac{c}{8}$ **67** $\frac{b^2 - a^2}{c^2}$ **68** $\frac{a}{b} - \frac{b}{c} \cdot \frac{c}{a}$

69 $|a - b - c|$ **70** $|a| - |b| - |c|$

C

CALCULATOR USAGE (71–80)

In Exercises 71–76 a student stored on a calculator the value of −5 for x. Determine the value that would result from the algebraic expressions shown.

71 X² − 5X + 7

72 X² − 5(X + 7)

73 (8 − X)(X − 9)

74 (8 − X)²

75 (X − 15)/(X + 1)

76 X − 15/X + 1

In Exercises 77–80 store the value of −2.25 for x in your calculator. Then evaluate each of these expressions.

77 $x^2 - 33x + 81$ **78** $x^3 - 5x^2 - 9x + 11$

79 $(x + 7.85)^2$ **80** $\frac{3x^2 - 4x - 8}{4x}$

Revenue and Cost Formulas (81–86)

In Exercises 81–86 the revenue per unit of producing x units is given by $R(x) = -2x^2 + 128x$ and the cost per unit is given by $C(x) = x^2 - 52x + 730$. Evaluate each of these expressions.

81 $R(5)$ **82** $R(10)$ **83** $C(5)$

84 $C(10)$ **85** $R(20)$ **86** $C(20)$

DISCUSSION QUESTIONS (87 AND 88)

87 Explain the role of the distributive property when you simplify an algebraic expression by combining like terms.

88 Use the distributive property to describe why removing the parentheses from $-(5a - 4b + 2c)$ yields the expression $-5a + 4b - 2c$.

CHALLENGE QUESTION

a. Calculate the value of

$$\frac{2}{3}\cdot\frac{3}{4}\cdot\frac{4}{5}\cdot\;\cdots\;\cdot\frac{97}{98}\cdot\frac{98}{99}\cdot\frac{99}{100}.$$

b. A soup can with a height of $4\frac{1}{2}$ inches has a radius of $1\frac{1}{2}$ inches. The can's label covers the vertical surface of the can except for the $\frac{1}{8}$-inch rims at top and bottom. Determine the area of the label.

SECTION SKILL REINFORCERS

1 The result of adding two real numbers is called their ________.
2 The result of subtracting two real numbers is called their ________.
3 The result of multiplying two real numbers is called their ________.
4 The result of dividing two real numbers is called their ________.
5 $2x + (3y + 5z) = (2x + 3y) + 5z$ illustrates the ________ property.

1-4 *Solving Linear Equations*

SECTION OBJECTIVES

9 Solve linear equations in one variable.

10 Transpose an equation to solve for a specified variable.

An **equation** is a statement that two quantities are equal. The left side of the equation is called the **left member,** and the right side is called the **right member.**

$$\underbrace{3x + 5}_{\text{Left member}} = \underbrace{5x + 1}_{\text{Right member}}$$

Left member — Equal symbol — Right member

A statement of equality may be either true or false. An equation that is always true is an **identity,** whereas an equation that is always false is a **contradiction.** If an equation contains a variable, it may be true for some values of the variable but false for other values. Such an equation is called a

A Mathematical Note

The first recorded use of = as the equal symbol was by Robert Recorde, an English author, in 1557. His justification for this notation was "No two thyngs coulde be more equalle than two straight lines."

conditional equation, because its truth depends on the conditions—that is, on the values of the variables.

Most of our work consists of determining values for variables that will make conditional equations true. These values are said to **satisfy the equation** and are called **roots,** or **solutions,** of the equation. The collection of all solutions is the **solution set.** (The solution set that has no members is called the **null set** and is denoted by $\varnothing$.) The process of finding the solution set is called **solving the equation.**

EXAMPLE 1 Classifying Equations

Identify each equation as either an identity, a contradiction, or a conditional equation.

	SOLUTIONS	
(a) $4x = 12$	Conditional equation	The value 3 is a solution, since $4(3) = 12$. Because 3 is the only solution, this equation is conditional.
(b) $2x + 3x = 5x$	Identity	This statement is true for each real number that can be substituted for x, so this equation is an identity. The solution set is the set of all real numbers.
(c) $w = w + 1$	Contradiction	No value of w will satisfy this equation. Since the equation is always false, it is a contradiction. The solution set is $\varnothing$.
(d) $w = 2w$	Conditional equation	The only solution of this equation is 0. Thus this equation is a conditional equation whose solution set is $\{0\}$. ▼

As Example 1 illustrates, the presence of a variable in an equation does not necessarily mean that the equation is conditional. Some equations containing variables are contradictions, whereas other equations containing variables are identities. Identities containing variables are frequently used to state important properties. For example, the identity $a(b + c) = ab + ac$ is used to state the distributive property.

A **linear equation in one variable** contains only constants and a single variable with an exponent of 1. A variable with an exponent of 1 is referred to as a first-degree term; thus these equations are called **first-degree equations.** Every linear equation in one variable can be simplified to the form $ax = b$, where x represents the real variable, a and b are real constants, and $a \neq 0$.

LINEAR EQUATION IN ONE VARIABLE

A **linear equation in one variable** x is an equation that can be written in the form $ax = b$, where a and b are real constants and $a \neq 0$.

EXAMPLE 2 Identifying Linear Equations

Determine which of these equations are linear equations in one variable.

	SOLUTIONS	
(a) $5x = 40$	Linear equation in the variable x	The exponent of x is 1.
(b) $y^2 = 25$	Not linear	The exponent of y is 2. Thus the equation is a second-degree equation, not a first-degree equation.
(c) $5z - 8 = 2(z - 11)$	Linear equation in the variable z	The only variable is z, and its degree is 1. Using the properties of equivalent equations (in the following box), we can rewrite this equation as $3z = -14$.
(d) $7(2x - 3y) = 4(x - 2y)$	Linear equation in two variables	This linear equation contains two variables, x and y.
(e) $5x - 3(x + 1)$	Not a linear equation	This is a linear expression, but it is not an equation.

Equivalent equations have the same solution set. To solve an equation whose solution is not obvious, we form simpler equivalent equations until we obtain an equation whose solution is obvious.

A Geometric Viewpoint: Using a Balance Scale and Solving Equations

When thinking of an equation, you may find it helpful to use the concept of a balance scale that has the left member in balance with the right member. We must always perform the same operation on both members to preserve this balance.

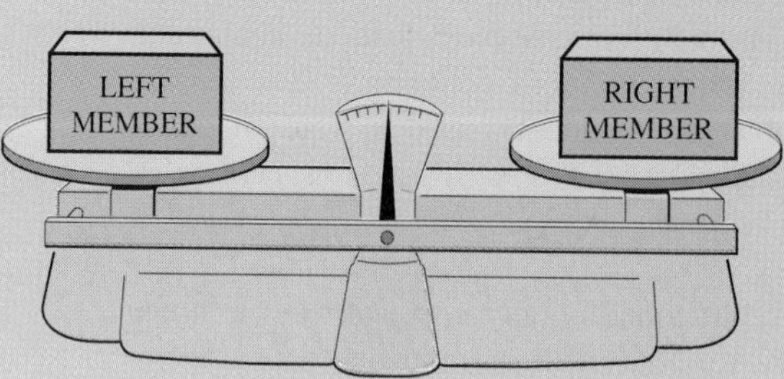

The general strategy for solving a linear equation, regardless of its complexity, is to isolate the variable whose value is to be determined on one side of the equation and place all other terms on the other side.

Two properties that are used to solve many equations are given in the following box; they are a direct result of the theorems of equality presented in Section 1-1.

FORMING EQUIVALENT EQUATIONS

If a, b, and c are real numbers, then

1 $a = b$ is equivalent to $a + c = b + c$. — Adding the same number to both members of an equation produces an equivalent equation.

2 $a = b$ is equivalent to $ac = bc$ for $c \neq 0$. — Multiplying both members of an equation by the same nonzero number produces an equivalent equation.

Subtracting c from a number is the same as adding $-c$; thus subtracting the same number from both members of an equation produces an equivalent equation. Similarly, dividing a number by c ($c \neq 0$) is the same as multiplying by $\frac{1}{c}$, so dividing both members of an equation by a nonzero number also produces an equivalent equation. Therefore $a = b$ is equivalent to $a - c = b - c$, and $a = b$ is also equivalent to $a \div c = b \div c$ for $c \neq 0$.

EXAMPLE 3 Solving Linear Equations

Solve each of these equations.

SOLUTIONS

(a) $w - 5 = 11$

$$w - 5 = 11$$

$$w - 5 + 5 = 11 + 5$$ Add 5 to both sides in order to isolate w in the left member of the equation.

$$w = 16$$ The solution set is {16}.

(b) $\frac{2t}{3} = 8$

$$\frac{2t}{3} = 8$$

$$\left(\frac{3}{2}\right)\left(\frac{2t}{3}\right) = \left(\frac{3}{2}\right)(8)$$ Multiply both sides by $\frac{3}{2}$ to obtain a coefficient of 1 for t and thus isolate t in the left member of the equation.

$$t = 12$$ The solution set is {12}. ▼

EXAMPLE 4 Solving a Linear Equation

Solve $5x - 6 = 3x + 2$.

SOLUTION

$$5x - 6 = 3x + 2$$

$$5x - 6 - 3x = 3x + 2 - 3x$$ Subtract $3x$ from both sides to isolate x in the left member of the equation.

$$2x - 6 = 2$$

$$2x - 6 + 6 = 2 + 6$$ Add 6 to both sides to isolate the constant terms in the right member of the equation.

$$2x = 8$$

$$\frac{2x}{2} = \frac{8}{2}$$ Divide both sides by 2 to obtain a coefficient of 1 for x.

$$x = 4$$

Check

$$5x - 6 = 3x + 2$$
$$5(4) - 6 \stackrel{?}{=} 3(4) + 2$$
$$20 - 6 \stackrel{?}{=} 12 + 2$$
$$14 = 14 \text{ checks.}$$

The solution set is {4}. ▼

All linear equations in one variable can be solved by using the procedure outlined in the box below. Complicated linear equations may involve more steps than shown in the examples in this section, but the method is the same. The Addition Theorem of Equality is generally used to isolate the variables on one side of the equation, and then the Multiplication Theorem of Equality is used to obtain a coefficient of 1.

Solving Linear Equations

Step 1 If the equation contains fractions, multiply both sides of the equation by the least common denominator (LCD) of all the fractions.

Step 2 If the equation created in Step 1 contains grouping symbols, first use the distributive property to remove the grouping symbols and then simplify each side of the equation by combining like terms.

Step 3 Use the Addition Theorem of Equality to isolate the variable terms on one side and the constant terms on the other side.

Step 4 Use the Multiplication Theorem of Equality to solve the equivalent equation produced in Step 3.

EXAMPLE 5 Solving a Linear Equation Containing Parentheses

Solve $2(3x - 5) = 7(2x + 5) - 5$ for x, and check the solution.

SOLUTION

$$2(3x - 5) = 7(2x + 5) - 5$$
$$6x - 10 = 14x + 35 - 5$$
$$6x - 10 = 14x + 30$$

Use the distributive property to distribute the factor 2 on the left side and the factor 7 on the right side. Then simplify each side of the equation separately.

$$6x - 10 + 10 = 14x + 30 + 10$$
$$6x = 14x + 40$$
$$6x - 14x = 14x + 40 - 14x$$
$$-8x = 40$$

After adding 10 to both members, isolate the variable terms in the left member by using the Addition Theorem of Equality to subtract $14x$ from (or add $-14x$ to) both members.

$$-8x\left(-\frac{1}{8}\right) = 40\left(-\frac{1}{8}\right)$$
$$x = -5$$

Then solve for x, using the Multiplication Theorem of Equality to multiply both members by $-\frac{1}{8}$. Note that multiplying by $-\frac{1}{8}$ is the same as dividing both sides by -8.

Check

$$2[3(-5) - 5] \stackrel{?}{=} 7[2(-5) + 5] - 5$$
$$2(-15 - 5) \stackrel{?}{=} 7(-10 + 5) - 5$$
$$2(-20) \stackrel{?}{=} 7(-5) - 5$$
$$-40 \stackrel{?}{=} -35 - 5$$
$$-40 = -40 \text{ checks.}$$

Caution: Use parentheses when substituting -5 for x to avoid an order-of-operations error.

▼

EXAMPLE 6 Solving a Linear Equation Containing Fractions

Solve $\dfrac{3v-3}{6} = \dfrac{4v+1}{15} + 2$.

SOLUTION

$\dfrac{3v-3}{6} = \dfrac{4v+1}{15} + 2$	$6 = 2 \cdot 3$ $15 = \quad 3 \cdot 5$ $\text{LCD} = 2 \cdot 3 \cdot 5 = 30$
$30\left(\dfrac{3v-3}{6}\right) = 30\left(\dfrac{4v+1}{15} + 2\right)$	Multiply both members by the LCD, 30.
$5(3v-3) = 30\left(\dfrac{4v+1}{15}\right) + 30(2)$	Use the distributive property to remove parentheses.
$15v - 15 = 2(4v+1) + 60$	
$15v - 15 = 8v + 62$	Add 15 to both members to isolate the constant term in the right member.
$15v = 8v + 77$	
$7v = 77$	Subtract $8v$ from both members to isolate the variable term in the left member. Then divide both members by 7.
$v = 11$	The solution set is {11}.

In the process of solving equations, we sometimes encounter instances where the variable is eliminated as we produce simpler equivalent equations. In Example 7(a) we obtain a contradiction; in Example 7(b) we obtain an identity.

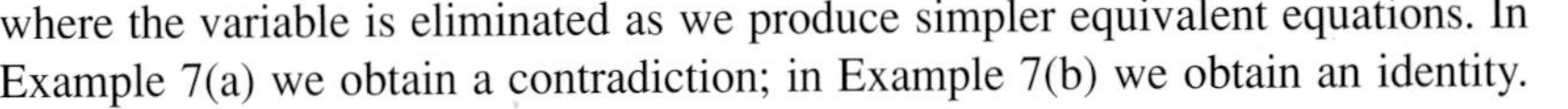

SELF-CHECK

1 Solve $(4k-3) - 2(k+4) = 3(k+7)$, and check the solution.

2 Solve $\dfrac{z}{6} + 2 = \dfrac{z}{4}$.

EXAMPLE 7 Solving Equations When the Variable Is Eliminated

Find the solution set for each of these equations.

(a) $2(x+1) = 2(x-1)$

(b) $3(x-4) = 2(x+3) - (18-x)$

SOLUTIONS

(a) $2(x+1) = 2(x-1)$

$2x + 2 = 2x - 2$	Remove parentheses. Then subtract $2x$ from both members.
$2 = -2$	This equation is a contradiction.

Since this contradiction is equivalent to the original equation, there is no solution to this equation. The solution set is the null set, ∅.

Answer ∅

(b) $3(x-4) = 2(x+3) - (18-x)$

$3x - 12 = 2x + 6 - 18 + x$	Remove parentheses, and combine like terms.
$3x - 12 = 3x - 12$	Add 12, and then subtract $3x$.
$0 = 0$	This equation is an identity.

Since this identity is equivalent to the original equation, the solution set is the set of all real numbers, $\mathbb{R}$.

Answer $\mathbb{R}$

SELF-CHECK ANSWERS 

1 $k = -32$ **2** $z = 24$

The equation in the next example, $\frac{2x + 3}{x - 5} = \frac{3x - 2}{x - 5}$, is not a linear equation. We can, however, form a linear equation by multiplying both members by the LCD, $x - 5$. Is this new linear equation equivalent to the original equation—that is, do both of these equations have the same solution set? The answer to this question depends on whether we multiplied by zero or by a nonzero number. If $x = 5$, then $x - 5 = 0$, and we have multiplied both members by zero. In this case the new equation may not be equivalent to the original equation. Any value that incorrectly occurs as a solution in the last step and is not a solution of the original equation is called an **extraneous value.** We will examine extraneous values in more detail in Chapter 6.

EXAMPLE 8 Solving an Equation with a Variable in the Denominator

Solve $\frac{2x + 3}{x - 5} = \frac{3x - 2}{x - 5}$.

SOLUTION

$$\frac{2x + 3}{x - 5} = \frac{3x - 2}{x - 5}$$

$$(x - 5)\left(\frac{2x + 3}{x - 5}\right) = (x - 5)\left(\frac{3x - 2}{x - 5}\right)$$ Multiply both members by $x - 5$ for $x \neq 5$.

$$2x + 3 = 3x - 2$$ Simplify both sides of the equation. Then add 2 and subtract $2x$.

$$5 = x$$

$$x = 5$$ This value cannot be a solution to the original equation because it causes division by zero.

There is no solution for this equation; the solution set is the null set, $\varnothing$.

Answer $\varnothing$

A **formula** is a general rule that is stated in mathematical language, usually as an equation with specific variables already selected to represent each quantity in the problem to which it applies. In formulas it is very helpful to use representative letters, such as A for area, V for volume, and I for interest. A list of well-known formulas, including those for the areas of common regions and the volumes of common solids, appears on the inside front cover of this book.

Formulas are sometimes called **literal equations** because they contain more than one letter of the alphabet as variables. The Addition and Multiplication Theorems of Equality can be used to create alternative forms of formulas. We can specify the variable that we wish to be the subject of the statement of equality and then transpose the literal equation to write it in the desired form. This is called **solving for a specified variable,** or **transposing an equation.** In the formula for the area of a trapezoid in Example 10, we will solve for the base, b.

In your own work you may want to highlight or underline the specified variable in your first step. This practice will help keep you focused on your goal of isolating a particular variable.

A Mathematical Note

Although mathematical terms are used very precisely, different terms may be used in different countries. Compare the textbook vocabulary used in two different English-speaking countries: In Australia textbooks use the term *transposing an equation,* whereas in the United States we use the term *solving for a specified variable.*

EXAMPLE 9 Solving an Equation for y

Solve each of these equations for y.

SOLUTION

(a) $2x + 3y = 6$

$2x + 3y = 6$ — The specified variable is y.

$3y = -2x + 6$ — Subtract $2x$ from both sides to isolate the specified variable on the left side.

$y = -\frac{2}{3}x + 2$ — Divide both sides by 3.

(b) $ay - by = c$, assuming $a \neq b$

$ay - by = c$ — The specified variable is y.

$(a - b)y = c$ — The y variable occurs in two terms, but we want it to occur only once in the final form. Thus we use the distributive property to factor out y.

$y = \frac{c}{a - b}$ — Divide both members by $a - b$. (Note that since $a \neq b$, $a - b$ is not zero and we can divide by $a - b$.) ▼

SELF-CHECK 

1 Solve $12x - 4y = 16$ for y.
2 Solve $P = 2w + 2l$ for w.
3 Solve $m = \frac{n}{3} - b$ for n.

EXAMPLE 10 Formula for the Area of a Trapezoid

Solve $A = \frac{1}{2}h(a + b)$, the formula for the area of a trapezoid, for b.

SOLUTION

$A = \frac{1}{2}h(a + b)$ — The specified variable is b.

$2A = h(a + b)$ — Multiply both members by the LCD, 2.

$2A = ha + hb$ — Distribute the multiplication of h.

$2A - ha = hb$ — Subtract ha from both members to isolate the term containing b.

$\frac{2A - ha}{h} = b$ — Divide both members by h.

$b = \frac{2A - ha}{h}$

Caution: Although this answer could also be written as $b = \frac{2A}{h} - a$, note that $b = \frac{2A - a}{h}$ is incorrect. ▼

SELF-CHECK ANSWERS ✓

1 $y = 3x - 4$ **2** $w = \frac{P - 2l}{2}$ **3** $n = 3b + 3m$

EXAMPLE 11 Transposing an Equation Containing Parentheses

Solve $2(3a - b) = 7(2a + b) - 5$ for a.

SOLUTION

$2(3a - b) = 7(2a + b) - 5$ The specified variable is a.

$6a - 2b = 14a + 7b - 5$ Use the distributive property to remove the parentheses.

$6a = 14a + 9b - 5$

$-8a = 9b - 5$ Isolate the specified variable in the left member by adding $2b$ to both sides and then subtracting $14a$ from both sides.

$\dfrac{-8a}{-8} = \dfrac{9b - 5}{-8}$ Obtain a coefficient of 1 for variable a by dividing both sides by -8.

$$a = \frac{-9b + 5}{8}$$

EXAMPLE 12 Temperature Conversions

Solve $F = \frac{9}{5}C + 32$ for C, and complete the following table showing equivalent Fahrenheit and Celsius temperatures.

F	−40°	32°	98.6°	212°
C				

SOLUTION

$F = \frac{9}{5}C + 32$ The specified variable is C.

$F - 32 = \frac{9}{5}C$ Subtract 32 from both members.

$C = \frac{5}{9}(F - 32)$ Multiply both members by $\frac{5}{9}$.

Substitute the given values in the equation, and complete the table (see the figure).

$$C = \frac{5}{9}(-40 - 32) = \frac{5}{9}(-72) = -40 \qquad C = \frac{5}{9}(32 - 32) = \frac{5}{9}(0) = 0$$

$$C = \frac{5}{9}(98.6 - 32) = \frac{5}{9}(66.6) = 37 \qquad C = \frac{5}{9}(212 - 32) = \frac{5}{9}(180) = 100$$

F	−40°	32°	98.6°	212°
C	−40°	0°	37°	100°

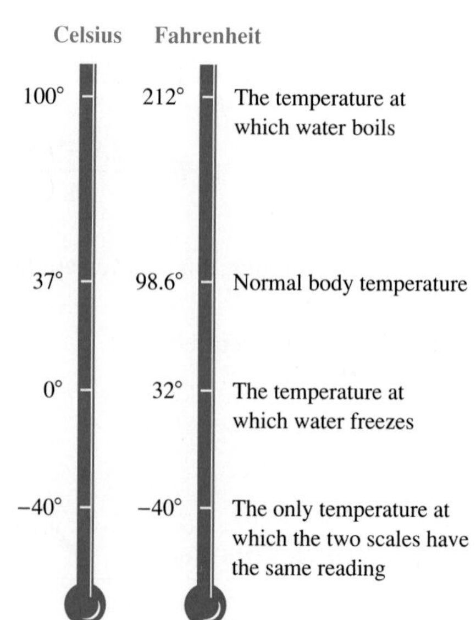

Exercises 1-4

A

In Exercises 1–6 fill in the blanks to complete each statement.

1 Equations that have the same solution set are called ________ equations.

2 A value of the variable that makes an equation true is called a(n) ________ of the equation.

3 An equation that has no solution is called a(n) ________.

4 An equation that is always true is called a(n) ________.

5 A first-degree equation in one variable is called a(n) ________ equation in one variable.

6 A linear equation that has exactly one solution is a(n) ________ equation.

In Exercises 7 and 8 solve each linear equation, and check your solution.

7 **a.** $v + 5 = 22$ **b.** $13 = m + 9$

c. $5x = 20$ **d.** $\frac{3n}{7} = 6$

e. $3m - 5 = 5m + 3$

8 **a.** $w - 11 = 28$ **b.** $78 = t - 2$

c. $4z = -20$ **d.** $\frac{5v}{11} = \frac{3}{22}$

e. $4m + 7 = 7m - 5$

In Exercises 9–34 solve each equation.

9 **a.** $-2n = 1$ **b.** $\frac{4z}{7} = \frac{28}{3}$

c. $-x = 0$ **d.** $2t = t + 5$

10 **a.** $-6r = 2$ **b.** $\frac{3y}{5} = \frac{9}{2}$

c. $-\frac{z}{2} = 7$ **d.** $5t + 6 = 4t$

11 $3a - 6 = 4a + 13$

12 $7k - 9 = 8k + 19$

13 $2t + 1 = 5t - 27 - 10t$

14 $4(z - 3) = 3z - 5$

15 $3n - 7 + (n + 1) = 11n + 1$

16 $2 - 6(s + 1) = 4(2 - 3s) + 6$

17 $(y - 6) + (y - 2) + y = -8$

18 $12x + 5x - 4 = 3x - 4 + 2x$

19 $2(z + 5) - 2(5 - z) = 8$

20 $26t + (t + 1) = 25t + (2t + 1)$

21 $183x = 183x - 2$

22 $3(2w - 5) - (3 - 6w) = 12$

23 $12 + 3(2w - 5) = -(3 - 6w)$

24 $2(v + 2) = 27 - 2(v + 3)$

25 $7(w - 1) - 4(2w + 3) = 2(w + 1) - 3(3 - w)$

26 $11(x - 3) + 4(2x + 1) = 5(3x - 2) + 13$

27 $7(m - 1) - 4(2m + 3) = 2(m + 1) - 3(4 - m)$

28 $4(s - 1) + 3(s + 2) = 7(s + 1) - 5$

29 $\frac{v}{3} - \frac{1}{6} = \frac{v + 1}{9}$

30 $\frac{2(t + 1)}{3} = \frac{2(t - 5)}{5}$

31 $\frac{v}{3} + 20 = \frac{v}{4} - \frac{v}{12} - \frac{v}{9}$

32 $\frac{s}{14} - \frac{2}{5} = \frac{s}{21} + \frac{4s}{175}$

33 $\frac{4x + 2}{5} + 2 = 3 - \frac{3 - 7x}{2}$

34 $\frac{24w - 67}{60} = \frac{3w - 8}{12}$

In Exercises 35–38 solve each linear equation for *y*.

35 $12x + 3y = 15$

36 $12x - 2y = 20$

37 $3x - 2y = 7$

38 $2x + 3y = -8$

In Exercises 39–54 solve each equation for the variable specified.

39 **Volume of a Parallelepiped** $V = lwh$ for h

40 **Perimeter of a Triangle** $P = a + b + c$ for b

41 **Amount of an Investment** $A = p + prt$ for r

42 **Area of a Triangle** $A = \frac{1}{2}bh$ for b

43 **Volume of a Cone** $V = \frac{1}{3}\pi r^2 h$ for h

44 **Interest Formula** $I = PRT$ for R

45 **Slope-Intercept Form of a Line** $y = mx + b$ for x

46 **Sum of an Arithmetic Sequence** $S = \frac{n}{2}(a + l)$ for a

47 **Area of a Trapezoid** $A = \frac{1}{2}h(a + b)$ for a

48 **Point-Slope Form of a Line** $y - y_1 = m(x - x_1)$ for x

49 **Last Term of an Arithmetic Sequence** $l = a + (n - 1)d$ for d

50 **Boyle's Law, in Chemistry** $\frac{P_1}{P_2} = \frac{V_2}{V_1}$ for V_1

51 **Sum of an Infinite Geometric Sequence** $S = \frac{a}{1 - r}$ for r

52 **Amount of an Investment** $A = p + prt$ for p

53 Surface Area of a Cylinder $S = 2\pi r^2 + 2\pi rh$ for h

54 Charles's Law, in Chemistry $\frac{V_1}{T_1} = \frac{V_2}{T_2}$ for T_1

B

Exercises 55–60 contrast the concept of solving an equation with that of simplifying an expression. In part a solve the equation, and in part b perform the indicated operations and simplify the result.

	Solve	*Simplify*
55	**a.** $5m - 3(m - 2) = 0$	**b.** $5m - 3(m - 2)$
56	**a.** $5m - 3 - (m - 2) = 0$	**b.** $5m - 3 - (m - 2)$
57	**a.** $4(x - 3) - 5(x + 2) = 0$	**b.** $4(x - 3) - 5(x + 2)$
58	**a.** $4(x - 3) - 5x + 2 = 0$	**b.** $4(x - 3) - 5x + 2$
59	**a.** $\frac{v - 1}{12} - \frac{1}{6} = 0$	**b.** $\frac{v - 1}{12} - \frac{1}{6}$
60	**a.** $\frac{v + 2}{15} + \frac{1}{3} = 0$	**b.** $\frac{v + 2}{15} + \frac{1}{3}$

ESTIMATION SKILLS (61–66)

61 The best mental estimate of the solution of $4.9984y = 39.973$ is

a. 65 **b.** 6.5 **c.** 7.5
d. 8 **e.** 8.5

62 The best mental estimate of the solution of $7.0186x = 631.0452$ is

a. 900 **b.** 90 **c.** 9
d. 0.9 **e.** 8.5

63 The best mental estimate of the solution of $w - 7.8934 = 4.10527$ is

a. 12 **b.** 11 **c.** −11
d. −12 **e.** 10

64 The best mental estimate of the solution of $m + 0.05931 = 0.04027$ is

a. 0.02 **b.** −0.02 **c.** 0.01
d. −0.01 **e.** −0.03

65 The best mental estimate of the solution of $10z - 0.01z = 21.99$ is

a. 0.12 **b.** 0.22 **c.** 2.2
d. 12.0 **e.** 220

66 The best mental estimate of the solution of $8.4t + 3.4 = 1.001 - 1.57t$ is

a. −0.24 **b.** −2.4 **c.** 2.4
d. 24 **e.** −24

In Exercises 67–70 solve each equation. Be careful not to include any extraneous values as solutions.

67 $\frac{2}{x - 5} + 6 = \frac{8}{x - 5}$

68 $\frac{3}{z - 1} + 2 = \frac{5}{z - 1}$

69 $\frac{5}{w - 4} = 3 + \frac{5}{w - 4}$

70 $\frac{z}{z - 4} = \frac{4}{z - 4} - 1$

In Exercises 71–74 solve each equation for *x*.

71 $ax = b - cx$

72 $2ax + bx = c$

73 $4(2w - 3x) = 5(3w - 2x) + 4$

74 $3(2x - 5b) + 7x = 4x - (x - 3b) + 1$

In Exercises 75–78 solve each equation for *y′*.

75 $y + 2y' = 3x$

76 $x + 2yy' - 5 = 0$

77 $2x + y + xy' + 2yy' = 0$

78 $x + xy' + y + y + y' = 0$

C

CALCULATOR USAGE (79–82)

In Exercises 79–82 solve each equation with the aid of a calculator. Round all answers to the nearest hundredth.

79 $4.15v + 29.35 = 7.29v + 189.21$

80 $4.17(3.1t + 57) = 6.23(2.2t - 16)$

81 $-0.139(2.8x - 1.39) = 6.71(3.1x + 4.67)$

82 $\frac{4.08x - 3.10}{2.41} = 6.85$

83 Give an example of real numbers a, b, and c for which $ac = bc$ but $a \neq b$.

84 Give an example of real numbers a, b, and c for which $a = b$ but it canot be said that $\frac{a}{c}$ equals $\frac{b}{c}$.

85 Give an example of real numbers a, b, c, and d for which $a + b = c + d$ but $a \neq c$ and $b \neq d$.

86 Is the equation that expresses the associative property of addition an identity, a contradiction, or a conditional equation?

87 If the equation $a = b$ is an identity, then the equation $a + c = b + c$ is a(n) ________.

88 If the equation $a = b$ is a conditional equation, then the equation $a - c = b - c$ is a(n) ________.

DISCUSSION QUESTIONS (89–90)

89 Discuss the procedure for solving a linear equation. In your discussion of equivalent equations, use the analogy of a balance scale and refer to the Addition and Multiplication Theorems of Equality.

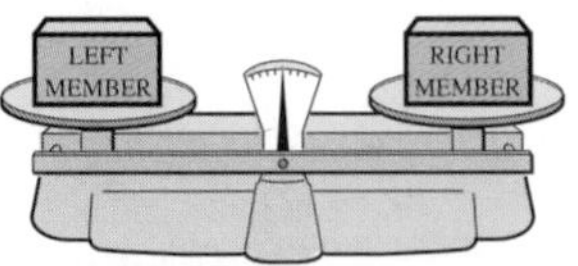

Figure for Exercise 89

90 Explain why the equation $a = b$ might not be equivalent to the equation $ac = bc$.

CHALLENGE QUESTION

A tub containing 100 pounds of saltwater that is 99% water is set out in the morning sun. At the end of the day the sun has evaporated some of the water, leaving saltwater that is 96% water. How many pounds of saltwater remain?

SECTION SKILL REINFORCERS

Simplify each of these expressions.

1 -10^2 **2** $(-10)^2$ **3** $1^2 + 1^2 + 1^2 + 1^2$ **4** $(1 + 1 + 1 + 1)^2$ **5** $4^3 - 4^2$

1-5 *Applications of Linear Equations*

SECTION OBJECTIVE

11 Solve word problems using the mixture and rate principles.

After college the problems that you will encounter involving mathematics will usually be stated in words. Thus you must be able to work word problems in order to fully benefit from any mathematics that you have learned.

In this section we will give a strategy for solving word problems. As you study, try to focus on learning and implementing this strategy rather than on solving individual problems.

STRATEGY FOR SOLVING WORD PROBLEMS

Step 1 Read the problem carefully to determine what you are being asked to find.

Step 2 Represent each unknown numerical quantity by using a representative variable, and specify precisely what the variable represents. If necessary, make a sketch.

Step 3 If necessary, condense the problem into a **word equation**. Then, using the variable chosen in Step 2, translate the word equation into an algebraic equation.

Step 4 Solve the algebraic equation, and answer the question asked by the problem.

Step 5 Check the reasonableness of your answer.

A Mathematical Note

My grandfather Albert Murphy taught for 31 years, from 1907 to 1938. His total salary for those 31 years of teaching in one-room schools was $14,238. Almost all of the arithmetic problems in one of the books he taught from consisted of word problems. Problem 61 at the end of this section was taken from a 1911 arithmetic text that he used.

An important part of this strategy is translating key words and phrases into their algebraic equivalents (see Figure 1-10). Although it is unlikely that you will encounter numeric word problems in your future job, these problems provide excellent practice in translating key words into their algebraic equivalents. As you work these problems, it is wise to read carefully with a pencil in hand to highlight key words and phrases. Examine each problem for words such as *find, determine,* or *evaluate* or for a sentence ending with a question mark. Also examine the problem for words that indicate one of the operations of addition, subtraction, multiplication, division, or exponentiation.

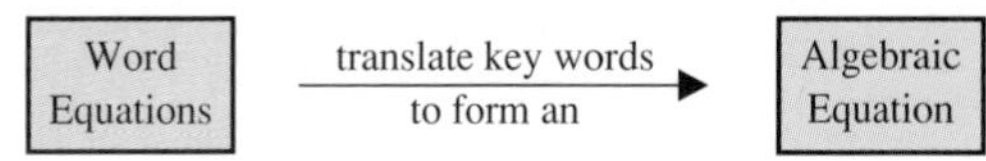

Figure 1-10

EXAMPLE 1 Numeric Word Problem

Five more than twice a number is thirty-seven. Find this number.

SOLUTION Let

$n =$ The unknown number — Represent the unknown number with a variable.

| 5 | more than | twice a number | is | 37 |

First write the equation in words.

$$2n + 5 = 37$$

Translate this **word equation** into an algebraic equation, using the variable n. Then solve this equation.

$$2n = 32$$

$$n = 16$$

Check Twice 16 plus 5 is 37.

Answer The number is 16.

The correct position for the equal sign in an equation can often be determined by locating a verb that means "equal to"—for example, *is, are, is equal to, is the same as,* or *equals.*

Consecutive integers are integers that are adjacent in the usual counting sequence, such as 16 and 17 or -8 and -7. If n represents an integer, then $n + 1$ and $n + 2$ represent the next two consecutive integers. Consecutive even integers, such as 10, 12, and 14, differ by 2. Consecutive odd integers, such as 7, 9, and 11, also differ by 2. If m is an odd integer, then the next three odd integers are $m + 2$, $m + 4$, and $m + 6$.

SELF-CHECK

Four less than five times a number is forty-one. Find this number.

REPRESENTING CONSECUTIVE INTEGERS

Consecutive integers differ by one. If n is an integer, the next two consecutive integers are $n + 1$ and $n + 2$.

Consecutive even integers differ by two. If n is an even integer, the next two consecutive even integers are $n + 2$ and $n + 4$.

Consecutive odd integers differ by two. If n is an odd integer, the next two consecutive odd integers are $n + 2$ and $n + 4$.

SELF-CHECK ANSWER 

The number is 9.

EXAMPLE 2 Consecutive Integers

Find two consecutive integers whose sum is 75.

SOLUTION Let

n = First integer

$n + 1$ = Second integer — Consecutive integers differ by 1.

First integer + Second integer = 75 — First write the equation, using a combination of words and symbols.

$n + (n + 1) = 75$ — Translate this **word equation** into an algebraic equation.

$2n + 1 = 75$ — Then simplify, and solve for the first integer, n, and the second integer, $n + 1$.

$2n = 74$

$n = 37$

$n + 1 = 38$

Answer The integers are 37 and 38. — Do these numbers check? ▼

EXAMPLE 3 Consecutive Odd Integers

If the first of two consecutive odd integers is tripled, the result is 24 more than the second integer. Determine these integers.

SOLUTION Let

n = First odd integer

$n + 2$ = Second odd integer — Remember that odd integers differ by 2.

Three times the first integer = Second integer + 24 — Write the equation in words and symbols.

$3n = (n + 2) + 24$ — Translate this **word equation** into an algebraic equation.

$3n = n + 26$ — Then simplify, and solve for the first integer, n, and the second integer, $n + 2$.

$2n = 26$

$n = 13$

$n + 2 = 15$

Answer The consecutive odd integers are 13 and 15. ▼

When two numbers are compared using the operation of division, we often use the term **ratio**.

SELF-CHECK 

Find two consecutive even integers whose sum is 98.

RATIO

The **ratio** of a to b means $a \div b$ and is denoted by $\frac{a}{b}$ or $a:b$.

SELF-CHECK ANSWER 

The integers are 48 and 50.

EXAMPLE 4 Ratio of Two Numbers

The ratio of two numbers is $\frac{3}{5}$. If the first number is 8 less than the second number, determine each of these numbers.

SOLUTION Let

$n =$ Second number

$n - 8 =$ First number

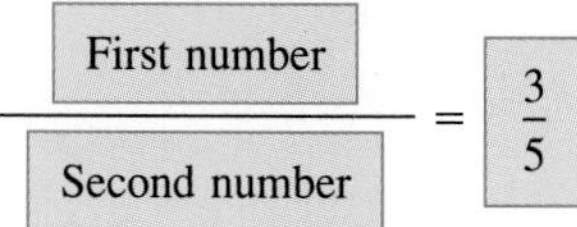

$$\frac{\boxed{\text{First number}}}{\boxed{\text{Second number}}} = \boxed{\frac{3}{5}}$$ This is the **word equation.**

$$\frac{n-8}{n} = \frac{3}{5}$$ Translate this word equation into an algebraic equation.

$$5n\left(\frac{n-8}{n}\right) = 5n\left(\frac{3}{5}\right)$$ Multiply both sides by the LCD, $5n$, for $n \neq 0$.

$$5(n-8) = 3n$$ Simplify, and then solve for n and $n - 8$.

$$5n - 40 = 3n$$

$$2n = 40$$

$$n = 20$$

$$n - 8 = 12$$

Answer The numbers are 12 and 20. Do these numbers check?

It has been shown that many students who have difficulty with word problems often either skip the step of identifying the variable or make the identification so vague that it is not useful. Therefore you should take the time to identify your unknown precisely with a representative variable, such as P for profit or C for cost. Try to avoid the common habit of always selecting the variable x, since this may prevent you from focusing precisely on the quantity you are trying to find. Also, using a representative variable will establish a good habit that will carry over into any computer programming you might do.

Two general principles that are used to form equations for a wide variety of applications are the mixture principle and the rate principle.

General Principles Used to Form Equations

Mixture principle for two ingredients:

Amount in first + Amount in second = Total amount in mixture

Rate principle:

Amount = Rate · Base

The following lists show how these principles can be used in several applications. These lists also illustrate the power of a general mathematical principle to solve seemingly unrelated problems.

MIXTURE PRINCIPLE

Amount in first + Amount in second = Total amount in mixture

Applications

1 Area in first region + Area in second region = Total area
2 Distance by car + Distance by plane = Total distance
3 Value of dimes + Value of quarters = Total value of coins
4 Interest on bonds + Interest on CDs = Total interest
5 Medicine in first solution + Medicine in second solution = Total medicine in mixture
6 Work by craftsperson + Work by apprentice = Total work

RATE PRINCIPLE

Amount = Rate · Base

Applications

1 Amount of work = Rate of work · Time worked ($W = RT$)
2 Amount of distance = Rate of travel · Time traveled ($D = RT$)
3 Amount of interest = Principal · Rate of interest · Time ($I = PRT$)
4 Amount of money = Value per coin · Number of coins
5 Amount of medicine in solution = Percent of medicine in solution · Quantity of solution

EXAMPLE 5 Area of a Window

In order to calculate heat loss in a new building, the architect must determine the total area of several Norman windows in this building. Determine the area of the Norman window shown in the figure. (A Norman window has a semicircular top on a rectangular base.)

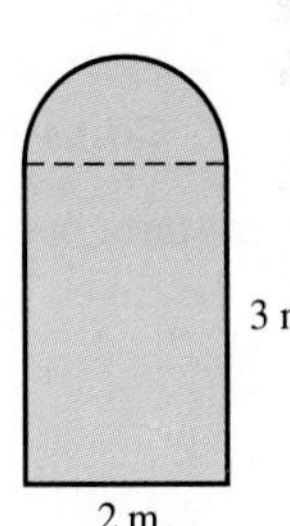

SOLUTION

Total area = Area of rectangular base + Area of semicircle

The **word equation** is based on the mixture principle.

$A = l \cdot w + \frac{1}{2}\pi r^2$

Write the area, using the formulas for the area of a rectangle and a semicircle.

$A = 3(2) + \frac{1}{2}\pi(1)^2$

Substitute the given values into the equation, and then simplify. Since the diameter of the semicircle is 2 m, the radius is 1 m.

$A = 6 + \frac{1}{2}\pi$

$A \approx 7.6$

Approximate using a calculator with a  key.

Answer The total area is approximately 7.6 m^2. 

EXAMPLE 6 Distance Traveled by Two Planes

Two planes depart in opposite directions. After 2 hours the faster plane has traveled three times as far as the slower plane. Find the distance that each plane has traveled if the distance between the planes is 840 kilometers (see the figure).

SOLUTION Let

d = Distance traveled by the slower plane

$3d$ = Distance traveled by the faster plane

The faster plane has traveled three times the distance the slower plane has traveled.

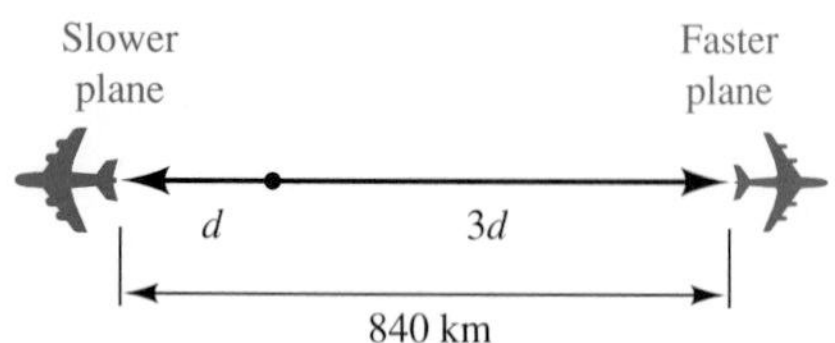

Physical applications can often be clarified by making a sketch to represent the problem.

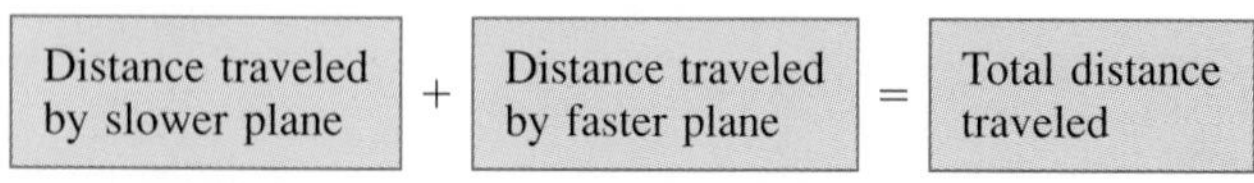

This **word equation** is based on the mixture principle.

$$d + 3d = 840$$
$$4d = 840$$
$$d = 210$$
$$3d = 630$$

Substitute the values into the word equation, and then solve for d.

Check $210 + 630 = 840$

Answer The slower plane has traveled 210 km, and the faster plane has traveled 630 km.

Do these distances seem reasonable?

EXAMPLE 7 Coins Needed for Change

A paper carrier starts out to make his collections with $4.00 worth of pennies and nickels. If he takes three times as many pennies as nickels, how many of each does he take?

SOLUTION Let

n = Number of nickels

$3n$ = Number of pennies

The problem is to determine how many pennies and how many nickels the paper carrier has. There are three times as many pennies as nickels.

Value of nickels	+	Value of pennies	=	Total value of the coins

The **word equation** is based on the mixture principle.

Rate Principle:	**Value per Coin**	·	**Number of Coins**	=	**Total Value of Coins**
Nickels	0.05	·	n	=	$0.05n$
Pennies	0.01	·	$3n$	=	$0.03n$

Use the rate principle and a table to organize the information for the different types of coins. The values needed in the word equation are in the right-hand column of the table. The values in the first two columns and the equation at the top of the table are used to fill in this third column.

$$0.05n + 0.03n = 4.00$$

Substitute the values from the table into the word equation.

$$0.08n = 4.00$$
$$8n = 400$$
$$n = 50$$
$$3n = 150$$

Multiply both members by 100, simplify, and solve for n.

Answer The paper carrier has 150 pennies and 50 nickels.

Are these numbers reasonable?

EXAMPLE 8 Interest on Two Savings Accounts

A retiree deposited a total of \$5000 in two different savings accounts. The interest rate on one account was 5.5%, and the interest rate on the other account was 6.0%. At the end of the first year the interest earned on the first account was \$102.50 more than the interest earned on the second account. Determine the amount deposited in each account.

SOLUTION Let

$$p = \text{Principal invested at } 5.5\%$$
$$5000 - p = \text{Principal invested at } 6.0\%$$

The total principal invested was \$5000.

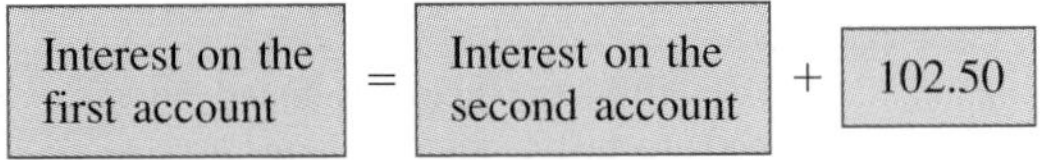

Word equation

Rate Principle:	Principal · Rate · Time = Interest
First account	$p \cdot 0.055 \cdot 1 = 0.055p$
Second account	$(5000 - p) \cdot 0.060 \cdot 1 = 0.060(5000 - p)$

Organize the information for each account in a table. The values in the word equation are in the right-hand column of this table. The values in the first three columns and the equation at the top of the table are used to fill in this fourth column.

$$0.055p = 0.060(5000 - p) + 102.50$$

Substitute the values from the table into the word equation.

$$0.055p = 300 - 0.060p + 102.50$$
$$0.115p = 402.50$$
$$p = 3500$$
$$5000 - p = 1500$$

Simplify, and then solve for p.

Answer \$3500 was invested in the first account at 5.5%, and \$1500 was invested in the second account at 6%.

Are these values reasonable? Do they check?

SELF-CHECK 

A total of \$1900 worth of dimes and nickels is collected from a toll machine. A typical collection contains nine times as many dimes as nickels. Assuming this collection is typical, determine the number of each in the collection.

The preceding examples have illustrated the use of tables with some word problems involving the mixture and rate principles. The tables summarize the information in a clear format that makes it easy to form the algebraic equation for each problem. This format is particularly useful for problems involving solutions of chemicals or medicines. For example, the table in Example 9 makes it easier to preserve the distinction between the amount of medicine and the amount of solution that contains the medicine.

SELF-CHECK ANSWER 

The collection contains 2000 nickels and 18,000 dimes.

EXAMPLE 9 Mixture of Two Solutions of Medicine

A nurse must administer 5 ounces of a 12% solution of medicine. The hospital has a 25% solution and a 5% solution of this medicine in stock. How many ounces of each must be mixed to obtain 5 ounces of a 12% solution? (See the figure.)

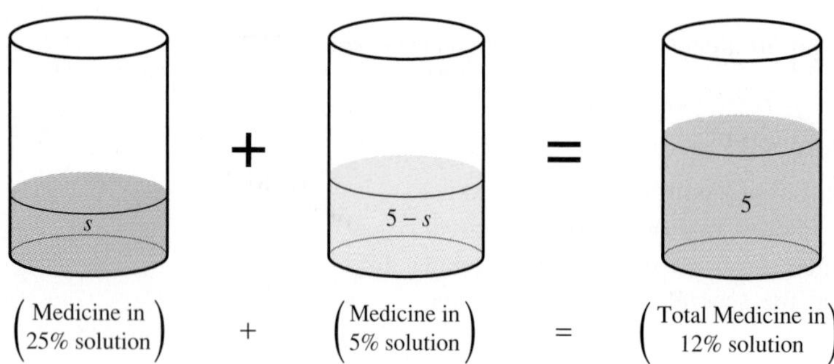

(Medicine in 25% solution) + (Medicine in 5% solution) = (Total Medicine in 12% solution)

SOLUTION Let

$$s = \text{Number of ounces of 25\% solution}$$
$$5 - s = \text{Number of ounces of 5\% solution}$$

Identify the unknowns, using the fact that the total number of ounces is 5.

Medicine in 25% solution	+	Medicine in 5% solution	=	Total medicine in 12% solution

The **word equation** is based on the mixture principle.

Rate Principle:	Percent of Solution as a Decimal	·	Number of Ounces of Solution	=	Amount of Medicine in Solution
25% solution	0.25	·	s	=	$0.25s$
5% solution	0.05	·	$5 - s$	=	$0.05(5 - s)$
Total mixture	0.12	·	5	=	$0.12(5)$

Use the rate principle and a table to organize the information for each solution. The values needed in the word equation are in the right-hand column of this table. The values in the first two columns and the equation at the top of the table are used to fill in the third column.

$$0.25s + 0.05(5 - s) = 0.12(5)$$

Substitute the values from the table into the word equation. Multiply both members by 100.

Simplify, and then solve for s.

$$25s + 5(5 - s) = 12(5)$$
$$25s + 25 - 5s = 60$$
$$20s = 35$$
$$s = 1.75$$
$$5 - s = 3.25$$

Answer The nurse should mix 1.75 oz of the 25% solution with 3.25 oz of the 5% solution.

Do these values check?

Many problems involve calculating the time required to complete a task or to do some work. In these problems it is often necessary to determine the rate of work. Since $W = RT$ (Work = Rate · Time), the rate of work is $R = \dfrac{W}{T}$. This formula is used to determine the rate of work in each part of the next example.

EXAMPLE 10 Rate of Work

Determine the rate of work in each of these situations.

	SOLUTIONS
(a) A painter can paint a house in 4 days.	$R = \frac{1}{4}$ house per day
(b) An outlet pipe can drain a pool in 7 hours.	$R = \frac{1}{7}$ pool per hour
(c) A printer can print 450 characters in 3 seconds.	$R = \frac{450 \text{ characters}}{3 \text{ seconds}} = 150$ cps
(d) A computer takes h hours to print 10,000 payroll checks.	$R = \frac{10{,}000}{h}$ checks per hour
(e) An accountant can complete a monthly report in t hours.	$R = \frac{1}{t}$ report per hour

SELF-CHECK

A pharmacist must prepare 40 milliliters of a 25% solution. She does not have this particular solution in stock, but she does have both a 40% solution and 20% solution on hand. Fill in the table below to determine how much of each solution she should mix to obtain the desired prescription.

	Percent of Solution as a Decimal	· mL of Solution	= mL of Drug
20%	0.20		
40%	0.40		
Mixture	0.25		

EXAMPLE 11 Work Done by Two Painters

Painter A can paint a small house in 9 hours. When painter B helps him, it takes the two working together only 6 hours. How many hours would it take painter B to paint the house working alone?

SOLUTION Let

t = Time in hours for B to paint the house if working alone

Be very specific about whose time t represents and whether the time is for the painter alone or with help.

Work A does	+	Work B does	=	Total work in painting 1 house

The **word equation** is based on the mixture principle.

	Rate	·	Time	=	Work
Painter A	$\frac{1}{9}$	·	6	=	$\frac{2}{3}$
Painter B	$\frac{1}{t}$	·	6	=	$\frac{6}{t}$

The table shows the work done by each painter when they work together for 6 hours. Since A's time for 1 house is 9 hours, his rate is $\frac{1}{9}$ house per hour. A paints $\frac{2}{3}$ of the house.

SELF-CHECK ANSWER

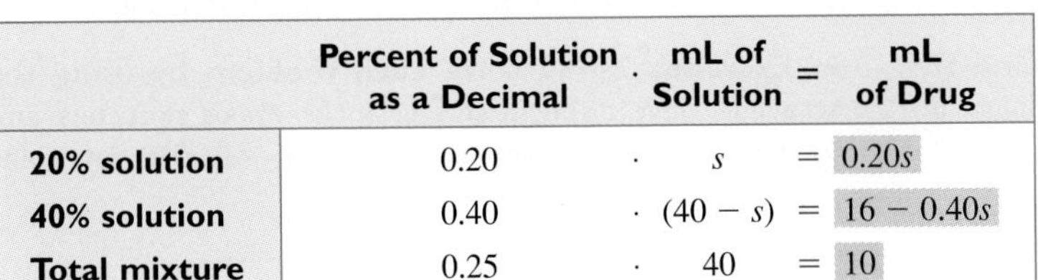

	Percent of Solution as a Decimal	·	mL of Solution	=	mL of Drug
20% solution	0.20	·	s	=	$0.20s$
40% solution	0.40	·	$(40 - s)$	=	$16 - 0.40s$
Total mixture	0.25	·	40	=	10

She should mix 30 mL of 20% solution and 10 mL of 40% solution.

$$\frac{2}{3} + \frac{6}{t} = 1$$

Substitute the values from the table into the word equation.

$$3t\left(\frac{2}{3}\right) + 3t\left(\frac{6}{t}\right) = 3t(1)$$

Multiply both members by the LCD, $3t$. Then solve for t.

$$2t + 18 = 3t$$

$$18 = t$$

Answer Working alone, painter B could paint the house in 18 h.

Does this value check?

SELF-CHECK

Machine A takes 6 hours to clean the streets in one sector of a city, whereas machine B can do the same job in 4 hours. How many hours will it take to clean this sector if the machines work simultaneously?

Exercises 1-5

A

Numeric Word Problems (1–20)

In Exercises 1–20 write an equation for each problem using the variable n, and then use this equation to solve the problem.

1 Seven more than twice a number is thirteen. Find this number.

2 Forty more than three times a number equals 19. Find this number.

3 Five times a number plus 2 equals three times the number added to 2. Find this number.

4 Six times a number decreased by four times the number is equal to ten. Find the number.

5 If the sum of a number and -3 is divided by 10, the result is equal to the quotient obtained when the number minus 2 is divided by 3. What is the number?

6 If the sum of a number and 1 is divided by 6, the result is equal to the quotient obtained when the number plus 3 is divided by 10. What is this number?

7 Find two consecutive integers whose sum is 85.

8 Find two consecutive integers whose sum is -41.

9 Find two consecutive odd integers whose sum is 112.

10 Find two consecutive even integers whose sum is 190.

11 If the first of two consecutive odd integers is multiplied by 5, the result is 26 more than the second integer. Find each integer.

12 If the smaller of two consecutive integers is decreased by 5, the result is four times the larger integer. Determine the smaller integer.

13 The sum of the smallest and the largest of three consecutive odd integers is 82. Determine the smallest of these integers.

14 The sum of the smallest and the largest of three consecutive even integers is 100. Determine all three of these integers.

15 The ratio of two numbers is $\frac{3}{7}$. If the first number is 20 less than the second number, determine each of these numbers.

16 The ratio of two numbers is $\frac{4}{5}$. If the first number is 4 less than the second number, determine each of these numbers.

17 The ratio of two numbers is $\frac{5}{6}$. The second number is 18 less than twice the first number. Find both numbers.

18 The ratio of two numbers is $\frac{3}{8}$. The second number is 15 less than three times the first number. Find both numbers.

19 What number is increased by 6 when it is tripled?

20 What number is decreased by 7 when it is doubled?

In Exercises 21–42 solve each problem by using the strategy developed in this chapter. Make sketches and form tables wherever appropriate. All the formulas

SELF-CHECK ANSWER

It will take 2.4 h to clean this sector if the machines work simultaneously.

needed to work Exercises 25–30 are included on the inside cover of this book. Assume all curved portions shown are either circles or semicircles.

21 Complementary Angles Angle A contains a degrees, and angle B is 12 degrees larger than angle A. Angles A and B are complementary. Find the number of degrees in each angle. (Two angles are complementary if the sum of their measures is 90°.)

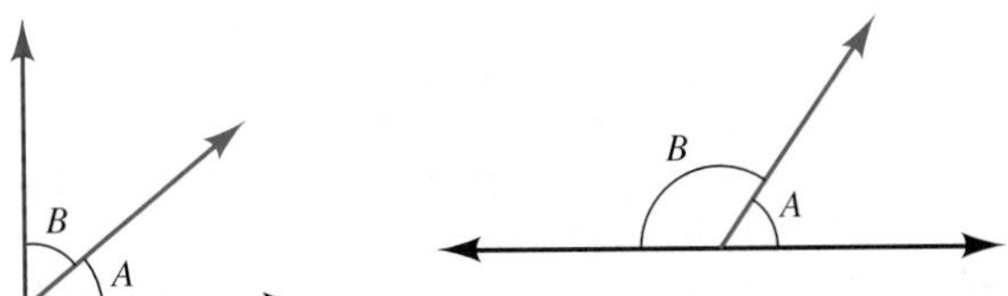

Figure for Exercise 21 **Figure for Exercise 22**

22 Supplementary Angles Angle A contains a degrees, and angle B is 36 degrees larger than angle A. Angles A and B are supplementary. Find the number of degrees in each angle. (Two angles are supplementary if the sum of their measures is 180°.)

23 GM Credit Card In 1992 General Motors started sponsoring a new credit card. For each charge to this card GM credits 5% of the purchase price toward an account that the customer can apply to the purchase of a new GM automobile. The maximum amount that can be accumulated in this account is \$3500. What amount would the customer have to charge with this card to accumulate a maximum credit of \$3500?

Figure for Exercise 23

24 Credit Card Interest Rate The monthly interest charge on a credit card balance of \$650 was \$9.75. What was the monthly interest rate charged by this credit card company?

In Exercises 25 and 26 find the perimeter of each figure.

25

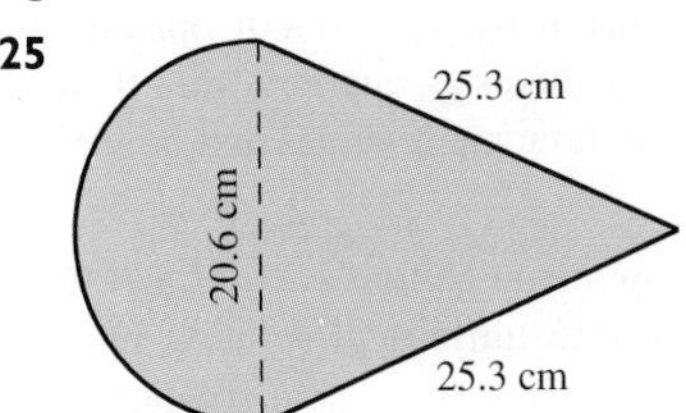

26

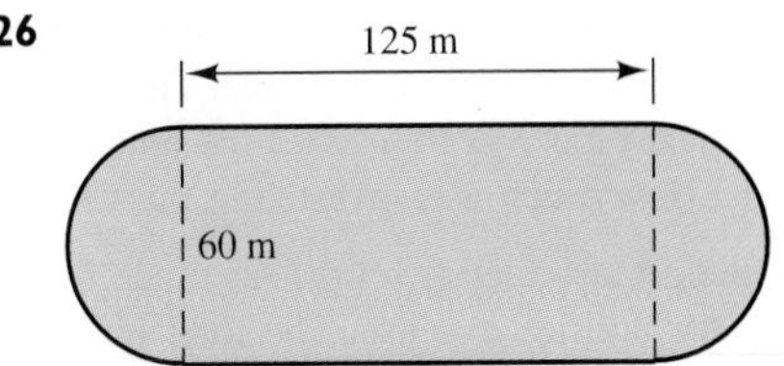

In Exercises 27 and 28 find the total area in each region.

27

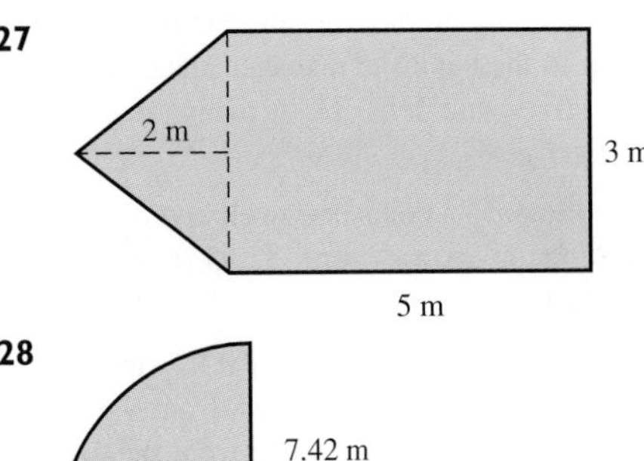

28

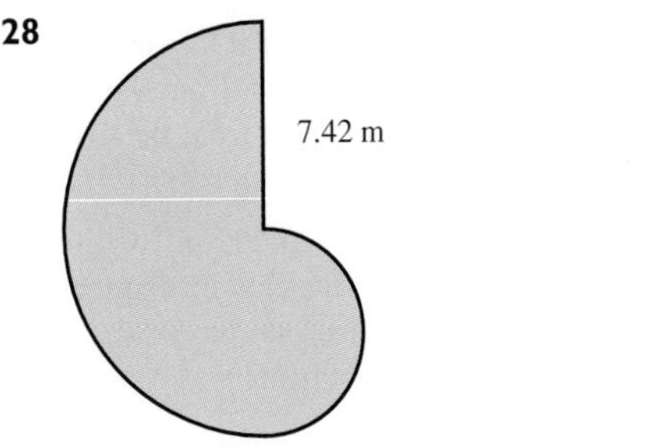

In Exercises 29 and 30 find the total volume of each solid.

29 (shaded portion only) **30** (shaded portion only)

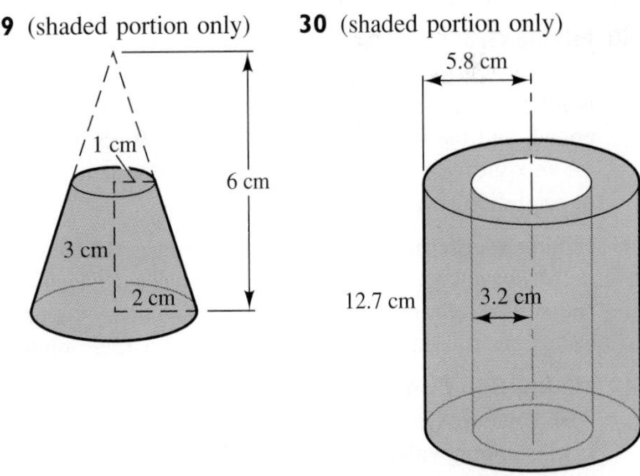

31 Distance by Ship Two boats depart simultaneously in opposite directions. Later they are 31 kilometers apart. If one boat traveled 5 kilometers farther than the other, how far did the slower boat travel?

32 Distance by Submarine Two submarines depart simultaneously in opposite directions. After 3 hours one submarine has traveled twice as far as the slower submarine. Find the distance each submarine has traveled if the distance between them is 63 kilometers.

33 Investments An investment of \$6250 was split between certificates of deposit and a utility mutual fund. The amount invested in the mutual fund was four times the amount invested in the CDs. How much was invested in the CDs?

34 Budgeting A budget allocated five times as much for interest expense as for utilities expense. The total budgeted for these expenses was \$6300. How much was budgeted for utilities?

35 Change A supermarket usually needs three times as many \$5 bills as \$10 bills in order to transact its daily business. On one trip to the bank the manager plans to pick up \$5000 worth of fives and tens. How many of each bill should the manager get?

36 Coins A girl collected 75 coins in nickels and dimes from her allowance. If the coins are worth \$5.95, how many of each did she collect?

37 Interest A family invested \$4500, part at 8% and the rest at 10% annual interest. The yearly interest from the 8% investment was \$90 more than the interest from the 10% investment. How much was invested at each rate?

38 Profit A man invested a total of \$25,000 in two investments. He made a profit of 5% on the first investment and 2% on the second investment. If his total profit was \$650, what was the amount of each investment?

39 Mixing Meats Meat worth \$2.05 per pound is mixed with meat worth \$1.25 per pound to produce 500 pounds of hamburger worth \$1.35 per pound. How much of each cut of meat is used in the hamburger?

40 Mixing Grass Seeds A homeowner purchased a mixture of grass seed composed of fine perennial bluegrass and rapid-growing annual rye. The bluegrass cost \$1.25 per pound, and the rye cost \$0.95 per pound. If 25 pounds of the mixture was purchased for \$29.15, how many pounds of each did the homeowner buy?

41 Mixing Medicines A nurse must administer 4 ounces of a 15% solution of medicine. In stock are a 10% solution and a 50% solution of this medicine. How many ounces of each should he mix to obtain 4 ounces of a 15% solution?

42 Mixing Solutions A druggist needs 20 milliliters of a 30% solution. To obtain this solution, she mixes an 80% stock solution with a diluent (0% solution). How many milliliters of the stock solution and how many milliliters of the diluent should be used?

In Exercises 43 and 44 determine the rate of work for each situation.

43 a. A roofer can complete a roof in 5 days.

b. A boy can mow a yard in 4 hours.

c. A manufacturer can produce 36 cars in 2 hours.

d. A secretary can type a report in t hours.

44 a. A repair person can fix a copy machine in 3 hours.

b. A girl can paint her room in 2 days.

c. A printer can produce 24 signs in 8 hours.

d. A hose can fill a swimming pool in t hours.

In Exercises 45 and 46 solve each problem using the strategy developed in this chapter.

45 Planting Shrubbery A landscape contractor has two employees plant shrubbery around a new office building. The more experienced employee could do the job alone in 2 days, whereas the new employee would need 3 days working alone. How many days will it take them to plant the shrubbery if they work together?

46 Filling a Trough One pipe can fill a watering trough in 50 minutes, and another can fill it in 75 minutes. If both pipes are turned on, how many minutes will it take to fill half of the trough?

B

In Exercises 47–61 solve each problem using the strategy developed in this chapter.

Numeric Word Problems (47–50)

47 The sum of a number, 2 more than three times the number, and 5 less than four times the number is 133. What is this number?

48 The sum of three numbers is 71. The second number is 8 less than the first, and the third is 3 more than twice the first. Find the three numbers.

49 The first of four whole numbers is twice the third, the second is 4 less than the first, and the fourth is 6 more than the second. If their sum is 82, find the four numbers.

50 The first of four whole numbers is 2 more than the third, the second is 4 more than the third, and the fourth is twice the first. If their sum is 65, find the four numbers.

51 Painting A painter can paint a room in 3 hours, whereas an apprentice would need 5 hours. How many hours would it take the two of them to paint the room if they worked together?

52 Fueling a Missile One hose can fuel a missile in 30 minutes, whereas a second hose takes 40 minutes. If both are turned on, how many minutes will it take to fuel the missile?

53 Coins At the end of her shift, a tollbooth operator removed four times as many dimes as quarters from the coin box. If the total was \$104, how many dimes and how many quarters were there?

54 Mixing Solutions How many gallons each of a 20% nitric acid solution and a 45% nitric acid solution must be used to make 6 gallons of a 30% nitric acid solution?

55 Mixing Alloys A goldsmith has two alloys that are 50% and 80% pure gold, respectively. How many grains of each must be used to make 300 grains of an alloy that is 72% pure gold?

56 Preparing an Herbicide A farmer is preparing an herbicide by mixing a 90% solution with pure water (0% solution). How much solution and how much water are needed to fill a 225-gallon tank with 2% solution?

Investments (57 and 58)

57 A retiree wants to invest \$20,000 so as to produce a total monthly income of \$144.50. He invests \$7000 at his local bank at $7\frac{1}{2}$%. What rate must he earn on the rest of the investment to reach his goal?

58 A school district planned to invest \$200,000 of its building fund in two kinds of 6-month certificates of deposit. If the rates are 8.5% and 9%, how much should be invested at each rate in order to earn \$8875.00 in 6 months?

59 Distance by Car Two cars start toward each other at the same time from two cities that are 382.5 kilometers apart. If one car averages 40 kilometers per hour and the other 45 kilometers per hour, how much time will elapse before they meet?

60 Distance by Plane A jet and a tanker plane are 1050 miles apart. The jet radios to set up a rendezvous for refueling in 1 hour and 15 minutes. If the tanker plane flies at 265 miles per hour, how fast must the jet fly?

61 Profit or Loss A man sold two farms for \$3000 each; on one he gained 10%, and on the other he lost 10%. Determine his total gain or loss.*

62 Concrete Pool Apron Approximate to the nearest tenth of a cubic yard the amount of concrete that should be ordered to pave the shaded apron of a swimming pool, as illustrated in the figure below. The concrete will be poured to a consistent depth of 4 inches.

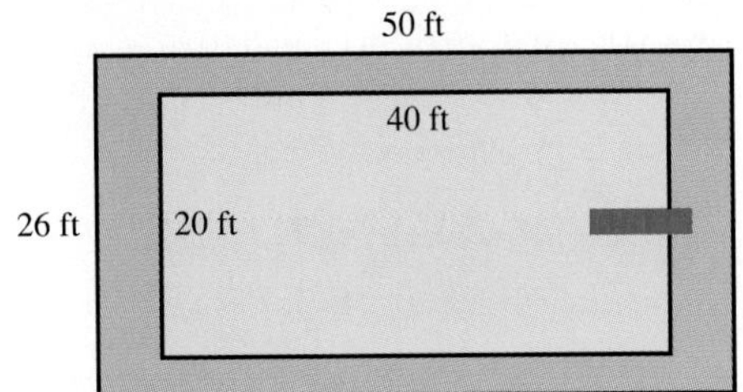

Figure for Exercise 62

63 Find the area of the shaded region shown in the figure.

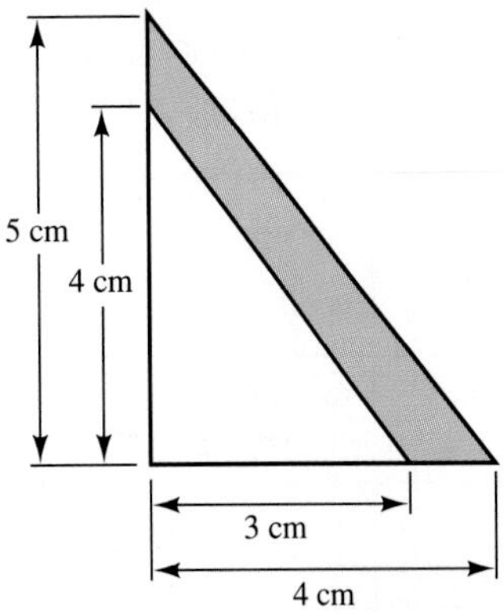

Figure for Exercise 63

64 Determine the area of the shaded region shown in the figure. Assume both curved portions are semicircles of the same radius.

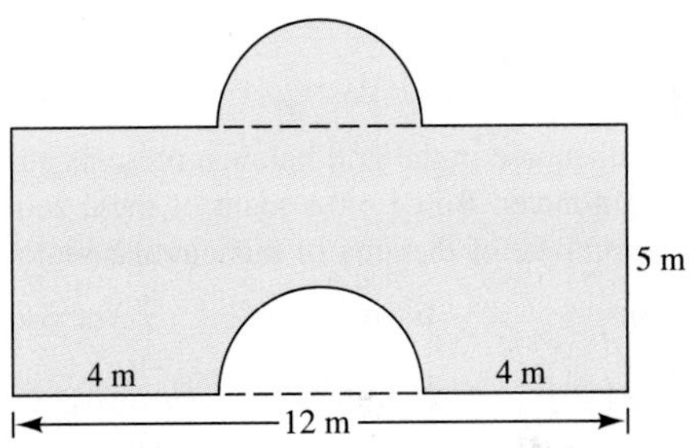

Figure for Exercise 64

65 Find the area of the shaded parallelogram plus the area of the square in the figure.†

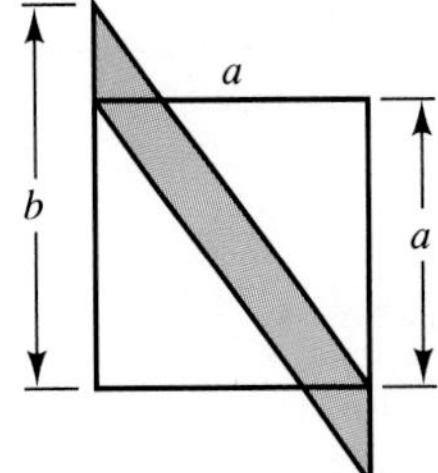

Figure for Exercise 65

---

*This problem comes from the 1911 text *Arithmetic by Analysis* by J. H. Diebel, published by the author. My grandfather Albert Murphy taught from this book for several years in one-room schools in southern Illinois.

†From Donald J. Albers, Stephen B. Rodi, and Ann E. Watkins, *New Directions in Two-Year College Mathematics,* Proceedings of the Sloan Foundation Conference on Two-Year College Mathematics (New York: Springer-Verlag, 1985), p. 357.

ESTIMATION SKILLS (66 AND 67)

66 Each small square in the grid in the figure below represents an area of 1 square centimeter. Based on a count of these square units, the best estimate of the area of this region is

a. 5 cm^2 **b.** 10 cm^2 **c.** 15 cm^2

d. 20 cm^2 **e.** 25 cm^2

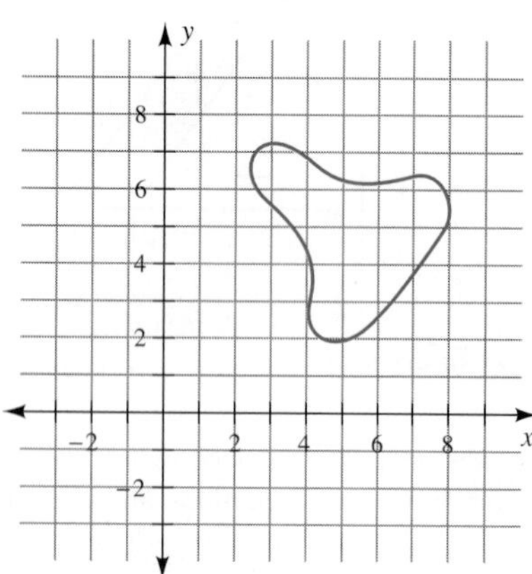

Figure for Exercise 66

67 Each small square in the grid below represents an area of 1 square centimeter. Based on a count of these square units, the best estimate of the area of this circular sector is

a. 12 cm^2 **b.** 10 cm^2 **c.** 8 cm^2

d. 6 cm^2 **e.** 4 cm^2

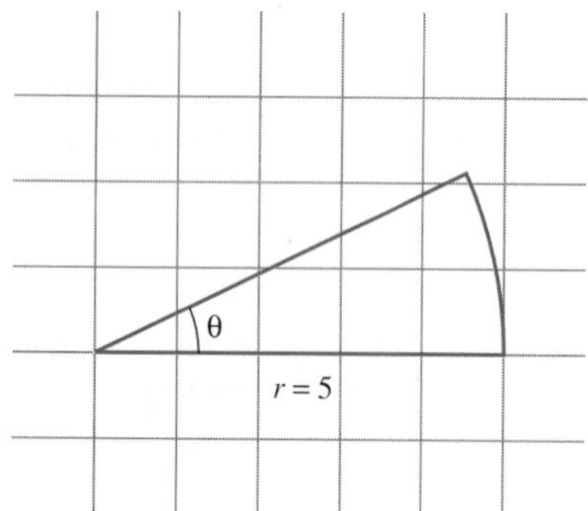

Figure for Exercise 67

DISCUSSION QUESTIONS (68 AND 69)

68 Suppose a classmate of yours is having some difficulty working with word problems. When you observe this student's work, you notice that the unknowns have not been identified. What would you do to help this student?

69 A student's answer to a question asking for the monthly interest charge on a credit card was \$5400. Explain why this answer is not reasonable.

CHALLENGE QUESTION

Mentally determine which is greater, the height or the circumference of the can shown in the figure.

SECTION SKILL REINFORCERS

Solve each of these equations.

1 $n + (n + 1) + (n + 2) = 66$ **2** $3(t + 1) = 5t + 1$ **3** $\frac{x - 3}{3} + \frac{2x - 7}{6} = \frac{5 - x}{2}$

4 $3(x + 4) = 3x + 4$ **5** $2(x + 2) = 3(x + 1) - (x - 1)$

1-6 Solving Linear Inequalities

SECTION OBJECTIVES

12 Use interval notation.

13 Solve linear inequalities in one variable.

An inequality may be thought of as a restriction on a variable. For example, if the cost of an item can be at most \$35, we can write $c \leq 35$. A compound inequality is formed when we place more than one restriction on a variable—for example, when we say that a golfer must hit a drive at least 230 yards, but not more than 245 yards, in order to be on a certain green. Two simple inequalities can be formed from this statement, $d \geq 230$ and $d \leq 245$, but a more concise notation is $230 \leq d \leq 245$.

The **compound inequality** $a \leq x \leq b$ is equivalent to $x \geq a$ *and* $x \leq b$. Read this inequality with x as the subject of the sentence: "x is greater than or equal to a and less than or equal to b." The graph of this inequality and other related inequalities is given in the box below. In the graphs parentheses denote that the endpoint is not included in the interval, and brackets denote that it is included in the interval. This convention for parentheses and brackets is also used in interval notation. **Interval notation** for $1 \leq x < 3$ is $[1, 3)$.

A Mathematical Note

The symbols > and < for greater than and less than are due to Thomas Harriot (1631). These symbols were not immediately accepted, as many mathematicians preferred the symbols ⊏ and ⊐.

INTERVAL NOTATION

Inequality Notation	*Meaning*	*Graph*	*Interval Notation*
$x > a$	x is greater than a	←——(——→ at a	$(a, +\infty)$
$x \geq a$	x is greater than or equal to a	←——[——→ at a	$[a, +\infty)$
$x < a$	x is less than a	←——)——→ at a	$(-\infty, a)$
$x \leq a$	x is less than or equal to a	←——]——→ at a	$(-\infty, a]$
$a < x < b$	x is greater than a and less than b	←(——)→ at a, b	(a, b)
$a < x \leq b$	x is greater than a and less than or equal to b	←(——]→ at a, b	$(a, b]$
$a \leq x < b$	x is greater than or equal to a and less than b	←[——)→ at a, b	$[a, b)$
$a \leq x \leq b$	x is greater than or equal to a and less than or equal to b	←[——]→ at a, b	$[a, b]$

The **infinity symbol,** ∞, is *not* a real number; rather it signifies that the values continue through extremely large values without any end or bound. The symbol $+\infty$ in $[a, +\infty)$ indicates that the interval continues unbounded to the right. The symbol $-\infty$ in $(-\infty, a]$ indicates that the interval continues unbounded to the left.

A Mathematical Note

The symbol ∞ was used to represent infinity by John Wallis in *Arithmetica Infinitorum* in 1655. The Romans had commonly used this symbol to represent one thousand. Likewise, we now use the word *myriad* to mean any large number, although the Greeks used it to mean ten thousand.

EXAMPLE 1 Translating Inequality Notation

Write a verbal statement for each of these inequalities, and then sketch the graph of the inequality.

SOLUTIONS

(a) $-2 < y \leq 4$ y is greater than -2 and less than or equal to 4. In interval notation this is the interval $(-2, 4]$.

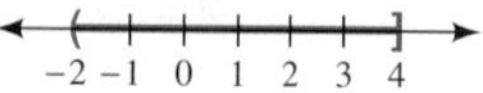

(b) $-3 \leq z < 2$ z is greater than or equal to -3 and less than 2. In interval notation this is the interval $[-3, 2)$.

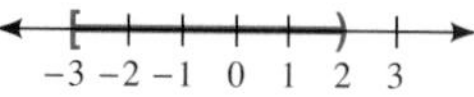

SELF-CHECK

1 Write $w > -3$ and $w \leq 4$ as a single compound inequality.
2 Graph $-2 \leq x < 1$.
3 Write the inequality corresponding to the interval $[-5, 1]$.
4 Write $x \geq 11$ using interval notation.

To solve an inequality is to find all the real values of the variable that make the inequality true. These values are the **roots,** or **solutions,** of the inequality. The methods used to solve inequalities rely on and are similar to the methods that we have used to solve equations.

Most of the inequalities that we will consider are conditional inequalities. **Conditional inequalities** contain a variable and are true for some, but not all, real values of the variable. An inequality that is always true is called an **absolute inequality,** whereas an inequality that is always false is called a **contradiction.**

EXAMPLE 2 Classifying Inequalities

Identify each inequality as an absolute inequality, a conditional inequality, or a contradiction.

SOLUTIONS

(a) $x < x + 1$ This inequality is true for every value of x and is therefore an *absolute inequality.*

The solution set is the set of all real numbers. The graph of the solution set is the entire real number line.

(b) $y < y - 1$ This inequality is false for every value of y and is therefore a *contradiction.*

The solution set is the null set. There are no solutions to plot on the number line.

(c) $z > 4$ This inequality is a conditional inequality, since it is true for real numbers greater than 4 but is false if 4 or any value less than 4 is substituted for z.

The point 5 satisfies this *conditional inequality,* since $5 > 4$ is a true statement. However, 4 and 3.5 are not solutions. The graph of the solution set is shown below.

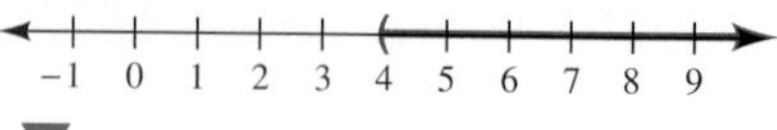

SELF-CHECK ANSWERS

1 $-3 < w \leq 4$ 2 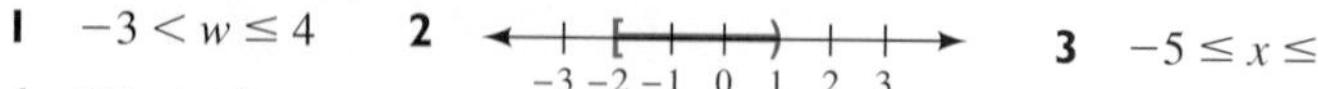 3 $-5 \leq x \leq 1$
4 $[11, +\infty)$

Most of our efforts involving inequalities are directed toward solving conditional inequalities. As we work toward the goal of isolating the variable on one side of the inequality, we form simpler inequalities that are equivalent to the given inequality. **Equivalent inequalities** are inequalities that have the same solution set. In addition to the properties of inequalities given in Section 1-1, we will frequently use the properties given in the following box.

FORMING EQUIVALENT INEQUALITIES

Given real numbers a, b, and c:

Order-Preserving Properties

$a < b$ is equivalent to $a + c < b + c$.	Adding the same number to both members of an inequality produces an equivalent inequality.
$a < b$ is equivalent to $ac < bc$ for $c > 0$.	Multiplying both members of an inequality by the same positive number produces an equivalent inequality.

Order-Reversing Property

$a < b$ is equivalent to $ac > bc$ for $c < 0$.	Multiplying both members of an inequality by the same negative number produces an equivalent inequality if the order of the inequality is reversed.

Similar statements can also be made about the order relations less than or equal to, greater than, and greater than or equal to. Statements about subtraction and division were omitted from the box, since these statements follow directly from the addition and multiplication principles. Illustrations of these properties follow the geometric viewpoint on the next page.

In the following examples, note that we solve the inequalities just as if they were equations *except* when we multiply or divide by a negative number. Remember to reverse the order relation if you multiply or divide both members of an inequality by a negative number.

EXAMPLE 3 Solving a Linear Inequality

Solve $2x + 4 > 12$. Then graph the solution set, and check one value from the solution set.

SOLUTION

$$2x + 4 > 12$$

$$2x + 4 - 4 > 12 - 4 \qquad \text{Subtracting 4 preserves the order.}$$

$$2x > 8$$

$$\frac{2x}{2} > \frac{8}{2} \qquad \text{Dividing by +2 preserves the order.}$$

$$x > 4$$

Answer $(4, +\infty)$

Remember, the parenthesis on 4 means that 4 is not in the solution set.

-1 0 1 2 3 4 5 6 7 8 9

3 does not check

5 checks

Test the value 5:

$$2(5) + 4 \overset{?}{>} 12$$
$$14 > 12 \text{ is true.}$$

Since 5 checks, it is in the solution set.

Test the value 3:

$$2(3) + 4 \overset{?}{>} 12$$
$$10 > 12 \text{ is false.}$$

Since 3 does not check, it is not in the solution set.

Caution: By checking one or two values you can sometimes catch a careless error. However, you cannot check all values in this infinite interval, and a check of only one or two values does not guarantee that the entire interval is correct.

A Geometric Viewpoint: The Order-Preserving and Order-Reversing Properties

Addition:

$$-2 < 2$$
$$-2 + 3 < 2 + 3$$
$$1 < 5$$

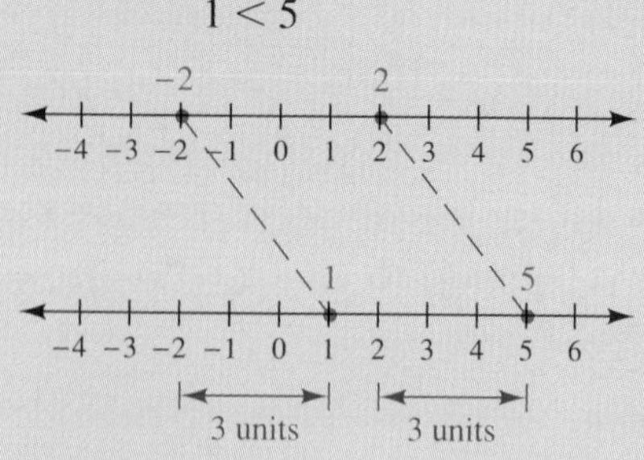

Adding 3 preserves the order relation, as both points are shifted to the right 3 units. Likewise, subtracting 3 would preserve the order relation by shifting both points to the left 3 units.

Multiplication by a positive number:

$$1 < 2$$
$$3(1) < 3(2)$$
$$3 < 6$$

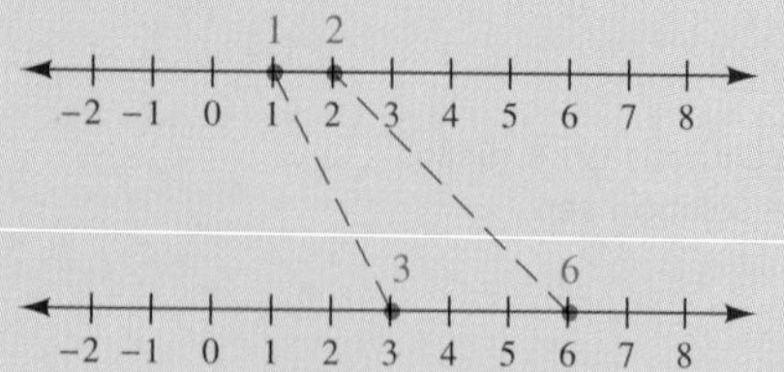

Multiplying by $+3$ preserves the order relation, although the distance between the points appears to stretch out. Likewise, dividing by $+3$ would preserve the order relation, although the distance between the points would appear to shrink.

Multiplication by a negative number:

$$-1 < 1$$
$$(-3)(-1) > (-3)(+1)$$
$$3 > -3$$

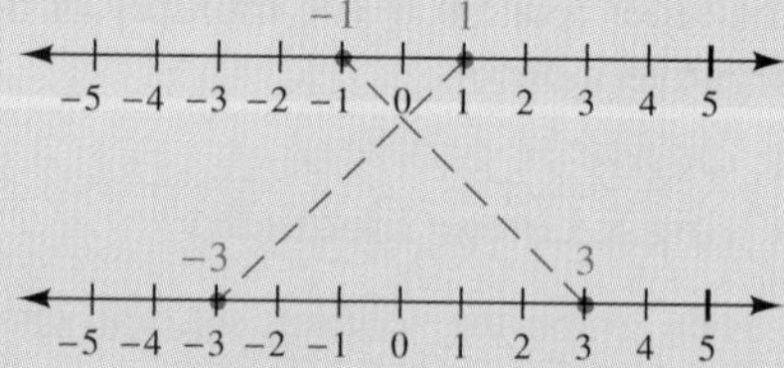

Multiplying by -3 changes the sign of each member of the inequality and thus reverses the order of the products on the number line. The order relation therefore is reversed when each side is multiplied by -3. Likewise, dividing by -3 would reverse the order relation.

EXAMPLE 4 Solving a Linear Inequality Containing Parentheses

Solve $5(3x - 2) \leq 18x + 5$, and sketch the graph of the solution set.

SOLUTION

$5(3x - 2) \leq 18x + 5$	
$15x - 10 \leq 18x + 5$	Use the distributive property to remove the parentheses.
$15x \leq 18x + 15$	Adding 10 to both sides preserves the order.
$-3x \leq 15$	Subtracting $18x$ from both sides preserves the order.
$x \geq -5$	Dividing by -3 reverses the order.

Answer $[-5, +\infty)$

−10 −9 −8 −7 −6 −5 −4 −3 −2 −1 0

EXAMPLE 5 Solving a Linear Inequality Containing Fractions

Solve $\dfrac{2x - 5}{2} > \dfrac{3x + 2}{3}$.

SOLUTION

$\dfrac{2x - 5}{2} > \dfrac{3x + 2}{3}$	
$6\left(\dfrac{2x - 5}{2}\right) > 6\left(\dfrac{3x + 2}{3}\right)$	Multiplying by the LCD, 6, preserves the order.
$3(2x - 5) > 2(3x + 2)$	
$6x - 15 > 6x + 4$	Use the distributive property, and simplify each side.
$-15 > 4$ is a contradiction.	Subtracting $6x$ from both members preserves the order.

Since this inequality is equivalent to the original inequality, there is no solution to this inequality. The solution set is the null set.

The same operations can often be performed on both of the inequalities forming a compound inequality. In such cases we can solve the compound inequality directly, as illustrated in the next example.

EXAMPLE 6 Solving a Compound Inequality

Determine the solution of $18 < 6 - 6t \leq 30$.

SOLUTION

$18 < 6 - 6t \leq 30$	
$12 < -6t \leq 24$	Subtracting 6 preserves the order.
$-2 > t \geq -4$	Dividing by -6 reverses the order.
$-4 \leq t < -2$	The usual form is to write the compound inequality with the smallest number on the left so that the endpoints of the interval are in the order in which they occur graphically.

Answer $[-4, -2)$

SELF-CHECK

Fill in the blank in each problem with the correct inequality symbol.

1 $x \geq y$
$x - 5$ ____ $y - 5$

2 $x \leq y$
$-3x$ ____ $-3y$

3 $x < y$
$x + 5$ ____ $y + 5$

4 $x > y$
$\dfrac{x}{2}$ ____ $\dfrac{y}{2}$

SELF-CHECK ANSWERS 

1 $\geq$ **2** $\geq$ **3** $<$ **4** $>$

For some compound inequalities the variable cannot be isolated between the inequalities as in the previous example. In such cases the compound inequality must be split into two simple inequalities that can be solved individually. The final solution is then formed by the intersection of these individual sets. The intersection of two sets is often indicated by the connecting word *and* because the intersection of set A and set B consists of the elements that are in set A *and* in set B.

SELF-CHECK

1 Solve $-3s + 5 \leq 26$.
2 Solve $-12 < -2(t - 3) \leq 4$.

INTERSECTION OF TWO SETS

The **intersection** of set A and set B is represented by $A \cap B$. $A \cap B$ is the set that contains the elements in *both* A *and* B.

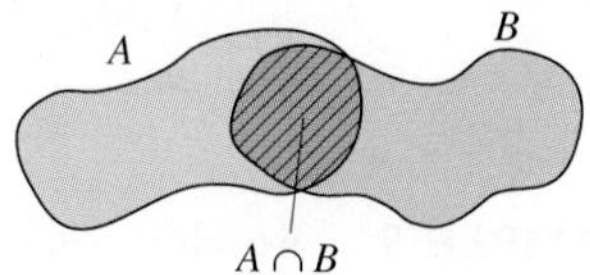

EXAMPLE 7 An Intersection of the Solution Sets of Two Inequalities

Solve $5v - 8 \leq 2v + 4 < 4v - 2$.

SOLUTION This compound inequality is equivalent to

$5v - 8 \leq 2v + 4$	and	$2v + 4 < 4v - 2$	Solve each of the inequalities individually.
$3v - 8 \leq 4$		$-2v + 4 < -2$	
$3v \leq 12$		$-2v < -6$	
$v \leq 4$	and	$v > 3$	

The intersection of $v \leq 4$ and $v > 3$ is $3 < v \leq 4$.

Answer $(3, 4]$

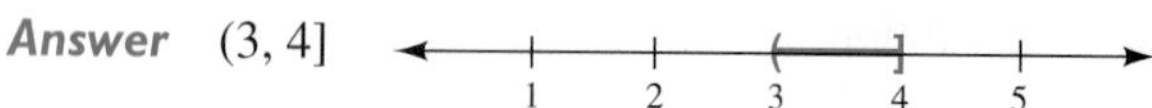

In many problems we are asked to form the *union* of the solution sets of two inequalities. The union of two sets is often indicated by the connecting word *or* since the union of set A and set B consists of the elements in set A *or* in set B. The graph of the solution set of $x < 2$ *or* $x > 5$ is sketched in Figure 1-11.

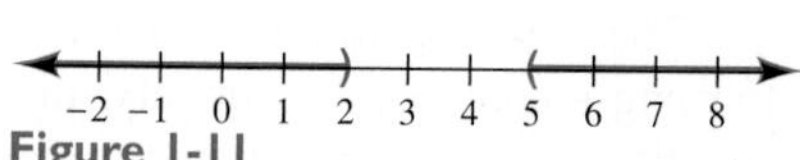

Figure 1-11

UNION OF TWO SETS

The **union** of set A and set B is represented by $A \cup B$. $A \cup B$ is the set that contains the elements in A *or* B (or both).

Note: Elements are listed only once in the union, even if they occur in both sets.

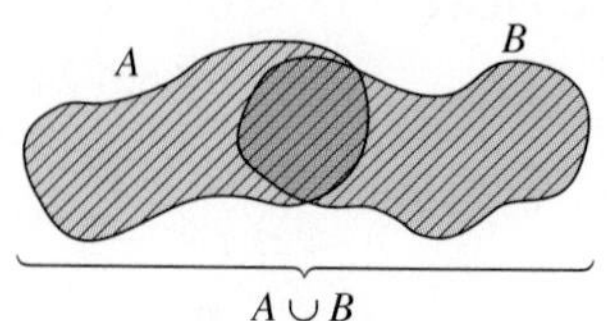

SELF-CHECK ANSWERS

1 $[-7, +\infty)$ 2 $1 \leq t < 9$

EXAMPLE 8 A Union of the Solution Sets of Two Inequalities

Solve $2(w-5) \geq 3w-8$ or $3(w+2) > w+8$, and graph the solution.

SOLUTION

$$2(w-5) \geq 3w-8 \quad \text{or} \quad 3(w+2) > w+8$$
$$2w-10 \geq 3w-8 \qquad 3w+6 > w+8$$
$$2w \geq 3w+2 \qquad 3w > w+2$$
$$-w \geq 2 \qquad 2w > 2$$
$$w \leq -2 \quad \text{or} \quad w > 1$$

Solve each of these inequalities individually. Then graph all points that are in one set or the other.

Answer $(-\infty, -2] \cup (1, +\infty)$

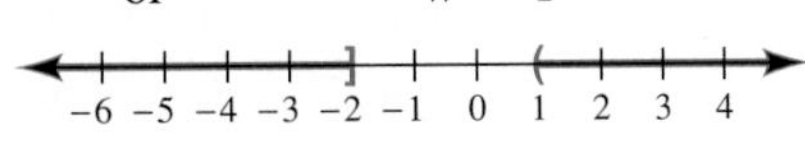

The union symbol, ∪, is used to indicate that the numbers can be in either the first interval or the second interval.

EXAMPLE 9 An Inequality in Verbal Form

If $5z$ minus 3 is either less than -8 or greater than or equal to 2, find z.

SOLUTION

$5z-3 < -8$ or $5z-3 \geq 2$ — Translate this verbal statement into the corresponding algebraic inequality.

$5z < -5$ $\quad 5z \geq 5$ — Adding 3 preserves the order.

$z < -1$ or $z \geq 1$ — Dividing by +5 preserves the order.

Answer $(-\infty, -1) \cup [1, +\infty)$

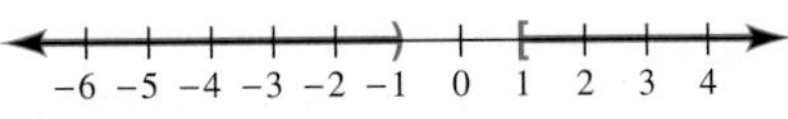

Caution: Answers should be expressed in the simplest form. The customary form for writing compound inequalities places the smallest number on the left and the largest on the right. The inequality will then match both the number line and inequality notation. For example, the inequalities $a < x$ *and* $x < b$ are generally written as $a < x < b$. The order in which a and b occur matches the interval notation (a, b). Be careful *not* to represent the inequality $a < x$ *or* $x < b$ with the notation $a < x < b$, since the latter notation represents $a < x$ *and* $x < b$.

Exercises 1-6

A

In Exercises 1 and 2 sketch the graph of each inequality.

1 a. $w > -2$ **b.** $-3 \leq y < 1$

c. $t < -1$ or $t \geq 2$

2 a. $x \leq 3$ **b.** $2 < z \leq 5$

c. $v \leq -5$ or $v > -2$

In Exercises 3 and 4 write an inequality for each graph. Use x as the variable.

3 a. −8 −7 −6 −5 −4 −3 −2 −1 0

b. −4 −3 −2 −1 0 1 2 3 4

c. −5 −4 −3 −2 −1 0 1 2 3

4 a.

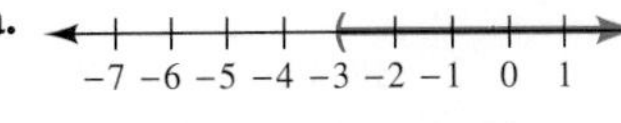

b.

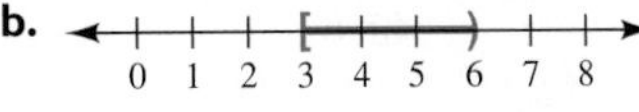

c. 2 3 4 5 6 7 8 9 10

In Exercises 5 and 6 use interval notation to represent each inequality.

5 a. $1 < x < 5$ **b.** $-3 \leq x \leq 7$

c. $x < 3$ **d.** $x \geq -2$

6 a. $-3 \leq x < 8$ **b.** $2 < x \leq 4$

c. $x \leq 4$ **d.** $x > 5$

In Exercises 7 and 8 write an inequality for each interval. Use x as the variable.

7 a. $[-2, 3)$ **b.** $(-4, 9)$ **c.** $[0, +\infty)$ **d.** $(-\infty, 6)$

8 **a.** $[-4, 4]$ **b.** $(5, 7)$ **c.** $(-\infty, -3]$ **d.** $[1, +\infty)$

In Exercises 9 and 10 test each of the following values to determine whether it is a solution of the given inequality: $-4, -3, -2, -1, \frac{1}{2}, 1, 2, 3,$ and 4. List the values that are solutions.

9 **a.** $w > -3$ **b.** $-2 < y \le 2$
c. $z \le -3$ or $z > -1$

10 **a.** $x \le -2$ **b.** $y < -2$ or $y \ge 2$
c. $-3 \le z < -1$

In Exercises 11–50 solve each inequality. In Exercises 11–18, 29–34, and 41–46, also sketch the graph of the solution set.

11 $t - 3 < 2$
12 $v + 2 \ge 1$
13 $7v - 2 \ge 6v - 1$
14 $3w + 4 < 2w + 3$
15 $4w > -8$
16 $5v < -25$
17 $-6z \le 18$
18 $-7t \le -21$
19 $\frac{-4x}{5} > -20$
20 $\frac{-6y}{7} > 42$
21 $2(3y + 4) \ge -4y + 24$
22 $3(x - 5) < 5x + 15$
23 $7(2t - 5) > 3(6t - 11)$
24 $5(7v - 3) \le 9(6 + 3v)$
25 $\frac{p}{4} - \frac{5}{4} > \frac{3p}{8} - \frac{3}{2}$
26 $\frac{7k}{8} - \frac{2}{3} \ge \frac{3k}{4} - \frac{1}{2}$
27 $3(x - 7) - 5(4 - x) - 3 > -4(x - 1)$
28 $4(x - 1) + 3(x + 3) \le 7(x + 1) - 5$
29 $7 < y - 5 \le 9$
30 $4 \le x + 3 < 6$
31 $-8 \le 4x < 44$
32 $-13 < 13m \le 26$
33 $-4 < -2m < 6$
34 $-35 \le -5w \le -20$
35 $4 \le \frac{-2r}{3} \le 16$
36 $-10 < \frac{-5t}{2} \le 20$
37 $11 \le 3z + 2 \le 17$
38 $15 < 2m - 7 \le 33$
39 $4 \le 2(5t - 3) < 7$
40 $6 < 3(2v + 4) < 18$
41 $-3x < 15$ and $-5x > -15$
42 $-4v > 28$ and $-7v < 70$
43 $2y + 1 \le -3$ or $-2y + 1 < -3$
44 $7x + 3 \ge 5x - 5$ or $3x - 2 \ge 5x + 8$
45 $-45 < -9w < 0$
46 $-15 < -5v < 0$
47 $2x + 1 < 3x - 3 \le 5(x - 1)$
48 $2x + 5 \le 4x + 1 < 3x + 4$
49 $\frac{3(t - 1)}{4} + 13 \ge \frac{5(9 - t)}{6}$
50 $\frac{3x - 5}{6} + \frac{4x - 9}{9} < \frac{5x - 8}{4}$

B

In Exercises 51 and 52 determine each intersection or union, and write the answer in interval notation.

51 **a.** $(2, 4) \cap (3, 5]$ **b.** $(2, 4) \cup (3, 5]$
c. $(2, +\infty) \cap (-\infty, 5]$

52 **a.** $[2, 4) \cap (3, +\infty)$ **b.** $[2, 4) \cup (3, +\infty)$
c. $(2, +\infty) \cup (-\infty, 5]$

53 Find a value of x for which
a. $7x > 2x$ **b.** $7x < 2x$ **c.** $7x = 2x$

54 Find a value of n for which
a. $\frac{1}{n} < 2$ and $1 < 2n$ **b.** $\frac{1}{n} < 2$ but $1 > 2n$

In Exercises 55 and 56 write an inequality for each statement.

55 **a.** x is at least 4. **b.** x is at most 9.
c. x is between 3 and 8.

56 **a.** x is at most 2. **b.** x is at least 7.
c. x is between -1 and 5.

In Exercises 57–68 solve each inequality.

57 $-4x - 6 < -2x - 1 \le -x - 3$
58 $-3(x - 2) \le -2x + 3 < -3x + 10$
59 $\frac{v - 1}{3} < \frac{v}{2} \le \frac{4v + 3}{10}$
60 $x - 1 < \frac{3x - 1}{2} < 1$
61 $5(v - 3) + 7 < 2v + 3(v - 2)$
62 $12 + 3(2w - 5) \le -(4 - 6w)$
63 $7d - 1 < 5d + 7 < 6d + 3$
64 $z - 1 < 2$ and $2 > z + 1$
65 $5(v + 4) < 28$ and $3(v - 4) > -11$
66 $z - 1 < \frac{2z + 1}{3}$ or $\frac{4z - 1}{3} \ge z - 1$
67 $2z - 1 < 3z + 1$ or $5z - 6 < 3z - 12$
68 $7(4m - 5) - 16 < 5(8m - 3)$ and $11 - 4(2m - 3) \ge -3(2 - 7m)$

ESTIMATION SKILLS (69–72)

69 The best mental estimate of the solution of $4.8v < 19.7$ is
a. $v < 15$ **b.** $v < 4$ **c.** $v > 4$ **d.** $v > 15$

70 The best mental estimate of the solution of $-2.97m < 11.96$ is
a. $m < 4$ **b.** $m < -4$ **c.** $m > -4$ **d.** $m > -15$

71 The best mental estimate of the solution of $-5.98 < -3.01x < 9.02$ is

a. $-3 < x < 2$ b. $-2 < x < 3$
c. $-4 < x < 6$ d. $-3 < x < 3$

72 The best mental estimate of the solution of $-19.8 < -9.94x < 29.7$ is

a. $-10 < x < 40$ b. $-3 < x < 2$
c. $-2 < x < 3$ d. $-20 < x < 30$

In Exercises 73–78 first write an algebraic inequality for the stated problem, and then solve the inequality.

73 Twice the sum of a and 3 is no more than 10.

74 Two-thirds of c is at least c minus 5.

75 Nine less than $2d$ is at most two times the quantity $d - 12$.

76 Four w plus 7 is greater than negative 5 and is less than or equal to 11.

77 Two minus $3m$ is greater than 5 or less than negative 1.

78 Six minus $2x$ is greater than 10 or less than or equal to 0.

79 Temperature Range The range of acceptable Celsius temperatures for a piece of equipment is from 15° to 30°. Solve $15 \le \frac{5}{9}(F - 32) \le 30$ to find the acceptable range of Fahrenheit temperatures.

80 Profit Interval The revenue R in dollars produced by selling x units of a product is given by $R(x) = 45x$. The cost C in dollars of producing and marketing x units is given by $C(x) = 23x + 3850$. In order to make a profit, revenue must exceed cost. Determine the profit interval; that is, determine the values of x that will generate a profit.

Calculator Usage (81 and 82)

In Exercises 81 and 82 solve each inequality with the aid of a calculator. Round all answers to the nearest hundredth.

81 $15.98x - 17.23 < 21.87(x - 11.05)$

82 $24.86(2.49x + 61.83) \ge -4.78(3.42 - 11.76x)$

Discussion Question

83 Describe why adding a negative number to both members of an inequality preserves the order of the inequality but multiplying both members of an inequality by a negative number reverses the order of the inequality.

Challenge Question

a. Which is larger, $25(\pi - 2)$ or $25(4 - \pi)$?

b. Which wastes more space, placing a container with a square-shaped base in a circular-shaped shipping crate or placing a container with a circular-shaped base in a square-shaped shipping crate? Use the figures and formulas shown below to answer this question.

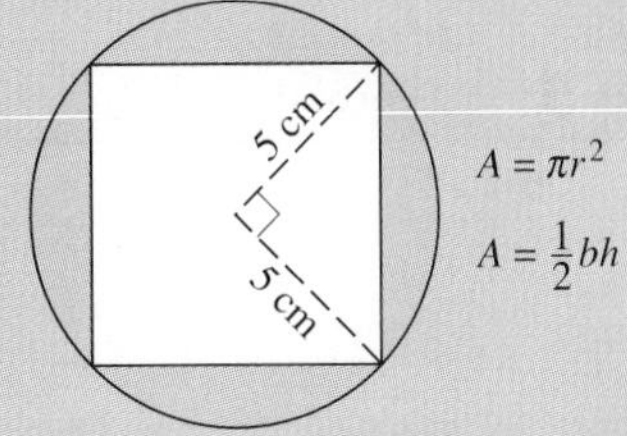

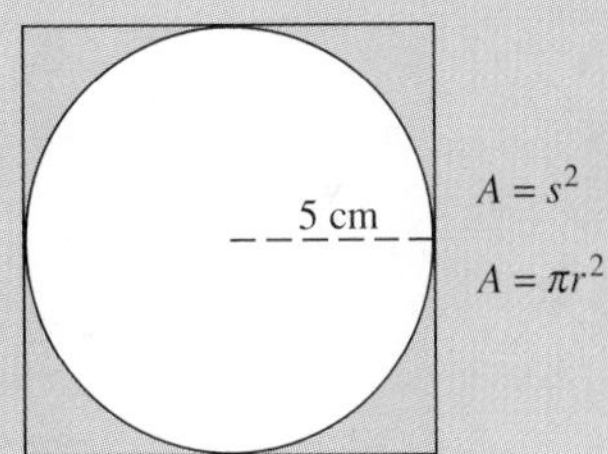

Section Skill Reinforcers

1 Twice 4 more than a number equals five times 3 less than the number. Find the number.
2 One-half a number is equal to the sum of three-fourths of the number and -2. Find the number.
3 What number is equal to two-thirds of itself?
4 What number is equal to 8 less than three-sevenths of itself?
5 Find four consecutive odd integers if the sum of the first three of these integers is 70 larger than the fourth integer.

1-7 *Solving Absolute-Value Equations and Inequalities*

SECTION OBJECTIVE

14 Solve absolute-value equations and inequalities.

As you work through problems involving absolute value, remember that the absolute value of a real number is merely its distance from the origin on the number line. Thus the absolute-value inequality $|x| < 4$ represents the interval of points for which the distance from the origin is less than 4 units. Likewise, the absolute-value inequality $|x| > 4$ represents the two intervals containing points for which the distance from the origin is more than 4 units. Table 1-2 illustrates these concepts.

TABLE 1-2 *Absolute-Value Expressions*

Absolute-Value Expression	Meaning of Expression	Graph
$\|x\| = 4$	$x = -4$ or $x = 4$	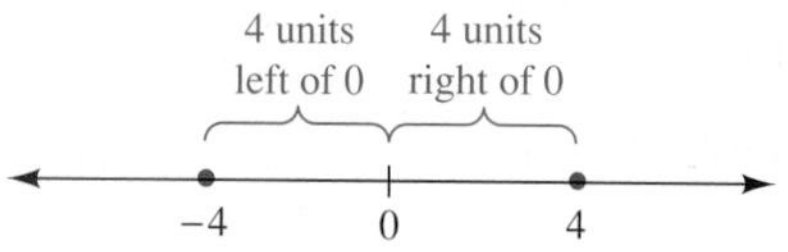
$\|x\| < 4$	$-4 < x < 4$	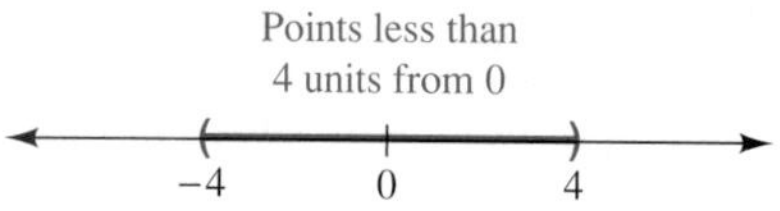
$\|x\| > 4$	$x < -4$ or $x > 4$	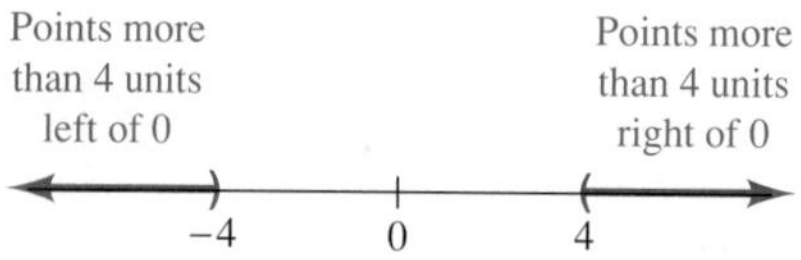

If a is larger than b, then the distance from a to b is given by the difference $a - b$. To indicate that the distance between a and b is always nonnegative, we can denote this distance by $|a - b|$. In particular, $|x - 0|$, or $|x|$, can be interpreted as the distance between x and the origin, as shown in Table 1-2. Likewise, $|a + b| = |a - (-b)|$ equals the distance between a and $-b$. This concept of distance is examined in the next two examples.

EXAMPLE 1 Representing Distance Using Absolute-Value Notation

Use absolute-value notation to represent the distance between each pair of points.

SOLUTIONS

(a) -3.7 and 4.3 $\quad |-3.7 - 4.3| = |-8|$

$$= 8$$

(b) $3x$ and $-4y$ $\quad |3x - (-4y)| = |3x + 4y|$

(c) $-5x$ and $8x$ $\quad$ $|-5x - 8x| = |-13x|$ $\quad$ One property of absolute value is that $|ab| = |a| \cdot |b|$.

$$= |-13| \cdot |x|$$
$$= 13|x|$$

▼

SELF-CHECK ✓

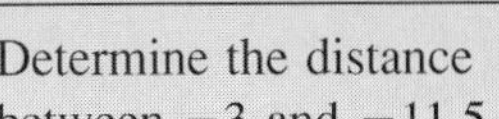

Determine the distance between -3 and -11.5.

EXAMPLE 2 Using Distance to Interpret an Absolute-Value Equation

Interpret $|x - 3| = 4$ using the geometric concept of distance.

SOLUTION The fact that the distance between x and 3 is 4 units means that x is either 4 units to the left of 3 or 4 units to the right of 3, as shown in the figure.

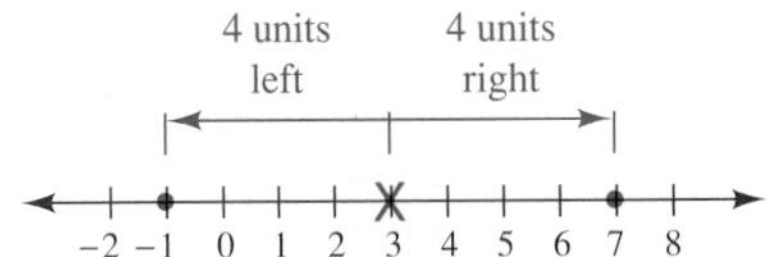

$$x - 3 = -4 \quad \text{or} \quad x - 3 = +4$$
$$x = -1 \quad \text{or} \quad x = 7$$

The solution set of $|x - 3| = 4$ is $\{-1, 7\}$. Do these values check? ▼

EXAMPLE 3 Using Distance to Interpret an Absolute-Value Inequality

Interpret $|x + 2| > 3$ using the geometric concept of distance.

SOLUTION The expression $|x + 2| > 3$ is equivalent to $|x - (-2)| > 3$. The fact that the distance between x and -2 is more than 3 units means that x is either more than 3 units to the left of -2 or more than 3 units to the right of -2, as shown in the figure.

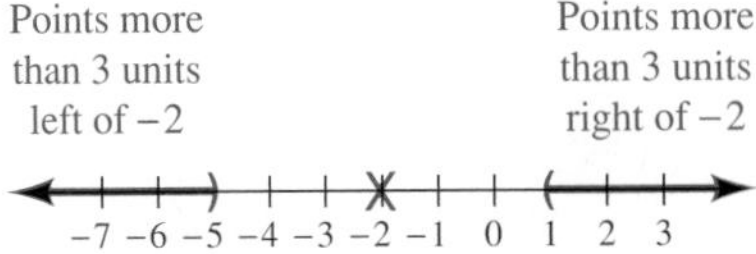

$$x - (-2) < -3 \quad \text{or} \quad x - (-2) > 3$$
$$x < -5 \quad \text{or} \quad x > 1$$

Thus $|x + 2| > 3$ is equivalent to $x < -5$ or $x > 1$. Check a value from each of these intervals. ▼

EXAMPLE 4 Absolute-Value Notation for an Interval

Write an absolute-value inequality to represent the interval shown in the figure.

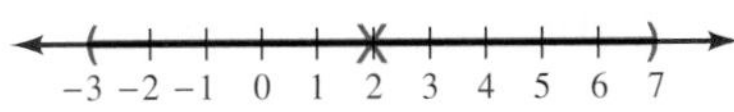

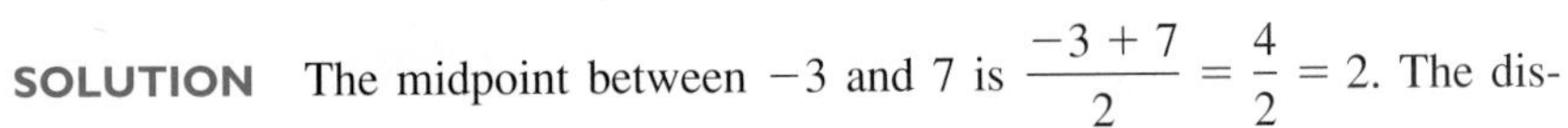

SOLUTION The midpoint between -3 and 7 is $\frac{-3 + 7}{2} = \frac{4}{2} = 2$. The distance from this midpoint to one end is $|7 - 2| = |5| = 5$. The interval of points between -3 and 7 is the interval of points less than 5 units from 2. In absolute-value notation this interval is $|x - 2| < 5$. ▼

Although we could continue to solve algebraic equations and inequalities through this geometric approach, it is much more efficient to use an algebraic approach. The algebraic method given in the following box generalizes the results of the observations we have made so far.

SELF-CHECK ANSWER 

8.5

SOLVING ABSOLUTE-VALUE EQUATIONS AND INEQUALITIES

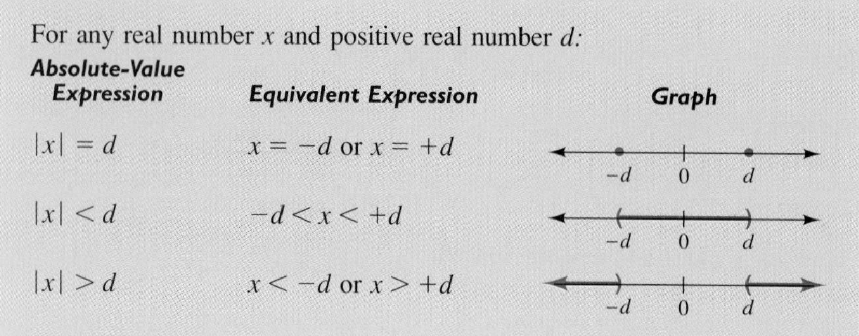

For any real number x and positive real number d:

Absolute-Value Expression	Equivalent Expression	Graph
$\lvert x\rvert = d$	$x = -d$ or $x = +d$	$-d$ 0 d
$\lvert x\rvert < d$	$-d < x < +d$	$-d$ 0 d
$\lvert x\rvert > d$	$x < -d$ or $x > +d$	$-d$ 0 d

Similar statements can also be made about the order relations less than or equal to ($\leq$) and greater than or equal to ($\geq$). Because these statements apply to any real expression within the absolute-value symbols, we will use them to solve the equations and inequalities in the following examples.

EXAMPLE 5 Solving an Absolute-Value Equation

Solve $\lvert 2y - 3\rvert = 31$.

SOLUTION $\lvert 2y - 3\rvert = 31$

$2y - 3 = -31$ or $2y - 3 = 31$ — Substitute $2y - 3$ for x in $\lvert x\rvert = 31$, and note that $\lvert x\rvert = 31$ is equivalent to $x = -31$ or $x = 31$.

$2y = -28$ $\quad$ $2y = 34$ — Solve each of the linear equations.

$y = -14$ or $y = 17$

Answer The solution set is $\{-14, 17\}$. — Do these values check?

EXAMPLE 6 Solving an Absolute-Value Inequality

Solve $\lvert 4v - 6\rvert > 10$, and graph the solution.

SOLUTION $4v - 6 < -10$ or $4v - 6 > +10$ — Substitute $4v-6$ for x in $\lvert x\rvert > 10$, and note that $\lvert x\rvert > 10$ is equivalent to $x < -10$ or $x > +10$.

$4v < -4$ $\quad$ $4v > 16$ — Adding 6 preserves the order.

$v < -1$ or $v > 4$ — Dividing by $+4$ preserves the order.

Answer $(-\infty, -1) \cup (4, +\infty)$

−4 −3 −2 −1 0 1 2 3 4 5 6

EXAMPLE 7 Solving an Absolute-Value Inequality

Solve $\lvert 4m - 3\rvert + 5 \leq 10$, and graph the solution.

SOLUTION $\lvert 4m - 3\rvert + 5 \leq 10$

$\lvert 4m - 3\rvert \leq 5$ — First isolate the absolute-value expression on the left side.

$-5 \leq 4m - 3 \leq 5$ — Substitute $4m - 3$ for x in $\lvert x\rvert \leq 5$, and note that $\lvert x\rvert \leq 5$ is equivalent to $-5 \leq x \leq 5$.

$-2 \le 4m \le 8$ Adding 3 preserves the order.

$-0.5 \le m \le 2$ Dividing by +4 preserves the order.

Answer $[-0.5, 2]$

−0.5
−1 0 1 2 3

▼

The absolute value of an expression is never negative, so each rule stated previously in the box was given for positive values of d. If an absolute-value expression is compared to a negative number, then the equation or inequality can be solved by inspection, as illustrated in the next example.

EXAMPLE 8 Special Cases Involving Absolute Value

Solve each of these equations and inequalities.

SOLUTIONS

(a) $|5w - 3(2w - 7)| = -2$ — The solution set is the null set, $\varnothing$. Since the absolute value of an expression is never negative, this equation has no solution.

(b) $|3v + 2(4v - 7)| > -3$ — The solution set is the set of all real numbers, $\mathbb{R}$. Since the absolute value of an expression is always greater than or equal to 0, this expression is greater than −3 for all real numbers.

(c) $|7(x - 3) + 9| < -1$ — The solution set is the null set, $\varnothing$. Since an absolute-value expression is always nonnegative, this absolute-value expression is never less than −1.

If $|a| = |b|$, then a and b are equal in magnitude but their signs can either agree or disagree. Thus $|a| = |b|$ implies that either $a = b$ or $a = -b$. This result is used to solve the next example.

$|a| = |b|$ is equivalent to $a = b$ or $a = -b$.

SELF-CHECK 

Solve each inequality.

1 $|5w + 7| < 8$

2 $|7v - 5| \ge 9$

EXAMPLE 9 Solving an Equation Involving Two Absolute-Value Expressions

Solve $|3x - 5| = |5x - 7|$.

SOLUTION $|3x - 5| = |5x - 7|$

$3x - 5 = 5x - 7$ or $3x - 5 = -(5x - 7)$ — $|a| = |b|$ is equivalent to $a = b$ or $a = -b$.

$-2x = -2$ $\quad$ $3x - 5 = -5x + 7$

$x = 1$ $\quad$ $8x = 12$

$x = 1$ or $x = \dfrac{3}{2}$

Answer The solution set is $\left\{1, \dfrac{3}{2}\right\}$. Do these values check? ▼

SELF-CHECK ANSWERS 

1 $-3 < w < \dfrac{1}{5}$ **2** $v \le \dfrac{-4}{7}$ or $v \ge 2$

In industrial applications there is generally an allowance for a small variation, or leeway, between the standard size for a part or component and the actual size. This acceptable variation is called the **tolerance.** For a 42-centimeter steel rod, for example, the tolerance may be 0.05 centimeter. A worker on an assembly line might merely lay the rod on a table that is marked to indicate the upper and lower limits of tolerance, as shown in Figure 1-12. However, an engineer doing calculations would need to describe this tolerance algebraically. The expression $|r - 42| \le 0.05$ is an algebraic statement that the length of the rod, r, and the desired length of 42 centimeters can differ by at most 0.05 centimeter.

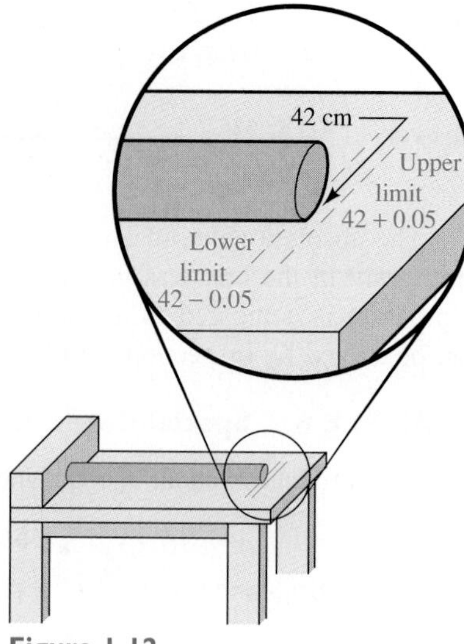

Figure 1-12

The next example illustrates the use of a TI-82 graphics calculator to evaluate absolute-value expressions. To indicate $|x|$, we use ABS(x) where ABS denotes absolute-value. Although these keystrokes are representative, keystrokes will vary to some extent, depending on the model of calculator you have.

EXAMPLE 10 Calculator Usage

Use a calculator to evaluate $f(-4.52)$ for the formula $f(x) = |8x - 50|$.

SOLUTION

Keystrokes

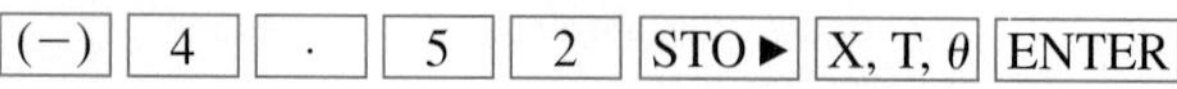

First store the value of −4.52 for x.

[2nd] [ABS] [( ] [8] [X, T, θ] [−] [5] [0] [ )] [ENTER] → 86.16

Then enter the formula into the calculator.

Answer $f(-4.52) = 86.16$

Exercises 1-7

In Exercises 1 and 2 calculate the distance between each pair of points.

1 a. 6 and −1 **b.** −19 and −7

c. $3a$ and $-4b$

2 a. −3 and 15 **b.** −21 and −45

c. $-7a$ and $-3b$

In Exercises 3–8 write an absolute-value equation or inequality to represent each set of points.

3 a.

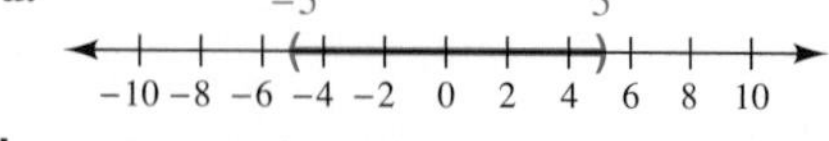

b.

−5 −4 −3 −2 −1 0 1 2 3 4 5

c.

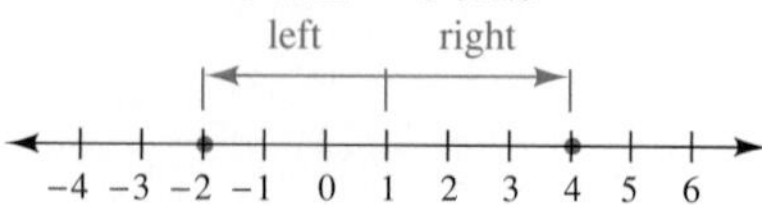

d. The points between −4 and 4

4 a.

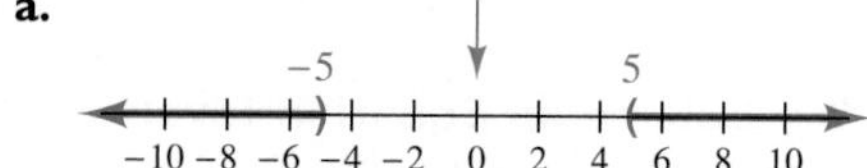

b.

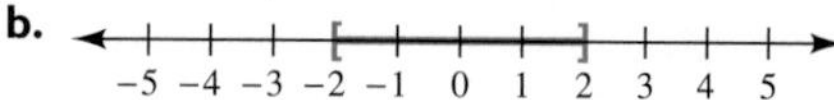

c.

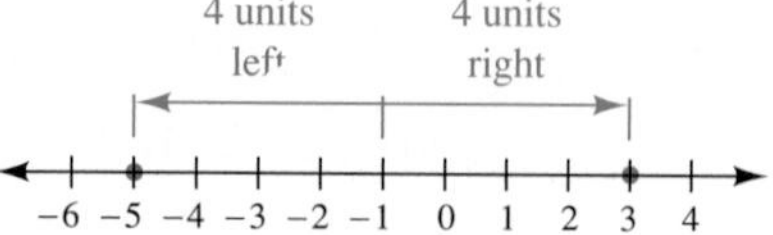

d. The points that are at least −3 and at most 3

5 a.

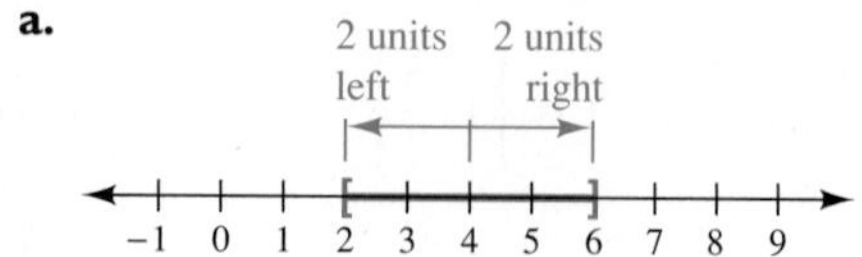

b.

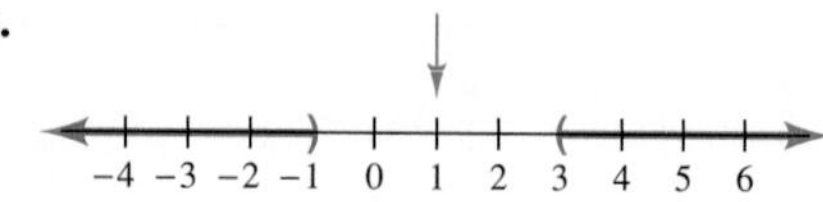

c. 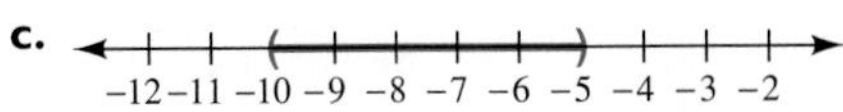

d. The points that are at least 0 and at most 4

6 a. 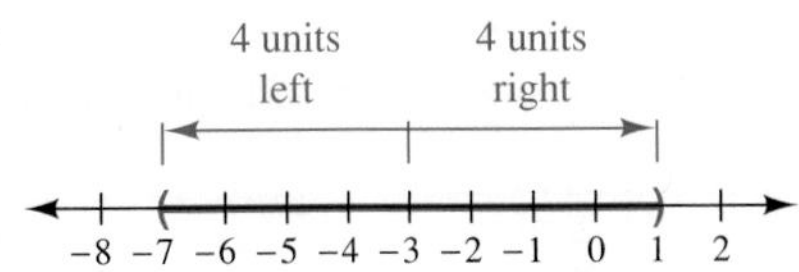

b. 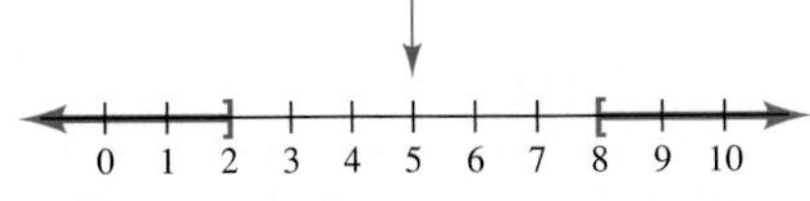

c. −4 −2 0 2 4 6 8 10 12 14 16

d. The points between −8 and 2

7 a. $[-3, 3]$ **b.** $(-\infty, -3) \cup (3, +\infty)$ **c.** $(-7, 7)$

8 a. $(-9, 9)$ **b.** $(-\infty, -9] \cup [9, +\infty)$ **c.** $[4, 8]$

In Exercises 9–12 solve each equation.

9 a. $|a| = 6$ **b.** $|m| = -1$

c. $|k + 3| = 2$ **d.** $|x| = 0$

10 a. $|-a| = 5$ **b.** $|m| = -2$

c. $|j - 2| = 5$ **d.** $|x - 1| = 0$

11 a. $|2n| = 8$ **b.** $\left|-\frac{3}{5}q\right| = 15$

c. $|2s - 6| = 0$

12 a. $|3m| = 6$ **b.** $\left|-\frac{1}{2}p\right| = 8$

c. $|2x - 5| = 10$

In Exercises 13–22 solve each inequality, and sketch the graph of the solution set.

13 $|2p - 1| < 5$ **14** $|3a + 4| \geq 2$

15 $|3t - 5| \geq 1$ **16** $|2(t - 1) + 3| < 5$

17 $\left|\frac{-7v}{2}\right| \leq 14$ **18** $\left|\frac{-5y}{11}\right| \leq 5.5$

19 $|2x + 3| - 4 > 1$ **20** $|2(x + 3) - 4| \geq 1$

21 $|4(x - 1) - 2(3x + 2)| < 3$

22 $|3(2x + 1) - 5(1 - x)| < 2$

In Exercises 23–42 solve each equation and inequality.

23 $|3d + 4| \leq -5$ **24** $|3d + 4| \geq -5$

25 $|2x + 1| > -1$ **26** $|2x + 1| < -1$

27 $6 + |2y - 1| \geq 10$ **28** $11 + |3v + 1| \leq 18$

29 $\left|\frac{v - 1}{7}\right| \leq 14$ **30** $\left|\frac{2t - 1}{3}\right| = 6$

31 $|2a + 1| + 4 = 11$ **32** $\left|\frac{2 - c}{3}\right| \geq 12$

33 $|4m - 3| = 9$ **34** $\left|8 - \frac{11}{7}d\right| > -\frac{2}{3}$

35 $5\left|\frac{k}{2} - 1\right| \geq 10$ **36** $2\left|\frac{k}{5} - 3\right| > 14$

37 $|2(2x - 1) - (x - 3)| < 11$

38 $|3(2x + 1) - 4(3x - 2)| < 5$

39 $|2x| = |x - 1|$ **40** $|3x| = |4 - x|$

41 $3\left|\frac{3h - 5}{2}\right| + 5 > 15$ **42** $2\left|5 - \frac{k}{3}\right| - 5 \geq -3$

B

In Exercises 43–48 graph each inequality, and then rewrite the inequality using absolute-value notation.

43 $-7 \leq x \leq 7$ **44** $x < -5$ or $x > 5$

45 $x < 1$ or $x > 5$ **46** $-7 \leq x \leq -1$

47 $-1 < x < 0$ **48** $6 < x < 9$

49 Monitoring Electrical Voltage A monitor at a power station continuously checks the voltage difference between two sources, x and y. If the difference between these two sources is greater than 5 volts, a warning light turns on. Express this inequality using absolute-value notation.

50 Mass of Engine Parts An engine has two parts, a and b, that counterbalance each other. The difference between their masses must be less than 0.5 gram or excessive vibration will result. Express this inequality using absolute-value notation.

ESTIMATION SKILLS (51–52)

In Exercises 51 and 52 select the best estimate for the solution of each equation. Make your estimates without using a calculator.

51 $|7.908342x| = 4.7448$

a. −5, 5 **b.** −0.5, 0.5 **c.** −6, 6

d. −0.6, 0.6 **e.** −3, 6

52 $|0.050908x| = 2.0741$

a. −4, 4 **b.** −40, 40 **c.** −2.5, 2.5

d. −25, 25 **e.** −4, 24

In Exercises 53–60 solve each equation or inequality.

53 $-|0.03y| = -1.5$ **54** $|-0.07w| = 2.1$

55 $\left|\frac{5 - 2b}{8}\right| = 7$ **56** $3|x - 2| - 5 = 4$

57 $\left|\frac{2v}{5} - \frac{1}{3}\right| + 1 \leq 2$ **58** $\left|\frac{3m}{7} - \frac{1}{4}\right| - 3 = 31$

59 $|t - 7| = |4t + 8|$ **60** $|3 - w| = |3w + 1|$

C

In Exercises 61–64 express the tolerance interval as an absolute-value inequality, and determine the lower and upper limits of the interval.

Tolerance (61–64)

61 The desired length is 15 meters, with a tolerance of ± 0.12 meter.

62 The desired length is 13.6 millimeters, with a tolerance of ± 0.4 millimeter.

63 The desired volume is 26.9 liters, with a tolerance of ± 0.9 liter.

64 The desired volume is 117.8 milliliters, with a tolerance of ± 4.3 milliliters.

CALCULATOR USAGE (65–68)

Some calculators use the ABS key to calculate the absolute value of an expression. Assume that a student has stored −5 for x prior to entering the expressions shown in Exercises 65 and 66. Determine the value of each expression.

65 abs (2X + 7)

66 abs (X^2 − 3X − 2)

67 Use a calculator to evaluate $f(-7.91)$ for the formula $f(x) = |7x - 34|$.

68 Use a calculator to evaluate $f(-12.53)$ for the formula $f(x) = |-x^2 + 2x - 5|$.

DISCUSSION QUESTION

69 A classmate in your algebra class says that you can take the absolute value of any expression by just removing the signs of each term. This is *not* correct! Use $|-x|$ and $|x - 2|$ to describe what is wrong with this student's logic.

CHALLENGE QUESTION

Given the formula $f(x) = \big|\,||x - 1| - 2| - 3\big|$, evaluate $f(-1)$, $f(1)$, and $f(10)$.

SECTION SKILL REINFORCERS

Simplify each expression.

1 -7^2 **2** $(-7)^2$ **3** $(7 - 1)^2$ **4** $7^2 - 1^2$ **5** $(-1)^{46}$

Key Concepts for Chapter 1

The key concepts reviewed in this first chapter are crucial to understanding the material that follows in the other chapters. You must be able to do the calculations in this chapter consistently. Even an occasional error in working with integers suggests possible confusion about one of the basic concepts. You can use the mastery test and chapter review to check your consistency and to eliminate any sources of confusion.

1 The relationship of important subsets of the real numbers is summarized in the tree diagram below.

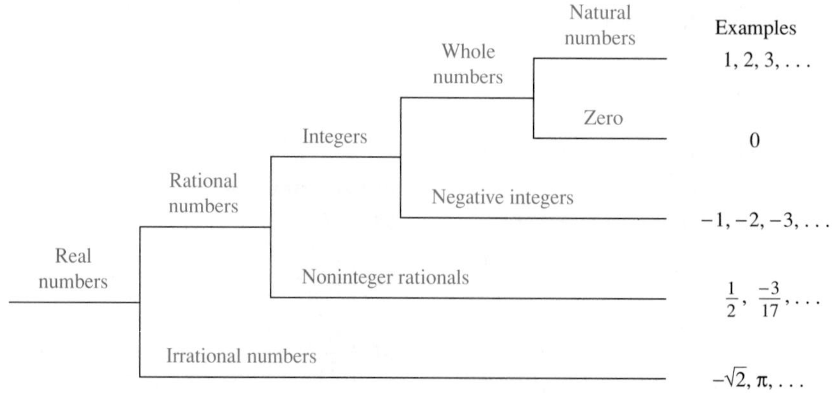

Figure for Key Concept 1

2 The absolute value of a, denoted by $|a|$, is the distance between the origin and the point with coordinate a on the real number line. For a real number a,

$$|a| = a \text{ if } a \geq 0$$
$$|a| = -a \text{ if } a < 0$$

3 Identities and inverses:

a. The additive identity is 0, and the additive inverse of a is $-a$.

b. The multiplicative identity is 1, and the multiplicative inverse of a is $\frac{1}{a}$ if $a \neq 0$.

4 For every real number a:

a. $\frac{a}{0}$ is undefined. **b.** $\frac{0}{a} = 0$ for $a \neq 0$

c. $\frac{a}{a} = 1$ for $a \neq 0$ **d.** $0 \cdot a = 0$

e. $-(-a) = a$

5 If a and b are added, then a and b are called *terms*. If a and b are multiplied, then a and b are called *factors*.

6 The $-$ symbol has three uses:

a. Negative: The $-$ symbol is used to indicate that a number to the left of the origin on the number line is negative.

b. Opposite: The $-$ symbol is used to indicate the opposite of a real number—that is, a number the same distance from the origin on the number line but on the opposite side of the origin.

c. Minus: The $-$ symbol is used to indicate the operation of subtraction.

7 Arithmetic of fractions:

a. Addition: $\frac{a}{c} + \frac{b}{c} = \frac{a+b}{c}$ for $c \neq 0$

b. Subtraction: $\frac{a}{c} - \frac{b}{c} = \frac{a-b}{c}$ for $c \neq 0$

c. Multiplication: $\frac{a}{b} \cdot \frac{c}{d} = \frac{a \cdot c}{b \cdot d}$ for $b \neq 0$ and $d \neq 0$

d. Division: $\frac{a}{b} \div \frac{c}{d} = \frac{a}{b} \cdot \frac{d}{c} = \frac{a \cdot d}{b \cdot c}$ for $b \neq 0$, $c \neq 0$, and $d \neq 0$

8 Three signs of a fraction:

$$-\frac{a}{b} = \frac{-a}{b} = \frac{a}{-b} = -\frac{-a}{-b}$$

$$\frac{a}{b} = \frac{-a}{-b} = -\frac{-a}{b} = -\frac{a}{-b}$$

9 Exponential notation: For any real number b and natural number n,

$$b^n = \underbrace{b \cdot b \cdot \cdots \cdot b}_{n \text{ factors}} \quad \text{with base } b \text{ and exponent } n$$

b^n is read as "b to the nth power."

10 Order of operations:

Step 1: Start with the expression within the innermost grouping symbols.
Step 2: Perform all exponentiations.
Step 3: Perform all multiplications and divisions, working from left to right.
Step 4: Perform all additions and subtractions, working from left to right.

11 Equivalent equations: If a, b, and c are real numbers, then

a. $a = b$ is equivalent to $a + c = b + c$.

b. $a = b$ is equivalent to $ac = bc$ for $c \neq 0$.

12 Consecutive integers:

a. If n is an integer, the next two consecutive integers are $n + 1$ and $n + 2$.

b. If n is an even integer, the next two consecutive even integers are $n + 2$ and $n + 4$.

c. If n is an odd integer, the next two consecutive odd integers are $n + 2$ and $n + 4$.

13 General principles used to form equations:

a. Mixture principle for two ingredients:

[Amount in first] + [Amount in second] = [Total amount in mixture]

b. Rate principle:

[Amount] = [Rate] · [Base]

14 Equivalent inequalities: If a, b, and c are real numbers, then

a. $a < b$ is equivalent to $a + c < b + c$.

b. $a < b$ is equivalent to $ac < bc$ if c is positive.

} Order-preserving properties

c. $a < b$ is equivalent to $ac > bc$ if c is negative. } Order-reversing property

15 Compound inequalities:
$a \leq x \leq b$ is equivalent to $x \geq a$ and $x \leq b$.

a. The values satisfying $x \geq a$ and $x \leq b$ satisfy *both* of these individual inequalities.

b. The values satisfying $x \geq a$ or $x \leq b$ are those values that satisfy either the first inequality or the second inequality (or both).

16 Interval notation:

Inequality Notation	*Interval Notation*
$x > a$	$(a, +\infty)$
$x \geq a$	$[a, +\infty)$
$x < a$	$(-\infty, a)$
$x \leq a$	$(-\infty, a]$
$a < x < b$	(a, b)
$a < x \leq b$	$(a, b]$
$a \leq x < b$	$[a, b)$
$a \leq x \leq b$	$[a, b]$

17 Absolute-value equations and inequalities: If x is a real number and d is a positive real number, then

a. $|x| = d$ is equivalent to $x = -d$ or $x = d$.

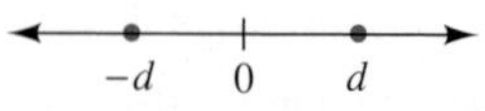

b. $|x| = -d$ is a contradiction and has no solution.

c. $|x| < d$ is equivalent to $-d < x < d$.

d. $|x| < -d$ is a contradiction and has no solution.

e. $|x| > d$ is equivalent to $x < -d$ or $x > d$.

f. $|x| > -d$ is an absolute inequality, and the solution set is the set of all real numbers.

18 Intersection and union:

a. Intersection: $A \cap B$ is the set that contains the elements in *both A and B*.

b. Union: $A \cup B$ is the set that contains the elements in A *or* B (or both).

Review Exercises for Chapter 1

In Exercises 1–20 perform the indicated operations, and simplify the results.

1 **a.** $15 + (-3)$ **b.** $15 - (-3)$ **c.** $15(-3)$ **d.** $15 \div (-3)$

2 **a.** -3^2 **b.** $(-3)^2$ **c.** -2^3 **d.** $(2-3)^{23}$

3 **a.** $-7 + 3 \cdot 5^2$ **b.** $(-7+3) \cdot 5^2$ **c.** $-7 + (3 \cdot 5)^2$ **d.** $(-7 + 3 \cdot 5)^2$

4 **a.** $-5(4)(-3)(2)$ **b.** $-5(-4)(-3)(2)$ **c.** $-5(-4)(-3)(-2)$ **d.** $-5(0)(3)(-2)$

5 **a.** $(13-3)^2$ **b.** $(13)^2 - 3^2$ **c.** $2(13-3)$ **d.** $(-1)^{13} - 13^1$

6 **a.** $2^5 - 5^2$ **b.** $(7-4)^2 - (7^2 - 4^2)$ **c.** $2 + (5)(5) \div (5)(5)$ **d.** $2 + [(5)(5)] \div [(5)(5)]$

7 **a.** $|-7|$ **b.** $|-4|$ **c.** $|-7-4|$ **d.** $|-7| - |-4|$

8 $7 - 5[6 + 2(9-6)]$

9 $-11 + 2[7 - 2(4-9) - 8]$

10 $\dfrac{7^2 - 3^2}{(7-3)^2}$

11 $-\dfrac{3}{5} - \dfrac{1}{2} + \dfrac{-3}{10}$

12 $\dfrac{3}{8} \div \left(-\dfrac{5}{12}\right)$

13 $\dfrac{27 - 4 \cdot 6}{5 - 8}$

14 $2(7-11) - 3(-4-1)$

15 $5^3 - 2^3 - (5-2)^3$

16 $-15 - 2^3 \cdot 3 + 5(11 - 3^2)$

17 $-34 \div 2 \cdot 3 + (5 - 9)$

18 $\dfrac{3}{8} \div \dfrac{-5}{12} \div \dfrac{-10}{21}$

19 $\dfrac{-4}{5} \div \left(\dfrac{3}{5} - \dfrac{5}{3}\right)$

20 $|-7 - (-3)| - (6 - |-4|)$

In Exercises 21–26 evaluate each expression for $w = -2$, $x = -3$, and $y = -4$.

21 $3w - x + 2y$ **22** $-5wxy$ **23** $(2w-5)(x-y)$

24 $wx^2 - y^2$ **25** $\dfrac{w}{x} - \dfrac{x}{y}$ **26** $\dfrac{wx - y}{2x - y}$

In Exercises 27–32 simplify each expression by combining like terms.

27 $4m - 2m + 7m$

28 $-a + 3b + 5a - 9b$

29 $2x - 7 - 3(4x - 2)$

30 $2(x - 3y + 1) - 3(2x - y - 4)$

31 $-3[x - 2(x - 2y)] - 4(x - 3y)$

32 $-2x - 3(4x - y) + 4(5x - 2y)$

In Exercises 33 and 34 write an algebraic expression for each verbal expression. Do not simplify these expressions.

33 Twice the square of the opposite of x

34 Negative seven minus the opposite of y

In Exercises 35 and 36 identify the property that justifies each statement.

35 a. $(2x + 2y) + z = 2x + (2y + z)$

b. $(2x + 2y) + z = (2y + 2x) + z$

c. $(2x + 2y) + z = 2(x + y) + z$

36 a. $(ab)(c + d) = (c + d)(ab)$

b. $(ab)(c + d) = (ab)(c) + (ab)(d)$

c. $(ab)(c + d) = (ab)(d + c)$

37 a. The additive inverse of $x - 3$ is ________.

b. The multiplicative inverse of $x - 3$ is ________ for $x \neq$ ________.

c. What number has no multiplicative inverse?

38 a. Numbers that are added are called ________.

b. Numbers that are multiplied are called ________.

c. The expression $x^2 - 5x + 7$ has ________ terms.

39 a. The additive inverse of $-\frac{4}{5}$ is ________.

b. The reciprocal of $-\frac{4}{5}$ is ________.

c. What number has an absolute value that is not positive?

40 a. What prime number is even?

b. What two integers are between $\sqrt{3}$ and π?

c. Which of these real numbers is an irrational number: $-\sqrt{25}, -4.14, 0, 2.12, \sqrt{7}, 4.\overline{15}$?

41 Perimeter and Area

a. Determine the perimeter of the given region to the nearest tenth of a centimeter. Assume the curved portions shown are semicircles.

b. Determine the area of this figure to the nearest tenth of a square centimeter.

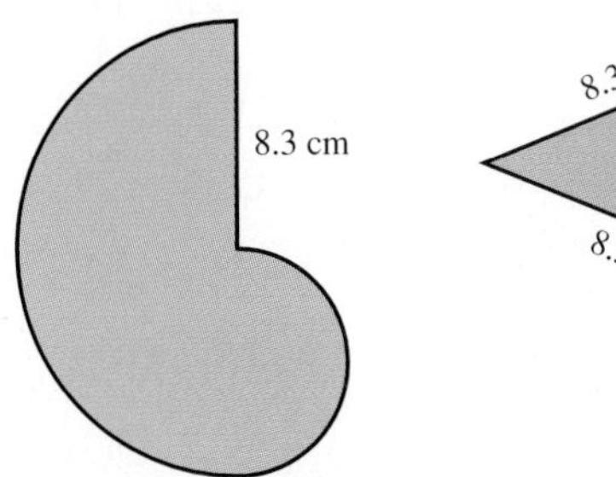

Figure for Exercise 41

8.3 cm

6.1 cm

8.3 cm

Figure for Exercise 42

42 Perimeter Determine the perimeter of the given region to the nearest tenth of a centimeter. Assume the curved portion shown is a semicircle.

In Exercises 43–45 simplify each expression in part a; solve the equation in part b.

	Simplify	*Solve*
43	**a.** $5m - 3 + m - 2m + 5$	**b.** $5m - 3 + m - 2m + 5 = 0$
44	**a.** $n - 8 - 5n - 2(n - 1)$	**b.** $n - 8 - 5n - 2(n - 1) = 0$
45	**a.** $\frac{w}{6} - \frac{w + 30}{15} - \frac{2w}{10} - 4$	**b.** $\frac{w}{6} - \frac{w + 30}{15} - \frac{2w}{10} - 4 = 0$

In Exercises 46–66 solve each equation or inequality. Graph the solution of each inequality for Exercises 50–61.

46 $5(x + 1) - 3(x - 1) = 2(x + 5)$

47 $\frac{3m + 1}{4} - \frac{1}{10} = \frac{4m - 3}{5}$

48 $\frac{x - 4}{x - 5} + 3 = \frac{11 - x}{x - 5}$

49 $\frac{x}{x + 3} + 2 = \frac{-3}{x + 3}$

50 $7x - 19 < 9x - 27$

51 $5(2y + 3) \geq 4(3y - 7)$

52 $1 < 2x - 3 \leq 9$

53 $14 \leq 5 - 3y < 17$

54 $3(2w - 8) < 7w + 5$ and $11w < 9w + 6$

55 $5w - 3 \geq 6w + 1$ or $3w - 2 \leq 6w + 4$

56 $|2r - 5| = 17$

57 $|2t - 5| \leq 3$

58 $|5t - 2| > 8$

59 $|y - 5| < -3$

60 $|y - 5| > -3$

61 $\frac{2z}{5} - \frac{14z}{10} < -5$

62 $\frac{5(z - 4)}{6} = \frac{2(z + 4)}{9} + \frac{1}{18}$

63 $5(2t - 4) - 3(7 - 2t) = 6(3t + 4) - 69$

64 $\frac{5(z - 4)}{6} = \frac{z + 5}{5} + \frac{z + 4}{7}$

65 $\frac{6r + 3}{11} + 2 \leq \frac{4r + 8}{7} + 1$

66 $1 - 3|x + 2| = -14$

67 Solve $W = FD$ for D.

68 Solve $V = \frac{1}{3}\pi r^2 h$ for h.

69 Solve $S = \frac{n}{2}[2a + (n - 1)d]$ for a.

70 Solve $A = P + PRT$ for R.

71 Numeric Equality One number is 2 more than three times another number. What are the numbers if their sum is 82?

72 Consecutive Odd Integers Find two consecutive odd integers whose sum is 348.

73 Numeric Inequality Five times a number is greater than 10 and less than or equal to 15. Find all such numbers.

74 Absolute-Value Inequality The absolute value of the difference of three times a number minus 4 is found. When 5 is subtracted from this absolute value, the result is 6. Find all such numbers.

75 Typing Speed How many words per minute must a typist average in order to type 2250 words in half an hour?

76 Running Speed What average speed must a runner maintain to have a 4-minute mile? Give the answer in miles per hour.

77 Dimensions of a Rectangle The length of a rectangle is 2 more than three times its width. If the perimeter is 92 centimeters, what is the width?

78 Hiking Speed Two military squads are 60 kilometers apart. They plan to rendezvous at some intermediate point. If one squad hikes 1 kilometer per hour faster than the other squad and they are to meet in 4 hours, what must the rate of each squad be in kilometers per hour?

79 Painting Painter A can paint a sign in 9 hours. When painter B helps him, it takes the two working together only 6 hours. How many hours would it take painter B working alone to paint the sign?

80 Draining a Pond The water level in a holding pond is being lowered so that repairs can be made. The outlet pipe can empty the full pond in 48 days if there is no water input. However, water is flowing into the pond at a rate such that the empty pond would be full in 100 days if no water were drained. If the outlet pipe is opened when the pond is full and water is flowing into the pond, how long will it take to drain the pond so that it is only one-fourth full?

81 Investment A woman invested a total of \$15,000 in two investments. She made a profit of 8% on the first investment and lost 6% on the second investment, for a net loss of \$480. How much did she invest at each rate?

82 Stamps A total of 100 stamps were purchased for \$26.90. Some of the stamps cost \$0.23 each, and the rest cost \$0.29 each. How many of each type were purchased?

83 Evaporation of Water How many pounds of water must be evaporated from 400 pounds of 4% salt solution to produce a 10% solution?

84 Economic Indicators The ratio of two readings from an economic indicator was $\frac{3}{5}$. The first reading was 9 units below normal, and the second reading was 5 units above normal. What is the normal reading?

85 Write an absolute-value inequality to represent each set of numbers.

a. 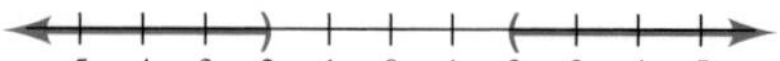

b. 

c. $[-7, -1]$ **d.** $(-\infty, -5) \cup (7, +\infty)$

86 Write an absolute-value inequality to represent these intervals.

a. $(-4, 4)$ **b.** $(-\infty, -1] \cup [1, +\infty)$

87 Simplify these sets, and write your answer in interval notation.

a. $(-2, 1) \cap (-1, 3]$ **b.** $(-2, 1) \cup (-1, 3]$

CALCULATOR USAGE (88 AND 89)

88 Use a calculator to evaluate $f(-8.25)$ for $f(x) = -3x^2 - 8x + 7$.

89 Use a calculator with an ABS key to evaluate $f(-4.44)$ for $f(x) = |-5x^2 - 7x + 3|$.

Error Analysis (90–92)

90 Explain why the notation $x/y - 2$ is an incorrect way of denoting the quotient of x divided by the quantity y minus 2.

91 Explain why the notation $x \cdot - y$ is an unacceptable format for denoting the product of x and the opposite of y.

92 Explain why the format

$$3x + 2 = 7x - 6$$
$$-4x = -8$$
$$x = 2$$

is acceptable, whereas the format

$$3x + 2 = 7x - 6$$
$$= -4x = -8$$
$$= \quad x = 2$$

is unacceptable.

Mastery Test for Chapter 1

Exercise numbers correspond to Section Objective numbers.

[1-1] **1** Given $S = \left\{-5, -\sqrt{5}, -1.5, 0, 2.\bar{5}, \sqrt{9}, \pi, 3.444\ldots, 4.131131113\ldots, 5\frac{2}{3}\right\}$,

a. List all natural numbers in S.

b. List all whole numbers in S.

c. List all integers in S.

d. List all rational numbers in S.

e. List all irrational numbers in S.

f. List all real numbers in S.

g. List all prime numbers between 32 and 40.

h. List all composite numbers between 79 and 87.

[1-1] **2** Simplify each of these expressions.

a. $|8|$ **b.** $|-8|$

c. $|7 - 12|$ **d.** $|7| - |12|$

[1-1] **3** Identify the property that justifies each statement.

a. $x(v + w) = xv + xw$

b. $x(v + w) = (v + w)x$

c. $x(v + w) = x(w + v)$

d. $(r + 2s) + 3t = r + (2s + 3t)$

[1-2] **4** Perform the operations indicated.

a. $-15 + 17$ **b.** $-15 - 17$

c. $1.23 + (-2.51)$ **d.** $\frac{1}{2} - \frac{1}{3}$

e. $-30(-5)$ **f.** $-30 \div 5$

g. $-2(5)(-10)$ **h.** $\left(\frac{-3}{4}\right) \div \left(\frac{5}{8}\right)$

[1-2] **5** Perform the operations indicated.

a. -6^2 **b.** $(-6)^2$

c. $(3 + 5)^2$ **d.** $3^2 + 5^2$

[1-2] **6** Perform the operations indicated, and simplify the results.

a. $-4 + 5(3)$ **b.** $-2 - 3 \cdot 4^2$

c. $-5 + 2[6 - (3 - 11)]$

d. $(7 - 2)^2 + 7^2 - 2^2$

[1-3] **7** Evaluate each of the following algebraic expressions for $x = -2$ and $y = -5$.

a. $3x - 4y$ **b.** $3xy - 4$

c. $x + y^2$ **d.** $(x + y)^2$

[1-3] **8** Simplify each of these algebraic expressions.

a. $-5x + 7x$

b. $3x - 4y - 2x + 9y$

c. $x - 3 - 2(x + 1)$

d. $2(3x + y) - (2x - 5y)$

[1-4] **9** Solve each of these equations.

a. $5(2v - 3) - 4(3v - 2) = 3(4 - 5v)$

b. $\frac{y + 1}{2} = \frac{y + 2}{3} + \frac{y - 1}{6}$

c. $\frac{z}{z - 3} = 2 + \frac{3}{z - 3}$

[1-4] **10** Solve each of these equations for x.

a. $z = xy$ **b.** $z = x + y$

c. $z = \frac{xy + a}{b}$ **d.** $z = \frac{w(x + y)}{v}$

[1-5] **11 a.** The sum of a number and 17 is 3 more than twice this number. Find the number.

b. The sum of two consecutive odd integers is 256. Find these integers.

c. Determine the area of the figure shown to the nearest tenth of a square centimeter.

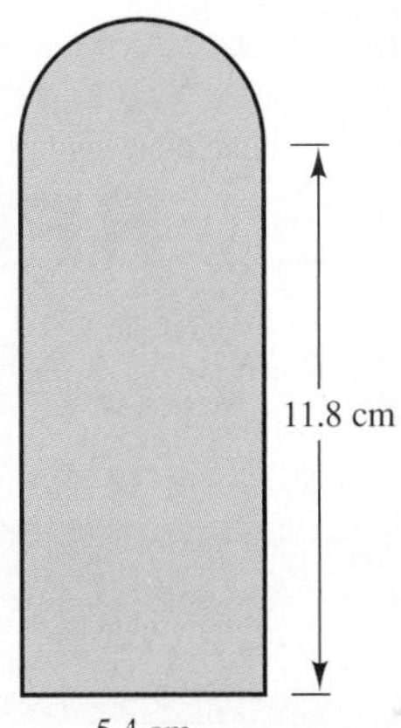

Figure for Exercise 11c

d. A movie theater charges an admission price of \$3 for an adult and \$1.50 for a child. If 700 tickets were sold and the total revenue received was \$1650, how many tickets of each type were sold?

e. Two airplanes depart simultaneously from parallel runways at the same airport and travel in the same direction. The first plane averages 640 kilometers per hour, and the second plane averages 768 kilometers per hour. How many hours will it be before they are 544 kilometers apart?

[1-6] **12 a.** Express $-3 \leq x < 11$ using interval notation.

b. Express $x > \pi$ using interval notation.

c. Use interval notation to represent the graph shown.

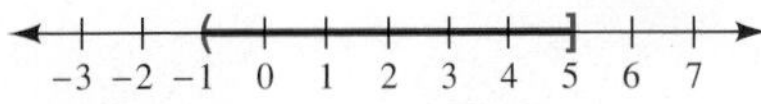

Figure for Exercise 12c

[1-6] **13** Solve each of these inequalities.

a. $7x - 19 < 9x - 27$

b. $2w + 1 < 3$ or $3w - 2 \geq 13$

c. $-28 \leq -3v - 7 < 8$

d. $3(2w - 8) < 7(w + 1) - 2$ and $11(w - 4) < 9(w + 1) - 41$

[1-7] **14** Solve each equation or inequality.

a. $|2x + 5| = 49$ **b.** $|2x - 3| < 19$

c. $\frac{2}{3}|x - 5| - 4 < 2$

Chapter 2

GRAPHS, RELATIONS, AND LINEAR FUNCTIONS

AN APPLICATION

Math in the Movies

Many of today's most popular movies use computers to work their visual magic. In a process called ray tracing a series of calculations converts an imaginary three-dimensional scene into a two-dimensional image by tracking the rays of light that would travel from the scene to the moviegoer's eye if the scene really existed. Depicted schematically, the process looks something like this:

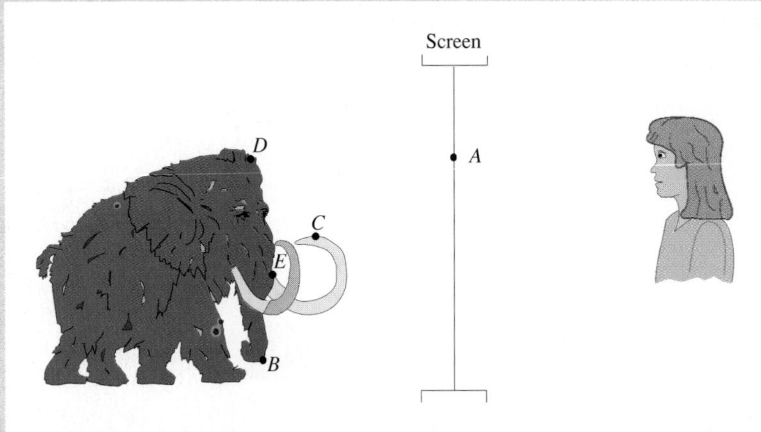

The image on the movie screen is created on a computer screen first, using millions of tiny bits of color called pixels. To find out what should appear at point A on the screen, we can follow a straight line from the viewer's eye through A to the first object the line hits. Using other, more complicated techniques to determine the effects of imaginary lighting, we can then determine precisely what color the pixel at point A should be. In this example should we be looking at the lighting of point B, C, D, or E? (A ruler is sufficient to find out; you do not need a pen or pencil.)

Over a million points have to be plotted to produce a single realistic-looking frame of film. Theoretically, this could be done by hand, but obviously computers are the only practical way to generate the thousands of frames making up a single movie sequence. Computers, however, do not use a pencil and ruler to trace rays. Instead, they use program routines based on the fact that a straight line can be represented mathematically by a linear equation. In this chapter you will learn more about linear equations, and at the end of the chapter you will have a chance to apply your new knowledge to some problems involving ray tracing.

CHAPTER 2 OBJECTIVES

1 **Plot points on a rectangular coordinate system (Section 2-1).**

2 **Graph a linear equation using the intercepts of the line (Section 2-1).**

3 **Draw a broken line graph through a set of points (Section 2-1).**

4 **Calculate the slope of a line (Section 2-2).**

5 **Use the slope-intercept form to graph a line (Section 2-2).**

6 **Use the special forms of horizontal and vertical lines (Section 2-3).**

7 **Use the point-slope form of a line (Section 2-3).**

8 **Determine whether a relation is a function (Section 2-4).**

9 **Determine the domain and the range of a function (Section 2-4).**

10 **Evaluate an expression using functional notation (Section 2-5).**

11 **Translate a statement of variation into an equation (Section 2-5).**

An important use of mathematics is to discover and explain patterns and relationships. Graphs are important because they allow us to use our visual sense to see patterns. To paraphrase an old saying, one graph may reveal as much as a thousand words. This chapter will use the two-dimensional rectangular coordinate system to examine linear functions. We will then examine other types of functions that relate two variables.

2-1 The Rectangular Coordinate System

SECTION OBJECTIVES

1 **Plot points on a rectangular coordinate system.**

2 **Graph a linear equation using the intercepts of the line.**

3 **Draw a broken line graph through a set of points.**

On the real number line each point represents a real number. On the real plane each point represents an ordered pair of real numbers. The **rectangular coordinate system** shown in Figure 2-1 is frequently called the **Cartesian coordinate system** in honor of René Descartes (see the Mathematical Note). The horizontal number line is usually called the **x-axis,** and the vertical number line is usually called the **y-axis.** The axes cross at the **origin** and separate the plane into four **quadrants,** which are labeled I, II, III, and IV, in a counterclockwise direction starting from the upper right. The points on the axes are not considered part of any of the quadrants.

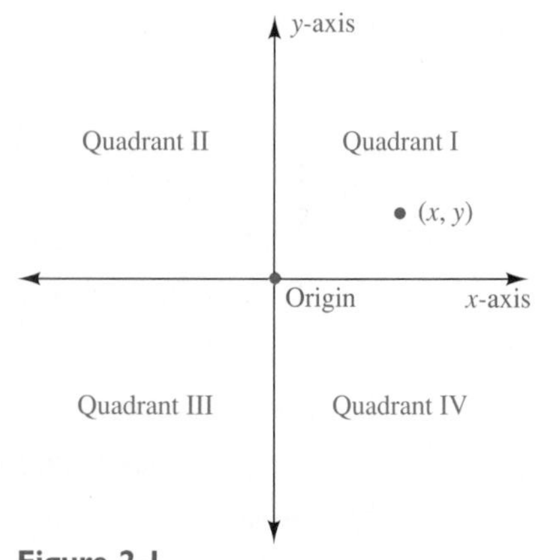

Figure 2-1
The Rectangular Coordinate System

Each point in the plane corresponds to a unique ordered pair of real numbers that identify its horizontal and vertical location with respect to the origin. The first coordinate of the ordered pair is called the **x-coordinate,** and the second coordinate is called the **y-coordinate.** The coordinates of the points A, B,

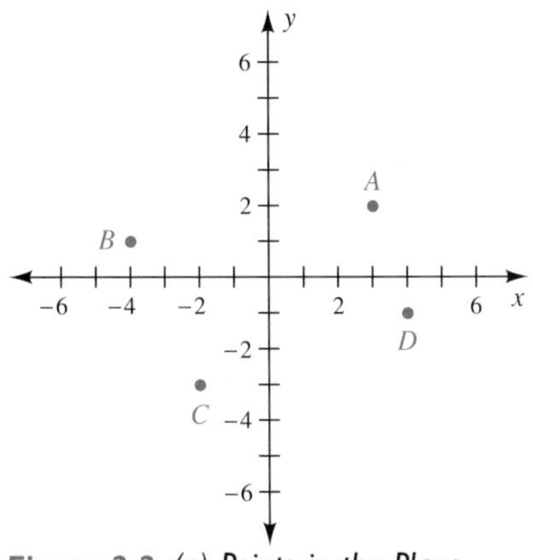

Figure 2-2 *(a) Points in the Plane* *(b) Coordinate Sign Pattern*

> **A Mathematical Note**
>
> René Descartes (1596–1650), born to a noble French family, was known for his studies in anatomy, astronomy, chemistry, physics, and philosophy, as well as mathematics. Prior to Descartes, algebra was concerned with numbers and calculations, and geometry was concerned with figures and shapes. Descartes merged the power of these two areas into analytic geometry—his most famous discovery.

C, and *D* in Figure 2-2(a) are

A: (3, 2), which is in quadrant I;

B: (−4, 1), which is in quadrant II;

C: (−2, −3), which is in quadrant III; and

D: (4, −1), which is in quadrant IV.

All points within the same quadrant have the same sign pattern. For example, in quadrant I both coordinates are positive, which is shown by (+, +). Knowledge of the sign pattern for each quadrant, as shown in Figure 2-2(b), is useful in trigonometry. These sign patterns can also be helpful in analyzing scatter diagrams in statistics.

A **scatter diagram** for a set of data points is simply a graph of these points that allows us to examine the data for some type of pattern. For example, if the data points all lie near a line, then the relationship is approximately a linear relationship.

The magnitude of the coordinates is used to determine the scale on each axis. If these coordinates differ significantly in size, then a different scale can be used on each axis. This practice is common in statistics, when the x and y variables often represent quantities measured in different units.

EXAMPLE 1 A Scatter Diagram from a Nature Study

As part of a study of koalas a naturalist observed the behavior of one koala for ten days. During the day she recorded both the mean temperature in degrees Celsius and the number of active minutes of this very sluggish animal. Determine whether a linear relationship exists between the temperature and the koala's active minutes.

SOLUTION Ordered pairs are often presented in a table format.

Temperature, x	15	16	18	20	17	14	19	21	23	22
Active Minutes, y	55	60	65	70	60	50	70	80	80	75

The first ordered pair from this table is (15, 55). The break in the x-axis shown in the figure denotes that the numbers from 0 to 13 have been omitted. Likewise, the break in the y-axis denotes omission of the numbers from 0 to 50.

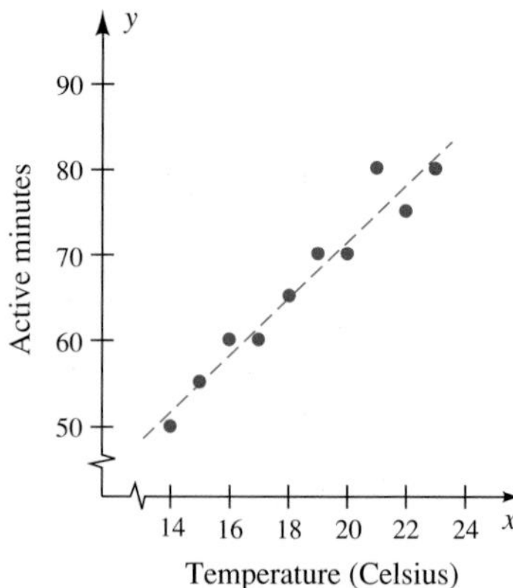

The dashed line on the graph shows that the data points all lie near a line. Because these points all lie near a line, it appears that there is approximately a linear relationship between the temperature and the active minutes of this animal. ▼

An equation of the form $Ax + By + C = 0$ is called a **first-degree equation** because the exponent on both x and y is understood to be 1. The graph of all ordered pairs that satisfy this equation is a straight line. Thus this equation is called a **linear equation.**

General Form of a Linear Equation

The general form of a linear equation is

$$Ax + By + C = 0$$

where A, B, and C are real constants and A and B are not both zero.

EXAMPLE 2 Identifying Linear Equations

Determine which of these equations are linear equations.

	SOLUTIONS	
(a) $2x + 3y - 1 = 0$	Linear equation	In the general form, $Ax + By + C = 0$, $A = 2, B = 3$, and $C = -1$.
(b) $y = 5x - 4$	Linear equation	This equation can be written in general form as $5x - y - 4 = 0$.
(c) $y = 7x^2 + 4$	Not a linear equation	This is a second-degree equation since the exponent on x is 2.
(d) $y = 4$	Linear equation	This equation can be written in general form as $0x + y - 4 = 0$.
(e) $x = 1$	Linear equation	This equation can be written in general form as $x + 0y - 1 = 0$.
(f) $x + 3y - 4$	Not a linear equation	This is a linear expression, but it is not an equation. ▼

EXAMPLE 3 Checking Possible Solutions of an Equation

Test the points (5, 0), (0, 2), and (1, 1) to determine whether each point is a solution of the linear equation $2x + 5y = 10$.

SOLUTION

$$\begin{aligned} 2x + 5y &= 10 \\ 2(5) + 5(0) &\stackrel{?}{=} 10 \\ 10 + 0 &\stackrel{?}{=} 10 \\ 10 &= 10 \text{ checks.} \end{aligned}$$

$$\begin{aligned} 2x + 5y &= 10 \\ 2(0) + 5(2) &\stackrel{?}{=} 10 \\ 0 + 10 &\stackrel{?}{=} 10 \\ 10 &= 10 \text{ checks.} \end{aligned}$$

$$\begin{aligned} 2x + 5y &= 10 \\ 2(1) + 5(1) &\stackrel{?}{=} 10 \\ 2 + 5 &\stackrel{?}{=} 10 \\ 7 &= 10 \text{ does not check.} \end{aligned}$$

Substitute the test points into the equation for x and y. Then simplify the left side of the equation.

Answer (5, 0) and (0, 2) are solutions; (1, 1) is not a solution. ▼

To graph a line, only two points need to be plotted. However, it is wise to plot a third point as a check. Points on a line can be determined by forming a table of values using the given equation. To prepare a table of values, we arbitrarily select values for the variable x and then calculate the corresponding values of y from the equation.

> **A Mathematical Note**
>
> W. W. Sawyer makes the point in his 1943 book *Mathematician's Delight* that the first mathematicians were practical men who built or made things. Some of our terminology can be traced to this source. For example, the word *straight* comes from Old English for "stretched," whereas the word *line* is the same as that for "linen thread." Thus a straight line is literally a stretched linen thread—as anyone who is planting potatoes or laying bricks knows.

EXAMPLE 4 Preparing a Table of Values

Graph $2x + 3y - 6 = 0$.

SOLUTION Prepare a table of values by arbitrarily selecting the x-coordinates -3, 0, and 3. Substitute these values into the equation, and then calculate the corresponding y-coordinates.

Set $x = -3$, and solve for y.

$$\begin{aligned} 2x + 3y - 6 &= 0 \\ 2(-3) + 3y - 6 &= 0 \\ -6 + 3y - 6 &= 0 \\ 3y &= 12 \\ y &= 4 \end{aligned}$$

Set $x = 0$, and solve for y.

$$\begin{aligned} 2x + 3y - 6 &= 0 \\ 2(0) + 3y - 6 &= 0 \\ 0 + 3y - 6 &= 0 \\ 3y &= 6 \\ y &= 2 \end{aligned}$$

Set $x = 3$, and solve for y.

$$\begin{aligned} 2x + 3y - 6 &= 0 \\ 2(3) + 3y - 6 &= 0 \\ 6 + 3y - 6 &= 0 \\ 3y &= 0 \\ y &= 0 \end{aligned}$$

x	y
-3	4
0	2
3	0

Select arbitrary x values, and calculate the y values; plot these points, and then draw the line through them.

Although the choice of coordinates used in forming a table of values is arbitrary, we usually include $x = 0$ and $y = 0$ because they result in relatively easy computations and because they give the points at which the line intercepts, or crosses, the axes. The graph crosses the x-axis at the **x-intercept** $(a, 0)$ and the y-axis at the **y-intercept** $(0, b)$. (See Figure 2-3.) Sometimes we abbreviate the terminology, using a instead of $(a, 0)$ for the x-intercept and b instead of $(0, b)$ for the y-intercept.

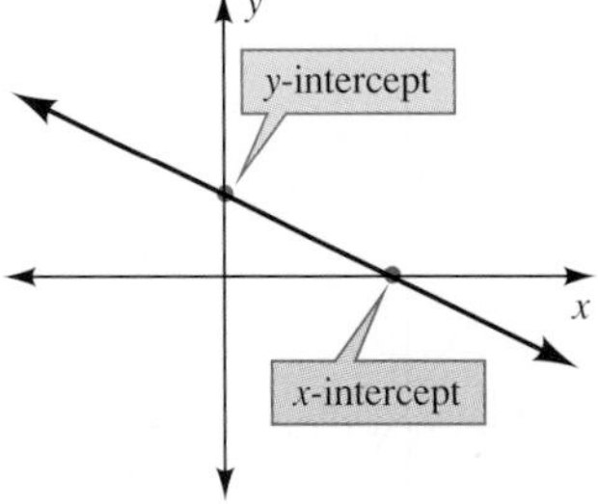

Figure 2-3

FINDING THE x- AND y-INTERCEPTS

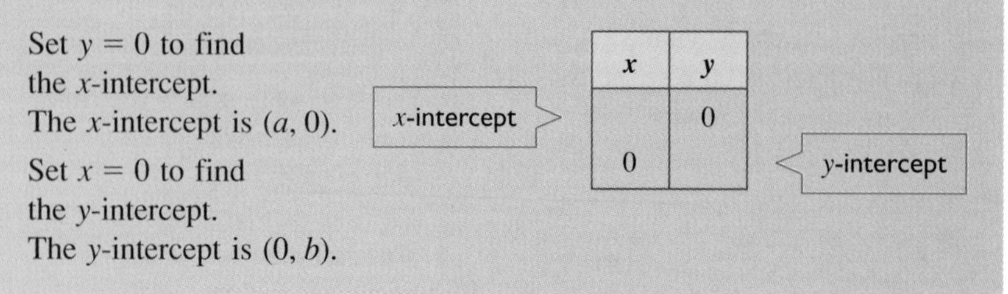

EXAMPLE 5 Calculating the Intercepts of a Line

Calculate the intercepts of $4x - 3y = 12$, and graph the line through these intercepts.

SOLUTION

x-intercept

Set $y = 0$.

$$\begin{aligned} 4x - 3y &= 12 \\ 4x - 3(0) &= 12 \\ 4x &= 12 \\ x &= 3 \end{aligned}$$

y-intercept

Set $x = 0$.

$$\begin{aligned} 4x - 3y &= 12 \\ 4(0) - 3y &= 12 \\ -3y &= 12 \\ y &= -4 \end{aligned}$$

x	y
3	0
0	−4

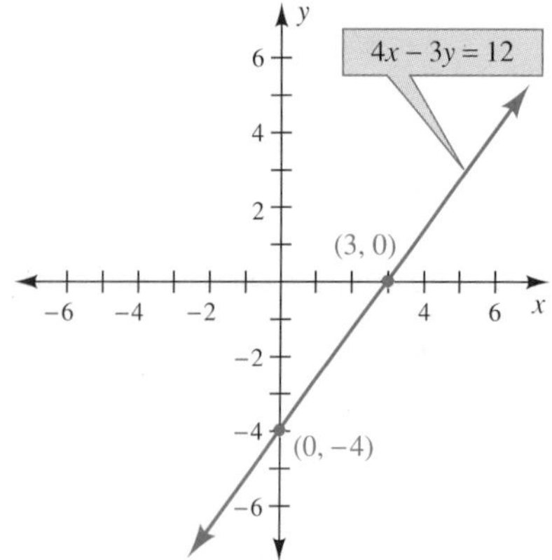

Plot the intercepts, and then draw a line through them.

If either A or B equals 0 in the general form $Ax + By + C = 0$, then one of the variables will be missing. Equations of this form will be graphed as either horizontal or vertical lines. If A and B are not 0 but $C = 0$, then the graph will

SELF-CHECK

Graph $4x + 3y = -12$.

SELF-CHECK ANSWER

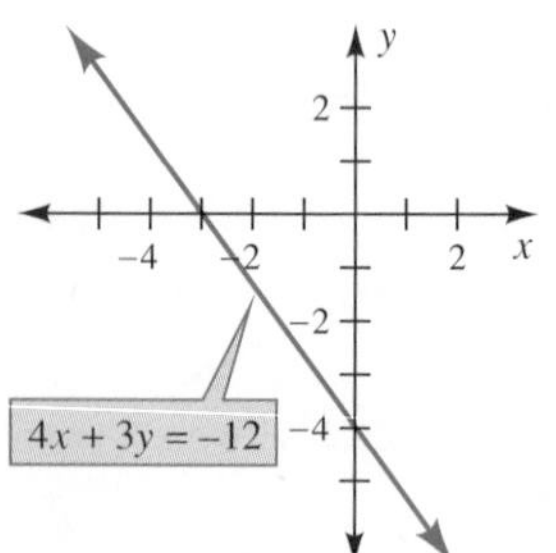

be an oblique (slanted) line through the origin. In each of these special cases there is only one intercept instead of distinct x- and y-intercepts. We will examine these special cases in the next two examples.

EXAMPLE 6 Graphs of Horizontal and Vertical Lines

Graph each of these linear equations.

SOLUTIONS

(a) $2y + 6 = 0$

$$2y + 6 = 0$$
$$2y = -6$$
$$y = -3$$

x	y
0	−3
−2	−3
2	−3

This equation could also be written in the general form $0x + 2y + 6 = 0$, or $0x + y + 3 = 0$. The y-coordinate is always -3, and the x-coordinate can be any real number since its coefficient is 0.

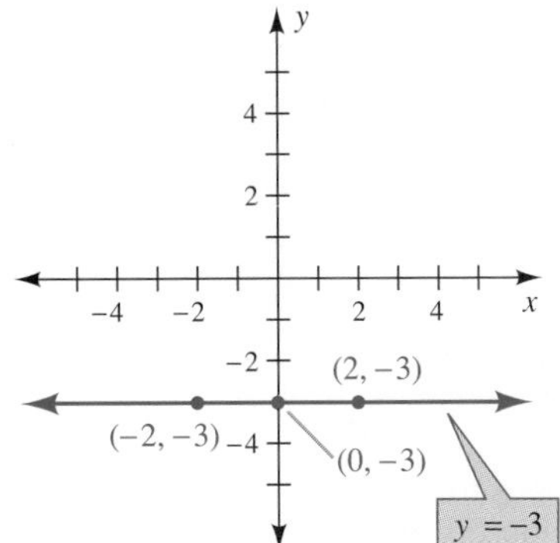

The graph of this equation is a horizontal line with y-intercept $(0, -3)$.

(b) $3x - 5 = 0$

$$3x - 5 = 0$$
$$3x = 5$$
$$x = \frac{5}{3}$$

x	y
$\frac{5}{3}$	1
$\frac{5}{3}$	0
$\frac{5}{3}$	2

This equation could also be written in the general form $3x + 0y - 5 = 0$. The x-coordinate is always $\frac{5}{3}$. The y-coordinate can be any real number.

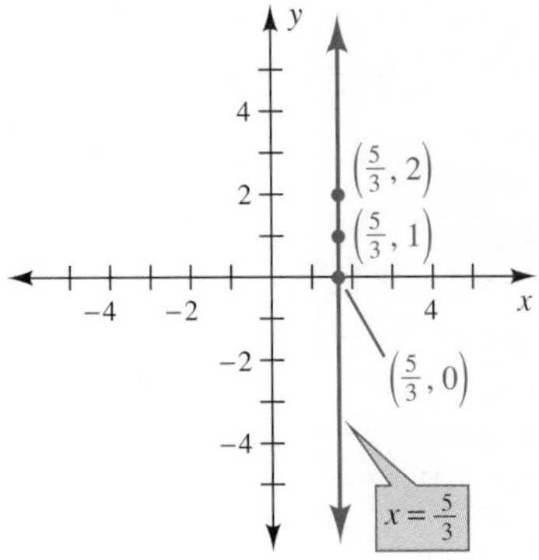

The graph of this equation is a vertical line with x-intercept $\left(\frac{5}{3}, 0\right)$.

▼

EXAMPLE 7 Graph of an Oblique Line Through the Origin

Graph $2x + 3y = 0$.

SOLUTION $2x + 3y = 0$

x	y
0	0
0	0
3	−2

The point (0, 0) is listed twice in the table to emphasize that it is both the x- and the y-intercept.

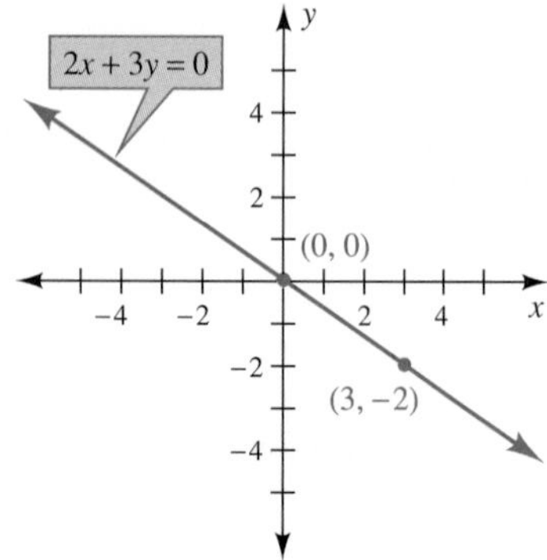

Plot the points in the table.

A summary of the method we have used to graph a linear equation is given in the following box. This summary refers to both oblique (slanted) lines and the special cases of horizontal and vertical lines. Remember that a third point is often plotted as a check.

GRAPHING A LINEAR EQUATION

To graph $Ax + By + C = 0$, plot the intercepts and one other point. Then draw the line through these points.

x	y
?	0
0	?
?	?

Possibilities

1 An oblique line: If $A \neq 0$ and $B \neq 0$, then the line is oblique. This line will pass through the origin if $C = 0$.

2 A horizontal line: If $A = 0$, the equation can be written in the form y equals some constant and the line will be horizontal.

3 A vertical line: If $B = 0$, the equation can be written in the form x equals some constant and the line will be vertical.

The emphasis in the first few examples has been to use known points to graph a line. It is just as important to be able to use a given graph to determine the points on the line. Newspapers and magazines often use graphs to present information. A linear graph not only shows the overall pattern between two variables, but also graphically gives the ordered pairs that relate the x and y variables. In the next example, we will use a graph to answer questions related to specific points on the graph.

SELF-CHECK

Graph each of these linear equations.

1 $2x + y = 4$ **2** $x = 4$
3 $y = 4$

EXAMPLE 8 Fuel Consumption

The engineers responsible for building and testing a new vehicle produced the following graph, which approximates the mileage in miles per gallon for the steady speed at which the vehicle is driven. Use this graph to

(a) Determine the mileage at 40 miles per hour

(b) Determine the speed at which the mileage is 25 miles per gallon

SOLUTION

(a) The mileage at 40 mi/h is approximately 43 mi/gal.

(b) The speed when the mileage is 25 mi/gal is approximately 63 mi/h.

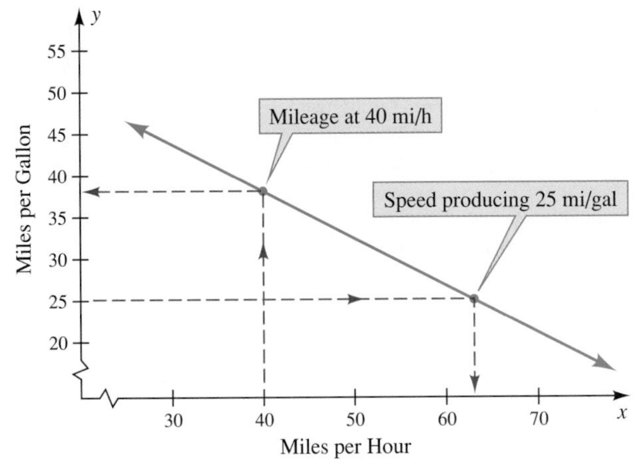

SELF-CHECK

4 Use the graph in Example 8 to determine the mileage at 50 miles per hour.

SELF-CHECK ANSWERS

1

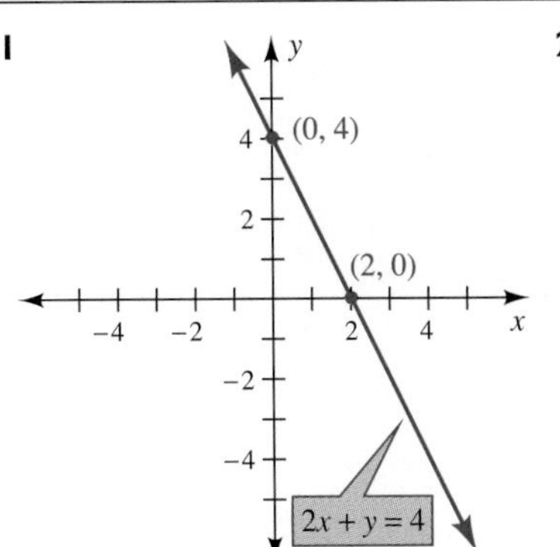

2

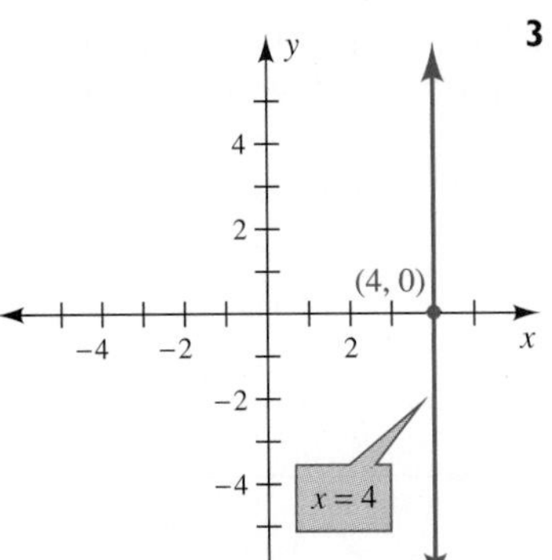

3

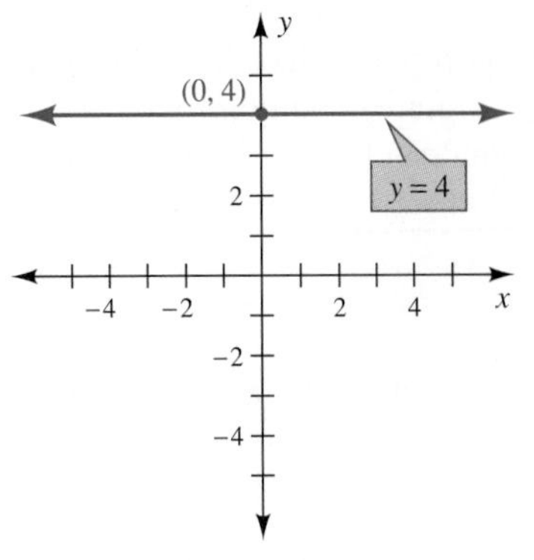

4 35 mi/gal

One important use of linear graphs is to approximate other, more complicated graphs. Over a limited range of values the points on a line can be used to closely approximate the points on another curve. By using a series of line segments, we can approximate curves as illustrated in Figure 2-4.

Spreadsheet programs and statistics frequently use line graphs to analyze data. A **line graph,** also called a **broken line graph,** is formed by using line segments to connect points with adjacent x-coordinates. Line graphs allow us to use known data points to approximate an unknown curve and unknown points on the curve. There are two types of linear approximations, **interpolation** and **extrapolation.** Linear interpolation uses two known points to approximate an unknown point between the known points. Linear extrapolation approximates an unknown point beyond the range of all the given points.

Use of a line graph is illustrated in the next example, in which a physics student has recorded the distance of a ball at eight different times.

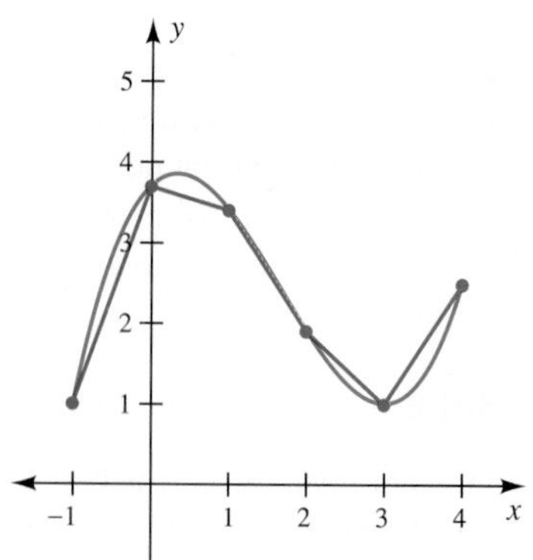

Figure 2-4
A Graph Approximated by Linear Segments

EXAMPLE 9 Approximations Using a Line Graph

Draw a line graph of the points shown in the table. Use the graph and linear interpolation to approximate the distance at 8 seconds. Then use the graph and linear extrapolation to approximate the distance at a time of 0 seconds.

x (seconds)	y (feet)
1.0	200
1.9	285
2.8	350
3.5	380
4.5	400
5.6	380
7.1	290
8.9	90

SOLUTION

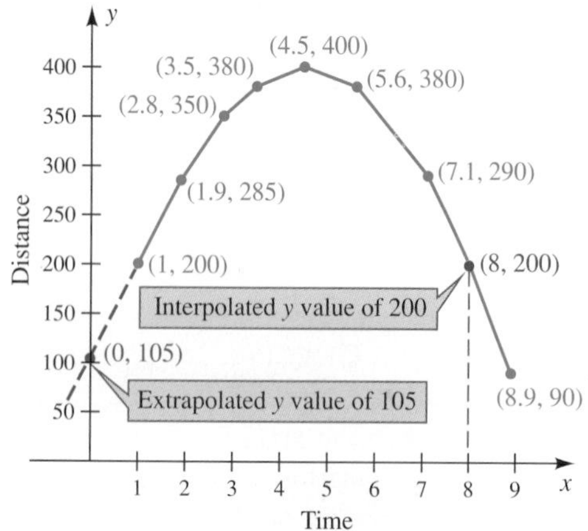

Interpolation: 8 sec is between the known values at 7.1 sec and 8.9 sec. Use the line segment connecting these known values to approximate the y-coordinate of the point on the segment whose x-coordinate is 8.
Extrapolation: 0 sec is earlier than any of the given times. Extend the line segment through the first pair of points to a point with an x-coordinate of 0. Then approximate the y-coordinate of this point.

Answer The distance at 8 sec is approximately 200 ft. The distance at 0 sec is approximately 105 ft.

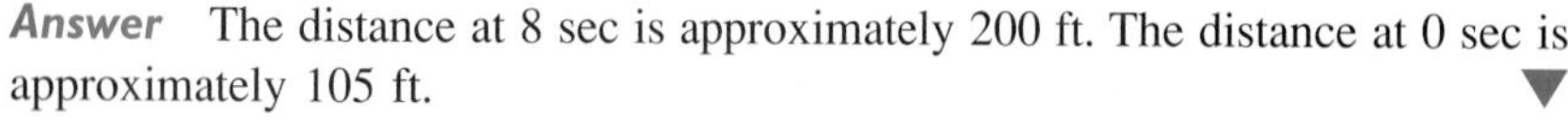

Caution: The accuracy of interpolations is dependent in part on the distance between the known points. Extrapolations into the future should be viewed with extreme care, as future conditions may well change and not follow past trends. A similar warning can be given about extrapolations into the past.

SELF-CHECK 

Use the graph in Example 9 to determine the distance at 6.5 seconds.

SELF-CHECK ANSWER 

330 ft

Exercises 2-1

A

In Exercises 1 and 2 give the coordinates of points A–D and the quadrant in which each point is located.

1

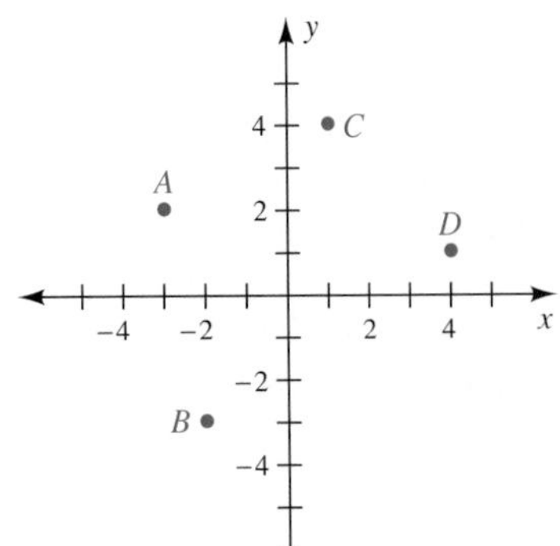

2

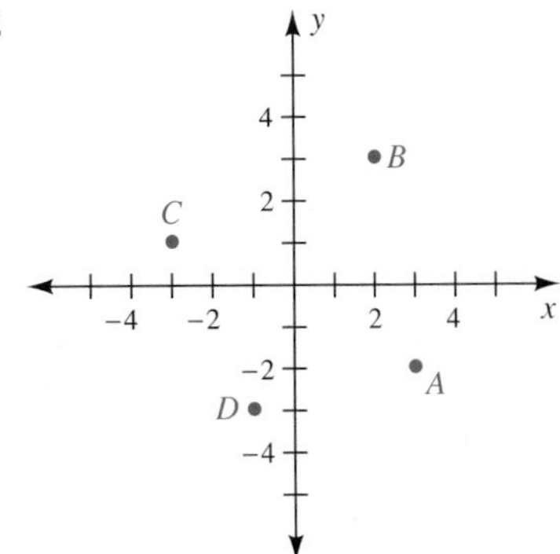

In Exercises 3 and 4 plot and label the points whose coordinates are given.

3 $A = (2, -3)$, $B = (-4, 1)$, $C = (-3, -1)$, $D = (5, 2)$, $E = (0, 0)$

4 $A = (-2, -2)$, $B = (-3, 1)$, $C = (4, 2)$, $D = (5, -2)$

5 Give the quadrants in which points A–E from Exercise 3 are located.

6 Give the quadrants in which points A–D from Exercise 4 are located.

In Exercises 7 and 8 test the points (1, 3), (−9, −3), and (−2, 5) in the given equations to determine whether each point is a solution of the equation.

7 $3x - 5y + 12 = 0$

8 $2x - 7y - 3 = 0$

In Exercises 9–16 complete the table shown for the given equation. Then use the points from the table to graph this linear equation.

x	0		2
y		0	

9 $x + 2y = 2$

10 $x - 2y = 6$

11 $2x - y = 6$

12 $2x + 5y = 10$

13 $3x - 2y = 12$

14 $3x + 2y = 12$

15 $5x - 4y = 20$

16 $5x - 4y = 0$

In Exercises 17–26 plot the intercepts of each equation, and then graph the linear equation.

17 $4x - 3y = 12$

18 $3x + 4y = 12$

19 $x + y = 5$

20 $x - y = 4$

21 $x = 7$

22 $y = -3$

23 $2y + 10 = 0$

24 $3x + 12 = 0$

25 $3x + 5y = 0$

26 $2x - 5y = 0$

In Exercises 27–30, use the given graph to complete the table of values shown.

x		0	2
y	0		

27

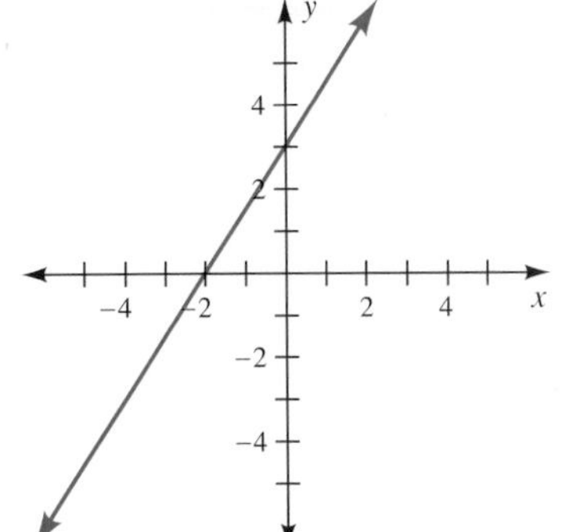

28

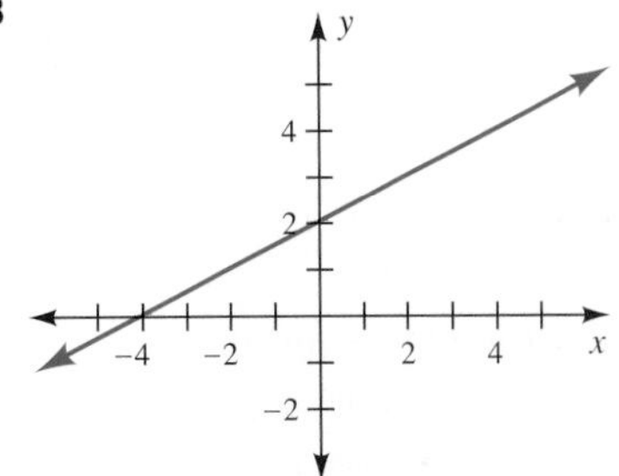

29

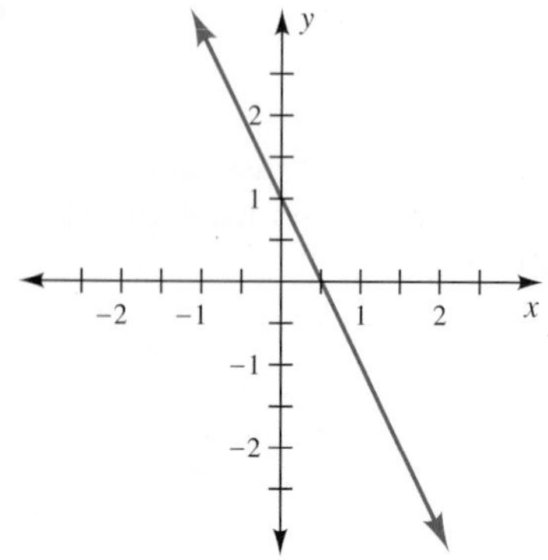

30

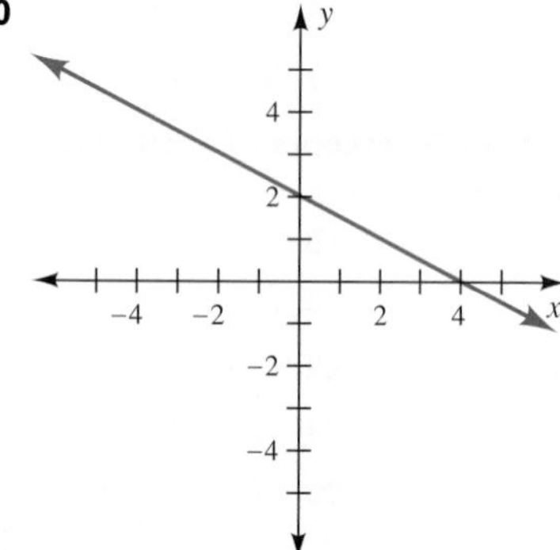

In Exercises 31 and 32 use the given line graph to complete the table of values shown.

x	1	2	3	4	5
y					

31

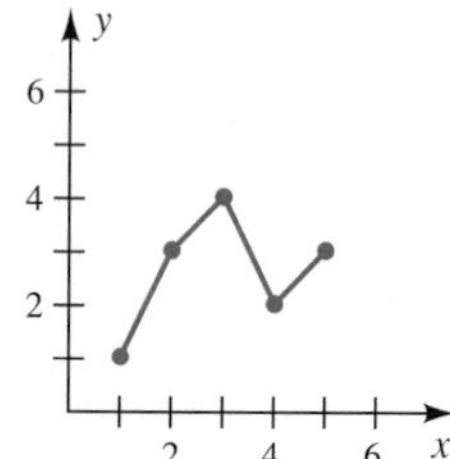

32

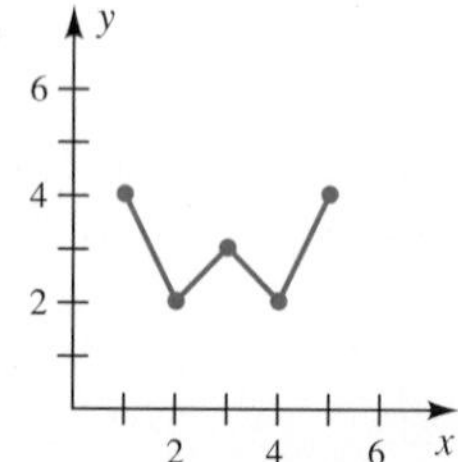

In Exercises 33–40 plot the points specified.

33 All points for which the x-coordinate is 2

34 All points for which the x-coordinate is -3

35 All points for which the y-coordinate is -2

36 All points for which the y-coordinate is 4

37 All points for which the x- and y-coordinates are equal

38 All points in the plane for which the y-coordinate is the opposite of the x-coordinate

39 The triangle with vertices at $(-2, 2)$, $(2, 2)$, and $(0, -3)$

40 The square with vertices at $(3, 3)$, $(-3, 3)$, $(-3, -3)$, and $(3, -3)$

41 Value of an Investment The graph below was published by an investment company to show the value of \$1000 invested with their company in 1985. Use this graph to determine

a. The value of the investment in 1989

b. The year that the value of the investment was \$1500

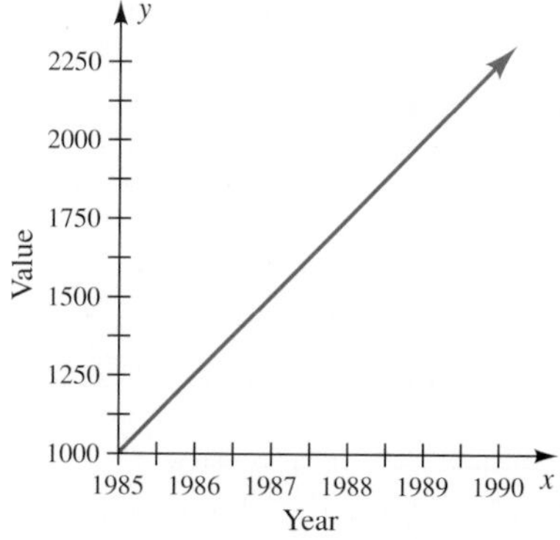

Figure for Exercise 41

42 Cost of a Telephone Call The graph below was distributed by a local phone cooperative to show the cost of a phone call from one city to another. Use this graph to determine

a. The cost of a 3-minute call

b. The length of a call that costs \$0.55

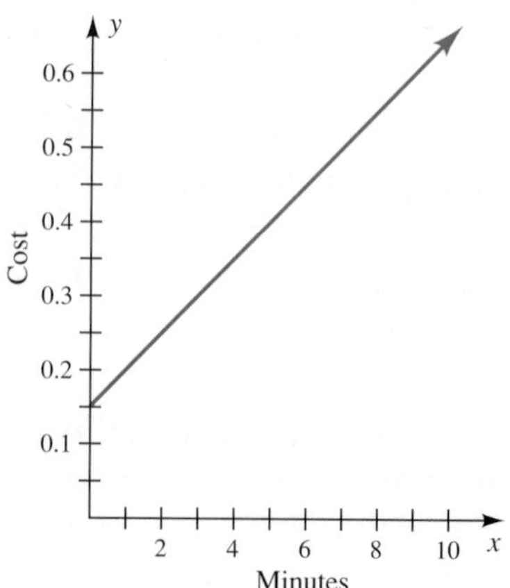

Figure for Exercise 42

43 Keyboard Sales The sales of computer keyboards by a keyboard manufacturer are given for selected months in the line graph below. The sales are in thousands of keyboards. Use this graph to

a. Determine the sales in July

b. Determine the sales in September

c. Interpolate to approximate the sales in August

d. Extrapolate to approximate the sales in December

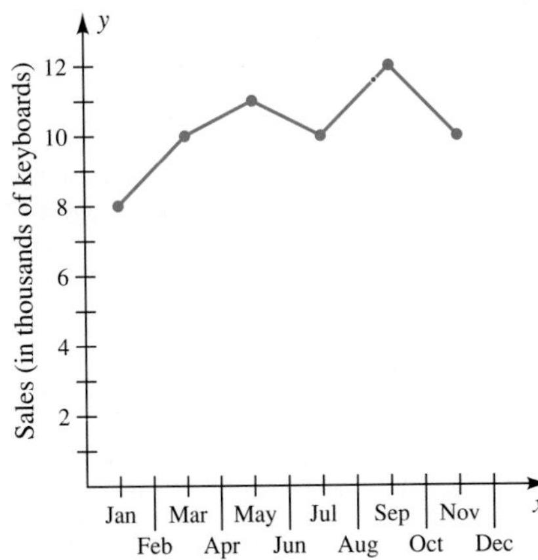

Figure for Exercise 43

44 Stock Prices An investor recorded the weekly stock prices for a share of a mutual fund. His incomplete records for 9 weeks are given in the line graph below. Use this graph to

a. Determine the price in week 3

b. Determine the price in week 5

c. Interpolate to approximate the price in week 6

d. Extrapolate to approximate the price in week 8

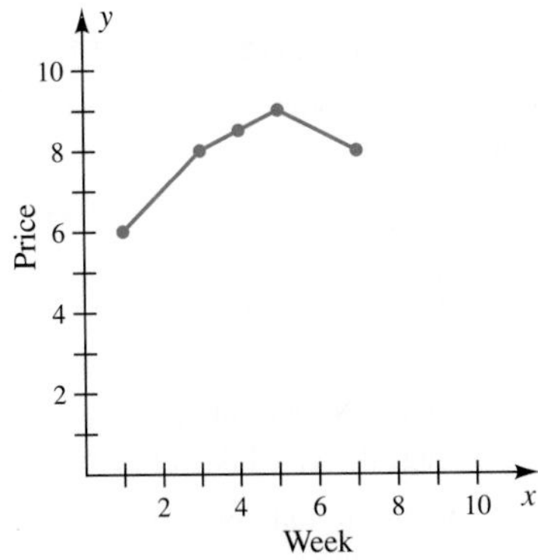

Figure for Exercise 44

B

Scatter Diagrams (45–51)

45 Draw a scatter diagram for the data below, which compare the number of cigarettes smoked per day by an expectant mother to the birthweight of her child. Does this appear to be a linear relationship?

Number of Cigarettes, x	Birthweight in Kilograms, y
5	3.20
10	3.15
15	3.10
20	3.06
25	3.09
30	3.05
35	3.01
40	3.00

46 Draw a scatter diagram for the data below, which compare the year to the annual cost (tuition) of attending the University of Illinois for that year. Does this appear to be a linear relationship?

Year	Cost
1945	80
1950	116
1955	180
1960	200
1965	270
1970	790
1975	850
1980	916
1985	1723
1990	2806

47 The data points graphed in a scatter diagram have positive x- and y-coordinates. In what quadrant are these points located?

48 The data points graphed in a scatter diagram have negative x-coordinates and positive y-coordinates. In what quadrant are these points located?

49 The data points graphed in a scatter diagram have negative x- and y-coordinates. In what quadrant are these points located?

50 The data points graphed in a scatter diagram have positive x-coordinates and negative y-coordinates. In what quadrant are these points located?

51 The y-coordinates of the data points graphed in a scatter diagram are neither positive nor negative. Where are these points located?

In Exercises 52–56 translate each statement into a linear equation.

52 Four times x is three more than twice y.

53 Seven times y is nine less than five times x.

54 Two more than y is three times the quantity x plus four.

55 Eight less than y is four times the quantity x minus seven.

56 One-half of y is two-thirds of the quantity x minus one.

In Exercises 57 and 58 draw a line graph for each table of values.

57

x	1	2	3	4	5
y	1	3	2	5	0

58

x	1	2	3	4	5
y	4	3	0	1	3

C

In Exercises 59–62 determine the distance between each pair of points.

59 (5, 7) and (5, −3)

60 (−2, 11) and (−2, 6)

61 (5, 7) and (2, 7)

62 (−2, 11) and (2, 11)

63 The following table gives the points scored by Shaquille O'Neal for five consecutive games. Draw a line graph for these data.

x	1	2	3	4	5
y	20	32	28	35	29

64 The following table gives the strikeouts recorded by Tom Glavine for five consecutive games. Draw a line graph for these data.

x	1	2	3	4	5
y	8	11	7	5	9

In Exercises 65 and 66 use the given equation to complete the table of values shown. Then draw a line graph for these points.

x	0	1	2	3	4	5	6
y							

65 $y = x^2 - 6x + 8$

66 $y = -x^2 + 6x - 7$

67 Utility Bills The utility bills for selected months were given by a homeowner to a prospective buyer of her house. Use this information to

a. Draw a line graph for these data

b. Interpolate to approximate the August bill

c. Extrapolate to approximate the December bill

x	Jan	Mar	Apr	Jun	Jul	Sep	Nov
y	\$250	\$175	\$350	\$100	\$175	\$150	\$200

68 Water Bills A homeowner analyzed water bills for selected months, as shown in the table, for budget purposes. Use this information to

a. Draw a line graph for these data

b. Interpolate to approximate the August bill

c. Extrapolate to approximate the December bill

x	Jan	Mar	Apr	Jun	Jul	Sep	Nov
y	\$20	\$25	\$40	\$45	\$40	\$35	\$25

Discussion Questions (69–71)

69 Determine whether the points in the table are all on the same line. Explain your answer.

x	1	2	3	4	5
y	5	8	12	15	18

70 Discuss some of the pitfalls of using interpolation to approximate the August water bill in Exercise 68.

71 Discuss some of the pitfalls of using extrapolation to approximate the December water bill in Exercise 68.

CHALLENGE QUESTION

Observe the visual pattern in each of these line graphs, and then use this pattern to predict what the next point will be if this pattern continues.

a.

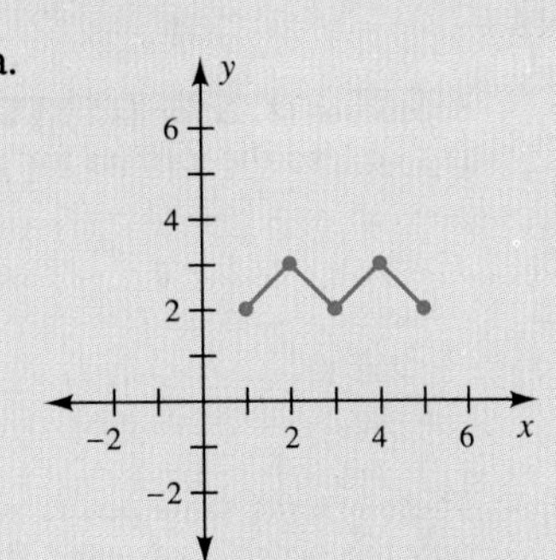

b.

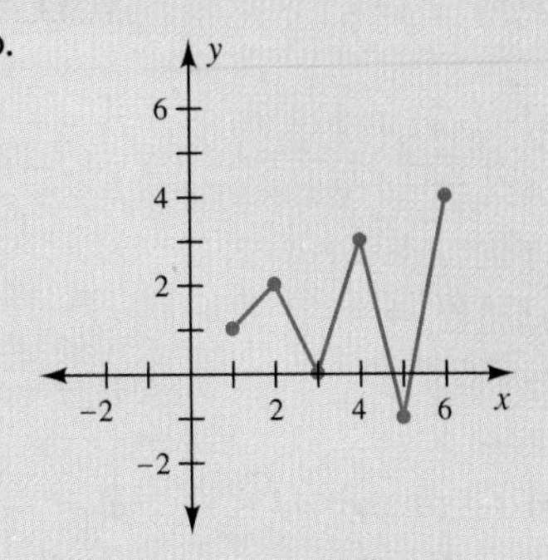

SECTION SKILL REINFORCERS

Simplify each of these expressions.

1 $\dfrac{5-3}{7-11}$ **2** $\dfrac{8-(-5)}{-1-2}$ **3** $\dfrac{8-1}{-2-(-5)}$ **4** $\dfrac{\frac{1}{2}-\frac{1}{3}}{\frac{1}{4}-\frac{1}{5}}$ **5** $\dfrac{\frac{1}{5}-\frac{1}{10}}{\frac{3}{4}-\frac{1}{2}}$

2-2 *Slope and Slope-Intercept Form of a Line*

SECTION OBJECTIVES

4 Calculate the slope of a line.

5 Use the slope-intercept form to graph a line.

Linear graphs are characterized by their constant rate of change. Visually, this rate of change is displayed by the steepness of the line. Other types of graphs will be steeper in some intervals than in others, but a line will have the same steepness between any two points on the line. The slope of a nonvertical line is a number that measures the steepness of the line, or how the line slants. To measure steepness, select any two points on the line and form the ratio of the change in y to the change in x, as shown in Figure 2-5.

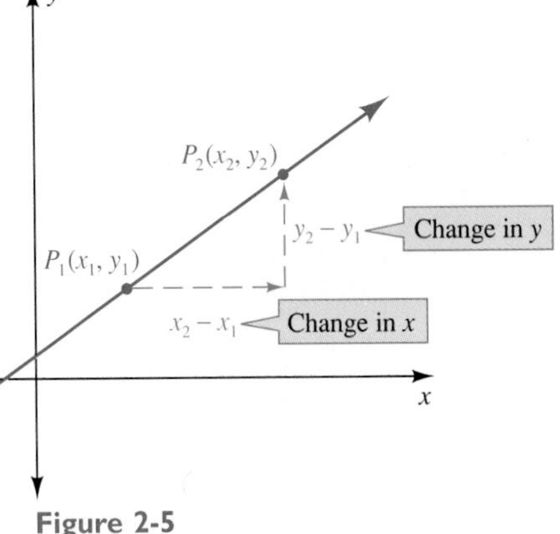

Figure 2-5
Slope of a Line

SLOPE

The slope m of a line through points (x_1, y_1) and (x_2, y_2) for $x_1 \neq x_2$ is

$$m = \frac{\text{Change in } y}{\text{Change in } x} = \frac{y_2 - y_1}{x_2 - x_1}$$

EXAMPLE 1 Calculating the Slope of a Line Through Given Points

Calculate the slope of a line through the given points.

SOLUTIONS

(a) (1, 2) and (4, 6)

$$m = \frac{y_2 - y_1}{x_2 - x_1}$$

Substitute (1, 2) for (x_1, y_1) and (4, 6) for (x_2, y_2) into the formula for slope.

$$m = \frac{6 - 2}{4 - 1}$$

$$m = \frac{4}{3}$$

(b) (4, 6) and (1, 2)

$$m = \frac{y_2 - y_1}{x_2 - x_1}$$

These are the same points as in part (a); only the order is reversed. Because they determine the same line, the slope must be the same.

$$m = \frac{2 - 6}{1 - 4}$$

$$m = \frac{-4}{-3} = \frac{4}{3}$$

A Mathematical Note

The origin of the use of m to designate slope is unknown. In his book *Mathematical Circles Revisited,* Howard Eves says that m might have been used because slopes were first studied with respect to mountains. Frank Mauz of Honolulu Community College suggests that m may be derived from the French word *monter,* which means "to mount, to climb, or to slope up."

The conclusion we can derive from Example 1 is that we are free to label either of two points as the first point when we calculate the slope of a line through these two points. However, once this decision has been made, we must use the same order for the subtraction in both the numerator and the denominator.

$$m = \frac{y_2 - y_1}{x_2 - x_1} \quad \text{or} \quad m = \frac{y_1 - y_2}{x_1 - x_2} \qquad m \neq \frac{y_2 - y_1}{x_1 - x_2}$$

Both correct

Incorrect: Note that these expressions would differ in sign.

EXAMPLE 2 Calculating the Slope of a Line Through Given Points

Calculate the slope of each line using the points labeled on the line.

SOLUTIONS

(a)

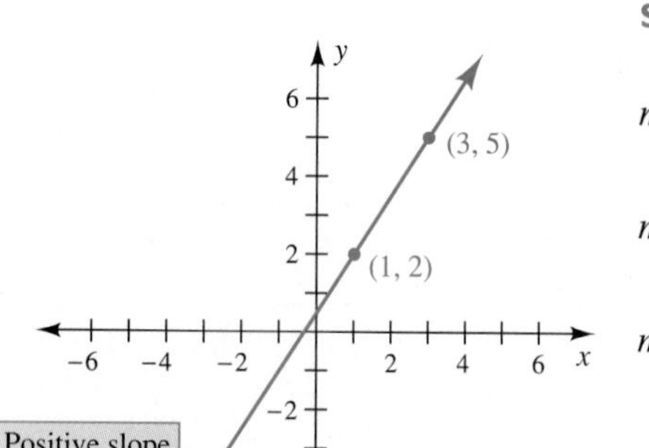

$$m = \frac{y_2 - y_1}{x_2 - x_1}$$

This line slopes upward to the right, and its slope is a positive number.

$$m = \frac{5 - 2}{3 - 1}$$

$$m = \frac{3}{2}$$

(b)

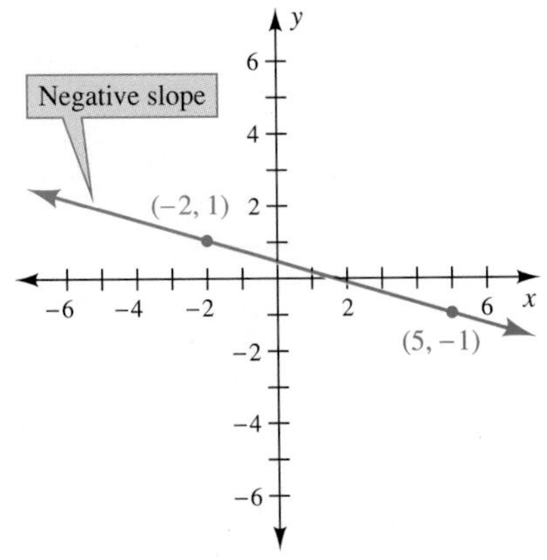

$$m = \frac{y_2 - y_1}{x_2 - x_1}$$

$$m = \frac{-1 - 1}{5 - (-2)}$$

$$m = -\frac{2}{7}$$

This line slopes downward to the right, and its slope is a negative number.

(c)

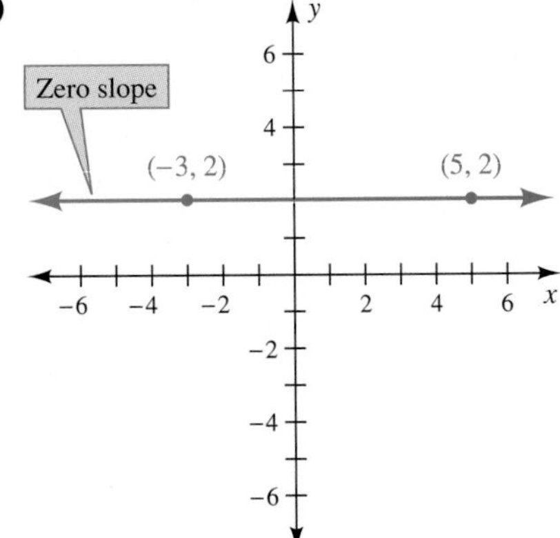

$$m = \frac{y_2 - y_1}{x_2 - x_1}$$

$$m = \frac{2 - 2}{5 - (-3)}$$

$$m = \frac{0}{8}$$

$$m = 0$$

This line is horizontal, and its slope is 0.

(d)

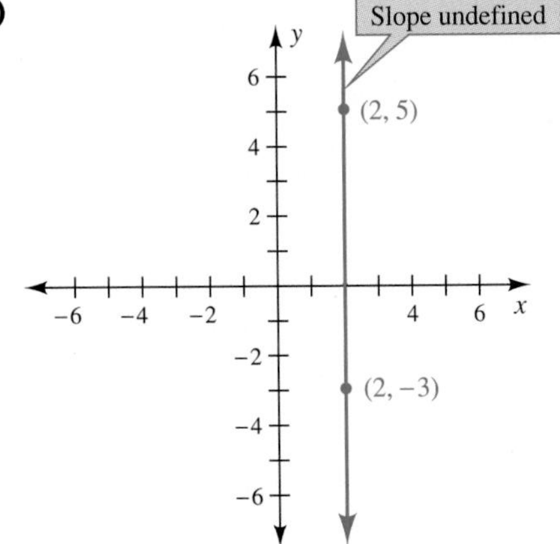

$$m = \frac{y_2 - y_1}{x_2 - x_1}$$

$$m = \frac{5 - (-3)}{2 - 2}$$

$$m = \frac{8}{0}$$

m is undefined.

This line is vertical, and its slope is undefined because division by 0 is undefined.

SELF-CHECK

Calculate the slope of the line through each of the following pairs of points.

1 (5, 6) and (6, 4)
2 (−3, −2) and (5, 2)
3 (−4, −5) and (4, −5)

SELF-CHECK ANSWERS

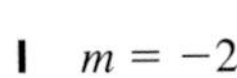

1 $m = -2$ **2** $m = \frac{1}{2}$ **3** $m = 0$

It is very important to be able to visualize the difference between positive and negative slopes. The ability to picture in your mind a line with a slope of $+1$ or -1 will help you estimate the slope of a line by inspection when both axes are labeled with the same scale. Note in Fig 2-6(a) that the steepest lines have slopes of magnitude greater than 1.

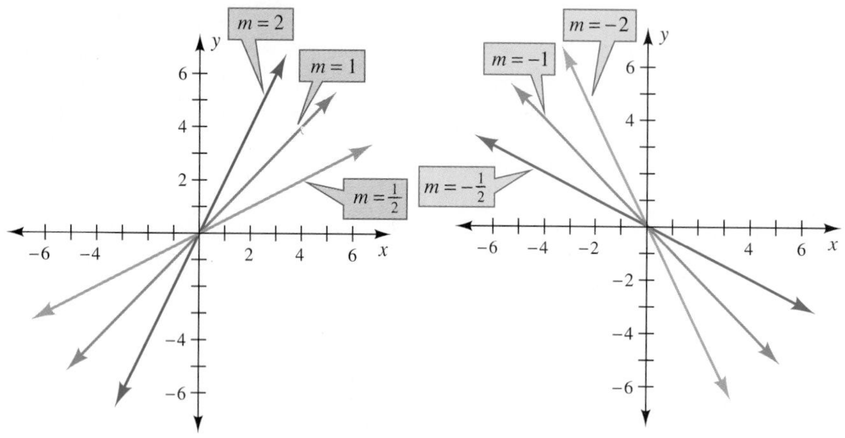

Figure 2-6 *(a) Lines with Positive Slope* *(b) Lines with Negative Slope*

CLASSIFYING LINES BY THEIR SLOPES

Slope	*Description of Line*	*Graph of Line*
Positive	Slopes upward to the right	
Negative	Slopes downward to the right	
Zero	Horizontal line	
Undefined	Vertical line	

We can calculate the slope of a line by selecting any two points on the line. We will illustrate this process in the next example and then show a shortcut for determining the slope of a line directly from its equation.

EXAMPLE 3 Calculating the Slope of a Line Defined by an Equation

Calculate the slope of the line defined by $2x + 3y = 6$.

SOLUTION First determine any two points on the line.

x-intercept

Set $y = 0$.

$$2x + 3y = 6$$
$$2x + 3(0) = 6$$
$$2x = 6$$
$$x = 3$$

y-intercept

Set $x = 0$.

$$2x + 3y = 6$$
$$2(0) + 3y = 6$$
$$3y = 6$$
$$y = 2$$

The intercepts are calculated.

x	y
3	0
0	2

$$m = \frac{y_2 - y_1}{x_2 - x_1}$$
$$m = \frac{2 - 0}{0 - 3}$$
$$m = -\frac{2}{3}$$

Then use these two points to calculate the slope.

Answer The slope of the line is $-\frac{2}{3}$. ▼

The strategy used in the previous example can be used to calculate the slope m of a line through the y-intercept $(0, b)$ and any other point (x, y) that satisfies the equation of the line. Using the definition of slope and these two points, we obtain

$$m = \frac{y_2 - y_1}{x_2 - x_1}$$

$$m = \frac{y - b}{x - 0}$$ Substitute the points $(0, b)$ and (x, y) into the slope formula.

$$mx = y - b$$ Multiply both sides by x.

$$y = mx + b$$ Add b to both sides of the equation.

The result of the work shown above is that the equation of a line with slope m and y-intercept $(0, b)$ can be written in the form $y = mx + b$.

SLOPE-INTERCEPT FORM

The slope-intercept form of the equation of a line with slope m and y-intercept b is

$$y = mx + b$$

We will now rework Example 3 using the slope-intercept form.

EXAMPLE 4 Using the Slope-Intercept Form

Determine the slope and the y-intercept of $2x + 3y = 6$.

SOLUTION $2x + 3y = 6$ — Solve this equation for y in order to express it in slope-intercept form.

$3y = -2x + 6$

$y = -\frac{2}{3}x + 2$

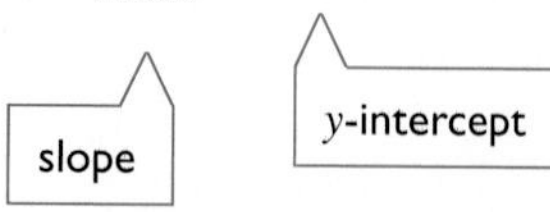

Answer The slope is $-\frac{2}{3}$, and the y-intercept is 2.

SELF-CHECK 

Determine the slope and the y-intercept of these lines.

1 $y = \frac{3}{4}x - 8$

2 $4x - 3y = 12$

We can draw a line if we know two points on the line. We can also draw a line if we know one point and the slope of the line. In the next example the y-intercept is used as the first point and the slope is used to change the x- and y-coordinates to produce a second point. A line is then drawn through the two points.

EXAMPLE 5 Using the Slope-Intercept Form to Graph a Line

Graph $y = \frac{5}{3}x - 1$.

SOLUTION $y = \frac{5}{3}x - 1$ — First determine the y-intercept, and plot this point.

y-intercept $(0, -1)$

$y = \frac{5}{3}x - 1$ — Then determine the slope, and use the relative changes of x and y, as given by the slope, to plot a second point.

$m = \frac{+5}{+3} = \frac{\text{Change in } y}{\text{Change in } x}$

SELF-CHECK ANSWERS

1 $m = \frac{3}{4}$, $(0, -8)$ 2 $m = \frac{4}{3}$, $(0, -4)$

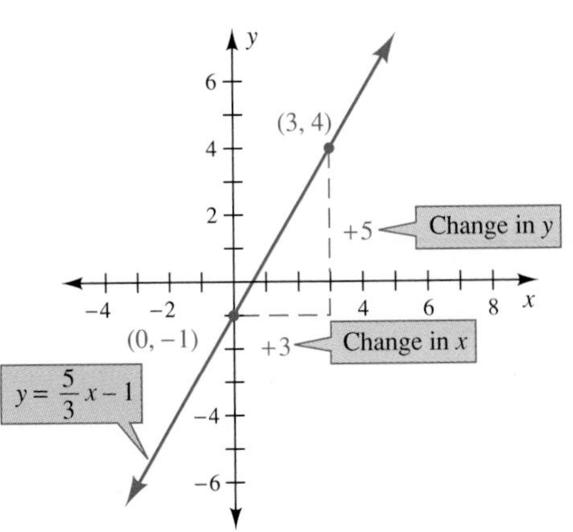

Finally, draw the line through these points.

▼

SELF-CHECK

Graph $y = -\frac{3}{4}x + 5$.

Graphics calculators are often used to graph straight lines and other curves. The slope-intercept form $y = mx + b$ is frequently required by graphics calculators and computer programs because it provides a formula for calculating the y-coordinate for a given x-coordinate. The next example illustrates one way to graph an equation in slope-intercept form on a TI-82 graphics calculator. Although these keystrokes are representative, keystrokes will vary to some extent depending on the model of calculator you have.

EXAMPLE 6 Calculator Usage

Use a graphics calculator to graph $y = \frac{x}{2} + 1$.

SOLUTION

Keystrokes

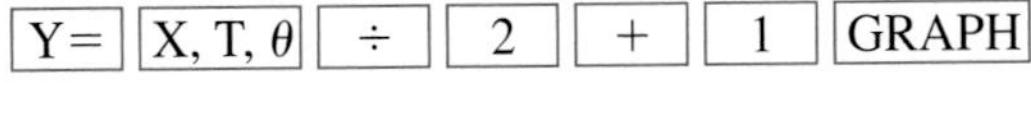

Graph this equation assuming that the window of values has already been set to [−4, 4] for x and [−4, 4] for y.

[−4, 4] for x, [−4, 4] for y

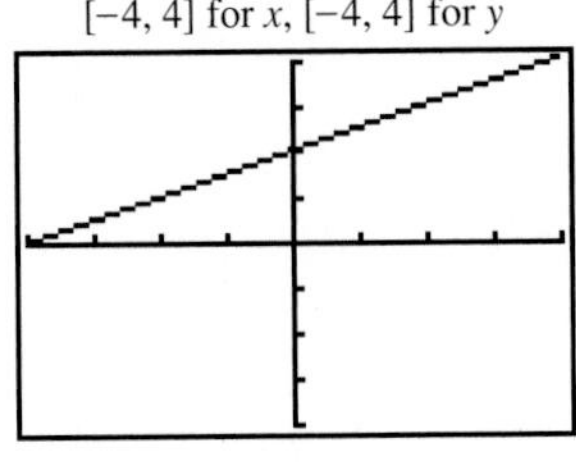

▼

SELF-CHECK ANSWER

[−2, 8] for x, [−2, 8] for y

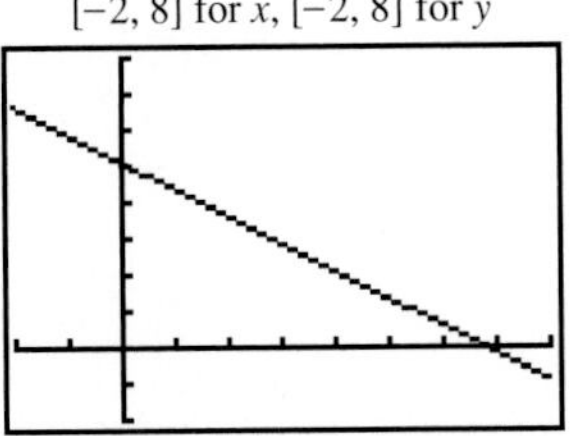

The slopes of two lines can be used to determine if the lines are parallel or perpendicular. Parallel lines have the same slope, since they rise or fall at the same rate. (Vertical lines are parallel to each other even though the slope of a vertical line is undefined.) Lines that are perpendicular to each other have slopes that are negative reciprocals.

PARALLEL AND PERPENDICULAR LINES

If L_1 and L_2 are distinct nonvertical lines with slopes m_1 and m_2, respectively, then

1 L_1 and L_2 are **parallel** if and only if

$$m_1 = m_2$$

2 L_1 and L_2 are **perpendicular** if and only if

$$m_1 = -\frac{1}{m_2}$$

That is,

$$m_1 \cdot m_2 = -1$$

The slope of a vertical line is undefined. All vertical lines are parallel to each other and are perpendicular to all horizontal lines.

Parallel Lines

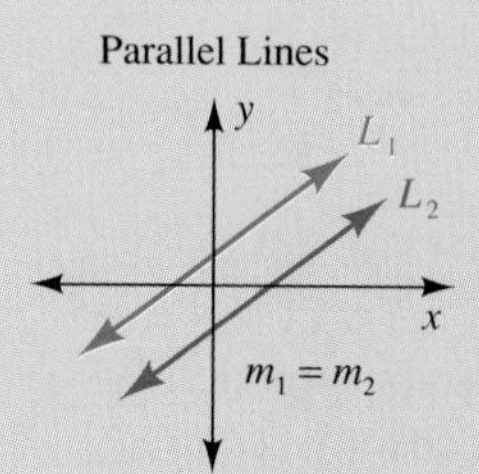

Perpendicular Lines

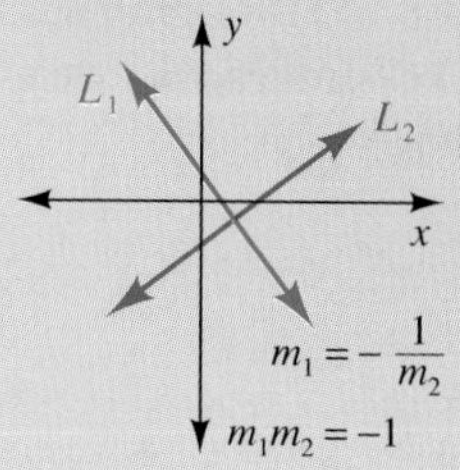

EXAMPLE 7 Determining Whether Two Lines Are Parallel or Perpendicular

Determine whether the lines $2x + 3y = 18$ and $3x - 2y = 12$ are parallel, perpendicular, or neither parallel nor perpendicular.

SOLUTION First determine the slope of each line.

First Line	***Second Line***	
$2x + 3y = 18$	$3x - 2y = 12$	Write each equation in slope-intercept form to determine the slope.
$3y = -2x + 18$	$-2y = -3x + 12$	
$y = -\frac{2}{3}x + 6$	$y = \frac{3}{2}x - 6$	
$m_1 = -\frac{2}{3}$	$m_2 = \frac{3}{2}$	

Since $m_1 \neq m_2$, the lines are not parallel. Since $m_1 m_2 = \left(-\frac{2}{3}\right)\left(\frac{3}{2}\right) = -1$, the lines are perpendicular.

Answer The lines are perpendicular.

EXAMPLE 8 An Application of Slope

Use the slopes of the sides to determine whether the vertices $A(-1, 1)$, $B(1, 3)$, and $C(3, 1)$ form a right triangle.

SOLUTION Triangle ABC is a right triangle if two sides are perpendicular.

Slope of $\overline{AB}$	Slope of $\overline{AC}$	Slope of $\overline{BC}$	
$m_1 = \frac{3-1}{1-(-1)}$	$m_2 = \frac{1-1}{3-(-1)}$	$m_3 = \frac{1-3}{3-1}$	First calculate the slope of each side of the triangle.
$m_1 = \frac{2}{2}$	$m_2 = \frac{0}{4}$	$m_3 = \frac{-2}{2}$	
$m_1 = 1$	$m_2 = 0$	$m_3 = -1$	Then compare these slopes.

Since $m_1 \cdot m_3 = -1$, AB is perpendicular to BC. Thus ABC is a right triangle, with $\angle B$ the right angle.

SELF-CHECK

1. Calculate the slope of a line parallel to $2x - 7y = 14$.
2. Calculate the slope of a line perpendicular to $2x - 7y = 14$.

The concept of slope is used in many applications in which two quantities are changing simultaneously along a straight line. Some of these applications use the terms *rise* and *run*. The slope of a line is the ratio of the vertical rise or fall to the horizontal run (see Figure 2-7). Other applications using slope may use the term *grade* or *angle of elevation*. Example 9 involves the grade of a drainage pipe.

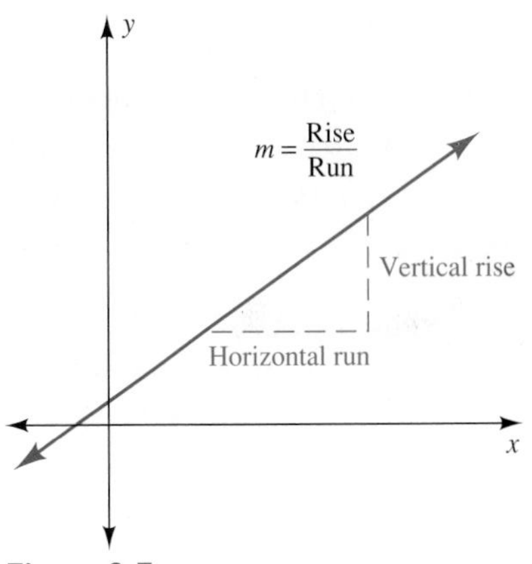

Figure 2-7

EXAMPLE 9 An Engineering Application of Slope

An engineer has specified that the drainage pipe for a building be laid so that it has 5% grade. Determine the change in elevation that a contractor must allow for 400 feet of this pipe.

SOLUTION Let

$y =$ Change in elevation — Sketch the problem using a convenient placement of the origin to simplify the computations (see the figure).

$m = \frac{y_2 - y_1}{x_2 - x_1}$ — Use the formula for slope.

$-\frac{5}{100} = \frac{y - 0}{400 - 0}$ — Substitute in the changes for y and x, and write -5% as $-\frac{5}{100}$ for a 5% drop in grade.

$-\frac{5}{100} = \frac{y}{400}$

$-100y = 5(400)$ — Solve this ratio for y.

$-100y = 2000$

$y = -20$

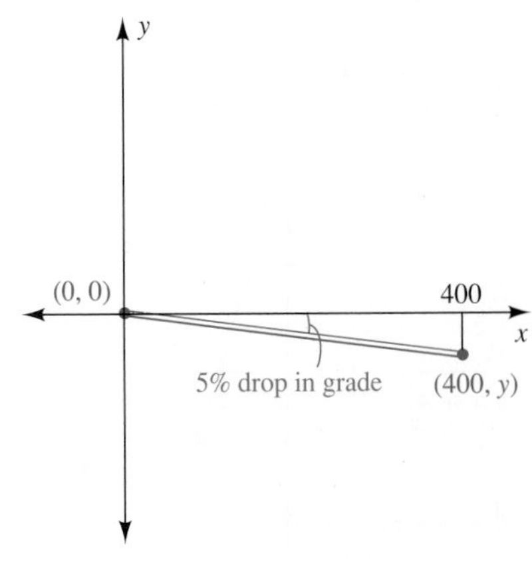

Answer This section of pipe would drop 20 ft in elevation.

SELF-CHECK ANSWERS 

1 $\frac{2}{7}$ 2 $-\frac{7}{2}$

Exercises 2-2

A

In Exercises 1–10 calculate the slope of the line through the given points.

1 (5, 8) and (3, 12)
2 (9, 1) and (3, 5)
3 (−2, 5) and (5, −3)
4 (0, −20) and (−9, 20)
5 $\left(-\frac{1}{2}, \frac{2}{3}\right)$ and $\left(\frac{1}{2}, -\frac{1}{3}\right)$
6 $\left(\frac{4}{5}, \frac{3}{2}\right)$ and $\left(-\frac{1}{5}, \frac{1}{2}\right)$
7 (−11, 7) and (−11, 3)
8 (2, −6) and (2, 6)
9 (−11, 7) and (6, 7)
10 (2, −6) and (−3, −6)

In Exercises 11–18 determine the slope of each line.

11 $y = 2x - 9$
12 $y = 8x + 3$
13 $5x + 3y = 15$
14 $2x - 7y = 9$
15 $y = 3$
16 $3y = 1$
17 $6x = 9$
18 $2x = -5$

In Exercises 19–24 determine the slope of each line.

19

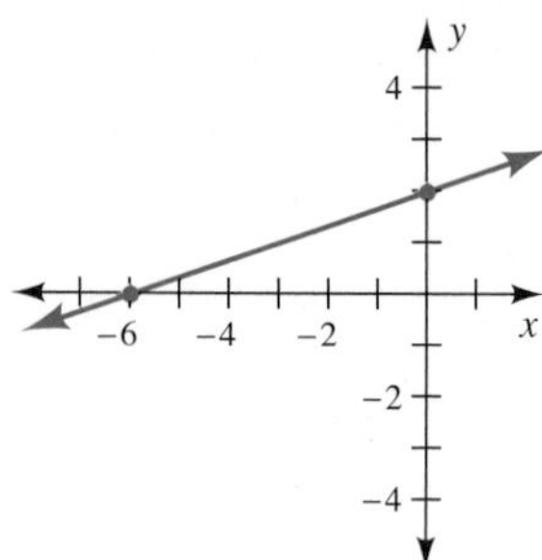

20

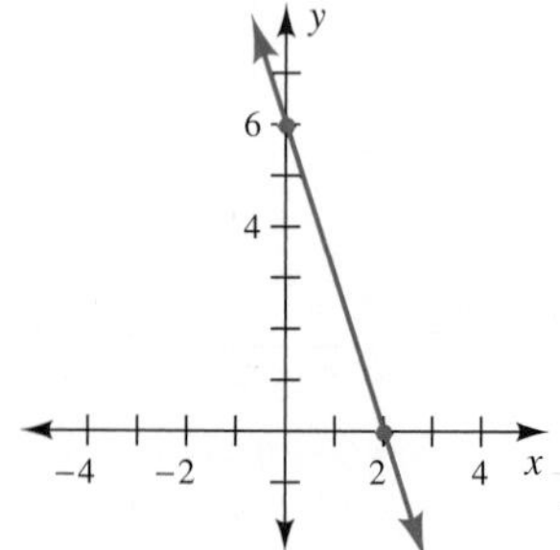

21

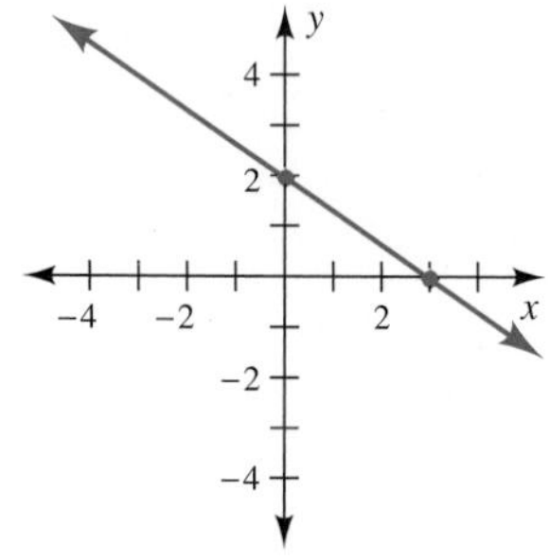

22

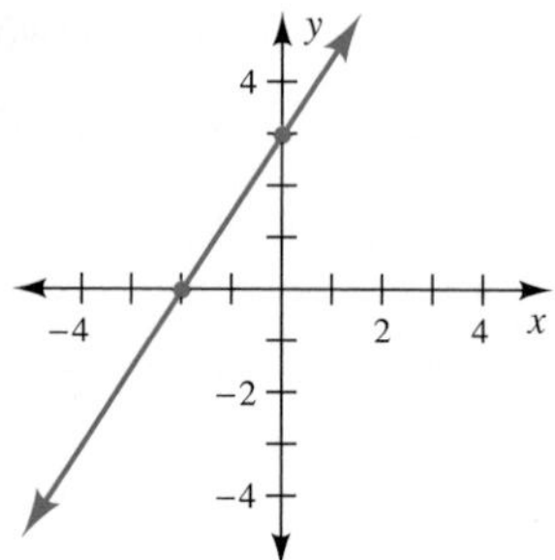

23

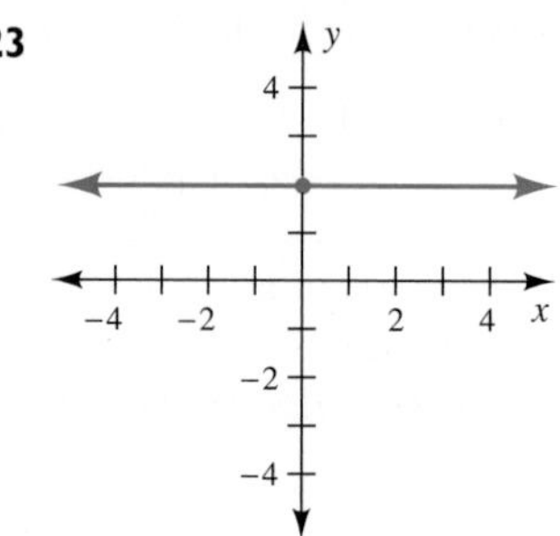

24

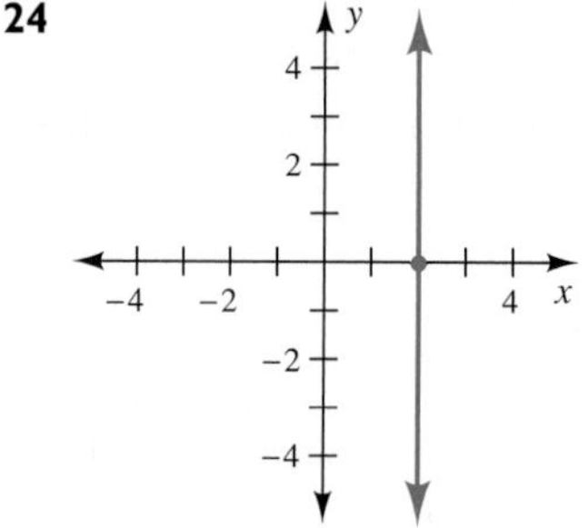

In Exercises 25–30 determine whether the line through the first pair of points is parallel to the line through the second pair of points, perpendicular to this line, or neither parallel nor perpendicular to it.

25 (−2, 8) and (3, −2)
(4, 1) and (7, −5)

26 (−4, −7) and (0, 5)
(−4, −10) and (6, 10)

27 (−3, 2) and (4, −1)
(4, −5) and (7, −11)

28 (9, −2) and (7, −2)
(0, 5) and (−5, 5)

29 (−4, 3) and (−1, −2)
(5, −6) and (10, −3)

30 (−7, −1) and (7, −1)
(5, −2) and (5, −4)

In Exercises 31–38 determine whether the lines given by the equations are parallel, perpendicular, or neither parallel nor perpendicular.

31 $y = 4x + 2$
$y = 3x + 2$

32 $x = 6$
$y = 11$

33 $x = 5$
$y = 5$

34 $x = 9$
$x = -9$

35 $y = -3$
$y = 17$

36 $3x - 2y = 6$
$5x + 4y = 20$

37 $2x + 3y = 4$
$6y = -12 - 4x$

38 $5x - y = 10$
$2y = 10x - 30$

39 **a.** What is the slope of a line parallel to the line defined by $y = 4x - 2$?

b. What is the slope of a line perpendicular to the line defined by $y = 4x - 2$?

40 **a.** What is the slope of a line parallel to the line defined by $-4x + 7y = 28$?

b. What is the slope of a line perpendicular to the line defined by $-4x + 7y = 28$?

In Exercises 41–50 graph the line that passes through the given point and has the given slope.

41 $(-1, -2)$, $m = \frac{3}{5}$

42 $(-2, -4)$, $m = \frac{4}{9}$

43 $(3, -4)$, $m = -\frac{2}{3}$

44 $(-5, 3)$, $m = -\frac{4}{7}$

45 $(0, 0)$, $m = 2$

46 $(0, 0)$, $m = -3$

47 $(2, 3)$, $m = 0$

48 $(3, 2)$, $m = 0$

49 $(2, 3)$, m undefined

50 $(3, 2)$, m undefined

In Exercises 51–54 graph each line.

51 $y = 2x - 3$

52 $y = -3x + 2$

53 $y = x$

54 $y = -x$

55 Draw a line parallel to the line shown in the figure and passing through the origin.

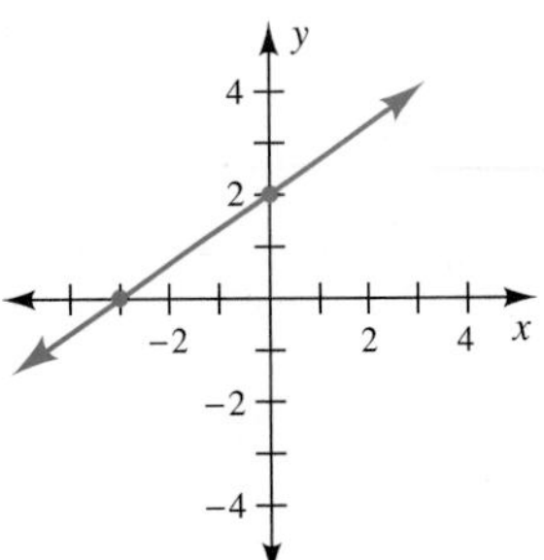

Figure for Exercises 55 and 56

56 Draw a line perpendicular to the line shown in the figure and passing through the origin.

B

CALCULATOR USAGE (57–62)

In Exercises 57–60 use a graphics calculator to graph the linear functions defined by y_1, y_2, and y_3 on the same standard viewing window of [−10, 10] for x and [−10, 10] for y. Then visually compare these slopes and y-intercepts.

57 $y_1 = x - 3$
$y_2 = 2x - 3$
$y_3 = \frac{x}{2} - 3$

58 $y_1 = -x + 2$
$y_2 = -4x + 2$
$y_3 = -\frac{x}{4} + 2$

59 $y_1 = -2x - 4$
$y_2 = -2x - 1$
$y_3 = -2x + 3$

60 $y_1 = \frac{2x}{3} + 1$
$y_2 = \frac{2x}{3} + 4$
$y_3 = \frac{2x}{3} - 2$

61 Graph $y = 0.5x + 25$ on your standard viewing window of $[-10, 10]$ for x and $[-10, 10]$ for y. Describe what you see. What can you do to obtain a better viewing window?

62 Graph $y = -2x - 40$ on your standard viewing window of $[-10, 10]$ for x and $[-10, 10]$ for y. Describe what you see. What can you do to obtain a better viewing window?

In Exercises 63–66 use the slopes of the sides of each triangle to determine whether the given vertices form a right triangle.

63 $(-3, -1)$, $(-4, 2)$, and $(-1, 5)$

64 $(4, 4)$, $(-2, -6)$, and $(-4, 2)$

65 $(-1, -4)$, $(3, -3)$, and $(-3, 4)$

66 $(-1, 5)$, $(1, -4)$, and $(4, 3)$

67 Choose the letter of the graph that best matches the description given.

a.

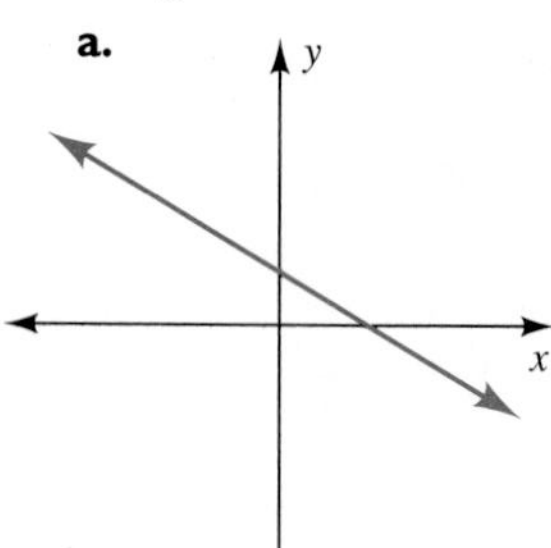

b.

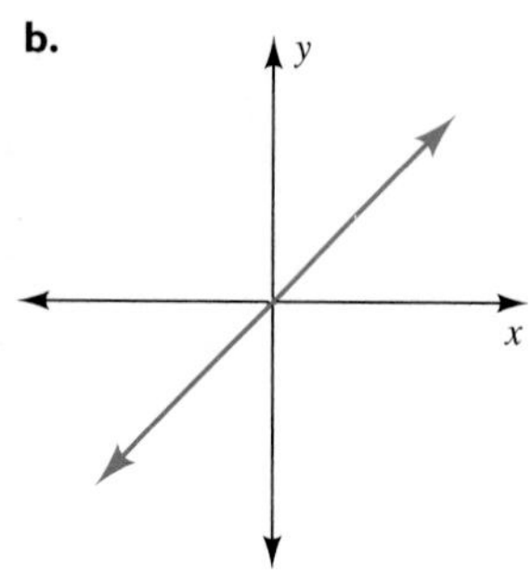

c.

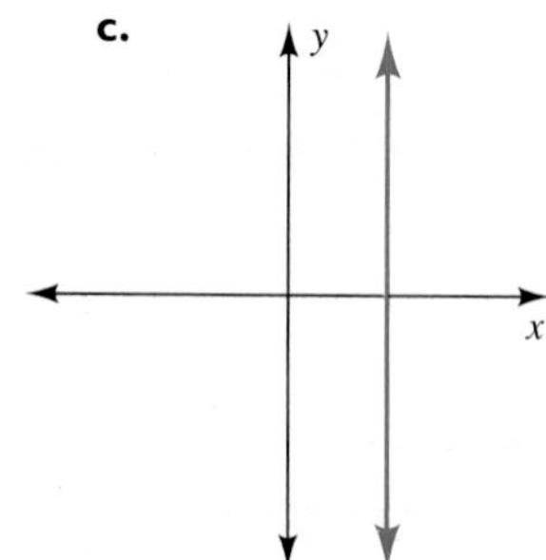

d.

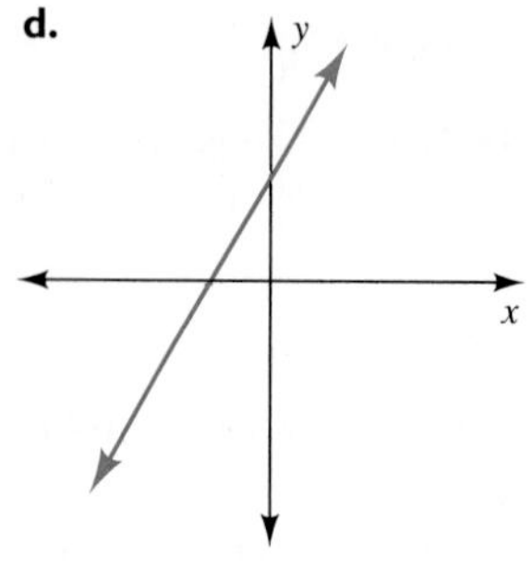

e.

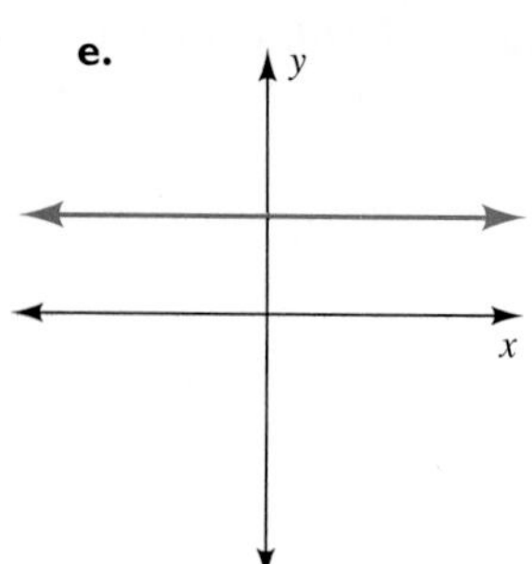

f.

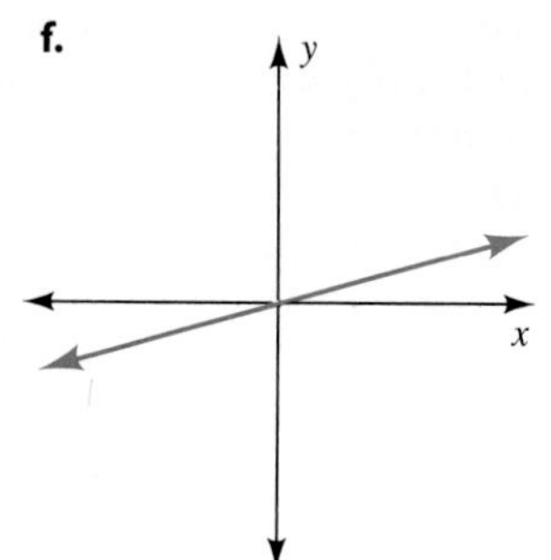

i. A slope of 0 **ii.** A positive slope less than 1

iii. A slope of 1

iv. A positive slope greater than 1

v. A negative slope between 0 and -1

vi. An undefined slope

ESTIMATION SKILLS (68 AND 69)

In Exercises 68 and 69 select the best estimate of the slope of the given line.

68

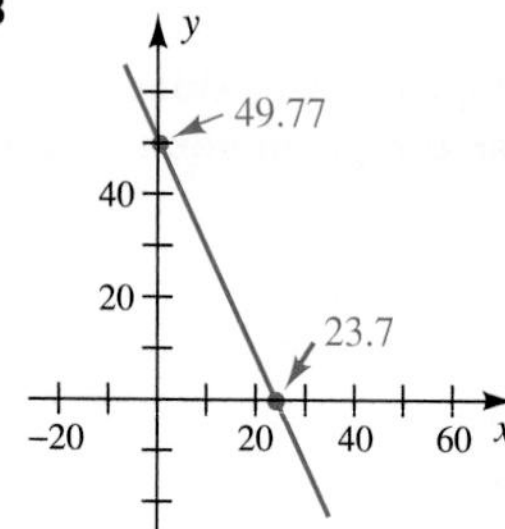

a. -2

b. $-\frac{1}{2}$

c. $\frac{1}{2}$

d. 2

e. $\frac{3}{2}$

69

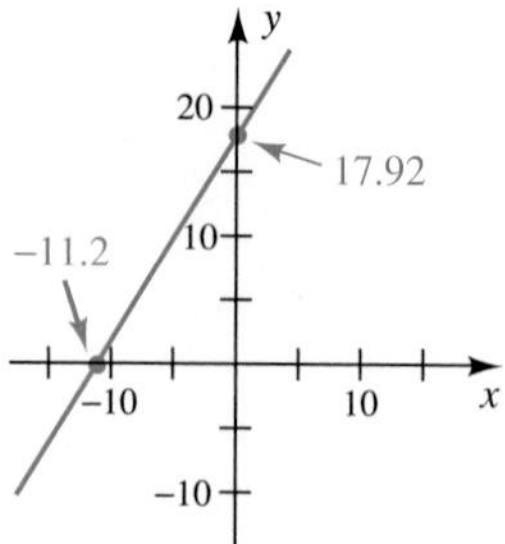

a. -2

b. $-\frac{1}{2}$

c. $\frac{1}{2}$

d. 2

e. $\frac{3}{2}$

Grade of a Highway (70 and 71)

70 An engineer has determined that for several reasons, including highway safety, the slope of a particular section of highway should not exceed a 4% grade. If this maximum grade is allowed on a section of 1800 meters, how much change in elevation will be permitted on this section? (The change of elevation can be controlled by topping hills and filling in low places. See the figure below.)

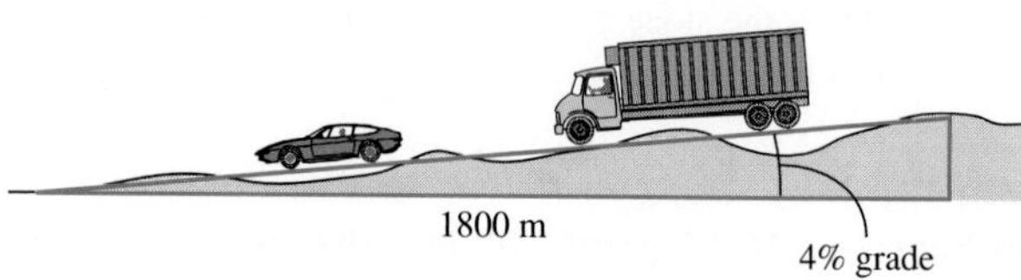

Figure for Exercise 70

71 How much change in elevation would be permitted on a 2400-meter section of the highway described in Exercise 70?

72 Sewage Pipe Elevation How much change in elevation must a contractor allow over 150 feet of right of way for a sewage pipe with a 4% grade?

73 Water Line Elevation A 250-meter section of water line has a 7% grade. What is the change in elevation from one end to the other?

74 A line passes through quadrants I, II, and III but not through quadrant IV. Is the slope of this line positive or negative?

75 A line passes through quadrants I, II, and IV but not through quadrant III. Is the slope of this line positive or negative?

76 A line passes through quadrants II, III, and IV but not through quadrant I. Is the slope of this line positive or negative?

77 A line passes through quadrants, I, III, and IV but not through quadrant II. Is the slope of this line positive or negative?

78 A line passes through quadrants I and III but not through quadrants II and IV. Is the slope of this line positive or negative?

79 A line passes through quadrants II and IV but not through quadrants I and III. Is the slope of this line positive or negative?

80 What is the slope of the x-axis?

81 What is the slope of the y-axis?

C

In Exercises 82 and 83 write the equation of each line in slope-intercept form.

82

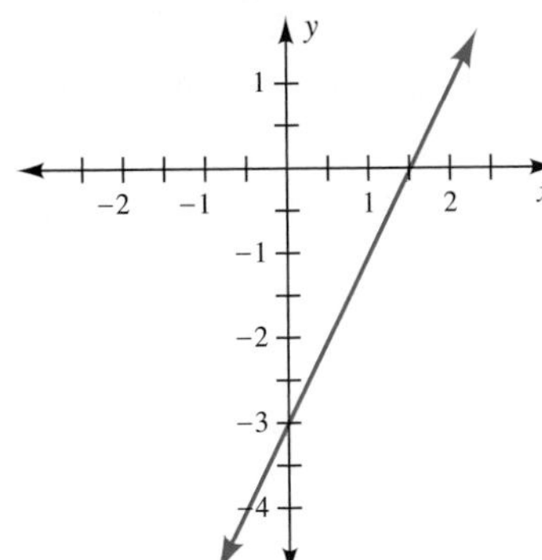

83

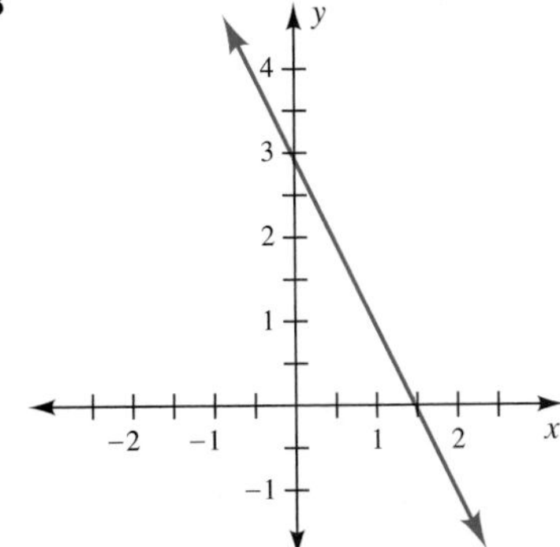

In Exercises 84 and 85 write the equation of each line.

84

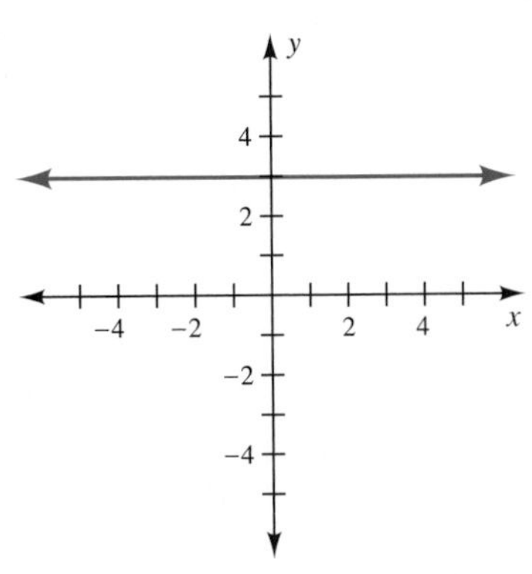

85

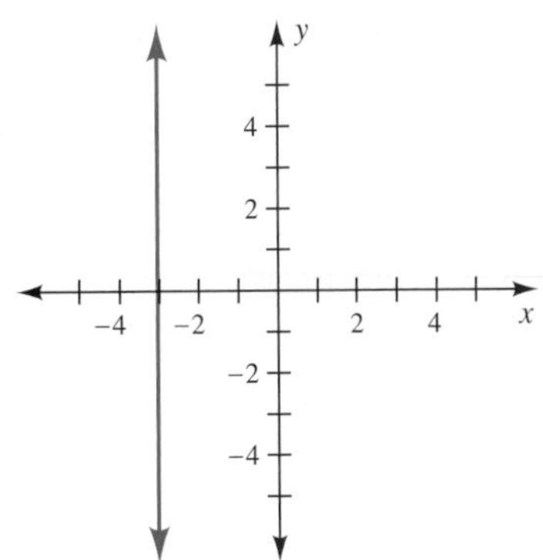

86 **Price of a Mutual Fund** The line graph shown gives the price of a share of a mutual fund stock for selected months. If the slope of the line graph is positive, the price is increasing. If the slope of the line graph is negative, the price is decreasing.

a. Is the price increasing or decreasing in February?

b. Is the price increasing or decreasing in October?

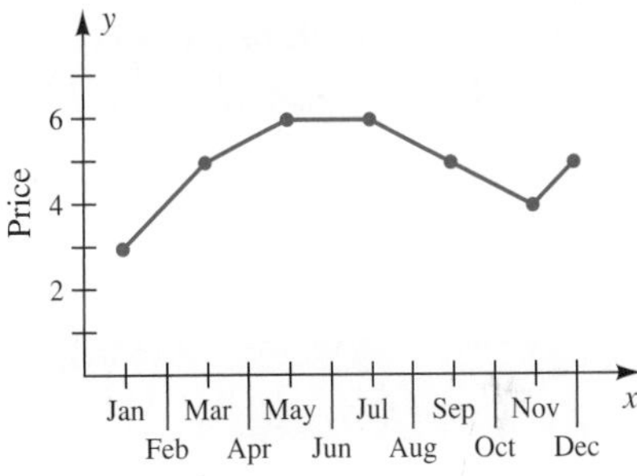

Figure for Exercise 86

87 **Price of Silver** The line graph shown gives the price per ounce of silver for selected months. If the slope of the line graph is positive, the price is increasing. If the slope of the line graph is negative, the price is decreasing.

a. Is the price increasing or decreasing in February?

b. Is the price increasing or decreasing in October?

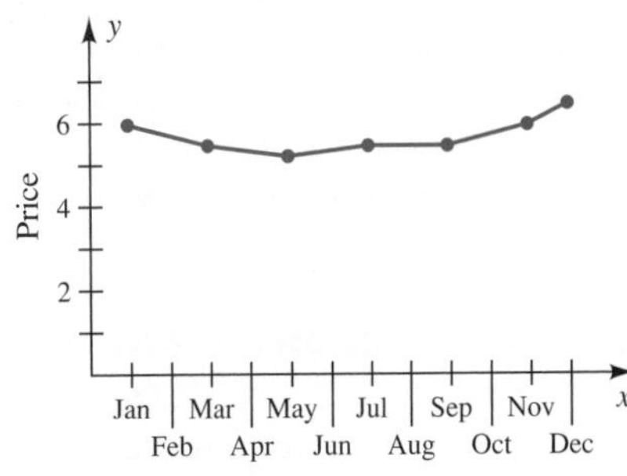

Figure for Exercise 87

DISCUSSION QUESTIONS (88–91)

88 Describe in your own words the steps to follow to graph a line using a given point on the line and the slope of the line.

89 Without graphing these data, explain why the relationship among these points is not a linear relationship.

x	-3	-2	-1	0	1	2	3
y	17	7	1	-1	1	7	17

90 Describe a procedure for determining without graphing whether three points are on the same line.

91 **Error Analysis** A student *incorrectly* calculated the slope of the line through the points in the table. Find the error in this student's work, and then explain the source of the error.

x	4	6	8	10
y	32	35.5	39	42.5

$$m = \frac{42.5 - 39}{8 - 10} = -1.75$$

CHALLENGE QUESTION

Determine the next term in each of the following sequences. (*Hint:* Consider each of these numbers as successive y-coordinates of a line graph; then use the visual pattern, as in the Challenge Question in Section 2-1.)

a. 2, 5, 8, 11, 14, ______
b. 2, 5, 2, 5, 2, ______
c. 2, 4, 1, 5, 0, 6, −1, ______

SECTION SKILL REINFORCERS

Write the solution of each of these inequalities in interval notation.

1 $|x| \le 3$ **2** $|x| > 4$ **3** $(-5, 4] \cap [-2, +\infty)$ **4** $(-5, 4] \cup [-2, 7)$ **5** $5 \le 2x + 3 < 9$

2-3 Special Forms of Linear Equations

SECTION OBJECTIVES

6 **Use the special forms of horizontal and vertical lines.**

7 **Use the point-slope form of a line.**

Because linear equations are frequently used to describe problems in many different subjects, it is constructive to study these equations in detail. We will examine different forms of linear equations so that you will learn to choose the form that is most appropriate for a given problem.

In Section 2-1 we used the general form $Ax + By + C = 0$ and observed the special forms for horizontal and vertical lines. The equation of every horizontal line is given by $y = k$ for some constant k, and the equation of every vertical line is given by $x = h$ for some constant h.

EQUATIONS OF HORIZONTAL AND VERTICAL LINES

Horizontal line: $y = k$ for a real constant k

Vertical line: $x = h$ for a real constant h

To graph the horizontal line given by $y = k$, we need only plot the intercept $(0, k)$ and draw a line parallel to the x-axis. The strategy is similar for vertical lines.

EXAMPLE 1 Graphing Horizontal and Vertical Lines

Graph these linear equations.

SOLUTIONS

(a) $y = 3$

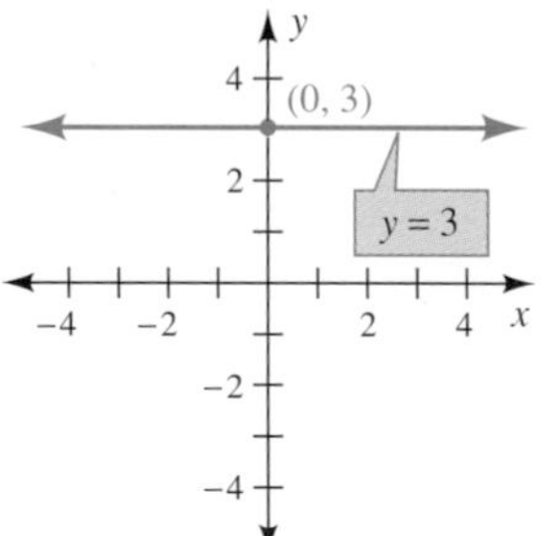

This is the equation of a horizontal line with y-intercept $(0, 3)$. The y-coordinate of each point on this line is 3.

(b) $x = -2$

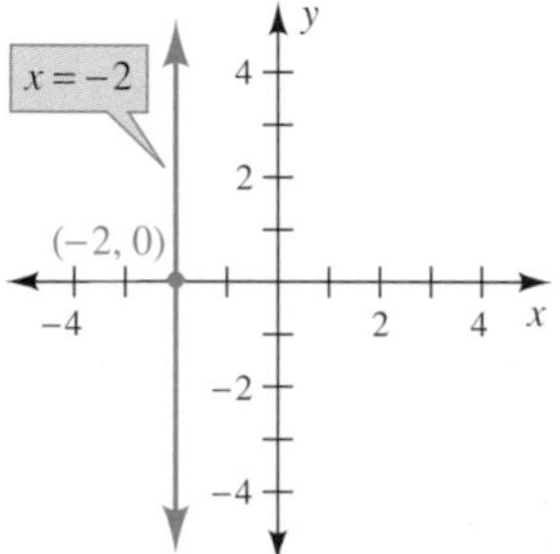

This is the equation of a vertical line with x-intercept $(-2, 0)$. The x-coordinate of each point on this line is -2.

▼

We need exactly two pieces of information to write the equation of a linear function or to draw its graph. Section 2-2 illustrated the procedure for drawing a line when we are given its slope and y-intercept. We will now combine these concepts to write the equation of a line given its slope and any point on the line.

Suppose we know that a line passes through point (x_1, y_1) and has slope m. To determine the equation relating x and y for any other point (x, y) on this line, we substitute this information into the slope formula to obtain

$$m = \frac{y - y_1}{x - x_1}$$

which simplifies to

$$y - y_1 = m(x - x_1)$$

This form of a linear equation is called the **point-slope form.**

POINT-SLOPE FORM

The point-slope form of the equation of a line through (x_1, y_1) with slope m is

$$y - y_1 = m(x - x_1)$$

EXAMPLE 2 Writing the Equation of a Line Given a Point and the Slope

Write the general form of the equation for the line passing through $(2, -1)$ with slope $\frac{2}{3}$.

SOLUTION

$y - y_1 = m(x - x_1)$	Point-slope form
$y - (-1) = \frac{2}{3}(x - 2)$	Substitute $(2, -1)$ for (x_1, y_1) and $\frac{2}{3}$ for m.
$3(y + 1) = 2(x - 2)$	Multiply both sides by the LCD, 3.
$3y + 3 = 2x - 4$	Simplify.
$7 = 2x - 3y$	
$2x - 3y - 7 = 0$	The general form is $Ax + By + C = 0$.

EXAMPLE 3 Using the Point-Slope Form to Graph a Line

Graph the line defined by equation $y - 2 = -\frac{2}{3}(x + 1)$.

SOLUTION From the point-slope form this line goes through the point $(-1, 2)$ with slope $-\frac{2}{3}$.

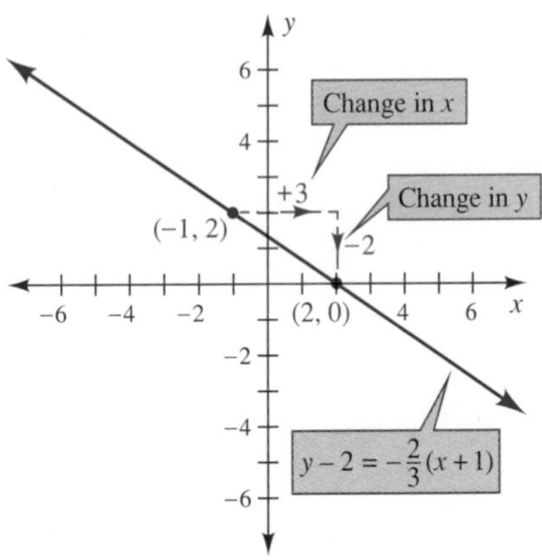

First plot the point $(-1, 2)$, and then use the relative changes of x and y as given by the slope to plot a second point. Finally, draw the line through these points.

EXAMPLE 4 Writing the Equation of a Line Given Two Points

Write the general form of the equation of the line passing through $(4, -2)$ and $(-1, -1)$.

SOLUTION

$m = \frac{y_2 - y_1}{x_2 - x_1} = \frac{-1 - (-2)}{-1 - 4} = \frac{1}{-5} = -\frac{1}{5}$	Since the slope is not given, first calculate m.
$y - y_1 = m(x - x_1)$	Point-slope form
$y - (-2) = -\frac{1}{5}(x - 4)$	Substitute $(4, -2)$ and $-\frac{1}{5}$ into the point-slope form.

$$5(y + 2) = -(x - 4)$$ Multiply both sides by the LCD, 5.

$$5y + 10 = -x + 4$$ Simplify.

$$x + 5y + 6 = 0$$ The general form is $Ax + By + C = 0$.

In Example 4 both points are used to calculate the slope. Either point can then be used with the point-slope form to determine the equation of the line. The equation will be the same in both cases.

SELF-CHECK

1 Write the equation of a horizontal line through $(3, -5)$.
2 Write the equation of a vertical line through $(3, -5)$.
3 Write the equation of the line passing through $(-3, 4)$ and $(2, 1)$.

Following is a summary of the forms of linear equations.*

1 General form: $Ax + By + C = 0$
2 Horizontal line: $y = k$ for a real constant k
3 Vertical line: $x = h$ for a real constant h
4 Point-slope form: $y - y_1 = m(x - x_1)$
5 Slope-intercept form: $y = mx + b$

The general form of a linear equation is often used to express answers, since this form can be written using only integers when A, B, and C are rational numbers. The point-slope form and the slope-intercept form are usually used to work problems, since the given information about the line often fits directly into these forms.

EXAMPLE 5 Using the Special Forms to Write the Equation of a Line

Write the general form of the line satisfying the given conditions.

SOLUTIONS

(a) $m = -\frac{3}{7}$ and the y-intercept is -4.

$$y = mx + b$$
$$y = -\frac{3}{7}x + (-4)$$
$$7y = -3x - 28$$
$$3x + 7y + 28 = 0$$

The slope-intercept is most appropriate. Substitute the given information into this form, and then rewrite the equation in the general form.

(b) $m = \frac{7}{5}$ and the x-intercept is -6.

$$y - y_1 = m(x - x_1)$$
$$y - 0 = \frac{7}{5}[x - (-6)]$$
$$5y = 7(x + 6)$$
$$5y = 7x + 42$$
$$7x - 5y + 42 = 0$$

Caution: Do not use the slope-intercept form, since it is the x-intercept that is given, not the y-intercept. Use the point-slope form, with the x-intercept $(-6, 0)$ as the given point.

*Other forms of linear equations are used in calculus. These forms can be derived as needed from the forms presented in this text.

SELF-CHECK ANSWERS 

1 $y = -5$ **2** $x = 3$ **3** $3x + 5y - 11 = 0$

EXAMPLE 6 Using the Special Forms to Write the Equation of a Line

Write the equation of the line satisfying the given conditions.

SOLUTIONS

(a) A line through $(-3, 4)$ and $(-3, -8)$

$x = -3$

This line is vertical, because both points have the same x-coordinate. Every vertical line has the form $x = h$.

(b) A line through $(-3, 4)$ and $(9, 4)$

$y = 4$

This line is horizontal, because both points have the same y-coordinate. Every horizontal line has the form $y = k$.

(c) A line through $(-3, 4)$ and parallel to $2x + 5y + 7 = 0$

$$2x + 5y + 7 = 0$$
$$5y = -2x - 7$$
$$y = -\frac{2}{5}x - \frac{7}{5}$$
$$m = -\frac{2}{5}$$

Parallel lines have the same slope, so write the given line in slope-intercept form in order to determine its slope.

$$y - y_1 = m(x - x_1)$$
$$y - 4 = -\frac{2}{5}[x - (-3)]$$
$$5(y - 4) = -2(x + 3)$$
$$5y - 20 = -2x - 6$$
$$2x + 5y - 14 = 0$$

Now substitute the slope and the given point into the point-slope form.

(d) A line through $(-3, 4)$ and perpendicular to $2x + 5y + 7 = 0$

$$m = \frac{5}{2}$$
$$y - y_1 = m(x - x_1)$$
$$y - 4 = \frac{5}{2}[x - (-3)]$$
$$2(y - 4) = 5(x + 3)$$
$$2y - 8 = 5x + 15$$
$$5x - 2y + 23 = 0$$

Perpendicular lines have slopes whose product is -1. Thus the slope here is the opposite of the reciprocal of the slope found in part (c). The opposite of the reciprocal of $-\frac{2}{5}$ is $\frac{5}{2}$.

SELF-CHECK

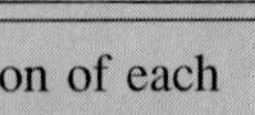

Write the equation of each line described.

1 A vertical line through $(2, 4)$

2 A horizontal line through $(2, 4)$

3 A line through $(-2, 3)$ and $(-2, -3)$

4 A line through $(-2, 3)$ and $(2, 3)$

SELF-CHECK ANSWERS

 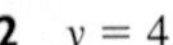 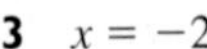 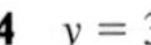

1 $x = 2$ **2** $y = 4$ **3** $x = -2$ **4** $y = 3$

In Section 2-1 we used line graphs and made graphical approximations using linear interpolation and linear extrapolation. We will expand on this concept now by using linear equations to make linear approximations.

EXAMPLE 7 Approximations Using a Line Graph

The height in inches and the corresponding weight in pounds of a young woman upon different visits to her physician are recorded on the line graph shown. Determine the equation of the line passing through *A* and *B*, and use this equation to extrapolate her weight when her height was 51 inches.

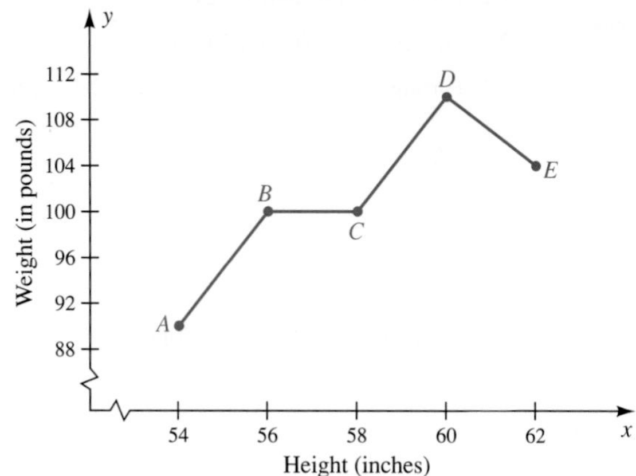

SOLUTION First use the points $A(54, 90)$ and $B(56, 100)$ to determine the equation of the line through *A* and *B*.

$$m = \frac{y_2 - y_1}{x_2 - x_1} = \frac{100 - 90}{56 - 54} = \frac{10}{2} = 5$$ Calculate the slope of $\overline{AB}$.

$$y - y_1 = m(x - x_1)$$

$$y - 90 = 5(x - 54)$$

$$y = 5x - 270 + 90$$

Then substitute the point (54, 90) and the slope 5 into the point-slope form.

$$y = 5x - 180$$

This is the slope-intercept form of the line through *A* and *B*.

$$y = 5(51) - 180$$

$$y = 255 - 180$$

$$y = 75$$

Then substitute in the height of 51 in. to calculate the weight.

Answer She weighed approximately 75 lb when she was 51 in. tall. ▼

SELF-CHECK 

Use the line graph in Example 7 to interpolate an approximation of her weight at a height of 59 inches.

SELF-CHECK ANSWER ✓

She weighed approximately 105 lb when she was 59 in. tall.

Exercises 2-3

A

In Exercises 1 and 2 write each equation in slope-intercept form, and then give the slope and the y-intercept.

1 a. $4x - 5y = 7$ **b.** $y - 4 = 2(x - 2)$

2 a. $-3x + 7y = 4$ **b.** $y - 11 = \frac{1}{2}(x + 4)$

In Exercises 3–30 write the general form of a line satisfying the conditions given.

3 a. A horizontal line through $(-4, 7)$

b. A vertical line through $(-4, 7)$

4 a. A vertical line through $(4, -7)$

b. A horizontal line through $(4, -7)$

5 a. $m = 2$, $b = -5$ **b.** $m = -\frac{3}{4}$, $b = \frac{1}{2}$

c. $m = 0$, $b = 7$

6 a. $m = 3$, $b = 8$ **b.** $m = -\frac{3}{7}$, $b = 2$

c. $m = 0$, $b = -7$

7 a. Parallel to the y-axis through $(-5, 9)$

b. Perpendicular to the y-axis through $(-5, 9)$

c. Perpendicular to the x-axis through $(2, \pi)$

8 a. Parallel to the y-axis through $(-2, 7)$

b. Perpendicular to the y-axis through $(-2, 7)$

c. Parallel to the x-axis through $(-\pi, -4)$

9 $m = -3$, through $(-2, 7)$

10 $m = -4$, through $(7, -2)$

11 $m = \frac{4}{7}$, through $(8, -1)$

12 $m = \frac{5}{8}$, through $(-4, -5)$

13 $m = 0$, through $(-6, 6)$

14 $m = 0$, through $(6, -6)$

15 $m = -\frac{3}{7}$, through $(0, 0)$

16 $m = -\frac{9}{4}$, through $(0, 0)$

17 $m = \frac{5}{8}$, with x-intercept -4

18 $m = -\frac{2}{7}$, with x-intercept 3

19 Through $(-2, 7)$ and $(2, 3)$

20 Through $(4, -9)$ and $(-4, -5)$

21 Through $(4, -6)$ and $(4, 6)$

22 Through $(-2, -3)$ and $(-2, 5)$

23 Through $(3, 0)$ and $(0, 2)$

24 Through $(-4, 0)$ and $(0, 5)$

25 Through $(-2, 6)$ and perpendicular to a line with slope $-\frac{5}{2}$

26 Through $(5, -7)$ and perpendicular to a line with slope $\frac{7}{3}$

27 Through $(5, 8)$ with slope undefined

28 Through $(-3, -4)$ with slope undefined

29

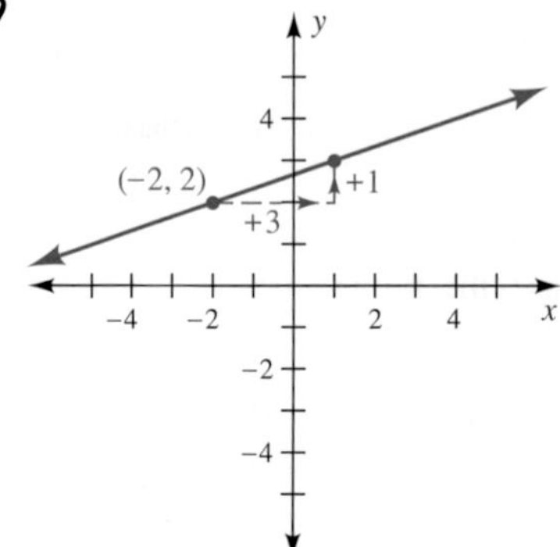

30

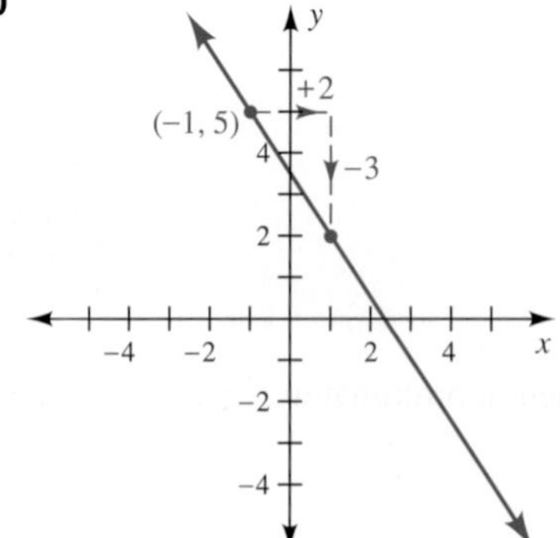

In Exercises 31–40 determine the slope of each line by inspection.

31 $y = 8$ **32** $y = -9$ **33** $y = 8x$

34 $y = -9x$ **35** $x = 8$ **36** $x = -9$

37 $y = -\frac{1}{8}x - 8$ **38** $y = \frac{1}{9}x - 9$

39 $y - 8 = -2(x + 9)$ **40** $y + 9 = 8(x - 2)$

In Exercises 41–48 write the general form of a line that is parallel to the given line and passes through the given point.

41 $y = \frac{4}{5}x - 7$; $(-2, 3)$ **42** $y = -\frac{7}{5}x + 9$; $(3, -2)$

43 $2x + 9y = 40$; $(-4, -9)$ **44** $7x - 5y = 30$; $(4, -10)$

45 $x = 11$; $(8, 3)$ **46** $x = -11$; $(5, 12)$

47 $y = -13$; $(-4, 7)$ **48** $y = 13$; $(-11, -9)$

In Exercises 49–56 write the general form of a line that is perpendicular to the given line and passes through the given point.

49 $y = -\frac{9}{5}x + 8$; $(4, -1)$ **50** $y = \frac{8}{3}x - 4$; $(-1, 4)$

51 $2x + 7y - 8 = 0$; $(-2, 5)$ **52** $3x - 5y + 9 = 0$; $(3, -5)$

53 $x = -4$; $(8, -8)$ **54** $x = 7$; $(9, -9)$

55 $y = \pi$; $(0, 0)$ **56** $y = 0$; $(-\pi, \pi)$

B

In Exercises 57–60 choose the equation that best describes each line.

57 a. $y = \frac{2}{3}x + 3$

b. $y = \frac{2}{3}x + 2$

c. $y = \frac{3}{2}x + 3$

d. $y = \frac{3}{2}x + 2$

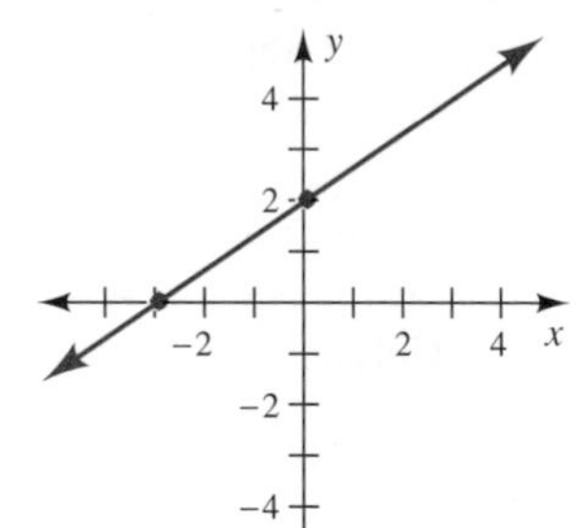

58 a. $y = \frac{2}{3}x + 3$

b. $y = \frac{3}{2}x + 3$

c. $y = -\frac{2}{3}x + 3$

d. $y = -\frac{3}{2}x + 3$

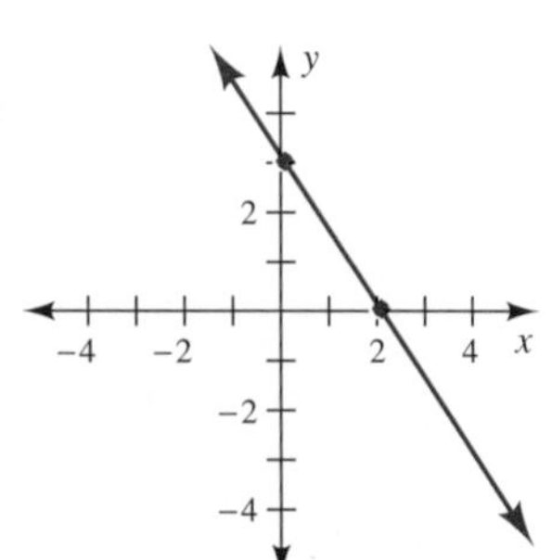

59 a. $y = \frac{x}{2} - 2$

b. $y = \frac{3}{2}x - 2$

c. $y = \frac{3}{2}x + 2$

d. $y = \frac{2}{3}x - 2$

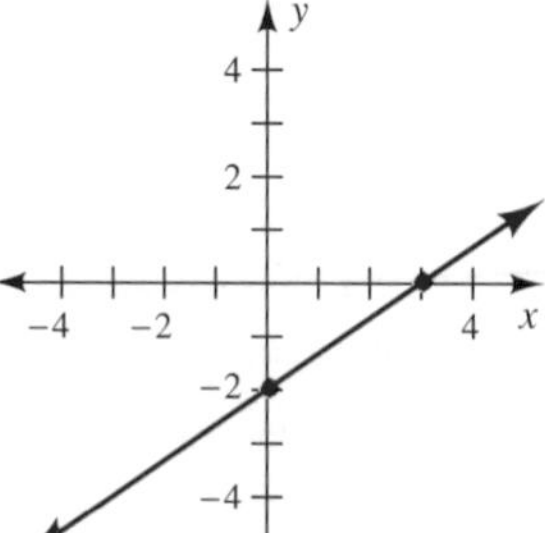

60 a. $y = -\frac{3}{2}x - 1$

b. $y = -\frac{3}{2}x - 2$

c. $y = -\frac{2}{3}x - 2$

d. $y = -\frac{2}{3}x - 3$

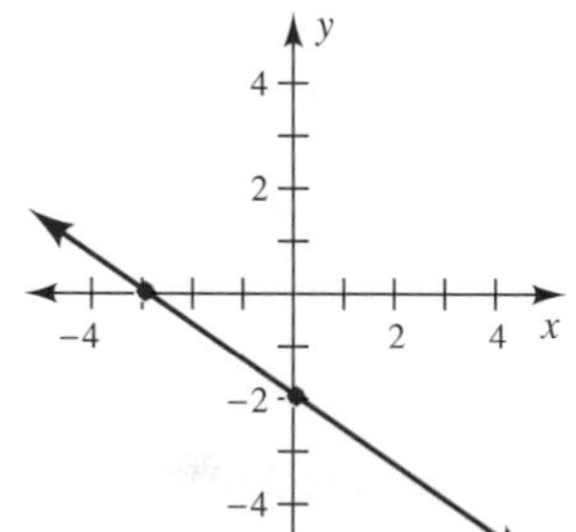

In Exercises 61–63 graph each line using the *y*-intercept and the slope of the line.

61 $y = 2x - 4$ **62** $y = -3x + 7$

63 $y = \frac{4}{9}x - 5$

In Exercises 64–66 use the point-slope to determine a point on the line and the slope of the line. Then use this point and the slope to graph the line.

64 $y - 3 = -2(x + 4)$

65 $y + 3 = \frac{3}{2}(x - 1)$

66 $y + 1 = \frac{5}{2}(x + 3)$

67 Temperature Conversions The freezing temperature of water is 0° Celsius and 32° Fahrenheit. The boiling temperature of water is 100° Celsius and 212° Fahrenheit. Using these pairs of data, determine the linear equation relating these two temperature scales, using the variables C and F.

68 Salary The salary for a clerk selling beauty supplies is \$35 per day, plus a 15% commission on all sales. Write a linear equation giving the salary S in terms of the daily sales d.

69 Scatter Diagram The table of values shown gives the height y in inches of a boy at an age of x years.

a. Form a scatter diagram by plotting these points.

b. Draw a line that seems to best fit these data—that is, a line that overall seems closest to these points.

c. Write the equation of the line you drew in part b.

d. Use the equation in part c to predict this boy's height at age fourteen.

x	0	4	6	8	10	12
y	20	33	37	46	48	56

70 Broken Line Graph Use the data in Exercise 69 to draw a broken line graph. Use this graph to interpolate the boy's height at age 11.

CALCULATOR USAGE (71–74)

In Exercises 71–74 use a graphics calculator to graph each equation. Use [−3, 3] for x and [−5, 5] for y.

71 $y = 2.58x + 4.17$ **72** $y = -2.58x + 4.17$

73 $y = -0.73x - 1.82$ **74** $y = 0.73x - 1.82$

In Exercises 75–80 use the line graph shown to answer each question. This line graph shows the relationship between the year and the average miles per gallon of all automobiles in the United States.

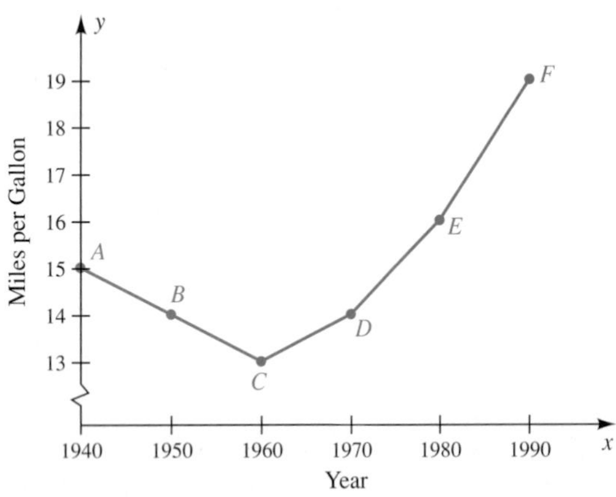

Figure for Exercises 75–80

75 Determine the slope of $\overline{AB}$.

76 Determine the slope of $\overline{BC}$.

77 Determine the slope of $\overline{EF}$.

78 Interpolate the average miles per gallon in 1975.

79 Determine the equation of the line through E and F.

80 Use the equation in Exercise 79 to extrapolate the average miles per gallon in the year 2000.

CHALLENGE QUESTION

When my wife and I arrived in Australia in 1989, our host from Dandenong College commented that it was a chilly day of 10°. When my wife looked surprised, he noted that this was the Celsius temperature and that she could obtain the approximate Fahrenheit temperature by doubling the Celsius temperature and then adding 30°. The actual conversion equation is $F = \frac{9}{5}C + 32$. Graph both the actual conversion equation and the approximation given by our host over Celsius temperatures ranging from 0° to 35°. Then compare these linear graphs.

a. When will the approximation be exactly correct?

b. Over what Celsius temperatures will the Fahrenheit approximation be too small?

c. Over what Celsius temperatures will the Fahreneit approximation be too large?

d. What is the largest error that this approximation produces over the range of 0° to 35° Celsius?

SECTION SKILL REINFORCERS

Simplify each of the following expressions.

1 $-1^8 + (-1)^7$ **2** $13 - 5[18 - 8(9 - 11)]$ **3** $12 \div 3 \cdot 2 - 3 \cdot 5^2$ **4** $(11 - 1)^2 - (11^2 - 1^2)$

5 $\dfrac{6^2 - 4^2}{(6 - 4)^2}$

2-4 Relations and Functions

Section Objectives

8 Determine whether a relation is a function.

9 Determine the domain and the range of a function.

Scientists, as well as those in government and business, often study relationships among different quantities. At first they may examine these relationships graphically, with a scatter diagram, a line graph, or a histogram. Once the relationship has become clear, it is often expressed by a formula in which the various quantities are represented mathematically by variables. Such formulas can then be used to make predictions and to project future results. For example, a nuclear scientist can predict the amount of radioactive material that will remain after 100 years, an engineer can predict the maximum load that a bridge can safely carry, and a banker can project the future value of a certificate of deposit. Because working with relationships is fundamental to many diverse subject areas, we will study the concept of a mathematical relationship in detail.

To a mathematician a **relation** is a correspondence that associates elements in one set with those in another set. Each of the correspondences shown below is a relation.

To each time of the day $\xrightarrow{\text{there corresponds}}$ a temperature.
To each room in a house $\xrightarrow{\text{there corresponds}}$ an area.
To each person $\xrightarrow{\text{there corresponds}}$ a weight.
To each real number $\xrightarrow{\text{there corresponds}}$ its square.
To each nonzero real number $\xrightarrow{\text{there correspond}}$ two square roots.

Relation

A relation is a correspondence from a domain set D to a range set R that pairs each element in D with one or more elements in R.

For many applications it is important to be able to project a specific unique result. A **function** is a relation with the additional property that it pairs each domain element with exactly one range element. Thus every function is a relation, but some relations are not functions. The last relation in the list above is not a function since nonzero real numbers have two different square roots.

Function

A function is a relation that pairs each element in a domain D with exactly one element in a range R.

A Mathematical Note

The word *function* was introduced by the German mathematician Gottfried Wilhelm Freiherr von Leibniz (c. 1682). Leibniz is credited as a developer of calculus, together with the English mathematician Sir Isaach Newton (c. 1680). Leibniz contributed both original terminology and notation, as well as his results in the areas of algebra and calculus.

EXAMPLE 1 Identifying Functions

Classify each correspondence as either a function, a relation that is not a function, or a correspondence that is not a relation.

SOLUTIONS

(a) D R
$3 \to 15$
$9 \to 39$
$13 \to 55$

This relation is a function.

The domain $D = \{3, 9, 13\}$ and the range $R = \{15, 39, 55\}$.

(b) D R
$16 \to 4$
$16 \searrow -4$
$0 \to 0$

This relation is not a function.

The element 16 is not paired with exactly one element in the range; it is paired with both 4 and −4.

(c) D R
$-4 \to 16$
$4 \nearrow$ (16)
$0 \to 0$

This relation is a function.

The domain $D = \{-4, 4, 0\}$ and the range $R = \{16, 0\}$. −4 is paired only with 16; 4 is paired only with 16; and 0 is paired only with 0. Thus each element in D is paired with exactly one element in R.

(d) D R
$3 \to \frac{1}{3}$
$2 \to \frac{1}{2}$
$1 \to 0$
0

This correspondence is not a relation.

The element 0 is not paired with any element in the range. Each element in D must be paired with an element in R for a correspondence to be a relation.

The mapping, or arrow, notation used in the previous example is helpful in illustrating that a function is just a special correspondence between two sets. However, **ordered-pair notation** can convey this same information more concisely. The first coordinate of an ordered pair is the input from the domain, and the second coordinate is the output in the range.

SELF-CHECK

Classify each correspondence as either a function, a relation that is not a function, or a correspondence that is not a relation.

1 D R
$0 \to 0$
$3 \to 3$
$-3 \nearrow$ (3)

2 D R
$0 \to 0$
$3 \to 3$
$3 \searrow -3$

3 D R
$-3 \to -2$
$0 \to -1$
3

EXAMPLE 2 Using Ordered-Pair Notation to Represent a Function

SOLUTION

D	R	Ordered pair
$-2 \to$	-1	$(-2, -1)$
$-1 \to$	3	$(-1, 3)$
$0 \to$	5	$(0, 5)$
$1 \to$	7	$(1, 7)$
$2 \to$	9	$(2, 9)$

Both notations indicate a function with $D = \{-2, -1, 0, 1, 2\}$ and $R = \{-1, 3, 5, 7, 9\}$, and both indicate how the elements are paired. This set of ordered pairs could also be written horizontally as $\{(-2, -1), (-1, 3), (0, 5), (1, 7), (2, 9)\}$.

SELF-CHECK ANSWERS

1 A function **2** A relation but not a function **3** Not a relation

EXAMPLE 3 Using Mapping Notation to Represent a Function

Rewrite the function $\{(-5, 5), (5, 5), (-7, 7), (7, 7)\}$ using the mapping notation.

SOLUTION

D		R
-5	$\rightarrow$	5
5	$\nearrow$	
-7	$\rightarrow$	7
7	$\nearrow$	

This relation is a function. Each element of the domain is paired with exactly one element in the range. The domain $D = \{-5, 5, -7, 7\}$ and the range $R = \{5, 7\}$. ▼

A relation can be defined by a set of **input-output** pairs, with the domain the set of first coordinates and the range the set of second coordinates. Thus the special type of correspondence required for a relation to be a function can also be defined in terms of ordered pairs.

ORDERED-PAIR DEFINITION OF A FUNCTION

A function is a set of ordered pairs with no pairs that have the same first coordinate and different second coordinates.

The Cartesian coordinate system provides a pictorial means of presenting the relationship between two variables. Since each point in a plane can be uniquely identified by an ordered pair (x, y), graphs can be used to represent mathematical relations. The x-coordinate of the ordered pair (x, y) is called the **independent variable,** and the y-coordinate is called the **dependent variable.**

EXAMPLE 4 Graphing Relations

Graph each of these relations.

SOLUTIONS

(a)

D		R
-2	$\rightarrow$	-3
-1	$\rightarrow$	-1
0	$\rightarrow$	1
1	$\rightarrow$	3
2	$\rightarrow$	5

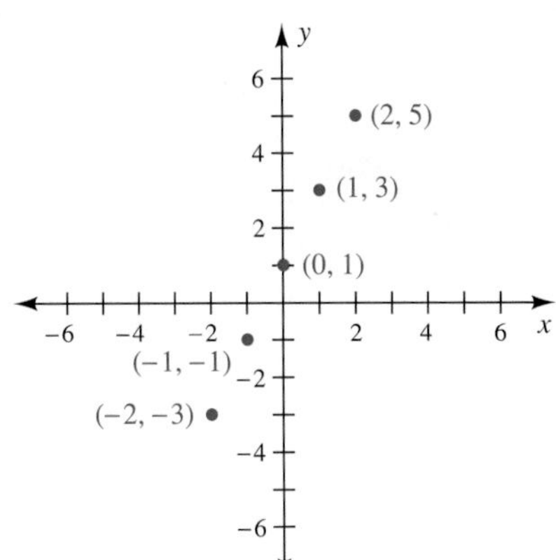

This relation is a function with domain $D = \{-2, -1, 0, 1, 2\}$ and range $R = \{-3, -1, 1, 3, 5\}$.

(b)

x	y
−2	2
−1	2
0	2
1	2
2	2

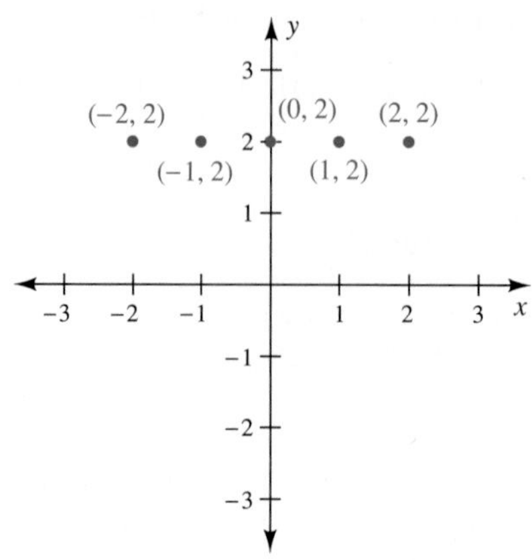

This relation is a function with domain $D = \{-2, -1, 0, 1, 2\}$ and range $R = \{2\}$.

(c) $\{(2, -2), (2, -1), (2, 0), (2, 1), (2, 2)\}$

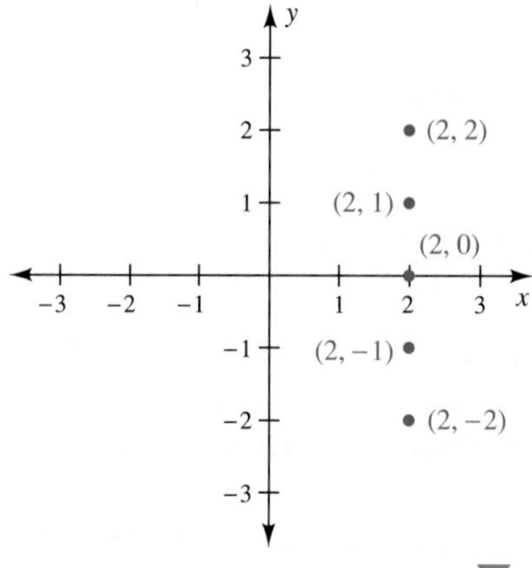

This relation is not a function because different ordered pairs have the same first coordinate.

SELF-CHECK

Determine which of these relations is a function.

1 $\{(8, 11), (9, 11), (12, 11)\}$
2 $\{(6, 4), (3, 8), (6, 5)\}$

If a function is defined by a graph, then the ordered pairs can be determined by examining the points that form the graph. The domain can be found by projecting these points onto the x-axis, and the range can be found by projecting these points onto the y-axis.

DOMAIN AND RANGE FROM THE GRAPH OF A FUNCTION

Domain: The domain is the projection of the graph onto the x-axis.
Range: The range is the projection of the graph onto the y-axis.

SELF-CHECK ANSWERS 

1 A function **2** A relation but not a function

EXAMPLE 5 Determining the Domain and Range from the Graph of a Relation

Write the domain and the range of the relations defined by these graphs.

SOLUTIONS

(a)

$D = \{-3, 1, 4\}$

$R = \{-2, -1, 2\}$

This relation is a function consisting of the ordered pairs $\{(-3, -1), (1, 2), (4, -2)\}$. The points project onto -3, 1, and 4 on the x-axis and -2, -1, and 2 on the y-axis.

(b)

$D = \{-2, -1, 0, 1, 2\}$

$R = \{0, 1, 4\}$

This relation is a function consisting of the ordered pairs $\{(-2, 4), (-1, 1), (0, 0), (1, 1), (2, 4)\}$.

(c)

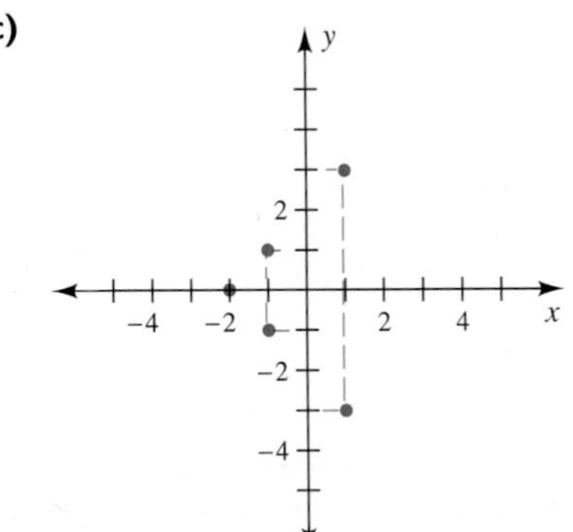

$D = \{-2, -1, 1\}$

$R = \{-3, -1, 0, 1, 3\}$

This relation, which consists of the ordered pairs $\{(-2, 0), (-1, 1), (-1, -1), (1, 3), (1, -3)\}$, is not a function.

Histograms and line graphs are also used to show the relationship between two sets of values. In the line graph shown in Example 6 we will consider the endpoints of the line segments as given data that have been collected. The line

SELF-CHECK 

Write the domain and the range of the functions defined by these graphs.

1

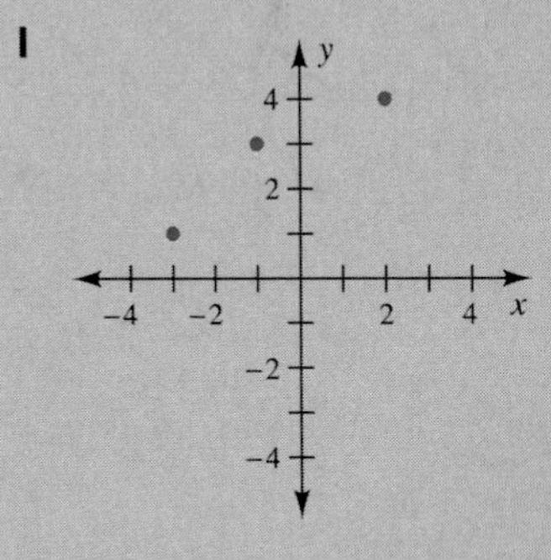

2

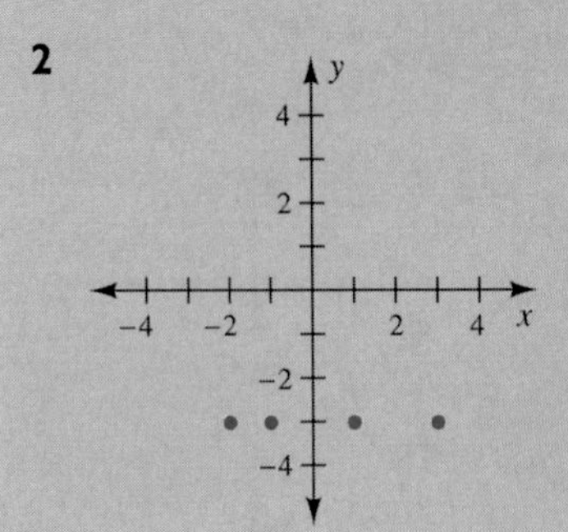

SELF-CHECK ANSWERS 

1 $D = \{-3, -1, 2\}$, $R = \{1, 3, 4\}$ **2** $D = \{-2, -1, 1, 3\}$, $R = \{-3\}$

segments contain many other points that are not part of the original domain. These line segments make the visual pattern more apparent and facilitate the interpolation of unknown points.

EXAMPLE 6 Determining the Domain and Range from Histograms and Line Graphs

Both the histogram and the line graph display the percent of world motor vehicle production of the United States for selected years. Write this function in ordered-pair notation. Then determine the domain and the range of this function.

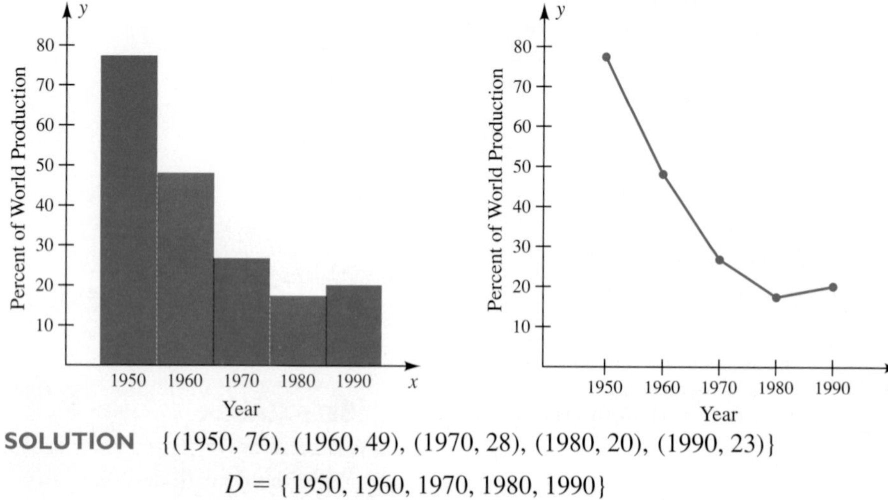

SOLUTION $\{(1950, 76), (1960, 49), (1970, 28), (1980, 20), (1990, 23)\}$

$$D = \{1950, 1960, 1970, 1980, 1990\}$$

$$R = \{76, 49, 28, 20, 23\}$$

If the domain set of a function is small, then we may be able to list all the ordered pairs of the function or define the function using a table or mapping notation. If the domain is infinite, we cannot possibly list all the ordered pairs. A graph can easily illustrate an infinite set of points, however, and thus define a function with an infinite domain.

EXAMPLE 7 Determining the Domain and Range from the Graph of a Function

Write the domain and the range of the relations defined by these graphs.

SOLUTIONS

(a)

The domain is the infinite set of points in the interval $[1, 5)$. The range is the infinite set of points in the interval $[-1, 4]$.

The domain, determined by projecting the graph onto the x-axis, includes the endpoint 1 but does not include the endpoint 5. The range is the projection of the graph onto the y-axis.

(b)

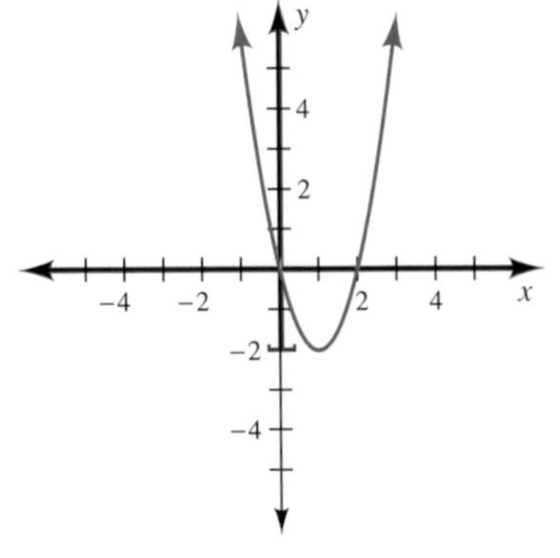

The domain is the set of all real numbers. The range is the interval $[-2, +\infty)$.

The graph continues to both the left and the right; thus the projection onto the x-axis includes all real numbers. The projection onto the y-axis includes all real numbers greater than or equal to -2.

SELF-CHECK 

Write the domain and the range of the function defined by this graph.

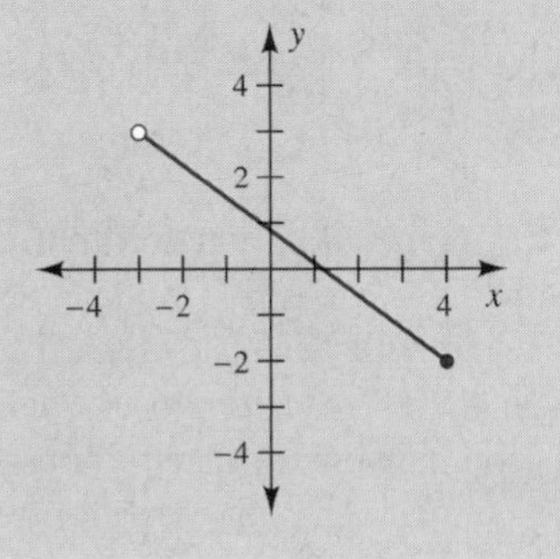

Not all graphs represent functions. Some useful relations assign the same x value to more than one y value. In such cases there will be more than one point on the graph for certain x values. We can visually determine whether a graph represents a function by using the **vertical line test** to inspect the graph, because any two points on the same vertical line must have the same x value.

VERTICAL LINE TEST

Imagine a vertical line placed on the same coordinate system as the given graph. If at any position this vertical line intersects the graph at more than one point, then the graph does *not* represent a function. Otherwise, the graph *does* represent a function.

EXAMPLE 8 Using the Vertical Line Test

Use the vertical line test to determine whether each relation is a function.

SOLUTIONS

(a)

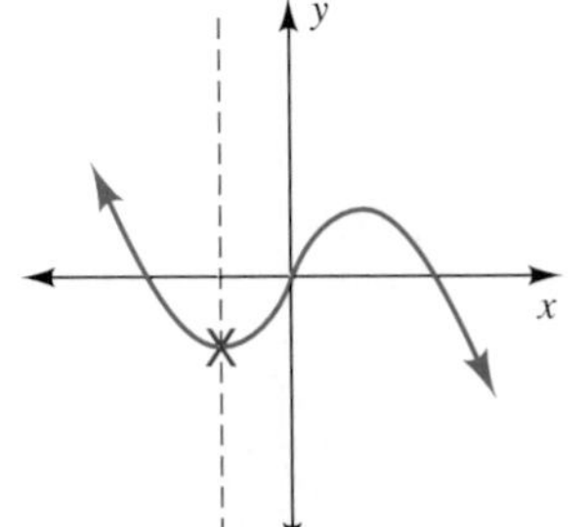

Function

SELF-CHECK ANSWER 

D: $(-3, 4]$, R: $[-2, 3)$

(b)

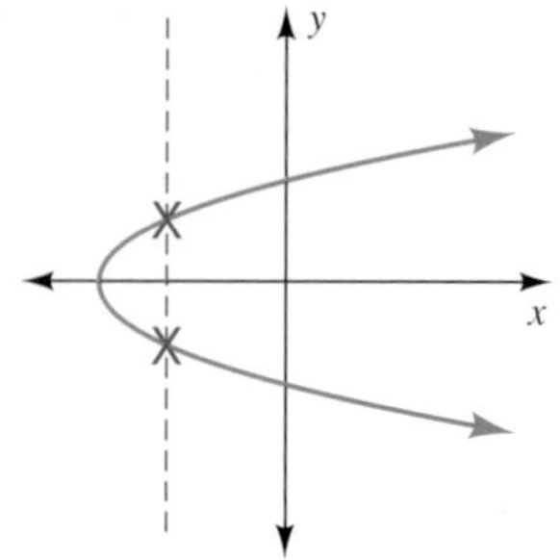

Not a function

The vertical line shown intersects the graph at two points.

(c)

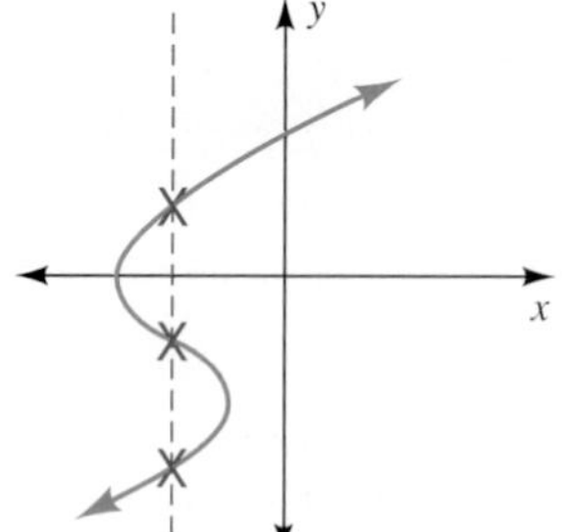

Not a function

The vertical line shown intersects the graph at three points.

SELF-CHECK

Use the vertical line test to determine whether each relation is a function.

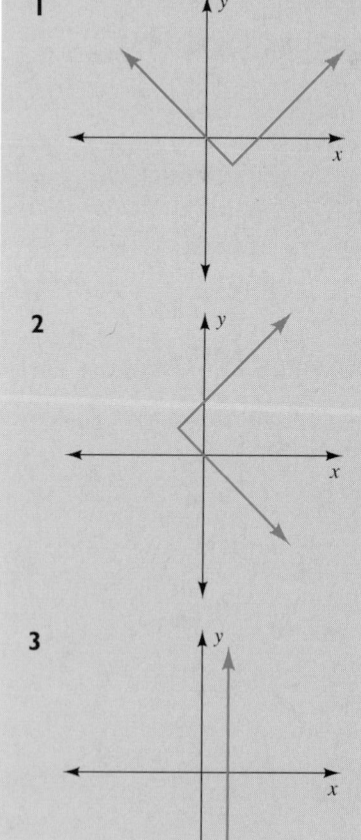

The graph of a function that is **increasing** will slope upward to the right and thus have a positive slope. On the other hand, a graph of a function that is **decreasing** will slope downward to the right and thus have a negative slope. We can therefore determine visually where a function is increasing and even make comparisons to determine where the function is increasing most rapidly. These comparisons allow doctors, engineers, accountants, and others to obtain important information about key events from data presented in graphical form. This is illustrated in the next example.

EXAMPLE 9 Identifying Important Events from a Graph

The function defined by the graph shown on the next page gives the water depth in feet behind the dam of a reservoir. Match each time period (1, 2, 3, and 4) to the event that occurred during that time period.

Event A: The dam gate is completely closed, and there is light runoff into the reservoir.

Event B: The dam gate is completely closed, and a heavy rain begins upstream from the reservoir.

Event C: The dam gate is set to slightly open as light runoff continues.

Event D: The dam gate is completely opened, and the rain lessens.

SELF-CHECK ANSWERS

1 A function **2** A relation but not a function
3 A relation but not a function

SOLUTION At time 1, event B occurs. The graph shows a rapid increase starting at 1 in the domain.

At time 2, event D occurs. The graph shows a decrease starting at 2 in the domain.

At time 3, event C occurs. The graph shows the reservoir level is constant starting at 3 in the domain.

At time 4, event A occurs. The graph is slowly increasing starting at 4 in the domain. ▼

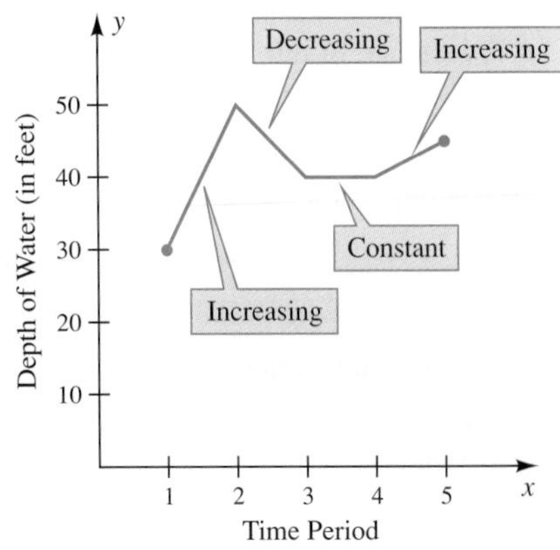

In this section we have used a variety of notations to specify functions, including mapping notation, tables, ordered pairs, and graphs. You should be familiar with all of these common ways of denoting functions, because no one way is best for all situations. In the next section functions will be defined using equations and functional notation.

Exercises 2-4

A

In Exercises 1–12 classify each correspondence as either a function, a relation that is not a function, or a correspondence that is not a relation. State the domain and the range of each relation.

1 *D* *R*
7 → 3
8 → 4
9 → 6

2 *D* *R*
7 → 3
8 ↗
9 ↗

3 *D* *R*
7 → 3
8 → 4
↘ 6

4 *D* *R*
7 → 3
8 → 4
9

5

x	y
−3	π
0	π
1	π
9	π

6

x	y
π	−3
π	0
π	1
π	9

7

x	y
−3	π
0	
1	5
9	13

8

x	y
−3	3
0	0
1	−1
9	−9

9 $\{(7, 11), (7, 2), (7, -4)\}$

10 $\{(5, 8), (-5, -3), (0, -3), (1, 8)\}$

11 $\{(1, 4), (2, 4), (3, 4), (\pi, 4)\}$

12 $\{(\pi, 1), (1, \pi), (2, 9), (9, 2)\}$

In Exercises 13–16 determine whether each graph represents a function or a relation that is not a function. State the domain and the range of each relation.

13 **14**

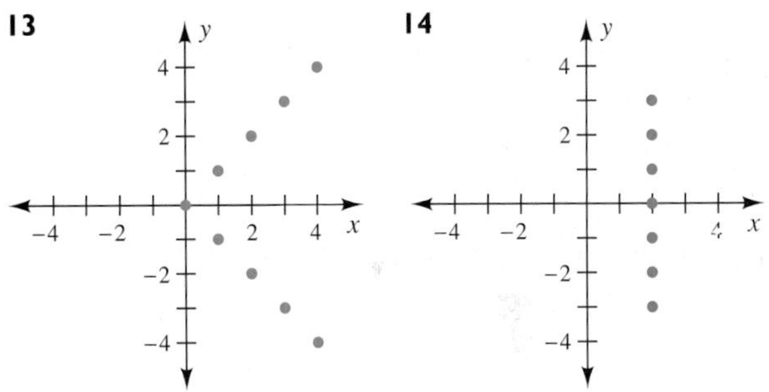

15 **16**

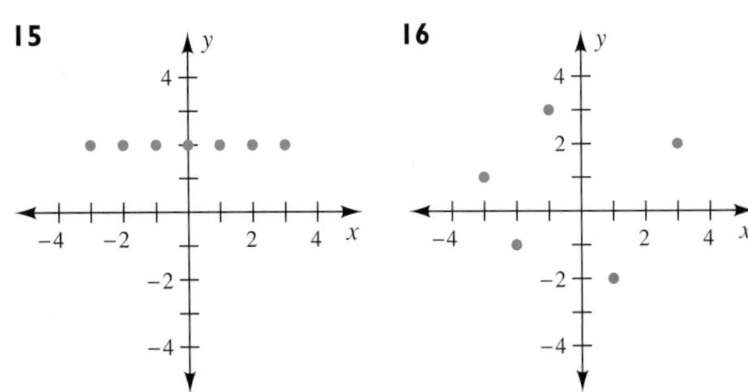

17 Use the function defined by the table below to complete each part of this exercise.

x	−5	−3	−2	0	1	4
y	4	2	0	−2	−3	−4

a. Express this function using mapping notation.

b. Express this function using ordered-pair notation.

c. Graph this function.

18 Use the function defined by the table below to complete each part of this exercise.

x	-5	-3	-2	0	1	4
y	3	3	3	3	3	3

a. Express this function using mapping notation.

b. Express this function using ordered-pair notation.

c. Graph this function.

19 Use the function defined by the graph below to complete each part of this exercise.

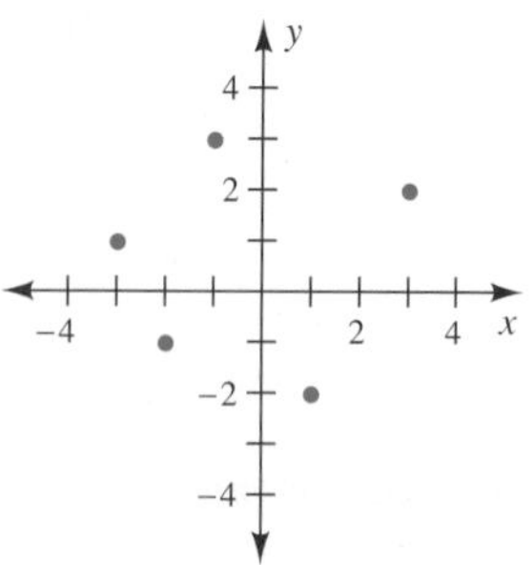

Figure for Exercise 19

a. Express this function using mapping notation.

b. Express this function using ordered-pair notation.

c. Express this function using a table format.

20 Use the function defined by the graph below to complete each part of this exercise.

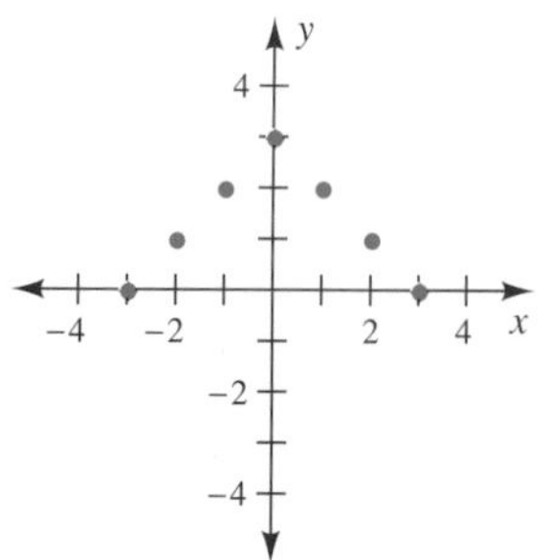

Figure for Exercise 20

a. Express this function using mapping notation.

b. Express this function using ordered-pair notation.

c. Express this function using a table format.

21 Use the function defined by the given mapping notation to complete each part of this exercise.

$$\begin{array}{rcr} D & & R \\ -1 & \to & 5 \\ 1 & \to & 3 \\ 2 & \to & -1 \\ 4 & \to & -2 \end{array}$$

a. Express this function using a table format.

b. Express this function using ordered-pair notation.

c. Graph this function.

22 Use the function defined by the given mapping notation to complete each part of this exercise.

$$\begin{array}{rcr} D & & R \\ -4 & \to & -4 \\ -2 & \to & -2 \\ 0 & \to & 0 \\ 1 & \to & 1 \\ 3 & \to & 3 \end{array}$$

a. Express this function using a table format.

b. Express this function using ordered-pair notation.

c. Graph this function.

23 Use the function $\{(-5, 4), (-3, 4), (1, 4), (2, 4), (3, 4)\}$ to complete each part of this exercise.

a. Express this function using a table format.

b. Express this function using mapping notation.

c. Graph this function.

24 Use the function $\{(-4, -2), (-2, 4), (4, -3), (3, 1)\}$ to complete each part of this exercise.

a. Express this function using a table format.

b. Express this function using mapping notation.

c. Graph this function.

Test Results (25 and 26)

25 Use the function defined by the histogram shown to complete each part of this exercise.

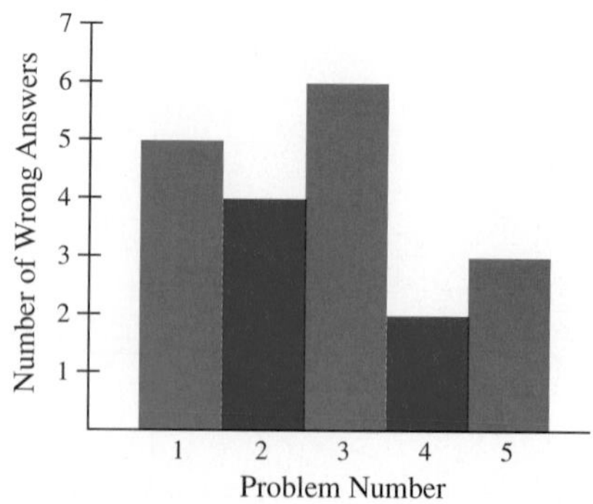

Figure for Exercise 25

a. Express this function using mapping notation.

b. Express this function using ordered-pair notation.

c. Express this function using a table format.

26 Use the function defined by the given histogram to complete each part of this exercise.

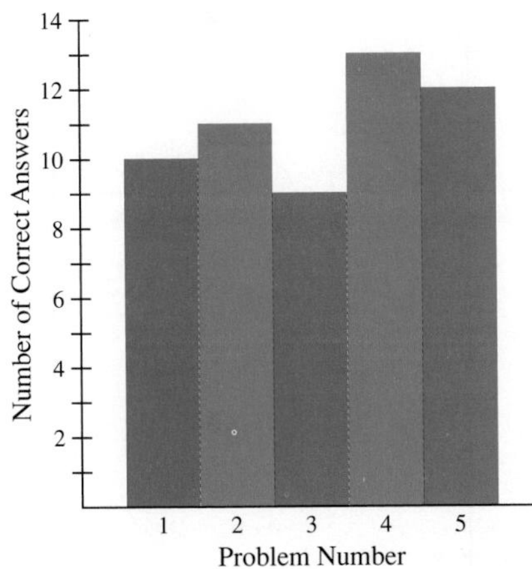

Figure for Exercise 26

a. Express this function using mapping notation.

b. Express this function using ordered-pair notation.

c. Express this function using a table format.

27 Tax Obligation Days Use the function defined by the line graph shown to complete each part of this exercise. The domain consists only of the x-coordinates of the endpoints of each line segment. Estimate the y-coordinates by inspection of the graph.

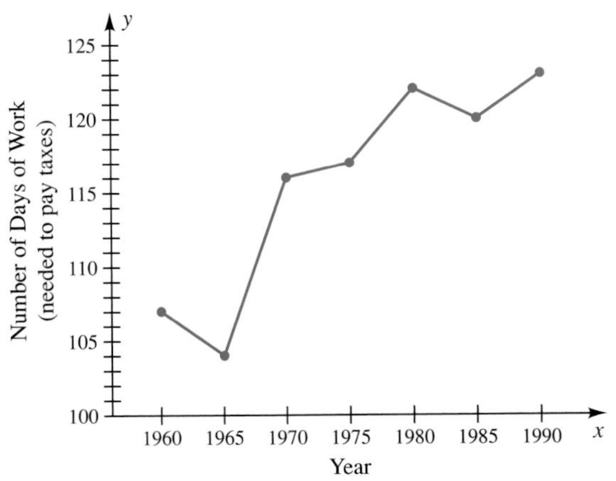

Figure for Exercise 27

a. Express this function using mapping notation.

b. Express this function using ordered-pair notation.

c. Express this function using a table format.

28 Average Baseball Salaries Use the function defined by the given line graph to complete each part of this exercise. The domain consists only of the x-coordinates of the endpoints of each line segment. Estimate the y-coordinates by inspection of the graph.

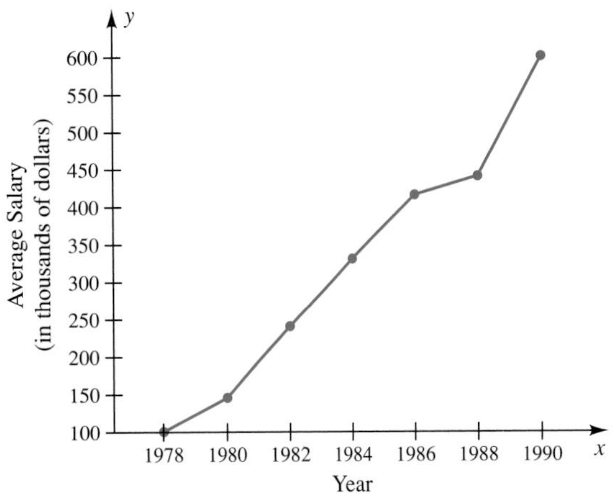

Figure for Exercise 28

a. Express this function using mapping notation.

b. Express this function using ordered-pair notation.

c. Express this function using a table format.

In Exercises 29–48 use the vertical line test to determine which graphs represent functions.

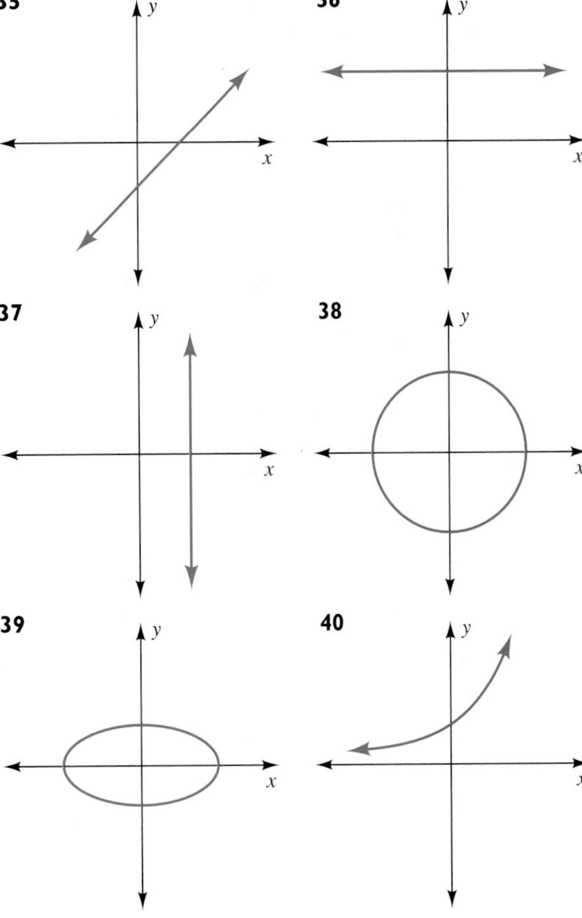

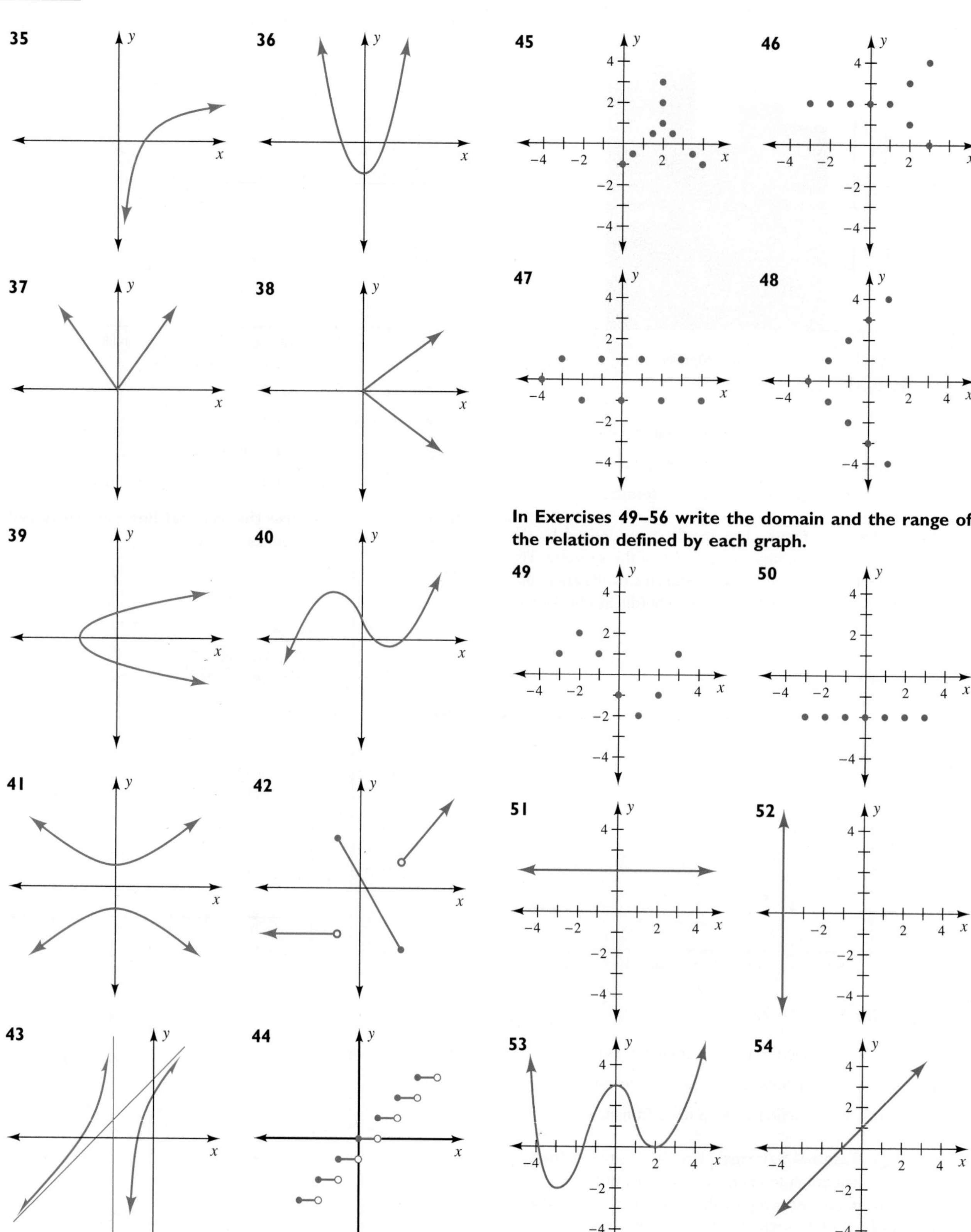
35
36
37
38
39
40
41
42
43
44
45
46
47
48
In Exercises 49–56 write the domain and the range of the relation defined by each graph.
49
50
51
52
53
54

55

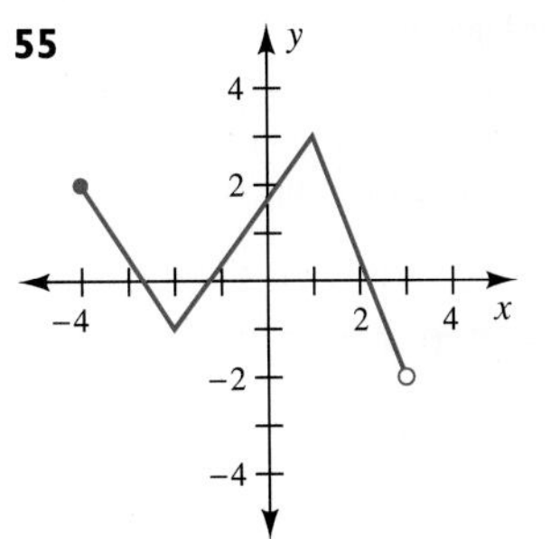

56

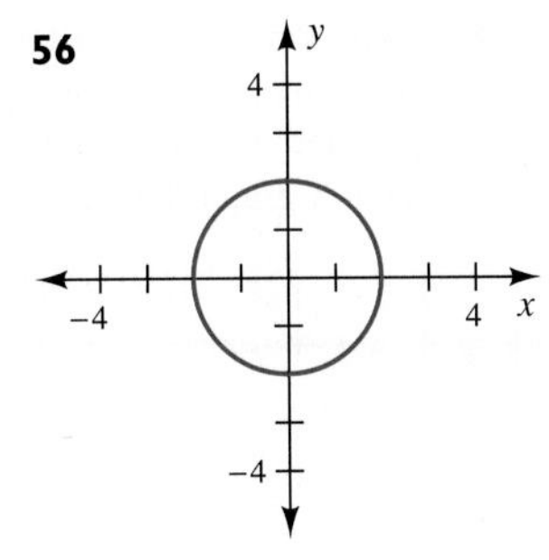

B

57 A function is defined by the graph shown on the domain $[1, 5]$. Use this graph to determine the following information.

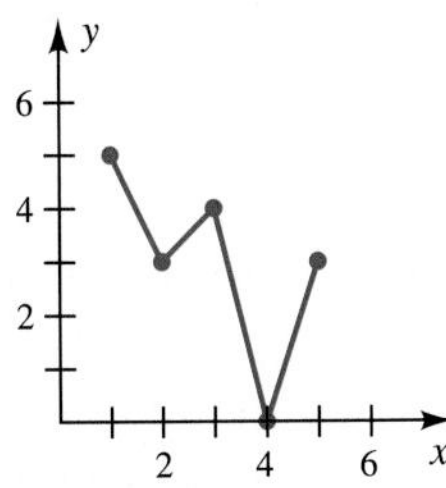

Figure for Exercise 57

a. Is the function increasing or decreasing on the interval $(1, 2)$?

b. Is the function increasing or decreasing on the inverval $(2, 3)$?

c. Is the function increasing or decreasing on the interval $(3, 4)$?

d. On which interval is the function increasing most rapidly?

58 A function is defined by the graph shown on the domain $[1, 5]$. Use this graph to determine the following information.

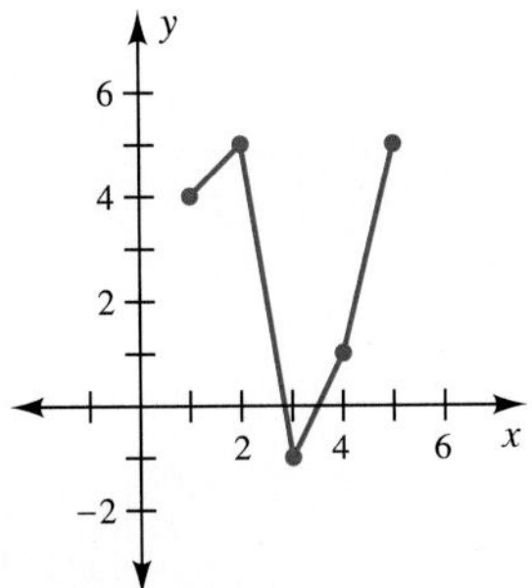

Figure for Exercise 58

a. Is the function increasing or decreasing on the interval $(1, 2)$?

b. Is the function increasing or decreasing on the interval $(2, 3)$?

c. Is the function increasing or decreasing on the interval $(3, 4)$?

d. On which interval is the function increasing most rapidly?

59 Bacterial Blood Count The function defined by the graph shown gives the bacterial blood count for a patient over a period of days. Match each time period (1, 2, 3, and 4) to the event that occurred during that time period.

Event A: The patient contracts a new bacterial infection.

Event B: The patient discontinues an antibiotic, and the bacterial count returns to a normal level.

Event C: The patient starts taking an antibiotic.

Event D: The bacterial level continues at a normal level.

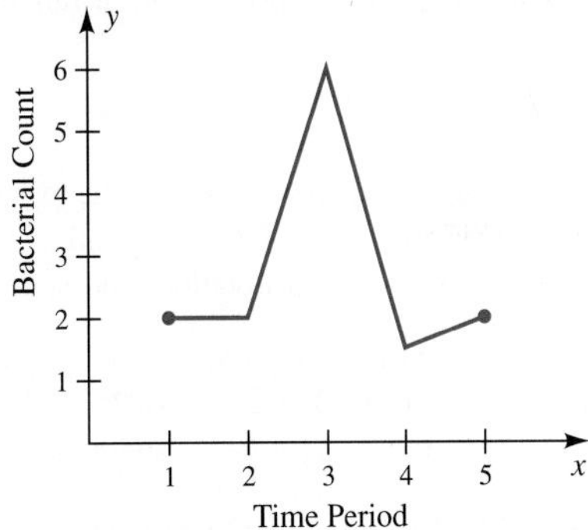

Figure for Exercise 59

60 Company Sales Revenue The function defined by the graph shown gives the gross revenue of a company over a period of months. Match each time period (1, 2, 3, and 4) to the event that occurred during that time period.

Event A: The company introduces a new model of its product, and its sales roar.

Event B: The economy improves, and sales improve slightly.

Event C: Sales of the current model peak and level off.

Event D: Sales of the current model start declining.

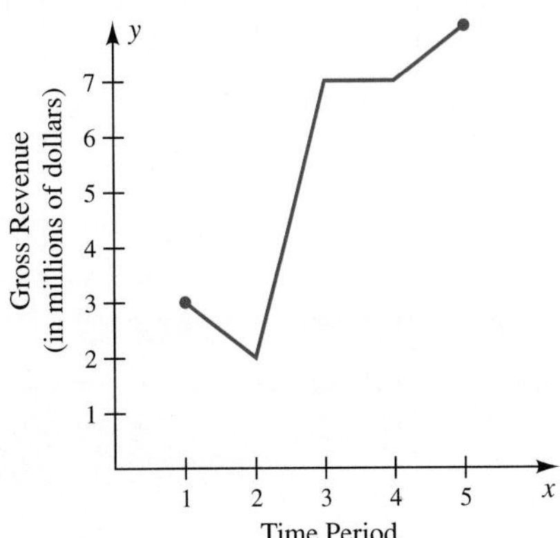

Figure for Exercise 60

61 Graph a straight line that does *not* represent a function.

62 Graph a semicircle that does *not* represent a function.

63 Graph a semicircle that represents a function.

64 Can a graph that is a circle ever represent a function? Explain why or why not.

65 Can a graph that is an ellipse ever represent a function? Explain why or why not.

66 Graph a function whose domain is the set of all real numbers and whose range is $\{2\}$.

67 Explain why the correspondence that associates each person with his or her biological parents is not a function.

68 Explain why the correspondence that associates a given sales tax with the purchase price is not a function.

C

69 A function consists of the ordered pairs (x, y), where each x is from the domain $\{-4, -2, 0, 1, 3\}$ and $y = |x|$. Write this function in ordered-pair notation, and state its range.

70 A function consists of the ordered pairs (x, y), where each x is from the domain $\{-3, -2, -1, 0, 2\}$ and $y = x^2 - 5$. Express this function using a table, and state its range.

71 A function consists of the ordered pairs (x, y), where each x is from the domain $\{0, 1, 4, 9\}$ and $y = \sqrt{x}$. Express this function using mapping notation, and state its range.

72 A function consists of the ordered pairs (x, y), where each x is from the domain $\{-8, -1, 0, 1, 8\}$ and $y = \sqrt[3]{x}$. Express this function using ordered-pair notation, and state its range.

73 A function consists of the ordered pairs (x, y), where x can be any real number and $y = x^2$. Complete the table below, plot ordered pairs, connect the points, and then sketch a graph that represents the entire function.

x	-3	-2	-1	$-\frac{1}{2}$	0	$\frac{1}{2}$	1	2	3
y									

74 A function consists of the ordered pairs (x, y), where x can be any real number and $y = -x^2$. Complete the table below, plot the ordered pairs, connect the points, and then sketch a graph that represents the entire function.

x	-3	-2	-1	$-\frac{1}{2}$	0	$\frac{1}{2}$	1	2	3
y									

CALCULATOR USAGE (75–78)

In Exercises 75–78 the domain of each function is the set of all real numbers. Graph each function on a graphics calculator, and use this graph to estimate the range of the function.

75 $y = x(x - 2)(x - 3)(x - 4)$

76 $y = x(x - 2)(x - 3)(x + 2)$

77 $y = -x^2(x + 1)(x - 2)$

78 $y = -x^2(x - 1)(x + 2)$

DISCUSSION QUESTION

79 Explain in your own words the vertical line test.

CHALLENGE QUESTION

A metal sphere is placed in a cylindrical beaker marked in cubic centimeters on the side. Approximate the radius of this sphere.

SECTION SKILL REINFORCERS

Evaluate each of these expressions for $x = -10$.

1 x^2 **2** $-x^2$ **3** $(-x)^2$ **4** $7x - 5$ **5** $-3x^2 + 8x - 4$

2-5 Functional Notation

SECTION OBJECTIVES

10 Evaluate an expression using functional notation.

11 Translate a statement of variation into an equation.

For a function with many domain elements, it is not practical to list all possible ordered pairs. Thus we often examine the relationship between the domain elements and the range elements to find a pattern or formula that describes how each element in the domain is paired with an element in the range.

EXAMPLE 1 Writing a Formula for a Function

Write a formula that describes how each domain element x is paired with a range element y.

D		R
-2	$\xrightarrow{\text{triple}}$	-6
-1	$\longrightarrow$	-3
0	$\longrightarrow$	0
1	$\longrightarrow$	3
2	$\longrightarrow$	6
3	$\longrightarrow$	9

SOLUTION

In this function each element of D is tripled to obtain the corresponding value in R. We can represent this relationship by $x \rightarrow 3x$, $(x, 3x)$, or the ordered pairs (x, y), where $y = 3x$.

▼

The function in Example 1 can be denoted by $y = 3x$ or by $f(x) = 3x$, where the domain is $\{-2, -1, 0, 1, 2, 3\}$ and the range is $\{-6, -3, 0, 3, 6, 9\}$. The formula notation $f(x)$ that we used in Section 1-3 is also called **functional notation.**

FUNCTIONAL NOTATION

The expression $f(x)$ is read as "f of x" or "the value of f at x." The variable x represents an element of the domain, and $f(x)$ is the corresponding element in the range.

If $y = f(x)$, the x-coordinate of (x, y) is the independent variable and the y-coordinate, which equals $f(x)$, is the dependent variable. Once a value of x in the domain has been selected, the formula f determines a unique functional value of y in the range. If a value of x is given for the function, we sometimes call this value the **argument of the function,** and the resulting y in the range is the **value of the function** for this argument.

The notations $y = 2x + 1$ and $f(x) = 2x + 1$ both define a linear function whose graph is a straight line with slope 2 and y-intercept $(0, 1)$. The function in Example 2 could also be expressed as $y = x^2 + 2x - 1$. The functional form $f(x) = x^2 + 2x - 1$ is used to clarify the evaluation of the function at each argument.

EXAMPLE 2 Evaluating a Function

Evaluate the function $f(x) = x^2 + 2x - 1$ for each of these arguments.

SOLUTIONS

(a) -1 $\quad f(-1) = (-1)^2 + 2(-1) - 1 = 1 - 2 - 1 = -2$

Substitute each argument for x in the formula $f(x) = x^2 + 2x - 1$. Use parentheses wherever necessary to avoid errors in sign or in the order of operations.

(b) 0 $\quad f(0) = (0)^2 + 2(0) - 1 = 0 + 0 - 1 = -1$

(c) 2 $\quad f(2) = (2)^2 + 2(2) - 1 = 4 + 4 - 1 = 7$

(d) h $\quad f(h) = (h)^2 + 2(h) - 1 = h^2 + 2h - 1$

(e) $h + 1$ $\quad f(h+1) = (h+1)^2 + 2(h+1) - 1 = h^2 + 2h + 1 + 2h + 2 - 1 = h^2 + 4h + 2$ ▼

EXAMPLE 3 Evaluating a Function

Evaluate each of these expressions for $g(x) = 3x^2 - 5$.

SOLUTIONS

(a) $g(h) - g(2)$ $\quad g(h) - g(2) = [3(h)^2 - 5] - [3(2)^2 - 5] = (3h^2 - 5) - (7) = 3h^2 - 12$

Compare the calculations in parts (a) and (b), and note the distinct meanings of these two notations.

(b) $g(h - 2)$ $\quad g(h-2) = 3(h-2)^2 - 5 = 3(h^2 - 4h + 4) - 5 = 3h^2 - 12h + 7$ ▼

SELF-CHECK

Evaluate each of these expressions for $f(x) = \sqrt{3x + 4}$.

1 $f(-1)$ **2** $f(0)$
3 $f(7)$ **4** $f(h + 2)$

Functions are frequently denoted by the letter f, g, or h and are usually defined by giving the formula that describes how the elements in the domain, D, are paired with those in the range, R. Usually the domain and the range are not listed. Instead, it is more convenient to have the domain and the range implied by the formula. In algebra the **implied domain** is usually understood to be the set of all possible real numbers so that the range also consists of real

SELF-CHECK ANSWERS

1 1 **2** 2 **3** 5 **4** $\sqrt{3h + 10}$

numbers. Any values that would cause division by zero must be excluded from the domain. Other values may be excluded to avoid imaginary numbers in the range.* A function that is restricted to real values in the range is called a **real-valued function.**

DOMAIN OF $y = f(x)$

The domain of a real-valued function $y = f(x)$ is understood to be the set of all real numbers for which the formula is defined and that yield real values in the range. In particular, this means that the domain *excludes* values that

1 cause division by zero or

2 cause negative values under a square root symbol.

EXAMPLE 4 Determining the Domain of a Function

Determine the domain of each of these functions.

	SOLUTIONS	
(a) $f(x) = 2x + 7$	$D = \mathbb{R}$	The formula says to "double x and then add 7." There are no values that need to be excluded because they cause division by 0 or result in square roots of negative values. Thus the domain is the set of all real numbers.
(b) $g(x) = \dfrac{3x - 5}{(x + 4)(x - 5)}$	$D = \mathbb{R} \sim \{-4, 5\}$ (all real numbers except -4 and 5)	The values -4 and 5 would cause division by 0 and thus must be excluded from the domain. (The symbol $\sim$ indicates exclusion.) No other values have to be excluded.
(c) $h(x) = \sqrt{2x + 12}$	D: $[-6, +\infty)$	The domain consists of the real numbers for which $2x + 12 \geq 0$. Solving for x gives $x \geq -6$. Values less than -6 are excluded to avoid negative numbers under the square root symbol. ▼

Determining the range of a function defined by functional notation is often not as crucial as determining the domain, and it can be a rather complicated process for some formulas. We will therefore postpone this topic until Section 10-1 on inverse functions.

SELF-CHECK 

Determine the domain for each of these functions.

1 $f(x) = \dfrac{2x + 9}{x^2 - x - 6}$

2 $g(x) = \sqrt{3x - 5}$

3 $h(x) = 2x^3 - 9x + 5$

*Imaginary numbers will be examined in Chapter 7.

SELF-CHECK ANSWERS 

1 $\mathbb{R} \sim \{-2, 3\}$ **2** $\left[\dfrac{5}{3}, +\infty\right)$ **3** $\mathbb{R}$

In order to emphasize that a function associates each element in the domain with a unique element in the range, some teachers use the concept of a "function machine." (See Figure 2-8.) This machine takes raw material from the domain, manipulates it according to the specified formula, and then produces an output that is a value in the range. In the past these machines were merely figments of the imagination, which teachers used for instructional purposes. Today, however, most students have access to genuine function machines in the forms of computers and calculators. Most scientific calculators can evaluate the functions $f(x) = x^2$, $f(x) = \sqrt{x}$, and $f(x) = \frac{1}{x}$.

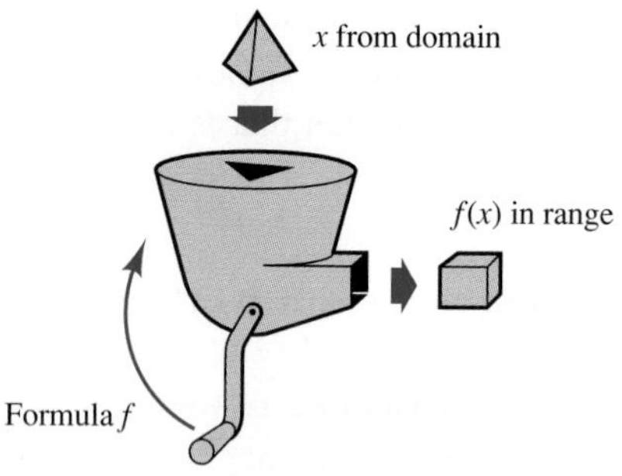

x from domain

f(x) in range

Formula f

Figure 2-8
A "Function Machine"

EXAMPLE 5 Calculator Usage

Determine the domain of each of these functions, and then use a calculator to evaluate $f(-9)$, $f(0)$, and $f(9)$.

SOLUTIONS

(a) $f(x) = x^2$ Domain $= \mathbb{R}$

Keystrokes

$f(-9)$:

Scientific [9] [+/−] [x^2] → 81

Graphics [(] [(−)] [9] [)] [x^2] [ENTER] → 81

This function is defined for all real numbers. The boxes show the correct key sequence to use on a typical scientific calculator and on a typical graphics calculator. The last value displayed to the right is the answer.

$f(0)$:

Scientific [0] [x^2] → 0

Graphics [0] [x^2] [ENTER] → 0

$f(9)$:

Scientific [9] [x^2] → 81

Graphics [9] [x^2] [ENTER] → 81

(b) $f(x) = \sqrt{x}$ Domain: $[0, +\infty)$

Keystrokes

$f(-9)$:

Scientific [9] [+/−] [$\sqrt{x}$] → Error

Graphics [$\sqrt{\ }$] [(−)] [9] [ENTER] → Error

For $x < 0$, this function produces imaginary values. Thus an error message may result on some calculators when we try to evaluate $f(-9)$.

$f(0)$:

Scientific [0] [$\sqrt{x}$] → 0

Graphics [$\sqrt{\ }$] [0] [ENTER] → 0

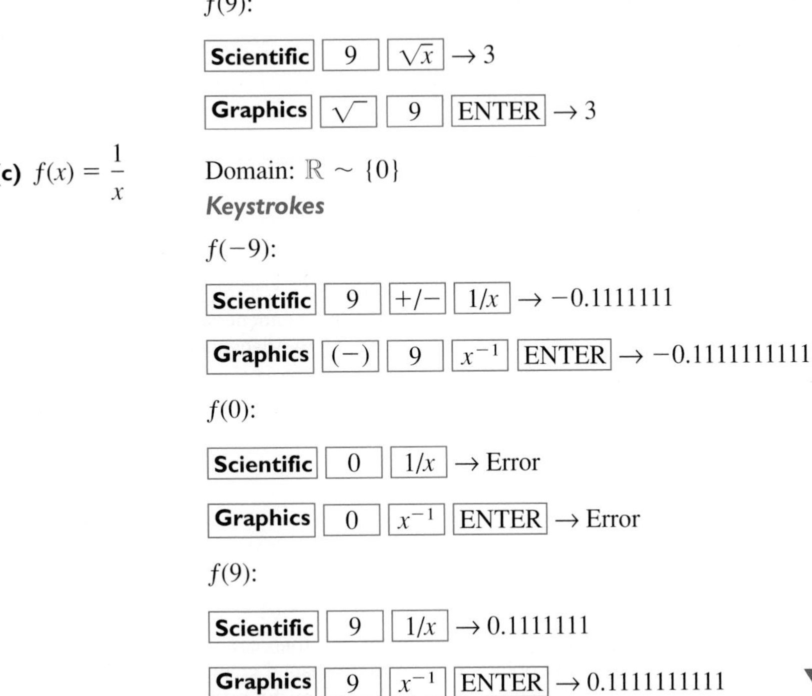

$f(9)$:

Scientific [9] [$\sqrt{x}$] $\to 3$

Graphics [$\sqrt{\ }$] [9] [ENTER] $\to 3$

(c) $f(x) = \dfrac{1}{x}$ Domain: $\mathbb{R} \sim \{0\}$

Keystrokes

$f(-9)$:

Scientific [9] [+/−] [$1/x$] $\to -0.1111111$

Graphics [(−)] [9] [x^{-1}] [ENTER] $\to -0.1111111111$

$f(0)$:

Scientific [0] [$1/x$] $\to$ Error

Graphics [0] [x^{-1}] [ENTER] $\to$ Error

$f(9)$:

Scientific [9] [$1/x$] $\to 0.1111111$

Graphics [9] [x^{-1}] [ENTER] $\to 0.1111111111$ ▼

The domain does not include zero because $x = 0$ would cause division by zero. Thus an error message results when we try to evaluate $f(0)$. The decimal 0.1111111 is approximately equal to $\frac{1}{9}$.

Functions of the form $f(x) = mx + b$ or the alternative form $y = mx + b$ are linear functions, which we know how to graph. We will now discuss graphing other types of functions. If the domain of the function is infinite, then it is obvious that we cannot graph the function by plotting the points one at a time. However, we can still sketch these functions. One procedure is to make a table of representative values and then plot these points. The rest of the graph can be approximated by connecting these points with a line graph. As we plot more points, the line graph will begin to look smoother and will more closely approximate the actual graph. In fact, this is basically the approach used by graphics calculators, which light up closely spaced pixels on the screen to approximate the graph. Line graph approximations of $f(x) = x^2$ are shown in Figure 2-9(a) and (b) for 7 points and for 13 points, respectively.

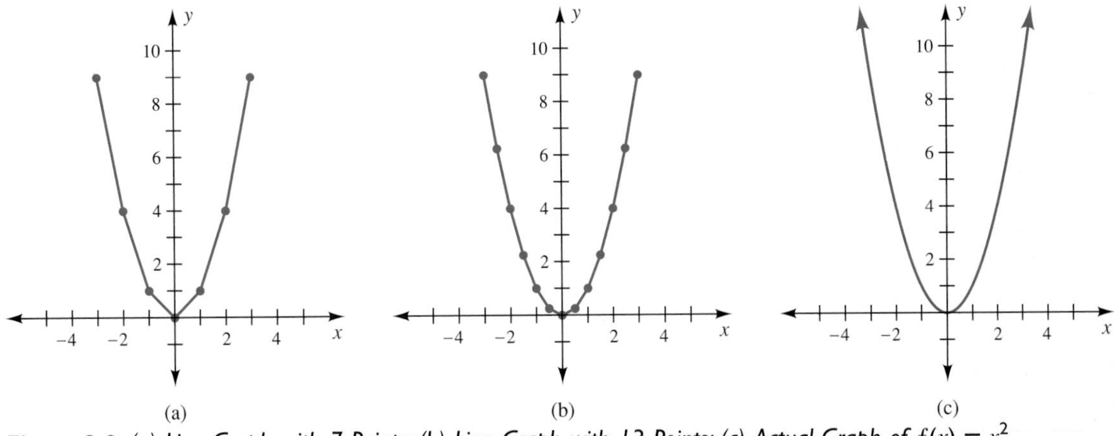

Figure 2-9 *(a) Line Graph with 7 Points; (b) Line Graph with 13 Points; (c) Actual Graph of* $f(x) = x^2$

As you progress in your study of algebra, you will develop an ability to recognize the shape of many functions from their defining equation. Then you can plot only a few representative points and quickly fill in the appropriate shape. The next example illustrates the characteristic V-shape of an absolute-value function.

EXAMPLE 6 Graphing an Absolute-Value Function

Graph the function defined by $f(x) = |x|$.

SOLUTION

$f(x) = |x|$
$f(-3) = |-3| = 3$
$f(-2) = |-2| = 2$
$f(-1) = |-1| = 1$
$f(0) = |0| = 0$
$f(1) = |1| = 1$
$f(2) = |2| = 2$
$f(3) = |3| = 3$

x	$y = f(x)$
−3	3
−2	2
−1	1
0	0
1	1
2	2
3	3

Prepare a table of values by arbitrarily selecting values for the independent variable x and then calculating the dependent values $f(x)$.

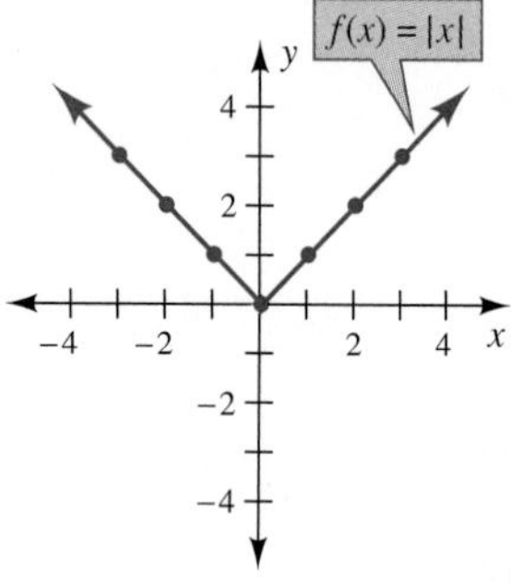

Plot these coordinates and then connect the points to form the V-shaped graph shown to the left. Note from the graph that the domain of $f(x) = |x|$ is the set of all real numbers and the range is the set of all nonnegative real numbers.

Graphics calculators can also be used to graph absolute-value functions. Note that the characteristic V-shape of $y = |x|$ has been shifted down 2 units in the graph in Example 7.

EXAMPLE 7 Calculator Usage

Graph $y = |x| - 2$.

SOLUTION

Keystrokes

[Y=] [2nd] [ABS] [X, T, θ] [−] [2] [GRAPH]

Assume that the window of values has already been set to [−5, 5] for x and [−3, 4] for y.

[−5, 5] for x, [−3, 4] for y

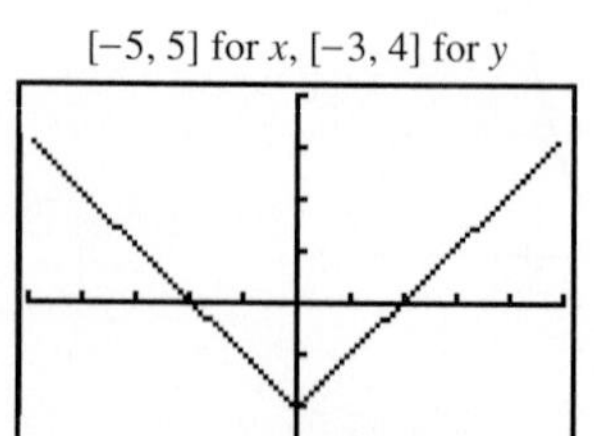

People use words as well as formulas to describe mathematical functions. In particular, the language of variation is often used to describe mathematical functions and relations. The three most common types of variation are given in the following box.

VARIATION

If x, y, and z are variables and k is a nonzero constant, then

Direct Variation: $y = kx$ means that "y varies directly as x" or that "y is directly proportional to x."

Inverse Variation: $y = \frac{k}{x}$ means that "y varies inversely as x" or that "y is inversely proportional to x."

Joint Variation: $z = kxy$ means that "z varies directly as the product of x and y" or that "z varies jointly with x and y."

The constant k is called the **constant of variation.**

EXAMPLE 8 Translating Statements of Variation into Algebraic Equations

Write the equation of variation for each statement of variation.

	SOLUTIONS
(a) z varies directly as x and inversely as y.	$z = \frac{kx}{y}$
(b) V varies jointly as h and A.	$V = khA$

▼

EXAMPLE 9 Hooke's Law

Hooke's Law states that the distance d a spring stretches is directly proportional to the mass m attached to the bottom of the spring. In algebraic form this statement is written as $d = km$. In an experiment with a 5-kilogram mass, a spring stretches 6 centimeters. If a second experiment is run with an 8-kilogram mass, how many centimeters should the spring stretch?

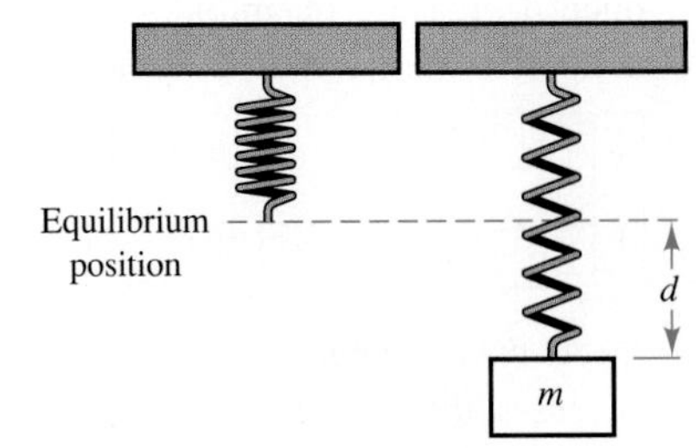

SOLUTION

$d = km$	Hooke's Law, with d for distance and m for mass
$6 = k(5)$	Substitute the data from the first experiment into the equation.
$k = 1.2$	Calculate the constant of variation.
$d = 1.2m$	Substitute this value of k into Hooke's Law.
$d = (1.2)(8)$	Substitute the mass from the second experiment.
$d = 9.6$	

Answer The spring should stretch 9.6 cm.

▼

EXAMPLE 10 Solving for the Constant of Variation

If y is inversely proportional to x, transpose the equation of variation to solve for the constant of variation k in terms of x and y and then determine k if $y = 12$ when $x = 4$.

SOLUTION y is inversely proportional to x.

$y = \frac{k}{x}$ Translate the word equation into an algebraic equation.

$xy = x\left(\frac{k}{x}\right)$ Multiply both sides of the equation by x.

$xy = k$ Then solve for k.

$k = xy$

$k = (4)(12)$ Substitute the given values for x and y.

$k = 48$

When y varies directly as x, the constant of variation will determine whether y increases or decreases as x increases. If the constant of variation is positive, as in $y = 2x$, then y increases as x increases. If the constant of variation is negative, as in $y = -2x$, then y decreases as x increases. However, in both of these cases increasing magnitudes of x yield increasing magnitudes of y. This relationship is discussed further in the geometric viewpoint below.

SELF-CHECK 

1 The pressure exerted on an object submerged in an experimental fluid is directly proportional to the depth at which the object is submerged. The pressure on the object at 4 meters is 9 kilograms per square centimeter. What pressure is exerted at 10 meters?

2 The variable a varies jointly as b and c and inversely as d. Write the equation of variation, and then transpose this equation to solve for the constant of variation k.

A Geometric Viewpoint: Direct and Inverse Variation

Variables that are in direct and inverse variation have quite distinctive behaviors that can readily be observed from a table of values for the variables. For $y = kx$, a statement of direct variation, increasing magnitudes of the x variable produce increasing magnitudes of the y variable. By contrast, for $y = \frac{k}{x}$, a statement of inverse variation, increasing magnitudes of the x variable produce decreasing magnitudes of the y variable. These behaviors are illustrated in the following tables.

x	1	2	3	4	6	12	24
y	2	4	6	8	12	24	48

(a) As the magnitude of x increases, the magnitude of y increases. In this case, $y = 2x$.

x	1	2	3	4	6	12	24
y	24	12	8	6	4	2	1

(b) As the magnitude of x increases, the magnitude of y decreases. In this case, $y = \frac{24}{x}$.

The distinctive behaviors of direct and inverse variation can also be readily observed from the graphs of the ordered pairs (x, y) on a rectangular coordinate system. This is illustrated by the graphs of $y = 1 \cdot x$ and $y = \frac{1}{x}$ shown below.

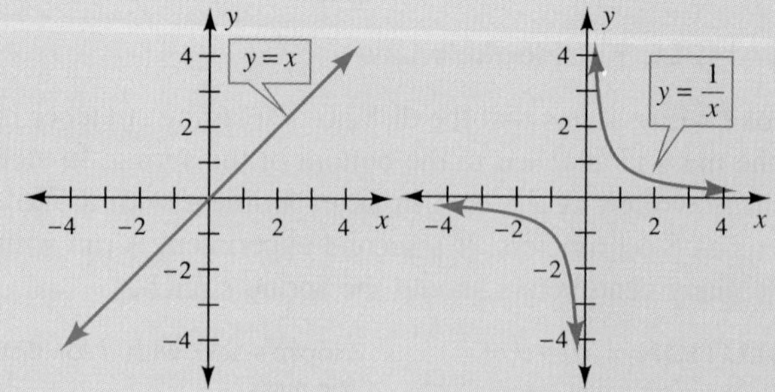

(a) As the magnitude of x increases, the magnitude of y increases.

(b) As the magnitude of x increases, the magnitude of y decreases.

SELF-CHECK ANSWERS 

1 22.5 kg/cm^2 2 $a = \frac{kbc}{d}$; $k = \frac{ad}{bc}$

Exercises 2-5

A

In Exercises 1 and 2 evaluate each expression, given $f(x) = 2x - 4$ and $g(x) = 3x + 5$.

1 a. $f(0)$ **b.** $f(-1)$ **c.** $f(10)$ **d.** $f(-10)$

2 a. $g(0)$ **b.** $g(-1)$ **c.** $g(10)$ **d.** $g(-10)$

In Exercises 3 and 4 evaluate each expression, given $f(x) = 2x^2 - 5x + 3$ and $g(x) = 3x^2 + 2x - 5$.

3 a. $f(0)$ **b.** $f(1)$ **c.** $f(2)$ **d.** $f(-2)$

4 a. $g(0)$ **b.** $g(1)$ **c.** $g(2)$ **d.** $g(-2)$

In Exercises 5 and 6 evaluate each expression, given $f(x) = \dfrac{x - 2}{x + 3}$ and $g(x) = \sqrt{2x - 7}$. If the given argument is not in the domain of the function, state that the function is undefined for this argument.

5 a. $f(-5)$ **b.** $f(2)$ **c.** $f(-3)$ **d.** $-f(1)$

6 a. $g(8)$ **b.** $g(4)$ **c.** $g\left(\dfrac{7}{2}\right)$ **d.** $-g(16)$

In Exercises 7 and 8 evaluate each expression, given $f(x) = 2x^2 - 3x + 1$.

7 a. $f(3)$ **b.** $f(2)$ **c.** $f(3) + f(2)$ **d.** $f(3 + 2)$

8 a. $f(6)$ **b.** $f(4)$ **c.** $f(6) - f(4)$ **d.** $f(6 - 4)$

In Exercises 9 and 10 evaluate each expression for $f(x) = 3x + 2$ and $g(x) = 2x - 3$.

9 a. $f(h)$ **b.** $f(h) - f(1)$ **c.** $f(h - 1)$

10 a. $g(h)$ **b.** $g(h) - g(1)$ **c.** $g(h - 1)$

In Exercises 11–24 determine the domain of each function.

11 $f(x) = 2x + 7$

12 $g(x) = 7x - 2$

13 $h(x) = 9x^2 - 3x + 1$

14 $f(x) = -x^2 + 9x$

15 $g(x) = |x + 8|$

16 $h(x) = |9 - 5x|$

17 $f(x) = 9$

18 $g(x) = -7$

19 $h(x) = \dfrac{5x - 4}{x + 2}$

20 $f(x) = \dfrac{x - 5}{2x + 6}$

21 $g(x) = \dfrac{(x + 3)(x - 3)}{(x - 2)(x + 2)}$

22 $h(x) = \dfrac{(2x + 1)(2x - 1)}{(x - 4)(x + 3)}$

23 $f(x) = \sqrt{x - 1}$

24 $g(x) = \sqrt{x + 3}$

In Exercises 25–34 write an equation for each statement of variation, using k as the constant of variation.

25 a varies directly as b.

26 m varies directly as n.

27 w varies inversely as z.

28 v varies inversely as w.

29 a varies jointly with b and c.

30 y varies jointly with w and x.

31 Pressure in a Balloon The pressure P inside a balloon is inversely proportional to the volume V of gas.

32 Distance by Car The distance d that a car travels varies jointly with its rate of speed r and the time t that it travels.

33 Volume of a Gas The volume V of a gas varies directly as the absolute temperature T and inversely as the pressure P.

34 Electricity from a Windmill The number of kilowatts W of electricity that can be produced by a windmill varies directly as the cube of the speed of the wind s.

In Exercises 35–38 write a statement of variation for each equation.

35 $y = \dfrac{k}{vw}$

36 $y = \dfrac{kvw}{x}$

37 $a = \dfrac{kbc}{mn}$

38 $a = \dfrac{kbcm}{n}$

In Exercises 39–44 find the constant of variation for each statement of variation.

39 y varies directly as x and $y = 12$ when $x = 8$.

40 y varies inversely as x and $y = 4$ when $x = 3$.

41 y varies inversely as x and $y = \dfrac{3}{2}$ when $x = 4$.

42 y varies jointly with w and x and $y = 30$ when $w = 2$ and $x = 5$.

43 y varies jointly with w and x and $y = 48$ when $w = \dfrac{1}{3}$ and $x = 4$.

44 y varies directly as x and inversely as w and $y = 50$ when $w = 20$ and $x = 25$.

In Exercises 45–48 use the given statement of variation to solve each problem.

45 y varies inversely as x and $y = 25$ when $x = 2$. Find y when $x = 10$.

46 a varies directly as b and $a = 12$ when $b = \dfrac{2}{3}$. Find a when $b = \dfrac{3}{2}$.

47 a varies directly as b and inversely as c and $a = 3$ when $b = 9$ and $c = 12$. Find a when $b = 15$ and $c = 6$.

48 a varies directly as b and inversely as c and $a = 6$ when $b = 2$ and $c = 24$. Find a when $b = 5$ and $c = 60$.

B

CALCULATOR USAGE (49–56)

In Exercises 49–52 use a calculator to approximate each expression to four significant digits. Let $f(x) = \sqrt{x + 1}$ and $g(x) = \dfrac{1}{x - 2}$. If you get an error message, explain why.

49 a. $f(6)$ **b.** $g(20)$

50 a. $g(\pi)$ **b.** $f(\pi)$

51 a. $f(-6)$ **b.** $g(-6)$

52 a. $f(2)$ **b.** $g(2)$

In Exercises 53–56 use a graphics calculator to graph all three functions on the same set of axes. How does the graph of $y = f(x) + c$ compare to the graph of $y = f(x)$?

53 a. $y = |x|$ **b.** $y = |x| + 2$ **c.** $y = |x| - 1$

54 a. $y = |x|$ **b.** $y = |x| - 3$ **c.** $y = |x| + 1$

55 a. $y = |x|$ **b.** $y = |x - 2|$ **c.** $y = |x + 1|$

56 a. $y = |x|$ **b.** $y = |x + 3|$ **c.** $y = |x - 1|$

In Exercises 57–60 use the given function to complete the tables of values in parts a and b. Then sketch a line graph approximation of the function based on each table.

57 $f(x) = -x^2$

58 $f(x) = -x^2 + 2$

59 $f(x) = x^2 - 2$

60 $f(x) = x^2 - 4$

Part a

x	y
−3	
−2	
−1	
0	
1	
2	
3	

Part b

x	y
−3	
−2.5	
−2	
−1.5	
−1	
−0.5	
0	
0.5	
1	
1.5	
2	
2.5	
3	

In Exercises 61–64 use the given statements of variation to solve each problem.

61 Hooke's Law Hooke's Law states that the distance a spring stretches is directly proportional to the mass attached to the bottom of the spring. If a 4-kilogram mass stretches a spring 3 centimeters, what mass would be required to stretch the spring 5 centimeters? (See the figure.)

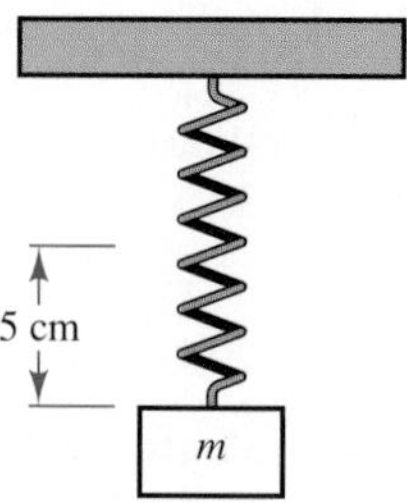

Figure for Exercise 61

62 Pressure The pressure exerted on a submerged object is directly proportional to the depth the object is submerged. If the pressure at 60 centimeters is 0.06 kilogram per square centimeter, find the pressure at 80 centimeters.

63 Electrical Current The current in an electrical circuit varies inversely as the resistance. When the resistance is 4 ohms, the current is 24 amps. Determine the current when the resistance is 6 ohms.

64 Paint Coverage The area covered by a paint varies inversely with the thickness with which it is applied. When the paint is applied 0.5 millimeter thick, a container of paint will cover 420 square meters. How many square meters will this container cover if the paint is applied 0.75 millimeter thick?

C

In Exercises 65–68 determine without graphing whether these points all satisfy a linear function. (*Hint:* Linear functions are characterized by a constant rate of change.)

65

x	1	2	3	4	5
y	−1	1	3	5	7

66

x	1	2	3	4	5
y	5	8	11	14	17

67

x	1	2	3	4	5
y	0	3	8	15	24

68

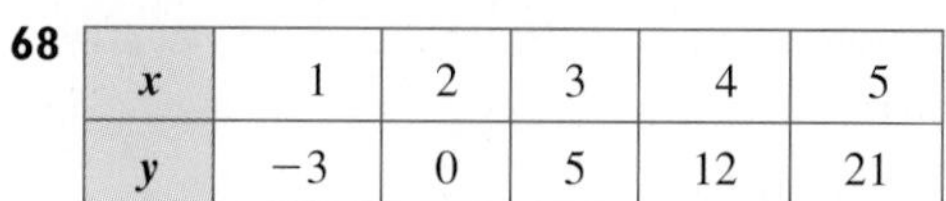

x	1	2	3	4	5
y	−3	0	5	12	21

69 Match each of these statements with the correct equation.

a. y varies directly as x, and y increases when x increases.

b. y varies directly as x, and y decreases when x increases.

c. y varies inversely as x, and y decreases when x increases.

d. y varies inversely as x, and y increases when x increases.

i. $y = \frac{6}{x}$ **ii.** $y = \frac{1}{2}x$

iii. $y = -\frac{4}{x}$ **iv.** $y = -6x$

In Exercises 70–74 each graph is either a graph of $y = kx$ (an equation of direct variation) or a graph of $y = \frac{k}{x}$ (an equation of inverse variation). Determine whether each graph illustrates direct or inverse variation, and then use selected points to determine the constant of variation k.

70

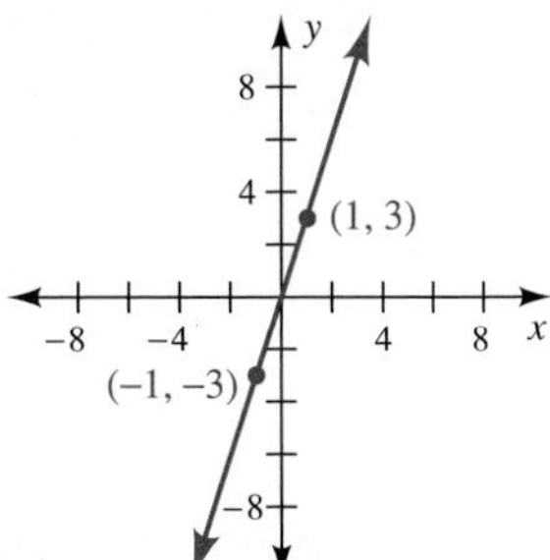

71

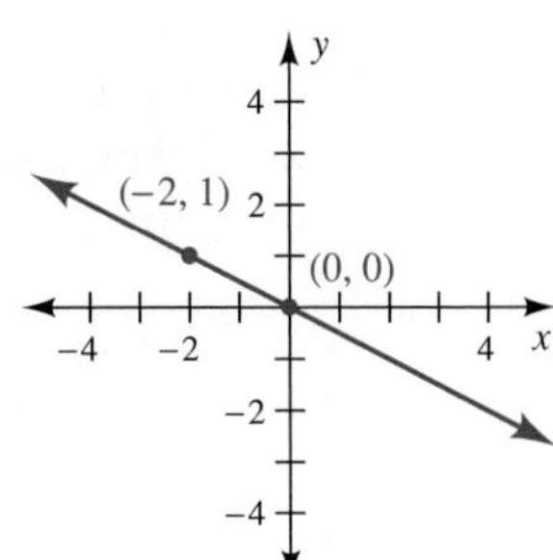

72

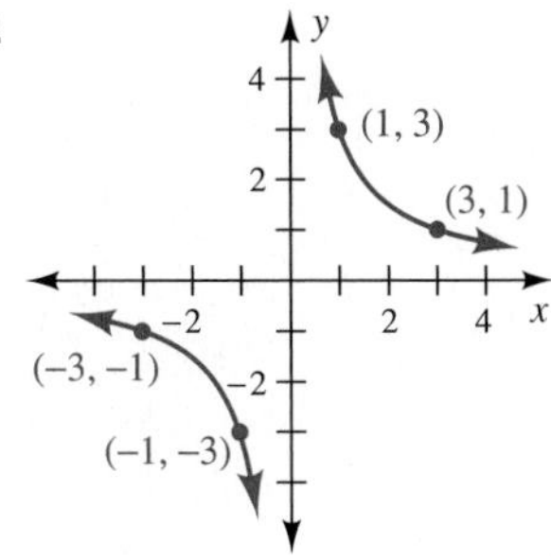

73

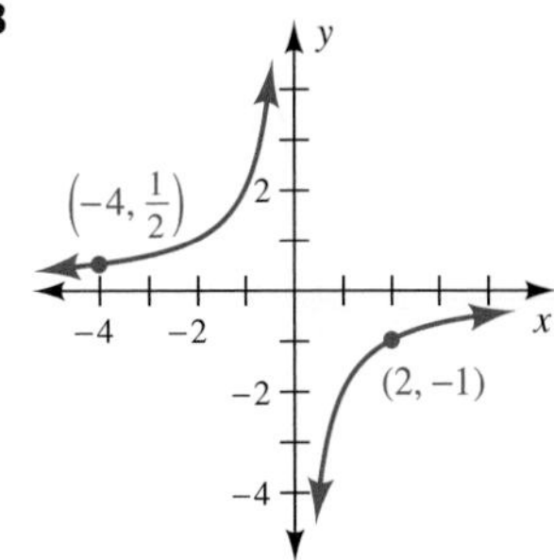

74

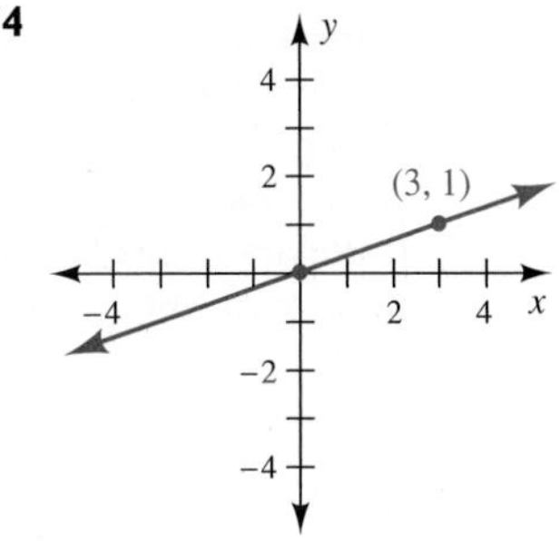

In Exercises 75 and 76 each graph illustrates an equation of direct variation of the form $y = kx^2$. Use selected points to determine the constant of variation k.

75

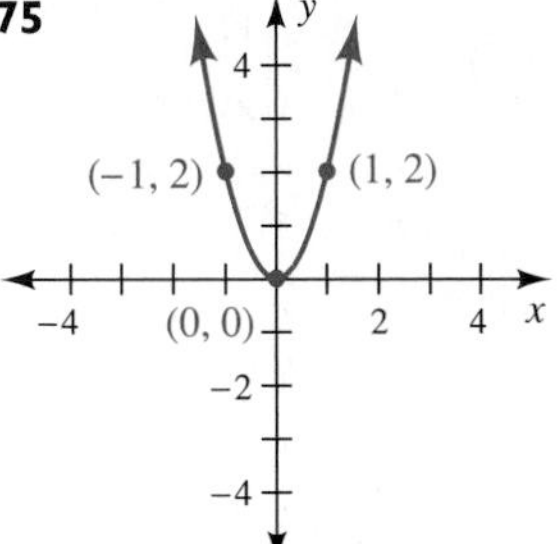

76

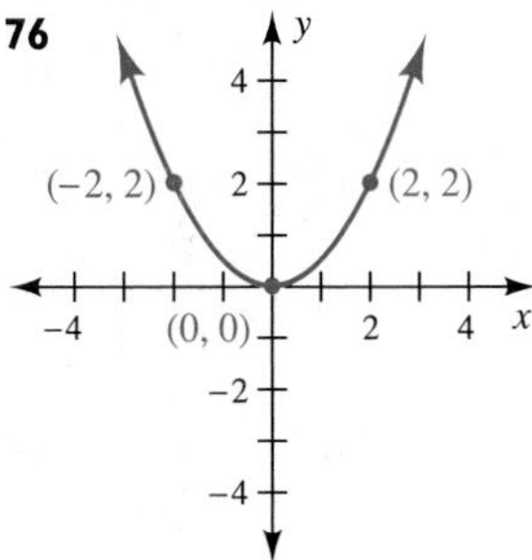

In Exercises 77–80 each table of values illustrates variation either of the form $y = kx$ or of the form $y = \frac{k}{x}$. Determine whether the variation illustrated is direct variation or inverse variation. Also determine the constant of variation k.

77

x	2	4	6	8	10	12
y	6	12	18	24	30	36

78

x	2	4	6	8	10	12
y	1	2	3	4	5	6

79

x	2	4	6	8	10	12
y	60	30	20	15	12	10

80

x	2	4	6	8	10	12
y	9	4.5	3	2.25	1.8	1.5

Profit Function (81–83)

In Exercises 81–83 use the profit function given by the equation $P(u) = -u^2 + 60u + 700$ to evaluate each expression. The variable u, representing the number of units produced, is the independent variable, and the dependent variable P is the profit in dollars of producing this number of units. This formula is valid only from 0 to 60 units of production.

81 $P(0)$ **82** $P(30)$ **83** $P(55)$

In Exercises 84 and 85 select the function that is represented by the line graph approximation.

84

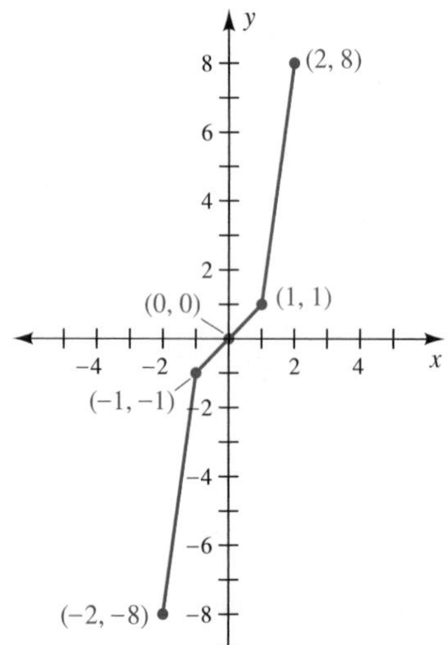

a. $f(x) = -x^2$
b. $f(x) = x^2$
c. $f(x) = -x^3$
d. $f(x) = x^3$
e. $f(x) = x^4$

85

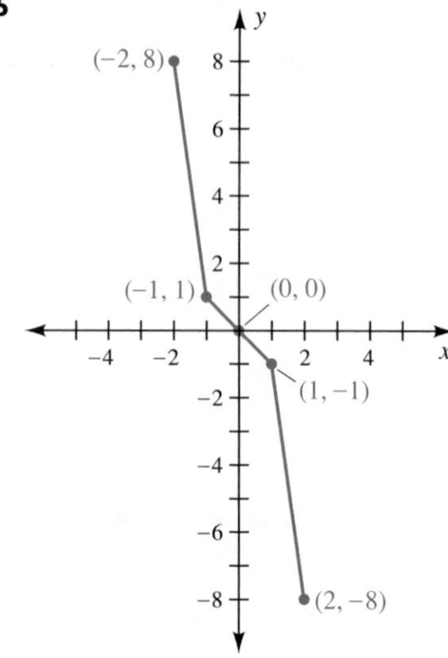

a. $f(x) = -x^2$
b. $f(x) = x^2$
c. $f(x) = -x^3$
d. $f(x) = x^3$
e. $f(x) = x^4$

Discussion Question

86 A politician claims that the tax revenue for one county is directly proportional to the tax rate. An opponent claims that the tax revenue is inversely proportional to the tax rate. Discuss what will happen to tax revenues if the tax rate is increased. Talk about both claims.

Challenge Question

Determine a function $y = f(x)$ that is satisfied by each of the points in this table of values. (*Hint:* If you have trouble doing this using only algebra, use a line graph to examine the geometric pattern.)

x	−2	−1	0	1	2
y	−4	−1	2	5	8

Section Skill Reinforcers

Determine the slope and the y-intercept of each of these linear functions.

1 $f(x) = 2x - 3$ **2** $f(x) = 4x + 2$ **3** $f(x) = \frac{2}{3}x + 6$ **4** $f(x) = -\frac{4}{3}x + 12$ **5** $f(x) = 8$

Key Concepts for Chapter 2

1 All points in the same quadrant of the rectangular coordinate system have the same sign pattern (see the figure).

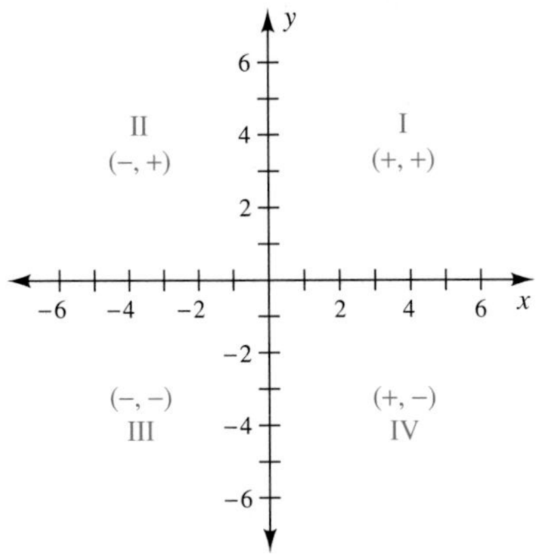

Figure for Key Concept 1

2 Forms of linear equations:

a. General form: $Ax + By + C = 0$

b. Horizontal line: $y = k$ for a real constant k

c. Vertical line: $x = h$ for a real constant h

d. Point-slope form: $y - y_1 = m(x - x_1)$

e. Slope-intercept form: $y = mx + b$

f. Oblique line through the origin: $Ax + By = 0$

3 The x-intercept of a line is the point where the line crosses the x-axis. The y-intercept of a line is the point where the line crosses the y-axis. To find the x-intercept, set $y = 0$. To find the y-intercept, set $x = 0$.

4 A line graph (or a broken line graph) is a graph formed by using line segments to connect points with adjacent x-coordinates.

5 Linear interpolation uses known points on a line to approximate an unknown point between these points. Linear extrapolation uses known points on a line to approximate an unknown point beyond the range of all the given points.

6 Slope:

a. The slope m of a line through points (x_1, y_1) and (x_2, y_2) for $x_1 \neq x_2$ is $m = \dfrac{y_2 - y_1}{x_2 - x_1}$.

b. The slope of $y = mx + b$ is m.

c. The slope of a horizontal line is zero.

d. The slope of a vertical line is undefined.

e. Parallel lines have the same slope.

f. Perpendicular lines have slopes whose product is -1.

g. A line with a positive slope slopes upward to the right.

h. A line with a negative slope slopes downward to the right.

7 A relation is a correspondence from a domain set D to a range set R that pairs each element in D with one or more elements in R. Thus a relation is a set of ordered pairs.

8 A function is a relation that pairs each element in a domain D with exactly one element in R. Thus a function is a set of ordered pairs with no pairs that have the same first coordinate and different second coordinates.

9 Methods of denoting functions:

a. Mapping notation

b. Ordered-pair notation

c. Tables of values

d. Graphs

e. Functional notation

10 If a function is graphed, the domain of the function is the projection of the graph onto the x-axis. The range of the function is the projection of the graph onto the y-axis.

11 The vertical line test can be used to determine whether a graph represents a function. The horizontal line test (Chapter 10) can be used to determine whether the graph of a function represents a one-to-one function.

12 Functional notation: $f(x)$ is read as "f of x" or "the value of f at x." Functions are frequently denoted by the letter f, g, or h. In algebra the implied domain is usually understood to be the set of all possible real numbers so that the range also consists of real numbers. In particular, this means that the domain *excludes* values that cause division by zero or cause negative values under a square root symbol.

13 Variation: If x, y, and z are variables and k is a constant,

a. Direct variation: $y = kx$ means that y varies directly as x.

b. Inverse variation: $y = \dfrac{k}{x}$ means that y varies inversely as x.

c. Joint variation: $z = kxy$ means that z varies jointly as x and y.

Review Exercises for Chapter 2

In Exercises 1–4 graph the line through the given points, and then calculate the slope of the line through these points.

1 $(-2, -5)$ and $(2, -2)$

2 $(-7, 11)$ and $(5, 11)$

3 $(4, -3)$ and $(-4, -3)$

4 $(-4, 8)$ and $(1, -4)$

In Exercises 5–10 graph each linear equation.

5 $3x - 5y - 15 = 0$

6 $x = -4$

7 $y = 3$

8 $y = -\frac{5}{3}x + 4$

9 $y + 3 = \frac{3}{4}(x + 5)$

10 $x = -y$

11 Draw a broken line graph through this set of points: $\{(1950, 40), (1960, 48), (1970, 40), (1980, 37), (1990, 37)\}$. These points give the average price in dollars for shares listed on the New York Stock Exchange for selected years. Use this graph to estimate the average share price in 1965.

In Exercises 12 and 13 determine whether the line through the first pair of points is parallel to the line through the second pair of points, perpendicular to this line, or neither parallel nor perpendicular to this line.

12 $(-2, 2)$ and $(2, 3)$; $(-2, -2)$ and $(6, 0)$

13 $(-3, 2)$ and $(1, 1)$; $(-5, 1)$ and $(2, -1)$

In Exercises 14 and 15 determine whether the lines given by the equations are parallel, perpendicular, or neither parallel nor perpendicular.

14 $y = \frac{1}{2}x + 5$, $y = -2x + 1$

15 $2x + 6y = 10$, $3x + 9y = -10$

16 Determine whether the vertices $(-4, 3)$, $(0, 5)$, and $(4, 3)$ form a right triangle.

In Exercises 17–27 write the general form of the line satisfying the given conditions.

17 Through $(-2, 3)$ with slope $-\frac{2}{3}$

18 Through $(3, -2)$ and $(4, -5)$

19 $m = -\frac{4}{5}$, $b = 2$

20 $m = -\frac{5}{2}$, $a = 2$

21 Through $(0, 0)$ and parallel to $y = 5x - 6$

22 Through $(0, 0)$ and perpendicular to $y = 6x + 5$

23 A vertical line through $(-5, 4)$

24 A horizontal line through $(-5, 4)$

25 A line parallel to the x-axis through $(1, 7)$

26 A line perpendicular to the x-axis through $(1, 7)$

27 Perpendicular to the line through $(1, 6)$ and $(5, 10)$ and passing through $(2, 3)$

In Exercises 28 and 29 write the equation of each line in general form.

28

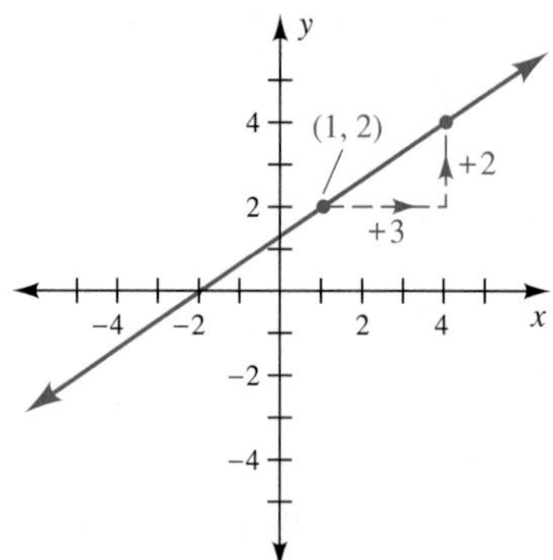

29

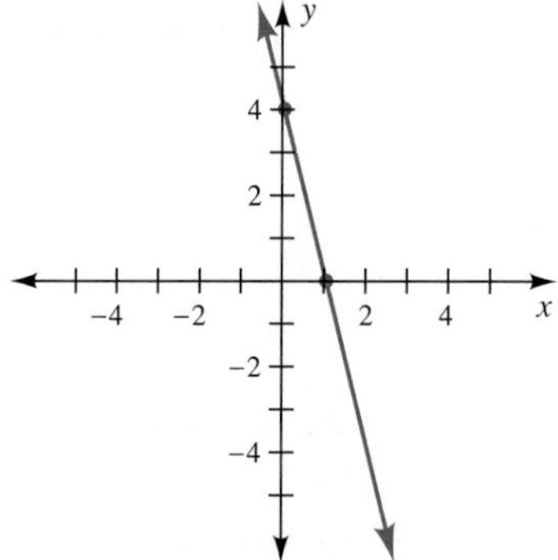

30 Determine the slope of each line segment in the line graph shown: $\overline{AB}$, $\overline{BC}$, $\overline{CD}$, and $\overline{DE}$. On which interval is this function increasing?

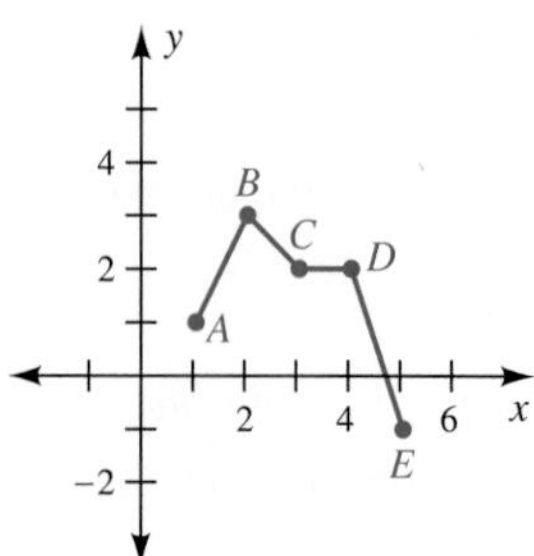

Figure for Exercise 30

In Exercises 31–42 classify each relation as either a function or a relation that is not a function. State the domain and the range of each relation.

31 $D \quad R$
$9 \to 1$
$8 \to 2$
$3 \to 5$
$6 \to 4$

32 $D \quad R$
$9 \to 1$
$8 \nearrow$
$3 \nearrow$

33 $D \quad R$
$1 \to 9$
$\searrow 8$
$6 \to 4$
$7 \to 5$

34

x	y
-5	25
$-\pi$	19
0	16
π	18
11	13

35

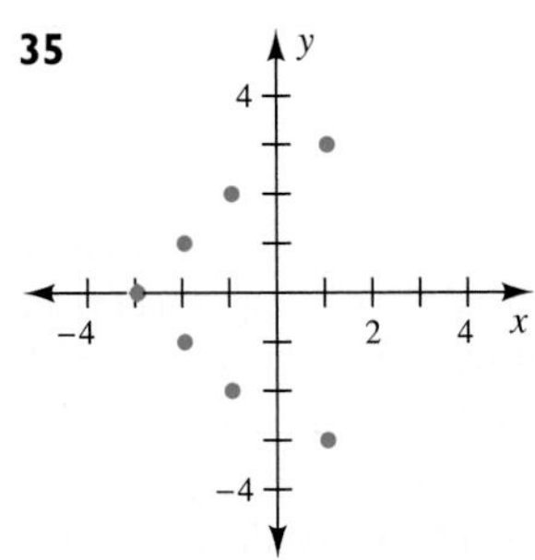

36

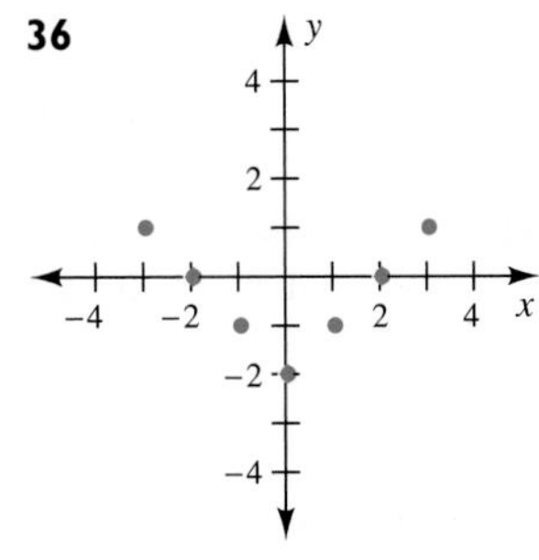

37 $f = \{(1, 2), (2, 1), (3, 7), (4, 7)\}$

38 $\{(5, 3), (4, 8), (7, 2), (4, -7)\}$

39

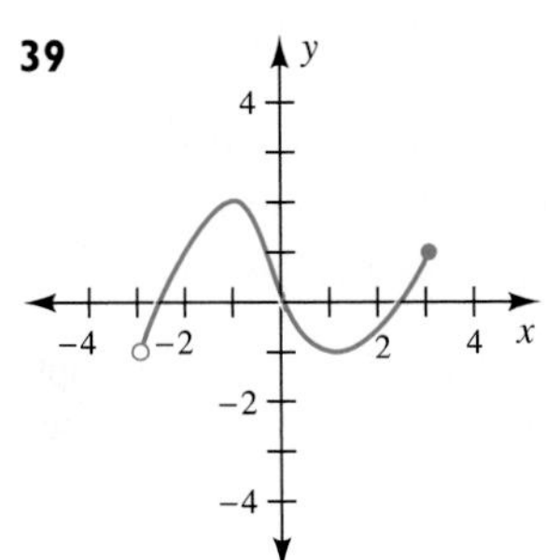

40

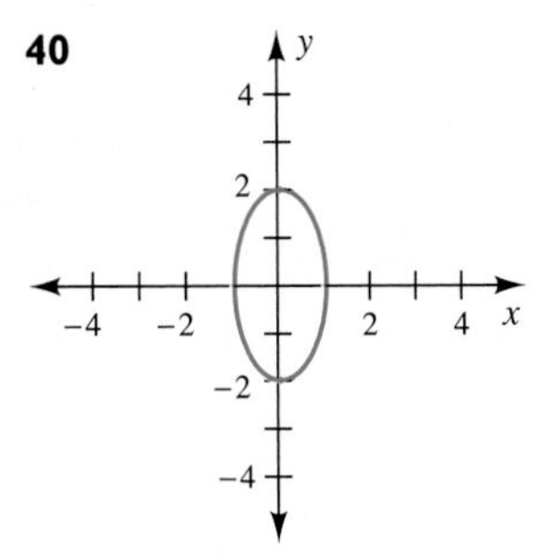

41

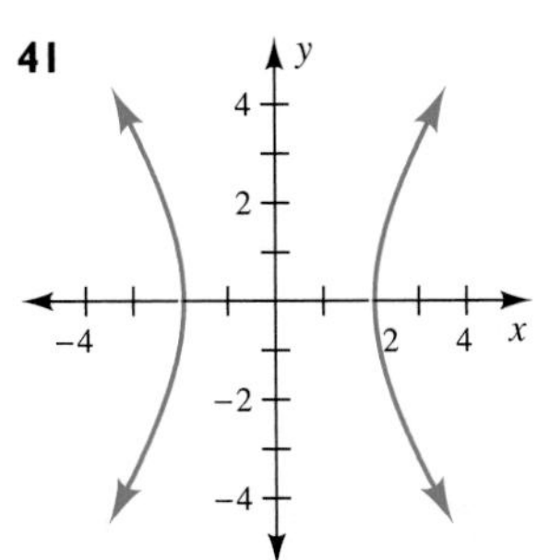

42

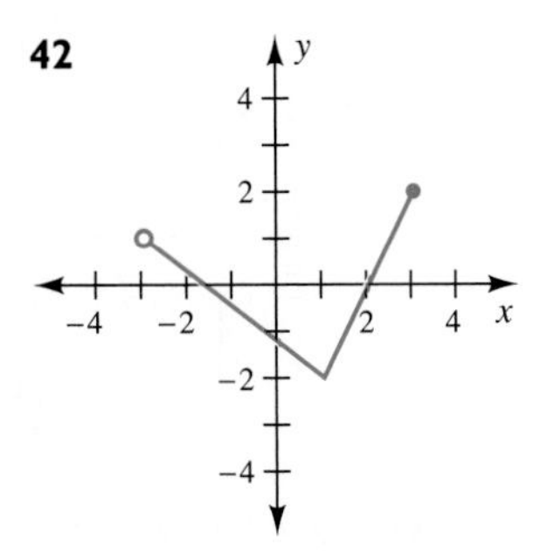

43 A function consists of the ordered pairs (x, y), where each x is from the domain $\{-3, -2, -1, 0, 1, 2, 3\}$ and $y = |x|$. Write this function using ordered-pair notation.

44 A function consists of the ordered pairs (x, y), where each x is from the domain $\{-4, -2, 0, 1, 3, 5\}$ and $y = -x^2 + 10$. Write this function using ordered-pair notation.

45 Acreage Burned by Fires Write the function defined by the histogram shown in ordered-pair notation. This histogram gives the acreage in the United States burned by forest and range fires for selected years.

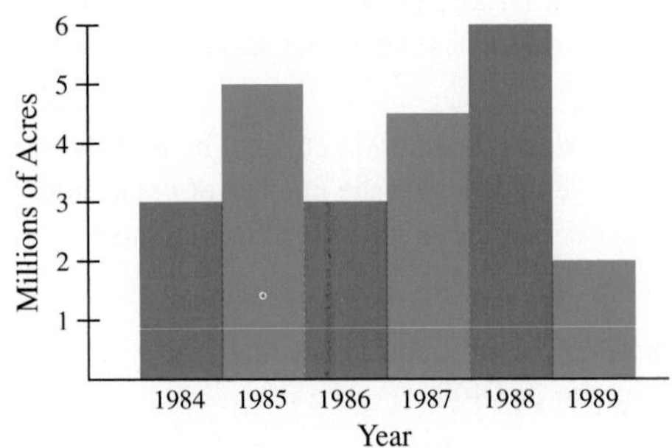

Figure for Exercise 45

46 Graduation Rates Use the given line graph to complete this table of values. The graph gives for selected years the percent of new U.S. college graduates who are female.

x	1940	1950	1960	1970	1980	1990
y						

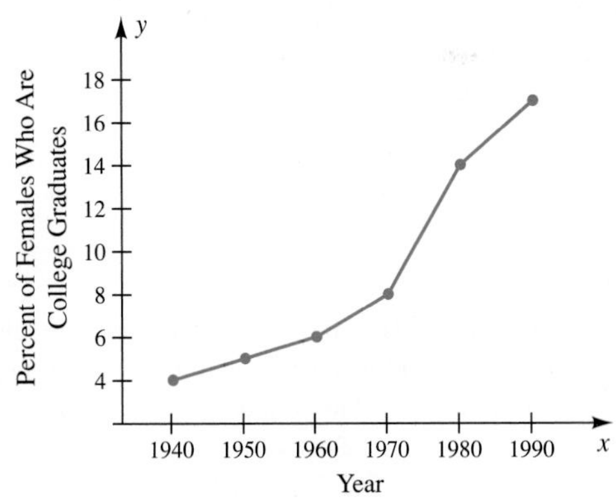

Figure for Exercise 46

In Exercises 47–50 evaluate each expression for $f(x) = 3x - 2$.

47 $f(0)$ **48** $f(5)$ **49** $f(-7)$ **50** $-f(7)$

In Exercises 51–56 evaluate each expression for $f(x) = 2x^2 - x + 3$.

51 $f(0)$ **52** $f(-2)$ **53** $-f(5)$

54 $f(h)$ **55** $f(h) - f(2)$ **56** $f(h - 2)$

57 Recipe Proportions The amount of sugar used in a recipe varies directly as the number of people for whom the recipe is prepared. Three cups of sugar are needed for 50 people. How many cups of sugar are needed for 75 people?

58 The variable R varies directly as L and inversely as r. Write this statement of a variation as an equation.

59 Gas Consumption The number of liters of gasoline used by a car traveling at a fixed speed varies directly as the distance traveled and inversely as the square of the speed. If a car uses 20 liters of gas during a 256-kilometer trip made at 64 kilometers per hour, how many liters will be used during a 150-kilometer trip made at 80 kilometers per hour?

60 Tuition Fees The tuition charged by a Midwest community college depends on the number of credit hours the student takes. Use the graph shown to determine

a. the tuition cost for 12 credit hours

b. the number of credit hours that can be taken for a tuition fee of $400

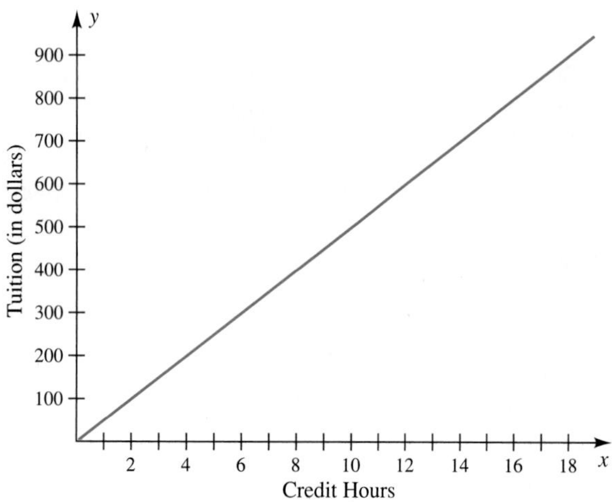

Figure for Exercise 60

61 Use the function $f(x) = x(x - 1)$ to complete the tables of values in parts a and b. Then sketch a line graph approximation of the function based upon each table.

Part a

x	y
-3	
-2	
-1	
0	
1	
2	
3	

Part b

x	y
-3	
-2.5	
-2	
-1.5	
-1	
-0.5	
0	
0.5	
1	
1.5	
2	
2.5	
3	

Calculator Usage

62 Use a graphics calculator to graph the functions defined by y_1, y_2, and y_3 on the same standard viewing window of $[-10, 10]$ for x and $[-10, 10]$ for y: $y_1 = x^2$, $y_2 = x^2 - 2$, and $y_3 = x^2 - 4$. Visually compare these graphs.

63 Trout Farming The function defined by the graph shown gives the total weight of fish at a trout pond over a period of months. Match each time period (1, 2, 3, and 4) to the event that occurred during that time period.

Event A: Harvesting is initiated in this pond.

Event B: A new hatching is added to the pond, and these trout begin to grow.

Event C: A predator migrates to this pond and reduces the growth rate of the fish in this pond.

Event D: Harvesting is discontinued in this pond.

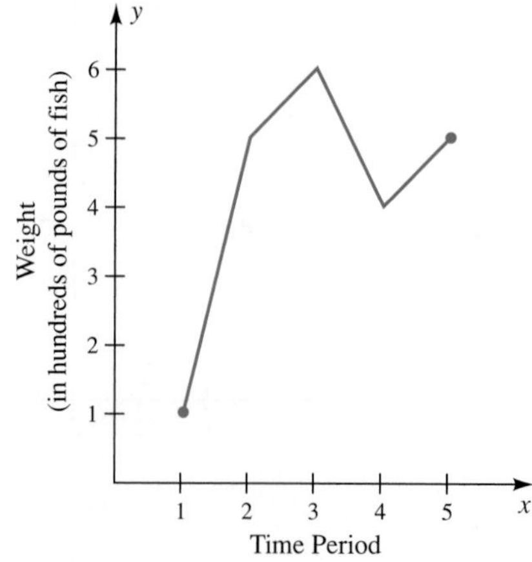

Figure for Exercise 63

In Exercises 64–66 determine the domain of each function.

64 $f(x) = 8x^3 - 9x - 11$

65 $g(x) = |3x + 7|$

66 $h(x) = \dfrac{x - 1}{x + 2}$

67 Grade of a Highway A highway is limited to a 4% slope. What is the maximum change in elevation in feet of a section covering 1 mile (5280 feet)?

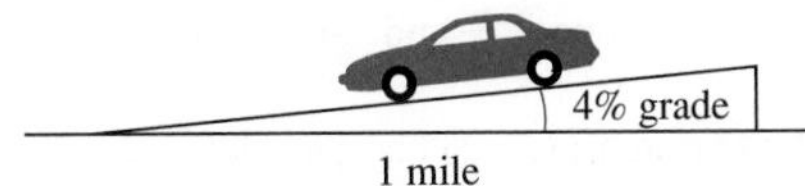

Figure for Exercise 67

68 Coin Sorting A coin-sorting machine shakes coins through a series of curved paths, channeling each denomination of coin into a separate stack. The stacks are labeled so that one can determine how many coins are in each stack by reading the mark next to the top coin in the stack. Thus the number of coins is a function of the height of the stack. Mark off your own label on a sheet of paper that could be glued to a stacking tray for nickels similar to the one shown in the figure.

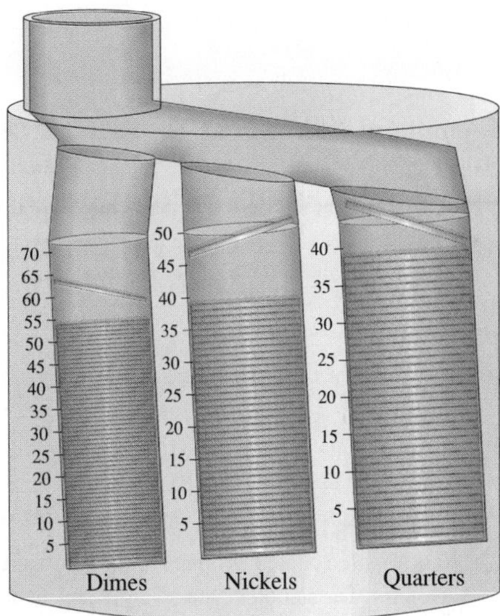

Figure for Exercise 68

Mastery Test for Chapter 2

Exercise numbers correspond to Section Objective numbers.

[2-1] **1** Plot each of these points on a rectangular coordinate system.

a. $(-3, 4)$ **b.** $(-2, -3)$ **c.** $(3, -2)$

d. $(4, 1)$ **e.** $(0, 2)$

[2-1] **2** Graph each of these linear equations.

a. $2x + 5y + 10 = 0$

b. $x = 4$ **c.** $y = -3$

[2-1] **3** Draw a broken line graph through each set of points.

a. The United States production of electricity by petroleum for selected years is given by the following points. The electricity production is given in billions of kilowatt-hours.

$\{(1960, 50), (1965, 65), (1970, 180), (1975, 290), (1980, 250), (1985, 100)\}$

b. The United States production of electricity by natural gas for selected years is given by the following points. The electricity production is given in billions of kilowatt-hours.

$\{(1960, 160), (1965, 220), (1970, 370), (1975, 300), (1980, 350), (1985, 290)\}$

[2-2] **4** Calculate the slope of each line.

a. $y = 7$

b. $x = 7$

c. $y = 2x - 5$

d. The line through $(2, 5)$ and $(8, -7)$

e. The line perpendicular to $2x + 3y + 4 = 0$

[2-2] **5** Use the slope-intercept form to graph each of these lines.

a. $y = 2x - 3$

b. $y = -\frac{3}{5}x + 4$

c. $y = \frac{4}{3}x$

[2-3] **6** Write the equation of a line satisfying the given conditions.

a. A horizontal line through $(7, -2)$

b. A vertical line through $(7, -2)$

c. A vertical line with x-intercept $(5, 0)$

d. A horizontal line with y-intercept $(0, 8)$

[2-3] **7** Write the general form of a line satisfying the given conditions.

a. Through $(4, -2)$ with slope $\frac{3}{7}$

b. Through $(-2, 5)$ and $(4, 8)$

c. With slope $-\frac{2}{3}$ and y-intercept -5

d. Through $(7, -2)$ and parallel to $2x + 3y + 8 = 0$

[2-4] **8** Determine which of these relations are functions.

a.

D *R*
$-3 \rightarrow 9$
$3 \nearrow$
$-4 \rightarrow 16$
$4 \nearrow$

b.

x	y
9	−3
9	3
16	−4
16	4

c.

d.

e.

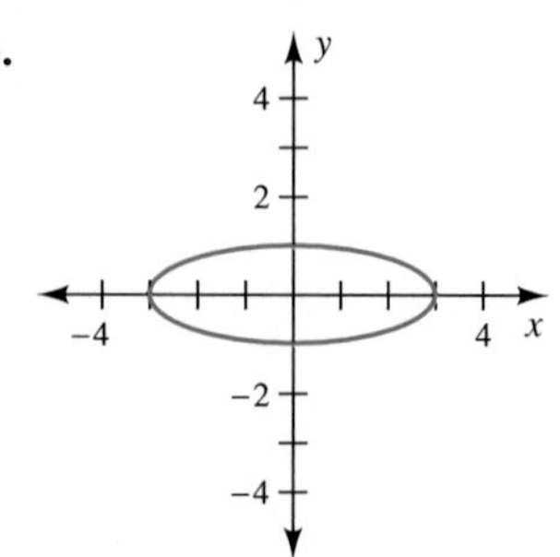

f. $\{(6, 6), (5, 6), (3, 6)\}$

[2-4] **9** Determine the domain and the range of each of these functions.

a. $\{(1, 8), (2, 4), (7, 11), (8, 13)\}$

b. $f(x) = 7$ **c.** $f(x) = \sqrt{x - 3}$

d.

x	y
−3	0
−2	1
−1	2
0	7

e.

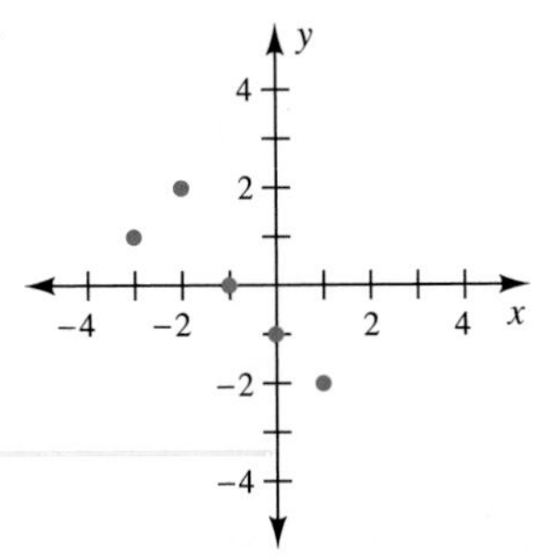

f.

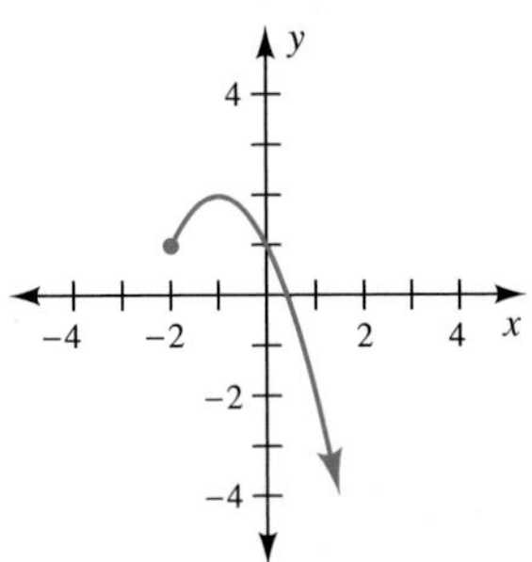

[2-5] **10** Evaluate each of the following expressions for $f(x) = 3x^2 + x - 12$.

a. $f(0)$ **b.** $f(2)$ **c.** $f(-2)$

d. $f(h)$ **e.** $f(h + 2)$ **f.** $f(h) + f(2)$

[2-5] **11** **a.** The variable y varies directly as x. When $x = 10$, $y = 15$. Find y when x is 24.

b. Write an equation for this statement of variation: z varies directly as x squared and inversely as y cubed.

c. Write a statement of variation for the equation $V = kr^2h$.

Active Application

Imagine that you are a computer programmer writing a new ray-tracing program for use in special effects. To make sure the program is running as intended, you want to conduct a trial run.

You first use the program on a movie scene involving a post that is standing in an otherwise empty grass field. You place the eye of the imaginary viewer at (40, 10) and the post with its top at (4, 5) and its bottom at (3, −1). (The post leans.) The movie screen, which is twenty feet tall, runs vertically from (10, 0) to (10, 20). (To keep things simple, we will keep all measurements in feet and ignore the difference in size between the movie screen and the computer screen on which you are working. We will also work with x- and y-coordinates only; a complete treatment of three dimensions would involve z-coordinates as well.) At what point on the screen should the top of the post appear? How about the bottom of the post? Give your answers in terms of x- and y-coordinates.

Chapter 3

Systems of Linear Equations and Inequalities

AN APPLICATION

When the Right Solution Counts

In most hospitals nurses are usually responsible for administering medication. One of the several means of administering medication is via intravenous drip (IV), which enables the medication to go directly into the patient's bloodstream.

One advantage of administering a drug by IV is that the drug is introduced into the patient's body slowly, over a long period of time, lessening the stress on the patient's system. Drugs given by IV are diluted to a far lower concentration than would be used for an injection. Dilution is simply a matter of adding saline solution (which is nothing more than water plus a little salt) to a concentrate of the drug. Drugs are often shipped in concentrated form, allowing custom mixing based on patient needs. A patient with malfunctioning kidneys, for example, may require restricted fluid intake, and that may mean a somewhat more concentrated IV solution than usual.

Mixing an IV solution requires a certain amount of math. In this chapter you will learn how to solve equations like those used in mixing solutions. Try your skills at this problem: A nurse has a solution that is 15% by volume of a certain drug and wants to dilute it so that the mixture is only 5% by volume of the drug. If the nurse starts with 100 milliliters of the concentrated drug, how much saline solution should be used?

CHAPTER 3 OBJECTIVES

1 **Solve a system of linear equations in two variables graphically (Section 3-1).**

2 **Graph the solution of a system of linear inequalities in two variables (Section 3-1).**

3 **Solve a system of linear equations by the substitution method (Section 3-2).**

4 **Solve a system of linear equations by the addition-subtraction method (Section 3-2).**

5 **Solve a system of three linear equations in three variables (Section 3-3).**

6 **Solve word problems using systems of linear equations (Section 3-4).**

A linear equation $Ax + By + C = 0$ is satisfied by an infinite number of ordered pairs (x, y), and thus its graph contains an infinite number of points. When we consider two or more of these linear equations simultaneously, the equations are referred to as a **system of linear equations.** A **solution** of a system of two equations in two variables is an ordered pair of numbers that satisfies both equations simultaneously. The process of finding the solution of a system of equations is referred to as **simultaneously solving the system.**

3-1 *Graphing Systems of Linear Equations and Inequalities*

SECTION OBJECTIVES

1 **Solve a system of linear equations in two variables graphically.**

2 **Graph the solution of a system of linear inequalities in two variables.**

We can determine whether an ordered pair of coordinates is a solution of a system by substituting these coordinates into each equation of the system. To be a solution of the system, this ordered pair must satisfy each equation.

EXAMPLE 1 Determining Whether an Ordered Pair Is a Solution of a System of Linear Equations

Determine whether the given ordered pair is a solution of the system $\begin{Bmatrix} 3x + y = 2 \\ 6x - y = -11 \end{Bmatrix}$.

SOLUTIONS

(a) (0, 11)

$$3x + y = 2 \qquad\qquad 6x - y = -11$$

$$3(0) + 11 \stackrel{?}{=} 2 \qquad\qquad 6(0) - 11 \stackrel{?}{=} -11$$

$$11 = 2 \text{ is false.} \qquad\qquad -11 = -11 \text{ is true.}$$

The ordered pair (0, 11) is not a solution of the system because it does not satisfy both equations.

(b) $(1, -1)$

$$3x + y = 2 \qquad 6x - y = -11$$
$$3(1) + (-1) \stackrel{?}{=} 2 \qquad 6(1) - (-1) \stackrel{?}{=} -11$$
$$2 = 2 \text{ is true.} \qquad 7 = -11 \text{ is false.}$$

The ordered pair $(1, -1)$ is not a solution of the system.

(c) $(-1, 5)$

$$3x + y = 2 \qquad 6x - y = -11$$
$$3(-1) + 5 \stackrel{?}{=} 2 \qquad 6(-1) - 5 \stackrel{?}{=} -11$$
$$2 = 2 \text{ is true.} \qquad -11 = -11 \text{ is true.}$$

The ordered pair $(-1, 5)$ is a solution of the system. ▼

The graphical method can be used to solve many systems of equations with two variables. To solve a system using this method, we graph each equation and then estimate the coordinates of each point of intersection. The graphical method is intuitive, since we can see the point where the graphs intersect. This method is also quite powerful, since we can use it with graphics calculators on systems of equations whose graphs are more complicated than straight lines. However, because the solutions are based on estimates, errors can occur. To avoid serious errors of estimation, it is important to check an estimated solution by substituting both coordinates into each equation of the system.

EXAMPLE 2 Solving a Linear System Graphically

Solve the linear system $\begin{Bmatrix} x + y = 1 \\ 2x - y = -4 \end{Bmatrix}$ graphically.

SOLUTION Graph both equations on the same coordinate system by plotting their intercepts, as shown in the figure.

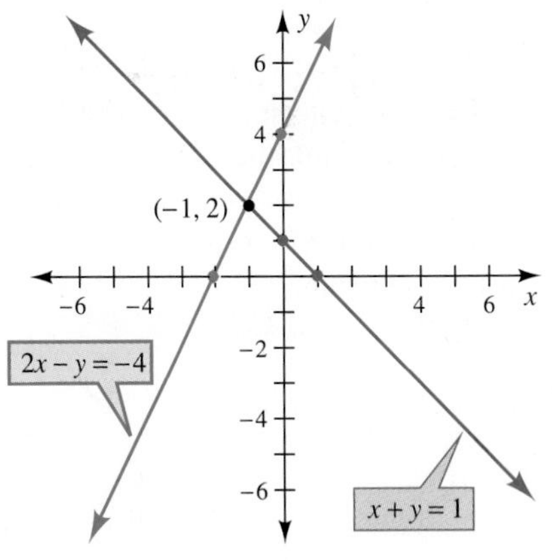

$x + y = 1$

x	$y = 1 - x$
0	1
1	0

$2x - y = -4$

x	$y = 2x + 4$
0	4
−2	0

When we inspect the graph, we find that the point of intersection is $(-1, 2)$. Check this value in both equations.

$$x + y = 1 \qquad 2x - y = -4$$
$$-1 + 2 \stackrel{?}{=} 1 \qquad 2(-1) - 2 \stackrel{?}{=} -4$$
$$1 = 1 \text{ is true.} \qquad -4 = -4 \text{ is true.}$$

Since $(-1, 2)$ checks in both equations, it is a solution of the system.

Answer $(-1, 2)$ ▼

There are three ways in which two lines in a plane can be related. Thus there are three types of linear systems that can be associated with a pair of lines in a plane. The three possibilities are as follows:

1 The lines are distinct and intersect in a single point, in which case there is exactly one simultaneous solution.

2 The lines are distinct and parallel, in which case there is no simultaneous solution.

3 The lines coincide (both equations represent the same line), in which case there are an infinite number of simultaneous solutions.

If a system of two equations in two variables has a solution, the system is called **consistent;** otherwise, it is called **inconsistent.** If the equations have distinct graphs, the equations are called **independent;** if the graphs coincide, the equations are called **dependent.** The following box illustrates the three possibilities listed above.

SELF-CHECK

Solve the linear system $\begin{cases} 2x - y = 5 \\ x + 2y = 0 \end{cases}$ graphically.

TYPES OF SOLUTION SETS FOR LINEAR SYSTEMS WITH TWO EQUATIONS

The linear system $\begin{cases} A_1x + B_1y + C_1 = 0 \\ A_2x + B_2y + C_2 = 0 \end{cases}$ can have

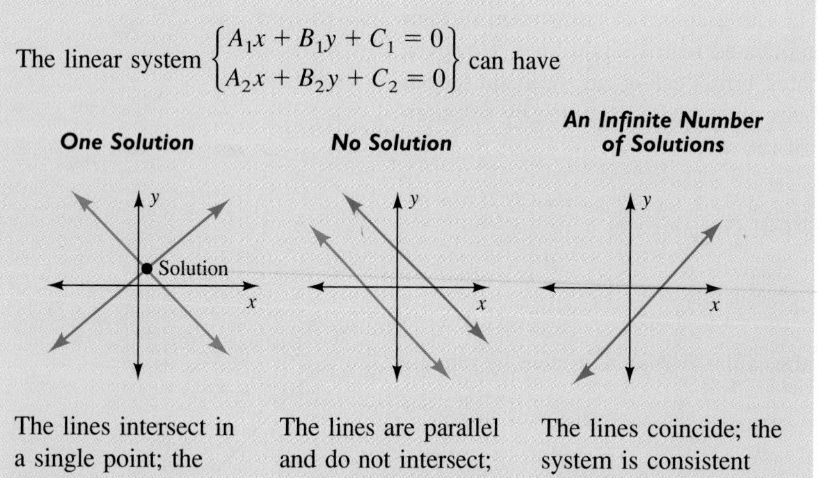

The lines intersect in a single point; the system is consistent and the equations are independent.	The lines are parallel and do not intersect; the system is inconsistent.	The lines coincide; the system is consistent and the equations are dependent.

SELF-CHECK ANSWER

(2, −1)

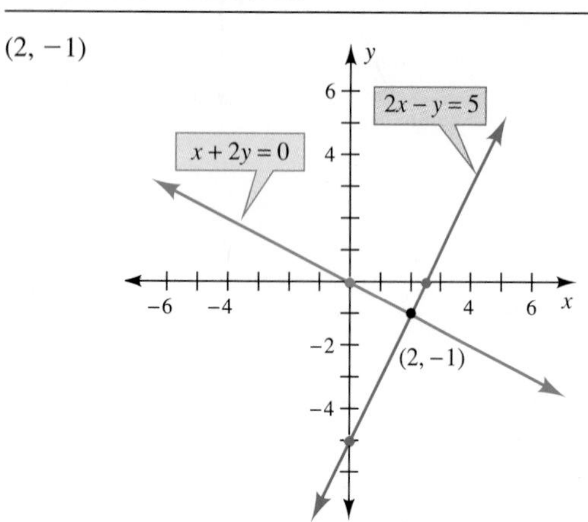

EXAMPLE 3 Solving Linear Systems Graphically

Solve each of these systems graphically.

SOLUTIONS

(a) $\begin{Bmatrix} x - y - 2 = 0 \\ -x + 2y - 6 = 0 \end{Bmatrix}$

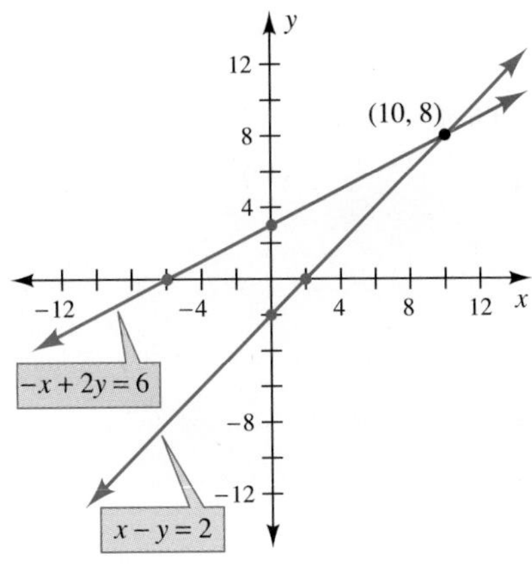

Answer (10, 8) (The check is left to you.)

(b) $\begin{Bmatrix} 6x + 3y - 12 = 0 \\ 4x + 2y + 2 = 0 \end{Bmatrix}$

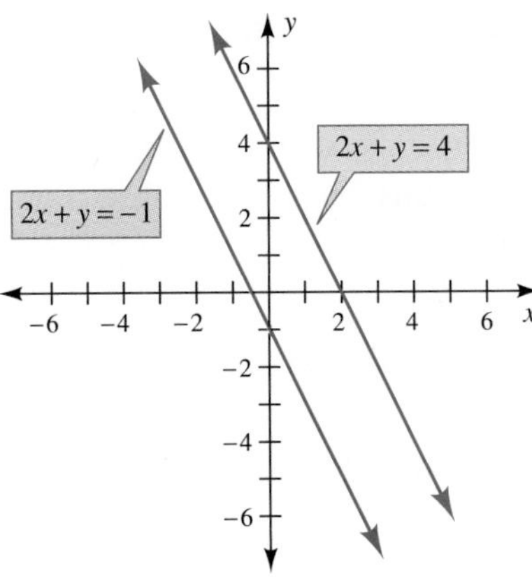

Answer No solution; an inconsistent system

(c) $\begin{Bmatrix} 3x - 9y - 18 = 0 \\ 2x - 6y - 12 = 0 \end{Bmatrix}$

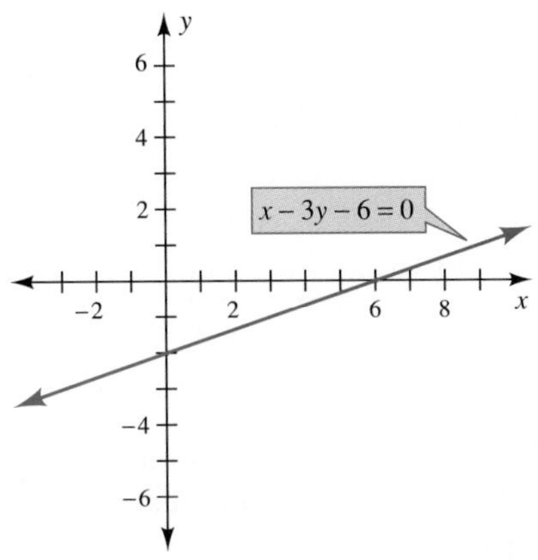

Answer The equations are dependent, and the system has an infinite number of solutions; every point on the line represents a solution of the system.

▼

Example 3 illustrates that (1) lines that intersect in a single point are not parallel and thus have different slopes; (2) lines that do not intersect are parallel and thus have the same slope but different y-intercepts; and (3) lines that coincide have the same slope and the same y-intercept. Therefore we can quickly determine how many solutions a system of linear equations has by first writing each equation in slope-intercept form.

NUMBER OF SOLUTIONS OF A LINEAR SYSTEM

The linear system $\left\{\begin{array}{l} y = m_1x + b_1 \\ y = m_2x + b_2 \end{array}\right\}$ has

1 one solution if $m_1 \neq m_2$.

2 no solution if $m_1 = m_2$ and $b_1 \neq b_2$.

3 an infinite number of solutions if $m_1 = m_2$ and $b_1 = b_2$.

The only linear equations for which this strategy will not work are those that represent vertical lines. Recall that a vertical line has no slope and its equation cannot be written in slope-intercept form. Systems that contain a vertical line can often be solved by inspection.

EXAMPLE 4 Determining the Number of Solutions of a System of Linear Equations

Determine the number of solutions of each of these systems.

SOLUTIONS

(a) $\left\{\begin{array}{r} x - y - 2 = 0 \\ -x + 2y - 6 = 0 \end{array}\right\}$

$$x - y - 2 = 0 \qquad\qquad -x + 2y - 6 = 0$$
$$y = 1x - 2 \qquad\qquad 2y = x + 6$$
$$y = \frac{1}{2}x + 3$$
$$m_1 = 1 \qquad\qquad m_2 = \frac{1}{2}$$

Since $m_1 \neq m_2$, the system has one solution. [See Example 3(a) for this solution.]

(b) $\left\{\begin{array}{r} 6x + 3y - 12 = 0 \\ 4x + 2y + 2 = 0 \end{array}\right\}$

$$6x + 3y - 12 = 0 \qquad\qquad 4x + 2y + 2 = 0$$
$$3y = -6x + 12 \qquad\qquad 2y = -4x - 2$$
$$y = -2x + 4 \qquad\qquad y = -2x - 1$$
$$m_1 = -2,\ b_1 = 4 \qquad\qquad m_2 = -2,\ b_2 = -1$$

Since $m_1 = m_2$ but $b_1 \neq b_2$, the system is inconsistent and has no solution. [Note in Example 3(b) that these lines are distinct and parallel.]

(c) $\begin{cases} 3x - 9y - 18 = 0 \\ 2x - 6y - 12 = 0 \end{cases}$

$$3x - 9y - 18 = 0 \qquad 2x - 6y - 12 = 0$$
$$9y = 3x - 18 \qquad 6y = 2x - 12$$
$$y = \frac{1}{3}x - 2 \qquad y = \frac{1}{3}x - 2$$

$$m_1 = \frac{1}{3},\ b_1 = -2 \qquad m_2 = \frac{1}{3},\ b_2 = -2$$

Since $m_1 = m_2$ and $b_1 = b_2$, the equations are dependent and the system has an infinite number of solutions. [Note in Example 3(c) that the lines coincide.]

SOLVING A SYSTEM OF LINEAR EQUATIONS USING A GRAPHICS CALCULATOR

Graphics calculators have the capability to simultaneously graph a variety of functions, including linear functions. The simultaneous solution of a system can then be observed by noting the coordinates of the point of intersection. Before we illustrate the process, consider the pair of graphs in Figure 3-1, obtained on a graphics calculator. Can you determine the solution of each system? Could these graphs represent the same system of equations?

SELF-CHECK 

Determine the number of solutions of each system.

1 $\begin{cases} 3x - 4y = 12 \\ -6x + 8y = 12 \end{cases}$

2 $\begin{cases} 5x - 10y - 20 = 0 \\ 4x = 8y + 16 \end{cases}$

3 $\begin{cases} 2x + y = -2 \\ x + 2y = -10 \end{cases}$

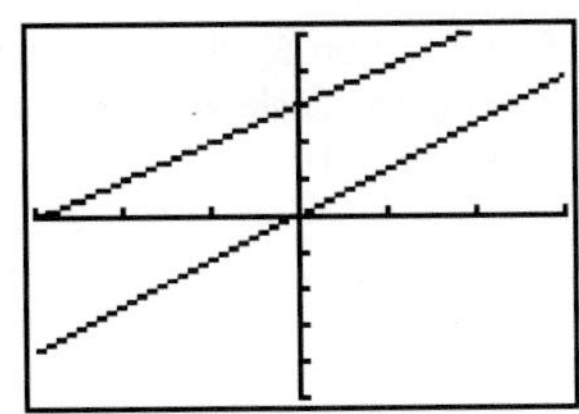
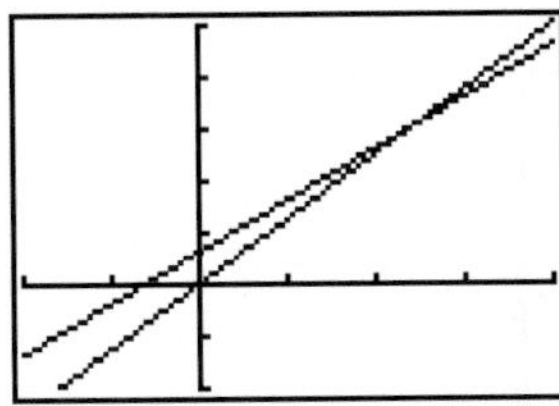

Figure 3-1

Although these windows display considerably different views, both graphs are of the same system of equations: $y = x + 3$ and $y = 1.25x$. (The viewing rectangle for the left window is $x = [-3, 3]$ and $y = [-5, 5]$, and the viewing rectangle for the right window is $x = [-10, 20]$ and $y = [-10, 25]$.) At first glance it might appear that the window on the left represents an inconsistent system. However, the scale of values and the range of values used can alter the view considerably. Lines that appear parallel may really intersect at a point outside the range of values shown on the window. By examining the equations as we did in Example 4, we can predict the number of solutions of the system before we graph the equations. We should then be careful to select an appropriate window that displays the solution of the system. Sometimes we may examine several windows before selecting an appropriate view. Figure 3-1 emphasizes the importance of labeling the scale on any graph that you make.

Although we can visually approximate the point of intersection on a calculator window, it is more accurate to use the TRACE feature and the arrow keys

SELF-CHECK ANSWERS 

1 No solution **2** An infinite number of solutions **3** One solution

to position the cursor approximately over the point of intersection and then read the coordinates from the display. Graphics calculators can even use the ZOOM feature to enlarge this point and refine the approximation. In Example 5 the TI-82 is used to illustrate features shared by several models of graphics calculators.

EXAMPLE 5 Calculator Usage

Approximate to the nearest tenth each coordinate of the solution of the system $\begin{Bmatrix} 2x + 3y = 9.3 \\ 5x - 4y = 3.7 \end{Bmatrix}$.

SOLUTION

$$2x + 3y = 9.3 \longrightarrow 3y = -2x + 9.3$$

$$y = -\frac{2x}{3} + 3.1$$

First write each equation in the form $y = f(x)$.

$$5x - 4y = 3.7 \longrightarrow -4y = -5x + 3.7$$

$$y = 1.25x - 0.925$$

1 [Y=] [(−)] [2] [X, T, θ] [÷] [3] [+] [3] [.] [1] [▼]

[1] [.] [2] [5] [X, T, θ] [−] [.] [9] [2] [5]

Enter the formulas for both functions into y_1 and y_2.

2 [ZOOM] [6 (standard option)]

[−10, 10] for x, [−10, 10] for y

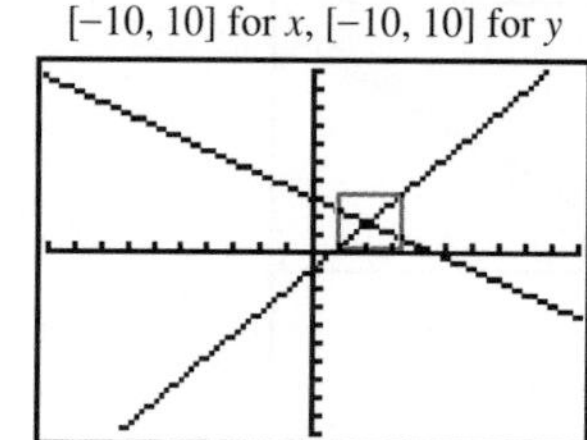

Select option 6, which is the standard viewing rectangle with $-10 \le x \le 10$ and $-10 \le y \le 10$.

3 [ZOOM] [1 (box option)]

Select approximately (0.2, 0.3) as one corner of the box by pressing [ENTER]. Then use the arrow keys to move the cursor to approximately (3.4, 3.2), and press [ENTER] to fix the diagonal corner of the box.

Select option 1, the box option. Then fix the corners of the new viewing rectangle, and zoom in on this rectangle.

[0.21, 3.40] for x, [0.32, 3.23] for y

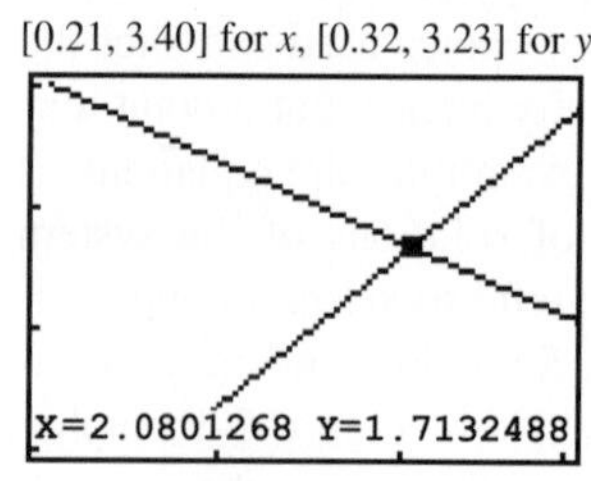

4 [TRACE]

Use the arrow keys to move the cursor over the point of intersection, which is approximately (2.1, 1.7). To check this approximate solution of this system of equations, substitute these coordinates into both of the given equations.

The accuracy of the approximation can be determined by observing how much the coordinates change as the cursor moves from one side of the point of intersection to the other side.

Check

$$2x + 3y = 9.3 \qquad\qquad 5x - 4y = 3.7$$
$$2(2.1) + 3(1.7) \stackrel{?}{=} 9.3 \qquad\qquad 5(2.1) - 4(1.7) \stackrel{?}{=} 3.7$$
$$4.2 + 5.1 \stackrel{?}{=} 9.3 \qquad\qquad 10.5 - 6.8 \stackrel{?}{=} 3.7$$
$$9.3 = 9.3 \text{ checks.} \qquad\qquad 3.7 = 3.7 \text{ checks.}$$

Answer (2.1, 1.7)

The line $Ax + By + C = 0$ in Figure 3-2 separates the plane into two regions called **half-planes.** One half-plane will satisfy $Ax + By + C < 0$, and the other will satisfy $Ax + By + C > 0$.

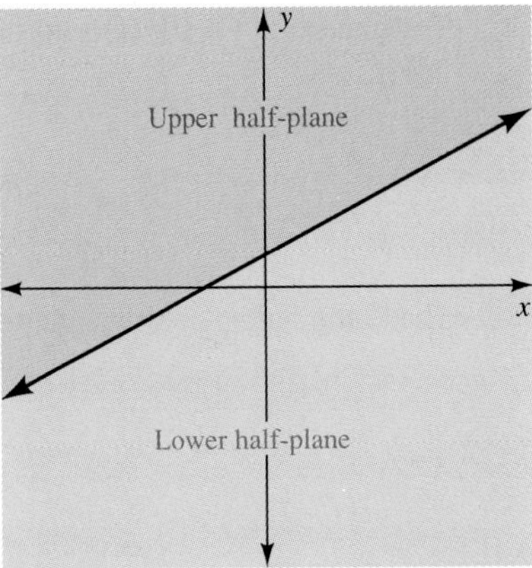

Figure 3-2
Half-Planes

GRAPHING A LINEAR INEQUALITY

Step 1 Graph the equality $Ax + By + C = 0$, using
- **a.** a solid line if the equality is included in the solution or
- **b.** a dashed line if the equality is not included in the solution.

Step 2 Choose an arbitrary test point not on the line; (0, 0) is often convenient. Substitute this test point into the inequality.

Step 3
- **a.** If the test point satisfies the inequality, shade the half-plane containing this point.
- **b.** If the test point does not satisfy the inequality, shade the other half-plane.

EXAMPLE 6 Graphing a Linear Inequality

Graph $3x - 5y + 15 < 0$.

SOLUTION

1 Draw a dashed line for $3x - 5y + 15 = 0$, since the equality is not part of the solution. The line passes through the intercepts $(-5, 0)$ and $(0, 3)$.

2 Test the origin:

$$3x - 5y + 15 < 0$$
$$3(0) - 5(0) + 15 \stackrel{?}{<} 0$$
$$15 < 0 \text{ is false.}$$

3 Shade the half-plane that does *not* include the test point (0, 0).

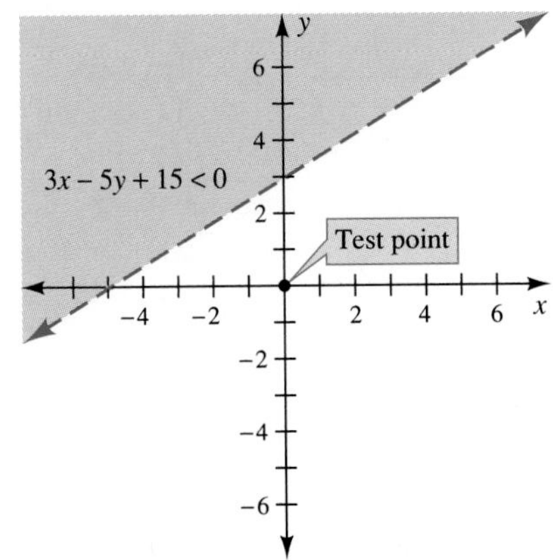

EXAMPLE 7 Graphing a Linear Inequality

Graph $5x \geq 3y$.

SOLUTION

1 Draw a solid line for $5x = 3y$, since the original statement includes the equality. The line passes through (0, 0) and (3, 5).

2 Test the point (1, 0). [Do not test (0, 0), since the origin lies on the line.]

$$5x \geq 3y$$
$$5(1) \stackrel{?}{\geq} 3(0)$$
$$5 \geq 0 \text{ is true.}$$

3 Shade the half-plane that includes the test point (1, 0).

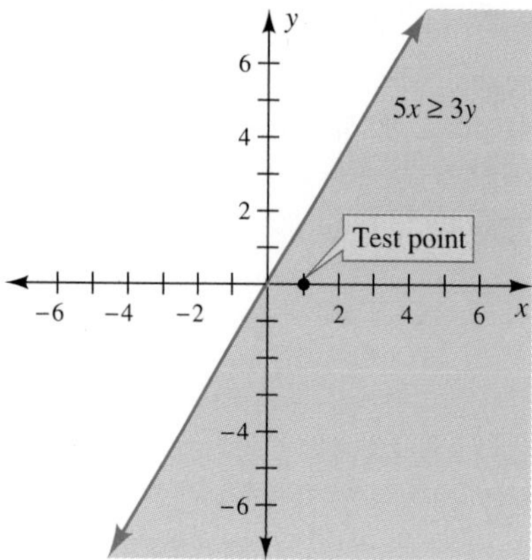

To graph a system of linear inequalities, we first graph each inequality on the same coordinate system. The solution of the system is the intersection of these two individual regions. To clarify which points satisfy each inequality, we will use horizontal and vertical lines to indicate the solutions of the individual inequalities. The solution of the system—the cross-hatched region where these lines intersect—will then be shaded for greater emphasis.

SELF-CHECK 

Graph $2x + y \leq 10$.

SELF-CHECK ANSWER

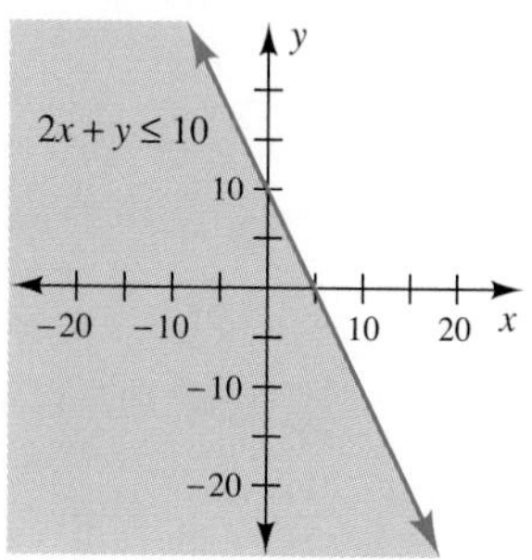

EXAMPLE 8 Graphing a System of Linear Inequalities

Graph the solution of $\begin{Bmatrix} x + y \geq 2 \\ 3x - 2y < 6 \end{Bmatrix}$.

SOLUTION Graph $x + y = 2$ with a solid line.

$x + y = 2$

x	y
2	0
0	2

$$x + y \geq 2$$
$$0 + 0 \overset{?}{\geq} 2$$
$$0 \geq 2 \text{ is false.}$$

Use the intercepts (2, 0) and (0, 2) to plot the line. Then test the point (0, 0).

Indicate the region that does *not* contain the test value (0, 0) with vertical lines.

Graph $3x - 2y = 6$ with a dashed line.

$3x - 2y = 6$

x	y
2	0
0	−3

$$3x - 2y < 6$$
$$3(0) - (0) \overset{?}{<} 6$$
$$0 < 6 \text{ is true.}$$

Use the intercepts (2, 0) and (0, −3) to plot the dashed line. Then test the point (0, 0).

Indicate the region that contains the test value (0, 0) with horizontal lines. Then use solid shading to emphasize that every point in the cross-hatched region is a solution of the system containing both of these inequalities.

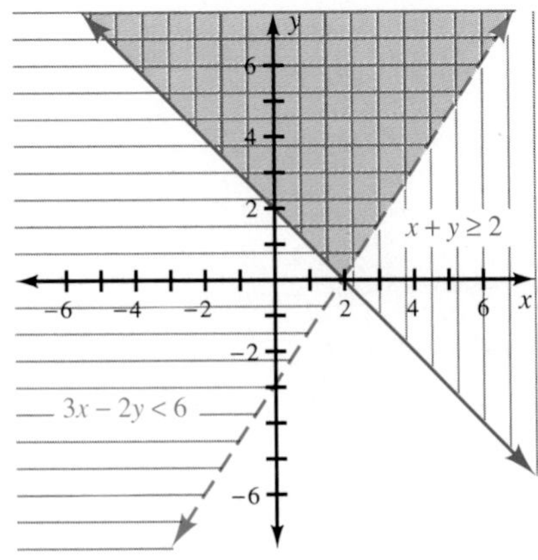

▼

EXAMPLE 9 Graphing a System of Linear Inequalities

Graph the solution of $\begin{Bmatrix} x \geq -3 \\ x \leq 2 \end{Bmatrix}$.

SOLUTION Inspection reveals that both $x = -3$ and $x = 2$ represent vertical lines. The region containing (0, 0) satisfies both inequalities. Thus the solution set is the strip between these solid lines.

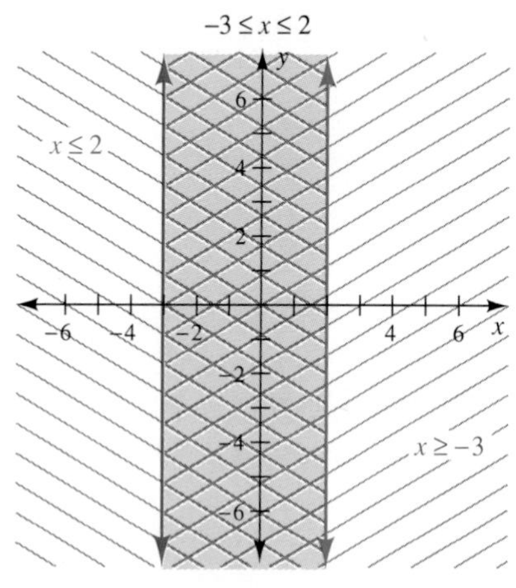

EXAMPLE 10 Graphing a System of Linear Inequalities

Graph the solution of $\begin{Bmatrix} y \leq 4 \\ x - y \leq 2 \\ y \geq -x \end{Bmatrix}$.

SOLUTION First graph each line on the same coordinate system. Then test the point (1, 1) in each inequality.

$y \leq 4$	$x - y \leq 2$	$y \geq -x$
$1 \leq 4$ is true.	$1 - 1 \stackrel{?}{\leq} 2$	$1 \geq -1$ is true.
	$0 \leq 2$ is true.	

SELF-CHECK 

Graph the solution of each of these systems of inequalities.

1 $\begin{Bmatrix} 2x - y \leq 4 \\ x + 2y > 6 \end{Bmatrix}$

2 $\begin{Bmatrix} y < 4 \\ y > -3 \end{Bmatrix}$

SELF-CHECK ANSWERS ✓

1

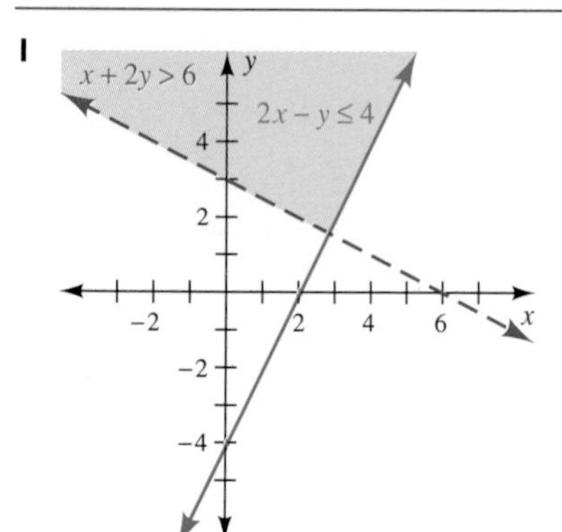

2

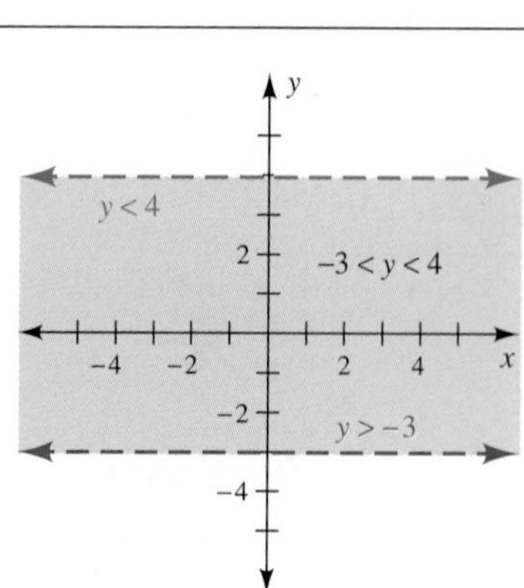

Shade each individual region with lines, and then use solid shading to indicate the triangular region that is the solution of the system.

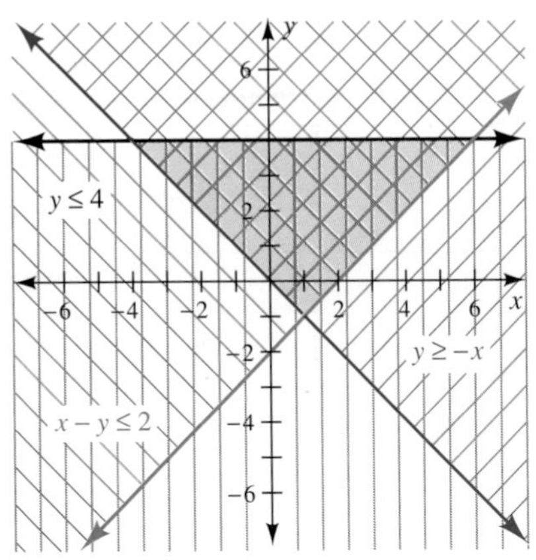

Exercises 3-1

A

1 Determine whether $\left(-2, \frac{1}{3}\right)$ is a solution of each of these systems.

a. $\begin{aligned} 4x - 3y &= -9 \\ -x + 3y &= 4 \end{aligned}$ **b.** $\begin{aligned} 3x + 6y &= 4 \\ x + 6y &= 0 \end{aligned}$

c. $\begin{aligned} x - 3y &= -3 \\ 2x + 3y &= -3 \end{aligned}$

2 Determine whether $\left(3, -\frac{1}{2}\right)$ is a solution of each of these systems.

a. $\begin{aligned} 2x + 4y &= 6 \\ x - 2y &= 4 \end{aligned}$ **b.** $\begin{aligned} x + 2y &= 2 \\ 3x - 4y &= 11 \end{aligned}$

c. $\begin{aligned} 5x + 2y &= 14 \\ 3x - 2y &= 7 \end{aligned}$

In Exercises 3–18 solve each linear system graphically.

3 $\begin{aligned} x - y &= 2 \\ x + 3y &= 6 \end{aligned}$ **4** $\begin{aligned} x + y &= 4 \\ 2x + y &= 5 \end{aligned}$

5 $\begin{aligned} 2x + y - 6 &= 0 \\ 2x - y - 6 &= 0 \end{aligned}$ **6** $\begin{aligned} 2x - y - 3 &= 0 \\ x + y - 6 &= 0 \end{aligned}$

7 $\begin{aligned} 3x - 3y - 9 &= 0 \\ 2x - 2y &= 4 \end{aligned}$ **8** $\begin{aligned} \frac{x}{2} + \frac{y}{2} - 2 &= 0 \\ \frac{x}{3} + \frac{y}{3} &= -1 \end{aligned}$

9 $\begin{aligned} 5x - 10y &= 15 \\ 2x - 4y &= 6 \end{aligned}$ **10** $\begin{aligned} y &= x - 1 \\ y &= \frac{1}{2}x + \frac{1}{2} \end{aligned}$

11 $\begin{aligned} y &= x + 6 \\ y &= 2x + 5 \end{aligned}$ **12** $\begin{aligned} 12x + 4y &= 8 \\ 9x + 3y &= 6 \end{aligned}$ **13** $\begin{aligned} x &= 3 \\ x + y &= 5 \end{aligned}$

14 $\begin{aligned} x + y &= 6 \\ y &= 2 \end{aligned}$ **15** $\begin{aligned} y &= 5x \\ 3y &= x \end{aligned}$ **16** $\begin{aligned} 4x &= y \\ 2y &= -x \end{aligned}$

17 $\begin{aligned} x &= -2 \\ y &= 3 \end{aligned}$ **18** $\begin{aligned} x &= 3 \\ y &= -4 \end{aligned}$

In Exercises 19–26 graph each linear inequality.

19 $x - y \geq 5$ **20** $2x + y \leq 6$

21 $3x - 2y - 12 < 0$ **22** $-5x + 2y + 10 < 0$

23 $x > 3y$ **24** $5x < -y$

25 $\frac{1}{2}x + \frac{1}{3}y \leq 1$ **26** $\frac{1}{4}x - \frac{1}{5}y \geq 1$

In Exercises 27–40 graph the solution of each system of linear inequalities.

27 $\begin{aligned} x - y &\geq 4 \\ 2x - y &< 6 \end{aligned}$ **28** $\begin{aligned} x + 2y &\leq 2 \\ 2x - y &\geq 4 \end{aligned}$

29 $\begin{aligned} 2x - 5y - 10 &< 0 \\ 2x - y - 7 &\geq 0 \end{aligned}$ **30** $\begin{aligned} 3x + 2y - 12 &> 0 \\ 2x + 5y - 10 &< 0 \end{aligned}$

31 $\begin{aligned} \frac{x}{2} - \frac{y}{2} &< 1 \\ \frac{x}{2} + \frac{y}{2} &> -1 \end{aligned}$ **32** $\begin{aligned} \frac{x}{3} - \frac{y}{3} &\geq -1 \\ \frac{x}{4} + \frac{y}{4} &\geq -1 \end{aligned}$

33 $\begin{aligned} x &\geq 1 \\ x &< 4 \end{aligned}$ **34** $\begin{aligned} x &> -5 \\ x &\leq -2 \end{aligned}$

35 $\begin{aligned} y &< -2 \\ y &\geq -6 \end{aligned}$ **36** $\begin{aligned} y &> 3 \\ y &< 4 \end{aligned}$

37 $x > -2$
$x \le 3$

38 $y < 5$
$y > 3$

39 $y \ge -2$
$y \le -4$

40 $x \le -4$
$x \ge -1$

B

ESTIMATION SKILLS (41–44)

41 Average Salaries for Professional Athletes The line graphs shown give the average salaries for professional baseball and basketball players for selected years. Use these graphs to determine the time period in which these averages were approximately the same. What was the approximate salary at this time?

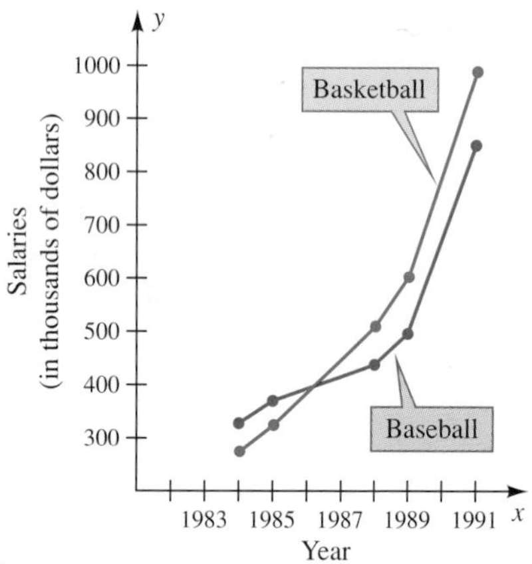

Figure for Exercise 41

42 Causes of Death An important part of setting national policy is understanding the scope of our problems. The two line graphs shown give the number of deaths due to suicide and AIDS in the United States in selected years. Use these graphs to determine the time period in which the number of deaths from AIDS was approximately the same as the number of deaths from suicide.

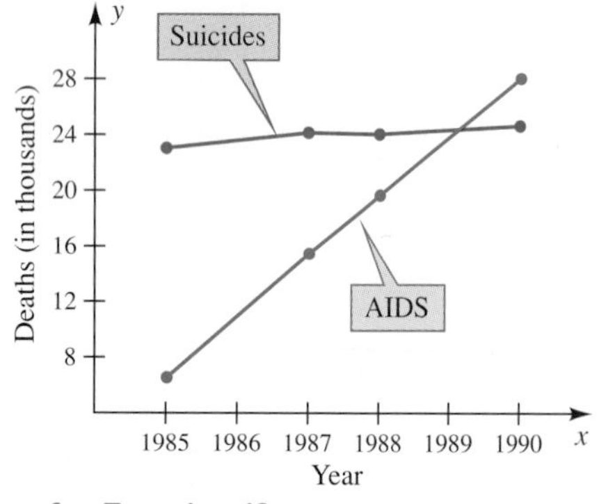

Figure for Exercise 42

In Exercises 43 and 44 use the graph shown on the calculator display to estimate the slope of each line. *Caution:* Be sure to read the range of values indicated above each graph.

43 a. [−3, 3] for x, [−10, 10] for y

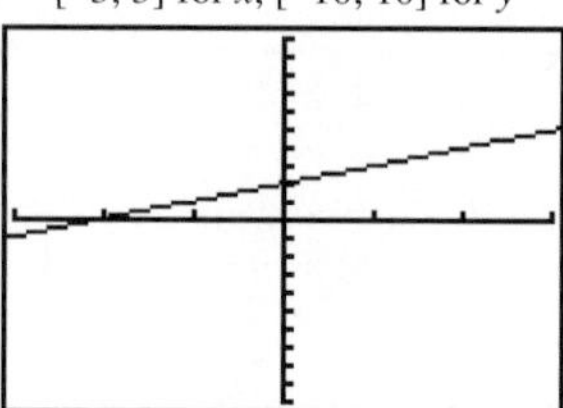

b. [−8, 8] for x, [−2, 3] for y

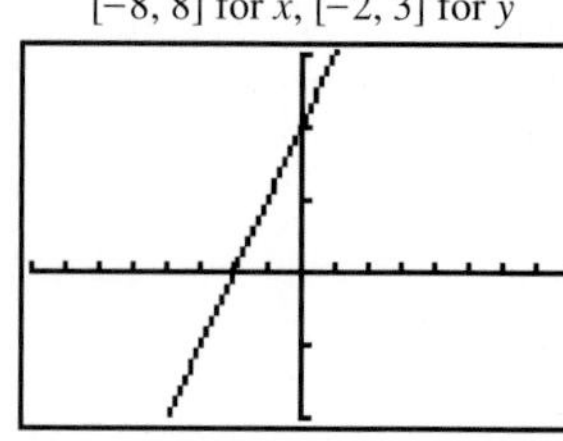

44 a. [−10, 10] for x, [−3, 3] for y

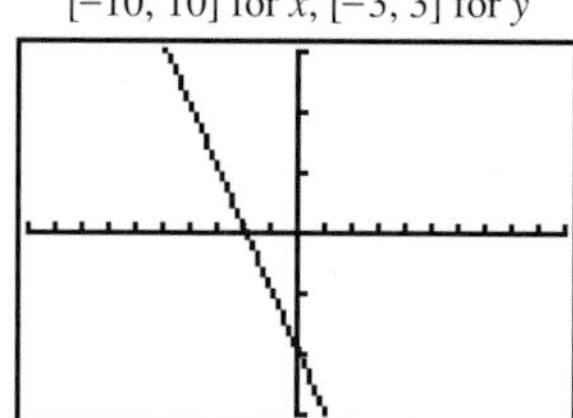

b. [−3, 3] for x, [−10, 10] for y

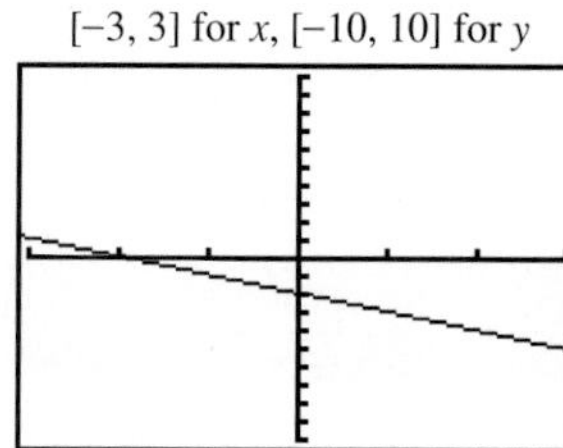

In Exercises 45–54 write each equation in slope-intercept form, if possible, and then determine the number of solutions of each system.

45 $2x - 7y = 14$
$3x + 5y + 11 = 0$

46 $6x = 15y - 9$
$8x - 20y + 12 = 0$

47 $6x - 2y = 8$
$-9x + 3y = 10$

48 $4x + 3y - 9 = 0$
$-2x + 7y + 8 = 0$

49 $\frac{x}{3} - \frac{y}{2} = 1$
$x = \frac{3y}{2} + 3$

50 $\frac{x}{3} + \frac{y}{2} + 1 = 0$
$x = 3 - \frac{3y}{2}$

51 $0.5x - 0.4y = 2.8$
$0.3x + 0.2y = 0.06$

52 $0.6x + 0.3y + 0.9 = 0$
$0.2x - 0.5y - 0.4 = 0$

53 $\frac{9x}{2} = 4\frac{1}{3}$
$2x - y = 5$

54 $-3x + 8 = 7$
$-x + 3y + 4 = 9$

C

In Exercises 55–62 graph the solution of each system of linear inequalities.

55 $x \ge 0$
$y \ge 0$
$2x + 3y < 6$

56 $x \ge 0$
$y \ge 0$
$5x + 2y < 10$

57 $x + y > 0$
$x - y < 0$
$y < 4$

58 $2x + 3y \le 0$
$3x - 2y \ge 0$
$y \ge -4$

59 $x \ge -2$
$x \le 2$
$y \ge -3$
$y \le 3$

60 $x > 1$
$x < 3$
$y > -5$
$y < -2$

61 $3x + 3y - 15 \le 0$
$6x + 2y - 18 \le 0$
$x \ge 0$
$y \ge 0$

62 $x - y - 2 \le 0$
$2x + 2y - 8 \le 0$
$x \ge 0$
$y \ge 0$

Calculator Usage (63–66)

In Exercises 63–66 use a graphics calculator to graph the solution of each system of equations; then approximate each coordinate to the nearest tenth.

63 $2x + y = 1$
$7x - y = 0$

64 $7x - 14y = -1$
$-7x + 7y = -2$

65 $2x - y = 1.1$
$3x + 2y = 11.4$

66 $x - 3y = 0$
$9x - 9y = 4$

67 Rope Lengths A rope encloses a rectangle whose perimeter is at most 10 meters. This rope is cut into two lengths, x and y. Assume these lengths are positive. Write the system of inequalities for this problem, and then graph this system of inequalities.

68 Salaries Employee Y makes at most $50 more per week than employee X. Together they are paid at most $800 per week. Assume the salaries of X and Y are nonnegative. Write the system of inequalities for this problem, and then graph this system of inequalities.

Discussion Questions (69 and 70)

69 Explain in your own words what is meant by an inconsistent system of equations.

70 Write your own word problem that the following system of inequalities algebraically models.

$$x + 2y \le 28$$
$$x \ge 0$$
$$y \ge 0$$

Section Skill Reinforcers

If x and y represent real numbers, write an algebraic equation for each of these statements.

1 Three more than the sum of these numbers is seven. **2** Five less than the sum of these numbers is two. **3** Twice the sum of these numbers is eight. **4** Twice x minus y is eleven. **5** Seven more than x equals y.

3-2 *Solving Systems of Linear Equations in Two Variables*

Section Objectives

3 Solve a system of linear equations by the substitution method.

4 Solve a system of linear equations by the addition-subtraction method.

Although the graphical method enables you to visualize a solution to a system of equations in two variables, it is not appropriate for three or more variables. Even with systems of two equations and two variables, this method can be time consuming, and it is sometimes inaccurate because it can be difficult to determine the exact point of intersection of the two lines. We will now develop two efficient algebraic methods that give exact solutions: the substitution method and the addition-subtraction method.

The substitution principle says that a quantity can be substituted into an expression for any quantity that it equals without changing the value of the expression. This principle is the basis for the **substitution method,** which is described in the following box.

THE SUBSTITUTION METHOD

Step 1 Solve one of the equations for one of the variables in terms of the other variable.

Step 2 Substitute the expression obtained in Step 1 into the other equation (eliminating one of the variables), and solve the resulting equation.

Step 3 Substitute the value obtained in Step 2 into the equation obtained in Step 1 to find the value of the other variable.

The ordered pair obtained in Steps 2 and 3 is the solution.

EXAMPLE 1 Solving a System of Linear Equations by the Substitution Method

Solve $\begin{Bmatrix} 2x + y = 1 \\ 3x - 6y = 4 \end{Bmatrix}$ by the substitution method.

SOLUTION

1 $2x + y = 1$ — Solve the first equation for y.

$$y = 1 - 2x$$

2

$$3x - 6y = 4$$

$$3x - 6(1 - 2x) = 4$$

Substitute for y in the second equation, and then solve for x.

$$15x = 10$$

$$x = \frac{10}{15}$$

$$x = \frac{2}{3}$$

3 $y = 1 - 2x$

$$y = 1 - 2\left(\frac{2}{3}\right)$$

Substitute $\frac{2}{3}$ for x in the equation that was solved for y.

$$y = 1 - \frac{4}{3}$$

$$y = -\frac{1}{3}$$

Answer $\left(\frac{2}{3}, -\frac{1}{3}\right)$ — The solution set is $\left\{\left(\frac{2}{3}, -\frac{1}{3}\right)\right\}$. Does this solution check?

The algebraic solution of a consistent system of independent equations will result in a unique value for each variable and thus a unique ordered pair satisfying the system. The algebraic solution of an inconsistent system will result in a contradiction and thus no solution. The algebraic solution of a consistent system of dependent equations will result in an identity and thus an infinite number of solutions.

EXAMPLE 2 Solving a Consistent System of Dependent Equations by the Substitution Method

Solve $\begin{Bmatrix} x - 3y + 2 = 0 \\ 2x - 6y + 4 = 0 \end{Bmatrix}$ by the substitution method.

SOLUTION

$x - 3y + 2 = 0$ Solve the first equation for x.

$x = 3y - 2$

$2x - 6y + 4 = 0$

$2(3y - 2) - 6y + 4 = 0$ Substitute for x in the second equation, and then simplify.

$6y - 4 - 6y + 4 = 0$

$0 = 0$ This equation is an identity, so the given equations are dependent.

Answer This consistent system of dependent equations has an infinite number of solutions. Every ordered pair (x, y) that satisfies $x - 3y + 2 = 0$ is an element of the solution set. ▼

EXAMPLE 3 Solving an Inconsistent System by the Substitution Method

Solve $\begin{Bmatrix} 2x - 6y = 4 \\ -3x + 9y = 6 \end{Bmatrix}$ by the substitution method.

SOLUTION

$2x - 6y = 4$ Solve the first equation for x.

$2x = 6y + 4$

$x = 3y + 2$

$-3x + 9y = 6$

$-3(3y + 2) + 9y = 6$ Substitute for x in the second equation, and then simplify.

$-9y - 6 + 9y = 6$

$-6 = 6$ is false. This equation is a contradiction, so the system is an inconsistent system.

Answer This inconsistent system has no solution. The solution set is the null set, ∅. ▼

SELF-CHECK 

Solve each of these systems by the substitution method.

1 $3x - y = 3$
$6x + 4y = -6$

2 $4x + 2y = 5$
$2x + y = 3$

The substitution method is well suited to systems that contain at least one coefficient of 1 or -1. Other systems may be solved more easily using the **addition-subtraction method,** which is described in the next box. This method is also called the **elimination method** because the procedure involves eliminating a variable in one of the equations by adding equals to equals.

SELF-CHECK ANSWERS 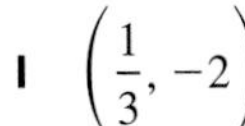

1 $\left(\frac{1}{3}, -2\right)$ **2** This inconsistent system has no solution.

ADDITION-SUBTRACTION METHOD

Step 1 If necessary, multiply each equation by a constant so that the equations have one variable whose coefficients are the same except for their signs.

Step 2 Add the new equations to eliminate a variable, and then solve the resulting equation.

Step 3 Substitute this value into one of the original equations (**back-substitution**), and solve for the other variable.

The ordered pair obtained in Steps 2 and 3 is the solution.

EXAMPLE 4 Solving a System of Linear Equations by the Addition-Subtraction Method

Solve $\begin{cases} 2x + 3y = 1 \\ 3x - 3y = -21 \end{cases}$ by the addition-subtraction method.

SOLUTION

$$\begin{aligned} 2x + 3y &= 1 \\ 3x - 3y &= -21 \\ \hline 5x \quad &= -20 \end{aligned}$$

Add the equations to eliminate y.

$$x = -4$$

Solve for x.

$$\begin{aligned} 2x + 3y &= 1 \\ 2(-4) + 3y &= 1 \\ -8 + 3y &= 1 \\ 3y &= 9 \\ y &= 3 \end{aligned}$$

Back-substitute the x value into the first equation of the original system, and solve for y.

Answer $(-4, 3)$

Does this value check?

EXAMPLE 5 Solving a System of Linear Equations by the Addition-Subtraction Method

Solve $\begin{cases} 3x + 5y = 1 \\ 6x + 13y = -1 \end{cases}$ by the addition-subtraction method.

SOLUTION

$$\begin{cases} 3x + 5y = 1 \\ 6x + 13y = -1 \end{cases} \xrightarrow[r_2' = r_2]{r_1' = -2r_1} \begin{aligned} -6x - 10y &= -2 \\ 6x + 13y &= -1 \\ \hline 3y &= -3 \\ y &= -1 \end{aligned}$$

Multiply both members of the original equation by -2, using the notation $r_1' = -2r_1$ to indicate how the new first equation was obtained. Then add these new equations, and solve for y.

$$\begin{aligned} 3x + 5y &= 1 \\ 3x + 5(-1) &= 1 \\ 3x - 5 &= 1 \\ 3x &= 6 \\ x &= 2 \end{aligned}$$

Back-substitute this y value into the first equation in the original system.

Answer $(2, -1)$

Does this solution check?

EXAMPLE 6 **Solving a System of Linear Equations by the Addition-Subtraction Method**

Solve $\begin{cases} 3x + 2y = -6 \\ 5x - 7y = -41 \end{cases}$ by the addition-subtraction method.

SOLUTION $\begin{cases} 3x + 2y = -6 \\ 5x - 7y = -41 \end{cases} \xrightarrow[r_2' = -3r_2]{r_1' = 5r_1} \begin{array}{r} 15x + 10y = -30 \\ -15x + 21y = 123 \\ \hline 31y = 93 \\ y = 3 \end{array}$

Multiply both members of the first equation by 5 and both members of the second equation by -3. Add these new equations, and then solve for y.

$$\begin{aligned} 3x + 2y &= -6 \\ 3x + 2(3) &= -6 \\ 3x + 6 &= -6 \\ 3x &= -12 \\ x &= -4 \end{aligned}$$

Back-substitute this y value into the first equation in the original system.

Answer $(-4, 3)$

Does this solution check?

EXAMPLE 7 **Solving a Consistent System of Dependent Equations by the Addition-Subtraction Method**

Solve $\begin{cases} 2x - 6y = 4 \\ 3x - 9y = 6 \end{cases}$ by the addition-subtraction method.

SOLUTION $\begin{cases} 2x - 6y = 4 \\ 3x - 9y = 6 \end{cases} \xrightarrow[r_2' = -2r_2]{r_1' = 3r_1} \begin{array}{r} 6x - 18y = 12 \\ -6x + 18y = -12 \\ \hline 0x + 0y = 0 \\ 0 = 0 \end{array}$

Multiply by 3.

Multiply by -2.

Add these new equations.

This equation is an identity, so the given equations are dependent.

Answer This consistent system of dependent equations has an infinite number of solutions. Every ordered pair (x, y) that satisfies $x - 3y = 2$ is an element of the solution set.

$x - 3y = 2$ is equivalent to both $2x - 6y = 4$ and $3x - 9y = 6$.

Word problems that contain more than one unknown are often easier to solve by using a different variable for each unknown than by expressing each unknown in terms of the same variable. Some of the word problems examined in this chapter are quite similar to those solved in earlier chapters using only one variable. The difference is that we now use *more than one* variable to solve these problems.

SELF-CHECK 

Solve the following systems of linear equations.

1 $\begin{cases} 3x + 2y = 2 \\ 6x - 6y = -1 \end{cases}$

2 $\begin{cases} 7x + 4y = -2 \\ 2x - 3y = -13 \end{cases}$

1 $\left(\frac{1}{3}, \frac{1}{2}\right)$ **2** $(-2, 3)$

EXAMPLE 8 A Numeric Word Problem

One number is three more than five times a second number. Twice the first number subtracted from three times the second number yields a difference of 1. Find these numbers.

SOLUTION Let

m = First of these two numbers

n = Second of these two numbers

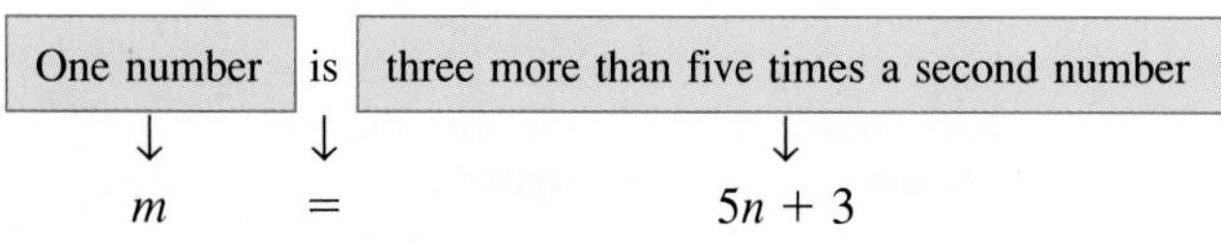

Substitute the variables into the **word equations.**

Twice the first number subtracted from three times the second number	yields	1
↓	↓	↓
$3n - 2m$	$=$	1

$$\begin{cases} m - 5n = 3 \\ -2m + 3n = 1 \end{cases} \xrightarrow[r_2' = r_2]{r_1' = 2r_1} \begin{array}{r} 2m - 10n = 6 \\ -2m + 3n = 1 \\ \hline -7n = 7 \\ n = -1 \end{array}$$

Multiply both members of the first equation by 2.

Add these new equations.

Solve for n.

$$\begin{aligned} m &= 5n + 3 \\ &= 5(-1) + 3 \\ &= -5 + 3 \\ m &= -2 \end{aligned}$$

Back-substitute this value of n into the first equation.

Answer The first number is -2, and the second number is -1. Do these values check?

The following system of equations is not a linear system, but it can still be solved by the addition-subtraction method. We will solve for $\frac{1}{x}$ and $\frac{1}{y}$ and then reciprocate to obtain x and y.

EXAMPLE 9 A System of Nonlinear Equations

Solve $\begin{cases} \frac{3}{x} - \frac{3}{y} = -7 \\ \frac{5}{x} + \frac{6}{y} = -8 \end{cases}$ by the addition-subtraction method.

SOLUTION

$$\begin{Bmatrix} \frac{3}{x} - \frac{3}{y} = -7 \\ \frac{5}{x} + \frac{6}{y} = -8 \end{Bmatrix} \xrightarrow[r_2' = r_2]{r_1' = 2r_1} \begin{aligned} \frac{6}{x} - \frac{6}{y} &= -14 \\ \frac{5}{x} + \frac{6}{y} &= -8 \\ \hline \frac{11}{x} &= -22 \\ \frac{1}{x} &= -2 \end{aligned}$$

Multiply both members of the first equation by 2.

Add these equations, and then solve for $\frac{1}{x}$.

$$\frac{3}{x} - \frac{3}{y} = -7$$
$$3\left(\frac{1}{x}\right) - \frac{3}{y} = -7$$
$$3(-2) - \frac{3}{y} = -7$$
$$-6 - \frac{3}{y} = -7$$
$$-\frac{3}{y} = -1$$
$$\frac{1}{y} = \frac{1}{3}$$

Back-substitute -2 for $\frac{1}{x}$ in the first equation of the original system, and solve for $\frac{1}{y}$.

Since $\frac{1}{x} = -2$, $x = -\frac{1}{2}$. Likewise, since $\frac{1}{y} = \frac{1}{3}$, $y = 3$.

Answer $\left(-\frac{1}{2}, 3\right)$

Does this answer check?

Exercises 3-2

A

In Exercises 1–12 solve each system of equations by the substitution method. Identify the systems that are inconsistent or contain dependent equations.

1 $x = y + 1$
$2x - 3y = -2$

2 $y = 3x + 7$
$2x + 5y = 1$

3 $x - 2y = -2$
$3x + 2y = -2$

4 $2x + y = 2$
$6x + y = 0$

5 $2x + 4y - 7 = 0$
$3x - y = 0$

6 $3x + 6y + 4 = 0$
$x - 2y = 0$

7 $2x + 4y = 2$
$x + 2y = 1$

8 $3x - y + 2 = 0$
$6x - 2y + 4 = 0$

9 $3x = 7y$
$11x = 5y$

10 $12x = 5y$
$9x = 2y$

11 $x + 5y = 3$
$2x + 10y = 8$

12 $2x + 6y = 5$
$x + 3y = 2$

In Exercises 13–24 solve each system of equations by the addition-subtraction method. Identify the systems that are inconsistent or contain dependent equations.

13 $2x - 3y = 1$
$4x + 3y = -7$

14 $2x + 5y = 11$
$2x - 3y = -13$

15 $2x - 7y = -8$
$4x + 5y = 60$

16 $11x - 3y = 46$
$5x + 6y = 43$

17 $4x - 3y = -38$
$9x + 2y = -33$

18 $4x - 9y = 48$
$5x + 13y = -37$

19 $8x - 10y = 4$
$12x - 15y = 6$

20 $20x - 70y = 30$
$4x - 14y = 20$

21 $\frac{x}{2} + \frac{y}{2} = -\frac{1}{12}$
$\frac{x}{3} - \frac{y}{4} = -\frac{1}{4}$

22 $\frac{x}{4} + \frac{y}{5} = -\frac{1}{20}$
$\frac{x}{8} - \frac{y}{10} = \frac{9}{40}$

23 $\frac{1}{2}x - \frac{1}{3}y = \frac{1}{6}$
$15x - 10y = 2$

24 $x - \frac{1}{4}y = \frac{3}{4}$
$12x - 3y = 9$

In Exercises 25–40 solve each system of equations by either the substitution method or the addition-subtraction method, whichever seems more appropriate.

25 $5x + 2y = 8$
$5x + 3y = 12$

26 $x - 7y = 14$
$2x - 7y = 21$

27 $2x + 2y = -1$
$3x - 2y = -4$

28 $3x + 2y = -1$
$-3x + 4y = 4$

29 $y = x - 5$
$4x + y = 5$

30 $x - 4y = 1$
$2x - 5y = -1$

31 $3x + 5y = -8$
$9x - 10y = 1$

32 $4x + 7y = 13$
$6x + 5y = 3$

33 $\frac{3}{4}x + \frac{2}{3}y = -1$
$\frac{1}{2}x - \frac{5}{3}y = -7$

34 $\frac{3}{5}x - \frac{3}{2}y = 6$
$\frac{4}{5}x + \frac{1}{2}y = 3$

35 $\frac{4x}{3} - \frac{2y}{7} = -2$
$\frac{5x}{3} + \frac{3y}{7} = -8$

36 $\frac{6x}{5} - \frac{y}{2} = -\frac{9}{10}$
$\frac{4x}{5} + \frac{y}{5} = 1$

37 $0.2x - 0.1y = -0.6$
$0.2x + 0.1y = -1.0$

38 $-0.4x + 0.2y = -0.7$
$0.3x + 0.1y = -0.1$

39 $2(x - y) - 3(x + y) = -3$
$5(x + 2y) - (2x - 3y) = 9$

40 $3(x - 2y) + 2(3x + y) = -29$
$4(x + y) - 5(x - y) = 46$

B

Numeric Word Problems (41–48)

In Exercises 41–48 solve each problem using a system of two equations in two variables.

41 Find two numbers whose sum is 142 and whose difference is 28.

42 Find two numbers whose sum is 17 and whose difference is 89.

43 One number is 5 more than three times another. Find these numbers if their sum is 77.

44 One number is 15 more than another. Find these numbers if their sum is 193.

45 One number is 4 less than three times a second number. The second number is 17 less than twice the first number. Find these numbers.

46 The second of two numbers is 5 less than twice the first number. The first number is 40 less than three times the second number. Find these numbers.

47 One number is 2 more than four times a second number. Three times the first number minus eleven times the second number is 9. Find these numbers.

48 One number is 7 more than five times a second number. Four times the first number plus six times the second number is 2. Find these numbers.

Insurance Costs (49 and 50)

49 A family has a choice of two health insurance plans through the husband's employer. The first option is an HMO that will cost the family \$150 per month and will cover 100% of their health care expenses. The second option is insurance that will cost \$110 per month and will cover 80% of the family's health care expenses. At what level of health care expenses will both options cost the family the same amount for one year?

50 A family has a choice of two health insurance plans through the wife's employer. The first option is an HMO, which will cost the family \$135 per month and will cover 100% of their health care expenses. The second option is insurance, which will cost \$105 per month and will cover 85% of the family's health care expenses. At what level of health care expenses will both options cost the family the same amount for one year?

In Exercises 51–56 solve each system of nonlinear equations.

51 $\frac{12}{x} + \frac{4}{y} = 5$
$\frac{2}{x} - \frac{1}{y} = 0$

52 $\frac{1}{x} - \frac{2}{y} = 7$
$\frac{2}{x} + \frac{1}{y} = -1$

53 $\frac{3}{x} + \frac{2}{y} = 10$
$\frac{2}{x} - \frac{5}{y} = -6$

54 $\frac{15}{x} + \frac{6}{y} = 16$
$\frac{5}{x} - \frac{3}{y} = 2$

55 $\frac{1}{x+1} - \frac{1}{y-2} = 1$
$\frac{1}{x+1} + \frac{1}{y-2} = 5$

56 $\frac{1}{x-3} - \frac{1}{y+2} = \frac{1}{6}$
$\frac{1}{x-3} + \frac{1}{y+2} = \frac{5}{6}$

C

CALCULATOR USAGE (57–60)

In Exercises 57–60 solve each system of equations using a calculator. Give answers accurate to the nearest hundredth.

57 $1.67x + 2.34y = 2.91$
$2.06x - 1.78y = 1.49$

58 $7.35x - 8.24y = 42.08$
$4.28x + 9.03y = -20.58$

59 $\sqrt{2}x - \sqrt{5}y = -31.70$
$\sqrt{3}x + \sqrt{7}y = 9.79$

60 $\pi x + \pi y = -2.325$
$\sqrt{2}x + \sqrt{11}y = -5.688$

In Exercises 61 and 62 solve each system of linear equations for the ordered pair (x_1, x_2) that satisfies the system.

61 $2x_1 + 3x_2 = -4$
$3x_1 - 5x_2 = 13$

62 $2x_1 + 3x_2 = 11$
$3x_1 + 5x_2 = 31$

In Exercises 63 and 64 find the values of a and b that make $(x, y) = (5, 3)$ a solution of the system of equations.

63 $ax + by = 2$
$2ax + by = 13$

64 $3ax - by = 0$
$ax + 2by = 70$

In Exercises 65–68 solve each system for (x, y) in terms of the nonzero constants a and b.

65 $x - ay = b$
$x + ay = 2b$

66 $3ax + y = b$
$2ax - y = 4b$

67 $2ax + by = 3$
$3ax - 2by = 4$

68 $5ax - 3by = 7$
$2ax + 4by = 11$

DISCUSSION QUESTION

69 Explain in your own words what happens when a system of dependent equations is solved by the addition-subtraction method.

CHALLENGE QUESTION

Solve $\begin{Bmatrix} a_1x + b_1y = c_1 \\ a_2x + b_2y = c_2 \end{Bmatrix}$ for (x, y) in terms of coefficients a_1, a_2, b_1, b_2, c_1, and c_2. Assume that $a_1b_2 - a_2b_1 \neq 0$.

SECTION SKILL REINFORCERS

Solve the following exercises, given points $P = (3, 7)$ and $Q = (7, -3)$.

1 Determine the slope of the line through P and Q.
2 Write the equation of the line through P and Q.
3 Write the equation of a vertical line through P.
4 Write the equation of a vertical line through Q.
5 Write the equation of a horizontal line through Q.

3-3 Systems of Linear Equations in Three Variables

SECTION OBJECTIVE

5 Solve a system of three linear equations in three variables.

A Geometric Viewpoint: A First-Degree Equation in Three Variables

A first-degree equation in two variables of the form $Ax + By + C = 0$ is called a linear equation, since its graph is a straight line if A and B are not both zero. Similarly, a first-degree equation in three variables of the form $Ax + By + Cz + D = 0$ is also called a linear equation. However, this name is misleading because if A, B, and C are not all zero, the graph of $Ax + By + Cz + D = 0$ is not a line but a plane in three-dimensional space.

The graph of a three-dimensional space on two-dimensional paper is limited in its portrayal of the third dimension. Nonetheless, we can give the viewer a feeling for planes in a three-dimensional space by orienting the x-, y-, and z-axes as shown in the figure. This graph illustrates the plane $2x + 3y + 4z = 12$, whose x-intercept is (6, 0, 0), whose y-intercept is (0, 4, 0), and whose z-intercept is (0, 0, 3).

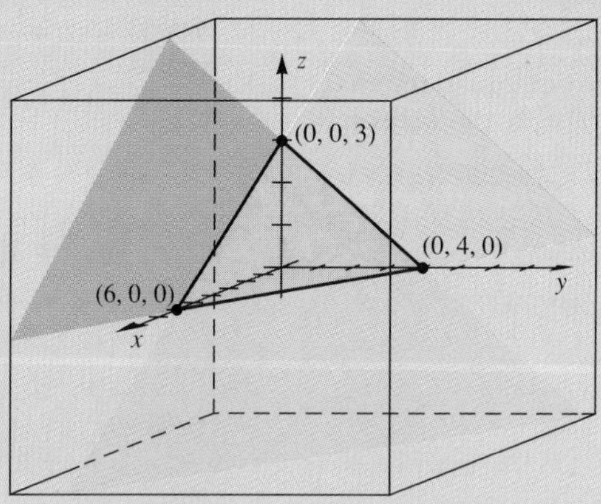

A system of three equations in three variables can be viewed geometrically as the intersection of a set of three planes. These planes may intersect in either one point, no points, or an infinite number of points. The illustrations in the following box show some of the ways we can obtain these solutions. Can you sketch other ways of obtaining these solution sets?

TYPES OF SOLUTION SETS FOR LINEAR SYSTEMS WITH THREE EQUATIONS

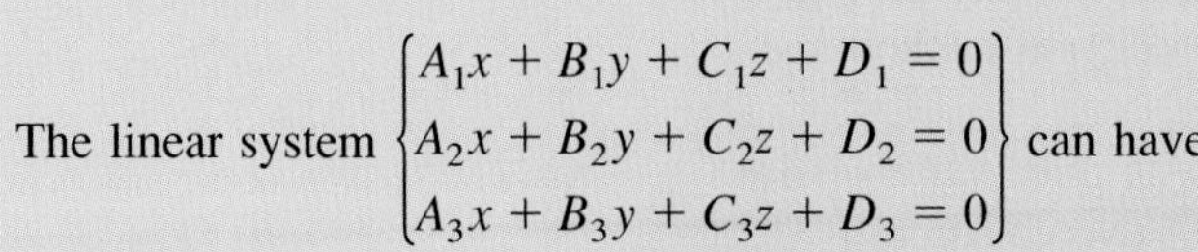

The linear system $\begin{Bmatrix} A_1x + B_1y + C_1z + D_1 = 0 \\ A_2x + B_2y + C_2z + D_2 = 0 \\ A_3x + B_3y + C_3z + D_3 = 0 \end{Bmatrix}$ can have

One Solution	***No Solution***	***An Infinite Number of Solutions***
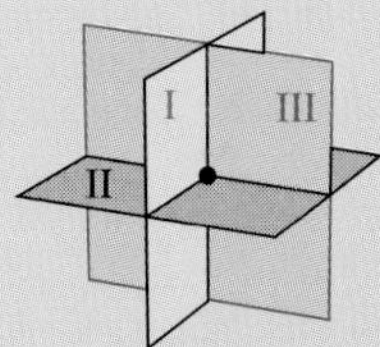	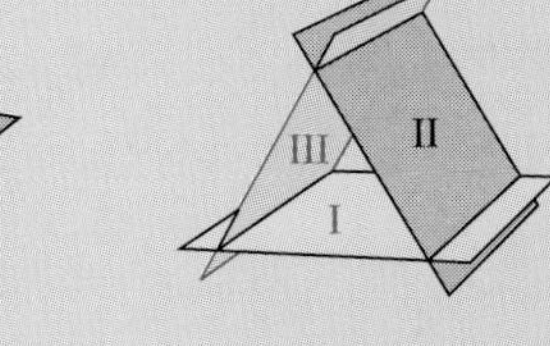	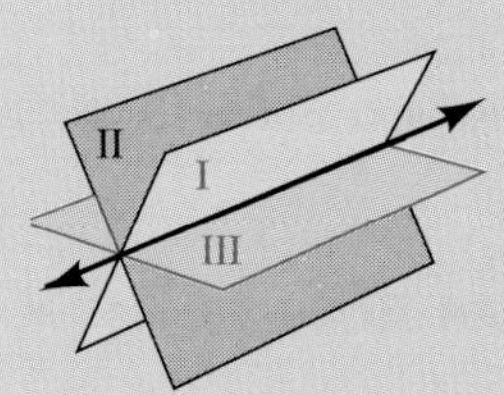
The planes intersect in a single point P; the system is consistent and the equations are independent.	The planes have no point in common; the system is inconsistent.	The planes intersect along a line and thus have an infinite number of common points; the system is consistent and the equations are dependent.

An ordered triple is a solution of a system of equations with three variables if and only if it is a solution of each equation in the system.

EXAMPLE 1 Determining Whether an Ordered Triple Is a Solution of a System of Linear Equations

Determine whether $(2, -3, 5)$ is a solution of $\begin{Bmatrix} x + y - z = -6 \\ 2x - y + z = 12 \\ 3x - 2y - 2z = 3 \end{Bmatrix}$.

SOLUTION

$$x + y - z = -6$$
$$2 + (-3) - 5 \stackrel{?}{=} -6$$
$$-6 = -6 \text{ is true.}$$

$$2x - y + z = 12$$
$$2(2) - (-3) + 5 \stackrel{?}{=} 12$$
$$4 + 3 + 5 \stackrel{?}{=} 12$$
$$12 = 12 \text{ is true.}$$

$$3x - 2y - 2z = 3$$
$$3(2) - 2(-3) - 2(5) \stackrel{?}{=} 3$$
$$6 + 6 - 10 \stackrel{?}{=} 3$$
$$2 = 3 \text{ is false.}$$

Answer No; $(2, -3, 5)$ must satisfy each equation to be a solution of the system. ▼

The figures in the box on the previous page can give us an intuitive understanding of the possible solutions to these systems; however, it is not practical to actually solve these systems graphically. Thus we will rely entirely on algebraic methods. In this section we will use the addition-subtraction method and the substitution method.

The strategy for solving a system of three equations with three variables is just an extension of the procedure used to solve systems of two equations:

$$\begin{Bmatrix}\text{3 equations in}\\ \text{3 variables}\end{Bmatrix} \longrightarrow \begin{Bmatrix}\text{2 equations in}\\ \text{2 variables}\end{Bmatrix} \longrightarrow \begin{Bmatrix}\text{1 equation in}\\ \text{1 variable}\end{Bmatrix}$$

$$\text{back-substitute} \longleftarrow \text{back-substitute} \longleftarrow$$

This strategy is described in the box below and illustrated by the following examples.

Solving Systems of Three Linear Equations in Three Variables

Step 1 Produce a system of two equations in two variables by using the addition-subtraction principle with two different pairs of equations to eliminate one of the variables.

Step 2 Use this system of equations in two variables to produce a single equation with only one variable. Then solve for this variable.

Step 3 Use the value determined in Step 2 to solve for the other two variables. First back-substitute this value into one of the two equations to solve for a second variable. Finally, back-substitute both of these values into one of the original equations to solve for the third variable.

The ordered triple obtained in Steps 2 and 3 is the solution.

If you obtain a contradiction in one of these steps, the system is inconsistent and has no solution. If you obtain an identity in any step, the system either contains dependent equations with infinitely many solutions or is inconsistent with no solution.

EXAMPLE 2 Solving a System of Three Linear Equations in Three Variables

Solve the system
$$\begin{matrix}(1)\\(2)\\(3)\end{matrix} \begin{Bmatrix} x + y + z = 2 \\ -x + y - 2z = 1 \\ x + y - z = 0 \end{Bmatrix}.$$

SOLUTION Eliminate one variable, x, by adding equations (1) and (2) and then (2) and (3).

$$\begin{array}{ll} (1) & x + y + z = 2 \\ (2) & \underline{-x + y - 2z = 1} \\ & 2y - z = 3 \end{array} \qquad \begin{array}{ll} (2) & -x + y - 2z = 1 \\ (3) & \underline{x + y - z = 0} \\ & 2y - 3z = 1 \end{array}$$

We now have two equations in two variables.

The resulting system is $\begin{Bmatrix} 2y - z = 3 \\ 2y - 3z = 1 \end{Bmatrix}$.

Eliminate the variable y from this system of two equations.

$$\begin{Bmatrix} 2y - z = 3 \\ 2y - 3z = 1 \end{Bmatrix} \xrightarrow[r_2' = -r_2]{r_1' = r_1} \begin{array}{r} 2y - z = 3 \\ -2y + 3z = -1 \\ \hline 2z = 2 \\ z = 1 \end{array}$$

Eliminate y to obtain a single equation in the variable z, and then solve for z.

Back-substitute this value of z into one of the equations with two variables.

$$\begin{aligned} 2y - z &= 3 \\ 2y - (1) &= 3 \\ 2y &= 4 \\ y &= 2 \end{aligned}$$

Then back-substitute the values of z and y into one of the original equations with all three variables.

$$\begin{aligned} (1) \quad x + y + z &= 2 \\ x + 2 + 1 &= 2 \\ x &= -1 \end{aligned}$$

Answer $(-1, 2, 1)$

Check Check $(-1, 2, 1)$.

(1) $x + y + z = 2$
$-1 + 2 + 1 \stackrel{?}{=} 2$
$2 = 2$ is true.

(2) $-x + y - 2z = 1$
$-(-1) + 2 - 2(1) \stackrel{?}{=} 1$
$1 = 1$ is true.

(3) $x + y - z = 0$
$-1 + 2 - 1 \stackrel{?}{=} 0$
$0 = 0$ is true.

The solution $(-1, 2, 1)$ checks in all three equations. ▼

Sometimes it is convenient to use both the substitution method and the addition-subtraction method to solve a system of three equations in three variables. This procedure is illustrated in the next example.

EXAMPLE 3 Solving a System of Three Linear Equations in Three Variables

Solve $\begin{Bmatrix} 2x - y = -1 \\ x + 2y - z = 19 \\ 3x - 2y + 4z = -32 \end{Bmatrix}$.

SOLUTION Use the substitution method to produce a system of two equations in two variables.

$$\begin{Bmatrix} 2x - y = -1 \\ x + 2y - z = 19 \\ 3x - 2y + 4z = -32 \end{Bmatrix} \longrightarrow \begin{Bmatrix} y = 2x + 1 \\ x + 2(2x + 1) - z = 19 \\ 3x - 2(2x + 1) + 4z = -32 \end{Bmatrix}$$

Solve the first equation for y, and substitute for y in the other two equations.

$$\longrightarrow \begin{Bmatrix} x + 4x + 2 - z = 19 \\ 3x - 4x - 2 + 4z = -32 \end{Bmatrix}$$

$$\longrightarrow \begin{Bmatrix} 5x - z = 17 \\ -x + 4z = -30 \end{Bmatrix}$$

Simplify the last two of these equations.

Then use the addition-subtraction method to produce a single equation with only one variable.

$$\begin{Bmatrix} 5x - z = 17 \\ -x + 4z = -30 \end{Bmatrix} \xrightarrow[r_2' = 5r_2]{r_1' = r_1} \begin{array}{r} 5x - z = 17 \\ \underline{-5x + 20z = -150} \\ 19z = -133 \\ z = -7 \end{array}$$

Multiply both members of the second equation by 5, and then add these new equations.

Solve for z.

Back-substitute to solve for the other two variables.

$$\begin{aligned} -x + 4z &= -30 \\ -x + 4(-7) &= -30 \\ -x - 28 &= -30 \\ -x &= -2 \\ x &= 2 \end{aligned}$$

Back-substitute for z, and then solve for x.

$$\begin{aligned} y &= 2x + 1 \\ y &= 2(2) + 1 \\ y &= 5 \end{aligned}$$

Back-substitute for x, and solve for y.

Answer $(2, 5, -7)$ Be sure to write the ordered triple in the correct order. Does this solution check? ▼

EXAMPLE 4 Solving an Inconsistent System

Solve the system
$$\begin{array}{l} \textbf{(1)} \\ \textbf{(2)} \\ \textbf{(3)} \end{array} \begin{Bmatrix} x + y - z = 3 \\ x - 5y + z = 4 \\ -4x + 5y + z = 5 \end{Bmatrix}.$$

SOLUTION Eliminate the variable z by adding equations (1) and (2) and then (1) and (3).

$$\begin{array}{lr} \textbf{(1)} & x + y - z = 3 \\ \textbf{(2)} & \underline{x - 5y + z = 4} \\ & 2x - 4y = 7 \end{array} \qquad \begin{array}{lr} \textbf{(1)} & x + y - z = 3 \\ \textbf{(3)} & \underline{-4x + 5y + z = 5} \\ & -3x + 6y = 8 \end{array}$$

Eliminate the variable y from this system of two equations in two variables.

$$\begin{Bmatrix} 2x - 4y = 7 \\ -3x + 6y = 8 \end{Bmatrix} \xrightarrow[r_2' = 2r_2]{r_1' = 3r_1} \begin{array}{r} 6x - 12y = 21 \\ \underline{-6x + 12y = 16} \\ 0 = 37 \end{array}$$

Add these equations.

This equation is a contradiction, so the system has no solution.

Answer This system is inconsistent and has no solution. ▼

SELF-CHECK ✓

Solve
$$\begin{Bmatrix} -x + y + z = 6 \\ x - 3y + 3z = 10 \\ x + y - 4z = -17 \end{Bmatrix}.$$

SELF-CHECK ANSWER ✓

(1, 2, 5)

The word problem in the next example uses three different variables to represent the three unknowns. Thus three separate *word equations* are necessary to solve for these variables. In general, a word problem with a unique solution must contain information that will yield as many equations as variables. For example, a problem with three variables should contain information that will yield three equations.

EXAMPLE 5 Dimensions of a Triangle

The perimeter of a triangle is 63 centimeters. The length of the longest side is 7 centimeters less than the sum of the lengths of the other two sides. The longest side is also 17 centimeters longer than the difference of the lengths of the other two sides. Find the length of each side.

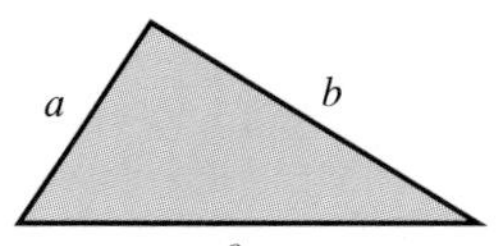

SOLUTION Draw a sketch like the one in the figure, and let

a = Length of the shortest side in centimeters

b = Length of the second side in centimeters

c = Length of the longest side in centimeters

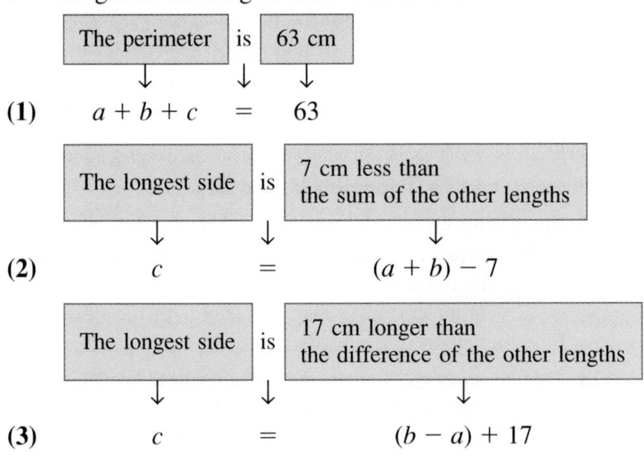

Translate each **word equation** into the corresponding algebraic equation.

Eliminate the variable c by substituting $a + b - 7$ for c in equations (1) and (3).

(1)
$$a + b + c = 63$$
$$a + b + (a + b - 7) = 63$$
$$2a + 2b - 7 = 63$$
$$2a + 2b = 70$$
$$a + b = 35$$

(3)
$$c = (b - a) + 17$$
$$a + b - 7 = (b - a) + 17$$
$$2a = 24$$
$$a = 12$$

In the equation obtained by substituting into equation (3), both b and c have been eliminated. Thus we can immediately back-substitute for a.

$$a + b = 35$$
$$12 + b = 35 \qquad \text{Back-substitute for } a.$$
$$b = 23$$

$$c = (a + b) - 7$$
$$c = (12 + 23) - 7 \qquad \text{Back-substitute for } a \text{ and } b.$$
$$c = 28$$

Answer The triangle has sides of 12 cm, 23 cm, and 28 cm. Do these values check?

▼

Exercises 3-3

A

In Exercises 1–20 solve each system of three linear equations in three variables.

1 $\begin{aligned} x + 2y + z &= 11 \\ -x - y + 2z &= 1 \\ 2x - y + z &= 4 \end{aligned}$

2 $\begin{aligned} x + 2y + 3z &= 7 \\ x - 3y - 2z &= -13 \\ 2x - y + 2z &= 5 \end{aligned}$

3 $\begin{aligned} x + y - z &= 1 \\ 2x + y + z &= 4 \\ x - y - 2z &= -2 \end{aligned}$

4 $\begin{aligned} 3x - y + 2z &= 4 \\ 2x + 2y - z &= 10 \\ x - y + 3z &= -4 \end{aligned}$

5 $\begin{aligned} 5x + y + 3z &= -1 \\ 2x - y + 4z &= -6 \\ 3x + y - 2z &= 7 \end{aligned}$

6 $\begin{aligned} 3x - y + z &= 8 \\ 2x + 3y - z &= 0 \\ 4x + 2y + z &= 7 \end{aligned}$

7 $\begin{aligned} x - 10y + 3z &= -5 \\ 2x - 15y + z &= 7 \\ 3x + 5y - 2z &= 8 \end{aligned}$

8 $\begin{aligned} x + 2y + 2z &= 4 \\ 2x + y + 2z &= 3 \\ 3x + y - 4z &= 2 \end{aligned}$

9 $\begin{aligned} x + y &= -2 \\ -y + z &= 2 \\ x - z &= -1 \end{aligned}$

10 $\begin{aligned} 2x + y &= 7 \\ y - z &= 2 \\ x + z &= 2 \end{aligned}$

11 $\begin{aligned} 2x + z &= 7 \\ y - 2z &= -5 \\ x + 2y &= 4 \end{aligned}$

12 $\begin{aligned} x + y &= 0 \\ x + 2z &= 5 \\ y + z &= 4 \end{aligned}$

13 $\begin{aligned} 2x - 4y + 2z &= 6 \\ 3x - 6y + 3z &= 10 \\ 4x - 8y + 4z &= 11 \end{aligned}$

14 $\begin{aligned} x + 2y + 2z &= 2 \\ 2x - y + z &= 1 \\ 4x + 3y + 5z &= 3 \end{aligned}$

15 $\begin{aligned} \frac{x}{6} + \frac{y}{3} + \frac{z}{2} &= 1 \\ x - \frac{y}{2} + \frac{z}{2} &= 1 \\ \frac{x}{6} - \frac{y}{2} + \frac{z}{3} &= 0 \end{aligned}$

16 $\begin{aligned} \frac{x}{2} + y - \frac{z}{2} &= -\frac{1}{2} \\ \frac{x}{2} + \frac{y}{2} - \frac{3z}{2} &= -1 \\ \frac{x}{2} - \frac{y}{2} + \frac{z}{2} &= 2 \end{aligned}$

17 $\begin{aligned} 0.2x + 0.1y + 0.1z &= 0.6 \\ 0.3x + 0.2y + 0.2z &= 1.0 \\ -0.1x + 0.3y + 0.1z &= 0 \end{aligned}$

18 $\begin{aligned} 0.2x + 0.1y - 0.5z &= 0.3 \\ 0.1x - 0.3y - 0.2z &= -0.1 \\ 0.1x + 0.1y + 0.3z &= 1.3 \end{aligned}$

19 $\begin{aligned} 121x &= -11 \\ -35y &= 70 \\ 48z &= 24 \end{aligned}$

20 $\begin{aligned} -33y &= 11 \\ 19x &= -38 \\ 17z &= -170 \end{aligned}$

B

Numeric Word Problems (21–24)

21 The largest of three numbers is 2 more than the sum of the other two numbers. The sum of the largest number and the smallest number is 32 more than the second number. Find each of these numbers if their sum is 198.

22 The largest of three numbers is three times the second number. The second number equals the sum of the smallest number and the largest number. Find each of these numbers if their sum is 30.

23 The sum of three numbers is 65. The sum of the first two numbers exceeds the third number by 3. The sum of the last two numbers exceeds the first number by 35. Find these numbers.

24 The sum of three numbers is 0. The sum of the first number and twice the second number is 5 less than the third number. The sum of the first number and three times the third number is 2 more than the second number. Find these numbers.

Dimensions of a Triangle (25 and 26)

25 The perimeter of a triangle is 168 centimeters. The length of the longest side is twice that of the shortest side. The sum of the lengths of the shortest side and the longest side is 48 centimeters more than the length of the other side. Find the length of each side.

26 The length of the longest side of a triangle is 12 centimeters less than the sum of the lengths of the other two sides. The length of the shortest side is 10 centimeters more than one-half the length of the third side. Find the length of each side if the perimeter is 188 centimeters.

Angles of a Triangle (27 and 28)

27 One angle of a triangle is twice as large as another. The third angle is 9° larger than the sum of the other two angles. Find the number of degrees in each angle. (*Hint:* The sum of the angles of a triangle is 180°.)

28 The smallest angle in a triangle is 78° less than the largest angle. The other angle is three times as large as the smallest angle. How many degrees are in each angle?

C

In Exercises 29 and 30 solve each system of linear equations for the ordered triple (x_1, x_2, x_3) that satisfies the system.

29 $\begin{aligned} x_1 + 2x_2 + 3x_3 &= 12 \\ x_1 - 2x_2 + 2x_3 &= 15 \\ x_1 + 3x_2 + 4x_3 &= 15 \end{aligned}$

30 $\begin{aligned} x_1 + x_2 + x_3 &= -3 \\ x_1 + 2x_2 - x_3 &= 4 \\ x_1 - 2x_2 - 2x_3 &= 3 \end{aligned}$

In Exercises 31 and 32 solve each nonlinear system of equations.

31 $\frac{1}{x} - \frac{2}{y} - \frac{1}{z} = 5$
$\frac{2}{x} - \frac{1}{y} + \frac{3}{z} = 11$
$-\frac{3}{x} - \frac{2}{y} + \frac{2}{z} = 7$

32 $\frac{1}{x} - \frac{2}{y} - \frac{2}{z} = 10$
$\frac{2}{x} + \frac{3}{y} - \frac{1}{z} = 4$
$\frac{3}{x} + \frac{1}{y} - \frac{1}{z} = 8$

In Exercises 33 and 34 find constants, *a*, *b*, and *c* such that (1, −3, 5) is a solution of the system of equations.

33 $ax + by + cz = 5$
$ax - by - cz = -1$
$2ax + 3by + 4cz = 13$

34 $ax - 2by + 3cz = 50$
$2ax + 3by - cz = -26$
$3ax - by + 2cz = 30$

In Exercises 35 and 36 solve the system for (*x*, *y*, *z*) in terms of the nonzero constants *a*, *b*, and *c*.

35 $2ax + cz = 2$
$by - cz = 1$
$ax - by = 0$

36 $ax + by + cz = 0$
$2ax - by + cz = 14$
$-ax + by + 2cz = -21$

Calculator Usage

37 a. Use the functions given by $y_1 = 2x^2 + x - 3$ and $y_2 = -x^4 + 3x^3 - 4x$ to complete the table. Then use these points to draw a line graph approximation of each function on the same coordinate system.

x	y_1	y_2
−2.5		
−2		
−1.5		
0		
0.5		
1.5		

b. Use a graphics calculator to simultaneously graph these functions, and then use the ZOOM and TRACE features to approximate to the nearest tenth the coordinates of the points of intersection of these graphs.

Discussion Question

38 Describe in geometric terms how it is possible to obtain an identity in the course of solving a system of three linear equations in three variables when the system is inconsistent and has no solution.

Challenge Question

The number on each line segment drawn in the figure represents the total area of the two squares connected by the segment. Determine the length of a side of each square.

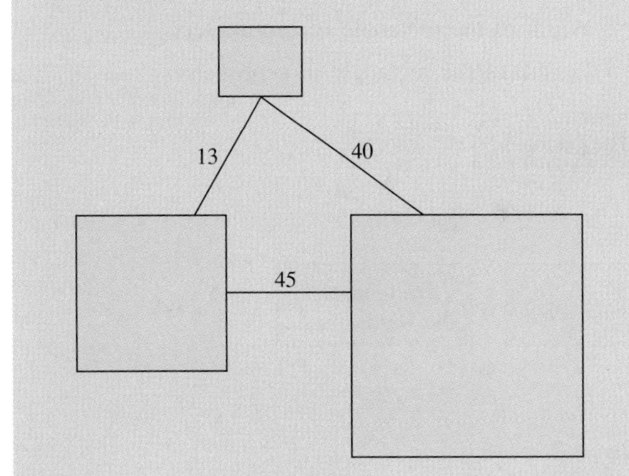

Section Skill Reinforcers

1 Write the equation of a horizontal line through (−4, 7). **2** Write the equation of the line through (−4, 7) and (1, −3). **3** Write the equation of a line through (−4, 7) and (0, 0). **4** Write the equation of a line through (−4, 7) and parallel to $y = -\frac{3x}{5}$. **5** Write the equation of a line through (−4, 7) and perpendicular to $y = -\frac{3}{5}$.

3-4 Applications of Systems of Linear Equations

SECTION OBJECTIVE

6 Solve word problems using systems of linear equations.

In keeping with the text's emphasis on problem solving, this section presents a number of applications. We will rely on the strategy that we first employed in Section 1-5 to solve problems with a single equation and only one variable. If a word problem involves two unknowns, we will generally use two variables and thus will need two equations. Similarly, if a word problem involves three unknowns, we will generally use a system of three equations with three variables. Remember to check all answers to verify that these numbers are meaningful in the given application.

EXAMPLE 1 Dimensions of a Rectangle

The perimeter of a rectangle is 78 centimeters. If the length is 7 centimeters more than the width, find the dimensions of the rectangle.

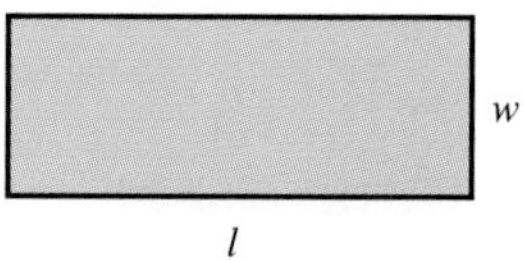

SOLUTION Draw a sketch like the one in the figure, and let

w = Width of the rectangle in centimeters
l = Length of the rectangle in centimeters

The perimeter	is	78 cm
↓	↓	↓
$2w + 2l$	$=$	78

Translate each **word equation** into the corresponding algebraic equation.

The length	is	7 cm more than the width
↓	↓	↓
l	$=$	$w + 7$

$$\begin{Bmatrix} 2w + 2l = 78 \\ l = w + 7 \end{Bmatrix}$$

$$\begin{aligned} 2w + 2(w + 7) &= 78 \\ 2w + 2w + 14 &= 78 \\ 4w &= 64 \\ w &= 16 \end{aligned}$$

Substitute for l in the first equation, and then solve for w.

$$\begin{aligned} l &= w + 7 \\ l &= 16 + 7 \\ l &= 23 \end{aligned}$$

Back-substitute for w, and then solve for l.

Answer The width is 16 cm, and the length is 23 cm. Do these dimensions check?

EXAMPLE 2 Food Rations

A zookeeper mixes two types of food for one of the animals in her care. Food A contains 20% protein and 50% carbohydrates, and food B contains 34% protein and 58% carbohydrates. How many grams of each of these foods should be mixed to obtain a ration containing 146 grams of protein and 284 grams of carbohydrates?

SOLUTION Let

a = Number of grams of food A

b = Number of grams of food B

	Amount of Ingredient in Food A	+	Amount of Ingredient in Food B	=	Total Amount of This Ingredient
Protein	$0.20a$	+	$0.34b$	=	146
Carbohydrates	$0.50a$	+	$0.58b$	=	284

The **word equation** is based on the mixture principle. The amount of each ingredient found in each food is determined by the rate principle.

$$\begin{cases} 0.20a + 0.34b = 146 \\ 0.50a + 0.58b = 284 \end{cases} \xrightarrow[r_2' = -2r_2]{r_1' = 5r_1} \begin{aligned} a + 1.70b &= 730 \\ -a - 1.16b &= -568 \\ \hline 0.54b &= 162 \end{aligned}$$

Use the addition-subtraction method to eliminate a.

$$b = 300$$

Then divide both members by 0.54.

$$\begin{aligned} a + 1.70b &= 730 \\ a + 1.70(300) &= 730 \\ a + 510 &= 730 \\ a &= 220 \end{aligned}$$

Back-substitute for b, and then solve for a.

Answer Mix 220 g of food A with 300 g of food B.

Do these values check?

EXAMPLE 3 Rate of a Boat in a Flowing Stream

A paddlewheel riverboat takes 30 minutes to go 10 kilometers downstream and $2\frac{1}{2}$ hours to return upstream. Determine the rate of the boat in still water and the rate of the current.

SOLUTION Let

b = Riverboat's rate in still water in kilometers per hour

c = Rate of the current in kilometers per hour

SELF-CHECK 

A coffee drinker buys a mixture of two types of coffee beans. One of the varieties costs \$6 per pound, and the other costs \$7 per pound. If 12 pounds of the mixture cost \$79.50, how many pounds of each variety are used?

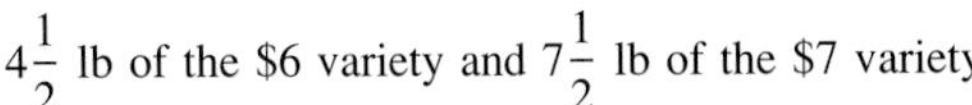

SELF-CHECK ANSWER

$4\frac{1}{2}$ lb of the \$6 variety and $7\frac{1}{2}$ lb of the \$7 variety

	Rate · Time = Distance
Downstream	$(b + c) \cdot \left(\frac{1}{2}\right) = 10$
Upstream	$(b - c) \cdot \left(\frac{5}{2}\right) = 10$

To maintain consistent units, convert 30 minutes to $\frac{1}{2}$ hour.

$$\begin{cases} \frac{1}{2}(b + c) = 10 \\ \frac{5}{2}(b - c) = 10 \end{cases} \xrightarrow[r_2' = 2r_2]{r_1' = 10r_1} \begin{aligned} 5b + 5c &= 100 \\ 5b - 5c &= 20 \\ \hline 10b \quad &= 120 \\ b &= 12 \end{aligned}$$

Use the addition-subtraction method to eliminate c, and then solve for b.

$$\begin{aligned} 5b + 5c &= 100 \\ 5(12) + 5c &= 100 \\ 60 + 5c &= 100 \\ 5c &= 40 \\ c &= 8 \end{aligned}$$

Back-substitute for b, and then solve for c.

Answer The rate of the boat in still water is 12 km/h, and the rate of the current is 8 km/h.

Do these values check?

EXAMPLE 4 Measure of the Angles of a Triangle

Angle A of a triangle is 68° more than the sum of the other two angles. This angle is also 31° more than the sum of angle B and twice angle C. Find the number of degrees in each angle.

SOLUTION Draw a sketch as in the figure shown, and let

a = Number of degrees in angle A

b = Number of degrees in angle B

c = Number of degrees in angle C

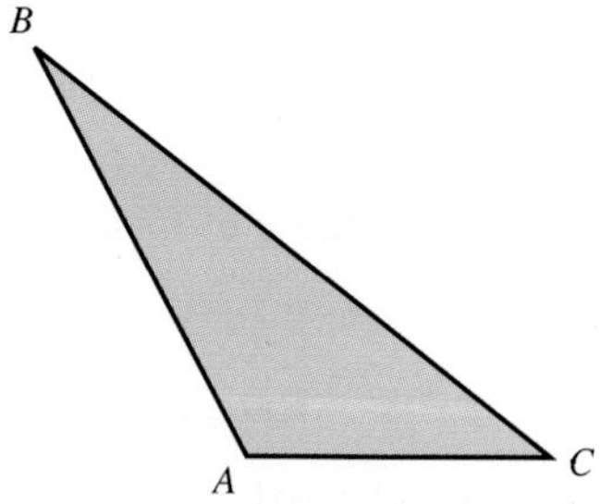

Although only two word equations are given directly, a third word equation is implied: The sum of the number of degrees of the angles of a triangle is always 180.

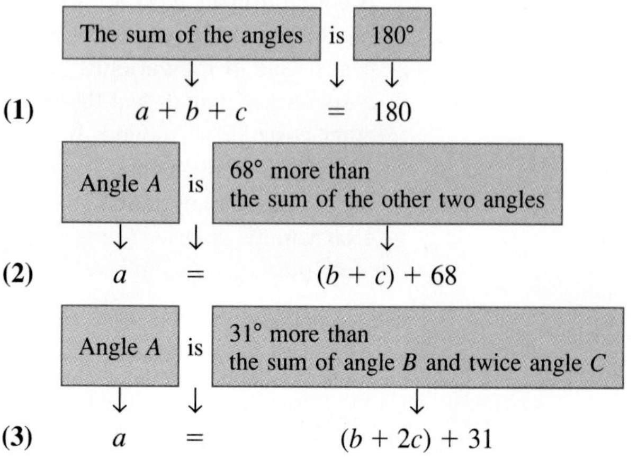

Translate each **word equation** into the corresponding algebraic equation.

Next rewrite these equations so that the constant terms are isolated in the right members of the equations.

$$\begin{array}{l} (1) \\ (2) \\ (3) \end{array} \begin{Bmatrix} a + b + c = 180 \\ a - b - c = 68 \\ a - b - 2c = 31 \end{Bmatrix}$$

Eliminate both b and c by adding equations (1) and (2).

$$\begin{array}{ll} (1) & a + b + c = 180 \\ (2) & \underline{a - b - c = 68} \\ & 2a = 248 \\ & a = 124 \end{array}$$

Now back-substitute this value of a into equations (1) and (3).

$$\begin{array}{llll} (1) & a + b + c = 180 & (3) & a - b - 2c = 31 \\ & 124 + b + c = 180 & & 124 - b - 2c = 31 \\ & b + c = 56 & & b + 2c = 93 \end{array}$$

Next eliminate b from this system of two equations in two variables.

$$\begin{Bmatrix} b + c = 56 \\ b + 2c = 93 \end{Bmatrix} \xrightarrow[r_2' = r_2]{r_1' = -r_1} \begin{array}{r} -b - c = -56 \\ \underline{b + 2c = 93} \\ c = 37 \end{array}$$

Back-substitute this value of c, and solve for b.

$$\begin{aligned} b + c &= 56 \\ b + 37 &= 56 \\ b &= 19 \end{aligned}$$

Thus $a = 124$, $b = 19$, and $c = 37$.

Answer Angle A is 124°, angle B is 19°, and angle C is 37°. ▼

EXAMPLE 5 Determining the Equation of a Parabola

The equation of a parabola that passes through points $(-1, -3)$, $(1, 3)$, and $(2, 12)$ is of the form $y = ax^2 + bx + c$. Find this equation by substituting the x and y values of each point into the equation $y = ax^2 + bx + c$ and then solving for a, b, and c.

SOLUTION

$$\begin{array}{llllll} (x, y) & & y = ax^2 & + bx & + c & \\ (-1, -3) & \rightarrow & -3 = a(-1)^2 & + b(-1) & + c \rightarrow (1) & a - b + c = -3 \\ (1, 3) & \rightarrow & 3 = a(1)^2 & + b(1) & + c \rightarrow (2) & a + b + c = 3 \\ (2, 12) & \rightarrow & 12 = a(2)^2 & + b(2) & + c \rightarrow (3) & 4a + 2b + c = 12 \end{array}$$

Eliminate both a and c by subtracting equation (2) from equation (1).

$$\begin{array}{ll} (1) & a - b + c = -3 \\ (2) & \underline{a + b + c = 3} \\ & -2b = -6 \\ & b = 3 \end{array}$$

Now back-substitute this value of b into equations (2) and (3).

$$\begin{aligned} &(2) \quad a + b + c = 3 & &(3) \quad 4a + 2b + c = 12 \\ & \quad a + 3 + c = 3 & & \quad 4a + 2(3) + c = 12 \\ & \quad a + c = 0 & & \quad 4a + c = 6 \end{aligned}$$

Next eliminate c from this system of two equations in two variables.

$$\begin{cases} a + c = 0 \\ 4a + c = 6 \end{cases} \xrightarrow[r_2' = r_2]{r_1' = -r_1} \begin{aligned} -a - c &= 0 \\ 4a + c &= 6 \\ \hline 3a &= 6 \\ a &= 2 \end{aligned}$$

Back-substitute this value of a, and solve for c.

$$\begin{aligned} a + c &= 0 \\ 2 + c &= 0 \\ c &= -2 \end{aligned}$$

Thus $a = 2$, $b = 3$, and $c = -2$.

Answer $y = 2x^2 + 3x - 2$

Exercises 3-4

A

In Exercises 1–22 solve each problem using a system of two linear equations in two variables.

Numeric Word Problems (1–4)

1 One number is 17 more than five times a second number. Four times the first number minus the second number is 11. Find these numbers.

2 The sum of twice one number and three times a second number is 41. Seven times the first number minus twice the second number is 6. Find these numbers.

3 The sum of two numbers is 30, and the difference of the larger minus the smaller is 16. Find these numbers.

4 The sum of two numbers is 3, and the difference of the larger minus the smaller is 2. Find these numbers.

5 Dimensions of a Lot The perimeter of a rectangular lot is 500 feet. The length of the lot is 20 feet less than twice the width. Find the dimensions of the lot (see the figure).

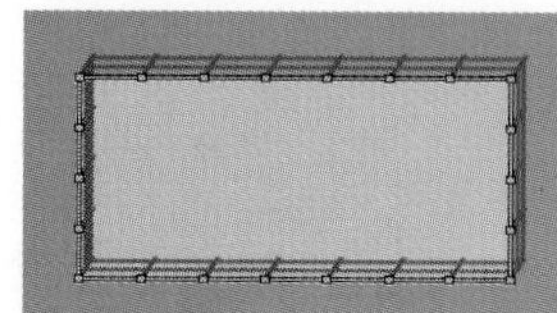

Figure for Exercise 5

6 Dimensions of a Room A rectangular room is three times as long as it is wide. If the perimeter is 26 meters, find the dimensions of the room.

7 Supplementary Angles Two angles are supplementary (their sum is 180°), and the larger is 12° more than three times the smaller. Find the number of degrees in each angle (see the figure).

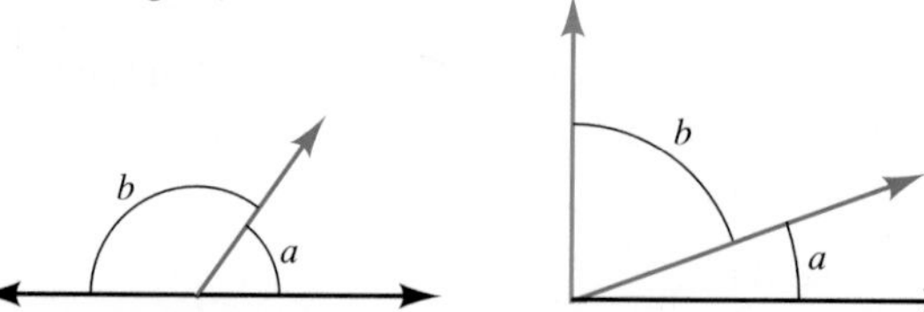

Figure for Exercise 7

Figure for Exercise 8

8 Complementary Angles Two angles are complementary (their sum is 90°), and the larger is 12° less than three times the smaller. Find the number of degrees in each angle (see the figure).

9 Cereal Mixture A dietitian is trying to create a nutritious breakfast cereal by combining two types of grain. Type A contains 25% protein and 40% carbohydrates, and type B contains 11% protein and 48% carbohydrates. How many grams of each of these grains should the dietitian use if the cereal is to contain 307 grams of protein and 856 grams of carbohydrates?

10 Feed Mixture A specialist in animal nutrition is experimenting with different feed rations by mixing feed from two sources. Source A contains 5% soybean meal and 60% ground corn. Source B contains 9% soybean meal and 55% ground corn. How many pounds of each should be mixed together to obtain a ration containing 88 pounds of soybean meal and 685 pounds of ground corn?

Fertilizer Mixture (11 and 12)

11 One fertilizer is 10% potash, and another is 20% potash. How many pounds of each fertilizer should be used to obtain a 100-pound mixture that is 16% potash?

12 One fertilizer is 18% nitrogen, and another is 12% nitrogen. A 200-pound mixture of these fertilizers is 14.55% nitrogen. How many pounds of each fertilizer were used to make the mixture?

13 Investment Income An investor received an income of \$365 from two investments. One investment earned interest at a rate of 8%, and the other earned interest at a rate of 11%. If the amounts invested at each rate had been switched, the investment would have generated \$395 in revenue. Find the amount invested at each rate.

14 Interest on Loans The total interest charged on two loans was \$667.50. The interest rate of one loan was 12%, and the other was 15%. If the amounts borrowed at each rate had been switched, the interest would have been \$750. Find the amount of each loan.

15 Number of Movie Tickets A theater that charges \$5 for adult tickets and \$2 for youth tickets collects receipts of \$4950 from the first showing of a movie. If all tickets had been priced at \$4, the income would have been \$5100. Find the number of adults and the number of youths at the first showing.

16 Photo Chemicals A photo lab paid \$235 for two chemicals. Chemical A costs \$5 per liter, and chemical B costs \$9 per liter. If the price of each chemical were to increase by \$1 per liter, the cost of this order would be \$270. How many liters of each chemical were ordered?

17 Machine Time A factory uses two machines to paint the cars produced on an assembly line. Each subcompact requires 5 minutes on machine A and 7 minutes on machine B. Each full-sized car requires 6 minutes on machine A and 8 minutes on machine B. If machine A can be operated 350 minutes per shift and machine B can be operated 480 minutes per shift, how many cars of each type could be painted if both machines were running at capacity?

18 Workers' Time One woodworking project requires 2 hours of a craftsman's time and 2 hours of his assistant's time. A second project requires 2 hours of the craftsman's time and 4 hours of his assistant's time. Find out how many of each of these two projects would be needed to occupy the craftsman for 32 hours and his assistant for 56 hours.

19 Speed of a Plane A plane can fly 1300 miles in 2 hours when it is flying with the wind. However, the plane takes $3\frac{1}{4}$ hours to fly this same distance when it is flying against the wind. Find the air speed of the plane and the speed of the wind.

20 Speed of a Barge A barge can go 90 miles downstream in 5 hours, but it takes $22\frac{1}{2}$ hours to make this trip back upstream. Determine the rate of the barge in still water and the rate of the current.

Break-Even Point (21 and 22)

21 The cost of producing x units of product includes fixed overhead costs of \$3000 and a variable cost of \$75 per unit. The income produced from selling x units of production is \$100 per unit. Determine the break-even value—that is, the number of units of production for which the cost and income are equal.

22 The cost of producing x units of product includes fixed overhead costs of \$5000 and a variable cost of \$125 per unit. The income produced from selling x units of production is \$150 per unit. Determine the break-even value—that is, the number of units of production for which the cost and income are equal.

B

In Exercises 23–30 solve each problem using a system of three linear equations with three variables.

Numeric Word Problems (23 and 24)

23 The sum of three numbers is 155. The largest number is 45 more than the sum of the other two numbers. The smallest number is twice the difference of the middle number minus the largest number. Find these numbers.

24 The sum of three numbers is three times as large as the middle number. Twice the largest number minus the smallest number is twice the middle number. When the sum of the largest and the smallest numbers is doubled and this product is increased by 2, the result is twelve times as large as the middle number. Find these numbers.

Angles of a Triangle (25 and 26)

25 One angle of a triangle is twice as large as the sum of the other two angles. The largest angle exceeds the smallest angle by 100°. Find the number of degrees in each angle (see the figure).

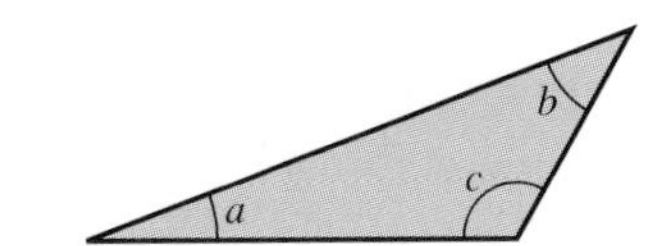

Figure for Exercise 25

26 The largest angle of a triangle is 20° less than the sum of the other two angles. Twice the sum of the largest angle and the medium-sized angle is seven times as large as the smallest angle. Find the number of degrees in each angle (see the figure).

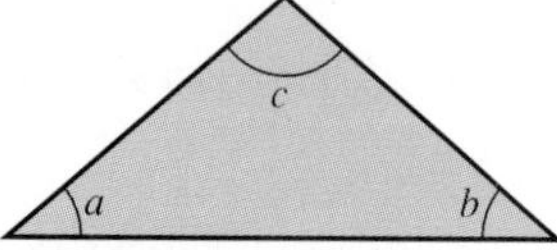

Figure for Exercise 26

Equation of a Parabola (27 and 28)

27 The equation of a parabola that opens vertically and passes through the points (1, 2), (−1, 12), and (2, 6) is of the form $y = ax^2 + bx + c$. Find this equation (see the figure).

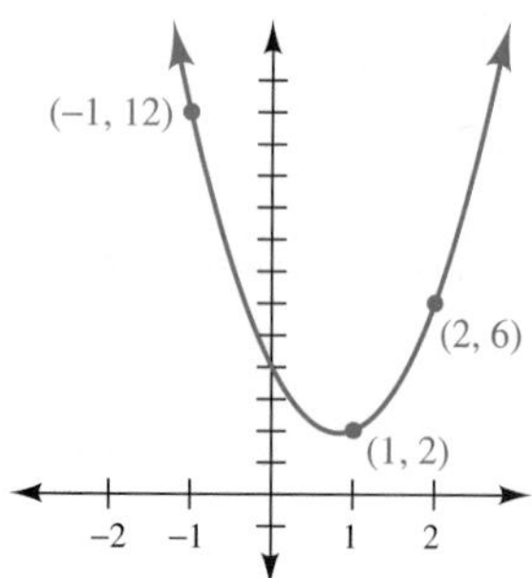

Figure for Exercise 27

28 The equation of a parabola that opens vertically and passes through the points (1, −5), (2, −2), and (−2, 10) is of the form $y = ax^2 + bx + c$. Find this equation (see the figure).

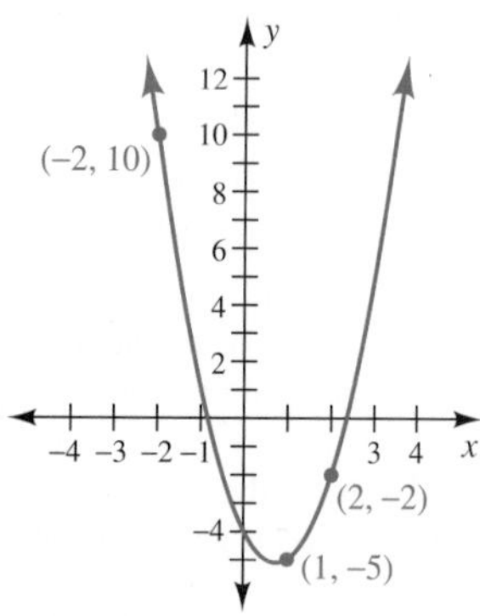

Figure for Exercise 28

Equation of a Circle (29 and 30)

29 Find constants D, E, and F such that the circle defined by the equation $x^2 + y^2 + Dx + Ey + F = 0$ passes through the points (1, 0), (−1, −2), and (3, −2).

30 Find constants D, E, and F such that the circle defined by the equation $x^2 + y^2 + Dx + Ey + F = 0$ passes through the points (−3, −4), (2, 1), and (0, 5).

C

In Exercises 31–36 solve each problem using a system of linear equations.

31 Cubic Equation Find constants A, B, and C for the equation $y = x^3 + Ax^2 + Bx + C$ so that the points (−1, −7), (1, 1) and (2, 17) satisfy the equation (see the figure).

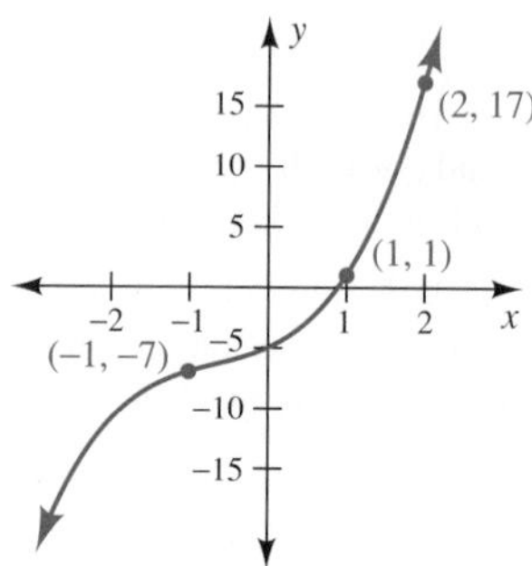

Figure for Exercise 31

32 Investment Income A teacher split \$18,000 of her retirement money among three investments. During the first year, a savings account provided income at the rate of 7% per year, an insurance company annuity earned 10% per year, and utility stocks earned 12% per year, for a total annual investment income of \$1770. During the second year, the same investments earned only \$1590. The savings account and the annuity earned the same amount as in the first year, but the utility stocks earned only 9%. Determine the amount invested in each of these investments.

33 Mixture of Foods A zookeeper mixes three foods, the contents of which are described in the table below. How many grams of each food are needed to produce a mixture with 133 grams of fat, 494 grams of protein, and 1700 grams of carbohydrates?

	A	B	C
Fat	6%	4%	5%
Protein	15%	18%	20%
Carbohydrates	45%	65%	70%

34 Production of Computer Chips Last month a small electronics company spent \$44,250 producing three types of special-purpose computer chips. The chips cost \$2.00, \$3.25, and \$4.50 to produce, and they were sold for \$2.50, \$3.50, and \$4.75, respectively. If a total of 11,500 chips were produced and the gross income was \$47,375, determine how many of each type of chip were produced.

35 Use of Farmland A farmer must decide how many acres of each of three crops to plant during this growing season. The farmer must pay a certain amount for seed and devote a certain amount of labor and water to each acre of crop planted, as shown in the table below.

	A	B	C
Seed Cost	\$120	\$85	\$80
Hours of Labor	4	12	8
Gallons of Water	500	900	700

The amount of money available to pay for seed is \$26,350. The farmer's family can devote 2520 hours to tending the crops, and the farmer has access to 210,000 gallons of water for irrigation. How many acres of each crop would use *all* of these resources?

36 Emptying a Tank A portable storage tank used by a fire department can be emptied in 15 minutes by three hoses running simultaneously. If the first two hoses operate for 10 minutes and then are turned off, the third hose takes an additional 20 minutes to empty the tank. If the second hose is replaced by another hose like the first hose, these three hoses running simultaneously can empty the tank in 18 minutes. How many minutes would it take each hose separately to empty the tank?

DISCUSSION QUESTION

37 Write your own word problem that the following system of equations algebraically models.

$$\begin{aligned} x + y + z &= 10{,}000 \\ 0.05x + 0.05y + 0.08z &= 650 \\ 0.04x + 0.06y + 0.07z &= 590 \end{aligned}$$

CHALLENGE QUESTION

A computer that is generating special effects for a movie is creating the image of a post, with the post's top at point (4, 5) and its bottom at (3, −1). For this scene the coordinate system being used places the movie screen's top at (1, 18) and its bottom at (13, 2). The viewer is located at (31, 28). Approximate to the nearest hundredth the coordinates of the points where the top and the bottom of the post will appear on the screen.

SECTION SKILL REINFORCERS

1 Evaluate $f(2)$ for $f(x) = 3x^2 - 2x + 4$. **2** Evaluate $f(-2)$ for $f(x) = 3x^2 - 2x + 4$. **3** Evaluate $-f(2)$ for $f(x) = 3x^2 - 2x + 4$. **4** For $f(x) = 0.25x^3$, complete a table of values using the x values of $-3, -2, -1, 0, 1, 2,$ and 3. **5** Sketch the line graph approximation of the graph of $f(x) = 0.25x^3$ using the points from the table of values completed above.

Key Concepts for Chapter 3

1 Systems of linear equations:

a. Consistent systems of independent equations have exactly one solution.

b. Inconsistent systems have no solution.

c. Consistent systems of dependent equations have an infinite number of solutions.

2 The linear system $\left\{\begin{aligned} y &= m_1x + b_1 \\ y &= m_2x + b_2 \end{aligned}\right\}$ has

a. one solution if $m_1 \neq m_2$.

b. no solution if $m_1 = m_2$ and $b_1 \neq b_2$.

c. an infinite number of solutions if $m_1 = m_2$ and $b_1 = b_2$.

3 Methods of solving systems of linear equations:

a. Graphical method

b. Substitution method

c. Addition-subtraction method

4 To graph the solution of a linear inequality:

a. Graph the corresponding equality using a solid line if the equality is included in the solution or a dashed line if the equality is not included in the solution.

b. Choose an arbitrary test point not on the line; (0, 0) is often convenient. Substitute this test point into the inequality.

c. If the test point satisfies the inequality, shade the region containing this point. If the test point does not satisfy the inequality, shade the other region.

Review Exercises for Chapter 3

In Exercises 1–6 graph the solution of each system of equations or inequalities.

1 $x - y = 1$
$5x - 2y = 4$

2 $x - 3y = 0$
$2x + 3y = -9$

3 $x - 2y = 4$
$x - 2y = -4$

4 $2x - 3y < 6$
$2x + 5y > 10$

5 $3x + 4y \le 12$
$3x - 4y \le 12$

6 $2x + y \le 4$
$x \ge 0$
$y \ge 0$

In Exercises 7–16 solve each system of equations by either the substitution method or the addition-subtraction method.

7 $3x + 4y - 9 = 0$
$4x + 3y + 2 = 0$

8 $y = -1.5x + 1$
$6x + 6y = 7$

9 $y = \frac{2}{5}x - 1$
$6x - 15y = 10$

10 $0.4x + 0.7y = -0.3$
$0.5x - 0.8y = 8.0$

11 $x - 3z = 16$
$2x - y + 2z = 10$
$3x + y - 2z = 25$

12 $2x + 3y + 5z = 1$
$2x - 3y - 5z = 1$
$2x - 3y + 5z = -1$

13 $\frac{x}{6} + \frac{y}{10} = 1$
$\frac{x}{3} - \frac{3y}{5} = -2$

14 $0.2x - 0.3y = -0.27$
$0.7x + 0.8y = 0.35$

15 $x + y + z = -3$
$2x + 3y - 5z = 22$
$3x - 5y + 2z = -5$

16 $\frac{6}{x} + \frac{6}{y} = 1$
$\frac{8}{x} - \frac{9}{y} = 7$

In Exercises 17–19 write each equation in slope-intercept form, and then identify the type of system and the number of solutions of the system.

17 $15x - 25y + 20 = 0$
$21x - 35y + 30 = 0$

18 $15x - 25y + 20 = 0$
$21x - 35y + 28 = 0$

19 $15x - 25y + 20 = 0$
$21x + 35y + 30 = 0$

CALCULATOR USAGE (20–22)

In Exercises 20–22 use a graphics calculator to graph the solution of each system of equations, and then approximate each coordinate to the nearest tenth.

20 $y = 0.25x - 0.45$
$y = -0.3x + 0.1$

21 $y = -\frac{5x}{3} + 10$
$y = \frac{5x}{9} + \frac{10}{3}$

22 $0.2x - 0.3y = -0.27$
$0.7x + 0.8y = 0.35$

23 Find constants a, b, and c such that $(-2, 1, 7)$ is a solution of the system of equations

$$\begin{aligned} ax + by + cz &= 35 \\ ax + by - cz &= -21 \\ 2ax - by + 3cz &= 89 \end{aligned}$$

24 Numeric Word Problem One number is 3 more than twice a second number. Three times the first number minus seven times the second number yields a difference of 5. Find these numbers.

25 Supplementary Angles Two angles are supplementary, and the larger angle is 12° more than six times the smaller angle. Find the number of degrees in each angle (see the figure).

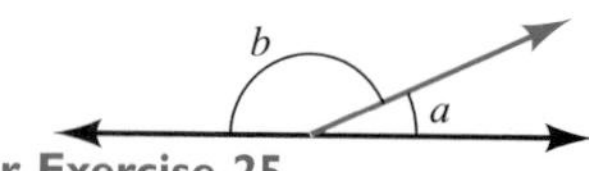

Figure for Exercise 25

26 Dietary Mixture A dietitian is creating a food supplement using two sources. Source A contains 15% protein and 40% carbohydrates, and source B contains 20% protein and 30% carbohydrates. How many grams of each source should be mixed to obtain a supplement containing 90 grams of protein and 170 grams of carbohydrates?

27 Dimensions of a Triangle The perimeter of the triangle shown in the figure is 170 centimeters. The length of the longest side is 10 centimeters less than the sum of the lengths of the other two sides. The length of the shortest side is one-half the length of the longest side. Determine the length of each side.

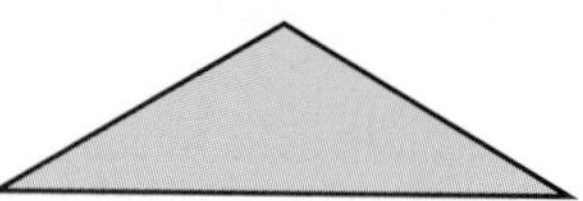

Figure for Exercise 27

28 Numeric Word Problem The sum of two numbers is 7 more than a third number. Twice the first number plus three times the second number is 44 more than the third number. Four times the first number plus the second number is 8 less than the third number. Find each of these numbers.

29 Equation of a Parabola Find the equation of a parabola of the form $y = ax^2 + bx + c$ that passes through the points $(0, 6)$, $(1, -1)$, and $(2, 4)$ (see the figure).

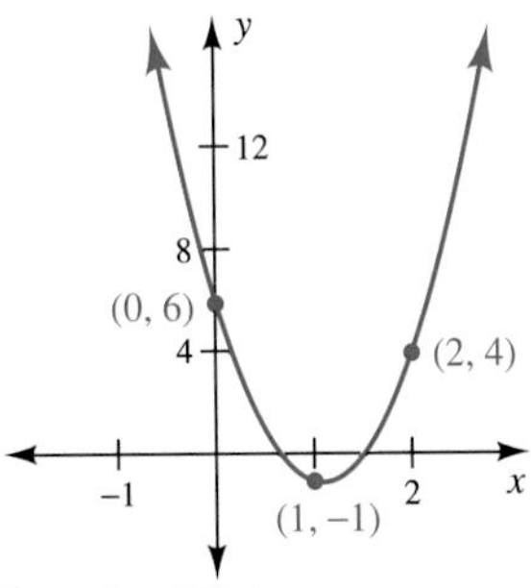

Figure for Exercise 29

30 Transportation Costs A traveler has two options for transportation from an airport to her hotel. One option is to take the airport shuttle for \$10. The second option is to use a taxi that costs \$4 plus \$0.75 per mile. Graph each of these cost options, and determine the distance that would result in the same cost for both options.

31 Sales of Sporting Equipment The line graphs shown in the figure give the sales in millions of dollars of golf shoes and of water skis for selected years. Determine the year(s) in which these sales were approximately the same.

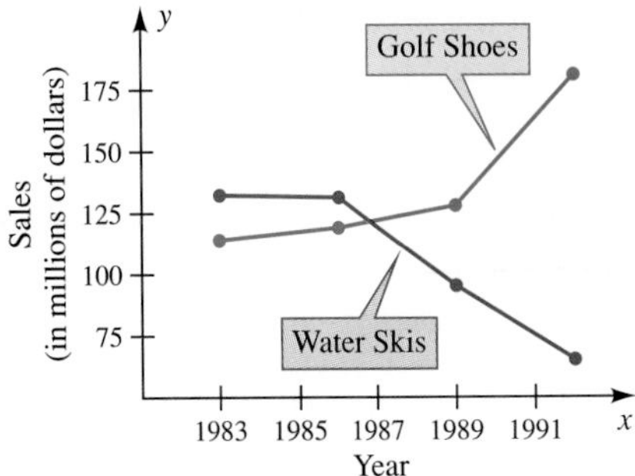

Figure for Exercise 31

Mastery Test for Chapter 3

Exercise numbers correspond to Section Objective numbers.

[3-1] 1 Solve each of these systems graphically.

a. $3x - 5y + 15 = 0$
$12x - 5y - 30 = 0$

b. $y = 3x - 4$
$y = -\frac{5}{2}x + \frac{3}{2}$

c. $y = \frac{3}{4}x - 2$
$y = \frac{3}{4}x + 2$

[3-1] 2 Graph the solution of each of these systems of inequalities.

a. $5x - 2y > 10$
$3x + 2y \leq 6$

b. $y \geq -3$
$y < 4$

c. $x \geq 0$
$y \geq 0$
$x + y \leq 2$

[3-2] 3 Solve each system of equations by the substitution method.

a. $y = 3x - 14$
$4x + 7y = -23$

b. $x - 3y = 11$
$5x + 4y = -2$

c. $y = 2x - 3$
$-3x + 1.5y = -4$

[3-2] 4 Solve each system of equations by the addition-subtraction method.

a. $8x + 9y = 1$
$2x - 3y = 2$

b. $2x - 8y = 10$
$-3x + 12y = -15$

c. $7x - 8y = 4$
$14x + 24y = -7$

[3-3] 5 Solve each of these systems of linear equations.

a. $x - 2y + 4z = 5$
$2x + y - 3z = 9$
$3x + y - 2z = 15$

b. $x + 2y + 3z = 17$
$3x + y - 2z = -21$
$-5x + 7y + 4z = 31$

[3-4] 6 Solve each problem using a system of equations.

a. Dimensions of a Triangle The perimeter of a triangle is 22 centimeters. The length of the shortest side is 12 centimeters less than the sum of the lengths of the other two sides. The longest side is twice as long as the shortest side. Find the length of each side.

b. Interest on Two Investments A man retired and invested some of his savings at 7.5% and the rest at 9%. His interest for one year was \$795. If he had switched the amounts invested at each rate, his interest for the year would have been \$855. How much did he invest at each rate?

c. Speed of a Barge A barge can go 230 kilometers downstream in 10 hours. However, the same barge takes 46 hours to make the trip back upstream. Determine the rate of the barge in still water and the rate of the current.

Cumulative Review of Chapters 1–3

The limited purpose of this review is to help you gauge your mastery of the first three chapters. It is not meant to examine each detail from these chapters, nor is it meant to focus on specific sections that may be emphasized at any particular school.

In Exercises 1–12 evaluate each expression for $x = -2$, $y = -3$, and $z = 4$.

1 $x + yz$ **2** $(x + y)z$ **3** $-5xyz$

4 $\dfrac{z - x}{y}$ **5** $|x + y + z|$ **6** $|x| + |y| + |z|$

7 $x^2 + y^2$ **8** $(x + y)^2$ **9** $(y + z)^3$

10 $y^3 + z^3$ **11** $x^2y^2 + xy^2$ **12** $\dfrac{x - 2(3y + z) + 2}{z - 3(2x + y)}$

In Exercises 13–16 simplify each expression in part a and solve each equation in part b.

	Simplify	*Solve*
13	**a.** $6m - 2(m - 1)$	**b.** $6m - 2(m - 1) = 0$
14	**a.** $3(x - 4) + 3(x + 2)$	**b.** $3(x - 4) + 3(x + 2) = 0$
15	**a.** $5(2x - 4) - 2(5x - 9)$	**b.** $5(2x - 4) - 2(5x - 9) = 0$
16	**a.** $9(4x - 2) - 6(6x - 3)$	**b.** $9(4x - 2) - 6(6x - 3) = 0$

In Exercises 17–20 simplify each expression.

17 $3x - 4y - 5z - (2x - 3y + 8z)$

18 $-2a - 3b + 8c - (3a - 2b - 7c)$

19 $18 - 6[2m - (3m - 5)]$

20 $-24 - 3[6r - (9 - 4r) - 11]$

In Exercises 21–26 solve each equation for x.

21 $3(5x - 2) - 4(3x - 1) = -11$

22 $\dfrac{x + 3}{2} + \dfrac{3x + 1}{4} = 8$

23 $a = bx$ **24** $c = \dfrac{2x - 3}{b}$

25 $|2x - 5| = 23$ **26** $\dfrac{|x - 3|}{2} = 7$

In Exercises 27–30 solve each inequality.

27 $2m - 5 < 4m - 11$ **28** $-13 < 3v - 1 < 8$

29 $|x + 1| < 5$ **30** $2x < 6$ or $3x > 21$

31 Write the interval notation for each of these inequalities.

a. $2 \le x < 4$ **b.** $-2 < x \le 5$

c. $-1 \le x \le 0$ **d.** $x \ge 3$

In Exercises 32–35 name the property that justifies each statement.

32 $5(x + z) = 5x + 5z$ **33** $5(x + z) = 5(z + x)$

34 $5(x + z) = (x + z)(5)$ **35** $5 + (x + z) = (5 + x) + z$

36 Identify all of the numbers from the set $\left\{-7, -4.73, -\pi, -\sqrt{4}, -\dfrac{3}{7}, 0, \sqrt{7}, 17\right\}$ that are

a. Real numbers **b.** Irrational numbers

c. Rational numbers **d.** Integers

e. Whole numbers **f.** Natural numbers

37 Calculate the slope of the line through $(-1, -8)$ and $(5, 4)$.

In Exercises 38–41 determine the slope of each line.

38 $y - 2 = 3(x + 4)$ **39** $y = 4$

40 $x = 4$ **41** $3x + 4y = 24$

In Exercises 42–45 graph the line satisfying the given conditions.

42 x-intercept $(2, 0)$ and y-intercept $(0, -3)$

43 $m = \dfrac{3}{5}$ with y-intercept $(0, -2)$

44 $x = 2$ **45** $y = 2$

In Exercises 46–50 write the equation of the line satisfying the given conditions.

46 A horizontal line through $(3, 7)$

47 A vertical line through $(3, 7)$

48 A line through $(3, 7)$ with slope -2

49 A line through $(3, 7)$ and $(7, 3)$

50 A line through $(3, 7)$ and parallel to $y = -\dfrac{2}{3}x + 5$

In Exercises 51–58 evaluate each expression, given $f(x) = 5x^3 - 3x^2 + x - 2$ and $g(x) = \dfrac{2x + 1}{x - 2}$.

51 $f(2)$ **52** $f(3)$ **53** $f(2) - f(3)$

54 $f(2 - 3)$ **55** $g(0)$ **56** $-g(0.5)$

57 $g(-0.5)$ **58** $g(2)$

In Exercises 59 and 60 determine the domain of each function.

59 $f(x) = 5x - 7$ **60** $f(x) = \dfrac{5}{x - 7}$

In Exercises 61–64 determine whether each relation is a function.

61 $\{(3, 4), (5, 7), (2, 4), (6, 7)\}$

62 $\{(2, 1), (3, 8), (4, 7), (2, 6)\}$

63 **64**

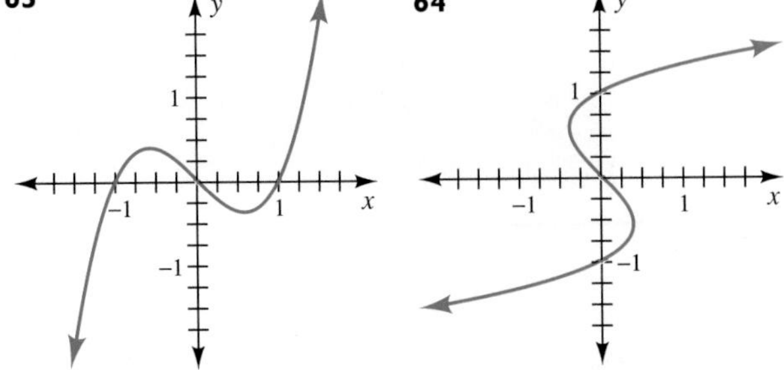

65 500-Meter Speed Skating The following table gives the women's record times in seconds for 500-meter speed skating for selected years. Draw a line graph for this function. The 1988 record was set by Bonnie Blair, a five-time gold medal winner in the Olympics and a former student at Parkland College.

Year	1964	1968	1972	1976	1980	1984	1988
Time	45.0	46.1	43.3	42.8	41.8	41.0	39.1

66 World Population The line graph shown in the figure gives the world population in billions for selected years starting with 1801.

a. In what year did the population reach 2 billion?

b. What was the population in 1986?

c. How many years did it take for the population to increase from 1 billion to 2 billion?

d. Was the population increasing faster in 1930 or in 1960?

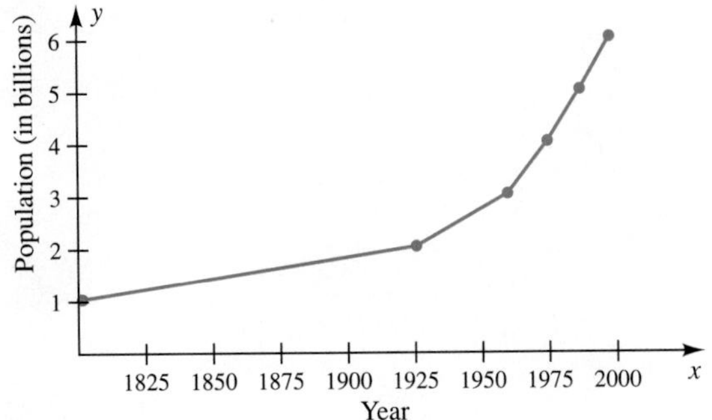

Figure for Exercise 66

67 Biological Infection The function defined by the graph shown gives the level of a biological infection in a person over a 3-week period. Match each time period 1, 2, 3, and 4 to the event that occurred in that time period.

Event A: An effective antibiotic is administered.

Event B: The person is exposed repeatedly to the infection in her work environment.

Event C: The antibiotic is discontinued, and the infection level returns to normal.

Event D: The body's immune system weakens, and the infection level begins to increase.

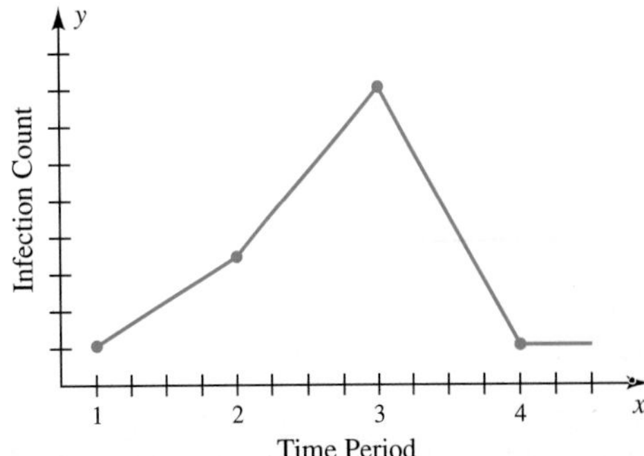

Figure for Exercise 67

In Exercises 68–70 graph the solution of each system of equations or inequalities.

68 $3x - y = -1$
$2x + y = 6$

69 $3x + 4y \leq 12$
$x \geq 2y$

70 $3x + 4y \leq 12$
$x \geq 0$
$y \leq 0$

In Exercises 71–73 solve each system of equations.

71 $x = 5y + 4$
$3x - 4y = -10$

72 $4x + 9y = -1$
$6x + 6y = 1$

73 $x + 2y + 3z = 3$
$x - 2y + 4z = 8$
$2x + 3y + 5z = 6$

74 Variation The variable v varies directly as x and inversely as y. Find v when $x = 5$ and $y = 6$, if $v = 2$ when $x = \frac{1}{2}$ and $y = 3$.

75 Electronic Detection The ratio of two readings from an electronic detection device was $\frac{5}{12}$. If the second reading was 28 units more than the first reading, determine the number of units for each reading.

76 Investment A high school student invested \$2000 of summer income to save for college. A portion was deposited in a savings account that earned only 4%. The rest was invested in a local electrical utility and earned 8%. The total income for 1 year was \$136. How much did the student put in the savings account?

77 Dog Food Mixture A mixture of dog food is created by mixing two sources. Source A contains 12% protein and 40% carbohydrates, and source B contains 25% protein and 30% carbohydrates. How many grams of each source should be mixed to obtain a supplement containing 272 grams of protein and 480 grams of carbohydrates?

78 Calculator Usage

a. Draw a line graph approximation of $y = (x - 2)^2 - 3$ by using the points with x-coordinates $-1, 0, 1, 2, 3, 4,$ and 5.

b. Then use a graphics calculator to graph this function.

Chapter 4

INTEGER EXPONENTS AND POLYNOMIALS

AN APPLICATION

Microeconomics

Economics is the study of finance and resource management. Microeconomics, as the name implies, focuses on the small and local—the operation of a single company, the finances of a single household, or the buying behavior of a single consumer. A husband and wife deciding which health insurance to buy may not think of themselves as engaging in microeconomics, but that is exactly what they are doing when they estimate how high a premium they can afford and what coverage they need.

Whereas a family often makes financial decisions based on a few calculations and lots of intuition, a business needs more reliable methods—methods based on mathematical models of the company's situation. Many of these models involve **polynomials,** which you will study in this chapter. At the end of the chapter you will be able to try your hand at advising a company by analyzing polynomial equations.

For now, let's consider a question involving no calculation. A company with five employees produces old-fashioned, push-powered lawn mowers. The company finds that it makes very little money selling 100 mowers a month, a good deal of money selling 300 a month, and not very much money selling 500 a month. We can picture the relationship between its sales and profits as shown in the following figure.

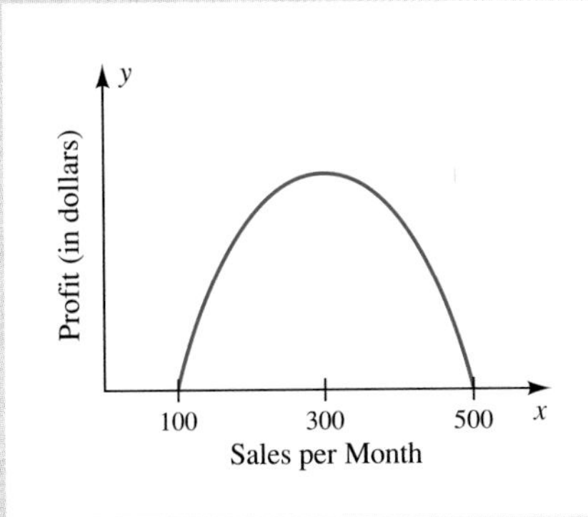

That low sales produce little profit is to be expected, but why might high sales produce little profit? What problems might arise in manufacturing and selling 500 push mowers a month? High sales levels are not always linked to reduced profits, but the phenomenon is quite common.

Chapter 4 Objectives

1 Use integer exponents (Section 4-1).

2 Use the properties of exponents (Section 4-2).

3 Use scientific notation (Section 4-2).

4 Determine the degree of a polynomial, classify a polynomial according to the number of terms, and write a polynomial in standard form (Section 4-3).

5 Evaluate a polynomial for given values of the variables (Section 4-3).

6 Add and subtract polynomials (Section 4-4).

7 Multiply polynomials using a vertical format (Section 4-5).

8 Multiply binomials by inspection (Section 4-5).

9 Divide polynomials (Section 4-6).

10 Use synthetic division (Section 4-7).*

11 Evaluate $P(a)$ using the Remainder Theorem (Section 4-7).*

This chapter will review the properties of natural number exponents. The concept of exponents will be extended to include zero and negative integers, and then integer exponents will be used to express numbers in scientific notation. Polynomials and operations with polynomials will also be examined. Polynomials, the simplest of algebraic expressions, play a fundamental role in the study of algebra and are encountered in many common problems in business and the sciences.

4-1 Integer Exponents

Section Objective

1 Use integer exponents.

Natural number exponents provide a concise way of indicating repeated multiplication. For example, it is much easier to write the exponential form x^{10} than the expanded form

$$x \cdot x \cdot x \cdot x \cdot x \cdot x \cdot x \cdot x \cdot x \cdot x$$

Note that the exponent of an algebraic expression refers only to the quantity immediately in front of it. As the next example illustrates, it is important to properly determine the base when using an exponential expression.

EXAMPLE 1 Identifying Bases and Exponents

Determine the base and the exponent of each of these exponential expressions.

SOLUTIONS

(a) x^2 — x^2 means $x \cdot x$, with the base x and exponent 2; it is read as "x squared" or "x to the second power."

(b) $-x^2$ — $-x^2$ means $-(x \cdot x)$, with base x and exponent 2.

---

*These objectives are part of an optional section.

(c) $(-x)^2$ — $(-x)^2$ means $(-x)(-x)$, with base $-x$ and exponent 2.

(d) $(2c)^3$ — $(2c)^3$ means $(2c)(2c)(2c)$, with base $2c$ and exponent 3.

(e) $2c^3$ — $2c^3$ means $2 \cdot c \cdot c \cdot c$, with base c (not $2c$) and exponent 3; it is read as "two times c cubed."

(f) z — z means the same as z^1, with base z. When no exponent is written, the exponent is understood to be 1. ▼

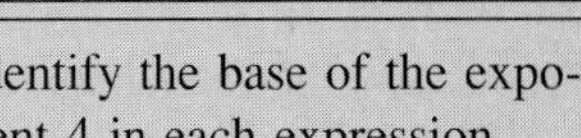

SELF-CHECK

Identify the base of the exponent 4 in each expression.

1	$5x^4$	**2**	$(5x)^4$
3	$(-y)^4$	**4**	$-y^4$
5	xy^4	**6**	$-(xy)^4$
7	-9^4	**8**	$(-9)^4$

Much of the usefulness of exponents lies in the fact that certain of their properties can be used to simplify many computations. These properties follow directly from the definition of natural number exponents. Let us start by examining the product $x^4 \cdot x^5$. Since x^4 means that x is used as a factor four times and x^5 means that x is used as a factor another five times, the product $x^4 \cdot x^5$ has x used as a factor a total of nine times.

$$x^4 \cdot x^5 = \overbrace{x \cdot x \cdot x \cdot x}^{\text{4 factors of } x} \cdot \overbrace{x \cdot x \cdot x \cdot x \cdot x}^{\text{5 factors of } x} = \overbrace{x \cdot x \cdot x \cdot x \cdot x \cdot x \cdot x \cdot x \cdot x}^{\text{9 factors of } x} = x^9$$

In general, in the product $x^m \cdot x^n$, the notation x^m means that x is used as a factor m times, and the notation x^n means that x is used as a factor another n times. Since x is used as a factor a total of $m + n$ times in this product, $x^m \cdot x^n = x^{m+n}$.

$$x^m \cdot x^n = \overbrace{x \cdot x \cdot \cdots \cdot x}^{m \text{ factors of } x} \cdot \overbrace{x \cdot x \cdot \cdots \cdot x}^{n \text{ factors of } x} = \overbrace{x \cdot x \cdot \cdots \cdot x}^{m+n \text{ factors of } x} = x^{m+n}$$

This theorem, known as the product rule, says that the product of two exponential expressions with the same base can be determined by adding the exponents.

PRODUCT RULE

For any real number x and natural numbers m and n,

$$x^m \cdot x^n = x^{m+n} \quad \text{Add the exponents.}$$

Be careful not to apply the product rule unless both of the factors have the same base.

EXAMPLE 2 Using the Product Rule

Use the product rule to simplify each of these expressions.

SOLUTIONS

(a) $(3x^4)(7x^8)$

$$(3x^4)(7x^8) = 21x^{4+8}$$
$$= 21x^{12}$$

Add the exponents with the common base.

SELF-CHECK ANSWERS 

1 x **2** $5x$ **3** $-y$ **4** y **5** y **6** xy **7** 9 **8** -9

(b) $(2x^2)(5y^3)$ $\quad$ $(2x^2)(5y^3) = 10x^2y^3$ $\quad$ Since the bases of x^2 and y^3 are not the same, this expression cannot be simplified further.

(c) $(-2a^2b^3)(6a^4b^7)$ $\quad$ $(-2a^2b^3)(6a^4b^7) = -12a^{2+4}b^{3+7}$
$= -12a^6b^{10}$ $\quad$ Only exponents on factors with the same base can be combined.

(d) $(x^{2n})(x^{n+5})$ $\quad$ $(x^{2n})(x^{n+5}) = x^{2n+(n+5)}$
$= x^{3n+5}$ $\quad$ Since the bases are the same, the exponents can be added.

The second basic theorem describing operations with exponents is the quotient rule. This time let us start by examining the quotient $\frac{x^7}{x^4}$, $x \neq 0$. Since x^7 means that x is used as a factor seven times and x^4 means that x is used as a factor four times, the fraction can be reduced to x^3, as illustrated below.

$$\frac{x^7}{x^4} = \frac{x \cdot x \cdot x \cdot x \cdot x \cdot x \cdot x}{x \cdot x \cdot x \cdot x} \begin{array}{l} \longleftarrow \text{7 factors of } x \\ \longleftarrow \text{4 factors of } x \end{array}$$

$$= x \cdot x \cdot x \qquad \longleftarrow \text{3 factors } x$$

$$= x^3$$

In general, if m and n are natural numbers and $m > n$, then the quotient $\frac{x^m}{x^n}$ $(x \neq 0)$ can be reduced by dividing out each of the n common factors of x. Once n of the m factors of x have been divided out, there are $m - n$ of these factors of x left. Thus the quotient rule for $x \neq 0$ and natural numbers m and n and $m > n$ states that

$$\frac{x^m}{x^n} = x^{m-n}$$

Likewise, if $m < n$, then

$$\frac{x^m}{x^n} = \frac{1}{x^{n-m}}$$

This rule can be applied only when the numerator and the denominator have the same base.

EXAMPLE 3 Using the Quotient Rule

Use the quotient rule to simplify each of these expressions. Assume that all variables are nonzero.

SOLUTIONS

(a) $\frac{15x^{11}}{3x^7}$ $\quad$ $\frac{15x^{11}}{3x^7} = 5x^{11-7}$
$= 5x^4$ $\quad$ Combine the exponents on the common base x. Subtract the smaller exponent from the larger exponent.

(b) $\frac{14x^3}{21x^8}$ $\quad$ $\frac{14x^3}{21x^8} = \frac{2}{3x^{8-3}}$
$= \frac{2}{3x^5}$ $\quad$ Subtract the smaller exponent from the larger exponent.

(c) $\dfrac{26x^5}{39y^2}$ $\quad \dfrac{26x^5}{39y^2} = \dfrac{2x^5}{3y^2}$ First divide both the numerator and the denominator by 13. Since the bases of x^5 and y^2 are not the same, this expression cannot be simplified further.

(d) $\dfrac{a^5b^3}{a^3b^2}$ $\quad \dfrac{a^5b^3}{a^3b^2} = a^{5-3}b^{3-2} = a^2b$ Only exponents on factors with the same base can be combined.

(e) $\dfrac{x^{2n+3}}{x^{n+1}}$ $\quad \dfrac{x^{2n+3}}{x^{n+1}} = x^{(2n+3)-(n+1)} = x^{n+2}$ Subtract the exponent in the denominator from the exponent in the numerator. ▼

For $x \neq 0$, we know that $\dfrac{x^m}{x^m} = 1$. If we apply the quotient rule and subtract exponents, we obtain $\dfrac{x^m}{x^m} = x^{m-m} = x^0$. Thus for $x \neq 0$, x^0 must equal 1 to be consistent with the other properties of mathematics.

DEFINITION OF x^0

For any nonzero real number x,

$$x^0 = 1$$

Note: 0^0 is undefined.

EXAMPLE 4 Simplifying Expressions with Zero Exponents

Simplify each of these expressions, assuming all bases are nonzero.

	SOLUTIONS	
(a) 5^0	$5^0 = 1$	The base is 5.
(b) $(-5)^0$	$(-5)^0 = 1$	The base is -5.
(c) -5^0	$-5^0 = -1$	The base is 5 (not -5); $-5^0 = -(5^0)$.
(d) $(x + 3y)^0$	$(x + 3y)^0 = 1$	Note the different bases in parts (d), (e),
(e) $x^0 + (3y)^0$	$x^0 + (3y)^0 = 1 + 1 = 2$	and (f). In part (e) the base of $(3y)^0$ is $3y$, whereas in part (f) the base of the
(f) $x^0 + 3y^0$	$x^0 + 3y^0 = 1 + 3(1) = 4$	term $3y^0$ is just y. ▼

To develop a definition for negative exponents, let us consider the quotient $\dfrac{x^2}{x^5}$ for $x \neq 0$. We know that

$$\frac{x^2}{x^5} = \frac{x \cdot x}{x \cdot x \cdot x \cdot x \cdot x} = \frac{1}{x^3}$$

If we apply the quotient rule, subtracting the exponent in the denominator from the exponent in the numerator, we obtain $\frac{x^2}{x^5} = x^{2-5} = x^{-3}$. Thus x^{-3} must equal $\frac{1}{x^3}$ to be consistent with the other properties of mathematics. The general definition is stated as follows.

A Mathematical Note

John Wallis, in *Arithmetica Infinitorum* in 1655, was the first writer to explain the use of zero and negative exponents.

DEFINITION OF x^{-n}

For any nonzero real number x and natural number n,

$x^{-n} = \frac{1}{x^n}$ Reciprocate the base, and then use a positive exponent n.

EXAMPLE 5 Simplifying Expressions with Negative Exponents

Simplify each of these expressions.

SOLUTIONS

(a) 3^{-2} $\quad 3^{-2} = \frac{1}{3^2} = \frac{1}{9}$ $\quad$ Reciprocate the base, and then use the positive exponent 2.

(b) -3^{-2} $\quad -3^{-2} = -\frac{1}{3^2} = -\frac{1}{9}$ $\quad$ The base is 3. To indicate a base of −3, we would have to use parentheses and write $(-3)^{-2}$.

(c) $2^{-1} + 5^{-1}$ $\quad 2^{-1} + 5^{-1} = \frac{1}{2} + \frac{1}{5} = \frac{7}{10}$

(d) $(2 + 5)^{-1}$ $\quad (2 + 5)^{-1} = 7^{-1} = \frac{1}{7}$

It is important to note the distinction in parts (c) and (d) between applying the exponent to each individual term and applying the exponent to the quantity as a whole. Failure to observe this important distinction leads to many order-of-operations errors.

Since we have defined $x^0 = 1$ and $x^{-n} = \frac{1}{x^n}$ to be consistent with the quotient rule, we can now restate the quotient rule for $\frac{x^m}{x^n}$ to include the cases where m is not greater than n.

SELF-CHECK

Simplify each expression.

1 $3^0 + 4^0$ **2** $(3 + 4)^0$
3 $(3 + 4)^{-2}$
4 $3^{-2} + 4^{-2}$

QUOTIENT RULE

For any nonzero real number x and integers m and n,

$\frac{x^m}{x^n} = x^{m-n}$ Subtract the exponents.

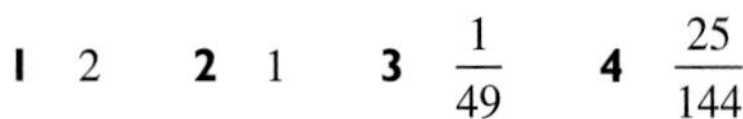

SELF-CHECK ANSWERS

1 2 **2** 1 **3** $\frac{1}{49}$ **4** $\frac{25}{144}$

It can be shown that both the product rule and the quotient rule apply to all integer exponents. After you apply the properties to simplify an expression, it is customary to write your answer using only positive exponents.

EXAMPLE 6 Simplifying Exponential Expressions

Simplify each of these expressions, assuming that all variables are nonzero.

SOLUTIONS

(a) $x^{-7} \cdot x^{-7}$

$$x^{-7} \cdot x^{-7} = x^{-7+(-7)} = x^{-14} = \frac{1}{x^{14}}$$

Product rule: Add the exponents. Then reciprocate, and express the result in terms of positive exponents.

(b) $x^{-7} + x^{-7}$

$$x^{-7} + x^{-7} = 2x^{-7} = \frac{2}{x^7}$$

Add similar terms.

Note the distinction between this expression and the expression in part (a).

(c) $\dfrac{18x^{-5}}{27x^{-7}}$

$$\frac{18x^{-5}}{27x^{-7}} = \frac{2}{3}x^{-5-(-7)} = \frac{2}{3}x^2$$

Quotient rule: Subtract the smaller exponent from the larger exponent to obtain an answer with a positive exponent.

(d) $\dfrac{x^2x^{-5}}{x^4}$

$$\frac{x^2x^{-5}}{x^4} = \frac{x^{-3}}{x^4} = \frac{1}{x^7}$$

Product rule: Add the exponents. Quotient rule: Subtract the smaller exponent from the larger exponent to obtain an answer with a positive exponent.

(e) $\left(\dfrac{x}{y}\right)^{-1}$

$$\left(\frac{x}{y}\right)^{-1} = \frac{1}{\left(\frac{x}{y}\right)}$$

Definition of negative exponents

$$= \frac{1}{1} \cdot \frac{y}{x} = \frac{y}{x}$$

To divide, invert and multiply. These steps are shown to confirm the accuracy of the result; however, you should follow the shorter method illustrated in the next example. ▼

The result from part (e) of Example 6 can be generalized to the useful result given in the following box.

QUOTIENT TO A NEGATIVE POWER

For nonzero real numbers x and y and any positive integer n,

$$\left(\frac{x}{y}\right)^{-n} = \left(\frac{y}{x}\right)^{n}$$

Reciprocate the base, and then use a positive exponent n.

EXAMPLE 7 Simplifying Expressions with Negative Exponents

Simplify each of these expressions.

SOLUTIONS

(a) $\left(\frac{3}{7}\right)^{-1}$ $\qquad \left(\frac{3}{7}\right)^{-1} = \frac{7}{3}$

Reciprocate the base of $\frac{3}{7}$, and use the positive exponent 1.

(b) $\left(\frac{1}{2}+\frac{1}{3}\right)^{-1}$ $\qquad \left(\frac{1}{2}+\frac{1}{3}\right)^{-1} = \left(\frac{5}{6}\right)^{-1}$

$$= \frac{6}{5}$$

It is important to note the distinction in parts (b) and (c) between applying an exponent to a whole quantity and applying this exponent to each individual term of a quantity. Failure to observe this distinction leads to many order-of-operations errors.

(c) $\left(\frac{1}{2}\right)^{-1} + \left(\frac{1}{3}\right)^{-1}$ $\qquad \left(\frac{1}{2}\right)^{-1} + \left(\frac{1}{3}\right)^{-1} = 2 + 3$

$$= 5$$

SELF-CHECK

Simplify $\dfrac{(18x^3y^{-2})(15x^{-2}y^5)}{27x^{-4}y^4}$, assuming that $x \neq 0$ and $y \neq 0$.

EXAMPLE 8 Evaluating Algebraic Expressions with Negative Exponents

Evaluate each of these expressions for $x = -2$ and $y = -3$.

SOLUTIONS

(a) $(2xy)^{-2}$ $\qquad [2(-2)(-3)]^{-2} = (12)^{-2}$

$$= \frac{1}{12^2}$$

$$= \frac{1}{144}$$

Substitute the given values, and then simplify the expression within the brackets. Remove the negative exponent by reciprocating the base.

(b) $2xy^{-2}$ $\qquad 2(-2)(-3)^{-2} = \frac{-2(2)}{(-3)^2}$

Substitute the given values, and then remove the negative exponent by reciprocating the base.

$$= \frac{-4}{9}$$

Simplify both the numerator and the denominator.

$$= -\frac{4}{9}$$

To be sure of the correct usage of your calculator, it is wise to go through a problem with simple values whose result is known before undertaking the calculation of similar problems whose results are unknown. The next example illustrates correct calculator usage for problems involving negative exponents.

SELF-CHECK ANSWER 

$\dfrac{10x^5}{y}$

EXAMPLE 9 Calculator Usage

(a) Store the value of -5 for x in your calculator.
(b) Evaluate x^{-2}.
(c) Evaluate $-x^{-2}$.
(d) Evaluate $(-x)^{-2}$.

SOLUTION

Keystrokes

(a) [(−)] [5] [STO] [X, T, θ] [ENTER] — First store the value of -5 for x.

(b) [X, T, θ] [∧] [(−)] [2] [ENTER] $\rightarrow 0.04$ — $x^{-2} = (-5)^{-2} = \frac{1}{25} = 0.04$

(c) [(−)] [X, T, θ] [∧] [(−)] [2] [ENTER] $\rightarrow -0.04$ — $-x^{-2} = -(-5)^{-2} = -0.04$

(d) [(] [(−)] [X, T, θ] [)] [∧] [(−)] [2] [ENTER] $\rightarrow 0.04$ — $(-x)^{-2} = [-(-5)]^{-2}$ $= (5)^{-2} = 0.04$

Exercises 4-1

A

In Exercises 1–18 simplify each expression completely. Assume that all bases are nonzero real numbers.

1 **a.** 5^2 **b.** -5^2 **c.** $(-5)^2$ **d.** $-(-5)^2$

2 **a.** 3^4 **b.** -3^4 **c.** $(-3)^4$ **d.** $-(-3)^4$

3 **a.** $(-1)^{41}$ **b.** $(-1)^{58}$ **c.** 0^{58} **d.** 58^0

4 **a.** $(7-8)^9$ **b.** $(10-8)^6$ **c.** $(12-12)^{12}$ **d.** $(99-100)^{101}$

5 **a.** 3^0 **b.** -3^0 **c.** $(-3)^0$ **d.** $-(-3)^0$

6 **a.** 7^0 **b.** -7^0 **c.** $(-7)^0$ **d.** $-(-7)^0$

7 **a.** 2^{-1} **b.** -2^1 **c.** -2^{-1} **d.** $(-2)^{-1}$

8 **a.** 3^{-1} **b.** -3^1 **c.** -3^{-1} **d.** $(-3)^{-1}$

9 **a.** $(3x-4y)^0$ **b.** $(3x)^0-(4y)^0$ **c.** $3x^0-4y^0$ **d.** $3x^0+(-4y)^0$

10 **a.** $(7x-5y)^0$ **b.** $(7x)^0-(5y)^0$ **c.** $7x^0-5y^0$ **d.** $(7x)^0+(-5y)^0$

11 **a.** 10^2 **b.** 10^5 **c.** 10^{-2} **d.** 10^{-5}

12 **a.** 10^3 **b.** 10^4 **c.** 10^{-3} **d.** 10^{-4}

13 **a.** $\left(\frac{5}{9}\right)^{-1}$ **b.** $\left(\frac{2}{3}\right)^{-2}$ **c.** $(2+3)^{-1}$ **d.** $2^{-1}+3^{-1}$

14 **a.** $\left(\frac{7}{9}\right)^{-1}$ **b.** $\left(\frac{2}{5}\right)^{-2}$ **c.** $(3+4)^{-1}$ **d.** $3^{-1}+4^{-1}$

15 **a.** $4^5 \cdot 4^{-3} + 5^9 \cdot 5^{-6}$ **b.** $\frac{6^0}{6^{-2}} + \frac{8^{-7}}{8^{-8}}$

16 **a.** $10^7 \cdot 10^{-5} - 3^6 \cdot 3^{-4}$ **b.** $\frac{7^{-4}}{7^{-6}} + \frac{9^{-5}}{9^{-6}}$

17 $(4a-9b)^0 + (4a)^0 + (-9b)^0 + 4a^0 - 9b^0$

18 $(5a+7b)^0 + (5a)^0 + (7b)^0 + 5a^0 + 7b^0$

In Exercises 19–54 simplify each expression. Write the answer using only positive exponents. Assume that all bases are nonzero real numbers.

19 x^7x^{11} **20** y^5y^{14} **21** $\frac{x^{11}}{x^7}$

22 $\frac{y}{y^5}$ **23** $\frac{x^7}{x^{11}}$ **24** $\frac{y^5}{y^{14}}$

25 $a \cdot a^0 \cdot a^8$ **26** $b \cdot b^0 \cdot b^7$ **27** $y^{-5} \cdot y^3 \cdot y^{-6}$

28 $y^4 \cdot y^0 \cdot y^{-6}$ **29** $(5x^7)(-7x^5)$ **30** $(4y^6)(-6y^4)$

31 $(-3a^2b^3)(-4a^5b^2)$ **32** $(-7a^3b^3)(-8a^4b^5)$

33 $\frac{48v^{17}}{16v^9}$ **34** $\frac{15w^{21}}{12w^{17}}$

35 $\frac{-12x^5y^5}{14x^4y^5}$ **36** $\frac{-121x^3y^4}{77x^3y^3}$

37 $(x-y)^{-5}(x-y)^4$ **38** $(x-y)^{-3}(x-y)^4$

39 $r^{-1} + s^{-1}$ **40** $v^{-1} - w^{-1}$

41 $(r+s)^{-1}$ **42** $(v-w)^{-1}$

43 $x^{-8} + x^{-8}$ **44** $x^{-5} + x^{-5}$

45 $\frac{6a^{-5}b^6}{8a^7b^{-8}}$ **46** $\frac{14a^7c^9}{21a^4c^{-5}}$

47 $(2x^3y^{-4})(-3x^{-2}y^2)$ **48** $(5x^{-6}y^{-4})(-7x^3y^{-3})$

49 $\frac{(8x^3y^{-2})(-4x^{-2}y^7)}{12x^{-1}y^{-8}}$ **50** $\frac{(-15x^4y^{-5})(14x^{-3}y^6)}{35x^{-5}y^3}$

51 $\frac{(6a^2b^3)(35a^{-4}b^{-2})}{(15a^4b^5)(7a^{-3}b^{-7})}$ **52** $\frac{(44a^{-3}b^5)(39a^2b^{-4})}{(26a^{-6}b^6)(55a^5b^{-5})}$

53 $\left(\frac{-8x^{-9}y^7}{13x^7y^{-11}}\right)^0$ **54** $\left(\frac{-73x^8y^{-11}}{87x^{-7}y^{13}}\right)^0$

B

In Exercises 55–58 simplify each expression. Write the answer using only positive exponents. Assume that all bases are nonzero real numbers.

55 **a.** 4^{-3} **b.** 3^{-4} **c.** -4^{-3} **d.** -3^{-4}

56 **a.** 2^{-5} **b.** 5^{-2} **c.** -2^{-5} **d.** -5^{-2}

57 **a.** $(x^{-1})(x^{-1})$ **b.** $x^{-1} + x^{-1}$

c. $\frac{1}{x^{-1}} + \frac{1}{x^{-1}}$ **d.** $\frac{1}{x^{-1}} \cdot \frac{1}{x^{-1}}$

58 **a.** $x^{-2}x^{-2}$ **b.** $x^{-2} + x^{-2}$

c. $\frac{1}{x^{-2}} + \frac{1}{x^{-2}}$ **d.** $\frac{1}{x^{-2}}\,\frac{1}{x^{-2}}$

In Exercises 59–62 evaluate each expression for $x = -1$, $y = -2$, and $z = -3$.

59 **a.** $x^2 + y^2$ **b.** $(x+y)^2$

c. $-y^2$ **d.** $(-y)^2$

60 **a.** xyz^2 **b.** $(xyz)^2$

c. $-z^2$ **d.** $(-z)^2$

61 **a.** $(x+y)^{-1}$ **b.** $x^{-1} + y^{-1}$

c. $-y^{-1}$ **d.** $(-y)^{-1}$

62 **a.** $(2x)^{-1} + (3y)^{-1}$ **b.** $2x^{-1} + 3y^{-1}$

c. $-z^{-1}$ **d.** $(-z)^{-1}$

In Exercises 63–68 simplify each expression. Assume that x and y are nonzero real numbers and that m is a natural number.

63 $x^5 \cdot x^m \cdot x^{-2}$ **64** $x^{3m+5}x^{4m-2}$ **65** $\frac{y^{6m+3}}{y^{5m-2}}$

66 $\frac{x^{7m-11}}{x^{2m-19}}$ **67** $\frac{x^{m+1}x^{2m+3}}{x^{3m+2}}$ **68** $\frac{x^{2m+1}x^{3m+4}}{x^{4m+3}}$

Estimation Skills (69–72)

69 The best mental estimate of $(2.568 \div 3.621 + 14.38)^0$ is ______.

a. -27 **b.** -2.7 **c.** -1

d. 1 **e.** 27

70 The best mental estimate of $\left(\frac{0.40183}{0.80301}\right)^{-1}$ is ______.

a. -2 **b.** -1 **c.** -0.5

d. 0.5 **e.** 2

71 The best mental estimate of $\left(\frac{361.895}{180.047}\right)^{-1}$ is ______.

a. 0.5 **b.** 0.8 **c.** 1.1

d. 1.3 **e.** 1.9

72 The best mental estimate of $[(1.998)(5.001)]^{-2}$ is ______________.

a. -0.1 **b.** -1.0 **c.** 1.0

d. 0.1 **e.** 0.01

C

In Exercises 73–76 determine which expressions are undefined, and simplify those that are defined.

73 **a.** 0^1 **b.** 1^0 **c.** $\frac{1}{0}$ **d.** $\frac{0}{1}$

74 **a.** $\frac{-2}{0}$ **b.** $\frac{0}{-2}$ **c.** 2^0 **d.** 0^2

75 **a.** -1^0 **b.** $(-1)^0$ **c.** 0^{-1} **d.** $(3^0 - 4^0)^0$

76 **a.** -2^0 **b.** $(-2)^0$ **c.** 0^{-2} **d.** $(5^0 - 6^0)^0$

Calculator Usage (77–80)

In Exercises 77 and 78 the value of 2 is stored for x; then the following expressions are entered into a calculator. Determine the value of each expression.

77 (5X − 8)^ −3

78 (4X + 2)^ −4

In Exercises 79 and 80 store the value of −2.5 for x and −0.5 for y in your calculator. Then evaluate each of these expressions.

79 a. $x^{-1} - y^{-1}$

b. $(x - y)^{-1}$

80 a. $9(x + y)^{-2}$

b. $9x^{-2} + y^{-2}$

DISCUSSION QUESTIONS (81 AND 82)

81 Write to a classmate explaining the different meanings of -3^2 and $(-3)^2$.

82 Write to a classmate explaining why it is incorrect to interpret expressions with negative exponents as expressions whose value is negative.

CHALLENGE QUESTION

a. Simplify each of these expressions, using the definition of x^n as denoting n factors of x.
i. $(x^2)^2$ ii. $(x^2)^3$ iii. $(x^2)^4$ iv. $(x^2)^5$

b. Can you use the results of part a to simplify $(x^9)^{11}$?

c. Can you generalize these results to simplify $(x^m)^n$ for natural numbers m and n?

d. Use the x values −2, −1, 0, 1, 2, and 3 to form a line graph approximation of the function $f(x) = 2^x$.

e. Use this line graph to approximate $2^{1/2}$.

SECTION SKILL REINFORCERS

1 a. Simplify $4x - (3x - 2)$. **b.** Solve $4x - (3x - 2) = 0$.

2 a. Simplify $2(x - 2) + 3(x + 4)$. **b.** Solve $2(x - 2) + 3(x + 4) = 0$.

3 a. Simplify $5(x - 3) + 3(x - 3)$. **b.** Solve $5(x - 3) + 3(x - 3) = 0$.

4 a. Simplify $\frac{x-2}{6} - \frac{2}{3}$. **b.** Solve $\frac{x-2}{6} - \frac{2}{3} = 0$.

5 a. Simplify $\frac{x+3}{10} - \frac{1}{2}$. **b.** Solve $\frac{x+3}{10} - \frac{1}{2} = 0$.

4-2 *Other Properties of Exponents and Scientific Notation*

SECTION OBJECTIVES

2 Use the properties of exponents.

3 Use scientific notation.

The expression $(x^3)^2$ can be interpreted as $x^3 \cdot x^3$ and rewritten as $(x \cdot x \cdot x)(x \cdot x \cdot x)$, which equals x^6. Thus $(x^3)^2 = x^{3 \cdot 2} = x^6$. Similarly, if m and n are natural numbers, $(x^m)^n = x^{mn}$. This power rule is illustrated below.

$$(x^m)^n = \overbrace{x^m \cdot x^m \cdot \cdots \cdot x^m}^{n \text{ factors of } x^m}$$

$$= \overbrace{(x \cdot x \cdot \cdots \cdot x)(x \cdot x \cdot \cdots \cdot x) \cdot \cdots \cdot (x \cdot x \cdot \cdots \cdot x)}^{n \text{ groups of } m \text{ factors of } x}$$

$$= \overbrace{(x \cdot x \cdot x \cdot x \cdot x \cdot x \cdot \cdots \cdot x \cdot x \cdot x)}^{m \cdot n \text{ factors of } x}$$

$$= x^{mn}$$

The properties given in the following box can be shown to be true for all integer exponents. The last two properties are generalizations of the product, quotient, and power rules.

PROPERTIES OF EXPONENTS

For any nonzero real numbers x and y and integer exponents m and n,

Product Rule: $x^m x^n = x^{m+n}$

Quotient Rule: $\dfrac{x^m}{x^n} = x^{m-n}$

Power Rule: $(x^m)^n = x^{mn}$

Product to a Power: $(xy)^m = x^m y^m$

Quotient to a Power: $\left(\dfrac{x}{y}\right)^m = \dfrac{x^m}{y^m}$

$$\left(\frac{x}{y}\right)^{-n} = \frac{y^n}{x^n}$$

EXAMPLE 1 Using the Power Rule for Exponents

Simplify each of these expressions, assuming that all variables are nonzero.

SOLUTIONS

(a) $(x^7)^3$

$(x^7)^3 = x^{7 \cdot 3} = x^{21}$ Power rule: Multiply exponents.

(b) $(xy)^5$

$(xy)^5 = x^5y^5$ Product to a power

(c) $(x^2y^3)^6$

$(x^2y^3)^6 = (x^2)^6(y^3)^6$ Product to a power

$= x^{2 \cdot 6}y^{3 \cdot 6}$ Power rule: Multiply exponents.

$= x^{12}y^{18}$

(d) $\left(\dfrac{x^2}{y^3}\right)^4$

$\left(\dfrac{x^2}{y^3}\right)^4 = \dfrac{(x^2)^4}{(y^3)^4}$ Quotient to a power

$= \dfrac{x^{2 \cdot 4}}{y^{3 \cdot 4}}$ Power rule: Multiply exponents.

$= \dfrac{x^8}{y^{12}}$

(e) $\left(\dfrac{x^4}{y^7}\right)^{-2}$

$\left(\dfrac{x^4}{y^7}\right)^{-2} = \left(\dfrac{y^7}{x^4}\right)^2$ Reciprocate the base.

$= \dfrac{(y^7)^2}{(x^4)^2}$ Quotient to a power

$= \dfrac{y^{7 \cdot 2}}{x^{4 \cdot 2}}$ Power rule: Multiply exponents.

$= \dfrac{y^{14}}{x^8}$

EXAMPLE 2 Simplifying Expressions with Negative Exponents

Simplify each of these expressions completely, leaving only positive exponents. Assume that all variables are nonzero.

SOLUTIONS

(a) $(2a^{-2}b^4)^{-3}$

$(2a^{-2}b^4)^{-3} = 2^{-3}(a^{-2})^{-3}(b^4)^{-3}$ Product to a power

$= 2^{-3}a^6b^{-12}$ Power rule: Multiply the exponents.

$= \dfrac{a^6}{2^3b^{12}}$ Express in terms of positive exponents by reciprocating the bases.

$= \dfrac{a^6}{8b^{12}}$

(b) $\left(\dfrac{12x^{-2}y^4}{15x^5y^{-6}}\right)^{-2}$

$\left(\dfrac{12x^{-2}y^4}{15x^5y^{-6}}\right)^{-2} = \left(\dfrac{4y^{10}}{5x^7}\right)^{-2}$ First simplify the expression inside the parentheses, observing carefully the order of operations. Use the quotient rule on each base by subtracting the smaller exponent from the larger exponent.

$= \left(\dfrac{5x^7}{4y^{10}}\right)^2$ Next reciprocate the base to remove the negative exponent.

$= \dfrac{5^2(x^7)^2}{4^2(y^{10})^2}$ Raise each factor to the second power.

$= \dfrac{25x^{14}}{16y^{20}}$ ▼

SELF-CHECK 

Simplify $\left(\dfrac{14x^{-3}y^2}{35x^2y^{-4}}\right)^{-3}$, assuming that $x \neq 0$ and $y \neq 0$.

EXAMPLE 3 Simplifying Expressions with Variable Exponents

Simplify $(x^{2n-3}x^{n-2})^2$, assuming that $x \neq 0$ and n is a positive integer.

SOLUTION $(x^{2n-3}x^{n-2})^2 = [x^{(2n-3)+(n-2)}]^2$ Product rule: Add exponents.

$= (x^{3n-5})^2$

$= x^{2(3n-5)}$ Power rule: Multiply exponents.

$= x^{6n-10}$ ▼

EXAMPLE 4 Evaluating Exponential Expressions

Evaluate $x - yz^2$ for $x = -2$, $y = -3$, and $z = -4$.

SOLUTION $x - yz^2 = (-2) - (-3)(-4)^2$ Substitute the given values, using parentheses to avoid order-of-operations errors.

$= -2 - (-3)(16)$

$= -2 + 48$

$= 46$ ▼

SELF-CHECK ANSWER 

$\dfrac{125x^{15}}{8y^{18}}$

One useful application of exponents is in scientific notation. Since our system of representing numbers is based on 10, powers of 10 are used in scientific notation. This notation is used extensively in the sciences and is also implemented by most calculators and computers. It is convenient for representing extremely large or extremely small numbers and is easy to use, since we can apply the properties of exponents to simplify calculations. Each of the powers of 10 from 10,000 to 0.001 is written below in exponential notation.

$$10{,}000 = 10^4$$
$$1000 = 10^3$$
$$100 = 10^2$$
$$10 = 10^1$$
$$1 = 10^0$$
$$0.1 = 10^{-1}$$
$$0.01 = 10^{-2}$$
$$0.001 = 10^{-3}$$

Note in the list above that the exponent determines the position of the decimal point. We use this fact in **scientific notation** to write any number as a product of a number between 1 and 10 (or -1 and -10 if negative) and an appropriate power of 10. Table 4-1 shows some representative numbers written in scientific notation.

A Mathematical Note

Extremely small numbers and extremely large numbers are difficult to comprehend. For example, try to grasp the value of each of the following amounts: a million dollars, a billion dollars, and a trillion dollars. Spent at the rate of a dollar a second, a million dollars would last almost 12 days, a billion dollars would last almost 32 years, and a trillion dollars would last almost 32,000 years.

TABLE 4-1 *Scientific Notation*

Number				Scientific Notation
4321	=	4.321×1000	=	4.321×10^3
43.21	=	4.321×10	=	4.321×10^1
0.4321	=	$4.321 \times (0.1)$	=	4.321×10^{-1}
0.004321	=	$4.321 \times (0.001)$	=	4.321×10^{-3}
-43.21	=	-4.321×10	=	-4.321×10^1

The steps for writing a number in scientific notation are given in the following box. Remember that positive exponents indicate larger magnitudes and negative exponents indicate smaller magnitudes.

Writing a Number in Scientific Notation

Step 1 Move the decimal point immediately to the right of the first nonzero digit of the number.

Step 2 Multiply by a power of 10 determined by counting the number of places the decimal point has been moved.

- **a.** The exponent on 10 is positive if the magnitude of the original number is greater than 10.
- **b.** The exponent on 10 is negative if the magnitude of the original number is less than 1.

EXAMPLE 5 Writing Numbers in Scientific Notation

Write each of these numbers in scientific notation.

SOLUTIONS

(a) 87,654,321 $87{,}654{,}321 = 8.7654321 \times 10^7$ (7 places to the left) — The decimal point is moved seven places. The exponent on 10 is positive 7, since the original number is greater than 10.

(b) 0.0000123 $0.0000123 = 1.23 \times 10^{-5}$ (5 places to the right) — The decimal point is moved five places. Since the original number is less than 1, the exponent on 10 is −5.

(c) 57.83×10^4

$$57.83 \times 10^4 = (5.783 \times 10) \times 10^4$$
$$= 5.783 \times (10^1 \times 10^4)$$
$$= 5.783 \times 10^5$$

▼

EXAMPLE 6 Writing Numbers in Standard Decimal Notation

Write each of these numbers in standard decimal notation.

SOLUTIONS

(a) 8.765×10^4 $8.765 \times 10^4 = 87650$ (4 places to the right)

(b) 8.765×10^{-3} $8.765 \times 10^{-3} = 0.008765$ (3 places to the left)

▼

The display on a calculator can show only a limited number of digits, so most scientific models are designed to accept input and to display answers in scientific notation. Test your calculator by performing the multiplication (0.000 000 5)(.000 03). With pencil and paper it is easy to compute the product, 0.000 000 000 015, which can also be written in scientific notation as 1.5×10^{-11}. Note that this number has too many digits for a calculator (which typically displays 8 to 10 digits) to show in standard decimal format. Therefore a scientific model will display something like

1.5 −11

which is interpreted to mean 1.5×10^{-11}.

Scientific Calculator

EXAMPLE 7 Interpreting a Calculator Display

A calculator display (as shown in the figures) shows

7.834 06

What number is being represented?

SOLUTION The number being represented is 7.834×10^6 or, in standard decimal form, 7,834,000. ▼

Graphics Calculator TI-82

Calculator manufacturers use different labels for the key used to enter the power of 10 of a number in scientific notation. This key is labeled [EE] in the next example; other calculators use an [EXP] key instead.

EXAMPLE 8 Calculator Keystrokes

Illustrate a typical sequence for entering into a calculator the value

0.000 000 000 123

SOLUTION In scientific notation this number is 1.23×10^{-10}. Thus a typical key sequence is

Keystrokes

Scientific

[1] [.] [2] [3] [EE] [1] [0] [+/−] [=]

On a scientific calculator, the number 10 is entered and then the change-of-sign key is used to make this exponent negative. This sequence is distinct from that used to enter a negative exponent on a graphics calculator.

Graphics

[1] [.] [2] [3] [EE] [(−)] [1] [0] [ENTER]

The display should now show

[1.23 −10]

We can also use scientific notation to perform pencil-and-paper calculations or to perform quick estimates of otherwise lengthy calculations.

SELF-CHECK ✓

Write in standard decimal form the number represented by the calculator display

[4.98 04]

EXAMPLE 9 Using Scientific Notation to Make Estimates

Estimate the value of $\dfrac{(99{,}894)(0.0000149987)}{29{,}987{,}423{,}116}$.

SOLUTION

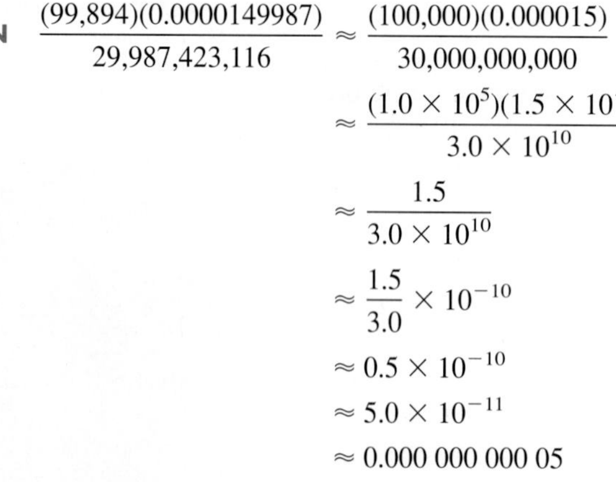

$$\frac{(99{,}894)(0.0000149987)}{29{,}987{,}423{,}116} \approx \frac{(100{,}000)(0.000015)}{30{,}000{,}000{,}000}$$

First approximate each factor.

$$\approx \frac{(1.0 \times 10^{5})(1.5 \times 10^{-5})}{3.0 \times 10^{10}}$$

Then express these approximations in scientific notation.

$$\approx \frac{1.5}{3.0 \times 10^{10}}$$
$$\approx \frac{1.5}{3.0} \times 10^{-10}$$
$$\approx 0.5 \times 10^{-10}$$
$$\approx 5.0 \times 10^{-11}$$
$$\approx 0.000\,000\,000\,05$$

Simplify the numerator, then reduce the fraction and express this result in standard decimal notation.

SELF-CHECK ANSWER ✓

49,800

Exercises 4-2

A

In Exercises 1–36 simplify each expression completely. Write each answer using only positive exponents. Assume that all bases are nonzero real numbers.

1 a. $(x^2)^3$ **b.** $(x^2)^{-3}$ **c.** $(xy)^2$ **d.** $(xy^2)^3$

2 a. $(m^3)^4$ **b.** $(m^3)^{-4}$ **c.** $(mn)^4$ **d.** $(mn^2)^4$

3 a. $(v^{-4})^{-5}$ **b.** $(v^2w^3)^4$ **c.** $(v^2w^{-3})^{-4}$ **d.** $(v^2w^{-3})^0$

4 a. $(a^{-2})^{-6}$ **b.** $(a^3b^4)^6$ **c.** $(a^3b^{-4})^{-6}$ **d.** $(a^3b^{-4})^0$

5 a. $\left(\frac{v^2}{w^3}\right)^4$ **b.** $\left(\frac{v^2}{w^3}\right)^{-4}$ **c.** $\left(\frac{v^{-2}}{w^3}\right)^4$ **d.** $\left(\frac{v^{-2}}{w^3}\right)^{-4}$

6 a. $\left(\frac{x^3}{y^4}\right)^5$ **b.** $\left(\frac{x^3}{y^4}\right)^{-5}$ **c.** $\left(\frac{x^3}{y^{-4}}\right)^5$ **d.** $\left(\frac{x^3}{y^{-4}}\right)^{-5}$

7 a. $(2x^2)^3$ **b.** $(3x^{-3})^2$ **c.** $(3x^{-3})^{-2}$ **d.** $(2x^2)^0$

8 a. $(4y^3)^2$ **b.** $(4y^{-3})^2$ **c.** $(4y^{-3})^{-2}$ **d.** $(4y^{-3})^0$

9 a. $\left(\frac{2a}{b^2}\right)^3$ **b.** $\left(\frac{2a}{b^2}\right)^{-3}$ **c.** $\left(\frac{2a^{-1}}{b^{-2}}\right)^3$ **d.** $\left(\frac{2a}{b^{-2}}\right)^{-3}$

10 a. $\left(\frac{5m^2}{n^3}\right)^2$ **b.** $\left(\frac{5m^2}{n^3}\right)^{-2}$ **c.** $\left(\frac{5m^{-2}}{n^{-3}}\right)^2$ **d.** $\left(\frac{5m^2}{n^{-3}}\right)^{-2}$

11 $(2^{-2})^{-3}$ **12** $(5^{-2})^{-2}$

13 $(2^{-7}\cdot 2^5)^{-2}$ **14** $(7^8\cdot 7^{-9})^{-2}$

15 $\left(\frac{10^{-5}}{10^{-3}}\right)^{-4}$ **16** $\left(\frac{4^{-7}}{4^{-6}}\right)^{-3}$

17 $(2x^3y^4)^5$ **18** $(3x^2y^6)^4$

19 $(3x^2y^{-3})^4$ **20** $(5x^3y^{-2})^3$

21 $\left(\frac{81x^2y^3z}{54x^2z^2}\right)^2$ **22** $\left(-\frac{121m^4n^7}{110m^7n^4}\right)^3$

23 $\left(\frac{2a^{-3}b^4}{3a^{-4}b^{-5}}\right)^{-2}$ **24** $\left(\frac{4a^5b^{-3}}{5a^{-2}b^{-4}}\right)^{-2}$

25 $\left[\frac{(a^2b^2c^3)^3}{(a^{-2}b^3c)^{-2}}\right]^{-4}$ **26** $\left[\frac{(a^{-3}b^4c^5)^{-2}}{(a^3bc^{-2})^{-3}}\right]^{-5}$

27 $(-3v^2w^{-3})^{-1}(-2v^{-2}w^4)^{-2}$ **28** $\frac{(-2xy^{-3})^{-2}}{(-10x^{-2}y^{-2})^{-4}}$

29 $\left[\frac{(373x^3y^{-5}z^{-17})^{-4}}{(11.74x^{-2}y^8z^{11})^{-2}}\right]^0$ **30** $\left[\frac{-18(39x^3y^{-4})^{-9}}{73(ab^{-4}c^{11})^7}\right]^0$

31 $\left[\frac{(-2ab^2c^{-3})^3(3a^{-2}b)^2}{36a^2b^2c^3}\right]^{-2}$

32 $\left[\frac{(-3xyz^{-1})^2(5x^{-3}y^{-1}z)}{90x^2y^{-2}z}\right]^{-3}$

33 $\left(\frac{63^0-7}{0^3-12}\right)^{-3}$ **34** $\left(\frac{-14}{-25}\cdot\frac{-15}{21}\right)^{-3}$

35 $\left(\frac{-22}{33}\cdot\frac{-21}{35}\right)^{-2}$ **36** $\frac{(-3v^2w^{-3})^{-1}(-2v^{-2}w^4)^{-2}}{(6v^{-4}w^{-5})^{-3}}$

In Exercises 37–40 evaluate each expression for $x = 2$, $y = -3$, and $z = -5$.

37 a. $x^2 - y^2 - z^2$ **b.** $(x - y - z)^2$

38 a. $(3x + y - z)^2$ **b.** $(3x)^2 + y^2 - z^2$

39 a. $3(x + y)^2 + (-z)^2$ **b.** $3x^2 + y^2 + (-z)^2$

40 a. $(x + y + z)^{-1}$ **b.** $x^{-1} + y^{-1} + z^{-1}$

In Exercises 41–44 convert each number in scientific notation to standard decimal form.

41 Speed of Light The speed of light is approximately 2.998×10^{10} centimeters per second.

42 Distance from the Sun The earth is approximately 1.49×10^8 kilometers from the sun.

43 Computer Speed One computer can complete an operation in approximately 3.0×10^{-9} second (3 nanoseconds).

44 Mass of an Atom The mass of an atom of hydrogen is approximately 1.673×10^{-24} gram.

In Exercises 45–52 express each number in scientific notation.

45 Avogadro's Number Avogadro's number is 602,300,000,000,000,000,000,000 molecules per mole.

46 Optical Fibers Optical fibers have transmitted data at the rate of 20 gigabits per second—that is, 20,000,000,000 bits per second.

47 Universal Gravitational Constant The universal gravitational constant is 0.000 000 000 066 73 Nm^2/kg^2.

48 Mass of an Electron The mass of an electron is 0.000 548 6 amu (atomic mass units).

49 9,870,000 **50** 0.000 000 987

51 0.0018×10^{-3}

52 -1249.63×10^{4}

B

ESTIMATION SKILLS (53–58)

In Exercises 53–58 use scientific notation to select the best estimate of the value of each expression. Make your estimates without using a calculator.

53 (0.000 000 799 8)(3,999,087,241)

a. 3000 **b.** 300 **c.** 30 **d.** 20 **e.** 400

54 (35,984,361,089)(0.000 199 99)

a. 70 **b.** 7000 **c.** 70,000 **d.** 700,000 **e.** 7,000,000

55 $\dfrac{4{,}897{,}634{,}010}{25{,}010{,}089}$

a. 2 **b.** 20 **c.** 200 **d.** 2000 **e.** 20,000

56 $\dfrac{0.000\,000\,645}{0.000\,015\,9}$

a. 0.000 4 **b.** 0.004 **c.** 0.04 **d.** 0.08 **e.** 0.009

57 $(0.0199)^3$

a. 0.000 008 **b.** 0.000 004 **c.** 0.008 **d.** 0.004 **e.** 0.06

58 $(0.300\,018)^2$

a. 0.06 **b.** 0.07 **c.** 0.09 **d.** 0.27 **e.** 2.39

In Exercises 59–66 simplify each expression. Assume that x and y are nonzero real numbers and that m is a natural number.

59 $(x^{m+3})^2$ **60** $(x^{2m+1})^3$ **61** $\dfrac{x^{m+4}}{x^{2m+6}}$

62 $\left(\dfrac{y^{m-4}}{y^{3m+6}}\right)^{-3}$ **63** $(x^m y^3)^3$ **64** $(x^{m+1}y^{m+1})^4$

65 $(x^{m+1}y^{-m})^{-2}$ **66** $(x^{-m}y^{m+4})^{-3}$

CALCULATOR USAGE (67–78)

In Exercises 67–70 give the standard decimal form of the number represented on each calculator display.

67 [7.89 −05] **68** [3.11 −08]

69 [−4.71 06] **70** [−9.34 05]

In Exercises 71–74 give the standard decimal form of the number represented by each graphics calculator display. The number following the letter E is the exponent on 10 in scientific notation.

71 [1.23 E −9] **72** [2.46 E − 11]

73 [1.2345 E 8] **74** [2.469 E 13]

C

In Exercises 75–78 store the value of 1.23 for x and 5.67 for y in your calculator. Then evaluate each of these expressions to the nearest thousandth.

75 a. $(x + y)^3$ **b.** $x^3 + y^3$

76 a. $(x - y)^{-1}$ **b.** $x^{-1} - y^{-1}$

77 $\dfrac{(5.489 \times 10^{11})(7.216 \times 10^{4})}{1.887 \times 10^{14}}$

78 $\dfrac{(2.643 \times 10^{-7})(14{,}829)^2}{(7.217 \times 10^{-1})^2}$

DISCUSSION QUESTION

79 Describe why scientific notation is called "scientific notation" and why this notation was invented.

CHALLENGE QUESTION

a. The speed of light is approximately 2.9979×10^8 meters per second. The radius of the Earth is approximately 6.378×10^6 meters. How many times can an electrical signal circle the Earth in 1 second?
b. The distance from the Earth to the sun is approximately 149,000,000,000 meters. Approximately how many seconds will it take light originating on the sun to reach the Earth?
c. A computer can complete an operation in 1 picosecond, which is one-trillionth (10^{-12}) of a second. How far will an electrical signal travel in 1 picosecond?
d. One light-year is not a unit of time but a unit of distance. A light-year is the distance that light travels in one year. The speed of light is approximately 2.9979×10^8 meters per second. Convert the distance of 1 light-year to meters. Express your answer accurate to three significant digits.
e. Determine how many seconds elapsed between the birth of Christ and the beginning of 1995. Express your answer in scientific notation accurate to two significant digits.

SECTION SKILL REINFORCERS

Simplify each of these expressions.

1 100^0 **2** 0^{100} **3** 100^{-1} **4** -1^{100} **5** $(-1)^{100}$

4-3 *Polynomials*

SECTION OBJECTIVES

4 Determine the degree of a polynomial, classify a polynomial according to the number of terms, and write a polynomial in standard form.

5 Evaluate a polynomial for given values of the variables.

Recall from Section 1-3 that a term of an algebraic expression can consist of a single constant, a variable, or a product of constants and variables. An algebraic expression consisting of a single term of the form ax^n, where a is a real number and n is a whole number, is called a **monomial.** Since repeated multiplication by a variable can occur, monomials can include positive integer exponents. Negative exponents are not permitted, however, because division by a variable is not allowed in a monomial.

EXAMPLE 1 Determine Whether Each Expression Is a Monomial

Determine which of these expressions are monomials.

	SOLUTIONS	
(a) $3x^4$, $-5x^3y$, $\frac{7}{3}x^2y^2$, $-1.95y^5$, and 18	All of these expressions are monomials.	In particular, $\frac{7}{3}x^2y^2$ is a monomial with constant factor $\frac{7}{3}$. There is no variable in the denominator of this monomial.
(b) $\frac{5}{w}$, or $5w^{-1}$	This expression is not a monomial.	A monomial cannot contain division by a variable.
(c) $x^{1/2}$, y^{-3}, and $x^3y^2z^{-4}$	These expressions are not monomials.	These expressions contain exponents that are not natural numbers.

▼

Polynomials are algebraic expressions that contain only monomials as terms (or addends). Polynomials containing one, two, and three terms are called, respectively, **monomials, binomials,** and **trinomials.** Example 2 illustrates these classifications.

EXAMPLE 2 Determine Whether Each Expression Is a Polynomial

Determine whether each expression is a polynomial. Classify each polynomial according to the number of terms it contains.

SOLUTIONS

(a) 10 — 10 is a constant monomial.

(b) $4y^2 + 10$ — $4y^2 + 10$ is a binomial, with terms $4y^2$ and 10.

(c) $5w^2 - 7w + 4$ — $5w^2 - 7w + 4$ is a trinomial in w, with terms $5w^2$, $-7w$, and 4.

(d) $-\frac{1}{3}x^4 + 7x^2 - 9x + 11$ — $-\frac{1}{3}x^4 + 7x^2 - 9x + 11$ is a polynomial in x, with four terms.

(e) $a^2 + 3ab + b^2$ — $a^2 + 3ab + b^2$ is a trinomial, with terms a^2, $3ab$, and b^2.

(f) $\dfrac{4x^2 - 7x + 9}{3x - 8}$ — $\dfrac{4x^2 - 7x + 9}{3x - 8}$ is not a polynomial. A polynomial cannot contain a variable in the denominator.

SELF-CHECK

Determine whether each expression is a polynomial. Classify each polynomial according to the number of terms it contains.

1 $5x^2yz - 17$

2 $\dfrac{5x^2y}{z - 197}$

3 $5x^2 - yz + 14$

4 $85x^2yx$

Recall from Section 1-3 that the constant factor of a monomial is called the **numerical coefficient**, or simply the **coefficient.** The coefficient of $-7v^2w^3$ is -7, not 7. The coefficient of x is understood to be 1, and the coefficient of $-x$ is understood to be -1. Although the word *coefficient* is usually not applied to a constant term, it can be. For example, the coefficient of 5 is 5 (note that $5 = 5x^0$ and the coefficient of $5x^0$ is 5).

The **degree of a monomial** is the sum of the exponents on the variables. A nonzero constant is understood to have degree 0, but no degree is assigned to the monomial 0.

EXAMPLE 3 Classifying Monomials by Degree

Give the degree of each of these monomials.

SOLUTIONS

(a) $-73v$ — $-73v$ has degree 1. The exponent on v is understood to be 1.

(b) π — π has degree 0. All nonzero constants have degree 0.

(c) $-4x^2y^3z$ — $-4x^2y^3z$ has degree 6. The sum of exponents 2, 3, and 1 is 6.

(d) 0 — 0 has no degree assigned to it.

The **degree of a polynomial** is the degree of the term of the highest degree. To find the degree of a polynomial, do *not* sum the degrees of the terms; instead, take only the degree of the term of highest degree.

SELF-CHECK ANSWERS 

1 Binomial **2** Not a polynomial **3** Trinomial **4** Monomial

EXAMPLE 4 Classifying Polynomials by Degree

Give the degree of each of these polynomials.

	SOLUTIONS	
(a) $4w - 13$	1	The first term of this binomial is of degree 1, and the constant term is of degree 0.
(b) $-2x^2y^3$	5	The exponents 2 and 3 on this monomial have a sum of 5.
(c) $x^3y + 2x^2y^2 + 4xy^3$	4	Each of these terms is of the fourth degree.
(d) $5x^3 - 3x^2y^2$	4	The degrees of the individual terms are 3 and 4, respectively. The largest of these numbers is 4.

SELF-CHECK 

Determine the degree of each polynomial.

1 $17a^2b^3c^4$ **2** 18

3 $18a^2 - b^3c^4$

4 $a + b + c + d$

Polynomials containing two or more terms are frequently written in a standard form. Use of this standard form makes it easier to compare polynomials and to perform operations on them. A polynomial is in **standard form** if (1) the variables are written in alphabetical order in each term and (2) the terms are arranged in order of decreasing powers of the first variable. If there is a constant term, it is written last. The arrangement of successive terms in decreasing powers is frequently referred to as **descending order.**

EXAMPLE 5 Writing Polynomials in Standard Form

Express each of these polynomials in standard form.

	SOLUTIONS	
(a) $3c^3a^2b$	$3a^2bc^3$	Write the factors in alphabetical order.
(b) $5 + a^4 - 3a^2 + 6a - 7a^3$	$a^4 - 7a^3 - 3a^2 + 6a + 5$	Arrange the terms in descending order.
(c) $y^2 + 4yx + x^2$	$x^2 + 4xy + y^2$	Write each term in alphabetical order, and then arrange the terms in decreasing powers of the first variable.

POLYNOMIAL IN x

An algebraic expression of the form

$$a_nx^n + a_{n-1}x^{n-1} + a_{n-2}x^{n-2} + \cdots + a_2x^2 + a_1x + a_0$$

with real numbers $a_n, a_{n-1}, a_{n-2}, \ldots, a_2, a_1, a_0$ and whole number n is called a **polynomial in x.**

A polynomial in x can be represented by the notation $P(x)$. For example, we can let $P(x) = 2x^2 - 4x + 11$. To evaluate this polynomial for $x = 3$, we then substitute 3 for x in the polynomial. Symbolically,

$$P(3) = 2(3)^2 - 4(3) + 11 = 17$$

SELF-CHECK ANSWERS 

1 9 **2** 0 **3** 7 **4** 1

EXAMPLE 6 Evaluating a Polynomial

Evaluate the polynomial $P(x) = x^3 - 9x^2 + 7x - 5$ for the values given.

SOLUTIONS

(a) $P(-1)$ $\quad P(-1) = (-1)^3 - 9(-1)^2 + 7(-1) - 5$
$= -22$

(b) $P(2)$ $\quad P(2) = (2)^3 - 9(2)^2 + 7(2) - 5$
$= -19$

(c) $P(k)$ $\quad P(k) = k^3 - 9k^2 + 7k - 5$

Note the use of parentheses in parts (a) and (b) to enclose the value of x for which the polynomial is evaluated. This notation should be used to prevent order-of-operations errors.

SELF-CHECK

Write each of these polynomials in standard form.

1 $-3w^2v^4$
2 $4s - 9 + 7s^2$
3 $7yx + 9x^2 - 8y^2$
4 Given $P(x) = 2x^3 - 5x + 11$, evaluate $P(-2)$.

EXAMPLE 7 Profit Polynomial

The profit in dollars made by selling t units is given by the polynomial $P(t) = -t^2 + 14t - 33$. Evaluate and interpret each expression.

SOLUTIONS

(a) $P(0)$ $\quad P(0) = -(0)^2 + 14(0) - 33 = -33$
Selling 0 units results in a loss of \$33.

(b) $P(3)$ $\quad P(3) = -(3)^2 + 14(3) - 33 = 0$
The seller breaks even if 3 units are sold.

(c) $P(10)$ $\quad P(10) = -(10)^2 + 14(10) - 33 = 7$
The seller makes \$7 if 10 units are sold.

It is sometimes useful to be able to evaluate a polynomial $P(x)$ for several different values of x. The next two examples show two different ways to use a graphics calculator to facilitate these calculations. In Example 8 the graph of the polynomial function is used to approximate the desired values. In Example 9 we use the TABLE feature available on a TI-82 calculator to evaluate the polynomial for the given values.

EXAMPLE 8 Calculator Usage

Use the graph of $y = -0.25x^2 + x + 2$ to evaluate $P(2)$, $P(4)$, and $P(6)$ for $P(x) = -0.25x^2 + x + 2$.

SOLUTION

Keystrokes

[Y=] [(−)] [.] [2] [5] [X, T, θ] [x^2] [+] [X, T, θ] [+] [2]

First enter the equation to be graphed.

[ZOOM] [8: Integer] [ENTER] [TRACE] [▶]

Selecting this option from the ZOOM options allows us to easily evaluate $P(x)$ for integer values of x. The TRACE feature displays both the x- and y- coordinates of the points on the graph.

1 $-3v^4w^2$ **2** $7s^2 + 4s - 9$ **3** $9x^2 + 7xy - 8y^2$ **4** $P(-2) = 5$

$P(2) = \ \ 3$	Use the left [◄] and right [►] arrow keys to select the x value of 2 and read the y value of 3. Thus $P(2) = 3$.
$P(4) = \ \ 2$	Use the arrow keys to move to the point $x = 4$ and $y = 2$. Thus $P(4) = 2$.
$P(6) = -1$	Finally, move to the point $x = 6$ and $y = -1$. Thus $P(6) = -1$.

▼

EXAMPLE 9 Calculator Usage

Use the TABLE feature on the TI-82 to evaluate $P(2)$, $P(4)$, and $P(6)$ for $P(x) = -0.25x^2 + x + 2$.

SOLUTION

Keystrokes

[Y=] [(−)] [·] [2] [5] [X, T, θ] [x^2] [+] [X, T, θ] [+] [2] — First enter the function to be evaluated.

[2nd] [TABLE]

X	Y1	
0	2	
1	2.75	
2	3	
3	2.75	
4	2	
5	.75	
6	−1	
X = 0		

Selecting the TABLE feature will cause the calculator to display a table of x- and y-coordinates of points that satisfy this function. If Tblset on the calculator has Tblmin = 0 and ▲Tbl = 1, then the table of values shown to the left will be displayed. The highlighted value on the display can be changed, and new values can be obtained on the display by using the up [▲] and down [▼] arrow keys.

Answer $P(2) = 3$, $P(4) = 2$, $P(6) = -1$

▼

EXAMPLE 10 Polynomials Representing Areas

Given the figure shown, use the variables x and y to write a polynomial describing each of the quantities indicated.

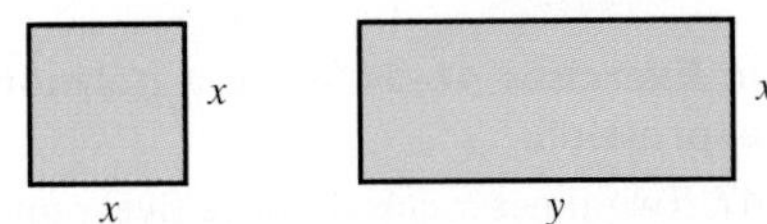

SOLUTIONS

(a) The area of the square	x^2	The area of a square is $A = s^2$. In this case the side is of length x, so $A = x^2$.
(b) The area of the larger rectangle	xy	The area of a rectangle is $A = l \cdot w$. In this case $A = xy$.
(c) The total area of the two regions	$x^2 + xy$	The total area is the area of the square plus the area of the rectangle.

▼

Exercises 4-3

A

In Exercises 1 and 2 determine whether each expression is a monomial. If the expression is a monomial, give its numerical coefficient.

1 a. $-7x^2y^5$ **b.** $\frac{x^3}{7}$ **c.** $\frac{7}{x^3}$ **d.** $-7x^3y^{-4}$

2 a. -13 **b.** $3x^7$ **c.** $3x^{-7}$ **d.** $-x^2y^3$

In Exercises 3 and 4 classify each polynomial as a monomial, binomial, or trinomial. Give the degree of each polynomial.

3 a. $3x^3 - y^2z^2$ **b.** $-3x^3y^2z^2$

c. $x^3 + y^2 - 3z^2$ **d.** 3

4 a. 11 **b.** $3x^3y^2 - z^2$

c. $3x^3 + y^2 - z^2$ **d.** $x^3y^2z^2 - 3$

In Exercises 5 and 6 write each polynomial in standard form.

5 a. $-7c^2a^5b^2$ **b.** $7 - x^2 + x$

c. $5y^2x - 3x^2y$ **d.** $3x^2y^2 + x^3y - 13xy^3$

6 a. $8y^2x^3z^4$ **b.** $14w - 9 + 11w^2$

c. $-4b^4a^3 + 9ba^5$ **d.** $y^2x - 17 + x^2y$

In Exercises 7–12 answer each question, referring to the polynomial $-x^4 - 2x^3 - 7x^2 + x - 52$.

7 What is the constant term of the polynomial?

8 What is the coefficient of the first term?

9 What is the coefficient of the third term?

10 What is the degree of the second term?

11 What is the degree of the polynomial?

12 What is the degree of the fifth term?

In Exercises 13–20 evaluate each expression, given that $P(x) = x^2 - x + 3$.

13 $P(0)$ **14** $P(-1)$ **15** $P(1)$

16 $P(2)$ **17** $P(-2)$ **18** $P(5)$

19 $P(k)$ **20** $P(-k)$

In Exercises 21–30 evaluate each expression, given that $P(x) = x^3 - 6x^2 - x + 10$.

21 $P(0)$ **22** $P(1)$ **23** $P(-1)$

24 $P(-2)$ **25** $P(2)$ **26** $P(k)$

27 $P(-k)$ **28** $P(-10)$ **29** $P(10)$

30 $P[P(0)]$

31 Write a first-degree binomial in x with a coefficient of 5 on the first term and a constant term of 7.

32 Write a second-degree trinomial in x with 3 as the coefficient of each term.

33 Write a fifth-degree monomial in x with a coefficient of -2.

34 Write a third-degree polynomial in x that has four terms with a coefficient of 1.

B

Profit Polynomials (35–40)

In Exercises 35–40 evaluate and interpret each expression. The profit in dollars made by selling t units is given by $P(t) = -t^2 + 22t - 40$.

35 $P(0)$ **36** $P(1)$ **37** $P(2)$

38 $P(10)$ **39** $P(20)$ **40** $P(22)$

Perimeter and Area Polynomials (41–46)

In Exercises 41–46, given the figure shown, write a polynomial to describe each of the quantities indicated.

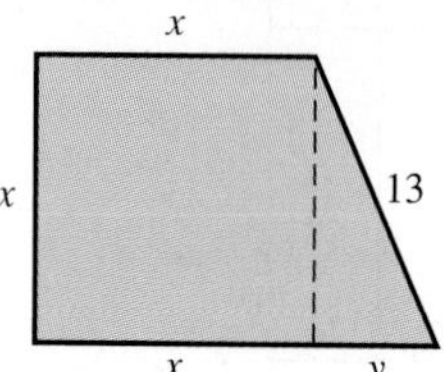

Figure for Exercises 41–46

41 The perimeter of the entire figure

42 The perimeter of the square portion of the figure

43 The perimeter of the triangular portion of the figure

44 The area of the square region

45 The area of the triangular region

46 The area of the entire region

In Exercises 47–54 write a polynomial for each verbal expression.

47 Two times x cubed minus five x plus eleven

48 Nine times y squared plus seven y minus nine

49 Three times the sixth power of w plus five times the fourth power of w

50 Eight times the fifth power of v minus v cubed plus eleven v

51 x squared minus y squared

52 a cubed plus b cubed

53 Seven more than twice w

54 Eight less than the opposite of four y

Estimation Skills (55–58)

In Exercises 55–58 select the best estimate of the value of each expression. Use $P(x) = x^3 + x^2 + x$, and make your estimates without using a calculator.

55 $P(9.9)$

a. 100 **b.** 200 **c.** 300 **d.** 1000 **e.** 1100

56 $P(1.1)$

a. 1.0 **b.** 2.4 **c.** 3.6 **d.** 6.8 **e.** 10.0

57 $P(-0.9)$

a. -1.8 **b.** -0.8 **c.** 0.8 **d.** 1.8 **e.** 3.8

58 $P(2.01)$

a. 114.2 **b.** 14.2 **c.** 10.2 **d.** 4.2 **e.** 0.2

59 Line Graph of Company Profits The function defined by the graph shown gives the total profit a company made by producing x units of a product. Match each number of units of production I, II, III, and IV to the event that occurred at this level of production.

Event A: To avoid a strike, the company agrees to a costly incentive program that proves to be unprofitable.
Event B: As production is started, the company covers its fixed costs quickly and moves into profitability.
Event C: To meet increased demand, the company hires another shift of workers, thereby increasing overall profit but also increasing its cost per unit.
Event D: To meet increased demand, the company decides to use expensive overtime and reduces its overall profit.

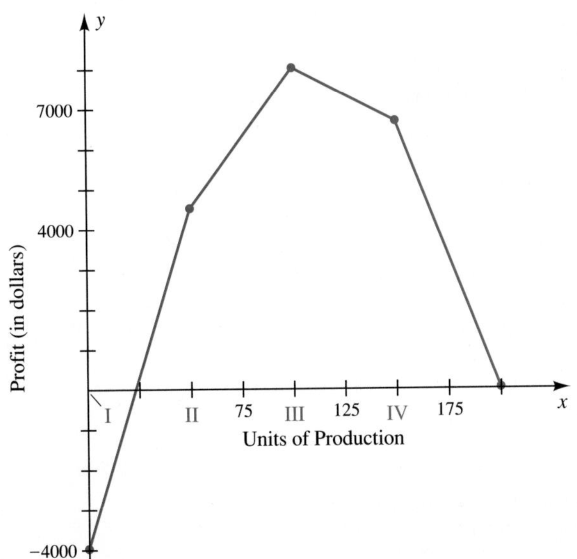

Figure for Exercise 59

C

Calculator Usage (60–65)

In Exercises 60–63 use a calculator to evaluate each expression to the nearest hundredth, given that

$$P(x) = 1.29x^2 - 0.281x - 11.7$$

and

$$Q(x) = 4.72x^2 + 6.83x - 27.29$$

60 $P(2.1)$

61 $Q(1.2)$

62 $P(5.1) - Q(1.5)$

63 $[P(2.2)][Q(1.1)]$

In Exercises 64 and 65 use a graphics calculator to graph $y = P(x)$; then use the TRACE feature to complete the table. If you have a TI-82 calculator, you can use the TABLE feature to complete this table.

x	-4	-3	-2	-1	0	1	2	3	4
$P(x)$									

64 $P(x) = x^2 - 2x - 8$

65 $P(x) = -x^2 + 5x + 4$

Discussion Questions (66–68)

66 Write in your own words a definition of a polynomial.

67 Rewrite the polynomial $5x^3 + 5x^2y + 7xy^2 + y^3$ twenty-four different ways by reordering the four terms of this polynomial all possible ways. Then explain why it is advantageous to write polynomials in standard form.

68 Determine the degree of each of these polynomials:

a. x^2 **b.** x^3 **c.** 5 **d.** 0 **e.** x^2x^3 **f.** $5x^2$ **g.** $0x^2 = 0$

69 If two monomials are multiplied, how is the degree of the product related to the degree of each factor? Using this information, describe why no degree is assigned to 0.

Challenge Question

Suppose one long rope is stretched taut around the equator of the Earth so that it touches each point on this equator. Another rope is placed in a similar fashion about the equator of the moon. Each rope is then increased in length so that each rope is exactly one meter above the surface all the way around the equator. Which rope will be increased more in length—the one around the Earth or the one around the moon?

SECTION SKILL REINFORCERS

Use interval notation to represent each of these inequalities.

1 $x \geq 3$ **2** $x < 4$ **3** $1 \leq x < 4$ **4** $-1 < x < 7$ **5** $-8 \leq x \leq -2$

4-4 Addition and Subtraction of Polynomials

SECTION OBJECTIVE

6 Add and subtract polynomials.

In Chapter 1 we saw that the distributive property can be used to rewrite $4x + 7x$ as $(4 + 7)x$, or $11x$. This means that we can add polynomials by adding the coefficients of like terms. **Like, or similar, terms** must have identical variable factors, including exponents; only their numerical coefficients can differ.

EXAMPLE 1 Classifying Terms as Like or Unlike

Classify the terms in each pair as like or unlike terms.

	SOLUTIONS	
(a) $3x^2y$ and $-5x^2y$	Like terms	The variable factors are identical.
(b) $3x^2y$ and $-5xy^2$	Unlike terms	The exponents on the variables are not identical.
(c) $3x^2y$ and $3yx^2$	Like terms	Although $3yx^2$ is not written in the standard form, the terms are similar.
(d) 21 and $7x^0$	Like terms	Since $7x^0 = 7$, both of these terms are constants.

Since subtraction of a term is the same as the addition of that term's opposite, we can also subtract like terms by subtracting their coefficients. Although the following example shows the step used in applying the distributive property, you should be able to perform this step mentally.

EXAMPLE 2 Adding and Subtracting Like Terms

Perform the operations indicated in each expression.

	SOLUTIONS	
(a) $4v^2 + 7v^2$	$4v^2 + 7v^2 = (4 + 7)v^2$	Distributive property
	$= 11v^2$	
(b) $3x^2y - 5x^2y$	$3x^2y - 5x^2y = (3 - 5)x^2y$	Distributive property
	$= -2x^2y$	
(c) $3x^2y - 5xy^2$	$3x^2y - 5xy^2$	This expression cannot be simplified further because the terms are not like terms.

Since the commutative and associative properties of addition allow us to add terms in any order, we can add polynomials by grouping similar terms together and then adding their coefficients. This process is generally described as **combining like terms.**

Two different formats are commonly used to add or subtract polynomials—the vertical format and the horizontal format. Whichever format you use, you should express the answer in standard form. Both formats involve combining coefficients of the like terms. The vertical format simplifies the computations by permitting alignment of similar terms in the same column. This alignment of similar terms is also useful in multiplying and dividing polynomials.

EXAMPLE 3 Adding Polynomials

Use the vertical format to add these polynomials.

(a) $(x^3 - x^2 + 4) + (3x - 7x^2 - 9)$

(b) $(4x + 13z - 3y) + (8y + 7x - 6z) + (-8x - 5z + 2y)$

SOLUTIONS

(a)

$$\begin{array}{r} x^3 - x^2 + 0x + 4 \\ - 7x^2 + 3x - 9 \\ \hline x^3 - 8x^2 + 3x - 5 \end{array}$$

Write each polynomial in standard form. Note that $x^3 - x^2 + 0x + 4$ equals $x^3 - x^2 + 4$. You may find it helpful to include this term with a zero coefficient in order to align like terms. Then add similar terms.

(b)

$$\begin{array}{r} 4x - 3y + 13z \\ 7x + 8y - 6z \\ -8x + 2y - 5z \\ \hline 3x + 7y + 2z \end{array}$$

▼

Remember that $-(x - y + z)$ means $-1 \cdot (x - y + z)$. Using the distributive property, we have

$$-(x - y + z) = -1(x) - 1(-y) - 1(z) = -x + y - z$$

Therefore the opposite of a polynomial is formed by changing the sign of each term of the polynomial.

When using the vertical format to subtract polynomials, place the polynomial being subtracted on the lower row and change the sign of each term. To prevent errors, you may write the new signs in circles above the original signs. This notation is also useful in dividing polynomials.

EXAMPLE 4 Subtracting Polynomials

Use the vertical format to subtract $(4x^2 + 3x - 2)$ from $(3x^2 - 7x + 4)$.

SOLUTION

$$\begin{array}{r} 3x^2 - 7x + 4 \\ (-)\ \underline{4x^2 + 3x - 2} \end{array} \quad \text{means} \quad \begin{array}{r} 3x^2 - 7x + 4 \\ (+)\ -4x^2 - 3x + 2 \\ \hline -x^2 - 10x + 6 \end{array} \quad \text{or} \quad \begin{array}{r} 3x^2 - 7x + 4 \\ \ominus \quad \ominus \quad \oplus \\ 4x^2 + 3x - 2 \\ \hline -x^2 - 10x + 6 \end{array}$$

▼

EXAMPLE 5 Subtracting Polynomials

Subtract $5y - 6y^2 - 8$ from $19 - 7y + 8y^2$.

SOLUTION

$$\begin{array}{rrr} 8y^2 & - \ 7y & + \ 19 \\ \oplus & \ominus & \oplus \\ -6y^2 & + \ 5y & - \ 8 \\ \hline 14y^2 & - \ 12y & + \ 27 \end{array}$$

First write each polynomial in standard form, aligning only similar terms. To subtract, change the sign of each term being subtracted, as noted in the circles.

The horizontal format requires less rewriting than the vertical format, but you must exercise more caution to ensure that you combine only coefficients of like terms. To make it easier to identify similar terms, you might want to identify all like terms with the same kind of underlining or the same color highlighter.

SELF-CHECK

Perform the operations indicated in each expression.

1
$$\begin{array}{r} 2r^2 - 4rs + 7s^2 \\ (+)\ \underline{3r^2 + 5rs - 3s^2} \end{array}$$

2
$$\begin{array}{r} 2r^2 - 4rs + 7s^2 \\ (-)\ \underline{3r^2 + 5rs - 3s^2} \end{array}$$

EXAMPLE 6 Adding Polynomials

Use the horizontal format to add

$$(5v^2 - 5 + 3v) + (7v + 9v^2 - 11) + (3v^2 - 8v + 7)$$

SOLUTION

$$(5v^2 - 5 + 3v) + (7v + 9v^2 - 11) + (3v^2 - 8v + 7)$$
$$= 17v^2 + 2v - 9$$

Use a common notation to identify all like terms. Write the answer in standard form.

For subtraction problems first remove the parentheses by changing the sign of each term being subtracted. Failure to do so is very likely to cause a sign error.

EXAMPLE 7 Subtracting Polynomials

Subtract $7w^2 - 4vw + 9v^2$ from $3vw - 5v^2 + w^2$.

SOLUTION

$$(3vw - 5v^2 + w^2) - (7w^2 - 4vw + 9v^2)$$
$$= 3vw - 5v^2 + w^2 - 7w^2 + 4vw - 9v^2$$
$$= -14v^2 + 7vw - 6w^2$$

Remove parentheses, changing the sign of each term being subtracted. Denote like terms, and then combine like terms. Write the answer in standard form.

EXAMPLE 8 Adding Polynomials Containing Parentheses

Simplify $[5x^2 + 7x - (3x - 5)] - [(8x - 9) - (3x^2 - 4x - 1)]$.

SOLUTION

$$[5x^2 + 7x - (3x - 5)] - [(8x - 9) - (3x^2 - 4x - 1)]$$
$$= [5x^2 + 7x - 3x + 5] - [8x - 9 - 3x^2 + 4x + 1]$$

First remove the parentheses inside the brackets.

$$= [5x^2 + 4x + 5] - [-3x^2 + 12x - 8]$$

Then combine like terms inside the brackets.

$$= 5x^2 + 4x + 5 + 3x^2 - 12x + 8$$
$$= 8x^2 - 8x + 13$$

Finally, remove the brackets, and combine like terms.

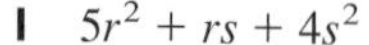

SELF-CHECK ANSWERS

1 $5r^2 + rs + 4s^2$ **2** $-r^2 - 9rs + 10s^2$

One of the reasons polynomials are so useful is that a single polynomial can represent the behavior of a whole set of data. In combination, polynomials can be used to describe relationships between various quantities, such as revenue, cost, and profit.

> **SELF-CHECK** ✓
>
> Using the horizontal format, simplify
> $(2r^3 - 3r + 4r^2 - 15) - (11 + 9r^2 - 7r + 4r^3)$.

EXAMPLE 9 Revenue, Cost, and Profit

The revenue in dollars produced by selling t gadgets is given by the polynomial $R(t) = 5t^2 - 2t$, and the cost of these t gadgets is given by $C(t) = 2t + 17$. Find the profit $P(t)$ when t gadgets are sold.

SOLUTION [Profit] = [Revenue] − [Cost] — Word equation

$P(t) = R(t) - C(t)$

$P(t) = (5t^2 - 2t) - (2t + 17)$ — Substitute the given expressions for $R(t)$ and $C(t)$, and then combine like terms.

$P(t) = 5t^2 - 2t - 2t - 17$

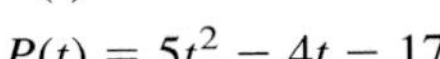

$P(t) = 5t^2 - 4t - 17$

EXAMPLE 10 Perimeter of a Figure

The polynomial that represents the perimeter of the figure shown is $4x^2 + 7x + 17$. Determine the length of the slanted side.

SOLUTION Let

l = Length of the slanted side — Identify the unknown length with an appropriate variable.

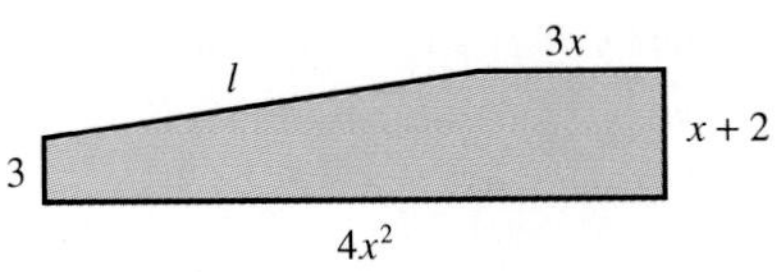

[Length of the slanted side] = [Total perimeter] − [Sum of the lengths of the other sides] — Word equation

$l = [4x^2 + 7x + 17] - [3 + 4x^2 + (x + 2) + 3x]$ — Substitute for the perimeter and the lengths of the given sides.

$l = (4dx^2 + 7x + 17) - (4x^2 + 4x + 5)$ — Simplify, and then subtract by combining like terms.

$l = 4x^2 + 7x + 17 - 4x^2 - 4x - 5$

$l = 3x + 12$ ▼

SELF-CHECK ANSWER 

$-2r^3 - 5r^2 + 4r - 26$

Exercises 4-4

A

In Exercises 1 and 2 classify the terms in each pair as like or unlike terms. Simplify the terms or put them in standard form first, if necessary. If the terms are like terms, find their sum.

1 **a.** $5x, 5y$ **b.** $-7x, 11x$

c. $3x^3y^2, 7x^2y^3$ **d.** $21x^4y^2, -12y^2x^4$

2 **a.** $5y, -9y$ **b.** $2v, -2w$

c. $-2a^2b^3, -5b^3a^2$ **d.** $-31v^2w, 13w^2v$

In Exercises 3–34 perform the indicated operations.

3
$$\begin{array}{r} 6y+5 \\ (+)\ \underline{2y+1} \end{array}$$

4
$$\begin{array}{r} 13a-7 \\ (+)\ \underline{-4a-9} \end{array}$$

5
$$\begin{array}{r} 12v-19 \\ (-)\ \underline{7v-8} \end{array}$$

6
$$\begin{array}{r} 17c-18 \\ (-)\ \underline{9c+20} \end{array}$$

7
$$\begin{array}{r} x^2+3x-5 \\ (+)\ \underline{2x^2-2x+9} \end{array}$$

8
$$\begin{array}{r} 3x^2-7x-11 \\ (+)\ \underline{4x^2+8x+6} \end{array}$$

9
$$\begin{array}{r} x^2+3x-5 \\ (-)\ \underline{2x^2-2x+9} \end{array}$$

10
$$\begin{array}{r} 3x^2-7x-11 \\ (-)\ \underline{4x^2+8x+6} \end{array}$$

11
$$\begin{array}{r} 11ab+5bc-6cd \\ (-)\ \underline{5ab-2bc-3cd} \end{array}$$

12
$$\begin{array}{r} 4t^3+4t^2+4t-7 \\ (-)\ \underline{-2t^3+3t^2-6t+8} \end{array}$$

13
$$\begin{array}{r} 7m^2-11m+8 \\ 5m^2+7m+14 \\ (+)\ \underline{-9m^2-8m-17} \end{array}$$

14
$$\begin{array}{r} x^2-2xy+y^2 \\ 3x^2-2xy-y^2 \\ (+)\ \underline{x^2+2xy+y^2} \end{array}$$

15 $(2a+5b)+(4a-9b)$ **16** $(3c-7d)+(8c-11d)$

17 $(2a+5b)-(4a-9b)$ **18** $(3c-7d)-(8c-11d)$

19 $(3x-2y+z)+(2x-4y+3z)$

20 $(r+3s-5t+7)+(6r-2s-9t-16)$

21 $(x^2+9y^2-5xy)-(-7xy-3y^2+2x^2)$

22 $(-4ab-b^2+3a^2)-(7b^2+a^2-5ab)$

23 $(-3a-5b+6c)-(-4a-2b-c)$

24 $(2a+3b-7c)-(8a-9b-c)$

25 $[(2y-3x-7z)+(-9z-10x+y)]-(-4x+7z+9y)$

26 $(-12a+7c-2b)-[(c-b+3a)+(4b-c-a)]$

27 $[(2v-3w)-(2w-4x)]-[(v-x)-(2v-5w)]$

28 $[(2m^2-3m-4)-(m^2-5)]-$
$[(m^2-4m)-(3m^2-7m+2)]$

29 $(a^2-2ab+b^2)-[(3a^2-b^2)-(a^2-ab-b^2)]$

30 $[(m^2-n^2)-(m^2-2m+n^2)]-$
$[(m^2+n^2)-(m^2+mn-n^2)]$

31 $3a-b-[a-(2a+3b)-(a-2b)]$

32 $[(-7y+5y^2+3)-(-3+y^2+8y)]-$
$[(7y-5)-(2y^2+8)-(3y-y^2)]$

33
$$\begin{array}{rrrrr} 5x^4 & -7x^3y & +12x^2y^2 & -3xy^3 & +4y^4 \\ 2x^4 & +4x^3y & & -7xy^3 & +9y^4 \\ (+)\ \underline{3x^4} & \underline{} & \underline{-9x^2y^2} & \underline{} & \underline{-y^4} \end{array}$$

34
$$\begin{array}{rrrrrr} 3a^5 & -4a^4 & & +2a^2 & -7a & +11 \\ 7a^5 & +2a^4 & -9a^3 & & +4a & -2 \\ 2a^5 & & -7a^3 & -a^2 & +a & -1 \\ (+)\ \underline{a^5} & \underline{+a^4} & \underline{+2a^3} & \underline{-2a^2} & \underline{+2a} & \underline{+4} \end{array}$$

In Exercises 35–44 perform the indicated operations.

35 Find the sum of $(2x^2-7x-11)$ and $(3x^2-8x+4)$.

36 Find the sum of $(5r+8-7s)$ and $(13s-4r+11)$.

37 Subtract $(-8a^2-4a+13)$ from $(-7a^2-17)$.

38 Subtract $(-7m+2n-9)$ from $(-4n-5+3m)$.

39 Subtract $(1.1x-2.2y-3.3)$ from $(4.4x-5.5y+6.6)$.

40 Subtract $\left(\frac{1}{2}x+\frac{2}{3}y\right)$ from $\left(\frac{3}{4}x-\frac{1}{3}y\right)$.

41 Add $(2a-3b)$ to the sum of $(4a+5b)$ and $(6a-7b)$.

42 Add $(3x-5y)$ to the difference of $(7x-9y)$ minus $(8x-4y)$.

43 Subtract (w^3-11w^4+13) from the sum of $(w+5w^2+7w^3-9)$ and $(-8-2w^2+7w+w^4)$.

44 Subtract $(7x^2-9xy-13y^2)$ from the sum of $(-8x^2-2xy+y^2)$ and $(11x^2+4xy+12y^2)$.

B

Perimeter of a Figure (45 and 46)

45 The polynomial that represents the perimeter of the figure shown is $3x^2-2x+11$. Determine the length of the slanted side.

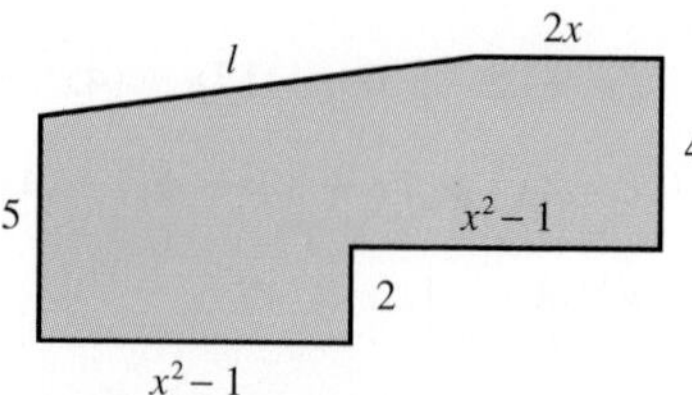

Figure for Exercise 45

46 The polynomial that represents the perimeter of the figure shown is $5x^2+3x-7$. Determine the length of the slanted side.

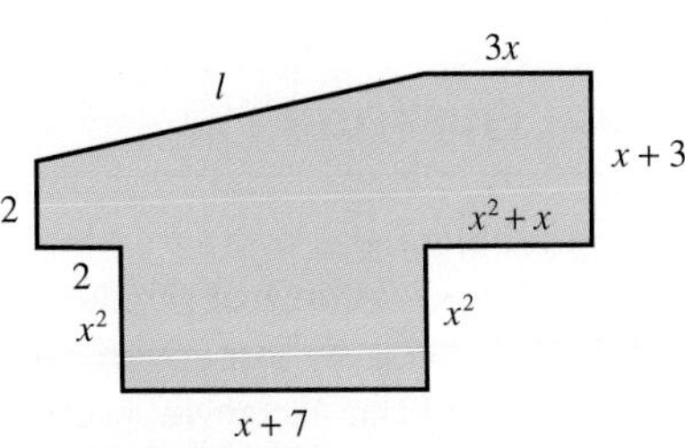

Figure for Exercise 46

Area of a Region (47 and 48)

47 Referring to the figure, write a polynomial for the shaded area inside the larger square with side x and outside the smaller square with side y.

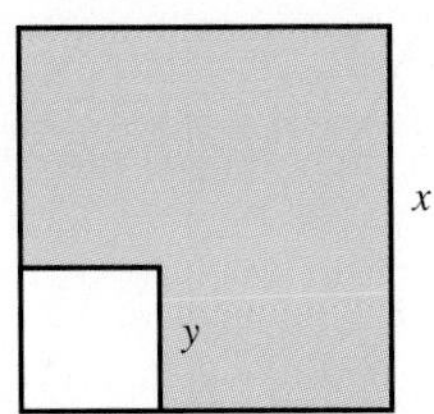

Figure for Exercise 47

48 Write a polynomial for the area enclosed between the concentric circles shown in the figure.

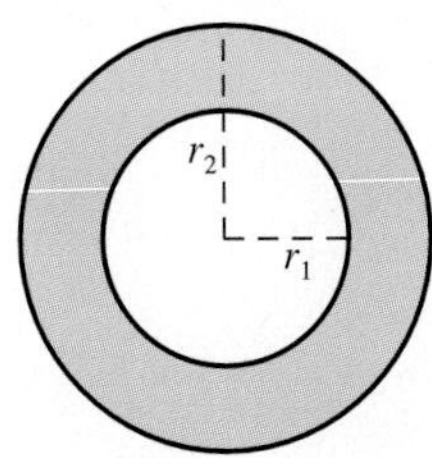

Figure for Exercise 48

49 Give an example of two third-degree polynomials whose sum is a constant.

50 Give an example of two second-degree polynomials whose sum is a first-degree polynomial.

51 Give an example of two binomials whose sum is a trinomial.

52 Give an example of two third-degree polynomials whose sum is x^2.

53 Give an example of two trinomials whose sum is a monomial.

54 Give an example of two second-degree monomials whose sum has no degree.

Estimation Skills (55 and 56)

In Exercises 55 and 56 select the best estimate for the simplified expression that results from performing the indicated operations. Make your estimates without using a calculator.

55 $(4.3051x - 2.4917y) - (-1.7064x + 5.5099y)$

a. $6.0x - 8.0y$ **b.** $6.0x + 8.0y$ **c.** $5.0x - 7.0y$
d. $5.0x + 7.0y$ **e.** $5.0x - 8.0y$

56 $(17.9854m - 29.0143) - (-109.1043m - 36.0348)$

a. $127m - 7$ **b.** $127m + 7$ **c.** $92m - 7$
d. $92m + 7$ **e.** $92m - 65$

C

Production, Marketing, and Total Costs (57–60)

In Exercises 57–60 use the given formulas for $P(t)$ and $M(t)$ and the formula $C(t) = P(t) + M(t)$ to find the total cost $C(t)$ of producing t units of a product. $P(t)$ is the production cost of these t units, and $M(t)$ is the marketing cost of these t units.

57 $P(t) = 2t^3 - t + 12,\ M(t) = 5t^2 + 8t + 11$

58 $P(t) = 4t^3 - t^2 + 8t + 20,\ M(t) = 2t^2 + 7t + 9$

59 $P(t) = 2t^4 - t^2 + 3t + 8,\ M(t) = 5t^2 - 3t + 24$

60 $P(t) = 3t^3 + t^2 - 8t + 30,\ M(t) = 2t^2 - 7t + 8$

Revenue, Cost, and Profit (61–64)

In Exercises 61–64 use the given formulas for $R(t)$ and $C(t)$ and the formula $P(t) = R(t) - C(t)$ to find the profit $P(t)$ from selling t units of a product. $R(t)$ is the revenue produced by selling t units, and $C(t)$ is the cost of t units.

61 $R(t) = 8t^2 - 3t,\ C(t) = 5t + 18$

62 $R(t) = 9t^2 - 12t,\ C(t) = 4t^2 + 3t + 11$

63 $R(t) = t^3 + t^2 - 7t,\ C(t) = 3t^2 + t + 35$

64 $R(t) = t^3 + t^2 - 8t + 30,\ C(t) = 2t^2 + 27$

65 Referring to the figure shown, answer these questions.

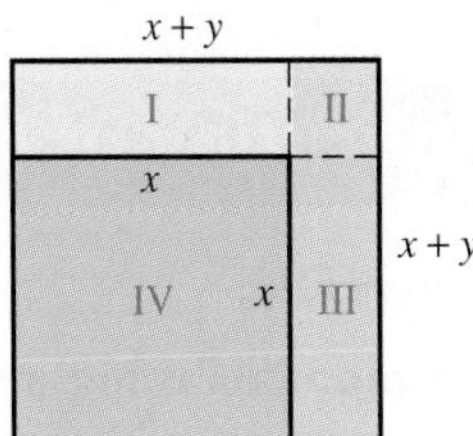

Figure for Exercise 65

a. Which rectangle has area $(x + y)(x + y)$?

b. Which rectangle has area x^2?

c. Which rectangle has area y^2?

d. Which rectangles have area xy?

e. Does $(x + y)(x + y) = x^2 + 2xy + y^2$?

Calculator Usage (66–68)

66 Use a calculator to simplify $(1.4139x^2 - 8.7353y^2) + (2.3245x^2 + 4.6537y^2)$.

67 Store the value of -4.78 for x in your calculator. Then approximate to the nearest thousandth $P(-4.78)$ for $P(x) = -2.39x^5 + 7.8x^4 + 3.6x^3 - 8.4x$.

68 Use a graphics calculator to graph $y_1 = x^2 + x - 2$, $y_2 = -x^2 + 2x - 3$, and $y_3 = 3x - 5$, and then use the TRACE feature to complete the table. If you have a TI-82 calculator, you can use the TABLE feature to complete this table. For each value of x, verify that $y_3 = y_1 + y_2$.

x	-2	-1	0	1	2
y_1					
y_2					
y_3					

69 Error Analysis A classmate subtracted $4x^2 - 5x - 9$ from $3x^2 + 9x + 13$ by writing

$$3x^2 + 9x + 13 - 4x^2 - 5x - 9$$

to obtain $-x^2 + 4x + 4$. Explain the *error* in this work, and make a suggestion as to how this classmate could organize the work to avoid this error.

Discussion Question

70 Describe in your own words the procedure for subtracting one polynomial from another polynomial.

Challenge Question

The following problem was presented by Martin Gardner, a well-known author of puzzles for *Scientific American.* Determine by inspection the length of the diagonal AC of the rectangle inscribed in the one quadrant of a circle shown in the figure. Explain your reasoning.

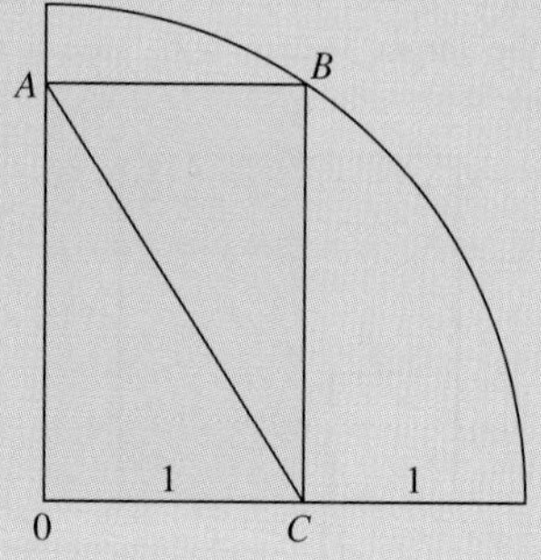

Section Skill Reinforcers

Simplify each of these expressions.

1 $24 - 4 \cdot 3^2$ **2** $(24 - 4) \cdot 3^2$ **3** $24 - (4 \cdot 3)^2$ **4** $24 \div 4 \cdot 3^2$ **5** $(24 - 4 \cdot 3)^2$

4-5 Multiplication of Polynomials

Section Objectives

7 Multiply polynomials using a vertical format.

8 Multiply binomials by inspection.

The following example illustrates how the distributive property is used to multiply a polynomial by a monomial. For example,

$$3x(2x^2 - 5x + 9) = 3x(2x^2) + 3x(-5x) + 3x(9)$$
$$= 6x^3 - 15x^2 + 27x$$

Distribute the multiplication of $3x$ to each term of the trinomial.

Thus to multiply a polynomial by a monomial, multiply each term of the polynomial by the monomial.

EXAMPLE 1 Distributing a Monomial Factor

Multiply these polynomials.

SOLUTIONS

(a) $6a(3a^2 - 5a - 4)$

$6a(3a^2 - 5a - 4)$
$= 6a(3a^2) + 6a(-5a) + 6a(-4)$
$= 18a^3 - 30a^2 - 24a$

Distribute the multiplication of the monomial to each term of the other factor.

(b) $(2x^2 - 3xy + 9y^2)(xy)$

$(2x^2 - 3xy + 9y^2)(xy)$
$= (2x^2)(xy) + (-3xy)(xy) + (9y^2)(xy)$
$= 2x^3y - 3x^2y^2 + 9xy^3$ ▼

SELF-CHECK ✓

Multiply $2x(7x^2 - x - 3)$.

Note the repeated use of the distributive property in the multiplication of the following binomials.

EXAMPLE 2 Multiplying Binomials

Multiply $(2x + y)$ by $(3x + 5y)$.

SOLUTION $(2x + y)(3x + 5y) = 2x(3x + 5y) + y(3x + 5y)$

Distribute the factor $3x + 5y$ to both terms inside the first set of parentheses.

$= (2x)(3x) + (2x)(5y) + (y)(3x) + (y)(5y)$

Then distribute the multiplication of each monomial factor.

$= 6x^2 + 10xy + 3xy + 5y^2$
$= 6x^2 + 13xy + 5y^2$

Combine like terms.

▼

A vertical format is a convenient form for ensuring that each term of the first polynomial is multiplied by each term of the second polynomial. To use the vertical format, first write each polynomial in standard form and then multiply the terms from left to right so that the product will also be in standard form.

EXAMPLE 3 Multiplying Binomials

Multiply $(2x + y)$ by $(3x + 5y)$ using the vertical format.

SOLUTION

$$\begin{array}{rl} 2x + y & \\ \underline{3x + 5y} & \\ 6x^2 + 3xy & \longleftarrow \text{The product of } 3x \text{ and } 2x + y \\ \underline{10xy + 5y^2} & \longleftarrow \text{The product of } 5y \text{ and } 2x + y \\ 6x^2 + 13xy + 5y^2 & \longleftarrow \text{The product in standard form} \end{array}$$

▼

SELF-CHECK ANSWER 

$14x^3 - 2x^2 - 6x$

EXAMPLE 4 Multiplying a Trinomial by a Binomial

Multiply $(3x + 6 + 4x^2)$ by $(7 + x)$.

SOLUTION

$$\begin{array}{rl} 4x^2 + 3x + 6 & \text{First write each factor in standard form.} \\ x + 7 & \\ \hline 4x^3 + 3x^2 + 6x \qquad & \longleftarrow x(4x^2 + 3x + 6) \\ 28x^2 + 21x + 42 & \longleftarrow 7(4x^2 + 3x + 6) \\ \hline 4x^3 + 31x^2 + 27x + 42 & \end{array}$$

Note that similar terms are aligned in the same column.

In algebra you will often need to find the product of two binomials, so it is important to be able to form these products quickly without using the vertical format. This is a basic skill needed to factor polynomials.

The steps used to form the product $(2x + y)(3x + 5y)$ in Example 2 are shown below. To remember the steps involved in forming the product of two binomials, we use the acronym *FOIL*.

$$\begin{aligned} (2x + y)(3x + 5y) &= 2x(3x + 5y) + y(3x + 5y) \\ &= \underbrace{(2x)(3x)}_{F} + \underbrace{(2x)(5y)}_{O} + \underbrace{(y)(3x)}_{I} + \underbrace{(y)(5y)}_{L} \\ &= 6x^2 + 10xy + 3xy + 5y^2 \\ &= 6x^2 + 13xy + 5y^2 \end{aligned}$$

First terms, $(2x)(3x)$
Outer terms, $+ (2x)(5y)$
Inner terms, $+ (y)(3x)$
Last terms, $+ (y)(5y)$

At first you may find it helpful to draw lines connecting each of the indicated products of the FOIL procedure. This technique is illustrated below. Once you feel comfortable using the procedure, you should omit this mental crutch.

EXAMPLE 5 Using the FOIL Procedure

Use the FOIL procedure to find each of the indicated products.

SOLUTIONS

(a) $(x - 7)(x - 8)$

$(x - 7)(x - 8) = x^2 - 15x + 56$ (F, O, I, L)

The middle term is the sum of $-8x$ and $-7x$.

(b) $(3m + 4n)(2m - 7n)$

$(3m + 4n)(2m - 7n) = 6m^2 - 13mn - 28n^2$ (F, O, I, L)

The middle term is the sum of $-21mn$ and $8mn$.

A Geometric Viewpoint: Using Graphs to Determine Whether Two Polynomials Are Equal

Two polynomials $P_1(x)$ and $P_2(x)$ are equal only if the graphs of $y_1 = P_1(x)$ and $y_2 = P_2(x)$ completely overlap. The graphs of $y_1 = (2x - 3)(3x + 1)$ and $y_2 = 6x^2 - 7x + 3$ are shown on the graphics calculator window to the right. By comparing these graphs as they are produced, we can note that there are two distinct graphs. We can conclude that $(2x - 3)(3x + 1) \neq 6x^2 - 7x + 3$. Correct multiplication produces $(2x - 3)(3x + 1) = 6x^2 - 7x - 3$.

[−10, 10] for x, [−10, 10] for y

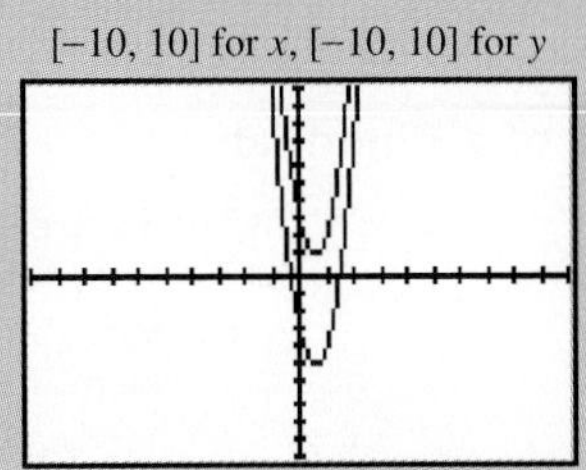

EXAMPLE 6 Calculator Usage

Use the graphs of $y_1 = (2x - 3)(3x + 1)$ and $y_2 = 6x^2 - 7x - 3$ to determine whether $(2x - 3)(3x + 1)$ equals $6x^2 - 7x - 3$.

SOLUTION

Keystrokes

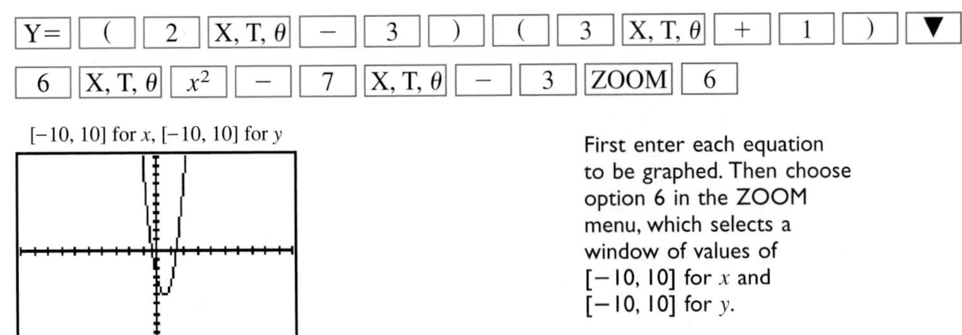

First enter each equation to be graphed. Then choose option 6 in the ZOOM menu, which selects a window of values of [−10, 10] for x and [−10, 10] for y.

Inspection suggests that the two graphs of $y_1 = (2x - 3)(3x + 1)$ and $y_2 = 6x^2 - 7x - 3$ overlap completely, a result that can be further examined by using the TRACE option on both graphs. Since roundoff error and an inappropriate scale can sometimes give a misleading visual impression, the apparent overlap does not prove that these polynomials are equal. However, graphing serves as a good check and can reveal careless errors, as noted in the geometric viewpoint above.

Answer $(2x - 3)(3x + 1) = 6x^2 - 7x - 3$ ▼

SELF-CHECK

Find each product.

1 $(2x - 3)(3x + 7)$

2 $(x^4 - 3x^2 + 7x)(x - 3)$

Determine each product by inspection.

3 $(5x - 6)(2x - 3)$

4 $(7x + 5)(3x + 2)$

SELF-CHECK ANSWERS

1 $6x^2 + 5x - 21$ **2** $x^5 - 3x^4 - 3x^3 + 16x^2 - 21x$

3 $10x^2 - 27x + 18$ **4** $21x^2 + 29x + 10$

Certain types of products occur so often that you should memorize their special forms. You may wish to verify each of the forms below by using the FOIL procedure.

SPECIAL PRODUCTS

The Square of a Binomial:

$(a + b)^2 = a^2 + 2ab + b^2$

$(a - b)^2 = a^2 - 2ab + b^2$

The square of a binomial is the sum of
a. the square of the first term of the binomial,
b. twice the product of the two terms of the binomial, and
c. the square of the last term of the binomial.

A Sum Times a Difference:

$(a + b)(a - b) = a^2 - b^2$

The product of a sum times a difference is the difference of their squares. (There is no middle term.)

EXAMPLE 7 Multiplying Special Products

Find each of the following special products.

SOLUTIONS

(a) $(x + 3)^2$

$(x + 3)^2 = x^2 + 6x + 9$

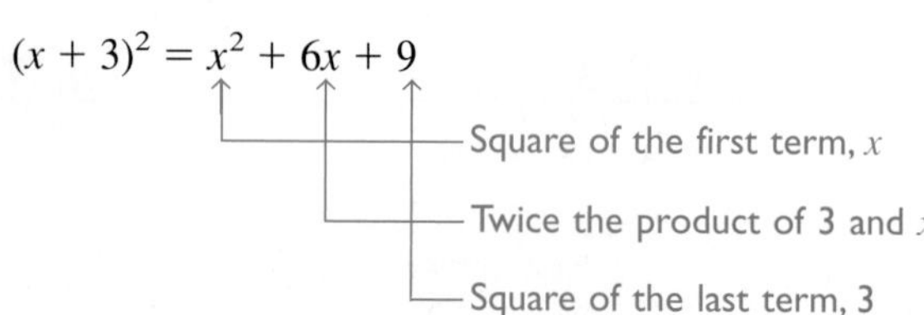

(b) $(6m - 5n)^2$

$(6m - 5n)^2 = 36m^2 - 60mn + 25n^2$

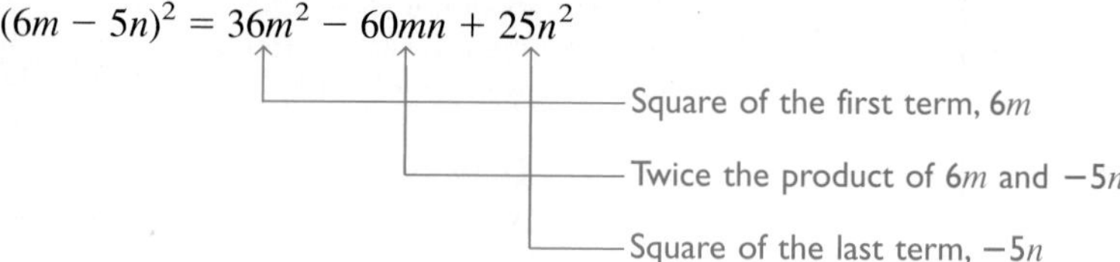

(c) $(2x + 11y)(2x - 11y)$

$(2x + 11y)(2x - 11y) = 4x^2 - 121y^2$

Square of the first term, $2x$
Minus the square of the second term, $11y$

EXAMPLE 8 Cubing a Binomial

Find $(2x - 3)^3$.

SOLUTION

$$\begin{aligned}(2x - 3)^3 &= (2x - 3)(2x - 3)^2 \\ &= (2x - 3)[(2x)^2 + 2(2x)(-3) + (-3)^2] \\ &= (2x - 3)(4x^2 - 12x + 9) \\ &= 2x(4x^2 - 12x + 9) - 3(4x^2 - 12x + 9) \\ &= 8x^3 - 24x^2 + 18x - 12x^2 + 36x - 27 \\ &= 8x^3 - 36x^2 + 54x - 27\end{aligned}$$

First square the binomial by inspection.

Distribute the factor $4x^2 - 12x + 9$ to each term of $2x - 3$, and then combine like terms.

The patterns we have used to multiply binomials can also be applied to expressions that are not binomials, as illustrated in the next example.

EXAMPLE 9 Using Special Products to Multiply Other Expressions

Multiply these algebraic expressions.

(a) $(x^{-1} + y^{-1})(x^{-1} - y^{-1})$

(b) $(x^m + y^n)^2$

(c) $(x + 2y + 3)(x + 2y - 3)$

SOLUTIONS

(a) $(x^{-1} + y^{-1})(x^{-1} - y^{-1}) = (x^{-1})^2 - (y^{-1})^2$
$= x^{-2} - y^{-2}$

A sum times a difference is the difference of the squares.

(b) $(x^m + y^n)^2 = (x^m)^2 + 2(x^m)(y^n) + (y^n)^2$
$= x^{2m} + 2x^my^n + y^{2n}$

Square of a sum

(c) $(x + 2y + 3)(x + 2y - 3) = [(x + 2y) + 3][(x + 2y) - 3]$ Group as a sum times a difference.

$= (x + 2y)^2 - (3)^2$ The product is the difference of the squares.

$= x^2 + 4xy + 4y^2 - 9$ Now square each term by inspection.

EXAMPLE 10 A Revenue Polynomial

The number of units demanded by consumers when each unit is priced at d dollars is given by the polynomial $N(d) = -d^2 + 400$. Find the revenue $R(d)$ generated if these units are sold at d dollars each.

SOLUTION | Revenue | = | Price per unit | · | Number of units sold | **Word equation**

$R(d) = d \cdot N(d)$

$R(d) = d(-d^2 + 400)$

$R(d) = -d^3 + 400d$

Substitute the given expression for $N(d)$, and then multiply.

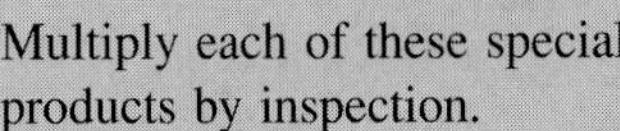

SELF-CHECK

Multiply each of these special products by inspection.

1 $(2m + 3n)^2$
2 $(5v - 7w)^2$
3 $(7x - 9y)(7x + 9y)$

Exercises 4-5

A

In Exercises 1–60 find each of the indicated products.

1 $2x(3x^3)$ **2** $-5x(2x^4)$ **3** $(-2xy^2)^3$
4 $(5x^2y)^2$ **5** $3x(x - 5)$ **6** $-2x(x - 3)$
7 $3x(x + y)$ **8** $5x(x - y)$ **9** $-4x(2x - 5)$
10 $-7x(3x - 2)$ **11** $-4a(a^2 - 2a - 3)$
12 $-5b(b^2 - 6b - 7)$ **13** $(2x^2 - 5x - 3)(2x)$
14 $(-3x^2 + 7x - 1)(3x)$ **15** $2x^2y(3x - 4y)$

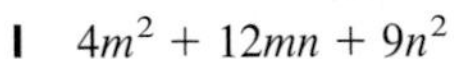

SELF-CHECK ANSWERS

1 $4m^2 + 12mn + 9n^2$ **2** $25v^2 - 70vw + 49w^2$ **3** $49x^2 - 81y^2$

16 $5xy^2(x^2 - y)$

17 $(-3xy^2)(2x - 5y)$

18 $(-2x^2y)(2x - 3y)$

19 $(x - 2y)(x + 3y)$

20 $(x - 3y)(x + 4y)$

21 $(x + 2y)(x + 3y)$

22 $(x + 3y)(x + 4y)$

23 $(x - 2y)(x - 3y)$

24 $(x - 3y)(x - 4y)$

25 $(x + 3)(x^2 - 5x - 1)$

26 $(x + 4)(x^2 - 2x - 4)$

27 $(5a - b)(2a + b)$

28 $(11x - 4y)(3x + y)$

29 $(3a^2 + 5a + 7)(2a - 3)$

30 $(2b^2 - 5b + 7)(3b - 4)$

31 $(x - 1)(x + 1)$

32 $(x - 3)(x + 3)$

33 $(2x + 5)(2x - 5)$

34 $(3x + 4)(3x - 4)$

35 $(x - 8y)(x + 8y)$

36 $(4x - y)(4x + y)$

37 $(ab - 3)(ab + 3)$

38 $(x^2 - 2)(x^2 + 3)$

39 $(m + 5)^2$

40 $(m + 6)^2$

41 $(2x + 1)^2$

42 $(c + 3d)^2$

43 $(7x + 5y)^2$

44 $(4x + 3y)^2$

45 $(x - 5)^2$

46 $(x - 7)^2$

47 $(w - z)^2$

48 $(x - 2y)^2$

49 $(3m - n)^2$

50 $(x - 3y)^2$

51 $(4x - 3y)^2$

52 $(5x - 4y)^2$

53 $(xy - 5)(xy + 5)$

54 $(ab + c)(ab - c)$

55 $(2x^2 - 9y)(2x^2 + 9y)$

56 $(7xy - 5z)(7xy + 5z)$

57 $(x^2 - y^2)(x^2 + y^2)$

58 $(x^3 + y^3)(x^3 - y^3)$

59 $[(x + y) + z][(x + y) - z]$

60 $[(2a + 3b) - c][(2a + 3b) + c]$

B

In Exercises 61–72 simplify each expression.

61 $(5x + y - 3z)(5x + y + 3z)$

62 $(7v + 3w + 2z)(7v + 3w - 2z)$

63 $(2a - b + c)(2a - b - c)$

64 $(m - 3n - 1)(m + 3n + 1)$

65 $-12x(x^2 + 3x - 11) - 5x(x^2 + 9)$

66 $(a - 1)(a + 3) - (a + 2)(a + 1)$

67 $(r - s)^2 - (r^2 - s^2)$

68 $(5v - 3z)(5v + 3z) - [(5v)^2 + (3z)^2]$

69 $(x - 2)^3$

70 $(3x + 1)^3$

71 $(2w - 1)(3w + 2)(4w - 5)$

72 $[(x + 1)(x - 1)]^2$

CALCULATOR USAGE (73–78)

In Exercises 73–77 use a graphics calculator to graph $y_1 = P_1(x)$ and $y_2 = P_2(x)$. Then use these graphs to determine whether $P_1(x) = P_2(x)$.

73 $P_1(x) = (3x - 4)(x + 2)$
$P_2(x) = 3x^2 - 2x - 8$

74 $P_1(x) = (4x - 3)(3x + 2)$
$P_2(x) = 12x^2 - x + 6$

75 $P_1(x) = (2.5x + 3.5)(2x + 2)$
$P_2(x) = 5x^2 + 12x + 7$

76 $P_1(x) = (1.4x + 1.5)(5x + 2)$
$P_2(x) = 7x^2 + 8.3x + 3$

77 $P_1(x) = (1.7x - 2.3)(1.7x + 3.4)$
$P_2(x) = 2.89x^2 - 1.87x - 7.82$

78 On the standard viewing window, $[-10, 10]$ for x and $[-10, 10]$ for y, only one graph appears when the equations $y_1 = (3x - 1)(4x + 5)$ and $y_2 = 12x^2 + 11x + 15$ are graphed. However, $(3x - 1)(4x + 5) \neq 12x^2 + 11x + 15$. Explain this.

79 Use a graphics calculator to complete the table below for $y_1 = x - 1$, $y_2 = x + 1$, and $y_3 = x^2 - 1$. If you have a TI-82 calculator, you can use the TABLE feature to complete this table. For each value of x, verify that $y_3 = y_1 \cdot y_2$.

x	−2	−1	0	1	2
y_1					
y_2					
y_3					

C

In Exercises 80–89 simplify each expression, assuming that all exponents are integers.

80 $(2x^{-3} + y^{-2})^2$

81 $(4a^{-2} - y^{-5})^2$

82 $-5m^{-1}n^{-1}(3m^3n - 2m^2n + 3mn)$

83 $-25s^{-2}(6r^3s^2 - 6r^2s^3 - 3rs^4 + s^5)$

84 $(x^m - 2)(x^m + 2)$

85 $(x^m + 3)(x^m - 4)$

86 $(x^m - y^n)^2$

87 $(x^m - y^n)(x^m + y^n)$

88 $(x^m + y^n)(x^{2m} - x^my^n + y^{2n})$

89 $(x^m - y^n)(x^{2m} + x^my^n + y^{2n})$

Revenue Polynomials (90 and 91)

In Exercises 90 and 91 find the revenue $R(t)$ from selling the units produced by an assembly line in t hours. Use the formula $R(t) = N(t)C(t)$, where $N(t)$ is the number of units produced in t hours and $C(t)$ is the cost per unit when the assembly line operates for t hours.

90 $N(t) = -3t^2 - 6t + 756$, $C(t) = 2t + 1$

91 $N(t) = -4t^2 + 7t + 203$, $C(t) = 3t + 5$

92 Determine the area of the rectangle shown in the figure.

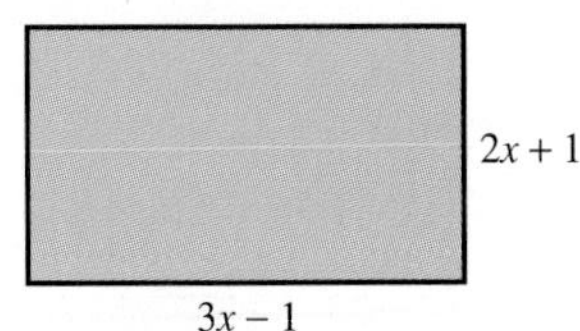

Figure for Exercise 92

93 Determine the area of the triangle shown in the figure.

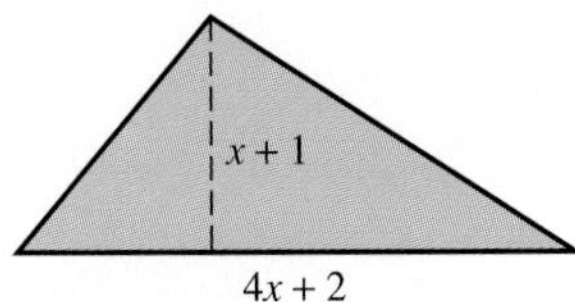

Figure for Exercise 93

94 Area of a Region Write, in simplified form, a polynomial for the area of the shaded region shown in the figure.

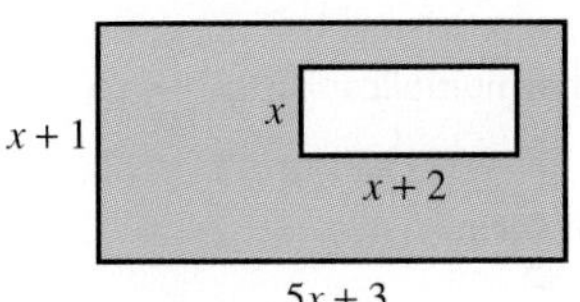

Figure for Exercise 94

DISCUSSION QUESTIONS (95 AND 96)

95 Compare the multiplication of the integers with the multiplication of the polynomials (shown below). In particular, note the value of each polynomial for $x = 10$.

Multiplication of Integers	***Multiplication of Polynomials***
Problem: Multiply 32 times 21.	**Problem:** Multiply $3x + 2$ times $2x + 1$.
21	$2x + 1$
(×) 32	(×) $3x + 2$
42	$4x + 2$
63	$6x^2 + 3x$
672	$6x^2 + 7x + 2$
$(32)(21) = 672$	$(3x + 2)(2x + 1) = 6x^2 + 7x + 2$

96 Multiply $(2x - 5y)(3x + 4y)$, and then describe the steps you used, including each time you used the distributive property.

CHALLENGE QUESTION

a. Simplify the product $(x - a)(x - b)(x - c) \cdot \cdots \cdot (x - z)$.

b. Use the figure shown to answer all parts of this question.

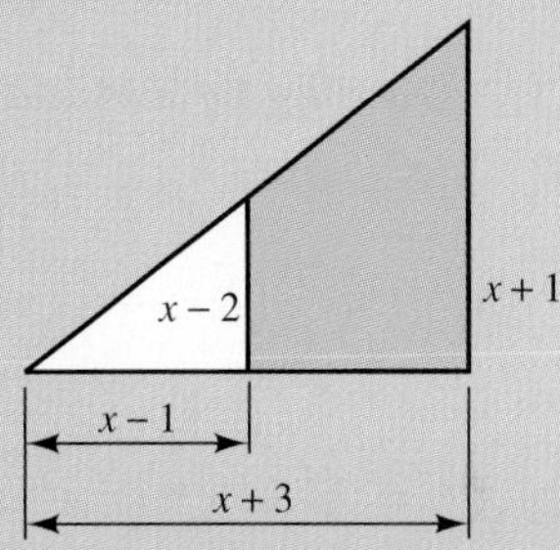

i. Use the formula for the area of a trapezoid to write in simplified form a polynomial for the area of the shaded region.

ii. Write, in simplified form, a polynomial for the area of the shaded region by considering this area as the difference of the areas of two triangles.

iii. Equate the two expressions from parts i and ii, and then solve for x.

iv. Determine the area of the shaded region.

SECTION SKILL REINFORCERS

Simplify each of these expressions.

1 $\dfrac{x^{18}}{x^2}$ **2** $x^{18}x^2$ **3** $(x^{18})^2$ **4** $\dfrac{x^{18}}{x^{-2}}$ **5** $\dfrac{x^{-18}}{x^2}$

4-6 *Division of Polynomials*

SECTION OBJECTIVE

9 Divide polynomials.

Example 1 illustrates the division of a polynomial by a monomial. Since dividing by $2x$ is the same as multiplying by $\frac{1}{2x}$, we can use the distributive property to simplify the expression. We will assume throughout this section that the variables are restricted to values that will avoid division by zero.

EXAMPLE 1 Dividing by a Monomial

Simplify $(6x^4 + 4x^2 + 2x) \div 2x$.

SOLUTION

$(6x^4 + 4x^2 + 2x) \div 2x = (6x^4 + 4x^2 + 2x) \cdot \frac{1}{2x}$ Multiply by the reciprocal of $2x$, $\frac{1}{2x}$.

$= \frac{6x^4}{2x} + \frac{4x^2}{2x} + \frac{2x}{2x}$ Distribute the multiplication of $\frac{1}{2x}$ to each term.

$= 3x^3 + 2x + 1$ Reduce each term.

The typical procedure for dividing a polynomial by a monomial is to divide each term of the polynomial by the monomial. This is illustrated by the fractional forms shown in the next example. Some of the steps are done mentally.

EXAMPLE 2 Dividing by a Monomial

Perform the following divisions.

SOLUTIONS

(a) $(8a^8 - 6a^6) \div 2a^2$

$\frac{8a^8 - 6a^6}{2a^2} = \frac{8a^8}{2a^2} - \frac{6a^6}{2a^2}$

$= 4a^6 - 3a^4$

Rewrite the quotient as the sum of fractions that share a common denominator, and then reduce these fractions.

(b) $\frac{15a^4b^2 - 10a^3b^3}{-5a^2b^2}$

$\frac{15a^4b^2 - 10a^3b^3}{-5a^2b^2} = \frac{15a^4b^2}{-5a^2b^2} - \frac{10a^3b^3}{-5a^2b^2}$

$= -3a^2 + 2ab$

Rewrite the quotient as the difference of two fractions that share a common denominator, and then reduce these fractions.

(c) $\frac{5v^2 - 10v + 25}{5v}$

$\frac{5v^2 - 10v + 25}{5v} = \frac{5v^2}{5v} - \frac{10v}{5v} + \frac{25}{5v}$

$= v - 2 + \frac{5}{v}$

Rewrite the quotient as three separate terms, and then reduce each individual term.

(d) $\frac{x^{2m} - x^{m+4}}{x^m}$

$\frac{x^{2m} - x^{m+4}}{x^m} = \frac{x^{2m}}{x^m} - \frac{x^{m+4}}{x^m}$

$= x^m - x^4$

Rewrite the expression as two separate terms. Then perform the indicated divisions by subtracting exponents: $2m - m = m$ and $(m + 4) - m = 4$.

The procedure for dividing a polynomial by another polynomial is very similar to the long division of whole numbers. To use this long-division algorithm, be sure to write both the divisor and the dividend in standard form. (An algorithm is a step-by-step procedure, generally repetitive in nature.)

SELF-CHECK

Perform each division.

1 $(5x^5 - 2x^4) \div x^3$

2 $(21w^4 - 35w^3 + 14w^2) \div (-7w)$

LONG DIVISION OF POLYNOMIALS

Step 1 Write the polynomials in the long-division format, expressing each in standard form.

Step 2 Divide the first term of the divisor into the first term of the dividend. The result of this division is the first term of the quotient.

Step 3 Multiply the first term of the quotient times every term in the divisor. Write this product under the dividend, aligning like terms.

Step 4 Subtract this product from the dividend, and bring down the next term.

Step 5 Use the result of Step 4 as a new dividend, and repeat Steps 2 through 4 until either the remainder is zero or the degree of the remainder is less than the degree of the divisor.

To illustrate the close relationship between the long division of whole numbers and the long division of polynomials, we will compare these procedures side by side. Before examining this comparison, note that if 10 is substituted for x in $(6x^2 + 7x + 2) \div (2x + 1)$, we obtain $[6(10)^2 + 7(10) + 2] \div [2(10) + 1]$, or $(600 + 70 + 2) \div (20 + 1) = 672 \div 21$. Thus for this value of x, both procedures denote the same thing.

Long Division of Whole Numbers	***Long Division of Polynomials***	
Problem: Divide 672 by 21.	**Problem:** Divide $6x^2 + 7x + 2$ by $2x + 1$.	

Step 1 Write the division in the long-division format.

$$21\overline{)672} \qquad\qquad 2x + 1\overline{)6x^2 + 7x + 2}$$

Step 2 Divide to obtain the first term in the quotient.

$$\begin{array}{r} 3 \\ 21\overline{)672} \end{array} \qquad\qquad \begin{array}{r} 3x \qquad\quad \\ 2x + 1\overline{)6x^2 + 7x + 2} \end{array}$$

$6x^2$ divided by $2x$ is $3x$.

Step 3 Multiply the first term in the quotient by the divisor.

$$\begin{array}{r} 3 \\ 21\overline{)672} \\ \underline{63}\; \end{array} \qquad\qquad \begin{array}{r} 3x \qquad\quad \\ 2x + 1\overline{)6x^2 + 7x + 2} \\ \underline{6x^2 + 3x} \qquad \end{array}$$

Note the alignment of similar terms.

SELF-CHECK ANSWERS 

1 $5x^2 - 2x$ **2** $-3w^3 + 5w^2 - 2w$

Step 4 Subtract this product from the dividend, and bring down the next term.

$$\begin{array}{r|l} & 3 \\ \hline 21 & 672 \\ & \underline{63} \\ & 42 \end{array} \qquad \begin{array}{r|l} & 3x \\ \hline 2x + 1 & 6x^2 + 7x + 2 \\ & \underline{6x^2 + 3x} \\ & 4x + 2 \end{array}$$

Step 5 Repeat Steps 2–4 to obtain the next term in the quotient.

$$\begin{array}{r|l} & 32 \\ \hline 21 & 672 \\ & \underline{63} \\ & 42 \\ & \underline{42} \\ & 0 \end{array} \qquad \begin{array}{r|l} & 3x + 2 \\ \hline 2x + 1 & 6x^2 + 7x + 2 \\ & \underline{6x^2 + 3x} \\ & 4x + 2 \\ & \underline{4x + 2} \\ & 0 \end{array}$$

Since the remainder is 0, the division is finished.

Answer $672 \div 21 = 32$ $\qquad (6x^2 + 7x + 2) \div (2x + 1) = 3x + 2$

EXAMPLE 3 Step-by-Step Use of the Long-Division Algorithm

Divide $3x^2 - 2x - 8$ by $x - 2$.

SOLUTION

1 $x - 2\overline{)3x^2 - 2x - 8}$

Set up the format for long division, writing both polynomials in standard form.

2 $$\begin{array}{r|l} & 3x \\ \hline x - 2 & 3x^2 - 2x - 8 \end{array}$$

Divide $3x^2$ (the first term of the dividend) by x (the first term of the divisor) to obtain $3x$ (the first term of the quotient). Align similar terms.

3 $$\begin{array}{r|l} & 3x \\ \hline x - 2 & 3x^2 - 2x - 8 \\ & \underline{3x^2 - 6x} \end{array}$$

Multiply $3x$ times every term in the divisor, aligning under similar terms in the dividend.

4 $$\begin{array}{r|l} & 3x \\ \hline x - 2 & 3x^2 - 2x - 8 \\ & \ominus \quad \oplus \\ & \underline{3x^2 - 6x} \\ & 4x - 8 \end{array}$$

Subtract $3x^2 - 6x$ from the dividend by changing the sign of each term and adding.

5 $$\begin{array}{r|l} & 3x + 4 \\ \hline x - 2 & 3x^2 - 2x - 8 \\ & \ominus \quad \oplus \\ & \underline{3x^2 - 6x} \\ & 4x - 8 \\ & \ominus \quad \oplus \\ & \underline{4x - 8} \\ & 0 \end{array}$$

Divide $4x$ (the first term in the last row) by x (the first term in the divisor) to obtain 4 (the next term in the quotient). Multiply each term of the divisor by 4, aligning under similar terms in the dividend. Then subtract to obtain the remainder of 0.

Answer $(3x^2 - 2x - 8) \div (x - 2) = 3x + 4$

↑ Dividend ↑ Divisor ↑ Quotient

Check The answer to this division problem can be checked by multiplying the divisor by the quotient. Multiply to verify that

$$\underset{\text{Divisor}}{(x - 2)}\underset{\text{Quotient}}{(3x + 4)} = \underset{\text{Dividend}}{3x^2 - 2x - 8}$$

▼

EXAMPLE 4 Dividing a Polynomial with Missing Terms

Divide $6x^4 + 45x - 6$ by $3x + 6$.

SOLUTION

$$\begin{array}{r} 2x^3 - 4x^2 + 8x - 1 \\ 3x+6\overline{)\,6x^4 + 0x^3 + 0x^2 + 45x - 6} \\ \ominus\ \ \ominus \qquad\qquad\qquad \\ \underline{6x^4 + 12x^3} \qquad\qquad\qquad \\ -12x^3 \qquad\qquad\qquad \\ \oplus\ \ \oplus \qquad\qquad \\ \underline{-12x^3 - 24x^2} \qquad\qquad \\ 24x^2 + 45x \qquad \\ \ominus\ \ \ominus \qquad \\ \underline{24x^2 + 48x} \qquad \\ -3x - 6 \\ \oplus\ \ \oplus \\ \underline{-3x - 6} \\ 0 \end{array}$$

Either leave space for the missing terms in the dividend or write in these terms with zero coefficients in order to keep similar terms aligned.

Answer $\dfrac{6x^4 + 45x - 6}{3x + 6} = 2x^3 - 4x^2 + 8x - 1$ ▼

EXAMPLE 5 Dividing a Polynomial That Is Not in Standard Form

Divide $19x^2 + 30 + x + 6x^3$ by $7 + 2x$, and check the answer.

SOLUTION

$$\begin{array}{r} 3x^2 - x + 4 \\ 2x+7\overline{)\,6x^3 + 19x^2 + x + 30} \\ \ominus\ \ \ominus \qquad\qquad \\ \underline{6x^3 + 21x^2} \qquad\qquad \\ -2x^2 + x \qquad \\ \oplus\ \ \oplus \qquad \\ \underline{-2x^2 - 7x} \qquad \\ 8x + 30 \\ \ominus\ \ \ominus \\ \underline{8x + 28} \\ 2 \end{array}$$

Write both the divisor and the dividend in standard form, and then divide using the long-division algorithm.

Note that the division is not exact, since the remainder is 2.

Answer $\dfrac{6x^3 + 19x^2 + x + 30}{2x + 7} = 3x^2 - x + 4 + \dfrac{2}{2x + 7}$

Check $(2x + 7)\left[3x^2 - x + 4 + \dfrac{2}{2x + 7}\right]$

$= (2x + 7)(3x^2 - x + 4) + (2x + 7)\left(\dfrac{2}{2x + 7}\right)$

$= (6x^3 + 19x^2 + x + 28) + 2$

$= 6x^3 + 19x^2 + x + 30$

Distribute the factor $2x + 7$, and then perform the indicated multiplications. Since the result is the dividend, the answer checks.

EXAMPLE 6 Dividing by a Second-Degree Divisor

Divide $43x^2 + 27 - 15x + 5x^3$ by $-2x + 5x^2 + 3$.

SOLUTION

$$\begin{array}{r} x + 9 \\ 5x^2 - 2x + 3\,\overline{)\,5x^3 + 43x^2 - 15x + 27} \\ \ominus\ \ \oplus\ \ \ominus\quad\ \ \\ \underline{5x^3 - 2x^2 + 3x}\qquad \\ 45x^2 - 18x + 27 \\ \ominus\ \ \oplus\ \ \ominus \\ \underline{45x^2 - 18x + 27} \\ 0 \end{array}$$

Write both divisor and dividend in standard form. Then divide $5x^2$ into $5x^3$ to obtain x.

Multiply x times $5x^2 - 2x + 3$ to obtain $5x^3 - 2x^2 + 3x$, and then subtract this quantity.

Divide $5x^2$ into $45x^2$ to obtain 9.

Multiply 9 times $5x^2 - 2x + 3$ to obtain $45x^2 - 18x + 27$, and then subtract this quantity to complete the division.

Thus

$(5x^3 + 43x^2 - 15x + 27) \div (5x^2 - 2x + 3) = x + 9$

You can multiply the divisor and the quotient to check this answer.

SELF-CHECK

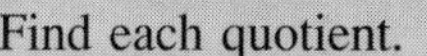

Find each quotient.

1 $(m^2 - 7m + 10) \div (m - 5)$

2 $(3r^3 + 7r^2 + 27r + 60) \div (3r + 7)$

EXAMPLE 7 Dividing a Polynomial with Missing Terms

Find $(a^4 - b^4) \div (a + b)$.

SOLUTION

$$\begin{array}{r} a^3 - a^2b + ab^2 - b^3 \\ a + b\,\overline{)\,a^4 + 0a^3b + 0a^2b^2 + 0ab^3 - b^4} \\ \ominus\ \ \ominus\qquad\qquad\qquad\qquad \\ \underline{a^4 + a^3b}\qquad\qquad\qquad\qquad \\ -a^3b + 0a^2b^2\qquad\qquad \\ \oplus\qquad\ \oplus\qquad\qquad \\ \underline{-a^3b - a^2b^2}\qquad\qquad \\ a^2b^2 + 0ab^3\qquad \\ \ominus\qquad\ \ominus\qquad \\ \underline{a^2b^2 + ab^3}\qquad \\ -ab^3 - b^4 \\ \oplus\qquad\ \oplus \\ \underline{-ab^3 - b^4} \\ 0 \end{array}$$

Place zero coefficients for the missing terms to facilitate the alignment of similar terms.

Answer $(a^4 - b^4) \div (a + b) = a^3 - a^2b + ab^2 - b^3$

SELF-CHECK ANSWERS 

1 $m - 2$ 2 $r^2 + 9 - \dfrac{3}{3r + 7}$

One way to check to determine whether two algebraic expressions are equal is to substitute a test value into each expression. If the value does not check, then the expressions are not equal. However, note that the fact that one or two values check does not mean that the expressions are equal for all values.

EXAMPLE 8 Checking an Answer

A student performed the division indicated by $\frac{6x^3 + 22x^2 - 58x + 24}{x^2 + 5x - 3}$ and obtained $6x - 7$. Check this answer by evaluating both expressions for $x = 1$.

SOLUTION At $x = 1$ the value of $\frac{6x^3 + 22x^2 - 58x + 24}{x^2 + 5x - 3}$ is

$$\frac{6(1)^3 + 22(1)^2 - 58(1) + 24}{(1)^2 + 5(1) - 3} = \frac{-6}{3} = -2$$ Substitute 1 for x in each expression.

At $x = 1$ the value of $6x - 7$ is

$$6(1) - 7 = -1$$

Since these values are not equal, the answer to the division is not correct.

$$\frac{6x^3 + 22x^2 - 58x + 24}{x^2 + 5x - 3} \neq 6x - 7$$

The correct division would yield $\frac{6x^3 + 22x^2 - 58x + 24}{x^2 + 5x - 3} = 6x - 8$

EXAMPLE 9 Dividing Polynomials with Variable Exponents

Find $(x^{3n} + 1) \div (x^n + 1)$.

SOLUTION

$$\begin{array}{r|l} & x^{2n} - x^n + 1 \\ x^n + 1 & x^{3n} + 0x^{2n} + 0x^n + 1 \\ & \ominus \quad \ominus \\ & x^{3n} + x^{2n} \\ \hline & -x^{2n} + 0x^n \\ & \oplus \quad \oplus \\ & -x^{2n} - x^n \\ \hline & x^n + 1 \\ & \ominus \quad \ominus \\ & x^n + 1 \\ \hline & 0 \end{array}$$

Place zero coefficients for the missing terms to facilitate the alignment of similar terms.

Answer $(x^{3n} + 1) \div (x^n + 1) = x^{2n} - x^n + 1$

EXAMPLE 10 An Average Cost Polynomial

The cost of producing t units is given by the polynomial $C(t) = t^3 + 3t^2 + t$. Find $A(t)$, the average cost per unit of producing these t units.

SOLUTION $\boxed{\text{Average cost}} = \frac{\boxed{\text{Total cost}}}{\boxed{\text{Number of units}}}$ Word equation

$$A(t) = \frac{C(t)}{t}$$

$$A(t) = \frac{t^3 + 3t^2 + t}{t}$$ Substitute the given expressions for $C(t)$, and then divide each term by t.

$$A(t) = t^2 + 3t + 1$$

Exercises 4-6

A

In Exercises 1–40 perform the indicated operations, and simplify the results.

1 $\frac{12x^3}{4x}$ 2 $\frac{6x^6}{2x^2}$ 3 $\frac{42a^4b^3}{-6a^2b^2}$

4 $\frac{65a^5b^6}{5a^3b^3}$ 5 $\frac{96x^{17}y^{14}}{-8x^{11}y^{11}}$ 6 $\frac{-125a^4b^4c^4}{-25a^4b^3c^2}$

7 $\left(\frac{34x^5y^0z^{-7}}{-17x^3y^{-4}z^{-5}}\right)^3$ 8 $\left(\frac{-45v^{-2}w^2z^0}{-15v^{-3}w^2z^3}\right)^4$

9 $\left(\frac{-121a^4b^{-3}c^2}{-77a^3b^0c^{-2}}\right)^{-2}$ 10 $\left(\frac{-63x^{-4}y^3z^2}{-42x^6y^3z^0}\right)^{-3}$

11 $\frac{(2a^3b^2c^4)^4}{(4ab^2c^3)^2}$ 12 $\frac{(-3ab^2c^{-3})^3}{(9a^3bc^2)^2}$

13 $\frac{12x^2 - 18x}{6x}$ 14 $\frac{20x^2 - 35x}{5x}$

15 $\frac{2x + 8y}{2}$ 16 $\frac{10v^2 - 15v}{5}$

17 $\frac{y^3 - 2y^2 + 3y}{y}$ 18 $\frac{z^3 + 5z^2 + 7z}{z}$

19 $\frac{-a^2x + 2ax}{-ax}$ 20 $\frac{10m^2n^2 - m^2n}{m^2n}$

21 $\frac{25a^3b^2 - 10a^2b^3 + 15ab^4}{-5a^2b^2}$

22 $\frac{39x^3y - 65xy^2 - 78x^2y^2}{13xy}$

23 Divide $18m^4 - 9m^3 + 6m^2$ by $-3m$.

24 Divide $4x^2y$ into $20x^5y - 16x^3y^3 + 12x^2y^4$.

25 $x + 2\overline{)x^2 - 5x - 14}$ 26 $x - 7\overline{)x^2 - 8x + 7}$

27 $\frac{b^2 + 4b - 12}{b + 6}$ 28 $\frac{y^2 - 7y + 10}{y - 5}$

29 $\frac{10n^2 - 11n + 3}{2n - 1}$ 30 $\frac{2k^2 - k - 6}{2k + 3}$

31 $\frac{15v^2 + 14v - 6}{3v + 4}$ 32 $\frac{6w^2 - 5w - 7}{2w + 1}$

33 $\frac{10x^2 + 23xy - 5y^2}{5x - y}$

34 $\frac{6a^2 + 9ab - 23b^2}{2a + 5b}$

35 $\frac{6v^3 + 9v^2 - 2v - 3}{3v^2 - 1}$

36 $\frac{4v^4 + 6v^3 + 3v - 1}{2v^2 - 1}$

37 $\frac{21x^4 - 7x^3 + 62x^2 - 9x + 45}{3x^2 - x + 5}$

38 $\frac{2z^5 + 4z^4 - z^3 + 2z^2 - z}{z^2 + 2z - 1}$

39 $\frac{a^5 - 1}{a - 1}$

40 $\frac{b^4 - 16}{b - 2}$

B

In Exercises 41–54 perform the indicated divisions, and simplify the quotient as completely as possible. Assume that all exponents are integers.

41 $2x - y\overline{)32x^5 \qquad - y^5}$

42 $5x - 2y\overline{)625x^4 \qquad - 16y^4}$

43 Divide $-8v^3 + 2v + 12v^4 - 3v^2$ by $-1 + 4v^2$.

44 Divide $-3 + 4x^2$ into $7x^3 + 8x^4 + 12x^5 - 12x + 3 - 10x^2$.

45 $\frac{4x^{4m} + 6x^{3m} - 10x^{2m}}{-2x^{2m}}$

46 $\frac{15y^{3n+6} - 18y^{2n+4} - 21y^{n+7}}{-3y^{n+4}}$

47 $\frac{7a^2b^3c^4}{14ab^2c^3} + \frac{3a^4b^2c^3}{6a^3bc^2}$

48 $\frac{14m^4n^5}{21m^3n^4} - \frac{5mn^2}{30n}$

49 $\dfrac{15c^3 - 10c^2}{5c^2} - \dfrac{7c^2 - 35c}{7c}$

50 $\dfrac{3t^3 - 5t}{t} + \dfrac{2t^4 - 6t^2}{2t^2}$

51 $17 + \dfrac{b^2 + 7b + 6}{b + 1}$

52 $\dfrac{6x^2 + 13x + 6}{2x + 3} - \dfrac{3x^2 - 19x + 28}{3x - 7}$

53 $\left(\dfrac{x^2 - 7x + 6}{x - 6}\right)(x + 1) - 5\left(\dfrac{x^2 - 6x - 7}{x - 7}\right)$

54 $\left(\dfrac{x^2 - y^2}{x - y}\right)^2$

C

In Exercises 55–60 complete each division problem to find the indicated quotient.

55 $3x^n - 1 \overline{)27x^{3n} - 1}$

56 $x^n + 5 \overline{)x^{4n} - 625}$

57 $2x - 4 \overline{)3x^2 - 5x - 2}$, quotient begins $\frac{3}{2}x$; partial work: $3x^2 - 6x$, remainder line $x - 2$

58 $3x + 6 \overline{)5x^2 + 9x - 2}$, quotient begins $\frac{5}{3}x$; partial work: $5x^2 + 10x$, remainder line $-\ x - 2$

59 $3x + 9 \overline{)4x^2 + 7x - 15}$

60 $4x - 2 \overline{)6x^2 + 7x - 5}$

61 Area of a Rectangle The area of a rectangle is $20m^2 + 7m - 6$. Find the width of the rectangle if the length is $5m - 2$ (see the figure).

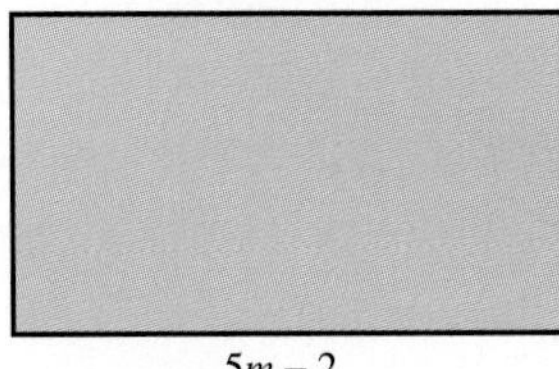

Figure for Exercise 61

62 Area of a Triangle The area of a triangle is $28n^2 + 17n - 3$. Find the height of the triangle if the base is $14n - 2$ (see the figure).

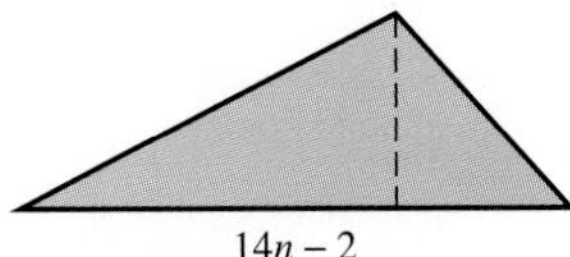

Figure for Exercise 62

63 Volume of a Box The volume of a rectangular box is $k^3 + 3k^2 - k - 3$. Find the length of the box if the width is $k + 1$ and the height is $k - 1$.

64 Entries in a Computer Array A two-dimensional array in a computer program has $30k^2 + 39k + 12$ entries. If there are $6k + 3$ rows, how many columns are there?

65 If a trinomial in x of degree 5 is divided by a trinomial in x of degree 3, what is the degree of the quotient?

66 If the dividend is a fourth-degree polynomial in x and the quotient is a first-degree polynomial in x, what is the degree of the divisor?

67 When a polynomial is divided by $2x + 3$, the quotient is $5x - 9$. Determine this polynomial.

68 When a polynomial is divided by $x^2 - x + 3$, the quotient is $2x + 7$. Determine this polynomial.

Average Cost Polynomials (69–72)

In Exercises 69–72 find the average cost per unit, $A(t)$, of the units produced by an assembly line in t hours. Use the formula $A(t) = C(t) \div N(t)$, where $C(t)$ is the total cost of operating the assembly line for t hours and $N(t)$ is the number of units produced in t hours.

69 $C(t) = 3t^2 - 5t - 2$, $N(t) = 3t + 1$

70 $C(t) = t^3 + 27$, $N(t) = t + 3$

71 $C(t) = t^3 + 64$, $N(t) = t + 4$

72 $C(t) = t^4 + 4t^2 - 5$, $N(t) = t^2 + 5$

In Exercises 73 and 74 determine whether the correct quotient for the division on the left is given by choice a or by choice b. Check both possibilities by evaluating all expressions for $x = 1$.

73 $\dfrac{10.8x^3 - 20.4x^2 - 8.24x + 11.36}{2.4x - 1.6}$

a. $4.5x^2 - 5.5x - 7.1$

b. $4.5x^2 + 5.5x - 7.1$

74 $\dfrac{3.45x^4 - 6.87x^3 - 26.43x^2 + 3.03x + 23.94}{1.5x + 2.1}$

a. $2.3x^3 - 7.8x^2 + 6.7x + 11.4$

b. $2.3x^3 - 7.8x^2 - 6.7x + 11.4$

Calculator Usage (75–77)

In Exercises 75 and 76 use a calculator to simplify each expression.

75 $\dfrac{0.57024x^2y^2 - 1.452xy^3}{0.132xy^2}$

76 $\dfrac{9.03x^5y^3 - 13.33x^4y^4 - 2.15x^3y^5}{4.3x^2y^3}$

77 Use a graphics calculator to complete this table for $y_1 = x^2 - 5x + 6$, $y_2 = x - 3$, and $y_3 = x - 2$. If you have a TI-82 calculator, you can use the TABLE feature to complete this table. For each value of x, verify that $y_3 = \dfrac{y_1}{y_2}$.

x	-2	-1	0	1	2
y_1					
y_2					
y_3					

CHALLENGE QUESTION

Work questions a–c, and then use these results to write your prediction of the answers to questions d and e.

a. $(x^2 - y^2) \div (x - y)$
b. $(x^3 - y^3) \div (x - y)$
c. $(x^4 - y^4) \div (x - y)$
d. $(x^8 - y^8) \div (x - y)$
e. $(x^n - y^n) \div (x - y)$

DISCUSSION QUESTION

78 Divide $6x^3 - 26x^2 + 44x - 32$ by $3x^2 - 7x + 8$, and write a description of each step you used in this long division.

SECTION SKILL REINFORCERS

Determine the rate of work in each of these situations.

1 A carpenter can roof a garage in 2 days.
2 A carpenter can roof a garage in one-half of a day.
3 A pump can fill a tank in 5 hours.
4 A computer printer can print 40 pages in 5 minutes.
5 A machine can produce a part in t minutes.

4-7* *Synthetic Division*

SECTION OBJECTIVES

10 Use synthetic division.

11 Evaluate $P(a)$ using the Remainder Theorem.

When the divisor is a binomial of the form $x - a$, **synthetic division** can be used to eliminate most steps in the long-division procedure and thus considerably shorten the process. In the long-division process several pieces of information are repeated. Synthetic division eliminates this duplication yet uses all the key information. Thus the answer is obtained "synthetically"—without doing the actual long division.

We will start with a long-division example and then, step by step, develop the logic that justifies shortening this procedure to synthetic division. After this logic has been developed, this section will illustrate how to use synthetic division.

STEPS IN SHORTENING THE LONG-DIVISION PROCEDURE

Step 1 *Omit the variables.* When both of the polynomials are in descending powers of x, the position of each term indicates the associated power of x. Thus we can omit the variables and write only the coefficients, using zero coefficients for any powers that are missing.

*This is an optional section.

Long Division

$$
\begin{array}{r|l}
 & 3x^2 + 6x + 7 \\ \hline
x - 2 & 3x^3 + 0x^2 - 5x - 10 \\
 & \ominus \quad \oplus \\
 & \underline{3x^3 - 6x^2} \\
 & \qquad 6x^2 - 5x \\
 & \qquad \ominus \quad \oplus \\
 & \qquad \underline{6x^2 - 12x} \\
 & \qquad\qquad 7x - 10 \\
 & \qquad\qquad \ominus \quad \oplus \\
 & \qquad\qquad \underline{7x - 14} \\
 & \qquad\qquad\qquad 4
\end{array}
$$

Variables Omitted

$$
\begin{array}{rr|rrrr}
 & & & 3 & 6 & 7 \\ \hline
1 & -2 & 3 & 0 & -5 & -10 \\
 & & 3 & -6 & & \\ \hline
 & & & 6 & -5 & \\
 & & & 6 & -12 & \\ \hline
 & & & & 7 & -10 \\
 & & & & 7 & -14 \\ \hline
 & & & & & 4
\end{array}
$$

Step 2 *Omit duplications in each column.* Each number that is shaded in Step 1 is a duplicate of the number in the column directly above it. No information will be lost if these duplicates are omitted.

$$
\begin{array}{rr|rrrr}
 & & & 3 & 6 & 7 \\ \hline
1 & -2 & 3 & 0 & -5 & -10 \\
 & & & -6 & & \\ \hline
 & & & 6 & & \\
 & & & & -12 & \\ \hline
 & & & & 7 & \\
 & & & & & -14 \\ \hline
 & & & & & 4
\end{array}
$$

Step 3 *Condense the notation.* Several spaces appear in the notation in Step 2. The notation can be written more concisely by compressing it vertically.

$$
\begin{array}{rr|rrrr}
 & & & 3 & 6 & 7 \\ \hline
1 & -2 & 3 & 0 & -5 & -10 \\
 & & & -6 & -12 & -14 \\ \hline
 & & 3 & 6 & 7 & 4
\end{array}
$$

Step 4 *Omit the top line.* Since we are developing this procedure only for divisors of the form $x - a$, it is possible to omit the coefficient 1 of x in the divisor. Since the numbers in the top line are duplicates of those in the bottom line, they too can be omitted.

$$
\begin{array}{r|rrrr}
-2 & 3 & 0 & -5 & -10 \\
 & & -6 & -12 & -14 \\ \hline
 & 3 & 6 & 7 & 4
\end{array}
$$

Step 5 *Change the sign and convert from subtraction to addition.* At each step in the long-division process, we subtract each new line from the line above. This procedure is simplified by changing the signs in the notation and adding. (Note that we have also simplified the notation by removing the long rule above the dividend and placing a short rule beneath the divisor instead.)

$$
\begin{array}{r|rrrr}
\underline{+2} & 3 & 0 & -5 & -10 \\
 & & 6 & 12 & 14 \\ \hline
 & 3 & 6 & 7 & 4
\end{array}
$$

The degree of the quotient is 1 less than the degree of the dividend, and the coefficients of the quotient's terms are given by the bottom line in the synthetic division. The last term in this bottom line is the remainder.

$$(3x^3 - 5x - 10) \div (x - 2) = 3x^2 + 6x + 7 + \frac{4}{x - 2}$$

If the divisor is an exact factor of the dividend, then this remainder will be 0.

Now that we have derived synthetic division from long division, let us apply this new method.

SYNTHETIC DIVISION

Step 1 Arrange the polynomials in standard form with a zero coefficient for any missing term.

Step 2 Write $+a$ for the divisor, $x - a$; write the coefficients from the dividend; and recopy the leading coefficient on the bottom row.

Step 3 Multiply a by the value just written on the bottom row, and write the product in the next column on the second row.

Step 4 Add the values in this new column, and write the result in the bottom row. Then repeat steps 3 and 4 until this zigzag pattern has filled all the columns.

Step 5 Write the quotient from the last row, using the last value as the remainder.

EXAMPLE 1 Using Synthetic Division

Use synthetic division to find the quotient of $\dfrac{x^3 - 11x + 15 - x^2}{x - 3}$.

SOLUTION

1 $(x^3 - x^2 - 11x + 15) \div (x - 3)$ Arrange the polynomials in standard form.

2 Write +3 for the divisor, write the coefficients from the dividend, and recopy the leading coefficient in the bottom row.

$$\begin{array}{r|rrrr} +3 & 1 & -1 & -11 & 15 \\ & & & & \\ \hline & 1 & & & \end{array}$$

3 Multiply 3 times the value on the bottom row: $+3 \cdot 1 = 3$. Write the product in the next column on the second row.

$$\begin{array}{r|rrrr} +3 & 1 & -1 & -11 & 15 \\ & \downarrow & 3 & & \\ \hline & 1 & & & \end{array}$$

4 Add the values in this column, and write the result on the bottom row.

$$\begin{array}{r|rrrr} +3 & 1 & -1 & -11 & 15 \\ & \downarrow & 3 & & \\ \hline & 1 & 2 & & \end{array}$$

$$\begin{array}{r|rrrr} +3 & 1 & -1 & -11 & 15 \\ & \downarrow & 3 & 6 & -15 \\ \hline & 1 & 2 & -5 & 0 \end{array}$$

Repeat Steps 3 and 4 until the zigzag pattern has filled all the columns.

5 Quotient: $x^2 + 2x - 5$
Remainder: 0

Write the quotient from the last row, using the last value as the remainder.

Thus

$$\frac{x^3 - x^2 - 11x + 15}{x - 3} = x^2 + 2x - 5$$

▼

SELF-CHECK 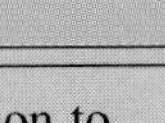

Use synthetic division to divide

$$x^4 + 4x^3 - x^2 - 3x + 4$$

by $x + 4$.

EXAMPLE 2 Using Synthetic Division to Divide a Polynomial Not in Standard Form

Use synthetic division to divide $(-7x^2 + 2x^3 - 16)$ by $(x - 4)$.

SOLUTION

$$\begin{array}{r|rrrr} +4 & 2 & -7 & 0 & -16 \\ & \downarrow & 8 & 4 & 16 \\ \hline & 2 & 1 & 4 & 0 \end{array}$$

Write the dividend in standard form, using a coefficient of 0 for the missing x term.

Thus

$$(2x^3 - 7x^2 - 16) \div (x - 4) = 2x^2 + x + 4$$

▼

EXAMPLE 3 Using Synthetic Division to Divide a Polynomial Not in Standard Form

Use synthetic division to divide $x + 2$ into $x + 65 - 7x^2 + 4x^3$.

SOLUTION

$$\begin{array}{r|rrrr} -2 & 4 & -7 & 1 & 65 \\ & \downarrow & -8 & 30 & -62 \\ \hline & 4 & -15 & 31 & 3 \end{array}$$

Write the dividend in the standard form $4x^3 - 7x^2 + x + 65$. Note that the divisor is $x - (-2)$.

Thus

$$\frac{4x^3 - 7x^2 + x + 65}{x + 2} = 4x^2 - 15x + 31 + \frac{3}{x + 2}$$

Note that the degree of the quotient is 1 less than that of the dividend.

▼

EXAMPLE 4 Using Synthetic Division to Divide a Polynomial with Missing Terms

Use synthetic division to divide $y^5 - 243$ by $y - 3$.

SOLUTION

$$\begin{array}{r|rrrrrr} 3 & 1 & 0 & 0 & 0 & 0 & -243 \\ & \downarrow & 3 & 9 & 27 & 81 & 243 \\ \hline & 1 & 3 & 9 & 27 & 81 & 0 \end{array}$$

Note the zero coefficients for the missing y^4, y^3, y^2, and y terms.

Thus

$$\frac{y^5 - 243}{y - 3} = y^4 + 3y^3 + 9y^2 + 27y + 81$$

▼

SELF-CHECK ANSWER 

$x^3 - x + 1$

EXAMPLE 5 Dividing a Polynomial by $x - k$

Divide $x^4 - kx^3 - 5x^2 + (5k + 4)x - 4k$ by $x - k$.

SOLUTION

$$\begin{array}{r|rrrrr} k & 1 & -k & -5 & (5k+4) & -4k \\ & \downarrow & k & 0 & (-5k+0) & 4k \\ \hline & 1 & 0 & -5 & 4 & 0 \end{array}$$

Since in this format each column contains the coefficients of terms of a specific degree, parentheses are used to emphasize that $5k + 4$ is a coefficient of a first-degree term.

Thus

$$\frac{x^4 - kx^3 - 5x^2 + (5k + 4)x - 4k}{x - k} = x^3 - 5x + 4$$

One of the primary uses of synthetic division is to evaluate polynomials. A brief examination of this topic follows. If a polynomial $P(x)$ is divided by $x - a$, resulting in a quotient polynomial $Q(x)$ and remainder R, then we can write

$$P(x) = (x - a)Q(x) + R$$

If $P(x)$ is evaluated for a, then

$$\begin{aligned} P(a) &= (a - a)Q(a) + R \\ &= 0 \cdot Q(a) + R \\ &= R \end{aligned}$$

Thus when $P(x)$ is divided by $x - a$, $P(a) = R$, the remainder. This result is known as the Remainder Theorem.

REMAINDER THEOREM

If a polynomial $P(x)$ is divided by $x - a$, then the remainder is equal to $P(a)$.

EXAMPLE 6 Evaluating a Polynomial for a Given Value

Given $P(x) = 4x^4 - 5x^3 + 7x^2 + 9$, evaluate $P(3)$ by

(a) substituting 3 for x in $P(x)$.

(b) determining the remainder when $P(x)$ is divided by $x - 3$.

SOLUTIONS

(a)
$$\begin{aligned} P(x) &= 4x^4 - 5x^3 + 7x^2 + 9 \\ P(3) &= 4(3)^4 - 5(3)^3 + 7(3)^2 + 9 \\ &= 4(81) - 5(27) + 7(9) + 9 \\ &= 324 - 135 + 63 + 9 \\ &= 261 \end{aligned}$$

Substitute 3 for x in $P(x)$, and then simplify, following the correct order of operations.

(b)
$$\begin{array}{r|rrrrr} 3 & 4 & -5 & 7 & 0 & 9 \\ & \downarrow & 12 & 21 & 84 & 252 \\ \hline & 4 & 7 & 28 & 84 & 261 \end{array}$$

Write +3 for the divisor, $x - 3$, and include a coefficient of 0 for the missing x term.

$P(3) = 261$

The remainder, 261, is the value of $P(3)$.

EXAMPLE 7 Using Synthetic Division and the Remainder Theorem

Use synthetic division to evaluate $P(1.1)$, given

$$P(x) = x^5 + 2.9x^4 - 1.4x^3 + 2.7x^2 - 4.6x - 1.2$$

SOLUTION

$$\begin{array}{r|rrrrrr} 1.1 & 1 & 2.9 & -1.4 & 2.7 & -4.6 & -1.2 \\ & & 1.1 & 4.4 & 3.3 & 6.6 & 2.2 \\ \hline & 1 & 4.0 & 3.0 & 6.0 & 2.0 & 1.0 \end{array}$$

Use synthetic division to find the remainder that equals $P(1.1)$.

Thus $P(1.1) = 1.0$.

SELF-CHECK

Use synthetic division to evaluate $P(2)$, given $P(x) = 2x^3 - 5x + 11$.

Exercises 4-7

A

In Exercises 1–20 use synthetic division to determine each quotient.

1 $(x^2 - 7x + 10) \div (x - 5)$

2 $(x^2 + x - 90) \div (x - 9)$

3 $(2y^2 - 3y - 20) \div (y - 4)$

4 $(3z^2 - 13z + 14) \div (z - 2)$

5 $\dfrac{15w^2 - 21 + 5w^3 - 7w}{w + 3}$

6 $\dfrac{6 - 17v - 27v^2 - 4v^3}{v + 6}$

7 $a + 1 \overline{)2a - a^2 + 3 + 4a^3 + 4a^4}$

8 $b + 2 \overline{)11b^3 + 2b^2 + 5b^4 + 14 + 7b}$

9 $\dfrac{7c + c^2 + 10}{c + 3}$

10 $\dfrac{4d - 10 + d^2}{d - 2}$

11 $(-7p - 3p^4 - 2 + p^6) \div (p - 2)$

12 $(5 - 3n^2 + 2n^4) \div (n - 1)$

13 $\dfrac{m^2 - 5m - 5m^3 + m^4}{m - 5}$

14 $\dfrac{r^2 - 6r - 6r^3 + r^4}{r - 6}$

15 $(3s^4 - s^3 + 9s^2 - 1) \div \left(s - \dfrac{1}{3}\right)$

16 $(4t^4 - 2t^3 + 6t^2 - 4t - 4) \div \left(t + \dfrac{1}{2}\right)$

17 $\dfrac{b^5 + b^3 - 2}{b - 1}$

18 $\dfrac{a^7 + a^5 - 22a^3 + 20}{a + 2}$

19 $\dfrac{z^4 - 256}{z - 4}$

20 $\dfrac{w^7 - 128}{w - 2}$

B

In Exercises 21–30 use the Remainder Theorem and synthetic division to evaluate $P(a)$.

21 $P(x) = x^5 + 5x^4 + 10x^3 + 10x^2 + 5x + 1,\ a = 2$

22 $P(x) = x^5 + 5x^4 + 10x^3 + 10x^2 + 5x + 1,\ a = -1$

23 $P(x) = x^5 + 5x^4 + 10x^3 + 10x^2 + 5x + 1,\ a = -3$

24 $P(x) = x^6 + x^4 + 4x^2 + 2x + 1,\ a = 3$

25 $P(x) = x^6 + x^4 + 4x^2 + 2x + 1,\ a = 4$

26 $P(x) = x^6 + x^4 + 4x^2 + 2x + 1,\ a = -2$

27 $P(x) = x^4 - 1.3x^3 + 1.7x^2 - 1.2x - 1.4,\ a = 2.3$

28 $P(x) = 2x^3 - 9.2x^2 + 4.5x + 16,\ a = 7.1$

29 $P(x) = 3x^3 - 9.7x^2 + 5.5x + 3,\ a = -4.9$

30 $P(x) = 5x^4 - 8.5x^3 + 4x^2 + 3.2x + 3,\ a = -1.7$

31 Divide $x^6 - (k^2 + 1)x^4 + kx^3 + 3x^2 + (-3k + 1)x - k$ by $x - k$.

32 Divide $2x^4 + 2kx^3 + 2x^2 + 2kx$ by $x + k$.

SELF-CHECK ANSWER

$P(2) = 17$

C

CALCULATOR USAGE (33–40)

In Exercises 33–40 use a calculator, the Remainder Theorem, and synthetic division to evaluate $P(a)$.

33 $P(y) = 2.7y^2 - 85y - 1.4$, $a = 2.5$

34 $P(w) = 3.4w^2 - 90w + 5.7$, $a = 3.5$

35 $P(x) = 0.014x^3 - 0.68x + 3.7$, $a = -6.2$

36 $P(t) = 0.041t^3 + 0.4t^2 - 8.9$, $a = -2.6$

37 $P(v) = 4.98v^4 - 3.1v^2 - 8.2$, $a = -1.1$

38 $P(y) = 2.2y^3 - 10.064y^2 - 40.1328y - 29.8656$, $a = -3.96$

39 $P(x) = x^3 - x^2 - 3x - 1$, $a = \sqrt{17}$
[Approximate $P(a)$ to the nearest thousandth.]

40 $P(x) = x^3 - x^2 - 3x - 1$, $a = 1 + \sqrt{2}$
[Approximate $P(a)$ to the nearest thousandth.]

DISCUSSION QUESTION

41 Describe in your own words the steps to follow to use synthetic division.

SECTION SKILL REINFORCERS

Expand each of these expressions.

1 $(3x + 5y)^2$ **2** $(5x - 3y)^2$ **3** $(4a + b)(4a - b)$ **4** $(10a - b)(10a + b)$ **5** $(3a + 2b)(2a - 3b)$

Key Concepts for Chapter 4

1 Exponents:

a. For any nonzero real number x and any natural number n,

$$x^n = x \cdot x \cdot x \cdot \cdots \cdot x \qquad n \text{ factors of } x$$

b. For any nonzero real number x,

$$x^0 = 1 \qquad \text{and} \qquad 0^0 \text{ is undefined}$$

c. For any nonzero real number x and any natural number n,

$$x^{-n} = \frac{1}{x^n}$$

2 Properties of exponents: For any nonzero real numbers x and y and integer exponents m and n,

Product rule: $x^m x^n = x^{m+n}$

Quotient rule: $\frac{x^m}{x^n} = x^{m-n}$

Power rule: $(x^m)^n = x^{mn}$

Product to a power: $(xy)^m = x^m y^m$

Quotient to a power: $\left(\frac{x}{y}\right)^m = \frac{x^m}{y^m}$

$\left(\frac{x}{y}\right)^{-n} = \frac{y^n}{x^n}$

3 A number is in scientific notation when it is expressed as the product of a number between 1 and 10 (or −1 and −10) and an appropriate power of 10. On many calculators an [EE] or [EXP] key is used to enter in the appropriate power of 10. The letter E often precedes the power of 10 on calculator screens or on computer printouts.

4 Polynomials can contain exponents that are natural numbers. They cannot contain variables in the denominator or under a radical. Exponents on the variables cannot be negative or fractional.

5 Monomials contain one term, binomials contain two terms, and trinomials contain three terms.

6 The degree of a monomial is the sum of the exponents on all the variables in the term. The degree of a polynomial is the same as the degree of the term of highest degree.

7 A polynomial is in standard form if the variables are written in alphabetical order in each term and the terms are arranged in order of decreasing powers of the first variable.

8 To add or subtract polynomials, combine like terms.

9 Products of polynomials:

The square of a sum: $(a + b)^2 = a^2 + 2ab + b^2$

The square of a difference: $(a - b)^2 = a^2 - 2ab + b^2$

A sum times a difference: $(a + b)(a - b) = a^2 - b^2$

To multiply other binomials by inspection, use the FOIL method. To multiply other polynomials, you may want to use the vertical format.

10 Quotients of polynomials:

a. To divide a polynomial by a monomial, divide each term by the monomial and reduce.

b. To divide a polynomial by a polynomial of more than one term, you may want to use the long-division algorithm.

***11** Synthetic division can be used to divide a polynomial by the binomial $x - a$.

***12** Remainder Theorem: If a polynomial $P(x)$ is divided by $x - a$, then the remainder is equal to $P(a)$.

Review Exercises for Chapter 4

In Exercises 1–32 simplify each expression. Assume that all bases are nonzero real numbers.

1 -8^2

2 $(2 + 8)^2 + 2^2 + 8^2$

3 $(5 + 7)^0 + 5^0 + 7^0$

4 $0^1 + 1^{10} + 100^0 - 0^{100}$

5 $\left(\frac{1}{5} - \frac{1}{10}\right)^{-1} + \left(\frac{1}{5}\right)^{-1} - \left(\frac{1}{10}\right)^{-1}$

6 $5 \cdot 3^2 + (5 \cdot 3)^2$

7 $11^7 \cdot 11^{-5}$

8 $\dfrac{13^{-5}}{13^{-7}}$

9 $(3ab^{-3})(2a^4b^4)$

10 $\dfrac{12vw^7}{4v^{-3}w^5}$

11 $(-2x^3y^{-4})^{-2}$

12 $\left(\dfrac{-25x^3y^{-3}}{15x^2y^{-5}}\right)^{-4}$

13 $\left(\dfrac{3v^{-2}}{2v}\right)^{-3}\left(\dfrac{4v^2}{3v^{-1}}\right)^{-2}$

14 $\left[\left(\dfrac{x^{-3}}{z^{-3}}\right)^{-2}\left(\dfrac{x^{-4}}{z^3}\right)^{-3}\right]^{-4}$

15 $(5x + 7y) - (2x - 9y) - (x + 2y)$

16 $(5x + 7y) - 2[x - 9y - 3(x + 4y)]$

17 $(y^2 - 7xy + 3x^2) - (8xy - 9y^2 + 11x^2)$

18 $(3x^2 - 7x - 9)(2x + 3)$

19 $(7x - 9)(7x + 9)$

20 $(4x + 5y)^2$

21 $(3v - 7w)^2$

22 $(2x + 7y)(3x - 5y)$

23 $(a + b)^2 - (a^2 + b^2)$

24 $(12y^3 - 28y - 21 + 11y^2) \div (3y + 2)$

25 $\dfrac{36m^4n^2 - 45m^3n^4}{-9m^3n^2}$

26 $\dfrac{27z^3 - 1}{3z - 1}$

27 $(x - y)^2 + (x + y)(x - y)$

28 $\dfrac{x^2 + x - 6}{x - 2} + \dfrac{x^2 - 9x + 14}{x - 7}$

29 $6x - 3(4x - 9) - 2(x + 1)(x - 1)$

30 $[(x - 5)(x + 6)^2 - (x - 7)]^0$

31 $(2v - 3w)^3$

32 $(3x - 5y)^0 + (3x)^0 - (5y)^0 + 3x^0 - 5y^0$

In Exercises 33–36 evaluate each expression for $w = -1$, $x = -2$, $y = -3$, and $z = -4$.

33 $wxy - z$

34 $-y(w - 2x)^2$

35 $\dfrac{w - 2x + 3y}{2z - (y - wx)}$

36 $w^2 + y^2 - (w + y)^2$

In Exercises 37–42 evaluate each expression for the polynomial $P(x) = -7x^3 + 2x^2 - x - 5$, using the values given.

37 $P(0)$

38 $P(1)$

39 $P(-2)$

40 $P(a)$

41 $P(-a)$

42 $2P(a) - P(2a)$

43 Write $2xy^3 - 5y^4 + 4x^2y^2 - 3x^3y + x^4$ in standard form, and give its degree.

44 Write a third-degree binomial in x whose leading coefficient is 7 and whose last term is the constant -3.

45 Perimeter of a Region Write a polynomial for the perimeter of the region shown in the figure.

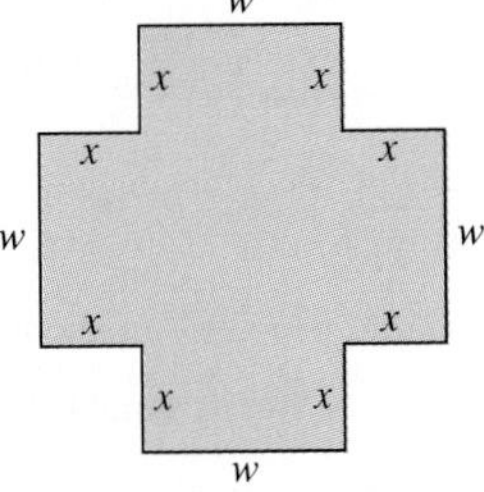

Figure for Exercises 45 and 46

46 Area of a Region Write a polynomial for the area of the region shown in the figure.

*These key concepts are part of an optional section.

In Exercises 47 and 48 simplify each expression in part a, and solve each equation in part b.

Simplify ***Solve***

47 a. $(x + 5)^2 - (x^2 + 15)$ **b.** $(x + 5)^2 - (x^2 + 15) = 0$

Simplify

48 a. $(x - 7)(4x + 1) - (2x + 3)^2 + 94$

Solve

b. $(x - 7)(4x + 1) - (2x + 3)^2 + 94 = 0$

In Exercises 49–52 assume that x and y are nonzero real numbers and m and n are natural numbers. Simplify each expression.

49 $\dfrac{x^{m+2}}{x^{m-4}}$ **50** $(x^{m+2})(x^{m+4})$

51 $(x^{m+2})^2$ **52** $(x^{m+1})^{m+2}$

CALCULATOR USAGE (53–55)

53 Use a graphics calculator to graph $y = x^2 - x - 2$ and the TRACE or TABLE feature to complete the table below.

x	−4	−3	−2	−1	0	1	2	3	4
y									

In Exercises 54 and 55 use a graphics calculator to graph $y_1 = P_1(x)$ and $y_2 = P_2(x)$. Then use these graphs to determine whether $P_1(x) = P_2(x)$.

54 $P_1(x) = (0.4x + 1.5)(0.4x - 1.5)$
$P_2(x) = 0.16x^2 - 1.25$

55 $P_1(x) = (1.7x - 2.3)^2$
$P_2(x) = 2.89x^2 + 5.29$

56 Explain why $(2x + 5)^2 = 7$ cannot be simplified to $4x^2 + 25 = 7$.

In Exercises 57 and 58 use synthetic division to find the indicated quotients.

***57** $\dfrac{2x^5 + 10x^4 + 3x^3 + 6x^2 - 45x}{x + 5}$ ***58** $\dfrac{x^4 + 6x^2 + 8x + 3}{x + 3}$

In Exercises 59 and 60 use synthetic division and the Remainder Theorem to evaluate each expression for $P(x) = 6x^5 - 3x^4 + 4x^3 + 3x + 5$.

59 $P(2)$ **60** $P(-3)$

61 The product of a polynomial and $3x - 4$ is $6x^2 + 13x - 28$. Determine this polynomial factor.

62 When $20x^2 - 31x - 9$ is divided by a polynomial, the quotient is $5x - 9$. Determine this polynomial divisor.

Mastery Test for Chapter 4

Exercise numbers correspond to Section Objective numbers.

[4-1] **1** Evaluate each of these expressions completely.

a. $-3^2 - 2^3$ **b.** $(2^{-1} + 5^{-1})^{-2}$

c. $(2^{-1})^{-2} + (5^{-1})^{-2}$

d. $(9x + 4y)^0 + (9x)^0 + (4y)^0 + 9x^0 + 4^0$

[4-2] **2** Simplify each of the following expressions as completely as possible, removing all negative exponents.

a. $(-7x^2y^3)(3x^5y^{-4})$ **b.** $\dfrac{-18x^{13}y^{-7}}{6x^{15}y^{-9}}$

c. $\left(\dfrac{-14x^{-3}y^4}{-49x^5y^{-4}}\right)^{-2}$

[4-2] **3 a.** Write 0.000 017 293 in scientific notation.

b. Write 5.98×10^6 in standard decimal form.

c. Use a calculator to evaluate $\dfrac{(2.98 \times 10^{51})(7.29 \times 10^{-28})}{8.93 \times 10^{24}}$ to the nearest thousandth.

[4-3] **4** Write each of these polynomials in standard form. Then classify each polynomial as a monomial, a binomial, or a trinomial, and give its degree.

a. $3y^2 - 7yx + 4x^2$

b. $-9 + 8x$

c. $-11b^4c^3a^2$

[4-3] **5** Given $P(x) = 3x^2 - 7x - 9$, evaluate each of the following expressions.

a. $P(0)$ **b.** $P(-1)$ **c.** $P(2)$

[4-4] **6** Simplify each of these expressions.

a. $(5v + 9w) + (7v - 13w)$

b. $(5v + 9w) - (7v - 13w)$

c. $(x^2 - 7x - 2) - (3x^2 - 8x + 5)$

[4-5] **7** Find each product.

a.
$$\begin{array}{r} 9x^2 + 8x - 4 \\ \times \quad -7x + 2 \\ \hline \end{array}$$

b.
$$\begin{array}{r} 5x^2 - x + 4 \\ \times\ 2x^2 + x - 3 \\ \hline \end{array}$$

*These exercises are part of an optional section.

[4-5] **8** Determine each of the following products by inspection.

a. $(3x - 5y)^2$ **b.** $(3x + 5y)^2$

c. $(11v + 7)(11v - 7)$ **d.** $(13x + 7y)(2x - 3y)$

[4-6] **9** Find each quotient.

a. $\dfrac{15a^2b^3 - 35ab^4}{5ab}$ **b.** $\dfrac{6x^2 - 11x - 10}{2x - 5}$

c. $\dfrac{z^4 - 81}{z - 3}$ **d.** $\dfrac{4v^4 - 6v^3 + 9}{2v^2 + 1}$

*[4-7] **10** Use synthetic division to find each quotient.

a. $\dfrac{y^3 - 3y^2 + 4y - 12}{y - 3}$

b. $\dfrac{w^3 - 64}{w - 4}$

c. $(4y^3 - 3y^2 - 2y + 1) \div (y + 1)$

d. $(3z^4 - 2z^2 - 5z) \div (z - 2)$

*[4-7] **11** Use synthetic division and the Remainder Theorem to evaluate each of these expressions for $P(x) = 7x^4 - 9x^3 + 11x^2 - 4$.

a. $P(2)$ **b.** $P(-2)$

c. $P(10)$ **d.** $P(-3)$

Active Application

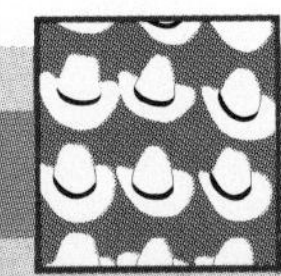

Suppose a record company wants to produce a new CD that they hope will be at least a minor hit for at least one season. The company wants you, an industry analyst and consultant, to advise them on how many disks to distribute.

Producing a CD involves both fixed and variable costs. The cost of the initial recording session is the same whether the company sells 500 disks or 5 million. How much is spent on manufacturing and packaging materials, on the other hand, depends on how many disks are made. You estimate the total fixed costs of producing the CD at $40,000. Data support a model of variable costs represented by $10x$, where x is thousands of disks produced and the costs are measured in thousands of dollars. For gross revenue from sales of these disks you use $\dfrac{x^3}{2} - 12x^2 + 96x - 6$, a polynomial reflecting the fact that as production increases, the additional disks get harder and harder to sell. (This model is only accurate for x below 10, but you are sure that production will be no higher than 10,000 disks.)

1 The company's gross profit (not including corporate taxes, etc.) is gross revenue minus fixed and variable expenses. What is the company's profit on the CD, expressed in terms of thousands of disks produced?

2 How many disks should the company produce to maximize its profit? Round your answer to the nearest hundred. If you take calculus later, you will learn how to derive this answer directly; in this case, use trial and error to determine the optimal value of x. If you have a graphing calculator, you may want to enter the appropriate polynomial and use the TRACE and/or ZOOM features to determine the x value that yields the highest profit.

3 Express the profit as a percent of the total cost. Write the answer as a polynomial plus a remainder; then determine the number of disks that will maximize this percent of profit. Does this level of production also maximize profits?

*These exercises are part of an optional section.

Chapter 5

FACTORING POLYNOMIALS

AN APPLICATION

The Mathematics of Flight

The science of aircraft flight, called *aeronautics,* is similar to *aerodynamics,* which is the study of how air and other gases act on solid bodies moving through them. Both aeronautics and aerodynamics make prominent use of equations containing polynomials. In this chapter you will learn how to solve such equations.

One intuitively simple application of polynomial equations to aircraft flight has to do with the conditions necessary for smooth, level flight. The various forces in the air that an aircraft must overcome to fly are collectively termed *drag.* If you have ridden a bicycle at high speed, you may have felt the wind tug at your clothing, and you certainly have felt the air press against your chest and face. These forces are called friction drag and air pressure drag. An airplane is also subject to a kind of drag created when the wings act on the air to keep the plane aloft.

To counteract drag, the airplane uses its engines, the total force of which is called the *thrust.* The plane can fly at a given speed only if the thrust is at least equal to the drag. For a jet aircraft we can picture the relationship between thrust and drag as shown in the following figure.

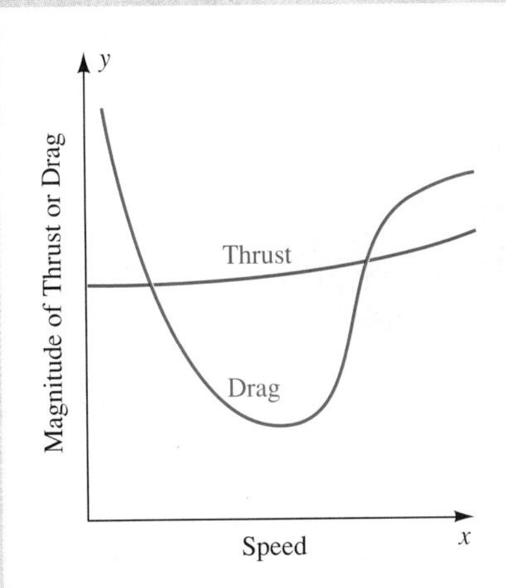

The two points where the thrust and drag curves cross—where drag exactly equals available thrust—represent the slowest and the fastest speeds attainable. To find these two speeds, we can set up an equation stating that drag equals available thrust and then solve this equation algebraically. This is the sort of thing you will be learning to do in this chapter. In the meantime ask yourself, Why is the drag very high when the air speed is low? (It may help to try to picture a slow-flying jet airplane.)

Chapter 5 Objectives

1 **Factor out the greatest common factor of a polynomial (Section 5-1).**

2 **By inspection, factor perfect square trinomials, the difference of two squares, and the sum or difference of two cubes (Section 5-2).**

3 **Factor quadratic trinomials (Section 5-3).**

4 **Factor polynomials by the method of grouping (Section 5-4).**

5 **Determine the most appropriate technique for factoring a polynomial (Section 5-4).**

6 **Use factoring to solve selected second- and third-degree equations (Section 5-5).**

7 **Use factorable quadratic equations to solve applied problems (Section 5-6).**

8 **Use the Pythagorean Theorem to solve applied problems (Section 5-6).**

Factoring is the process of rewriting an algebraic expression as a product of its factors. In Chapter 3 we multiplied $(2x + 1)(x + 3)$ to obtain the product $2x^2 + 7x + 3$. We will now reverse the process; given the polynomial $2x^2 + 7x + 3$, we will produce the factored form $(2x + 1)(x + 3)$.

Multiplying ↓ $(2x + 1)(x + 3)$
$= 2x^2 + 7x + 3$ ↑ Factoring

Factoring is extremely useful throughout algebra. It is used in solving many equations, and it is necessary for working with fractions. We will limit our discussion to factoring polynomials using integral coefficients, which is called **factoring over the integers.** A polynomial is **prime over the integers** if its only factorization is the trivial factorization involving 1 or -1.

5-1 *Factoring Out the Greatest Common Factor*

Section Objective

1 **Factor out the greatest common factor of a polynomial.**

To factor a polynomial, you must be able to recognize or calculate at least one factor other than 1 or -1. Usually the greatest common factor (GCF) of a polynomial is the easiest factor to recognize. The best way to start factoring a polynomial is to determine the GCF. If this GCF is not 1, then factor out the GCF before looking for other factors. Loosely speaking, the GCF of a polynomial is the "largest" factor common to each term. To be more precise, the **GCF of a polynomial** is the common factor that contains

1 the largest possible numerical coefficient and

2 the largest possible exponent on each variable factor.

FINDING THE GREATEST COMMON FACTOR OF A POLYNOMIAL

The greatest common factor (GCF) of a polynomial is the product of

1. the GCF of the coefficients of all the terms of the polynomial and
2. the variable factors common to all terms. (The exponent on each variable factor will be the smallest exponent that occurs on that variable factor in any of the terms.)

Remember that the GCF of a set of integers is the largest integer that will exactly divide each integer in the set. If the GCF is not obvious upon inspection, then it can be determined by examining the prime factorization of each of the integers. The GCF is the product of all the factors common to each of the integers.

EXAMPLE 1 Determining the GCF of a Binomial

Find the GCF of $12x^3y + 18x^2y^2$.

SOLUTION

$$12x^3y = 2^2 \cdot 3 \cdot x^3y$$
$$18x^2y^2 = 2 \cdot 3^2 \cdot x^2y^2$$
$$\text{GCF} = 2 \cdot 3 \cdot x^2y$$
$$\text{GCF} = 6x^2y$$

The only prime factors common to each of these coefficients are 2 and 3. The variable factors common to both terms are x^2 and y.

▼

The polynomial $4x^4 - 8x^3 + 12x^2$ can be written as $4x^2(x^2 - 2x + 3)$. Expressing this polynomial as the product of the GCF $4x^2$ and $x^2 - 2x + 3$ is referred to as **factoring out the greatest common factor.** If you know the GCF of a polynomial, you can find the other factor by division. This division often can be performed by inspection.

EXAMPLE 2 Factoring the GCF Out of a Binomial

Factor out the greatest common factor of $20x^2y - 30xy^2$.

SOLUTION Inspection reveals that the GCF of $20x^2y - 30xy^2$ is $10xy$. Thus

$$20x^2y - 30xy^2 = 10xy(2x - 3y)$$

$20x^2y = 2^2 \cdot 5x^2y$
$30xy^2 = 2 \cdot 3 \cdot 5xy^2$
$\text{GCF} = 2 \cdot 5xy$
$\text{GCF} = 10xy$
The factor $2x - 3y$ is obtained by dividing $20x^2y - 30xy^2$ by $10xy$.

▼

EXAMPLE 3 Factoring the GCF Out of a Trinomial

Factor out the greatest common factor of $26x^2y - 39xy - 65y^2$.

SOLUTION Inspection reveals that the GCF of $26x^2y - 39xy - 65y^2$ is $13y$. Thus

$$26x^2y - 39xy - 65y^2 = 13y(2x^2 - 3x - 5y)$$

$26x^2y = 2 \cdot 13x^2y$
$39xy = 3 \cdot 13xy$
$65y^2 = 5 \cdot 13y^2$
$\text{GCF} = 13y$
Divide by $13y$ to obtain the other factor:
$$\frac{26x^2y - 39xy - 65y^2}{13y} = 2x^2 - 3x - 5y$$

▼

EXAMPLE 4 Factoring the GCF Out of a Trinomial Whose Leading Coefficient Is Negative

Factor out the greatest common factor of $-12a^3b + 18a^2b^2 - 30ab^3$.

SOLUTION Inspection reveals that the GCF of $-12a^3b + 18a^2b^2 - 30ab^3$ is $6ab$. Thus

$$-12a^3b + 18a^2b^2 - 30ab^3 = 6ab(-2a^2 + 3ab - 5b^2)$$
$$= -6ab(2a^2 - 3ab + 5b^2)$$

$12a^3b = 2^2 \cdot 3a^3b$
$18a^2b^2 = 2 \cdot 3^2a^2b^2$
$30ab^3 = 2 \cdot 3 \cdot 5ab^3$
$\text{GCF} = 6ab$

Divide by the GCF, $6ab$, to obtain the other factor. When the coefficient of the first term of the second factor is negative, we sometimes rewrite the factorization to make this coefficient positive.

Sometimes the GCF is a polynomial with two or more terms. It may be easier to understand these problems if you substitute a single variable for this GCF. With practice you will no longer need to use this step and can eliminate it. The strategy is illustrated in the next two examples.

SELF-CHECK

Factor out the GCF from each polynomial.

1 $18y^6 - 9y^4$
2 $6x^3 - 9x^2 + 12x$

EXAMPLE 5 Factoring Out a Binomial GCF

Factor $(x + 2y)b + (x + 2y)(3c)$.

SOLUTION

$(x + 2y)b + (x + 2y)(3c)$ is of the form $ab + a(3c)$. — Substitute a for the GCF, $x + 2y$.

$ab + a(3c) = a(b + 3c)$ — Factor out the GCF, a.

$(x + 2y)b + (x + 2y)(3c) = (x + 2y)(b + 3c)$ — Substitute $x + 2y$ back in for a.

EXAMPLE 6 Factoring Out a Binomial GCF

Factor $(a - 3b)y^2 + (a - 3b)y + (3b - a)$.

SOLUTION

$(a - 3b)y^2 + (a - 3b)y + (3b - a)$ is of the form $xy^2 + xy - x$. — Substitute x for the GCF, $a - 3b$. Since $x = a - 3b$, substitute $-x$ for $-(a - 3b) = 3b - a$ in the last term.

$xy^2 + xy - x = x(y^2 + y - 1)$ — Factor out the GCF, x.

$(a - 3b)y^2 + (a - 3b)y + (3b - a) = (a - 3b)(y^2 + y - 1)$ — Substitute $a - 3b$ back in for x.

Note the last term of -1 in the factor above. This term is needed to make the product of the factors produce all the terms in the original expression.

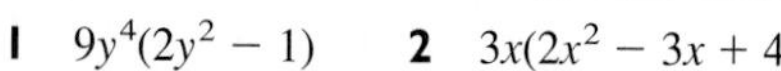
SELF-CHECK ANSWERS

1 $9y^4(2y^2 - 1)$ 2 $3x(2x^2 - 3x + 4)$

A Geometric Viewpoint: Factoring Out the GCF

The area of the first rectangle, $x(2x - 1)$, plus the area of the second rectangle, $8(2x - 1)$, yields a total area of $(x + 8)(2x - 1)$.

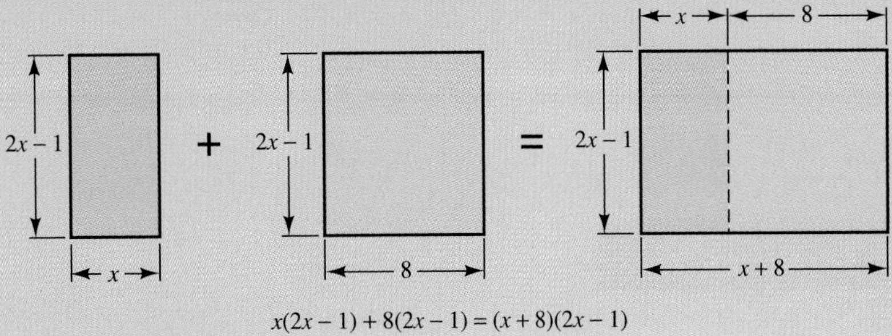

Factoring out the GCF is the easiest method of factoring, so you should always try it before trying other types of factoring. Some polynomials with four or more terms may not have any common factor other than 1 (or -1). Nonetheless, we can sometimes still take advantage of common factors by grouping terms, as illustrated in Example 7. This type of factoring is called **factoring by grouping.**

EXAMPLE 7 Factoring a Polynomial by Grouping

Factor $3x^2 - 6x + 5x - 10$ by grouping.

SOLUTION $3x^2 - 6x + 5x - 10 = 3x(x - 2) + 5(x - 2)$
$= (3x + 5)(x - 2)$

The first two terms have a common factor of $3x$, and the last two terms have a common factor of 5. Next factor out the common factor $x - 2$.

Factoring by grouping will be examined in more detail in Section 5-4.

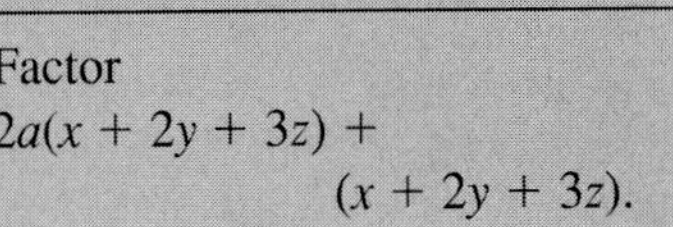

SELF-CHECK

Factor $2a(x + 2y + 3z) + (x + 2y + 3z)$.

EXAMPLE 8 Factoring a Polynomial with Variable Exponents

Factor $5x^{2m+1} - 7x^{2m}$.

SOLUTION The GCF of $5x^{2m+1} - 7x^{2m}$ is x^{2m}, because x^{2m} is the highest power of x that is a factor of both terms.

$5x^{2m+1} - 7x^{2m} = x^{2m}(5x - 7)$ Divide by x^{2m} to obtain the other factor:

$$\frac{5x^{2m+1} - 7x^{2m}}{x^{2m}} = 5x - 7$$

Remember to subtract exponents in order to divide these terms by x^{2m}.

The greatest common factor of $x^4 + x^3 + x^2$ is x^2. Because the smallest exponent on the terms x^4, x^3, and x^2 is 2 (the exponent on x^2), the greatest common factor must be x^2. Similarly, when examining the terms x^{-1}, $-x^{-2}$, and $-x^{-3}$, we note that the smallest exponent is -3. In the next example we will factor the GCF x^{-3} out of the expression $x^{-1} - x^{-2} - x^{-3}$.

SELF-CHECK ANSWER 

$(x + 2y + 3z)(2a + 1)$

EXAMPLE 9 Factoring Out a Factor with a Negative Exponent

Factor x^{-3} out of $x^{-1} - x^{-2} - x^{-3}$.

SOLUTION $x^{-1} - x^{-2} - x^{-3} = x^{-3}(x^2 - x - 1)$ Divide $x^{-1} - x^{-2} - x^{-3}$ by x^{-3} to obtain the other factor:

$$\frac{x^{-1} - x^{-2} - x^{-3}}{x^{-3}} = x^2 - x - 1$$

Exercises 5-1

A

In Exercises 1–10 determine the greatest common factor of each polynomial.

1 $10x - 2$
2 $15x + 10$
3 $3a^3 - 3a^2$
4 $7a^5 + 7a^3$
5 $9x^2 - 18xy - 15y^2$
6 $12x^2 - 15xy + 9y^2$
7 $12a^2b^3 - 20ab^2$
8 $14x^3y^5 - 21x^3y^5$
9 $x(x - 2y) - 3(x - 2y)$
10 $x(3x - y) + 4(3x - y)$

In Exercises 11–14 complete each factorization so that the first term in the second factor has a positive coefficient.

11 $-5x + 7 = -1(?)$
12 $-8v + 9 = -1(?)$
13 $-12m + 20 = -4(?)$
14 $-15a + 35 = -5(?)$

In Exercises 15–48 factor the greatest common factor from each polynomial.

15 $5y - 35$
16 $13z - 26$
17 $7x - 7y$
18 $24a - 16b$
19 $x^3 - x^2$
20 $m^6 + m^4$
21 $15m^6 - 20m^4$
22 $12x^7 - 21x^5$
23 $-22x^5y^3 + 33x^3y^4$
24 $-9a^3b + 6a^2b$
25 $4x^3 - 12x^2 - 6x$
26 $12m^5 - 18m^4 + 30m^3$
27 $-14x^{10} + 21x^8 + 35x^5$
28 $-3m^4 - 9m^3 - 3m^2$
29 $15a^3b^2 - 25a^2b^3 - 5ab$
30 $12a^2b^3 - 20ab^2 + 4ab$
31 $x(4a - b) - 2(4a - b)$
32 $a(3x - 1) + b(3x - 1)$
33 $-35x^3y^3 + 49x^2y^4 - 77xy^5$
34 $-55x^3y^4z^7 - 121x^2y^6z^9 + 33xy^5z^{11}$
35 $a(x + 3y) - 4c(x + 3y)$
36 $(x + y)(7a) - (x + y)(3b)$
37 $(14x - 3y + z)a - (14x - 3y + z)(2b)$
38 $3x(2a - 3b + c) - 2y(2a - 3b + c)$
39 $15x(2a - 7c) - 21(2a - 7c)$
40 $14x(3a + 5b) - 21(3a + 5b)$
41 $(a + 11b)(17x) - (a + 11b)(9y) + (a + 11b)(4z)$
42 $3a(19x - y) + 5b(19x - y) - 7c(19x - y)$
43 $(2x - 3y)a - b(3y - 2x)$
44 $2y(5x - 7y) + 3z(7y - 5x)$
45 $11z(7v - 4w) + (4w - 7v)$
46 $13a(3m - 5n) - (5n - 3m)$
47 $(117x + 31y)(54a^3b^2c) - 90a^2b^3c^3(117x + 31y)$
48 $x(5a - 4b - 3c - d) - 2y(5a - 4b - 3c - d)$

In Exercises 49–56 factor each polynomial using the technique of grouping illustrated in this section.

49 $x^2 - 2x + 5x - 10$
50 $x^2 + 3x + 2x + 6$
51 $2ax - 3a - 14x + 21$
52 $3am - 4a - 6m + 8$
53 $4ax - 3ay + 4bx - 3by$
54 $5ax + 2ay - 5bx - 2by$
55 $4a^2 - 8ab + 12ac + 3a - 6b + 9c$
56 $2x^3 - x^2 - 5x + 2x^2y - xy - 5y$

B

57 **a.** Factor $3vw^2$ out of $-6v^2w^2 + 15vw^3$.
b. Factor $-3vw^2$ out of $-6v^2w^2 + 15vw^3$.

58 **a.** Factor $11ab^2c$ out of $-33a^3b^2c + 121a^2b^3c^2 - 143ab^3c^4$.
b. Factor $-11ab^2c$ out of $-33a^3b^2c + 121a^2b^3c^2 - 143ab^3c^4$.

59 Factor x^3 out of $x^{n+3} + x^3$.

60 Factor w^m out of $w^{m+2} - 5w^m$.

In Exercises 61–66 factor the greatest common factor from each polynomial.

61 $x^{2m+1} + 7x^{2m}$
62 $4y^{n+3} - 5y^{n+2}$
63 $2a^{4n} - 3a^{3n}$

64 $3v^{5m} - 6v^{4m} + 9v^{3m}$

65 $15w^{7n} - 25w^{5n} + 20w^{4n}$

66 $2x^{2m}y^n - 3x^my^{n+1}$

C

67 $x - y$ is a factor of $x^3 - y^3$. Use long division to determine the other factor.

68 $x + y$ is a factor of $x^3 + y^3$. Use long division to determine the other factor.

69 The area of the rectangle shown in the figure is $4x^2 + 6x$, and its width is $2x$. Find its length.

$4x^2 + 6x$	$2x$

Figure for Exercise 69

70 The volume of the rectangular box shown in the figure is $x^3 + 4x^2 + 3x$. Find the area of the base if the height is $x + 1$.

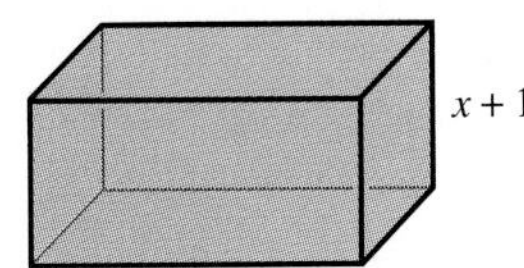

Figure for Exercise 70

71 Factor x^{-2} out of $3x^{-1} + 2x^{-2}$.

72 Factor y^{-3} out of $4y^{-1} - 5y^{-2} - 7y^{-3}$.

73 Factor a^{-m} out of $3a^2 - 5a^{-m}$.

74 Factor b^{-n} out of $2b^n - 4 + 3b^{-n}$.

75 Factor x^{-2n} out of $x^n - 3 - 2x^{-n} + 7x^{-2n}$.

76 Factor $x^{-1}y^{-1}$ out of $xy^{-1} + x^{-1}y$.

Calculator Usage (77–80)

In Exercises 77–80 fill in the missing factor.

77 $6.28x^2 + 9.42x + 15.7 = 3.14(\qquad)$

78 $-14.76x^3 + 27.06x = -2.46x(\qquad)$

79 $-13.7a(\qquad) = -41.1ax^2 + 27.4ax + 123.3a$

80 $9.3ax(\qquad) = 37.2ax^3 - 46.5ax^2 + 55.8ax$

81 Complete the factorization of $x(x + 1) + 3(x + 1)$, which is illustrated by the figure.

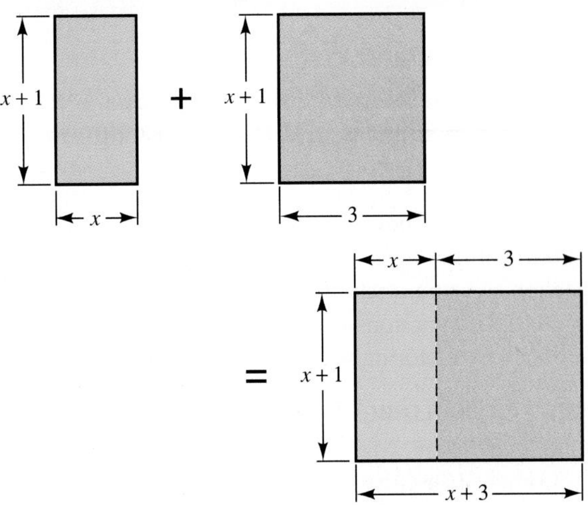

Figure for Exercise 81

Discussion Question

82 State in your own words a procedure for determining the GCF of a polynomial.

Challenge Question

A proper factor of a natural number n is a factor of n less than n. Natural numbers can be classified as deficient, perfect, or abundant. A perfect number is a natural number that equals the sum of all of its proper factors. For example, 6 is a perfect number because $6 = 1 + 2 + 3$, the sum of its proper factors. If the sum of the proper factors of n is less than n, the number is called a deficient number. If the sum of the proper factors of n is greater than n, n is called an abundant number. Classify each of the natural numbers 4 through 30 as deficient, perfect, or abundant.

Section Skill Reinforcers

Expand each of these expressions.

1 $(5v + 6w)(5v - 6w)$ **2** $(5v + 6w)^2$ **3** $(5v - 6w)^2$ **4** $(5a - b)(25a^2 + 5ab + b^2)$
5 $(2m + n)(4m^2 - 2mn + n^2)$

5-2 Factoring Special Forms

SECTION OBJECTIVE

2 By inspection, factor perfect square trinomials, the difference of two squares, and the sum or difference of two cubes.

Certain forms of polynomials occur so frequently that you should memorize the factors of these forms. Polynomials that fit these forms can be factored by inspection. As you study the special forms of polynomials, pay particular attention to the distinctions between them so that you can identify them quickly. We will start by factoring the special products covered in Section 4-5.

PERFECT SQUARE TRINOMIALS

Square of a Sum:

$a^2 + 2ab + b^2 = (a + b)^2$

Square of a Difference:

$a^2 - 2ab + b^2 = (a - b)^2$

Both the first term and the last term must be perfect squares. The middle term must be either plus or minus twice the product of the square roots of the first and last terms.

EXAMPLE 1 Identifying Special Forms

Determine which of these polynomials is a perfect square trinomial.

SOLUTIONS

(a) $8x^2 - 3x + 1$ This trinomial is not the square of any binomial that has integral coefficients.

The coefficient of the first term, 8, is not the square of any integer. Thus the first term cannot be a perfect square.

(b) $25x^2 + 10x + 1$ This trinomial fits the form of the square of a sum, as illustrated below:

$$25x^2 + 10x + 1 = (5x)^2 + 2(5x)(1) + (1)^2$$
$$= (5x + 1)^2$$

Note that $25x^2 = (5x)^2$ and $1 = (1)^2$. Thus the first and last terms are perfect squares. The middle term, $10x$, can also be written as $2(5x)(1)$.

(c) $64x^2 - 32x + 2$ This trinomial is not a perfect square.

The last term, 2, is not a perfect square of any integer.

(d) $81x^2 + 100y^2$ This binomial is not a perfect square.

The square of a binomial is always a trinomial, never a binomial. Note that $81x^2 + 100y^2 \neq (9x + 10y)^2$; $(9x + 10y)^2 = 81x^2 + 180xy + 100y^2$.

(e) $4w^2 + 5w + 1$ This trinomial is not a perfect square.

Although both the first term and the last term are perfect squares, the middle term is not twice the product of the numbers that are squared to obtain the first and last terms. The square of $2w + 1$ is $4w^2 + 4w + 1$, not $4w^2 + 5w + 1$.

(f) $4x^2 - 12xy - 9y^2$ This trinomial is not a perfect square.

The last term has a negative coefficient; therefore this trinomial is not the square of a binomial. For a trinomial to be a perfect square, both the first term and the last term must have positive coefficients.

EXAMPLE 2 Factoring Perfect Square Trinomials

Show that each trinomial fits the form of a perfect square trinomial, and then factor it.

SOLUTIONS

(a) $49v^2 - 14v + 1$

$$49v^2 - 14v + 1 = (7v)^2 + 2(7v)(-1) + (-1)^2 = (7v - 1)^2$$

(b) $100s^2 + 60st + 9t^2$

$$100s^2 + 60st + 9t^2 = (10s)^2 + 2(10s)(3t) + (3t)^2 = (10s + 3t)^2$$

(c) $x^{2m} + 12x^m + 36$

$$x^{2m} + 12x^m + 36 = (x^m)^2 + 2(x^m)(6) + (6)^2 = (x^m + 6)^2$$

Each of these trinomials is first written in the form $a^2 \pm 2ab + b^2$. Then the values of a and b from this form are used to write the factored form.

Each of the products shown below contains terms with zero coefficients; thus these products can be written as binomials. Because these special binomials occur frequently, you should memorize their factors.

$$\begin{array}{rrr} a + b & & \\ a - b & & \\ \hline a^2 + ab & & \\ -ab & - b^2 & \\ \hline a^2 & - b^2 & \end{array} \qquad \begin{array}{rrrr} a^2 + ab + b^2 & & & \\ a - b & & & \\ \hline a^3 + a^2b + ab^2 & & & \\ -a^2b - ab^2 & - b^3 & & \\ \hline a^3 & - b^3 & & \end{array} \qquad \begin{array}{rrrr} a^2 - ab + b^2 & & & \\ a + b & & & \\ \hline a^3 - a^2b + ab^2 & & & \\ +a^2b - ab^2 & + b^3 & & \\ \hline a^3 & + b^3 & & \end{array}$$

SELF-CHECK

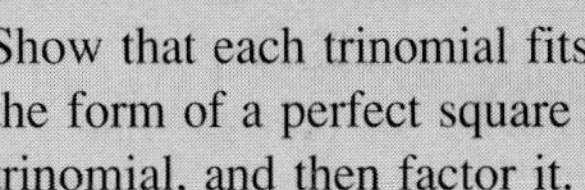

Show that each trinomial fits the form of a perfect square trinomial, and then factor it.

1 $y^2 + 6y + 9$

2 $4m^2 - 20mn + 25n^2$

SPECIAL BINOMIALS

$a^2 - b^2 = (a - b)(a + b)$	The difference of two squares
$a^2 + b^2$ is prime.*	The sum of two squares
$a^3 + b^3 = (a + b)(a^2 - ab + b^2)$	The sum of two cubes
$a^3 - b^3 = (a - b)(a^2 + ab + b^2)$	The difference of two cubes

*If a^2 and b^2 are only second-degree terms and have no common factor other than 1

EXAMPLE 3 Identifying Special Forms

Determine which of these polynomials is the difference of two squares.

SOLUTIONS

(a) $x^2 - xy - y^2$ This trinomial is not the difference of two squares.

The difference of two squares is always a binomial, never a trinomial.

(b) $3x^2 - 4$ This binomial is not the difference of two squares.

Since the coefficient 3 is not a perfect square, the first term is not a perfect square.

SELF-CHECK ANSWERS

1 $y^2 + 2(3)(y) + 3^2 = (y + 3)^2$ 2 $(2m)^2 + 2(2m)(-5n) + (-5n)^2 = (2m - 5n)^2$

(c) $169x^2 - 64$ This binomial fits the form of the difference of two squares:

$$169x^2 - 64 = (13x)^2 - (8)^2$$
$$= (13x - 8)(13x + 8)$$

Note that $169x^2 = (13x)^2$ and $64 = 8^2$.

(d) $25x^2 + 36y^2$ This binomial is not the difference of two squares.

This binomial is the sum of two squares, not the difference of two squares. It is prime.

(e) $100x^2 - 37y^2$ This binomial is not the difference of two squares.

Since 37 is not a perfect square, the last term is not a perfect square.

A Geometric Viewpoint: Factoring the Difference of Two Squares

The difference between the areas of the two squares with sides of x and y is $x^2 - y^2$. This is illustrated by the L-shaped red region in the figure to the right of the first equal symbol. This region consists of two rectangular regions, both of width $x - y$. One has length x, and the other has length y.

These two rectangles can be combined to form a rectangle with width $x - y$ and length $x + y$. The area of this rectangle is $(x - y)(x + y)$.

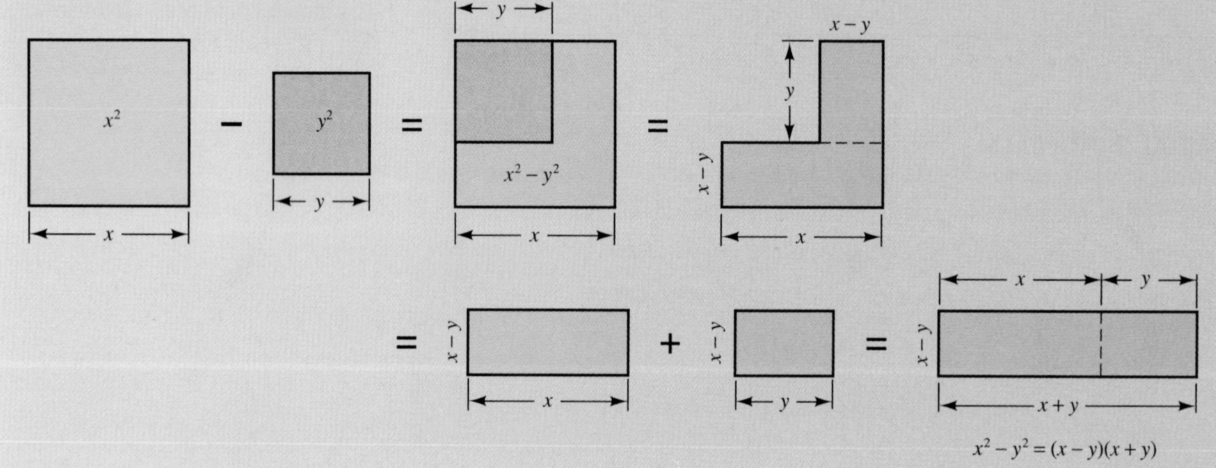

EXAMPLE 4 Factoring the Difference of Two Squares

Show that each binomial fits the form of the difference of two squares, and then factor it.

SOLUTIONS

(a) $49m^2 - 36n^2$

$$49m^2 - 36n^2 = (7m)^2 - (6n)^2$$
$$= (7m - 6n)(7m + 6n)$$

(b) $81x^2 - 25$

$$81x^2 - 25 = (9x)^2 - 5^2$$
$$= (9x - 5)(9x + 5)$$

Each of these binomials is first written in the form $a^2 - b^2$. Then the values of a and b from this form are used to write the factored form.

(c) $121v^4 - 25w^6$ $\quad$ $121v^4 - 25w^6 = (11v^2)^2 - (5w^3)^2$

$$= (11v^2 + 5w^3)(11v^2 - 5w^3)$$

(d) $x^{2m} - y^{2n}$ $\quad$ $x^{2m} - y^{2n} = (x^m)^2 - (y^n)^2$

$$= (x^m + y^n)(x^m - y^n)$$

▼

In the previous example note that the factored form of $a^2 - b^2$ can be written as either $(a - b)(a + b)$ or $(a + b)(a - b)$ because of the commutative property of multiplication.

EXAMPLE 5 Factoring a Fourth-Degree Binomial

Factor $v^4 - 16$ as completely as possible.

SOLUTION $v^4 - 16 = (v^2)^2 - (4)^2$ $\quad$ Factor this binomial as a difference of two squares.

$= (v^2 - 4)(v^2 + 4)$

$= (v - 2)(v + 2)(v^2 + 4)$ $\quad$ Then factor $v^2 - 4$, which is also a difference of two squares. $v^2 + 4$ is a sum of two squares and therefore is prime.

▼

It is helpful to be able to recognize the perfect squares, 4, 9, 16, 25, etc., when you are examining a binomial to determine if it is a difference of two squares. Likewise, it is helpful to be able to recognize the perfect cubes, 8, 27, 64, 125, etc., when you are examining a binomial to determine if it is a sum or difference of two cubes.

SELF-CHECK 

Factor each binomial.

1 $16m^2 - 169$

2 $100s^2 - 81t^4$

FACTORING THE SUM OR DIFFERENCE OF TWO CUBES

$$a^3 + b^3 = (a + b)(a^2 - ab + b^2)$$

$$a^3 - b^3 = (a - b)(a^2 + ab + b^2)$$

Step 1 Express the polynomial as a sum (or difference) of two perfect cubes.

Step 2 Write the binomial factor that is a sum (or difference) of the two cube roots.

Step 3 Use the binomial factor to obtain each term of the trinomial factor as follows:

a. The square of the first term of the binomial is the first term of the trinomial factor.

b. The opposite of the product of the two terms of the binomial is the second term of the trinomial factor.

c. The square of the last term of the binomial is the third term of the trinomial factor.

Caution: Be careful when writing the trinomial factor for the sum or difference of two cubes. A common error is to write the wrong coefficient for the middle term.

SELF-CHECK ANSWERS 

1 $(4m + 13)(4m - 13)$ $\quad$ **2** $(10s + 9t^2)(10s - 9t^2)$

EXAMPLE 6 Identifying Special Forms

Determine which of these binomials is the sum or difference of two cubes.

SOLUTIONS

(a) $x^3 + 25$ — This binomial is not the sum of two cubes. The last term, 25, is not a perfect cube.

(b) $a^3 + 27a^2 + 1$ — This trinomial is not the sum or difference of two cubes. The sum or difference of two cubes is always a binomial, never a trinomial.

(c) $8s^4 - 27t^3$ — This binomial is not the difference of two cubes. The first term is not a perfect cube because s^4 is not a perfect cube.

(d) $27x^3 + y^3$ — This binomial fits the form of the sum of two cubes, as illustrated below:

$$27x^3 + y^3 = (3x)^3 + (y)^3$$

Note that $27x^3 = (3x)^3$ and $y^3 = (y)^3$. This factorization is completed in Example 7(a).

EXAMPLE 7 Factoring the Sum or Difference of Two Cubes

Factor each of these polynomials.

SOLUTIONS

(a) $27x^3 + y^3$

$$27x^3 + y^3 = (3x)^3 + (y)^3$$

First express the binomial in the form of the sum of two cubes.

$$= (3x + y)[(\ )^2 - (\ )(\ ) + (\ )^2]$$

Then write the binomial factor, the sum of the two cube roots.

$$= (3x + y)[(3x)^2 - (3x)(y) + (y)^2]$$
$$= (3x + y)(9x^2 - 3xy + y^2)$$

Then use the binomial factor to obtain each term of the trinomial factor.

Square of the first term, $3x$
Opposite of the product of the two terms
Square of the last term, y

(b) $8w^3 - 27z^3$

$$8w^3 - 27z^3 = (2w)^3 - (3z)^3$$

Express this binomial as the difference of two cubes.

$$= (2w - 3z)[(\ )^2 + (\ )(\ ) + (\ )^2]$$

Write the binomial factor, the difference of the two cube roots.

$$= (2w - 3z)[(2w)^2 + (2w)(3z) + (3z)^2]$$
$$= (2w - 3z)(4w^2 + 6wz + 9z^2)$$

Then use the binomial factor to obtain each term of the trinomial factor.

Square of the first term, $2w$
Opposite of the product of the two terms
Square of the last term, $-3z$

(c) $8a^6 + 125$

$$8a^6 + 125 = (2a^2)^3 + (5)^3$$

Express this binomial as the sum of two cubes.

$$= (2a^2 + 5)[(\ )^2 - (\ )(\ ) + (\ )^2]$$

Write the binomial factor, the sum of the two cube roots.

$$= (2a^2 + 5)[(2a^2)^2 - (2a^2)(5) + (5)^2]$$
$$= (2a^2 + 5)(4a^4 - 10a^2 + 25)$$

Then use the binomial factor to fill in each term of the trinomial factor.

(d) $x^{3m} - y^{3n}$

$$x^{3m} - y^{3n} = (x^m)^3 - (y^n)^3$$
Express this binomial as the difference of two cubes.

$$= (x^m - y^n)[(\ )^2 + (\ )(\ ) + (\ )^2]$$
Write the binomial factor.

$$= (x^m - y^n)[(x^m)^2 + (x^m)(y^n) + (y^n)^2]$$
$$= (x^m - y^n)(x^{2m} + x^m y^n + y^{2n})$$
Use the binomial factor to obtain each term of the trinomial factor.

Caution: This coefficient is 1, not 2.

In Section 5-4 we will summarize a general factoring strategy; however, it is wise to begin now to distinguish among the various types of factoring as you work through the examples and exercises. Work on recognizing patterns based on the number of terms and on identifying perfect squares and perfect cubes. With more complicated polynomials it is sometimes easier to follow the logic if you use a substitution to clarify the basic structure of the problem.

EXAMPLE 8 Factoring the Difference of Two Squares

Factor $(2x + 3y)^2 - 16z^2$.

SOLUTION $(2x + 3y)^2 - (4z)^2$ is of the form $a^2 - b^2$.

$$a^2 - b^2 = (a + b)(a - b)$$
Substitute a for $(2x + 3y)$ and b for $4z$. Then factor this difference of two squares.

$$(2x + 3y)^2 - (4z)^2 = [(2x + 3y) + 4z][(2x + 3y) - 4z]$$
$$(2x + 3y)^2 - 16z^2 = (2x + 3y + 4z)(2x + 3y - 4z)$$
Substitute $2x + 3y$ back in for a and $4z$ back in for b, and then simplify.

Although a polynomial may be factorable by more than one method, you will always obtain the same factorization if you continue factoring until all factors are prime. It is generally most efficient, however, to begin by factoring out the GCF.

SELF-CHECK

Factor each binomial.

1 $64s^3 + t^3$

2 $27y^6 - 1000$

EXAMPLE 9 Factoring a Polynomial into Prime Factors

Factor $2v^4 - 2v$ as completely as possible.

SOLUTION $2v^4 - 2v = 2v(v^3 - 1)$ First factor out the GCF, $2v$.

$= 2v(v^3 - 1^3)$ Then note that $v^3 - 1$ can be written as the difference of two cubes.

$= 2v(v - 1)(v^2 + v + 1)$ Factor this difference of two cubes.

Caution: Do not forget to recopy this factor. The factorization is incorrect if this factor is omitted.

SELF-CHECK ANSWERS

1 $(4s + t)(16s^2 - 4st + t^2)$ 2 $(3y^2 - 10)(9y^4 + 30y^2 + 100)$

Exercises 5-2

A

In Exercises 1–60 factor each of these special forms as completely as possible.

1 $9a^2 - 1$
2 $b^2 - 16$
3 $v^2 - 81w^2$
4 $169a^2 - 4b^2$
5 $a^2 + 6a + 9$
6 $z^2 - 10z + 25$
7 $x^2 - 16xy + 64y^2$
8 $121a^2 + 22ab + b^2$
9 $w^3 + z^3$
10 $m^3 - n^3$
11 $8x^3 - 1$
12 $27a^3 + 1$
13 $16s^2 - 121t^2$
14 $9a^2 - 49b^2$
15 $25a^4 - 36b^6$
16 $9m^6n^4 - 121p^8$
17 $49s^2 + 28s + 4$
18 $4w^2 + 12wz + 9z^2$
19 $x^3 - 125y^3$
20 $x^3 + 64y^3$
21 $27m^3 + 8n^3$
22 $216x^3 + 125y^3$
23 $9x^2 - 12xy + 4y^2$
24 $25s^4 - 20s^2t + 4t^2$
25 $v^2 + 9$
26 $m^2 + 1$
27 $y^4 - 4y^2 + 4$
28 $36w^2 - 60w + 25$
29 $x^4 - 1$
30 $m^4 - 16$
31 $9x^6y^2 - 1$
32 $4x^6y^2 - 9$
33 $x^3 - 64$
34 $x^2 - 64$
35 $x^2 + 64$
36 $x^6 - 64$
37 $m^4 + 100$
38 $y^4 + 4$
39 $(a + 2b)^2 - 4$
40 $(3m - 2n)^2 - 25$
41 $81 - (5v - 7w)^2$
42 $144 - (3a + 5b)^2$
43 $(x + y)^2 - (a - b)^2$
44 $(2v - w)^2 - (3c + d)^2$
45 $(x + y)^2 + 2(x + y) + 1$
46 $(x - y)^2 - 4(x - y) + 4$
47 $(2a - b)^2 - 6(2a - b) + 9$
48 $(v + 3w)^2 + 10(v + 3w) + 25$
49 $(4v - 3w)^3 - 27$
50 $(7a + 2b)^3 + 125$
51 $(x + 5)^3 + x^3$
52 $(y + 4)^3 - y^3$
53 $6ab + 9b^2 + a^2$
54 $-10mn + m^2 + 25n^2$
55 $w^6 - 22w^3 + 121$
56 $36a^4 - 49c^4$
57 $169x^6 - 144z^2$
58 $81a^8 + 18a^4d^2 + d^4$
59 $25m^{10} + 20m^5p + 4p^2$
60 $125x^9 - z^9$

B

In Exercises 61–66 first factor out the greatest common factor, and then use the special forms to complete the factorization of each polynomial.

61 $3x^3 - 363x$
62 $12(a - b)x^2 - 75(a - b)$
63 $5ax^3 - 40a$
64 $7a^4b^2 - 7ab^5$
65 $6ax^2 + 36ax + 54a$
66 $2x^5 - 60x^4 + 450x^3$

67 $x^2 + y^2$ is the sum of two squares and is prime. $x^6 + y^6$ is the sum of two squares but is not prime. Factor $x^6 + y^6$ after first expressing it as the sum of two cubes.

68 $4x^2 + 9$ is the sum of two squares and is prime. $36x^2 + 9$ is the sum of two squares but is not prime. Factor $36x^2 + 9$.

In Exercises 69–76 factor each polynomial completely. Assume all exponents are positive integers.

69 $(x + y)^3 - x^3$
70 $x^{4m} - y^2$
71 $v^{6n} - 9$
72 $v^{6n} - 27$
73 $w^{2m} - 24w^m + 144$
74 $4r^{2n} + 20r^n + 25$
75 $x^{6m} - y^{8n}$
76 $x^{6m} - y^{9n}$

C

CALCULATOR USAGE (77–82)

Completely factor each of these polynomials.

77 $2.86x^2 - 25.74 = 2.86(\quad)$
$= 2.86(\quad)(\quad)$

78 $3.44x^3 - 27.52 = 3.44(\quad)$
$= 3.44(\quad)(\quad)$

79 $-21.8x^2 - 21.8x - 5.45 = -5.45(\quad)$
$= -5.45(\quad)^2$

80 $-6.5x^2 + 65x - 162.5 = -6.5(\quad)$
$= -6.5(\quad)^2$

In Exercises 81 and 82 use a graphics calculator to graph $y_1 = P_1(x)$ and $y_2 = P_2(x)$. Then use these graphs to determine whether $P_1(x)$ will factor into the polynomial given by $P_2(x)$.

81 $P_1(x) = 5.61x^2 + 2.49x - 0.63$
$P_2(x) = (1.1x - 2.4)(5.1x + 1.3)$

82 $P_1(x) = 6.25x^2 - 3x - 4.05$
$P_2(x) = (2.5x + 2.3)(2.5x - 3.5)$

83 Write a polynomial to represent the area enclosed between the concentric circles shown in the figure. Then factor this polynomial. The radius of the smaller circle is r_1, and the radius of the larger circle is r_2.

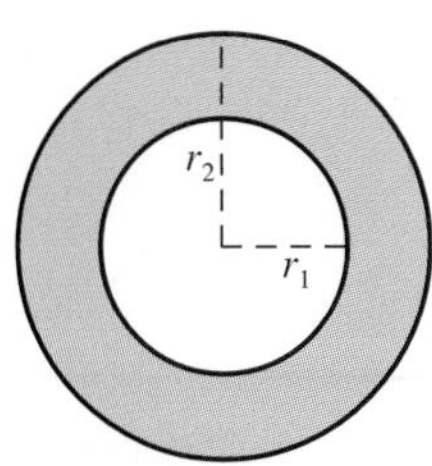

Figure for Exercise 83

84 Verify that $(125s^3 - 64t^3)$ factors as $(5s - 4t) \cdot (25s^2 + 20st + 16t^2)$ by multiplying these factors.

85 Completely factor $16x^4 - 72x^2y^2 + 81y^4$.

86 Completely factor $(a^2 - b^2)x^2 + 6(a^2 - b^2)x + 9(a^2 - b^2)$.

87 Complete the factorization of $x^2 - 25$, which is illustrated in the figure.

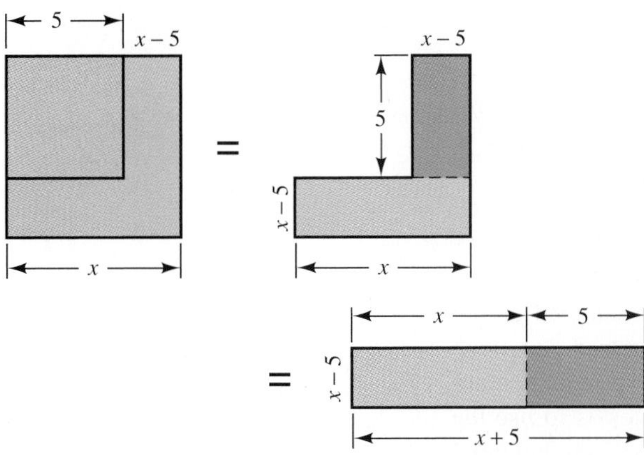

Figure for Exercise 87

Discussion Question

88 Explain why it is incorrect to say that a sum of two squares is always prime.

Challenge Question

a. Show that the difference between the volume of the larger cube and that of the smaller cube in the figure can be written as $(x + 1)^2 + x(x + 1) + x^2$.

b. Illustrate with figures, as in Exercise 87, the factorization $x^2 + 2xy + y^2 = (x + y)^2$.

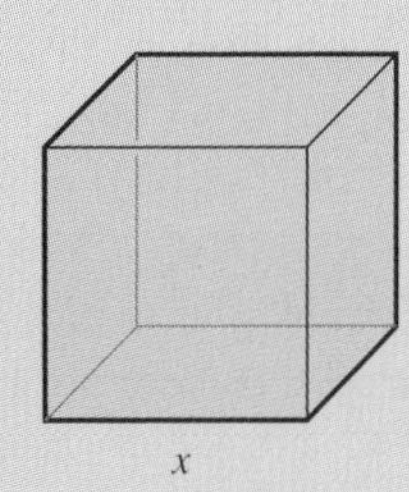

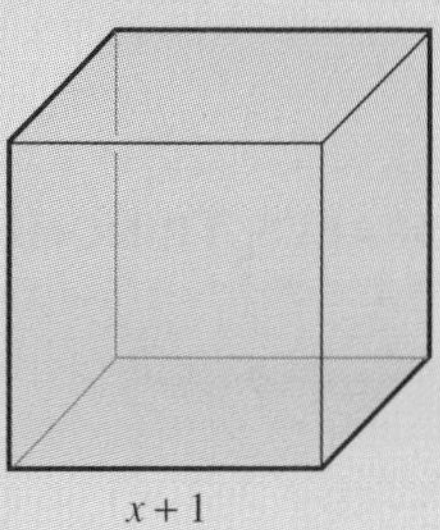

Section Skill Reinforcers

Factor each of these expressions.

1 $2a(x + y) - 3b(x + y)$ **2** $3x(a + b) - (a + b)$ **3** $(3ax - 15a) + (2x - 10)$
4 $12bx - 4by - 21x + 7y$ **5** $12x^2 - 3xy + 20x - 5y$

5-3 Factoring Trinomials

Section Objective

3 Factor quadratic trinomials.

In the last section we examined only trinomials that were perfect squares. We will now present two methods for factoring general trinomials of the form $ax^2 + bx + c$ or the form $ax^2 + bxy + cy^2$. The procedure in this section will enable you either to factor a trinomial or to determine that it is prime.

The trinomial $ax^2 + bx + c$ is referred to as a **quadratic trinomial.** The ax^2 term is called the **second-degree term,** or the **quadratic term;** bx is the **first-degree term,** or the **linear term;** and c is the **constant term.** For example, $6x^2 - 7x - 5$ is a quadratic trinomial with $a = 6$, $b = -7$, and $c = -5$. The quadratic term is $6x^2$, the linear term is $-7x$, and the constant term is -5.

METHOD I: FACTORING TRINOMIALS BY TRIAL AND ERROR

The easiest quadratic trinomials to factor are those in which the leading coefficient, a, is equal to 1, so we will begin with these trinomials. Either $x^2 + bx + c$ is prime or it can be factored as $(x + c_1)(x + c_2)$. Because

$$(x + c_1)(x + c_2) = x^2 + (c_1 + c_2)x + c_1c_2$$

a factorization of this form must have

$$c_1 + c_2 = b \qquad \text{and} \qquad c_1c_2 = c$$

FACTORABLE TRINOMIALS

A trinomial $x^2 + bx + c$ is factorable into a pair of binomial factors with integer coefficients if and only if there are two integers whose product is c and whose sum is b.

Since several possible factors c_1 and c_2 may have to be examined before the correct factors of c are found, this process is referred to as trial and error. Nonetheless, you should always examine the possibilities in a systematic manner rather than through random guessing. An important first step is to use the signs of b and c from $x^2 + bx + c$ to determine the sign pattern of the binomial factors.

SIGN PATTERN FOR THE FACTORS OF $x^2 + bx + c$

If the constant c is positive, the factors of c must have the same sign. These factors will share the same sign as the linear coefficient, b.

If the constant c is negative, the factors of c must be opposite in sign. The sign of the constant factor with the larger absolute value will be the same as that of the linear coefficient, b.

EXAMPLE 1 Factoring a Trinomial

Factor $x^2 - 10x + 24$.

SOLUTION First set up a form with the correct sign pattern.

$x^2 - 10x + 24 = (x - ?)(x - ?)$ Since the constant term, 24, is positive and the linear coefficient, -10, is negative, both constant terms must be negative.

Then use the factors of 24 to examine all possible factors of this form.

Possible Factors	Resulting Linear Term
$(x-1)(x-24)$	$-25x$
$(x-2)(x-12)$	$-14x$
$(x-3)(x-8)$	$-11x$
$(x-4)(x-6)$	$-10x$

	Factors of 24		*Sum of Factors*
	-1	-24	-25
	-2	-12	-14
	-3	-8	-11
	-4	-6	-10

The only factors of 24 that yield a sum of -10 are -4 and -6.

Answer $x^2 - 10x + 24 = (x-4)(x-6)$ ▼

EXAMPLE 2 Factoring a Trinomial

Factor $y^2 + 7y - 60$.

SOLUTION First set up a form with the correct sign pattern.

$y^2 + 7y - 60 = (y - ?)(y + ?)$ The constant term, -60, is negative, so the constant terms are of opposite sign. Also, the linear coefficient, $+7$, is positive, so the positive constant must be larger in absolute value than the negative constant.

Now use the factors of -60 to examine all possible factors of this form.

Possible Factors	Resulting Linear Term
$(y-1)(y+60)$	$59y$
$(y-2)(y+30)$	$28y$
$(y-3)(y+20)$	$17y$
$(y-4)(y+15)$	$11y$
$(y-5)(y+12)$	$7y$
$(y-6)(y+10)$	$4y$

	Factors of −60		*Sum of Factors*
	-1	60	59
	-2	30	28
	-3	20	17
	-4	15	11
	-5	12	7
	-6	10	4

The only factors of -60 that yield a sum of $+7$ are -5 and 12.

Answer $y^2 + 7y - 60 = (y-5)(y+12)$ ▼

EXAMPLE 3 A Prime Trinomial

Factor $w^2 + 9w + 12$.

SOLUTION First set up a form with the correct sign pattern.

$w^2 + 9w + 12 = (w + ?)(w + ?)$ The constant term, 12, is positive and the linear coefficient, 9, is positive, so both constants must be positive.

Now use the factors of 12 to examine all possible factors of this form.

Possible Factors	Resulting Linear Term
$(w + 1)(w + 12)$	$13w$
$(w + 2)(w + 6)$	$8w$
$(w + 3)(w + 4)$	$7w$

Factors of 12		*Sum of Factors*
1	12	13
2	6	8
3	4	7

None of the factors of 12 have a sum of 9. Since we have systematically examined and eliminated all integral factors of 12, we know that this polynomial is prime over the integers.

Answer $w^2 + 9w + 12$ is prime over the integers.

One way to determine if $ax^2 + bx + c$ is prime over the integers is to systematically eliminate all possibilities, as in Example 3. Another way to determine if $ax^2 + bx + c$ is prime is given by the test in the following box. The justification for this test is provided in Chapter 8 when the discriminant is discussed. This test uses the relationship that exists between the factors of $ax^2 + bx + c$ and the solutions of $ax^2 + bx + c = 0$.

TEST FOR FACTORING $ax^2 + bx + c$

The trinomial $ax^2 + bx + c$ with integer coefficients a, b, and c will factor over the integers if $b^2 - 4ac$ is a perfect square. If $b^2 - 4ac$ is not a perfect square, then $ax^2 + bx + c$ is prime.

Note in Example 3 that we know $w^2 + 9w + 12$ is prime since $b^2 - 4ac = 9^2 - 4(1)(12) = 33$ is not a perfect square. In Example 2, $y^2 + 7y - 60$ is factorable since $b^2 - 4ac = 7^2 - 4(1)(-60) = 289$ is the perfect square of 17.

Next we will use this trial-and-error process to factor trinomials in which the leading coefficient is not 1. We will continue to determine the sign pattern first in order to limit the possibilities that must be examined.

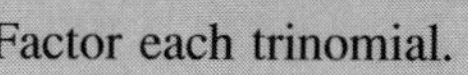

SELF-CHECK

Factor each trinomial.

1 $v^2 + 8v + 12$

2 $m^2 - 7m + 12$

3 $n^2 - 11n - 12$

FACTORING TRINOMIALS BY SYSTEMATIC TRIAL AND ERROR

Step 1 Make a form for the two binomial factors, and fill in the obvious information, such as the sign pattern.

Step 2 List the possible factors of the first and last terms that fit this pattern.

Step 3 Select the factors that yield the correct middle term. If all possibilities fail, the polynomial is prime.

SELF-CHECK ANSWERS 

1 $(v + 6)(v + 2)$ 2 $(m - 4)(m - 3)$ 3 $(n + 1)(n - 12)$

EXAMPLE 4 Factoring a Trinomial Whose Leading Coefficient Is Not 1

Factor $6x^2 - 7x - 5$.

SOLUTION The form with the correct sign pattern is

$6x^2 - 7x - 5 = (?x - ?)(?x + ?)$ The blank form has opposite signs, since the constant term, −5, is negative. All possible combinations of the factors of 6 and −5 are used to fill in the sign pattern.

Possible Factors	Resulting Linear Terms
$(6x + 1)(x - 5)$	$-29x$
$(6x - 1)(x + 5)$	$29x$
$(3x + 1)(2x - 5)$	$-13x$
$(3x - 1)(2x + 5)$	$13x$
$(2x + 1)(3x - 5)$	$-7x$
$(2x - 1)(3x + 5)$	$7x$
$(x + 1)(6x - 5)$	x
$(x - 1)(6x + 5)$	$-x$

Factors of 6			*Factors of −5*	
6	1	☞	1	−5
3	2		−1	5
☞ 2	3			
1	6			

Obviously, when you work an actual problem, you can stop listing factors as soon as you determine the correct factors.

Answer $6x^2 - 7x - 5 = (2x + 1)(3x - 5)$ ▼

EXAMPLE 5 Factoring Trinomials with Two Variables

Factor $12a^2 - 29ab + 15b^2$.

SOLUTION The form with the correct sign pattern is

$$12a^2 - 29ab + 15b^2 = (?a - ?b)(?a - ?b)$$

Both coefficients of b must be negative in this form, since 15 is positive and −29 is negative.

The possibilities that can be used to fill in this form will come from the factors of 12 and 15, listed below.

Factors of 12		*Factors of 15*	
1	12	1	15
2	6	☞ 3	5
☞ 3	4		

Try to determine the correct factorization by inspection. If necessary, start listing these factors on scrap paper. The correct factors are pointed out here.

The correct factorization obtained from these possibilities is

$$(3a - 5b)(4a - 3b)$$

Answer $12a^2 - 29ab + 15b^2 = (3a - 5b)(4a - 3b)$

To check the factorization, multiply these factors using the FOIL procedure. ▼

With practice you should be able to factor a trinomial simply by thinking of the possibilities without listing them. Remember to be systematic, however. You must use some trial and error, but there is no need to make wild guesses. If you eliminate the possibilities in an orderly manner, you can be certain when a second-degree trinomial is prime.

EXAMPLE 6 A Prime Trinomial

Factor $11x^2 + 3xy - 10y^2$.

SOLUTION $b^2 - 4ac = 3^2 - 4(11)(-10)$
$= 449$

Since $\sqrt{449} \approx 21.19$, 449 is not a perfect square. Thus $11x^2 + 3xy - 10y^2$ is prime by the test for factoring trinomials.

The test for factoring $ax^2 + bx + c$ extends to factoring trinomials with two variables.

Answer $11x^2 + 3xy - 10y^2$ is prime over the integers.

EXAMPLE 7 Factoring a Trinomial Whose Leading Coefficient Is Negative

Factor $-6x^2 + 29xy - 13y^2$.

SOLUTION $-6x^2 + 29xy - 13y^2$ is equal to $-(6x^2 - 29xy + 13y^2)$. The form with the correct sign pattern is

$$-6x^2 + 29xy - 13y^2 = -(6x^2 - 29xy + 13y^2)$$
$$= -(?x - ?y)(?x - ?y)$$

Since the leading coefficient is negative, we first factor out -1. The sign pattern within the parentheses is then determined from the coefficients -29 and 13.

Factors of 6	Factors of 13
1 6	☞ 1 13
☞ 2 3	

The correct factorization can be determined by trial and error, using these possibilities.

Answer $-6x^2 + 29xy - 13y^2 = -(2x - y)(3x - 13y)$

Check the factorization by multiplying the factors.

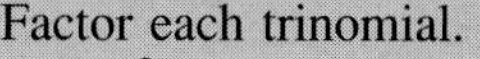

SELF-CHECK

Factor each trinomial.

1 $3m^2 + 5m + 2$
2 $5n^2 - 17n + 6$

EXAMPLE 8 Using Substitution to Factor Trinomials

Factor $(a + b)^2 - 3(a + b) - 28$.

SOLUTION $(a + b)^2 - 3(a + b) - 28$ is of the form $x^2 - 3x - 28$. The form with the correct sign pattern for this polynomial is

$$x^2 - 3x - 28 = (?x + ?)(?x - ?)$$

Substitute x for $(a + b)$. This form has opposite signs for the constants, since -28 is negative.

Factors of −28		Sum of Factors
1	−28	−27
2	−14	−12
☞ 4	−7	−3

The correct factorization obtained from these possibilities is

$$x^2 - 3x - 28 = (x + 4)(x - 7)$$
$$(a + b)^2 - 3(a + b) - 28 = [(a + b) + 4][(a + b) - 7]$$

Substitute $(a + b)$ back in for x, and then simplify.

Answer $(a + b)^2 - 3(a + b) - 28 = (a + b + 4)(a + b - 7)$

SELF-CHECK ANSWERS

1 $(3m + 2)(m + 1)$ 2 $(5n - 2)(n - 3)$

Remember that it is generally most efficient to begin factoring by factoring out the GCF.

EXAMPLE 9 Completely Factoring a Trinomial

Factor $-20m^3n^2 + 52m^2n^3 - 32mn^4$.

SOLUTION

$-20m^3n^2 + 52m^2n^3 - 32mn^4 = -4mn^2(5m^2 - 13mn + 8n^2)$

First factor out the opposite of the GCF, $-4mn^2$. Factor out a negative so that the trinomial factor's leading coefficient will be positive.

$-20m^3n^2 = -2^2 \cdot 5m^3n^2$
$52m^2n^3 = 2^2 \cdot 13m^2n^3$
$-32mn^4 = -2^5mn^4$
$\text{GCF} = 2^2mn^2 = 4mn^2$

$= -4mn^2(?m - ?n)(?m - ?n)$

Write the form, and try factors of 5 and 8 until the correct factors are found.

$= -4mn^2(5m - 8n)(m - n)$

METHOD 2: FACTORING TRINOMIALS BY GROUPING

An alternative to the trial-and-error method for factoring trinomials of the form $ax^2 + bx + c$ is a method known as the AC method or the grouping method. To understand the logic of this method, first consider $x^2 + bx + c$ factored as $(x + c_1)(x + c_2)$. Multiplying these factors, we obtain $x^2 + c_1x + c_2x + c_1c_2 = x^2 + bx + c$. Thus c must have two factors, c_1 and c_2, whose sum is b. This is the basis for the trial-and-error method for factoring $x^2 + bx + c$.

This logic can be extended to factor $ax^2 + bx + c$ as $(a_1x + c_1)(a_2x + c_2)$. Multiplying these factors, we obtain

$$a_1x(a_2x + c_2) + c_1(a_2x + c_2) = a_1a_2x^2 + a_1c_2x + a_2c_1x + c_1c_2$$

This means that if we can factor ac into two factors (a_1c_2) and (a_2c_1) whose sum is b, we can then factor out the common factor $a_2x + c_2$ to obtain

$$a_1x(a_2x + c_2) + c_1(a_2x + c_2) = (a_1x + c_1)(a_2x + c_2)$$

FACTORING TRINOMIALS BY THE GROUPING METHOD

To factor $ax^2 + bx + c$, list all possible factors of ac and select factors b_1 and b_2 whose sum is b. Rewrite $ax^2 + bx + c$ as $ax^2 + b_1x + b_2x + c$, and then factor this polynomial by the grouping strategy from Section 5-1.

We will now factor trinomials like those in Examples 4 and 5 by the grouping method. This will allow you to compare the method of grouping to the trial-and-error method.

EXAMPLE 10 Factoring a Trinomial by the Grouping Method

Factor $6x^2 - 7x - 5$. (This is the same trinomial factored in Example 4.)

SOLUTION

Factors of -30		Sum of Factors
1	-30	-29
2	-15	-13
3	-10	-7
5	-6	-1
6	-5	1
10	-3	7
15	-2	13
30	-1	29

The form $ax^2 + bx + c$ has $a = 6$, $b = -7$, and $c = -5$. Therefore $ac = (6)(-5) = -30$, and the two factors of -30 must be of opposite sign. The two factors highlighted have a sum of -7, which is the coefficient b. (You can stop listing possibilities as soon as you determine the correct factors.)

$$6x^2 - 7x - 5 = 6x^2 + 3x - 10x - 5$$

Rewrite the linear term $-7x$ as the sum of the two terms found above.

$$= 3x(2x + 1) - 5(2x + 1)$$

$$= (2x + 1)(3x - 5)$$

Then factor this polynomial by grouping.

EXAMPLE 11 Factoring a Trinomial by the Grouping Method

Factor $12x^2 - 29xy + 15y^2$. (This is similar to the trinomial factored in Example 5.)

SOLUTION

Factors of 180		Sum of Factors
-1	-180	-181
-2	-90	-92
-3	-60	-63
-4	-45	-49
-5	-36	-41
-6	-30	-36
-9	-20	-29

The form $ax^2 + bxy + cy^2$ has $a = 12$, $b = -29$, and $c = 15$. Therefore $ac = (12)(15) = 180$, and the two factors of 180 must both be negative. Stop your list when the sum is -29, the coefficient b.

$$12x^2 - 29xy + 15y^2 = 12x^2 - 9xy - 20xy + 15y^2$$

Rewrite the middle term $-29xy$ as the sum of the two terms found above.

$$= 3x(4x - 3y) - 5y(4x - 3y)$$

$$= (3x - 5y)(4x - 3y)$$

Then factor this polynomial by grouping.

A Geometric Viewpoint: Determining Factors from a Graph

The factors of $x^2 + bx + c$ (over the set of real numbers) correspond exactly to the x-intercepts of the graph of $y = x^2 + bx + c$. The graph of $y = x^2 - x - 6$, as exhibited on a graphics calculator, is shown in the figure. This graph crosses the x-axis at $x = -2$ and $x = 3$, which corresponds to the factorization $x^2 - x - 6 = (x + 2)(x - 3)$. In general, if $(a, 0)$ is an x-intercept of the graph of $y = x^2 + bx + c$, then $x - a$ is a factor of $x^2 + bx + c$. Graphs of $y = ax^2 + bx + c$ are examined in detail in Section 9-2.

[−4, 4] for x, [−8, 4] for y

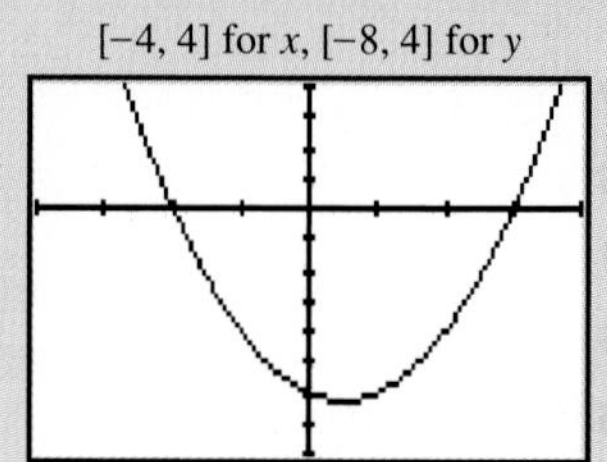

EXAMPLE 12 Calculator Usage

Use the graph of $y = x^2 - 3x - 4$ to factor $x^2 - 3x - 4$.

SOLUTION

Keystrokes

First enter the equation to be graphed. Assume that the window of values has already been set to $[-5, 6]$ for x and $[-8, 10]$ for y.

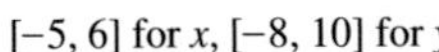

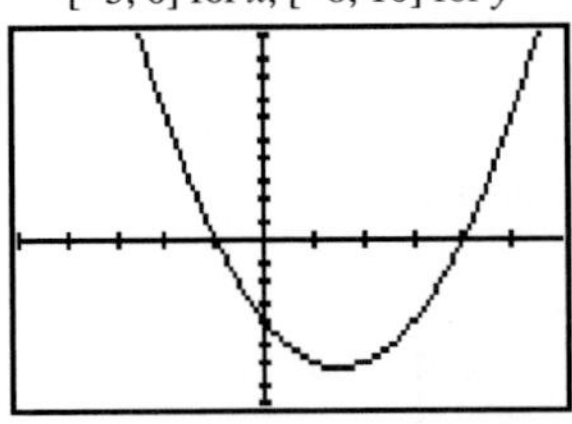

By inspection (or by using the TRACE option), we find that the x-intercepts are approximately -1 and $+4$. Thus the factors are $[x - (-1)](x - 4)$.

Answer $x^2 - 3x - 4 = (x + 1)(x - 4)$ ▼

Exercises 5-3

A

In Exercises 1–44 factor each of these polynomials completely.

1 a. $y^2 - 7y + 6$ **b.** $y^2 - 5y - 6$

c. $y^2 - y - 6$ **d.** $y^2 + y - 6$

2 a. $a^2 + 2a - 35$ **b.** $a^2 + 12a + 35$

c. $a^2 + 36a + 35$ **d.** $a^2 - 2a - 35$

3 a. $6x^2 - 13x - 5$ **b.** $6x^2 - 17x + 5$

c. $6x^2 + 13x - 5$ **d.** $6x^2 + x - 5$

4 a. $10x^2 + 11x + 3$ **b.** $10x^2 - x - 3$

c. $10x^2 - 13x - 3$ **d.** $10x^2 - 29x - 3$

5 $t^2 + 18t + 77$ **6** $w^2 - 11w + 28$

7 $b^2 + 9b - 36$ **8** $t^2 + 2t - 48$

9 $x^2 - 5x - 36$ **10** $z^2 + 29z + 100$

11 $5v^2 + 12v + 7$ **12** $3w^2 - 5w + 2$

13 $x^2 - 9x - 36$ **14** $y^2 + 2y - 48$

15 $2x^2 + 5x + 3$ **16** $3v^2 - 7v - 6$

17 $-11x^2 - 6x + 5$ **18** $-7y^2 + 6y + 13$

19 $6c^2 + 35cd + 11d^2$ **20** $25m^2 + 20mn - 21n^2$

21 $-a^2 + ab + 6b^2$ **22** $-a^2 + 8ab - 15b^2$

23 $2v^2 + vw - 21w^2$ **24** $2v^2 + 13vw + 21w^2$

25 $2v^2 + 11vw - 21w^2$ **26** $2v^2 - 17vw + 21w^2$

27 $x^4 - 5x^2 - 14$ **28** $x^4 - 15x^2 + 14$

29 $x^4 - 2x^2y^2 - 99y^4$ **30** $x^4 + 20x^2y^2 + 99y^4$

31 $-42a^2 + 5ab + 25b^2$ **32** $-24c^2 + 94cd - 35d^2$

33 $(a - 2b)^2 - 2(a - 2b) - 15$

34 $(a + 3b)^2 + 8(a + 3b) + 15$

35 $6(x - y)^2 + 5(x - y) + 1$

36 $6(x + y)^2 - (x + y) - 1$

37 $25x^2 + 4xy - 21y^2$

38 $45m^2 - 88mn + 35n^2$

39 $11(a + b)^2 + 100(a + b) + 9$

40 $15(x - y)^2 + 31(x - y)z + 14z^2$

41 $x^6 - 2x^3 - 8$ **42** $x^6 + 6x^3 + 8$

43 $x^6 + x^3y^2 - 20y^4$ **44** $x^6 - 8x^3y^2 - 20y^4$

B

In Exercises 45–52 first factor out the greatest common factor, and then completely factor each of these polynomials.

45 $5v^3 - 15v^2 - 20v$ **46** $2x^3 + 6x^2 + 4x$

47 $-6a^3b + 12a^2b^2 + 144ab^3$

48 $-35v^2w^3 - 7v^2w^2 + 42v^2w$

49 $24x^3yz - 20x^2y^2z - 156xy^3z$

50 $165m^3n^2 + 22m^2n^3 - 264mn^4$

51 $8(a - b)c^2 + 6(a - b)c - 9(a - b)$

52 $6(c - d)v^2 - 16(c - d)vw - 12(c - d)w^2$

In Exercises 53–56 some of the polynomials are factorable, and some are prime over the integers. Factor those that are factorable, and label the others as prime.

53 **a.** $m^2 - 81$ **b.** $m^2 + 81$

c. $m^2 + 18m + 81$ **d.** $m^2 + m + 81$

54 **a.** $m^2 - 21$ **b.** $m^2 + 21$

c. $m^2 - 4m - 21$ **d.** $m^2 + 4m + 21$

55 **a.** $x^2 + 12x + 36$ **b.** $x^2 - 12x + 36$

c. $-x^2 + 12x - 36$ **d.** $x^2 + 36$

56 **a.** $m^2 - n^2$ **b.** $m^2 + n^2$

c. $m^2 + mn + n^2$ **d.** $m^2 + 2mn + n^2$

C

In Exercises 57–62 factor each polynomial. Assume that all exponents are integers.

57 $x^{2n} + x^n - 20$ **58** $4z^{2m} - 9z^m + 5$

59 $5x^{2m} + 14x^m - 3$ **60** $12v^{4m} - 19v^{2m} + 5$

61 $x^2(x - y)^2 + xy(x - y) - 6y^2$

62 $m^2(m + n)^2 - mn(m + n) - 12n^2$

CALCULATOR USAGE (63–70)

In Exercises 63 and 64 use the given graph of $y = P(x)$ to factor $P(x)$.

63 $P(x) = x^2 + 2x - 15$ **64** $P(x) = x^2 + 2x - 8$

[−8, 6] for x, [−20, 25] for y

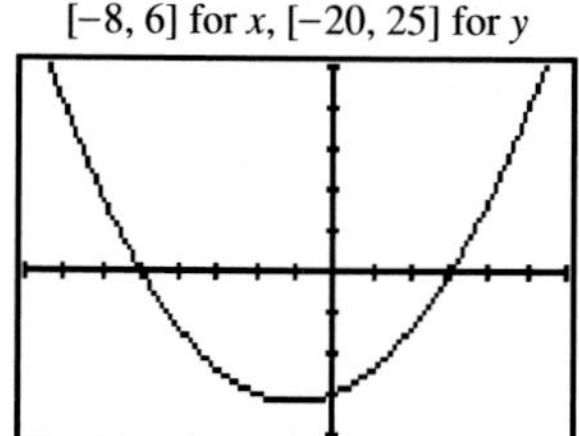

[−8, 6] for x, [−15, 20] for y

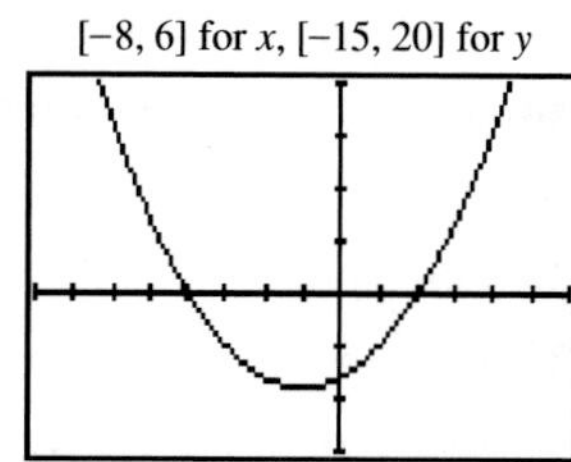

In Exercises 65 and 66 use a graphics calculator to graph $y = P(x)$, and then use this graph to factor $P(x)$. (*Hint:* Use the window of values shown.)

WINDOW FORMAT
Xmin = −10
Xmax = 28
Xscl = 2
Ymin = −250
Ymax = 175
Yscl = 25

65 $P(x) = x^2 - 18x - 144$ **66** $P(x) = x^2 - 16x - 80$

In Exercises 67–70 use a calculator to determine whether each polynomial is prime or factorable. (The polynomial $ax^2 + bx + c$ is prime over the integers if $b^2 - 4ac$ is not a perfect square.)

67 $6x^2 + 55x + 119$ **68** $6x^2 + 55x + 19$

69 $6x^2 + 55x + 25$ **70** $6x^2 + 55x + 49$

71 **Length of a Rectangle** The width of the rectangle shown in the figure is $5a - 7$, and its area is $15a^2 - a - 28$. Factor the polynomial representing the area, and then determine the length of the rectangle.

Figure for Exercise 71

72 **Width of a Rectangle** The length of the rectangle in the figure is $4x + 3y$, and its area is $28x^2 + 13xy - 6y^2$. Determine its width.

Figure for Exercise 72

73 **Investment** If a principal P is invested at a simple interest rate r for t years, the value of this investment after t years will be $P + Prt$. Factor this expression.

74 **Consecutive Integers** If m is an integer, show that $m^3 + 3m^2 + 2m$ is the product of three consecutive integers. (*Hint:* Factor the expression.)

75 **Consecutive Even Integers** If m is an even integer, show that $m^3 + 6m^2 + 8m$ is the product of three consecutive even integers.

76 **Price per Unit** The total revenue generated by selling $4t + 5$ units is $36t^2 + 17t - 35$ dollars. Determine the price per unit.

77 **Cost per Item** The total cost of producing items over a period of t hours is given by $10t^2 - t - 3$ dollars. The number of items produced during this same period is given by $5t - 3$. Determine the cost per item during this time period.

DISCUSSION QUESTION

78 Describe why the test for factoring trinomials of the form $ax^2 + bx + c$ can be used to show that the binomial $x^2 + k^2$ is prime.

CHALLENGE QUESTION

Describe how you can use a graphics calculator and the graph of $y = x^2 + bx + c$ to determine whether $x^2 + bx + c$ is prime over the real numbers. $x^2 + bx + c$ is prime over the real numbers if it will not factor as $(x - r_1)(x - r_2)$ for real numbers r_1 and r_2.

SECTION SKILL REINFORCERS

Write each of these numbers in scientific notation.

1 12,300 **2** 789,000 **3** 0.078 **4** 0.0032 **5** 0.000071

5-4 Factoring by Grouping and a General Factoring Strategy

SECTION OBJECTIVES

4 Factor polynomials by the method of grouping.

5 Determine the most appropriate technique for factoring a polynomial.

The factoring techniques developed in the previous sections are used primarily with binomials and trinomials. For polynomials with four or more terms a technique called **factoring by grouping** is often used. Instead of examining the entire polynomial at one time, we first group together terms that we already know how to factor. Thus factoring by grouping relies heavily on the ability to spot common factors and special forms.

EXAMPLE 1 Factoring by Grouping

Factor $3ac - bc + 6a - 2b$.

SOLUTION

$$\begin{aligned}3ac - bc + 6a - 2b &= (3ac - bc) + (6a - 2b)\\ &= c(3a - b) + 2(3a - b)\\ &= (3a - b)(c + 2)\end{aligned}$$

Group together the first two terms, with a GCF of c, and the last two terms, with a GCF of 2. Then factor $3a - b$ out of both terms.

EXAMPLE 2 Factoring by Grouping

Factor $x^2 - y^2 + x - y$.

SOLUTION

$$\begin{aligned}x^2 - y^2 + x - y &= (x^2 - y^2) + (x - y)\\ &= (x - y)(x + y) + (x - y)(1)\\ &= (x - y)(x + y + 1)\end{aligned}$$

Note that the first two terms form the difference of two squares. Group them together, and factor $x^2 - y^2$. Then factor $x - y$ out of both terms.

Caution: This term of 1 must be included to account for the term $(x - y)(1)$ in the previous step.

The terms of a polynomial can be grouped several ways. It is possible that one grouping will lead to a factorization, whereas other groupings will prove useless. For example, it is tempting to group $z^2 - a^2 + 4ab - 4b^2$ as $(z^2 - a^2) + (4ab - 4b^2)$, which would result in $(z + a)(z - a) + 4b(a - b)$. However, this grouping, which has a GCF of 1, is not useful. A more useful grouping of this polynomial is shown in the next example.

EXAMPLE 3 **Factoring by Grouping**

Factor $z^2 - a^2 + 4ab - 4b^2$.

SOLUTION $z^2 - a^2 + 4ab - 4b^2 = z^2 - (a^2 - 4ab + 4b^2)$ — Group the last three terms together, noting the sign change of each term within the parentheses.

$= z^2 - (a - 2b)^2$ — Then factor the perfect square trinomial within the parentheses.

$= [z + (a - 2b)][z - (a - 2b)]$

$= (z + a - 2b)(z - a + 2b)$ — This polynomial fits the form of the difference of two squares. Factor this special form, and then simplify.

EXAMPLE 4 **Factoring by Grouping**

Factor $y^3 + x + x^3 + y$.

SOLUTION $y^3 + x + x^3 + y = (x^3 + y^3) + (x + y)$ — Group together the sum of the two perfect cubes, and factor this special form.

$= (x + y)(x^2 - xy + y^2) + (x + y)(1)$ — Then factor out the GCF, $x + y$.

$= (x + y)[(x^2 - xy + y^2) + 1]$ — Be sure to include the last term of 1.

$= (x + y)(x^2 - xy + y^2 + 1)$

It is important to practice factoring a variety of polynomials so that you can quickly select the appropriate technique. The exercises at the end of this section contain a mix of problems intended to promote the development of a general factoring strategy. The following box outlines the methods of factoring covered earlier in this chapter.

SELF-CHECK ✓

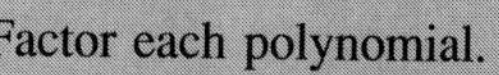

Factor each polynomial.

1 $rt - st + rv - sv$

2 $x^2 - 10x + 25 - 36y^2$

FACTORING A POLYNOMIAL OVER THE INTEGERS

After factoring out the GCF (greatest common factor), proceed as follows.

Binomials: Factor special forms:

$a^2 - b^2 = (a + b)(a - b)$	Difference of two squares
$a^3 - b^3 = (a - b)(a^2 + ab + b^2)$	Difference of two cubes
$a^3 + b^3 = (a + b)(a^2 - ab + b^2)$	Sum of two cubes
$a^2 + b^2$ is prime.	The sum of two squares is prime if a^2 and b^2 are only second-degree terms and have no common factor other than 1.

Trinomials: Factor forms that are perfect squares:

$a^2 + 2ab + b^2 = (a + b)^2$	Square of a sum
$a^2 - 2ab + b^2 = (a - b)^2$	Square of a difference

Otherwise, factor by trial and error or by grouping.

Polynomials of Four or More Terms: Factor by grouping.

SELF-CHECK ANSWERS ✓

1 $(r - s)(t + v)$ 2 $(x + 6y - 5)(x - 6y - 5)$

EXAMPLE 5 Completely Factoring a Binomial

Factor $5x^3y - 5xy^3$.

SOLUTION $5x^3y - 5xy^3 = 5xy(x^2 - y^2)$ Factor out the GCF, $5xy$.

$= 5xy(x - y)(x + y)$ Noting that $x^2 - y^2$ is the difference of two perfect squares, factor this special form.

EXAMPLE 6 Completely Factoring a Binomial

Factor $6a^4b^2 - 6ab^5$.

SOLUTION $6a^4b^2 - 6ab^5 = 6ab^2(a^3 - b^3)$ Factor out the GCF, $6ab^2$.

$= 6ab^2(a - b)(a^2 + ab + b^2)$ Next factor the difference of two perfect cubes.

EXAMPLE 7 Completely Factoring a Polynomial

Factor $3x^3y^2 - 3xy^4 - 3x^2y^2 + 3xy^3$.

SOLUTION $3x^3y^2 - 3xy^4 - 3x^2y^2 + 3xy^3 = 3xy^2(x^2 - y^2 - x + y)$ Factor out the GCF, $3xy^2$.

$= 3xy^2[(x^2 - y^2) - (x - y)]$ Group together the difference of two squares. Note the sign change of each term within the last set of parentheses.

$= 3xy^2[(x - y)(x + y) - (x - y)]$ Factor the difference of two squares as a sum times a difference.

$= 3xy^2(x - y)(x + y - 1)$ Then factor out the common factor, $x - y$. Be sure to include the last term of -1 in the last factor.

The substitution shown in the next example is not necessary and may be skipped. However, the substitution is shown below to emphasize that this polynomial can be considered a trinomial of the form $a^2 - 8ab + 15b^2$. In order to apply the special forms and the methods of factoring presented here, it is important to recognize the variety of expressions that fit these forms.

EXAMPLE 8 Using Substitution to Factor a Polynomial

Factor $(x + 3y)^2 - 8b(x + 3y) + 15b^2$.

SOLUTION

$(x + 3y)^2 - 8b(x + 3y) + 15b^2$ is of the form $a^2 - 8ab + 15b^2$.

$a^2 - 8ab + 15b^2 = (a - ?b)(a - ?b)$ Substitute a for $x + 3y$. Then factor by trial and error, after first writing the form showing the sign pattern.

$a^2 - 8ab + 15b^2 = (a - 3b)(a - 5b)$ Test the possible factors of 15 until you find the correct factors.

$(x + 3y)^2 - 8b(x + 3y) + 15b^2 = [(x + 3y) - 3b][(x + 3y) - 5b]$ Substitute $x + 3y$ back for a. Then simplify.

$= (x + 3y - 3b)(x + 3y - 5b)$

Answer $(x + 3y)^2 - 8b(x + 3y) + 15b^2 = (x + 3y - 3b)(x + 3y - 5b)$ 

A polynomial is prime over the integers if its only factorization is the trivial factorization involving 1 or -1. You should continue all factorizations until each factor, other than a monomial factor, is prime. The polynomials presented in the examples and exercises in this section either are prime or can be factored by the strategy given in this section. Other methods of factoring are usually considered in College Algebra courses.

SELF-CHECK

Factor $7m^3 - 35m^2 - 42m$.

Exercises 5-4

A

In Exercises 1–26 factor each polynomial completely using the technique of grouping.

1 $ac + bc + ad + bd$
2 $xy + xz + 2y + 2z$
3 $3a - 6b + 5ac - 10bc$
4 $6a^2 + 3ab + 2a + b$
5 $x^2 - xy + 5x - 5y$
6 $ac + bc + a + b$
7 $ab + bc - ad - cd$
8 $xy + xz - 2y - 2z$
9 $v^2 - vw - 7v + 7w$
10 $mn - 7m - n + 7$
11 $4a^2 + 12a + 9 - 16b^2$
12 $16x^2 - a^2 - 2a - 1$
13 $3mn + 15m - kn - 5k$
14 $az^2 + bz^2 + aw^2 + bw^2$
15 $x^2 - y^2 + 2y - 1$
16 $x^2 - 5xy - 6y^2 + x - 6y$
17 $x^3 - y^3 - x + y$
18 $x^3 + y^3 + x^2 - y^2$
19 $a^2 + 2a + 1 + ab + b$
20 $x^3 + y^3 - 3x - 3y$
21 $az^3 + bz^3 + aw^2 + bw^2$
22 $s^3 + 11s^2 + s + 11$
23 $9b^2 - 24b + 16 - a^2$
24 $ax - ay - az + bx - by - bz$
25 $ay^2 + 2ay - y + a - 1$
26 $x^2 - 14x + 49 - 16y^2$

B

In Exercises 27–74 completely factor each polynomial using the strategy outlined in this section.

27 $64y^2 - 9z^2$
28 $25a^2 - 144b^2$
29 $16x^2 + 49y^2$
30 $3ax^2 + 33ax + 72a$
31 $12x^2 - 27x + 15$
32 $121m^2 + 49n^2$
33 $49a^2 - 28a + 4$
34 $25a^2 - 10a + 1$
35 $x(a - b) + y(a - b)$
36 $a(x - y) - b(x - y)$
37 $10w^2 - 6w - 21$
38 $3s^2 + 3s + 3t - 3t^2$
39 $25v^2 - vw + 36w^2$
40 $bw^3 - b$
41 $4x^{10} + 12x^5y^3 + 9y^6$
42 $9s^2 - 63$
43 $12x^3y - 12xy^3$
44 $25y^2 - 30yz + 9z^2$
45 $cx + cy + dx + dy$
46 $5a^2bc - 5b^3c$
47 $ax^2 + ax + bxy + by$
48 $20x^3y - 245xy^3$
49 $x^6 + 4x^3y + 4y^2$
50 $3ax^2 + 3ay^2$
51 $4bx^3 - 32b$
52 $100s^4 + 120s^3t + 36s^2t^2$
53 $5x^2 - 55$
54 $(4x^2 - 12xy + 9y^2) + (72ay - 48ax) - 25a^2$
55 $(25x^2 - 10xy + y^2) + (10xz - 2yz) - 24z^2$
56 $63a^3b - 175ab$
57 $49b^2 + 126bc + 81c^2$
58 $5ab^3 - 5a$
59 $-71ax^4 + 71a$
60 $-12x^2 + 12xy - 3y^2$
61 $8x^6 - y^3$
62 $x^3 - y^3 + x - y$
63 $x^2 + 2xy + y^2 - 16z^2$
64 $9x^2 - 6x + 1 - 25y^2$
65 $2x^2 + 2x + 2y - 2y^2$
66 $12ax^2 - 10axy - 12ay^2$
67 $8ax^2 - 648ay^4$
68 $7s^5t - 7st^5$
69 $-6ax^3 + 24ax$
70 $3ax^2 - 3ay^2 + 6ay - 3a$
71 $x^3 + 4x^2y + 4xy^2$
72 $200x^2 + 2$
73 $(a - 2b)x^2 - 2(a - 2b)x - 24(a - 2b)$
74 $36(a - b)x^2 - 6(a - b)xy - 20(a - b)y^2$

C

In Exercises 75–78 factor each polynomial completely. Assume all exponents are natural numbers.

75 $81y^{2n} - 16$
76 $x^{2m} - y^{2n}$
77 $4x^{2m} + 20x^my^n + 25y^{2n}$
78 $10x^{2m} + 4x^my^n - 6y^{2n}$

SELF-CHECK ANSWER

$7m(m + 1)(m - 6)$

79 Factor $2x^{-1} - 10x^{-2} - 28x^{-3}$ by first factoring out $2x^{-3}$.

80 Factor $-45 + 60y^{-1} - 20y^{-2}$ by first factoring out $-5y^{-2}$.

CALCULATOR USAGE (81 AND 82)

In Exercises 81 and 82 use a graphics calculator to graph $y = P(x)$, and then use this graph to factor $P(x)$. (*Hint:* Use the window of values shown.)

WINDOW FORMAT
Xmin = −24
Xmax = 10
Xscl = 2
Ymin = −200
Ymax = 175
Yscl = 25

81 $P(x) = x^2 + 14x - 120$ **82** $P(x) = x^2 + 14x - 72$

DISCUSSION QUESTION

83 Outline a procedure for factoring a polynomial over the integers. In this outline include all of the methods covered in this chapter.

CHALLENGE QUESTION

Factor $2025x^4 + 2700x^2y + 900y^2$ first by factoring this perfect square trinomial as the square of a sum. Then start over and factor the GCF from this trinomial. Complete each of these factorizations, and compare these results, stating which factorization you find easier.

SECTION SKILL REINFORCERS

Solve each of these equations for x.

1 $3x - 5 = 0$ **2** $5x + 3 = 0$ **3** $5(x + 3) = 0$ **4** $3(2x - 3) = 4x + 5$ **5** $3x + xy = 4y + 5$

5-5 *Solving Equations by Factoring*

SECTION OBJECTIVE

6 Use factoring to solve selected second- and third-degree equations.

A **quadratic equation** in x is a second-degree equation that can be written in the **standard form** $ax^2 + bx + c = 0$, where a, b, and c represent real constants and $a \neq 0$. (If we allowed a to equal zero, the equation would not be quadratic; instead it would be the linear equation $bx + c = 0$.)

QUADRATIC EQUATION

If a, b, and c are real constants and $a \neq 0$, then

$$ax^2 + bx + c = 0$$

(ax^2: Quadratic term; bx: Linear term; c: Constant term)

is the standard form of a quadratic equation in x.

A Mathematical Note

The importance of quadratic equations has been noted by many ancient civilizations. A surviving document from Egypt (c. 2000 B.C.) contains quadratic equations. The Hindu mathematician Brahmagupta (c. 628) included a version of the quadratic formula for solving quadratic equations in his works.

EXAMPLE 1 Identifying Quadratic Equations

Determine which of these equations are quadratic equations in one variable. Write each of the quadratic equations in standard form.

SOLUTIONS

(a) $5y - 6 = -y^2$

This equation is a quadratic equation in y and can be written in standard form as

$$y^2 + 5y - 6 = 0$$

In this standard form $a = 1$, $b = 5$, and $c = -6$. It is customary to write the equation so that a is positive.

(b) $5w = 12$

This equation is a linear equation in w; it is not a quadratic equation.

(c) $5v^2 = 3v - 4$

This equation is a quadratic equation in v and can be written in standard form as

$$5v^2 - 3v + 4 = 0$$

The quadratic term is $5v^2$, the linear term is $-3v$, and the constant term is 4.

(d) $x^2 + 7x = x^3 - 5$

This equation is a third-degree equation; it is not a quadratic equation.

(e) $x^2 + 3x + 7$

This is not an equation, just a polynomial expression.

(f) $x^2 = 6x$

This equation is a quadratic equation and can be written in standard form as

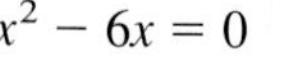

$$x^2 - 6x = 0$$

In standard form, $a = 1$, $b = -6$, and c is understood to be 0.

The easiest method for solving many quadratic equations is to use factoring and the zero-factor principle. (Because some quadratic polynomials are not factorable over the integers, some quadratic equations cannot be solved by factoring over the integers. We will cover methods for solving these quadratic equations in Chapter 8.)

ZERO-FACTOR PRINCIPLE

If a and b are real numbers, then $ab = 0$ if and only if

$a = 0$ or $b = 0$ — The word "or" is used in the inclusive sense to mean $a = 0$ or $b = 0$ (or both a and b are zero).

Proof

Part a If either a or b equals zero, then $ab = 0$ by the Zero-Factor Theorem in Section 1-1.

Part b Assume $ab = 0$. If both a and b are zero, then the statement is obviously true. Thus we will examine the case in which one factor, say a, is not zero. Since $a \neq 0$, $\frac{1}{a}$ exists. Replacing ab with zero, we obtain $\frac{1}{a}(ab) = \frac{1}{a}(0)$. This is equivalent to $\left(\frac{1}{a} \cdot a\right)b = 0$. Hence $(1)b = 0$, so $b = 0$, which is what we wanted to prove.

EXAMPLE 2 Solving a Quadratic Equation in Factored Form

Solve $(v - 2)(v + 3) = 0$.

SOLUTION $(v - 2)(v + 3) = 0$

$v - 2 = 0$ or $v + 3 = 0$ By the zero-factor principle, this is equivalent to stating that either the first factor or the second factor is zero.

$v = 2$ $\quad v = -3$ Notice that the solution set $\{-3, 2\}$ contains two numbers. ▼

SOLVING QUADRATIC EQUATIONS BY FACTORING

Step 1 Write the equation in standard form, with the right member zero.

Step 2 Factor the left member of the equation.

Step 3 Set each factor equal to zero.

Step 4 Solve the resulting first-degree equations.

EXAMPLE 3 Solving a Quadratic Equation Not in Standard Form

Solve $x^2 - 7x + 12 = 2$.

SOLUTION $x^2 - 7x + 12 = 2$

$x^2 - 7x + 10 = 0$ Subtract 2 from each member to put the equation in standard form.

$(x - 2)(x - 5) = 0$ Factor the left member of the equation.

$x - 2 = 0$ or $x - 5 = 0$ Set each factor equal to zero.

$x = 2$ $\quad x = 5$ Solve each of the linear equations.

Check For $x = 2$,

$$(2)^2 - 7(2) + 12 \stackrel{?}{=} 2$$
$$4 - 14 + 12 \stackrel{?}{=} 2$$
$$2 = 2 \text{ checks.}$$

For $x = 5$,

$$(5)^2 - 7(5) + 12 \stackrel{?}{=} 2$$
$$25 - 35 + 12 \stackrel{?}{=} 2$$
$$2 = 2 \text{ checks.}$$

▼

EXAMPLE 4 Solving a Quadratic Equation Containing Fractions

Solve $\frac{1}{8}y^2 - \frac{1}{4}y - 1 = 0$.

SOLUTION $\frac{1}{8}y^2 - \frac{1}{4}y - 1 = 0$

$y^2 - 2y - 8 = 0$ Multiply both sides by the LCD, 8, to obtain integer coefficients.

$(y - 4)(y + 2) = 0$ Factor.

$y - 4 = 0$	or	$y + 2 = 0$	Set each factor equal to zero.
$y = 4$		$y = -2$	Solve each of these linear equations.

EXAMPLE 5 Solving a Quadratic Equation

Solve $(6z + 1)(z - 2) = 8$.

SOLUTION

$$(6z + 1)(z - 2) = 8$$
$$6z^2 - 11z - 2 = 8$$
$$6z^2 - 11z - 10 = 0$$
$$(2z - 5)(3z + 2) = 0$$

Caution: The zero-factor principle works only if the product is zero. The first step is to write the equation in standard form so that the right member is zero. Then factor, using the trial-and-error method.

$2z - 5 = 0$	or	$3z + 2 = 0$	Set each factor equal to zero. Then solve each of the linear equations for z.
$2z = 5$		$3z = -2$	
$z = \frac{5}{2}$		$z = -\frac{2}{3}$	

In Example 6 both solutions of the quadratic equation are the same. In this case we call the root a **double root**, or a **root of multiplicity two.**

SELF-CHECK

Solve each quadratic equation by factoring.

1 $(m - 4)(m + 11) = 0$

2 $y^2 - y = 12$

EXAMPLE 6 Solving a Quadratic Equation with a Double Root

Solve $9x^2 = 30x - 25$.

SOLUTION

$9x^2 = 30x - 25$	
$9x^2 - 30x + 25 = 0$	Write the equation in standard form.
$(3x - 5)^2 = 0$	Factor the perfect square trinomial.

$3x - 5 = 0$	or	$3x - 5 = 0$	Set each factor equal to zero.
$3x = 5$		$3x = 5$	Solve each of the linear equations.
$x = \frac{5}{3}$		$x = \frac{5}{3}$	$\frac{5}{3}$ is a root of multiplicity two.

The zero-factor principle can also be extended to solve equations of higher degree if they can be factored.

EXAMPLE 7 Solving a Cubic Equation by Factoring

Solve $w^3 - w^2 - 12w = 0$.

SOLUTION

$w^3 - w^2 - 12w = 0$	
$w(w^2 - w - 12) = 0$	Factor out the GCF, w.
$w(w - 4)(w + 3) = 0$	Then factor the trinomial by trial and error.

$w = 0$	or	$w - 4 = 0$	or	$w + 3 = 0$	Set each factor equal to zero by using the zero-factor principle.
$w = 0$		$w = 4$		$w = -3$	Solve each of the linear equations. Note that this third-degree equation has three solutions.

SELF-CHECK ANSWERS

1 $m = 4$ or $m = -11$ 2 $y = 4$ or $y = -3$

A Geometric Viewpoint: Estimating Solutions from a Graph

The real solutions of the equation $ax^2 + bx + c = 0$ correspond exactly to the x-intercepts of the graph of $y = ax^2 + bx + c$. The graph of $y = x^2 - x - 6$, as exhibited on the display of a T1-82 calculator, is shown. This graph crosses the x-axis at $x = -2$ and $x = 3$. These x values are also the solutions of $x^2 - x - 6 = 0$, which factors as $(x + 2)(x - 3) = 0$. Graphs of $y = ax^2 + bx + c$ are examined in detail in Section 9-2.

[−6, 6] for x, [−6.5, 1.5] for y

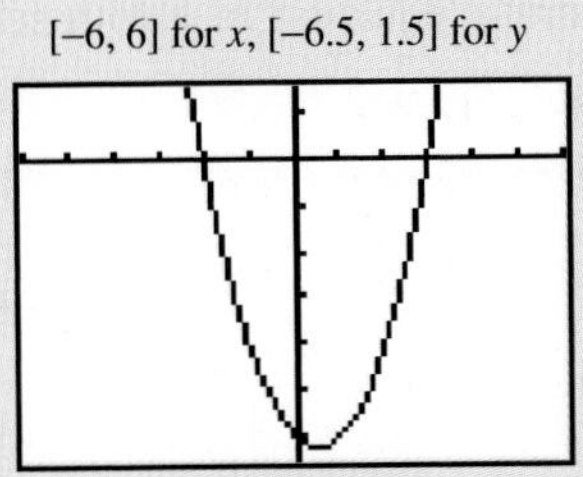

EXAMPLE 8 Calculator Usage

Use the graph of $y = 4x^2 - 4x - 15$ to solve $4x^2 - 4x - 15 = 0$ and to factor $4x^2 - 4x - 15$.

SELF-CHECK

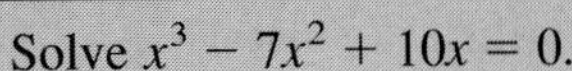

Solve $x^3 - 7x^2 + 10x = 0$.

SOLUTION

Keystrokes

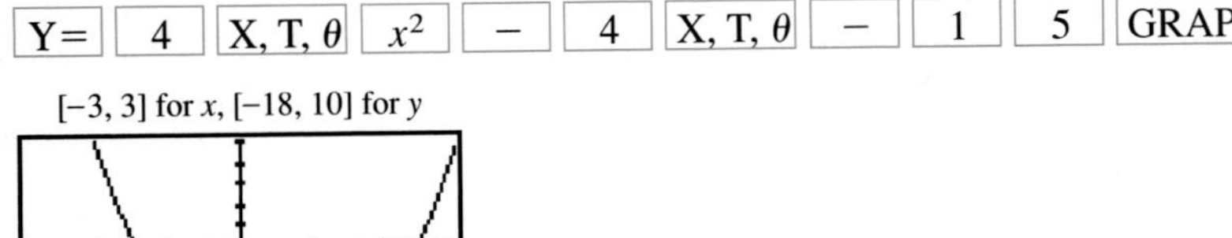

Y= | 4 | X, T, θ | x^2 | − | 4 | X, T, θ | − | 1 | 5 | GRAPH

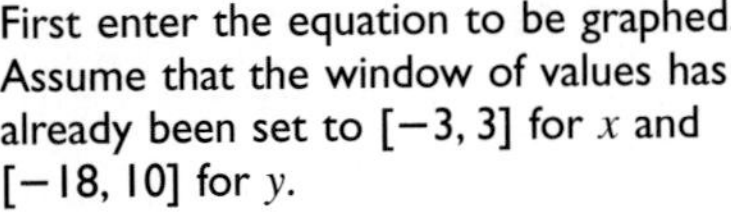

First enter the equation to be graphed. Assume that the window of values has already been set to [−3, 3] for x and [−18, 10] for y.

[−3, 3] for x, [−18, 10] for y

By inspection or by using the TRACE feature, we see that the graph crosses the x-axis at $x = -\frac{3}{2}$ and $x = \frac{5}{2}$. The solutions of $4x^2 - 4x - 15 = 0$ are $-\frac{3}{2}$ and $\frac{5}{2}$.

The x-intercepts correspond exactly to the solutions of the equation.

$$x = -\frac{3}{2} \quad \text{or} \quad x = \frac{5}{2}$$

$$2x = -3 \qquad\qquad 2x = 5$$

$$2x + 3 = 0 \qquad\qquad 2x - 5 = 0$$

$$(2x + 3)(2x - 5) = 0$$

$$4x^2 - 4x - 15 = 0$$

Next use these solutions to produce the factors of $4x^2 - 4x - 15$.

Answer The solutions are $-\frac{3}{2}$ and $\frac{5}{2}$; $4x^2 - 4x - 15 = (2x + 3)(2x - 5)$. ▼

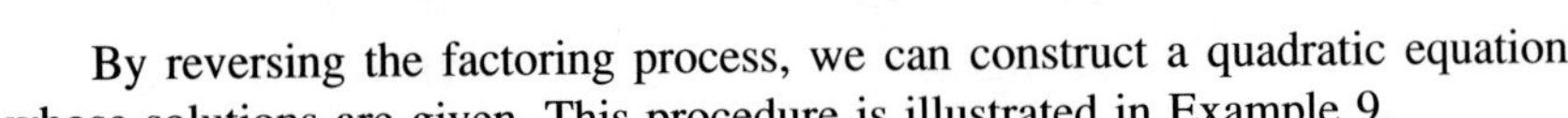

By reversing the factoring process, we can construct a quadratic equation whose solutions are given. This procedure is illustrated in Example 9.

SELF-CHECK ANSWER 

$x = 0$ or $x = 2$ or $x = 5$

EXAMPLE 9 Constructing a Quadratic Equation Whose Solutions Are Given

Construct a quadratic equation in x with solutions of -1 and $\frac{2}{3}$.

SOLUTION $x = -1$ or $x = \frac{2}{3}$ The solutions are -1 and $\frac{2}{3}$.

$x + 1 = 0$ $\quad 3x = 2$

$x + 1 = 0$ $\quad 3x - 2 = 0$ Rewrite the equation so that the right member is zero.

$(x + 1)(3x - 2) = 0$ These factors equal zero, so their product is zero.

$3x^2 + x - 2 = 0$

Answer $3x^2 + x - 2 = 0$ is a quadratic equation with solutions of -1 and $\frac{2}{3}$.

EXAMPLE 10 Solving a Quadratic Equation for a Specified Variable

Solve $x^2 - 3xy - 4y^2 = 0$ for x.

SOLUTION $x^2 - 3xy - 4y^2 = 0$

$(x + y)(x - 4y) = 0$ Factor the left member of the equation.

$x + y = 0$ or $x - 4y = 0$ Using the zero-factor principle, set each factor equal to zero, and then solve for x.

$x = -y$ $\quad x = 4y$

Exercises 5-5

A

In Exercises 1 and 2 identify each choice as a quadratic equation, a linear equation, or neither a quadratic equation nor a linear equation.

1 a. $5x - 3 = 7$ **b.** $5x^2 - 3x = 7$

c. $5x^2 - 3x - 7$ **d.** $\frac{x^2}{x - 1} = x^2$

2 a. $\frac{1}{x^2 - 1} = 2x$ **b.** $\frac{x}{x - 1} = 2x^2$

c. $x^2 - 1 = 2x$ **d.** $-2(x - 3) + 4$

In Exercises 3 and 4 write each quadratic equation in standard form, and identify *a*, *b*, and *c*.

3 a. $2x^2 - 7x + 3 = 0$ **b.** $8x^2 = 3x$

c. $(x - 1)(2x + 1) = 3$

4 a. $3m^2 = 17$ **b.** $4y = 6 - y^2$

c. $(3x + 1)(x - 2) = 3$

In Exercises 5–48 solve each equation.

5 $(m - 8)(m + 17) = 0$ **6** $(m + 15)(m - 3) = 0$

7 $(2n - 5)(3n + 1) = 0$ **8** $(5n + 2)(n - 6) = 0$

9 $(z - 1)(z + 2)(2z - 7) = 0$ **10** $z(z + 6)(6z + 1) = 0$

11 $v^2 - 121 = 0$ **12** $v^2 - 169 = 0$

13 $x^2 + 3x + 2 = 0$ **14** $x^2 + 5x + 4 = 0$

15 $y^2 - 3y = 18$ **16** $t^2 + 4t = 12$

17 $3v^2 = -v$ **18** $5v^2 = v$

19 $2w^2 = 7w + 15$ **20** $3w^2 = 17w + 6$

21 $6x^2 + 19x + 10 = 0$ **22** $10x^2 - 17x + 3 = 0$

23 $y^2 = -13y - 42$ **24** $y^2 = 15y - 54$

25 $9m^2 = 42m - 49$ **26** $25m^2 = 90m - 81$

27 $70z^2 = 5z + 15$ **28** $4x^2 = 9$

29 $9x^2 = 25$ **30** $(r + 6)(r - 1) = -10$

31 $r(r + 3) = 10$ **32** $\frac{m^2}{18} - \frac{m}{6} - 1 = 0$

33 $\frac{t^2}{20} - \frac{t}{4} + \frac{1}{5} = 0$

34 $8v^2 + 24v = 0$

35 $(v - 12)(v + 1) = -40$

36 $(v + 5)(v + 3) = 5v + 25$

37 $(2w - 3)^2 = 25$

38 $(3w + 2)^2 = 16$

39 $(3x - 8)(x + 1) = (x + 1)(x - 3)$

40 $3x(x - 1) = 2x(x + 1) + 6$

41 $(t - 5)(t + 3)(2t + 3) = 0$

42 $(t + 1)(t - 1)(3t + 1) = 0$

43 $v(v^2 - 5v - 24) = 0$

44 $v(v^2 + 14v + 24) = 0$

45 $w(6w^2 + 5w - 6) = 0$

46 $w(5w^2 + w - 4) = 0$

47 $14y^3 = 3y - 19y^2$

48 $10y^3 = 29y^2 - 10y$

In Exercises 49 and 50 construct a quadratic equation in x with the solutions specified.

49 a. -3 and 3 **b.** $\frac{3}{7}$ and -2

c. 0 and -4 **d.** A double root of 3

50 a. -4 and 4 **b.** $-\frac{2}{5}$ and $\frac{5}{2}$

c. 0 and 6 **d.** A double root of -2

B

Exercises 51–54 have two parts. In part a solve the equation, and in part b perform the indicated operations and simplify the result.

	Solve	*Simplify*
51	**a.** $(5m - 3)(m - 2) = 0$	**b.** $(5m - 3)(m - 2)$
52	**a.** $(4x - 3)(5x + 2) = 0$	**b.** $(4x - 3)(5x + 2)$
53	**a.** $4(x - 3)(5x + 2) = 0$	**b.** $4(x - 3)(5x + 2)$
54	**a.** $x(x - 3)(5x + 2) = 0$	**b.** $x(x - 3)(5x + 2)$

ESTIMATION SKILLS (55–58)

55 Use the graph of $y = x^2 - x - 2$ shown to estimate the solutions of $x^2 - x - 2 = 0$. Do these solutions check?

[−6, 6] for x, [−4, 4] for y

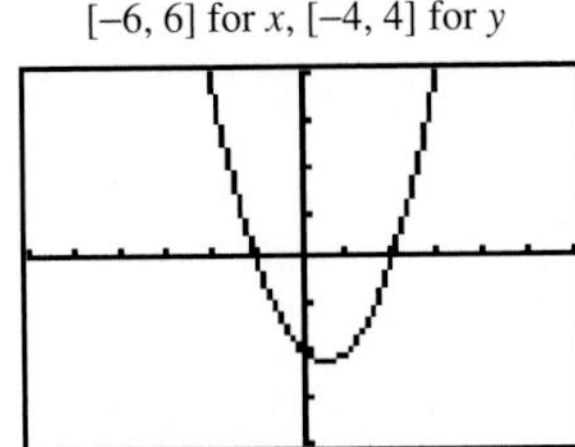

Figure for Exercise 55

56 Use the graph of $y = x^2 + x - 6$ shown to estimate the solutions of $x^2 + x - 6 = 0$. Do these solutions check?

[−9, 9] for x, [−7, 5] for y

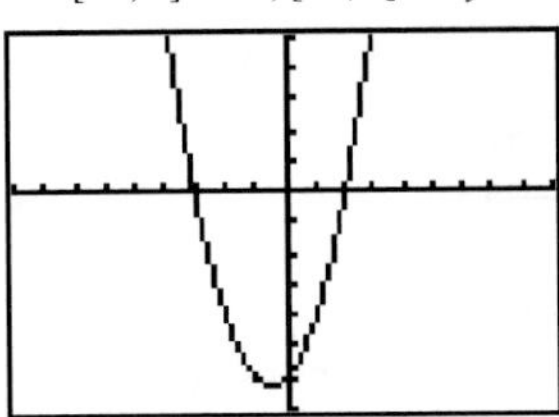

Figure for Exercise 56

57 Use the graph of $y = -x^3 - x^2 + 2x$ shown to estimate the solutions of $-x^3 - x^2 + 2x = 0$. Do these solutions check?

[−6, 6] for x, [−4, 4] for y

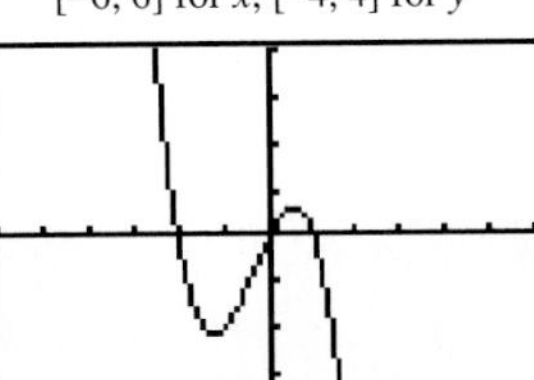

Figure for Exercise 57

58 Use the graph of $y = 4x^3 - 4x^2 - 5x + 3$ shown to estimate the solutions of $4x^3 - 4x^2 - 5x + 3 = 0$. Do these solutions check?

[−2, 2] for x, [−3, 5] for y

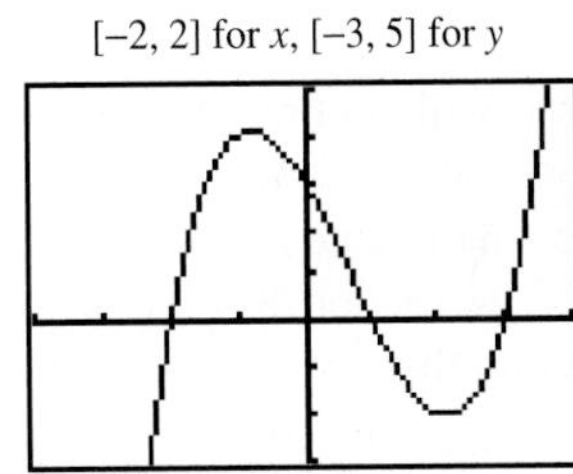

Figure for Exercise 58

C

In Exercises 59–62 construct a cubic (third-degree) equation in x with the solutions specified.

59 0, 1, and 2

60 -1, 0, and 1

61 2, -3, and 4

62 -5, 6, and 10

CALCULATOR USAGE (63–66)

Use a graphics calculator to graph $y = P(x)$ and to estimate the solutions of $P(x) = 0$. Then use the graph to factor $P(x)$.

63 $P(x) = 2x^2 - 15x + 28$

64 $P(x) = 2x^2 + 21x - 36$

65 $P(x) = 2x^2 - 25x + 72$

66 $P(x) = 4x^2 + 4x - 35$

67 Solve $x^2 - 5xy - 6y^2 = 0$ for x. (*Hint:* Factor the left member.)

68 Solve $6x^2 - xy = 2y^2$ for x.

69 Solve $(x')^2 + 2xx' - 3x^2 = 0$ for x'.

70 Solve $(x')^2 + 2xx' - 15x^2 = 0$ for x'.

DISCUSSION QUESTIONS (71 AND 72)

71 Explain in your own words the zero-factor principle.

72 Explain the procedure to follow to solve a quadratic equation by factoring.

SECTION SKILL REINFORCERS

If n represents an odd integer, write an expression to represent each of the following numbers.

1 The next even integer **2** The next odd integer **3** The previous odd integer
4 The previous even integer **5** Five less than three times this integer

5-6 *Applications Yielding Quadratic Equations*

SECTION OBJECTIVES

7 Use factorable quadratic equations to solve applied problems.

8 Use the Pythagorean Theorem to solve applied problems.

In this section we will continue to use the principles and strategy developed in Chapter 1. We will also examine applications of the Pythagorean Theorem. The applications are intentionally limited to problems resulting in factorable quadratic equations. In Section 8-4 we will examine some problems resulting in quadratic equations that are not factorable over the integers. The primary emphasis of this section is not on introducing new material but on continuing the development of your problem-solving skills.

Following is a summary of the strategy developed in Chapter 1:

1 Read the problem carefully to determine what you are being asked to find, and then identify your variables.

2 Apply well-known formulas to the problem.

3 Translate key phrases into algebraic statements.

4 Use tables and sketches to clarify the relationships among variables and to form equations.

5 Use the rate principle and the mixture principle to form equations.

6 Check your solutions.

After solving the equations that you form, always inspect the solutions to see if they are appropriate for the original problem. Check not only for extraneous values but also for answers that may not be meaningful in the application, such as negative lengths.

EXAMPLE 1 Dimensions of a Rectangle

The length of a rectangle is 1 centimeter less than four times the width. Find the dimensions of this rectangle if the area is 60 cm^2.

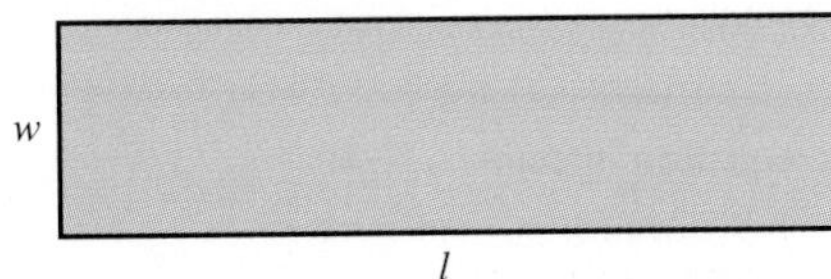

SOLUTION Drawing a sketch as in the figure shown, let

w = Width of the rectangle in centimeters

$4w - 1$ = Length of the rectangle in centimeters

Area	=	Length	·	Width

The **word equation** is based on the formula for the area of a rectangle.

$60 = (4w - 1)w$ — Substitute the area, the length, and the width into the word equation.

$60 = 4w^2 - w$ — Simplify, and rewrite the quadratic equation in standard form.

$4w^2 - w - 60 = 0$

$(4w + 15)(w - 4) = 0$ — Factor the left member by the trial-and-error method or by the grouping method.

$4w = -15$ or $w - 4 = 0$ — Set each factor equal to zero.

$w = -\frac{15}{4}$ $\qquad w = 4$ — Solve the linear equations for the width, and then determine the length.

This negative value is not a meaningful answer for this problem.

$4w - 1 = 4(4) - 1$

$4w - 1 = 15$

Answer The rectangle has a width of 4 cm and a length of 15 cm.

Do these dimensions check?

EXAMPLE 2 An Application to Construction

A string 140 feet long is used to outline the foundation of a rectangular house with an area of 1200 square feet. Determine the dimensions of this house.

SOLUTION Drawing a sketch as in the figure shown, let

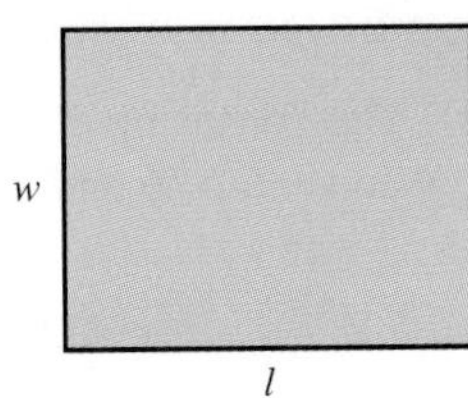

w = Width of the house in feet

l = Length of the house in feet

The representation for the length is developed next.

$P = 2w + 2l$ — Formula for the perimeter of a rectangle

$140 = 2w + 2l$ — Substitute the given perimeter.

$70 = w + l$ — Divide both members by 2.

$l = 70 - w$ — Solve for the length, l.

Thus w equals the width of the house in feet and $70 - w$ equals the length of the house in feet.

Area	=	Length	·	Width

The **word equation** is based on the formula for the area of a rectangle.

$1200 = (70 - w)(w)$ — Substitute the area, the length, and the width into the word equation.

$1200 = 70w - w^2$ — Simplify.

$w^2 - 70w + 1200 = 0$ — Rewrite the quadratic equation in standard form.

$(w - 30)(w - 40) = 0$ — Factor by the trial-and-error method or by the grouping method.

$w - 30 = 0$ or $w - 40 = 0$ — Set each factor equal to zero.

$w = 30$ $\quad$ $w = 40$

$70 - w = 40$ $\quad$ $70 - w = 30$ — The dimensions are either 30 ft × 40 ft or 40 ft × 30 ft—in either case a rectangle with one side 30 ft and the other side 40 ft.

Answer One side of the house is 30 ft and the other side is 40 ft.

SELF-CHECK

The length of a rectangle is 3 meters more than twice the width. Find the dimensions of this rectangle if the area is 65 square meters.

EXAMPLE 3 Consecutive Even Integers

Find two consecutive even integers whose product is 48.

SOLUTION Let

n = Smaller of the two even integers

$n + 2$ = Larger of the two even integers

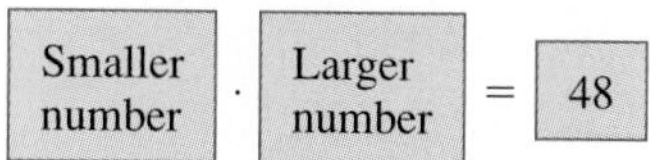

Word equation

$n(n + 2) = 48$ — Form the algebraic equation by substituting into the word equation.

$n^2 + 2n - 48 = 0$ — Write the quadratic equation in standard form.

$(n + 8)(n - 6) = 0$ — Factor the left member by the trial-and-error method.

$n + 8 = 0$ or $n - 6 = 0$ — Set each factor equal to zero.

$n = -8$ $\quad$ $n = 6$ — Solve the linear equations for the smaller integer.

$n + 2 = -6$ $\quad$ $n + 2 = 8$ — Determine the larger integer.

Answer The consecutive even integers are either −8 and −6 or 6 and 8. — Do these values check?

SELF-CHECK ANSWER

Width = 5 m, length = 13 m

EXAMPLE 4 Consecutive Integers

The sum of the squares of two consecutive integers is 61. Find these integers.

SOLUTION Let

$n =$ Smaller of the two integers

$n + 1 =$ Larger of the two integers

Square of the first integer	+	Square of the second integer	=	61	Word equation

$n^2 + (n + 1)^2 = 61$ — Algebraic equation

$n^2 + n^2 + 2n + 1 = 61$ — Be sure to note the middle term, $2n$, when the binomial is expanded.

$2n^2 + 2n + 1 = 61$

$2n^2 + 2n - 60 = 0$ — Subtract 61 from each member to write the quadratic equation in standard form.

$n^2 + n - 30 = 0$ — Divide both members by 2.

$(n + 6)(n - 5) = 0$ — Factor.

$n + 6 = 0$ or $n - 5 = 0$ — Set each factor equal to zero.

$n = -6$ $n = 5$ — Solve for the first integer.

$n + 1 = -5$ $n + 1 = 6$ — Solve for the second integer.

Answer The consecutive integers are either -6 and -5 or 5 and 6. — Do these values check?

SELF-CHECK 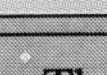

The sum of two numbers is 13, and their product is 40. Find these numbers.

One of the theorems most often used in mathematics is the Pythagorean Theorem, which states an important relationship among the sides of a right triangle. A **right triangle** is a triangle containing a 90° angle; that is, two sides of the triangle are perpendicular.

PYTHAGOREAN THEOREM

If triangle ABC is a right triangle, then

$$a^2 + b^2 = c^2$$

(The converse of this theorem is also true. If $a^2 + b^2 = c^2$, then triangle ABC is a right triangle.)

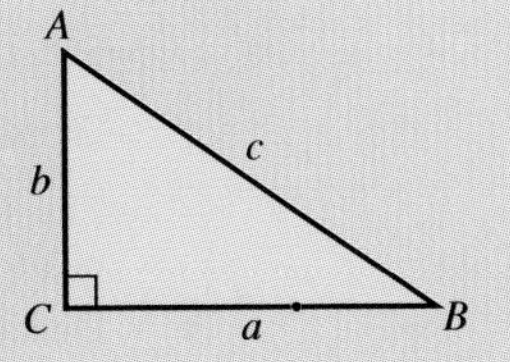

A Mathematical Note

The Greek mathematician Pythagoras (c. 500 B.C.) taught orally and required secrecy of his initiates. Thus records of the society formed by Pythagoras are anecdotal. His society produced a theory of numbers that was part science and part numerology. It assigned numbers to many abstract concepts—for example, 1 for reason, 2 for opinion, 3 for harmony, and 4 for justice. The star pentagon was the secret symbol of the Pythagoreans.

SELF-CHECK ANSWER 

5 and 8

A Geometric Viewpoint: The Pythagorean Theorem

Geometrically, the Pythagorean Theorem says that the sum of the areas of the squares formed on the legs of a right triangle is equal to the area of the square formed on the hypotenuse. The figure shown illustrates this for a 3-4-5 right triangle.

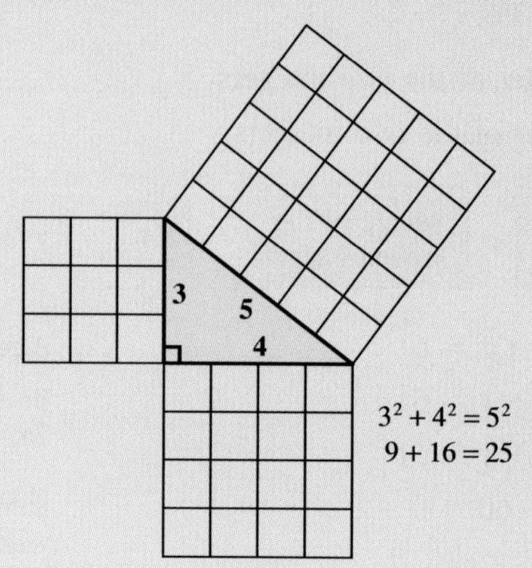

The area of the square formed on the shorter leg is $3^2 = 9$.

The area of the square formed on the longer leg is $4^2 = 16$.

The area of the square formed on the hypotenuse is $5^2 = 25$.

By the Pythagorean formula, $a^2 + b^2 = c^2$; this figure illustrates that $3^2 + 4^2 = 5^2$ since $9 + 16 = 25$.

EXAMPLE 5 An Application of the Pythagorean Theorem

The hypotenuse of a right triangle is 3 centimeters more than twice the length of the shortest side. The longer leg is 3 centimeters less than three times the length of the shortest side. Find the length of each side.

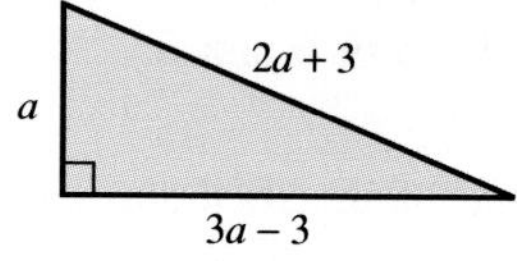

SOLUTION Referring to the figure shown, let

a = Length of the shorter leg in centimeters

$3a - 3$ = Length of the longer leg in centimeters

$2a + 3$ = Length of the hypotenuse in centimeters

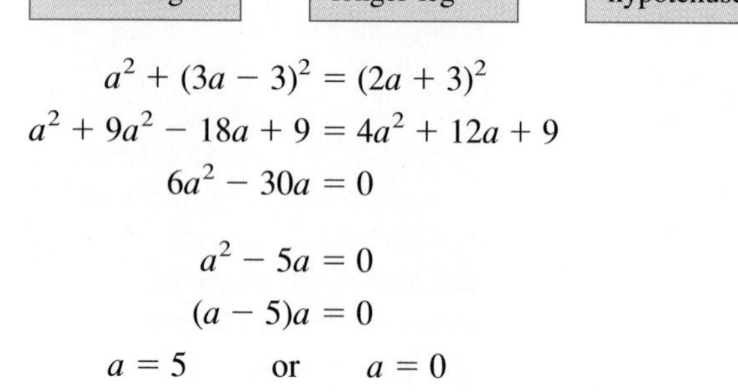

The word equation is based on the Pythagorean Theorem, $a^2 + b^2 = c^2$.

$a^2 + (3a - 3)^2 = (2a + 3)^2$ — Substitute for a, b, and c.

$a^2 + 9a^2 - 18a + 9 = 4a^2 + 12a + 9$ — Simplify.

$6a^2 - 30a = 0$ — Combine like terms, and write the quadratic equation in standard form.

$a^2 - 5a = 0$ — Divide both sides by 6.

$(a - 5)a = 0$ — Factor.

$a = 5$ or $a = 0$ — Set each factor equal to zero.

$3a - 3 = 12$

$2a + 3 = 13$

A length of zero would not yield a triangle and is not a meaningful answer for this problem.

Answer The legs are 5 cm and 12 cm, and the hypotenuse is 13 cm.

EXAMPLE 6 An Application of the Pythagorean Theorem

Two airplanes depart simultaneously from an airport. One flies due south; the other flies due east at a rate 50 miles per hour faster than that of the first airplane. After 2 hours radar indicates that the airplanes are 500 miles apart. What is the ground speed of each airplane?

SELF-CHECK

The length of a rectangle is 3 meters more than three times the width. A diagonal of the rectangle is 1 meter more than the length. Find the dimensions of the rectangle. (*Hint:* Use the Pythagorean Theorem.)

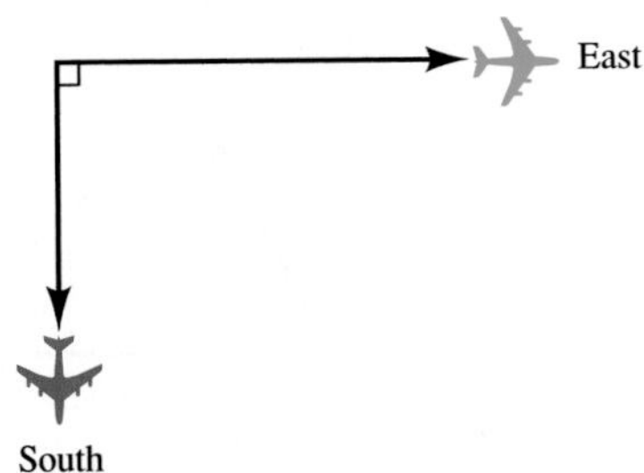

SOLUTION Drawing a sketch as in the figure shown, let

r = Ground speed of the first plane in miles per hour

$r + 50$ = Ground speed of the second plane in miles per hour

Rate principle:	Rate · Time = Distance
First plane	$r \cdot 2 = 2r$
Second plane	$(r + 50) \cdot 2 = 2r + 100$

Use the rate principle and a table to organize the information for each plane.

$$\left[\text{Distance first plane travels}\right]^2 + \left[\text{Distance second plane travels}\right]^2 = \left[\text{Distance between the two planes}\right]^2$$

The word equation is based on the Pythagorean Theorem.

$$(2r)^2 + (2r + 100)^2 = (500)^2$$

Substitute the distances into the word equation.

$$4r^2 + (4r^2 + 400r + 10{,}000) = 250{,}000$$

Simplify.

$$8r^2 + 400r - 240{,}000 = 0$$

Write the equation in standard form.

$$r^2 + 50r - 30{,}000 = 0$$

Divide both members by 8.

$$(r - 150)(r + 200) = 0$$

Factor by the trial-and-error method.

$$r - 150 = 0 \quad \text{or} \quad r + 200 = 0$$

Then set each factor equal to zero.

$$r = 150 \qquad r = -200$$

$$r + 50 = 200$$

A negative rate is not meaningful for this problem.

Answer The first plane is flying south at 150 mi/h, and the second is flying east at 200 mi/h.

Do these rates check?

SELF-CHECK ANSWER 

The width is 7 m, and the length is 24 m.

Exercises 5-6

In Exercises 1–19 solve each problem.

Numeric Word Problems (1–12)

1 Find two consecutive integers whose product is 156.

2 Find two consecutive integers whose product is 240.

3 Find two consecutive even integers whose product is 288.

4 Find two consecutive odd integers whose product is 99.

5 The sum of two numbers is 10, and their product is -75. Find these numbers.

6 The difference of two numbers is 5, and their product is 66. Find these numbers.

7 The difference of two numbers is 2, and their product is $\frac{5}{4}$. Find these numbers.

8 The sum of two numbers is 4, and their product is $\frac{15}{4}$. Find these numbers.

9 The sum of the squares of two consecutive integers is 113. Find these integers.

10 The sum of the squares of three consecutive integers is 50. Find these integers.

11 One number is 4 more than three times another number. Find these numbers if their product is 175.

12 One number is 3 less than four times another number. Find these numbers if their product is 10.

Dimensions of a Rectangle (13–15)

13 The length of a rectangle (see the figure) is 2 centimeters more than three times the width. Find the dimensions of this rectangle if the area is 21 square centimeters.

Figure for Exercises 13–15

14 The length of a rectangle (see the figure) is 10 meters less than twice the width. Find the dimensions of this rectangle if the area is 72 square centimeters.

15 A rectangular pad (see the figure) is outlined by a 44-meter stripe of paint. Find the dimensions of this pad if its surface area is 96 square meters.

Dimensions of a Triangle (16–19)

16 The base of a triangle (see the figure) is 3 meters longer than the height. Find the base if the area of this triangle is 77 square meters.

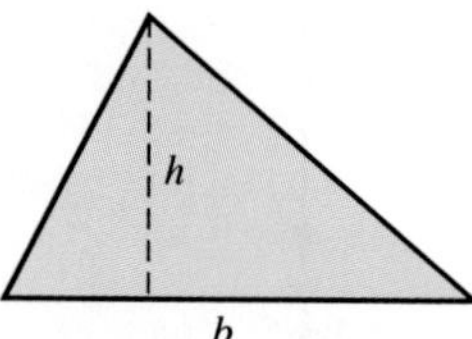

Figure for Exercise 16

17 The base of the triangle shown in the figure is 5 centimeters longer than the height. Determine the height of the triangle if its area is 12 square centimeters.

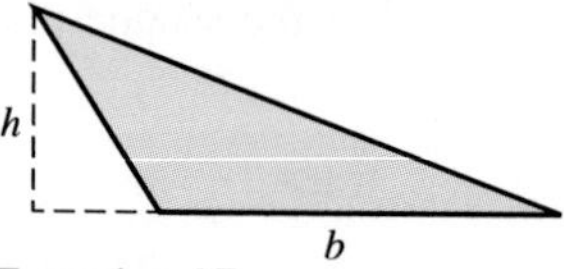

Figure for Exercise 17

18 The hypotenuse of the right triangle shown in the figure is 1 centimeter less than twice the length of the shortest side. The longer leg is 1 centimeter longer than the length of the shortest side. Find the length of each side.

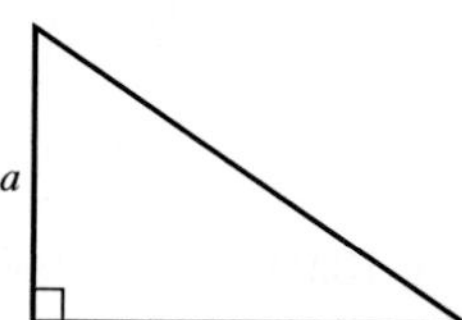

Figure for Exercises 18 and 19

19 The hypotenuse of the right triangle shown in the figure is 2 centimeters less than three times the length of the shortest side. The longer leg is 2 centimeters longer than twice the length of the shortest side. Find the length of each side.

In Exercises 20–29 solve each problem.

Altitude of a Parallelogram (20 and 21)

20 The base of a parallelogram (see the figure) is 6 centimeters more than the altitude. If the area of this parallelogram is 16 square centimeters, determine the altitude.

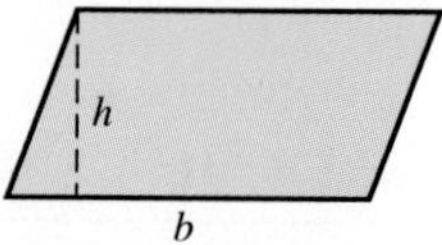

Figure for Exercises 20 and 21

21 The base of a parallelogram (see the figure) is 2 centimeters less than twice the altitude. If the area of this parallelogram is 24 square centimeters, determine the altitude.

22 Length of a Square If each side of a square is increased by 2 centimeters, the total area of both the new square and the original square will be 34 square centimeters. What is the length of each side of the original square? (See the figure.)

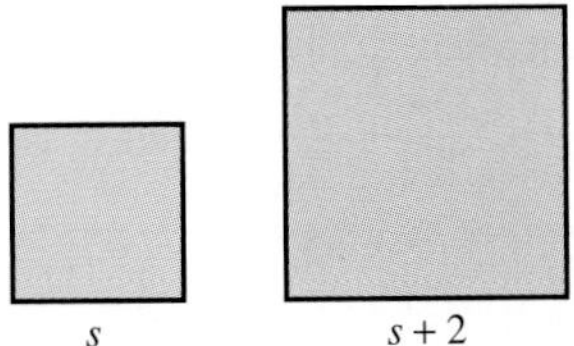

Figure for Exercise 22

23 Diameter of a Storage Bin The length of the diagonal brace in the cylindrical storage bin shown in the figure is 17 meters. If the height of the cylindrical portion of the bin is 7 meters more than the diameter, determine the diameter.

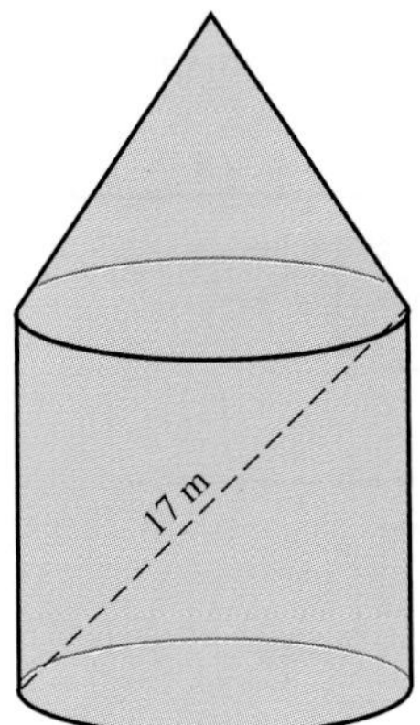

Figure for Exercise 23

Distance by Plane (24 and 25)

24 Upon leaving an airport, an airplane flew due south and then due east. After it had flown 17 miles farther east than it had flown south, it was 25 miles from the airport. How far south had it flown? (See the figure.)

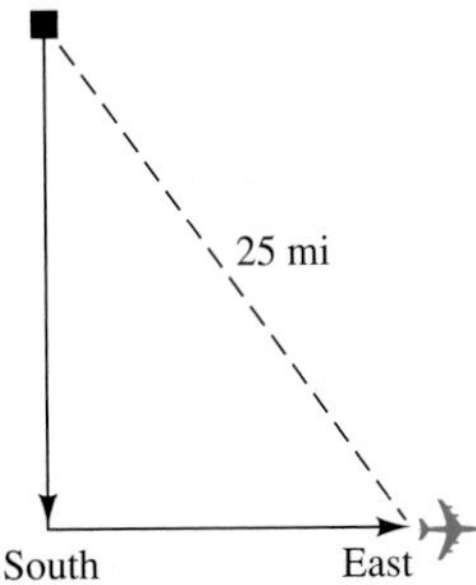

Figure for Exercise 24

25 Upon leaving an airport, an airplane flew due west and then due north. After it had flown 1 mile farther north than it had flown west, it was 29 miles from the airport. How far west had it flown? (See the figure.)

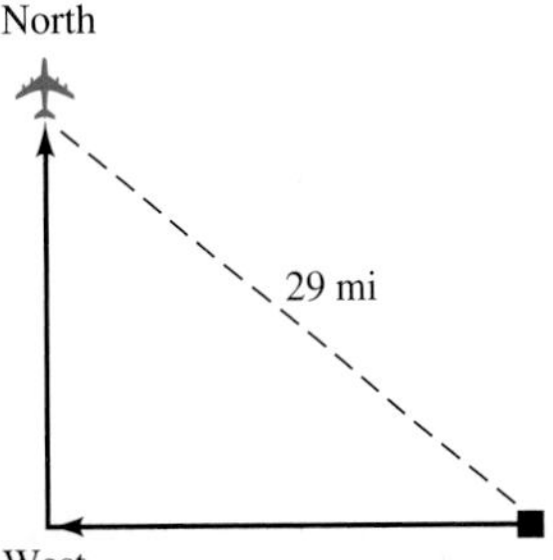

Figure for Exercise 25

26 Position of a Ladder A 17-foot ladder is leaning against a wall. The vertical distance from the bottom of the wall to the top of the ladder is 7 feet longer than the horizontal distance from the bottom of the wall to the bottom of the ladder. How far is the bottom of the ladder from the bottom of the wall? (See the figure.)

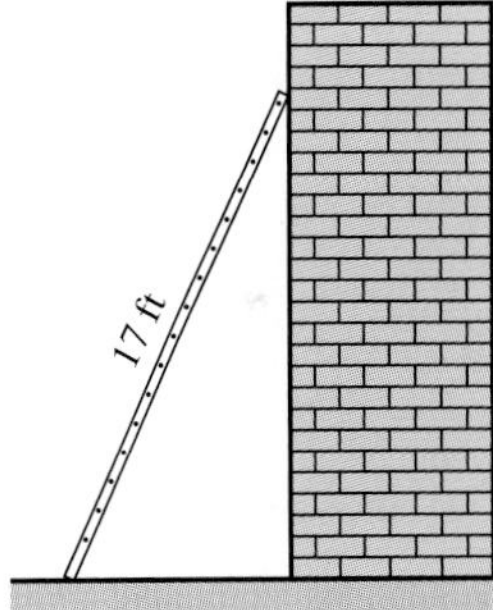

Figure for Exercise 26

27 Room Width Examining the blueprints for a rectangular room that is 17 feet longer than it is wide, an electrician determines that a wire run diagonally across this room will be 53 feet long. What is the width of the room? (See the figure.)

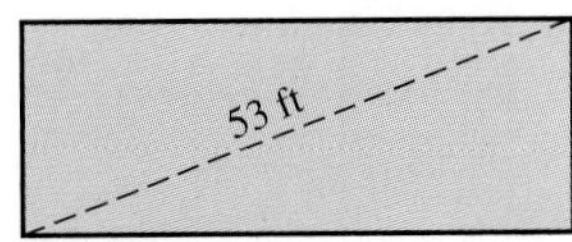

Figure for Exercise 27

28 Ground Speed of Planes Two airplanes depart simultaneously from an airport. One flies due south; the other flies due east at a rate 30 miles per hour faster than that of the first airplane. After 3 hours, radar indicates that the airplanes are 450 miles apart. What is the ground speed of each airplane? (See the figure.)

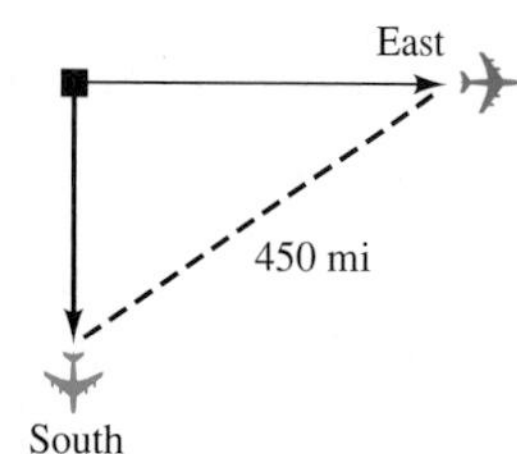

Figure for Exercise 28

29 Speed of an Airplane Two airplanes depart simultaneously from an airport. One flies due south; the other flies due east at a rate 10 miles per hour faster than that of the first airplane. After 1 hour radar indicates that the airplanes are 290 miles apart. What is the ground speed of each airplane? (See the figure.)

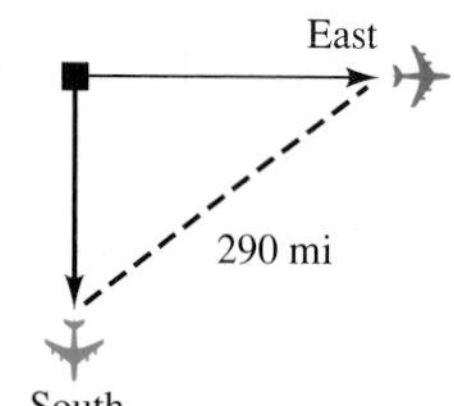

Figure for Exercise 29

C

In Exercises 30–34 solve each problem.

30 Consecutive Integers The sum of the squares of three consecutive integers is 5. Find these integers.

31 Constructing a Box A rectangular sheet of metal is three times as long as it is wide. A 2-centimeter square is cut from each corner of the sheet, and the sides are folded up to form an open box with a volume of 312 cubic centimeters. Find the width of the sheet. (See the figure.)

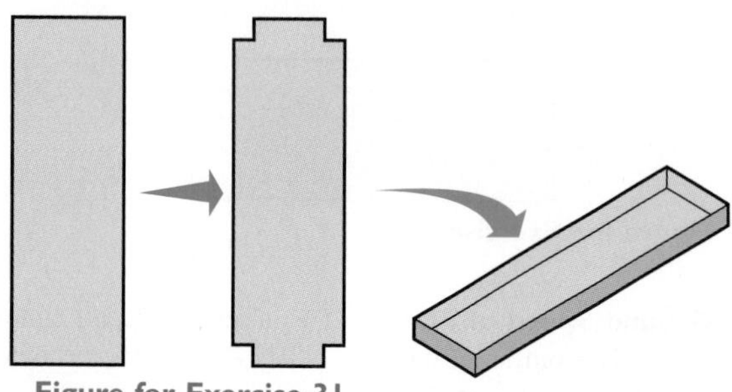
Figure for Exercise 31

Aircraft Drag (32 and 33)

32 For subsonic speeds the drag on a particular airplane is equal to $k(v^2 - 6v + 19)$, where v is velocity in hundreds of miles per hour and k is a constant. The available thrust is $k(15 - v)$. Determine the minimum and maximum speeds at which the aircraft can fly. (*Hint:* Recall from the application at the beginning of this chapter that these speeds are the points where the thrust and drag curves cross—in other words, where thrust = drag.)

33 For subsonic speeds the drag on a particular airplane is equal to $k\left(v^2 - \frac{35v}{6} + \frac{55}{3}\right)$, where v is velocity in hundreds of miles per hour and k is a constant. The available thrust is $k\left(\frac{v^2}{6} + 10\right)$. Determine the minimum and maximum speeds at which the aircraft can fly.

34 Rope Length The length of one piece of rope is 8 meters less than twice the length of another piece of rope. Each rope is used to enclose a square region. The area of the region enclosed by the longer rope is 279 square meters more than the area of the region enclosed by the shorter rope. Determine the length of the shorter rope. (See the figure.)

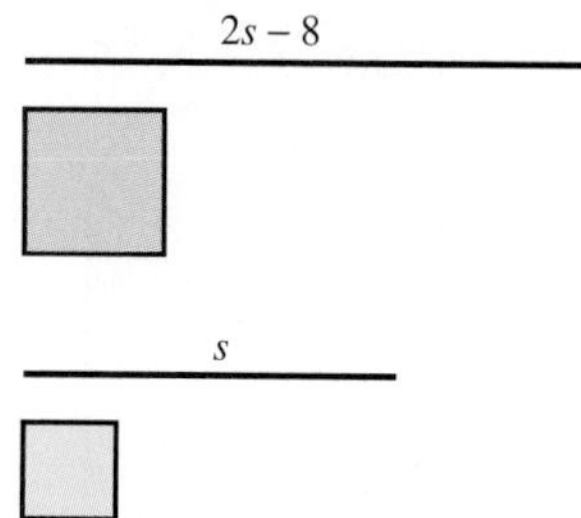

Figure for Exercise 34

35 Radius of a Circle The area enclosed between two concentric circles is 56π square centimeters. The radius of the larger circle is 1 centimeter less than twice the radius of the smaller circle. Determine the length of the shorter radius. (See the figure.)

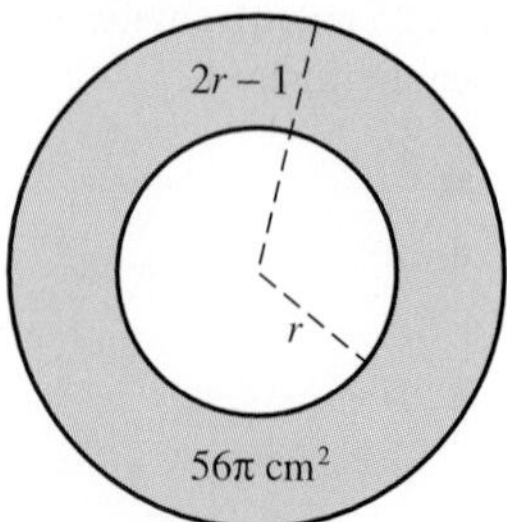

Figure for Exercise 35

36 Dimensions of a Mat A square mat used for athletic exercises has a uniform red border on all four sides. The rest of the mat is blue. The width of the blue square is three-fourths the width of the entire square. If the area colored red is 28 square meters, determine the length of each side of the mat. (See the figure.)

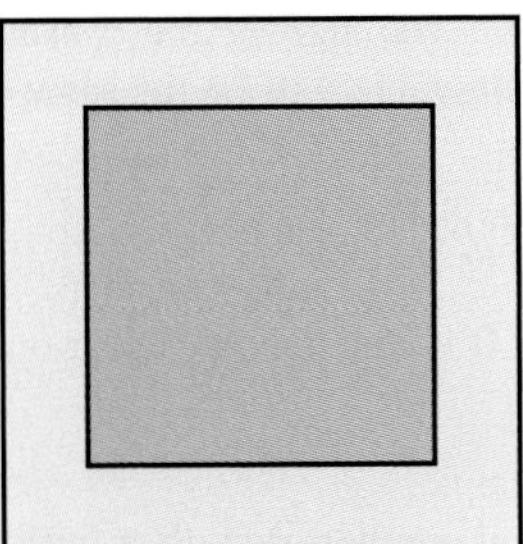

Figure for Exercise 36

DISCUSSION QUESTIONS (37 AND 38)

37 The first twelve exercises of this exercise set can be classified as numeric word problems. These problems are criticized by some because their "practical" application is not apparent. Although these problems do have some practical applications, these applications are somewhat limited. Discuss whether you think it is appropriate to practice your problem-solving skills by working these exercises.

38 Write your own problem that the equation $x^2 + (x + 1)^2 = (x + 9)^2$ algebraically models.

CHALLENGE QUESTION

The square shown in the figure was etched in stone, along with the single word "BEHOLD," by a Hindu mathematician, who meant that this figure offers visual proof of the Pythagorean Theorem. Describe how this square can be used to prove the Pythagorean Theorem.

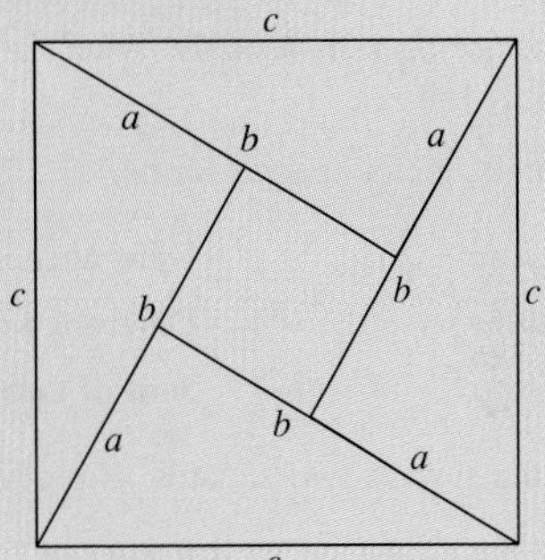

SECTION SKILL REINFORCERS

Factor each of these polynomials completely over the integers.

1 $x^2 - 64$ **2** $x^2 + 64$ **3** $x^3 - 64$ **4** $x^3 + 64$ **5** $x^2 + 16x + 64$

Key Concepts for Chapter 5

1 Prime polynomial:
A polynomial is prime over the integers if its only factorization with integer coefficients is the trivial factorization involving 1 or -1.

2 The GCF of a polynomial:
The greatest common factor of a polynomial is the factor common to each term that has

a. the largest possible numerical coefficient and

b. the largest possible exponent on each variable factor.

3 Strategy for factoring polynomials: After factoring out the GCF, proceed as follows.

a. For binomials, factor special forms:

$a^2 - b^2 = (a + b)(a - b)$	Difference of two squares
$a^3 - b^3 = (a - b)(a^2 + ab + b^2)$	Difference of two cubes
$a^3 + b^3 = (a + b)(a^2 - ab + b^2)$	Sum of two cubes

($a^2 + b^2$, the sum of two squares, is prime.)

b. For trinomials, factor forms that are perfect squares.

$a^2 + 2ab + b^2 = (a + b)^2$	Square of a sum
$a^2 - 2ab + b^2 = (a - b)^2$	Square of a difference

Otherwise, factor by trial and error or by the grouping method.

c. For polynomials of four or more terms, factor by grouping.

4 Sign pattern for the factors of $x^2 + bx + c$:

a. If the constant c is positive, the factors of c must have the same sign. These factors will share the same sign as the linear coefficient, b.

b. If the constant c is negative, the factors of c must be opposite in sign. The sign of the constant factor with the larger absolute value will be the same as the linear coefficient, b.

5 Quadratic equation:
If a, b, and c are real constants and $a \neq 0$, then $ax^2 + bx + c = 0$ is the standard form of a quadratic equation in x.

6 Zero-factor principle:
If a and b are real numbers and $ab = 0$, then $a = 0$ or $b = 0$.

7 Solving quadratic equations by factoring:

a. Write the equation in standard form, with the right member zero.

b. Factor the left member of the equation.

c. Set each factor equal to zero.

d. Solve the resulting first-degree equations.

8 Pythagorean Theorem:
If triangle ABC is a right triangle, then $a^2 + b^2 = c^2$. (The converse of this theorem is also true. If $a^2 + b^2 = c^2$, then triangle ABC is a right triangle.)

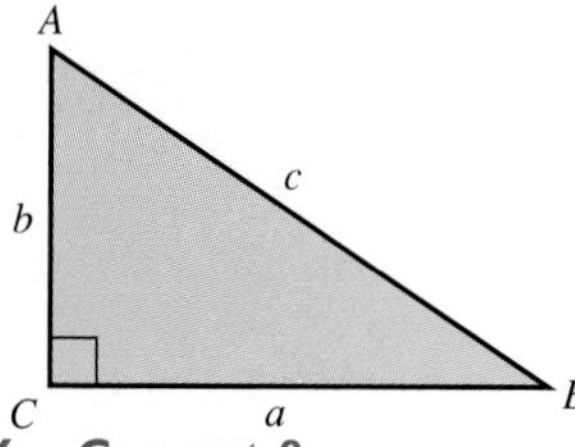

Figure for Key Concept 8

Review Exercises for Chapter 5

In Exercises 1–32 factor each polynomial completely over the integers.

1 $4x - 36$ **2** $2x^2 - 10x$

3 $12ax^2 - 24ax$ **4** $x^2 - 4$

5 $4x^2 - 1$ **6** $x^2 + 4x + 4$

7 $7x^2 - 28$ **8** $m^2 - 64$

9 $x^2 - 11x + 18$ **10** $m^2 + 10m + 21$

11 $ax + bx + ay + by$ **12** $ax + 2ay - 3x - 6y$

13 $x^2 - xy - 42y^2$ **14** $v^2 + 2v + 1 - w^2$

15 $9y^2 + 30y + 25$ **16** $x^2 - 16$

17 $8x^3 - y^3$ **18** $2x^4 - 16x$

19 $11x^2 - 44y^2$ **20** $11x^2 - 11y^2 + 33x + 33y$

21 $4ax^2 - 9ay^2$ **22** $5a(2x - 3y) + 3b(3y - 2x)$

23 $x^3y - 5x^2y + 4xy$ **24** $6x^2 + 61xy + 10y^2$

25 $4x^2 + 12xy + 9y^2$ **26** $x^2(a + b) - 25(a + b)$

27 $a^2(x + 5y) + 2ab(x + 5y) + b^2(x + 5y)$

28 $v^2 + 9w^2$ **29** $20x^2 + 50x - 30$

30 $v^4 - v^3 - v + 1$ **31** $x^{2m} - 1$

32 $x^{3m} - 1$

In Exercises 33–44 solve each equation.

33 $(x - 5)(x + 1) = 0$ **34** $(2x - 3)(3x + 2) = 0$

35 $x(x - 7)(7x - 2) = 0$ **36** $v^2 - 4v - 21 = 0$

37 $10y^2 + 13y - 3 = 0$ **38** $6w^2 = 11w + 21$

39 $\frac{v^2}{30} = \frac{v}{15} + \frac{1}{2}$ **40** $(x + 6)(x - 2) = 9$

41 $x(x^2 - 36) = 0$ **42** $2x(x^2 - 10x + 25) = 0$

43 $(3w + 2)(w + 1) = (2w + 3)(w - 2)$

44 $\frac{z^2}{12} = \frac{z + 3}{3}$

In Exercises 45–49 construct a quadratic equation in x with the solutions specified.

45 $-7, 7$ **46** $2, -11$ **47** $\frac{3}{5}, \frac{-1}{2}$ **48** $0, 8$

49 A double root of -2

50 Construct a cubic equation in x with solutions of 0, 5, and 7.

51 Consecutive Integers Find two consecutive integers whose product is 55 more than their sum.

52 Dimensions of a Pen Thirty meters of fencing encloses the rectangular pen shown in the figure. Find the dimensions of the pen.

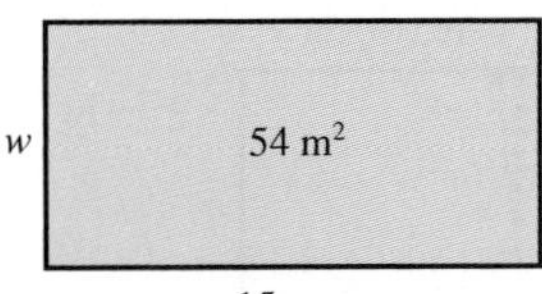

Figure for Exercise 52

53 Dimensions of a Pool As shown in the figure, a rectangular swimming pool with an area of 60 square meters is surrounded by a concrete apron of uniform width. If the distance from one end of the apron to the other is 4 meters longer than the distance from side to side, find the dimensions of the pool. (*Hint:* Let w be the width of the pool; the width of the apron is not needed.)

Figure for Exercise 53

54 Distance by Plane Upon leaving an airport, an airplane flies due west and then due north. After it has flown 31 kilometers farther north than it few west, it is 41 kilometers from the airport. How far west did it fly?

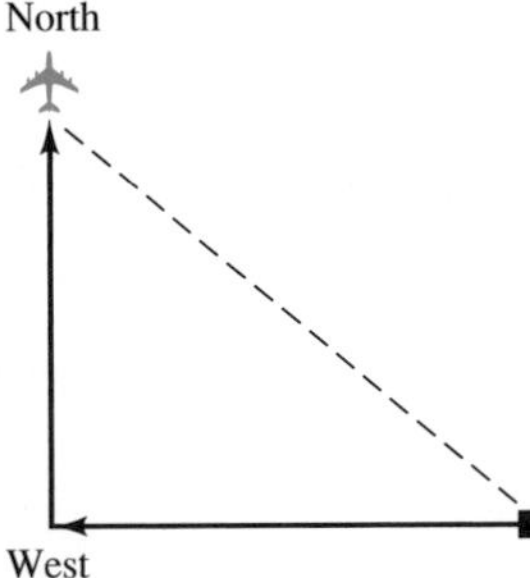

Figure for Exercise 54

55 Revenue The price charged per unit for t units of sales is $400 - t^2$ dollars.

a. Find the polynomial representing the revenue generated by selling t units.

b. Find the values of t that will generate a revenue of 0 dollars.

56 Factor $2x^{-3}$ out of $12x^{-1} - 14x^{-2} - 40x^{-3}$, and then factor this expression completely.

CALCULATOR USAGE (57–66)

In Exercises 57 and 58 use the graph of $y = P(x)$ shown on the graphics calculator display to factor the third-degree polynomial $P(x)$.

57 $[-10, 10]$ for x, $[-10, 10]$ for y

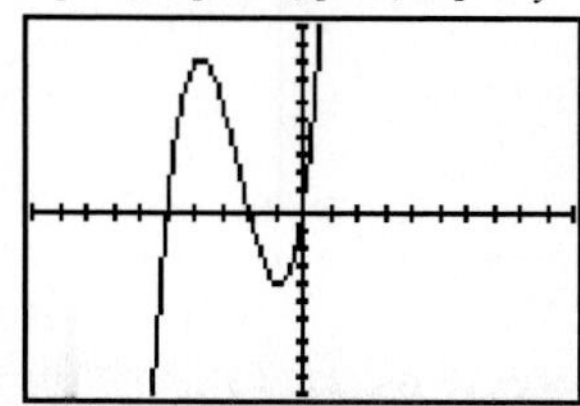

58 $[-10, 10]$ for x, $[-10, 10]$ for y

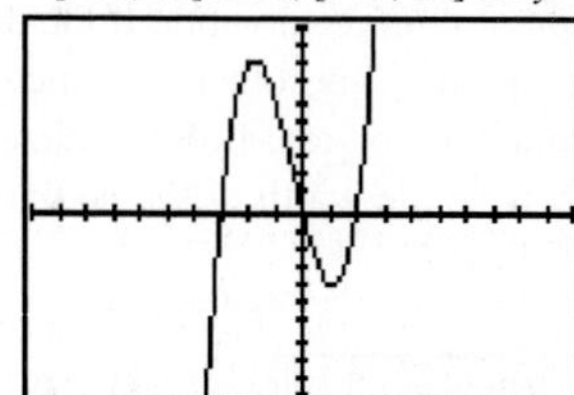

Use a graphics calculator to graph $y = P(x)$. Then use this graph to factor $P(x)$.

59 $P(x) = 2x^3 + 3x^2 - 23x - 12$

60 $P(x) = x^4 + 2x^3 - 9x^2 - 2x + 8$

61 $P(x) = 4x^4 + 4x^3 - 49x^2 - x + 12$

62 $P(x) = 2x^4 - x^3 - 8x^2 + x + 6$

63 Use the graph of $y = 4x^2 - 1$ shown on the graphics calculator window to solve $4x^2 - 1 = 0$.

$[-2, 2]$ for x, $[-2, 4]$ for y

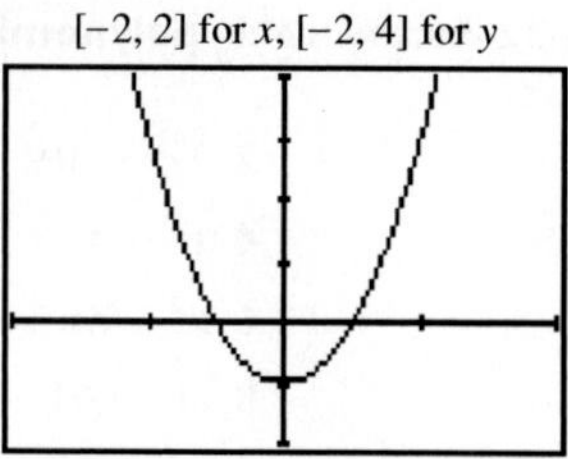

64 Use the graph of $y = x^3 - 4x^2 - x + 4$ shown on the graphics calculator window to solve $x^3 - 4x^2 - x + 4 = 0$.

$[-5, 5]$ for x, $[-9, 9]$ for y

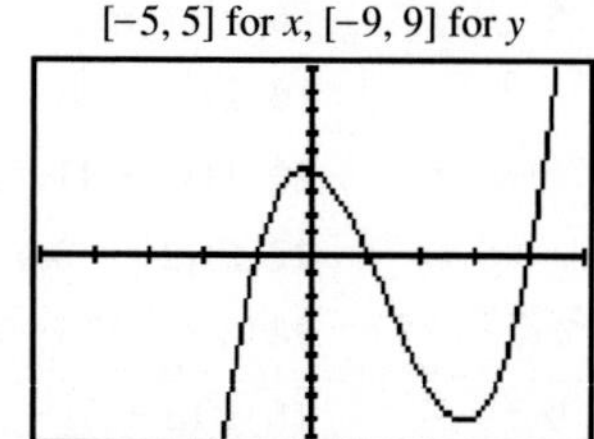

In Exercises 65 and 66 use a graphics calculator to graph $y_1 = P_1(x)$ and $y_2 = P_2(x)$. Then use these graphs to determine whether $P_1(x) = P_2(x)$.

65 $P_1(x) = (1.7x - 2.3)(2.1x - 1.4)$
$P_2(x) = 3.57x^2 - 7.21x + 3.22$

66 $P_1(x) = (1.3x + 2.4)(2.2x - 1.7)$
$P_2(x) = 2.86x^2 - 3.07x - 4.08$

Mastery Test for Chapter 5

Exercise numbers correspond to Section Objective numbers.

[5-1] 1 Factor the greatest common factor from each of these polynomials.

a. $7a - 42$ **b.** $5x^2y - 15y$

c. $x(x - 2y) + 3(x - 2y)$ **d.** $6abc - 12ab$

[5-2] 2 Factor each of these special forms as completely as possible.

a. $x^2 - 4xy + 4y^2$ **b.** $x^2 - 4y^2$

c. $27x^3 - 1$ **d.** $25x^2 + 10x + 1$

[5-3] 3 Factor each of these trinomials as completely as possible.

a. $w^2 - 4w - 45$ **b.** $w^2 + 14w + 45$

c. $x^2 - xy - 12y^2$ **d.** $x^2 - 13xy + 12y^2$

[5-4] 4 Factor each of these polynomials as completely as possible.

a. $2ax + 3a + 2bx + 3b$

b. $14ax - 6bx - 35ay + 15by$

c. $a^2 - 4b^2 + a - 2b$

d. $x^2 + 2xy + y^2 - 4$

[5-4] 5 Factor each of these polynomials as completely as possible.

a. $5x^2 - 20$ **b.** $5x^3 - 5$

c. $2ax^2 + 20ax + 50a$

d. $3bx^2 - 3bxy - 60by^2$

[5-5] 6 Solve each of these equations.

a. $(2x + 1)(x - 3) = 0$ **b.** $x^2 - 2x - 99 = 0$

c. $x^2 - 5x = 84$

d. $(x + 3)(x - 2)(x - 5) = 0$

[5-6] 7 Solve each of these problems.

a. Find two consecutive even integers whose product is 168.

b. The sum of the squares of two consecutive odd integers is 74. Find these integers.

c. The base of the triangle shown in the figure is 5 centimeters longer than the height. Determine the height of this triangle if its area is 63 square centimeters.

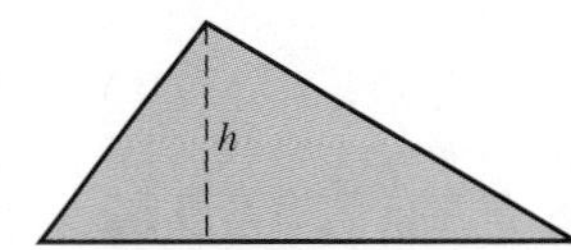

Figure for Exercise 7c

[5-6] 8 Solve each of these problems.

a. The hypotenuse of the right triangle shown in the figure is 4 centimeters less than three times the length of the shortest side. The longer leg is 4 centimeters longer than twice the length of the shortest side. Determine the length of each side of this triangle.

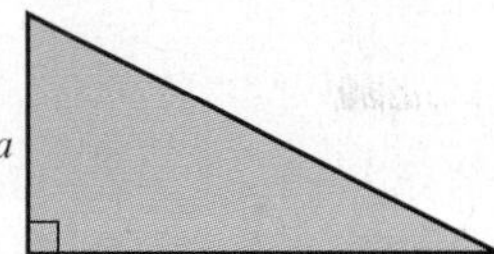

Figure for Exercise 8a

b. A round stock of metal that is $18\sqrt{2}$ centimeters in diameter is milled into a square piece of stock. How long are the sides of the largest square that can be milled? (See the figure.)

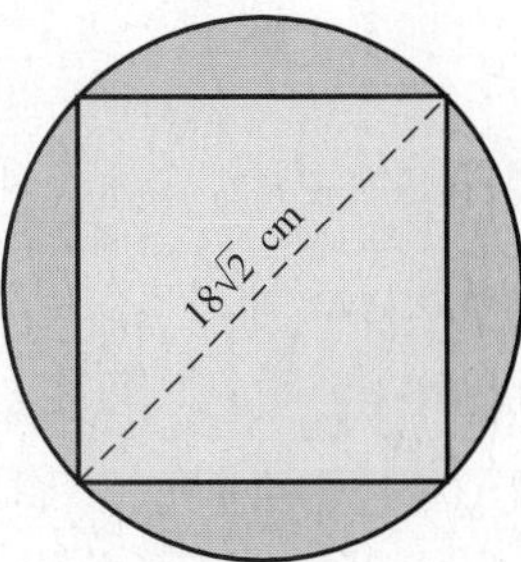

Figure for Exercise 8b

Chapter 6

RATIONAL EXPRESSIONS

AN APPLICATION

When Math Is a Matter of Minutes

At the beginning of the last chapter we discussed the aerodynamics of airplane flight. From a more commercial aspect time is a concern second only to safety in the airline industry. Airline schedules are woven together in an intricate web, and a late flight can have a domino effect on the schedules of both passengers and airlines. A late flight into Chicago may affect several connecting flights out of the city, causing those flights to leave later than planned or to leave with less than full capacity. This, in turn, can cause further delays for other airplanes taking off and landing, costing each passenger and airline time and money.

Occasionally, a plane arrives earlier than scheduled. This may be because the pilot had the advantage of a tailwind. A tailwind saves time without requiring the pilot to increase the plane's air speed—its speed relative to the surrounding air. A headwind, on the other hand, slows the plane down (relative to the ground) and may necessitate increased air speed. Solutions to some of the more complex problems involving air speed require the use of equations containing **rational expressions.** This chapter will introduce you to the algebra of such equations and give you the tools necessary to solve the problems in the active application at the end of the chapter. Let's try an easy problem first. Suppose a plane is about to fly 1851 miles from San Francisco to Chicago. The flight is scheduled to take 4 hours and 4 minutes. If the plane has a 20-mile-per-hour tailwind, what must its air speed be?

CHAPTER 6 OBJECTIVES

1 **Reduce a rational expression to lowest terms (Section 6-1).**

2 **Multiply and divide rational expressions (Section 6-2).**

3 **Add and subtract rational expressions (Section 6-3).**

4 **Simplify rational expressions in which the order of operations must be determined (Section 6-4).**

5 **Simplify complex fractions (Section 6-4).**

6 **Solve equations containing rational expressions (Section 6-5).**

7 **Solve applied problems that yield equations with fractions (Section 6-6).**

A fraction that contains at least one variable in the denominator is frequently called an **algebraic fraction.** A fraction that is the ratio of two polynomials is called a **rational expression.** The following are rational expressions:

$$\frac{4}{3}, \quad \frac{7x}{5y}, \quad \frac{x^2 - 1}{x^3 - 3}, \quad \text{and} \quad \frac{x^2 + 3xy + 2y^2}{x - y}$$

The algebraic fraction $\frac{4}{\sqrt{x}}$ is not a rational expression because the denominator, $\sqrt{x}$, is not a polynomial. Neither is the expression $\frac{3x - 5}{0}$ a rational expression; it is a meaningless expression because division by zero is undefined.

Since division by zero is undefined, we must never permit the variables to assume values that would cause division by zero. Any value that would cause division by zero is called an **excluded value.**

An **excluded value** of an expression is a value of a variable that would cause division by zero in the expression.

The replacement set for the variables in a rational expression is understood to include the set of all real numbers except those excluded because they would result in division by zero. The rational expression $\frac{4x - 7}{x - 5}$, for example, is defined for all real values of x except 5; the value 5 is an excluded value, since 5 would cause division by zero. The excluded values for $\frac{2x + 1}{x^2 - 9}$ are -3 and 3, because either of these values would cause division by zero.

Unless the problem states otherwise, the domain set for variables is assumed to be the set of all real numbers except for the excluded values.

6-1 Reducing Rational Expressions

SECTION OBJECTIVE

1 Reduce a rational expression to lowest terms.

The rules for simplifying rational expressions are identical to those for simplifying rational numbers, which we reviewed in Chapter 1. A rational expression is in lowest terms when the numerator and the denominator have no common factor other than -1 or 1. The basic principle for reducing rational expressions is as follows: If a, b, and c are real algebraic expressions and $b \neq 0$ and $c \neq 0$, then

$$\frac{ac}{bc} = \frac{a}{b}$$

That is, we can divide both the numerator and the denominator of a rational expression by any nonzero factor.

A Mathematical Note

Nobel prizes were established by Alfred Nobel, the inventor of dynamite, because he felt guilty about creating a product that had caused so many injuries and deaths. These prizes are given in chemistry, physics, physiology or medicine, literature, world peace, and economics. Noticeably missing from this list is the field of mathematics. Howard W. Eves, in his book *In Mathematical Circles,* writes that Nobel was greatly antagonized by the mathematician G. M. Mittag-Leffler (1846–1927) and did not award a prize in mathematics because he was concerned that Mittag-Leffler might win such an award.

EXAMPLE 1 Reducing a Rational Expression with a Monomial Denominator

Reduce $\frac{15x^3y^2}{25x^4y^2}$ to lowest terms.

SOLUTION

$$\frac{15x^3y^2}{25x^4y^2} = \frac{5x^3y^2\,(3)}{5x^3y^2\,(5x)}$$

Factor the GCF out of the numerator and the denominator. Note that x and y must both be nonzero to avoid division by zero.

$$= \frac{3}{5x}$$

This is the same answer you would get using the properties of exponents given in Chapter 4.

REDUCING A RATIONAL EXPRESSION TO LOWEST TERMS

Step 1 Factor both the numerator and the denominator of the rational expression.

Step 2 Divide the numerator and the denominator by any common nonzero factors.

EXAMPLE 2 Reducing Rational Expressions with Binomial Denominators

Reduce each of these rational expressions to lowest terms.

SOLUTIONS

(a) $\frac{3x-6}{x^2-2x}$

$$\frac{3x-6}{x^2-2x} = \frac{3(x-2)}{x(x-2)}$$

Factor the GCF, 3, out of the numerator and the GCF, x, out of the denominator. Divide both the numerator and the denominator by the common factor, $x - 2$. We must assume that $x \neq 2$ to avoid division by zero.

$$= \frac{3}{x}$$

(b) $\dfrac{x^2 - y^2}{5x - 5y}$

$$\frac{x^2 - y^2}{5x - 5y} = \frac{(x + y)\cancel{(x - y)}}{5\cancel{(x - y)}} = \frac{x + y}{5}$$

Factor the numerator as the difference of two squares, and factor the GCF, 5, out of the denominator. Then divide both by the common factor, $x - y$. Note that we must assume that $x \neq y$ to avoid division by zero. ▼

SELF-CHECK ✓

Reduce each rational expression to lowest terms.

1 $\dfrac{22a^2b^3}{33a^3b^2}$ **2** $\dfrac{x^2 - 9}{4x - 12}$

EXAMPLE 3 A Rational Expression Already in Reduced Form

Reduce $\dfrac{3x - 7}{3x + 11}$ to lowest terms.

SOLUTION $\dfrac{3x - 7}{3x + 11}$ is already in reduced form, since the greatest common factor of the numerator and the denominator is 1. ▼

A classic error in Example 3 would be to cancel the $3x$ terms in the numerator and the denominator. Always remember to divide *each* term in the numerator and the denominator by the same factor. To divide only some of these terms by a common factor is to commit an error in the order of operations. $\dfrac{3x - 7}{3x + 11}$ does *not* equal $\dfrac{3x}{3x} - \dfrac{7}{11}$. **You can divide both the numerator and the denominator by a common factor, but you *cannot* cancel terms.**

Two polynomials are **opposites** if *every* term of the first polynomial is the opposite of a corresponding term in the second polynomial. The ratio of opposites is -1.

EXAMPLE 4 Reducing a Rational Expression When the Denominator Is the Opposite of the Numerator

Reduce each of these rational expressions to lowest terms.

SOLUTIONS

(a) $\dfrac{2x - 3y}{3y - 2x}$

$$\frac{2x - 3y}{3y - 2x} = \frac{\overset{1}{\cancel{2x - 3y}}}{\underset{-1}{-\cancel{(2x - 3y)}}} = -1$$

The denominator is the opposite of the numerator. The ratio of these opposites is -1.

(b) $\dfrac{-4a + 3b + c}{4a - 3b - c}$

$$\frac{-4a + 3b + c}{4a - 3b - c} = \frac{\overset{-1}{-\cancel{(4a - 3b - c)}}}{\underset{1}{\cancel{4a - 3b - c}}} = -1$$

This expression equals -1, because the numerator and the denominator are opposites.

SELF-CHECK ANSWERS ✓

1 $\dfrac{2b}{3a}$ **2** $\dfrac{x + 3}{4}$

(c) $\dfrac{7r + 3s - t}{-7r + 3s + t}$ $\qquad \dfrac{7r + 3s - t}{-7r + 3s + t}$ is already in reduced form.

The numerator and the denominator are not opposites, because the coefficients of the s terms are not opposites. Again, be careful not to make the classic error of canceling individual terms common to the numerator and the denominator. ▼

EXAMPLE 5 Reducing a Rational Expression by Factoring

Reduce $\dfrac{x^2 - 4y^2}{x^2 - 5xy - 14y^2}$ to lowest terms.

SOLUTION

$$\frac{x^2 - 4y^2}{x^2 - 5xy - 14y^2} = \frac{(x + 2y)(x - 2y)}{(x + 2y)(x - 7y)}$$

Factor the numerator as a difference of two squares. Factor the denominator by trial and error.

$$= \frac{x - 2y}{x - 7y}$$

Then divide both by the common factor, $x + 2y$. ▼

Two rational expressions $R_1(x)$ and $R_2(x)$ are equal only if the graphs of $y_1 = R_1(x)$ and $y_2 = R_2(x)$ completely overlap for all points where both are defined. Whereas we expect the graphs to be identical at all points except excluded values, errors of approximation may cause graphs produced on graphics calculators to differ somewhat near the excluded values. In the next example we will use the graphs of $y_1 = \dfrac{x^2 - 4x - 21}{x - 7}$ and $y_2 = x - 3$ to determine whether $\dfrac{x^2 - 4x - 21}{x - 7}$ reduces to $x - 3$.

EXAMPLE 6 Calculator Usage

Use the graphs of $y_1 = \dfrac{x^2 - 4x - 21}{x - 7}$ and $y_2 = x - 3$ to determine whether $\dfrac{x^2 - 4x - 21}{x - 7}$ equals $x - 3$.

SOLUTION

Keystrokes

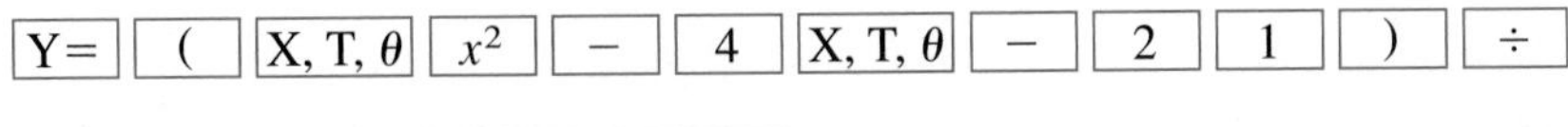

(X, T, θ − 7) ▼

X, T, θ − 3 ZOOM 6

First enter each equation to be graphed. Choosing option 6 in the ZOOM menu selects a window of values of $[-10, 10]$ for x and $[-10, 10]$ for y.

$[-10, 10]$ for x, $[-10, 10]$ for y

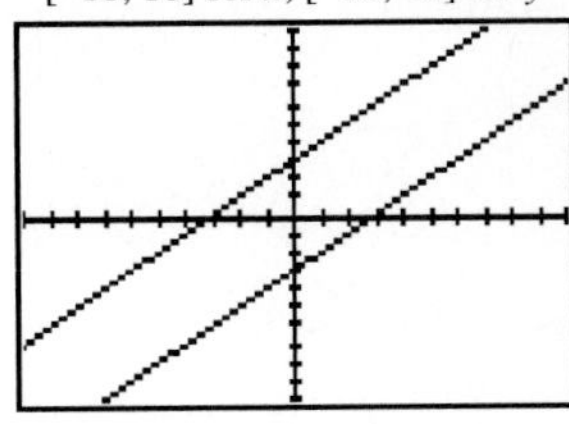

Inspection reveals that the graphs of $y_1 = \dfrac{x^2 - 4x - 21}{x - 7}$ and $y_2 = x - 3$ are different. Therefore $\dfrac{x^2 - 4x - 21}{x - 7} \neq x - 3$. The correct simplification is $\dfrac{x^2 - 4x - 21}{x - 7} = x + 3$. ▼

SELF-CHECK

Reduce each rational expression to lowest terms.

1 $\dfrac{r - 3s}{-r + 3s}$ 2 $\dfrac{5r - 10s}{4s^2 - r^2}$

The fundamental principle of fractions, the property that $\dfrac{ac}{bc} = \dfrac{a}{b}$ for $b \neq 0$ and $c \neq 0$, is used both to reduce fractions to lowest terms and to express fractions so that they have a common denominator. In the next example both the numerator and the denominator are multiplied by the same nonzero value, called the **building factor.** This step is important in adding rational expressions.

EXAMPLE 7 Converting from One Denominator to Another

Determine the missing numerator in $\dfrac{3}{7} = \dfrac{?}{28}$.

SOLUTION

$$\frac{3}{7} = \frac{?}{28}$$

The building factor, 4, can be determined either by examining the factors of 28 or by directly dividing 28 by 7.

$$= \frac{?}{4 \cdot 7}$$

$$\frac{3}{7} = \frac{4}{4} \cdot \frac{3}{7}$$

Multiply $\frac{3}{7}$ by the multiplicative identity, 1, in the form $\frac{4}{4}$.

$$= \frac{12}{28}$$

▼

EXAMPLE 8 Converting from One Denominator to Another

Determine the missing numerator in $\dfrac{5}{3} = \dfrac{?}{6x - 3y}$.

SOLUTION

$$\frac{5}{3} = \frac{?}{6x - 3y}$$

The building factor, $2x - y$, can be determined either by examining the factors of $6x - 3y$ or by directly dividing $6x - 3y$ by 3.

$$= \frac{?}{3(2x - y)}$$

$$\frac{5}{3} = \frac{5}{3} \cdot \frac{2x - y}{2x - y}$$

Multiply $\frac{5}{3}$ by the multiplicative identity, 1, in the form $\frac{2x - y}{2x - y}$.

$$= \frac{10x - 5y}{6x - 3y}$$

▼

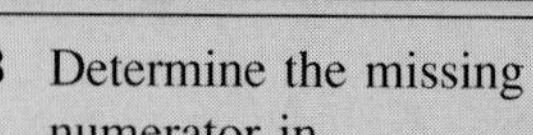

SELF-CHECK

3 Determine the missing numerator in

$$\frac{3x - 2}{2x - 3} = \frac{?}{10x^2 - 7x - 12}$$

SELF-CHECK ANSWERS

1 -1 2 $\dfrac{-5}{r + 2s}$ 3 $\dfrac{3x - 2}{2x - 3} = \dfrac{15x^2 + 2x - 8}{10x^2 - 7x - 12}$

EXAMPLE 9 Determining a Missing Denominator of Equal Fractions

Determine the missing denominator in $\frac{x+3}{x-2} = \frac{x^2+2x-3}{?}$.

SOLUTION

$$\frac{x+3}{x-2} = \frac{x^2+2x-3}{?} = \frac{(x+3)(x-1)}{?}$$

The building factor, $x - 1$, can be determined by examining the factors of $x^2 + 2x - 3$ or by dividing $x^2 + 2x - 3$ by $x + 3$.

$$\frac{x+3}{x-2} = \frac{x+3}{x-2} \cdot \frac{x-1}{x-1} = \frac{x^2+2x-3}{x^2-3x+2}$$

Multiply $\frac{x+3}{x-2}$ by the multiplicative identity, 1, in the form $\frac{x-1}{x-1}$. ▼

As noted earlier, we assume throughout this chapter that the replacement set for each variable excludes all values that would cause division by zero. In analyzing some mathematical expressions, such as the one in the next example, we must be very aware of the excluded values.

EXAMPLE 10 Determining Excluded Values

Determine the excluded values for $\frac{2x^2+7x+8}{3x^2+7x-6}$.

SOLUTION Solve

$3x^2 + 7x - 6 = 0$ — The only excluded values are those that cause the denominator to be 0.

$(3x - 2)(x + 3) = 0$ — Factor the quadratic equation, set the factors equal to 0, and solve the resulting linear equations.

$3x - 2 = 0$ or $x + 3 = 0$

$x = \frac{2}{3}$ $\qquad x = -3$

Answer The excluded values are -3 and $\frac{2}{3}$. — The given rational expression is defined for all other real numbers. ▼

Exercises 6-1

A

In Exercises 1–44 reduce each rational expression to lowest terms. Assume that all values of the variables that cause division by zero are excluded.

1 $\frac{10x^2}{15x^3}$ **2** $\frac{18m^5}{30m^3}$ **3** $\frac{22a^2b^3}{33a^3b}$

4 $\frac{21x^7y^5}{35x^4y^8}$ **5** $\frac{30x^2-45x}{5x}$ **6** $\frac{46x^4-69x^3}{23x^2}$

7 $\frac{7x}{14x^2-21x}$ **8** $\frac{6m^2}{9m^4-15m^3}$

9 $\frac{a^2b(2x-3y)}{-ab^2(2x-3y)}$ **10** $\frac{ab(2x-3y)^2}{ab(2x-3y)^3}$

11 $\frac{7x-8y}{8y-7x}$ **12** $\frac{3a-5b}{5b-3a}$

13 $\frac{ax-ay}{by-bx}$ **14** $\frac{5x-10}{8-4x}$

15 $\frac{(x-2y)(x+5y)}{(x+2y)(x+5y)}$

16 $\frac{(x-4y)(x+4y)}{(x-4y)(x+7y)}$

17 $\frac{x^2-y^2}{3x+3y}$

18 $\frac{v^2-w^2}{7v-7w}$

19 $\frac{25x^2-4}{14-35x}$

20 $\frac{4x^2-9}{9-6x}$

21 $\frac{3x^2-2xy}{3x^2-5xy+2y^2}$

22 $\frac{x^2-2xy+y^2}{x^2-y^2}$

23 $\frac{9x^2-6xy+y^2}{9x^2-y^2}$

24 $\frac{x^2-y^2}{x^3-y^3}$

25 $\frac{4x^2+12xy+9y^2}{14x+21y}$

26 $\frac{x^2+9x+14}{x^2+5x-14}$

27 $\frac{14ax+21a}{35ay+42a}$

28 $\frac{3ax-3ay}{3ax+3ay}$

29 $\frac{ax-y-z}{y+z-ax}$

30 $\frac{x^2-x-3}{x+3-x^2}$

31 $\frac{2a^2-ab-b^2}{a^2-b^2}$

32 $\frac{ax+bx-ay-by}{5x-5y}$

33 $\frac{vx+vy-wx-wy}{v^2-w^2}$

34 $\frac{2x^2-9x-5}{4x^2-1}$

35 $\frac{x^2-25}{3x^2+14x-5}$

36 $\frac{2x^2+xy-3y^2}{3x^2-5xy+2y^2}$

37 $\frac{5a^2+4ab-b^2}{5a^2-6ab+b^2}$

38 $\frac{4a^2-4ab+b^2}{2a^2+ab-b^2}$

39 $\frac{b^2-2b+1-a^2}{5ab-5a+5a^2}$

40 $\frac{a^2-4ab+4b^2-z^2}{5a-10b+5z}$

41 $\frac{x^3-y^3}{-x^2+2xy-y^2}$

42 $\frac{x^4-y^4}{x^2+y^2}$

43 $\frac{12x^2+24xy+12y^2}{16x^2-16y^2}$

44 $\frac{6y^2+11yz-7z^2}{9y^2+42yz+49z^2}$

B

In Exercises 45–54 fill in the missing numerator or denominator.

45 $\frac{7}{144}=\frac{21}{?}$

46 $\frac{35}{36}=\frac{385}{?}$

47 $\frac{7x-8y}{3a-5b}=\frac{?}{10b-6a}$

48 $\frac{11x-13y}{5x+7y}=\frac{65y-55x}{?}$

49 $\frac{a+b}{10}=\frac{a^2-b^2}{?}$

50 $\frac{6}{x-y}=\frac{?}{x^2-y^2}$

51 $\frac{5}{x-6}=\frac{?}{x^2-7x+6}$

52 $\frac{2x-y}{x+3y}=\frac{?}{x^2+5xy+6y^2}$

53 $\frac{2x-y}{x+3y}=\frac{2x^2+xy-y^2}{?}$

54 $\frac{3}{s+t}=\frac{?}{s^3+t^3}$

In Exercises 55–64 reduce each rational expression to lowest terms. Assume that all values of the variables that cause division by zero are excluded.

55 $\frac{y^2-169}{y^2+12y-13}$

56 $\frac{33ay-88ax}{64x^2-9y^2}$

57 $\frac{14x^2-9xy+y^2}{y^2-7xy}$

58 $\frac{(a+b)^2+7(a+b)+6}{(a+b)^2-1}$

59 $\frac{x^2+x+y-y^2}{2x^3+2y^3}$

60 $\frac{a^3+b^3}{(a+b)^3}$

61 $\frac{a^2+2a+1+ab+b}{9a+9b+9}$

62 $\frac{(a-2b)^2-9}{(a-2b)^2-(a-2b)-6}$

63 $\frac{(m-n)^3}{m^3-n^3}$

64 $\frac{10x^2+13xy-3y^2}{5x^2-26xy+5y^2}$

C

CALCULATOR USAGE (65–70)

In Exercises 65–68 use a graphics calculator to graph $y_1 = R_1(x)$ and $y_2 = R_2(x)$. Then use these graphs to determine whether $R_1(x) = R_2(x)$.

65 $R_1(x) = \frac{10.8x^2-15.8x-10.6}{5.4x-10.6}$
$R_2(x) = 2x+1$

66 $R_1(x) = \frac{10.8x^2-15.8x-10.6}{2x+1}$
$R_2(x) = 5.4x-10.6$

67 $R_1(x) = \frac{8.5x^2+9.1x-29.4}{1.7x+4.2}$
$R_2(x) = 5x+7$

68 $R_1(x) = \frac{8.5x^2+9.1x-29.4}{5x-7}$
$R_2(x) = 1.7x-4.2$

In Exercises 69 and 70 use a graphics calculator to graph $y_1 = \frac{1}{x}$ and $y_2 = \frac{1}{x^2}$. Then use these graphs to match these functions with the given broken-line graphs.

69

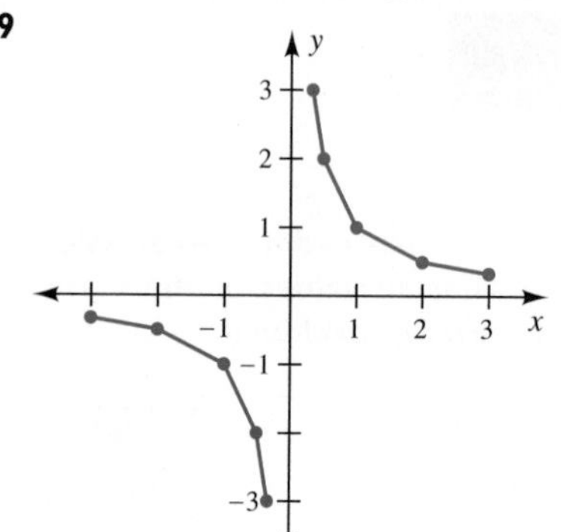

70

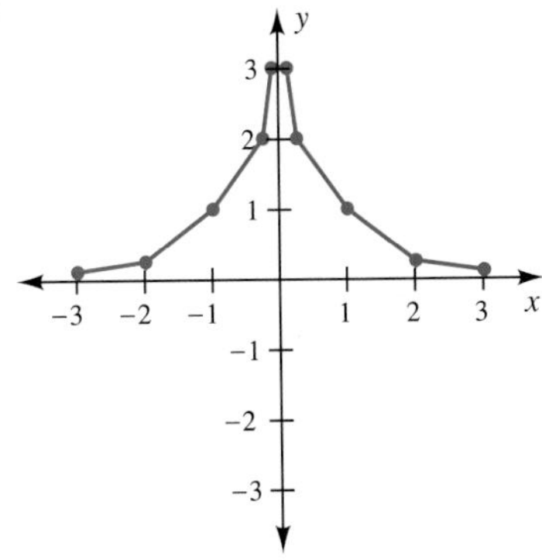

In Exercises 71–76 determine the excluded values for each rational expression.

71 $\dfrac{x}{x+7}$ **72** $\dfrac{2x+1}{2x-1}$

73 $\dfrac{5x-11}{7x-14}$ **74** $\dfrac{3x-7}{x^2-81}$

75 $\dfrac{x^2-6x-7}{x^2+6x+9}$ **76** $\dfrac{4x+19}{6x^2-13x-28}$

In Exercises 77 and 78 reduce each rational expression to lowest terms. Assume that m is a natural number and that all values of the variables that cause division by zero are excluded.

77 $\dfrac{x^{2m}-25}{7x^m+35}$ **78** $\dfrac{x^{m+1}-4x}{x^{2m}-16}$

79 Factor x^{-2} out of both the numerator and the denominator of $\dfrac{2+x^{-1}y-15x^{-2}y^2}{1+6x^{-1}y+9x^{-2}y^2}$, and then reduce the expression to lowest terms.

80 Factor $x^{-1}y^{-1}$ out of both the numerator and the denominator of $\dfrac{4xy^{-1}+12+9x^{-1}y}{6xy^{-1}+5-6x^{-1}y}$, and then reduce the expression to lowest terms.

81 Plant Breakdowns After t years the estimated number of breakdowns at one plant is given by $N = 2.5t + 4.5$. The total cost in thousands of dollars of these breakdowns is estimated to be $T = t^2 + 1.8t$. Determine the average cost of these breakdowns.

Error Analysis (82 and 83)

In Exercises 82 and 83 examine each student's work, and correct any mistakes you find.

82 Student A

$$\frac{x-3}{x+5} = \frac{3}{5}$$

83 Student B

$$\begin{aligned}\frac{x^2-9}{x^2-4x+1} &= \frac{(x-3)(x+3)}{(x-3)(x-1)}\\ &= \frac{x+3}{x-1}\\ &= \frac{3}{-1}\\ &= -3\end{aligned}$$

Discussion Questions (84 and 85)

84 Explain in your own words what an excluded value is.

85 Explain in your own words the procedure for reducing a rational expression to lowest terms.

Challenge Question

Both squares shown in the figure have sides of 2 centimeters. The second square is positioned so that one of its vertices is at the center of the first square. Determine the area of the shaded region enclosed between the squares.

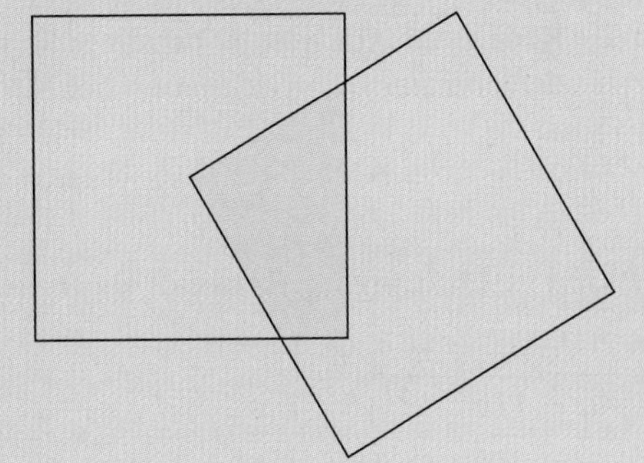

Section Skill Reinforcers

Simplify each of these expressions.

1 $2^{-1} + 2^{-1}$ **2** $(2 + 2)^{-1}$ **3** $\dfrac{14}{15} \div \dfrac{10}{21}$ **4** $\dfrac{14}{15} \div \dfrac{21}{10}$ **5** $\dfrac{2xy}{z^2} \div \dfrac{4x}{3z}$

6-2 Multiplication and Division of Rational Expressions

SECTION OBJECTIVE

2 Multiply and divide rational expressions.

Since the variables in a rational expression represent real numbers, the rules and procedures for performing multiplication and division with rational expressions are the same as those used in Chapter 1 for performing these operations with arithmetic fractions. (Multiplication and division will be covered before addition and subtraction so that you can develop some of the skills needed to add rational expressions with different denominators.) The rule for multiplying real expressions is as follows: If a, b, c, and d are real expressions and $b \neq 0$ and $d \neq 0$, then

$$\frac{a}{b} \cdot \frac{c}{d} = \frac{a \cdot c}{b \cdot d}$$

MULTIPLYING RATIONAL EXPRESSIONS

Step 1 Factor the numerators and the denominators. Then write the product as a single fraction, indicating the product of the numerators and the product of the denominators.

Step 2 Reduce this fraction by dividing the numerator and the denominator by any common nonzero factors.

EXAMPLE 1 Multiplying Rational Expressions

Multiply $\dfrac{5xy}{4a - 12b} \cdot \dfrac{3a - 9b}{15x^2}$.

SOLUTION

$$\frac{5xy}{4a - 12b} \cdot \frac{3a - 9b}{15x^2} = \frac{5xy}{4(a - 3b)} \cdot \frac{3(a - 3b)}{15x^2}$$

Factor the numerators and the denominators. Then write the product as a single fraction.

$$= \frac{\overset{y}{\cancel{5xy}}\cancel{(3)}\cancel{(a - 3b)}}{4\cancel{(a - 3b)}\underset{x}{\cancel{(15x^2)}}}$$

Reduce this fraction by dividing out the common factors, $(a - 3b)$ and $15x$.

$$= \frac{y}{4x}$$

EXAMPLE 2 **Multiplying Rational Expressions**

Multiply $\dfrac{2x-y}{4x^2-4y^2} \cdot \dfrac{x^2+2xy+y^2}{2x^2+xy-y^2}$.

SOLUTION

$$\frac{2x-y}{4x^2-4y^2} \cdot \frac{x^2+2xy+y^2}{2x^2+xy-y^2} = \frac{2x-y}{4(x^2-y^2)} \cdot \frac{(x+y)(x+y)}{(2x-y)(x+y)}$$

Factor the numerators and the denominators. Then write the product as a single fraction.

$$= \frac{\cancel{(2x-y)}\cancel{(x+y)}\cancel{(x+y)}}{4(x-y)\cancel{(x+y)}\cancel{(2x-y)}\cancel{(x+y)}}$$

Reduce this fraction by dividing by the common factors.

$$= \frac{1}{4(x-y)}$$

▼

EXAMPLE 3 **Multiplying Rational Expressions**

Multiply $\dfrac{x^2-y^2}{x^3-y^3} \cdot (x^2+xy+y^2)$.

SOLUTION

$$\frac{x^2-y^2}{x^3-y^3} \cdot (x^2+xy+y^2) = \frac{(x-y)(x+y)}{(x-y)(x^2+xy+y^2)} \cdot \frac{x^2+xy+y^2}{1}$$

Factor the first numerator and the first denominator. Note that the second factor has a denominator of 1.

$$= \frac{\cancel{(x-y)}(x+y)\cancel{(x^2+xy+y^2)}}{\cancel{(x-y)}\cancel{(x^2+xy+y^2)}(1)}$$

Reduce by dividing out the common factors.

$$= \frac{x+y}{1}$$

$$= x+y$$

▼

The rule for dividing rational expressions is the same one used in Chapter 1 for dividing arithmetic fractions: If a, b, c, and d are real expressions and $b \neq 0$, $c \neq 0$, and $d \neq 0$, then

$$\frac{a}{b} \div \frac{c}{d} = \frac{a}{b} \cdot \frac{d}{c} = \frac{a \cdot d}{b \cdot c}$$

SELF-CHECK 

Find each product.

1 $\dfrac{2x-5y}{7xy} \cdot \dfrac{35x^2}{25y-10x}$

2 $\dfrac{10x^2-6x}{7xy-7y} \cdot \dfrac{5x^2-2x-3}{25x^2-9}$

DIVIDING RATIONAL EXPRESSIONS

Step 1 Rewrite the division problem as the product of the dividend and the reciprocal of the divisor.

Step 2 Perform the multiplication using the rule for multiplying rational expressions.

SELF-CHECK ANSWERS 

1 $-\dfrac{x}{y}$ **2** $\dfrac{2x}{7y}$

EXAMPLE 4 Dividing Rational Expressions

Divide $\dfrac{10ab}{x-3y} \div \dfrac{-5a^3}{7x-21y}$.

SOLUTION

$$\frac{10ab}{x-3y} \div \frac{-5a^3}{7x-21y} = \frac{10ab}{x-3y} \cdot \frac{7x-21y}{-5a^3}$$

Rewrite the division problem as the product of the dividend and the reciprocal of the divisor.

$$= \frac{\overset{2b}{\cancel{(10ab)}}(7)\cancel{(x-3y)}}{\cancel{(x-3y)}\underset{-a^2}{\cancel{(-5a^3)}}}$$

Factor, and then reduce by dividing out the common factors.

$$= \frac{14b}{-a^2} = -\frac{14b}{a^2}$$

SELF-CHECK

Divide $\dfrac{10x-15}{x^2-5x} \div \dfrac{14x-21}{10-2x}$.

EXAMPLE 5 Dividing Rational Expressions

Divide $\dfrac{x^2-9}{x^2-25}$ by $\dfrac{x-3}{x+5}$.

SOLUTION

$$\frac{x^2-9}{x^2-25} \div \frac{x-3}{x+5} = \frac{x^2-9}{x^2-25} \cdot \frac{x+5}{x-3}$$

Rewrite the division problem as the product of the dividend and the reciprocal of the divisor.

$$= \frac{(x+3)\cancel{(x-3)}\cancel{(x+5)}}{\cancel{(x+5)}(x-5)\cancel{(x-3)}}$$

Factor, and then reduce by dividing out the common factors.

$$= \frac{x+3}{x-5}$$

EXAMPLE 6 Simplifying a Rational Expression Involving Both Multiplication and Division

Simplify $\dfrac{12x}{x+y} \div \left(\dfrac{5x-5}{x^2-y^2} \cdot \dfrac{3xy}{xy-y}\right)$.

SOLUTION

$$\frac{12x}{x+y} \div \left(\frac{5x-5}{x^2-y^2} \cdot \frac{3xy}{xy-y}\right) = \frac{12x}{x+y} \div \left[\frac{(5x-5)(3xy)}{(x^2-y^2)(xy-y)}\right]$$

The grouping symbols indicate that the multiplication inside the brackets has priority over the division. Multiply the numerators and the denominators within the brackets first.

$$= \frac{12x}{x+y} \cdot \frac{(x^2-y^2)(xy-y)}{(5x-5)(3xy)}$$

Then invert the divisor within the brackets and multiply.

$$= \frac{\overset{4}{\cancel{(12x)}}(x-y)\cancel{(x+y)}\cancel{(y)}\cancel{(x-1)}}{\cancel{(x+y)}(5)\cancel{(x-1)}\underset{\cancel{x}}{\cancel{(3xy)}}}$$

Factor the numerator and the denominator, and then reduce by dividing out the common factors.

$$= \frac{4(x-y)}{5}$$

SELF-CHECK ANSWER

$-\dfrac{10}{7x}$

EXAMPLE 7 **Simplifying a Rational Expression Involving Both Multiplication and Division**

Simplify $\frac{x^2 - 25}{x^2 - x - 12} \div \frac{x^2 - x - 20}{3x - 3} \cdot \frac{x^2 - 16}{x^2 + 4x - 5}$.

SOLUTION

$$\frac{x^2 - 25}{x^2 - x - 12} \div \frac{x^2 - x - 20}{3x - 3} \cdot \frac{x^2 - 16}{x^2 + 4x - 5}$$

Following the correct order of operations from left to right, invert only the fraction in the middle.

$$= \frac{x^2 - 25}{x^2 - x - 12} \cdot \frac{3x - 3}{x^2 - x - 20} \cdot \frac{x^2 - 16}{x^2 + 4x - 5}$$

$$= \frac{\cancel{(x + 5)}\cancel{(x - 5)}(3)\cancel{(x - 1)}\cancel{(x + 4)}\cancel{(x - 4)}}{(x + 3)\cancel{(x - 4)}\cancel{(x - 5)}\cancel{(x + 4)}\cancel{(x + 5)}\cancel{(x - 1)}}$$

Factor, and then reduce by dividing out the common factors.

$$= \frac{3}{x + 3}$$

Exercises 6-2

A

In Exercises 1–30 perform the indicated operations, and reduce all results to lowest terms.

1 $\frac{-3}{14} \div \frac{12}{7}$

2 $\frac{15}{11} \div \left(-\frac{25}{22}\right)$

3 $\frac{-4}{5} \cdot \frac{35}{6} \div \frac{-14}{9}$

4 $\frac{-3}{7} \div \frac{-33}{21} \div \frac{-15}{9}$

5 $\frac{20x^2}{6w} \div \frac{5x^3}{14w^2}$

6 $\frac{18w^3}{28y^3} \cdot \frac{8y^2}{15w}$

7 $\frac{15a - 5b}{xy^2} \cdot \frac{x^2y}{6a - 2b}$

8 $\frac{5x - 20}{10x} \div \frac{7x - 28}{14x^2}$

9 $\frac{14x - 49y}{a^2 - b^2} \cdot \frac{a - b}{35y - 10x}$

10 $\frac{3x^2 - 12}{14x - 28} \cdot \frac{7x}{11x + 22}$

11 $\frac{x^2 - 5x + 6}{10x - 20} \cdot \frac{5x - 15}{x^2 - 9}$

12 $\frac{5x^2y - 15xy}{x^2 - 4} \div \frac{x^2 - 9}{10x^2 - 20x}$

13 $\frac{x^2 - y^2}{x^2 - 2xy + y^2} \div \frac{3x + 3y}{7x - 21}$

14 $\frac{4x^2 + 12x + 9}{5x^3 - 2x^2} \div \frac{14x^2 + 21x}{10x^2y^2}$

15 $\frac{x^2 - 3x + 2}{x^2 - 4x + 4} \div \frac{x^2 - 2x + 1}{3x^2 - 12}$

16 $\frac{a^2 - 9b^2}{4a^2 - b^2} \cdot \frac{4a^2 - 4ab + b^2}{2a^2 - 7ab + 3b^2}$

17 $\frac{7(c - y) - a(c - y)}{2c^2 - 7cy + 5y^2} \cdot \frac{2c - 5y}{c - y}$

18 $\frac{5(a + b) - x(a + b)}{2a^2 - 2b^2} \cdot \frac{6ax}{15 - 3x}$

19 $\frac{10y - 14x}{x - 1} \cdot \frac{x^2 - 2x + 1}{21x - 15y}$

20 $\frac{3w - 5z}{6w^2z} \cdot \frac{18w^2z + 30wz^2}{9w^2 - 25z^2}$

21 $\frac{9x^2 - 9xy + 9y^2}{5x^2y + 5xy^2} \cdot \frac{2xy}{3x^3 + 3y^3}$

22 $\frac{4x^2 - 1}{18xy} \div \frac{6x - 3}{16x^2 + 8x}$

23 $(3x^2 - 14x - 5) \cdot \frac{x^2 - 2x - 35}{3x^2 - 20x - 7}$

24 $(6x^2 - 19x + 10) \div \frac{6x^2 - 11x - 10}{9x^2 + 12x + 4}$

25 $\frac{20x^2 + 3x - 9}{21x - 35x^2} \div (12x^2 - 11x - 15)$

26 $\frac{4a^2 - ab - 5b^2}{ax + by + ay + bx} \div (8a - 10b)$

27 $\frac{-4}{7} \cdot \left(\frac{15}{6} \div \frac{-10}{14}\right)$

28 $-\frac{2}{9} \div \left(\frac{22}{12} \cdot \frac{-20}{33}\right)$

29 $\frac{18x^2}{5a} \cdot \frac{15ax}{81a^2} \cdot \frac{44ax}{24x^3}$

30 $\frac{14ab^2}{18x} \cdot \frac{27x^3}{15a^3b^3} \div \frac{7a^2x}{25ab}$

B

In Exercises 31–42 perform the indicated operations, and reduce all results to lowest terms.

31 $\dfrac{x^3 - y^3}{6x^2 - 6y^2} \div \dfrac{2x^2 + 2xy + 2y^2}{9xy}$

32 $\dfrac{x^2 + x - y^2 - y}{3x^2 - 3y^2} \div \dfrac{5x + 5y + 5}{7x^2y + 7xy^2}$

33 $\dfrac{a(2x - y) - b(2x - y)}{a^2 - b^2} \div \dfrac{6x - 3y}{4a + 4b}$

34 $\dfrac{2x - 5}{2x^2 - x - 15} \cdot \dfrac{-2x^2 - x + 10}{3x + 4} \cdot \dfrac{-15x - 20}{2x^2 - 9x + 10}$

35 $\dfrac{x^4 - y^4}{-2x - 3y} \cdot \dfrac{2x + 3y}{7x^2 + 7y^2} \cdot \dfrac{12xy}{8x^2 + 8xy}$

36 $\dfrac{x^3 - y^3}{18x^2y^3} \cdot \dfrac{9x^2y + 9xy^2}{x^2 - y^2} \div \dfrac{x^2 + xy + y^2}{36y^4}$

37 $\dfrac{5a - b}{a^2 - 5ab + 4b^2} \div \left(\dfrac{6ab}{3a - 12b} \cdot \dfrac{b^2 - 5ab}{4a - 4b}\right)$

38 $\dfrac{x - y}{x^2 - y^2} \div \dfrac{6x}{x + y} \cdot \dfrac{15x^2}{4}$

39 $\dfrac{x^2 - xy}{5x^2y^2} \div \dfrac{4x - 4y}{3xy} \cdot \dfrac{20y}{11}$

40 $\dfrac{4v^2 + 11v + 6}{2v^2 - v - 10} \div \left(\dfrac{4v^2 - 21v - 18}{2v^2 - 3v - 5} \cdot \dfrac{4v^2 - 7v + 3}{4v^2 - 27v + 18}\right)$

41 $\dfrac{x(a - b + 2c) - 2y(a - b + 2c)}{x^2 - 4xy + 4y^2} \div \dfrac{2a^2 - 2ab + 4ac}{7xy^2 - 14y^3}$

42 $\dfrac{a(2x - y) - b(2x - y)}{a^2 - b^2} \div \dfrac{6x - 3y}{4a + 4b}$

ESTIMATION SKILLS (43–46)

In Exercises 43–46 mentally estimate the value of each expression for $x = 10.01$, and then select the answer closest to your estimate.

43 $\dfrac{(x + 6)(x - 7)}{(x - 7)(x - 6)}$

a. -4 **b.** -1 **c.** 1 **d.** 4

44 $\dfrac{(x - 2)(x + 5)}{(x - 2)(x - 5)}$

a. 3 **b.** 2 **c.** 1 **d.** -1

45 $\dfrac{x^2 + 50}{x^2 - 50}$

a. 25 **b.** 5 **c.** 3 **d.** -5

46 $\dfrac{x^2 - 80}{x - 6}$

a. 16 **b.** 12 **c.** 5 **d.** 1.5

C

CALCULATOR USAGE (47–50)

In Exercises 47–50 store the value of -2.5 for x in your calculator, and then approximate to the nearest thousandth the value of each of these expressions.

47 $\dfrac{x^2 - 8}{x^2 + 10}$

48 $\dfrac{x^2 + 4x - 7}{x^2 + x - 8}$

49 $\dfrac{(x + 6)(x - 7)}{(x + 3)(x + 5)}$

50 $\dfrac{(2x - 1)(x - 2)}{(3x + 1)(x - 3)}$

In Exercises 51–54 perform the indicated operations, and reduce all results to lowest terms.

51 $\dfrac{a^3 + a^2 + a + 1}{a^3 + a^2 + ab^2 + b^2} \div \dfrac{a^3 + a^2b + a + b}{2a^2 + 2ab - ab^2 - b^3}$

52 $\dfrac{av + bv - aw - bw}{av + aw - bv - bw} \div \dfrac{av + aw - 5bv - 5bw}{av - aw - 5bv + 5bw} \div \dfrac{a + b}{b - a}$

53 $\dfrac{3x - 6y}{7x^2y} \div \dfrac{17x^3 - 17y^3}{34xy^2} \div \dfrac{3x}{2x^2 + 2xy + 2y^2}$

54 $\dfrac{2x^2 - 3x - 2}{7x - 3} \cdot \dfrac{49x^2 - 42x + 9}{4x^2 - 1} \cdot \dfrac{2x^2 - 7x + 3}{10 - 5x}$

In Exercises 55–58 perform the indicated operations, and reduce all results to lowest terms. Assume that m is a natural number and that all values of the variables that cause division by zero are excluded.

55 $\dfrac{x^{2m+1}}{x^m - 1} \cdot \dfrac{3x^m - 3}{x}$

56 $\dfrac{x^{2m} - 1}{x^{2m} + x^m} \div \dfrac{x^m - 1}{x^{2m}}$

57 $\dfrac{x^{2m} - 1}{x^m - 1} \cdot \dfrac{3x^m + 3}{(x^m + 1)^2}$

58 $\dfrac{x^{3m} - x^m}{x^{2m} + x^m} \div \dfrac{x^{3m} - 1}{x^{2m} + x^m + 1}$

DISCUSSION QUESTIONS (59 AND 60)

59 Describe in your own words the procedure for multiplying two rational expressions.

60 Discuss whether the expressions $\dfrac{x^2 - 4}{x^2 - x - 6}$ and $\dfrac{x - 2}{x - 3}$ are equal for all values of x. Are there any values for which one expression is defined and the other expression is not?

CHALLENGE QUESTION

Reduce

$\dfrac{x}{x - 1} \cdot \dfrac{x - 1}{x - 2} \cdot \dfrac{x - 2}{x - 3} \cdot \cdots \cdot \dfrac{x - 8}{x - 9} \cdot \dfrac{x - 9}{x - 10}$, and list all the excluded values.

SECTION SKILL REINFORCERS

Simplify each expression.

1 $\frac{3}{7}+\frac{2}{5}$ **2** $\frac{3}{7}-\frac{2}{5}$ **3** $\frac{9}{20}-\frac{4}{15}+\frac{5}{12}$ **4** $\frac{9}{20}-\left(\frac{4}{15}+\frac{5}{12}\right)$ **5** $\frac{9}{20}-\frac{4}{15}\cdot\frac{5}{12}$

6-3 Addition and Subtraction of Rational Expressions

SECTION OBJECTIVE

3 Add and subtract rational expressions.

The rules for adding and subtracting rational expressions are the same rules given in Chapter 1 for adding and subtracting rational numbers: If a, b, and c are real expressions and $c \neq 0$, then

$$\frac{a}{c}+\frac{b}{c}=\frac{a+b}{c} \quad \text{and} \quad \frac{a}{c}-\frac{b}{c}=\frac{a-b}{c}$$

The result should always be reduced to lowest terms.

EXAMPLE 1 Combining Rational Expressions with the Same Denominator

Simplify each of these expressions.

SOLUTIONS

(a) $\frac{5}{24x}-\frac{7}{24x}$

$$\frac{5}{24x}-\frac{7}{24x}=\frac{5-7}{24x}$$

The denominators of these fractions are the same, so subtract the numerators.

$$=\frac{-2}{24x}$$

$$=\frac{-1}{12x}$$

Then reduce the sum to lowest terms by dividing both the numerator and the denominator by 2.

(b) $\frac{2x-1}{x^2}+\frac{4x+1}{x^2}$

$$\frac{2x-1}{x^2}+\frac{4x+1}{x^2}=\frac{(2x-1)+(4x+1)}{x^2}$$

These fractions have the same denominators, so add the numerators of these like terms.

$$=\frac{6x}{x^2}$$

$$=\frac{6}{x}$$

Then reduce this sum to lowest terms by dividing both the numerator and the denominator by x.

It is easy to make errors in the order of operations if you are not careful when adding or subtracting rational expressions. Example 2 illustrates how parentheses can be used to avoid these errors.

EXAMPLE 2 Subtracting Rational Expressions with the Same Denominator

Simplify each of these expressions.

SOLUTIONS

(a) $\frac{5x+7}{2x+2} - \frac{3x-1}{2x+2}$

$$\frac{5x+7}{2x+2} - \frac{3x-1}{2x+2} = \frac{(5x+7)-(3x-1)}{2x+2}$$

Note: A common error is to write the numerator as $5x + 7 - 3x - 1$. To avoid this error, enclose the terms of the numerator in parentheses before combining like terms.

$$= \frac{5x+7-3x+1}{2x+2}$$

$$= \frac{2x+8}{2x+2}$$

$$= \frac{\cancel{2}(x+4)}{\cancel{2}(x+1)}$$

Factor, and then reduce to lowest terms.

$$= \frac{x+4}{x+1}$$

(b) $\frac{x^2}{x^2-1} - \frac{2x-1}{x^2-1}$

$$\frac{x^2}{x^2-1} - \frac{2x-1}{x^2-1} = \frac{x^2-(2x-1)}{x^2-1}$$

Again, use parentheses in the numerator to avoid an error in sign.

$$= \frac{x^2-2x+1}{x^2-1}$$

Simplify the numerator.

$$= \frac{\cancel{(x-1)}(x-1)}{\cancel{(x-1)}(x+1)}$$

Factor, and then reduce to lowest terms.

$$= \frac{x-1}{x+1}$$

▼

Remember that $-\frac{a}{b} = \frac{-a}{b} = \frac{a}{-b}$. These alternative forms are often useful when you are working with two fractions whose denominators are opposites of each other.

EXAMPLE 3 Subtracting Rational Expressions Whose Denominators Are Opposites

Simplify $\frac{5x}{3x-6} - \frac{4x}{6-3x}$.

SOLUTION $\frac{5x}{3x-6} - \frac{4x}{6-3x} = \frac{5x}{3x-6} + \frac{4x}{3x-6}$

Noting that the denominators $3x - 6$ and $6 - 3x$ are opposites, change $-\frac{4x}{6-3x}$ to the equivalent form $\frac{4x}{3x-6}$.

$$= \frac{5x+4x}{3x-6}$$

Now that the denominators are the same, you can add the numerators of these like terms.

$$= \frac{\overset{3}{\cancel{9}}x}{\cancel{3}(x-2)}$$

Then factor the denominator, and reduce this result by dividing both the numerator and the denominator by 3.

$$= \frac{3x}{x-2}$$

▼

Fractions can be added or subtracted only if they are expressed in terms of a common denominator. Although any common denominator can be used, we can simplify our work considerably by using the least common denominator (LCD). Sometimes the LCD can be determined by inspection. In other cases it may be necessary to use the procedure given in the following box—the same procedure that was used in Chapter 1 for arithmetic fractions.

FINDING THE LCD OF TWO OR MORE FRACTIONS

Step 1 Factor each denominator completely, including constant factors. Express repeated factors in exponential form.

Step 2 List each factor to the highest power to which it occurs in any single factorization.

Step 3 Form the LCD by multiplying the factors listed in Step 2.

EXAMPLE 4 Determining the LCD for Two Fractions

The denominators of two fractions are $6x^2 - 30x + 36$ and $9x^2 + 9x - 108$. Find their LCD.

SOLUTION

$$6x^2 - 30x + 36 = 6(x^2 - 5x + 6)$$
$$= 2 \cdot 3(x - 2)(x - 3)$$

First factor each denominator completely. The factor 2 and the three binomial factors each occur once.

$$9x^2 + 9x - 108 = 9(x^2 + x - 12)$$
$$= 3^2(x - 3)(x + 4)$$

The factor 3 is used twice in the second factorization.

$$\text{LCD} = 2 \cdot 3^2(x - 2)(x - 3)(x + 4)$$
$$= 18(x - 2)(x - 3)(x + 4)$$

The LCD is the product of each factor to the highest power to which it occurs in any of these factorizations. It is generally best to leave the LCD in the factored form shown in the last line of this example. ▼

SELF-CHECK ✓

1 Simplify $\dfrac{9x - 2}{2x + 1} - \dfrac{5x - 4}{2x + 1}$.

2 Simplify $\dfrac{2y - 1}{7y - 3} - \dfrac{5y - 2}{3 - 7y}$.

3 Find the LCD of the denominators $x^2 - xy - 2y^2$ and $6xy - 3x^2$.

ADDING (SUBTRACTING) RATIONAL EXPRESSIONS

Step 1 Express the denominator of each rational expression in factored form, and then find the LCD.

Step 2 Convert each term to an equivalent rational expression whose denominator is the LCD.

Step 3 Retaining the LCD as the denominator, add (subtract) the numerators to form the sum (difference).

Step 4 Reduce the expression to lowest terms.

SELF-CHECK ANSWERS

1 2 **2** 1 **3** $3x(x + y)(x - 2y)$ (The LCD is generally expressed so that the constant coefficient is positive rather than negative.)

EXAMPLE 5 **Adding Two Rational Expressions with Different Denominators**

Simplify $\frac{7}{6xy^2} + \frac{2}{15x^2y}$.

SOLUTION $\frac{7}{6xy^2} + \frac{2}{15x^2y} = \frac{?}{30x^2y^2} + \frac{?}{30x^2y^2}$

First determine the LCD:
$6xy^2 = 2 \cdot 3xy^2$
$15x^2y = 3 \cdot 5x^2y$
$\text{LCD} = 2 \cdot 3 \cdot 5 \cdot x^2y^2$
$\text{LCD} = 30x^2y^2$

$= \frac{7}{6xy^2}\left(\frac{5x}{5x}\right) + \frac{2}{15x^2y}\left(\frac{2y}{2y}\right)$

Then convert each fraction so that it has this LCD as its denominator.

$= \frac{35x}{30x^2y^2} + \frac{4y}{30x^2y^2}$

Finally, add like terms.

$= \frac{35x + 4y}{30x^2y^2}$

▼

EXAMPLE 6 **Subtracting Two Rational Expressions with Different Denominators**

Simplify $\frac{5}{v+4} - \frac{4}{v-5}$.

SOLUTION $\frac{5}{v+4} - \frac{4}{v-5} = \frac{5}{v+4} \cdot \frac{v-5}{v-5} - \frac{4}{v-5} \cdot \frac{v+4}{v+4}$

The LCD is $(v + 4)(v - 5)$. Convert each fraction so that it has this LCD as its denominator.

$= \frac{5(v-5) - 4(v+4)}{(v+4)(v-5)}$

$= \frac{5v - 25 - 4v - 16}{(v+4)(v-5)}$

Add like terms, and simplify the numerator.

$= \frac{v - 41}{(v+4)(v-5)}$

▼

EXAMPLE 7 **Adding Rational Expressions with Different Denominators**

Add $\frac{x}{6x^2 - 30x + 36} + \frac{7}{9x^2 + 9x - 108}$.

SOLUTION $\frac{x}{6x^2 - 30x + 36} + \frac{7}{9x^2 + 9x - 108}$

$= \frac{x}{6(x-2)(x-3)} + \frac{7}{9(x-3)(x+4)}$

First factor each denominator. The LCD of these two fractions, $18(x - 2)(x - 3)(x + 4)$, was calculated in Example 4.

$= \frac{x}{6(x-2)(x-3)} \cdot \frac{3(x+4)}{3(x+4)} + \frac{7}{9(x-3)(x+4)} \cdot \frac{2(x-2)}{2(x-2)}$

Convert each term to a fraction with this LCD as its denominator.

$$= \frac{3x(x+4) + 14(x-2)}{18(x-2)(x-3)(x+4)}$$ Leaving the denominator in factored form, simplify the numerator by multiplying and then combining like terms.

$$= \frac{3x^2 + 12x + 14x - 28}{18(x-2)(x-3)(x+4)}$$

$$= \frac{3x^2 + 26x - 28}{18(x-2)(x-3)(x+4)}$$ The numerator is prime, so this expression is in reduced form. ▼

Some of the steps illustrated in Example 7 can be combined in order to decrease the amount of writing necessary in these problems. Although most of the steps are still shown in the following examples, some have been combined. When combining steps, be careful not to make an error in sign.

EXAMPLE 8 Combining Three Rational Expressions with Different Denominators

Simplify $\frac{2v + 11}{v^2 + v - 6} - \frac{2}{v+3} + \frac{3}{2-v}$.

SOLUTION $\frac{2v + 11}{v^2 + v - 6} - \frac{2}{v+3} + \frac{3}{2-v}$

$$= \frac{2v+11}{(v+3)(v-2)} - \frac{2}{v+3} + \frac{-3}{v-2}$$ Factor each denominator. Noting that the factor $v - 2$ in the first denominator is the opposite of the denominator $2 - v$ in the last term, change $\frac{3}{2-v}$ to the equivalent form $\frac{-3}{v-2}$.

$$= \frac{2v+11}{(v+3)(v-2)} - \frac{2}{v+3} \cdot \frac{v-2}{v-2} + \frac{-3}{v-2} \cdot \frac{v+3}{v+3}$$ Now convert each term so that it has the LCD, $(v + 3)(v - 2)$, as its denominator.

$$= \frac{(2v+11) - 2(v-2) - 3(v+3)}{(v+3)(v-2)}$$ Simplify the numerator by multiplying and then combining like terms.

$$= \frac{2v + 11 - 2v + 4 - 3v - 9}{(v+3)(v-2)}$$

$$= \frac{-3v + 6}{(v+3)(v-2)}$$

$$= \frac{-3\cancel{(v-2)}}{(v+3)\cancel{(v-2)}}$$ Factor the numerator, and reduce by dividing out the common factor, $v - 2$.

$$= \frac{-3}{v+3}$$ ▼

Note that when we multiply or divide rational expressions, we first reduce as much as possible and then perform the multiplication or division. On the other hand, when we add or subtract rational expressions, we add or subtract first, using a common denominator, and then reduce afterwards.

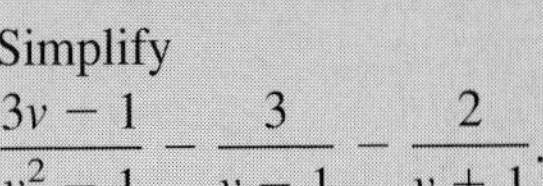

SELF-CHECK

Simplify $\frac{3v - 1}{v^2 - 1} - \frac{3}{v-1} - \frac{2}{v+1}$.

SELF-CHECK ANSWER 

$\frac{-2}{v-1}$

EXAMPLE 9 Combining Three Rational Expressions with Different Denominators

Simplify $3x - 5 - \dfrac{17}{x-1} + \dfrac{x}{1-x}$.

SOLUTION $3x - 5 - \dfrac{17}{x-1} + \dfrac{x}{1-x} = \dfrac{3x-5}{1} - \dfrac{17}{x-1} - \dfrac{x}{x-1}$ Change $3x - 5$ to the equivalent form $\dfrac{3x-5}{1}$ and $\dfrac{x}{1-x}$ to the equivalent form $-\dfrac{x}{x-1}$.

$= \dfrac{(3x-5)(x-1) - 17 - x}{x-1}$ Express each fraction in terms of the LCD, $x - 1$.

$= \dfrac{3x^2 - 8x + 5 - 17 - x}{x-1}$ Simplify the numerator by multiplying and then combining like terms.

$= \dfrac{3x^2 - 9x - 12}{x-1}$

$= \dfrac{3(x-4)(x+1)}{x-1}$ Factor the numerator to determine whether this rational expression can be reduced. Since the GCF of the numerator and the denominator is 1, this fraction is in lowest terms. Leave the numerator in factored form in the answer.

Exercises 6-3

A

In Exercises 1–8 perform the indicated operations, and reduce the results to lowest terms.

1 $\dfrac{5b+13}{3b^2} + \dfrac{b-4}{3b^2}$

2 $\dfrac{3x^2+1}{8x^3} - \dfrac{1-3x^2}{8x^3}$

3 $\dfrac{7}{x-7} - \dfrac{x}{x-7}$

4 $\dfrac{a-1}{a+7} + \dfrac{8}{a+7}$

5 $\dfrac{3s+7}{s^2-9} + \dfrac{s+5}{s^2-9}$

6 $\dfrac{5t-22}{t^2-5t+6} + \dfrac{4t-5}{t^2-5t+6}$

7 $\dfrac{x+a}{x(a+b)+y(a+b)} - \dfrac{x-b}{x(a+b)+y(a+b)}$

8 $\dfrac{2y^2+1}{2y^2-5y-12} - \dfrac{4-y}{2y^2-5y-12}$

In Exercises 9–20 find the least common denominator for the denominators listed.

9 18 and 45

10 16 and 40

11 6, 10, and 15

12 6, 12, and 18

13 $15x^2y^3$ and $27x^3y$

14 $35s^2t^3w^4$ and $55s^7t^3w^2$

15 $12(x+y)$ and $15(x+y)$

16 $21(3a-b)$ and $14(3a-b)$

17 $18x + 18y$ and $6x^2 + 6xy$

18 $4x^2y - 4xy^2$ and $26x^2 - 26y^2$

19 $x^2 - 36$ and $x^2 - 5x - 6$

20 $x^2 - 5x + 6$ and $x^2 - 9$

In Exercises 21–50 perform the indicated operations, and reduce the results to lowest terms.

21 $\dfrac{4}{9w} - \dfrac{7}{6w}$

22 $\dfrac{4}{5z} - \dfrac{1}{2z}$

23 $\dfrac{3v-1}{7v} - \dfrac{v-2}{14v}$

24 $\dfrac{12a-7}{35a} - \dfrac{a-1}{5a}$

25 $\dfrac{4}{b} + \dfrac{b}{b+4}$

26 $\dfrac{5}{c} - \dfrac{c}{c-7}$

27 $5 - \dfrac{1}{x}$

28 $7 + \dfrac{3}{y}$

29 $\dfrac{1}{x} - \dfrac{2}{x^2} + \dfrac{3}{x^3}$

30 $\dfrac{3}{y} + \dfrac{5}{y^2} - \dfrac{7}{y^3}$

31 $\dfrac{3}{x-2} - \dfrac{2}{x+3}$

32 $\dfrac{2}{x-5} - \dfrac{3}{x+6}$

33 $\dfrac{x}{x-1} - \dfrac{x}{x+1}$

34 $\dfrac{x}{x+2} - \dfrac{x}{x-2}$

35 $\frac{x+5}{x-4} - \frac{x+3}{x+2}$

36 $\frac{x-3}{x-2} - \frac{x-2}{x-3}$

37 $\frac{x-1}{x+2} + \frac{x+2}{x-1}$

38 $\frac{x+1}{x-3} - \frac{x-3}{x+1}$

39 $\frac{x}{77x-121y} - \frac{y}{49x-77y}$

40 $\frac{a}{6a-9b} - \frac{b}{4a-6b}$

41 $\frac{2}{(m+1)(m-2)} + \frac{3}{(m-2)(m+3)}$

42 $\frac{5}{(m-1)(m+2)} + \frac{2}{(m+2)(m-3)}$

43 $\frac{m+2}{m^2-6m+8} - \frac{8-3m}{m^2-5m+6}$

44 $\frac{3n+8}{n^2+6n+8} - \frac{4n+2}{n^2+n-12}$

45 $\frac{1}{a-3b} + \frac{b}{a^2-7ab+12b^2} + \frac{1}{a-4b}$

46 $\frac{5x}{5x-y} + \frac{2y^2}{25x^2-y^2} - \frac{5x}{5x+y}$

47 $\frac{x^2+5}{x^2-3x+4} - 2$

48 $\frac{2x^2-5x+3}{3x^2+7x+9} + 2$

49 $\frac{5}{2x+2} + \frac{x+5}{2x^2-2} - \frac{3}{x-1}$

50 $\frac{p+6}{p^2-4} - \frac{4}{p+2} - \frac{2}{p-2}$

B

In Exercises 51–60 perform the indicated operations, and reduce the results to lowest terms.

51 $\frac{1}{x-5} + \frac{1}{x+5} - \frac{10}{x^2-25}$

52 $\frac{42}{x^2-49} + \frac{3}{x+7} + \frac{3}{x-7}$

53 $\frac{3}{s-5t} + \frac{7}{s-2t} - \frac{9t}{s^2-7st+10t^2}$

54 $\frac{y+8}{y^2-2y-8} + \frac{1}{y+2} - \frac{4}{4-y}$

55 $\frac{2z+11}{z^2+z-6} + \frac{2}{z+3} - \frac{3}{z-2}$

56 $\frac{3}{2w-1} + \frac{4}{2w+1} - \frac{10w-3}{4w^2-1}$

57 $\frac{v+w}{vw} + \frac{w}{v^2-vw} - \frac{1}{w}$

58 $\frac{5v}{v^3-5v^2+6v} - \frac{4}{v-2} - \frac{3v}{3-v}$

59 $\frac{9w+2}{3w^2-2w-8} - \frac{7}{4-w-3w^2}$

60 $\frac{2x+y}{(x-y)(x-2y)} - \frac{x+4y}{(x-3y)(x-y)} - \frac{x-7y}{(x-3y)(x-2y)}$

C

In Exercises 61–64 perform the indicated operations, and reduce the results to lowest terms.

61 $\frac{2w-7}{w^2-5w+6} - \frac{2-4w}{w^2-w-6} + \frac{5w+2}{4-w^2}$

62 $\frac{2x^2-5y^2}{x^3-y^3} + \frac{x-y}{x^2+xy+y^2} + \frac{1}{x-y}$

63 $\frac{4m}{1-m^2} + \frac{2}{m+1} - 2$

64 $\frac{y-5}{x^2+5x+xy+5y} + \frac{1}{x+y} - \frac{2}{x+5}$

CALCULATOR USAGE (65 AND 66)

In Exercises 65 and 66 the sum obtained by performing the addition is given either by a or by b. Check both possibilities by evaluating all expressions for $x = 1$.

65 $\frac{1.5x-1.4}{4.25x^2-0.31x-7.36} + \frac{0.2x-0.9}{4.25x^2-0.31x-7.36}$

a. $\frac{1}{2.5x+3.2}$ **b.** $\frac{1}{2.5x-3.2}$

66 $\frac{2.1x-2.4}{12.25x^2-22.09} + \frac{1.4x-2.3}{12.25x^2-22.09}$

a. $\frac{2}{3.5x+4.7}$ **b.** $\frac{1}{3.5x+4.7}$

In Exercises 67–70 simplify each expression. Assume that m is a natural number and that all values of the variables that cause division by zero are excluded.

67 $\frac{x^m}{x^{2m}-4} - \frac{2}{x^{2m}-4}$

68 $\frac{2x^m}{x^{2m}-6x^m+9} - \frac{6}{x^{2m}-6x^m+9}$

69 $\frac{x^m}{x^{3m}-27} - \frac{3}{x^{3m}-27}$

70 $\frac{3x^m}{x^{2m}-x^m-20} - \frac{15}{x^{2m}-x^m-20}$

71 Total Work by Two Painters One painter can paint a house in t hours, while a second painter would take $t + 5$ hours to do the same house. In 8 hours the fractional portion of the job done by each is $\frac{8}{t}$ and $\frac{8}{t+5}$. What is the total work done by these painters in 8 hours?

72 Total Area Find the sum of the areas of the rectangle and the triangle shown.

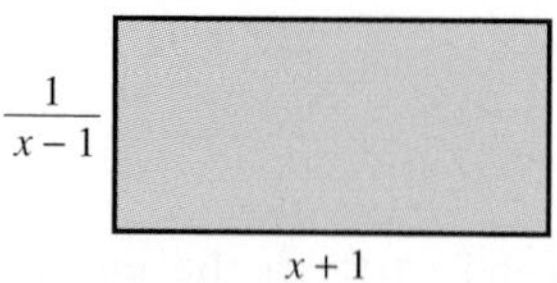

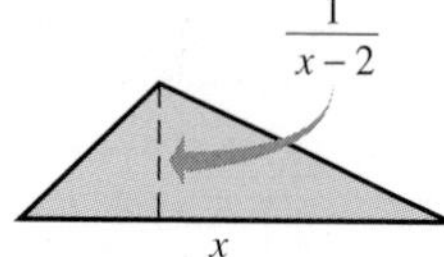

Figure for Exercise 72

73 The sum of a rational expression and $\frac{1}{x+5}$ is $\frac{4x}{2x^2 + 5x - 25}$. Determine this rational expression.

74 The difference formed by a rational expression minus $\frac{1}{x+5}$ is $\frac{27}{4x^2 + 13x - 35}$. Determine this rational expression.

DISCUSSION QUESTIONS (75 and 76)

75 Error Analysis Discuss the error made by the student whose work is shown below.

$$\frac{x}{x+1} - \frac{1}{x+1} \cdot \frac{2x}{x-1} = \frac{x-1}{x+1} \cdot \frac{2x}{x-1}$$
$$= \frac{2x}{x+1}$$

76 Describe in your own words the procedure for adding two rational expressions with different denominators.

CHALLENGE QUESTION

A rectangle has perimeter of 36 centimeters. Determine the dimensions that will give the following areas.

a. 32 cm^2 b. 56 cm^2
c. 72 cm^2 d. the maximum possible

SECTION SKILL REINFORCERS

Simplify each of these expressions.

1 $2^{-1} + 4^{-1}$ **2** $\left(\frac{1}{2} + \frac{1}{5}\right)^{-1}$ **3** $\left(\frac{1}{2}\right)^{-1} + \left(\frac{1}{3}\right)^{-1}$ **4** $-2^5 + 10^{-2}$ **5** $\left(\frac{2}{3}\right)^{-2} + \left(\frac{4}{5}\right)^{-1}$

6-4 *Combining Operations and Simplifying Complex Rational Expressions*

SECTION OBJECTIVES

4 **Simplify rational expressions in which the order of operations must be determined.**

5 **Simplify complex fractions.**

This section shows how to combine the operations already covered in the earlier sections of this chapter and provides further exercises involving rational expressions. The correct order of operations, from Section 1-4, is as follows:

1 Start with the expression within the innermost grouping symbols.

2 Perform all exponentiations.

3 Perform all multiplications and divisions, working from left to right.

4 Perform all additions and subtractions, working from left to right.

EXAMPLE 1 Using the Correct Order of Operations When Simplifying Rational Expressions

Simplify $\frac{2x}{5} + \frac{x^3}{15} \cdot \frac{9}{x^2}$.

SOLUTION

$$\frac{2x}{5} + \frac{x^3}{15} \cdot \frac{9}{x^2} = \frac{2x}{5} + \frac{9x^3}{15x^2}$$

Multiplication has higher priority than addition, so perform the indicated multiplication first.

$$= \frac{2x}{5} + \frac{3x}{5}$$

Reduce the product to lowest terms by dividing both the numerator and the denominator by $3x^2$.

$$= \frac{2x + 3x}{5}$$

Then perform the indicated addition, and reduce this sum to lowest terms.

$$= \frac{5x}{5}$$

$$= x$$

▼

EXAMPLE 2 Using the Correct Order of Operations When Simplifying Rational Expressions

Simplify $\left(\frac{2x}{5} + \frac{x^3}{15}\right) \cdot \frac{9}{x^2}$.

SOLUTION

$$\left(\frac{2x}{5} + \frac{x^3}{15}\right) \cdot \frac{9}{x^2} = \left(\frac{2x}{5} \cdot \frac{3}{3} + \frac{x^3}{15}\right) \cdot \frac{9}{x^2}$$

First add the terms inside the parentheses using a common denominator, 15.

$$= \frac{6x + x^3}{15} \cdot \frac{9}{x^2}$$

$$= \frac{\cancel{x}(x^2 + 6)}{\underset{5}{\cancel{15}}} \cdot \frac{\overset{3}{\cancel{9}}}{\underset{x}{\cancel{x^2}}}$$

Then multiply these factors. Reduce the product to lowest terms.

$$= \frac{3(x^2 + 6)}{5x}$$

▼

EXAMPLE 3 Using the Correct Order of Operations When Simplifying Rational Expressions

Simplify $\left(\frac{2}{x} - \frac{1}{3}\right) \div \left(\frac{2}{x} + \frac{1}{3}\right)$.

SOLUTION

$$\left(\frac{2}{x} - \frac{1}{3}\right) \div \left(\frac{2}{x} + \frac{1}{3}\right) = \left[\frac{2(3) - 1(x)}{3x}\right] \div \left[\frac{2(3) + 1(x)}{3x}\right]$$

First simplify the terms inside each set of parentheses. The LCD inside each set of parentheses is $3x$. Simplify each numerator.

$$= \frac{6 - x}{3x} \div \frac{6 + x}{3x}$$

$$= \frac{6 - x}{3\cancel{x}} \cdot \frac{3\cancel{x}}{6 + x}$$

To divide, invert the divisor and then multiply. Reduce by dividing out the common factors.

$$= \frac{6 - x}{6 + x}$$

$$= \frac{-(x - 6)}{x + 6}$$

$$= -\frac{x - 6}{x + 6}$$

SELF-CHECK

Simplify

$$\left(\frac{2}{x} - \frac{1}{y}\right) \div \left(\frac{6}{x} - \frac{3}{y}\right).$$

The original expression in Example 3 can be written as either

$$\left(\frac{2}{x} - \frac{1}{3}\right) \div \left(\frac{2}{x} + \frac{1}{3}\right) \quad \text{or} \quad \frac{\frac{2}{x} - \frac{1}{3}}{\frac{2}{x} + \frac{1}{3}}$$

The second form above is called a complex fraction. A **complex rational expression** is a rational expression whose numerator or denominator, or both, is also a rational expression. Each of the following fractions contains more than one fraction bar and is therefore a complex fraction:

$$\frac{7 + \frac{1}{x}}{y}, \quad \frac{8}{19 + \frac{1}{2}}, \quad \text{and} \quad \frac{\frac{m - n}{2m + 3}}{\frac{m^2 - 5n}{m^2 - 2mn + n^2}}$$

When expressing complex fractions as the numerator divided by the denominator, make sure the fraction bars are of different lengths so that you do not perform the operations in an incorrect order. As shown below, the expression $\dfrac{2}{\frac{3}{4}}$ does not equal $\dfrac{\frac{2}{3}}{4}$.

$$\frac{2}{\frac{3}{4}} = 2 \div \frac{3}{4} \quad \text{whereas} \quad \frac{\frac{2}{3}}{4} = \frac{2}{3} \div 4$$

$$= \frac{2}{1} \cdot \frac{4}{3} \qquad\qquad = \frac{2}{3} \cdot \frac{1}{4}$$

$$= \frac{8}{3} \qquad\qquad = \frac{1}{6}$$

There are two useful methods of simplifying complex fractions. Both are worth learning, since some problems can be worked more easily by the first method and others can be worked more easily by the second method. Example 4 illustrates Method I, and Example 5 illustrates Method II.

SELF-CHECK ANSWER

$\frac{1}{3}$

SIMPLIFYING COMPLEX FRACTIONS

Method I Express both the numerator and the denominator of the complex fraction as single terms; then multiply the numerator by the reciprocal of the denominator.

Method II Multiply both the numerator and the denominator of the complex fraction by the LCD of all the fractions that occur in the numerator and the denominator of the complex fraction.

EXAMPLE 4 Simplifying a Complex Fraction

Simplify $\dfrac{a^2 - b^2}{\dfrac{5a + 5b}{15a}}$ using Method I.

SOLUTION

$$\frac{a^2 - b^2}{\dfrac{5a + 5b}{15a}} = \frac{a^2 - b^2}{1} \div \frac{5a + 5b}{15a}$$

Rewrite this complex fraction as a division problem. A denominator of 1 is understood for $a^2 - b^2$.

$$= \frac{a^2 - b^2}{1} \cdot \frac{15a}{5a + 5b}$$

Then rewrite the division problem as the product of the dividend and the reciprocal of the divisor.

$$= \frac{\cancel{(a + b)}(a - b)}{1} \cdot \frac{\overset{3}{\cancel{15}}a}{\underset{1}{\cancel{5}}\cancel{(a + b)}}$$

Factor, and reduce by dividing out the common factors.

$$= \frac{3a(a - b)}{1} = 3a^2 - 3ab$$

▼

EXAMPLE 5 Simplifying a Complex Fraction

Simplify $\dfrac{1 - \dfrac{1}{x^3}}{1 + \dfrac{1}{x} + \dfrac{1}{x^2}}$ using Method II.

SOLUTION

$$\frac{1 - \dfrac{1}{x^3}}{1 + \dfrac{1}{x} + \dfrac{1}{x^2}} = \frac{1 - \dfrac{1}{x^3}}{1 + \dfrac{1}{x} + \dfrac{1}{x^2}} \cdot \frac{x^3}{x^3}$$

The LCD of all the fractions within the numerator and the denominator is x^3. Multiply both the numerator and the denominator by this LCD.

$$= \frac{x^3(1) - x^3\left(\dfrac{1}{x^3}\right)}{x^3(1) + x^3\left(\dfrac{1}{x}\right) + x^3\left(\dfrac{1}{x^2}\right)}$$

Caution: Be careful not to multiply the numerator and the denominator by different values.

$$= \frac{x^3 - 1}{x^3 + x^2 + x}$$

Simplify the numerator and the denominator.

$$= \frac{(x - 1)\cancel{(x^2 + x + 1)}}{x\cancel{(x^2 + x + 1)}}$$

$$= \frac{x - 1}{x}$$

Factor the numerator as the difference of two cubes. Factor the GCF, x, out of the denominator. Then reduce by dividing out the common factor, $x^2 + x + 1$.

▼

The next example shows both methods side by side so that you can compare them.

EXAMPLE 6 Simplifying a Complex Fraction

Simplify $\dfrac{\dfrac{x^2+b^2}{x^2-b^2}}{\dfrac{x-b}{x+b}+\dfrac{x+b}{x-b}}$.

SOLUTION

Method I

Invert the denominator and multiply.

$$\frac{\dfrac{x^2+b^2}{x^2-b^2}}{\dfrac{x-b}{x+b}+\dfrac{x+b}{x-b}}$$

$$=\frac{\dfrac{x^2+b^2}{x^2-b^2}}{\dfrac{(x-b)^2+(x+b)^2}{(x+b)(x-b)}}$$

$$=\frac{x^2+b^2}{x^2-b^2}\cdot\frac{(x+b)(x-b)}{x^2-2xb+b^2+x^2+2xb+b^2}$$

$$=\frac{x^2+b^2}{(x+b)(x-b)}\cdot\frac{(x+b)(x-b)}{2x^2+2b^2}$$

$$=\frac{(x^2+b^2)}{2\,(x^2+b^2)}$$

$$=\frac{1}{2}$$

Method II

Multiply the numerator and the denominator by the LCD, $(x-b)(x+b)$.

$$\frac{\dfrac{x^2+b^2}{x^2-b^2}}{\dfrac{x-b}{x+b}+\dfrac{x+b}{x-b}}$$

$$=\frac{\dfrac{x^2+b^2}{(x+b)(x-b)}}{\dfrac{x-b}{x+b}+\dfrac{x+b}{x-b}}\cdot\frac{(x+b)(x-b)}{(x+b)(x-b)}$$

$$=\frac{\dfrac{x^2+b^2}{(x+b)(x-b)}\cdot\dfrac{(x+b)(x-b)}{1}}{\dfrac{x-b}{x+b}\cdot\dfrac{(x+b)(x-b)}{1}+\dfrac{x+b}{x-b}\cdot\dfrac{(x+b)(x-b)}{1}}$$

$$=\frac{x^2+b^2}{(x-b)^2+(x+b)^2}$$

$$=\frac{x^2+b^2}{x^2-2xb+b^2+x^2+2xb+b^2}$$

$$=\frac{x^2+b^2}{2\,(x^2+b^2)}$$

$$=\frac{1}{2}$$

Since $a^{-n}=\dfrac{1}{a^n}$, fractions that involve negative exponents can be expressed as complex fractions. One way to simplify some expressions involving negative exponents is to first rewrite these expressions in terms of positive exponents. Then you can apply the methods just described to these complex fractions.

SELF-CHECK

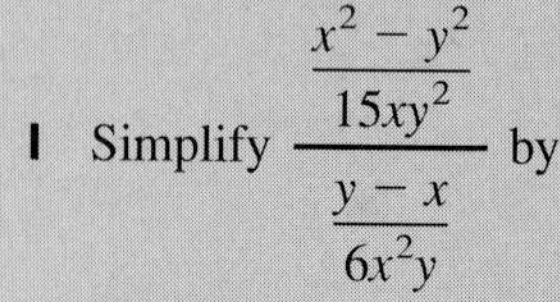

1 Simplify $\dfrac{\dfrac{x^2-y^2}{15xy^2}}{\dfrac{y-x}{6x^2y}}$ by Method I.

2 Simplify $\dfrac{x-\dfrac{1}{y}}{y-\dfrac{1}{x}}$ by Method II.

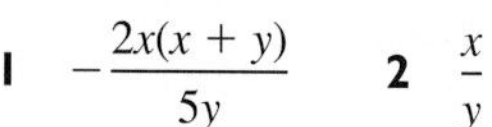

SELF-CHECK ANSWERS

1 $-\dfrac{2x(x+y)}{5y}$ **2** $\dfrac{x}{y}$

EXAMPLE 7 Simplifying a Fraction with Negative Exponents

Simplify $\dfrac{x^{-1} - x^{-2}}{x^{-1} + x^{-2}}$.

SOLUTION

$$\frac{x^{-1} - x^{-2}}{x^{-1} + x^{-2}} = \frac{\frac{1}{x} - \frac{1}{x^2}}{\frac{1}{x} + \frac{1}{x^2}}$$

Rewrite this expression as a complex fraction, using the definition of negative exponents.

$$= \frac{\frac{1}{x} - \frac{1}{x^2}}{\frac{1}{x} + \frac{1}{x^2}} \cdot \frac{x^2}{x^2}$$

Using Method II for simplifying complex fractions, multiply both the numerator and the denominator of the complex fraction by x^2, the LCD of all the fractions that occur in the numerator and the denominator of the complex fraction.

$$= \frac{\frac{1}{x}(x^2) - \frac{1}{x^2}(x^2)}{\frac{1}{x}(x^2) + \frac{1}{x^2}(x^2)}$$

$$= \frac{x - 1}{x + 1}$$

▼

In Example 7 the LCD of all the fractions within the numerator and the denominator of the complex fraction is x^2. Thus to simplify this complex fraction by Method II, we multiplied both the numerator and the denominator by x^2. This is exactly what we will do in Example 8, but without going through the work of writing the expression as a complex fraction. The alternative strategy illustrated in Example 8 is particularly appropriate when the negative exponents are applied only to monomials. The key is to select the appropriate expression to multiply times both the numerator and the denominator of the given fraction. The appropriate LCD to use as a factor can be determined by inspecting all the negative exponents in the original expression.

SELF-CHECK 

1 Determine the lowest power of x by which both the numerator and the denominator can be multiplied so as to eliminate all of the negative exponents in

$$\frac{x^{-1} - 4x^{-2} - 21x^{-3}}{x^{-1} + 6x^{-2} + 9x^{-3}}$$

2 Reduce the expression above to lowest terms.

EXAMPLE 8 Using an Alternative Method for Simplifying a Fraction with Negative Exponents

Simplify $\dfrac{x^{-1} - x^{-2}}{x^{-1} + x^{-2}}$.

SOLUTION

$$\frac{x^{-1} - x^{-2}}{x^{-1} + x^{-2}} = \frac{x^{-1} - x^{-2}}{x^{-1} + x^{-2}} \cdot \frac{x^2}{x^2}$$

$$= \frac{x^{-1}(x^2) - x^{-2}(x^2)}{x^{-1}(x^2) + x^{-2}(x^2)}$$

Multiply both the numerator and the denominator by x^2. Note that this is the lowest power of x that will eliminate all of the negative exponents on x in both the numerator and the denominator.

$$= \frac{x - 1}{x + 1}$$

Note that this result is the same as the answer in Example 7. ▼

SELF-CHECK ANSWERS 

1 x^3 **2** $\dfrac{x - 7}{x + 3}$

Exercises 6-4

A

In Exercises 1–44 simplify each expression.

1 $\frac{2x}{3} + \frac{x^2}{15} \cdot \frac{5}{x}$

2 $\frac{3x}{4} - \frac{x^2}{12} \cdot \frac{3}{x}$

3 $\left(\frac{2x}{3} + \frac{x^2}{15}\right) \cdot \frac{5}{x}$

4 $\left(\frac{3x}{4} - \frac{x^2}{12}\right) \cdot \frac{3}{x}$

5 $\frac{m}{2} + \frac{m^2}{10} \cdot \frac{15}{m}$

6 $\frac{3m}{7} + \frac{m^3}{14} \cdot \frac{7}{m^2}$

7 $\frac{3y}{5} - \frac{2y^5}{4} \div \frac{y^4}{7}$

8 $\frac{2y}{3} - \frac{y^4}{6} \div \frac{y^3}{9}$

9 $\frac{v}{v-6} - \frac{3v}{v+2} \cdot \frac{2v+4}{v^2-6v}$

10 $\frac{v}{v+10} + \frac{5v}{v-3} \cdot \frac{2v-6}{v^2+10v}$

11 $\frac{2-3x}{2x-3} + \frac{8-2x}{3x-6} \div \frac{4x-6}{3x-6}$

12 $\frac{6z+3}{3z-1} - \frac{5}{9z^2-1} \div \frac{1}{3z+2}$

13 $\left(\frac{2}{v+3} + v\right)\left(v - \frac{3}{v+2}\right)$

14 $\left(v - \frac{6v+35}{v+4}\right)\left(v - \frac{44}{v-7}\right)$

15 $\left(x - 2 - \frac{3}{x}\right) \div \left(1 + \frac{1}{x}\right)$

16 $\left(\frac{w+3}{2w+1} - 2w\right) \div \left(\frac{w-1}{2w+1} + 2w\right)$

17 $1 - \frac{4}{x} - \left(1 - \frac{2}{x}\right)^2$

18 $4 - \frac{12}{x} - \left(2 - \frac{3}{x}\right)^2$

19 $\left(1 + \frac{3}{5x}\right)^2 - \left(1 - \frac{3}{5x}\right)^2$

20 $\left(1 + \frac{2}{3x}\right)^2 - \left(1 - \frac{2}{3x}\right)^2$

21 $\dfrac{\frac{3}{5}}{\frac{7}{10}}$

22 $\dfrac{\frac{5}{6}}{\frac{2}{3}}$

23 $\dfrac{1 + \frac{1}{5}}{1 - \frac{1}{5}}$

24 $\dfrac{2 - \frac{3}{7}}{6 + \frac{2}{7}}$

25 $\dfrac{\frac{3}{5} - \frac{5}{3}}{\frac{1}{3} + \frac{1}{5}}$

26 $\dfrac{\frac{1}{2} + \frac{2}{3}}{\frac{3}{2} + 2}$

27 $\dfrac{\frac{12x^2}{5y}}{\frac{16x^2}{15y^2}}$

28 $\dfrac{\frac{18a^3}{25x}}{\frac{24a^3}{35x^2}}$

29 $\dfrac{\frac{w-z}{x^2y^2}}{\frac{w^2-z^2}{2xy}}$

30 $\dfrac{\frac{a^2-b^2}{6a^2b^3}}{\frac{a+b}{9a^3b}}$

31 $\dfrac{\frac{x^2-9x+14}{34x^4}}{\frac{5x^2-20}{17x^5}}$

32 $\dfrac{\frac{5s^2+5st+5t^2}{8s^3-8t^3}}{\frac{7s^2+28st+7t^2}{14s^2-14t^2}}$

33 $\dfrac{2 - \frac{1}{x}}{4 - \frac{1}{x^2}}$

34 $\dfrac{\frac{1}{x^2} - 49}{7 - \frac{1}{x}}$

35 $\dfrac{vw}{\frac{1}{v} + \frac{1}{w}}$

36 $\dfrac{\frac{1}{v} - \frac{1}{w}}{vw}$

37 $\dfrac{\frac{3}{x^2} - \frac{3}{x} + 3}{18 + \frac{18}{x^3}}$

38 $\dfrac{\frac{1}{x} - \frac{8}{x^2} + \frac{15}{x^3}}{1 - \frac{5}{x}}$

39 $\dfrac{3 + \frac{9}{x}}{\frac{15}{x^3} + \frac{8}{x^2} + \frac{1}{x}}$

40 $\dfrac{x + 2 - \frac{6}{2x+3}}{x + \frac{8x}{2x-1}}$

41 $\dfrac{\frac{w-a}{w+a} - \frac{w+a}{w-a}}{\frac{w^2+a^2}{w^2-a^2}}$

42 $\dfrac{w - 3 + \frac{2}{w-6}}{w - 2 - \frac{22}{w+7}}$

43 $\dfrac{\frac{12}{v^2} + \frac{1}{v} - 1}{\frac{24}{v^2} - \frac{2}{v} - 1}$

44 $\dfrac{\frac{12}{v^2} - \frac{1}{v} - 1}{\frac{8}{v^2} + \frac{6}{v} + 1}$

B

ESTIMATION SKILLS (45–48)

In Exercises 45–48 mentally estimate the value of each expression, and then select the answer closest to your estimation.

45 $\dfrac{\frac{198}{399}}{\frac{111}{443}}$ **a.** $\frac{1}{4}$ **b.** $\frac{1}{2}$ **c.** 1 **d.** 2 **e.** 4

46 $\dfrac{\frac{97}{299}}{\frac{101}{511}}$ **a.** $\frac{1}{3}$ **b.** $\frac{1}{5}$ **c.** $\frac{3}{5}$ **d.** $\frac{5}{3}$ **e.** 3

47 $\dfrac{\frac{99}{501}}{\frac{502}{98}}$ **a.** $\frac{1}{25}$ **b.** $\frac{1}{5}$ **c.** 1 **d.** 5 **e.** 25

48 $\dfrac{\frac{19}{101}}{\frac{24}{99}}$ **a.** $\frac{1}{5}$ **b.** $\frac{1}{4}$ **c.** $\frac{4}{5}$ **d.** 1 **e.** $\frac{5}{4}$

In Exercises 49–60 simplify each expression.

49 $\dfrac{x^{-2}}{x^{-2}+y^{-2}}$

50 $\dfrac{x^{-2}-y^{-2}}{y^{-2}}$

51 $\dfrac{v^{-1}+w^{-1}}{v^{-1}-w^{-1}}$

52 $\dfrac{m^2-n^2}{m^{-1}+n^{-1}}$

53 $\dfrac{m^3-n^3}{m^{-1}-n^{-1}}$

54 $\dfrac{x^3+y^3}{x^{-1}+y^{-1}}$

55 $\dfrac{a^{-2}-b^{-2}}{a^{-1}-b^{-1}}$

56 $\dfrac{b-a^{-1}}{a-b^{-1}}$

57 $\dfrac{xy^{-1}+x^{-1}y}{x^{-2}+y^{-2}}$

58 $\dfrac{x^{-1}y^{-2}+x^{-2}y^{-1}}{x^{-3}+y^{-3}}$

59 $\dfrac{\dfrac{\frac{1}{a}-\frac{1}{b}}{ab}}{\dfrac{1-\frac{a}{b}}{1+\frac{a}{b}}}$

60 $\dfrac{\dfrac{\frac{1}{x}+\frac{1}{y}}{\frac{1}{x}}}{\frac{y}{x}+1}$

C

Calculator Usage (61–64)

In Exercises 61–64 store the value of −2.85 for x in your calculator; then use your calculator to approximate to the nearest thousandth the value of each of these expressions.

61 $\dfrac{5x+\frac{5}{x}}{4x-\frac{4}{x}}$

62 $\dfrac{3x-\frac{1}{x}}{2x+\frac{1}{x}}$

63 $\dfrac{\frac{x+1}{x}}{\frac{x^2-1}{x-1}}$

64 $\dfrac{\frac{3x-1}{x+1}}{\frac{3x+1}{x-1}}$

In Exercises 65–68 simplify each expression.

65 $\dfrac{\dfrac{1-\frac{2}{x}}{1-\frac{4}{x^2}}}{\dfrac{1+\frac{1}{x}}{1+\frac{1}{x^3}}}$

66 $\dfrac{\dfrac{1+\frac{x}{y}}{1-\frac{x}{y}}}{\dfrac{x-\frac{y^2}{x}}{x+y}}$

67 $\dfrac{a-b}{ab}+\dfrac{a^3+b^3}{a^3b^3}\cdot\dfrac{a^2b^2}{4a^2-4ab+4b^2}$

68 $\dfrac{a+2}{a+3}-\dfrac{a^3-8}{a^2-9}\div\dfrac{a^2+2a+4}{a-3}$

In Exercises 69–72 simplify each expression.

69 $\dfrac{x^{-m}+y^{-n}}{x^{-m}y^{-n}}$

70 $\dfrac{x^{-m}-y^{-n}}{x^m-y^n}$

71 $\dfrac{x^{-m}-x^{-2m}}{x^{-m}+x^{-2m}}$

72 $\dfrac{x^{-m}+3x^{-2m}}{x^{-m}-2x^{-2m}}$

73 The total area of the rectangle and triangle shown is $\dfrac{3x}{x+1}$. Determine the height of the triangle.

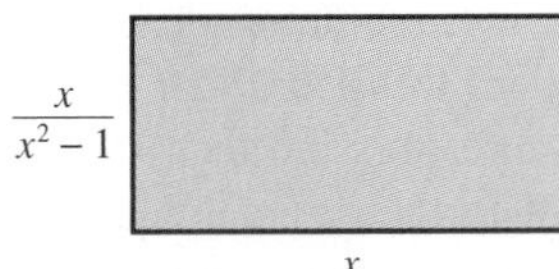

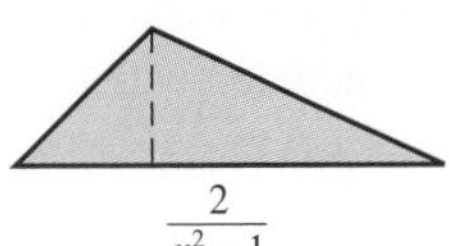

Figure for Exercise 73

Discussion Questions (74–76)

74 Describe in your own words what a complex fraction is.

75 Simplify $\dfrac{1-\frac{1}{x^2}}{1+\frac{3}{x}+\frac{2}{x^2}}$ using Method I, and then beside each step write an explanation of that step.

76 Simplify $\dfrac{1 - \dfrac{1}{x^2}}{1 + \dfrac{3}{x} + \dfrac{2}{x^2}}$ using Method II, and then beside each step write an explanation of that step.

CHALLENGE QUESTION

Using each of the digits 0, 1, 2, 3, 4, 5, 6, 7, 8, and 9 exactly once, write two fractions whose sum is 1.

SECTION SKILL REINFORCERS

1 Simplify $4(3x - 2) + 3(5x - 4)$. **2** Solve $4(3x - 2) + 3(5x - 4) = 7$. **3** Simplify $\dfrac{x}{12} - \dfrac{x}{15}$.

4 Solve $\dfrac{x}{12} - \dfrac{x}{15} = 0$. **5** Solve $\dfrac{x}{12} - \dfrac{x}{15} = 2$.

6-5 *Solving Equations Containing Rational Expressions*

SECTION OBJECTIVE

6 Solve equations containing rational expressions.

In Section 1-3 we solved fractional equations by multiplying both sides of the equation by the least common denominator (LCD). This method produces an equation equivalent to the original one as long as we do not multiply by zero. Multiplying by an expression that is equal to zero can produce an equation that is *not* equivalent to the original equation and thus can produce an extraneous value. An extraneous value is not a solution of the original equation, so when solving an equation with a variable in the denominator you must check that the solution does not include a value excluded from the domain of the variable because of division by zero.

SOLVING AN EQUATION CONTAINING RATIONAL EXPRESSIONS

Step 1 Multiply both sides of the equation by the LCD.

Step 2 Solve the resulting equation.

Step 3 Check the solution to determine whether it is an excluded value and therefore extraneous.

EXAMPLE 1 Solving an Equation Containing Rational Expressions

Solve $\dfrac{3}{x-7} + 5 = \dfrac{8}{x-7}$.

SOLUTION

$$\frac{3}{x-7} + 5 = \frac{8}{x-7}$$

Note that $x = 7$ is an excluded value.

$$\cancel{(x-7)}\left(\frac{3}{\cancel{x-7}}\right) + (x-7)(5) = \cancel{(x-7)}\left(\frac{8}{\cancel{x-7}}\right)$$

Multiply both members of the equation by the nonzero LCD, $x - 7$. The LCD is nonzero since 7 is an excluded value.

$$3 + 5x - 35 = 8$$

Solve the resulting equation.

$$5x = 40$$

$$x = 8$$

Since $x = 8$ is not an excluded value, this solution should check.

Answer The solution set is $\{8\}$.

EXAMPLE 2 Solving an Equation with an Extraneous Value

Solve $\dfrac{v}{v-3} = 4 - \dfrac{3}{3-v}$.

SOLUTION

$$\frac{v}{v-3} = 4 - \frac{3}{3-v}$$

$$\frac{v}{v-3} = 4 + \frac{3}{v-3}$$

Note that the only excluded value is $v = 3$. Noting that the denominators are opposites, change $-\dfrac{3}{3-v}$ to $\dfrac{3}{v-3}$.

$$\cancel{(v-3)}\left(\frac{v}{\cancel{v-3}}\right) = (v-3)(4) + \cancel{(v-3)}\left(\frac{3}{\cancel{v-3}}\right)$$

Multiply both sides by the nonzero LCD, $v - 3$. The LCD is nonzero since 3 is an excluded value.

$$v = 4v - 12 + 3$$

Solve the resulting equation for v.

$$-3v = -9$$

$$v = 3$$

$v = 3$ is the excluded value noted above, so this value is extraneous.

This value causes division by zero in the original equation, so there is no solution.

Answer The solution set is $\varnothing$.

EXAMPLE 3 Solving an Equation with an Extraneous Value

Solve $\dfrac{48}{z^2 - 2z - 15} + \dfrac{6}{z+3} = \dfrac{7}{z-5}$.

SOLUTION

$$\frac{48}{z^2 - 2z - 15} + \frac{6}{z+3} = \frac{7}{z-5}$$

$$\frac{48}{(z-5)(z+3)} + \frac{6}{z+3} = \frac{7}{z-5}$$

First factor the denominators in order to find the LCD and more easily determine the excluded values, 5 and -3.

$$(z-5)(z+3)\left[\frac{48}{(z-5)(z+3)} + \frac{6}{z+3}\right] = \cancel{(z-5)}(z+3)\left(\frac{7}{\cancel{z-5}}\right)$$

Multiply both sides by the nonzero LCD, $(z - 5)(z + 3)$. The LCD is nonzero since -3 and 5 are excluded values.

$$\frac{(z-5)(z+3)(48)}{(z-5)(z+3)} + \frac{(z-5)(z+3)(6)}{z+3} = (z+3)(7)$$

Use the distributive property to perform the multiplication in the left member.

$$48 + 6(z-5) = 7(z+3)$$
$$48 + 6z - 30 = 7z + 21$$
$$-z = 3$$
$$z = -3$$

Simplify, and solve the resulting equation.

$z = -3$ is one of the excluded values noted above, so it is an extraneous value.

This value causes division by zero in the original equation, so there is no solution.

Answer No solution

The solution set is $\varnothing$.

EXAMPLE 4 Solving an Equation with an Extraneous Value

Solve $\frac{y+1}{y+2} + \frac{y}{y-2} + 1 = \frac{8}{y^2-4}$.

SOLUTION

$$\frac{y+1}{y+2} + \frac{y}{y-2} + 1 = \frac{8}{y^2-4}$$
$$\frac{y+1}{y+2} + \frac{y}{y-2} + 1 = \frac{8}{(y+2)(y-2)}$$

Factor the denominator of the right member in order to determine the LCD, $(y+2)(y-2)$, and the excluded values, -2 and 2.

$$(y+2)(y-2)\left[\frac{y+1}{y+2} + \frac{y}{y-2} + 1\right] = (y+2)(y-2)\left[\frac{8}{(y+2)(y-2)}\right]$$
$$\frac{(y+2)(y-2)(y+1)}{y+2} + \frac{(y+2)(y-2)y}{y-2} + \frac{(y+2)(y-2)(1)}{1} = 8$$

Multiply both members by the nonzero LCD, $(y+2)(y-2)$. The LCD is nonzero since -2 and 2 are excluded values.

$$(y-2)(y+1) + (y+2)(y) + (y+2)(y-2)(1) = 8$$
$$y^2 - y - 2 + y^2 + 2y + y^2 - 4 = 8$$
$$3y^2 + y - 14 = 0$$
$$(3y+7)(y-2) = 0$$
$$3y + 7 = 0 \quad \text{or} \quad y - 2 = 0$$
$$y = -\frac{7}{3} \qquad\qquad y = 2$$

Write the resulting quadratic equation in standard form, and then solve it by factoring.

$y = 2$ is an excluded value; the value $y = -\frac{7}{3}$ should check.

The value $y = 2$ causes division by zero in the original equation and is therefore an extraneous value.

Answer $y = -\frac{7}{3}$ The solution set is $\left\{-\frac{7}{3}\right\}$.

SELF-CHECK 

Solve each of these equations.

1 $\frac{x-12}{x-4} = \frac{2x}{4-x} + 2$

2 $\frac{2y}{y+2} = \frac{y}{y+2} - 3$

SELF-CHECK ANSWERS 

1 No solution **2** $y = -\frac{3}{2}$

If the steps in the solution of an equation produce a contradiction, then the original equation has no solution. Example 5 shows how to interpret the answer when an identity is produced by the steps in the solution of an equation.

EXAMPLE 5 **An Equation That Simplifies to an Identity**

Solve $\dfrac{3}{x-4} - \dfrac{2}{x-2} = \dfrac{x+2}{x^2-6x+8}$.

SOLUTION

$$\frac{3}{x-4} - \frac{2}{x-2} = \frac{x+2}{x^2-6x+8}$$

$$\frac{3}{x-4} - \frac{2}{x-2} = \frac{x+2}{(x-4)(x-2)}$$

Factor the denominator of the right member in order to determine the LCD, $(x-4)(x-2)$, and the excluded values, 2 and 4.

$$(x-4)(x-2)\left[\frac{3}{x-4} - \frac{2}{x-2}\right] = (x-4)(x-2)\left[\frac{x+2}{(x-4)(x-2)}\right]$$

Multiply both members by the nonzero LCD, $(x-4)(x-2)$. The LCD is nonzero since 2 and 4 are excluded values. Simplify, and then solve the equation.

$$3(x-2) - 2(x-4) = x+2$$

$$3x - 6 - 2x + 8 = x + 2$$

$$x + 2 = x + 2$$

$$0 = 0 \text{ (an identity)}$$

Since the equation is an identity, its solution set is the set of all real numbers. However, it is not equivalent to the original equation, which has excluded values of 2 and 4. Thus the original equation is true for all real numbers except 2 and 4, for which it is meaningless.

Answer $\mathbb{R} \sim \{2, 4\}$ (This notation denotes a solution set consisting of all real numbers except 2 and 4.) ▼

In Chapter 2 we transposed formulas to solve for a specified variable. Example 6 involves transposing a formula that includes a rational fraction.

EXAMPLE 6 **Solving an Equation for a Specified Variable**

Solve $\dfrac{1}{R} = \dfrac{1}{r_1} + \dfrac{1}{r_2}$ for R.

SOLUTION

$$\frac{1}{R} = \frac{1}{r_1} + \frac{1}{r_2}$$

The variable r_1 is read as "r sub-one," and r_2 is read as "r sub-two." R, r_1, and r_2 are distinct variables.

$$Rr_1r_2\left(\frac{1}{R}\right) = Rr_1r_2\left(\frac{1}{r_1}\right) + Rr_1r_2\left(\frac{1}{r_2}\right)$$

Multiply both sides of the equation by the LCD, Rr_1r_2.

$$r_1r_2 = Rr_2 + Rr_1$$

$$r_1r_2 = R(r_2 + r_1)$$

Factor out the common factor of R on the right side of the equation.

$$\frac{r_1r_2}{r_1 + r_2} = R$$

Divide both sides of the equation by $r_1 + r_2$. (Note that $r_1 + r_2 = r_2 + r_1$.)

$$R = \frac{r_1r_2}{r_1 + r_2}$$

▼

Exercises 6-5

A

In Exercises 1 and 2 determine the values excluded from the domain of the variable because they would cause division by zero.

1 a. $\frac{3}{m-3}+5=\frac{2}{m-2}$

b. $\frac{3y}{(2y+3)(3y-2)}-\frac{7}{2y+3}=\frac{y-1}{3y-2}$

c. $\frac{4y-5}{2y^2+5y-3}=\frac{5y-4}{6y^2-y-1}$

2 a. $\frac{2}{n+2}+3=\frac{2}{n-5}$

b. $\frac{y+2}{4y^2-13y+3}=\frac{y}{4y-1}-\frac{5}{y-3}$

c. $\frac{7y}{6y^2-5y-6}=\frac{9y-1}{6y^2-11y-10}$

In Exercises 3–30 solve each equation.

3 $\frac{3}{z-1}+2=\frac{5}{z-1}$

4 $\frac{5}{z+3}-2=\frac{4}{z+3}$

5 $\frac{6w-1}{2w-1}-5=\frac{2w-3}{1-2w}$

6 $\frac{m}{m-2}-5=\frac{2}{m-2}$

7 $\frac{-3}{p+2}=\frac{-8}{p-3}$

8 $\frac{7}{p-4}=\frac{2}{p+1}$

9 $\frac{7}{3n-1}=\frac{2}{n+2}$

10 $\frac{10}{r-3}=\frac{34}{2r+1}$

11 $\frac{4}{k+2}=\frac{1}{3k+6}+\frac{11}{9}$

12 $\frac{5}{4k+1}=\frac{3}{8k+2}+1$

13 $\frac{3y}{(y+4)(y-2)}=\frac{5}{y-2}+\frac{2}{y+4}$

14 $\frac{y^2+18}{(2y+3)(y-3)}=\frac{5}{y-3}-\frac{1}{2y+3}$

15 $1-\frac{14}{y^2+4y+4}=\frac{7y}{y^2+4y+4}$

16 $1+\frac{6w}{w^2-6w+9}=\frac{18}{w^2-6w+9}$

17 $\frac{4}{x-5}+\frac{5}{x-2}=\frac{x+6}{3x-6}$

18 $\frac{w+1}{w}+\frac{14}{w-7}=\frac{3w-7}{w^2-7w}$

19 $\frac{1}{(t-1)^2}-3=\frac{2}{1-t}$

20 $\frac{8}{(t+4)^2}+\frac{2}{t+4}=3$

21 $\frac{z}{(z-2)(z+1)}-\frac{z}{(z+1)(z+3)}=\frac{3z}{(z+3)(z-2)}$

22 $\frac{4}{(z+1)(z-1)}=\frac{6-z}{(z+1)(z-2)}-\frac{8}{(z-2)(z-1)}$

23 $\frac{2v-5}{3v^2-v-14}+\frac{7}{3v-7}=\frac{8}{v+2}$

24 $\frac{v^2-2v+2}{6v^2+23v-4}+\frac{2}{v+4}=\frac{v}{6v-1}$

25 $\frac{m+4}{6m^2+5m-6}=\frac{m}{3m-2}-\frac{m}{2m+3}$

26 $\frac{2m+17}{2m^2+11m+14}+\frac{m-2}{m+2}=\frac{m-3}{2m+7}$

27 $\frac{x+1}{3x^2-4x+1}-\frac{x+1}{2x^2+x-3}=\frac{2}{6x^2+7x-3}$

28 $\frac{4y}{6y^2-7y-3}+\frac{2}{3y^2-2y-1}=\frac{y+2}{2y^2-5y+3}$

29 $\frac{2}{m+2}-\frac{1}{m+1}=\frac{1}{m}$

30 $\frac{m^2-1}{2m+1}=\frac{1-m}{3}$

In Exercises 31–38 transpose each equation to solve for the specified variable.

31 $I=\frac{k}{d}$ for d

32 $F=\frac{k}{r}$ for r

33 $\frac{1}{x}=\frac{1}{y}+\frac{1}{z}$ for x

34 $\frac{1}{x}=\frac{1}{y}-\frac{1}{z}$ for z

35 $h=\frac{2A}{B+b}$ for B

36 $h=\frac{2A}{B+b}$ for b

37 $\frac{1}{R}=\frac{1}{r_1}+\frac{1}{r_2}$ for r_1

38 $I=\frac{E}{r_1+r_2}$ for r_1

B

Exercises 39–44 have two parts. In part a solve the equation, and in part b perform the indicated operations and simplify the result.

	Solve		*Simplify*
39	a. $\frac{1}{p-1}=\frac{3}{p+1}$		b. $\frac{1}{p-1}-\frac{3}{p+1}$
40	a. $\frac{m-1}{m+1}=\frac{m-3}{m-2}$		b. $\frac{m-1}{m+1}-\frac{m-3}{m-2}$

Solve / **Simplify**

41 a. $\frac{x-1}{x+1} - 1 = \frac{x-6}{x-2}$ **b.** $\frac{x-1}{x+1} - 1 - \frac{x-6}{x-2}$

42 a. $\frac{x^2}{x+2} + \frac{x-1}{x-3} = 0$ **b.** $\frac{x^2}{x+2} + \frac{x-1}{x-3}$

Solve

43 a. $\frac{2x-8}{6x^2+x-2} = \frac{4}{3x+2} - \frac{2}{2x-1}$

Simplify

b. $\frac{2x-8}{6x^2+x-2} - \frac{4}{3x+2} + \frac{2}{2x-1}$

Solve

44 a. $\frac{w^2-w-3}{2w^2-9w+9} + \frac{1}{3-w} = \frac{w}{2w-3}$

Simplify

b. $\frac{w^2-w-3}{2w^2-9w+9} + \frac{1}{3-w} - \frac{w}{2w-3}$

In Exercises 45–50 solve each equation.

45 $\frac{z-2}{2z^2-5z+3} + \frac{3}{3z^2-2z-1} = \frac{3z}{6z^2-7z-3}$

46 $\frac{n-3}{n^2+5n+4} + \frac{n-2}{n^2+3n+2} = \frac{n^2-12}{(n+1)(n+2)(n+4)}$

47 $\frac{1}{n^2-5n+6} - \frac{1}{n^2-n-2} + \frac{3}{n^2-2n-3} = 0$

48 $\frac{12}{2w^2-13w+6} - \frac{7w-2}{2w^2+w-1} + \frac{w-20}{w^2-5w-6} = 0$

49 $\frac{x^2}{x^2-x-2} = \frac{2x}{x^2+x-6}$

50 $\frac{6v+6}{2v^2+7v-4} = \frac{3v}{v^2+2v-8} - \frac{5v-7}{2v^2-5v+2}$

C

In Exercises 51–60 solve each equation.

51 $\frac{5v+4}{2v^2+v-15} + \frac{3}{5-2v} - \frac{1}{v+3} = 0$

52 $\frac{x}{x-3} + \frac{1}{x-2} - \frac{1}{x+2} = \frac{x-12}{x^3-3x^2-4x+12}$

53 $\frac{x^2}{2x^2+9x-5} + \frac{2x}{x^2+2x-15} = \frac{4x}{(2x-1)(x+5)(x-3)}$

54 $\frac{x-2}{x^2+5x+6} = \frac{3}{x+3} - \frac{2}{x+2}$

55 $\frac{z-2}{4z^2-29z+30} - \frac{z+2}{5z^2-27z-18} = \frac{z+1}{20z^2-13z-15}$

56 $\frac{x^2}{4x^2-1} - \frac{3x}{2x^2+11x+5} = \frac{12x-x^2}{(4x^2-1)(x+5)}$

57 $\frac{w}{3w^2-11w+10} + \frac{5}{3w-5} = \frac{2}{w-2}$

58 $\frac{3n-7}{n^2-5n+6} + \frac{2n+8}{9-n^2} - \frac{n+2}{n^2+n-6} = 0$

59 $\frac{z-1}{z^2-2z-3} + \frac{z+1}{z^2-4z+3} = \frac{z+8}{z^2-1} + \frac{20}{(z-1)(z+1)(z-3)}$

60 $\frac{3w-8}{w^2-5w+6} + \frac{w+2}{w^2-6w+8} = \frac{5-2w}{w^2-7w+12} + \frac{12}{(w-4)(w-3)(w-2)}$

Discussion Question

61 Describe in your own words what an extraneous value is. Also describe what causes an extraneous value to occur in solving *some* equations containing rational expressions, but not in solving *all* of these equations.

Challenge Question

A rectangle has an area of 100 square centimeters. Determine the dimensions that will yield the following perimeters.

a. 202 cm b. 58 cm

c. 50 cm d. the minimum possible value

Section Skill Reinforcers

If m represents a positive real number, write an expression to represent each of these numbers.

1 The additive inverse of m **2** The multiplicative inverse of m **3** The reciprocal of m **4** The reciprocal of the number that is 3 more than m **5** The additive inverse of the number that is 3 more than m

6-6 Applications Yielding Equations with Fractions

SECTION OBJECTIVE

7 Solve applied problems that yield equations with fractions.

The problems in this section are specifically designed to yield equations involving algebraic fractions. You have already solved such equations in the previous section. The real key to working these problems is to form the correct equation using the problem-solving skills you learned in Chapter 1. Write down the important steps in order to avoid making careless errors. At this point it is important to master the material; you can use shortcuts later.

Perhaps the most important step in solving word problems is to determine the quantity you are to solve for and then to carefully identify this quantity with an appropriate variable. The strategy used to form equations is summarized below.

1 Read the problem carefully to determine what you are asked to find.
2 Use any well-known formula that is applicable to the problem.
3 Translate key phrases into algebraic statements.
4 Use sketches or tables to clarify the relationships among variables and to form equations.
5 Use the rate principle and the mixture principle to form equations.

A Mathematical Note

The problem-solving strategy given in Chapter 1 is used throughout this book, but this strategy is not new or unique. George Poyla (1888–1985) is well known for teaching problem-solving techniques. His best-selling book, *How to Solve It*, includes four steps to problem solving:

1 Understand the problem.
2 Devise a plan.
3 Carry out the plan.
4 Check back.

Compare these steps to those given in Chapter 1.

EXAMPLE 1 Numeric Word Problem

What number can be added to both the numerator and the denominator of $\frac{3}{11}$ to produce a fraction equal to $\frac{1}{2}$?

SOLUTION Let

n = Number added to the numerator and the denominator of the given fraction

$3 + n$ = New numerator

$11 + n$ = New denominator

$$\boxed{\text{New fraction}} = \boxed{\frac{1}{2}}$$ **Word equation**

$$\frac{3+n}{11+n} = \frac{1}{2}$$ Substitute the variables into the word equation.

$$2(11+n)\frac{3+n}{11+n} = 2(11+n)\frac{1}{2}$$ Multiply both sides of the equation by the LCD, $2(11 + n)$.

$$2(3 + n) = 11 + n$$

$$6 + 2n = 11 + n \qquad \text{Then simplify, and solve for } n.$$

$$n = 5$$

Answer The number 5 added to both the numerator and the denominator of $\frac{3}{11}$ yields

$$\frac{3 + 5}{11 + 5} = \frac{8}{16} = \frac{1}{2}$$

▼

EXAMPLE 2 Ratio of Gauge Readings

The ratio of two readings from a gauge is $\frac{1}{2}$. The reading in the numerator is 2 units below normal, whereas the reading in the denominator is 3 units above normal. What is the normal reading?

SOLUTION Let

n = Normal number of units

$n - 2$ = Number of units in the numerator

$n + 3$ = Number of units in the denominator

$$\boxed{\text{Ratio of readings}} = \boxed{\frac{1}{2}} \qquad \text{Word equation}$$

$$\frac{n - 2}{n + 3} = \frac{1}{2} \qquad \text{Substitute the values identified above into the word equation.}$$

$$2(n + 3)\left(\frac{n - 2}{n + 3}\right) = 2(n + 3)\left(\frac{1}{2}\right) \qquad \text{Multiply both members by the LCD, } 2(n + 3).$$

$$2(n - 2) = n + 3 \qquad \text{Simplify, and then solve for } n.$$

$$2n - 4 = n + 3$$

$$n = 7$$

Answer The normal reading is 7 units. ▼

SELF-CHECK

Twice the reciprocal of the larger of two consecutive integers minus the reciprocal of the smaller integer equals $\frac{3}{28}$. Find these integers.

The answers to many problems can be found by solving equations formed by translating key phrases into algebraic form. Remember to underline or highlight these key words as you read the problem.

SELF-CHECK ANSWER 

The integers are 7 and 8. (These are the only integral answers.)

EXAMPLE 3 Reciprocals of Consecutive Even Integers

The sum of the reciprocals of two consecutive even integers is $\frac{13}{84}$. Find these integers.

SOLUTION Let

n = Smaller integer

$\frac{1}{n}$ = Reciprocal of the smaller integer

$n + 2$ = Larger integer — Consecutive even integers differ by 2.

$\frac{1}{n+2}$ = Reciprocal of the larger integer

$$\boxed{\text{Sum of reciprocals}} = \boxed{\frac{13}{84}}$$ **Word equation**

$$\frac{1}{n} + \frac{1}{n+2} = \frac{13}{84}$$ Substitute the values identified above into the word equation.

$$84n(n+2)\left[\frac{1}{n} + \frac{1}{n+2}\right] = 84n(n+2)\left(\frac{13}{84}\right)$$ Multiply by the LCD, $84n(n + 2)$.

$$84(n+2) + 84n = 13n(n+2)$$
$$84n + 168 + 84n = 13n^2 + 26n$$ Simplify, and write the quadratic equation in standard form.

$$13n^2 - 142n - 168 = 0$$

$$(n - 12)(13n + 14) = 0$$ Factor the left member.

$$n - 12 = 0 \quad \text{or} \quad 13n + 14 = 0$$ Set each factor equal to zero.

$$n = 12 \qquad n = -\frac{14}{13}$$ Then solve for the smaller integer, n, and the larger integer, $n + 2$.

$$n + 2 = 14$$

This value is not an integer.

Answer The integers are 12 and 14.

Sometimes the equation used in a word problem is based on a well-known formula. For example, a formula used in analyzing the resistance of parallel electrical circuits is

$$\frac{1}{r_1} + \frac{1}{r_2} = \frac{1}{R}$$

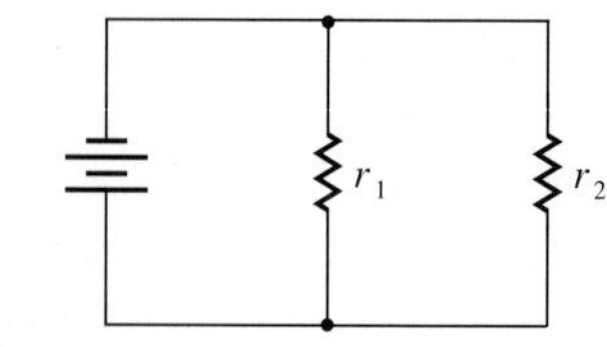

Figure 6-1

The total resistance in the circuit illustrated in Figure 6-1 is R; the individual resistances are r_1 and r_2.

EXAMPLE 4 Resistance in a Parallel Circuit

The total resistance of two resistors in a parallel circuit is 30 ohms. If one resistor has three times the resistance of the other, what is the resistance of each?

SOLUTION Draw a sketch as in Figure 6-1, and let

r = Resistance of the first resistor in ohms

$3r$ = Resistance of the second resistor in ohms

30 = Total resistance in ohms

$$\frac{1}{r_1} + \frac{1}{r_2} = \frac{1}{R}$$ The resistance formula

$$\frac{1}{r} + \frac{1}{3r} = \frac{1}{30}$$ Substitute the values specified in this problem.

$$30r\left(\frac{1}{r} + \frac{1}{3r}\right) = 30r\left(\frac{1}{30}\right)$$ Multiply by the LCD, $30r$.

$$30 + 10 = r$$ Simplify, and then solve for r.

$$40 = r$$

$$r = 40$$

$$3r = 120$$

Answer The resistance of the first resistor is 40 ohms, and the resistance of the second is 120 ohms. ▼

We will now examine a problem that uses both the mixture principle and the rate principle. When these principles are applied to problems involving the time it takes to complete a task or to do some work, they take the following form:

Work A does + Work B does = Total work   Mixture principle

Work done = Rate of work · Time worked   Rate principle

$$W = R \cdot T$$

Since $R = \frac{W}{T}$, if one job can be done in t hours, the rate of work is $\frac{1}{t}$ job per hour.

EXAMPLE 5 Work Done by Two Painters

Working alone, painter A can paint a sign in 6 hours less time than it would take painter B working alone. When A and B work together, the job takes only 4 hours. How many hours would it take each painter working alone to paint the sign?

SOLUTION Let

t = Time in hours for B to paint the sign, working alone

$t - 6$ = Time in hours for A to paint the sign, working alone

Identify specifically what you are trying to determine.

Work A does	+	Work B does	=	Total work of painting one sign

The **word equation** is based on the mixture principle.

	Rate · Time = Work
Painter A	$\frac{1}{t-6} \cdot 4 = \frac{4}{t-6}$
Painter B	$\frac{1}{t} \cdot 4 = \frac{4}{t}$

The right-hand column of the table is labeled "Work," since the word equation involves work that both A and B do together. Since painter A would take $t - 6$ hours, the rate is $\frac{1}{t-6}$. Since painter B would take t hours, the rate is $\frac{1}{t}$.

$$\frac{4}{t-6} + \frac{4}{t} = 1$$

Substitute the values from the table into the word equation. The total work done is 1 job.

$$t(t-6)\frac{4}{t-6} + t(t-6)\frac{4}{t} = t(t-6)(1)$$

Multiply by the LCD, $t(t - 6)$, and then solve for t.

$$4t + 4(t-6) = t^2 - 6t$$
$$4t + 4t - 24 = t^2 - 6t$$
$$8t - 24 = t^2 - 6t$$
$$0 = t^2 - 14t + 24$$
$$0 = (t-12)(t-2)$$

$t - 12 = 0$ or $t - 2 = 0$

$t = 12$ $\quad$ $t = 2$

$t - 6 = 6$ $\quad$ $t - 6 = -4$

This value is not appropriate.

Answer Working alone, painter A could paint the sign in 6 hours and painter B could paint the sign in 12 hours.

SELF-CHECK

A welder can weld a metal framework 5 hours faster than an apprentice can. How long will it take the apprentice to do the job alone if the welder and the apprentice can do it together in 6 hours?

The place value of each digit of a two-digit number, such as 26, determines the value of this number, as shown by $2(10) + 6 = 26$. In general, the value of a two-digit integer with a ten's digit t and a unit's digit u is $10t + u$. Example 6 is a problem involving the digits of a two-digit integer. One of the purposes of this type of problem is to give you further practice in carefully reading a paragraph and translating the words into a concise algebraic statement.

SELF-CHECK ANSWER 

The apprentice can do the job alone in 15 hours. (No other values check in this problem.)

EXAMPLE 6 Numeric Word Problem Involving Place Value

The ten's digit of a two-digit integer is 4 more than the unit's digit. The ratio of the number to the sum of the digits is $\frac{17}{2}$. Find this number.

SOLUTION Let

$$
\begin{aligned}
u &= \text{Unit's digit of this two-digit integer}\\
u + 4 &= \text{Ten's digit of this two-digit integer}\\
10(u + 4) + u &= \text{Value of this two-digit integer}\\
(u + 4) + u &= \text{Sum of the two digits of this number}
\end{aligned}
$$

$$\boxed{\frac{\text{Value of number}}{\text{Sum of digits}}} = \boxed{\frac{17}{2}}$$

Word equation

$$\frac{10(u + 4) + u}{(u + 4) + u} = \frac{17}{2}$$

Substitute the values identified above into the word equation.

$$\frac{11u + 40}{2u + 4} = \frac{17}{2}$$

Simplify by combining like terms.

$$2(u + 2)\left(\frac{11u + 40}{2(u + 2)}\right) = 2(u + 2)\left(\frac{17}{2}\right)$$

Multiply both members by the LCD, $2(u + 2)$.

$$
\begin{aligned}
11u + 40 &= 17(u + 2)\\
11u + 40 &= 17u + 34\\
-6u &= -6\\
u &= 1\\
u + 4 &= 5
\end{aligned}
$$

Simplify, and solve for the unit's digit u and the ten's digit $u + 4$.

Answer The number is 51.

Check

The value of this two-digit integer $= 10(5) + 1 = 51$

The sum of the two digits of this number $= 5 + 1 = 6$

The ratio of the number to the sum of the digits $= \frac{51}{6} = \frac{17}{2}$ ▼

Exercises 6-6

In Exercises 1–20 solve each problem.

Numeric Word Problems (1–14)

1 What number can be added to both the numerator and the denominator of $\frac{17}{25}$ to produce a fraction equal to $\frac{5}{7}$?

2 What number can be subtracted from both the numerator and the denominator of $\frac{17}{25}$ to produce a fraction equal to $\frac{3}{5}$?

3 The sum of the reciprocals of two consecutive integers is $\frac{11}{30}$. Find these integers.

4 The sum of the reciprocals of two consecutive even integers is $\frac{7}{24}$. Find these integers.

5 The sum of the reciprocals of two consecutive odd integers is sixteen times the reciprocal of their product. Find these integers.

6 The sum of the reciprocals of two consecutive even integers is ten times the reciprocal of their product. Find these integers.

7 The denominator of a fraction is an integer that is 6 more than the square of the numerator. The fraction equals $\frac{4}{35}$. Find the numerator.

8 The denominator of a fraction is an integer that is 6 less than the square of the numerator. The fraction equals $\frac{3}{25}$. Find the numerator.

9 Find two consecutive integers such that the sum of the reciprocal of the smaller number and the reciprocal of the square of the larger number is the same as the reciprocal of the product of the smaller number and the square of the larger number.

10 If the square of an integer is added to the numerator of $\frac{17}{35}$ and the integer is added to the denominator, the resulting fraction equals $\frac{11}{7}$. Find this integer.

11 The denominator of a fraction is an integer that is 4 more than the square of the numerator. If the fraction is reduced, it equals $\frac{3}{20}$. Find this numerator.

12 The denominator of a fraction is an integer that is 3 more than the square of the numerator. If the fraction is reduced, it equals $\frac{1}{4}$. Find this numerator.

13 The sum of a number and its reciprocal is $\frac{13}{6}$. Find this number.

14 The difference of a number minus its reciprocal is $\frac{9}{20}$. Find this number.

15 Ratio of Gauge Readings The ratio of two readings from a gauge is $\frac{4}{5}$. The first reading is 3 units above normal, and the second reading is 5 units above normal. What is the normal reading?

16 Ratio of Temperatures The ratio of two temperature readings is $\frac{7}{8}$. The first reading is 7 degrees below normal, whereas the second reading is 2 degrees above normal. What is the normal reading?

17 Wire Length A wire 16 meters long is cut into two pieces whose lengths have a ratio of 3 to 1. Find the length of each piece.

18 Rope Length A rope 20 meters long is cut into two pieces whose lengths have a ratio of 4 to 1. Find the length of each piece.

19 Investment An investor invests $12,000 in a combination of secure funds and high-risk funds. The ratio of dollars invested in secure funds to dollars invested in high-risk funds is 7 to 3. Find the amount invested in secure funds.

20 Complementary Angles The ratio of the measures of two complementary angles is 3 to 2. Find the measure of each angle.

Electrical Resistance (21 and 22)

In Exercises 21 and 22 use the formula $\frac{1}{r_1} + \frac{1}{r_2} = \frac{1}{R}$ to solve each problem. R represents the total resistance in a circuit with individual resistances r_1 and r_2.

21 The total resistance of two resistors in the parallel circuit shown in the figure is 40 ohms. If one resistor has twice the resistance of the other, what is the resistance of each?

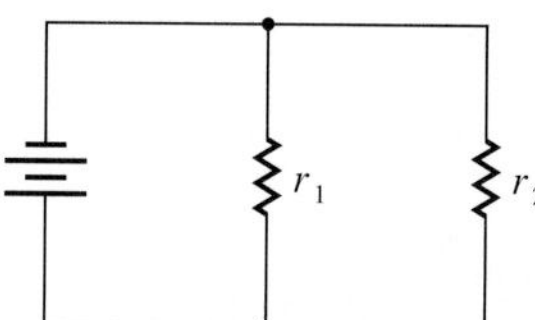

Figure for Exercises 21 and 22

22 The resistance of one resistor in a parallel circuit is 3 ohms greater than that of the second resistor in the circuit. Find the resistance of each resistor if the total resistance in the circuit is $5\frac{1}{7}$ ohms.

Place Value (23 and 24)

23 The unit's digit of a two-digit integer is 5 more than the ten's digit. The ratio of the sum of the digits to the integer is $\frac{1}{3}$. Find this integer.

24 The unit's digit of a two-digit integer is 4 more than the ten's digit. The ratio of the integer to the number formed by reversing the digits is $\frac{5}{17}$. Find this integer.

B

In Exercises 25–36 solve each problem.

Place Value (25 and 26)

25 The unit's digit of a two-digit integer is 3 more than the ten's digit. The ratio of the product of the digits to the integer is $\frac{1}{2}$. Find this integer.

26 The ten's digit of a two-digit integer is 6 more than the unit's digit. The ratio of the product of the digits to the integer is $\frac{9}{31}$. Find this integer.

Assets and Liabilities (27 and 28)

27 The assets of a small automobile dealership are approximated by $500x + 10{,}000$, and its liabilities are approximated by $100x + 14{,}000$, where x represents the number of vehicles sold. One measure of the strength of this business is the ratio of its assets to its liabilities. For the month of January the ratio was 5:4. How many vehicles were sold in January?

28 The assets of a heating and air-conditioning business are approximated by $300x + 5{,}000$, and its liabilities are approximated by $200x + 6{,}000$, where x represents the number of furnaces sold. One measure of the strength of this business is the ratio of its assets to its liabilities. For the month of February the ratio was $\frac{7}{6}$. How many furnaces were sold in February?

29 Production of Computer Chips A computer chip company can produce u units of a computer chip for \$48 per chip, plus an initial start-up investment of \$120,000. To compete on the market, it must be able to achieve an average cost of only \$72 per unit. How many units must the company produce to reach an average cost of \$72 per unit?

30 Kitchen Repairs A carpenter contracted to repair a kitchen for \$360. The work took him 6 hours more than he had estimated, and as a consequence he earned \$3 per hour less than he had estimated. How many hours had he estimated?

31 Painting A painter contracted to paint a duplex for \$672. The job took her 6 hours less than she had estimated, and as a consequence she earned \$2 per hour more than she had estimated. How many hours had she estimated?

32 Water Current Two canoes depart from the same point at the same time. After a period of time one has traveled 24 miles downstream and the other has traveled 10 miles upstream. If both canoes have a speed of 8.5 miles per hour in still water, determine the rate of the current.

33 Speed of a Car A trip of 255 miles includes 90 miles on gravel roads and the rest on paved roads. The driver averages 10 miles per hour faster on the paved roads than on the gravel roads. The total time for this trip is 5 hours. Determine the average speed on both types of road.

Length of a Square (34 and 35)

34 If the length of a side of a square is decreased by 8 centimeters, the area of the resulting square is only one-twenty-fifth of the area of the original square. Find the length of each side of the original square. (See the figure shown.)

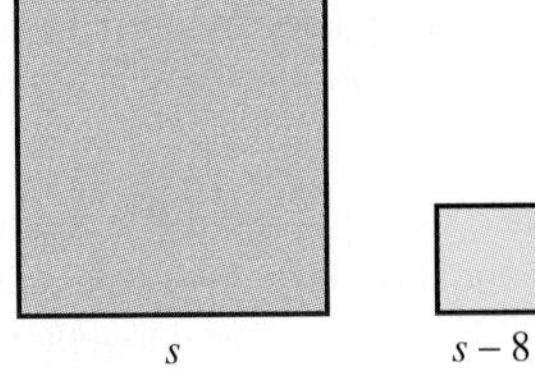

Figure for Exercise 34

35 If each side of a square is increased by 2 centimeters, the area of the new square is nine-fourths of the area of the original square. What is the length of each side of the original square? (See the figure shown.)

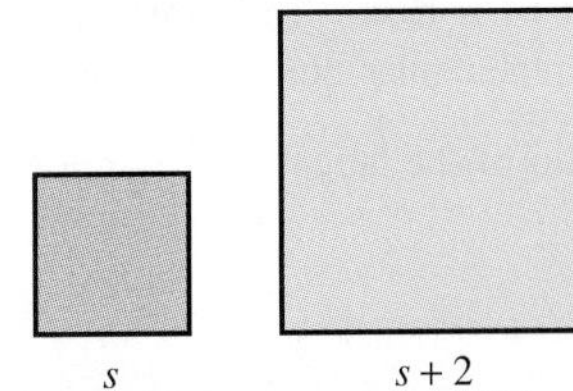

Figure for Exercise 35

36 Dimensions of a Metal Sheet A rectangular piece of metal is 3 centimeters longer than it is wide. A 1-centimeter strip is cut off each of the sides, leaving an area that is nine-twentieths of the original area. Find the original dimensions. (Refer to the figure shown.)

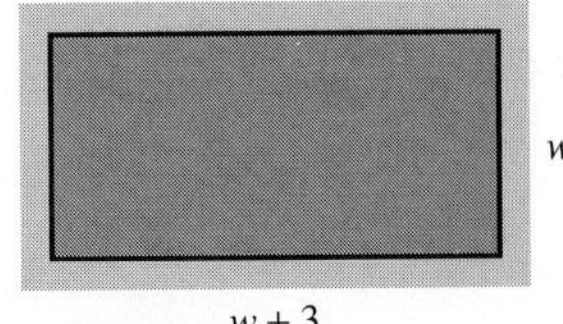

Figure for Exercise 36

C

In Exercises 37–46 solve each problem.

37 Consecutive Integers The reciprocal of the smaller of two consecutive integers minus the reciprocal of the larger integer is equal to the reciprocal of their product. Find all integers that satisfy this condition.

38 Search Time for Planes Search plane A can search an area for a crash victim in 50 hours. Planes A and B can jointly search the area in 30 hours. How many hours would it take plane B to search the area alone?

39 Generating a Mailing List Computer A takes twice as long as computer B to generate a mailing list. If a particular list can be generated in 9 hours when both computers are used, how many hours would it take each computer to generate the list working separately?

40 Processing Billings A new computer can process a firm's monthly billings in two-thirds of the time required by the old computer. If the billings can be completed in 9 hours when both computers are used, how many hours does it take each computer to process the billings working separately?

41 Filling a Tank One pipe can fill a cooling tank in 3 hours less than a second pipe. If the two pipes can fill seven-ninths of the tank in 4 hours when used together, how many hours would it take each pipe alone to fill the tank?

42 Wages In one week two workers each grossed $360 in hourly wages. The employee with more seniority made $1 per hour more than the other employee but worked 4 hours less. What is each employee's hourly rate of pay?

43 Dimensions of a Box The three dimensions (in centimeters) of a rectangular box are consecutive integers. The numerical ratio of the height to the area of the base is $\frac{2}{15}$. Find these dimensions if the height is the shortest dimension. (See the figure shown.)

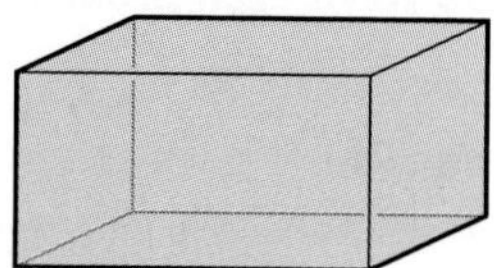

Figure for Exercise 43

Discussion Questions (44 and 45)

44 Discuss why it is important for you to practice writing the answers to word problems as full sentences. Address in your answer the broader objective that you are trying to accomplish when you practice problem-solving skills.

45 Write your own problem that the equation $\frac{6}{t} + \frac{6}{t-5} = 1$ algebraically models.

Challenge Question

A square sheet of metal with an area of 100 square centimeters has a circle of maximum size cut from it. Then a square of maximum area is cut from this circle. Only this square piece of metal is used. How much waste is there?

Section Skill Reinforcers

Simplify each of these expressions.

1 $\dfrac{\frac{1}{2}}{\frac{2}{3}}$ **2** $\dfrac{\frac{x}{y}}{\frac{3x}{4y}}$ **3** $\dfrac{1 - \frac{1}{x}}{1 - \frac{1}{x^2}}$ **4** $\dfrac{1 - \frac{2}{x} - \frac{3}{x^2}}{1 - \frac{5}{x} + \frac{6}{x^2}}$ **5** $\dfrac{1 - \frac{1}{x^3}}{1 + \frac{1}{x} + \frac{1}{x^2}}$

Key Concepts for Chapter 6

1 Excluded values:
The domain set for the variables in a rational expression is understood to include all real numbers except those excluded because they would result in division by zero.

2 Lowest terms:
To reduce a rational expression to lowest terms, factor both the numerator and the denominator and divide by any common nonzero factors. The expression is in lowest terms when the numerator and the denominator have no common factor other than -1 or 1.

3 Least common denominator (LCD):
The least common denominator of two or more fractions is the product formed by using each factor the greatest number of times it occurs in any of the denominators.

4 Operations with rational expressions:
If a, b, c, and d are real expressions, then

a. $\dfrac{a}{c} + \dfrac{b}{c} = \dfrac{a+b}{c}$ for $c \neq 0$

b. $\dfrac{a}{c} - \dfrac{b}{c} = \dfrac{a-b}{c}$ for $c \neq 0$

c. $\dfrac{a}{b} \cdot \dfrac{c}{d} = \dfrac{a \cdot c}{b \cdot d}$ for $b \neq 0$ and $d \neq 0$

d. $\dfrac{a}{b} \div \dfrac{c}{d} = \dfrac{a}{b} \cdot \dfrac{d}{c} = \dfrac{a \cdot d}{b \cdot c}$ for $b \neq 0$, $d \neq 0$, and $c \neq 0$

5 Signs of fractions:

a. $-\dfrac{a}{b} = \dfrac{-a}{b} = \dfrac{a}{-b}$

b. $\dfrac{1}{b-a} = -\dfrac{1}{a-b}$

c. $\dfrac{a-b}{b-a} = -1$

6 Order of operations:
When simplifying rational expressions, as with all other real expressions, perform operations in the following order:

a. Start with the expression within the innermost grouping symbols.

b. Perform all exponentiations.

c. Perform all multiplications and divisions, working from left to right.

d. Perform all additions and subtractions, working from left to right.

7 Simplifying complex fractions:

a. Method I: Express both the numerator and the denominator of the complex fraction as single terms, then multiply the numerator by the reciprocal of the denominator.

b. Method II: Multiply both the numerator and the denominator of the complex fraction by the LCD of all the fractions that occur in the numerator and the denominator of the complex fraction.

8 Simplifying fractions that involve negative exponents: To eliminate negative exponents on x from a fraction, multiply both the numerator and the denominator of this fraction by an appropriate power of x.

9 Solving equations containing rational expressions: When solving an equation with a variable in the denominator, check to make sure that the solution does not include an excluded value that would cause division by zero, thereby making the value extraneous.

10 Strategies for forming equations from word problems:

a. Read the problem carefully to determine what you are asked to find.

b. Use any well-known formula that is applicable to the problem.

c. Translate key phrases into algebraic statements.

d. Use sketches or tables to clarify relationships among variables and to form equations.

e. Use the rate principle and the mixture principle to form equations.

11 Rate of work:
If one job can be done in a time t, then the rate of work is $\dfrac{1}{t}$ of the job per unit of time.

Review Exercises for Chapter 6

In Exercises 1–7 reduce each expression to lowest terms.

1 $\dfrac{36x^2y}{12xy^3}$

2 $\dfrac{6a - 18b}{12b - 4a}$

3 $\dfrac{x^2 - y^2}{x^3 - y^3}$

4 $\dfrac{x^2 + x - 30}{2x^2 + 11x - 6}$

5 $\dfrac{9x^2 - 24xy + 16y^2}{12x^2 - 25xy + 12y^2}$

6 $\dfrac{cx - cy}{ax - ay + bx - by}$

7 $\dfrac{125v^3 + 343w^3}{10av + 14aw - 5bv - 7bw}$

In Exercises 8–10 determine the values excluded from the domain of the variable because they would cause division by zero in these rational expressions.

8 $\dfrac{7x - 9}{4x - 2}$

9 $\dfrac{5x^2 - 7x + 11}{x^2 - 36}$

10 $\dfrac{6x^2 - 23x + 21}{20x^2 - 23x - 7}$

In Exercises 11 and 12 find the missing numerators.

11 $\dfrac{2x + 1}{x - 3} = \dfrac{?}{x^2 - 6x + 9}$

12 $\dfrac{y - 3}{y + 4} = \dfrac{?}{2y^2 + 9y + 4}$

In Exercises 13 and 14 find the least common denominator for the denominators listed.

13 $18a^3b$ and $42a^2b^2$

14 $2x - 6y$ and $x^2 - 9y^2$

In Exercises 15–34 simplify each expression as completely as possible.

15 $\dfrac{36x^2 - 24x}{6x}$

16 $\dfrac{6xy}{2x - y} \cdot \dfrac{4x^2 - y^2}{3x^2}$

17 $\dfrac{9t^2 - 4}{16st} \div \dfrac{3t + 2}{16st^2 + 8st}$

18 $\dfrac{v^2 + 9vw + 8w^2}{v^2 - w^2} \div \dfrac{v^2 + 7vw - 8w^2}{v^2 + 5vw - 6w^2}$

19 $\dfrac{3x}{6x^2 + x - 1} - \dfrac{1}{6x^2 + x - 1}$

20 $\dfrac{2y}{27y^3 - 1} + \dfrac{y - 1}{27y^3 - 1}$

21 $\dfrac{1}{w + 1} - \dfrac{w}{w - 2} + \dfrac{w^2 + 2}{w^2 - w - 2}$

22 $\dfrac{3v}{3v^2 - 5v + 2} - \dfrac{2v}{2v^2 - v - 1}$

23 $\dfrac{6x}{2x - 3} - \dfrac{9x + 18}{4x^2 - 9} \cdot \dfrac{2x^2 - x - 6}{x^2 - 4}$

24 $\dfrac{2w + 4}{w^2 + 4w - 12} + \dfrac{w^2 - 169}{2w^2 - 13w + 21} \div \dfrac{w^2 - 15w + 26}{2w - 7}$

25 $\dfrac{6v^2 - 25v + 4}{6v^2 + 5v - 6} \div \dfrac{2v^2 - 3v - 20}{4v^2 + 16v + 15}$

26 $\dfrac{z^2 - 2z + 1}{z^5 - z^4} \cdot \dfrac{2z^4}{z^3 - 1} + \dfrac{2z^2 + 2z}{z^2 + z + 1}$

27 $\left(\dfrac{y}{4} - \dfrac{4}{y}\right)\left(y - \dfrac{y^2}{y + 4}\right)$

28 $\dfrac{5z + 5}{6z^2 + 13z + 6} - \dfrac{1 - 4z}{3z^2 - 7z - 6} - \dfrac{3z}{2z^2 - 3z - 9}$

29 $\dfrac{\dfrac{1}{a} + \dfrac{1}{a + 1}}{\dfrac{1}{a} + \dfrac{1}{a^2}}$

30 $\dfrac{x + \dfrac{44}{x + 5} - 10}{x + \dfrac{33}{x + 5} - 9}$

31 $\left(\dfrac{3v^2 + 11v + 6}{3v^2 + 5v + 2}\right)^2 \div \dfrac{v^2 - 9}{v^2 + 2v + 1}$

32 $\dfrac{36w^{-2} + 23w^{-1} - 8}{24 - 5w^{-1} - 36w^{-2}}$

33 $\left(m - \dfrac{15}{m + 2}\right) \div \left(m - 1 - \dfrac{10}{m + 2}\right)$

34 $\dfrac{6y}{3y + 2} + \dfrac{20y + 4}{15y^2 + 7y - 2} \div \dfrac{5y^2 - 24y - 5}{5y^2 - 26y + 5}$

In Exercises 35–42 solve each equation.

35 $\dfrac{7}{2x} = \dfrac{2}{x} - \dfrac{3}{2}$

36 $\dfrac{15}{w^2 + 5w} + \dfrac{w + 4}{w + 5} = \dfrac{w + 3}{w}$

37 $\dfrac{1}{y^2 + 5y + 6} - \dfrac{2}{y + 3} = \dfrac{7}{y + 2}$

38 $1 - \dfrac{14}{y^2 + 4y + 4} = \dfrac{7y}{y^2 + 4y + 4}$

39 $\dfrac{w}{w^2 - 9w + 20} + \dfrac{14}{w^2 - 3w - 4} = \dfrac{18}{w^2 - 4w - 5}$

40 $\dfrac{4}{v^2 + 7v + 10} - \dfrac{3}{v + 2} + \dfrac{8}{v + 5} = 0$

41 $\dfrac{5z + 11}{2z^2 + 7z - 4} = \dfrac{1}{z + 4} - \dfrac{3}{1 - 2z}$

42 $\dfrac{1}{2z^2 - 9z - 5} - \dfrac{1}{2z^2 - 6z - 20} = \dfrac{3}{(2z + 1)(2z + 4)(z - 5)}$

Calculator Usage (43 and 44)

43 The sum obtained by performing the addition below is given either by a or by b. Check both possibilities by evaluating all expressions for $x = 1$.

$$\frac{3.2x + 2.7}{25.38x^2 - 9.35x - 21.73} + \frac{2.2x + 1.4}{25.38x^2 - 9.35x - 21.73}$$

a. $\dfrac{1}{4.7x - 5.3}$

b. $\dfrac{1}{4.7x + 5.3}$

44 Use a graphics calculator to graph $y_1 = \dfrac{2.3x^2 - 12.0x + 11.2}{2.3x - 2.8}$ and $y_2 = x - 4$. Then use these graphs to determine whether $\dfrac{2.3x^2 - 12.0x + 11.2}{2.3x - 2.8}$ equals $x - 4$.

Exercises 45 and 46 have two parts. In part a solve the equation, and in part b perform the indicated operations and simplify the result.

	Solve	*Simplify*
45	**a.** $\dfrac{1}{m + 2} = \dfrac{3}{m + 4}$	**b.** $\dfrac{1}{m + 2} - \dfrac{3}{m + 4}$
46	**a.** $\dfrac{x - 2}{x + 1} = \dfrac{x - 4}{x - 6}$	**b.** $\dfrac{x - 2}{x + 1} - \dfrac{x - 4}{x - 6}$

In Exercises 47–54 solve each problem.

Numeric Word Problems (47–49)

47 The denominator of a fraction is 6 more than the numerator, and the fraction equals -1. Find the numerator.

48 The ratio of one number to another is $\frac{5}{2}$. If the first number is 5 less than three times the second, find both numbers.

49 The sum of the reciprocals of two consecutive odd integers is twelve times the reciprocal of their product. Find these integers.

50 Filling a Bathtub A hot-water faucet takes 7 minutes longer than the cold-water faucet to fill a bathtub. If both faucets were turned on, the tub would be half full in 6 minutes. How many minutes would it take to fill an empty tub with cold water if the hot-water faucet were turned off?

51 Average Cost A furniture company can produce u units of a desk for \$54 per desk, plus an initial start-up investment of \$20,000. To compete on the market, it must be able to achieve an average cost of only \$74 per unit. How many units must the company produce to reach an average cost of \$74 per unit?

52 Dimensions of a Metal Sheet A rectangular piece of metal is 6 centimeters longer than it is wide. A 0.5-centimeter strip is cut off of each side, leaving an area thirteen-sixteenths of the original area. Find the original dimensions. (See the figure shown.)

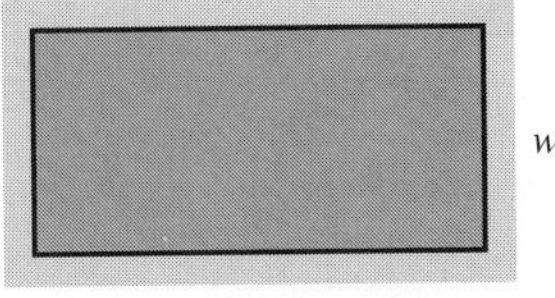

Figure for Exercise 52

53 Place Value The unit's digit of a two-digit integer is 2 more than the ten's digit. The ratio of the sum of the digits to the value of the number is $\frac{4}{19}$. Find this number.

54 Speed of Boats Two boats having the same speed in still water depart simultaneously from a dock, traveling in opposite directions in a river that has a current of 6 miles per hour. After a period of time one boat is 54 miles downstream and the other boat is 30 miles upstream. What is the speed of each boat in still water?

In Exercises 55–57 simplify each expression as completely as possible.

55 $\dfrac{x^{2m} - 10x^m + 25}{x^{2m} - x^m - 20}$

56 $\dfrac{4x^m - 5y^n}{x^m - y^n} + \dfrac{3y^n - 2x^m}{x^m - y^n}$

57 $\dfrac{2x^m - 4}{x^{2m} - 2x^m - 35} - \dfrac{x^{2m} - 4x^m - 5}{x^{2m} - 25} \cdot \dfrac{x^m + 3}{x^{2m} - 6x^m - 7}$

Error Analysis (58 and 59)

58 Explain why it is permissible to multiply the equation $\dfrac{x - 3}{x - 2} = \dfrac{2x - 5}{x - 2}$ by $x - 2$ to obtain $x - 3 = 2x - 5$, which does not contain fractions, but it is not permissible to multiply the expression $\dfrac{3x + 2}{x - 2}$ by $x - 2$ to obtain $3x + 2$.

59 Examine the student's work shown below, and correct any mistakes you find.

$$\frac{5x - 7}{2x - 12} - \frac{3x - 5}{2x - 12} = \frac{5x - 7 - 3x - 5}{2x - 12}$$
$$= \frac{2x - 12}{2x - 12}$$
$$= 1$$

Mastery Test for Chapter 6

Exercise numbers correspond to Section Objective numbers.

[6-1] 1 Reduce each rational expression to lowest terms.

a. $\dfrac{2x-6}{x^2-9}$

b. $\dfrac{2x^2-7x-15}{x^2-25}$

c. $\dfrac{4x^2-12xy+9y^2}{3ay-6by-2ax+4bx}$

[6-2] 2 Simplify each of the following expressions.

a. $\dfrac{ax-bx}{x^2}\cdot\dfrac{5x}{4a-4b}$

b. $\dfrac{x^2-1}{x+1}\div\dfrac{x^2-3x+2}{x-2}$

c. $\dfrac{v^2-v-20}{v^2-16}\div\dfrac{3v-15}{2v-8}$

[6-3] 3 Simplify each of the following expressions.

a. $\dfrac{3x-4}{2x-3}+\dfrac{x-2}{2x-3}$

b. $\dfrac{3x+7}{x^2+x-12}-\dfrac{2x+3}{x^2+x-12}$

c. $\dfrac{7}{w^2+w-12}+\dfrac{2}{w^2-8w+15}$

[6-4] 4 Simplify each of the following expressions.

a. $\dfrac{4x}{5}+\dfrac{x^2}{15}\cdot\dfrac{3}{x}$ **b.** $\left(\dfrac{4x}{5}+\dfrac{x^2}{15}\right)\cdot\dfrac{3}{x}$

c. $\dfrac{x-y}{x+y}+\dfrac{3x-21y}{x^2-y^2}\div\dfrac{15x^2-105xy}{10x^2y-10xy^2}$

[6-4] 5 Simplify each of the following expressions.

a. $\dfrac{4}{\frac{5}{6}}$ **b.** $\dfrac{\frac{x}{3}-2+\frac{3}{x}}{1-\frac{3}{x}}$

c. $\dfrac{36x^{-2}-3x^{-1}-18}{12x^{-2}-25x^{-1}+12}$

[6-5] 6 Solve each of the following equations.

a. $\dfrac{z-4}{z-2}=\dfrac{1}{z-2}$ **b.** $\dfrac{z-2}{z-1}+\dfrac{z-3}{2z-5}=1$

c. $\dfrac{z+11}{z^2-5z+4}=\dfrac{5}{z-4}+\dfrac{3}{1-z}$

[6-6] 7 Solve each of the following problems.

a. Ratio of Economic Indicators The ratio of two monthly economic indicators is $\frac{2}{3}$. The first monthly indicator is 2 units below normal, and the second monthly indicator is 2 units above normal. What is the normal number of units for this indicator?

b. Consecutive Even Integers The sum of the reciprocals of two consecutive even integers is $\frac{9}{40}$. Find these integers.

c. Work by Two Machines When members of a construction crew use two end loaders at once, they can move a pile of sand in 6 hours. If they use only the larger end loader, they can do the job in 5 hours less time than it would take if they used the smaller machine. How many hours would it take to do the job using the larger machine?

ACTIVE APPLICATION

1 A wilderness tour company flies its clients in pontoon planes to remote locations for day trips and also for extended expeditions. The planes travel at an airspeed of 110 mi/h, and wind along the flight route normally blows parallel to the flight path. On a day trip clients are supposed to get 8 full hours of fishing and other recreational activities. Day trips are supposed to take no more than 10 hours, including flying time. Using w as the speed of the wind on a given day, write an expression for the maximum distance that a day trip's destination can be from the airfield.

2 A day trip to a lake 100 miles away is scheduled on a day on which the wind is blowing at 30 mi/h. Given the information in problem 1, can the trip still take place?

3 A flight deep into the wilderness consists of three legs. The first leg is 100 miles long and normally involves no headwind or tailwind. The second leg is 85 miles long and usually has a headwind of variable speed. The third leg is 90 miles long and usually has a headwind about twice as strong as that of the second leg. Write an expression for the total time that the trip would take a plane flying at 110 mi/h.

4 A pilot flying the route in problem 3 encounters a headwind of 25 mi/h on the second leg. How long can the pilot expect the entire trip to take? Express your answer in hours and minutes.

Cumulative Review of Chapters 4–6

The limited purpose of this review is to help you gauge your mastery of these three chapters. It is not meant to examine each detail from these chapters, nor is it meant to focus on specific sections that may be emphasized at any one particular school.

In Exercises 1–23 simplify each expression.

1 $15x^2 - 5x^2$ **2** $15x^2 + 5x^2$ **3** $(15x^2)(5x^2)$

4 $\dfrac{15x^2}{5x^2}$ **5** -3^2 **6** $(-3)^2$

7 3^{-2} **8** 2^{-3} **9** $(2x + 3y)^0$

10 $(2x)^0 + (3y)^0$ **11** $2x^0 + 3y^0$ **12** $\left(\dfrac{1}{4} + \dfrac{1}{5}\right)^{-2}$

13 $3x + 4y - (2x - 5y)$ **14** $7a - 3b - 2(5a - b)$

15 $2x(3x^2 - 4x + y)$ **16** $(3v - w)(v + 2w)$

17 $(5x - 1)(5x + 1)$ **18** $(2a + 3b)^2$

19 $(2a^2b^3)(5a^3b^6)$ **20** $\dfrac{20x^7y^4}{5x^5y^5}$

21 $\dfrac{14a^3b^4 - 21a^2b^5}{7a^2b^3}$ **22** $\dfrac{z^4 - 1}{z - 1}$

23 $\left(\dfrac{15a^3b^3}{25a^2b^4}\right)^2$

24 Divide $15x^2 + 14x - 8$ by $5x - 2$.

25 What is the degree of $5x^3 + 2x^2 + x$?

26 Evaluate the expressions that are defined, and identify the expressions that are undefined.

a. $\dfrac{0}{5}$ **b.** $\dfrac{5}{0}$ **c.** 5^0 **d.** 0^5

e. 0^0 **f.** -1^0 **g.** 0^{-1}

27 Write 0.00078 in scientific notation.

28 What is the name of the property that justifies $3x(4x + 5) = (3x)(4x) + (3x)(5)$?

In Exercises 29–34 factor each polynomial over the integers.

29 $8x^2 - 72y^2$ **30** $18ax^2 - 60axy + 50ay^2$

31 $3x^2 + 20xy - 7y^2$ **32** $5a^3 - 40b^3$

33 $2x^2 + 10x + ax + 5a$ **34** $x^2 + 4xy + 4y^2 - 9$

In Exercises 35–42 perform the indicated operations, and express the answer in simplest form.

35 $\dfrac{x^2 - 4y^2}{x^2 + xy - 6y^2}$

36 $\dfrac{27x^3 - 1}{6x^2 + 13x - 5}$

37 $\dfrac{m^2 - n^2}{5m^2n + 5mn^2} \div \dfrac{m^3 - n^3}{3m^2 + 3mn + 3n^2}$

38 $\dfrac{a}{a - 2b} - \dfrac{b}{a - b}$

39 $\dfrac{2z + 11}{z^2 + z - 6} + \dfrac{2}{z + 3} - \dfrac{3}{z - 2}$

40 $\dfrac{x + 2}{x + 3} - \dfrac{x^3 - 8}{x^2 - 9} \cdot \dfrac{x - 3}{x^2 + 2x + 4}$

41 $\dfrac{\dfrac{1}{x^2} - 49}{7 - \dfrac{1}{x}}$ **42** $\dfrac{6x^{-2} - 5x^{-1} + 1}{1 - 4x^{-2}}$

In Exercises 43–48 solve each equation.

43 $(2x - 3)(x + 4) = 0$ **44** $6v^2 + 11v - 10 = 0$

45 $(2x - 3)(x + 4) = -12$ **46** $2w^3 - 8w = 0$

47 $\dfrac{12}{x + 2} = \dfrac{6}{x - 1}$

48 $\dfrac{4}{m^2 + 2m - 3} - \dfrac{3}{m + 2} = \dfrac{1}{m - 1}$

49 Solve $x^2 + 3xy - 10y^2 = 0$ for x.

50 Construct a quadratic equation in x with solutions -3 and $\dfrac{4}{5}$.

Exercises 51 and 52 have two parts. In part a perform the indicated operations and simplify the result, and in part b solve the equation.

Simplify

51 a. $\dfrac{5}{(x - 1)^2} - 3 - \dfrac{5x}{(x - 1)^2}$

Solve

b. $\dfrac{5}{(x - 1)^2} = 3 + \dfrac{5x}{(x - 1)^2}$

Simplify

52 a. $\dfrac{x - 1}{x + 2} + \dfrac{x}{x - 2} + 1 - \dfrac{8}{x^2 - 4}$

Solve

b. $\dfrac{x - 1}{x + 2} + \dfrac{x}{x - 2} + 1 = \dfrac{8}{x^2 - 4}$

CALCULATOR USAGE (53 AND 54)

53 The correct difference obtained by performing the addition below is given either by a or by b. Check both possibilities by evaluating all expressions for $x = 1$.

$$\frac{5.6x - 1.7}{7.35x^2 - 21.28x + 10.92} - \frac{1.4x + 1.1}{7.35x^2 - 21.28x + 10.92}$$

a. $\dfrac{1}{3.5x - 7.8}$

b. $\dfrac{2}{3.5x - 7.8}$

54 Use a graphics calculator to graph $y_1 = \dfrac{7.36x^2 - 32.87x + 36.4}{3.2x - 6.5}$ and $y_2 = 2.3x - 5.6$. Then use these graphs to determine whether $\dfrac{7.36x^2 - 32.87x + 36.4}{3.2x - 6.5}$ equals $2.3x - 5.6$.

55 Length of a Rafter Find r, the length in feet of the rafter shown in the figure. The lengths of the two braces are $r - 2$ and $r - 9$.

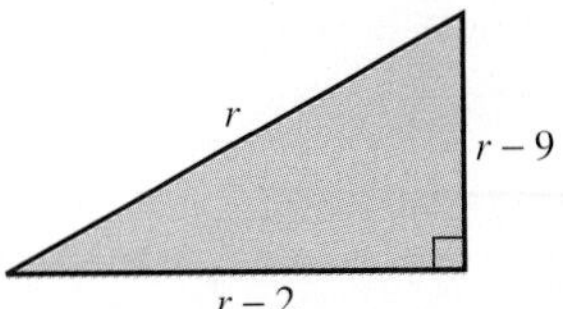

Figure for Exercise 55

56 Average Speed An automobile completed a 480-kilometer trip. If the average rate had been increased by 10 kilometers per hour, the trip would have taken 2 hours less than if the average rate had been decreased by 10 kilometers per hour. What was the average rate for this trip?

Chapter 7

EXPONENTS, ROOTS, AND RADICALS

AN APPLICATION

A Problem for Engineers

A structural engineer designing a bridge is faced with the challenge of creating a structure that will not collapse under its own weight. This is harder than it sounds, and the odd thing about the problem is that it depends on scale. Suppose that the engineer built a scale model of a bridge, using real concrete and real scaled-down steel girders, and found that the structure was solid. If she then tried to build a functioning bridge four, six, or ten times as large as the model, she might find that it could not hold up its own weight, let alone carry any traffic. The bigger the bridge, the worse the problem.

Why is this so? The answer lies in the relationship between an object's *volume* and the *area* of its cross-section. Recall from Chapter 1 that the volume of a rectangular box equals its length times its width times its height. If we think of a bridge girder as a long, thin rectangular box, we can see that quadrupling all its dimensions will make it $4 \times 4 \times 4 = 64$ times as large in volume, and hence 64 times as heavy. The trouble is that most measures of the the strength of the girder—its ability to withstand stresses of various kinds—vary directly as the area of the girder's cross-section, which is only $4 \times 4 = 16$ times as large. Therefore the girder is under 4 times the stress from its own weight as it was in the model.

One way to express the relationship between volume and strength directly (without reference to length, width, and height) is to say that for a beam *of a given shape,* strength varies directly as volume *to the two-thirds power.* That is, strength is proportional to $V^{2/3}$, where two-thirds is a *fractional exponent.* Fractional exponents will be explained in this chapter. In the meantime, note the qualifier "of a given shape." The familiar I-beam is a standard shape for steel girders because it provides maximum strength for minimum weight.

Let's apply the point about volume and area to a topic seemingly far removed from engineering. In *Gulliver's Travels* the hero voyages to the land of Brobdingnag, where he meets a race of people just like human beings but ten times as tall. What is wrong with this scenario?

Chapter 7 Objectives

1. **Interpret and use rational exponents (Section 7-1).**
2. **Interpret and use radical notation (Section 7-2).**
3. **Simplify radical expressions (Section 7-2).**
4. **Add and subtract radical expressions (Section 7-3).**
5. **Reduce the order of a radical (Section 7-3).**
6. **Multiply radical expressions (Section 7-4).**
7. **Divide and rationalize radical expressions (Section 7-5).**
8. **Solve equations involving radical expressions (Section 7-6).**
9. **Express complex numbers in standard form (Section 7-7).**
10. **Add, subtract, multiply, and divide complex numbers (Section 7-7).**

In Chapter 4 the meaning and the properties of integer exponents were introduced. This chapter will broaden the concept of exponents to include fractional exponents. It will also explain the meaning of rational number exponents and show that these exponents obey all the properties of integral exponents.

A fractional exponent is just another way of indicating the root of a number; thus we will also examine roots and radicals. This chapter will conclude with an examination of the meaning of the square root of a negative number.

Some important definitions and properties of exponents that were covered in Chapter 4 are summarized below.

Definitions:

$$x^n = \underbrace{x \cdot x \cdot x \cdot \cdots \cdot x}_{n \text{ factors}} \quad \text{for } n = 1, 2, 3, \ldots$$

$$x^0 = 1 \quad \text{for } x \neq 0$$

$$x^{-n} = \frac{1}{x^n} \quad \text{for } x \neq 0, n = 1, 2, 3, \ldots$$

Properties: For $x \neq 0$, $y \neq 0$, and integers m and n,

$$x^m \cdot x^n = x^{m+n} \quad \text{Product rule}$$

$$\frac{x^m}{x^n} = x^{m-n} \quad \text{Quotient rule}$$

$$(x^m)^n = x^{mn} \quad \text{Power rule}$$

$$(xy)^m = x^m y^m \quad \text{Product to a power}$$

$$\left(\frac{x}{y}\right)^m = \frac{x^m}{y^m} \quad \text{Quotient to a power}$$

$$\left(\frac{x}{y}\right)^{-m} = \left(\frac{y}{x}\right)^m \quad \text{Negative exponent}$$

7-1 Rational Exponents

SECTION OBJECTIVE

1 Interpret and use rational exponents.

Since fractional exponents are defined in terms of the principal root of a number, we will begin this section with a review of nth roots. A **square root** of a real number x is a number r for which $r^2 = x$. For example, -2 and 2 are both square roots of 4 since $(-2)^2 = 4$ and $2^2 = 4$. Similarly, a **cube root** of a real number x is a number r for which $r^3 = x$. For example, 5 is a cube root of 125 since $5^3 = 125$, and -2 is a cube root of -8 since $(-2)^3 = -8$. In general, an **nth root** of a real number x is a number r for which $r^n = x$. For example, $\frac{2}{3}$ is a square root of $\frac{4}{9}$ since $\left(\frac{2}{3}\right)^2 = \frac{4}{9}$. Likewise,

$-\frac{2}{3}$ is a square root of $\frac{4}{9}$ since $\left(-\frac{2}{3}\right)^2 = \frac{4}{9}$.

$-\frac{3}{5}$ is a cube root of $-\frac{27}{125}$ since $\left(-\frac{3}{5}\right)^3 = -\frac{27}{125}$.

5 is a fourth root of 625 since $5^4 = 625$.

-5 is a fourth root of 625 since $(-5)^4 = 625$.

1 is an nth root of 1 since $1^n = 1$ for any natural number n.

0 is an nth root of 0 since $0^n = 0$ for any natural number n.

Note that 4 has two real square roots, -2 and 2, whereas 8 has only one real cube root, 2. The characteristics of nth roots are described in the nth Root Theorem, which is usually not proven until more advanced courses.

*n*th ROOT THEOREM

If x is a real number and n is a natural number, then

For $x < 0$: If n is odd, x has exactly one real nth root. This nth root is negative.
If n is even, x has no real nth root.*

For $x = 0$: Zero is the only nth root of zero for all natural numbers n.

For $x > 0$: If n is odd, x has exactly one real nth root. This nth root is positive.
If n is even, x has two real nth roots. One nth root is negative, and the other (its opposite) is positive.

*Every nonzero real number has n nth roots that are complex numbers. Some of these roots may not be real numbers, however. We will restrict our discussion to roots that are real numbers until after we have covered imaginary numbers in Section 7-7.

A Mathematical Note

Pierre deFermat's last theorem is one of the most famous theorems in mathematics. It states that there do *not* exist positive integers x, y, z, and n such that $x^n + y^n = z^n$ for $n > 2$.

Two of the features that make this theorem so interesting are (1) it is easily understood—it resembles the Pythagorean Theorem $a^2 + b^2 = c^2$ and thus shows that the Pythagorean theorem does not generalize to a higher-degree equation—and (2) Fermat (c. 1635) wrote beside this theorem the marginal note "I have assuredly found an admirable proof of this [theorem], but the margin is too narrow to contain it."

After more than 300 years, a proof of this theorem was announced by Andrew Wiles in June of 1993. Since then a flaw has been discovered in the proof. At the time of the writing of this book the flaw has yet to be fixed, and Fermat's last theorem is still unproven.

To avoid ambiguity when using symbols to represent roots, we often use the term **principal *n*th root.** If there is only one real *n*th root of a number, this root is the principal root. If a number has both a positive and a negative *n*th root, the positive root is defined to be the principal root.

EXAMPLE 1 Determining Principal *n*th Roots

Give each principal *n*th root described below.

	SOLUTIONS	
(a) The principal square root of 4	2	-2 is a square root of 4, but it is not the principal square root.
(b) The principal fourth root of 625	5	-5 is a fourth root of 625, but it is not the principal fourth root.
(c) The principal cube root of 8	2	2 is the only real cube root of 8, so it is the principal cube root.
(d) The principal cube root of -8	-2	-2 is the only real cube root of -8, so it is the principal cube root.
(e) The principal square root of -9	This number has no real number as a principal square root.	By definition, the principal root must be a real number, and there is no real number whose square is -9.

According to the power rule for exponents,

$$(x^{1/n})^n = x^{n/n} = x$$

Thus it is logical to interpret $x^{1/n}$ as an *n*th root of x. For example, $x^{1/2}$ is a square root of x since $(x^{1/2})^2 = x$. Because we wish $x^{1/n}$ to have a unique value, we use $x^{1/n}$ to represent only the principal root of x.

THE PRINCIPAL *n*th ROOT

The principal *n*th root of the real number x is denoted by either $x^{1/n}$ or $\sqrt[n]{x}$. (See Section 7-2 for a discussion of the radical notation.)

For $x < 0$: If n is odd, the principal root is negative.
If n is even, there is no real *n*th root.

For $x = 0$: The principal root is 0.

For $x > 0$: The principal root is positive for all natural numbers n.

EXAMPLE 2 Writing *n*th Roots in Exponential Form

Simplify each of the following expressions.

	SOLUTIONS	
(a) $25^{1/2}$	$25^{1/2} = 5$	The principal square root of 25 is 5, not -5.
(b) $-25^{1/2}$	$-25^{1/2}$ means $-(25^{1/2}) = -5$.	Be careful not to confuse the notations in parts (b) and (c); their meanings are quite distinct. The negative base in part (c), -25, has no real square root.
(c) $(-25)^{1/2}$	$(-25)^{1/2}$ is not a real number.	

(d) $\left(\frac{8}{27}\right)^{1/3}$	$\left(\frac{8}{27}\right)^{1/3} = \frac{2}{3}$	The principal cube root of $\frac{8}{27}$ is $\frac{2}{3}$ since $\left(\frac{2}{3}\right)^3 = \frac{8}{27}$.
(e) $(-0.008)^{1/3}$	$(-0.008)^{1/3} = -0.2$	The principal cube root of -0.008 is -0.2 since $(-0.2)^3 = -0.008$.
(f) $32^{1/5}$	$32^{1/5} = 2$	The principal fifth root of 32 is 2 since $2^5 = 32$. ▼

To extend the definition of exponents to include any rational number $\frac{m}{n}$, we will again use the power rule for exponents. According to the power rule, both $(x^{1/n})^m$ and $(x^m)^{1/n}$ are equal to $x^{m/n}$. Thus $x^{2/3}$ can be interpreted as either $(x^{1/3})^2$ or $(x^2)^{1/3}$. Both the square of the cube root of x and the cube root of x squared have the same value. We will usually use the first form for arithmetic computations and the second form for algebraic results. In keeping with our earlier work with negative exponents, we will interpret $x^{-m/n}$ as $\frac{1}{x^{m/n}}$. These new definitions are summarized in the following box.

SELF-CHECK 

Simplify each of the following expressions.

1 $64^{1/2}$ **2** $64^{1/3}$
3 $(-64)^{1/2}$ **4** $-64^{1/2}$
5 $64^{1/6}$

RATIONAL EXPONENTS, $x^{m/n}$

For a real number x and natural numbers m and n,

$$x^{m/n} = (x^{1/n})^m = (x^m)^{1/n} \quad \text{if } x^{1/n} \text{ is a real number}$$

$$x^{-m/n} = \frac{1}{x^{m/n}} \quad \text{if } x \neq 0 \text{ and } x^{1/n} \text{ is a real number}$$

If $x < 0$ and n is even, then $x^{1/n}$ is not a real number.

To simplify computations, we generally assume that the rational exponent m/n is expressed in reduced form.

EXAMPLE 3 Evaluating Expressions with Rational Exponents

Simplify each of the following expressions.

	SOLUTIONS	
(a) $8^{2/3}$	$8^{2/3} = (8^{1/3})^2 = (2)^2 = 4$ or $8^{2/3} = (8^2)^{1/3} = (64)^{1/3} = 4$	We usually use the first form because it results in easier calculations.
(b) $-8^{2/3}$	$-8^{2/3} = -(8^{2/3}) = -(4) = -4$	Again note that you must be careful to correctly interpret the meaning of the negative signs in parts (b) and (c).
(c) $(-8)^{2/3}$	$(-8)^{2/3} = [(-8)^{1/3}]^2 = (-2)^2 = 4$	
(d) $8^{-2/3}$	$8^{-2/3} = \frac{1}{8^{2/3}} = \frac{1}{4}$	When you see negative exponents, think "reciprocate."

SELF-CHECK ANSWERS 

1 8 **2** 4 **3** No real principal root **4** -8 **5** 2

(e) $32^{3/5}$ $\quad 32^{3/5} = (32^{1/5})^3 = 2^3 = 8$

(f) $(-16)^{3/4}$ $\quad (-16)^{3/4}$ is not a real number. $\quad (-16)^{3/4}$ is not a real number since $(-16)^{1/4}$ is not a real number.

(g) $0^{-3/7}$ $\quad 0^{-3/7}$ is not a real number. $\quad 0^{-3/7} = \frac{1}{0^{3/7}} = \frac{1}{0}$, which is undefined.

(h) $5^{2/3}$ $\quad 5^{2/3}$ is a positive real number, but it cannot be simplified further. $\quad 5^{2/3}$ is an irrational number.

We have defined fractional exponents in such a way that all the properties of integral exponents now apply to all rational exponents. For now, however, we will restrict our variables to positive values; the next section will discuss the special care that must be exercised when the bases are negative.

EXAMPLE 4 Using the Properties of Exponents

Simplify each of the following expressions. Express all answers in terms of positive exponents. Assume that all variables represent positive real numbers.

SOLUTIONS

(a) $9^{3/8} \cdot 9^{1/8}$

$9^{3/8} \cdot 9^{1/8} = 9^{(3/8)+(1/8)}$ — Product rule—add exponents:

$= 9^{1/2}$ — $\frac{3}{8} + \frac{1}{8} = \frac{4}{8} = \frac{1}{2}$

$= 3$ — The principal square root of 9 is 3.

(b) $8^{1/5} \cdot 4^{1/5}$

$8^{1/5} \cdot 4^{1/5} = (8 \cdot 4)^{1/5}$ — Product to a power: $a^n b^n = (ab)^n$

$= 32^{1/5}$

$= 2$ — The principal fifth root of 32 is 2.

(c) $(z^{-2/3})^{-3}$

$(z^{-2/3})^{-3} = z^{(-2/3)(-3)}$ — Power rule—multiply exponents:

$= z^2$ — $\left(-\frac{2}{3}\right)(-3) = 2$

SELF-CHECK

Simplify each of the following expressions.

1 $64^{3/2}$ **2** $64^{2/3}$

3 $(-8)^{5/3}$ **4** $\left(\frac{27}{8}\right)^{-2/3}$

EXAMPLE 5 Using the Properties of Exponents

Simplify $\left(\frac{16x^4}{25}\right)^{-3/2}$. Express the answer in terms of positive exponents. Assume that x is a positive real number.

SOLUTION

$$\left(\frac{16x^4}{25}\right)^{-3/2} = \left[\left(\frac{16x^4}{25}\right)^{1/2}\right]^{-3} \qquad x^{m/n} = (x^{1/n})^m$$

$$= \left[\frac{16^{1/2}x^{4/2}}{25^{1/2}}\right]^{-3}$$

$$= \left[\frac{4x^2}{5}\right]^{-3}$$

Take the $\frac{1}{2}$ power of each factor in the numerator and the denominator, and then simplify. Note that $16^{1/2} = 4$ and $25^{1/2} = 5$.

SELF-CHECK ANSWERS

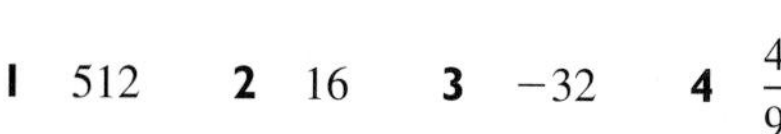

1 512 **2** 16 **3** −32 **4** $\frac{4}{9}$

$= \left[\dfrac{5}{4x^2}\right]^3$ Reciprocate the base for negative exponents.

$= \dfrac{5^3}{4^3(x^2)^3}$ Raise each factor to the third power. To raise a power to a power, multiply exponents.

$= \dfrac{125}{64x^6}$

The product of exponential expressions with two or more terms can be found using some of the skills developed for multiplying polynomials. In particular, we will use special forms, such as the product of a sum times a difference: $(a + b)(a - b) = a^2 - b^2$.

SELF-CHECK

Simplify each expression.

1 $\dfrac{y^{3/5}}{y^{1/4}}$ **2** $(m^{1/2})^{2/3}$

3 $(a^{-10}b^4)^{-1/2}$

EXAMPLE 6 Multiplying a Sum by a Difference

Simplify $(x^{1/2} + y^{1/2})(x^{1/2} - y^{1/2})$. Express the answer in terms of positive exponents. Assume that all variables represent positive real numbers.

SOLUTION $(x^{1/2} + y^{1/2})(x^{1/2} - y^{1/2}) = (x^{1/2})^2 - (y^{1/2})^2$ By inspection, observe that this expression fits the special form $(a + b)(a - b) = a^2 - b^2$, with $a = x^{1/2}$ and $b = y^{1/2}$.

$= x - y$ To raise a power to a power, multiply exponents.

EXAMPLE 7 Multiplying Using the Distributive Property

Simplify $2x^{2/3}(3x^{1/3} - 5x^{-2/3})$. Express the answer in terms of positive exponents. Assume that all variables represent positive real numbers.

SOLUTION $2x^{2/3}(3x^{1/3} - 5x^{-2/3}) = (2x^{2/3})(3x^{1/3}) - (2x^{2/3})(5x^{-2/3})$ First use the distributive property.

$= 6x^{3/3} - 10x^0$ Then simplify each term, adding the exponents of the factors with the same base.

$= 6x - 10$ Note that $x^0 = 1$.

EXAMPLE 8 Multiplying the Factors of the Sum of Two Cubes

Simplify $(x^{1/3} + y^{1/3})(x^{2/3} - x^{1/3}y^{1/3} + y^{2/3})$. Express the answer in terms of positive exponents. Assume that all variables represent positive real numbers.

SOLUTION $(x^{1/3} + y^{1/3})(x^{2/3} - x^{1/3}y^{1/3} + y^{2/3}) = (x^{1/3})^3 + (y^{1/3})^3$ By inspection, observe that this expression fits the special form $(a + b)(a^2 - ab + b^2) = a^3 + b^3$, with $a = x^{1/3}$ and $b = y^{1/3}$.

$= x + y$ To raise a power to a power, multiply exponents.

SELF-CHECK ANSWERS

1 $y^{7/20}$ **2** $m^{1/3}$ **3** $\dfrac{a^5}{b^2}$

EXAMPLE 9 Approximating Roots with a Calculator

Use a calculator to approximate the principal sixth root of 34 to five significant digits.

SOLUTION The sixth root of 34 is the same as $34^{1/6}$.

Keystrokes

Scientific

[3] [4] [x^y] [(] [1] [÷] [6] [)] [=] → 1.799892164

Graphics

[3] [4] [^] [(] [1] [÷] [6] [)] [ENTER] → 1.799892164

Answer The principal sixth root of 34, $34^{1/6}$, is approximately equal to 1.7999.

Exercises 7-1

A

In Exercises 1 and 2 write an algebraic expression for each phrase, using rational exponents to denote all *n*th roots.

1 a. The fifth root of x

b. The cube root of the quantity 2 plus z

c. Two plus the cube root of z

2 a. The sixth root of x

b. Three times the seventh root of y

c. The seventh root of $3y$

In Exercises 3–12 simplify each expression as completely as possible. Identify the expressions that are not real numbers.

3 a. $36^{1/2}$ **b.** $-36^{1/2}$ **c.** $(-36)^{1/2}$ **d.** $36^{-1/2}$

4 a. $27^{1/3}$ **b.** $-27^{1/3}$ **c.** $(-27)^{1/3}$ **d.** $27^{-1/3}$

5 a. $\left(\frac{8}{125}\right)^{1/3}$ **b.** $-\left(\frac{8}{125}\right)^{1/3}$ **c.** $\left(-\frac{8}{125}\right)^{1/3}$ **d.** $\left(\frac{8}{125}\right)^{-1/3}$

6 a. $0.09^{1/2}$ **b.** $-0.09^{1/2}$ **c.** $(-0.09)^{1/2}$ **d.** $0.09^{-1/2}$

7 a. $16^{1/2}$ **b.** $16^{1/4}$ **c.** $-16^{1/2}$ **d.** $16^{-1/4}$

8 a. $81^{1/2}$ **b.** $81^{1/4}$ **c.** $-81^{-1/2}$ **d.** $(-81)^{-1/4}$

9 a. $\left(\frac{8}{125}\right)^{2/3}$ **b.** $\left(\frac{-8}{125}\right)^{2/3}$ **c.** $\left(\frac{8}{125}\right)^{-2/3}$ **d.** $\left(\frac{-8}{125}\right)^{-2/3}$

10 a. $0.25^{3/2}$ **b.** $-0.25^{3/2}$ **c.** $0.25^{-3/2}$ **d.** $(-0.25)^{-3/2}$

11 a. $(25 + 144)^{1/2}$ **b.** $25^{1/2} + 144^{1/2}$ **c.** $(9 + 16)^{1/2}$ **d.** $9^{1/2} + 16^{1/2}$

12 a. $(81 + 1600)^{1/2}$ **b.** $81^{1/2} + 1600^{1/2}$ **c.** $(289 - 225)^{1/2}$ **d.** $289^{1/2} - 225^{1/2}$

In Exercises 13–36 simplify each expression. Express all answers in terms of positive exponents. Assume that all variables represent positive real numbers.

13 $5^{1/2} \cdot 5^{3/2}$ **14** $7^{5/3} \cdot 7^{1/3}$ **15** $(8^{5/3})^{2/5}$

16 $(32^{7/10})^{2/7}$ **17** $\frac{11^{4/3}}{11^{1/3}}$ **18** $\frac{9^{5/4}}{9^{3/4}}$

19 $(27^{1/12} \cdot 27^{-5/12})^{-2}$ **20** $(4^{1/5} \cdot 4^{2/5})^{5/2}$

21 $x^{1/3} \cdot x^{1/2}$ **22** $y^{1/4} \cdot y^{1/5}$

23 $\frac{x^{1/2}}{x^{1/3}}$ **24** $\frac{y^{1/4}}{y^{1/5}}$

25 $(z^{3/4})^{2/7}$ **26** $(z^{5/12})^{4/15}$

27 $\frac{w^{-2/3}}{w^{-5/3}}$ **28** $\frac{w^{3/4}}{w^{-3/8}}$

29 $(v^{-10}w^{-15})^{-1/5}$

30 $(v^{-26}w^{-39})^{-1/13}$

31 $(16v^{-2/5})^{3/2}$

32 $(25v^{-4/9})^{3/2}$

33 $\left(\dfrac{16n^{2/3}}{81n^{-2/3}}\right)^{-3/4}$

34 $\left(\dfrac{27n^{3/5}}{n^{-3/5}}\right)^{-2/3}$

35 $\dfrac{(27x^2y)^{1/2}(3xy)^{1/2}}{5x^{1/2}y^2}$

36 $\dfrac{(25x^2y^3z)^{1/3}(5xy^2z^2)^{1/3}}{3x^2y^{-1/3}z^{-2}}$

B

In Exercises 37–54 perform the indicated multiplications, and express the answers in terms of positive exponents. Assume that all variables represent positive real numbers.

37 $x^{3/5}(x^{2/5} - x^{-3/5})$

38 $x^{2/3}(x^{4/3} + x^{-2/3})$

39 $y^{-7/4}(2y^{11/4} - 3y^{7/4})$

40 $y^{-9/7}(5y^{23/7} - 6y^{16/7})$

41 $3w^{5/11}(2w^{17/11} - 5w^{6/11} - 9)$

42 $4w^{-5/3}(3w^{11/3} - 7w^{8/3} + 2w^{7/3})$

43 $(a^{1/2} + 3)(a^{1/2} - 3)$

44 $(2a^{1/2} - 3b^{1/2})(2a^{1/2} + 3b^{1/2})$

45 $(b^{3/5} - c^{5/3})(b^{3/5} + c^{5/3})$

46 $(b^{3/5} + c^{5/3})^2$

47 $(b^{3/5} - c^{5/3})(b^{3/5} - c^{5/3})$

48 $(x^{1/2} - x^{-1/2})^2$

49 $(x^{-1/2} + x^{1/2})^2$

50 $(x^{2/3} + x)(x^{-2/3} - x)$

51 $(y^{1/3} + 2)(y^{2/3} - 2y^{1/3} + 4)$

52 $(3y^{1/3} - 5)(9y^{2/3} + 15y^{1/3} + 25)$

53 $[(3^{1/3} + 5^{1/3})(3^{2/3} - 15^{1/3} + 5^{2/3})]^{5/3}$

54 $[(14^{1/3} - 5^{1/3})(14^{2/3} + 70^{1/3} + 5^{2/3})]^{3/2}$

Estimation Skills (55–58)

55 The best mental estimate of $145^{1/2}$ is ____.

a. 73 **b.** 72 **c.** 14 **d.** 13 **e.** 12

56 The best mental estimate of $170^{1/2}$ is ____.

a. 96 **b.** 97 **c.** 14 **d.** 13 **e.** 12

57 The best mental estimate of $1003^{1/3}$ is ____.

a. 333 **b.** 300 **c.** 133 **d.** 100 **e.** 10

58 The best mental estimate of $124^{1/3}$ is ____.

a. 42 **b.** 43 **c.** 41 **d.** 5 **e.** 10

C

In Exercises 59–61 simplify each expression. Assume that *x* and *y* represent positive real numbers and that *m* is a natural number.

59 $x^{m/3}x^{m/2}$

60 $\dfrac{16x^{m/2}}{2x^{m/3}}$

61 $(x^{m/3}y^{m/2})^6$

62 Factor $x^{5/2} + 3x^{3/2} - 4x^{1/2}$ by first factoring out $x^{1/2}$.

63 Factor $x^{3/2} - 25x^{-1/2}$ by first factoring out $x^{-1/2}$.

64 Complete the following table for $y = x^{1/2}$. Then use these points to draw a line graph of $y = x^{1/2}$.

x	0	0.25	0.81	1.00	1.44	2.25	4	9
y								

65 Complete the following table for $y = x^{1/3}$. Then use these points to draw a line graph of $y = x^{1/3}$.

x	0	0.027	0.125	1	8
y					

Calculator Usage (66–70)

In Exercises 66–70 use a calculator with a power key labeled $\boxed{y^x}$ or $\boxed{\wedge}$ to approximate each number to five significant digits.

66 $37^{1/3}$

67 $73^{1/4}$

68 $12^{2/3}$

69 $128^{2/7}$

70 The third power of the fifth root of 86

71 Give an example of a real number x for which $x^{1/2}$ is defined but $x^{-1/2}$ is undefined.

Discussion Questions (72 and 73)

72 Explain why $x^{1/2}$ is defined to be $\sqrt{x}$.

73 Give an example of a real number x for which $(x^2)^{1/2} \neq x$. Discuss why this does not violate the power rule which states $(x^a)^b = x^{ab}$.

Challenge Question

a. Which is larger: $(2^2)^2$ or $2^{(2^2)}$?
b. Which is larger: $(3^3)^3$ or $3^{(3^3)}$?
c. Which is larger: $2^{(3^4)}$ or $4^{(3^2)}$?

Section Skill Reinforcers

1 Which is a rational number: $\sqrt{9}$ or $\sqrt{10}$? **2** Which is an irrational number: $\sqrt{63}$ or $\sqrt{64}$? **3** List one rational number between $\sqrt{5}$ and $\sqrt{6}$. **4** Simplify $|-7| + |7|$. **5** Simplify $|23| - |-23|$.

7-2 Radicals

SECTION OBJECTIVES

2 Interpret and use radical notation.

3 Simplify radical expressions.

In Section 7-1 the principal nth root of a real number x was represented by $x^{1/n}$. This way of representing roots is very useful because it enables us to apply all the properties of exponents to simplify problems involving nth roots. However, another notation for nth roots that you may be more familiar with is radical notation. First we will examine this notation, and then we will relate it to the properties covered in Section 7-1.

> **A Mathematical Note**
>
> The radical sign, $\sqrt{}$, is composed of two parts: $\surd$ and $\overline{}$. The symbol $\surd$ comes from the letter r, the first letter of the Latin word *radix,* which means root. Thus the $\surd$ indicates that a root is to be taken of the quantity underneath the bar (also called the vinculum).
>
> On many calculators we can access the symbol $\surd$ but not the bar. To indicate the square root of a quantity, we can use parentheses instead of the bar. For example, $\sqrt{2x+1}$ can be represented on some calculators by $\surd\ (2x + 1)$.

RADICAL NOTATION

For a real number x and a natural number n, if $x^{1/n}$ is a real number,

$$\sqrt[n]{x} = x^{1/n}, \text{ the principal } n\text{th root of } x$$

- x is the radicand.
- $\sqrt{}$ is the radical sign, or the radical.
- n is the index, or the order of the radical.

If no index is written, $\sqrt{x}$ is interpreted as the square root of x, with index 2.

If $x > 0$ and n is even, $\sqrt[n]{x}$ always represents the positive nth root. For example, $\sqrt{25}$ denotes $+5$, the principal square root of 25. To denote the negative square root of 25, we write $-\sqrt{25} = -5$. If n is odd, then $\sqrt[n]{x}$ has the same sign as x. For example, $\sqrt[3]{8} = 2$ and $\sqrt[3]{-8} = -2$. If $x < 0$ and n is even, then $\sqrt[n]{x}$ is not a real number. For example, $\sqrt{-1}$ is not a real number since no real number squared yields -1. (We will examine this case in Section 7-7.)

EXAMPLE 1 Using Radical Notation

Write each radical in exponential notation, express the meaning of the notation in words, and then evaluate the expression.

SOLUTIONS

	Exponential Notation	*Meaning*	*Value*	*Check*
(a) $\sqrt{9}$	$9^{1/2}$	The principal square root of 9	3	$3^2 = 9$
(b) $\sqrt[3]{8}$	$8^{1/3}$	The principal cube root of 8	2	$2^3 = 8$
(c) $\sqrt[5]{-243}$	$(-243)^{1/5}$	The principal fifth root of -243	-3	$(-3)^5 = -243$
(d) $\sqrt[6]{0}$	$0^{1/6}$	The principal sixth root of 0	0	$0^6 = 0$

If $x^{1/n}$ is a real number,* then, as noted in the previous section,

$$x^{m/n} = (x^{1/n})^m = (x^m)^{1/n}$$

---

*$(-1)^{1/2}$ is one example of an expression of this form that is not a real number.

In radical notation, this statement becomes

$$x^{m/n} = (\sqrt[n]{x})^m = \sqrt[n]{x^m} \quad \text{if } \sqrt[n]{x} \text{ is a real number}$$

SELF-CHECK

Evaluate each expression.
1 $\sqrt{64}$ **2** $-\sqrt{64}$
3 $\sqrt[3]{64}$

EXAMPLE 2 Writing Expressions with Rational Exponents in Radical Form

Write each of the following expressions in radical notation.

SOLUTIONS

(a) $7^{3/4}$ $\quad 7^{3/4} = (\sqrt[4]{7})^3 = \sqrt[4]{7^3}$

(b) $(-9)^{7/5}$ $\quad (-9)^{7/5} = (\sqrt[5]{-9})^7 = \sqrt[5]{(-9)^7}$ ▼

EXAMPLE 3 Evaluating Radical Expressions

Evaluate each of these radical expressions.

SOLUTIONS

(a) $(\sqrt[3]{8})^2$ $\quad (\sqrt[3]{8})^2 = (2)^2 = 4$

(b) $\sqrt[3]{8^2}$ $\quad \sqrt[3]{8^2} = \sqrt[3]{64} = 4$

Compare parts (a) and (b), and note that for arithmetic expressions it is easier to carry out the steps as illustrated in part (a).

(c) $(\sqrt[5]{-32})^4$ $\quad (\sqrt[5]{-32})^4 = (-2)^4 = 16$

First evaluate the radical expression within the parentheses. Then raise this value to the appropriate power.

(d) $\sqrt{(-5)^2}$ $\quad \sqrt{(-5)^2} = \sqrt{25} = 5$

(e) $(\sqrt{-5})^2$ $\quad \sqrt{-5}$ is not a real number.

Compare parts (d) and (e), and note the important distinction between these two cases. Square roots of negative numbers are defined in Section 7-7. ▼

EXAMPLE 4 Calculator Usage

Use a calculator to approximate each of these numbers to five significant digits.

SOLUTIONS

(a) $\sqrt{83}$ *Keystrokes*

Scientific

[8] [3] [√] → 9.110433579

Graphics

[√] [8] [3] [ENTER] → 9.110433579

Answer $\sqrt{83} \approx 9.1104$

SELF-CHECK ANSWERS

1 8 **2** −8 **3** 4

(b) $\sqrt[5]{30}$ First express $\sqrt[5]{30}$ in exponential notation as $30^{1/5}$, then use these calculator keystrokes.

Keystrokes

Scientific

Graphics

Answer $\sqrt[5]{30} \approx 1.9744$

Graphics calculators make it very convenient to determine a whole table of square roots or cube roots. The TI-82 graphics calculator will either graph a function or directly display a table of values for the function. The table is obtained by using the TABLE key, as illustrated in the next example. Another method for obtaining these values is to graph $y = \sqrt{x}$ and then use the TRACE key to move along the graph. The calculator window display $x = 4$ and $y = 2$ means that $\sqrt{4} = 2$.

EXAMPLE 5 Calculator Usage

Use a graphics calculator and $y = \sqrt{x}$ to complete this table of values.

x	1	2	3	4	5	6	7
$\sqrt{x}$							

SOLUTION

Keystrokes (using the TABLE feature)

[Y=] [2nd] [√] [X, T, θ] [2nd] [TABLE]

X	Y1	
1	1	
2	1.4142	
3	1.7321	
4	2	
5	2.2361	
6	2.4495	
7	2.6458	
X = 1		

The window displayed assumes that the table has been set to the following values using the TblSet key:

TABLE SETUP
TblMin = 1
Δ Tbl = 1
Indpnt: Auto
Depend: Auto

Keystrokes (using the GRAPH feature)

[Y=] [2nd] [√] [X, T, θ] [ZOOM] [4]

[−4.7, 4.7] for x, [−3.1, 3.1] for y

The [4th] option on the ZOOM menu of the TI-82 will select appropriate decimal values for x. The TRACE key can be used to determine the y-coordinates corresponding to the x values in the table.

Answer

x	1	2	3	4	5	6	7
$\sqrt{x}$	1.000	1.4142	1.7321	2.0000	2.2361	2.4495	2.6458

As Example 3 showed, $\sqrt{(-5)^2} \neq (\sqrt{-5})^2$. The restriction that $(\sqrt[n]{x})^m = \sqrt[n]{x^m}$ if and only if $\sqrt[n]{x}$ is a real number is an important fact of algebra that is easily overlooked. This means that it is wise to be very careful when using this formula with variables. If $x < 0$ and n is an even number, then $\sqrt[n]{x}$ is not a real number and $(\sqrt[n]{x})^m \neq \sqrt[n]{x^m}$. In particular, if $x < 0$, then $\sqrt{x^2}$ will not be x. For example, if $x = -5$, then $\sqrt{x^2} = \sqrt{(-5)^2} = \sqrt{25} = 5$, not -5. Since $\sqrt{x^2}$ is positive whether x is positive or negative, we can use absolute-value notation to correctly denote the result in all cases:

$$\sqrt{x^2} = |x|$$

SELF-CHECK ✓

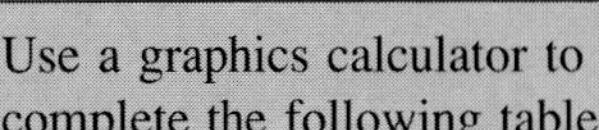

Use a graphics calculator to complete the following table.

x	10	11	12	13	14	15	16
$\sqrt{x}$							

There are two correct ways to handle $\sqrt[n]{x^n}$. The first option is the simpler one: we merely avoid the difficulty by restricting x to positive values, as we did in the exercises in Section 7-1. However, since odd roots pose no difficulty, the second option is to allow x to be negative and use absolute-value notation for the restricted case $\sqrt[n]{x^n}$ when n is even. This option is illustrated in the next example.

$\sqrt[n]{x^n}$

For any real number x,

$$\sqrt[n]{x^n} = |x| \quad \text{if } n \text{ is an even natural number}$$
$$\sqrt[n]{x^n} = x \quad \text{if } n \text{ is an odd natural number}$$

In particular, $\sqrt[2]{x^2} = |x|$.

EXAMPLE 6 Evaluating Radical Expressions

Simplify each of these radical expressions. Use absolute-value notation wherever necessary.

SOLUTIONS

(a) $\sqrt{(-5)^2}$ $\quad \sqrt{(-5)^2} = |-5| = 5$ $\quad$ $\sqrt[n]{x^n} = |x|$ if n is even.

(b) $\sqrt[3]{(-2)^3}$ $\quad \sqrt[3]{(-2)^3} = -2$ $\quad$ $\sqrt[n]{x^n} = x$ if n is odd: $\sqrt[3]{(-2)^3} = \sqrt[3]{-8} = -2$

SELF-CHECK ANSWER ✓

x	10	11	12	13	14	15	16
$\sqrt{x}$	3.1623	3.3166	3.4641	3.6056	3.7417	3.8730	4.0000

(c) $\sqrt{9x^2y^6}$

$$\sqrt{9x^2y^6} = \sqrt{(3xy^3)^2} = |3xy^3| = |3| \cdot |xy^3| = 3|xy^3|$$

First rewrite $9x^2y^6$ as the perfect square of $3xy^3$. Since the index of the radical is even (whenever no index appears the index is understood to be 2), absolute-value notation is needed for the answer.

(d) $\sqrt{9x^2y^4}$

$$\sqrt{9x^2y^4} = \sqrt{(3xy^2)^2}$$

First rewrite $9x^2y^4$ as the perfect square of $3xy^2$.

$$= |3xy^2|$$

Use absolute-value notation since the index of the radical is even (2).

$$= 3|x|y^2$$

Simplify the absolute-value expression, noting that $|3| = 3$ and $|y^2| = y^2$ since $y^2 \geq 0$ for all values of y.

(e) $\sqrt[3]{-125x^3y^6}$

$$\sqrt[3]{-125x^3y^6} = \sqrt[3]{(-5xy^2)^3}$$

First rewrite $-125x^3y^6$ as the perfect cube of $-5xy^2$.

$$= -5xy^2$$

The index (3) is odd, so absolute-value notation should not be used for this answer.

SELF-CHECK

Simplify each of these radical expressions. Use absolute-value notation wherever necessary.

1 $\sqrt[3]{y^6}$ **2** $\sqrt{y^6}$
3 $\sqrt[5]{-32y^5}$

Since radicals are just an alternative way of expressing rational exponents, the properties of radicals follow directly from the properties of exponents. Note that if $x^{1/n}$ or $\sqrt[n]{x}$ is not a real number, then these properties do not hold. For positive values of x and y, these properties are true for both even and odd nth roots.

PROPERTIES OF RADICALS

If $\sqrt[n]{x}$ and $\sqrt[n]{y}$ are both real numbers, then

Radical Form	***Exponential Form***
$\sqrt[n]{xy} = \sqrt[n]{x}\,\sqrt[n]{y}$	$(xy)^{1/n} = x^{1/n}y^{1/n}$
$\sqrt[n]{\frac{x}{y}} = \frac{\sqrt[n]{x}}{\sqrt[n]{y}}$ for $y \neq 0$	$\left(\frac{x}{y}\right)^{1/n} = \frac{x^{1/n}}{y^{1/n}}$ for $y \neq 0$

One way to simplify a radical expression is to use the property for the radical of a product to make the radicand as small as possible. The key is to recognize perfect nth powers. When working with square roots, look for perfect square factors, such as 4, 9, 16, and 25. Similarly, when working with cube roots, look for perfect cube factors, such as 8, 27, 64, and 125.

SELF-CHECK ANSWERS

1 y^2 **2** $|y^3|$ **3** $-2y$

EXAMPLE 7 Simplifying Radicals with Constant Radicands

Simplify each of these radical expressions.

SOLUTIONS

(a) $\sqrt{50}$ 　 $\sqrt{50} = \sqrt{25 \cdot 2}$ 　 25 is a perfect square factor of 50.

$= \sqrt{25}\sqrt{2}$ 　 $\sqrt{xy} = \sqrt{x}\sqrt{y}$

$= 5\sqrt{2}$

(b) $\sqrt{45}$ 　 $\sqrt{45} = \sqrt{9 \cdot 5}$ 　 9 is a perfect square factor of 45.

$= \sqrt{9}\sqrt{5}$ 　 $\sqrt{xy} = \sqrt{x}\sqrt{y}$

$= 3\sqrt{5}$

(c) $\sqrt[3]{40}$ 　 $\sqrt[3]{40} = \sqrt[3]{8 \cdot 5}$ 　 8 is a perfect cube factor of 40.

$= \sqrt[3]{8}\sqrt[3]{5}$ 　 $\sqrt[3]{xy} = \sqrt[3]{x}\sqrt[3]{y}$

$= 2\sqrt[3]{5}$ 　 *Caution:* Do not make the classic error of writing the answer incorrectly as $2\sqrt{5}$, leaving off the index of 3 on the radical symbol. ▼

A second way to simplify a radical expression is to use the property for the radical of a quotient to remove radicals from the denominator. Sometimes this property can also be used to remove fractions from the radicand.

EXAMPLE 8 Simplifying Radicals with Constant Radicands

Simplify each of these radical expressions.

SOLUTIONS

(a) $\sqrt{\frac{4}{9}}$ 　 $\sqrt{\frac{4}{9}} = \frac{\sqrt{4}}{\sqrt{9}}$ 　 $\sqrt{\frac{x}{y}} = \frac{\sqrt{x}}{\sqrt{y}}$

$= \frac{2}{3}$

(b) $\sqrt[3]{\frac{25}{8}}$ 　 $\sqrt[3]{\frac{25}{8}} = \frac{\sqrt[3]{25}}{\sqrt[3]{8}}$ 　 $\sqrt[3]{\frac{x}{y}} = \frac{\sqrt[3]{x}}{\sqrt[3]{y}}$

$= \frac{\sqrt[3]{25}}{2}$ 　 The numerator $\sqrt[3]{25}$ cannot be reduced further since it has no perfect cube factor other than 1.

(c) $\frac{\sqrt{75}}{\sqrt{3}}$ 　 $\frac{\sqrt{75}}{\sqrt{3}} = \sqrt{\frac{75}{3}}$ 　 $\frac{\sqrt{x}}{\sqrt{y}} = \sqrt{\frac{x}{y}}$

$= \sqrt{25}$ 　 Reduce the fraction in the radicand, and then take the square root.

$= 5$ ▼

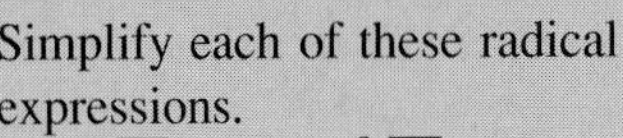

SELF-CHECK

Simplify each of these radical expressions.

1 $\sqrt{27}$ 　 **2** $\sqrt[3]{16}$

3 $\sqrt{\frac{36}{169}}$

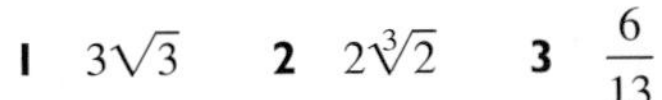

SELF-CHECK ANSWERS

1 $3\sqrt{3}$ 　 **2** $2\sqrt[3]{2}$ 　 **3** $\frac{6}{13}$

In the next example we restrict the variables to positive real numbers so that the product and quotient properties will be true. This restriction makes absolute-value notation unnecessary for the even nth roots.

EXAMPLE 9 Simplifying Radicals with Variable Radicands

Simplify each of these radical expressions. Assume that all variables represent positive real numbers.

SOLUTIONS

(a) $\sqrt{24v^2w^3}$

$\sqrt{24v^2w^3} = \sqrt{(4v^2w^2)(6w)}$ — $4v^2w^2$ is a perfect square factor of $24v^2w^3$.

$= \sqrt{4v^2w^2}\sqrt{6w}$ — $\sqrt{xy} = \sqrt{x}\sqrt{y}$

$= 2vw\sqrt{6w}$

(b) $\sqrt[3]{54v^5w^7}$

$\sqrt[3]{54v^5w^7} = \sqrt[3]{(27v^3w^6)(2v^2w)}$ — $27v^3w^6$ is a perfect cube factor of $54v^5w^7$.

$= \sqrt[3]{27v^3w^6}\sqrt[3]{2v^2w}$ — $\sqrt[3]{xy} = \sqrt[3]{x}\sqrt[3]{y}$

$= 3vw^2\sqrt[3]{2v^2w}$

(c) $\sqrt{\dfrac{a^3b^4}{c^2}}$

$\sqrt{\dfrac{a^3b^4}{c^2}} = \dfrac{\sqrt{a^3b^4}}{\sqrt{c^2}}$ — $\sqrt{\dfrac{x}{y}} = \dfrac{\sqrt{x}}{\sqrt{y}}$

$= \dfrac{\sqrt{(a^2b^4)(a)}}{c}$ — a^2b^4 is a perfect square factor of a^3b^4.

$= \dfrac{\sqrt{a^2b^4}\sqrt{a}}{c}$ — $\sqrt{xy} = \sqrt{x}\sqrt{y}$

$= \dfrac{ab^2\sqrt{a}}{c}$

The conditions that a radical expression must meet in order to be in simplified form are summarized in the following box. We have examined the first three conditions in this section; the fourth condition will be examined in Section 7-3.

SIMPLIFIED FORM FOR RADICAL EXPRESSIONS

A radical expression $\sqrt[n]{x^m}$ is in simplified form if and only if all of the following conditions are satisfied.

1 The radicand is as small as possible; that is, the exponent of m is less than the index of n.

2 There are no fractions in the radicand.

3 There are no radicals in the denominator.

4 The index is as small as possible: $\dfrac{m}{n}$, the fractional power, is not reducible.

Since we will use the quadratic formula frequently to solve quadratic equations, we will now examine a radical expression that comes from the quadratic formula.

EXAMPLE 10 A Radical Expression from the Quadratic Formula

Evaluate the radical expression $\frac{-b + \sqrt{b^2 - 4ac}}{2a}$ for $a = 1$, $b = -4$, and $c = -3$.

SOLUTION

$$\frac{-b + \sqrt{b^2 - 4ac}}{2a} = \frac{-(-4) + \sqrt{(-4)^2 - 4(1)(-3)}}{2(1)}$$ Substitute the given values.

$$= \frac{4 + \sqrt{16 + 12}}{2}$$ Then simplify the radical term in the numerator.

$$= \frac{4 + \sqrt{28}}{2}$$

$$= \frac{4 + \sqrt{4}\sqrt{7}}{2}$$ Note that 4 is a perfect square factor of 28.

$$= \frac{4 + 2\sqrt{7}}{2}$$

$$= \frac{2(2 + \sqrt{7})}{2}$$ Factor the GCF, 2, out of the numerator.

$$= 2 + \sqrt{7}$$ Then reduce by dividing the numerator and the denominator by the common factor, 2.

Exercises 7-2

A

In Exercises 1 and 2 write each radical in exponential notation, write the meaning of the notation in words, and then evaluate the expression.

1 a. $\sqrt{16}$ **b.** $\sqrt[3]{-1}$ **c.** $\sqrt[4]{16}$ **d.** $\sqrt[5]{0}$

2 a. $\sqrt{36}$ **b.** $\sqrt[3]{-8}$ **c.** $\sqrt[5]{32}$ **d.** $\sqrt[22]{1}$

In Exercises 3 and 4 write each expression in radical notation.

3 a. $5^{1/2}$ **b.** $5^{1/3}$ **c.** $7^{2/3}$ **d.** $11^{3/5}$

4 a. $3^{1/2}$ **b.** $2^{1/3}$ **c.** $5^{2/3}$ **d.** $13^{4/7}$

In Exercises 5–34 simplify each radical expression. Identify the expressions that are not real numbers.

5 a. $\sqrt{100}$ **b.** $-\sqrt{100}$

c. $\sqrt{-100}$ **d.** $\frac{1}{\sqrt{100}}$

6 a. $\sqrt{121}$ **b.** $-\sqrt{121}$

c. $\sqrt{-121}$ **d.** $\frac{1}{\sqrt{121}}$

7 a. $\sqrt[3]{27}$ **b.** $-\sqrt[3]{27}$

c. $\sqrt[3]{-27}$ **d.** $-\sqrt[3]{-27}$

8 a. $\sqrt[3]{125}$ **b.** $-\sqrt[3]{125}$

c. $\sqrt[3]{-125}$ **d.** $-\sqrt[3]{-125}$

9 a. $\sqrt[6]{64}$ **b.** $\sqrt[3]{64}$

c. $\sqrt[3]{-64}$ **d.** $\sqrt{64}$

10 a. $\sqrt[4]{81}$ **b.** $\sqrt{81}$

c. $\sqrt{-81}$ **d.** $-\sqrt{81}$

11 a. $\sqrt{\frac{16}{49}}$ **b.** $\sqrt[3]{-\frac{8}{125}}$

c. $\sqrt[5]{\frac{1}{32}}$ **d.** $\sqrt[3]{\left(\frac{8}{27}\right)^2}$

12 a. $\sqrt{\frac{144}{169}}$ **b.** $\sqrt[3]{-\frac{27}{1000}}$

c. $\sqrt[4]{\frac{1}{625}}$ **d.** $\sqrt{\left(\frac{4}{25}\right)^3}$

13 a. $\sqrt{0.25}$ **b.** $\sqrt{0.0001}$

c. $-\sqrt[3]{0.008}$ **d.** $\sqrt[4]{0.0001}$

14 a. $\sqrt{0.36}$ **b.** $\sqrt{0.09}$
c. $\sqrt[3]{-0.027}$ **d.** $\sqrt[3]{0.125}$

15 a. $\sqrt{9+16}$ **b.** $\sqrt{9}+\sqrt{16}$
c. $\sqrt{289-64}$ **d.** $\sqrt{289}-\sqrt{64}$

16 a. $\sqrt{64+225}$ **b.** $\sqrt{64}+\sqrt{225}$
c. $\sqrt{169-25}$ **d.** $\sqrt{169}-\sqrt{25}$

17 $\sqrt{75}$ **18** $\sqrt{8}$ **19** $\sqrt{28}$ **20** $\sqrt{63}$

21 $\sqrt{72}$ **22** $\sqrt{98}$ **23** $\sqrt[3]{24}$ **24** $\sqrt[3]{40}$

25 $\sqrt[3]{54}$ **26** $\sqrt[3]{135}$ **27** $\sqrt[4]{405}$ **28** $\sqrt[4]{48}$

29 $\sqrt{\frac{3}{64}}$ **30** $\sqrt{\frac{7}{81}}$ **31** $\sqrt{\frac{14}{200}}$ **32** $\sqrt{\frac{6}{75}}$

33 $\frac{\sqrt[3]{135}}{\sqrt[3]{5}}$ **34** $\frac{\sqrt[3]{1250}}{\sqrt[3]{10}}$

In Exercises 35–42 simplify each radical expression. The variables can represent any real number, so use absolute-value notation wherever necessary.

35 $\sqrt{25x^2}$ **36** $\sqrt{36w^2}$ **37** $\sqrt{144x^6}$

38 $\sqrt{121y^{10}}$ **39** $\sqrt[3]{8a^3}$ **40** $\sqrt[3]{-125b^3}$

41 $\sqrt[6]{x^{12}y^{18}}$ **42** $\sqrt[5]{x^{15}y^{25}}$

B

ESTIMATION SKILLS (43–46)

43 The best mental estimate of $\sqrt[3]{63}$ is ____.

a. 4 **b.** 3 **c.** 35
d. 30 **e.** 20

44 The best mental estimate of $\sqrt[4]{80}$ is ____.

a. 4 **b.** 3 **c.** 35
d. 30 **e.** 20

45 The best mental estimate of the length of each side of a square of area 26 cm^2 (see the figure) is ____.

a. 3 cm **b.** 4 cm **c.** 5 cm
d. 12 cm **e.** 13 cm

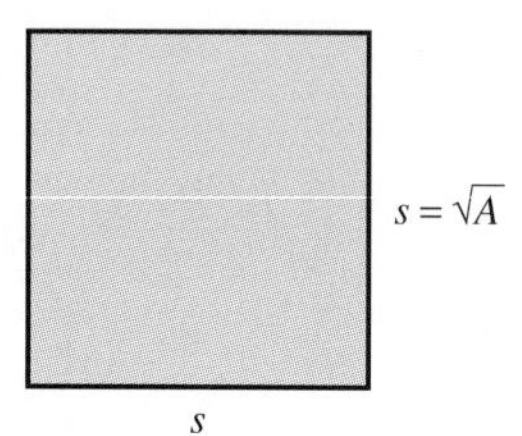

Figure for Exercise 45

46 The best mental estimate of the length of each side of a cubical box whose volume is 26 cm^3 (see the figure) is ____.

a. 3 cm **b.** 4 cm **c.** 5 cm
d. 12 cm **e.** 13 cm

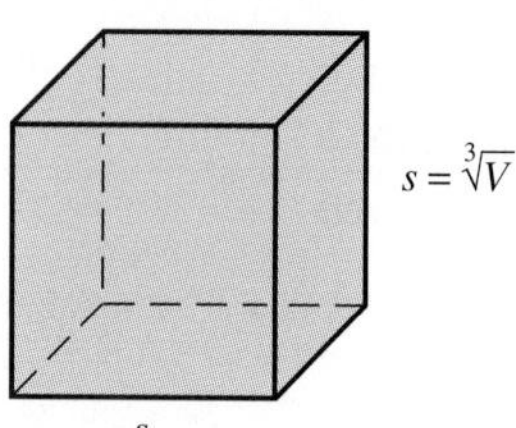

Figure for Exercise 46

In Exercises 47 and 48 write each expression using radical notation.

47 a. $(2v)^{3/4}$ **b.** $2(a^2b^3)^{1/7}$

48 a. $(4v)^{5/7}$ **b.** $3(a^2b^5)^{1/9}$

In Exercises 49–66 simplify each radical expression. Assume that all variables represent positive real numbers so that absolute-value notation is not necessary.

49 $\sqrt[4]{162}$ **50** $\sqrt[4]{70{,}000}$ **51** $\sqrt{50xy^3}$

52 $\sqrt{20a^5b^7}$ **53** $\sqrt{96v^{11}w^{13}}$ **54** $\sqrt[3]{40x^4y^7}$

55 $\sqrt[3]{24a^5b^{22}}$ **56** $\sqrt[5]{-64v^6w^{12}}$ **57** $\sqrt{\frac{a}{b^2}}$

58 $\sqrt{\frac{v^3w^5}{z^2}}$ **59** $\sqrt[3]{\frac{m^4n^3}{s^6}}$ **60** $\sqrt{\frac{5x}{20x^3}}$

61 $\sqrt[3]{\frac{15a^7b^8}{40a^2b^4}}$ **62** $\frac{\sqrt{18v^3w^5}}{\sqrt{98v^2w^2}}$ **63** $\frac{\sqrt[3]{3000v^5w^7}}{\sqrt[3]{3v^2w^2}}$

64 $\sqrt{y^2+2y+1}$ **65** $\sqrt[3]{(2v+3)^3}$ **66** $\sqrt[4]{(3v+5)^4}$

In Exercises 67–70 evaluate $\frac{-b+\sqrt{b^2-4ac}}{2a}$ for the given values of a, b, and c.

67 $a=1, b=-4, c=-7$ **68** $a=1, b=-6, c=4$

69 $a=2, b=-2, c=-1$ **70** $a=9, b=-12, c=-1$

C

CALCULATOR USAGE (71–78)

In Exercises 71–74 use a calculator with a power key labeled $\boxed{y^x}$ or $\boxed{\wedge}$ to approximate each number to five significant digits.

71 $\sqrt[4]{89}$ **72** $\sqrt[5]{129.6}$

73 $\sqrt{\sqrt{12}}$ **74** $\sqrt[3]{\sqrt{17}}$

In Exercises 75 and 76 approximate the length c of the hypotenuse of a right triangle, given the lengths a and b of the other two sides. Use a calculator, and approximate c to the nearest tenth of a centimeter. (See the figure shown.)

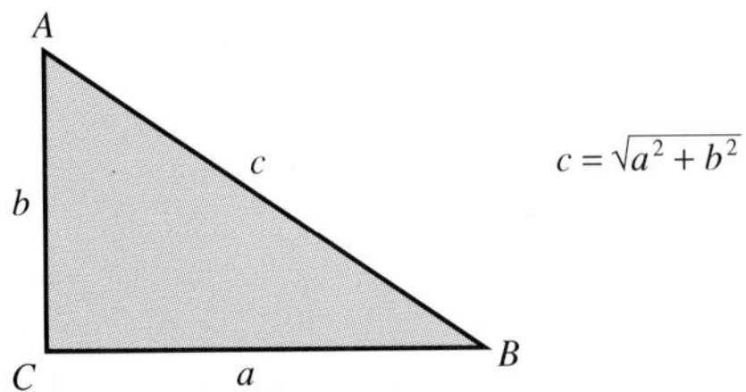

$c = \sqrt{a^2 + b^2}$

Figure for Exercises 75 and 76

75 $a = 29.5$ cm, $b = 35.7$ cm

76 $a = 43.6$ cm, $b = 57.3$ cm

77 Use a graphics calculator to graph $y = \sqrt[3]{x}$, using a window of [0, 30] for x and [0, 4] for y.

78 Use a graphics calculator to complete the following table. Give the missing values to the nearest hundredth.

x	1	2	3	4	5	6	7
$\sqrt[3]{x}$							

79 Free-Fall Time The time t in seconds for a hammer to fall a distance of d feet is given by the formula $t = \frac{\sqrt{d}}{4}$.

Complete the following table, and then use these points to complete the line graph. (See the figure.)

d	0	1	4	9	16	25
t	0	0.25	0.50			

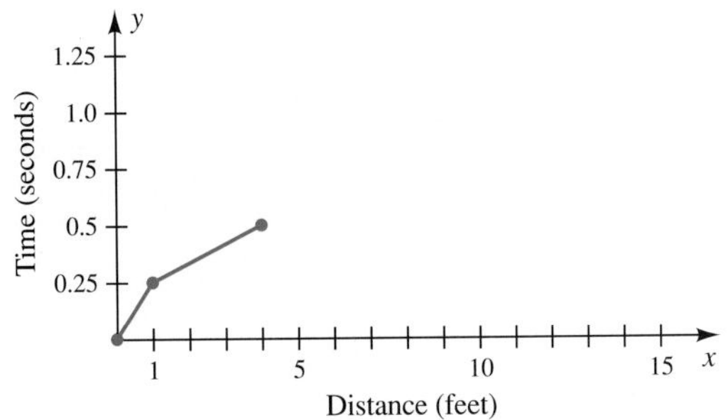

Figure for Exercise 79

80 Length of a Brace The length d of a diagonal brace from the lower corner of a rectangular storage box to the opposing upper corner is given by

$$L = \sqrt{w^2 + l^2 + h^2}$$

where w, l, and h are, respectively, the width, length, and height of the box. Determine the length of a diagonal of a box with width 5.2 meters, length 9.4 meters, and height 6.5 meters.

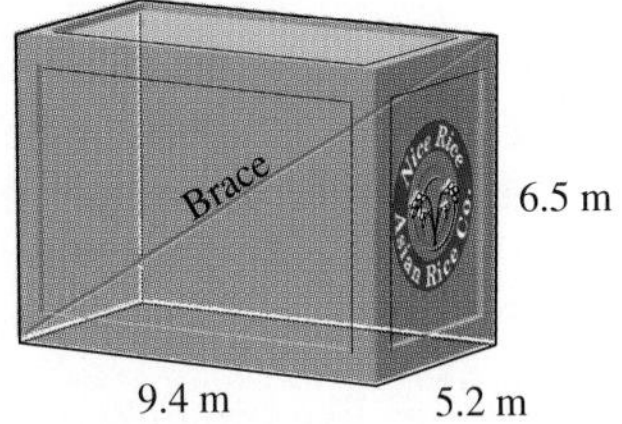

Figure for Exercise 80

Discussion Question

81 Explain why $\sqrt{\sqrt{256}} = \sqrt[4]{256}$.

Challenge Question

a. Using your calculator, evaluate $\sqrt{10}$. Then take the square root of this result, then the square root of this result, and so on, until you have taken ten square roots.

b. Using your calculator, evaluate $\sqrt{0.1}$. Then take the square root of this result, then the square root of this result, and so on, until you have taken ten square roots.

c. Make a conjecture based on your observations in parts a and b.

Section Skill Reinforcers

Simplify each expression.

1 $64^{1/3}$ **2** $64^{1/2}$ **3** 64^0 **4** $64^{-1/2}$ **5** 64^{-1}

7-3 Addition and Subtraction of Radical Expressions

SECTION OBJECTIVES

4 Add and subtract radical expressions.

5 Reduce the order of a radical.

We add like radicals in exactly the same way that we add like terms of a polynomial. Like radicals must have the same index and the same radicand; only the coefficients of the like terms can differ.

EXAMPLE 1 Classifying Radicals As Like or Unlike

Classify these radicals as like or unlike.

	SOLUTIONS	
(a) $-3\sqrt{2}$, $5\sqrt{2}$, and $\frac{\sqrt{2}}{3}$	Like radicals	The radicand of each of these square roots is 2.
(b) $-\sqrt{xy^2}$ and $4\sqrt{xy^2}$ for $x > 0$ and $y > 0$	Like radicals	The radicand of each of these square roots is xy^2. The coefficients of the radicals are -1 and 4.
(c) $7\sqrt{5}$ and $7\sqrt[3]{5}$	Unlike radicals	These radicals are unlike because they do not have the same index. One is a square root, and the other is a cube root.
(d) $11\sqrt{7}$ and $11\sqrt{8}$	Unlike radicals	These square roots are unlike because the radicands 7 and 8 are different.
(e) $\sqrt[3]{xy^2}$ and $\sqrt[3]{x^2y}$	Unlike radicals	These radicals are unlike because the radicands are not equal ($xy^2 \neq x^2y$).

Using the distributive property, we can rewrite $3\sqrt{2} + 4\sqrt{2}$ as $(3 + 4)\sqrt{2}$, which equals $7\sqrt{2}$. Note that we can add like radicals by adding their coefficients, just as we add like terms of a polynomial. This similarity is illustrated in Table 7-1.

SELF-CHECK

Classify each pair of radicals as either like or unlike.

1 $-\frac{\sqrt{7}}{2}$ and $\frac{\sqrt{7}}{3}$

2 $\sqrt[3]{7}$ and $-\sqrt{7}$

3 $\sqrt[7]{2}$ and $\sqrt[7]{3}$

TABLE 7-1 *Comparison of Polynomials and Radical Terms*

Addition of Polynomials	Addition of Radicals
$3x + 4x = 7x$	$3\sqrt{2} + 4\sqrt{2} = 7\sqrt{2}$
$9x + 5y - 6x = 3x + 5y$	$9\sqrt{7} + 5\sqrt{11} - 6\sqrt{7} = 3\sqrt{7} + 5\sqrt{11}$

SELF-CHECK ANSWERS 

1 Like **2** Unlike **3** Unlike

EXAMPLE 2 Adding and Subtracting Radicals

Perform the following additions and subtractions.

SOLUTIONS

(a) $\sqrt{3} - 5\sqrt{3}$

$\sqrt{3} - 5\sqrt{3} = (1 - 5)\sqrt{3}$
$= -4\sqrt{3}$

Subtract the coefficients of these like terms.

(b) $\sqrt[3]{7} - 5\sqrt[3]{7} + 12\sqrt[3]{7}$

$\sqrt[3]{7} - 5\sqrt[3]{7} + 12\sqrt[3]{7} = (1 - 5 + 12)\sqrt[3]{7}$
$= 8\sqrt[3]{7}$

Use the distributive property, and then combine like terms.

(c) $4\sqrt{5} + 7\sqrt[3]{5}$

$4\sqrt{5} + 7\sqrt[3]{5}$

This expression cannot be simplified since the terms are unlike.

(d) $3\sqrt[3]{x^2y} - 4\sqrt[3]{x^2y}$

$3\sqrt[3]{x^2y} - 4\sqrt[3]{x^2y} = (3 - 4)\sqrt[3]{x^2y}$
$= -\sqrt[3]{x^2y}$

Subtract the coefficients of these like terms.

(e) $(5\sqrt{11} + 6\sqrt[3]{11}) - (2\sqrt[3]{11} - 7\sqrt{11})$

$(5\sqrt{11} + 6\sqrt[3]{11}) - (2\sqrt[3]{11} - 7\sqrt{11})$
$= 5\sqrt{11} + 6\sqrt[3]{11} - 2\sqrt[3]{11} + 7\sqrt{11}$
$= (5 + 7)\sqrt{11} + (6 - 2)\sqrt[3]{11}$
$= 12\sqrt{11} + 4\sqrt[3]{11}$ ▼

First remove the parentheses, and then group like terms together. Then add the coefficients of the like terms.

One reason for always writing radicals in simplified form is to identify like terms. Radicals that appear to be unlike can sometimes be simplified so that the terms are alike. Remember to look for perfect square factors when working with square roots and for perfect cube factors when working with cube roots.

SELF-CHECK 

Perform the indicated operations.

1 $4\sqrt{6} - 9\sqrt{6}$
2 $8\sqrt[3]{xy} - 4\sqrt[3]{xy} + \sqrt[3]{xy}$

EXAMPLE 3 Simplifying Radicals and Then Combining Like Radicals

Simplify these radicals first, and then combine like radicals.

SOLUTIONS

(a) $7\sqrt{20} - 2\sqrt{45}$

$7\sqrt{20} - 2\sqrt{45} = 7\sqrt{4 \cdot 5} - 2\sqrt{9 \cdot 5}$
$= 7\sqrt{4}\sqrt{5} - 2\sqrt{9}\sqrt{5}$
$= 7(2\sqrt{5}) - 2(3\sqrt{5})$
$= 14\sqrt{5} - 6\sqrt{5}$
$= 8\sqrt{5}$

Note that 4 is a perfect square factor of 20 and that 9 is a perfect square factor of 45. Simplify using the property that $\sqrt{xy} = \sqrt{x}\sqrt{y}$.

Then combine like terms.

(b) $11\sqrt[3]{16} - 2\sqrt[3]{54}$

$11\sqrt[3]{16} - 2\sqrt[3]{54}$
$= 11\sqrt[3]{8 \cdot 2} - 2\sqrt[3]{27 \cdot 2}$
$= 11\sqrt[3]{8}\sqrt[3]{2} - 2\sqrt[3]{27}\sqrt[3]{2}$
$= 11(2\sqrt[3]{2}) - 2(3\sqrt[3]{2})$
$= 22\sqrt[3]{2} - 6\sqrt[3]{2}$
$= 16\sqrt[3]{2}$

Note that 8 is a perfect cube factor of 16 and that 27 is a perfect cube factor of 54. Simplify these radicals.

Then combine like terms.

SELF-CHECK ANSWERS 

1 $-5\sqrt{6}$ 2 $5\sqrt[3]{xy}$

(c) $3\sqrt{80x} - 2\sqrt{500x}$ for $x > 0$

$$\begin{aligned} 3\sqrt{80x} - 2\sqrt{500x} &= 3\sqrt{16 \cdot 5x} - 2\sqrt{100 \cdot 5x} \\ &= 3\sqrt{16}\sqrt{5x} - 2\sqrt{100}\sqrt{5x} \\ &= 3(4\sqrt{5x}) - 2(10\sqrt{5x}) \\ &= 12\sqrt{5x} - 20\sqrt{5x} \\ &= -8\sqrt{5x} \end{aligned}$$

16 is a perfect square factor of $80x$, and 100 is a perfect square factor of $500x$. Use this observation to simplify these radicals.

Then combine like terms.

(d) $\sqrt{18} + \sqrt{12}$

$$\begin{aligned} \sqrt{18} + \sqrt{12} &= \sqrt{9 \cdot 2} + \sqrt{4 \cdot 3} \\ &= \sqrt{9}\sqrt{2} + \sqrt{4}\sqrt{3} \\ &= 3\sqrt{2} + 2\sqrt{3} \end{aligned}$$

9 is a perfect square factor of 18, and 4 is a perfect square factor of 12.

These radicals are unlike, so the expression cannot be simplified further.

SELF-CHECK

Simplify each expression.

1 $2\sqrt{50} - 3\sqrt{200}$

2 $5\sqrt[3]{24x^4} + 7\sqrt[3]{81x^4}$

Reducing the order of a radical simplifies a radical expression by making the index as small as possible. We will rely on rational exponents and the properties of exponents to explain this simplification. You may perform some of these steps mentally.

EXAMPLE 4 Reducing the Order of a Radical

Reduce the order of each of these radicals. Assume that the variables represent positive real numbers.

SOLUTIONS

(a) $\sqrt[4]{x^2}$

$\sqrt[4]{x^2} = x^{2/4}$ — First express this radical in exponential notation.

$= x^{1/2}$ — Then reduce the rational exponent: $\frac{2}{4} = \frac{1}{2}$.

$= \sqrt{x}$ — Express this simplified result in radical notation.

(b) $\sqrt[5]{x^3}$

$\sqrt[5]{x^3} = x^{3/5}$
$= \sqrt[5]{x^3}$ — The fraction $\frac{3}{5}$ cannot be reduced, so this radical is already in reduced form.

(c) $\sqrt[9]{8x^3y^6}$

$\sqrt[9]{8x^3y^6} = (2^3x^3y^6)^{1/9}$ — First express this radical in exponential notation.

$= [(2xy^2)^3]^{1/9}$
$= (2xy^2)^{3/9}$ — Then rewrite it, noting that $2^3x^3y^6$ is a perfect cube of $2xy^2$. Use the power rule for exponents: $(x^m)^n = x^{mn}$.

$= (2xy^2)^{1/3}$ — Reduce the fractional exponent: $\frac{3}{9} = \frac{1}{3}$.

$= \sqrt[3]{2xy^2}$ — Express this result in radical notation.

SELF-CHECK ANSWERS

1 $-20\sqrt{2}$ 2 $31x\sqrt[3]{3x}$

The above examples show that $\sqrt[pn]{x^{pm}}$ can be simplified to $\sqrt[n]{x^m}$ if the index and the exponent on the radicand share some common factor p. This fact can be used to shorten the steps shown in Example 4. In Example 4(a),

$$\sqrt[4]{x^2} = \sqrt[2\cdot 2]{x^{2\cdot 1}} = \sqrt[2]{x^1} = \sqrt{x}$$

Example 5 also shows that radicals that appear unlike should be simplified before they are compared. In this case we simplify the radicals by reducing the order.

SELF-CHECK 

Simplify each of these radicals by reducing the order of the radical.

1 $\sqrt[6]{x^4}$ **2** $\sqrt[15]{v^5w^{10}}$

EXAMPLE 5 Reducing the Order of a Radical Before Adding

Add $2\sqrt[4]{25} + 7\sqrt[6]{125}$, after first simplifying each term.

SOLUTION $2\sqrt[4]{25} + 7\sqrt[6]{125} = 2\left[\sqrt[2\cdot 2]{5^2}\right] + 7\left[\sqrt[2\cdot 3]{5^3}\right]$ Express the radicands in exponential form.

$= 2\sqrt{5} + 7\sqrt{5}$ Then reduce the order of each term, and add the like terms.

$= 9\sqrt{5}$

EXAMPLE 6 Simplifying Radical Expressions

Simplify $\sqrt{x^2 + 2xy + y^2} + \sqrt{x^2 - 2xy + y^2}$, assuming that $x > y > 0$.

SOLUTION $\sqrt{x^2 + 2xy + y^2} + \sqrt{x^2 - 2xy + y^2} = \sqrt{(x + y)^2} + \sqrt{(x - y)^2}$ Express the radicands as perfect squares.

$= (x + y) + (x - y)$ If $x > y > 0$, then $x + y > 0$ and $x - y > 0$; thus absolute-value notation is not needed for these square roots. Then combine the like terms.

$= 2x$

Exercises 7-3

In Exercises 1–40 find the sum or difference. Assume that all variables represent positive real numbers so that absolute-value notation is not necessary.

1 $\sqrt{25} + \sqrt{49}$ **2** $\sqrt{36} + \sqrt{64}$

3 $2\sqrt{9} - 3\sqrt{16}$ **4** $5\sqrt{100} - 3\sqrt{4}$

5 $19\sqrt{2} + 11\sqrt{2}$ **6** $7\sqrt{3} + 8\sqrt{3}$

7 $6\sqrt{7x} - 9\sqrt{7x}$ **8** $4\sqrt{11y} - 7\sqrt{11y}$

9 $2\sqrt[3]{5} + 7\sqrt[3]{5} - \sqrt[3]{5}$ **10** $4\sqrt[3]{7} - 8\sqrt[3]{7} - 3\sqrt[3]{7}$

11 $7\sqrt[4]{17w} - 8\sqrt[4]{17w} + \sqrt[4]{17w}$

12 $6\sqrt[7]{5w} - 11\sqrt[7]{5w} + 5\sqrt[7]{5w}$

13 $7\sqrt{5} - 5\sqrt{7} - \sqrt{5} + \sqrt{7}$

14 $7\sqrt{2} - \sqrt{3} + 3\sqrt{2} + 2\sqrt{3}$

15 $\sqrt{28} + \sqrt{63}$ **16** $\sqrt{12} - \sqrt{27}$

17 $\sqrt{75} - \sqrt{48}$ **18** $\sqrt{44} + \sqrt{99}$

19 $3\sqrt{40} + 2\sqrt{90}$ **20** $7\sqrt{50} - 4\sqrt{18}$

21 $3\sqrt{50v} - 7\sqrt{32v}$ **22** $6\sqrt{45v} - 7\sqrt{320v}$

23 $5\sqrt{28w} - 4\sqrt{63w}$ **24** $3\sqrt{27b} - \sqrt{75b}$

25 $\sqrt[3]{24} - \sqrt[3]{375}$ **26** $\sqrt[3]{54} - \sqrt[3]{128}$

27 $2\sqrt{6} - \sqrt{54} + 5\sqrt{24}$ **28** $2\sqrt{45} - 3\sqrt{20} - \sqrt{80}$

29 $20\sqrt[3]{-81t^3} - 7\sqrt[3]{24t^3}$ **30** $11\sqrt[3]{-16t^3} - 5\sqrt[3]{54t^3}$

31 $9\sqrt[3]{40z^2} - 2\sqrt[3]{5000z^2}$

32 $7\sqrt{1210w^3} - 5\sqrt{1440w^3}$

SELF-CHECK ANSWERS 

1 $\sqrt[3]{x^2}$ **2** 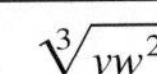 $\sqrt[3]{vw^2}$

33 $\sqrt{\frac{5}{4}} + \sqrt{\frac{5}{9}}$ **34** $\sqrt{\frac{7}{9}} - \sqrt{\frac{7}{25}}$

35 $2\sqrt{\frac{11x}{25}} - \sqrt{\frac{11x}{49}}$ **36** $3\sqrt{\frac{13x}{49}} - 5\sqrt{\frac{13x}{4}}$

37 $7\sqrt{0.98} + 2\sqrt{0.75} - \sqrt{0.12} + 5\sqrt{0.72}$

38 $4\sqrt{0.20} - 3\sqrt{0.90} - 7\sqrt{0.80} + \sqrt{1.60}$

39 $\sqrt{x^2 + 2xy + y^2} - \sqrt{x^2} - \sqrt{y^2}$

40 $\sqrt{x^2} + \sqrt{y^2} - \sqrt{x^2 - 2xy + y^2}$ for $x > y > 0$

In Exercises 41–46 simplify each radical by reducing the order of the radical. Assume that all variables represent positive real numbers so that absolute-value notation is not necessary.

41 $\sqrt[6]{x^3}$ **42** $\sqrt[6]{x^2}$ **43** $\sqrt[12]{81y^4z^8}$

44 $\sqrt[12]{64y^6z^6}$ **45** $\sqrt[9]{-27x^{27}}$ **46** $\sqrt[4]{25v^2w^4x^6}$

B

Estimation Skills (47–50)

47 The best mental estimate of $\frac{1 + \sqrt{9.05}}{2}$ is ____.

a. 1 **b.** 2 **c.** 3 **d.** 4 **e.** 5

48 The best mental estimate of $\frac{1 - \sqrt{9.05}}{2}$ is ____.

a. −4 **b.** −3 **c.** −2 **d.** −1 **e.** 2

49 The best mental estimate of $\frac{-4 - \sqrt{35.97}}{2}$ is ____.

a. −10 **b.** −6 **c.** −5 **d.** 6 **e.** 5

50 The best mental estimate of $\frac{-4 + \sqrt{35.97}}{2}$ is ____.

a. 1 **b.** 2 **c.** 3 **d.** 4 **e.** 5

Calculator Usage (51–60)

In Exercises 51–54 determine the value of each expression shown on the graphics calculator display. Assume that the values of 144 for x and 25 for y have been stored in your calculator.

51 √ X + Y

52 √ (X + Y)

53 √ (X + 52)

54 √ X + 52

In Exercises 55–58 store in your calculator the values of 23.6 for x and 31.5 for y, and then approximate the value of each expression to the nearest thousandth.

55 $(\sqrt{x} + \sqrt{y})^2$ **56** $(\sqrt{x} - \sqrt{y})^3$

57 $\sqrt{x^2 + y^2} - \sqrt{(x + y)^2}$ **58** $\sqrt[3]{x^2 + y^2} - \sqrt[3]{(x + y)^2}$

59 Use a graphics calculator to graph $y = \sqrt{x}$, $y = \sqrt{x} + 2$, and $y = \sqrt{x} + 4$. Use a window with [0, 25] for x and [0, 10] for y. Then compare these graphs.

60 Use a graphics calculator to graph $y = \sqrt{x}$, $y = \sqrt{x + 2}$, and $y = \sqrt{x + 4}$. Use a window with [−5, 20] for x and [0, 5] for y. Then compare these graphs.

C

In Exercises 61–66 simplify each radical term, and then combine like terms. Assume that all variables represent positive real numbers so that absolute-value notation is not necessary.

61 $2\sqrt{147a^2b} - 3\sqrt{363a^2b}$ **62** $5\sqrt[6]{27} - 7\sqrt[4]{9}$

63 $9\sqrt[6]{4} - 5\sqrt[9]{8}$ **64** $4\sqrt[15]{8} - 11\sqrt[20]{16}$

65 $2\sqrt[6]{x^2y^4} - 12\sqrt[9]{-x^3y^6}$

66 $4\sqrt[10]{x^6y^8} - 3\sqrt[20]{x^{12}y^{16}}$

In Exercises 67 and 68 simplify each radical expression. Assume that x and y represent positive real numbers and that m and n represent natural numbers.

67 $3\sqrt{x^{2m}y^{4n}} + 5\sqrt[3]{x^{3m}y^{6n}}$ **68** $7\sqrt[n]{x^ny^{n+1}}$

Error Analysis (69 and 70)

In Exercises 69 and 70 examine each student's work, and correct any mistakes you find.

69 Student A

$$\begin{aligned}\sqrt{64 + 225} &= \sqrt{64} + \sqrt{225}\\ &= 8 + 15\\ &= 23\end{aligned}$$

70 Student B

$$\begin{aligned}\sqrt{24} &= \sqrt{25 - 1}\\ &= \sqrt{25} - \sqrt{1}\\ &= 5 - 1\\ &= 4\end{aligned}$$

Discussion Question

71 Discuss the role of the distributive property in adding like radicals.

Section Skill Reinforcers

Simplify each of these expressions.

1 a^3a^4 **2** $a^{-5}a^8$ **3** $\frac{8x^4}{4x^3}$ **4** $\frac{12x^6}{-3x^{-3}}$ **5** $(12x^6)(-3x^{-3})$

7-4 Multiplication of Radical Expressions

SECTION OBJECTIVE

6 Multiply radical expressions.

To multiply radicals, we use the product property:

$$\sqrt[n]{x}\sqrt[n]{y} = \sqrt[n]{xy} \quad \text{for } x > 0 \text{ and } y > 0$$

When you are multiplying radicals, it is important to remember two things:

1 The product property does not hold true if n is even and the radicands are negative.*

2 The radicals must be of the same order before the radicands can be multiplied.

EXAMPLE 1 Multiplying Radicals

Perform each indicated multiplication, and then simplify the product.

SOLUTIONS

(a) $\sqrt{9x}\sqrt{x}$ for $x > 0$

$\sqrt{9x}\sqrt{x} = \sqrt{9x^2}$ — $\sqrt[n]{x}\sqrt[n]{y} = \sqrt[n]{xy}$

$= 3x$ — Absolute-value notation is not needed for the answer since $x > 0$.

(b) $(\sqrt[3]{20b^2})(\sqrt[3]{50b^2})$

$(\sqrt[3]{20b^2})(\sqrt[3]{50b^2}) = \sqrt[3]{1000b^4}$ — Multiply the radicands. Note that $1000b^3$ is a perfect cube factor of $1000b^4$.

$= \sqrt[3]{10^3b^4}$ — Simplify the radical.

$= \sqrt[3]{10^3b^3}\sqrt[3]{b}$

$= 10b\sqrt[3]{b}$

(c) $(11\sqrt[5]{8a^3})(4\sqrt[5]{4a^2})$

$(11\sqrt[5]{8a^3})(4\sqrt[5]{4a^2}) = 44\sqrt[5]{32a^5}$ — Multiply the coefficients and the radicands, respectively.

$= 44(2a)$ — Note that $32a^5$ is a perfect fifth power of $2a$.

$= 88a$ ▼

The multiplication of some radical expressions is similar to the multiplication of polynomials. Watch for special forms that can be multiplied by inspection, and use the FOIL method to multiply binomials when appropriate.

EXAMPLE 2 Multiplying Radicals with More Than One Term

Perform each indicated multiplication, and then simplify the product. Assume that the variables represent positive real numbers so that absolute-value notation is not necessary.

SOLUTIONS

(a) $2\sqrt{3}(3\sqrt{3} - 5)$

$2\sqrt{3}(3\sqrt{3} - 5)$

$= (2\sqrt{3})(3\sqrt{3}) - (2\sqrt{3})(5)$ — Distribute the multiplication of $2\sqrt{3}$, and then simplify.

$= 6(\sqrt{3})^2 - 10\sqrt{3}$

$= 18 - 10\sqrt{3}$

---

*To multiply $\sqrt{-2}\sqrt{-3}$, see Section 7-7.

(b) $(\sqrt{2x} + \sqrt{3y})(\sqrt{2x} - \sqrt{3y})$

$$\begin{aligned}(\sqrt{2x} + \sqrt{3y})(\sqrt{2x} - \sqrt{3y}) &= (\sqrt{2x})^2 - (\sqrt{3y})^2 \\ &= 2x - 3y\end{aligned}$$

Note that this expression is of the form $(a + b)(a - b) = a^2 - b^2$, with $a = \sqrt{2x}$ and $b = \sqrt{3y}$.

(c) $(4\sqrt{11} - 2)(3\sqrt{11} + 7)$

$$\begin{aligned}(4\sqrt{11} - 2)(3\sqrt{11} + 7) &= 12(\sqrt{11})^2 + 28\sqrt{11} - 6\sqrt{11} - 14 \\ &= 132 + 22\sqrt{11} - 14 \\ &= 118 + 22\sqrt{11}\end{aligned}$$

Multiply using the FOIL method, and then simplify by adding like terms.

(d) $(\sqrt[3]{x} - 2)(\sqrt[3]{x^2} + 2\sqrt[3]{x} + 4)$

$$\begin{aligned}(\sqrt[3]{x} - 2)(\sqrt[3]{x^2} + 2\sqrt[3]{x} + 4) &= (\sqrt[3]{x})^3 - (2)^3 \\ &= x - 8\end{aligned}$$

Note that this expression is of the form $(a - b)(a^2 + ab + b^2) = a^3 - b^3$, with $a = \sqrt[3]{x}$ and $b = 2$.

▼

The special products play an important role in simplifying many algebraic expressions. Once again we will use the product $(a + b)(a - b) = a^2 - b^2$. The radical expressions $\sqrt{a} + \sqrt{b}$ and $\sqrt{a} - \sqrt{b}$ are called **conjugates** of each other. Knowing that $(\sqrt{a} + \sqrt{b})(\sqrt{a} - \sqrt{b}) = a - b$ is important, since forming the product $a - b$ can be a way to simplify radicals.

SELF-CHECK 

Find each product.

1. $7\sqrt{10}(3\sqrt{2} - 2\sqrt{5})$
2. $(5\sqrt{3} + 2\sqrt{7})^2$

CONJUGATE RADICALS

The radical expressions $\sqrt{a} + \sqrt{b}$ and $\sqrt{a} - \sqrt{b}$ are conjugates of each other.

EXAMPLE 3 Identifying Conjugates

Write the conjugate of each of these radical expressions. Assume that x and y represent positive real numbers.

SOLUTIONS

(a) $\sqrt{2} + \sqrt{3}$ — The conjugate is $\sqrt{2} - \sqrt{3}$.

(b) $3 - 5\sqrt{11}$ — The conjugate is $3 + 5\sqrt{11}$.

(c) $2\sqrt{x} - 7\sqrt{y}$ — The conjugate is $2\sqrt{x} + 7\sqrt{y}$.

(d) $5\sqrt{3x} + 1$ — The conjugate is $5\sqrt{3x} - 1$. ▼

EXAMPLE 4 Multiplying Conjugates

Find the product of $\sqrt{7} + \sqrt{5}$ and its conjugate.

SOLUTION

$$\begin{aligned}(\sqrt{7} + \sqrt{5})(\sqrt{7} - \sqrt{5}) &= (\sqrt{7})^2 - (\sqrt{5})^2 \\ &= 7 - 5 \\ &= 2\end{aligned}$$

The conjugate of $\sqrt{7} + \sqrt{5}$ is $\sqrt{7} - \sqrt{5}$.

The product of these two irrational numbers is the natural number 2. ▼

SELF-CHECK ANSWERS 

1 $42\sqrt{5} - 70\sqrt{2}$ **2** $103 + 20\sqrt{21}$

EXAMPLE 5 Checking a Solution of a Quadratic Equation

Determine whether or not $1 + \sqrt{3}$ is a solution of $x^2 - 2x - 2 = 0$.

SOLUTION

$$x^2 - 2x - 2 = 0$$
$$(1 + \sqrt{3})^2 - 2(1 + \sqrt{3}) - 2 \stackrel{?}{=} 0$$
$$1 + 2\sqrt{3} + 3 - 2 - 2\sqrt{3} - 2 \stackrel{?}{=} 0$$
$$0 = 0$$

Substitute the given value of x into the equation to determine if it checks.

Since $1 + \sqrt{3}$ checks, it is a solution of the equation. ▼

To raise the order of a radical, we reverse the strategy used to reduce the order of a radical. Such conversions allow us to create radicals of the same order so that we can perform operations on them.

EXAMPLE 6 Changing the Order of a Radical

Convert each of these radical expressions to an equivalent sixth root.

SOLUTIONS

(a) $\sqrt[3]{5}$

$\sqrt[3]{5} = 5^{1/3}$ — The exponential form is shown to illustrate the logic of the solution; the steps could also be done in radical form.

$= 5^{2/6}$ — Rewrite the exponent $\frac{1}{3}$ as $\frac{2}{6}$.

$= \sqrt[6]{5^2}$ — Convert back to radical form.

$= \sqrt[6]{25}$

(b) $\sqrt{2y}$ for $y > 0$

$\sqrt{2y} = (2y)^{1/2}$ — Write the expression in exponential form.

$= (2y)^{3/6}$ — Rewrite the exponent $\frac{1}{2}$ as $\frac{3}{6}$ so that the result can be expressed as a sixth root.

$= \sqrt[6]{(2y)^3}$

$= \sqrt[6]{8y^3}$ ▼

EXAMPLE 7 Multiplying Radicals of Different Orders

Use the results from Example 6 to multiply $\sqrt[3]{5}\sqrt{2y}$.

SOLUTION

$$\sqrt[3]{5}\sqrt{2y} = \sqrt[6]{25}\sqrt[6]{8y^3}$$
$$= \sqrt[6]{25(8y^3)}$$
$$= \sqrt[6]{200y^3}$$

Convert both factors to the sixth order, which is the least common order $(2 \cdot 3 = 6)$. (See Example 6 for the steps required to change the order of these radicals.) Then multiply these sixth-order radicals. ▼

SELF-CHECK 

Convert each of these radical expressions to an equivalent fifteenth root.

1 $\sqrt[3]{2x}$ **2** $\sqrt[5]{3y}$

3 Multiply $\sqrt[3]{2x}$ by $\sqrt[5]{3y}$.

Assume that x and y represent positive real numbers.

1 $\sqrt[15]{32x^5}$ **2** $\sqrt[15]{27y^3}$ **3** $\sqrt[15]{864x^5y^3}$

Exercises 7-4

A

In Exercises 1–44 perform the indicated multiplications, and then simplify each product. Assume that the variables represent positive real numbers so that absolute-value notation is unnecessary.

1 $\sqrt{2}\sqrt{6}$

2 $\sqrt{3}\sqrt{6}$

3 $\sqrt{8w}\sqrt{2w}$

4 $\sqrt{3w^3}\sqrt{27w}$

5 $(2\sqrt{3})(4\sqrt{5})$

6 $(3\sqrt{7})(4\sqrt{2})$

7 $\sqrt[3]{9v}\sqrt[3]{-3v^2}$

8 $\sqrt[3]{-4v}\sqrt[3]{2v^2}$

9 $(2\sqrt{6z})(4\sqrt{3z})$

10 $(5\sqrt{14z})(11\sqrt{7z})$

11 $\sqrt{2}(5\sqrt{6})(8\sqrt{3})$

12 $-\sqrt{7}(3\sqrt{14})(\sqrt{8})$

13 $(\sqrt[3]{xy^2})^2$

14 $(\sqrt[3]{4ab^2})^2$

15 $(11\sqrt{6v})(-2\sqrt{3w})(\sqrt{2vw})$

16 $(-5\sqrt{15v})(-2\sqrt{5vw})(9\sqrt{3w})$

17 $\sqrt[3]{2ab^2}\sqrt[3]{4a^2c}\sqrt[3]{bc^2}$

18 $\sqrt[3]{a^2b}\sqrt[3]{3ac^2}\sqrt[3]{9b^2c}$

19 $\sqrt{2}(5\sqrt{2} - 1)$

20 $\sqrt{7}(2\sqrt{7} + 1)$

21 $-5\sqrt{3}(2\sqrt{3} - 7)$

22 $-4\sqrt{11}(2\sqrt{11} + 6)$

23 $\sqrt{3}(\sqrt{2} + \sqrt{5})$

24 $\sqrt{5}(\sqrt{3} + \sqrt{7})$

25 $3\sqrt{5}(2\sqrt{15} - 7\sqrt{35})$

26 $2\sqrt{7}(5\sqrt{14} - \sqrt{21})$

27 $-\sqrt[3]{4}(2\sqrt[3]{2} + \sqrt[3]{5})$

28 $-\sqrt[3]{6}(3\sqrt[3]{4} - 2\sqrt[3]{9})$

29 $3\sqrt{x}(\sqrt{6x} - 5\sqrt{x})$

30 $\sqrt{5x}(5\sqrt{x} - \sqrt{10x})$

31 $(\sqrt{7} + \sqrt{13})(\sqrt{7} - \sqrt{13})$

32 $(\sqrt{17} - \sqrt{2})(\sqrt{17} + \sqrt{2})$

33 $(\sqrt{3x} - \sqrt{y})(\sqrt{3x} + \sqrt{y})$

34 $(\sqrt{19z} + \sqrt{5y})(\sqrt{19z} - \sqrt{5y})$

35 $(2\sqrt{3} - \sqrt{2})^2$

36 $(3\sqrt{2} - \sqrt{3})^2$

37 $(\sqrt{a} + 5\sqrt{3b})^2$

38 $(\sqrt{3a} + 3\sqrt{5b})^2$

39 $(2\sqrt{6} - \sqrt{15})(\sqrt{6} + 2\sqrt{15})$

40 $(3\sqrt{10} + \sqrt{14})(2\sqrt{10} - 3\sqrt{14})$

41 $(\sqrt{v-2} + 3)(\sqrt{v-2} - 3)$

42 $(\sqrt{2v+3} - 5)(\sqrt{2v+3} + 5)$

43 $(\sqrt{7} - \sqrt{5})^2(\sqrt{7} + \sqrt{5})$

44 $(\sqrt{11} - \sqrt{17})(\sqrt{11} + \sqrt{17})^2$

B

In Exercises 45–52 multiply each radical expression by its conjugate, and simplify the result. Assume that the variables represent positive real numbers so that absolute-value notation is unnecessary.

45 $\sqrt{5} - \sqrt{2}$

46 $7 + \sqrt{6}$

47 $\sqrt{11} - 5$

48 $\sqrt{11} - 3$

49 $\sqrt{x} + \sqrt{3y}$

50 $\sqrt{2x} - \sqrt{y}$

51 $\sqrt{2v+1} + \sqrt{3v-1}$

52 $\sqrt{5v+2} - \sqrt{4v-3}$

In Exercises 53–60 perform the indicated operations, and simplify the result. Assume that the variables represent positive real numbers so that absolute-value notation is unnecessary.

53 $(7\sqrt{6})(2\sqrt{5}) - (3\sqrt{10})(2\sqrt{3})$

54 $(11\sqrt{10})(4\sqrt{7}) - (5\sqrt{14})(3\sqrt{5})$

55 $(5\sqrt[3]{14x^2})(3\sqrt[3]{-5y}) - (2\sqrt[3]{-7xy})(7\sqrt[3]{10x})$

56 $(2 + \sqrt[3]{2})(4 - 2\sqrt[3]{2} + \sqrt[3]{4})$

57 $(\sqrt[3]{2} - 5)(\sqrt[3]{4} + 5\sqrt[3]{2} + 25)$

58 $(\sqrt[3]{3} + 5)(\sqrt[3]{9} - 5\sqrt[3]{3} + 25)$

59 **a.** $(\sqrt{x} + \sqrt{y})(\sqrt{x} - \sqrt{y})$
b. $(\sqrt[3]{x} + \sqrt[3]{y})(\sqrt[3]{x} - \sqrt[3]{y})$
c. $(\sqrt[3]{x} + \sqrt[3]{y})(\sqrt[3]{x^2} - \sqrt[3]{xy} + \sqrt[3]{y^2})$

60 **a.** $(\sqrt{a} - \sqrt{b})(\sqrt{a} + \sqrt{b})$
b. $(\sqrt[3]{a} - \sqrt[3]{b})(\sqrt[3]{a} + \sqrt[3]{b})$
c. $(\sqrt[3]{a} - \sqrt[3]{b})(\sqrt[3]{a^2} + \sqrt[3]{ab} + \sqrt[3]{b^2})$

C

61 Determine whether $1 + \sqrt{2}$ is a solution of $x^2 - 2x - 1 = 0$.

62 Determine whether $1 - \sqrt{2}$ is a solution of $x^2 - 2x - 1 = 0$.

In Exercises 63–66 assume that the variables represent positive real numbers.

63 a. Convert $\sqrt{x}$ to an equivalent sixth root.

b. Convert $\sqrt[3]{y}$ to an equivalent sixth root.

c. Multiply $\sqrt{x}\sqrt[3]{y}$.

64 a. Convert $\sqrt{2}$ to an equivalent sixth root.

b. Convert $\sqrt[3]{5}$ to an equivalent sixth root.

c. Multiply $\sqrt{2} \cdot \sqrt[3]{5}$.

65 Multiply $\sqrt[3]{2} \cdot \sqrt[4]{3}$.

66 Multiply $\sqrt[3]{a}\sqrt[5]{b}$.

CALCULATOR USAGE (67–69)

In Exercises 67 and 68 approximate these expressions to the nearest hundredth for $a = 2.93$, $b = 7.21$, and $c = 1.22$.

67 $\dfrac{-b - \sqrt{b^2 - 4ac}}{2a}$

68 $\dfrac{-b + \sqrt{b^2 - 4ac}}{2a}$

69 Approximations of π Use a calculator to determine which of these approximations is closest to π:

a. $\dfrac{22}{7}$ **b.** $\dfrac{355}{113}$ **c.** $\sqrt{\sqrt{\dfrac{2143}{22}}}$

DISCUSSION QUESTION

70 Describe in your own words the procedure for multiplying two radicals of the same order.

CHALLENGE QUESTION

A formula* for determining the wind chill factor is

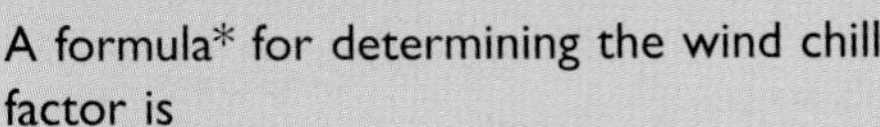

$$T_w = 91.4 - (T - 91.4)(0.02V - 0.30\sqrt{V} - 0.47)$$

where

T_w = wind chill temperature in degrees Fahrenheit
T = air temperature in degrees Fahrenheit
V = velocity of the wind in miles per hour

This formula is used for wind velocities of 5 mi/h or greater. For a temperature of 30° F and wind conditions of less than 5 mi/h, the wind chill temperature is approximately 30° F. Use this formula to complete the following table for $T = 30°$. Then use this table to complete the line graph of this function for $T = 30°$.

V	0	5	10	15	20	25	30	35	40
T_w	30	27	17						

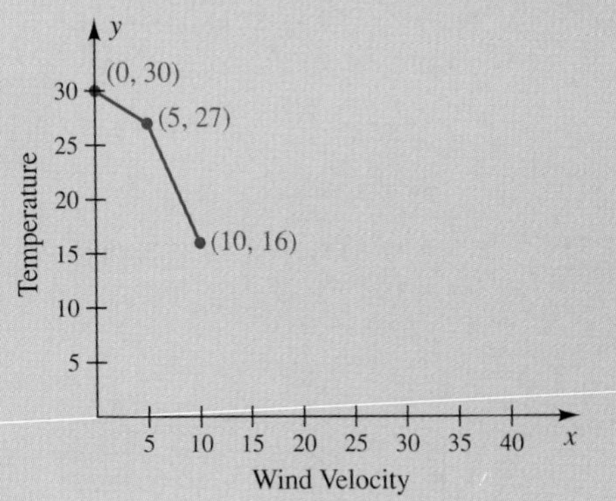

*A form of this formula was given to the author by Robert Scott of the Illinois State Water Survey.

SECTION SKILL REINFORCERS

Simplify each of these expressions.

1 $\sqrt{8 + 8}$ **2** $\sqrt{8} + \sqrt{8}$ **3** $\sqrt{8}\sqrt{8}$ **4** $\dfrac{\sqrt{8}}{\sqrt{8}}$ **5** $\sqrt{8} - \sqrt{8}$

7-5 Division of Radical Expressions

SECTION OBJECTIVE

7 Divide and rationalize radical expressions.

In Section 7-2 we simplified selected radicals of the form $\frac{\sqrt{x}}{\sqrt{y}}$ $(x > 0, y > 0)$ by first changing them to the form $\sqrt{\frac{x}{y}}$, as follows:

$$\frac{\sqrt{50}}{\sqrt{18}} = \sqrt{\frac{50}{18}} = \sqrt{\frac{25}{9}} = \frac{5}{3}$$

We do not always obtain a perfect square when we reduce, however, so we need a more general strategy for dealing with the problem of radicals in the denominator. The process of simplifying a radical expression by removing the radical from the denominator is called **rationalizing the denominator.** Relying on the property that $(\sqrt[n]{x^n}) = x$ for $x > 0$, we can multiply a square root by itself to obtain its square and thus remove the radical. Similarly, for any nth root, we can first multiply by an appropriate factor to produce a perfect nth power in the radicand.

EXAMPLE 1 Dividing Radicals

Perform the indicated operations. Assume a and b are positive real numbers.

SOLUTIONS

(a) $\sqrt{7} \div \sqrt{11}$

$\sqrt{7} \div \sqrt{11} = \frac{\sqrt{7}}{\sqrt{11}}$ Write the quotient in fractional form.

$= \frac{\sqrt{7}}{\sqrt{11}} \cdot \frac{\sqrt{11}}{\sqrt{11}}$ Multiply both the numerator and the denominator by $\sqrt{11}$ to make the radicand in the denominator a perfect square.

$= \frac{\sqrt{77}}{11}$ Note that the simplified form has the denominator rationalized; that is, the denominator is now the rational number 11 instead of the irrational number $\sqrt{11}$.

(b) $\frac{\sqrt{2a}}{\sqrt{3b}}$

$\frac{\sqrt{2a}}{\sqrt{3b}} = \frac{\sqrt{2a}}{\sqrt{3b}} \cdot \frac{\sqrt{3b}}{\sqrt{3b}}$ Multiply both the numerator and the denominator by $\sqrt{3b}$ to make the radicand in the denominator a perfect square.

$= \frac{\sqrt{6ab}}{3b}$

For a radicand to be a perfect nth power, every factor in the radicand must be a perfect nth power. When you are trying to determine the powers on each factor of the radicand, it is helpful to express the numerical coefficient as a product of the powers of its prime factors. The following box summarizes the strategy for rationalizing a radical expression whose denominator consists of a single radical term.

RATIONALIZING A RADICAL EXPRESSION WITH A SINGLE TERM IN THE DENOMINATOR

Step 1 Write the expression as a single radical, and reduce the fraction in the radicand to lowest terms.

Step 2 Multiply both the numerator and the denominator of this fraction by the radical that will make the radicand in the denominator a perfect *n*th power.

EXAMPLE 2 Rationalizing a Denominator with One Term

Simplify each radical expression by rationalizing the denominator. In part (c), assume that x and y are positive real numbers.

SOLUTIONS

(a) $\dfrac{\sqrt[3]{5v}}{\sqrt[3]{4w}}$

$$\frac{\sqrt[3]{5v}}{\sqrt[3]{4w}} = \frac{\sqrt[3]{5v}}{\sqrt[3]{2^2w}}$$

First write the numerical coefficient of the radicand in the denominator as a product of its prime factors: $4 = 2^2$.

$$= \frac{\sqrt[3]{5v}}{\sqrt[3]{2^2w}} \cdot \frac{\sqrt[3]{2w^2}}{\sqrt[3]{2w^2}}$$

Since $(2^2w)(2w^2) = 2^3w^3$, a perfect cube, multiply both the numerator and the denominator by $\sqrt[3]{2w^2}$.

$$= \frac{\sqrt[3]{10vw^2}}{\sqrt[3]{2^3w^3}}$$

$$= \frac{\sqrt[3]{10vw^2}}{2w}$$

(b) $\dfrac{\sqrt[5]{3d}}{\sqrt[5]{8b^2}}$

$$\frac{\sqrt[5]{3d}}{\sqrt[5]{8b^2}} = \frac{\sqrt[5]{3d}}{\sqrt[5]{2^3b^2}}$$

Write the coefficient 8 of the radicand in the denominator as a power of the prime factor 2.

$$= \frac{\sqrt[5]{3d}}{\sqrt[5]{2^3b^2}} \cdot \frac{\sqrt[5]{2^2b^3}}{\sqrt[5]{2^2b^3}}$$

Since $2^3b^2(2^2b^3) = 2^5b^5$, a perfect fifth power, multiply both the numerator and the denominator by $\sqrt[5]{2^2b^3}$.

$$= \frac{\sqrt[5]{12b^3d}}{\sqrt[5]{2^5b^5}}$$

$$= \frac{\sqrt[5]{12b^3d}}{2b}$$

(c) $\dfrac{\sqrt{15x^2y^2}}{\sqrt{35xy^3}}$

$$\frac{\sqrt{15x^2y^2}}{\sqrt{35xy^3}} = \sqrt{\frac{15x^2y^2}{35xy^3}}$$

Since the radicals share some factors, first write the expression as a single radical.

$$= \sqrt{\frac{3x(5xy^2)}{7y(5xy^2)}}$$

Then reduce the radicand by dividing out the common factor, $5xy^2$.

$$= \sqrt{\frac{3x}{7y}}$$

$$= \frac{\sqrt{3x}}{\sqrt{7y}} \cdot \frac{\sqrt{7y}}{\sqrt{7y}}$$

Multiply both the numerator and the denominator by $\sqrt{7y}$ to produce a perfect square in the radicand in the denominator.

$$= \frac{\sqrt{21xy}}{7y}$$

▼

We will now examine selected radical expressions with binomial denominators involving square roots. The key to simplifying these expressions is to use conjugates, because the product of the conjugates $\sqrt{x} + \sqrt{y}$ and $\sqrt{x} - \sqrt{y}$ contains no radical. For $x > 0$ and $y > 0$,

$$(\sqrt{x} + \sqrt{y})(\sqrt{x} - \sqrt{y}) = x - y$$

SELF-CHECK

Simplify each of these radical expressions. Assume that the variables represent positive real numbers.

1 $\dfrac{1}{\sqrt{6}}$ **2** $\dfrac{1}{\sqrt[3]{6}}$

3 $\sqrt{\dfrac{x}{y}}$ **4** $\sqrt[3]{\dfrac{y^2}{x}}$

RATIONALIZING A RADICAL EXPRESSION WITH A BINOMIAL DENOMINATOR INVOLVING SQUARE ROOTS

Step 1 Reduce the fraction to lowest terms.

Step 2 Multiply both the numerator and the denominator of the fraction by the conjugate of the denominator.

EXAMPLE 3 Rationalizing a Binomial Denominator

Simplify $\dfrac{6}{\sqrt{7} - \sqrt{5}}$ by rationalizing the denominator.

SOLUTION

$$\frac{6}{\sqrt{7} - \sqrt{5}} = \frac{6}{\sqrt{7} - \sqrt{5}} \cdot \frac{\sqrt{7} + \sqrt{5}}{\sqrt{7} + \sqrt{5}}$$

Multiply the numerator and the denominator by $\sqrt{7} + \sqrt{5}$, the conjugate of the denominator: $(\sqrt{7} - \sqrt{5})(\sqrt{7} + \sqrt{5}) = 7 - 5$

$$= \frac{6(\sqrt{7} + \sqrt{5})}{7 - 5}$$

$$= \frac{6(\sqrt{7} + \sqrt{5})}{2}$$

Divide the numerator and the denominator by their common factor, 2.

$$= 3(\sqrt{7} + \sqrt{5})$$

$$= 3\sqrt{7} + 3\sqrt{5}$$

EXAMPLE 4 Rationalizing a Binomial Denominator

Divide 30 by $2\sqrt{17} + 2\sqrt{14}$.

SOLUTION

$$30 \div (2\sqrt{17} + 2\sqrt{14}) = \frac{30}{2(\sqrt{17} + \sqrt{14})}$$

Express the quotient in fractional form, and factor the GCF of 2 out of the denominator.

$$= \frac{15}{\sqrt{17} + \sqrt{14}}$$

Reduce the fraction by dividing both the numerator and the denominator by 2.

$$= \frac{15}{\sqrt{17} + \sqrt{14}} \cdot \frac{\sqrt{17} - \sqrt{14}}{\sqrt{17} - \sqrt{14}}$$

$$= \frac{15(\sqrt{17} - \sqrt{14})}{17 - 14}$$

Multiply the numerator and the denominator by the conjugate of the denominator to make the denominator the product of the conjugates.

SELF-CHECK ANSWERS

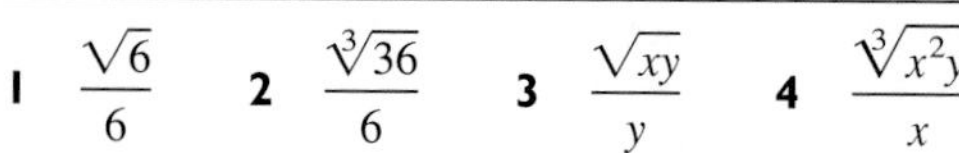

1 $\dfrac{\sqrt{6}}{6}$ **2** $\dfrac{\sqrt[3]{36}}{6}$ **3** $\dfrac{\sqrt{xy}}{y}$ **4** $\dfrac{\sqrt[3]{x^2y^2}}{x}$

$$= \frac{15(\sqrt{17} - \sqrt{14})}{3}$$ Simplify the denominator, and then reduce the fraction to lowest terms.

$$= 5(\sqrt{17} - \sqrt{14})$$
$$= 5\sqrt{17} - 5\sqrt{14}$$

SELF-CHECK

Simplify each expression.

1 $\dfrac{3}{2 - \sqrt{7}}$

2 $\dfrac{6}{\sqrt{11} + \sqrt{23}}$

EXAMPLE 5 Rationalizing a Denominator with Variables

Simplify $\dfrac{\sqrt{x}}{\sqrt{x} + \sqrt{y}}$ by rationalizing the denominator. Assume that x and y represent positive real numbers and that $x \neq y$.

SOLUTION $$\frac{\sqrt{x}}{\sqrt{x} + \sqrt{y}} = \frac{\sqrt{x}}{\sqrt{x} + \sqrt{y}} \cdot \frac{\sqrt{x} - \sqrt{y}}{\sqrt{x} - \sqrt{y}}$$ Multiply both the numerator and the denominator by $\sqrt{x} - \sqrt{y}$, the conjugate of the denominator.

$$= \frac{\sqrt{x}(\sqrt{x} - \sqrt{y})}{(\sqrt{x} + \sqrt{y})(\sqrt{x} - \sqrt{y})}$$ Distribute the factor of $\sqrt{x}$ in the numerator, and multiply the conjugates in the denominator.

$$= \frac{x - \sqrt{xy}}{x - y}$$

Exercises 7-5

A

In Exercises 1–32 perform each indicated division by rationalizing the denominator and then simplifying. Assume that all variables represent positive real numbers.

1 $\dfrac{2}{\sqrt{6}}$ **2** $\dfrac{5}{\sqrt{15}}$ **3** $\dfrac{\sqrt{5}}{\sqrt{8}}$ **4** $\dfrac{\sqrt{11}}{\sqrt{13}}$

5 $\dfrac{15}{\sqrt{3x}}$ **6** $\dfrac{14}{\sqrt{7y}}$ **7** $\dfrac{4}{\sqrt[3]{2}}$ **8** $\dfrac{12}{\sqrt[3]{3}}$

9 $\dfrac{\sqrt[3]{2}}{\sqrt[3]{25}}$ **10** $\dfrac{\sqrt[3]{3}}{\sqrt[3]{4}}$ **11** $\sqrt{\dfrac{6a^2b^3}{10ab^4}}$

12 $\sqrt{\dfrac{6ab}{15a^2b}}$ **13** $18 \div \sqrt{6}$ **14** $25 \div \sqrt{5}$

15 $\dfrac{\sqrt[3]{5v}}{\sqrt[3]{5w^2}}$ **16** $\dfrac{\sqrt[4]{v}}{\sqrt[4]{27w^3}}$ **17** $\dfrac{\sqrt[3]{4x^2y}}{\sqrt[3]{18xy^2}}$

18 $\sqrt[5]{\dfrac{10vx^3}{15v^4x^2}}$ **19** $\dfrac{\sqrt[5]{6(2x - y)}}{\sqrt[5]{6x^3(2x - y)}}$

20 $\dfrac{\sqrt[4]{12(3x - 4y)}}{\sqrt[4]{12x^3(3x - 4y)}}$ **21** $\dfrac{2}{1 - \sqrt{3}}$

22 $\dfrac{3}{1 + \sqrt{7}}$ **23** $\dfrac{12}{\sqrt{5} - 3}$

24 $\dfrac{36}{\sqrt{13} + 5}$ **25** $\dfrac{-6}{3\sqrt{5} - 3\sqrt{3}}$

26 $\dfrac{-15}{3\sqrt{7} - 3\sqrt{2}}$ **27** $\dfrac{\sqrt{7} - \sqrt{3}}{\sqrt{7} + \sqrt{3}}$

28 $\dfrac{\sqrt{7} + \sqrt{3}}{\sqrt{7} - \sqrt{3}}$ **29** $\dfrac{\sqrt{x} + \sqrt{y}}{\sqrt{x} - \sqrt{y}}$

30 $\dfrac{\sqrt{x} - \sqrt{y}}{\sqrt{x} + \sqrt{y}}$ **31** $\dfrac{\sqrt{a}}{\sqrt{a} - \sqrt{b}}$

32 $\dfrac{\sqrt{b}}{\sqrt{a} + \sqrt{b}}$

B

In Exercises 33–46 simplify each radical expression by rationalizing the denominator. Assume that all variables represent positive real numbers.

33 $\dfrac{-34\sqrt{2}}{2\sqrt{7} - 3\sqrt{5}}$ **34** $\dfrac{-19\sqrt{5}}{5\sqrt{3} - 3\sqrt{2}}$

SELF-CHECK ANSWERS

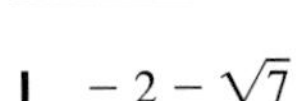

1 $-2 - \sqrt{7}$ **2** $\dfrac{\sqrt{23} - \sqrt{11}}{2}$

35 $\dfrac{\sqrt{27}-\sqrt{5}}{\sqrt{3}+\sqrt{5}}$

36 $\dfrac{\sqrt{125}+\sqrt{7}}{\sqrt{5}-\sqrt{7}}$

37 $\dfrac{5}{\sqrt{3x}+\sqrt{2y}}$

38 $\dfrac{7}{\sqrt{5x}-\sqrt{3y}}$

39 $\dfrac{5}{3\sqrt{x}-\sqrt{2y}}$

40 $\dfrac{9}{5\sqrt{x}+\sqrt{3y}}$

41 $\dfrac{2\sqrt{3x}+5\sqrt{2y}}{2\sqrt{3x}-5\sqrt{2y}}$

42 $\dfrac{4\sqrt{5x}-5\sqrt{3y}}{4\sqrt{5x}+5\sqrt{3y}}$

43 $\dfrac{\sqrt{x+4}-\sqrt{x}}{\sqrt{x+4}+\sqrt{x}}$

44 $\dfrac{\sqrt{a+b}-\sqrt{a}}{\sqrt{a+b}+\sqrt{b}}$

45 $(\sqrt{2}-\sqrt{3}) \div (\sqrt{2}+\sqrt{3})$

46 $(\sqrt{6}-\sqrt{15}) \div (\sqrt{10}-\sqrt{6})$

ESTIMATION SKILLS (47–50)

In Exercises 47–50 mentally estimate each of these expressions, and then select the answer that best matches your estimate.

47 $\dfrac{\sqrt{143}}{\sqrt{15}}$

a. 1 **b.** 2 **c.** 3 **d.** 4 **e.** 5

48 $\dfrac{\sqrt{101.7}}{\sqrt{24.96}}$

a. 1 **b.** 2 **c.** 3 **d.** 4 **e.** 5

49 $\dfrac{50}{\sqrt{35}-\sqrt{0.98}}$

a. 2 **b.** 4 **c.** 6 **d.** 8 **e.** 10

50 $\dfrac{10}{\sqrt{50}-\sqrt{26}}$

a. 5 **b.** $\dfrac{2}{3}$ **c.** 1 **d.** $\dfrac{3}{2}$ **e.** 2

C

CALCULATOR USAGE (51–54)

In Exercises 51–54 store in your calculator the value of 7.85 for x, and then approximate to the nearest hundredth the value of each of these expressions.

51 $\dfrac{\sqrt{x}+11}{\sqrt{x}-11}$

52 $\dfrac{\sqrt{2x}-3}{\sqrt{2x}+3}$

53 $\dfrac{\sqrt[3]{x}-4}{\sqrt[3]{x}+4}$

54 $\dfrac{\sqrt[3]{x}+5}{\sqrt[3]{x}-5}$

55 Pulse Rates The line graph in the figure gives the predicted pulse rate in beats per minute for men with a height of H inches.

a. Determine the approximate pulse rate for a man with a height of 68 inches.

b. Determine the approximate height of a man with a pulse rate of 75 beats per minutes.

c. Use a graphics calculator to graph $y = \dfrac{590}{\sqrt{H}}$, and compare this graph to the given line graph.

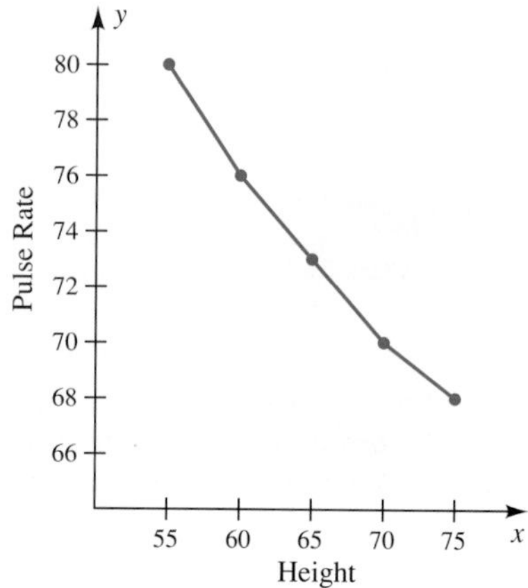

Figure for Exercise 55

56 Rationalize $\dfrac{1}{\sqrt[3]{2}-1}$ by simplifying

$$\frac{1}{\sqrt[3]{2}-1} \cdot \frac{\sqrt[3]{4}+\sqrt[3]{2}+1}{\sqrt[3]{4}+\sqrt[3]{2}+1}.$$

57 Rationalize $\dfrac{1}{\sqrt[3]{x}+\sqrt[3]{y}}$ by simplifying

$$\frac{1}{\sqrt[3]{x}+\sqrt[3]{y}} \cdot \frac{\sqrt[3]{x^2}-\sqrt[3]{xy}+\sqrt[3]{y^2}}{\sqrt[3]{x^2}-\sqrt[3]{xy}+\sqrt[3]{y^2}}.$$

58 a. Use the Pythagorean Theorem to determine the length of $\overline{AC}$ in the figure on the next page.

b. Use the Pythagorean Theorem to determine the length of $\overline{AD}$.

c. Write the radical expression that represents the difference of the lengths of $\overline{AD}$ and $\overline{AC}$.

d. Write the radical expression that represents the sum of the lengths of $\overline{AD}$ and $\overline{AC}$.

e. Write the radical expression that represents the ratio of the expressions in parts c and d of this exercise. Then simplify this radical expression.

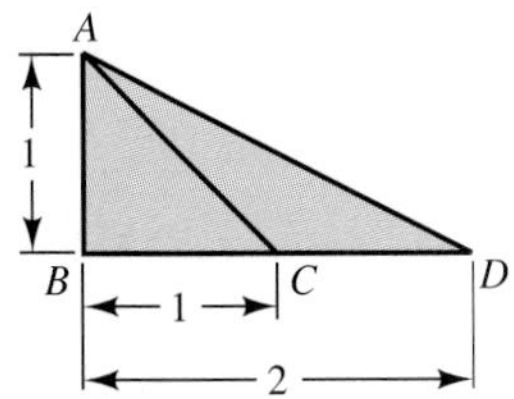

Figure for Exercise 58

Discussion Question

59 Describe in your own words the procedure for dividing two radicals of the same order.

Challenge Question

a. A circle and a square have the same area. Compare their perimeters.

b. A circle and a square have the same perimeter. Compare their areas.

Section Skill Reinforcers

Solve each of these equations for x.

1 $5(x - 2) + 3 = 2(x + 2) - 17$ **2** $2xy + 3 = 7$ **3** $2xy + 3x = 7$ **4** $2x^2 - 5x - 3 = 0$
5 $(x + 2)(x - 3) = 14$

7-6 *Equations with Radicals*

Section Objective

8 Solve equations involving radical expressions.

Equations that contain variables in a radicand are called **radical equations.** Since these equations occur frequently in various disciplines, it is important to know how to solve them. In this section we will examine radical equations that result in either linear equations or factorable quadratic equations. An example of such a radical equation is $\sqrt{x} = 3$. It is easy to verify that $x = 9$ is a solution of this equation because $\sqrt{9} = 3$. The key to solving radical equations is to raise both sides of the equation to the same power.

A Mathematical Note

Niels Abel, 1802–1829, was a Norwegian mathematician who was raised in extreme poverty and died of tuberculosis at the age of twenty-six. Nonetheless, he established several new mathematical concepts. Among his accomplishments was his proof, using radicals, that the general fifth-degree polynomial equation is impossible to solve exactly.

Power Theorem

For any real numbers a and b and natural number n,

$$\text{if } a = b \qquad \text{then } a^n = b^n$$

Caution: The equations $a = b$ and $a^n = b^n$ are not always equivalent. The equation $a^n = b^n$ may have a solution that is not a solution of $a = b$.

For example, $(-3)^2 = 3^2$, but $-3 \neq 3$. Since $a^n = b^n$ can be true when $a \neq b$, we must check all possible solutions in the original equation to determine whether they are really solutions or extraneous values. Recall from Section 1-4 that an *extraneous value* is a value that incorrectly occurs as a solution in the last step of the solution process but is *not* a solution of the original equation.

SOLVING RADICAL EQUATIONS CONTAINING A SINGLE RADICAL

Step 1 Isolate a radical term on one side of the equation.

Step 2 Raise both sides to the nth power.

Step 3 Solve the resulting equation. (If this equation contains a radical, repeat Steps 1 and 2.)

Step 4 Check the possible solutions in the original equation to determine whether they are really solutions or are extraneous.

EXAMPLE 1 Solving an Equation with One Radical Term

Solve $\sqrt{7-y}=4$.

SOLUTION

$$\sqrt{7-y}=4$$

$$(\sqrt{7-y})^2=4^2$$ Square both sides of the equation.

$$7-y=16$$ Simplify, and then solve the equation for y.

$$-y=9$$

$$y=-9$$

Check $\sqrt{7-(-9)}\stackrel{?}{=}4$

$\sqrt{16}=4$ is true.

Answer The solution set is $\{-9\}$.

EXAMPLE 2 Solving an Equation with One Radical Term

Solve $\sqrt{2z+3}=-5$.

SOLUTION

$$\sqrt{2z+3}=-5$$ This equation cannot have a solution since a principal root is always nonnegative. However, we will continue the solution process in order to see what happens.

$$(\sqrt{2z+3})^2=(-5)^2$$ Square both members of the equation.

$$2z+3=25$$ Simplify, and then solve for z.

$$2z=22$$

$$z=11$$

Check $\sqrt{2(11)+3}\stackrel{?}{=}-5$

$\sqrt{25}\stackrel{?}{=}-5$

$5=-5$ does not check.

The principal root of 25 is +5, not −5. Thus 11 is an extraneous value of the original equation.

Answer No solution

The solution set is the null set, $\varnothing$.

EXAMPLE 3 Solving an Equation with One Radical Term

Solve $\sqrt[3]{a-1} = 4$.

SOLUTION

$$\sqrt[3]{a-1} = 4$$

$$(\sqrt[3]{a-1})^3 = 4^3 \quad \text{Cube both members of the equation.}$$

$$a - 1 = 64 \quad \text{Simplify, and then solve the linear equation.}$$

$$a = 65$$

Check $\sqrt[3]{65 - 1} \stackrel{?}{=} 4$

$\sqrt[3]{64} = 4$ is true.

Answer $a = 65$ The solution set is $\{65\}$.

EXAMPLE 4 Solving an Equation with One Radical Term

Solve $x = \sqrt{x+6}$.

SOLUTION

$$x = \sqrt{x+6}$$

$$x^2 = x + 6 \quad \text{Square both members of the equation.}$$

$$x^2 - x - 6 = 0 \quad \text{Write the quadratic equation in standard form.}$$

$$(x-3)(x+2) = 0 \quad \text{Factor the left member.}$$

$$x - 3 = 0 \quad \text{or} \quad x + 2 = 0 \quad \text{Set each factor equal to zero.}$$

$$x = 3 \qquad x = -2 \quad \text{Solve for } x.$$

Check $x = 3$: $3 \stackrel{?}{=} \sqrt{3+6}$; $3 \stackrel{?}{=} \sqrt{9}$; $3 = 3$ is true. $x = -2$: $-2 \stackrel{?}{=} \sqrt{-2+6}$; $-2 \stackrel{?}{=} \sqrt{4}$; $-2 = 2$ is false.

Answer $x = 3$ (-2 is an extraneous value.) The solution set $\{3\}$ does not contain the extraneous value -2.

Whenever a radical equation contains another term on the same side of the equation as the radical, we begin by isolating the radical term on one side of the equation. Then we simplify both sides by squaring if the radicals are square roots, cubing if the radicals are cube roots, etc.

SELF-CHECK

Solve each equation.

1 $\sqrt{t-3} = 5$

2 $\sqrt{1-p} = -1$

EXAMPLE 5 Solving an Equation with One Radical Term

Solve $\sqrt{2w-3} + 9 = w$.

SOLUTION

$$\sqrt{2w-3} + 9 = w$$

$$\sqrt{2w-3} = w - 9 \quad \text{First isolate the radical term in the left member by subtracting 9 from both members.}$$

$$2w - 3 = w^2 - 18w + 81 \quad \text{Square both sides of the equation. (Be careful not to omit the middle term when you square the binomial in the right member.)}$$

SELF-CHECK ANSWERS 

1 $t = 28$ **2** No solution

$$w^2 - 20w + 84 = 0$$ Write the quadratic equation in standard form.

$$(w - 14)(w - 6) = 0$$ Factor, and solve for w.

$$w - 14 = 0 \quad \text{or} \quad w - 6 = 0$$

$$w = 14 \qquad\qquad w = 6$$

Check $w = 14$: $\sqrt{2(14) - 3} + 9 \stackrel{?}{=} 14$

$$\sqrt{25} + 9 \stackrel{?}{=} 14$$

$$5 + 9 \stackrel{?}{=} 14$$

$$14 = 14 \text{ is true.}$$

$w = 6$: $\sqrt{2(6) - 3} + 9 \stackrel{?}{=} 6$

$$\sqrt{9} + 9 \stackrel{?}{=} 6$$

$$3 + 9 \stackrel{?}{=} 6$$

$$12 = 6 \text{ is false.}$$

Answer $w = 14$ (6 is an extraneous value.) The solution set {14} does not contain the extraneous value 6.

EXAMPLE 6 Solving an Equation with Two Radical Terms

Solve $\sqrt[3]{3t + 1} - \sqrt[3]{5t - 9} = 0$.

SOLUTION $\sqrt[3]{3t + 1} - \sqrt[3]{5t - 9} = 0$

$$\sqrt[3]{3t + 1} = \sqrt[3]{5t - 9}$$ Isolate a radical on one side of the equation by adding $\sqrt[3]{5t - 9}$ to both members.

$$3t + 1 = 5t - 9$$ Cube both members of the equation.

$$-2t = -10$$ Then solve for t.

$$t = 5$$

Check $\sqrt[3]{3(5) + 1} - \sqrt[3]{5(5) - 9} \stackrel{?}{=} 0$

$$\sqrt[3]{16} - \sqrt[3]{16} \stackrel{?}{=} 0$$

$$0 = 0 \text{ is true.}$$

Answer $t = 5$

The same strategy used to solve radical equations for constant values of the variable can also be used to transform radical equations to solve for a specified variable.

SELF-CHECK

Solve each radical equation.

1 $\sqrt{6t + 5} - \sqrt{5 - 7t} = 0$

2 $\sqrt{6t + 5} + \sqrt{5 + 7t} = 0$

EXAMPLE 7 Solving for a Specified Variable

Solve $y = \dfrac{\sqrt[3]{2x - z}}{3}$ for x.

SOLUTION $y = \dfrac{\sqrt[3]{2x - z}}{3}$

$$3y = \sqrt[3]{2x - z}$$ Multiply both members by 3 to isolate the radical term in the right member of the equation.

$$27y^3 = 2x - z$$ Cube both members.

$$27y^3 + z = 2x$$ Isolate the term involving x by adding z to both members.

$$x = \frac{27y^3 + z}{2}$$ Then divide both members by 2.

SELF-CHECK ANSWERS

1 $t = 0$ **2** No solution

EXAMPLE 8 An Application of the Pythagorean Theorem

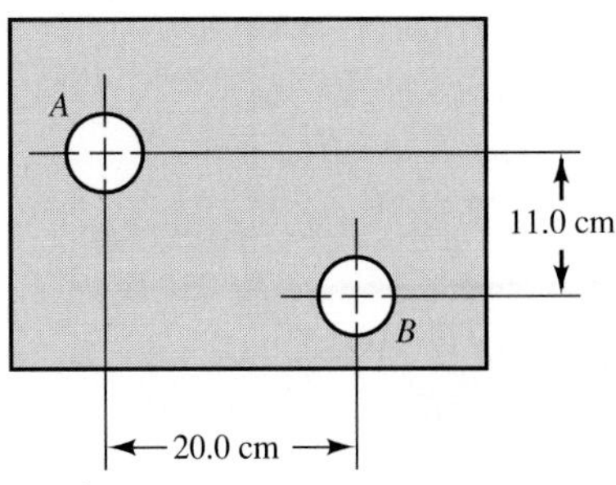

A machinist is measuring a metal block that is part of an automobile engine. Use the measurements shown on the drawing in the figure to determine the distance from the center of hole A to the center of hole B.

SOLUTION The information from the sketch is shown in condensed form in the right triangle labeled ABC (see the figure). The distance desired is the length of the side labeled c.

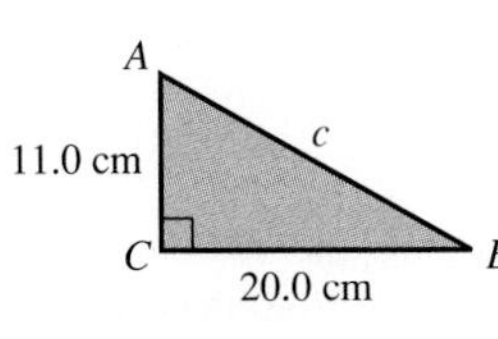

$$\begin{aligned} c &= \sqrt{a^2 + b^2} \\ &= \sqrt{20^2 + 11^2} \\ &= \sqrt{400 + 121} \\ &= \sqrt{521} \\ c &\approx 22.8 \end{aligned}$$

Since c is the hypotenuse of this right triangle, the Pythagorean Theorem can be used to calculate c with the given values of $a = 20$ cm and $b = 11$ cm.

Calculator approximation

Answer The distance from the center of hole A to the center of hole B is approximately 22.8 cm.

Remember to write your answers as full sentences. One of the most important skills required by employers is the ability to communicate information clearly.

EXAMPLE 9 Calculator Usage

Use the graphs of $y_1 = x$ and $y_2 = \sqrt{x + 6}$ to solve $x = \sqrt{x + 6}$.

SOLUTION

Keystrokes

[Y=] [X, T, θ] [▼] [2nd] [√‾] [(] [X, T, θ] [+] [6] [)] [GRAPH]

Graph the first equation as $y_1 = x$ and the second equation as $y_2 = \sqrt{x + 6}$. Assume that the window of x values and y values has already been set to $[-7, 5]$ for x and $[-1, 4]$ for y.

$[-7, 5]$ for x, $[-1, 4]$ for y

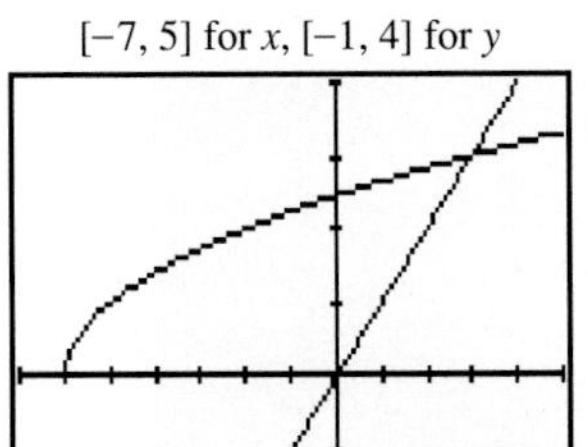

The point where the curves cross gives the x value for which $x = \sqrt{x + 6}$. By using the TRACE key (and perhaps the ZOOM key), we can approximate this value to be $x = 3$.

This is the same answer obtained in Example 4 by algebraic methods.

Answer $x = 3$

Exercises 7-6

In Exercises 1–25 solve each equation.

1 $\sqrt{t - 4} = 3$

2 $\sqrt{t - 2} = 12$

3 $\sqrt{c + 7} + 23 = 43$

4 $\sqrt{c + 5} - 3 = 6$

5 $\sqrt{3x - 21} + 7 = 2$

6 $\sqrt{25 + 2x} + 11 = 5$

7 $\sqrt[3]{2w + 1} = -2$

8 $\sqrt[3]{5w + 2} = -3$

9 $\sqrt[4]{6v - 2} = 2$

10 $\sqrt[4]{7v - 3} = 3$

11 $\sqrt[5]{w^2 - 4w} = 2$

12 $\sqrt[5]{2w^2 + w - 16} = -1$

13 $\sqrt{y^2 - y + 13} - y = 1$

14 $\sqrt{y^2 + y + 11} - y = 1$

15 $\sqrt{7t + 2} = 2t$

16 $\sqrt{-9t - 2} = 2t$

17 $\sqrt{w^2 - 2w + 1} = 2w$

18 $\sqrt{2w + 1} = w + 1$

19 $\sqrt{2x + 1} + 5 = 2x$

20 $\sqrt{7x - 3} + 1 = 3x$

21 $\sqrt{6u + 7} = \sqrt{11u + 7}$

22 $\sqrt{14u - 12} = \sqrt{10u - 8}$

23 $\sqrt[3]{2w^2 + 3w} = \sqrt[3]{2 + 2w - w^2}$

24 $\sqrt[3]{4w^2 + 13w + 9} = \sqrt[3]{w^2 + 2w - 1}$

25 $\sqrt{5z - 10} + \sqrt{3z + 8} = 0$

B

In Exercises 26–29 solve each problem for x.

26 The square root of the quantity x plus 5 is equal to 3.

27 The cube root of the quantity x plus 2 is equal to -2.

28 Four times the square root of x equals three times the quantity x minus 5.

29 If the square root of the quantity x plus 4 is added to 2, the result equals x.

In Exercises 30–33 solve each equation for the variable indicated. Assume that variables are restricted to meaningful values.

30 $b = \sqrt{c^2 - a^2 + h}$ for h

31 $\sqrt{\dfrac{a + b}{c - b}} = 5$ for b

32 $\sqrt{\dfrac{m - n}{2m + n}} = 3$ for n

33 $\sqrt{5 + \dfrac{m}{n}} = m$ for n

In Exercises 34–37 use the Pythagorean Theorem ($a^2 + b^2 = c^2$) or one of its alternative forms ($a = \sqrt{c^2 - b^2}$, $b = \sqrt{c^2 - a^2}$, and $c = \sqrt{a^2 + b^2}$) to solve each problem.

34 Squaring a Corner A carpenter is trying to "square" the corner of a building. She makes a mark 6 meters along one wall and another mark 8 meters along the other wall. What will the distance between the marks be if the walls are square? (See the figure.)

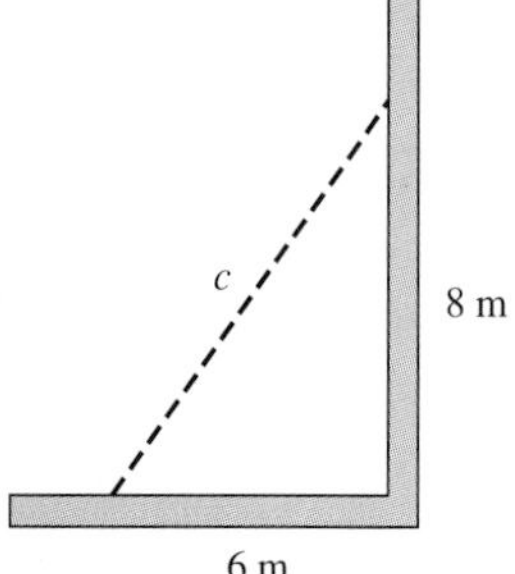

Figure for Exercise 34

35 Wiring a Room An electrician plans to run wiring diagonally across a rectangular room. The room is 12 feet wide and 16 feet long. How much wiring will be needed to cross this room? (See the figure.)

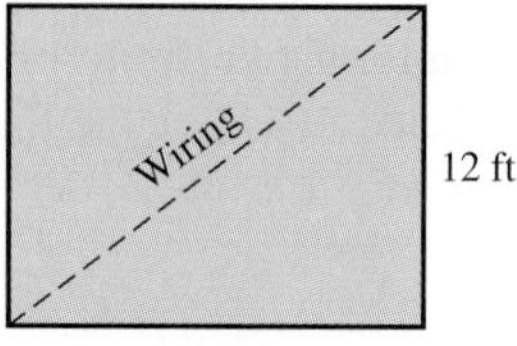

Figure for Exercise 35

36 Machinist's Measurements Using the measurements made by the machinist, find the distance from the center of a hole A to the center of the hole B shown in the figure.

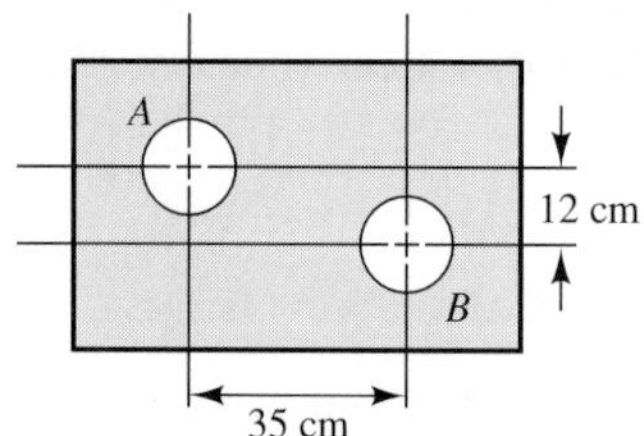

Figure for Exercise 36

37 Height of a Wall A 17-foot ladder is leaning against a chimney. If the bottom of the ladder is 8 feet from the base of the wall, how far is it from the bottom of the wall to the top of the ladder? (See the figure.)

Figure for Exercise 37

C

Calculator Exercises (38–42)

38 Use the graphs of $y_1 = \sqrt{x+3}$ and $y_2 = 1$ shown on the calculator window to solve $\sqrt{x+3} = 1$.

[−5, 5] for x, [−1, 4] for y

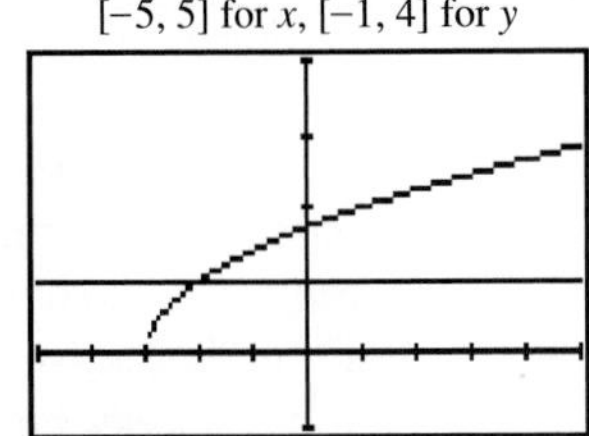

Figure for Exercise 38

In Exercises 39–42 use a graphics calculator to solve each equation.

39 Use the graphs of $y_1 = \sqrt{4x+1}$ and $y_2 = x - 1$ to solve $\sqrt{4x+1} = x - 1$.

40 Use the graphs of $y_1 = \sqrt{2x+1}$ and $y_2 = x - 1$ to solve $\sqrt{2x+1} = x - 1$.

41 Use the graphs of $y_1 = \sqrt{7x+2}$ and $y_2 = \sqrt{9x-2}$ to solve $\sqrt{7x+2} = \sqrt{9x-2}$.

42 Use the graphs of $y_1 = \sqrt{x^2+x+4}$ and $y_2 = 2x - 2$ to solve $\sqrt{x^2+x+4} = 2x - 2$.

In Exercises 43–48 solve each equation.

43 $\sqrt[3]{x^3 - 6x^2 + 12x} = x$

44 $\sqrt[3]{x^3 + 9x^2 + 27x} = x$

45 $\sqrt{4w^2 + 12w + 6} = 2w$

46 $\sqrt{9w^2 - 12w + 8} = 3w$

47 $\sqrt[3]{x^3 - 2x^2 + 3x - 2} = x - 1$

48 $\sqrt[3]{x^3 + 2x^2 + 3x + 5} = x + 1$

Production of Solar Cells (49–52)

In Exercises 49–52 use the fact that the dollar cost C of producing n solar cells per shift is given by the formula $C = 18\sqrt[3]{n^2} + 450$.

49 Find the overhead cost for one shift. That is, find the cost of producing 0 solar cells.

50 Find the cost of producing 27 solar cells.

51 Find the number of solar cells produced when the cost is \$738.

52 Find the number of solar cells produced when the cost is \$1098.

Discussion Question

53 Describe which step in the solution of radical equations introduces the extraneous values that occur in some problems.

Section Skill Reinforcers

Perform each of the indicated operations.

1 $(2x - 3y) - (4x - 5y)$ **2** $(2x + 3y) - 4(x - 5y)$ **3** $(2x + 3y)(4x - 5y)$ **4** $\dfrac{5x - 10y}{x - 2y}$ **5** $(x - 2y)(x + 2y)$

7-7 Complex Numbers

Section Objectives

9 **Express complex numbers in standard form.**

10 **Add, subtract, multiply, and divide complex numbers.**

Every real number is either negative, zero, or positive. Because the square of a negative number is positive, the square of zero is zero, and the square of a positive number is positive, there is no real number x for which $x^2 = -1$. The desire to solve equations of the form $x^2 = -1$ led mathematicians to define the number i so that $i^2 = -1$. We use i to represent $\sqrt{-1}$, and $-i$ to represent $-\sqrt{-1}$. When these numbers were first developed in the seventeenth century, they were called imaginary numbers, because mathematicians were not familiar with them and did not know a concrete application for them. One of the first concrete applications of imaginary numbers was developed in 1892, when Charles P. Steinmetz used them in his theory of alternating currents. We often use imaginary numbers in the computation of problems whose final answers are real numbers, just as we use fractions in the computation of problems whose answers are natural numbers.

The square root of any negative number is imaginary and can be expressed in terms of the imaginary unit i. For $x > 0$, we define $\sqrt{-x}$ to be $i\sqrt{x}$.

A Mathematical Note

Jean-Victor Poncelet, 1788–1867, was a lieutenant in Napoleon's army. After being left for dead on the Russian front, he was captured and marched for months across frozen plains in subzero weather. To survive the boredom of captivity, he decided to reproduce as much mathematics as he could. When he returned to France after the war, he had developed projective geometry and also a geometric interpretation of imaginary numbers.

THE IMAGINARY NUMBER i

$$i = \sqrt{-1} \qquad \text{so} \qquad i^2 = -1$$

For any positive real number x, $\sqrt{-x} = i\sqrt{x}$.

EXAMPLE 1 Writing Imaginary Numbers Using the i Notation

Write each imaginary number in terms of i.

SOLUTIONS

(a) $\sqrt{-25}$ $\quad \sqrt{-25} = i\sqrt{25}$ $\quad \sqrt{-x} = i\sqrt{x}$

$= i(5)$

$= 5i$

(b) $\sqrt{-3}$ $\quad \sqrt{-3} = i\sqrt{3}$

Using i, the real numbers, and the operations of addition, subtraction, multiplication, and division, we obtain numbers that can be written in the form $a + bi$, where a and b are real numbers. A combination of a real number and an imaginary number—that is, any number that can be written in the standard form $a + bi$—is called a **complex number.** If $b = 0$, then $a + bi$ is just the real number a. If $b \neq 0$, then $a + bi$ is called **imaginary.** If $a = 0$ and $b \neq 0$, then bi is called **pure imaginary.** Thus the complex numbers include both the real numbers and the pure imaginary numbers.

COMPLEX NUMBERS

If a and b are real numbers and $i = \sqrt{-1}$, then

$$a + bi$$

Real term ↑ ↑ Imaginary term

is a complex number.

A complex number is either a real number or an imaginary number, but not both. The relationships among the real, complex, and imaginary numbers are shown in Figure 7-1 and illustrated by the examples in Table 7-2.

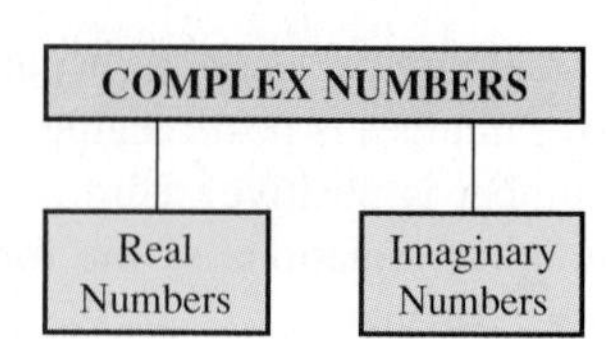

Figure 7-1

TABLE 7-2 *Complex Numbers*

	Real Term	Coefficient of the Imaginary Term	Classification
$3 - 4i$	3	-4	Imaginary
$6 = 6 + 0i$	6	0	Real
$-7i = 0 - 7i$	0	-7	Pure imaginary
$0 = 0 + 0i$	0	0	Real
$-\sqrt{25} = -5 + 0i$	-5	0	Real
$\sqrt{-25} = 0 + 5i$	0	5	Pure imaginary

Two complex numbers are equal if and only if both their real terms and their imaginary terms are equal.

EXAMPLE 2 Equality of Complex Numbers

Determine a and b such that the complex numbers are equal.

SOLUTIONS

(a) $a + bi = 6 + 11i$ $\quad a = 6, b = 11$ $\quad$ Equate the real parts and the imaginary parts.

(b) $-5 + bi = a + 9i$ $\quad a = -5, b = 9$

(c) $(a + 3) - 8i = 4 + 2bi$ $\quad a + 3 = 4, 2b = -8$

$$a = 1, \quad b = -4$$

▼

Since complex numbers consist of two terms, a real term and an imaginary term, the arithmetic of complex numbers is very similar to the arithmetic of binomials. The similarity is illustrated in Table 7-3 using the operation of addition.

SELF-CHECK 

Determine a and b such that the following complex numbers are equal.

1 $3 + bi = a - 17i$

2 $-\sqrt{36} + \sqrt{-49} = a + bi$

TABLE 7-3 *Comparison of Binomials and Complex Numbers*

Addition of Binomials	Addition of Complex Numbers
$2x + 3y$	$2 + 3i$
$+(4x + 7y)$	$+(4 + 7i)$
$6x + 10y$	$6 + 10i$

The addition of complex numbers is carried out by adding the real terms and the imaginary terms separately. That is,

$$(a + bi) + (c + di) = (a + c) + (b + d)i$$

Subtraction is performed similarly.

SELF-CHECK ANSWERS 

1 $a = 3, b = -17$ $\quad$ 2 $a = -6, b = 7$

EXAMPLE 3 Adding and Subtracting Complex Numbers

Perform the indicated operations.

SOLUTIONS

(a) $(2 + 5i) + (8 + 4i)$

$$(2 + 5i) + (8 + 4i) = (2 + 8) + (5 + 4)i$$
$$= 10 + 9i$$

Group like terms together, and then add like terms.

(b) $6 + (11 - 4i)$

$$6 + (11 - 4i) = (6 + 11) + (0 - 4)i$$
$$= 17 - 4i$$

(c) $(8 - 5i) - 6i$

$$(8 - 5i) - 6i = (8 + 0) + (-5 - 6)i$$
$$= 8 - 11i$$

Since $i^2 = -1$, higher powers of i can always be simplified to the **standard form,** $a + bi$. The first four powers of i are keys to simplifying higher powers to standard form and should therefore be memorized.

FIRST FOUR POWERS OF i

$$i^1 = i$$
$$i^2 = i \cdot i = -1$$
$$i^3 = i^2 \cdot i = (-1)i = -i$$
$$i^4 = i^2 \cdot i^2 = (-1)(-1) = 1$$

The powers of i repeat in cycles of four: i^4, i^8, i^{12}, etc., all of which are equal to 1. We will use this fact to simplify i^n, where n is any integer exponent, to either i, -1, $-i$, or 1.

EXAMPLE 4 Powers of i

Simplify each power of i.

SOLUTIONS

(a) i^7

$$i^7 = i^4 \cdot i^3$$
$$= (1)(-i)$$
$$= -i$$

First extract the largest multiple of 4 from each exponent. Then simplify, replacing i^4, i^8, i^{12}, etc., by 1.

(b) i^{13}

$$i^{13} = i^{12}i^1$$
$$= (1)(i)$$
$$= i$$

(c) i^{406}

$$i^{406} = i^{404} \cdot i^2$$
$$= (i^4)^{101} \cdot (-1)$$
$$= (1)^{101}(-1)$$
$$= (1)(-1)$$
$$= -1$$

$$4\overline{)406} = 101,\quad 404,\quad \text{remainder } 2$$

$406 = 4(101) + 2$

SELF-CHECK

Simplify each of these expressions.

1 $(21 + 4i) - (9 - 7i)$
2 i^{33}

SELF-CHECK ANSWERS

1 $12 + 11i$ **2** i

In working with complex numbers, it is important to remember the restriction on the formula given in Section 7-2: $(\sqrt[n]{x})^m = \sqrt[n]{x^m}$ if and only if $\sqrt[n]{x}$ is a real number. Specifically, we noted that $\sqrt{(-5)^2} \neq (\sqrt{-5})^2$:

$$\sqrt{(-5)^2} = \sqrt{25} \qquad \text{whereas} \qquad (\sqrt{-5})^2 = (i\sqrt{5})^2$$
$$= 5 \qquad\qquad\qquad\qquad = 5i^2$$
$$= -5$$

It is also crucial to remember that $\sqrt{x}\sqrt{y} \neq \sqrt{xy}$ for negative values of x and y. For example,

$$\sqrt{-2}\sqrt{-3} = (i\sqrt{2})(i\sqrt{3}) \qquad \text{whereas} \qquad \sqrt{(-2)(-3)} = \sqrt{+6}$$
$$= i^2\sqrt{6} \qquad\qquad\qquad = \sqrt{6}$$
$$= -\sqrt{6}$$

Note that, in the equation above, we wrote $i\sqrt{2}$ rather than $\sqrt{2}i$. The latter form could easily be confused with $\sqrt{2i}$, which is not the same as $i\sqrt{2}$. To make it clear that the factor i is not in the radicand, it is best to put the i in front of the radical.

To avoid the potential errors noted above, we recommend that you write every complex number in standard form before proceeding with any operations. This procedure is illustrated in Example 5.

STANDARD FORM FOR COMPLEX NUMBERS

Before performing any operations with complex numbers, express them in the standard form $a + bi$. Also express your answers in this form.

In the next example we will multiply complex numbers as if they were binomials with a real term and an imaginary term. Applying the FOIL method to the multiplication of the complex numbers, we have

$$(a + bi)(c + di) = \underset{\boxed{F}}{ac} + \underset{\boxed{O}}{adi} + \underset{\boxed{I}}{bci} + \underset{\boxed{L}}{bdi^2}$$

Replacing i^2 by -1, we have

$$(a + bi)(c + di) = (ac - bd) + (ad + bc)i$$

It is not necessary to memorize this result; simply multiply complex numbers as you do binomials.

EXAMPLE 5 Multiplying Complex Numbers

Simplify each of these expressions.

SOLUTIONS

(a) $3(4 - 5i)$ $\qquad 3(4 - 5i) = 3(4) - 3(5i)$ Distribute the factor 3.

$$= 12 - 15i$$

(b) $2i(7 + 3i)$

$2i(7 + 3i) = 2i(7) + 2i(3i)$ — Distribute the factor $2i$.
$= 14i + 6i^2$
$= 14i + 6(-1)$ — Replace i^2 by -1.
$= -6 + 14i$ — Write the answer in the standard $a + bi$ form.

(c) $\sqrt{-2}(\sqrt{3} - \sqrt{-2})$

$\sqrt{-2}(\sqrt{3} - \sqrt{-2}) = i\sqrt{2}(\sqrt{3} - i\sqrt{2})$ — First write each factor in the standard $a + bi$ form.
$= i\sqrt{2}(\sqrt{3}) - i\sqrt{2}(i\sqrt{2})$ — Then distribute the factor $i\sqrt{2}$.
$= i\sqrt{6} - 2i^2$ — Simplify each term, noting that $\sqrt{2}\sqrt{2} = 2$.
$= i\sqrt{6} - 2(-1)$ — Replace i^2 by -1.
$= 2 + i\sqrt{6}$ — Reorder the terms to write the answer in the standard form.

(d) $(5 - 6i)(5 + 6i)$

$(5 - 6i)(5 + 6i) = (5)^2 - (6i)^2$ — Note that this expression is of the form $(x - y)(x + y) = x^2 - y^2$.
$= 25 - 36i^2$
$= 25 - 36(-1)$ — Replace i^2 by -1.
$= 61$

(e) $(2 - 7i)(5 + 3i)$

$(2 - 7i)(5 + 3i) = 10 + 6i - 35i - 21i^2$ — Multiply, using the FOIL method.
$= 10 - 29i - 21(-1)$ — Replace i^2 by -1.
$= 10 - 29i + 21$
$= 31 - 29i$ ▼

Since $i = \sqrt{-1}$, any expression with i in the denominator can be considered to have a radical in the denominator. Thus dividing complex numbers is similar to rationalizing radicals. In keeping with our earlier definition of conjugates, we define the conjugate of $a + bi$ to be $a - bi$. The product of a complex number and its conjugate uses the special product of a sum times a difference to yield the difference of the squares. This product is always a real number, as illustrated in Example 6.

A Mathematical Note

The term *conjugates* for $a + bi$ and $a - bi$ was suggested by Augustin-Louis Cauchy (1789–1857). Cauchy was a prolific writer in mathematics, perhaps second only to Euler. In mathematical circles, Cauchy is also known for having reportedly lost important papers of Niels Abel (see page 391), which were sent to him by the French Academy for review.

Conjugate Complex Numbers

The complex numbers $a + bi$ and $a - bi$ are conjugates of each other.

EXAMPLE 6 Multiplying Complex Conjugates

Multiply each complex number by its conjugate.

SOLUTIONS

	Conjugate	*Product*	
(a) $3 + 4i$	$3 - 4i$	$(3 + 4i)(3 - 4i) = 9 - 16i^2$ $= 25$	Multiply, using the fact that this expression is of the form $(x + y)(x - y) = x^2 - y^2$.
(b) $5 - 2i$	$5 + 2i$	$(5 - 2i)(5 + 2i) = 25 - 4i^2$ $= 29$	
(c) 11	11	$(11)(11) = 121$	$11 = 11 + 0i$ $11 = 11 - 0i$ These numbers are conjugates.

(d) $8i$	$-8i$	$(8i)(-8i) = -64i^2$ $= 64$	$8i = 0 + 8i$ $-8i = 0 - 8i$ These numbers are conjugates.
(e) $a + bi$	$a - bi$	$(a + bi)(a - bi) = a^2 - b^2i^2$ $= a^2 + b^2$	The product $a^2 + b^2$ is a real number.

▼

SELF-CHECK 

Simplify each of these expressions.

1 $(5 + i)(6 + 2i)$
2 $(11 - 8i)(11 + 8i)$
3 $(1 + i)^2$

As illustrated in Example 6(e), the product of complex conjugates is always a real number. This fact plays a key role in the division of complex numbers; a formal description of this role is given in the following box.

DIVISION OF COMPLEX NUMBERS

Step 1 Write the division problem as a fraction.

Step 2 Multiply both the numerator and the denominator by the conjugate of the denominator.

Step 3 Simplify the result, and express it in standard $a + bi$ form.

EXAMPLE 7 Dividing Complex Numbers

Simplify $(8 - i) \div (1 - 2i)$.

SOLUTION

$(8 - i) \div (1 - 2i) = \dfrac{8 - i}{1 - 2i}$ — Write the division problem as a fraction.

$= \dfrac{8 - i}{1 - 2i} \cdot \dfrac{1 + 2i}{1 + 2i}$ — Multiply the numerator and the denominator by $1 + 2i$, the conjugate of the denominator.

$= \dfrac{8 + 15i - 2i^2}{1 - 4i^2}$ — Multiply the numerators using the FOIL method, and multiply the conjugates in the denominators.

$= \dfrac{10 + 15i}{5}$ — Simplify, replacing i^2 by -1.

$= \dfrac{10}{5} + \dfrac{15}{5}i$ — Write the result in standard $a + bi$ form.

$= 2 + 3i$

▼

EXAMPLE 8 Checking a Complex Number Solution of a Quadratic Equation

Verify that $1 + 2i$ is a solution of $x^2 - 2x + 5 = 0$.

SOLUTION

$x^2 - 2x + 5 = 0$

$(1 + 2i)^2 - 2(1 + 2i) + 5 \stackrel{?}{=} 0$ — Substitute $1 + 2i$ for x.

$(1 + 4i + 4i^2) - 2 - 4i + 5 \stackrel{?}{=} 0$ — Simplify, and then add like terms.

$1 + 4i - 4 - 2 - 4i + 5 \stackrel{?}{=} 0$

$0 = 0$ is true.

This value checks, so it is a solution of the equation. ▼

SELF-CHECK ANSWERS 

1 $28 + 16i$ 2 185 3 $2i$

Complex numbers are sometimes represented by ordered pair notation. An ordered pair representation of $a + bi$ is (a, b). For example, $(2, 3)$ represents $2 + 3i$, $(4, -5)$ represents $4 - 5i$, and $(0, -1)$ represents $-i$. Some calculators, such as the TI-85, use this ordered pair notation, as illustrated in Example 9.

SELF-CHECK

Simplify each of these expressions.

1 $\dfrac{2}{1 + i}$ **2** $\dfrac{3 + 2i}{2 - 3i}$

EXAMPLE 9 Calculator Usage

Use a calculator with complex number capability to evaluate $(2 - i)^3$.

SOLUTION

Keystrokes

(2 , (−) 1) ^ 3 ENTER → (2, −11)

You can check this answer with pencil and paper calculations.

Answer $(2 - i)^3 = 2 - 11i$

Exercises 7-7

A

In Exercises 1–10 simplify each expression, and write the result in the standard $a + bi$ form.

1 a. $-\sqrt{36}$ **b.** $\sqrt{-36}$ **c.** $-\sqrt{-36}$ **d.** $\sqrt{36}$

2 a. $-\sqrt{49}$ **b.** $\sqrt{-49}$ **c.** $-\sqrt{-49}$ **d.** $\sqrt{49}$

3 a. $\sqrt{-9} + \sqrt{16}$ **b.** $\sqrt{9} + \sqrt{-16}$ **c.** $-\sqrt{9} - \sqrt{-16}$ **d.** $\sqrt{-9} - \sqrt{16}$

4 a. $\sqrt{4} + \sqrt{-25}$ **b.** $\sqrt{-4} - \sqrt{25}$ **c.** $-\sqrt{4} - \sqrt{-25}$ **d.** $-\sqrt{-4} + \sqrt{-25}$

5 a. $\sqrt{-9 - 16}$ **b.** $-\sqrt{-9 - 16}$ **c.** $\sqrt{-6}\sqrt{-6}$ **d.** $\sqrt{(-6)(-6)}$

6 a. $\sqrt{-25 - 144}$ **b.** $\sqrt{-25} + \sqrt{-144}$ **c.** $\sqrt{-7}\sqrt{-7}$ **d.** $\sqrt{(-7)(-7)}$

7 a. $\sqrt{-4}\sqrt{-25}$ **b.** $\sqrt{(-4)(-25)}$ **c.** $-\sqrt{4}\sqrt{-25}$ **d.** $(-\sqrt{4})(-\sqrt{25})$

8 a. $\sqrt{-9}\sqrt{-100}$ **b.** $\sqrt{(-9)(-100)}$ **c.** $-\sqrt{9}\sqrt{-100}$ **d.** $(-\sqrt{9})(-\sqrt{100})$

9 a. $\sqrt{-\dfrac{25}{9}}$ **b.** $\dfrac{\sqrt{-25}}{\sqrt{9}}$ **c.** $\dfrac{\sqrt{25}}{\sqrt{-9}}$ **d.** $\sqrt{\dfrac{-25}{-9}}$

10 a. $\sqrt{\dfrac{-36}{-49}}$ **b.** $\dfrac{\sqrt{36}}{\sqrt{-49}}$ **c.** $\sqrt{-\dfrac{36}{49}}$ **d.** $\dfrac{\sqrt{-36}}{\sqrt{49}}$

In Exercises 11 and 12 identify the real values of a and b that will make each statement true.

11 a. $a + bi = 18 - 5i$ **b.** $a - 2i = -6 + bi$ **c.** $5a - 9i = 20 + (b + 2)i$ **d.** $3a - bi = 72$

12 a. $a + bi = -6 + 17i$ **b.** $-3 + bi = a + 11i$ **c.** $14a - 3bi = (5a - 1) + 2i$ **d.** $3a - bi = 72i$

In Exercises 13–42 perform the indicated operations, and express the result in the standard $a + bi$ form.

13 $(1 + 2i) + (8 - 3i)$ **14** $(6 - 7i) + (13 - 4i)$

15 $(5 + 3i) - (2 + 2i)$ **16** $(7 + i) - (5 - 2i)$

17 $6(10 + 4i)$ **18** $-9(3 - 12i)$

19 $(3 + 3i) + \dfrac{1}{2}(8 - 6i)$ **20** $(3 + 3i) - \dfrac{1}{3}(6 - 9i)$

21 $(2i)(3i)$ **22** $(5i)(7i)$

23 $2i(3 - 5i)$ **24** $-3i(4 - 2i)$

SELF-CHECK ANSWERS 

1 $1 - i$ **2** i

25 $i(3 + 2i) + 2(5 - 3i)$

26 $4(3 - 2i) - i(5 + 3i)$

27 $(2 - 7i)(2 + 7i)$

28 $(6 + 9i)(6 - 9i)$

29 $(5 - 2i)(4 + 7i)$

30 $(6 + i)(3 - 5i)$

31 $(5 + i)^2$

32 $(6 - i)^2$

33 $(4 - 3i)^2$

34 $(3 + 5i)^2$

35 $(4 + 7i)(7 - 4i)$

36 $(11 - 3i)(2 + 5i)$

37 $\sqrt{-3}(\sqrt{2} + \sqrt{-3})$

38 $(\sqrt{2} - \sqrt{-5})(\sqrt{-2} - 3\sqrt{-5})$

39 $2\sqrt{-75} + \sqrt{-27}$

40 $3\sqrt{-8} - 5\sqrt{-98}$

41 $\sqrt{4} + \sqrt{-9} - \sqrt{9} - \sqrt{-25}$

42 $\sqrt{64} + \sqrt{-36} - \sqrt{9} - \sqrt{-1}$

In Exercises 43–46 multiply each complex number by its conjugate.

43 $2 + 5i$ **44** $3 - 8i$ **45** $13i$ **46** 13

In Exercises 47–52 perform the indicated operations, and express the result in the standard $a + bi$ form.

47 $\dfrac{4}{1 + i}$ **48** $\dfrac{6}{1 - i}$ **49** $\dfrac{4 - i}{4 + i}$

50 $\dfrac{5 + i}{5 - i}$ **51** $85 \div (7 - 6i)$ **52** $185 \div (11 + 8i)$

In Exercises 53–56 simplify each power of i.

53 i^9 **54** i^{11} **55** i^{58} **56** i^{81}

B

57 Given $a = 1$, $b = 2$, and $c = 5$, evaluate both $\dfrac{-b - \sqrt{b^2 - 4ac}}{2a}$ and $\dfrac{-b + \sqrt{b^2 - 4ac}}{2a}$.

58 Given $a = 2$, $b = -3$, and $c = 2$, evaluate both $\dfrac{-b - \sqrt{b^2 - 4ac}}{2a}$ and $\dfrac{-b + \sqrt{b^2 - 4ac}}{2a}$.

59 **a.** Is 3 a real number?

b. Is 3 a complex number?

60 **a.** Is $3i$ a real number?

b. Is $3i$ a complex number?

61 If $z^2 = -1$ and $z \neq i$, what is the value of z?

62 What can you say about a complex number that is equal to its conjugate?

63 Give an example of two imaginary numbers whose sum is a real number.

64 Give an example of two imaginary numbers whose product is a real number.

65 Name a complex number by which you cannot divide.

66 Given $z = 3 - i$:

a. What is the conjugate of z?

b. What is the additive inverse of z?

c. What is the multiplicative inverse of z?

In Exercises 67 and 68 determine whether the given complex number is a solution of the quadratic equation.

67 **a.** $x^2 - 6x + 13 = 0$; $3 - 2i$

b. $x^2 - 6x + 13 = 0$; $3 + 2i$

68 **a.** $x^2 - 4x + 13 = 0$; $2 - 3i$

b. $x^2 - 4x + 13 = 0$; $2 + 3i$

ESTIMATION SKILLS (69–72)

In Exercises 69–72 mentally estimate the value of each expression, and select the answer closest to your estimate.

69 $\dfrac{4.02 + \sqrt{-3.99}}{1.99}$

a. $3 - 2i$ **b.** $2 + i$

c. $2 - i$ **d.** $2 + 2i$

70 $\dfrac{4.99 - \sqrt{-25.02}}{5.01}$

a. $1 - i$ **b.** $1 + i$

c. $-i$ **d.** i

71 $\dfrac{7.93 - \sqrt{-64.03}}{4.007}$

a. $2 + 4i$ **b.** $2 - 60i$

c. $2 - 2i$ **d.** $2 + 2i$

72 $\dfrac{8.02 + \sqrt{-4.03}}{3.987}$

a. $4 - i$ **b.** $4 + i$

c. $2 + i$ **d.** $2 + 0.5i$

C

CALCULATOR USAGE (73–76)

In Exercises 73–76 use a calculator with complex number capability to evaluate each expression.

73 $(1 - 2i)^3$ **74** $(1 - i)^4$

75 $\left(-\dfrac{1}{2} + \dfrac{\sqrt{3}}{2}i\right)^3$ **76** $\left(-\dfrac{1}{2} - \dfrac{\sqrt{3}}{2}i\right)^3$

In Exercises 77–82 simplify each expression, and write the result in the standard $a + bi$ form.

77 $(-10 + 22i) \div 2i$ **78** $\dfrac{58}{\sqrt{4} - \sqrt{-25}}$

79 i^{-15}

80 $\dfrac{25i}{\sqrt{-9}+\sqrt{-16}}$

81 $\sqrt[3]{-64}-\sqrt{-64}$

82 $\sqrt{-1{,}000{,}000}-\sqrt[3]{1{,}000{,}000}$

83 Show that i is a solution of $x^4 - 2x^3 - 2x^2 - 2x - 3 = 0$.

DISCUSSION QUESTIONS (84–87)

84 Does $x^2 + 4$ factor over the real numbers?

85 Does $(x + 2i)(x - 2i) = x^2 + 4$?

86 Does $x^2 + 4$ factor over the complex numbers?

87 Discuss some of the advantages of using complex numbers (rather than only real numbers) in factoring.

CHALLENGE QUESTION

Simplify

$$i + i^2 + i^3 + i^4 + \cdots + i^{99} + i^{100}$$

SECTION SKILL REINFORCERS

Factor each of these polynomials.

1 $x^2 + 20x + 100$ **2** $x^2 + 8xy + 16y^2$ **3** $4x^2 + 20xy + 25y^2$ **4** $x^4 + 6x^2y + 9y^2$ **5** $x^4 - 1$

Key Concepts for Chapter 7

1 Principal nth root:
The principal nth root of the real number x is denoted by either $x^{1/n}$ or $\sqrt[n]{x}$.

a. For $x < 0$: if n is odd, the principal root is negative. If n is even, there is no real nth root.

b. For $x = 0$: The principal root is 0.

c. For $x > 0$: The principal root is positive for all natural numbers n.

2 Rational exponents, $x^{m/n}$:
For a real number x and natural numbers m and n,

$$x^{m/n} = (x^{1/n})^m = (x^m)^{1/n} \quad \text{if } x^{1/n} \text{ is a real number}$$

$$x^{-m/n} = \frac{1}{x^{m/n}} \quad \text{if } x \neq 0 \text{ and } x^{1/n} \text{ is a real number}$$

If $x < 0$ and n is even, then $x^{1/n}$ is not a real number and the equalities listed above are not true. For example, $\sqrt{-5}$ is not a real number, and $\sqrt{(-5)^2} \neq (\sqrt{-5})^2$.

3 $\sqrt[n]{x^n}$:
For any real number x,

$$\sqrt[n]{x^n} = |x| \quad \text{if } n \text{ is even and } x \text{ is a real number}$$

$$\sqrt[n]{x^n} = x \quad \text{if } n \text{ is odd and } x \text{ is a real number}$$

In particular, $\sqrt[2]{x^2} = |x|$.

4 Properties of radicals:
If x and y are both positive real numbers, then

$$\sqrt[n]{xy} = \sqrt[n]{x}\sqrt[n]{y} \quad \text{and} \quad \sqrt[n]{\frac{x}{y}} = \frac{\sqrt[n]{x}}{\sqrt[n]{y}}$$

5 Reducing the order of a radical:
If x is a positive real number and m, n, and p are natural numbers, then

$$\sqrt[pn]{x^{pm}} = \sqrt[n]{x^m}$$

6 Simplified form for radical expressions:
A radical expression $\sqrt[n]{x^m}$ is in simplified form if and only if all of the following conditions are satisfied.

a. The radicand is as small as possible; that is, the exponent of m is less than the index of n.

b. There are no fractions in the radicand.

c. There are no radicals in the denominator.

d. The index is as small as possible: $\frac{m}{n}$, the fractional power, is not reducible.

7 Like radicals:
Like radicals have the same index and the same radicand; only the coefficients of like terms can differ.

8 Conjugates:
The radical expressions $\sqrt{a} + \sqrt{b}$ and $\sqrt{a} - \sqrt{b}$ are called conjugates of each other.

9 Operations with radical expressions:
The addition, subtraction, and multiplication of radical expressions are similar to the corresponding operations with polynomials. Division by a radical expression is accomplished by rationalizing the denominator.

10 Rationalizing a radical expression with a single term in the denominator:

   a. Write the expression as a single radical, and reduce the fraction in the radicand to lowest terms.

   b. Multiply both the numerator and the denominator of this fraction by the radical that will make the radicand in the denominator a perfect nth power.

11 Rationalizing a radical expression with a binomial denominator involving square roots:

   a. Reduce the fraction to lowest terms.

   b. Multiply both the numerator and the denominator of the fraction by the conjugate of the denominator.

12 Solving radical equations:

   a. Isolate a radical term on one side of the equation.

   b. Raise both sides to the nth power.

   c. Solve the resulting equation. (If this equation contains a radical, repeat Steps a and b.)

   d. Check each possible solution in the original equation to determine whether it is a solution or an extraneous value.

13 The imaginary number i:

$$i = \sqrt{-1}$$

For any positive real number x, $\sqrt{-x} = i\sqrt{x}$. The first four powers of i are $i^1 = i$, $i^2 = -1$, $i^3 = -i$, and $i^4 = 1$.

14 Complex numbers:
If a and b are real numbers and $i = \sqrt{-1}$, then

$$a + bi$$

Real term ↑ ↑ Imaginary term

is a complex number.

15 Operations with complex numbers:
The addition, subtraction, and multiplication of complex numbers are similar to the corresponding operations with binomials. Division by a complex number is accomplished by multiplying both the numerator and the denominator by the conjugate of the denominator and then simplifying the result.

16 Subsets of the complex numbers:
The relationships of important subsets of the set of complex numbers are summarized in the tree diagram, where a and b are real numbers and $i = \sqrt{-1}$.

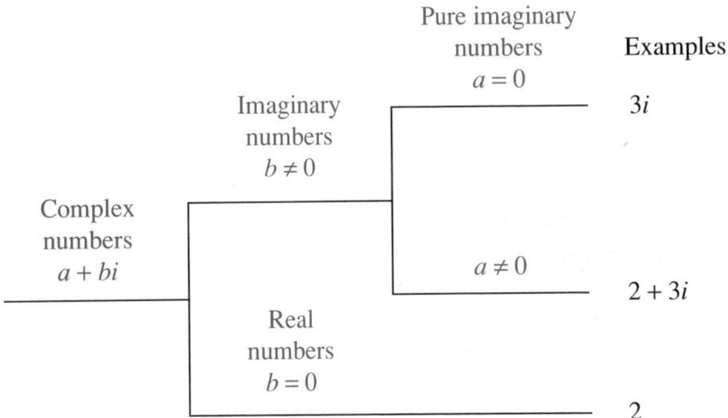

Figure for Key Concept 16

Review Exercises for Chapter 7

In Exercises 1–4 simplify each radical expression. The variables can represent any real number, so use absolute-value notation wherever necessary.

1 $\sqrt{64x^2}$ **2** $\sqrt[3]{64x^3}$

3 $\sqrt{(-7)^2}$ **4** $\sqrt[5]{-32x^5y^{10}}$

In Exercises 5–50 simplify each expression. Assume that the variables represent positive real numbers.

5 $49^{1/2}$ **6** $-49^{1/2}$ **7** $49^{-1/2}$

8 $\left(\frac{27}{125}\right)^{2/3}$ **9** $\left(\frac{-27}{125}\right)^{2/3}$ **10** $\left(\frac{27}{125}\right)^{-2/3}$

11 $(9)^{5/2}$ **12** $(625 - 576)^{1/2}$ **13** $625^{1/2} - 576^{1/2}$

14 $\sqrt{81x^4}$ **15** $\sqrt[4]{81x^4}$ **16** $\sqrt[3]{0}$

17 $\sqrt[3]{1}$ **18** $\sqrt[3]{-1}$ **19** $\sqrt[3]{\left(\frac{8}{27}\right)^2}$

20 $\sqrt[5]{-100{,}000x^5}$ **21** $\sqrt{\frac{18xy^2}{50xz^2}}$

22 $\frac{\sqrt{32x^3}}{\sqrt{98x}}$ **23** $\sqrt{6x}\sqrt{15x}$

24 $\sqrt[3]{18x}\sqrt[3]{12x^2}$ **25** $\sqrt{100x^2} - \sqrt{64x^2}$

26 $\sqrt{100x^2 - 64x^2}$ **27** $\sqrt{98x^3}$

28 $\sqrt[3]{40x^4}$ **29** $\sqrt{6x}\sqrt{15x^3}$

30 $\frac{\sqrt{75x^3}}{\sqrt{27x}}$ **31** $\frac{14}{\sqrt{7}}$

32 $\frac{14}{\sqrt{7} - \sqrt{5}}$ **33** $(5\sqrt{2} - 7\sqrt{3}) - (\sqrt{2} - 4\sqrt{3})$

34 $3\sqrt{8} - 7\sqrt{50}$ **35** $3\sqrt{3}(2\sqrt{12} - 9\sqrt{75})$

36 $(2\sqrt{2} - 5\sqrt{3})(2\sqrt{2} + 5\sqrt{3})$

37 $(2\sqrt{2} - 5\sqrt{3})^2$

38 $\frac{12}{2\sqrt{5} - 2\sqrt{3}}$ **39** $2\sqrt{45z} - 6\sqrt{20z}$

40 $(2\sqrt{14x} - 3\sqrt{6y})(3\sqrt{14x} + 2\sqrt{6y})$

41 $(2\sqrt{5x})(3\sqrt[3]{2y})$ **42** $\sqrt{9y^2 + 12y + 4}$

43 $(2x^{3/5}y^{1/3})(3x^{2/5}y^{-4/3})$

44 $(8x^{-6}y^9)^{2/3}$

45 $(-32x^{5/3}y^{3/2})^{2/5}$

46 $(5v^{1/2} - 2w^{1/2})(5v^{1/2} + 2w^{1/2})$

47 $\dfrac{(x^2y^2z^2)^{1/3}(xyz)^{2/3}}{(xyz)^{-2/3}}$

48 $(-8)^{-4/3}$

49 $2\sqrt{7}(3\sqrt{14} - \sqrt{35})$

50 $2v^{1/2}w^{1/3}(3v^{1/2}w^{2/3} - 5v^{-1/2}w^{4/3})$

In Exercises 51–54 solve each equation.

51 $\sqrt{4v + 24} = 2$

52 $\sqrt{z^2 + z - 2} - 5 = z$

53 $\sqrt{2t + 11} + 2 = t$

54 $\sqrt{5w - 6} + \sqrt{4w - 3} = 0$

55 Length of a Rafter What length rafter is needed to span a horizontal distance of 24 feet if the roof must rise 7 feet? (See the figure.)

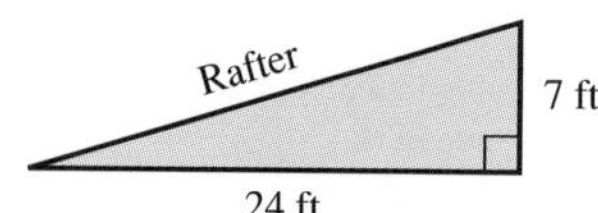

Figure for Exercise 55

56 Height of a Gate A 5-meter diagonal brace is used to reinforce a rectangular gate. The gate is 3 meters wide. How tall is the gate? (See the figure.)

Figure for Exercise 56

In Exercises 57–64 perform the indicated operations, writing the answer in standard $a + bi$ form.

57 $\sqrt{-25} - \sqrt{25}$

58 $\sqrt{-36} - \sqrt{-4}$

59 $2(4 - 3i) - 5(2 + 6i)$

60 $(5 - 7i)(6 + 3i)$

61 $(7 - 3i)^2$

62 $\dfrac{58}{2 + 5i}$

63 $\dfrac{5 + 3i}{5 - 3i}$

64 $i^6 + 2i^9$

CALCULATOR USAGE (65–67)

65 Approximate $\sqrt[5]{72}$ to the nearest thousandth.

66 Use the graph of $y = \sqrt{x + 1} - 2$, shown on the calculator window, to solve $\sqrt{x + 1} - 2 = 0$.

[−3, 8] for x, [−2, 2] for y

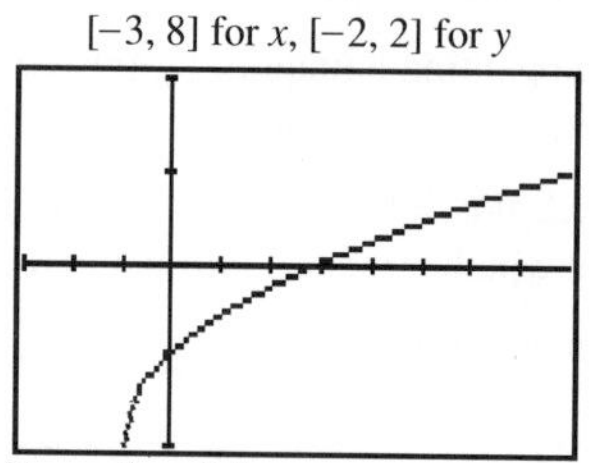

Figure for Exercise 66

67 Use the graphs of $y_1 = \sqrt{2v + 1}$ and $y_2 = \sqrt{v} + 1$ to solve $\sqrt{2v + 1} = \sqrt{v} + 1$.

68 Error Analysis Examine the student's work shown below, and correct all errors that you find.

$$\sqrt{3x - 1} = x - 1$$
$$3x - 1 = x^2 + 1$$
$$0 = x^2 - 3x + 2$$
$$0 = (x - 1)(x - 2)$$
$$x - 1 = 0 \quad \text{or} \quad x - 2 = 0$$
$$x = 1 \qquad\qquad x = 2$$

Mastery Test for Chapter 7

Exercise numbers correspond to Section Objective numbers.

[7-1] 1 Simplify each of these expressions as completely as possible. Assume that x and y are positive real numbers.

a. $36^{1/2}$ **b.** $125^{1/3}$

c. $(4x^2y^6)^{1/2}$ **d.** $(8x^9)^{2/3}$

[7-2] 2 Simplify each of these radical expressions. Use absolute-value notation wherever necessary.

a. $\sqrt{169 - 25}$ **b.** $\sqrt[3]{-125}$

c. $\sqrt{64x^2y^4}$ **d.** $\sqrt[3]{64x^{24}}$

[7-2] 3 Simplify each of these radical expressions. Assume that x is a positive real number.

a. $\sqrt{18x^2}$ **b.** $\sqrt[3]{40x^4}$

c. $\sqrt{\dfrac{5x^3}{20x}}$ **d.** $\dfrac{\sqrt[3]{x^5y^4}}{\sqrt[3]{x^2y}}$

[7-3] 4 Perform the operations indicated.

a. $8\sqrt{7} - 3\sqrt{7}$

b. $(5\sqrt{2} - 3\sqrt{5}) - (2\sqrt{2} - 7\sqrt{5})$

c. $3\sqrt{28} - 5\sqrt{63}$

d. $2\sqrt[3]{54x^4} - 5\sqrt[3]{16x^4}$

[7-3] **5** Reduce the order of each radical. Assume that x is a positive real number.

a. $\sqrt[4]{x^2}$ **b.** $\sqrt[6]{8}$

c. $\sqrt[15]{32x^5}$ **d.** $\sqrt[9]{8x^3y^6}$

[7-4] **6** Perform the operations indicated, and simplify the result. Assume that x is a positive real number.

a. $(3\sqrt{2})(5\sqrt{2})$ **b.** $(8\sqrt{6x})(5\sqrt{3x})$

c. $2\sqrt{5}(3\sqrt{5} - 2\sqrt{10})$

d. $(2\sqrt{3x} - \sqrt{5})(2\sqrt{3x} + \sqrt{5})$

[7-5] **7** Perform the indicated operations by rationalizing the denominators and then simplifying the results.

a. $\dfrac{\sqrt{18}}{\sqrt{2}}$ **b.** $\dfrac{18}{\sqrt{6}}$

c. $\dfrac{40}{\sqrt{7} - \sqrt{2}}$ **d.** $\dfrac{10}{\sqrt{9} + \sqrt{7}}$

[7-6] **8** Solve each of these equations.

a. $\sqrt{5v + 1} = 4$

b. $\sqrt{5w + 11} + 7 = 1$

c. $4 + \sqrt{4x - x^2} = x$

d. $\sqrt[3]{x + 1} = 2$

[7-6] **9** Simplify each of these complex numbers, expressing the answer in standard $a + bi$ form.

a. $\sqrt{-81}$ **b.** $\sqrt{16 - 25}$

c. $\sqrt{-16} - \sqrt{25}$ **d.** $i^2 + i^3$

[7-7] **10** Perform the operations indicated, expressing the answer in standard $a + bi$ form.

a. $2(4 - 5i) - 3(2 - 4i)$ **b.** $(4 - 5i)(2 - 4i)$

c. $\dfrac{4 - 5i}{2 - 4i}$ **d.** $(3 - i)^2$

ACTIVE APPLICATION

This application continues the theme of bridge building, as discussed in the application at the beginning of the chapter. These problems are typical of those an engineer might encounter. Suppose that you are designing a bridge that looks like the following illustration.

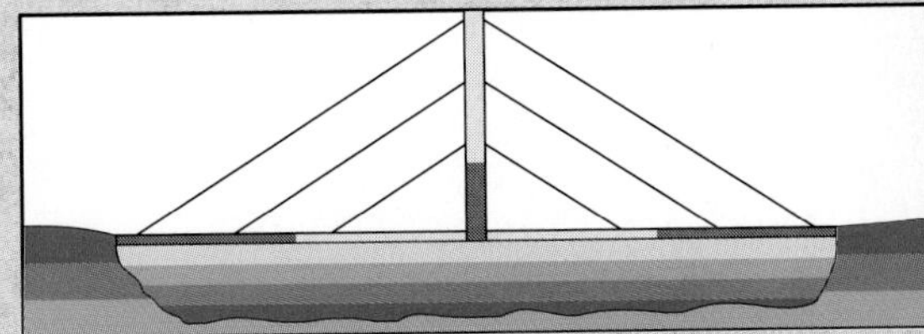

1 One of the first things you have to do is order the support cables from a supplier. The 2-inch-diameter cables attach to the vertical support at heights of 20, 30, and 40 feet above the roadbed, and they attach to the horizontal section at distances of 30, 45, and 60 feet from the vertical support on each side. How many total feet of cable will be needed?

2 The tensile strength of a cable varies directly as the area of its cross-section. You have just learned that the traffic across the bridge will include large trucks, which you had not expected. The bridge has to be redesigned from the ground up. After some calculations you determine that the tension the cable will be under is $1\frac{1}{3}$ times as large as the tension that the cable you ordered can safely handle. What is the minimum diameter for the cable you need?

3 Assume that the overall cost of building a bridge of this type varies directly as the maximum load capacity to the $\frac{3}{5}$th power. By what factor is the cost of the bridge increased if its maximum load capacity ends up being $1\frac{1}{3}$ times as great as originally planned? Write your answer as a percent.

Chapter 8

QUADRATIC EQUATIONS AND INEQUALITIES

An Application

Power Curves

Graphs, such as those in Chapter 2, provide a quick visual overview of the relationships among variables and have many applications. One such application involves the power output of engines. The amount of power an engine produces depends on the speed at which it is running. The graph relating power to speed, called the power curve, has a curved shape, as illustrated below.

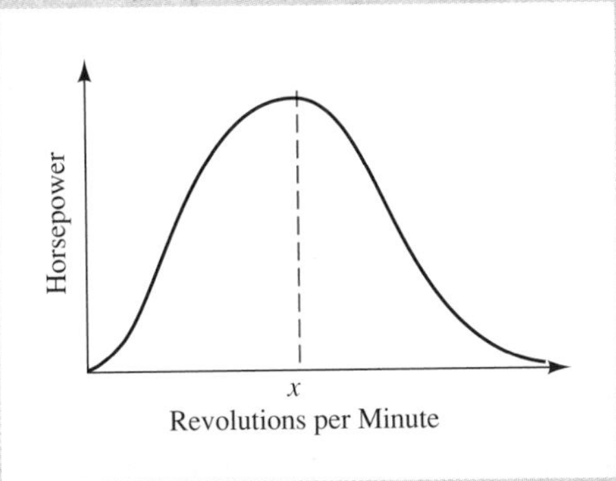

The point of maximum power, x, is important because engines usually perform better running above this point than they do running below it. Suppose that the engine in question belongs to an ocean-going powerboat. When the engine is temporarily put under increased load—because the boat must climb up the side of a wave or because a piece of kelp is caught in the propeller—the engine briefly slows down. An engine running above the point of maximum power will then produce *more* power and smoothly overcome the extra load. An engine running below the point of maximum power, on the other hand, will produce *less* power and tend to become sluggish. This is a prime consideration for a boat designer.

The idea of a power curve has become a popular metaphor in the business community. What do you think "staying ahead of the power curve" means in a production context?

CHAPTER 8 OBJECTIVES

1 **Solve a quadratic equation by extraction of roots (Section 8-1).**

2 **Solve a quadratic equation by completing the square (Section 8-1).**

3 **Solve a quadratic equation using the quadratic formula (Section 8-2).**

4 **Use the discriminant to determine the nature of the solutions of a quadratic equation (Section 8-2).**

5 **Solve equations of quadratic form (Section 8-3).**

6 **Solve equations with radicals and equations with rational expressions that can be simplified to quadratic equations (Section 8-3).**

7 **Use quadratic equations to solve applied problems (Section 8-4).**

8 **Solve quadratic and rational inequalities (Section 8-5).**

A quadratic equation in x is a second-degree equation that can be written in the standard form $ax^2 + bx + c = 0$, where a, b, and c represent real constants and $a \neq 0$. The importance of quadratic equations was recognized by ancient civilizations; we know that the Babylonians were able to solve these equations as early as 2000 B.C. In Section 5-5 we solved selected quadratic equations by factoring over the integers. This chapter will examine methods for solving quadratic equations that cannot be solved by factoring.

8-1 *Extraction of Roots and Completing the Square*

SECTION OBJECTIVES

1 **Solve a quadratic equation by extraction of roots.**

2 **Solve a quadratic equation by completing the square.**

In Section 5-5 we solved the equation $6z^2 - 11z - 10 = 0$ by rewriting it as $(2z - 5)(3z + 2) = 0$ and then setting each factor equal to zero. This method of factoring the left member of $ax^2 + bx + c = 0$ over the integers will not work for $x^2 - 5 = 0$ because $x^2 - 5$ is not factorable over the integers. The easiest way to solve this equation is by the method of **extraction of roots.**

The method of extraction of roots can be applied to any equation that can be written in the form $x^2 = k$; that is, it can be applied to any equation for which $b = 0$ when the equation is in standard form $ax^2 + bx + c = 0$. The procedure is to first solve the equation for x^2, obtaining $x^2 = k$, and then take the square roots of k. Depending on the value of k, the equation may have either two distinct roots or one double root. For example, $x^2 = 9$ has two distinct roots, -3 and 3, whereas $x^2 = 0$ has 0 as a double root.

To denote both possible square roots of k, we write $\pm\sqrt{k}$. The notation $\pm$ is read as "plus or minus." A radical sign alone denotes the principal square root, which is always positive; a radical sign preceded by a negative sign always denotes the negative square root. For example, $x = \pm\sqrt{9}$ means that $x = \sqrt{9}$ (in which case $x = 3$) or that $x = -\sqrt{9}$ (in which case $x = -3$).

The notation $\pm\sqrt{9}$ is read as "plus or minus the square root of nine."

EXTRACTION OF ROOTS

If k is a real number, then the equation $x^2 = k$ has solution(s) denoted by

$$x = \sqrt{k} \quad \text{and} \quad x = -\sqrt{k}$$

If $k > 0$ in $x^2 = k$, the equation will have two distinct real solutions. If $k = 0$, $x^2 = k$ will have a double root of 0. If $k < 0$, the two solutions of $x^2 = k$ will be imaginary. Each of these cases is examined in Example 1. The solution in part (b) of that example contains irrational numbers, and the solution in part (d) contains imaginary numbers.

EXAMPLE 1 Solving Quadratic Equations by Extraction of Roots

Solve each of these equations.

SOLUTIONS

(a) $5x^2 - 45 = 0$

$5x^2 - 45 = 0$

$x^2 = \frac{45}{5}$ — Solve for x^2 to write the equation in the form $x^2 = k$.

$x^2 = 9$

$x = \pm\sqrt{9}$ — To solve for x, take both square roots of the right member.

$x = -3$ or $x = 3$ — Remember to specify the two distinct roots; a common error is to write only one of these roots. The solution set is $\{-3, 3\}$.

(b) $2w^2 - 38 = 0$

$2w^2 - 38 = 0$

$w^2 = \frac{38}{2}$ — Solve for w^2.

$w^2 = 19$

$w = \pm\sqrt{19}$ — Extract the roots.

$w = -\sqrt{19}$ or $w = \sqrt{19}$ — *Caution:* Be sure to write the two distinct roots. The solution set is $\{-\sqrt{19}, \sqrt{19}\}$.

(c) $7z^2 + 13 = 13$

$7z^2 + 13 = 13$

$7z^2 = 0$ — Solve for z^2.

$z^2 = 0$

$z = \pm\sqrt{0}$ — Extract the roots.

$z = 0$ — Since $+\sqrt{0} = 0$ and $-\sqrt{0} = 0$, this equation has a double root. The solution set is $\{0\}$.

(d) $y^2 + 4 = 0$

$y^2 + 4 = 0$

$y^2 = -4$ — Solve for y^2.

$y = \pm\sqrt{-4}$ — Extract the roots.

$y = \pm 2i$ — Simplify this imaginary number.

$y = -2i$ or $y = 2i$ — Note that these two distinct roots are complex conjugates. The solution set is $\{-2i, 2i\}$. ▼

The technique of extraction of roots can be extended to solve quadratic equations for which $b \neq 0$ in the equation $ax^2 + bx + c = 0$. We will illustrate this extension by examining a quadratic equation whose left member is a perfect square and whose right member is a constant.

EXAMPLE 2 Solving Quadratic Equations by Extraction of Roots

Solve each of these equations.

SOLUTIONS

(a) $(x-3)^2 = 36$

$(x-3)^2 = 36$

$x - 3 = \pm 6$ Extract the roots.

$x = 3 \pm 6$ Solve for x.

$x = 3 - 6$ or $x = 3 + 6$ In this case there are two distinct roots.

$x = -3$ $x = 9$

(b) $(2y-5)^2 = 0$

$(2y-5)^2 = 0$

$2y - 5 = \pm\sqrt{0}$ Extract the roots.

$2y = 5 \pm 0$ Add 5 to both members, and then divide both members by 2.

$y = \frac{5 \pm 0}{2}$ Since $\frac{5+0}{2} = \frac{5}{2}$ and $\frac{5-0}{2} = \frac{5}{2}$, this equation has a double root.

$y = \frac{5}{2}$

(c) $(3v-2)^2 = 5$

$(3v-2)^2 = 5$

$3v - 2 = \pm\sqrt{5}$ Extract the roots.

$3v = 2 \pm \sqrt{5}$ Add 2 to both members.

$v = \frac{2 \pm \sqrt{5}}{3}$ Then divide by 3.

$v = \frac{2-\sqrt{5}}{3}$ or $v = \frac{2+\sqrt{5}}{3}$ These distinct roots are both real numbers. ▼

In Example 3 we begin by writing the left member of the equation as a perfect square so that the equation can be solved by extraction of roots.

EXAMPLE 3 Extending the Method of Extraction of Roots

Solve $y^2 - 10y + 25 = 9$.

SOLUTION $y^2 - 10y + 25 = 9$

$(y-5)^2 = 9$ The left member is a perfect square.

$y - 5 = \pm 3$ Extract the roots.

$y = 5 \pm 3$ Solve for y.

$y = 2$ or $y = 8$ The solution set is {2, 8}. ▼

SELF-CHECK ✓

Solve each of these quadratic equations by the method of extraction of roots.

1 $3s^2 - 147 = 0$
2 $(y-4)^2 = 25$
3 $(z-8)^2 = 0$

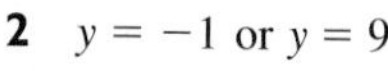

SELF-CHECK ANSWERS ✓

1 $s = -7$ or $s = 7$ 2 $y = -1$ or $y = 9$ 3 $z = 8$ (a double root)

The left member of a quadratic equation can always be written as a perfect square of the form

$$(x + k)^2 = x^2 + 2kx + k^2 \qquad \text{or} \qquad (x - k)^2 = x^2 - 2kx + k^2$$

In either case, the constant term needed, k^2, is the square of one-half the coefficient of x:

$$k^2 = \left(\frac{2k}{2}\right)^2 \qquad \text{and} \qquad k^2 = \left(\frac{-2k}{2}\right)^2$$

The process of writing the left member of an equation as a perfect square is called **completing the square.**

EXAMPLE 4 Constructing Perfect Square Trinomials

Find the constant term needed to make each of these expressions a perfect square trinomial.

SOLUTIONS

(a) $x^2 + 6x + ?$ $\qquad x^2 + 6x + \left(\frac{6}{2}\right)^2 = x^2 + 6x + 9$

$= (x + 3)^2$

In each case the constant term needed is the square of one-half the coefficient of x.

(b) $x^2 - 8x + ?$ $\qquad x^2 - 8x + \left(-\frac{8}{2}\right)^2 = x^2 - 8x + 16$

$= (x - 4)^2$

(c) $x^2 + 2ax + ?$ $\qquad x^2 + 2ax + \left(\frac{2a}{2}\right)^2 = x^2 + 2ax + a^2$

$= (x + a)^2$ ▼

The process of adding 9 to $x^2 + 6x$ to obtain $x^2 + 6x + 9$ is called completing the square, because it produces the perfect square $(x + 3)^2$. Completing the square is illustrated geometrically as follows.

A Geometric Viewpoint: Completing the Square

Adding the 9-square-unit square (each side of 3 units) to the upper right corner of the figure on the left produces the complete square shown to the right.

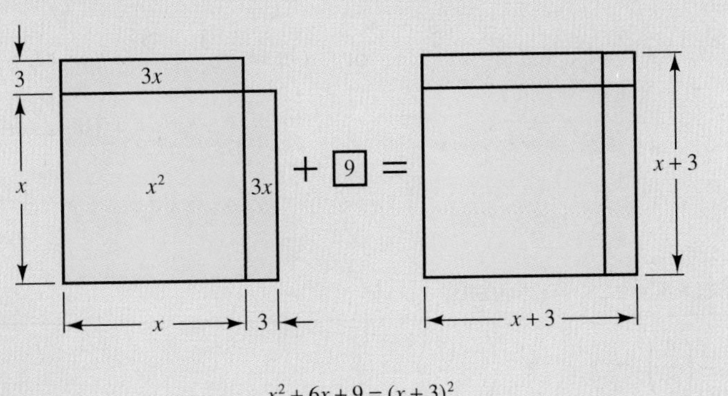

The process of completing the square can be used to solve any quadratic equation. The key steps in this process are shown in the following box.

A Mathematical Note

Many methods that we now use quickly and efficiently in algebraic form have their roots in ancient geometric methods. In particular, the method of completing the square was used by both Greek and Arab mathematicians in the geometric form shown in the preceding Geometric Viewpoint. Al-Khowârizmî (c. 825) was a noted Arab mathematician, astronomer, and author who illustrated this method in his writings.

SOLVING QUADRATIC EQUATIONS BY COMPLETING THE SQUARE

Step 1 Write the equation with the constant term on the right side.

Step 2 Divide both sides of the equation by the coefficient of x^2.

Step 3 Take one-half of the coefficient of x, square this number, and add the result to both sides of the equation.

Step 4 Write the left side of the equation as a perfect square.

Step 5 Solve this equation by extraction of roots.

EXAMPLE 5 Completing the Square: Two Rational Solutions

Solve $x^2 + 3x - 4 = 0$.

SOLUTION

$$x^2 + 3x - 4 = 0$$

$$x^2 + 3x = 4$$

Shift the constant term to the right side. (Note that the coefficient of x^2 is already 1.)

$$x^2 + 3x + \left(\frac{3}{2}\right)^2 = 4 + \left(\frac{3}{2}\right)^2$$

Take one-half of the coefficient of x; $\frac{1}{2}(3) = \frac{3}{2}$. Square this number: $\left(\frac{3}{2}\right)^2$. Then add this result to both sides of the equation.

$$\left(x + \frac{3}{2}\right)^2 = \frac{16}{4} + \frac{9}{4}$$

Write the left side as a perfect square.

$$\left(x + \frac{3}{2}\right)^2 = \frac{25}{4}$$

$$x + \frac{3}{2} = \pm\frac{5}{2}$$

Extract the roots.

$$x = -\frac{3}{2} \pm \frac{5}{2}$$

$$x = -\frac{3}{2} - \frac{5}{2} \quad \text{or} \quad x = -\frac{3}{2} + \frac{5}{2}$$

Simplify both solutions.

$$x = -4 \qquad\qquad x = 1$$

The solution set is $\{-4, 1\}$.

SELF-CHECK 

Determine the value that must be added to each expression to produce a perfect square trinomial.

1 $x^2 + 2x + ?$

2 $z^2 - 12z + ?$

SELF-CHECK ANSWERS

1 $\left(\frac{2}{2}\right)^2 = 1$ 2 $\left(\frac{-12}{2}\right)^2 = 36$

EXAMPLE 6 Completing the Square: Two Irrational Solutions

Solve $4x^2 - 12x + 7 = 0$ by completing the square.

SOLUTION

$$4x^2 - 12x + 7 = 0$$

$$4x^2 - 12x = -7$$ Shift the constant to the right side.

$$x^2 - 3x = -\frac{7}{4}$$ Divide both sides of the equation by the coefficient of x^2, which is 4.

$$x^2 - 3x + \left(-\frac{3}{2}\right)^2 = -\frac{7}{4} + \left(-\frac{3}{2}\right)^2$$ Take one-half of the coefficient of x: $\frac{1}{2}(-3) = -\frac{3}{2}$. Square this number: $\left(-\frac{3}{2}\right)^2$. Then add this result to both sides of the equation.

$$\left(x - \frac{3}{2}\right)^2 = \frac{1}{2}$$ Write the left side as a perfect square.

$$x - \frac{3}{2} = \pm\sqrt{\frac{1}{2}}$$ Extract the roots.

$$x = \frac{3}{2} \pm \frac{\sqrt{2}}{2}$$ Simplify, noting that $\frac{1}{\sqrt{2}} = \frac{1}{\sqrt{2}} \cdot \frac{\sqrt{2}}{\sqrt{2}} = \frac{\sqrt{2}}{2}$.

$$x = \frac{3 - \sqrt{2}}{2} \quad \text{or} \quad x = \frac{3 + \sqrt{2}}{2}$$ The solution set is $\left\{\frac{3 - \sqrt{2}}{2}, \frac{3 + \sqrt{2}}{2}\right\}$.

EXAMPLE 7 Completing the Square: Two Imaginary Solutions

Solve $x^2 - 4x + 5 = 0$.

SOLUTION

$$x^2 - 4x + 5 = 0$$

$$x^2 - 4x = -5$$ Shift the constant to the right side.

$$x^2 - 4x + (-2)^2 = -5 + (-2)^2$$ Take one-half of the coefficient of x: $\frac{1}{2}(-4) = -2$. Square this number: $(-2)^2$. Then add this result to both sides of the equation.

$$(x - 2)^2 = -1$$ Write the left side as a perfect square.

$$x - 2 = \pm\sqrt{-1}$$ Extract the roots.

$$x = 2 \pm i$$ Add 2 to both members, and replace $\sqrt{-1}$ by i.

$$x = 2 + i \quad \text{or} \quad x = 2 - i$$

We have now solved quadratic equations by factoring, by extraction of roots, and by completing the square. In cases where they are applicable, factoring and extraction of roots are generally easier to use; however, the process of complet-

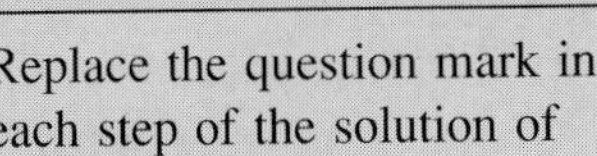

SELF-CHECK

Replace the question mark in each step of the solution of $5x^2 - 3x - 2 = 0$.

1 $5x^2 - 3x = ?$

2 $x^2 + ?x = \frac{2}{5}$

3 $x^2 - \frac{3}{5}x + \frac{9}{100} = \frac{2}{5} + ?$

4 $\left(x - \frac{3}{10}\right)^2 = ?$

5 **a.** $x - \frac{3}{10} = \pm ?$

$x = \frac{3}{10} \pm \frac{7}{10}$

$x = \frac{3}{10} - \frac{7}{10}$ or

$x = \frac{3}{10} + \frac{7}{10}$

b. $x = -\frac{2}{5}$ or $x = ?$

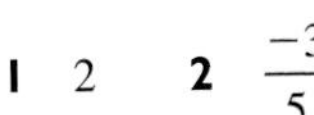

SELF-CHECK ANSWERS

1 2 2 $\frac{-3}{5}$ 3 $\frac{9}{100}$ 4 $\frac{49}{100}$ 5 **a.** $\frac{7}{10}$ **b.** 1

ing the square can be used to solve equations that cannot be solved by factoring over the integers. The quadratic formula, another method for solving quadratic equations, is covered in Section 8-2.

A solution produced by completing the square can be any complex number, real or imaginary. In Section 5-5 we reversed the factoring process to construct a quadratic equation whose rational solutions were given. We now use the same process to construct a quadratic equation given any complex solutions.

EXAMPLE 8 **Constructing Quadratic Equations from Given Solutions**

	SOLUTIONS	
(a) 3 and $-\frac{2}{5}$	$x = 3$ or $x = -\frac{2}{5}$	The solutions are 3 and $-\frac{2}{5}$.
	$x - 3 = 0$ or $5x = -2$	Rewrite these equations so that the right side is zero.
	$x - 3 = 0$ or $5x + 2 = 0$	
	$(x - 3)(5x + 2) = 0$	If either factor is zero, then their product is zero.
	$5x^2 - 13x - 6 = 0$	This equation has the given solutions.
(b) $2 + \sqrt{3}$ and $2 - \sqrt{3}$	$x = 2 + \sqrt{3}$ or $x = 2 - \sqrt{3}$	The solutions are $2 + \sqrt{3}$ and $2 - \sqrt{3}$.
	$x - (2 + \sqrt{3}) = 0$ or $x - (2 - \sqrt{3}) = 0$	Rewrite these equations so that the right side is zero.
	$(x - 2 - \sqrt{3})(x - 2 + \sqrt{3}) = 0$	If either of these factors is zero, their product is zero.
	$[(x - 2) - \sqrt{3}][(x - 2) + \sqrt{3}] = 0$	Noting the special form, multiply by inspection.
	$(x - 2)^2 - (\sqrt{3})^2 = 0$	
	$(x^2 - 4x + 4) - 3 = 0$	
	$x^2 - 4x + 1 = 0$	This equation has the given solutions.
(c) $3 + i$ and $3 - i$	$x = 3 + i$ or $x = 3 - i$	The solutions are $3 + i$ and $3 - i$.
	$x - (3 + i) = 0$ or $x - (3 - i) = 0$	Rewrite these equations so that the right side is zero.
	$(x - 3 - i)(x - 3 + i) = 0$	If either factor is zero, their product is zero.
	$[(x - 3) - i][(x - 3) + i] = 0$	
	$(x - 3)^2 - (i)^2 = 0$	Noting the special form, multiply by inspection.
	$(x^2 - 6x + 9) + 1 = 0$	$-i^2 = +1$
	$x^2 - 6x + 10 = 0$	This equation has the given solutions.

SELF-CHECK

Write a quadratic equation in x that has the given roots.

1 $5 - \sqrt{2}$ and $5 + \sqrt{2}$

2 $2 + 3i$ and $2 - 3i$

SELF-CHECK ANSWERS

1 $x^2 - 10x + 23 = 0$ **2** $x^2 - 4x + 13 = 0$

Exercises 8-1

A

In Exercises 1–20 solve each quadratic equation by extraction of roots.

1 $v^2 = 81$

2 $v^2 = 169$

3 $5w^2 - 20 = 0$

4 $7w^2 - 175 = 0$

5 $36x^2 = 49$

6 $144x^2 = 25$

7 $(y - 7)^2 - 9 = 0$

8 $(y + 5)^2 - 25 = 0$

9 $(2w + 1)^2 = 36$

10 $(3w - 4)^2 = 100$

11 $z^2 = 18$

12 $z^2 = 50$

13 $(5t - 3)^2 = 2$

14 $(3t + 5)^2 = 7$

15 $x^2 + 16 = 0$

16 $x^2 + 36 = 0$

17 $-7v^2 = 84$

18 $-5v^2 = 90$

19 $(2s - 1)^2 = -9$

20 $(5w + 2)^2 = -36$

In Exercises 21–24 determine the value of c so that each expression will be a perfect square trinomial.

21 $x^2 + 10x + c$

22 $x^2 - 14x + c$

23 $y^2 - 18y + c$

24 $y^2 + 20y + c$

In Exercises 25–28 determine the value of c so that the trinomial on the left side of the equation will be the square of a binomial.

25 $x^2 + 2x + c = 8 + c$

26 $x^2 - 2x + c = 8 + c$

27 $x^2 - \frac{2}{5}x + c = \frac{3}{5} + c$

28 $x^2 + \frac{2}{3}x + c = \frac{1}{3} + c$

In Exercises 29–42 solve each quadratic equation by completing the square.

29 $z^2 - 4z = 0$

30 $z^2 - 8z = 0$

31 $x^2 + 4x - 5 = 0$

32 $x^2 + 2x = 4$

33 $z^2 + 2z + 2 = 0$

34 $z^2 + 6z + 10 = 0$

35 $x^2 - \frac{3}{2}x = \frac{7}{16}$

36 $x^2 - \frac{2}{3}x = \frac{8}{9}$

37 $2v^2 - 10v + 12 = 0$

38 $3w^2 = -6w - 2$

39 $-2y^2 - 2 = 8y$

40 $-5y^2 - 20 = 30y$

41 $4t(t - 2) - 1 = 0$

42 $3t^2 = 6t(5t - 9)$

B

In Exercises 43–48 construct a quadratic equation in x that has the given solutions.

43 5 and $-\frac{1}{2}$

44 -6 and $\frac{2}{3}$

45 $-\sqrt{7}$ and $\sqrt{7}$

46 $-\sqrt{11}$ and $\sqrt{11}$

47 $-4i$ and $4i$

48 $-7i$ and $7i$

ESTIMATION SKILLS (49–52)

In Exercises 49–52 you are given the solution to a quadratic equation in radical form. Mentally estimate the value of each solution, and select the pair that is closest to your mental estimate.

49 $3 \pm \sqrt{50}$

a. $-8, 8$ b. $-7, 7$

c. $-7, 4$ d. $-4, 10$

50 $7 \pm \sqrt{63}$

a. $-15, 15$ b. $-14, 14$

c. $-1, 15$ d. $1, -9$

51 $-11 \pm \sqrt{10}$

a. $-21, 21$ b. $-14, -8$

c. $-14, 14$ d. $-8, 8$

52 $-13 \pm \sqrt{82}$

a. $-22, -4$ b. $-4, 4$

c. $-4, 22$ d. $-2, 24$

In Exercises 53–62 solve each quadratic equation either by factoring, by extraction of roots, or by completing the square.

53 $6v^2 + v - 15 = 0$

54 $(2v - 3)^2 = 25$

55 $(3x + 4)^2 = 49$

56 $-10x^2 - 11x + 6 = 0$

57 $-2y^2 - 4y + 1 = 0$

58 $y^2 - 8y + 13 = 0$

59 $w^2 + 12 = 0$

60 $-2w^2 + 12w + 19 = 0$

61 $49z^2 + 25 = 70z$

62 $25z^2 + 16 = -40z$

63 One number is 6 more than another number. Find these numbers if their product is 8.

64 The sum of a number and its reciprocal is 4. Find this number.

65 The rectangle shown in the figure is 4 centimeters longer than it is wide. Find the dimensions if the area is 8 square centimeters.

Figure for Exercise 65

66 The perimeter of the square shown in the figure is numerically 2 more than the area. Find the length in meters of one side of the square.

Figure for Exercise 66

C

In Exercises 67–70 solve each equation.

67 $x(x^2 - 5x + 4) = 0$

68 $-y^3 = 25y$

69 $(2y - 3)(y^2 - 6y + 13) = 0$

70 $x^3 - 2x^2 + 26x = 0$

CALCULATOR USAGE (71–74)

71 Use the graph of $y = 4x^2 - 8x - 5$ shown in the figure to estimate the solutions of $4x^2 - 8x - 5 = 0$. Then check your estimates.

[−2, 4] for x, [−10, 10] for y

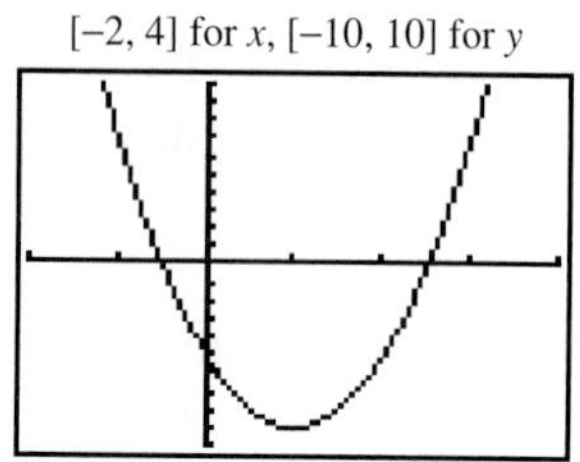

Figure for Exercise 71

In Exercises 72–74 use a graphics calculator to graph $y = P(x)$, and then use this graph to approximate to the nearest tenth the solutions of $P(x) = 0$.

72 $P(x) = 2x^2 + 4x + 1$

73 $P(x) = 5x^2 + 5x + 1$

74 $P(x) = 2x^2 - 5x - 1$

75 Use the equation $y = -x^2 - 5x + 1$ to complete the table.

x	−5	−4	−3	−2	−1	0	1
y	1	5					

Then use this table to complete the line graph started in the figure.

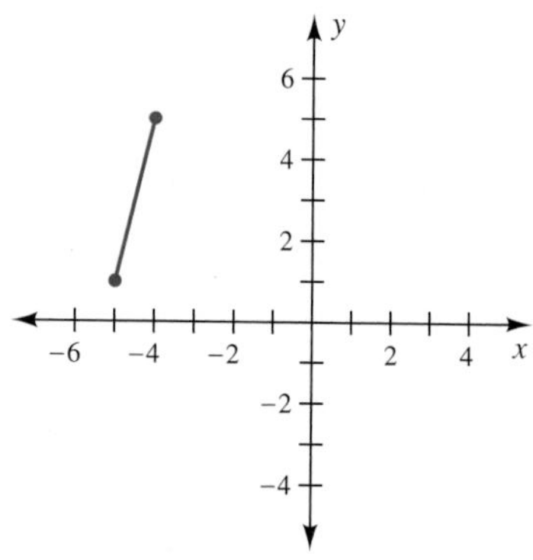

Figure for Exercise 75

In Exercises 76 and 77 construct a quadratic equation in x that has the given solutions.

76 $2 - \sqrt{5}$ and $2 + \sqrt{5}$

77 $3 - 4i$ and $3 + 4i$

DISCUSSION QUESTIONS (78 AND 79)

78 Compare the relative advantages and disadvantages of solving quadratic equations by factoring and by completing the square.

79 Write your own problem that the equation $x(x + 5) = 6$ algebraically models.

CHALLENGE QUESTION

A worker discovered that the sugar cubes at his office had dimensions that were one-half those of the sugar cubes at his home. He takes one cube of sugar in his coffee at home. How many cubes will he want in his coffee at work?

SECTION SKILL REINFORCERS

Simplify each of these radical expressions.

1 $\sqrt{12}$ **2** $\sqrt{50}$ **3** $\dfrac{4 + \sqrt{8}}{2}$ **4** $\dfrac{6 - \sqrt{18}}{3}$ **5** $\dfrac{-12 - \sqrt{300}}{2}$

8-2 The Quadratic Formula and the Discriminant

SECTION OBJECTIVES

3 Solve a quadratic equation using the quadratic formula.

4 Use the discriminant to determine the nature of the solutions of a quadratic equation.

The Babylonians were able to solve quadratic equations as early as 2000 B.C. They not only knew how to use completing the square, they also knew a general formula for solving a quadratic equation. This formula, which we know in its current form as the **quadratic formula,** is derived by completing the square. We could continue to solve each individual quadratic equation by completing the square. Rather than repeating these steps for each problem, however, we can solve the general quadratic equation $ax^2 + bx + c = 0$ by completing the square and then use this general solution as a formula that can be applied to any quadratic equation.

DERIVATION OF THE QUADRATIC FORMULA

$$ax^2 + bx + c = 0$$

Start with a quadratic equation in standard form with a, b, and c real numbers and $a > 0$.

$$ax^2 + bx = -c$$

Shift the constant to the right side.

$$x^2 + \frac{b}{a}x = -\frac{c}{a}$$

Divide both sides by the coefficient of x^2, a.

$$x^2 + \frac{b}{a}x + \left(\frac{b}{2a}\right)^2 = -\frac{c}{a} + \left(\frac{b}{2a}\right)^2$$

Add the square of one-half the coefficient of x.

$$\left(x + \frac{b}{2a}\right)^2 = -\frac{4ac}{4a^2} + \frac{b^2}{4a^2}$$

Write the left side as a perfect square.

$$\left(x + \frac{b}{2a}\right)^2 = \frac{b^2 - 4ac}{4a^2}$$

Simplify by combining the terms in the right member.

$$x + \frac{b}{2a} = \pm\sqrt{\frac{b^2 - 4ac}{4a^2}}$$

Extract the roots.

$$x = -\frac{b}{2a} \pm \frac{\sqrt{b^2 - 4ac}}{2a}$$

Simplify the radical.

$$x = \frac{-b \pm \sqrt{b^2 - 4ac}}{2a}$$

This is the quadratic formula.

$$x = \frac{-b + \sqrt{b^2 - 4ac}}{2a} \quad \text{or} \quad x = \frac{-b - \sqrt{b^2 - 4ac}}{2a}$$

You should memorize the quadratic formula, given in the following box. Note that the symbol $\pm$ is used to express the two solutions concisely.

QUADRATIC FORMULA

The solutions of the quadratic equation $ax^2 + bx + c = 0$ with real coefficients a, b, and c, when $a \neq 0$, are

$$x = \frac{-b \pm \sqrt{b^2 - 4ac}}{2a}$$

Before using this formula, be sure to write the quadratic equation you are trying to solve in standard form; otherwise, it is easy to make an error in the sign of a, b, or c.

EXAMPLE 1 The Quadratic Formula: Two Rational Solutions

Solve $x^2 - 5x - 6 = 0$, and then check the solutions.

SOLUTION

$$x = \frac{-(-5) \pm \sqrt{(-5)^2 - 4(1)(-6)}}{2(1)}$$

Substitute $a = 1$, $b = -5$, and $c = -6$ into $x = \dfrac{-b \pm \sqrt{b^2 - 4ac}}{2a}$.

$$x = \frac{5 \pm \sqrt{25 + 24}}{2} = \frac{5 \pm \sqrt{49}}{2} = \frac{5 \pm 7}{2}$$

Simplify, and find both solutions.

$$x = \frac{5 - 7}{2} \quad \text{or} \quad x = \frac{5 + 7}{2}$$

$$x = -1 \qquad\qquad x = 6$$

The solution set is $\{-1, 6\}$.

Check $x = -1$: $(-1)^2 - 5(-1) - 6 \stackrel{?}{=} 0$

$1 + 5 - 6 \stackrel{?}{=} 0$

$0 = 0$ checks.

$x = 6$: $(6)^2 - 5(6) - 6 \stackrel{?}{=} 0$

$36 - 30 - 6 \stackrel{?}{=} 0$

$0 = 0$ checks.

Note that we could also have solved the equation in Example 1 by factoring it as $(x + 1)(x - 6) = 0$. The primary advantage of the quadratic formula is that it can be used to solve problems that cannot be solved by factoring over the integers. This is illustrated by Example 2.

EXAMPLE 2 The Quadratic Formula: Two Irrational Solutions

Solve $4y^2 - 4y = 1$.

SOLUTION

$$4y^2 - 4y = 1$$

$$4y^2 - 4y - 1 = 0$$

First write the equation in standard form.

$$y = \frac{-(-4) \pm \sqrt{(-4)^2 - 4(4)(-1)}}{2(4)}$$

Substitute $a = 4$, $b = -4$, and $c = -1$ into the quadratic formula.

$$= \frac{4 \pm \sqrt{16 + 16}}{8}$$

$$= \frac{4 \pm \sqrt{32}}{8}$$

$$= \frac{4 \pm 4\sqrt{2}}{8}$$ Note that $\sqrt{32} = \sqrt{16} \cdot \sqrt{2} = 4\sqrt{2}$.

$$= \frac{4(1 \pm \sqrt{2})}{8}$$

$$= \frac{1 \pm \sqrt{2}}{2}$$ Simplify, and write both solutions separately.

$$y = \frac{1 - \sqrt{2}}{2} \quad \text{or} \quad y = \frac{1 + \sqrt{2}}{2}$$ The solution set is $\left\{\frac{1 - \sqrt{2}}{2}, \frac{1 + \sqrt{2}}{2}\right\}$.

$$y \approx -0.2071 \qquad y \approx 1.2071$$ Calculator approximation

▼

EXAMPLE 3 The Quadratic Formula: Two Imaginary Solutions

Solve $3w^2 = 4w - 2$.

SOLUTION

$$3w^2 = 4w - 2$$

$$3w^2 - 4w + 2 = 0$$ First write the equation in standard form.

$$w = \frac{-(-4) \pm \sqrt{(-4)^2 - 4(3)(2)}}{2(3)}$$ Substitute $a = 3$, $b = -4$, and $c = 2$ into the quadratic formula.

$$= \frac{4 \pm \sqrt{16 - 24}}{6}$$

$$= \frac{4 \pm \sqrt{-8}}{6}$$

$$= \frac{4 \pm 2i\sqrt{2}}{6}$$ Note that $\sqrt{-8} = \sqrt{-4}\sqrt{2} = 2i\sqrt{2}$.

$$= \frac{2 \pm i\sqrt{2}}{3}$$ Reduce this fraction by dividing both the numerator and the denominator by 2.

$$w = \frac{2}{3} - \frac{\sqrt{2}}{3}i \quad \text{or} \quad w = \frac{2}{3} + \frac{\sqrt{2}}{3}i$$ The solution set is $\left\{\frac{2}{3} - \frac{\sqrt{2}}{3}i, \frac{2}{3} + \frac{\sqrt{2}}{3}i\right\}$.

▼

Every quadratic equation has either two distinct roots or a double root. These roots may be either real numbers or imaginary numbers. The nature of the roots can be determined by examining the radicand, $b^2 - 4ac$, of the quadratic formula,

$$x = \frac{-b \pm \sqrt{b^2 - 4ac}}{2a}$$

Since $b^2 - 4ac$ can be used to discriminate between real solutions and imaginary solutions, it is called the **discriminant.**

SELF-CHECK 

1 Solve $z^2 = 2z + 2$ using the quadratic formula.
2 Solve $x^2 + 8 = 2x$ using the quadratic formula.

SELF-CHECK ANSWERS 

1 $z = 1 - \sqrt{3}$ or $z = 1 + \sqrt{3}$ 2 $x = 1 - i\sqrt{7}$ or $x = 1 + i\sqrt{7}$

Nature of the Solutions of a Quadratic Equation

Assume $ax^2 + bx + c = 0$ is a quadratic equation.

	Discriminant, $b^2 - 4ac$	**Nature of the Solutions**
For real numbers a, b, and c	Positive	Two distinct real solutions
	Zero	A double real solution
	Negative	Two imaginary solutions that are complex conjugates
For rational numbers a, b, and c	Positive and a perfect square	Two distinct rational solutions
	Positive and not a perfect square	Two distinct irrational solutions
	Zero	A rational solution of multiplicity two
	Negative	Two imaginary solutions that are complex conjugates

A Mathematical Note

James Joseph Sylvester (1814–1897) was born in England as James Joseph. He changed his last name to Sylvester when he moved to the United States. At Johns Hopkins University he led efforts to establish graduate work in mathematics in the United States. He also founded the *American Journal of Mathematics.* Among his lasting contributions to mathematics are the many new terms he introduced, including the term *discriminant.*

EXAMPLE 4 Determining the Nature of the Solutions of a Quadratic Equation

SOLUTIONS

(a) $x^2 - 6x + 8 = 0$

$$b^2 - 4ac = (-6)^2 - 4(1)(8) = 36 - 32 = 4$$

Substitute $a = 1$, $b = -6$, and $c = 8$ into the discriminant.

Since the discriminant, 4, is positive and a perfect square, the solutions are distinct rational numbers.

(b) $x^2 - 6x + 7 = 0$

$$b^2 - 4ac = (-6)^2 - 4(1)(7) = 36 - 28 = 8$$

Substitute $a = 1$, $b = -6$, and $c = 7$ into the discriminant.

This discriminant, 8, is positive and not a perfect square, so the solutions are distinct and irrational.

(c) $x^2 - 6x + 9 = 0$

$$b^2 - 4ac = (-6)^2 - 4(1)(9) = 36 - 36 = 0$$

Substitute $a = 1$, $b = -6$, and $c = 9$ into the discriminant.

This discriminant is zero, so the solution is a rational solution of multiplicity two.

(d) $x^2 - 6x + 10 = 0$

$$b^2 - 4ac = (-6)^2 - 4(1)(10) = 36 - 40 = -4$$

Substitute $a = 1$, $b = -6$, and $c = 10$ into the discriminant.

This discriminant, -4, is negative, so the solutions are imaginary and are complex conjugates.

A Geometric Viewpoint: The Nature of Solutions from a Graph

The real solutions of $ax^2 + bx + c = 0$ correspond exactly to the values of x at which the graph of $y = ax^2 + bx + c$ crosses the x-axis. The graph of $y = ax^2 + bx + c$ crosses the x-axis twice if $ax^2 + bx + c = 0$ has two distinct real solutions. The graph of $y = ax^2 + bx + c$ is tangent to the x-axis if $ax^2 + bx + c = 0$ has a real solution of multiplicity two. The graph of $y = ax^2 + bx + c$ will not touch the real x-axis if $ax^2 + bx + c = 0$ has imaginary solutions. Three graphs from the display of a TI-82 calculator are shown below. Graphs of $y = ax^2 + bx + c$ are examined in detail in Section 9-2.

$[-4, 5]$ for x, $[-7, 10]$ for y

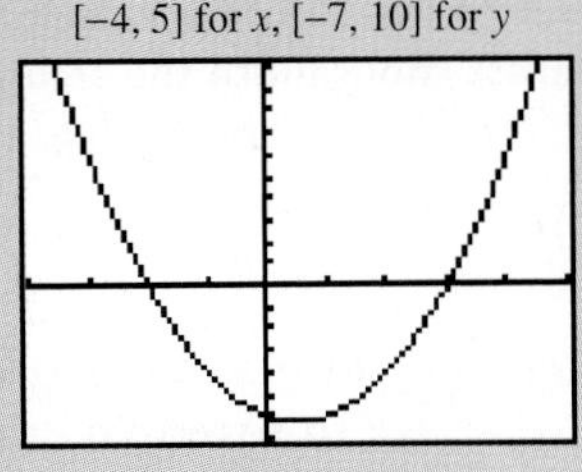

$[-2, 6]$ for x, $[-2, 10]$ for y

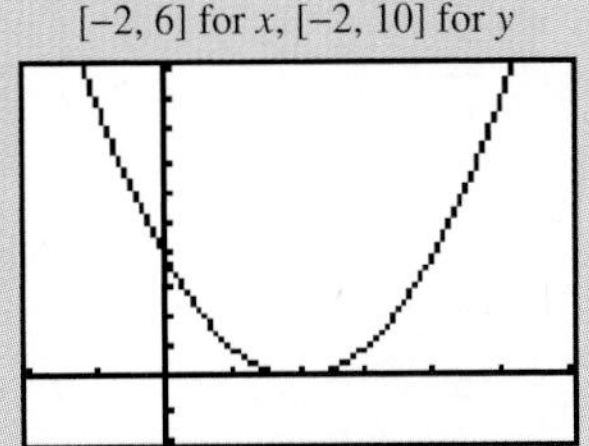

$[-2, 6]$ for x, $[0, 12]$ for y

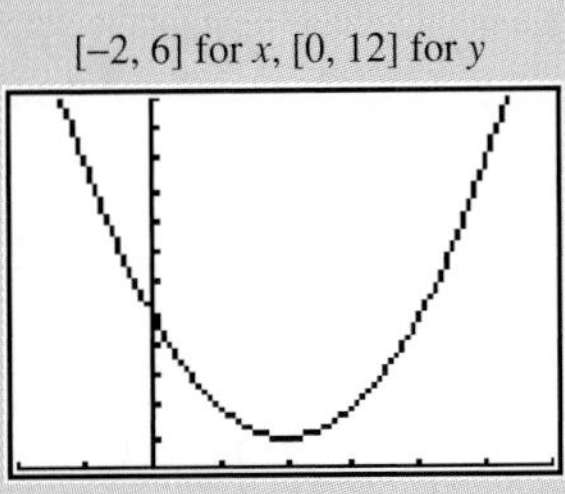

Many computer languages are not designed to evaluate the square root of a negative number. Programs written in these languages must first evaluate the discriminant before computing the square root of $b^2 - 4ac$; otherwise, an error message would be generated the first time the program encountered an equation with imaginary roots. (See the appendix for a program for solving quadratic equations on a TI-82 graphics calculator.)

SELF-CHECK 

Use the discriminant to determine the nature of the roots of each quadratic equation.

1. $4x^2 - 8x + 4 = 0$
2. $4x^2 - 8x + 5 = 0$
3. $4x^2 - 8x - 5 = 0$
4. $x^2 + x - 1 = 0$

EXAMPLE 5 The Discriminant and a Solution of Multiplicity Two

Determine k such that $4x^2 - 12x + k = 0$ will have a rational solution of multiplicity two.

SOLUTION

$b^2 - 4ac = 0$	Set the discriminant equal to zero to produce a rational solution of multiplicity two.
$(-12)^2 - 4(4)(k) = 0$	Substitute $a = 4$, $b = -12$, and $c = k$ into the discriminant.
$144 - 16k = 0$	Simplify, and solve for k.
$144 = 16k$	
$k = 9$	

SELF-CHECK ANSWERS 

1 Rational solution of multiplicity two 2 Imaginary solutions that are complex conjugates 3 Distinct rational solutions 4 Distinct irrational solutions

Exercises 8-2

A

In Exercises 1–20 solve each quadratic equation using the quadratic formula.

1 $8t^2 = 2t + 1$ **2** $2t^2 + 5t = -3$

3 $2z^2 - 3 = -5z$ **4** $9z^2 - 3z = 20$

5 $w^2 - 8 = 0$ **6** $5w^2 - 4 = 0$

7 $3v^2 + 7 = 0$ **8** $v^2 + 50 = 0$

9 $-5m^2 = -6m$ **10** $-6m^2 = 5m$

11 $-2w^2 + 6w = 5$ **12** $w^2 = 4w - 5$

13 $x(x - 6) = -13$ **14** $3v(v + 1) = -2 - v$

15 $4t^2 + 7 = 12t$ **16** $2t^2 = 3t + 1$

17 $9v^2 + 12v + 4 = 0$ **18** $25v^2 - 30v + 9 = 0$

19 $-5w^2 + 2w + 1 = 0$ **20** $-3w^2 - 7w - 3 = 0$

In Exercises 21–32 use the discriminant to determine the nature of the solutions of each equation. Identify the solutions as distinct rational solutions, distinct irrational solutions, a rational solution of multiplicity two, or imaginary solutions.

21 $x^2 + 8 = 0$ **22** $x^2 + 11 = 0$

23 $y^2 - 8 = 0$ **24** $y^2 - 11 = 0$

25 $z^2 = 10z - 25$ **26** $z^2 = 22z - 121$

27 $-3w^2 + 4w = 0$ **28** $-5w^2 + 2w = 0$

29 $4t^2 + 12t + 15 = 0$ **30** $3t^2 - 7t - 4 = 0$

31 $4v^2 - 3v = 0$ **32** $2v^2 - 5v = 0$

In Exercises 33–50 solve each equation by the most appropriate method. Use either factoring, extraction of roots, completing the square, or the quadratic formula.

33 $6v^2 - v = 35$ **34** $-15v^2 + 11v + 14 = 0$

35 $-y^2 = -225$ **36** $y^2 + 2y + 2 = 0$

37 $2w^2 - 3w + 2 = 0$ **38** $w^2 + 169 = 0$

39 $49x^2 - 70x + 25 = 0$ **40** $9x^2 + 66x + 121 = 0$

41 $3z^2 + 15 = 0$ **42** $7z^2 - 77 = 0$

43 $-17t^2 = -34t$ **44** $-23t^2 = 115t$

45 $-10m^2 + 11m = -6$ **46** $-2x^2 - 5x = 1$

47 $4x^2 - 8x + 5 = 0$ **48** $3x^2 + x + 3 = 0$

49 $(v - 1)^2 + (v + 3)^2 = 0$

50 $(2v + 1)^2 - (v - 4)^2 = 2v - 14$

In Exercises 51–56 determine the value of k for which the equation has a rational solution of multiplicity two.

51 $x^2 - 4x + k = 0$ **52** $x^2 - 10x + k = 0$

53 $kx^2 + 6x - 2 = 0$ **54** $kx^2 + 12x + 9 = 0$

55 $9x^2 + kx + 25 = 0$ **56** $25x^2 + kx + 4 = 0$

C

CALCULATOR USAGE (57–66)

In Exercises 57–60 use a calculator to approximate to the nearest thousandth the solutions of each equation.

57 $73y^2 - 85y + 13 = 0$ **58** $523w^2 - 372w - 208 = 0$

59 $\sqrt{3}x^2 + \sqrt{31}x = \sqrt{29}$ **60** $\sqrt{5}x^2 + \sqrt{23}x = \sqrt{17}$

In Exercises 61–63 use the given graphs of $y = ax^2 + bx + c$ to determine the nature of the solutions of $ax^2 + bx + c = 0$. Identify the solutions as two distinct real solutions, a rational solution of multiplicity two, or imaginary solutions.

61

[−5, 5] for x, [−2, 10] for y

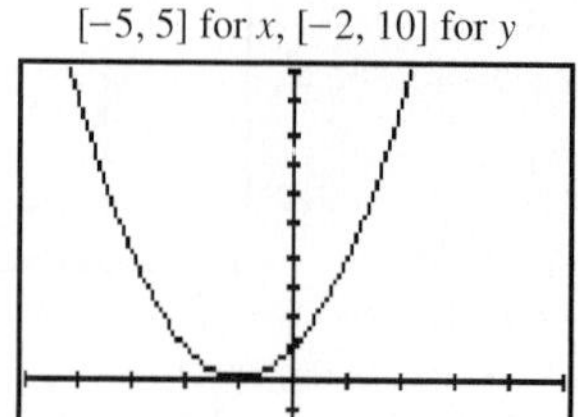

62

[−5, 5] for x, [−4, 9] for y

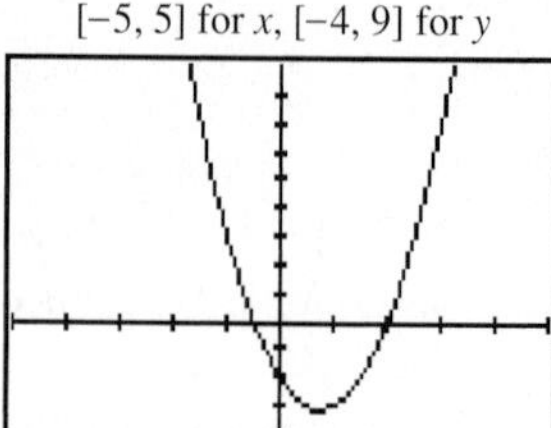

63

[−5, 5] for x, [−2, 10] for y

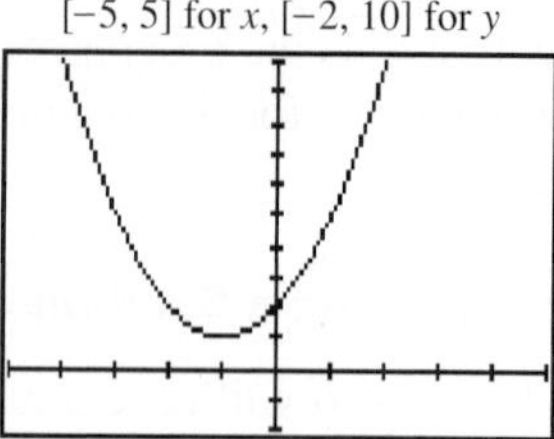

In Exercises 64–66 use a graphics calculator to graph $y = ax^2 + bx + c$, and then use this graph to determine the nature of the solutions of $ax^2 + bx + c = 0$.

64 $y = x^2 + 6$ **65** $y = x^2 + 6x + 9$ **66** $y = x^2 - 6$

Numeric Word Problems (67–69)

67 One number is 7 more than another number. Find these numbers if their product is −5.

68 The sum of a number and its reciprocal is 5. Find this number.

69 The difference of a number minus its reciprocal is 4. Find this number.

CHALLENGE QUESTION

One solution of $x^3 + ax + b = 0$ is given by Tartaglia's formula,

$$x = \sqrt[3]{\left(-\frac{b}{2}\right) + \sqrt{\left(\frac{b}{2}\right)^2 + \left(\frac{a}{3}\right)^3}} + \sqrt[3]{\left(-\frac{b}{2}\right) - \sqrt{\left(\frac{b}{2}\right)^2 + \left(\frac{a}{3}\right)^3}}$$

Use this formula to find one of the solutions of $x^3 - 2x + 4 = 0$.

SECTION SKILL REINFORCERS

1 A car travels 150 miles in 3 hours. What is its average rate? **2** A car travels a distance D in T hours. What is its average rate? **3** A repair person can do a job in 2 hours. What is the rate of work? **4** A machine can make 20 units of a product in 8 hours. What is the rate of work? **5** A person can do a job in T hours. What is the rate of work?

8-4 *Applications of Quadratic Equations*

SECTION OBJECTIVE

7 Use quadratic equations to solve applied problems.

In this section we will continue to use the strategy for word problems first given in Chapter 1. It is still very important to read the problem, determine the quantity you are to solve for, and then describe this quantity precisely with an appropriate variable. Some of the applications will use the Pythagorean Theorem and will produce quadratic equations with irrational solutions. The primary purpose of this section is not to introduce new material but to help you continue to develop skills for solving word problems.

EXAMPLE 1 A Numeric Word Problem

One number is 4 more than another number. Find these numbers if the ratio of their squares is $\frac{3}{5}$.

SOLUTION Let

n = Smaller number

$n + 4$ = Larger number

| Ratio of their squares | = | $\frac{3}{5}$ | **Word equation** |

$$\frac{n^2}{(n+4)^2} = \frac{3}{5}$$ Substitute the values identified above.

$$5(n+4)^2\left[\frac{n^2}{(n+4)^2}\right] = 5(n+4)^2\left(\frac{3}{5}\right)$$

Multiply both members by the LCD, $5(n + 4)^2$.

$$5n^2 = 3(n+4)^2$$
$$5n^2 = 3(n^2 + 8n + 16)$$
$$5n^2 = 3n^2 + 24n + 48$$
$$2n^2 - 24n - 48 = 0$$
$$n^2 - 12n - 24 = 0$$

Simplify, and then write the quadratic equation in standard form.

$$n = \frac{-(-12) \pm \sqrt{(-12)^2 - 4(1)(-24)}}{2(1)}$$

Since the left member is not factorable, use the quadratic formula with $a = 1$, $b = -12$, and $c = -24$.

$$= \frac{12 \pm \sqrt{144 + 96}}{2}$$
$$= \frac{12 \pm \sqrt{240}}{2}$$
$$= \frac{12 \pm \sqrt{16}\sqrt{15}}{2}$$
$$= \frac{12 \pm 4\sqrt{15}}{2}$$

Simplify the radical term, and then reduce by dividing the numerator and the denominator by 2.

$$n = 6 \pm 2\sqrt{15}$$
$$n + 4 = 10 \pm 2\sqrt{15}$$

There are two possibilities for the smaller number and two possibilities for the larger number.

Answer The numbers are either $6 - 2\sqrt{15}$ and $10 - 2\sqrt{15}$ or $6 + 2\sqrt{15}$ and $10 + 2\sqrt{15}$.

You could approximate these values with a calculator and then check them.

It is important to be able to relate the language of mathematics to your everyday language. The problems you encounter at work, for example, will likely be expressed orally or in written form. Your answers should use words similar to those used in the problems. It is important to practice expressing your ideas in words—what we call a *word equation*—and to practice writing your answers as full sentences so that you can communicate effectively in the workplace.

EXAMPLE 2 Electrical Resistance

The total resistance in a parallel circuit, such as the one shown in the figure, is given by the formula

$$\frac{1}{R_1} + \frac{1}{R_2} + \frac{1}{R_3} = \frac{1}{R_T}$$

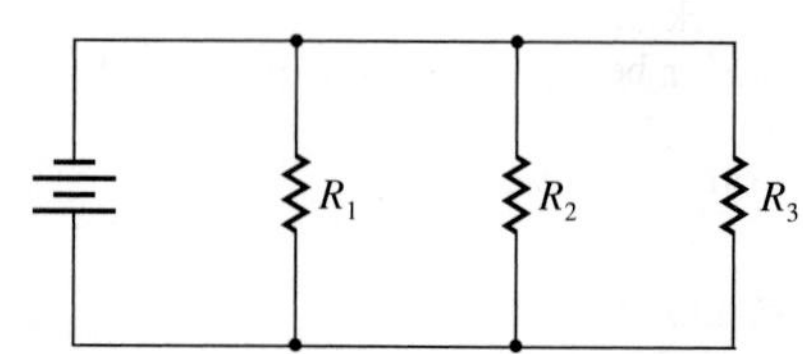

If R_2 is five times R_1 and R_3 is 1 more than R_1, find each of these values in ohms, given that the total resistance is 2 ohms.

SOLUTION Let

R_1 = Resistance of the first resistor

$5R_1$ = Resistance of the second resistor

$R_1 + 1$ = Resistance of the third resistor

$$\frac{1}{R_1} + \frac{1}{R_2} + \frac{1}{R_3} = \frac{1}{R_T}$$

The resistance formula for a parallel circuit

$$\frac{1}{R_1} + \frac{1}{5R_1} + \frac{1}{R_1 + 1} = \frac{1}{2}$$

Substitute the values identified above.

$$10R_1(R_1 + 1)\left[\frac{1}{R_1} + \frac{1}{5R_1} + \frac{1}{R_1 + 1}\right] = 10R_1(R_1 + 1)\left(\frac{1}{2}\right)$$

Multiply both members by the LCD, $10R_1(R_1 + 1)$.

$$10(R_1 + 1) + 2(R_1 + 1) + 10R_1 = 5R_1^{\,2} + 5R_1$$

$$-5R_1^{\,2} + 17R_1 + 12 = 0$$

$$5R_1^{\,2} - 17R_1 - 12 = 0$$

Simplify, and then write the quadratic equation in standard form.

$$(R_1 - 4)(5R_1 + 3) = 0$$

Solve by factoring.

$R_1 - 4 = 0$ or $5R_1 + 3 = 0$

$R_1 = 4$ $\qquad R_1 = \dfrac{-3}{5}$ (Not appropriate)

$5R_1 = 20$

$R_1 + 1 = 5$

Caution: Always check your answers to make sure they are reasonable and appropriate for the given problem.

Answer The resistances are 4 ohms, 20 ohms, and 5 ohms.

SELF-CHECK 

The length of a rectangle is 3 meters more than twice the width. Find the dimensions of this rectangle if the area is 152 square meters.

EXAMPLE 3 Work Done by Two Conveyor Belts

Two conveyor belts can unload a shipment of tomatoes in 6 hours. (See the figure.) Working alone, the slower belt would take 9 hours longer than the faster belt to do the job. How many hours would it take each belt working alone to do the job?

SOLUTION Let

t = Time in hours for the faster belt

$\frac{1}{t}$ = Rate of the faster belt

$t + 9$ = Time in hours for the slower belt

$\frac{1}{t + 9}$ = Rate of the slower belt

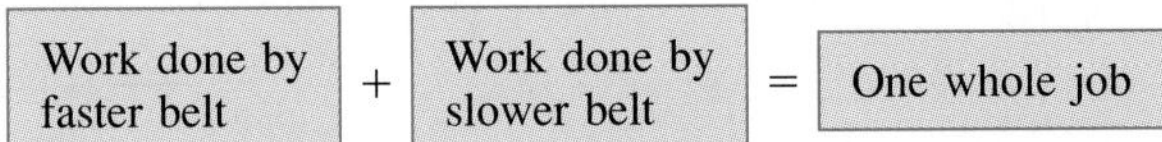

The **word equation** is based on the mixture principle.

	$R \cdot T = W$
Faster belt	$\frac{1}{t} \cdot 6 = \frac{6}{t}$
Slower belt	$\frac{1}{t + 9} \cdot 6 = \frac{6}{t + 9}$

Rate · Time = Work is one application of the rate principle. Use this equation and a table to organize the given information.

SELF-CHECK ANSWER 

The width of the rectangle is 8 m, and the length is 19 m.

$$\frac{6}{t} + \frac{6}{t+9} = 1$$

Substitute the values from the table into the word equation.

$$t(t+9)\left(\frac{6}{t} + \frac{6}{t+9}\right) = t(t+9)(1)$$

Multiply both members by the LCD, $t(t + 9)$.

$$6(t+9) + 6t = t(t+9)$$

$$6t + 54 + 6t = t^2 + 9t$$

$$0 = t^2 - 3t - 54$$

Simplify, and write the quadratic equation so that one member is zero.

$$0 = (t-9)(t+6)$$

Factor, and solve for t.

$t - 9 = 0$ or $t + 6 = 0$

$t = 9$ $\quad$ $t = -6$ (Not a meaningful answer)

$t + 9 = 18$

Answer The faster belt could do the job in 9 h, and the slower belt could do the job in 18 h.

SELF-CHECK

A secretary and a clerical assistant working together can stuff a batch of envelopes in 2 hours. Working alone, the secretary could do the job 3 hours faster than the assistant could. How long would it take each person working alone to do the job?

EXAMPLE 4 Boats in a Flowing River

Two boats that have the same speed in still water travel in opposite directions in a river with a current of 5 kilometers per hour. The boat going upstream departs 1 hour before the boat going downstream. A period of time after the downstream boat has departed, a radio conversation between the boats indicates that one boat is 44 kilometers upstream and the other boat is 75 kilometers downstream. Approximate to the nearest tenth of a kilometer per hour the speed of the boats in still water.

SOLUTION Let

r = Rate of each boat in still water in km/h

$r + 5$ = Rate of the boat going downstream in km/h

$r - 5$ = Rate of the boat going upstream in km/h

First identify the quantity being sought, and then describe this quantity with an appropriate variable.

[Time of boat going upstream] = [Time of boat going downstream] + [1]

Word equation

	$D \div R = T$
Boat going upstream	$44 \div (r-5) = \frac{44}{r-5}$
Boat going downstream	$75 \div (r+5) = \frac{75}{r+5}$

Since $D = RT$, $T = \frac{D}{R}$. Use this equation and a table to organize the given information.

$$\frac{44}{r-5} = \frac{75}{r+5} + 1$$

Substitute the values from the table into the word equation.

SELF-CHECK ANSWER 

3 h for the secretary, 6 h for the assistant

$$(r-5)(r+5)\left(\frac{44}{r-5}\right) = (r-5)(r+5)\left(\frac{75}{r+5}+1\right)$$

Multiply both members by the LCD, $(r-5)(r+5)$.

$$44(r+5) = 75(r-5) + (r-5)(r+5)$$

$$44r + 220 = 75r - 375 + r^2 - 25$$

$$0 = r^2 + 31r - 620$$

Simplify, and then write the quadratic equation so that one member is zero.

$$r = \frac{-31 \pm \sqrt{31^2 - 4(1)(-620)}}{2(1)}$$

Since the right member is not factorable, use the quadratic formula, with $a = 1$, $b = 31$, and $c = -620$.

$$r = \frac{-31 \pm \sqrt{3441}}{2}$$

$$r \approx \frac{-31 \pm 58.660038}{2}$$

Use a calculator to approximate the rate to the nearest tenth of a kilometer per hour.

$r \approx 13.8$ or $r \approx -44.8$

Not an appropriate solution

Answer The speed of each boat in still water is approximately 13.8 km/h.

EXAMPLE 5 Investment Income

An investment of \$4000 was split between two stocks on the New York Stock Exchange. At the end of 1 year, the first stock had gained \$300 in value and the second stock had gained \$110 in value. If the rate of gain on the second stock was 1% higher than the rate of gain on the first stock, find each rate of gain.

SOLUTION Let

r = Rate of gain on the first stock

$r + 0.01$ = Rate of gain on the second stock

Identify the quantities sought with an appropriate variable.

[Amount invested in first stock] + [Amount invested in second stock] = [\$4000]

This **word equation** is based on the mixture principle.

	$I \div R = P$
First stock	$300 \div r = \frac{300}{r}$
Second stock	$110 \div (r + 0.01) = \frac{110}{r+0.01}$

Since $I = PRT$ (with $T = 1$ year), $P = \frac{I}{R}$. Use this equation and a table to organize the given information.

$$\frac{300}{r} + \frac{110}{r+0.01} = 4000$$

Substitute the values from the table into the word equation.

$$r(r+0.01)\left(\frac{300}{r}\right) + r(r+0.01)\left(\frac{110}{r+0.01}\right) = r(r+0.01)(4000)$$

Multiply both members by the LCD, $r(r + 0.01)$.

$$300(r+0.01) + 110r = 4000r(r+0.01)$$

$$300r + 3 + 110r = 4000r^2 + 40r$$

Simplify, and then write the quadratic equation in standard form.

$$410r + 3 = 4000r^2 + 40r$$
$$4000r^2 - 370r - 3 = 0$$
$$(10r - 1)(400r + 3) = 0$$

Factor the left member, and then solve for the rate of gain r.

$$10r - 1 = 0 \quad \text{or} \quad 400r + 3 = 0$$
$$10r = 1 \qquad 400r = -3$$
$$r = 0.10 \qquad r = -0.0075$$
$$r + 0.01 = 0.11$$

Not an appropriate solution

Answer The rate of gain for the first stock was 10%, and the rate of gain for the second stock was 11%.

Do these values seem reasonable?

Exercises 8-4

A

In Exercises 1–22 solve each problem.

Numeric Word Problems (1–8)

1 Find two consecutive integers whose product is 132.

2 Find two consecutive even integers whose product is 440.

3 The sum of the squares of two consecutive odd integers is 202. Find these integers.

4 The sum of the squares of three consecutive integers is 110. Find these integers.

5 One number is 7 more than another number. Find these numbers if the sum of their reciprocals is 1.

6 If a number is subtracted from its reciprocal, the difference is −2. Find this real number.

7 Find two real numbers whose sum is 12 and whose product is 34.

8 Find two real numbers whose sum is 9 and whose product is 19.

9 Dimensions of a Family Room The length of the rectangular family room shown in the figure is 3 yards more than the width. If it takes 40 square yards of carpeting to cover this room, what are its dimensions?

Figure for Exercise 9

10 Dimensions of a Pen A 54-meter section of fencing encloses the rectangular pen shown in the figure. If the area of the pen is 170 square meters, find the length and the width of the pen.

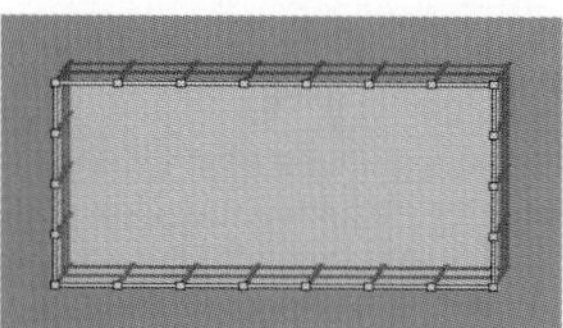

Figure for Exercise 10

11 Dimensions of a Pad A rectangular pad is outlined by a 48-meter stripe of paint. Find the dimensions of this pad if its surface area is 143 square meters.

12 Dimensions of a Rectangle The length of a rectangle is 3 meters less than twice the width. Find the dimensions of this rectangle if the area is 135 square meters.

13 Height of a Triangle The base of the triangle shown in the figure is 7 centimeters longer than the height. Determine the height of the triangle if its area is 30 square centimeters.

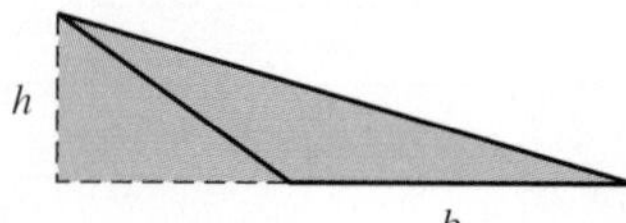

Figure for Exercise 13

14 Base of a Triangle The base of the triangle shown in the figure is 3 meters longer than the height. Find the base if the area of this triangle is 15 square meters.

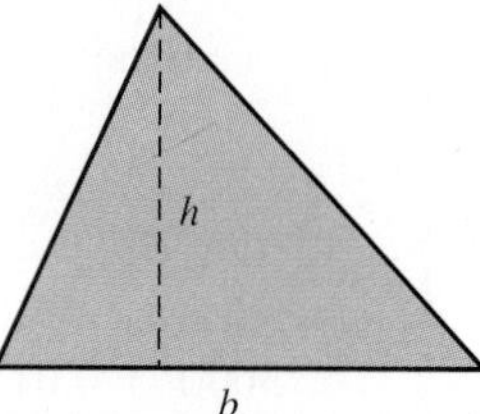

Figure for Exercise 14

15 Width of a Room An electrician examining the blueprints for a rectangular room that is 4 feet longer than it is wide determines that a wire run diagonally across this room will be 20 feet long. What is the width of the room? (See the figure.)

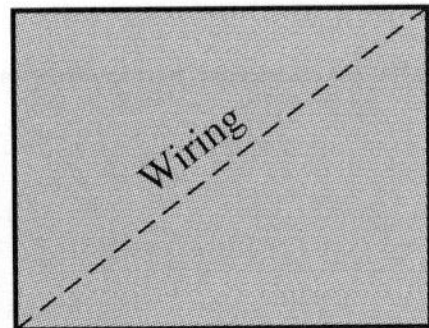

Figure for Exercise 15

16 Airplane Mileage Upon leaving an airport, an airplane flies due south and then due east. After it has flown 31 miles farther east than it flew south, it is 41 miles from the airport. How far south did it fly? (See the figure.)

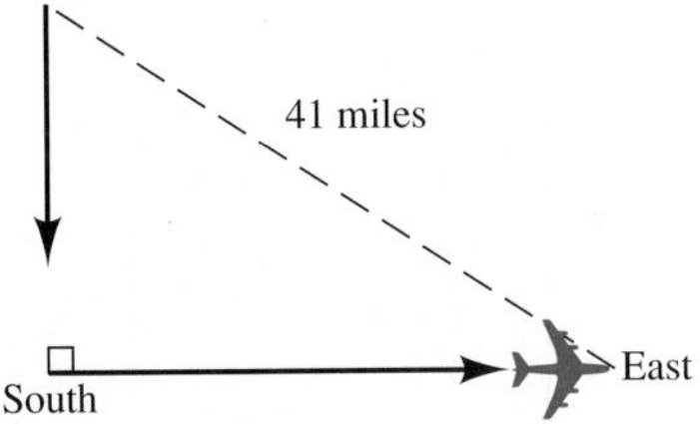

Figure for Exercise 16

Interest Rates (17 and 18)

17 An amount of $10,000 was invested in two 1-year certificates of deposit. In 1 year, the first CD earned $280 in interest and the second earned $480. If the interest rate on the second CD was 1% higher than that on the first CD, find each interest rate.

18 An investment of $6300 was split between two 1-year certificates of deposit. In 1 year, the first CD earned $200 in interest and the second earned $361. If the interest rate on the second CD is $1\frac{1}{2}$% higher than that on the first CD, find each interest rate.

19 Pumping Water Working together, two pumps can pump all the water from a service tunnel in 5 hours. Working alone, the smaller pump would take 3 hours longer than the larger pump to do the job. Approximate to the nearest tenth of an hour the time it would take each pump to do the job working alone.

20 Hours Worked Working together, two workers can finish a concrete floor in 40 minutes. Working alone, the worker with the smaller machine would take 25 minutes longer than the worker with the larger machine. Approximate to the nearest minute the time it would take the worker with the larger machine to finish the floor working alone.

21 Airplane Speed Two planes with the same air speed depart 1 hour apart and travel in opposite directions. The plane that departs first flies directly into a 40-mile-per-hour wind, and the second plane flies with this wind. After a period of time, radar indicates that the first plane has traveled 700 miles and the second plane has traveled 540 miles. Determine the air speed of each plane.

22 Boat Speed Two boats that have the same speed in still water travel in opposite directions on a river with a current of 7 kilometers per hour. The boat going downstream departs 1 hour after the boat going upstream. After a while, a person on shore contacts the captain of each boat by radio. One boat is 21 kilometers upstream, and the other boat is 42 kilometers downstream. What is the speed of each boat in still water?

B

In Exercises 23–32 solve each problem.

Electrical Resistance (23 and 24)

23 Suppose that the total resistance of the parallel circuit shown in the figure, $\frac{1}{R_1} + \frac{1}{R_2} + \frac{1}{R_3} = \frac{1}{R_T}$, is 3 ohms. If R_2 is three times R_1 and R_3 is 2 more than R_1, approximate to the nearest hundredth of an ohm the value of R_1.

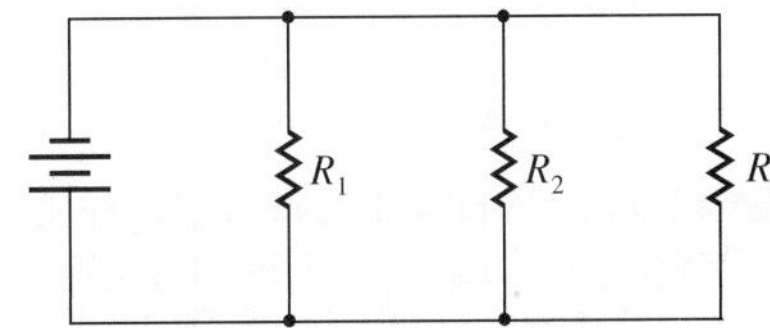

Figure for Exercises 23 and 24

24 Suppose that the total resistance of the parallel circuit shown in the figure, $\frac{1}{R_1} + \frac{1}{R_2} + \frac{1}{R_3} = \frac{1}{R_T}$, is 8 ohms. If R_2 is four times R_1 and R_3 is 5 more than R_1, approximate to the nearest hundredth of an ohm the value of R_1.

25 Area Between Two Circles The area enclosed between two concentric circles is 16π square centimeters. The radius of the larger circle is 1 centimeter less than twice the radius of the smaller circle. Determine the length of the shorter radius. (See the figure.)

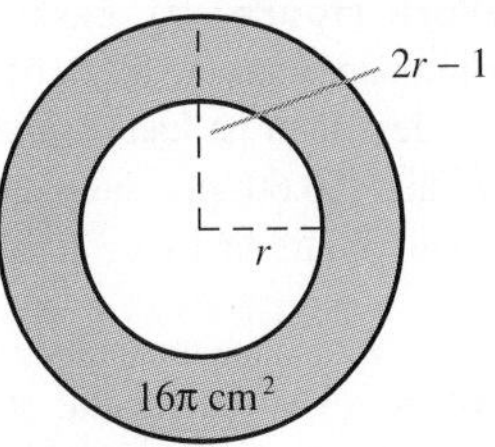

Figure for Exercise 25

26 Dimensions of a Mat A square mat used for athletic exercises has a red border on all four sides. The rest of the mat is blue. The width of the blue square is three-fourths the width of the entire square. If the area colored red is 112 square feet, determine the length of each side of the mat. (See the figure.)

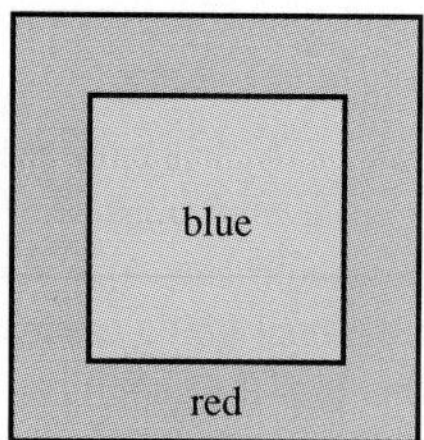

Figure for Exercise 26

27 Helicopter Ground Speed Two helicopters depart simultaneously from an airport. One flies due south; the other flies due east at a rate 70 miles per hour faster that that of the first helicopter. After 1 hour radar indicates that the helicopters are 170 miles apart. What is the ground speed of each helicopter? (See the figure.)

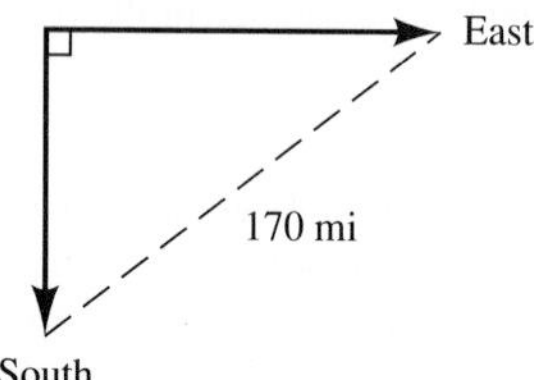

Figure for Exercise 27

28 Width of a Field A park district employee stopped for a break after mowing a strip around the 50-meter by 80-meter rectangular field shown in the figure. If she had completed one-half the field when she took her break, determine to the nearest tenth of a meter the width of this strip when she stopped.

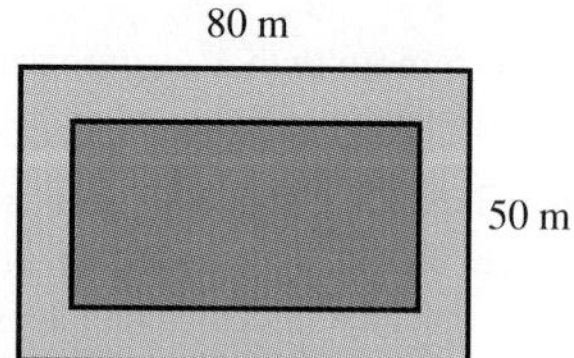

Figure for Exercise 28

29 Estimating Work Hours An electrician contracted to wire a small maintenance garage for $1612. The actual work took him 6 hours less than he had estimated, and as a consequence he earned $0.60 per hour more than he had planned. How many hours had he estimated this job would take?

30 Emptying a Tank A supplemental storage tank can be drained by an outlet pipe in 3 hours less than it takes the inlet pipe to fill an empty tank. If the tank is three-fourths full and both pipes are open, it takes 12 hours to empty the tank. Approximate to the nearest tenth of an hour the time it would take the outlet to drain a full tank if the inlet were closed.

31 Dimensions of Posterboard A rectangular piece of posterboard is 6.0 centimeters longer than it is wide. A 1.5-centimeter strip is cut off each of the sides, leaving an area $\frac{35}{48}$ the original area. Find the original dimensions. (See the figure shown.)

Figure for Exercise 31

32 Diameter of a Bin The length of the diagonal brace on the cylindrical storage bin shown in the figure is 25 feet. If the height of the bin is 4 feet more than the diameter, determine the diameter of this bin to the nearest tenth of a foot.

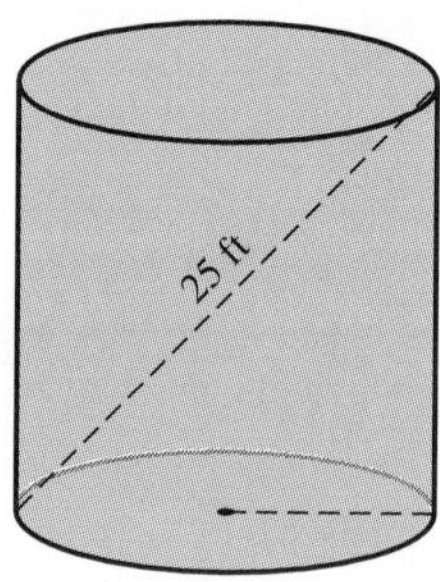

Figure for Exercise 32

C

In Exercises 33–40 solve each problem.

33 Distances on a Baseball Diamond The bases on a baseball diamond are placed at the corners of a square whose sides are 90 feet long. How much farther does a catcher have to throw the ball to get it from home plate to second base than from home plate to third base? (See the figure.)

Figure for Exercise 33

34 Speed of a Baseball If a catcher on a baseball team throws a baseball at 120 feet per second (over 80 miles per hour), approximately how long will it take his throw to go from home plate to second base? Approximately how long will it take his throw to go from home plate to third base? (*Hint:* See Exercise 33.)

Figure for Exercise 34

35 Length of a Rope The length of one piece of rope is 4 meters more than twice the length of another piece of rope. Each rope is used to enclose a square region. If the area enclosed by the longer rope is 480 square meters more than the area enclosed by the shorter rope, determine the length of the shorter rope to the nearest tenth of a meter.

36 Interest Rates The formula for computing the amount A of an investment of principal P invested at interest rate r for 1 year and compounded semiannually is $A = P\left(1 + \frac{r}{2}\right)^2$. Approximately what interest rate is necessary for \$1000 to grow to \$1180 in 1 year if the interest is compounded semiannually?

37 Distance to the Horizon The radius of the earth is approximately 4000 miles. Approximate to within 10 miles the distance from the horizon to a plane flying at an altitude of 4 miles. (See the figure.)

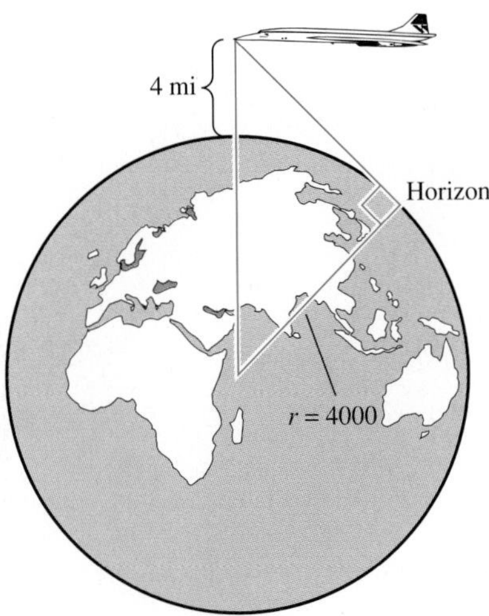

Figure for Exercise 37

38 Constructing a Box A metal box with an open top is to be formed by cutting 5-centimeter squares from each corner of a rectangular sheet of metal and then folding the sides up. The length of the box is to be 5 centimeters more than the width, and its capacity is to be 3700 cubic centimeters. Determine to the nearest tenth of a centimeter how wide the piece of sheet metal should be. (See the figure.)

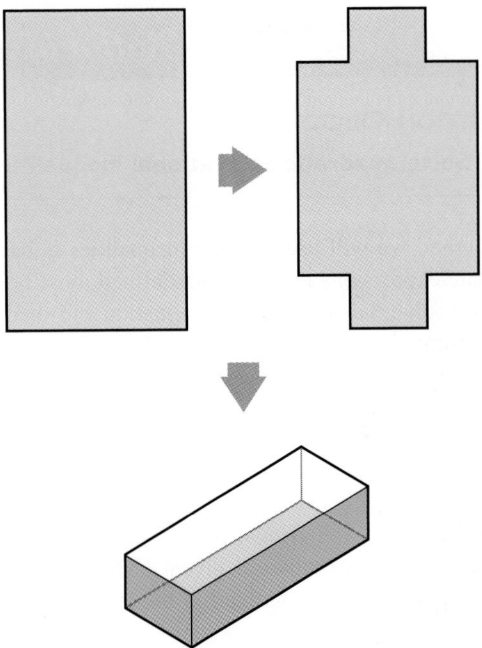

Figure for Exercise 38

39 Hourly Wages In one week a worker at a factory grossed \$342. Working in another area of the factory, his wife grossed \$378 in the same one-week period. The wife made \$1.50 more per hour than her husband did but worked 2 hours less. What is each spouse's hourly wage?

40 Height of a Box The length of the steel box shown in the figure is three times the height, and the width is 3 meters more than the height. If the length of the diagonal brace is 11 meters, determine the height of this box to the nearest tenth of a meter.

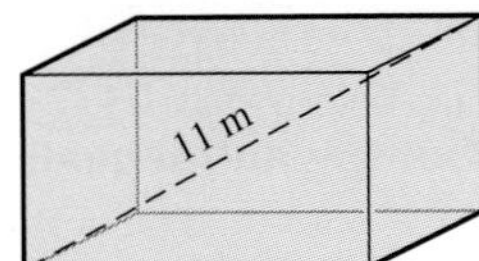

Figure for Exercise 40

Discussion Question

41 Write your own problem that the equation $(x + 7)^2 - x^2 = 119$ algebraically models.

SECTION SKILL REINFORCERS

Graph each of these intervals.

1 (3, 5) **2** [−2, 3] **3** (−1, 4] **4** (−2, +∞) **5** [−3, 1) ∪ (2, 5]

8-5 Quadratic and Rational Inequalities

SECTION OBJECTIVE

8 Solve quadratic and rational inequalities.

The method we will use to solve inequalities is based on the fact that two real algebraic expressions that are both defined must be either equal or unequal. We will first solve the corresponding equation and then use these points of equality to determine the solution of the inequality. For example, to graph the inequality $x > 2$, we first locate the point of equality $x = 2$. This point of equality subdivides the number line into two regions. In Figure 8-1, region A contains the numbers less than 2 and region B contains the numbers greater than 2.

In general, the points that form the boundaries of these regions of inequality are called **critical points** of the inequality. Since all of the numbers in a region formed by the critical points satisfy the same inequality, we can determine which regions satisfy a given inequality by testing a number from each region. This strategy is outlined in the following box.

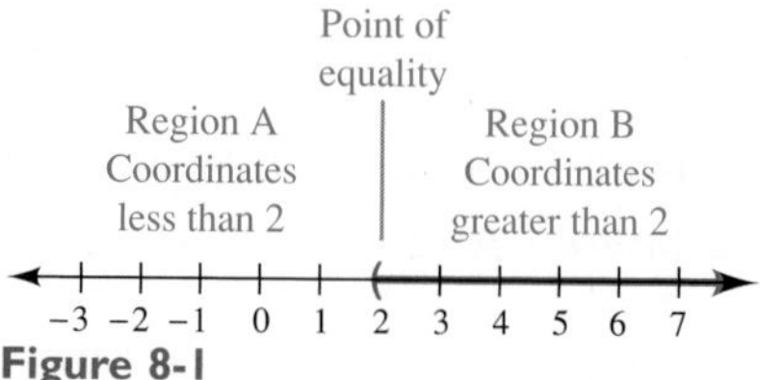

Figure 8-1

SOLVING A QUADRATIC INEQUALITY

Step 1 Find the points of equality, and plot these critical points on the number line.*

Step 2 Select an arbitrary test number from each region formed by the critical points.

Step 3 Substitute these test numbers into the inequality to determine which regions satisfy the inequality.

Step 4 Graph the solution set, and indicate this set algebraically.

*If there are no real critical points, treat the entire number line as a single region to be tested.

EXAMPLE 1 Solving an Inequality with Rational Critical Points

Solve $2x^2 - x < 15$.

SOLUTION Solve the corresponding equality.

$$2x^2 - x = 15$$

$$2x^2 - x - 15 = 0 \qquad \text{Put the quadratic equation in standard form, and then solve it by factoring.}$$

$$(2x + 5)(x - 3) = 0$$

$$2x + 5 = 0 \quad \text{or} \quad x - 3 = 0$$
$$2x = -5 \qquad\qquad x = 3$$
$$x = \frac{-5}{2}$$

Plot the critical points.

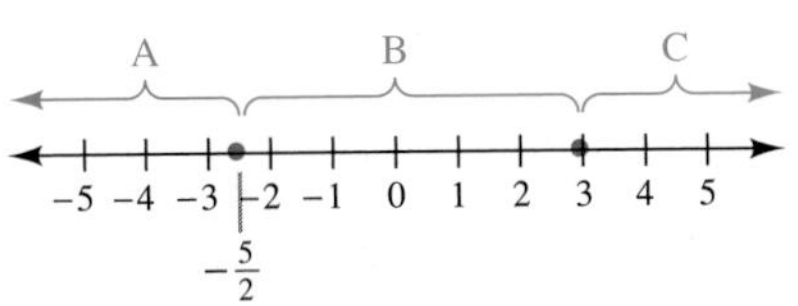

The critical points $\frac{-5}{2}$ and 3 separate the number line into three regions.

Select a test number from each region.

A: −3 is in region A.

B: 0 is in region B.

C: 4 is in region C.

The test numbers in each region are arbitrary, so it is wise to select numbers that simplify the computations.

Test these values.

Region A:

$$2(-3)^2 - (-3) \overset{?}{<} 15$$
$$2(9) + 3 \overset{?}{<} 15$$
$$21 < 15 \text{ is false.}$$

Region B:

$$2(0)^2 - (0) \overset{?}{<} 15$$
$$0 < 15 \text{ is true.}$$

Region C:

$$2(4)^2 - (4) \overset{?}{<} 15$$
$$2(16) - 4 \overset{?}{<} 15$$
$$28 < 15 \text{ is false.}$$

Graph the solution.

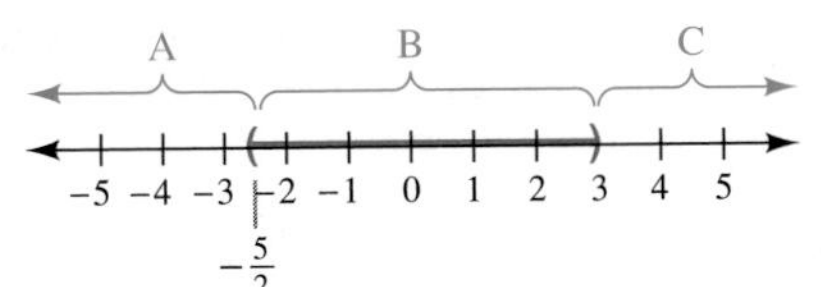

After graphing all of the points in the region that satisfies the original inequality, indicate this solution set algebraically.

Answer $\left(-\frac{5}{2}, 3\right)$

This is interval notation for $-\frac{5}{2} < x < 3$.

In Example 1 we would have used brackets instead of parentheses in the graph of the solution if the inequality had been less than or equal to instead of less than. If the inequality in Example 1 had been greater than instead of less than, the solution process would have produced the same regions to test, but a test of these regions would have yielded $x < \frac{-5}{2}$ or $x > 3$ as a solution of $2x^2 - x > 15$. The graph of the latter inequality is shown in Figure 8-2.

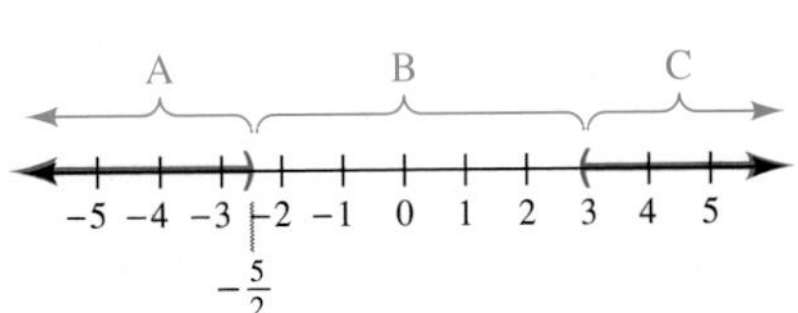

Figure 8-2

EXAMPLE 2 Solving an Inequality with Irrational Critical Points

Solve $x^2 - 2x \geq 6$.

SOLUTION Solve the corresponding equality.

$$x^2 - 2x = 6$$
$$x^2 - 2x - 6 = 0$$
$$x = \frac{-(-2) \pm \sqrt{(-2)^2 - 4(1)(-6)}}{2(1)}$$
$$= \frac{2 \pm \sqrt{4 + 24}}{2}$$

Since the left member is not factorable over the integers, we will use the quadratic formula, with $a = 1, b = -2$, and $c = -6$.

$$= \frac{2 \pm \sqrt{28}}{2}$$ Note that $\sqrt{28} = \sqrt{4}\sqrt{7} = 2\sqrt{7}$.

$$= \frac{2 \pm 2\sqrt{7}}{2}$$ Reduce the fraction by dividing the numerator and the denominator by 2.

$$= 1 \pm \sqrt{7}$$

Plot the critical points.

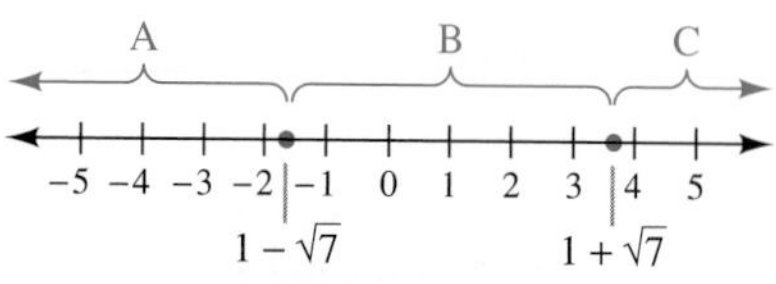

Use the approximations $1 - \sqrt{7} \approx -1.6$ and $1 + \sqrt{7} \approx 3.6$ to help locate the critical points.

Test a selected number from each region.

Region A: Test -2

$(-2)^2 - 2(-2) \overset{?}{\geq} 6$

$4 + 4 \overset{?}{\geq} 6$

$8 \geq 6$ is true.

Region B: Test 0

$(0)^2 - 2(0) \overset{?}{\geq} 6$

$0 \geq 6$ is false.

Region C: Test 4

$(4)^2 - 2(4) \overset{?}{\geq} 6$

$16 - 8 \overset{?}{\geq} 6$

$8 \geq 6$ is true.

Graph the solution.

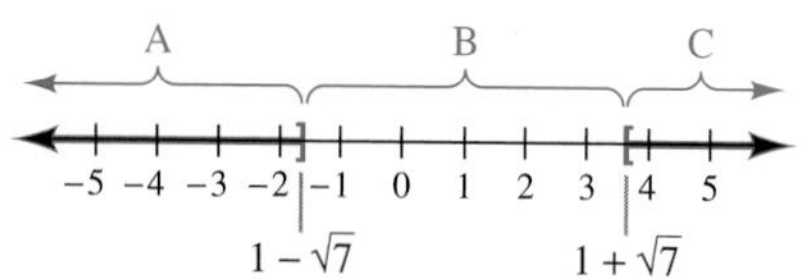

Graph the points from the regions that satisfy the original inequality, using brackets to indicate that the points of equality are in the solution set. Then indicate this solution set algebraically.

Answer $(-\infty, 1 - \sqrt{7}] \cup [1 + \sqrt{7}, +\infty)$

The union of these two intervals contains the x values $x \leq 1 - \sqrt{7}$ or $x \geq 1 + \sqrt{7}$.

EXAMPLE 3 Solving an Inequality with No Critical Point

Solve $(x + 3)^2 > 6x - 16$.

SOLUTION Solve the corresponding equality.

$$(x + 3)^2 = 6x - 16$$

$$x^2 + 6x + 9 = 6x - 16$$

$$x^2 = -25$$

$$x = \pm 5i$$

Solve the quadratic equation by the method of extraction of roots.

There are no real critical values. Test any real number, such as 0, to determine the solution.

Test 0:

$(0 + 3)^2 \overset{?}{>} 6(0) - 16$

$9 > -16$ is true.

The test number is a solution to the equation, so all real numbers are solutions. The inequality is an absolute inequality.

Answer The solution set is the set of all real numbers.

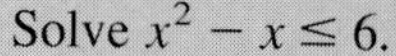

SELF-CHECK

Solve $x^2 - x \leq 6$.

If the inequality in Example 3 had been less than instead of greater than, the solution process would have been nearly identical except that the test value would have failed to satisfy the inequality. In that case there would have been no solution, and the solution set would have been the null set, ∅.

SELF-CHECK ANSWER

$[-2, 3]$

The procedure demonstrated above for solving quadratic inequalities can also be used to solve higher-degree inequalities. With inequalities of higher degree the only part of the procedure that may be more difficult is finding the critical points. The next example is already factored, so the critical values can easily be determined using the zero-factor principle.

EXAMPLE 4 Solving an Inequality with Three Critical Points

Solve $(x + 2)(x - 1)(x - 2) \geq 0$.

SOLUTION Solve the corresponding equality.

$(x + 2)(x - 1)(x - 2) = 0$

$x + 2 = 0$ or $x - 1 = 0$ or $x - 2 = 0$ Set each factor equal to zero.

$x = -2$ $\quad x = 1$ $\quad x = 2$

Plot the critical points.

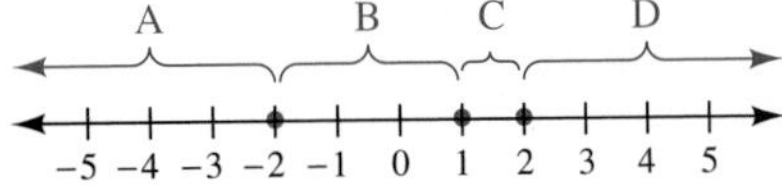

Test a number from each region.

Region A: Test -3

$(-3 + 2)(-3 - 1)(-3 - 2) \overset{?}{\geq} 0$

$(-1)(-4)(-5) \overset{?}{\geq} 0$

$-20 \geq 0$ is false.

Region B: Test 0

$(0 + 2)(0 - 1)(0 - 2) \overset{?}{\geq} 0$

$(2)(-1)(-2) \overset{?}{\geq} 0$

$4 \geq 0$ is true.

Region C: Test $\frac{3}{2}$

$\left(\frac{3}{2} + 2\right)\left(\frac{3}{2} - 1\right)\left(\frac{3}{2} - 2\right) \overset{?}{\geq} 0$

$\left(\frac{7}{2}\right)\left(\frac{1}{2}\right)\left(\frac{-1}{2}\right) \overset{?}{\geq} 0$

$-\frac{7}{8} \geq 0$ is false.

Region D: Test 3

$(3 + 2)(3 - 1)(3 - 2) \overset{?}{\geq} 0$

$(5)(2)(1) \overset{?}{\geq} 0$

$10 \geq 0$ is true.

Graph the solution.

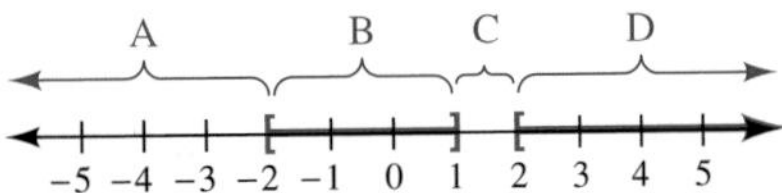

After graphing all the points from the regions that satisfy the original inequality, indicate the solution set for $-2 \leq x \leq 1$ or $x \geq 2$, using interval notation.

Answer $[-2, 1] \cup [2, +\infty)$

Inequalities involving rational expressions can be worked by a method similar to the one used above. When an inequality contains a variable in the denominator, it is much easier to solve the corresponding equality than to try to solve the given inequality directly. If both sides of the inequality $\frac{1}{2 - x} > x$ are multiplied by $2 - x$, we cannot tell whether the order should be preserved or

SELF-CHECK

Solve
$(x + 3)(x - 2)(x - 3) < 0$.

SELF-CHECK ANSWER 

$(-\infty, -3) \cup (2, 3)$

reversed since we do not know whether $2 - x$ is positive or negative. By solving the corresponding equality, $\frac{1}{2 - x} = x$, we can avoid the problem of having to consider both of these possibilities.

To determine the regions on the number line that satisfy an inequality, we first find the critical points that form the boundaries of these regions. These critical points are not points of inequality, so they must be either points of equality or points that cause the expression to be undefined because of division by zero.

SOLVING AN INEQUALITY CONTAINING RATIONAL EXPRESSIONS

Step 1 Find the critical points,* including
- **a.** the points that cause division by zero and
- **b.** the points of equality.

Step 2 Choose an arbitrary test number from each region formed by these critical points.

Step 3 Substitute these test numbers into the inequality to determine which regions satisfy the inequality.

Step 4 Graph the solution set, and indicate this set algebraically.

*If there are no real critical points, treat the entire number line as a single region to be tested.

EXAMPLE 5 Solving a Rational Inequality

Solve $\frac{6}{x - 2} \geq 3$.

SOLUTION

$$\frac{6}{x - 2} \geq 3$$

$$\frac{6}{x - 2} - 3 \geq 0$$ Start by setting the right side of the inequality equal to zero.

$$\frac{6 - 3(x - 2)}{x - 2} \geq 0$$ Then rewrite the left side of the inequality as a single fraction.

$$\frac{-3x + 12}{x - 2} \geq 0$$ Simplify the numerator, and express it in factored form.

$$\frac{-3(x - 4)}{x - 2} \geq 0$$

Find the critical point that causes division by zero.

$$x - 2 = 0$$ Set the denominator equal to zero.

$$x = 2$$

Find the points of equality.

$$-3(x - 4) = 0$$ If the fraction is equal to zero, then the numerator must equal zero.

$$x - 4 = 0$$ Divide both sides by -3.

$$x = 4$$

Plot the critical points 2 and 4.

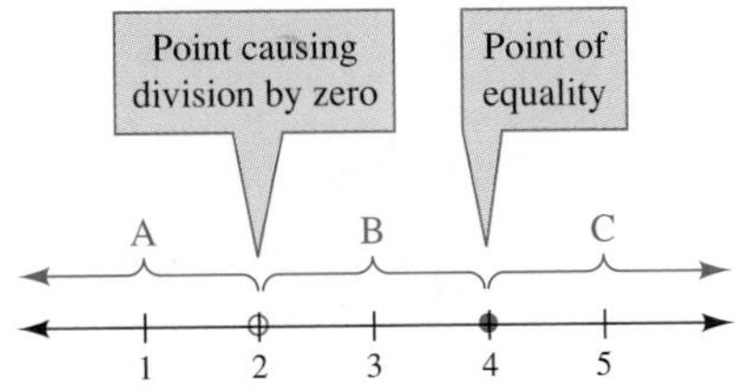

Test a number from each region.

Region A: Test 0

$$\frac{6}{0 - 2} \overset{?}{\geq} 3$$

$-3 \geq 3$ is false.

Region B: Test 3

$$\frac{6}{3 - 2} \overset{?}{\geq} 3$$

$6 \geq 3$ is true.

Region C: Test 5

$$\frac{6}{5 - 2} \overset{?}{\geq} 3$$

$2 \geq 3$ is false.

Graph the solution.

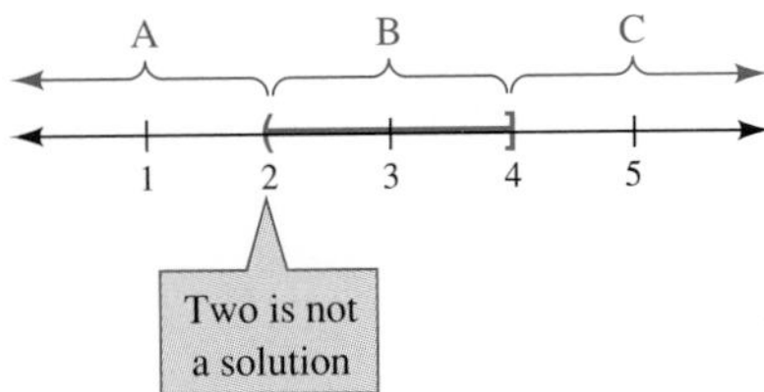
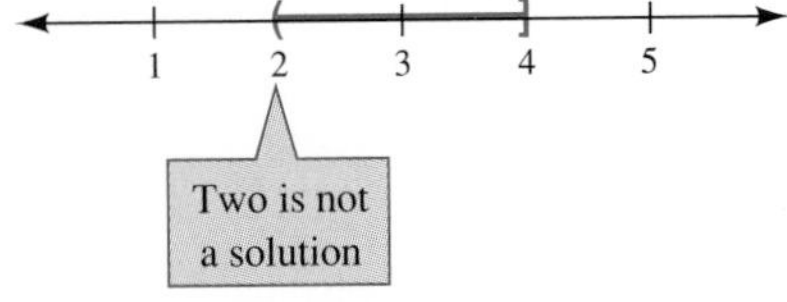

Note that 2 is not in the solution set, since the left side of the inequality is undefined at 2. The number 4 is a solution, since the left and right members are equal for $x = 4$.

Answer (2, 4]

$2 < x \leq 4$ ▼

SELF-CHECK

Solve $\dfrac{6}{y - 3} \geq 2$.

The solution of an inequality with the symbol $\leq$ or $\geq$ can include critical points that are points of equality, but it can never include critical points that cause division by zero, because these numbers produce expressions that are undefined.

EXAMPLE 6 Height of Fireworks

A faulty fireworks rocket is launched vertically with an initial velocity of 96 feet per second and then falls to the earth unexploded. (See the figure.) Its height, h, in feet after t seconds is given by $h = -16t^2 + 96t$. During what time interval after launch will the height of the rocket exceed 80 feet?

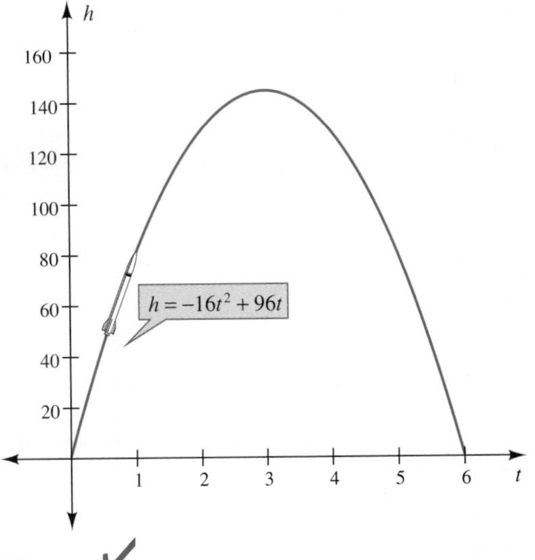

SELF-CHECK ANSWER

(3, 6]

SOLUTION Height > 80 — Word inequality

$-16t^2 + 96t > 80$ — Algebraic inequality

$-16t^2 + 96t - 80 > 0$ — Subtract 80 from both sides of the inequality.

$t^2 - 6t + 5 < 0$ — Dividing by -16 reverses the inequality.

Solve the corresponding equality for the critical points.

$t^2 - 6t + 5 = 0$

$(t - 1)(t - 5) = 0$ — Factor the left number.

$t - 1 = 0 \quad \text{or} \quad t - 5 = 0$ — Set each factor equal to zero.

$t = 1 \qquad\qquad t = 5$

Plot the critical points.

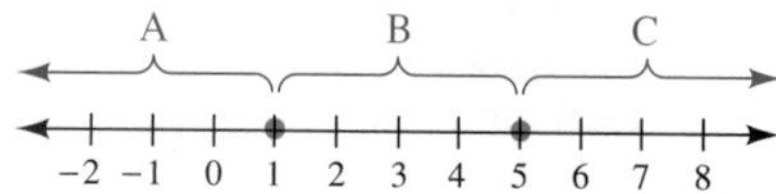

Test a value from each region.

Region A: Test 0

$0^2 - 6(0) + 5 \overset{?}{<} 0$

$5 < 0$ is false.

Region B: Test 2

$2^2 - 6(2) + 5 \overset{?}{<} 0$

$4 - 12 + 5 \overset{?}{<} 0$

$-3 < 0$ is true.

Region C: Test 6

$6^2 - 6(6) + 5 \overset{?}{<} 0$

$36 - 36 + 5 \overset{?}{<} 0$

$5 < 0$ is false.

Graph the solution.

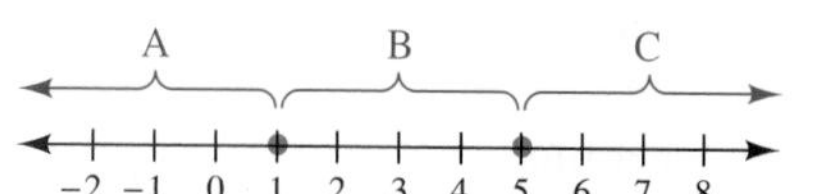

$1 < t < 5$

Answer The height of the fireworks rocket will exceed 80 feet between 1 second and 5 seconds after launch.

A Geometric Viewpoint: Using a Graph to Solve Inequalities

The real solutions of $ax^2 + bx + c < 0$ correspond to the values of x for which the graph of $y = ax^2 + bx + c$ is below the x-axis. The graph of $y = x^2 - x - 6$ on the display of a TI-82 calculator is shown to the right. This graph is below the x-axis on the interval $(-2, 3)$. This interval of values is also the solution of $x^2 - x - 6 < 0$.

Graphs of $y = ax^2 + bx + c$ are examined in detail in Section 9-2.

[−4, 5] for x, [−7, 10] for y

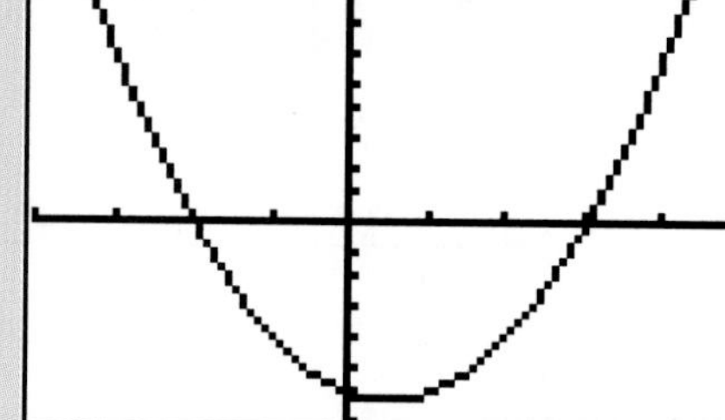

EXAMPLE 7 **Calculator Usage**

Use the graph of $y = 2x^2 + 3x - 2$ to solve $2x^2 + 3x - 2 < 0$.

SOLUTION

Keystrokes

[Y=] [2] [X, T, θ] [x^2] [+] [3] [X, T, θ] [−] [2] [GRAPH]

Assume that the window of values has already been set to $[-4, 3]$ for x and $[-5, 4]$ for y.

$[-4, 3]$ for x, $[-5, 4]$ for y

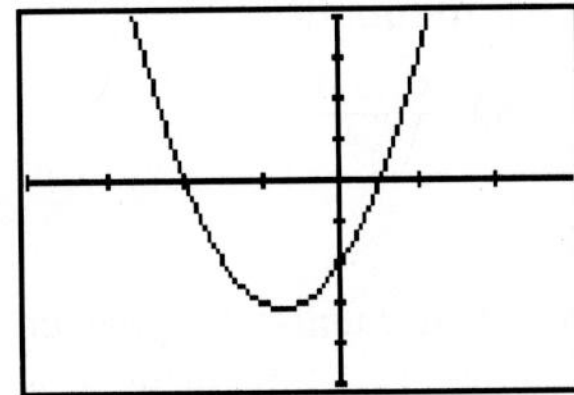

The solution of $2x^2 + 3x - 2 < 0$ corresponds to the interval for which the graph of $y = 2x^2 + 3x - 2$ is below the x-axis. By inspection or by using the TRACE feature, we can determine that the x-intercepts are -2 and $\frac{1}{2}$. For $-2 < x < \frac{1}{2}$, the graph is below the x-axis.

Answer $\left(-2, \frac{1}{2}\right)$

Exercises 8-5

A

In Exercises 1–36 solve each inequality.

1 $(x - 2)(x + 1) < 0$

2 $(x + 3)(x - 1) > 0$

3 $v(2 - v) \le 0$

4 $-v(v - 3) \ge 0$

5 $w^2 - 2w \ge 24$

6 $4w^2 < 4w + 3$

7 $y^2 > 2$

8 $y^2 \le 3$

9 $-3x^2 > 17x - 6$

10 $-x^2 < 15 - 8x$

11 $2x^2 - 3x \ge 2$

12 $2x^2 - 5x \le 3$

13 $6y^2 < 20 - 7y$

14 $6y^2 \ge 14 - 17y$

15 $v^2 + v - 1 \ge 0$

16 $v^2 + 2v > 1$

17 $(w + 1)(w - 1)(w - 3) > 0$

18 $(w + 2)(w - 1)(w - 4) \le 0$

19 $m(m - 3)(m + 3) \le 0$

20 $m(m - 2)(m + 4) > 0$

21 $(m - 2)^2 > 0$

22 $(m + 1)^2 > 0$

23 $\frac{6}{x - 3} < 2$

24 $\frac{6}{x - 1} < 2$

25 $\frac{2y - 6}{y} < 0$

26 $\frac{2y + 4}{y} > 0$

27 $\frac{m - 1}{m + 2} \ge 0$

28 $\frac{m + 2}{m - 3} \le 0$

29 $\frac{2x}{x - 2} < 6$

30 $\frac{3x}{x + 1} > 2$

31 $\frac{x - 1}{x + 1} \le 3$

32 $\frac{x - 2}{x + 2} < 3$

33 $\frac{2}{3 - r} < r$

34 $\frac{6}{5 - r} > r$

35 $\frac{6}{t - 5} \ge t$

36 $\frac{5}{t - 4} \le t$

B

37 Select the inequality whose solution is

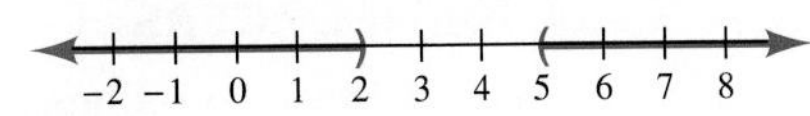

a. $(x + 2)(x + 5) > 0$

b. $(x - 2)(x - 5) > 0$

c. $(x + 2)(x + 5) < 0$

d. $(x - 2)(x - 5) < 0$

38 Select the inequality whose solution is

−5 −4 −3 −2 −1 0 1 2 3 4 5

a. $(x + 2)(x - 3) \leq 0$ **b.** $(x + 2)(x - 3) \geq 0$

c. $(x - 2)(x + 3) \leq 0$ **d.** $(x - 2)(x + 3) \geq 0$

39 Select the inequality whose solution is $[-3, 3]$.

a. $x^2 < 9$ **b.** $x^2 > 9$ **c.** $x^2 \leq 9$ **d.** $x^2 \geq 9$

40 Select the inequality whose solution is $(-\infty, -2) \cup (2, +\infty)$.

a. $x^2 > 4$ **b.** $x^2 < 4$ **c.** $x^2 > 2$ **d.** $x^2 < 2$

In Exercises 41–44 determine the values of k for which the given equation has real roots. A quadratic equation has real solutions if the discriminant $b^2 - 4ac$ is greater than or equal to zero.

41 $x^2 - kx + 9 = 0$ **42** $x^2 + kx + 25 = 0$

43 $x^2 - 2kx - 3k = 0$ **44** $x^2 - 2kx + 11 = 0$

In Exercises 45–54 solve each inequality.

45 $\frac{z^2 - 4}{3z} < 1$ **46** $\frac{z^2 - 6}{5z} \geq 1$

47 $v + 1 > \frac{2}{v}$ **48** $v < \frac{2v - 1}{v}$

49 $\frac{1}{x + 1} \geq \frac{x + 3}{x + 7}$ **50** $\frac{x}{x + 2} > \frac{2}{x - 1}$

51 $x^3 \leq 9x$ **52** $x^2 \geq 16x$

53 $v^3 + 2v^2 - 15v \geq 0$ **54** $v^3 + 2v^2 - 8v \leq 0$

C

CALCULATOR USAGE (55–60)

55 Use the graph of $y = x^2 - x - 6$ shown on the calculator window to solve $x^2 - x - 6 < 0$.

[−10, 10] for x, [−10, 10] for y

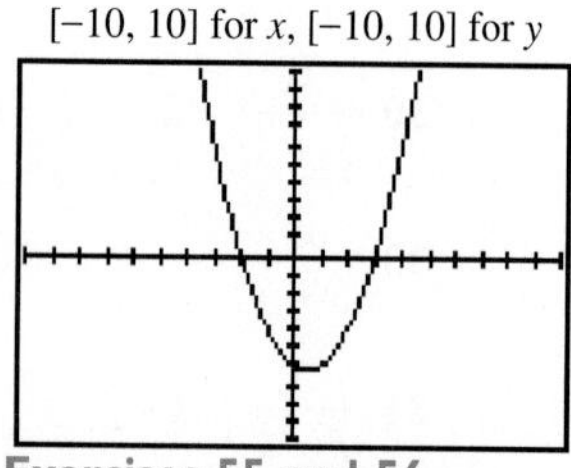

Figure for Exercises 55 and 56

56 Use the graph of $y = x^2 - x - 6$ shown on the calculator window to solve $x^2 - x - 6 > 0$.

In Exercises 57–60 use a graphics calculator to graph each equation and to solve the corresponding inequality.

57 Use the graph of $y = x^2 - 2x$ to solve $x^2 - 2x < 0$.

58 Use the graph of $y = 4x^2 - 1$ to solve $4x^2 - 1 < 0$.

59 Use the graph of $y = x^2 - x$ to solve $x^2 - x > 0$.

60 Use the graph of $y = x^3 - 4x$ to solve $x^3 - 4x > 0$.

In Exercises 61 and 62 solve each inequality.

61 $-4v^3 + 8v^2 > -v$ **62** $\frac{x^2 - 4x - 5}{x^2 - 9} < 0$

Profit (63 and 64)

63 The net income in dollars from selling t units of a product is given by $I = -t^2 + 120t - 1100$. If $I > 0$, there is a profit. Determine the values of t that will generate a profit.

64 The net income in dollars from selling t units of a product is given by $I = -t^2 + 190t - 925$. If $I > 0$, there is a profit. Determine the values of t that will generate a profit.

65 Height of a Golfball The height, h, in feet of a golfball after t seconds is given by $h = -16t^2 + 80t$. During what time interval after being initially struck with an eight iron will the height of the golfball exceed 64 feet?

66 Determine the values of v for which $\sqrt{8v^2 - 14v - 15}$ is a real number. This expression will be a real number when the radicand is greater than or equal to zero.

DISCUSSION QUESTION

67 The graph represents the profit y in dollars made by a company producing x units of a product. Discuss when the company is making a profit, when it is breaking even, when it is losing money, and when it is making the most money.

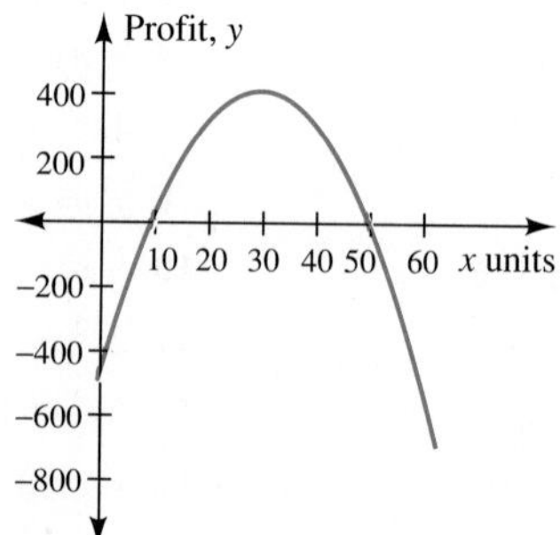

Figure for Exercise 67

SECTION SKILL REINFORCERS

Given $x = 5$, solve each of these equations for y.

1 $x + y = 11$ **2** $2x - y = 20$ **3** $2x + 3y = 1$ **4** $-x + 2y = 7$ **5** $\frac{x}{2} + \frac{y}{2} = 4$

Key Concepts for Chapter 8

1 The standard form of a quadratic equation in x is $ax^2 + bx + c = 0$, where a, b, and c represent real constants and $a \neq 0$.

2 A quadratic equation has either two distinct solutions or a solution of multiplicity two. These solutions may be either real or imaginary.

3 Methods of solving quadratic equations:

a. Factoring

b. Extraction of roots

c. Completing the square

d. The quadratic formula

$$x = \frac{-b \pm \sqrt{b^2 - 4ac}}{2a}$$

4 Nature of the solutions of a quadratic equation if a, b, and c are rational constants and $a \neq 0$:

Discriminant, $b^2 - 4ac$	***Nature of the Solutions***
Positive and a perfect square	Two distinct rational solutions
Positive and not a perfect square	Two distinct irrational solutions
Zero	A rational solution of multiplicity two
Negative	Two imaginary solutions that are complex conjugates

5 An equation in x is of quadratic form if it can be written as $az^2 + bz + c = 0$, where z is an algebraic expression in x.

6 The solution process for equations with rational expressions and for equations with radical expressions must include a check for extraneous values.

7 The critical points of an inequality are the points of equality and the points for which the expression is undefined.

8 The critical points of an inequality separate the number line into distinct regions. Either all the numbers in a region satisfy the inequality or none of the numbers in the region satisfy the inequality. Thus a test number from each region can be used to determine the solution of the inequality.

Review Exercises for Chapter 8

In Exercises 1–22 solve each equation by the most appropriate method.

1 $y^2 = 144$

2 $w^2 = 9w - 20$

3 $x^2 - 4x + 2 = 0$

4 $(v - 3)^2 = 36$

5 $10m^2 = 21m + 10$

6 $2y^2 = 6y - 9$

7 $-9v^2 + 24v = 16$

8 $(6x - 3)(x + 5) = 26x$

9 $y^4 = 81$

10 $z^4 = 14z^2 + 32$

11 $(m^2 - 3m)^2 - 44(m^2 - 3m) + 160 = 0$

12 $(5x^2 + 1) - 22\sqrt{5x^2 + 1} = -21$

13 $\left(\frac{2w - 1}{w - 2}\right)^2 + 2\left(\frac{2w - 1}{w - 2}\right) = 3$

14 $15m^{-2} = 17m^{-1} + 4$

15 $(n + 5)(2n - 3)(3n - 10) = 0$

16 $y^3 - 8 = 0$

17 $\sqrt{x + 12} - x = 0$

18 $\sqrt{2x - 1} = x - 2$

19 $m - 3 = \frac{4}{m}$

20 $\frac{x}{2x - 9} = \frac{x + 2}{x - 2}$

21 $v^{2/3} - 26v^{1/3} = 27$

22 $\frac{5y - 1}{3y + 5} - \frac{4y + 1}{2y - 3} = \frac{18}{6y^2 + y - 15}$

In Exercises 23–25 use the discriminant to determine the nature of the solutions of each equation. Identify the solutions as two distinct rational solutions, two distinct irrational solutions, a rational solution of multiplicity two, or two imaginary solutions.

23 $3x^2 = 30x - 75$

24 $4v^2 + 2v + 1 = 0$

25 $x^2 + \sqrt{11}x = 2$

In Exercises 26 and 27 determine the value of k such that the equation has a rational solution of multiplicity two.

26 $x^2 + kx + 25 = 0$

27 $kx^2 - 3kx + 9 = 0$

In Exercises 28 and 29 determine the value of the constant c so that the trinomial is the square of a binomial.

28 $y^2 - 18y + c$

29 $w^2 + 12w + c$

In Exercises 30 and 31 solve each quadratic equation by completing the square.

30 $v^2 = v + 3$

31 $3z^2 - 2z + 7 = 0$

In Exercises 32–34 solve each equation for x.

32 $x^2 - 3xy = 10y^2$

33 $5x^2 = 45y^2$

34 $x^2 + 4xy + y^2 = 0$ for $y \geq 0$

35 Select the inequality whose solution is

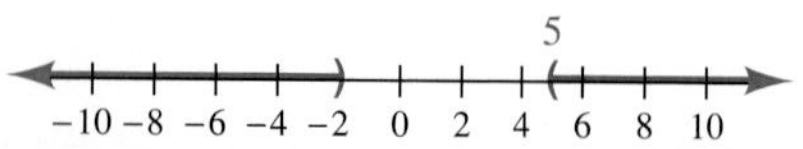

a. $(x + 2)(x - 5) < 0$ **b.** $(x + 2)(x - 5) > 0$
c. $(x - 2)(x - 5) < 0$ **d.** $(x - 2)(x - 5) > 0$

36 Select the inequality whose solution is $[-3, 4]$.

a. $(x + 3)(x - 4) \leq 0$ **b.** $(x - 3)(x + 4) \leq 0$
c. $(x + 3)(x - 4) \geq 0$ **d.** $(x - 3)(x + 4) \geq 0$

37 Select the inequality whose solution is $(-1, 1) \cup (2, +\infty)$.

a. $(x + 1)(x - 1)(x - 2) < 0$
b. $(x + 1)(x - 1)(x - 2) > 0$
c. $(x + 1)(x - 1)(x + 2) < 0$
d. $(x + 1)(x - 1)(x + 2) > 0$

In Exercises 38–43 solve each inequality.

38 $y^2 - 2y \geq 35$ **39** $x^2 - 7 < 0$

40 $v^2 - 4v < 1$ **41** $z(z^2 - 6z + 9) > 0$

42 $\frac{6}{m + 5} \geq m$ **43** $\frac{1 + w}{w + 4} \leq \frac{w - 1}{2}$

CALCULATOR USAGE (44–47)

44 Use the graph of $y = -x^2 - 3x + 10$ shown on the calculator window to solve $-x^2 - 3x + 10 > 0$.

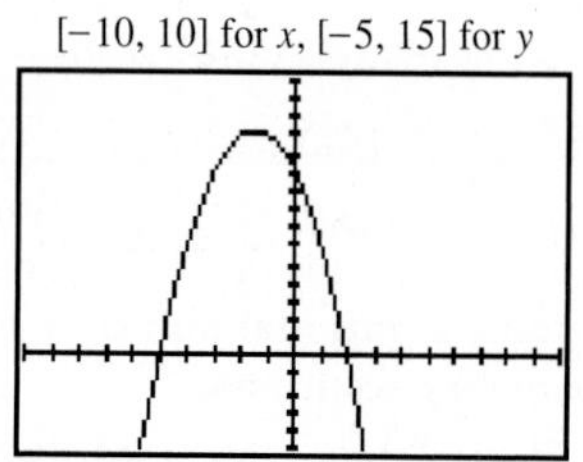

Figure for Exercise 44

In Exercises 45–47 use a graphics calculator to graph each equation and to solve the corresponding inequality.

45 Use the graph of $y = 2x^2 + 5x - 3$ to solve $2x^2 + 5x - 3 < 0$.

46 Use the graph of $y = -2x^2 + 7x + 4$ to solve $-2x^2 + 7x + 4 < 0$.

47 Use the graph of $y = x^3 + 3x^2 - x - 3$ to solve $x^3 + 3x^2 - x - 3 > 0$.

48 Determine the values of x for which $\sqrt{10x^2 - x - 3}$ is a real number.

49 Determine the value of k for which the solutions of $x^2 + kx + 1 = 0$ are real numbers.

Numeric Word Problems (50 and 51)

50 Find two numbers whose difference is 7 and whose product is -8.

51 One number is 4 more than another number. Find the smaller number if the sum of their reciprocals is 1.

52 Dimensions of a Right Triangle The length of the hypotenuse of the right triangle shown in the figure is 2 centimeters more than the length of the longer leg. If the longer leg is 7 centimeters longer than the shorter leg, determine the length of each side.

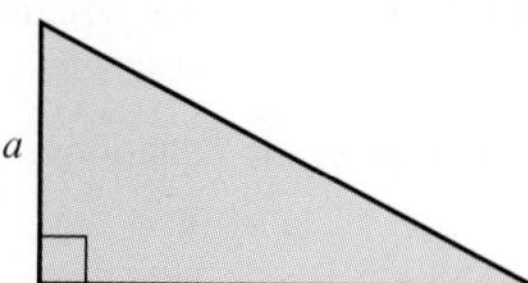

Figure for Exercise 52

53 Hours Worked Working together, two machine tool operators can complete a job in 8 hours. Working alone, the worker with the older machine would take 5 hours longer than the worker with the newer machine. Approximate to the nearest tenth of an hour the time it would take the worker with the newer machine to do the job working alone.

Interest Rates (54 and 55)

54 An amount of \$7000 was invested in two 1-year certificates of deposit. In 1 year, the first CD earned \$400 in interest and the second earned \$180. If the interest rate on the second CD was 1% higher than that on the first, find each interest rate.

55 The formula for computing the amount A of an investment of principal P at interest rate r for 1 year compounded semiannually is $A = P\left(1 + \frac{r}{2}\right)^2$. Approximately what interest rate is necessary for \$1000 to grow to \$1200 in 1 year if the interest is compounded semiannually?

56 Profit The net income in dollars produced by selling t units of a product is given by $I = -t^2 + 45t - 200$. If $I > 0$, there is a profit. Determine the values of t that will generate a profit.

57 Height of a Baseball The height, h, in feet of a baseball t seconds after being hit by a batter is given by $h = -16t^2 + 80t + 3$. During what time interval after the baseball is hit will its height exceed 67 feet? (See the figure.)

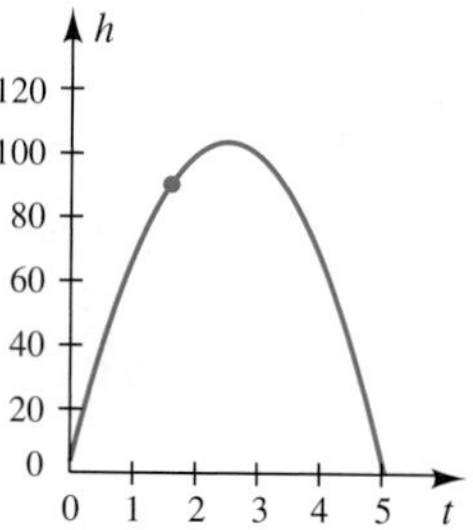

Figure for Exercise 57

58 Construct a quadratic equation in x that has the given solutions.

a. $-\frac{3}{4}$ and 5

b. $\frac{1+\sqrt{3}}{2}$ and $\frac{1-\sqrt{3}}{2}$

c. $2-5i$ and $2+5i$

59 Construct a cubic (third-degree) equation in x that has solutions 0, 4, and 7.

60 Error Analysis Examine the student's work shown below, and correct all of the errors that you find.

$$(x-2)(2x+3) = 2x-6$$
$$2x^2 - x - 6 = 2x - 6$$
$$2x^2 - x = 2x$$
$$2x - 1 = 2$$
$$2x = 3$$
$$x = \frac{3}{2}$$

Mastery Test for Chapter 8

Exercise numbers correspond to Section Objective numbers.

[8-1] 1 Solve each of these quadratic equations by extraction of roots.

a. $v^2 = 196$ **b.** $w^2 = -25$

c. $(2x-3)^2 = 49$

[8-1] 2 Solve each of these quadratic equations by completing the square.

a. $m^2 + 4m = 5$ **b.** $y^2 - 2y - 1 = 0$

c. $-4v^2 - 8v = 3$

[8-2] 3 Solve each of these quadratic equations using the quadratic formula.

a. $6x^2 - 19x = -10$ **b.** $w^2 + 2w = 4$

c. $3v^2 + 1 = 2v$

[8-2] 4 Use the discriminant to determine the nature of the solutions of each of these quadratic equations. Identify the solutions as two distinct rational solutions, two distinct irrational solutions, a rational solution of multiplicity two, or two imaginary solutions.

a. $5x^2 + 5x + 1 = 0$ **b.** $7y^2 = 84y - 252$

c. $-3w^2 = 2w + 1$

[8-3] 5 Solve each of these equations of quadratic form.

a. $x^4 - 34x^2 + 225 = 0$ **b.** $y - 8\sqrt{y} = 9$

c. $\left(\frac{v}{v-2}\right)^2 + \left(\frac{v}{v-2}\right) = 2$

[8-3] 6 Solve each of these equations.

a. $\sqrt{x+42} + x = 0$ **b.** $\sqrt{x+4} = x + 1$

c. $\frac{6}{w-1} + 1 = \frac{5}{w-3}$

[8-4] 7 Solve each of these problems.

a. The product of 1 more than a number and 1 less than a number is 1. Find this number.

b. The hypotenuse of the right triangle shown in the figure is 3 centimeters more than twice the length of the shortest side. The longer leg is 6 centimeters longer than the shortest side. Determine the length of each side of this triangle to the nearest tenth of a centimeter.

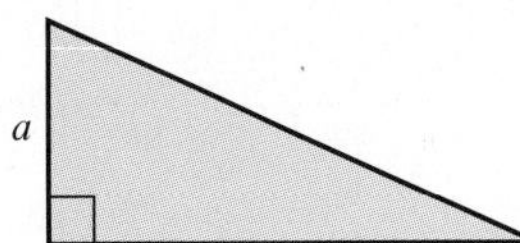

Figure for Exercise 7b

c. Two hoses together can fill a tank in 30 hours. The larger hose alone fills the tank in 32 hours less time than the smaller hose does. How many hours does it take the larger hose to fill the tank?

[8-5] 8 Solve each of these inequalities.

a. $x^2 - 3x > 10$ **b.** $3z^2 \leq -5z - 1$

c. $\frac{x^2 - 3x - 10}{x-8} \geq 0$ **d.** $\frac{5}{v-2} < \frac{3}{v-4}$

Chapter 9

INTRODUCTION TO CONIC SECTIONS

AN APPLICATION

Time Out for Calculating?

Many professional athletes and coaches now use sophisticated mathematics, implemented on computers, to gain a competitive edge over their opponents. Baseball managers use computers with statistics programs to determine which pitches batters are poorest at hitting. Marathon runners use computer-based motion analysis to fine tune their strides for maximum efficiency. And in mid-race yacht captains use laptop computers to compute optimal sail configurations, based on changing weather conditions.

But not all mathematical applications in sports require the use of computers. The motion of a ball in flight, such as a basketball shot toward a hoop or a football kicked toward a goalpost, can be described fairly well using a quadratic equation. The same sort of equation can describe the motion of a javelin or shot put. Analyzing such an equation can yield useful information. Suppose, for example, that you are a track-and-field coach looking at films of one of your shot-putters at practice; you are watching for the angle at which the shot put leaves the thrower's hand. Using techniques from calculus, it is possible to derive the angle that gives maximum distance. Intuitively, what do you think that angle would be?

The actual path of a thrown or kicked object can be illustrated by the graph of its quadratic function. The graph provides a means of analyzing certain aspects of the flight of the object, including the hang time of a punted football, given its initial upward velocity.

CHAPTER 9 OBJECTIVES

1 **Calculate the distance between two points (Section 9-1).**
2 **Determine the midpoint between two points (Section 9-1).**
3 **Graph a parabola (Section 9-2).**
4 **Determine the maximum or the minimum of a quadratic function (Section 9-2).**
5 **Graph a circle and an ellipse (Section 9-3).**
6 **Graph a hyperbola (Section 9-4).**
7 **Identify the type of conic section from its equation (Section 9-4).**

René Descartes developed analytic geometry in 1619 (see Section 2-1). Prior to Descartes, algebra was concerned with numbers and calculations, and geometry was concerned with figures and shapes. By associating points in the plane with ordered pairs of numbers, Descartes merged the power of these areas into analytic geometry. In this chapter we will use algebra to calculate both the distance between two points and the midpoint between two points. The distance and midpoint formulas are important algebraic tools for analyzing geometric shapes.

Some of the geometric shapes that we will examine in this chapter are parabolas, circles, ellipses, and hyperbolas. These shapes are referred to collectively as *conic sections,* because each of these figures can be obtained by intersecting a plane and a cone.

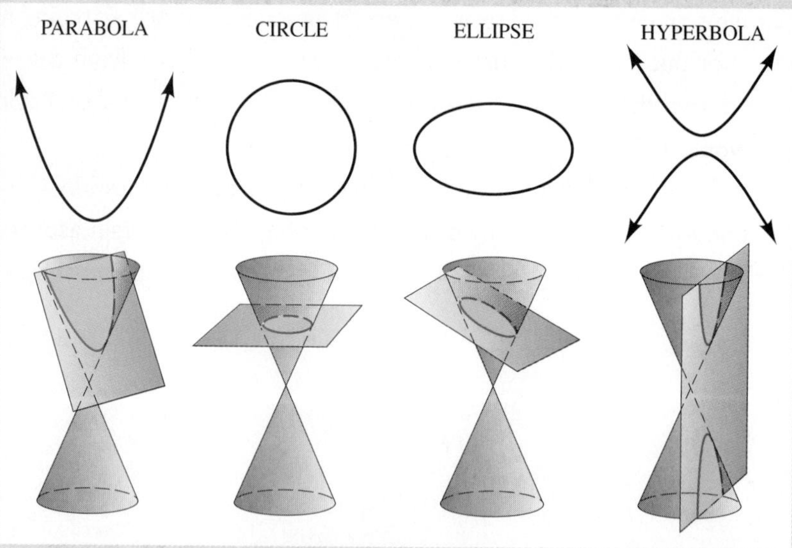

9-1 The Distance and Midpoint Formulas

SECTION OBJECTIVES

1 **Calculate the distance between two points.**
2 **Determine the midpoint between two points.**

An important measurement along a straight line is the distance between two points. To develop the formula for the distance between two points, we will first consider the horizontal and vertical changes between these points and then use the Pythagorean Theorem to compute the distance along the line between the points. The distance between points P and Q is denoted by $\overline{PQ}$.

Special Case
Calculate the distance between $P(2, 2)$ and $Q(5, 6)$.

General Case
Calculate the distance between $P(x_1, y_1)$ and $Q(x_2, y_2)$.

Step 1 Find the horizontal change from P to Q.

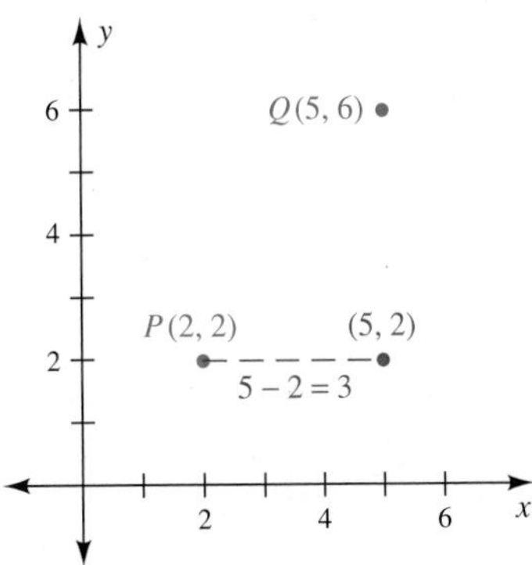

Horizontal distance = 5 − 2 = 3

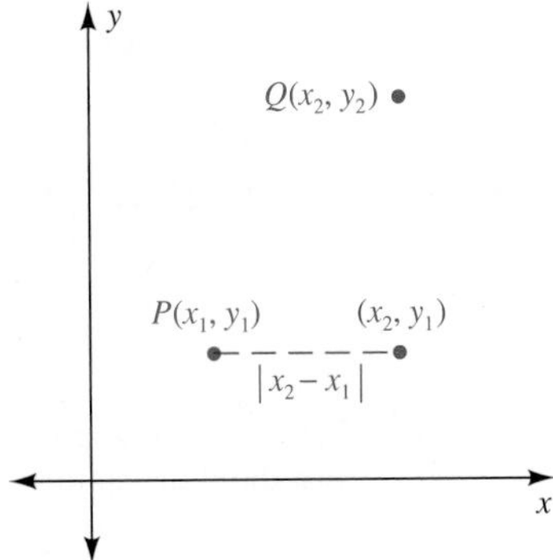

Horizontal distance = $|x_2 - x_1|$

Step 2 Find the vertical change from P to Q.

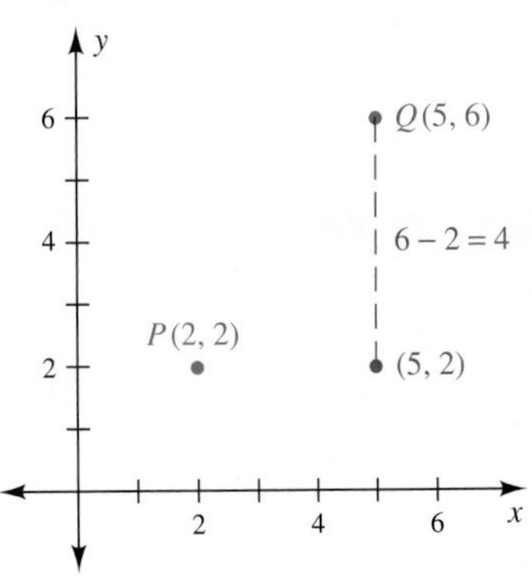

Vertical distance = 6 − 2 = 4

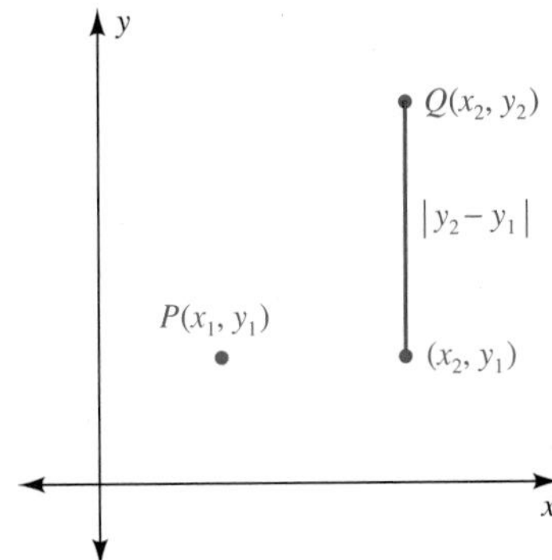

Vertical distance = $|y_2 - y_1|$

Step 3 Use the Pythagorean Theorem to find the length of hypotenuse PQ.

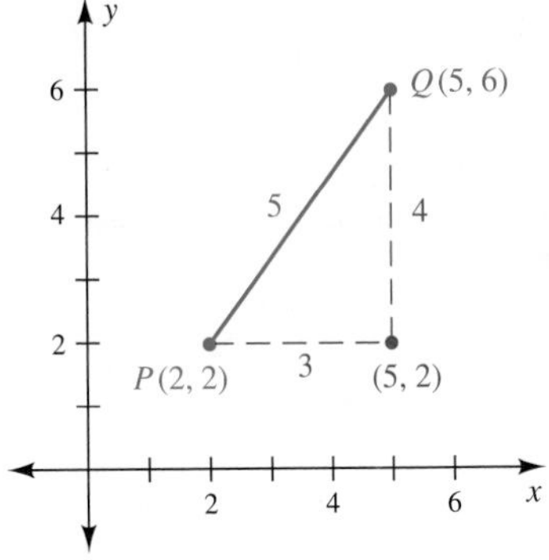

$$d = \sqrt{3^2 + 4^2}$$
$$= \sqrt{9 + 16}$$
$$= \sqrt{25} = 5$$

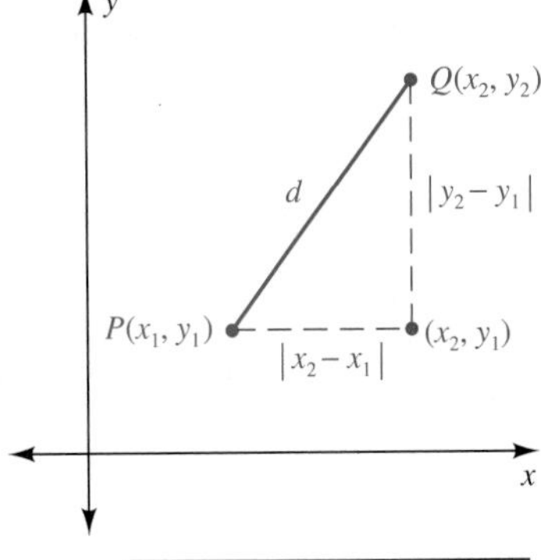

$$d = \sqrt{|x_2 - x_1|^2 + |y_2 - y_1|^2}$$
$$d = \sqrt{(x_2 - x_1)^2 + (y_2 - y_1)^2}$$

DISTANCE FORMULA

The distance d from (x_1, y_1) to (x_2, y_2) is given by

$$d = \sqrt{(x_2 - x_1)^2 + (y_2 - y_1)^2}$$

Absolute-value notation is not needed in the distance formula because the squares are always nonnegative. The formula is applicable in all cases, even if P and Q are on the same horizontal or vertical line.

EXAMPLE 1 Calculating the Distance Between Two Points

Calculate the distance between $(-3, 1)$ and $(5, -1)$.

SOLUTION $d = \sqrt{(x_2 - x_1)^2 + (y_2 - y_1)^2}$

$= \sqrt{[5 - (-3)]^2 + (-1 - 1)^2}$ Substitute the given values into the distance formula.

$= \sqrt{8^2 + (-2)^2}$

$= \sqrt{64 + 4}$

$= \sqrt{68}$

$= \sqrt{4}\sqrt{17}$

$d = 2\sqrt{17}$

The graph is shown in the figure.

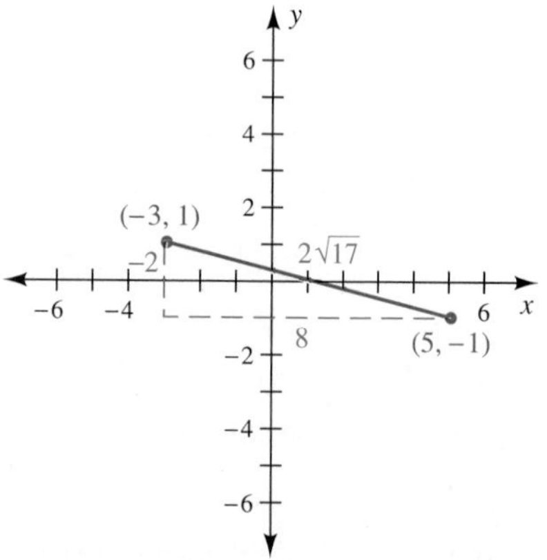

SELF-CHECK

Calculate the distance between $(2, -3)$ and $(-13, 5)$.

EXAMPLE 2 An Application of the Distance Formula

Determine whether the vertices $A(-4, 3)$, $B(5, 0)$, and $C(2, -3)$ form a right triangle.

SOLUTION ABC is a right triangle if $a^2 + b^2 = c^2$.

$a = \overline{BC} = \sqrt{(2 - 5)^2 + (-3 - 0)^2} = \sqrt{18}$

$b = \overline{AC} = \sqrt{[2 - (-4)]^2 + (-3 - 3)^2} = \sqrt{72}$

$c = \overline{AB} = \sqrt{[5 - (-4)]^2 + (0 - 3)^2} = \sqrt{90}$

Substitute the given values into the distance formula in order to calculate the length of each side.

The graph is shown in the figure.

Now check to see whether $a^2 + b^2 = c^2$.

$$(\sqrt{18})^2 + (\sqrt{72})^2 \stackrel{?}{=} (\sqrt{90})^2$$

$$18 + 72 \stackrel{?}{=} 90$$

$$90 = 90 \text{ checks.}$$

Answer ABC is a right triangle, with angle C the 90° angle and c the length of the hypotenuse.

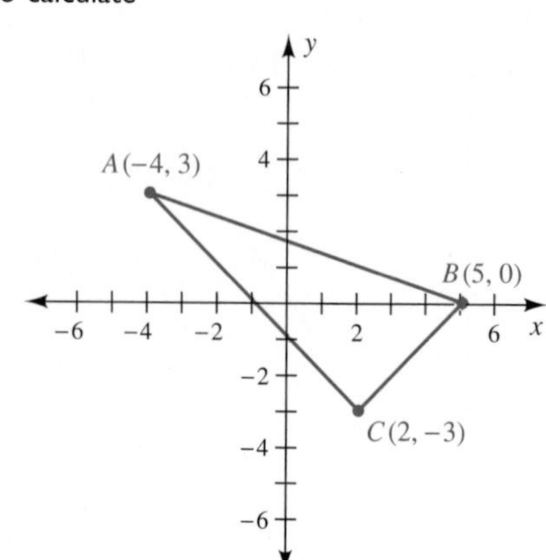

SELF-CHECK ANSWER 

17

EXAMPLE 3 An Application of the Distance Formula

Find all points with an x-coordinate of 6 that are 5 units from (3, 3).

SOLUTION Identify the desired point(s) by $(6, y)$.

$$d = \sqrt{(x_2 - x_1)^2 + (y_2 - y_1)^2}$$

$$5 = \sqrt{(6 - 3)^2 + (y - 3)^2}$$

$$5 = \sqrt{9 + y^2 - 6y + 9}$$

Substitute the given values into the distance formula. Then simplify the radicand.

$$5 = \sqrt{y^2 - 6y + 18}$$

Solve this radical equation for y.

$$25 = y^2 - 6y + 18$$

Square both sides of the equation.

$$0 = y^2 - 6y - 7$$

Subtract 25 from both sides.

$$(y + 1)(y - 7) = 0$$

Factor the trinomial.

$$y + 1 = 0 \quad \text{or} \quad y - 7 = 0$$

Set each factor equal to 0.

$$y = -1 \quad \text{or} \quad y = 7$$

Both values check.

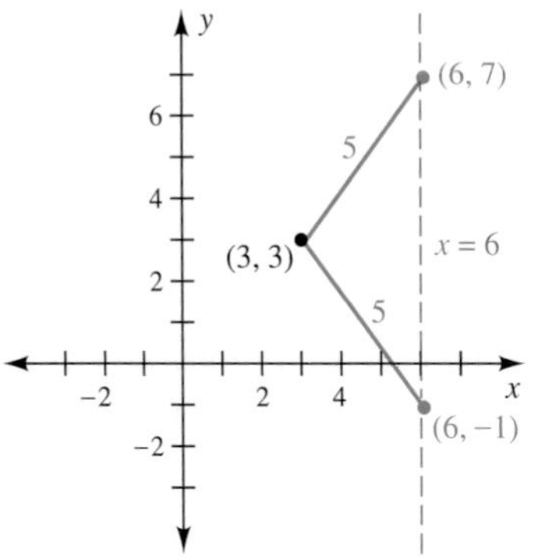

The graph is shown in the figure.

Answer The points (6, −1) and (6, 7) are both 5 units from (3, 3). ▼

The last formula that we will examine in this section is the midpoint formula. A student's average grade for two tests—the score midway between the two grades—is found by dividing the sum of the two scores by 2. Likewise, the midpoint of the line segment from P to Q can be found by averaging the coordinates of these points. The following midpoint formula can be established by using similar triangles.

MIDPOINT FORMULA

The midpoint (x, y) between $P(x_1, y_1)$ and $Q(x_2, y_2)$ is

$$(x, y) = \left(\frac{x_1 + x_2}{2}, \frac{y_1 + y_2}{2}\right)$$

EXAMPLE 4 Calculating the Midpoint Between Two Points

Determine the midpoint between (−8, −3) and (4, 1).

SOLUTION

$$(x, y) = \left(\frac{x_1 + x_2}{2}, \frac{y_1 + y_2}{2}\right)$$

$$(x, y) = \left(\frac{-8 + 4}{2}, \frac{-3 + 1}{2}\right)$$

Substitute the given values into the midpoint formula.

$$= \left(\frac{-4}{2}, \frac{-2}{2}\right)$$

$$(x, y) = (-2, -1)$$

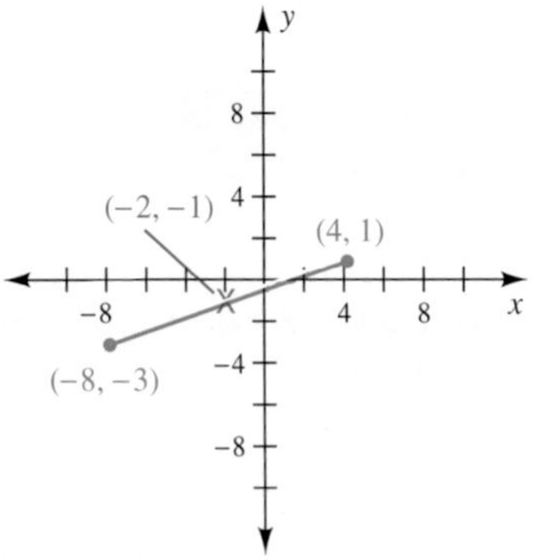

The graph is shown in the figure.

Answer The midpoint is (−2, −1). ▼

EXAMPLE 5 Determining the Equation of a Perpendicular Bisector

Determine the equation of the perpendicular bisector of the line segment connecting (4, −2) and (2, 6).

SOLUTION A bisector goes through the midpoint between (4, −2) and (2, 6).

$$(x, y) = \left(\frac{x_1 + x_2}{2}, \frac{y_1 + y_2}{2}\right)$$

$$(x, y) = \left(\frac{4 + 2}{2}, \frac{-2 + 6}{2}\right)$$ Substitute the given points into the midpoint formula.

$(x, y) = (3, 2)$ is the midpoint.

Perpendicular lines have slopes that are negative reciprocals.

$$m = \frac{y_2 - y_1}{x_2 - x_1}$$

$$m = \frac{6 - (-2)}{2 - 4}$$ Substitute the given points into the slope formula.

$$m = \frac{8}{-2}$$

$m = -4$ is the slope of the line segment.

$m = \dfrac{1}{4}$ is the slope of the perpendicular bisector.

Use the point-slope form of a line.

$$y - y_1 = m(x - x_1)$$

$$y - 2 = \frac{1}{4}(x - 3)$$ Substitute the midpoint and the slope into the point-slope formula.

$$4y - 8 = x - 3$$

$$0 = x - 4y + 5$$

$$x - 4y + 5 = 0$$ Write the equation in general form.

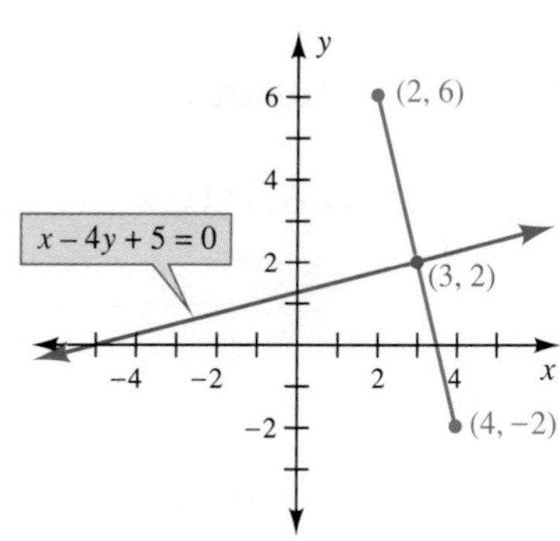

The graph is shown in the figure.

Answer The equation of the perpendicular bisector is $x - 4y + 5 = 0$.

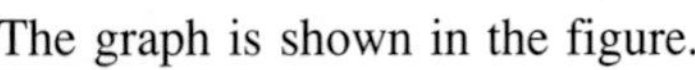

Exercises 9-1

A

In Exercises 1–8 calculate the distance between each pair of points.

1 (−2, −6) and (3, 6)

2 (−3, 2) and (1, −1)

3 (2, −7) and (−6, 8)

4 (0, −20) and (−9, 20)

5 $\left(-\frac{1}{2}, \frac{2}{3}\right)$ and $\left(\frac{1}{2}, -\frac{1}{3}\right)$

6 $\left(\frac{4}{5}, \frac{3}{2}\right)$ and $\left(-\frac{1}{5}, \frac{1}{2}\right)$

7 $(0, 0)$ and $(-\sqrt{2}, \sqrt{7})$

8 $(\sqrt{11}, -\sqrt{14})$ and $(0, 0)$

In Exercises 9–14 determine the midpoint between each pair of points.

9 (5, 11) and (7, 23)

10 (7, 2) and (3, 8)

11 (−4, −9) and (−16, 5)

12 (−3, −7) and (−7, 15)

13 $\left(\frac{1}{2}, \frac{1}{5}\right)$ and $\left(\frac{1}{3}, -\frac{1}{2}\right)$

14 $\left(-\frac{1}{4}, \frac{1}{3}\right)$ and $\left(\frac{1}{6}, \frac{2}{7}\right)$

In Exercises 15–19 use the distance formula to calculate the perimeter of each figure described.

15 The triangle formed by connecting the points (1, 2), (4, 6), and (4, 2)

16 The triangle formed by connecting the points (−2, −2), (22, −2), and (22, 5)

17 A square with one side connecting (0, 1) and (3, 5)

18 A rhombus with one side connecting (−1, 2) and (1, 4) (A rhombus is a parallelogram with all sides of equal length.)

19 An equilateral triangle with one side connecting (2, −5) and (−3, 7)

20 Calculate the length of the line segment connecting the midpoints D and E of the triangle in the figure.

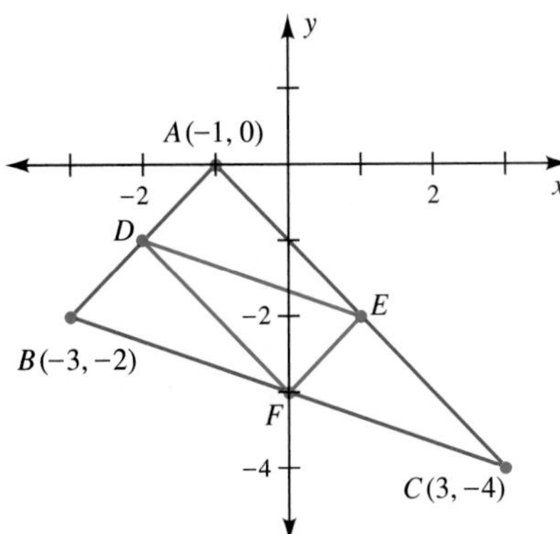

Figure for Exercises 20 and 21

21 Calculate the length of the line segment connecting the midpoints D and F of the triangle in the figure.

In Exercises 22–25 use the distance formula and the Pythagorean Theorem to determine whether the given vertices form a right triangle.

22 (−3, −1), (4, −4), and (−1, 1)

23 (−6, −1), (2, −1), and (0, 1)

24 (−5, −1), (2, −2), and (4, 4)

25 (3, −3), (−2, −1), and (5, 2)

B

26 Measuring a Buried Pipe A pipe-cleaning firm contracted to clean a pipe buried in a lake. Access points to the pipe are at points A and B on the edge of the lake, as shown in the figure. The contractor placed a stake as a reference point and then measured the coordinates in meters from this reference point to A and B. The coordinates of A and B are, respectively, (3.0, 5.2) and (37.8, 29.6). Approximate to the nearest tenth of a meter the distance between A and B.

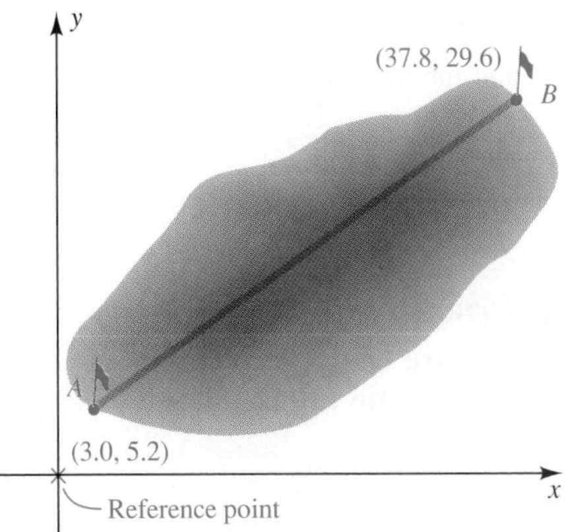

Figure for Exercise 26

27 Measuring an Engine Block A machinist is measuring a metal block that is part of an automobile engine. To find the distance from the center of hole A to the center of hole B, the machinist determines the coordinates shown on the drawing with respect to a reference point at the lower left corner of the metal block. Calculate the distance from the center of A to the center of B.

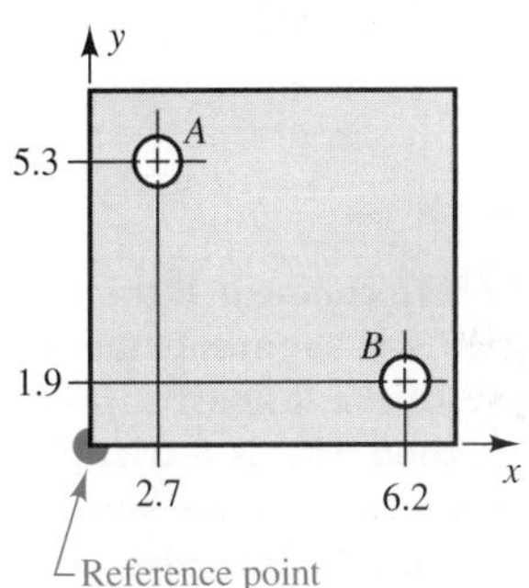

Figure for Exercise 27

28 Determine the equation of the perpendicular bisector of the line segment connecting (−4, 4) and (2, 6).

29 Determine the equation of the perpendicular bisector of the line segment connecting (−5, −3) and (3, 1).

30 Write the equation of the line through (2, 1) and through the midpoint between (7, 3) and (5, −1).

31 Write the equation of the line through (−5, −2) and through the midpoint between (3, 5) and (5, 1).

32 Write the equation of a line parallel to the x-axis and passing through the midpoint between $\left(\frac{1}{2}, \frac{1}{3}\right)$ and $\left(-\frac{1}{2}, \frac{5}{3}\right)$.

33 Write the equation of a line parallel to the y-axis and passing through the midpoint between $\left(\frac{1}{2}, \frac{1}{3}\right)$ and $\left(-\frac{1}{2}, \frac{5}{3}\right)$.

34 Write the equation of a line perpendicular to the y-axis and passing through the midpoint between $\left(-\frac{3}{2}, \frac{7}{3}\right)$ and $\left(\frac{1}{2}, -\frac{4}{3}\right)$.

CALCULATOR USAGE (35–40)

In Exercises 35–38 use a calculator to approximate to the nearest hundredth the distance between each pair of points.

35 $(\sqrt{2}, \sqrt{3})$ and $(\sqrt{5}, \sqrt{6})$

36 $(\sqrt{3}, \sqrt{7})$ and $(2\sqrt{5}, -\sqrt{6})$

37 $(3\sqrt{11}, -2\sqrt{7})$ and $(11\sqrt{3}, 7\sqrt{2})$

38 $(\sqrt{13}, \sqrt{17})$ and $(5\sqrt{2}, 2\sqrt{5})$

In Exercises 39–40 use a calculator to approximate to the nearest hundredth the midpoint between each pair of points.

39 $(\sqrt{2}, \sqrt{3})$ and $(\sqrt{5}, \sqrt{6})$

40 $(\sqrt{3}, \sqrt{7})$ and $(2\sqrt{5}, -\sqrt{6})$

C

Points $A, B,$ and C are collinear if the sum of the lengths of the shorter two line segments formed by the points equals the length of the longest segment. If the points are not collinear, then ABC is a triangle and the length of any one side is less than the sum of the lengths of the other two sides. In Exercises 41–44 determine whether the three points are collinear.

41 $(-2, 5)$, $(1, 2)$, and $(6, -3)$

42 $(-2, 4)$, $(1, 3)$, and $(6, 0)$

43 $(-3, -3)$, $(1, -2)$, and $(3, -1)$

44 $(-4, 4)$, $(2, 2)$, and $(5, 1)$

45 Use the distance formula to express the distance from the point (x, y) to the point (h, k).

46 Use the distance formula to express the distance from the point $(2a, 7a)$ to the point $(-a, 3a)$. Assume $a \geq 0$.

47 The points A and B are on the function defined by $f(x) = x^2 - 3x + 2$. If $A = (1, f(1))$ and $B = (5, f(5))$, determine the distance between A and B.

48 The points A and B are on the function defined by $f(x) = |2x - 3|$. If $A = (1, f(1))$ and $B = (5, f(5))$, determine the distance between A and B.

49 Draw all points that are 3 units from the origin.

50 Draw all points that are 2 units from the point $(0, 2)$.

DISCUSSION QUESTIONS (51 AND 52)

51 Describe in your own words the logic of the development of the distance formula.

52 Before Descartes, latitude and longitude were used to identify locations on Earth. Then Descartes introduced the use of x- and y-coordinates to identify points in the plane. Discuss any relationships that you see between these ideas.

CHALLENGE QUESTION

A delivery route is established between towns located at points $A, B, C, D, E,$ and F. The route starts at A and continues through each of the other towns before returning to A. Assume that there is a straight-line road connecting each pair of towns and that all coordinates are given in miles. Determine the shortest possible circuit.

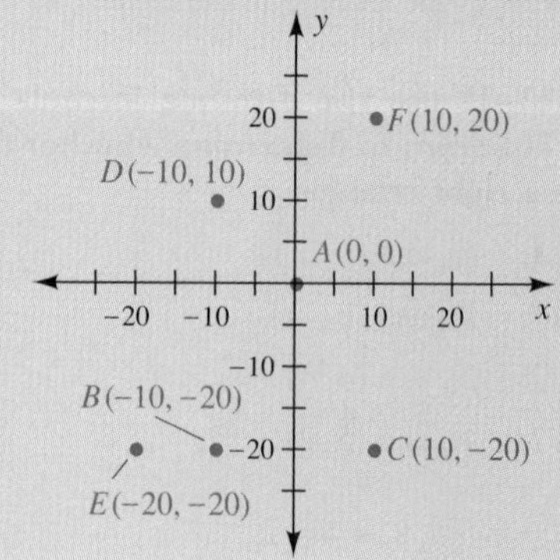

SECTION SKILL REINFORCERS

Evaluate each of these expressions given $f(x) = x^2 - 2x - 5$.

1 $f(3)$ **2** $-f(3)$ **3** $f(-3)$ **4** $f(3 + h)$ **5** $\frac{f(3 + h) - f(3)}{h}$

9-2 Parabolas and Quadratic Functions

SECTION OBJECTIVES

3 Graph a parabola.

4 Determine the maximum or the minimum of a quadratic function.

The equation $y = ax^2 + bx + c$ (or the alternative form $f(x) = ax^2 + bx + c$) is called a **quadratic function** if $a \neq 0$. The graph of a quadratic function of this form is always a parabola that is either concave upward or concave downward. A parabola is a smooth curve of the shape illustrated in Figure 9-1.

Parabolas opening upward

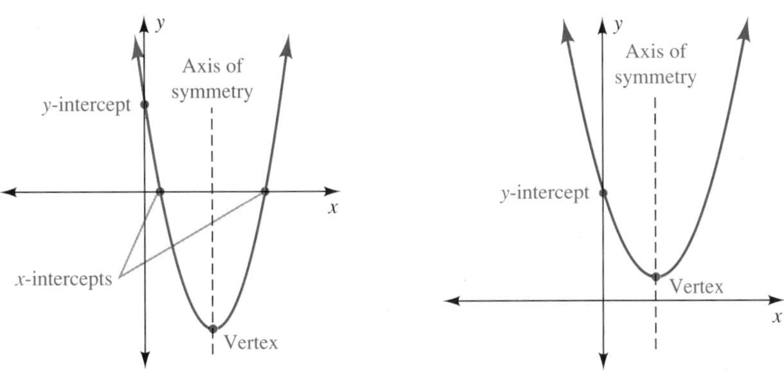

Parabolas opening downward

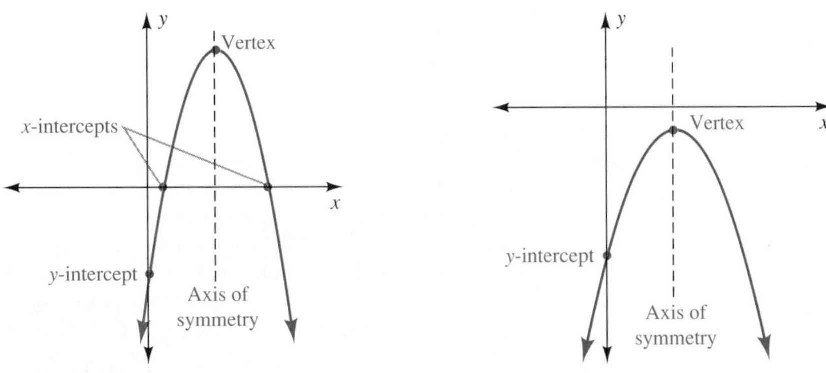

Figure 9-1

The **vertex** of a parabola that opens upward is its lowest point. The vertex of a parabola that opens downward is its highest point. The **axis of symmetry** of a parabola passes through the vertex. The portion of the parabola to the left of the axis of symmetry is a mirror image of the portion to the right of this axis of symmetry.

The parabola is used in the design of many products because of its useful geometric properties. Parabolic reflectors are used for spotlights, on microphones at sporting events, and in satellite receiver dishes. (See Figure 9-2.) The cables

Figure 9-2
Parabolic Satellite Dish

supporting some bridges hang in a parabolic shape, and the path of a thrown ball is parabolic (assuming that air resistance is negligible).

The parabola $y = x^2$ was approximated by line graphs in Section 2-5. These approximations are repeated below for your reference.

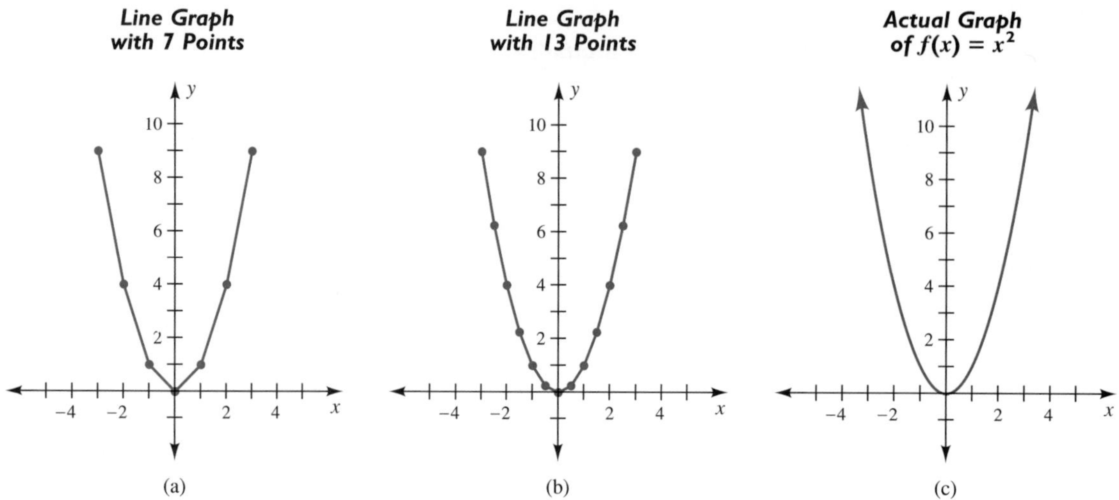

Each parabola from the family of parabolas defined by $f(x) = ax^2 + bx + c$ is similar to the reference parabola $y = x^2$. Our goal is to graph these parabolas quickly by recognizing their shapes rather than by tediously plotting points. We will first consider equations of the form $y = x^2 + k$. As illustrated in Example 1, the graph of $y = x^2 + k$ is identical to that of $y = x^2$, except that it is shifted (or translated) either up or down. By locating the most important point on a parabola—the vertex—we can locate the whole parabola.

EXAMPLE 1 Translating a Parabola Downward Two Units

Graph $y = x^2$ and $y = x^2 - 2$ on the same coordinate system.

SOLUTION

x	x^2	$x^2 - 2$
−3	9	7
−2	4	2
−1	1	−1
0	0	−2
1	1	−1
2	4	2
3	9	7

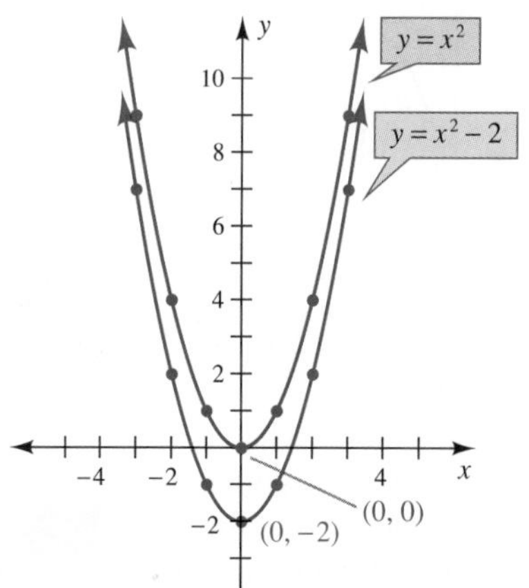

Each point (x, y) on $y = x^2$ is shifted down 2 units to obtain a corresponding point on $y = x^2 - 2$. In particular, the vertex shifts from (0, 0) on $y = x^2$ to (0, −2) on $y = x^2 - 2$. Both graphs are symmetric about $x = 0$, the y-axis.

In general, the graph of $y = x^2 + k$ is a parabola that is identical to the graph of $y = x^2$ except that it is shifted up k units if k is positive and down $|k|$ units if k is negative. The vertex is $(0, k)$.

EXAMPLE 2 Translating a Parabola Upward One Unit

Graph $y = x^2$ and $y = x^2 + 1$ on the same coordinate system.

SOLUTION

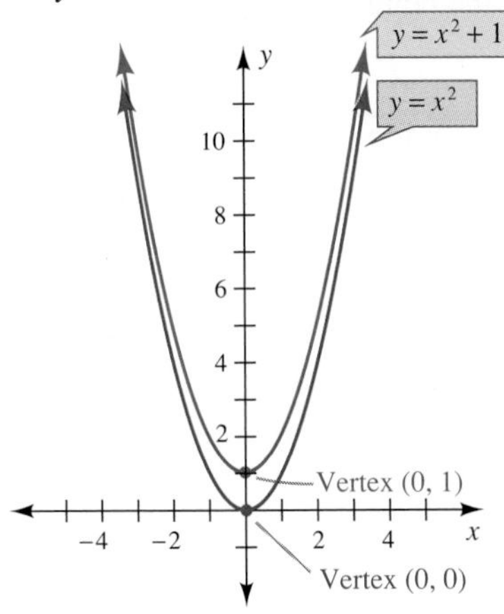

Draw the graph of $y = x^2$. Then shift this graph up 1 unit to form the graph $y = x^2 + 1$. The vertex shifts from (0, 0) to (0, 1). Both graphs are symmetric about $x = 0$, the y-axis.

▼

SELF-CHECK ✓

Use a graphics calculator to graph (or graph by hand) the parabolas $y = x^2 - 3$ and $y = x^2 + 2$.

We will now examine the graphs of $y = x^2$ and $y = (x - 2)^2$ in order to illustrate the general case $y = (x - h)^2$.

EXAMPLE 3 Translating a Parabola Two Units to the Right

Graph $y = x^2$ and $y = (x - 2)^2$ on the same coordinate system.

SOLUTION

x	x^2	$(x - 2)^2$
−3	9	25
−2	4	16
−1	1	9
0	0	4
1	1	1
2	4	0
3	9	1
4	16	4
5	25	9

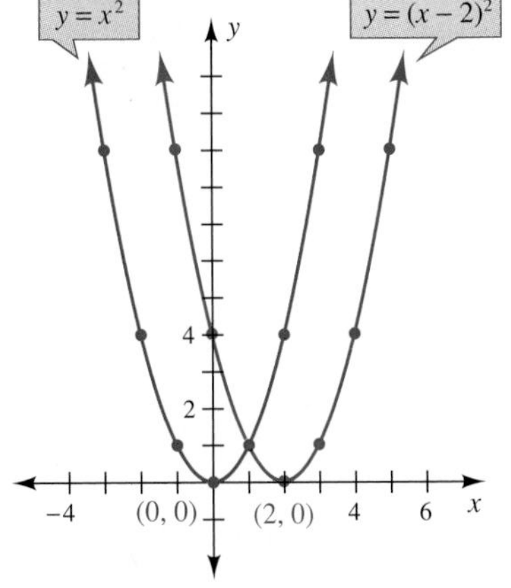

Each point (x, y) on $y = x^2$ is shifted 2 units to the right to obtain a corresponding point on $y = (x - 2)^2$. In particular, the vertex shifts from (0, 0) on $y = x^2$ to (2, 0) on $y = (x - 2)^2$. The axis of symmetry for $y = (x - 2)^2$ is the vertical line $x = 2$.

▼

SELF-CHECK ANSWER ✓

[−5, 5] for x, [−4, 8] for y

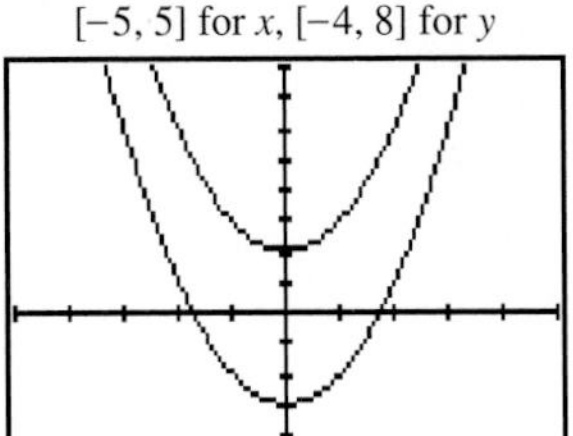

The expression $(x - h)^2$ is always nonnegative since it is the square of a real number. Thus the minimum y value of $y = (x - h)^2$ is $y = 0$, which will only occur when $x = h$. Therefore the vertex of $y = (x - h)^2$ will be at the point with the lowest y value, the point $(h, 0)$.

In general, the graph of $y = (x - h)^2$ is a parabola that is identical to the graph of $y = x^2$ except that it is shifted to the right h units if h is positive and to the left $|h|$ units if h is negative. The vertex is $(h, 0)$.

By combining vertical and horizontal translations, we can shift the parabola $y = x^2$ to obtain the parabola $y = (x - h)^2 + k$.

GRAPHING $y = (x - h)^2 + k$

The graph of $y = (x - h)^2 + k$ is a parabola that is identical to the graph of $y = x^2$ except that it is shifted so that the vertex is (h, k).

EXAMPLE 4 Combining Vertical and Horizontal Translations

Graph $y = x^2$ and $y = (x + 2)^2 - 1$ on the same coordinate system.

SOLUTION

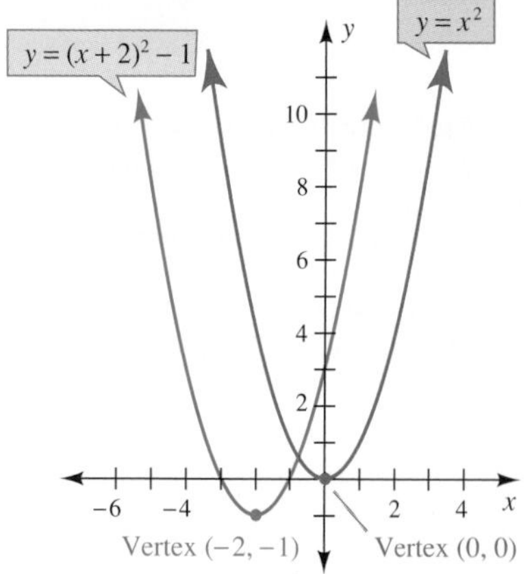

In the form $y = (x - h)^2 + k$, $y = [x - (-2)]^2 - 1$ has $h = -2$ and $k = -1$. First draw the graph of $y = x^2$, then shift this graph to the left 2 units and down 1 unit to form the graph of $y = [x - (-2)]^2 - 1$. The vertex shifts from $(0, 0)$ to $(-2, -1)$.

The next generalization that can be made about the parabola defined by $y = ax^2 + bx + c$ concerns the coefficient a of the x^2 term. If $a > 0$, the parabola opens upward (as in $y = x^2$). If $a < 0$, the parabola opens downward (as in $y = -x^2$). The magnitude of a also determines the "spread," or "span," of the parabola. The parabola $y = ax^2$ is wider than the reference parabola $y = x^2$ if $|a| < 1$ or narrower than $y = x^2$ if $|a| > 1$.

SELF-CHECK 

Determine the vertex of each of these parabolas.

1 $y = x^2 - 5$
2 $y = (x - 5)^2$
3 $y = (x - 3)^2 + 4$

SELF-CHECK ANSWERS 

1 $(0, -5)$ **2** $(5, 0)$ **3** $(3, 4)$

EXAMPLE 5 Graphing a Parabola with a Wider Span than $y = x^2$

Graph $y = x^2$ and $y = \frac{1}{4}x^2$ on the same coordinate system.

SOLUTION

x	x^2	$\frac{1}{4}x^2$
−3	9	2.25
−2	4	1
−1	1	0.25
0	0	0
1	1	0.25
2	4	1
3	9	2.25

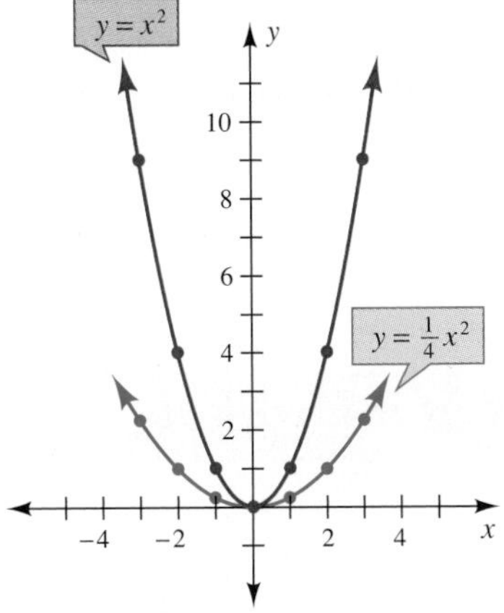

Since $\frac{1}{4}$, the coefficient of x^2, is positive, the parabola opens upward. Since $\left|\frac{1}{4}\right| < 1$, the parabola opens with a wider span than the reference parabola $y = x^2$.

The key information that has been developed in this section is summarized in the following box.

GRAPHING PARABOLAS

The graph of $y = a(x - h)^2 + k$, $a \neq 0$, is a parabola that

1 has its vertex at (h, k);

2 opens upward if $a > 0$ or downward if $a < 0$; and

3 is wider than the parabola $y = x^2$ if $|a| < 1$ or narrower if $|a| > 1$.

EXAMPLE 6 Determining the Vertex of a Parabola

Determine the vertex of each parabola, and determine whether the parabola opens upward or downward.

SOLUTIONS

(a) $y = 2(x + 3)^2 - 4$

The vertex is $(-3, -4)$, and the parabola opens upward. It is narrower than $y = x^2$.

In the form $y = a(x - h)^2 + k$, $y = 2[x - (-3)]^2 - 4$ has $a = 2$, $h = -3$, and $k = -4$.

(b) $y = -\frac{1}{3}(x - 5)^2 + 1$

The vertex is $(5, 1)$, and the parabola opens downward. It is wider than $y = x^2$.

In the form $y = a(x - h)^2 + k$, $a = -\frac{1}{3}$, $h = 5$, and $k = 1$.

If the equation of a parabola is given in the form $y = ax^2 + bx + c$, we can use the process of completing the square (Section 8-1) to put the equation in the form $y = a(x - h)^2 + k$; the vertex can then be determined by inspection, as illustrated in Example 7.

EXAMPLE 7 Using Completing the Square to Determine the Vertex of a Parabola

Determine the vertex of $y = x^2 + 4x - 1$.

SOLUTION

$$\begin{aligned} y &= x^2 + 4x - 1 \\ &= (x^2 + 4x + \boxed{c}\,) - 1 - \boxed{c} \\ &= (x^2 + 4x + \boxed{4}\,) - 1 - \boxed{4} \\ &= (x + 2)^2 - 5 \\ y &= [x - (-2)]^2 - 5 \end{aligned}$$

Adding and then subtracting c does not change the value of the right side of the equation. The value of c needed to complete the square is the square of half the coefficient of x:

$$c = \left(\frac{4}{2}\right)^2 = 2^2 = 4.$$

Answer The vertex is $(-2, -5)$.

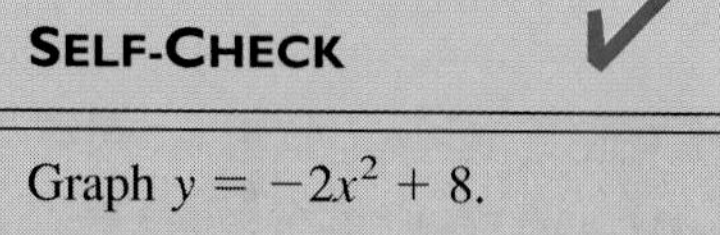

SELF-CHECK

Graph $y = -2x^2 + 8$.

A Geometric Viewpoint: The Vertex of a Parabola

The x-intercepts of a parabola that is defined by $y = ax^2 + bx + c$ are symmetric about the vertical axis of symmetry, which passes through the vertex. This is illustrated by the x-intercepts $(1, 0)$ and $(5, 0)$ of the parabola $y = x^2 - 6x + 5$, which is shown in the figure. These x-intercepts are symmetric about the axis of symmetry $x = 3$, which passes through the vertex $(3, -4)$.

In fact, if a parabola intersects any horizontal line, the two points of intersection will be symmetric about the axis of symmetry. Note that the points of intersection of this parabola with the line $y = 5$ are $(0, 5)$ and $(6, 5)$, which are also symmetric about the axis of symmetry $x = 3$. This fact can be used to obtain the vertex of the parabola. You will be asked to explain the details of the procedure in the Discussion Question at the end of this section.

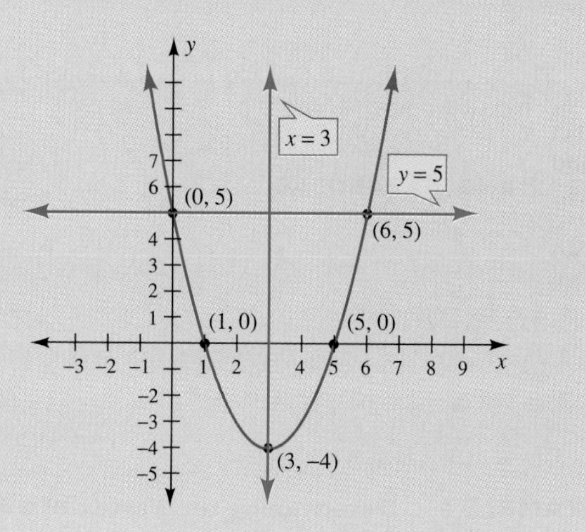

SELF-CHECK ANSWER

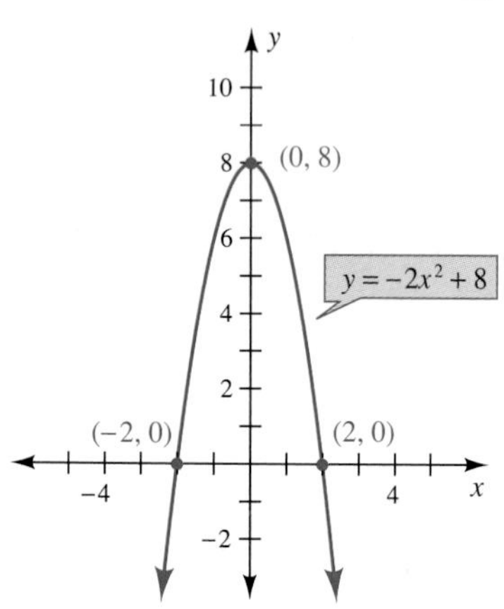

The value of $f(x)$ at the vertex of the parabola defined by the equation $f(x) = ax^2 + bx + c$ is either the maximum or the minimum of this quadratic function. We can use this fact to solve some maximum and minimum problems.

EXAMPLE 8 Maximum Area Enclosed by a Fence

A homeowner plans to enclose a rectangular playing area for the children. One side of the playing area will be formed by the wall of an existing building, and the other three sides will be formed by the fencing. What is the maximum area that can be enclosed with 16 meters of fencing?

SOLUTION Draw a sketch like the one in the figure, and let

w = Width of the rectangular area in meters

$16 - 2w$ = Length of the rectangular area in meters

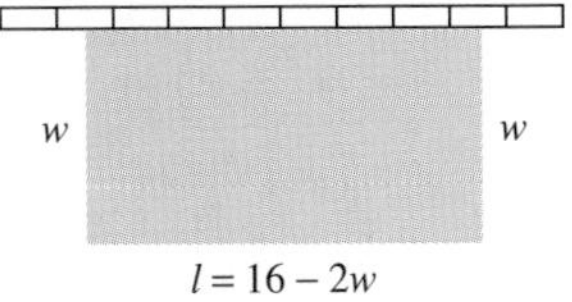

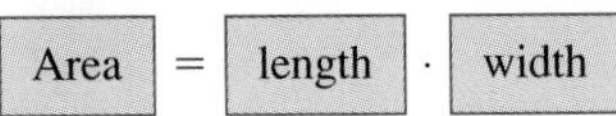

The **word equation** is based on the formula for the area of a rectangle.

$A = l \cdot w$

$A = (16 - 2w)w$ Substitute the width and the length into the word equation.

Since the area is a function of the width, we denote this relationship in functional notation as

$A(w) = w(16 - 2w)$ This is the equation of a parabola that opens downward, so the vertex will produce the maximum possible area.

$= -2w^2 + 16w$

$= -2(w^2 - 8w)$

$= -2(w^2 - 8w + c) + 2c$ Set up the equation to complete the square. Subtracting $2c$ and then adding $2c$ does not change the value of the right side of the equation.

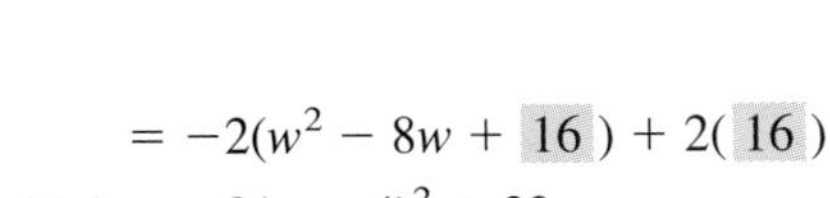

$= -2(w^2 - 8w + 16) + 2(16)$ The value of c needed to complete the square is the square of half the coefficient of x:

$A(w) = -2(w - 4)^2 + 32$

$$c = \left(\frac{-8}{2}\right)^2 = (-4)^2 = 16.$$

Vertex: (4, 32) The maximum value of A is the A-coordinate of (w, A) at the vertex. (See the figure.)

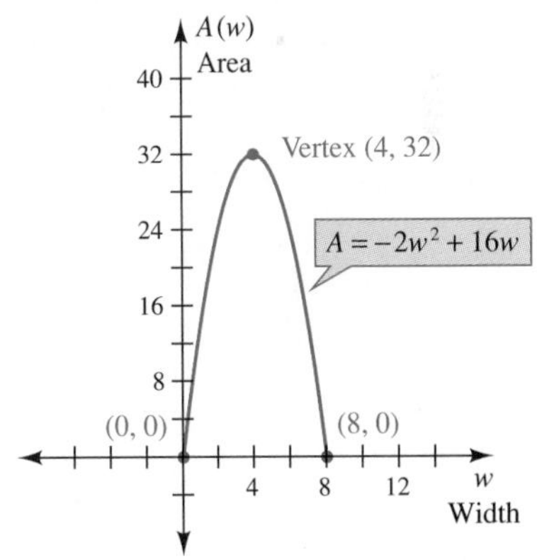

Answer The maximum area is 32 m^2 (attained when the width is 4 m). ▼

Equations of the form $x = ay^2 + by + c$, $a \neq 0$, define parabolas that open either to the left or to the right (instead of up or down, as in $y = ax^2 + bx + c$). The key information used to graph these parabolas can be obtained by the same method used to graph $y = ax^2 + bx + c$, except that the roles of x and y are interchanged.

SELF-CHECK 

Determine the minimum value of $f(x) = 7(x - 3)^2 + 8$ and the value of x for which $f(x)$ is a minimum.

SELF-CHECK ANSWER 

The minimum of 8 occurs when $x = 3$.

EXAMPLE 9 Graphing a Parabola That Opens to the Right

Graph $x = y^2$ by first selecting some arbitrary y values.

SOLUTION

$x = y^2$	y
4	−2
1	−1
0	0
1	1
4	2

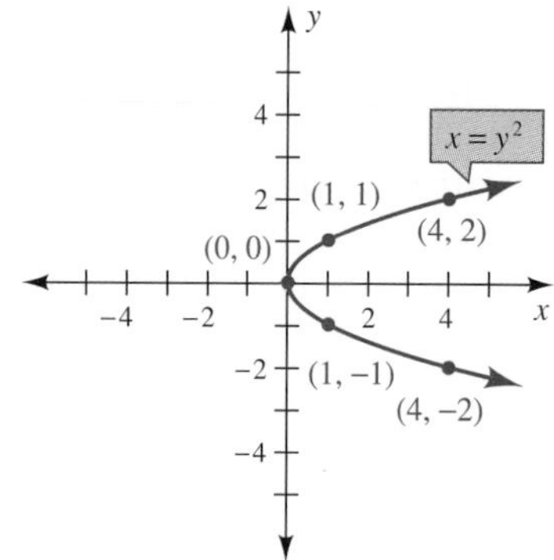

In the form $x = ay^2 + by + c$, this equation has $a = 1$, $b = 0$, and $c = 0$.

Use the vertical line test to verify that the parabola in this example is not a function of x.

Exercises 9-2

A

In Exercises 1–6 determine the vertex of each parabola and whether the parabola opens upward or downward.

1 a. $y = x^2 + 7$ **b.** $y = x^2 - 7$ **c.** $y = (x + 7)^2$ **d.** $y = (x - 7)^2$

2 a. $y = x^2 + 8$ **b.** $y = x^2 - 8$ **c.** $y = (x + 8)^2$ **d.** $y = (x - 8)^2$

3 a. $y = (x - 6)^2 - 1$ **b.** $y = (x + 6)^2 + 1$ **c.** $y = 6x^2 - 1$ **d.** $y = -6x^2 + 1$

4 a. $y = (x - 9)^2 + 2$ **b.** $y = (x + 9)^2 - 2$ **c.** $y = 2x^2 + 9$ **d.** $y = -9x^2 + 2$

5 a. $y = 2(x - 3)^2 + 4$ **b.** $y = -2(x + 3)^2 + 4$

6 a. $y = 4(x + 2)^2 - 3$ **b.** $y = -4(x + 2)^2 + 3$

In Exercises 7–12 graph $y = x^2$ and the parabola(s) defined by the given equation(s) on the same coordinate system. Label the vertex of each parabola.

7 a. $y = x^2 - 3$ **b.** $y = x^2 + 1$

8 a. $y = x^2 - 4$ **b.** $y = x^2 + 2$

9 a. $y = (x - 3)^2$ **b.** $y = (x + 1)^2$

10 a. $y = (x - 4)^2$ **b.** $y = (x + 2)^2$

11 $y = (x - 2)^2 - 3$ **12** $y = (x + 1)^2 - 4$

In Exercises 13 and 14 graph the parabolas defined by these equations on the same coordinate system.

13 a. $y = -2x^2$ **b.** $y = \frac{1}{2}x^2$

14 a. $y = -\frac{1}{2}x^2$ **b.** $y = 2x^2$

In Exercises 15 and 16 determine the maximum value of each function and the value of x for which $f(x)$ is a maximum.

15 $f(x) = -4(x + 7)^2 + 11$ **16** $f(x) = -7(x - 11)^2 + 4$

In Exercises 17 and 18 determine the minimum value of each function and the value of x for which $f(x)$ is a minimum.

17 $f(x) = 3(x - 8)^2 + 5$ **18** $f(x) = 8(x - 5)^2 + 3$

In Exercises 19–24 determine the vertex of each parabola by completing the square, and then graph each parabola.

19 $y = x^2 - 6x + 10$ **20** $y = x^2 + 6x + 8$

21 $y = x^2 + 2x - 3$ **22** $y = x^2 - 2x + 2$

23 $y = -x^2 - 2x$ **24** $y = -x^2 + 2x + 1$

In Exercises 25–28 complete the square to determine the minimum value of each function and the value of x for which $f(x)$ is a minimum.

25 $y = x^2 - 10x + 40$

26 $y = x^2 - 8x + 18$

27 $y = x^2 + 12x + 40$

28 $y = x^2 + 6x + 18$

B

In Exercises 29–32 graph the parabolas satisfying the given conditions.

29 Concave upward; vertex: $(4, -1)$; y-intercept: $(0, 15)$; x-intercepts: $(3, 0)$, $(5, 0)$; other points: $(1, 8)$

30 Concave downward; vertex: $(0, 4)$; y-intercept: $(0, 4)$; x-intercepts: $(2, 0)$, $(-2, 0)$; other points: $(1, 3)$, $(-1, 3)$

31 Concave downward; vertex: $(4, 1)$; y-intercept: $(0, -15)$; x-intercepts: $(3, 0)$, $(5, 0)$; other points: $(1, -8)$

32 Concave upward; vertex: $(2, -4)$; y-intercept: $(0, 0)$; x-intercepts: $(0, 0)$, $(4, 0)$; other points: $(-1, 5)$, $(1, -3)$

The graph of $y = ax^2 + bx + c$ has a y-intercept of $(0, c)$ that can be found by letting $x = 0$. The x-intercepts can be found by letting $y = 0$ and then solving $ax^2 + bx + c = 0$. In Exercises 33–36 determine the y-intercept and the x-intercepts of each parabola.

33 $y = x^2 - 5x + 6$

34 $y = x^2 - 9$

35 $y = x^2 - 7x - 8$

36 $y = -x^2 + x + 12$

37 Path of a Baseball The path of a baseball is given by $y = -16x^2 + 96x$, where y represents the height of the ball in feet and x is the number of seconds that have elapsed since the ball was released. Determine the highest point that the ball reaches.

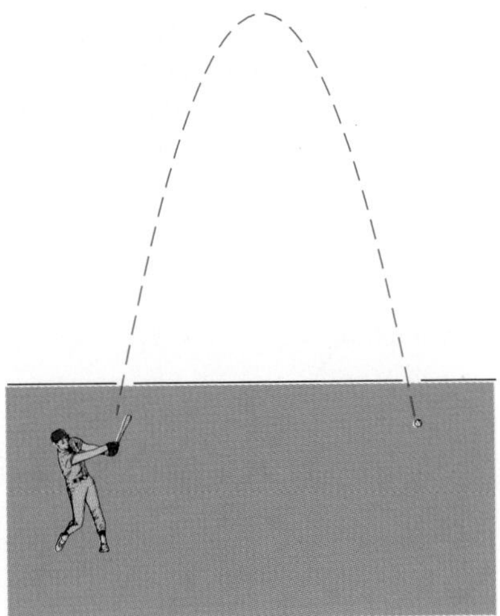

Figure for Exercise 37

38 Path of a Softball The path of a certain softball is given by $y = -16x^2 + 48x + 8$, where y represents the height of the ball in feet and x is the number of seconds that have elapsed since the ball was released. Determine the highest point that the ball reaches and the time it takes to reach this height.

39 Fencing an Area Determine the maximum rectangular area that can be enclosed by 40 meters of fencing.

40 Maximum Product

a. Find two numbers with a sum of 10 whose product is a maximum.

b. What is this maximum product?

41 Minimum Product

a. Find two numbers with a difference of 6 whose product is a minimum.

b. What is this minimum product?

42 Hiring Employees The manager of a small print shop estimates the daily profit in dollars from hiring x employees to be $P(x) = -8x^2 + 80x - 60$.

a. How many employees should be hired to maximize profit?

b. What is this maximum profit?

43 Grazing Cattle A farmer estimates the profit in dollars from grazing x cattle on a pasture to be $P(x) = -6.4x^2 + 320x - 3000$.

a. How many cattle should be grazed on this pasture to maximize profit?

b. What is this maximum profit?

44 Production of Computer Disks The average cost of producing x units of computer disks is given by $C(x) = x^2 - 40x + 576$.

a. How many disks should be produced to minimize the average cost?

b. What is the minimum average cost?

C

CALCULATOR USAGE (45–54)

In Exercises 45 and 46 approximate the domain and the range of the functions shown on the graphics calculator window.

45 $[-5, 5]$ for x, $[-7, 5]$ for y

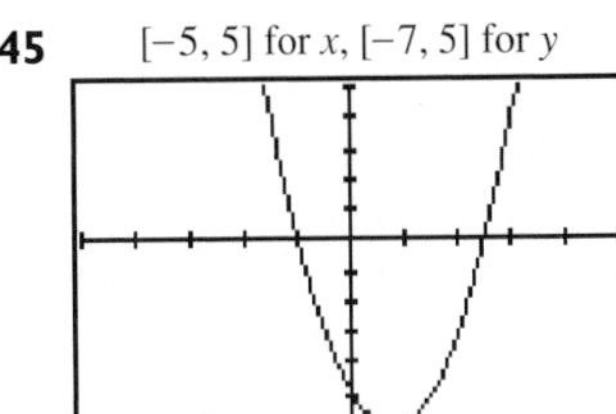

46 $[-6, 6]$ for x, $[-3, 7]$ for y

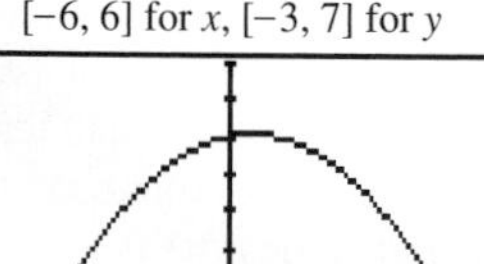

In Exercises 47 and 48 assume that a classmate has used a graphics calculator to graph the portion of a parabola shown. Use this graph, along with the superimposed dashed horizontal line, to determine for your classmate the x-coordinate of the missing vertex.

47 $[-2, 10]$ for x, $[-15, 6]$ for y

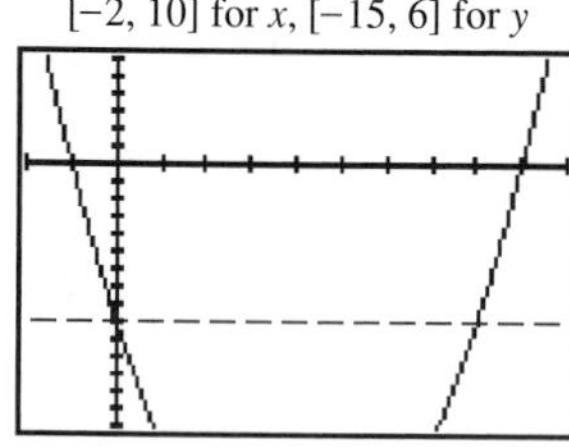

48 $[-8, 4]$ for x, $[-4, 14]$ for y

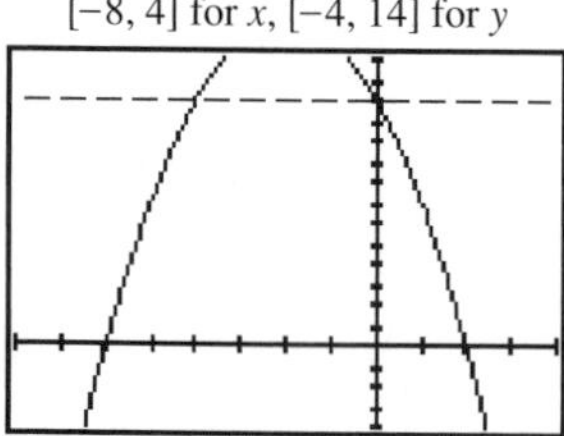

In Exercises 49–52 use a graphics calculator to graph $y = P(x)$, and then use this graph to factor $P(x)$.

49 $y = 2x^2 + 7x - 4$

50 $y = x^2 + 3x - 40$

51 $y = x^2 - x - 42$

52 $y = 2x^2 - 9x - 5$

In Exercises 53 and 54 use a graphics calculator to graph each function, using an appropriate window of values; then use the TRACE feature to approximate the maximum value of the function.

Power Curve (53 and 54)

53 A power boat engine has a power curve approximated by

$$y = -\frac{x^2}{50{,}000} + \frac{21x}{250} - \frac{16}{5}$$

where x is the number of revolutions per minute and y is the horsepower generated. At what number of revolutions per minute is the engine putting out maximum horsepower? What is the maximum horsepower? (See the Application at the beginning of this chapter.)

54 An engine in a stunt plane has a power curve approximated by

$$y = -\frac{x^2}{60{,}000} + \frac{x}{10} - 3$$

where x is the number of revolutions per minute and y is the horsepower generated. At what number of revolutions per minute is the engine putting out maximum horsepower? What is this maximum horsepower?

In Exercises 55–58 graph the parabola defined by each equation.

55 $x = y^2 - 1$

56 $x = (y + 1)^2$

57 $x = (y + 2)^2 - 1$

58 $x = (y - 3)^2 + 2$

DISCUSSION QUESTIONS (59 AND 60)

59 **a.** Discuss the effect of the coefficient a on the graph of $y = ax^2$.

b. Discuss the effect of the constant h on the graph of $y = (x - h)^2$.

c. Discuss the effect of the constant k on the graph of $y = x^2 + k$.

60 Using the Geometric Viewpoint in this section, describe how to obtain the x-coordinate of the vertex of a parabola defined by $y = ax^2 + bx + c$. Assume that you know where the parabola intersects a horizontal line. If you know the equation of the parabola, how can you find the y-coordinate of the vertex?

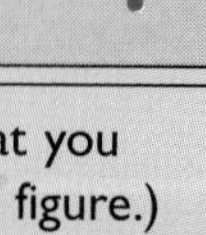

CHALLENGE QUESTION

What is the height of the tallest tower that you can brace using a 101-foot cable? (See the figure.) The top of the brace is approximately at the top of the tower, and the bottom of the brace must be anchored at least 20 feet from the bottom of the tower. Explain your reasoning.

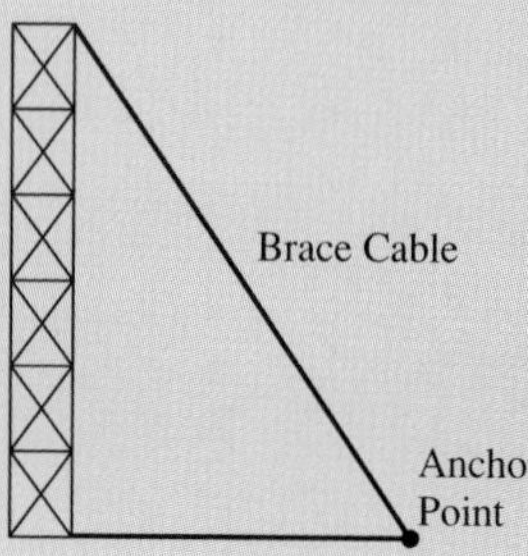

SECTION SKILL REINFORCERS

Choose the letter of the graph that represents each equation given.

a.

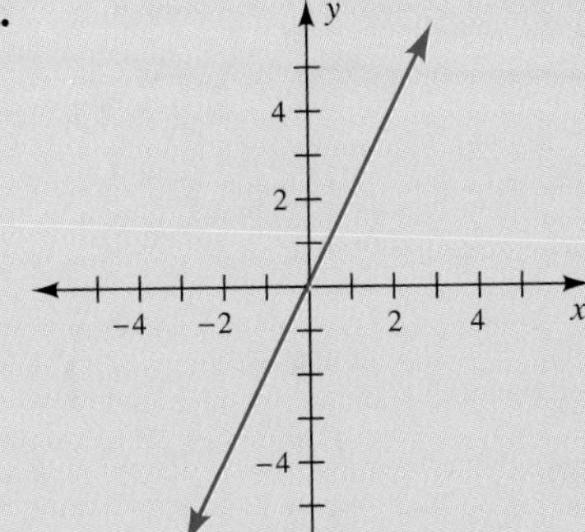

b.

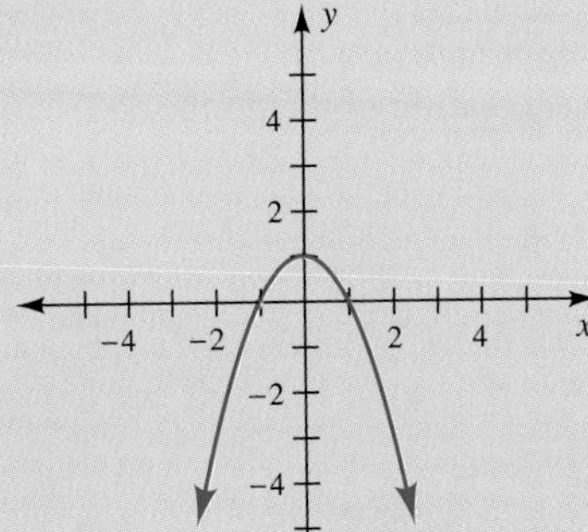

c.

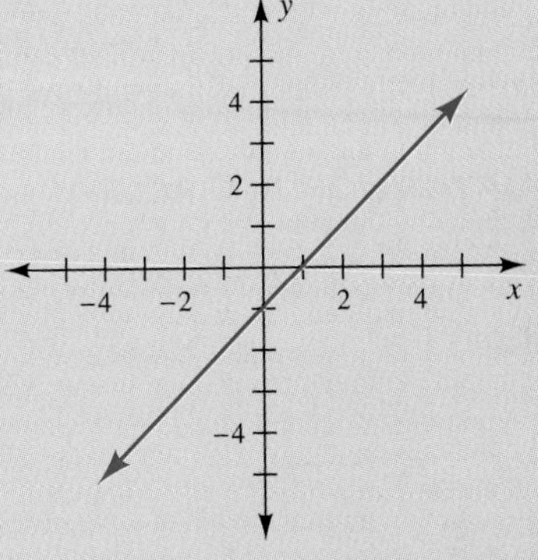

d.

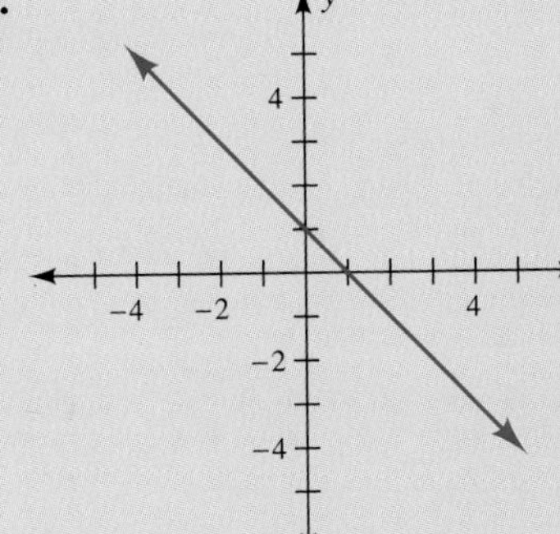

e.

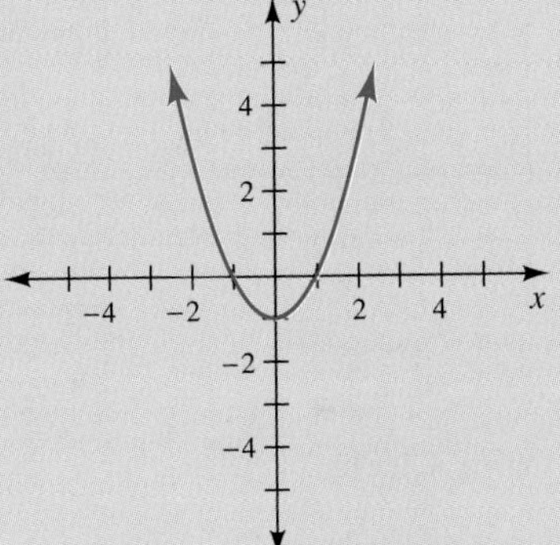

1 $y = x - 1$ **2** $y = x^2 - 1$ **3** $y = -x + 1$ **4** $y = -x^2 + 1$ **5** $y = 2x$

9-3 An Introduction to Conic Sections

SECTION OBJECTIVE

5 Graph a circle and an ellipse.

Circles, ellipses, and hyperbolas are relations. They are not functions, however, because each x-coordinate is not paired with exactly one y-coordinate—a fact that is easily determined by applying the vertical line test. These relations have many interesting properties that make them both useful and aesthetically pleasing. Our study of circles, ellipses, and hyperbolas will emphasize graphing these relations from their standard equations.

A **circle** is the set of all points in a plane that are the same distance from a fixed point. The fixed point is called the **center** of the circle, and the distance from the center to the points on the circle is called the **radius.** To draw a circle, place a loop of string over a tack and draw the circle as shown in Figure 9-3.

If the center of a circle is the point (h, k) and the radius is r, then the distance from any point (x, y) on the circle to (h, k) is r. From the distance formula in Section 9-1 we have

$$r = \sqrt{(x - h)^2 + (y - k)^2}$$

Squaring both sides of this equation gives the equation of the points on the circle.

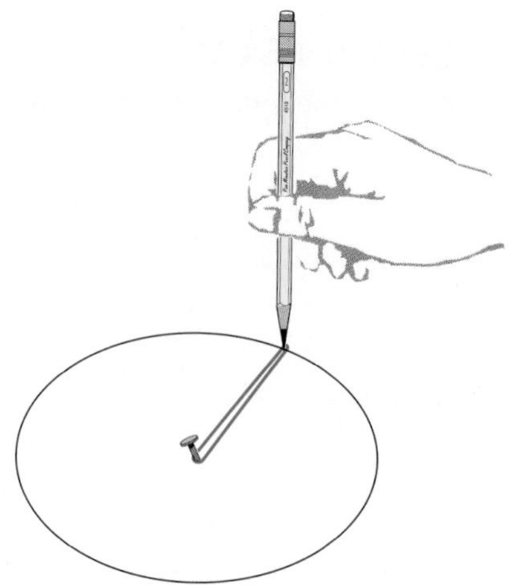

Figure 9-3
Drawing a Circle

EQUATION OF A CIRCLE

The standard form of the equation of a circle with center (h, k) and radius r is

$$(x - h)^2 + (y - k)^2 = r^2$$

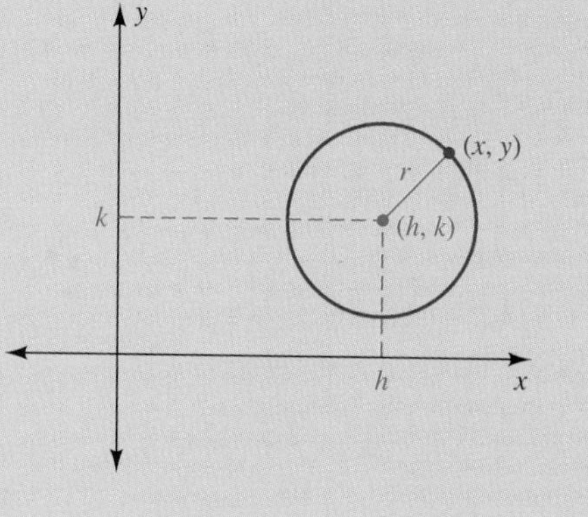

EXAMPLE 1 Writing the Equation of a Circle

Write the equation of a circle with its center at $(2, -4)$ and a radius of 3.

SOLUTION

$$(x - h)^2 + (y - k)^2 = r^2$$

$$(x - 2)^2 + [y - (-4)]^2 = 3^2$$

$$(x - 2)^2 + (y + 4)^2 = 9$$

Substitute the given values into the standard form of the equation of a circle.

The graph is shown in the figure.

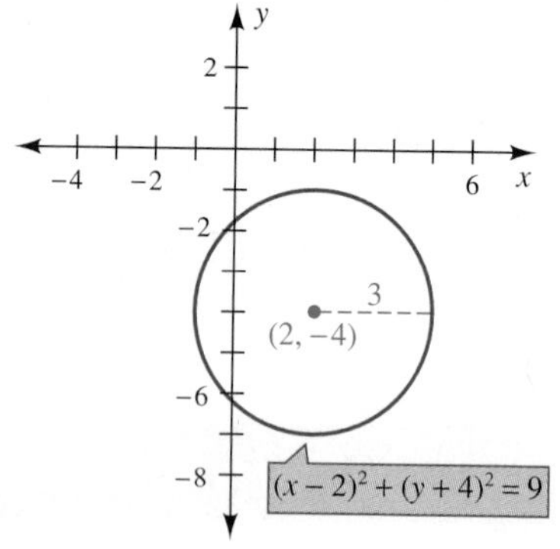

EXAMPLE 2 Determining the Center and Radius of a Circle

Determine the center and the radius of each of these circles.

SOLUTIONS

First write each equation in standard form, and then determine the center and the radius by inspection.

(a) $(x - 1)^2 + (y + 3)^2 = 4$

$$(x - 1)^2 + (y + 3)^2 = 4$$

$$(x - 1)^2 + [y - (-3)]^2 = 2^2$$

Center: $(1, -3)$

Radius: $r = 2$

The graph of the solution is shown in the figure.

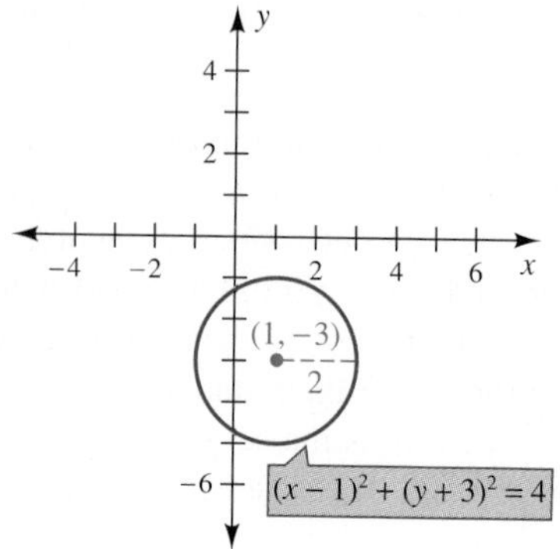

(b) $x^2 + y^2 = 16$

$$x^2 + y^2 = 16$$
$$(x - 0)^2 + (y - 0)^2 = 4^2$$

Center: $(0, 0)$

Radius: $r = 4$

The graph of the solution is shown in the figure.

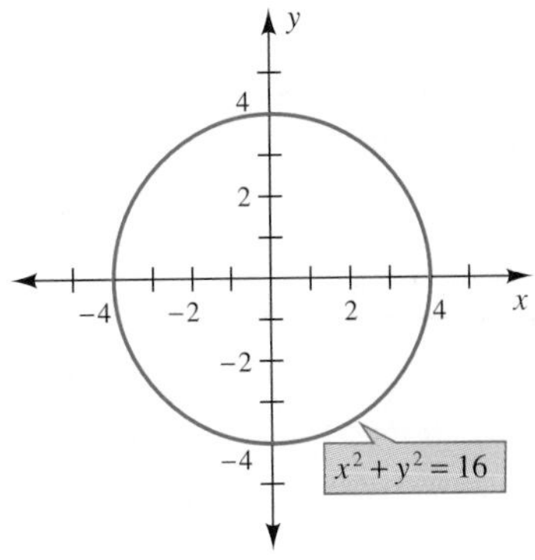

In Example 1 the general form $x^2 + y^2 - 4x + 8y + 11 = 0$ can be derived from the standard form of the circle, $(x - 2)^2 + (y + 4)^2 = 9$, by expansion. On the other hand, if we are given the general form, then we can use the process of completing the square to rewrite the equation in standard form so that the center and the radius will be obvious.

EXAMPLE 3 Completing the Square to Determine the Center and Radius of a Circle

Determine the center and the radius of $x^2 + y^2 + 4x - 6y + 12 = 0$.

SOLUTION

$$x^2 + y^2 + 4x - 6y + 12 = 0$$
$$(x^2 + 4x) + (y^2 - 6y) = -12$$
$$(x^2 + 4x + 4) + (y^2 - 6y + 9) = -12 + 4 + 9$$

First regroup the terms, and then complete the square.

$$(x + 2)^2 + (y - 3)^2 = 1$$
$$[x - (-2)]^2 + (y - 3)^2 = 1^2$$

By inspection determine the center and the radius from the standard form of a circle.

Answer Center: $(-2, 3)$; radius: $r = 1$

SELF-CHECK

Determine the center and the radius of each of these circles.

1. $(x + 3)^2 + (y - 4)^2 = 25$
2. $x^2 + y^2 = \frac{1}{4}$
3. $x^2 + y^2 + 10x - 22y + 110 = 0$

EXAMPLE 4 Graphing a Semicircle

Graph $y = \sqrt{9 - x^2}$.

SOLUTION Square both sides of the equation

$$y = \sqrt{9 - x^2}$$
$$y^2 = 9 - x^2 \quad \text{for } y \geq 0$$
$$x^2 + y^2 = 9 \quad \text{for } y \geq 0$$

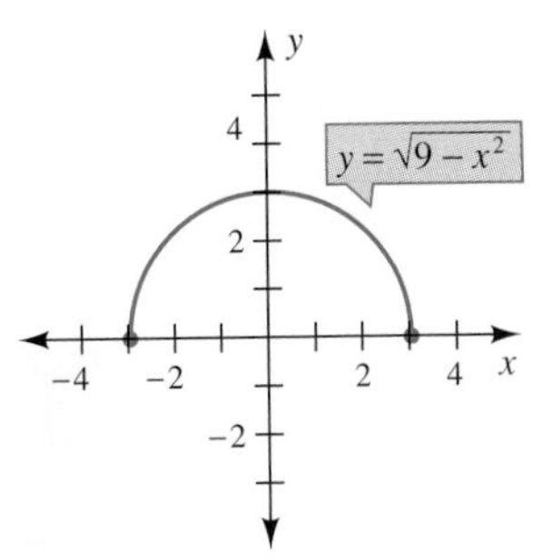

The resulting equation is of the form of a circle with center $(0, 0)$ and radius 3. However, since y is the principal square root, y is never negative. Thus the graph is only the upper semicircle. This equation is a function.

SELF-CHECK ANSWERS 

1. Center: $(-3, 4)$ Radius: $r = 5$
2. Center: $(0, 0)$ Radius: $r = \frac{1}{2}$
3. Center: $(-5, 11)$ Radius: $r = 6$

Graphics calculators can graph functions expressed in the explicit form $y = f(x)$. Current models are not designed to graph relations that are not functions. However, we can indirectly use a graphics calculator to graph circles by first decomposing the circle into two explicit functions. This is illustrated in the next example.

EXAMPLE 5 Using a Calculator to Graph a Circle

Use a graphics calculator to graph $x^2 + y^2 = 49$.

SOLUTION Solve $x^2 + y^2 = 49$ for y.

$$x^2 + y^2 = 49$$

$$y^2 = 49 - x^2 \qquad \text{Subtract } x^2 \text{ from both sides.}$$

$$y = \pm\sqrt{49 - x^2} \qquad \text{Solve for } y \text{ by extraction of roots.}$$

For $y \geq 0$, $y_1 = \sqrt{49 - x^2}$. This is the equation of the upper semicircle. For $y < 0$, $y_2 = -\sqrt{49 - x^2}$. This is the equation of the lower semicircle. Now use a graphics calculator to graph y_1 and y_2.

Keystrokes

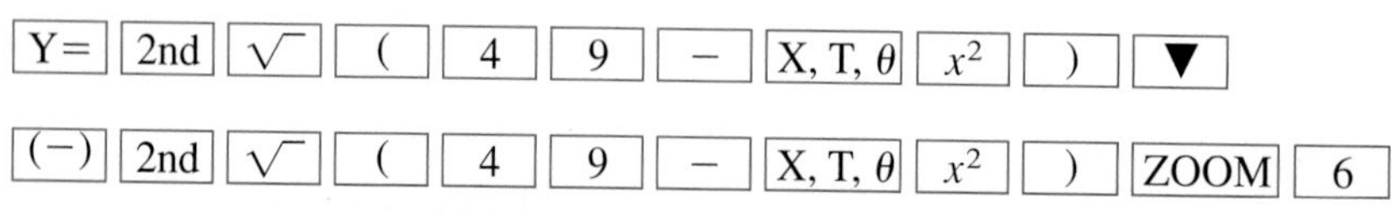

Graph the upper semicircle and the lower semicircle using the standard viewing window of $[-10, 10]$ for x and $[-10, 10]$ for y.

$[-10, 10]$ for x, $[-10, 10]$ for y

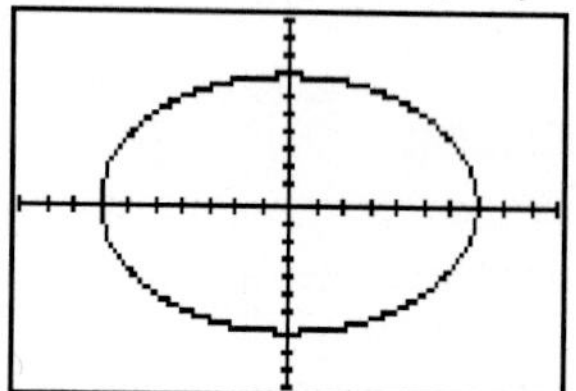

Note that the circle appears somewhat distorted because of the different numbers of pixels used horizontally and vertically. This distortion can be removed by pressing ZOOM 5 on a TI-82.

An **ellipse** is the set of all points in a plane the sum of whose distances from two fixed points is the same. The fixed points are called the **foci** of the ellipse. To draw an ellipse, place a loop of string over two tacks and draw the ellipse as illustrated in Figure 9-4. In this introduction to conics we will limit our discussion to ellipses centered at the origin.

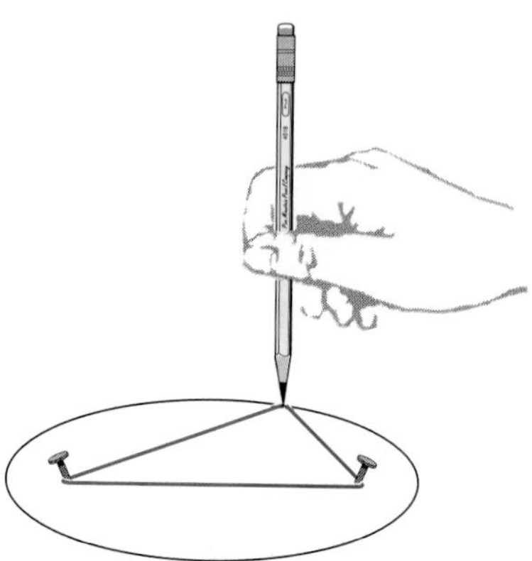

Figure 9-4 *Drawing an Ellipse*

A Mathematical Note

Ellipses have been studied and written about since 340 B.C. In the seventeenth century Johannes Kepler (c. 1610) discovered that the planets in our solar system have elliptical orbits with the sun as one focus. Today many machines are built with circular and elliptical gears.

EQUATION OF AN ELLIPSE

The standard form of the equation of an ellipse with its center at the origin, x-intercepts $(-a, 0)$ and $(a, 0)$, and y-intercepts $(0, -b)$ and $(0, b)$ is

$$\frac{x^2}{a^2} + \frac{y^2}{b^2} = 1$$

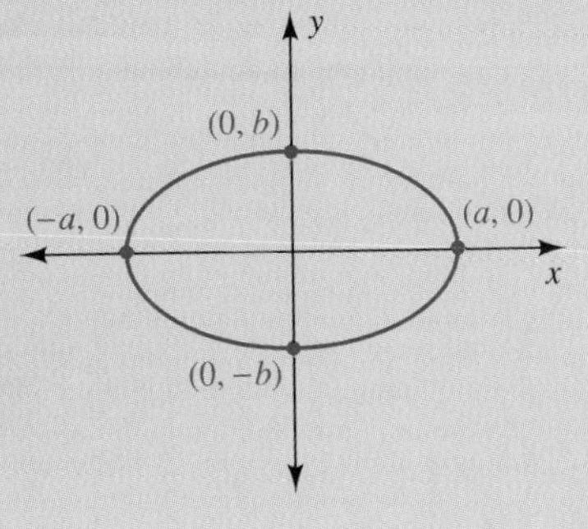

EXAMPLE 6 Writing the Equation of an Ellipse

Write the equation of an ellipse with center $(0, 0)$, x-intercepts $(-3, 0)$ and $(3, 0)$, and y-intercepts $(0, -5)$ and $(0, 5)$.

SOLUTION

$$\frac{x^2}{a^2} + \frac{y^2}{b^2} = 1$$

$$\frac{x^2}{3^2} + \frac{y^2}{5^2} = 1$$ Using the standard form of an ellipse, substitute 3 for a and 5 for b.

$$\frac{x^2}{9} + \frac{y^2}{25} = 1$$

$$25x^2 + 9y^2 = 225$$

The graph is shown in the figure.

Ellipses centered at the origin can be graphed by plotting only the four intercepts and then connecting these points with a smooth elliptical curve.

EXAMPLE 7 Graphing Ellipses

Graph each of these ellipses.

SOLUTIONS

(a) $\frac{x^2}{16} + \frac{y^2}{25} = 1$

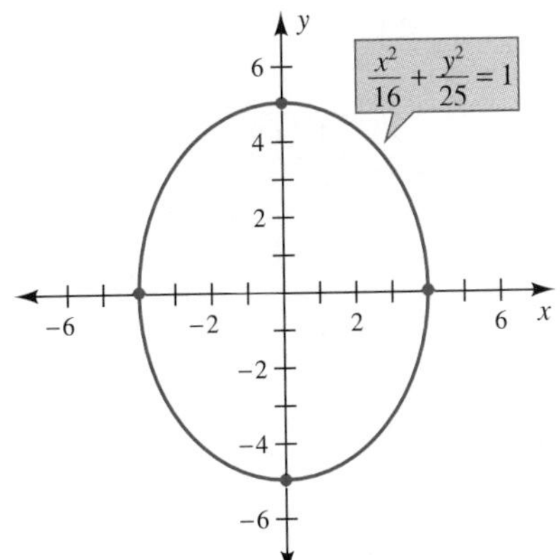

This ellipse is centered at the origin, with $a^2 = 16$ and $b^2 = 25$. Thus the x-intercepts are $(-4, 0)$ and $(4, 0)$, and the y-intercepts are $(0, -5)$ and $(0, 5)$. Plot these points, and then connect them with a smooth elliptical shape.

(b) $9x^2 + 25y^2 = 225$

$$\frac{x^2}{25} + \frac{y^2}{9} = 1$$

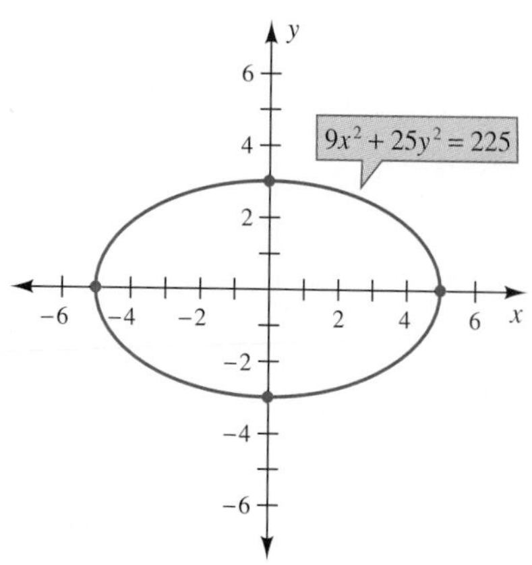

First divide both members of the equation by 225 so that the right member will be 1. This ellipse is centered at the origin, with x-intercepts $(-5, 0)$ and $(5, 0)$ and y-intercepts $(0, -3)$ and $(0, 3)$. Plot these points, and then connect them with a smooth elliptical shape.

SELF-CHECK

Graph each ellipse.

1 $\frac{x^2}{1} + \frac{y^2}{25} = 1$

2 $4x^2 + 9y^2 = 36$

EXAMPLE 8 Using a Calculator to Graph an Ellipse

Use a graphics calculator to graph $\frac{x^2}{36} + \frac{y^2}{4} = 1$.

SOLUTION Solve $\frac{x^2}{36} + \frac{y^2}{4} = 1$ for y.

$$\frac{x^2}{36} + \frac{y^2}{4} = 1$$

$x^2 + 9y^2 = 36$ Multiply both sides by 36.

$9y^2 = 36 - x^2$ Subtract x^2 from both sides.

$3y = \pm\sqrt{36 - x^2}$ Solve for y by extraction of roots.

$$y = \pm\frac{1}{3}\sqrt{36 - x^2}$$

For $y \geq 0$, $y_1 = \frac{1}{3}\sqrt{36 - x^2}$. This is the equation of the upper semiellipse.

SELF-CHECK ANSWERS

1

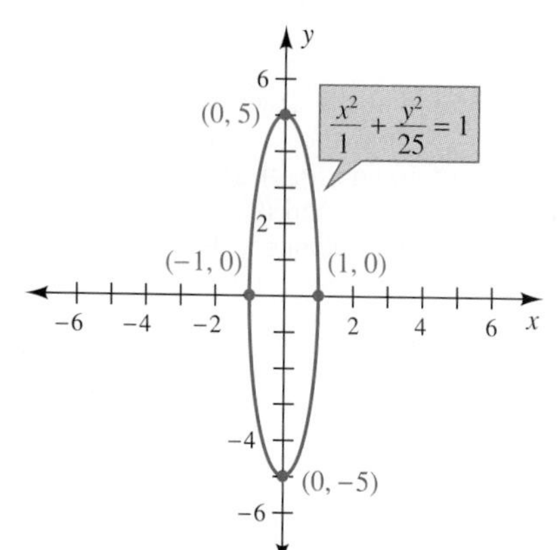

2

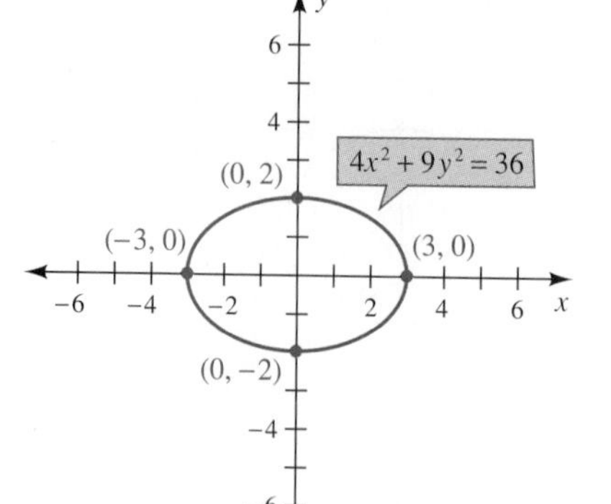

For $y < 0$, $y_2 = -\frac{1}{3}\sqrt{36 - x^2}$. This is the equation of the lower semiellipse.

Now use a graphics calculator to graph y_1 and y_2.

Keystrokes

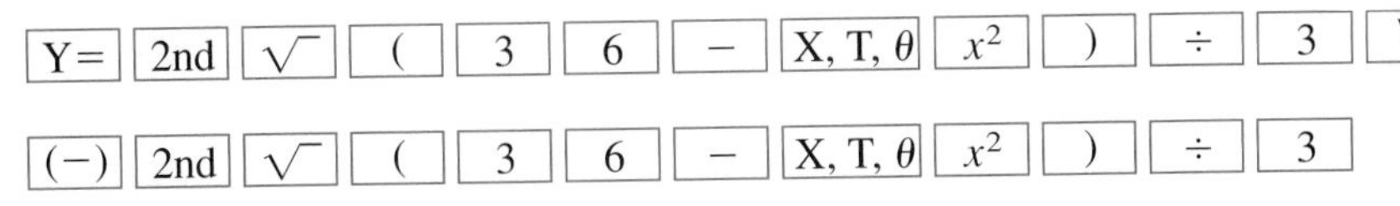

Graph the upper semiellipse and the lower semiellipse using the standard viewing window of $[-10, 10]$ for x and $[-10, 10]$ for y.

ZOOM 6

$[-10, 10]$ for x, $[-10, 10]$ for y

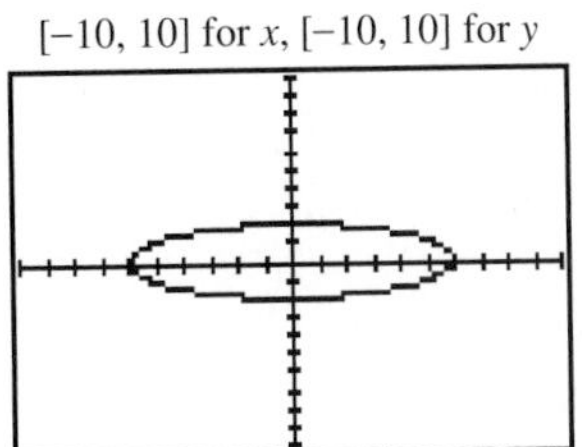

Note that the ellipse appears somewhat distorted because of the different numbers of pixels used horizontally and vertically. This distortion can be removed by pressing ZOOM 5 on a TI-82.

▼

Exercises 9-3

A

In Exercises 1–10 write the standard form of the equation of a circle satisfying the given conditions.

1 Center (3, 7); radius 2

2 Center (4, −5); radius 3

3 Center (−2, 1); radius $\frac{1}{2}$

4 Center (−3, −2); radius 0.4

5 Center (−5, −2); radius $\sqrt{3}$

6 Center (5, 0); radius $2\sqrt{5}$

7 Center (0, 0); passing through (4, 0)

8 Center (0, 0); passing through (0, −5)

9 Center (0, 0); passing through (−3, 4)

10 Center (0, 0); passing through (5, −12)

In Exercises 11–16 write the standard form of the equation of an ellipse with its center at the origin and with the given intercepts.

11 (−7, 0), (7, 0), (0, −10), and (0, 10)

12 (−5, 0), (5, 0), (0, −9), and (0, 9)

13 $\left(-\frac{1}{2}, 0\right)$, $\left(\frac{1}{2}, 0\right)$, (0, −2), and (0, 2)

14 (−3, 0), (3, 0), $\left(0, -\frac{1}{3}\right)$, and $\left(0, \frac{1}{3}\right)$

15 $(-\sqrt{2}, 0)$, $(\sqrt{2}, 0)$, $(0, -\sqrt{5})$, and $(0, \sqrt{5})$

16 $(-\sqrt{3}, 0)$, $(\sqrt{3}, 0)$, $(0, -\sqrt{7})$, and $(0, \sqrt{7})$

In Exercises 17–28 graph each conic section.

17 $x^2 + y^2 = 25$

18 $(x - 2)^2 + (y + 1)^2 = 4$

19 $(x + 1)^2 + (y - 3)^2 = 9$

20 $\frac{x^2}{36} + \frac{y^2}{25} = 1$

21 $\frac{x^2}{25} + \frac{x^2}{36} = 1$

22 $\frac{x^2}{25} + \frac{y^2}{25} = 1$

23 A circle with center (−4, 2) and radius 3

24 A circle with center (−2, −3) and radius 4

25 An ellipse with center (0, 0) and intercepts (−5, 0), (5, 0), (0, −3), and (0, 3)

26 An ellipse with center (0, 0) and intercepts (−2, 0), (2, 0), (0, −7), and (0, 7)

27 A circle that has a diameter with endpoints (4, −3) and (−2, 5)

28 A circle with center at (−1, −3) and passing through the point (2, 1)

In Exercises 29–34 choose the letter of the graph that represents the given equation.

a.

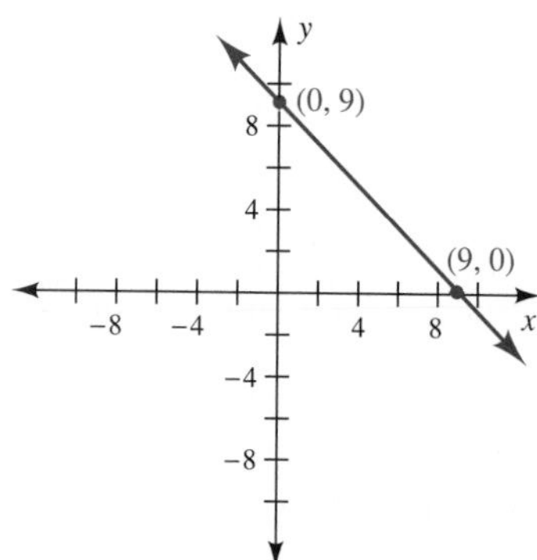

b.

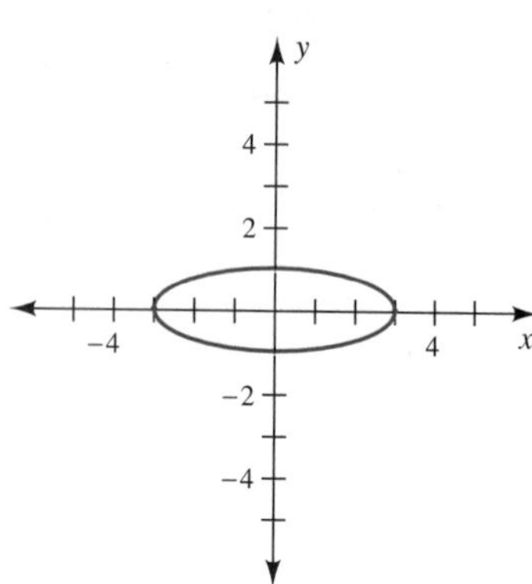

c.

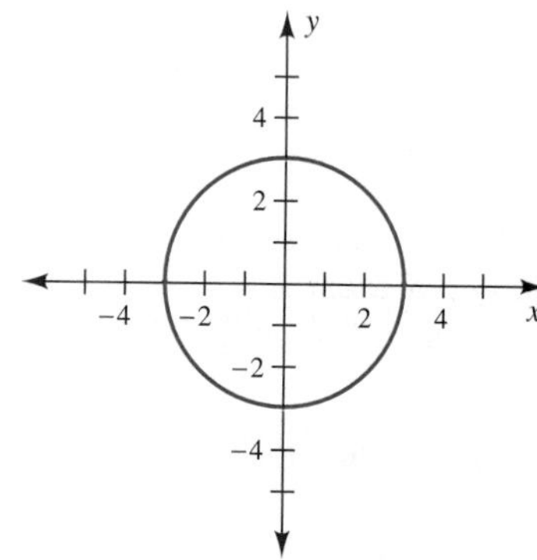

d.

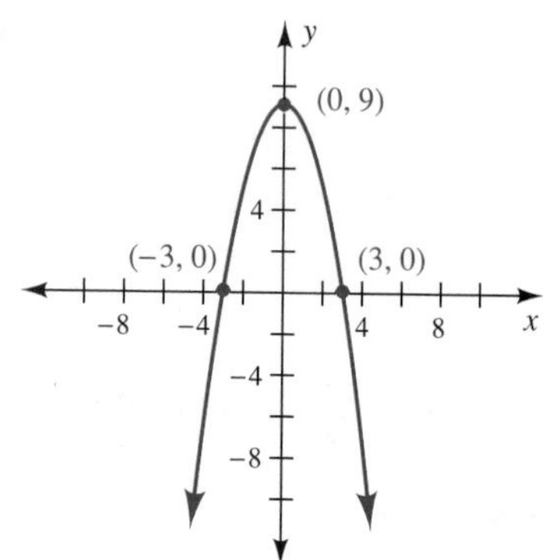

e.

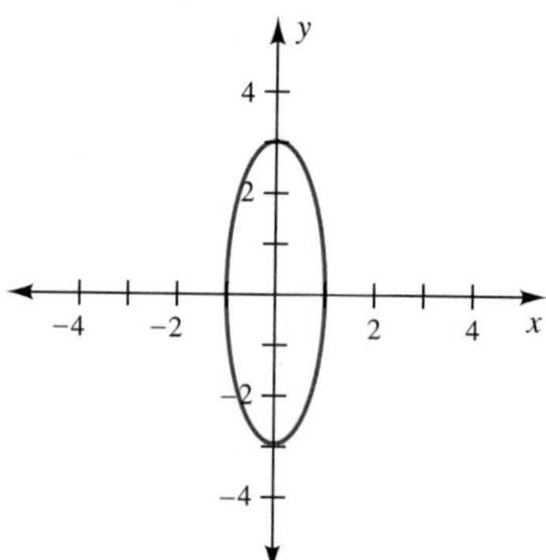

f.

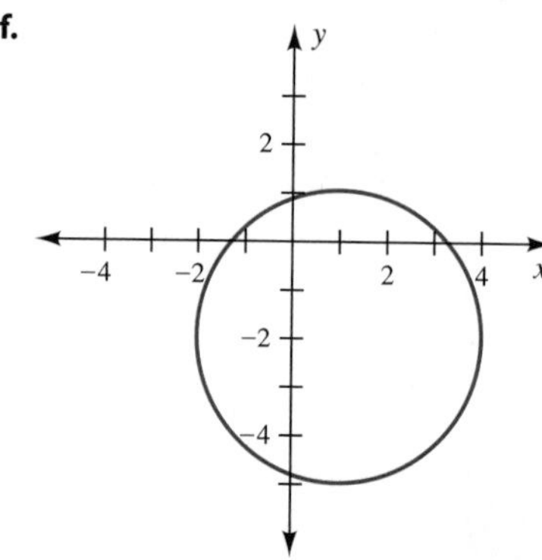

29 $x^2 + y^2 = 9$

30 $x^2 + y = 9$

31 $x + y = 9$

32 $9x^2 + y^2 = 9$

33 $(x - 1)^2 + (y + 2)^2 = 9$

34 $x^2 + 9y^2 = 9$

B

In Exercises 35–50 write each equation in standard form, determine the type of conic section, and graph the conic section.

35 $(x + 4)^2 + (y - 5)^2 = 25$

36 $(x - 5)^2 + (y + 4)^2 = 9$

37 $\frac{x^2}{4} + \frac{y^2}{25} = 1$

38 $\frac{x^2}{36} + \frac{y^2}{9} = 1$

39 $\frac{x^2}{25} + \frac{y^2}{81} = 1$

40 $\frac{x^2}{81} + \frac{y^2}{25} = 1$

41 $y = \frac{1}{4}x^2$

42 $y = -x^2 + 4x + 5$

43 $x^2 + y^2 = 64$

44 $16x^2 + y^2 = 64$

45 $x^2 + 16y^2 = 64$

46 $x^2 + y^2 - 4x + 8y + 19 = 0$

47 $x^2 + y^2 - 2x + 10y + 22 = 0$

48 $4x^2 + 36y^2 = 144$

49 $x^2 + y^2 + 10y + 16 = 0$

50 $x^2 + y^2 + 6x = 0$

51 Elliptical Window A window manufacturer sells a window that can be installed above another window. It markets this window as elliptical, although it is really a semiellipse with equation $y = \frac{14}{35}\sqrt{35^2 - x^2}$. What are the width and the height of this semiellipse?

Figure for Exercise 51

65 $\frac{(x-1)^2}{16} + \frac{(y-2)^2}{25} = 1$

66 $\frac{(x+1)^2}{16} + \frac{(y+2)^2}{25} = 1$

67 $\frac{(x+1)^2}{25} + \frac{(y+2)^2}{16} = 1$

68 Write the equation of a circle if the endpoints of a diameter are $(-3, -4)$ and $(3, 4)$.

69 Write the equation of a circle if the endpoints of a diameter are $(-5, 1)$ and $(3, 7)$.

70 Write the equation of an ellipse with intercepts $(-3, 0)$, $(3, 0)$, $(0, -5)$, and $(0, 5)$.

C

Calculator Usage (52–63)

In Exercises 52–59 use a graphics calculator to graph each function. Then approximate the domain and the range of the function.

52 $y = \sqrt{9 - x^2}$

53 $y = -\sqrt{9 - x^2}$

54 $y = -\sqrt{25 - x^2}$

55 $y = \sqrt{25 - x^2}$

56 $y = \sqrt{9 - 4x^2}$

57 $y = -\sqrt{9 - 4x^2}$

58 $y = x^2 - 2$ for $x \geq 0$

59 $y = x^2 - 2$ for $x \leq 0$

In Exercises 60–63 use a graphics calculator to graph each conic section.

60 $x^2 + y^2 = 81$

61 $(x - 1)^2 + y^2 = 64$

62 $\frac{x^2}{49} + \frac{y^2}{25} = 1$

63 $\frac{x^2}{25} + \frac{y^2}{64} = 1$

The standard equation of an ellipse with center (h, k) is $\frac{(x-h)^2}{a^2} + \frac{(y-k)^2}{b^2} = 1$. In Exercises 64–67 graph each ellipse.

64 $\frac{(x-1)^2}{25} + \frac{(y-2)^2}{16} = 1$

Discussion Question

71 Discuss the relationship that exists among the family of circles defined by $x^2 + y^2 = a$ for the following values of a: 16, 9, 4, 1, $\frac{1}{4}$, $\frac{1}{9}$, and $\frac{1}{100}$. What happens if $a = 0$?

Challenge Question

The ellipse shown in the figure is a mathematical model of the interrelationship between the number of aphids and the number of ladybugs in a monitored test plot. All numbers are given in hundreds. A description of event A is given below. Give a description of events B, C, and D.

Event A: The ladybug population is so large that it is not only reducing the aphid population, it is running out of food to support its own population.

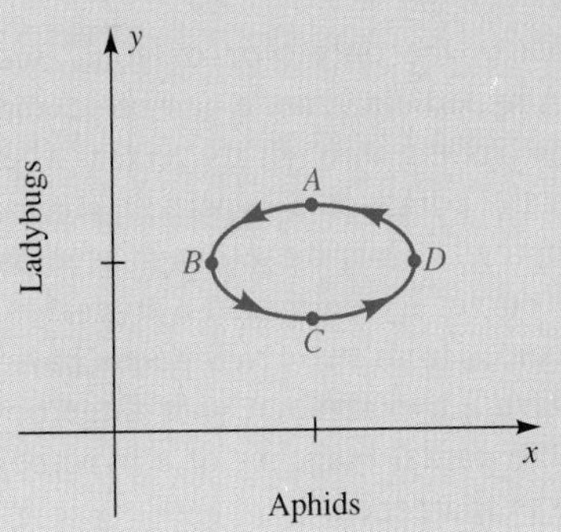

SECTION SKILL REINFORCERS

Determine the domain of each of these functions.

1

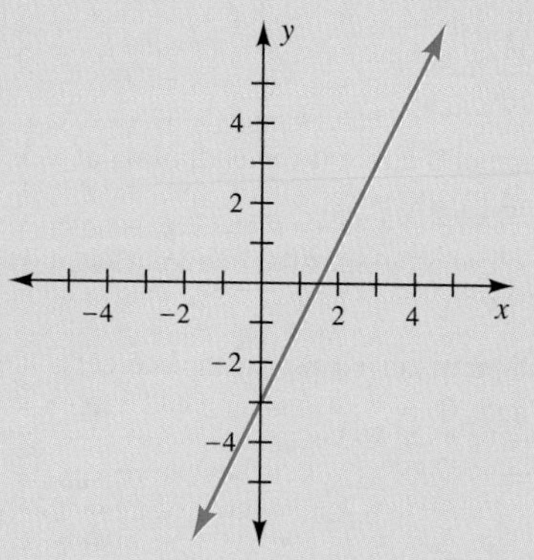

2

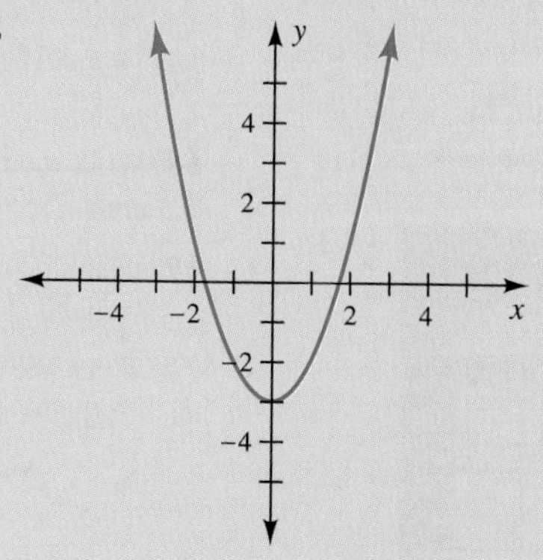

3

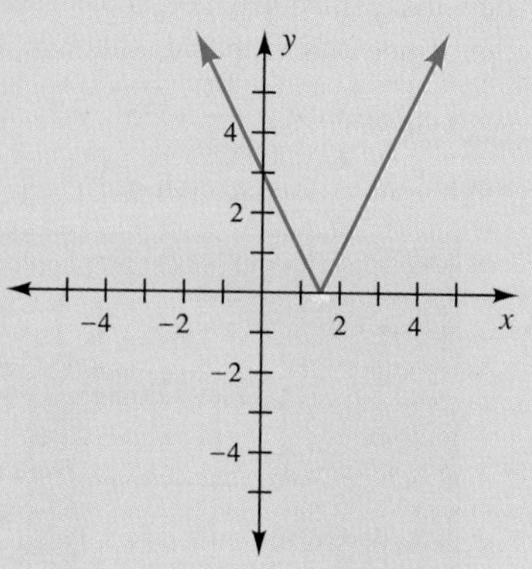

4

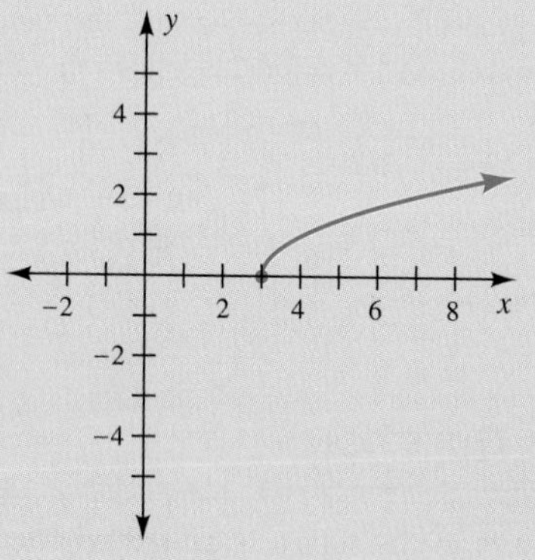

5

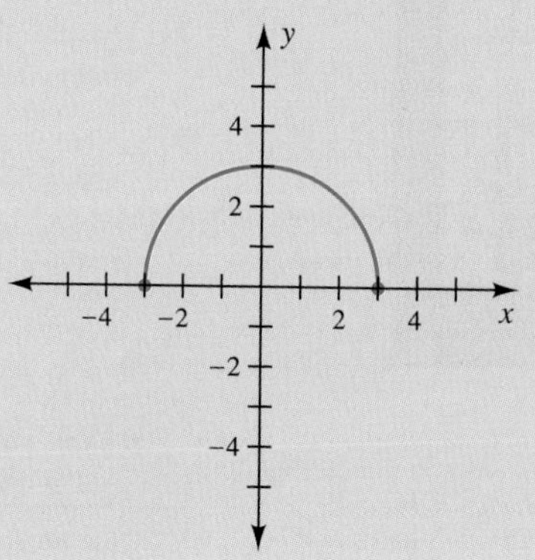

9-4 *Hyperbolas*

SECTION OBJECTIVES

6 Graph a hyperbola.

7 Identify the type of conic section from its equation.

A hyperbola is the only type of conic section that consists of two separate branches. A hyperbola is *not* a pair of parabolas; a hyperbola has a pair of linear **asymptotes** passing through its center, whereas a parabola has no asymptotes. The farther the points on a hyperbola are from the center, the closer they are to these asymptotes. Thus these linear asymptotes are a valuable aid in approximating the curve. By definition a **hyperbola** is the set of all points in a plane whose distances from two fixed points have a constant difference. Surprisingly, you may have a hyperbola at your fingertips right now. The figure to the right illustrates the partial branches of a hyperbola that are formed when a standard lead pencil is sharpened.

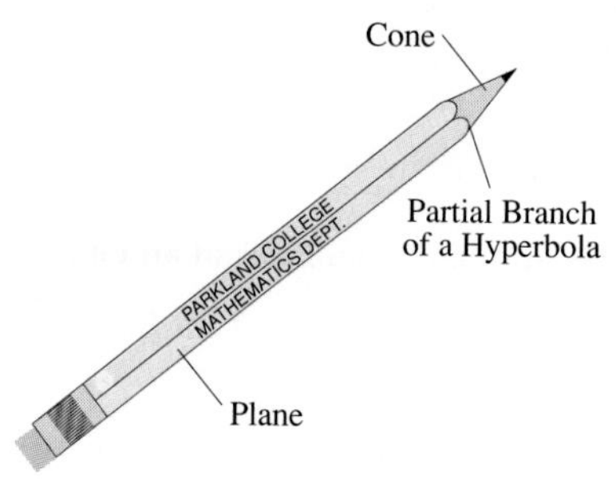

EQUATION OF A HYPERBOLA

Hyperbola Opening Horizontally:
The standard form of the equation of a hyperbola with its center at the origin and vertices $(-a, 0)$ and $(a, 0)$ is

$$\frac{x^2}{a^2} - \frac{y^2}{b^2} = 1$$

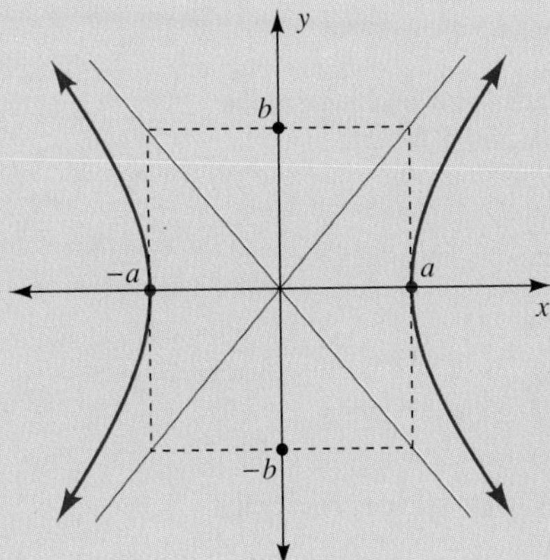

Hyperbola Opening Vertically:

The standard form of the equation of a hyperbola with its center at the origin and vertices $(0, -b)$ and $(0, b)$ is

$$\frac{y^2}{b^2} - \frac{x^2}{a^2} = 1$$

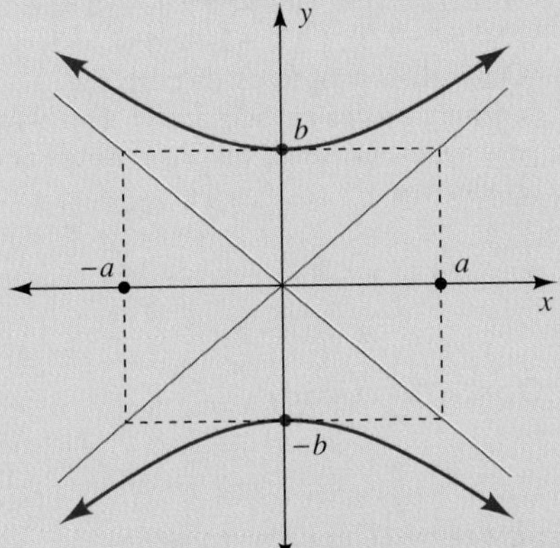

The linear asymptotes for both of these hyperbolas are $y = \frac{-b}{a}x$ and $y = \frac{b}{a}x$.

The linear asymptotes of a hyperbola pass through the corners of the rectangle formed by (a, b), $(-a, b)$, $(-a, -b)$, and $(a, -b)$. This rectangle, called the **fundamental rectangle,** can be used to quickly sketch the linear asymptotes. Use of the asymptotes then simplifies sketching the hyperbola, for it allows us to quickly approximate a portion of this complicated curve.

EXAMPLE 1 Writing the Equation of a Hyperbola

Write the equation of a hyperbola opening horizontally with center (0, 0), vertices (−2, 0) and (2, 0), and linear asymptotes $y = -\frac{5}{2}x$ and $y = \frac{5}{2}x$.

SOLUTION

$$\frac{x^2}{a^2} - \frac{y^2}{b^2} = 1$$

$$\frac{x^2}{2^2} - \frac{y^2}{5^2} = 1 \qquad \text{Using the standard form of a hyperbola, substitute 2 for } a \text{ and 5 for } b.$$

$$\frac{x^2}{4} - \frac{y^2}{25} = 1$$

$$25x^2 - 4y^2 = 100$$

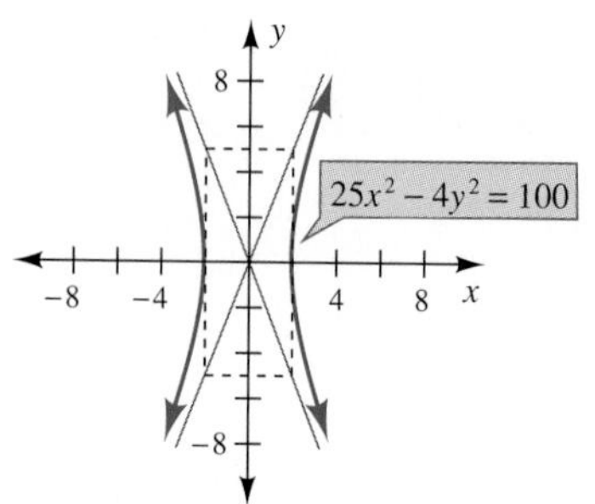

The graph is shown in the figure. ▼

EXAMPLE 2 Graphing a Hyperbola That Opens Horizontally

Graph $\frac{x^2}{4} - \frac{y^2}{9} = 1$.

SOLUTION

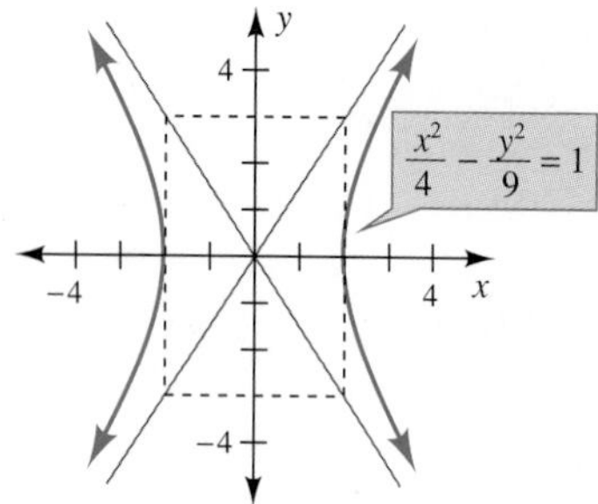

This hyperbola is centered at the origin, with $a^2 = 4$ and $b^2 = 9$. It opens horizontally, with vertices (−2, 0) and (2, 0). The corners of the fundamental rectangle are (2, 3), (−2, 3), (−2, −3), and (2, −3). Draw the asymptotes through the corners of this rectangle, and then sketch the hyperbola using the asymptotes as guidelines. ▼

EXAMPLE 3 Graphing a Hyperbola That Opens Vertically

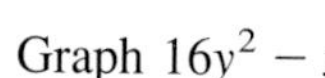

Graph $16y^2 - x^2 = 16$.

SOLUTION

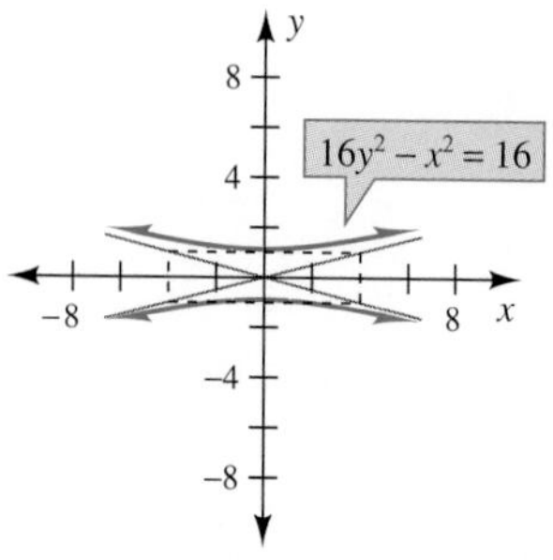

First divide by 16 in order to write the equation in standard form as $\frac{y^2}{1} - \frac{x^2}{16} = 1$. This hyperbola is centered at the origin. It opens vertically, with vertices (0, −1) and (0, 1). The corners of the fundamental rectangle are (4, 1), (−4, 1), (−4, −1), and (4, −1).

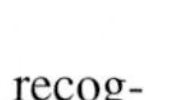

Our strategy for graphing conic sections is primarily dependent on recognizing shapes rather than on plotting points. Thus it is important to be able to recognize the standard forms of the various conic sections.

CONIC SECTIONS

Parabola:	$y = ax^2 + bx + c$	First degree in y, second degree in x. This relation is a function.
Circle:	$(x - h)^2 + (y - k)^2 = r^2$	Second degree in both x and y. The coefficients of x^2 and y^2 are the same.
Ellipse:	$\frac{x^2}{a^2} + \frac{y^2}{b^2} = 1$	Second degree in both x and y. The coefficients of x^2 and y^2 have the same sign but different magnitudes.
Hyperbola:	$\frac{x^2}{a^2} - \frac{y^2}{b^2} = 1$ $\frac{y^2}{b^2} - \frac{x^2}{a^2} = 1$	Second degree in both x and y. The coefficients of x^2 and y^2 have opposite signs.

SELF-CHECK 

Graph $4x^2 - 25y^2 = 100$.

EXAMPLE 4 Identifying Conic Sections

Identify the type of conic section represented by each of these equations.

SOLUTIONS

(a) $5x^2 - 7y^2 = 35$

This equation represents a hyperbola because the coefficients of x^2 and y^2 are of opposite sign. In standard form the equation is $\frac{x^2}{7} - \frac{y^2}{5} = 1$.

(b) $x^2 + y^2 + 4x + 16y + 19 = 0$

This equation represents a circle because x^2 and y^2 have the same coefficient. In standard form the equation is $(x + 2)^2 + (y + 8)^2 = 49$.

(c) $y = x^2 - 3$

This equation represents a parabola because the degree of y is 1 and the degree of x is 2.

(d) $8x^2 + 6y^2 = 48$

This equation represents an ellipse because the coefficients of x^2 and y^2 are positive but of different magnitudes. In standard form the equation is $\frac{x^2}{6} + \frac{y^2}{8} = 1$.

SELF-CHECK ANSWER

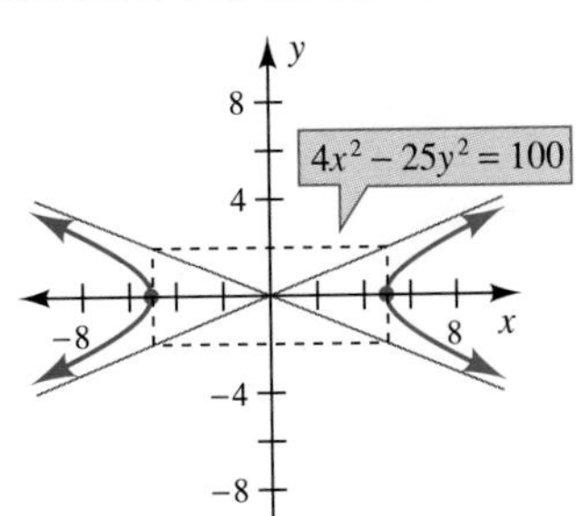

Every conic section can be defined by a second-degree equation in x and y of the general form $Ax^2 + Bxy + Cy^2 + Dx + Ey + F = 0$, where A, B, and C are not all zero. If A, B, and C are all zero, then this equation is the linear function $Dx + Ey + F = 0$. If $A \neq 0$, $B = 0$, and $C = 0$, then the equation $Ax^2 + Dx + Ey + F = 0$ is a quadratic function, which we examined in Section 9-3. The following table shows how the type of conic section can be determined from $B^2 - 4AC$. (The proof of this test is given in more advanced courses.)

$B^2 - 4AC$	Type of Conic Section
Negative	Ellipse
Zero	Parabola
Positive	Hyperbola

SELF-CHECK

Determine the type of conic section represented by each equation.

1 $2x^2 + y^2 = 1$
2 $2x^2 - y^2 = 1$
3 $y = x^2 + 1$
4 $x^2 + y^2 = 11$

We will now rework Example 4, using this test to identify each conic section.

EXAMPLE 5 Identifying Conic Sections

Identify the type of conic section represented by each of these equations.

SOLUTIONS

(a) $5x^2 - 7y^2 - 35 = 0$

$B^2 - 4AC = 0^2 - 4(5)(-7)$ Substitute $A = 5$, $B = 0$, and $C = -7$.

$= 140$

Since $B^2 - 4AC > 0$, this equation represents a hyperbola.

(b) $x^2 + y^2 + 4x + 16y + 19 = 0$

$B^2 - 4AC = 0^2 - 4(1)(1)$ Substitute $A = 1$, $B = 0$, and $C = 1$.

$= -4$

Since $B^2 - 4AC < 0$, this equation represents an ellipse. The fact that the coefficients of x^2 and y^2 are equal indicates that this ellipse is a circle.

(c) $x^2 - y - 3 = 0$

$B^2 - 4AC = 0^2 - 4(1)(0)$ Substitute $A = 1$, $B = 0$, and $C = 0$.

$= 0$

Since $B^2 - 4AC = 0$, this equation represents a parabola.

SELF-CHECK ANSWERS 

1 An ellipse 2 A hyperbola 3 A parabola 4 A circle

(d) $8x^2 + 6y^2 - 48 = 0$

$$B^2 - 4AC = 0^2 - 4(8)(6) = -192$$

Substitute $A = 8$, $B = 0$, and $C = 6$.

Since $B^2 - 4AC < 0$, this equation represents an ellipse.

(e) $xy - 1 = 0$

$$B^2 - 4AC = 1^2 - 4(0)(0) = 1$$

Substitute $A = 0$, $B = 1$, and $C = 0$. This hyperbola is centered at the origin, with the axes as asymptotes, and opens in quadrants I and III.

Since $B^2 - 4AC > 0$, this equation represents a hyperbola.

▼

Exercises 9-4

A

In Exercises 1–6 determine whether the hyperbola opens horizontally or vertically, and then determine its vertices.

1 $\frac{x^2}{49} - \frac{y^2}{36} = 1$　　**2** $\frac{y^2}{25} - \frac{x^2}{49} = 1$

3 $\frac{y^2}{4} - \frac{x^2}{5} = 1$　　**4** $\frac{x^2}{1} - \frac{y^2}{3} = 1$

5 $4x^2 = 36y^2 + 144$　　**6** $4x^2 = 36y^2 - 144$

In Exercises 7–12 write an equation of each hyperbola that is centered at the origin.

7 A hyperbola opening horizontally with vertices at $(-4, 0)$ and $(4, 0)$ and linear asymptotes $y = -\frac{5}{4}x$ and $y = \frac{5}{4}x$

8 A hyperbola opening horizontally with vertices at $(-5, 0)$ and $(5, 0)$ and linear asymptotes $y = -\frac{4}{5}x$ and $y = \frac{4}{5}x$

9 A hyperbola opening vertically with vertices at $(0, -4)$ and $(0, 4)$ and linear asymptotes $y = -\frac{4}{5}x$ and $y = \frac{4}{5}x$

10 A hyperbola opening vertically with vertices at $(0, -5)$ and $(0, 5)$ and linear asymptotes $y = -\frac{5}{4}x$ and $y = \frac{5}{4}x$

11 A hyperbola opening vertically, with corners of the fundamental rectangle at $(3, 2)$, $(-3, 2)$, $(-3, -2)$, and $(3, -2)$

12 A hyperbola opening horizontally, with corners of the fundamental rectangle at $(5, 1)$, $(-5, 1)$, $(-5, -1)$, and $(5, -1)$

In Exercises 13–20 graph each hyperbola, and label the vertices and corners of the fundamental rectangle.

13 A hyperbola opening horizontally, with center $(0, 0)$ and corners of the fundamental rectangle at $(1, 4)$, $(-1, 4)$, $(-1, -4)$, and $(1, -4)$

14 A hyperbola opening vertically, with center $(0, 0)$ and corners of the fundamental rectangle at $(5, 2)$, $(-5, 2)$, $(-5, -2)$, and $(5, -2)$

15 A hyperbola opening horizontally, with vertices $(-4, 0)$ and $(4, 0)$ and linear asymptotes $y = -\frac{3}{4}x$ and $y = \frac{3}{4}x$

16 A hyperbola opening vertically, with vertices $(0, -7)$ and $(0, 7)$ and linear asymptotes $y = -\frac{7}{5}x$ and $y = \frac{7}{5}x$

17 $\frac{x^2}{16} - \frac{y^2}{36} = 1$

18 $\frac{y^2}{25} - \frac{x^2}{4} = 1$

19 $\frac{y^2}{4} - \frac{x^2}{4} = 1$

20 $\frac{x^2}{9} - \frac{y^2}{9} = 1$

In Exercises 21–32 identify the type of conic section defined by each equation.

21 $y = x^2 + 1$　　**22** $y^2 = x^2 + 2$

23 $x^2 + y^2 = 2$　　**24** $x^2 + 2y^2 = 1$

25 $\frac{x^2}{2} + \frac{y^2}{3} = 1$　　**26** $\frac{x^2}{2} + \frac{y^2}{2} = 1$

27 $\frac{x^2}{2} + \frac{y}{2} = 1$

28 $\frac{x}{2} + \frac{y}{2} = 1$

29 $xy = 9$

30 $x^2 - 3y^2 + 2x - 5y + 10 = 0$

31 $x^2 + 3y^2 + 2x - 15y + 10 = 0$

32 $3x^2 + 3y^2 + 2x - 15y + 10 = 0$

In Exercises 33–40 choose the letter of the graph that represents the given equation.

a.

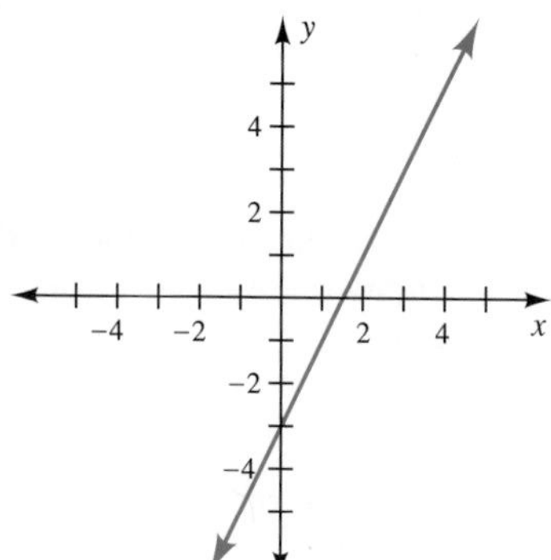

b.

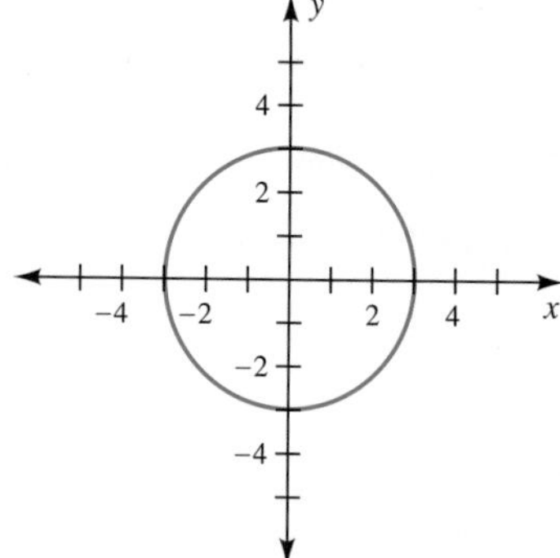

c.

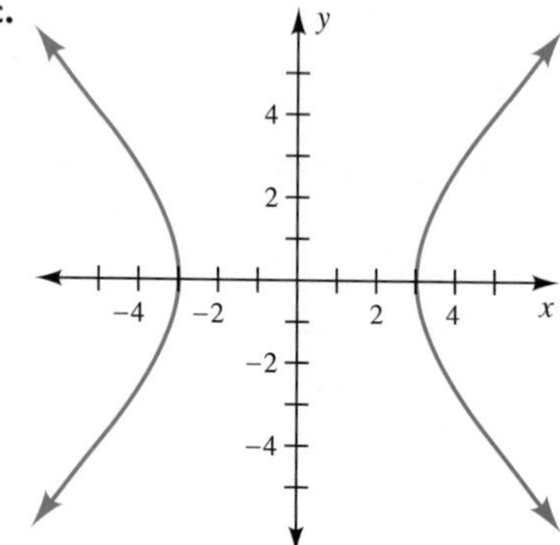

d.

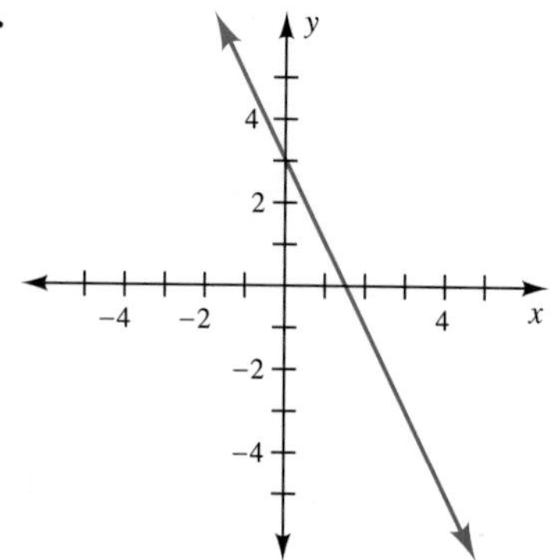

e.

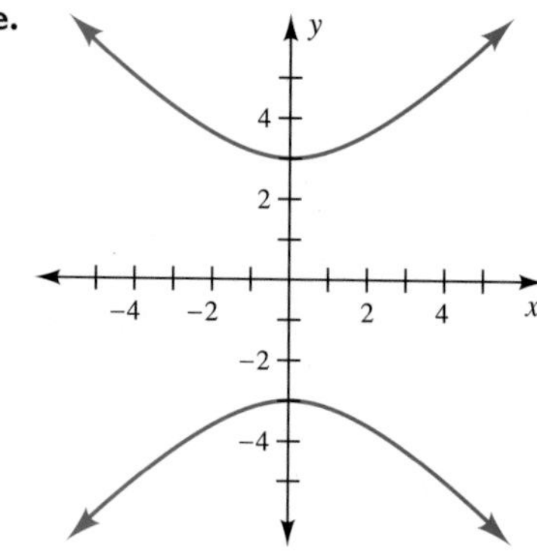

f.

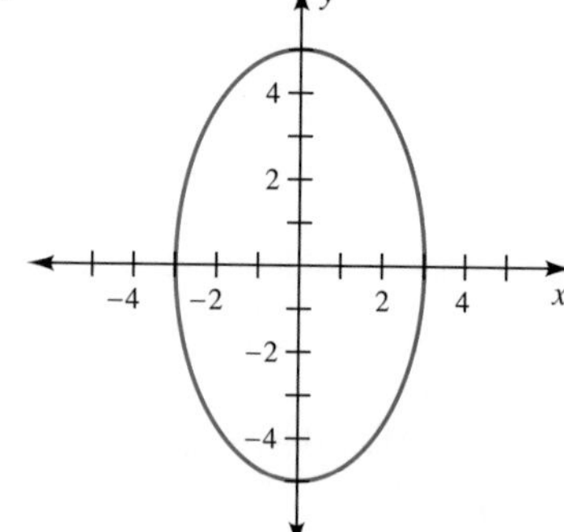

g.

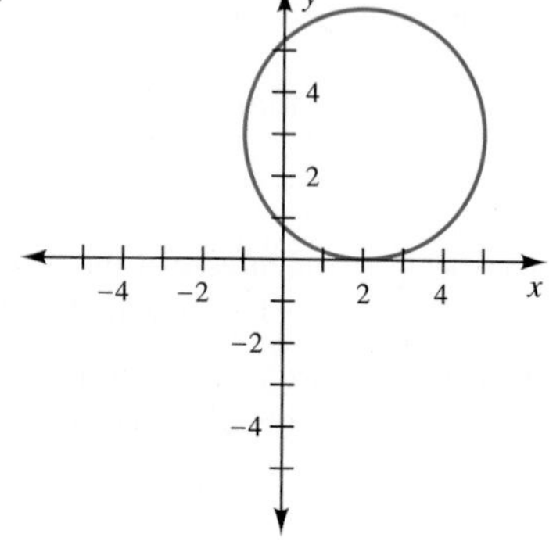

h.

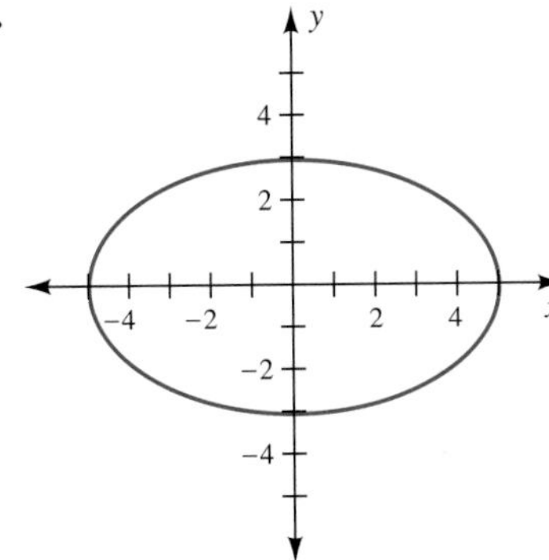

33 $2x + y - 3 = 0$

34 $y = 2x - 3$

35 $(x - 2)^2 + (y - 3)^2 = 9$

36 $\frac{x^2}{9} + \frac{y^2}{25} = 1$

37 $\frac{x^2}{9} + \frac{y^2}{9} = 1$

38 $\frac{x^2}{25} + \frac{y^2}{9} = 1$

39 $\frac{y^2}{9} - \frac{x^2}{9} = 1$

40 $\frac{x^2}{9} - \frac{y^2}{9} = 1$

CALCULATOR USAGE (41–50)

In Exercises 41–50 use a graphics calculator to graph each of these relations.

41 $y = |x - 2| - 3$

42 $y = 3x - 2$

43 $y = \sqrt{3x - 2}$

44 $y = \sqrt{16 - x^2}$

45 $y = -\sqrt{16 - x^2}$

46 $y = \sqrt{x^2 + 16}$

47 $y = -\sqrt{x^2 + 16}$

48 $y = 2\sqrt{1 - x^2}$

49 $\frac{x^2}{16} - \frac{y^2}{25} = 1$

50 $\frac{y^2}{25} - \frac{x^2}{16} = 1$

The standard equation of a hyperbola with center (h, k) and opening horizontally is $\frac{(x - h)^2}{a^2} - \frac{(y - k)^2}{b^2} = 1$. In Exercises 51 and 52 graph each hyperbola.

51 $\frac{(x - 1)^2}{25} - \frac{(y - 2)^2}{16} = 1$

52 $\frac{(x - 1)^2}{16} - \frac{(y - 2)^2}{25} = 1$

DISCUSSION QUESTIONS (53 AND 54)

53 Describe how each of the conic sections can be formed using cones and planes.

54 Describe a real-life example of each of the conic sections.

CHALLENGE QUESTION

A linear function $y = mx + b$ is characterized by the fact that equal changes in the independent variable x will result in constant changes in the dependent variable y. Complete the table for $y = 2x + 3$, which illustrates this concept. Then complete the second table for $y = x^2 + 3$, and describe what you have observed. Can you generalize based on what you have observed?

Table for $y = 2x + 3$

x	y	Change in y
1	5	
2	7	2
3	9	2
4	11	
5		
6		
7		
8		

Table for $y = x^2 + 3$

x	y	First-Order Change in y	Second-Order Change in y
1	4		
2	7	3	
3	12	5	2
4	19	7	2
5			
6			
7			
8			

SECTION SKILL REINFORCERS

Write in general form the equation of the line described in each exercise.

1 A vertical line through (2, 3) **2** A horizontal line through (2, 3) **3** A line through (2, 3) and (4, −4) **4** A line through (2, 3) and parallel to $y = 4x - 1$ **5** The perpendicular bisector of the line segment connecting (2, 3) and (8, 7)

Key Concepts for Chapter 9

1 Distance formula:
The distance d between (x_1, y_1) and (x_2, y_2) is given by

$$d = \sqrt{(x_2 - x_1)^2 + (y_2 - y_1)^2}$$

2 Midpoint formula:
The midpoint (x, y) between $P(x_1, y_1)$ and $Q(x_2, y_2)$ is given by

$$(x, y) = \left(\frac{x_1 + x_2}{2}, \frac{y_1 + y_2}{2}\right)$$

3 Parabolas:
The graph of $y = a(x - h)^2 + k$, $a \neq 0$, is a parabola that has its vertex at (h, k), opens upward if $a > 0$ or downward if $a < 0$, and is wider than the parabola $y = x^2$ if $|a| < 1$ or narrower if $|a| > 1$.

4 Maximum and minimum values:
A parabola that is concave downward has a maximum point at the vertex. A parabola that is concave upward has a minimum point at the vertex.

5 Conic sections:

a. Parabola: $y = ax^2 + bx + c$ is the equation of a parabola.

b. Circle: $(x - h)^2 + (y - k)^2 = r^2$ is the standard form of a circle with center (h, k) and radius r.

c. Ellipse: $\frac{x^2}{a^2} + \frac{y^2}{b^2} = 1$ is the standard form of an ellipse with center $(0, 0)$, x-intercepts $(-a, 0)$ and $(a, 0)$, and y-intercepts $(0, -b)$ and $(0, b)$.

d. Hyperbola: $\frac{x^2}{a^2} - \frac{y^2}{b^2} = 1$ is the standard form of a hyperbola that opens horizontally, with center $(0, 0)$ and vertices $(-a, 0)$ and $(a, 0)$. $\frac{y^2}{b^2} - \frac{x^2}{a^2} = 1$ is the standard form of the equation of a hyperbola that opens vertically, with center $(0, 0)$ and vertices $(0, -b)$ and $(0, b)$. The corners of the fundamental rectangle are (a, b), $(-a, b)$, $(-a, -b)$, and $(a, -b)$.

e. A conic section of the general form

$$Ax^2 + Bxy + Cy^2 + Dx + Ey + F = 0$$

is either an ellipse, a parabola, or a hyperbola, depending on whether $B^2 - 4AC$ is negative, zero, or positive, respectively.

Review Exercises for Chapter 9

In Exercises 1–4 calculate the distance between each pair of points, and determine the midpoint between these points.

1 $(-2, -5)$ and $(2, -2)$

2 $(-7, 11)$ and $(5, 11)$

3 $(4, -3)$ and $(-4, -3)$

4 $(-4, 8)$ and $(1, -4)$

5 Determine whether the points $(-2, 4)$, $(4, 6)$, and $(2, 2)$ form a right triangle.

6 Determine the equation of the perpendicular bisector of the line segment connecting $(4, 6)$ and $(2, 2)$.

In Exercises 7–10 determine the vertex of each parabola, and determine whether the parabola is concave upward or downward.

7 $y = -x^2 + 7$

8 $f(x) = (x + 11)^2$

9 $f(x) = (x + 10)^2 - 8$

10 $y = -x^2 + 6x + 5$

In Exercises 11–14 graph the parabola defined by each equation.

11 $y = -x^2 + 5$

12 $y = 3x^2$

13 $y = (x - 1)^2 + 2$

14 $y = x^2 - 4x + 5$

15 Fencing an Area Find the maximum rectangular area that can be enclosed by 128 meters of fencing if one side of the rectangle borders a long warehouse and does not require any fencing.

16 Minimum Product Find the two numbers with a difference of 10 whose product is a minimum. What is this minimum product?

In Exercises 17–28 determine the letter of the graph that represents the given equation.

17 $x = 3$

18 $y = 3$

19 $y = |x - 2|$

20 $y = -|x + 2|$

21 $y = x^2 - 3$

22 $y = -x^2 + 3$

23 $y = \sqrt{25 - x^2}$

24 $\frac{x^2}{1} + \frac{y^2}{9} = 1$

25 $(x - 3)^2 + y^2 = 9$

26 $\frac{x^2}{1} - \frac{y^2}{9} = 1$

27 $\frac{x^2}{9} + \frac{y^2}{1} = 1$

28 $x^2 + (y - 3)^2 = 9$

a.

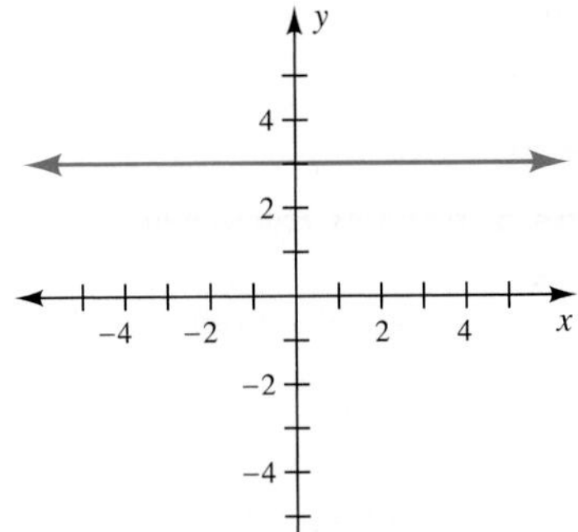

b.

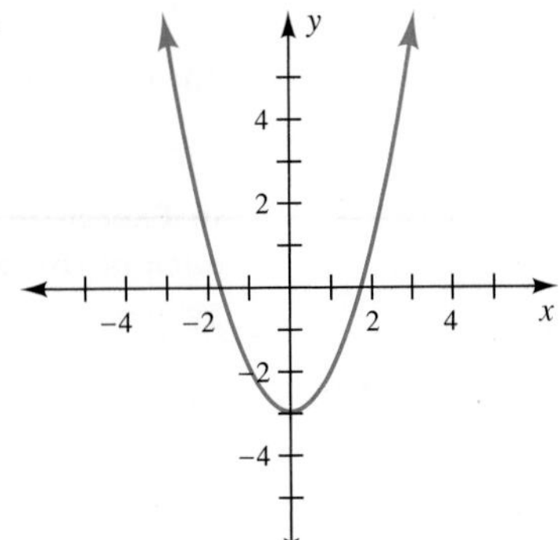

c.

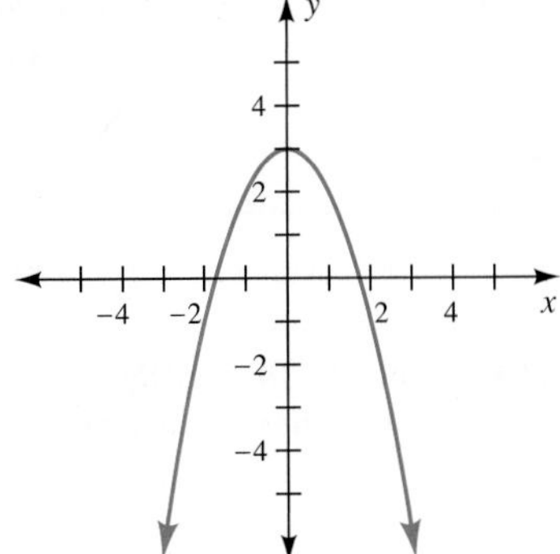

d.

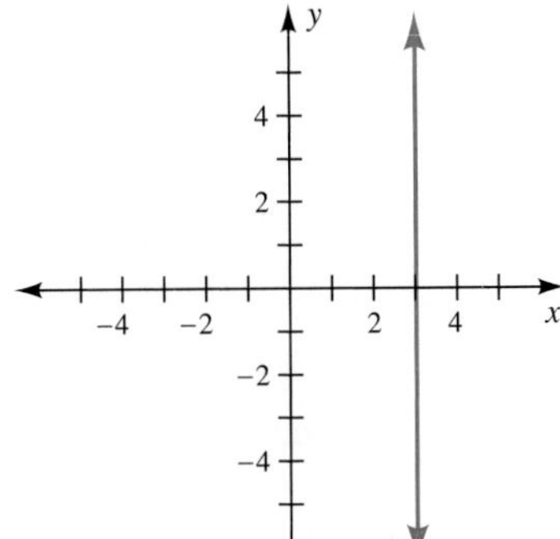

e.

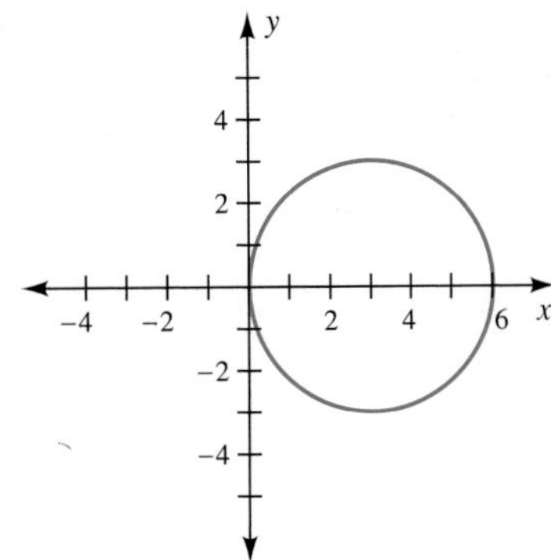

f.

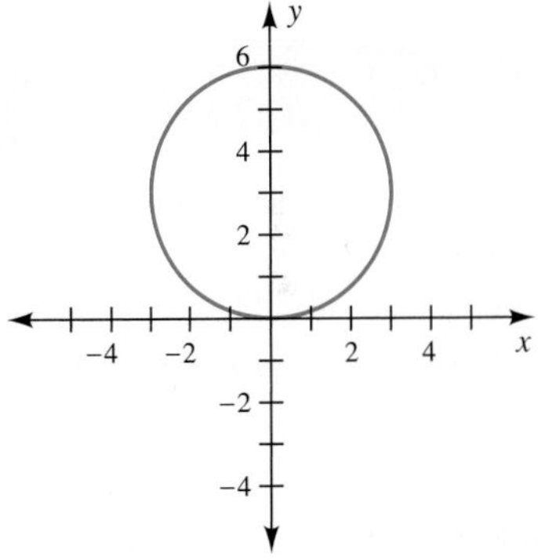

g.

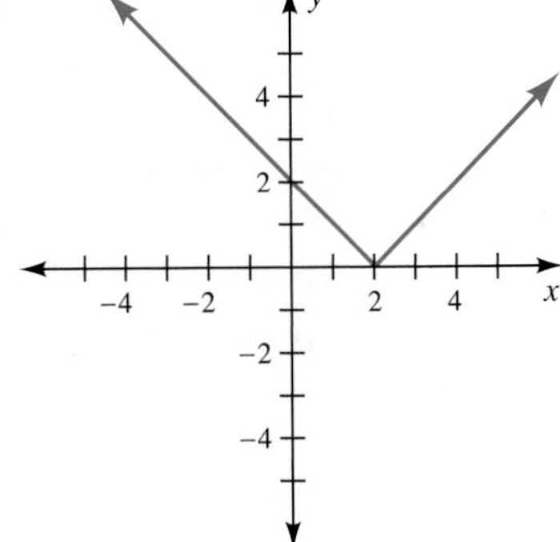

h.

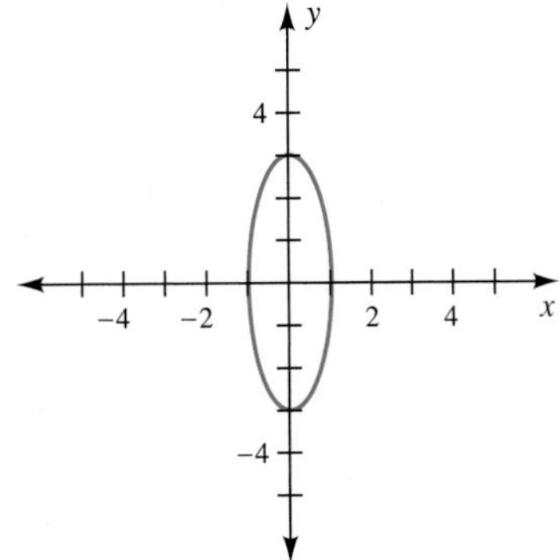

i.

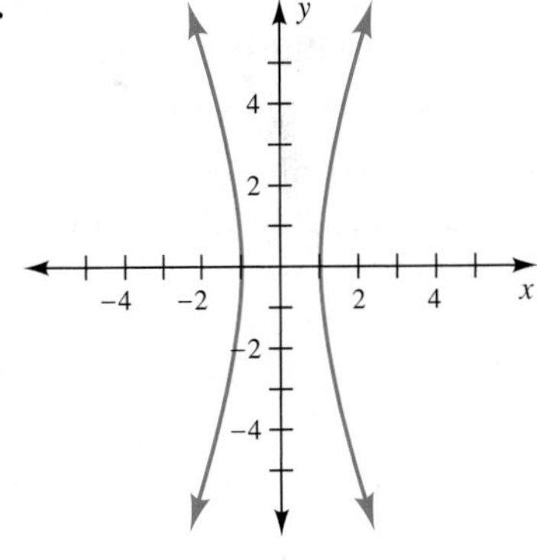

j.

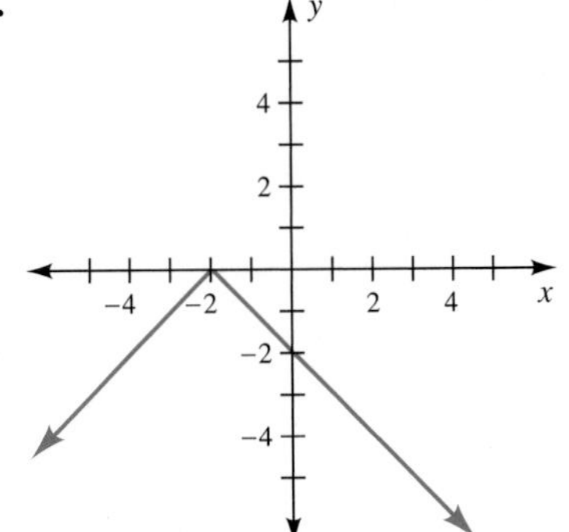

k.

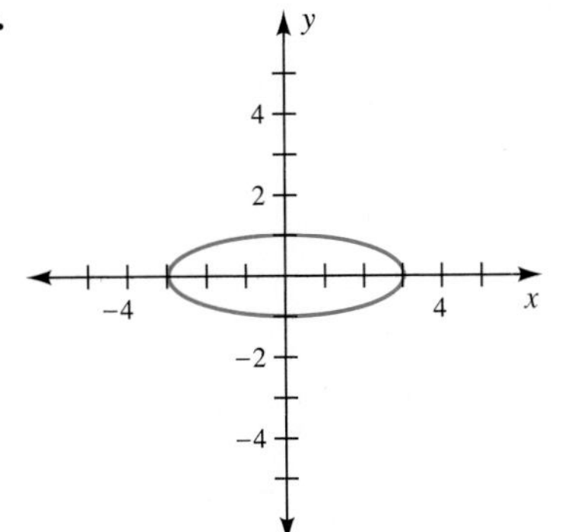

l.

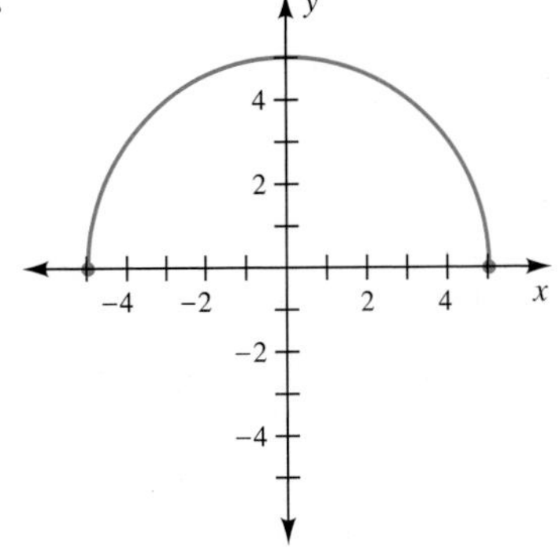

In Exercises 29–34 graph each relation.

29 $y = (x - 2)^2 - 4$

30 $(x + 2)^2 + (y - 3)^2 = 9$

31 $\frac{x^2}{64} + \frac{y^2}{25} = 1$

32 $\frac{x^2}{64} - \frac{y^2}{25} = 1$

33 $y = \sqrt{36 - x^2}$

34 $y = -\frac{1}{2}\sqrt{16 - x^2}$

In Exercises 35–38 identify each of these conic sections. Do not graph these equations.

35 $x^2 + y^2 + 12x - 6y - 4 = 0$

36 $49x^2 + 100y^2 = 4900$

37 $100x^2 + 49y^2 = 4900$

38 $4x^2 - 49y^2 = 196$

CALCULATOR USAGE (39–44)

In Exercises 39–44 use a graphics calculator to graph each of these relations.

39 $y = -(x + 3)^2 + 4$

40 $y = \sqrt{49 - x^2}$

41 $x^2 + y^2 = 8$

42 $y = \frac{1}{4}\sqrt{25 - x^2}$

43 $x^2 + 4y^2 = 4$

44 $x^2 - 4y^2 = 4$

Mastery Test for Chapter 9

Exercise numbers correspond to Section Objective numbers.

[9-1] 1 Calculate the distance between each pair of points.

a. (3, 9) and (−5, 9) **b.** (1, −2) and (7, 6)

c. (−2, 1) and (3, −11)

[9-1] 2 Calculate the midpoint between each pair of points.

a. (3, 9) and (9, −1) **b.** (4, −2) and (12, 6)

c. (−12, 13) and (0, −11)

[9-2] 3 Graph each of these parabolas.

a. $f(x) = (x - 1)^2$ **b.** $y = x^2 - 1$

c. $y = (x + 3)^2 - 4$ **d.** $y = x^2 + 8x + 11$

[9-2] 4 The profit P produced by the sale of an item is a function of the price p at which the item is sold. If $P(p) = -2p^2 + 80p - 600$, determine the maximum number of dollars of profit that can be made selling this item.

[9-3] 5 Graph each of these circles and ellipses.

a. $(x + 3)^2 + (y - 1)^2 = 9$

b. $x^2 + y^2 + 12x - 6y - 4 = 0$

c. $\frac{x^2}{9} + \frac{y^2}{36} = 1$ **d.** $4x^2 + 49y^2 = 196$

[9-4] 6 Graph each of these hyperbolas.

a. $\frac{x^2}{9} - \frac{y^2}{36} = 1$ **b.** $\frac{y^2}{9} - \frac{x^2}{36} = 1$

c. $\frac{y^2}{16} - \frac{x^2}{16} = 1$

[9-4] 7 Identify the type of conic section defined by each of these equations, but do not graph the equation.

a. $y = -5x^2 - 7x + 13$ **b.** $5x^2 + 7y^2 = 13$

c. $7x^2 + 7y^2 = 13$ **d.** $5x^2 - 7y^2 = 13$

ACTIVE APPLICATION

We will now analyze the motion of a baseball pitch—the familiar curve ball. Place the pitcher's mound at the origin of a coordinate system, and center the front edge of home plate 60 feet away on the positive x-axis. A curve ball thrown by a right-handed pitcher will have a counterclockwise spin, as viewed by the batter, and the pitch will curve toward the first quadrant. A curve ball's motion is actually quite complicated, but the curving path has several reasonable approximations.

1 We can treat the ball's path as a part of a parabola. Suppose the ball leaves the pitcher's hand at $(0, -1)$, initially traveling parallel to the x-axis and reaching home plate over the exact center of its front edge. If the ball's departure point is the vertex of the parabola, what is an equation for the ball's path?

2 If, as many baseball people say, a curve ball hooks more sharply toward the end of its flight than at the beginning, then an ellipse with major axis parallel to the x-axis may be a still better way to describe a curve ball's flight. Suppose the ball travels along the path $\dfrac{x^2}{100^2} + \dfrac{(y-9)^2}{10^2} = 1$. Where will it be when it passes the front edge of home plate? If the ball has to pass the front edge of home plate within $\frac{3}{4}$ ft of its center for a strike, will the pitch be a strike or a ball?

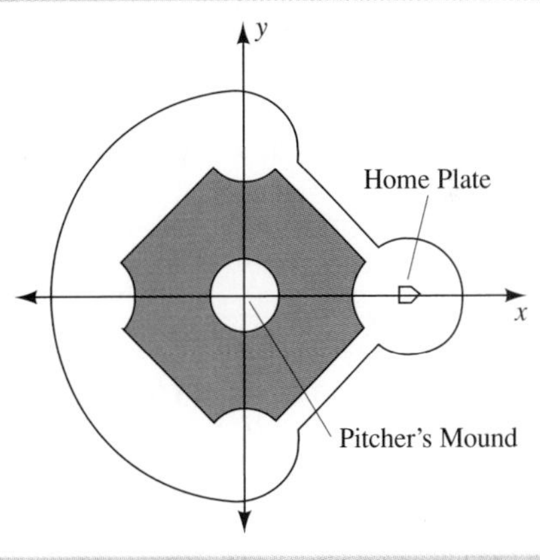

Cumulative Review of Chapters 7–9

The limited purpose of this review is to help you gauge your mastery of Chapters 7, 8, and 9. It is not meant to examine each detail from these chapters, nor is it meant to focus on specific sections that may be emphasized at any one particular school.

In Exercises 1–20 perform the indicated operations, and express the answer in simplest form.

1 $\left(\frac{8}{125}\right)^{2/3}$

2 $(36)^{-3/2}$

3 $\sqrt[2]{1{,}000{,}000} - \sqrt[3]{1{,}000{,}000}$

4 $(\sqrt{5} + \sqrt{2})^2$

5 $\dfrac{14}{\sqrt{7} + \sqrt{5}}$

6 $3\sqrt{8} - 5\sqrt{18}$

7 $(3x^{2/3}y^{1/4})(5x^{1/3}y^{-5/4})$

8 $\dfrac{12x^{5/3}y^{3/5}}{6x^{2/3}y^{-2/5}}$

9 $(16x^{8/3}y^{4/3})^{3/4}$

10 $\dfrac{(x^3y^3z^3)^{1/4}}{(xyz)^{1/2}}$

11 $\sqrt[3]{x^3y^5}$

12 $\sqrt{x^3y^5}$

13 $5(3 - 7i) + 2(4 + 2i)$

14 $(3 - 7i)(4 + 2i)$

15 $\dfrac{\sqrt{3x} + \sqrt{5y}}{\sqrt{3x} - \sqrt{5y}}$

16 $\dfrac{3 + 5i}{3 - 5i}$

17 $\sqrt{-36} - \sqrt{16}$

18 $\dfrac{130}{4 + 7i}$

19 $\dfrac{x^3 - y^3}{6x^2 - 6y^2} \div \dfrac{2x^2 + 2xy + 2y^2}{9xy}$

20 $\dfrac{2z + 11}{z^2 + z - 6} + \dfrac{2}{z + 3} - \dfrac{3}{z - 2}$

In Exercises 21–30 solve each equation.

21 $(2x - 3)(x + 4) = 0$

22 $6v^2 + 11v - 10 = 0$

23 $(2x - 3)(x + 4) = -12$

24 $2w^3 - 8w = 0$

25 $\sqrt{4a - 6} = 8$

26 $\sqrt[3]{4m + 12} = 2$

27 $\sqrt{5x + 6} = x$

28 $\dfrac{12}{x + 2} = \dfrac{6}{x - 1}$

29 $\dfrac{4}{m^2 + 2m - 3} - \dfrac{3}{m + 2} = \dfrac{1}{m - 1}$

30 $x^4 = 29x^2 - 100$

31 Solve $x^2 + 3xy - 10y^2 = 0$ for x.

32 Construct a quadratic equation in x with solutions -3 and $\frac{4}{5}$.

33 Length of a Rafter Find the length in feet of the rafter r shown in the figure. The lengths of the two braces are $r - 2$ and $r - 9$.

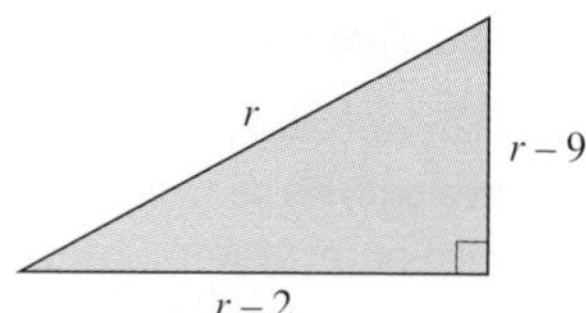

Figure for Exercise 33

34 Average Speed An automobile completed a 480-kilometer trip. If the average rate had been increased by 10 kilometers per hour, the trip would have taken 2 hours less than if the average rate had been decreased by 10 kilometers per hour. What was the average rate for this trip?

35 Solve $x^2 - 6x \geq -8$.

36 Solve $\dfrac{x^2 - x - 20}{x - 2} < 0$.

37 Calculate the distance between $(-2, -2)$ and $(3, 10)$.

38 Determine the midpoint between $(-2, 5)$ and $(6, -7)$.

39 Calculate the slope of the line through $(-1, -8)$ and $(5, 4)$.

40 Write the equation of a line through $(3, 7)$ and perpendicular to $y = -\frac{1}{5}x + 3$.

41 Write the equation of the perpendicular bisector of the line segment connecting $(1, -2)$ and $(5, 4)$.

42 Write the value of the discriminant of $2x^2 + 4x + 3 = 0$, and determine the nature of the solutions of this equation.

43 Average Cost The average cost of producing x units of a radio is given by $C(x) = x^2 - 85x + 1850$. Determine the minimum average cost.

44 Dog Food Mixture A mixture of dog food is created by mixing two sources. Source A contains 12% protein and 40% carbohydrates, and source B contains 25% protein and 30% carbohydrates. How many grams of each source should be mixed to obtain a supplement containing 272 grams of protein and 480 grams of carbohydrates?

45 Speed of an Airplane Two planes with the same air speed depart 1 hour apart and travel in opposite directions. The plane that departs first flies directly into a 50-mile-per-hour wind, and the second plane flies with this wind. After a period of time radar indicates that the first plane has traveled 375 miles and the second plane has traveled 450 miles. Determine the air speed of each plane.

46 Determine the vertex of the parabola $y = 2(x - 3)^2 + 4$.

47 Determine the center of the circle $x^2 + y^2 + 8x + 10y + 40 = 0$.

48 Determine the x- and y-intercepts of the ellipse $\dfrac{x^2}{100} + \dfrac{y^2}{49} = 1$.

49 Determine which way the hyperbola $\frac{x^2}{100} - \frac{y^2}{49} = 1$ opens, and determine its vertices.

50 Write the equation of a circle centered at $(2, -2)$ and passing through the point $(7, 10)$.

CALCULATOR USAGE (51–54)

In Exercises 51–54 use a graphics calculator to solve each equation or inequality.

51 $\sqrt{3x + 19} - 5 = 0$

52 $x^2 - x - 12 \geq 0$

53 $\frac{4x}{3} = \sqrt{25 - x^2}$

54 $(x - 1)(x - 4) = x - 1$

In Exercises 55–62 select the letter of the graph that represents the given equation.

55 $x^2 + y^2 = 4$

56 $x^2 + 4y^2 = 16$

57 $y = x^2 - 4$

58 $4x^2 + y^2 = 4$

59 $y = 4 - x^2$

60 $x^2 - y^2 = 4$

61 $y = 3x - 2$

62 $y = -2x + 3$

a.

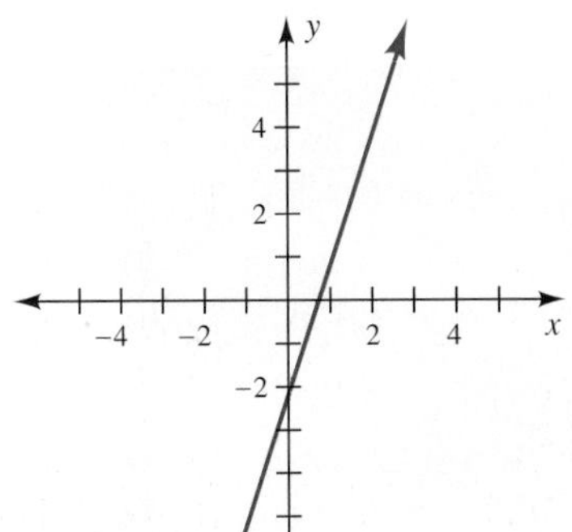

b.

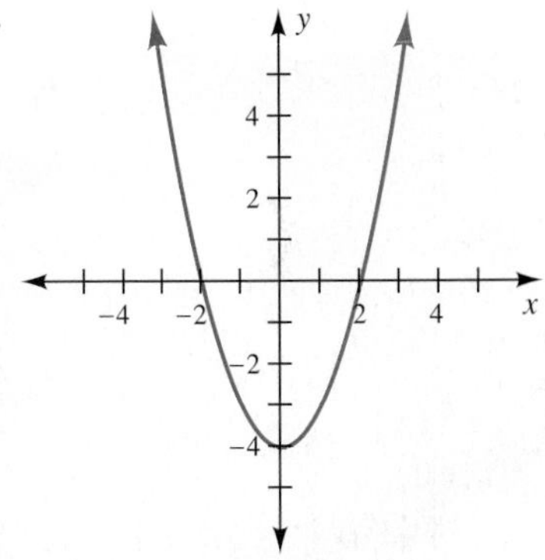

c.

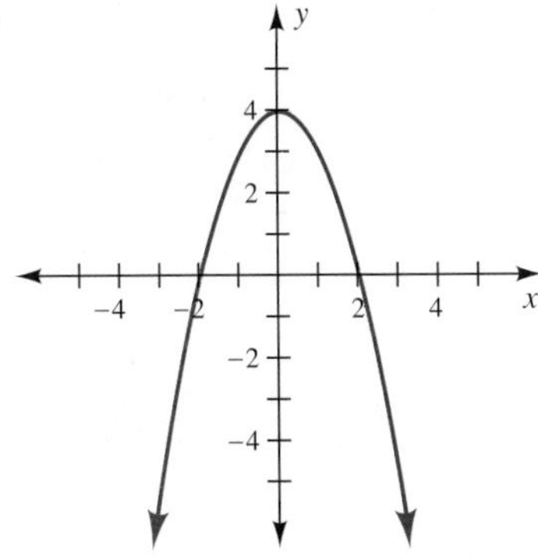

d.

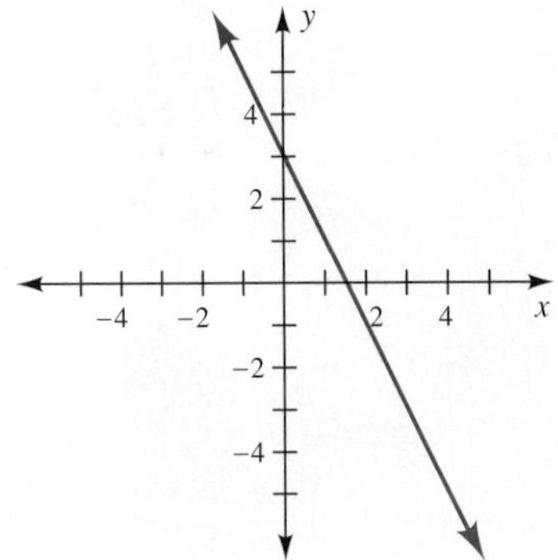

e.

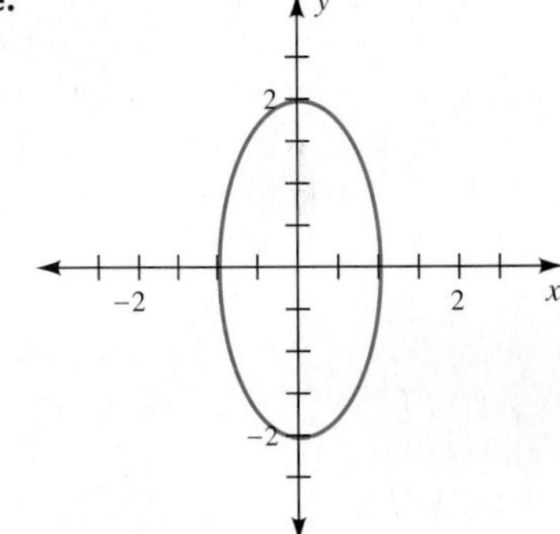

f.

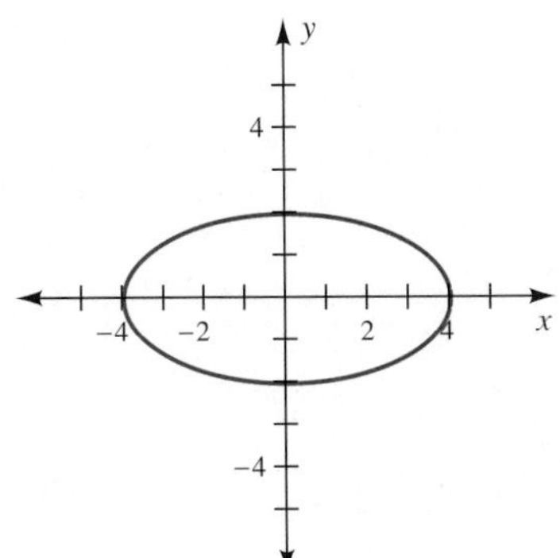

g.

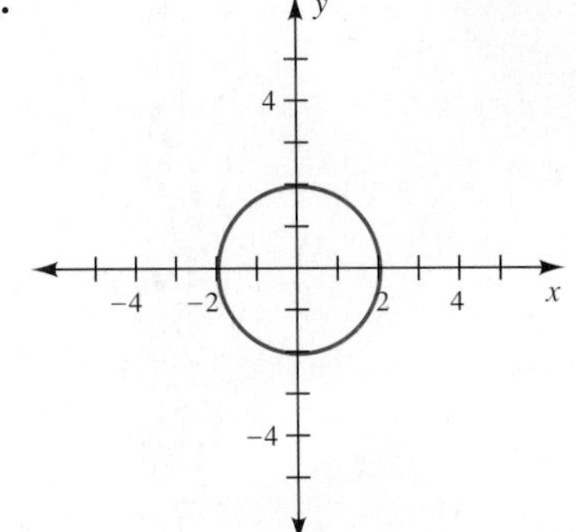

h.

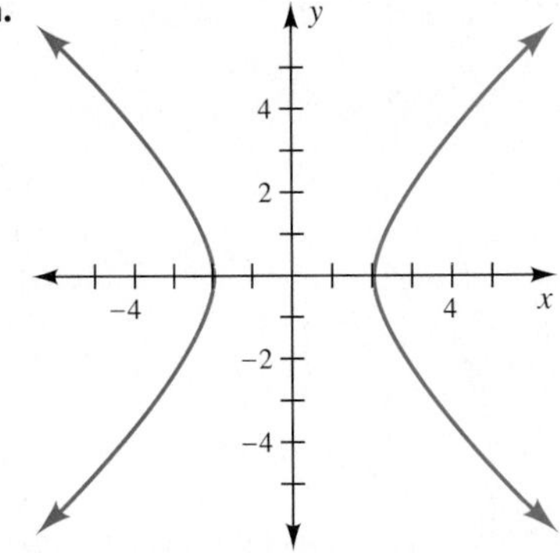

Chapter 10

Inverse, Exponential, and Logarithmic Functions

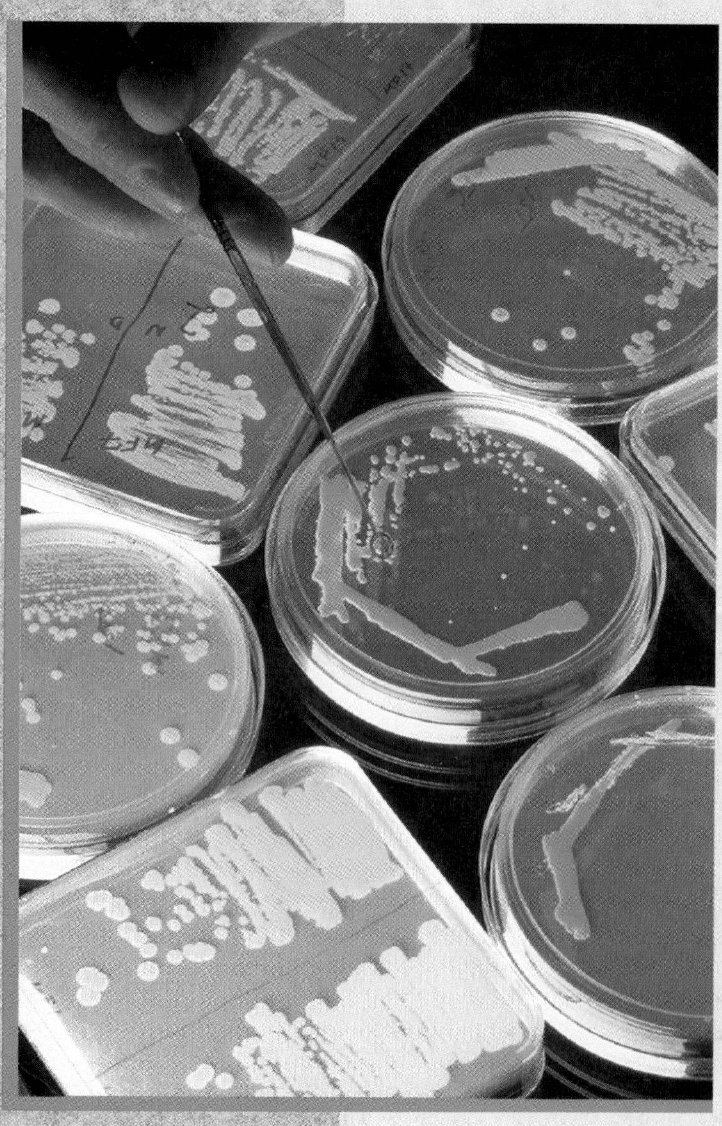

AN APPLICATION

Math for Microbes (and Microbe Watchers)

Like many microorganisms, amoebas reproduce by fission—after an amoeba has grown to a certain size, it splits into two new ones. Suppose that amoebas split once every minute. If a jar into which some amoebas are placed becomes full in exactly one hour, when is the jar half full?

The answer is 59 minutes. Surprised? Actually, the question is somewhat artificial. It takes longer than a minute for an amoeba to split (more on the order of half an hour); exactly how long it takes depends on various factors, including the food supply. Once the amoebas in the jar start to compete for food, their rate of multiplying tapers off. But the puzzle does illustrate an important point—in the early stages, at least, the growth of a population of organisms tends to be exponential, not linear.

If the number of amoebas in the jar grew linearly, we could use a linear model of the form

$$A = A_0 + kt$$

where A_0 is the original amount, t is time in minutes, and k is a constant.

With exponential growth, we must use a model of the form

$$A = A_0(2^t)$$

where, again, A_0 is the original amount and t is time in minutes. This model should help explain why our jar is half full one minute before it is completely full. When is the jar one-sixteenth full?

In this chapter you will learn more about exponents and the equations in which they appear. These equations often have application to far more significant phenomena than the growth of amoebas in a jar. For instance, in an outbreak of a disease the number of victims tends to rise exponentially at the start. This makes early detection and prevention of an epidemic difficult. Let's use the jar of amoebas again as an example. Suppose that you were looking through a room full of jars, trying to find the one jar containing amoebas. By the time there were enough amoebas in the jar for you to see them, the jar would be almost full. This is roughly the same problem that health officials face in attempting to identify and contain a new flu virus before it becomes epidemic.

CHAPTER 10 OBJECTIVES

1 **Write the inverse of a relation (Section 10-1).**

2 **Graph the inverse of a function (Section 10-1).**

3 **Graph exponential and logarithmic functions (Section 10-2).**

4 **Interpret and use logarithmic notation (Section 10-2).**

5 **Evaluate common and natural logarithms with a calculator (Section 10-3).**

6 **Use the product rule, the quotient rule, and the power rule for logarithms (Section 10-4).**

7 **Use the change-of-base formula for logarithms (Section 10-4).**

8 **Solve exponential and logarithmic equations (Section 10-5).**

9 **Use exponential and logarithmic equations to solve applied problems (Section 10-6).**

Exponential and logarithmic functions are used to solve many types of growth and decay problems. These functions are used by bankers to compute compound interest, by sociologists to predict population growth, and by archaeologists to compute the age of ancient objects through carbon-14 dating. All of these applications will be examined in this chapter. Because exponential and logarithmic functions are inverses of each other, we will start this chapter with an examination of inverse relations and inverse functions.

10-1 Inverse Relations and Inverse Functions

SECTION OBJECTIVES

1 **Write the inverse of a relation.**

2 **Graph the inverse of a function.**

Since a relation f is a set of ordered pairs (x, y), we can think of f as pairing each x value with a y value. For example, consider the relation that associates each number of books ordered with the dollar cost of these books. From the opposite perspective, if we know the cost of the books, we can determine the number of books that were ordered.

$$\text{number of books} \xrightarrow{f} \text{cost in dollars}$$

$$\text{number of books} \xleftarrow{\text{inverse of } f} \text{cost in dollars}$$

In the study of mathematical relationships, it is often valuable to be able to view both the relationship between x and y and the opposite relationship between y and x. In this section we will start a study of this topic by examining inverse functions.

THE INVERSE RELATION f^{-1}

If f is the set of ordered pairs (x, y), then f^{-1} is the set of ordered pairs (y, x) formed by reversing the ordered pairs of f.

Be careful not to misinterpret the notation f^{-1}, which represents the inverse relation of f. Although this notation appears identical to that used for negative exponents, f^{-1} does **not** represent $\frac{1}{f}$. The context of the notation with functions should make the intent of the notation clear.

EXAMPLE 1 Determining Inverses of Relations

Determine the inverse of each of these relations.

SOLUTIONS

(a) $f = \{(3, 4), (5, 8), (9, 4)\}$

$f^{-1} = \{(4, 3), (8, 5), (4, 9)\}$

Note that f is a function, but f^{-1} is only a relation because 4 is paired with two different elements.

(b)

$D \quad R$
$7 \to 1$
$8 \to 3$
$9 \to 5$

$D \quad R$
$1 \to 7$
$3 \to 8$
$5 \to 9$

Both the relation and its inverse are functions.

(c)

x	y
-3	8
0	4
0	6

x	y
8	-3
4	0
6	0

This relation is not a function, but its inverse is.

(d)

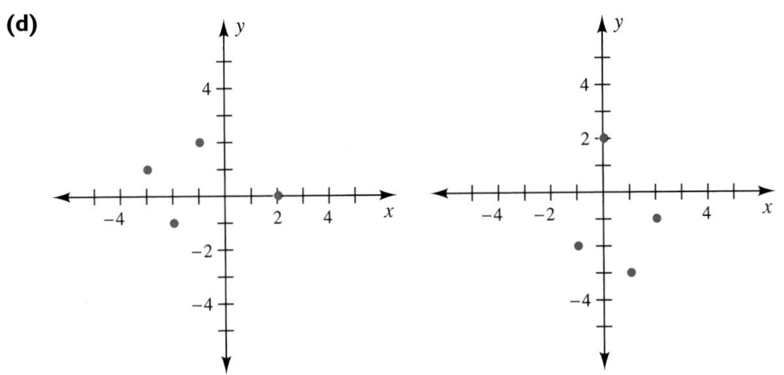

By the vertical line test both the relation and its inverse are functions. The set of points in the function is $\{(-3, 1), (-2, -1), (-1, 2), (2, 0)\}$. The set of points in the inverse of this function is $\{(1, -3), (-1, -2), (2, -1), (0, 2)\}$.

In Example 1(a), f was a function, but the inverse relation f^{-1} was not a function. Since f has two ordered pairs with the same second component—$(3, 4)$ and $(9, 4)$—its inverse has two ordered pairs with the same first component—$(4, 3)$ and $(4, 9)$. If distinct ordered pairs of a function f have distinct second components, then f^{-1} will also be a function. In this case f is a **one-to-one function** and f^{-1} is an **inverse function.**

ONE-TO-ONE FUNCTION

A function is one-to-one if different domain values produce different range values.

EXAMPLE 2 Identifying One-to-One Functions

Determine whether these functions are one-to-one and whether the inverse of the function is an inverse function.

SOLUTIONS

(a) $f = \{(-5, 6), (2, 7), (8, 9), (9, 3)\}$

This function is one-to-one, and thus the inverse, $f^{-1} = \{(6, -5), (7, 2), (9, 8), (3, 9)\}$, is an inverse function.

(b) $f = \{(-3, 9), (0, 0), (3, 9)\}$

This function is not one-to-one because both -3 and 3 are paired with 9. Thus the inverse relation, $f^{-1} = \{(9, -3), (0, 0), (9, 3)\}$, is not a function.

SELF-CHECK

Determine the inverse of $f = \{(3, \pi), (7, 11), (10, 14)\}$.

The vertical line test covered in Section 2-4 is a quick method for determining whether a graph represents a function. This test allows us to determine visually whether each x in the domain is paired with exactly one y-element in the range. Similarly, the horizontal line test is a quick method for determining whether the graph of a function represents a one-to-one function.

HORIZONTAL LINE TEST

Imagine a horizontal line placed on the same coordinate system as the given graph. If at any position this horizontal line intersects the graph of a function more than once, then the graph does *not* represent a one-to-one function.

EXAMPLE 3 Using the Horizontal Line Test

Use the horizontal line test to determine whether each graph represents a one-to-one function.

SOLUTIONS

(a)

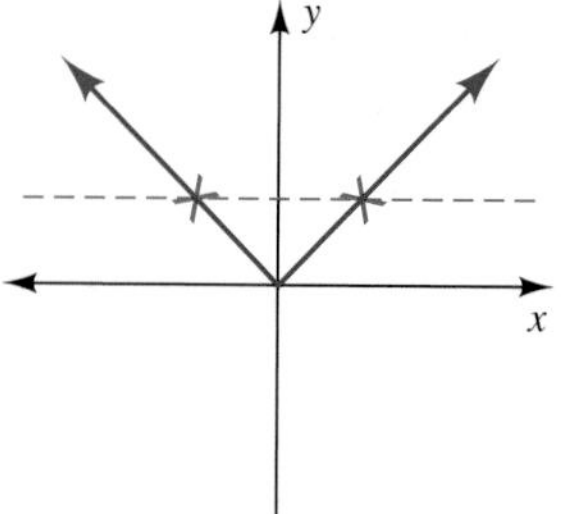

Not a one-to-one function

The horizontal line passes through more than one point of the graph of this function.

(b)

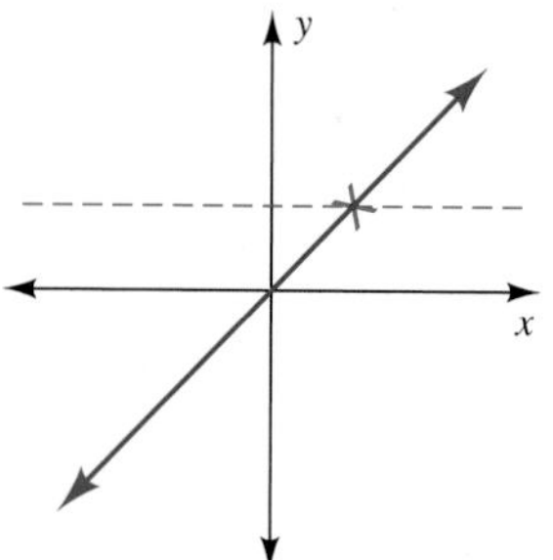

A one-to-one function

Any horizontal line will pass through exactly one point of the graph of this function.

SELF-CHECK ANSWER

$f^{-1} = \{(\pi, 3), (11, 7), (14, 10)\}$

If f is a one-to-one function defined by an equation, then the inverse function f^{-1} is also indirectly given by this equation. For example, if f is the doubling function given by $f(x) = 2x$, then the inverse function is the halving function, $f^{-1}(x) = \frac{x}{2}$. Generally, we use x to represent the domain element of a function. Although it may be slightly confusing at first, we will use x in the formulas for both f and f^{-1}. Since the inverse of a function is formed by reversing the ordered pairs of the function, the inverse of a function can be determined by interchanging x and y in the equation that defines the function.

SELF-CHECK 

In Exercises 1 and 2 determine whether each function is one-to-one.

1 $f = \{(5, 9), (7, 6), (9, 1)\}$

2

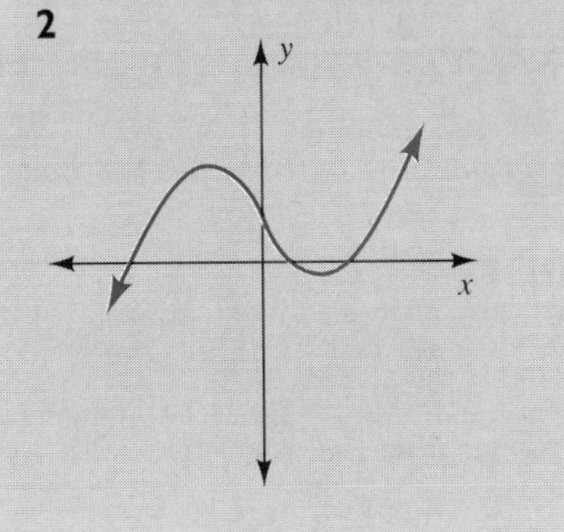

INVERSE OF A FUNCTION

To find the inverse of a function $y = f(x)$:

Step 1 Replace each x with y and each y with x.

Step 2 Solve the resulting equation for y, if possible. This inverse can then be written in functional notation by replacing y with $f^{-1}(x)$.

EXAMPLE 4 Determining the Inverse of a Function

Determine the inverse of $f(x) = 2x - 5$.

SOLUTION

$f(x) = 2x - 5$ First replace $f(x)$ by y to express the function in x-and-y notation.

$y = 2x - 5$

$x = 2y - 5$ Then interchange x and y in the equation to form the inverse.

$y = \frac{x + 5}{2}$ Solve for y.

$f^{-1}(x) = \frac{x + 5}{2}$ Rewrite the inverse using functional notation.

Check $f(3) = 2(3) - 5 = 1$ Thus (3, 1) is an ordered pair of $f(x)$.

$f^{-1}(1) = \frac{1 + 5}{2} = 3$ Thus (1, 3) is an ordered pair of $f^{-1}(x)$.

The check illustrates that $f^{-1}(x)$ reverses the ordered pairs of f. ▼

Remember from Section 2-5 that the domains and ranges of the functions considered in this text are assumed to be restricted to real numbers. Accordingly, when a function is defined by functional notation, it is usually understood that the domain is the set of all possible real numbers that yield real numbers in the range. Thus we specifically exclude from the domain values that would cause division by zero or that would result in imaginary values in the range. We will

SELF-CHECK 

3 If $f(x) = 3x + 2$, find $f^{-1}(x)$.

SELF-CHECK ANSWERS 

1 One-to-one **2** Not one-to-one **3** $f^{-1}(x) = \frac{x - 2}{3}$

now examine one method that can sometimes be used to determine the range of a function defined by functional notation.

The domain of a function f is the range of its inverse f^{-1}, and the range of f is the domain of f^{-1}. Consider $f = \{(-1, 3), (4, 7), (5, 6)\}$ and its inverse, $f^{-1} = \{(3, -1), (7, 4), (6, 5)\}$. The domain of f and the range of f^{-1} is $\{-1, 4, 5\}$. Similarly, the range of f and the domain of f^{-1} is $\{3, 7, 6\}$. Sometimes the easiest way to determine the range of f is by finding the domain of f^{-1}.

EXAMPLE 5 Determining the Domain and Range of a Function

Determine the domain and the range of $f(x) = \dfrac{x-2}{x-3}$.

SOLUTION

The domain is $\mathbb{R} \sim \{3\}$. — The value of 3 must be excluded from the domain to avoid division by zero.

To determine the range, we will first find f^{-1}.

$f(x) = \dfrac{x-2}{x-3}$	
$y = \dfrac{x-2}{x-3}$	Replace $f(x)$ by y to express the function in x-and-y notation.
$x = \dfrac{y-2}{y-3}$	Interchange x and y in the equation to form the inverse.
$x(y-3) = y-2$	Multiply both members by $y - 3$.
$xy - 3x = y - 2$	
$xy - y = 3x - 2$	Collect the y terms in the left member, and then factor out y.
$y(x-1) = 3x - 2$	
$y = \dfrac{3x-2}{x-1}$	Solve for y by dividing both members by $x - 1$.
$f^{-1}(x) = \dfrac{3x-2}{x-1}$	Rewrite the inverse using functional notation.

The domain of f^{-1} is $\mathbb{R} \sim \{1\}$.
Thus the range of f is also $\mathbb{R} \sim \{1\}$. — The value of 1 is excluded from this domain to avoid division by zero.

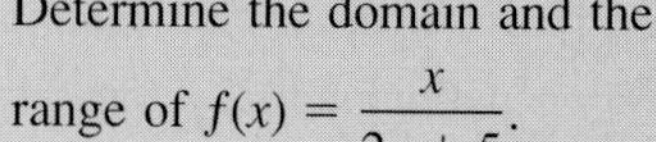

SELF-CHECK

Determine the domain and the range of $f(x) = \dfrac{x}{2x+5}$.

EXAMPLE 6 Graphing a Function and Its Inverse

Graph $f(x) = \dfrac{1}{3}x + 2$ and its inverse, $f^{-1}(x) = 3x - 6$, on the same coordinate system.

SELF-CHECK ANSWER 

Domain: $\mathbb{R} \sim \{-2.5\}$; range: $\mathbb{R} \sim \{0.5\}$

SOLUTION

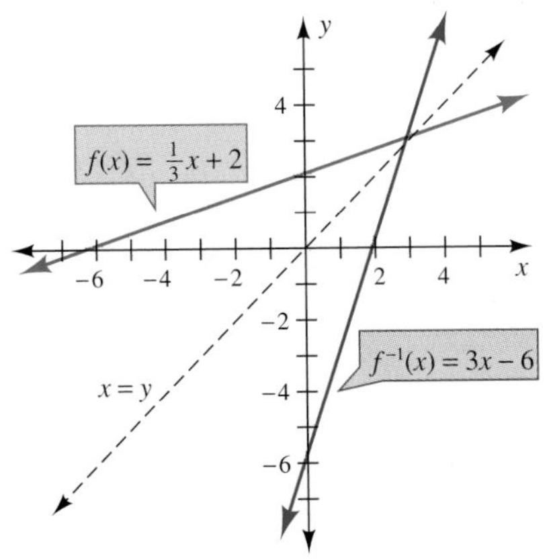

The graphs of both f and f^{-1} are straight lines. Inspection of the slope-intercept form reveals that f has slope $\frac{1}{3}$ and y-intercept 2 and f^{-1} has slope 3 with y-intercept -6.

Note that these graphs are symmetric about the line $x = 7$.

▼

A Geometric Viewpoint: Symmetry of f and f^{-1}

The graphs of $y = f(x)$ and $y = f^{-1}(x)$ are symmetric about the line $x = y$. One way to observe this symmetry is to make identical copies of $y = f(x)$ on separate sheets of clear plastic. To graph $y = f^{-1}(x)$, flip and rotate the second copy so as to interchange the x-axis and the y-axis, as shown in the figure.

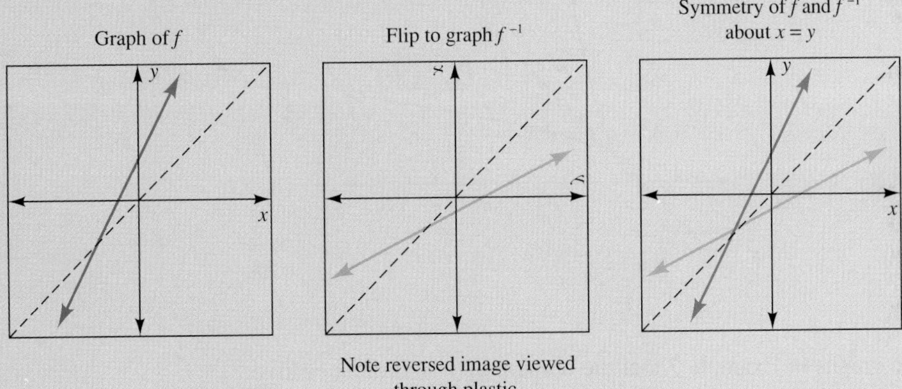

The graphs of f and f^{-1} are always symmetric about the line $x = y$ because the points (a, b) and (b, a) are mirror images about the graph of $x = y$. The reason for this symmetry is illustrated in Figure 10-1.

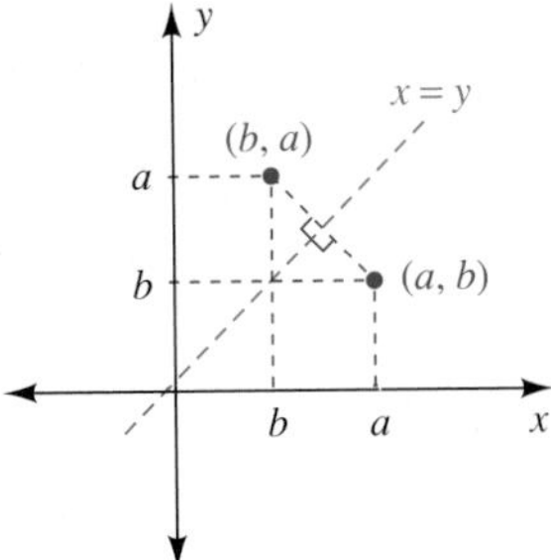

Figure 10-1 *Symmetry of (a, b) and (b, a) about $x = y$*

EXAMPLE 7 Graphing a Function and Its Inverse

Graph $f(x) = \sqrt{x}$ and its inverse on the same coordinate system.

SOLUTION

x	$f(x)$
0	0
1	1
4	2
9	3

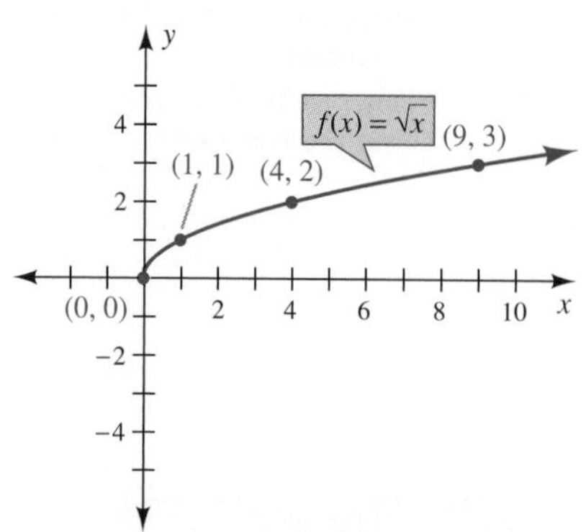

First graph $f(x) = \sqrt{x}$ by plotting a few points and then connecting them with a smooth curve.

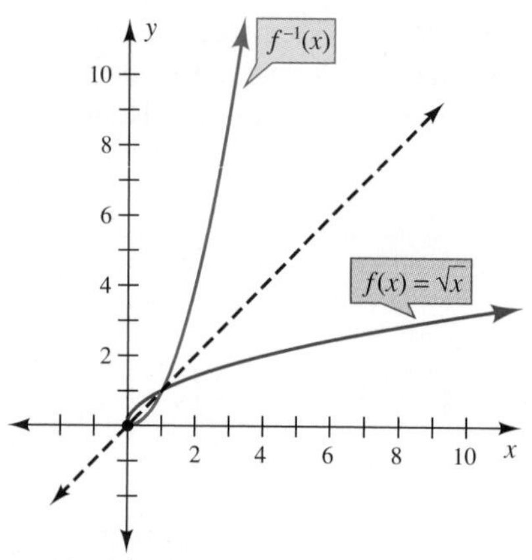

Then use the symmetry of f and f^{-1} about $x = y$ to graph $f^{-1}(x)$.

Note from the graphs in Example 7 that the domain of f is all nonnegative real numbers and the range of f is all nonnegative real numbers. Thus the domain and the range of f^{-1} are also all nonnegative real numbers. The inverse of $f(x) = \sqrt{x}$ is determined below.

$$f(x) = \sqrt{x}$$

$$y = \sqrt{x}$$ Express in x-and-y notation. Note that $x \geq 0$ and $y \geq 0$.

$$x = \sqrt{y}$$ Interchange x and y in the equation to form the inverse for $x \geq 0$ and $y \geq 0$.

$$x^2 = y$$ Then solve for y.

$$f^{-1}(x) = x^2 \quad \text{for } x \geq 0$$ Rewrite the inverse using functional notation.

The domain of $f^{-1}(x) = x^2$ is restricted to $x \geq 0$, as noted in the second graph in Example 7. This restriction forces the function to be one-to-one, so its inverse is also a function. The graph of $y = x^2$ for all real numbers x would be a parabola, which is not one-to-one; the restriction to $x \geq 0$ yields a graph that is only the right branch of this parabola.

Some models of graphics calculators have the ability to graph both a function and its inverse when you enter only the original function $y = f(x)$. The keystrokes for accomplishing this on a TI-82 graphics calculator are illustrated in the next example.

EXAMPLE 8 Using a Calculator to Graph a Function and Its Inverse

Use a graphics calculator to graph $y = \sqrt{5x + 25} - 5$ and its inverse on the same display window.

SOLUTION

Keystrokes

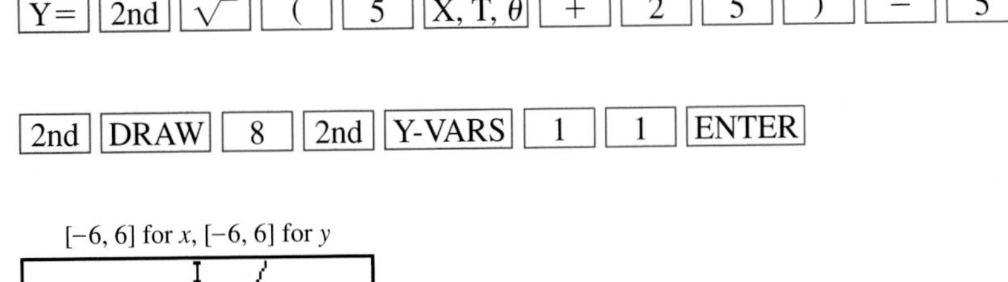

First enter the function. Then select option 8, the DrawInv option, from the DRAW menu. From the Y-VARS menu, choose to draw the inverse of Y_1, the function just defined. The window displayed to the left assumes that the window of values has already been set to $[-6, 6]$ for x and $[-6, 6]$ for y.

$[-6, 6]$ for x, $[-6, 6]$ for y

SELF-CHECK 

Graph $f(x) = 2x - 1$ and $f^{-1}(x) = \dfrac{x + 1}{2}$ on the same coordinate system.

SELF-CHECK ANSWER

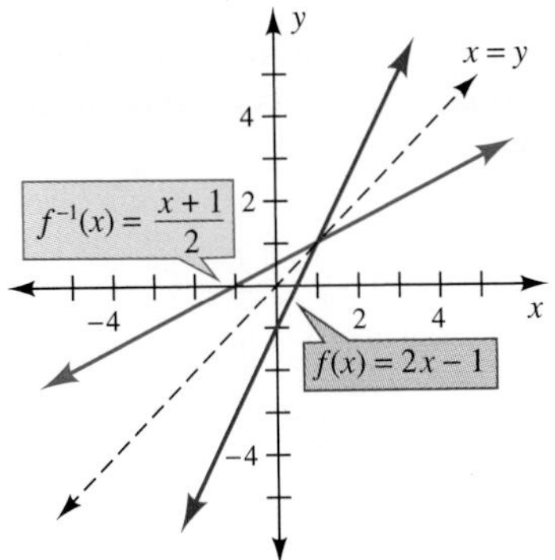

Exercises 10-1

A

In Exercises 1–12 write the inverse of each relation using ordered-pair notation.

1 $\{(1, 4), (3, 11), (8, 2)\}$

2 $\{(2, -1), (0, 3), (-4, 6)\}$

3 $\{(-3, 2), (-1, 2), (0, 2), (2, 2)\}$

4 $\{(-0.7, \pi), (-0.5, \pi), (1.3, \pi)\}$

5 $\{(a, b), (c, d)\}$

6 $\{(w, x), (y, z)\}$

7 $D \quad R$

$5 \to -5$

$5 \to 0$

$5 \to 5$

8 $D \quad R$

$8 \to 1$

$9 \to 1$

$10 \to 1$

9

x	y
−11	−1
−3	−1
3	1
11	1

10

x	y
0.7	0
0.8	0
0.9	0
1.0	1

11 **12**

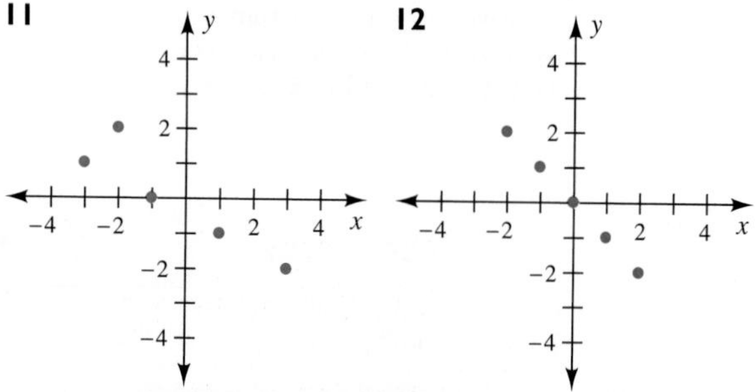

In Exercises 13–22 determine whether each function is one-to-one. Use the horizontal line test to determine whether the graphs represent one-to-one functions.

13 $\{(-\pi, 4), (3, \pi), (7, -9)\}$

14 $\{(-8, 9), (7, 13), (8, 9)\}$

15 $\{(7.5, 7), (7.8, 7), (8.3, 8)\}$

16 $\{(-5, 2), (-4, -1), (3, 7), (8, 8)\}$

17 **18**

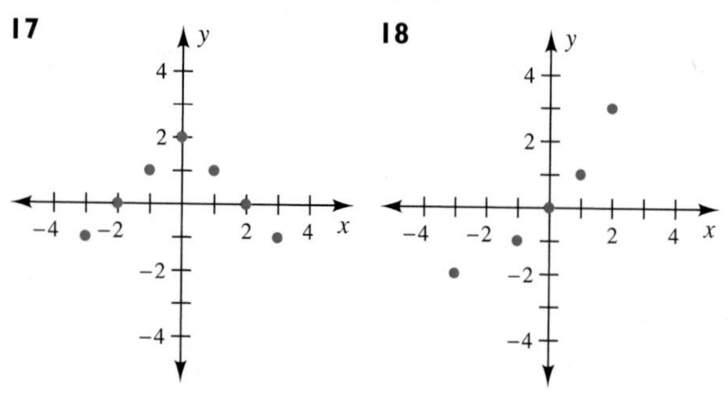

19 **20**

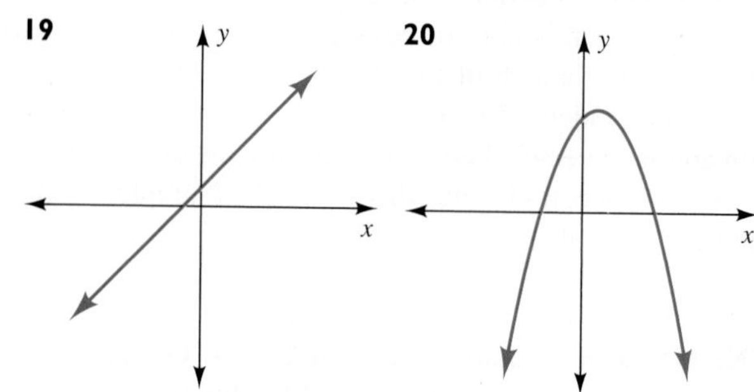

21 **22**

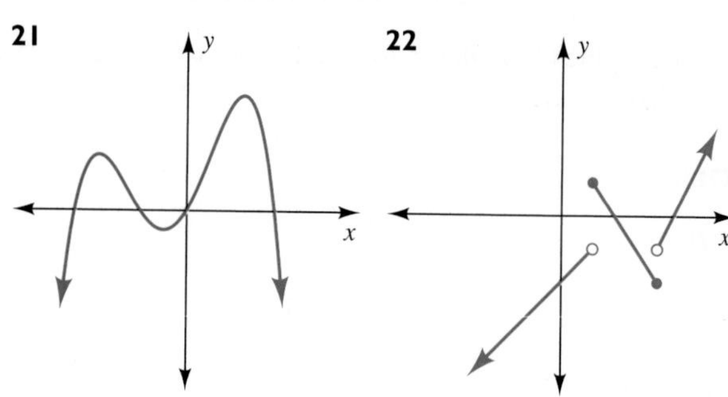

In Exercises 23–30 write the inverse of each function using functional notation.

23 $f(x) = 5x + 2$

24 $f(x) = 2x - 5$

25 $g(x) = x - 3$

26 $g(x) = x + 4$

27 $h(x) = \frac{1}{3}x - 1$

28 $h(x) = -\frac{1}{4}x + 1$

29 $f(x) = -x$

30 $f(x) = x$

In Exercises 31–40 sketch the inverse of each relation graphed. Show both the relation and its inverse on the same coordinate system.

31 **32**

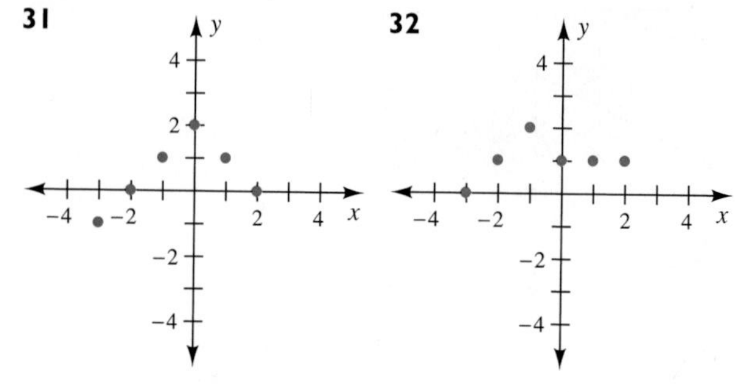

33 **34**

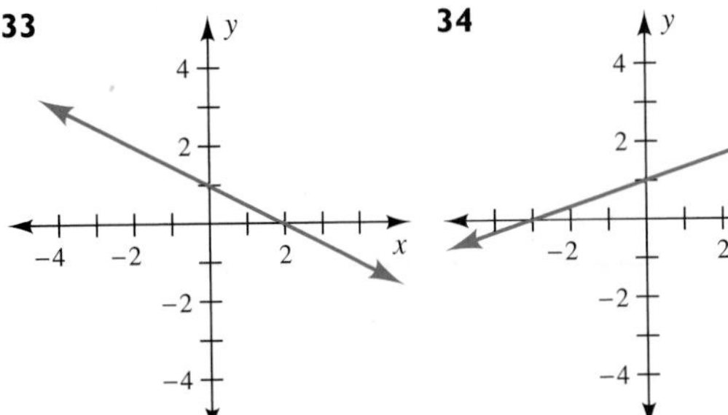

35 **36**

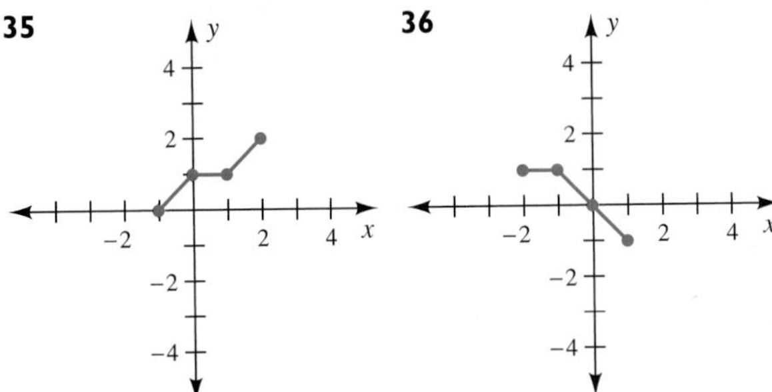

37 **38**

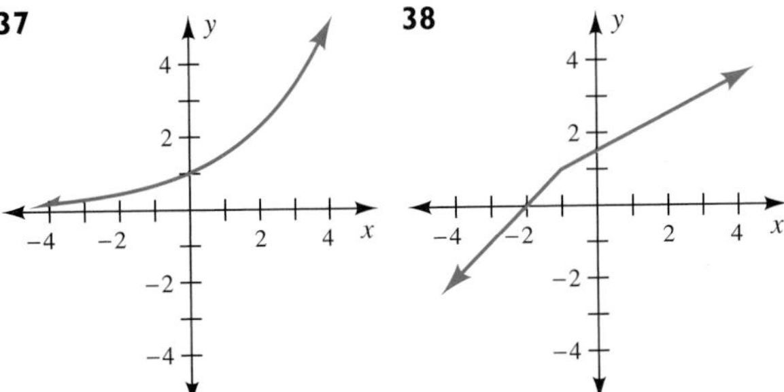

39 **40**

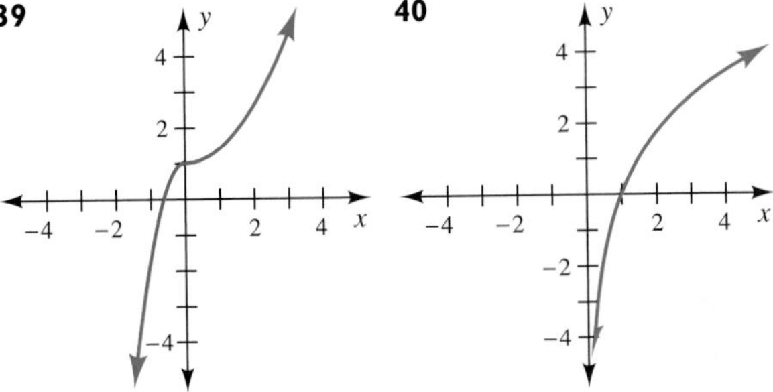

In Exercises 41–48 use $f(x) = 5x + 3$ and $f^{-1}(x) = \dfrac{x-3}{5}$ to evaluate each expression.

41 $f(5)$ **42** $f(2)$ **43** $f^{-1}(28)$

44 $f^{-1}(13)$ **45** $f(h)$ **46** $f^{-1}(h)$

47 $f^{-1}(5h + 3)$ **48** $f\left(\dfrac{h-3}{5}\right)$

B

Calculator Usage (49–56)

In Exercises 49–56 use a graphics calculator to graph the given function and its inverse on the same coordinate system.

49 $f(x) = -3x + 1$ **50** $f(x) = -2x + 3$

51 $g(x) = \dfrac{1}{4}x - 2$ **52** $g(x) = \dfrac{1}{5}x - 1$

53 $f(x) = \sqrt{2x + 4}$ **54** $f(x) = \sqrt{3x - 6}$

55 $g(x) = \sqrt[3]{x}$ **56** $g(x) = \sqrt[4]{x}$

In Exercises 57–62, given $f(x) = 5x - 2$, evaluate each of the following expressions for the given value of x:

a. $f(x)$ **b.** $f^{-1}(x)$ **c.** $\dfrac{1}{f(x)}$

57 $x = 3$ **58** $x = -2$ **59** $x = 18$

60 $x = 0$ **61** $x = a$ **62** $x = \dfrac{a+2}{5}$

In Exercises 63–68 determine the inverse of each function, and then determine the domain and the range of both the function and its inverse.

63 $f(x) = \dfrac{1}{x-1}$ **64** $f(x) = \dfrac{1}{x+2}$

65 $g(x) = \dfrac{x+3}{x-4}$ **66** $g(x) = \dfrac{x-5}{x+3}$

67 $h(x) = \dfrac{x-2}{2x+1}$ **68** $h(x) = \dfrac{x+5}{3x-1}$

69 a. Graph $f(x) = x^3$ and $f^{-1}(x)$ on the same coordinate system.

b. Write the inverse using functional notation.

c. State the domain and the range of both f and f^{-1}.

70 Production Cost The cost of producing x units of a product is given by the function $C(x) = 12x + 350$.

a. Determine the formula for $C^{-1}(x)$.

b. What does the variable x represent in $C^{-1}(x)$?

c. Determine the number of units that can be produced for \$1934.

Discussion Question

71 The equation $C = \dfrac{5}{9}(F - 32)$ is a formula for converting a given Fahrenheit temperature to Celsius. Find the formula that can be used to convert Celsius temperatures to Fahrenheit. Describe inverse functions, using these two formulas to illustrate your description.

CHALLENGE QUESTION

The pieces of the tangram shown can be used to form three different squares. Pieces I–IV can be used to form one square, and Piece V is itself a square. If all five pieces are used, a larger third square can be formed. Use the drawings in the text as a template to carefully make your own pieces on rigid paper. Then arrange these pieces to form the squares.

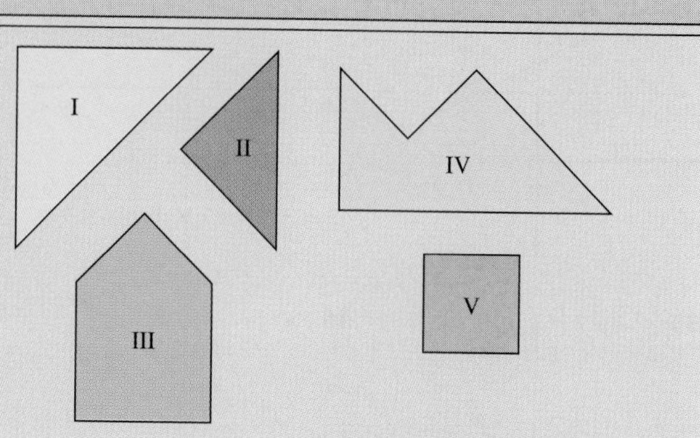

SECTION SKILL REINFORCERS

Simplify each of these expressions.

1 0^{10} **2** 10^0 **3** 10^{-1} **4** -1^{10} **5** $(-1)^{10}$

10-2 Exponential and Logarithmic Functions

SECTION OBJECTIVES

1 **Graph exponential and logarithmic functions.**

2 **Interpret and use logarithmic notation.**

An **exponential function** is a function defined by an equation of the form $y = b^x$ or $f(x) = b^x$; it has a variable rather than a constant in the exponent. Whereas $y = 2^x$ is an exponential function, $y = x^2$ is not an exponential function because the exponent is the constant 2.

EXPONENTIAL FUNCTION

If $b > 0$ and $b \neq 1$, then $f(x) = b^x$ is an exponential function with base b.

EXAMPLE 1 Identifying Exponential Functions

Determine which of these equations define exponential functions.

	SOLUTIONS	
(a) $y = 3^x$	An exponential function	This function has a base of 3.
(b) $y = x^3$	Not an exponential function	The exponent is the constant 3, not a variable.

(c) $f(x) = \left(\frac{4}{5}\right)^x$ An exponential function This function has a base of $\frac{4}{5}$.

(d) $f(x) = (-5)^x$ Not an exponential function The base of this exponential equation is -5. Since the base is negative, this equation does not represent an exponential function.

(e) $y = 1^x$ Not an exponential function $y = 1^x$ simplifies to $y = 1$, since 1 to any power equals 1. Thus it is a constant function. ▼

The domain of an exponential function $y = b^x$ is understood to be the set of all real numbers. To graph $y = 2^x$, we can plot some selected points and then connect these points with a smooth curve to illustrate the entire function, as shown below. A line graph approximation of $y = 2^x$ is shown side by side with the actual graph of the function.

Table of Values

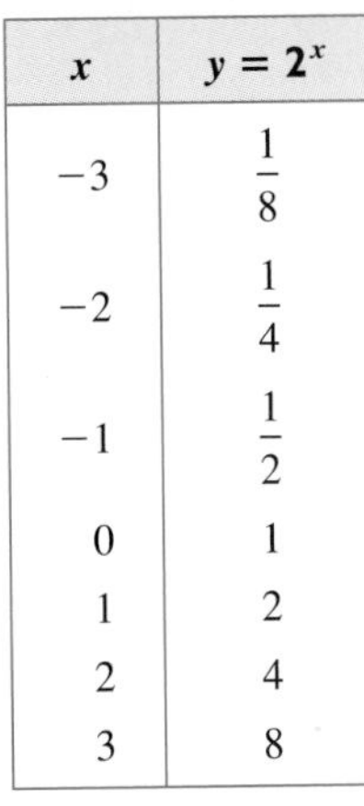

x	$y = 2^x$
-3	$\frac{1}{8}$
-2	$\frac{1}{4}$
-1	$\frac{1}{2}$
0	1
1	2
2	4
3	8

Line Graph Approximation of $y = 2^x$

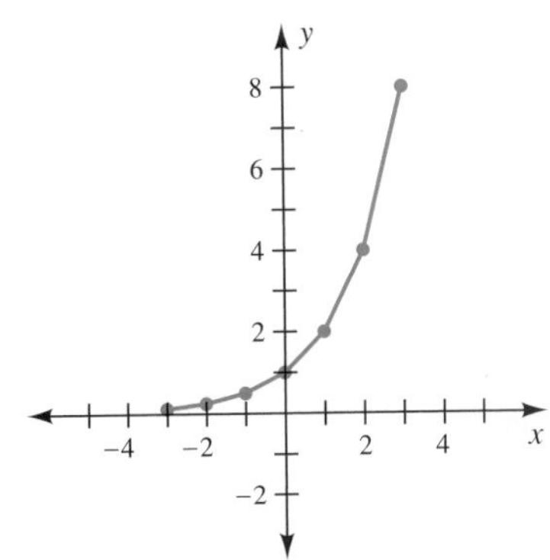

Actual Graph of $y = 2^x$

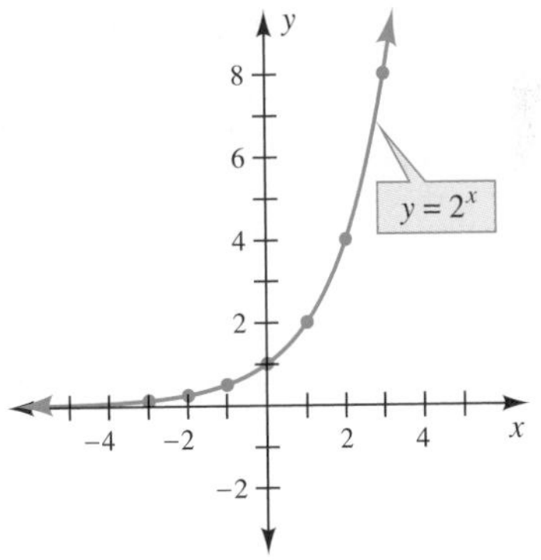

The actual graph of $y = 2^x$ shown above is based on the assumption that 2^x is defined for all real number exponents, when in fact the only exponents defined so far in this text have been rational numbers. A precise definition of expressions with irrational exponents, such as 2^π, is usually postponed until calculus. For now it is enough to know that the actual graph of $y = 2^x$ properly locates the point $(\pi, 2^\pi)$ and that for rational values of x close to π the value of 2^x will also be close to 2^π. We will assume throughout the rest of this text that $y = b^x$ is defined for any real number x. Values such as 2^π can be approximated with a calculator as needed.

EXAMPLE 2 Calculator Approximation

Use a calculator to approximate 2^π to five significant digits.

SOLUTION

Keystrokes

Scientific

[2] [y^x] [π] [=] $\rightarrow$ 8.8249778

Graphics

[2] [∧] [2nd] [π] [ENTER] $\rightarrow 8.824977827$

Answer $2^{\pi} \approx 8.8250$

SELF-CHECK

1 Approximate $5^{\sqrt{2}}$ to five significant digits.

The equations in the next example can all be solved easily without the use of a calculator because they involve familiar powers of 2, 3, 4, or other small integers. The properties we will use are given in the following box.

PROPERTIES OF EXPONENTIAL FUNCTIONS

For real exponents x and y and bases $a > 0$, $b > 0$,

1 for $b \neq 1$, $b^x = b^y$ if and only if $x = y$.

2 for $x \neq 0$, $a^x = b^x$ if and only if $a = b$.

EXAMPLE 3 Solving Exponential Equations

Solve each of these equations.

SOLUTIONS

(a) $3^{x-4} = 9$

$3^{x-4} = 9$

$3^{x-4} = 3^2$ — Substitute 3^2 for 9 in order to express both members in terms of the common base 3.

$x - 4 = 2$ — The exponents are equal since the bases are the same.

$x = 6$

(b) $4^x = \dfrac{1}{8}$

$4^x = \dfrac{1}{8}$

$(2^2)^x = 2^{-3}$ — Express both 4 and $\frac{1}{8}$ in terms of the common base 2.

$2^{2x} = 2^{-3}$ — Use the power rule for exponents to simplify the left member of this equation.

$2x = -3$ — The exponents are equal since the bases are the same.

$x = -\dfrac{3}{2}$

(c) $(b + 2)^4 = 625$

$(b + 2)^4 = 625$

$(b + 2)^4 = 5^4$ — Substitute 5^4 for 625.

$b + 2 = 5$ — The bases are equal since the exponents are the same.

$b = 3$

SELF-CHECK

Solve each of these equations.

2 $2^{x+7} = 16$ **3** $27^x = 9$

4 $b^{-3} = \dfrac{1}{64}$

SELF-CHECK ANSWERS 

1 $5^{\sqrt{2}} \approx 9.7385$ **2** $x = -3$ **3** $x = \dfrac{2}{3}$ **4** $b = 4$

If the base b is greater than 1, as in $y = 2^x$, then the graph of the function rises rapidly to the right as the base is used repeatedly as a factor. Thus an exponential function $y = b^x$ with $b > 1$ is an increasing function, called an **exponential growth function.** If $b < 1$, then the graph of $y = b^x$ declines rapidly to the right and is called an **exponential decay function.** A line graph approximation of the exponential decay function $y = \left(\frac{1}{2}\right)^x$ is shown side by side with the actual graph of the function.

Table of Values

x	$y = \left(\frac{1}{2}\right)^x$
-3	8
-2	4
-1	2
0	1
1	$\frac{1}{2}$
2	$\frac{1}{4}$
3	$\frac{1}{8}$

Line Graph Approximation of $y = \left(\frac{1}{2}\right)^x$

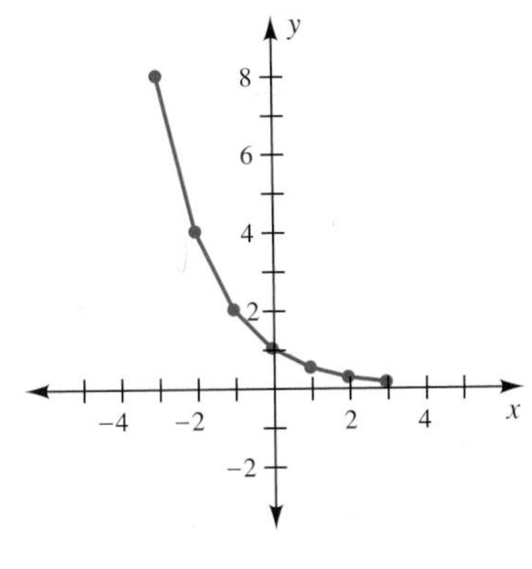

Actual Graph of $y = \left(\frac{1}{2}\right)^x$

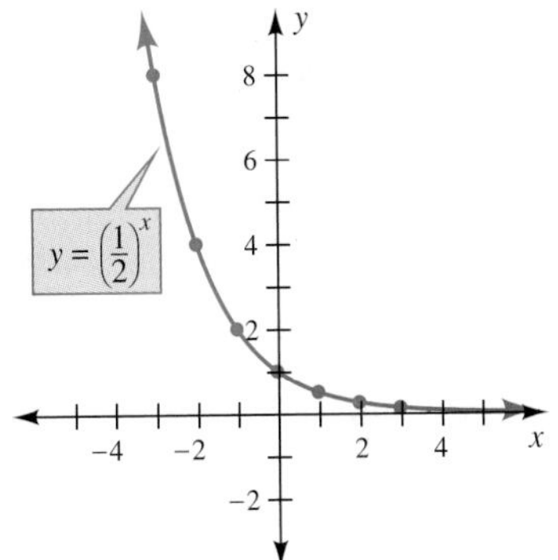

Note that both the graph of the exponential growth function $y = 2^x$ and the graph of the exponential decay function $y = \left(\frac{1}{2}\right)^x$ pass through the point (0, 1). Since $b^0 = 1$ (for $b \neq 0$), the graph of every exponential function $y = b^x$ will pass through the point (0, 1). Also note that both the graph of $y = 2^x$ and the graph of $y = \left(\frac{1}{2}\right)^x$ approach but never touch the x-axis. These characteristics are common to all growth and decay functions.

GROWTH AND DECAY FUNCTIONS

If $b > 1$, then $f(x) = b^x$ is called a growth function.

If $0 < b < 1$, then $f(x) = b^x$ is called a decay function.

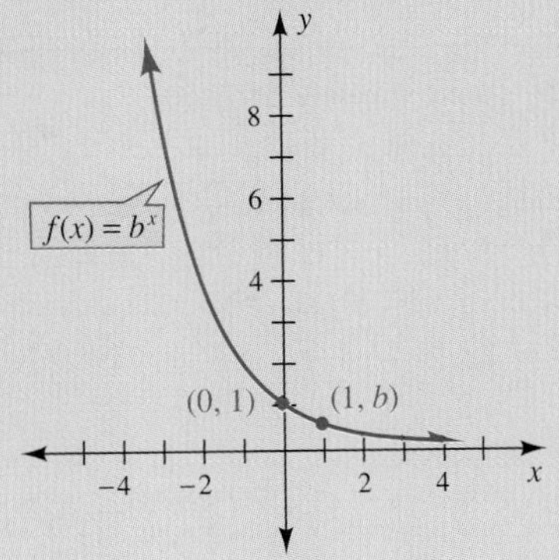

EXAMPLE 4 Compound Interest

The formula $A = P(1 + r)^t$ can be used to compute the total amount of money A that accumulates when a principal P is invested at a yearly interest rate r and left to compound for t years. Using this growth formula, compute the value of \$5000 invested at 9.25% and left to compound for 8 years.

SOLUTION

$A = P(1 + r)t$

$= 5000\,(1 + 0.0925)^8$ — Substitute the given values into the formula. Now use a calculator with a power key.

$= 5000\,(1.0925)^8$

$\approx 5000(2.029\,418\,3)$ — The answer may vary slightly depending on your calculator.

$A \approx \$10{,}147$ — This answer was rounded to five significant digits.

We will now consider the function defined by $x = 2^y$, where x is the independent variable. This function is formed by interchanging the roles of x and y in the function $y = 2^x$. Thus, in the language of Section 10-1, $x = 2^y$ is the inverse of $y = 2^x$. (Although you can master this material without having studied inverse functions, the additional insight provided by a knowledge of inverses is quite valuable.)

We cannot solve the function defined by $x = 2^y$ for y by using the basic operations and notation covered so far in this text. Thus the use of logarithmic notation to represent this useful function is introduced in the following box.

SELF-CHECK

1 Graph $y = 3^x$ and $y = \left(\frac{1}{3}\right)^x$ on the same rectangular coordinate system.

2 Use the formula $A = P(1 + r)^t$ and a calculator to determine the value of \$450 invested at 8.75% and left to compound for 7 years.

LOGARITHMIC FUNCTION

For $x > 0$, $b > 0$, and $b \neq 1$,

$$y = \log_b x \quad \text{if and only if} \quad b^y = x$$

Base: b
Logarithm or exponent*: y

The notation $\log_b x = y$ is read as "the logarithm of x base b is equal to y."

**Think:* Logarithms are exponents.

SELF-CHECK ANSWERS

1

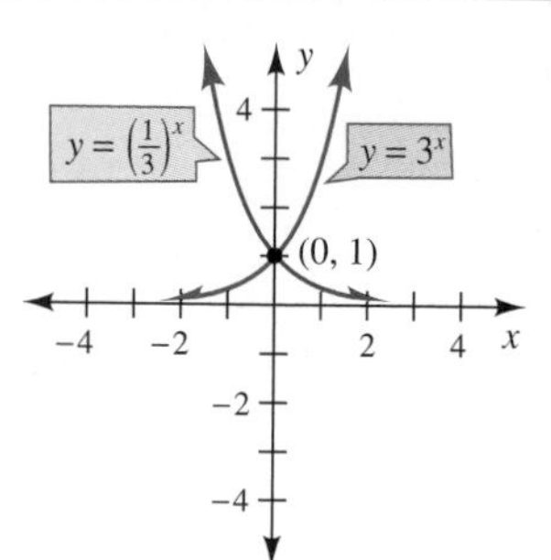

2 Approximately \$809.50

The relationship between the logarithmic form and the exponential form is illustrated in Example 5.

EXAMPLE 5 Translating Logarithmic Equations

Write the word equation for each of these logarithmic equations, and then write the corresponding exponential form.

SOLUTIONS

Logarithmic Form	*Word Equation*	*Exponential Form*
(a) $\log_5 25 = 2$	The log of 25 base 5 is 2.	$5^2 = 25$
(b) $\log_2 8 = 3$	The log of 8 base 2 is 3.	$2^3 = 8$
(c) $\log_3\left(\frac{1}{3}\right) = -1$	The log of $\frac{1}{3}$ base 3 is -1.	$3^{-1} = \frac{1}{3}$
(d) $\log_7\sqrt{7} = \frac{1}{2}$	The log of $\sqrt{7}$ base 7 is $\frac{1}{2}$.	$7^{1/2} = \sqrt{7}$

A Mathematical Note

Logarithms were invented by John Napier (1550–1617), a Scottish mathematician. His first publication on logarithms was published in 1614.

Napier's other inventions include "Napier's Bones," a mechanical calculating device that is an ancestor of the slide rule, and a hydraulic screw and revolving axle for controlling the water level in coal pits. He also worked on plans to use mirrors to burn and destroy enemy ships.

To evaluate $\log_b x$, think "What exponent on b is needed to obtain x?" Problems involving familiar powers of small integers can often be solved easily without use of a calculator. For problems involving more complicated values, a calculator is generally used.

EXAMPLE 6 Evaluating Logarithms by Inspection

Determine the value of each of these logarithms by inspection.

SOLUTIONS

(a) $\log_5 125$	$\log_5 125 = 3$	In exponential form, $5^3 = 125$.
(b) $\log_2\frac{1}{8}$	$\log_2\frac{1}{8} = -3$	In exponential form, $2^{-3} = \frac{1}{8}$.
(c) $\log_{49} 7$	$\log_{49} 7 = \frac{1}{2}$	In exponential form, $49^{1/2} = 7$.
(d) $\log_5 0$	$\log_5 0$ is undefined.	The argument of a logarithmic function cannot be zero.
(e) $\log_5(-25)$	$\log_5(-25)$ is undefined.	The argument of a logarithmic function cannot be negative.

Four properties that result from the definition of a logarithm are given in the following box. These important properties are used often in solving the logarithmic equations that arise in the solution of some word problems. We will use these properties to evaluate the logarithms in the next example.

SELF-CHECK 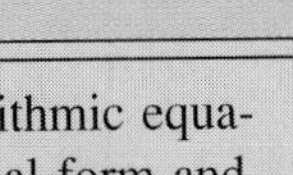

Write each logarithmic equation in exponential form and each exponential equation in logarithmic form.

1 $\log_7 49 = 2$

2 $\log_6\sqrt[5]{6} = \frac{1}{5}$

3 $3^{-4} = \frac{1}{81}$ **4** $6^0 = 1$

SELF-CHECK ANSWERS 

1 $7^2 = 49$ **2** $6^{1/5} = \sqrt[5]{6}$ **3** $\log_3\frac{1}{81} = -4$ **4** $\log_6 1 = 0$

PROPERTIES OF LOGARITHMIC FUNCTIONS

For $b > 0$ and $b \neq 1$,

1 $\log_b 1 = 0$ since $b^0 = 1$

2 $\log_b b = 1$ since $b^1 = b$

3 $\log_b \frac{1}{b} = -1$ since $b^{-1} = \frac{1}{b}$

4 $\log_b b^x = x$ since $b^x = b^x$

EXAMPLE 7 Using Properties to Evaluate Logarithms

Determine the value of each of these logarithms by inspection.

	SOLUTIONS	
(a) $\log_{19} 1$	$\log_{19} 1 = 0$	In exponential form, $19^0 = 1$.
(b) $\log_7 7$	$\log_7 7 = 1$	In exponential form, $7^1 = 7$.
(c) $\log_5 \frac{1}{5}$	$\log_5 \frac{1}{5} = -1$	In exponential form, $5^{-1} = \frac{1}{5}$.
(d) $\log_8 8^{2y}$	$\log_8 8^{2y} = 2y$	In exponential form, $8^{2y} = 8^{2y}$.

SELF-CHECK

Determine the value of each logarithm.

1 $\log_2 64$ **2** $\log_4 64$

3 $\log_8 64$ **4** $\log_7 \frac{1}{7}$

The equations in the next example can be solved without use of a calculator since the numbers involve well-known powers. We will examine equations of this type in more detail in Section 10-5.

EXAMPLE 8 Solving Equations

Solve each of these equations for x.

SOLUTIONS

(a) $\log_3(x - 1) = 2$

$\log_3(x - 1) = 2$

$x - 1 = 3^2$ First rewrite this equation in exponential form.

$x - 1 = 9$ Substitute 9 for 3^2, and then solve for x.

$x = 10$

(b) $\log_x 9 = -2$

$\log_x 9 = -2$

$x^{-2} = 9$ First rewrite this equation in exponential form.

$x^{-2} = 3^2$ Now work toward expressing each side in terms of the same exponent.

SELF-CHECK ANSWERS

1 6 **2** 3 **3** 2 **4** −1

$$x^{-2} = \left(\frac{1}{3}\right)^{-2}$$

$$x = \frac{1}{3}$$ The bases are equal since the exponents are equal.

(c) $\log_4 8 = x$

$$\log_4 8 = x$$

$$4^x = 8$$ First rewrite this equation in exponential form.

$$(2^2)^x = 2^3$$ Then express each side in terms of the common base of 2.

$$2^{2x} = 2^3$$ Use the power rule for exponents to rewrite the left member.

$$2x = 3$$ Then equate the exponents since the bases are the same.

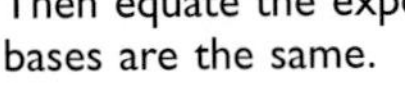

$$x = \frac{3}{2}$$

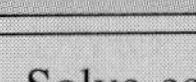

SELF-CHECK

Solve each of these equations for x.

1 $\log_2 x = 3$
2 $\log_{11}\sqrt{11} = x$
3 $\log_x 125 = 3$

From the definition of $y = \log_b x$ we know that this logarithmic function is defined only for positive values of x. A line graph approximation of $y = \log_2 x$ is shown side by side with the actual graph of the function.

Table of Values

x	$y = \log_2 x$
$\frac{1}{8}$	-3
$\frac{1}{4}$	-2
$\frac{1}{2}$	-1
1	0
2	1
4	2
8	3

Line Graph Approximation of $y = \log_2 x$

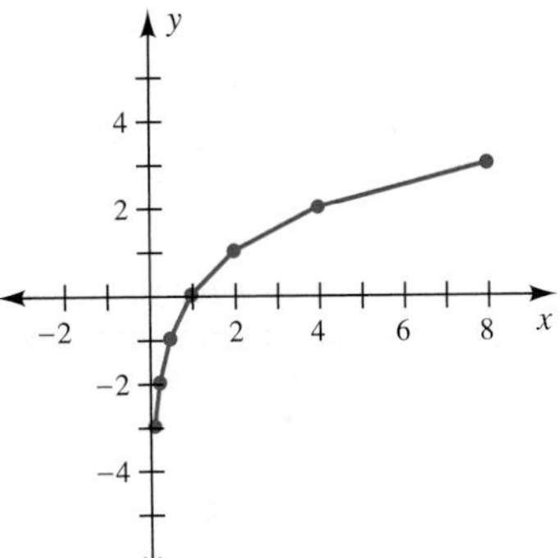

Actual Graph of $y = \log_2 x$

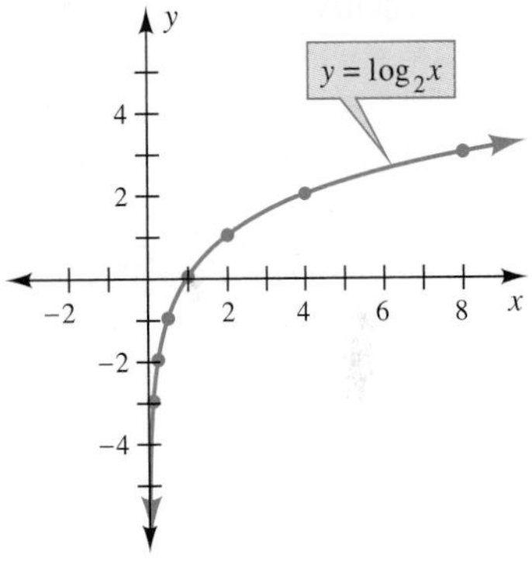

All logarithmic functions $y = \log_b x$ pass through the points (1, 0) and (b, 1). The actual graph of $y = \log_2 x$ shown above illustrates the characteristic shape of $y = \log_b x$ for $b > 1$: The graph approaches but never touches the negative portion of the y-axis as it extends to the left. It rises slowly to the right, extending continuously so that it contains a point above each positive x-coordinate. There are no points corresponding to zero or to negative arguments of x.

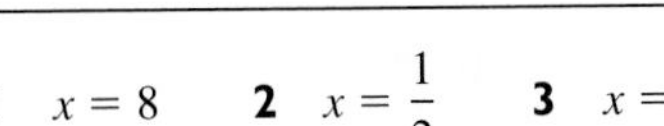

SELF-CHECK ANSWERS

1 $x = 8$ **2** $x = \frac{1}{2}$ **3** $x = 5$

GRAPH OF A LOGARITHMIC FUNCTION

If $b > 1$, then the graph of $y = \log_b x$ is

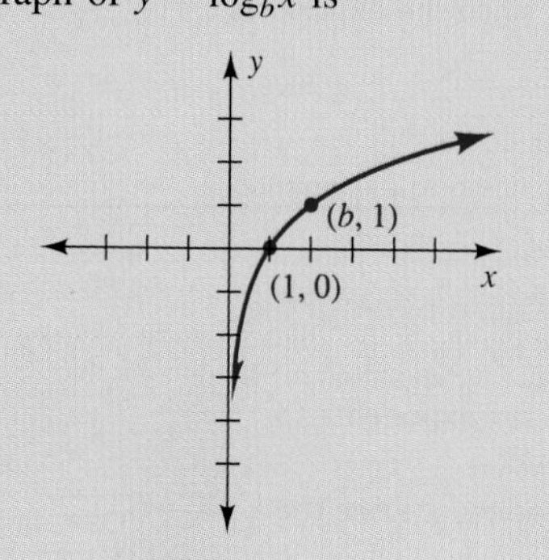

The next example illustrates the use of a TI-82 graphics calculator to graph $y = 2^x$ and its inverse on the same display window. Note that the graph of the inverse of $y = 2^x$ is the same as the graph of $y = \log_2 x$. The graphs of $y = 2^x$ and $y = \log_2 x$ are symmetric about the line $x = y$.

EXAMPLE 9 **Using a Calculator to Graph an Exponential Function and Its Inverse**

Use a graphics calculator to graph $y = 2^x$ and its inverse on the same display window.

SOLUTION

Keystrokes

[Y=] [2] [∧] [X, T, θ] [2nd] [DRAW] [8] [2nd] [Y-VARS]

[1] [1] [ENTER]

[−3, 6] for x, [−3, 6] for y

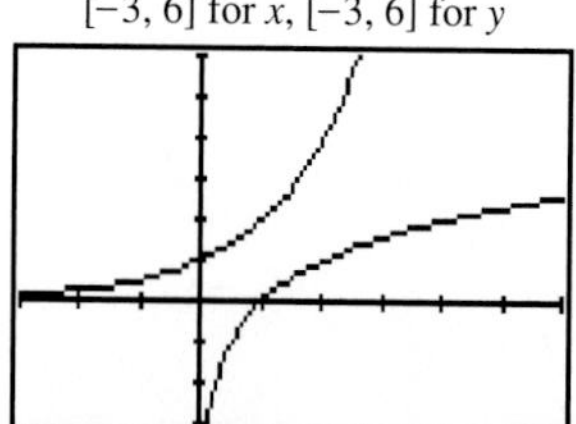

First enter the function. Then choose option 8, the DrawInv option, from the DRAW menu. From the Y-VARS menu, choose to draw the inverse of Y_1, the function just defined. The window displayed to the left assumes that the window of values has already been set to [−3, 6] for x and [−3, 6] for y.

Exercises 10-2

A

In Exercises 1–8 solve each equation without using a calculator.

1 a. $4^x = 16$ **b.** $2^y = 32$ **c.** $5^{v+1} = 125$ **d.** $\left(\frac{2}{3}\right)^w = \frac{8}{27}$

2 a. $2^x = 8$ **b.** $3^y = 81$ **c.** $7^{v+5} = 49$ **d.** $\left(\frac{3}{4}\right)^w = \frac{9}{16}$

3 a. $2^m = \frac{1}{2}$ **b.** $5^n = 1$ **c.** $25^x = 5$ **d.** $27^{2y} = 3$

4 a. $5^m = \frac{1}{5}$ **b.** $2^n = 1$ **c.** $49^x = 7$ **d.** $8^{5y} = 2$

5 a. $32^{3n-5} = 2$ **b.** $7^x = \frac{1}{49}$

c. $\left(\frac{2}{5}\right)^w = \frac{125}{8}$ **d.** $9^x = 27$

6 a. $64^{2n+9} = 2$ **b.** $5^x = \frac{1}{25}$

c. $\left(\frac{2}{7}\right)^w = \frac{49}{4}$ **d.** $16^x = 8$

7 a. $3^v = \sqrt{3}$ **b.** $5^w = \sqrt[3]{5}$

c. $10^x = 10{,}000$ **d.** $10^m = 0.01$

8 a. $7^v = \sqrt{7}$ **b.** $3^w = \sqrt[5]{3}$

c. $10^x = 100{,}000$ **d.** $10^m = 0.001$

In Exercises 9–12 one of the following three forms of the equation is given: logarithmic form, verbal form, or exponential form. For each equation write the other two forms.

9 a. Logarithmic form: $\log_6 36 = 2$

b. Verbal form: The log of 625 base 5 is 4.

c. Exponential form: $7^0 = 1$

10 a. Logarithmic form: $\log_3 81 = 4$

b. Verbal form: The log of 64 base 8 is 2.

c. Exponential form: $5^{-1} = \frac{1}{5}$

11 a. Logarithmic form: $\log_{15}\sqrt{15} = \frac{1}{2}$

b. Verbal form: The log of 17 base 13 is x.

c. Exponential form: $19^x = 23$

12 a. Logarithmic form: $\log_3\sqrt{3} = \frac{1}{2}$

b. Verbal form: The log of x base 8 is 1.3.

c. Exponential form: $7^v = 9$

In Exercises 13–16 write each logarithmic equation in exponential form.

13 $\log_{10}\frac{1}{100} = -2$ **14** $\log_2\sqrt[3]{4} = \frac{2}{3}$

15 $\log_b m = k$ **16** $\log_x y = z$

In Exercises 17 and 18 write each exponential equation in logarithmic form.

17 $16^{-1/2} = \frac{1}{4}$ **18** $m^p = n$

In Exercises 19–34 determine the value of each logarithm by inspection. Indicate which expressions are undefined.

19 $\log_{10}100$ **20** $\log_{10}100{,}000$ **21** $\log_{10}0.1$

22 $\log_{10}0.001$ **23** $\log_2 32$ **24** $\log_{11}121$

25 $\log_3\frac{1}{3}$ **26** $\log_2\frac{1}{8}$ **27** $\log_{11}\sqrt{11}$

28 $\log_{13}\sqrt[3]{13}$ **29** $\log_7\frac{1}{\sqrt[3]{7}}$ **30** $\log_{73}1$

31 $\log_{18}1$ **32** $\log_{3/4}\frac{16}{9}$ **33** $\log_5(-5)$

34 $\log_7 0$

In Exercises 35 and 36 graph each pair of functions on the same rectangular coordinate system. Lightly sketch the line of symmetry, $x = y$.

35 $y = 3^x$
$y = \log_3 x$

36 $y = \left(\frac{1}{3}\right)^x$
$y = \log_{1/3} x$

B

In Exercises 37–62 solve each equation for x without using a calculator.

37 $5^{x+7} = 25$ **38** $7^{x-3} = 49$ **39** $7^{3x} = \sqrt{7}$

40 $11^{2x-1} = \sqrt[3]{11}$ **41** $\left(\frac{3}{7}\right)^x = \frac{49}{9}$ **42** $\left(\frac{5}{4}\right)^{-x} = \frac{64}{125}$

43 $x^2 = 64$ **44** $x^3 = 64$

45 $2^{x^2} = 16$ **46** $3^{x^2-1} = 27$

47 $\log_3 1 = 2x + 1$ **48** $\log_3 3 = 4x - 1$

49 $\log_b b^{13} = x - 2$ **50** $\log_b b^{-11} = x + 5$

51 $\log_b\sqrt[5]{b} = x$ **52** $\log_b\sqrt[5]{b^2} = x$

53 $\log_5 x = 2$ **54** $\log_4 x = \frac{1}{2}$

55 $\log_5 x = -2$ **56** $\log_4 x = -2$

57 $\log_5 x = \frac{1}{2}$ **58** $\log_x 4 = \frac{1}{2}$

59 $\log_x 7 = \frac{1}{2}$ **60** $\log_x 11 = -1$

61 $\log_x 5 = 2$ **62** $\log_6 x = 1$

C

In Exercises 63–66 determine the value of b such that the graph of $y = b^x$ will pass through the points labeled on each graph.

63 **64**

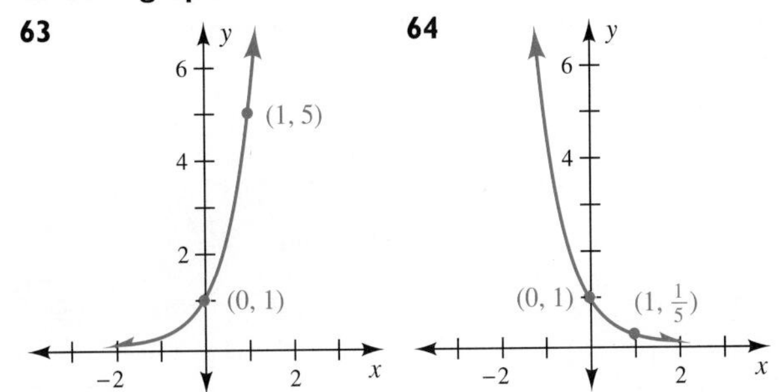

65 **66**

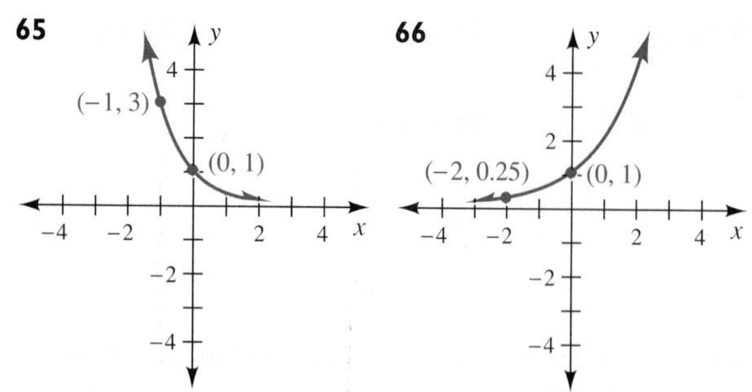

In Exercises 67–70 determine the value of b such that the graph of $y = \log_b x$ will pass through the points labeled on each graph.

67 **68**

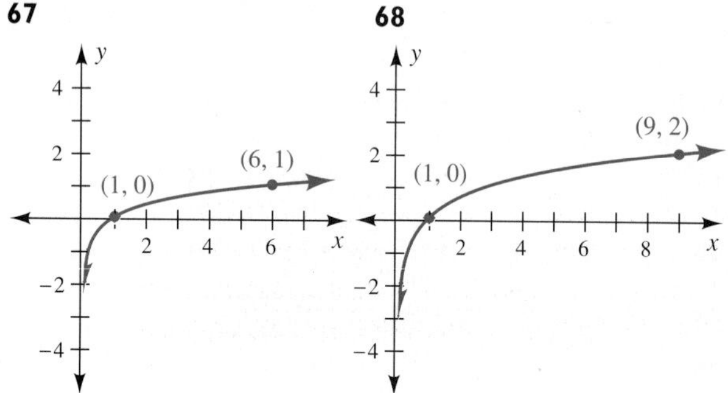

69 **70**

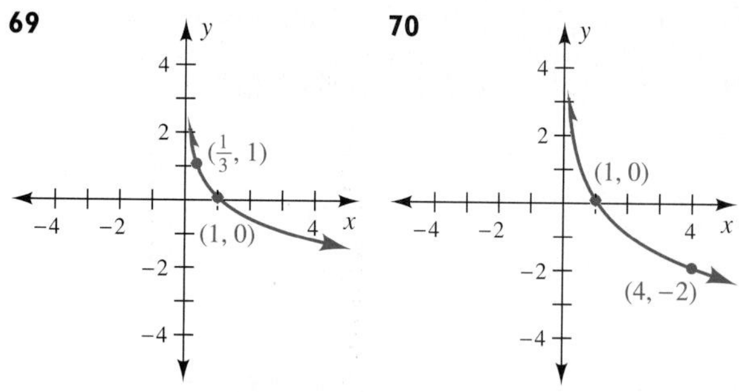

CALCULATOR USAGE (71–78)

In Exercises 71–74 use a calculator to approximate each expression to five significant digits.

71 3^{π} **72** $2^{\sqrt{3}}$ **73** $(5 + 7)^{\pi}$ **74** $5^{\pi} + 7^{\pi}$

In Exercises 75–78 use a graphics calculator to graph each pair of functions on the same rectangular coordinate system. Also graph the line of symmetry, $x = y$.

Use the ln and log keys to graph the logarithmic functions.

75 $y = e^x$
$y = \ln x$

76 $y = 10^x$
$y = \log x$

77 $y = 1.5^x$ and its inverse

78 $y = 0.5^x$ and its inverse

In Exercises 79 and 80 use the formula $A = P(1 + r)^t$ to determine the value A of a principal P invested at interest rate r for t years.

Compound Interest (79 and 80)

79 Determine the value of $700 invested at a 9.5% annual interest rate and left to compound for 10 years.

80 Determine the value of $1000 invested at an 11.25% annual interest rate and left to compound for 5 years.

DISCUSSION QUESTION

81 Explain why an exponential function $y = b^x$ does not use a base of $b = 1$.

CHALLENGE QUESTION

A linear function $y = mx + b$ is characterized by the fact that equal changes in the independent variable x will result in constant changes in the dependent variable y. Complete the following table for $y = 3x - 2$, which illustrates this concept. Then complete the second table for $y = 2^x$, and describe what you have observed. Can you generalize based on what you have observed?

Table for $y = 3x - 2$

x	y	Change in y
1	1	
2	4	3
3	7	3
4	10	
5		
6		
7		
8		

Table for $y = 2^x$

x	y	Ratio of y Values
1	2	
2	4	2
3	8	2
4	16	
5		
6		
7		
8		

SECTION SKILL REINFORCERS

Simplify each of these expressions.

1 x^3x^8 **2** $\frac{x^8}{x^3}$ **3** $(x^8)^3$ **4** $(x^5y^{-4})^5$ **5** $\frac{x^8}{x^{-3}}$

10-3 Using Calculators to Determine Logarithms

SECTION OBJECTIVE

3 Evaluate common and natural logarithms with a calculator.

Although $\log_b x$ is defined for any base $b > 0$ and $b \neq 1$, the only two bases commonly used are base 10 and base e. Since our number system is based upon powers of 10, the most convenient base for many computations is base 10. Logarithms to the base 10 are called **common logarithms.** The abbreviated form $\log x$ is used to denote $\log_{10} x$.

Mathematicians use base e extensively because many formulas are most easily stated using this base. The number e is irrational:

$$e \approx 2.718\ 281\ 828\ 459\ 045\ldots$$

You should remember that e is approximately equal to 2.718. Logarithms to the base e are called **natural logarithms.** The abbreviated form $\ln x$ is used to denote $\log_e x$.

A Mathematical Note

Swiss-born Leonard Euler (1701–1783) was hired by Catherine the Great of Russia to write the elementary mathematics textbooks for Russian schools. He wrote prolifically on various mathematical topics. From Euler's textbook *Introducio* came many symbols, such as i for $\sqrt{-1}$, π for the ratio of the circumference of a circle to its diameter, and e for the base of natural logarithms.

COMMON AND NATURAL LOGARITHMS

Common Logarithms: $\log x$ means $\log_{10} x$.

Natural Logarithms: $\ln x$ means $\log_e x$. $e \approx 2.718$

Only relatively simple logarithms, such as $\log 100 = 2$, can be determined by inspection. Thus common and natural logarithms have historically been determined through use of tables, slide rules, and other devices. Today calculators are usually used to determine these values. The examples in this section illustrate the key sequence for a typical calculator—the labels on the keys and the number of significant digits displayed will vary from model to model.

Logarithmic functions are also available in many computer languages; however, these functions may not mean what you expect. Note the following warning.

WARNING FOR COMPUTER USERS

Some computer languages provide only the natural logarithmic function, in which case this function may be called LOG instead of ln, even though the base is e. If you are unsure of the meaning of LOG on a particular computer, test a value such as LOG 10 to determine which base is being used.

EXAMPLE 1 Using a Calculator to Evaluate Logarithms

Use a calculator to determine these logarithmic values to five significant digits.

SOLUTIONS

(a) log 748

Keystrokes

Scientific

[7] [4] [8] [log] → 2.8739016

Graphics

[LOG] [7] [4] [8] [ENTER] → 2.873901598

Answer $\log 748 \approx 2.8739$

(b) ln 748

Keystrokes

Scientific

[7] [4] [8] [ln] → 6.617403

Graphics

[LN] [7] [4] [8] [ENTER] → 6.617402978

Answer $\ln 748 \approx 6.6174$

(c) log 0.0034

Keystrokes

Scientific

[0] [.] [0] [0] [3] [4] [log] → −2.4685211

Graphics

[LOG] [0] [.] [0] [0] [3] [4] [ENTER] → −2.468521083

Answer $\log 0.0034 \approx -2.4685$

(d) ln 0.0034

Keystrokes

Scientific

[0] [.] [0] [0] [3] [4] [ln] → −5.6839798

Graphics

[LN] [0] [.] [0] [0] [3] [4] [ENTER] → −5.683979847

Answer $\ln 0.0034 \approx -5.6840$

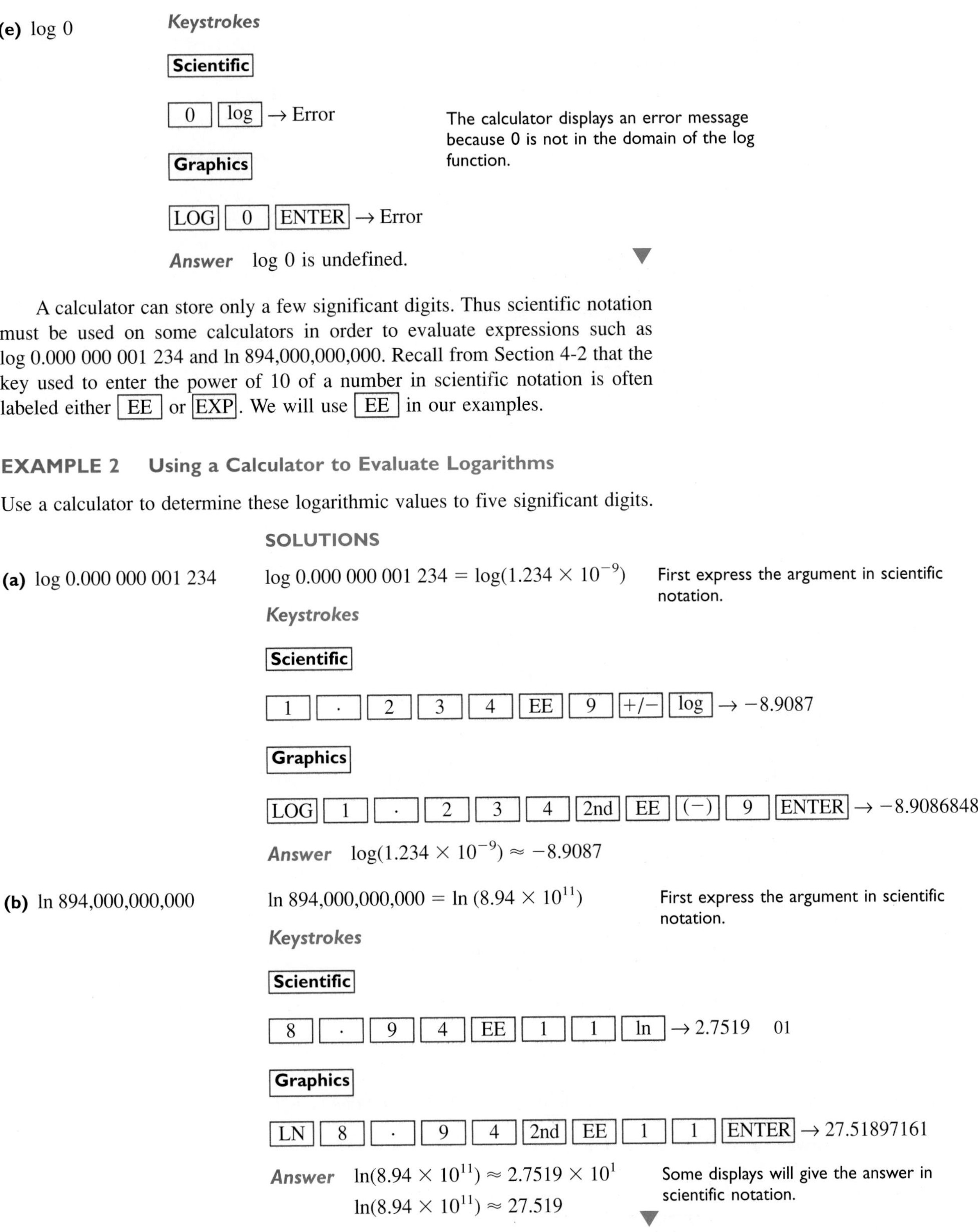

(e) log 0

Keystrokes

[Scientific]

[0] [log] → Error

The calculator displays an error message because 0 is not in the domain of the log function.

[Graphics]

[LOG] [0] [ENTER] → Error

Answer log 0 is undefined. ▼

A calculator can store only a few significant digits. Thus scientific notation must be used on some calculators in order to evaluate expressions such as log 0.000 000 001 234 and ln 894,000,000,000. Recall from Section 4-2 that the key used to enter the power of 10 of a number in scientific notation is often labeled either [EE] or [EXP]. We will use [EE] in our examples.

EXAMPLE 2 Using a Calculator to Evaluate Logarithms

Use a calculator to determine these logarithmic values to five significant digits.

SOLUTIONS

(a) log 0.000 000 001 234

$\log 0.000\ 000\ 001\ 234 = \log(1.234 \times 10^{-9})$

First express the argument in scientific notation.

Keystrokes

[Scientific]

[1] [.] [2] [3] [4] [EE] [9] [+/−] [log] → −8.9087

[Graphics]

[LOG] [1] [.] [2] [3] [4] [2nd] [EE] [(−)] [9] [ENTER] → −8.90868484

Answer $\log(1.234 \times 10^{-9}) \approx -8.9087$

(b) ln 894,000,000,000

$\ln 894{,}000{,}000{,}000 = \ln(8.94 \times 10^{11})$

First express the argument in scientific notation.

Keystrokes

[Scientific]

[8] [.] [9] [4] [EE] [1] [1] [ln] → 2.7519 01

[Graphics]

[LN] [8] [.] [9] [4] [2nd] [EE] [1] [1] [ENTER] → 27.51897161

Answer $\ln(8.94 \times 10^{11}) \approx 2.7519 \times 10^{1}$

$\ln(8.94 \times 10^{11}) \approx 27.519$ ▼

Some displays will give the answer in scientific notation.

Since logarithms and exponential functions are inverses of each other, many calculators incorporate both the $\ln x$ and the e^x functions into one key. In this case another key, such as an inverse key, is used to indicate which of the two functions is being selected. This design also permits use of the same key for both $\log x$ and 10^x.

SELF-CHECK ✓

Use a calculator to determine these logarithmic values to five significant digits.

1 $\log 0.47912$
2 $\ln 53.78$
3 $\log 5.81 \times 10^4$
4 $\ln 3.96 \times 10^{-5}$

BASE 10 AND BASE e

For $x > 0$,

1 $y = \log x$ is equivalent to $x = 10^y$.
2 $y = \ln x$ is equivalent to $x = e^y$.

EXAMPLE 3 Using a Calculator to Evaluate Exponential Expressions

Use a calculator to evaluate each of these exponential expressions to five significant digits.

SOLUTIONS

(a) $10^{2.1345}$ *Keystrokes*

Scientific

[2] [.] [1] [3] [4] [5] [10^x] → 136.3013 Method A

[2] [.] [1] [3] [4] [5] [INV] [log] → 136.3013 Method B

Graphics

[2nd] [10^x] [2] [.] [1] [3] [4] [5] [ENTER] → 136.3013006

Answer $10^{2.1345} \approx 136.30$

(b) $e^{-4.98}$ *Keystrokes*

Scientific

[4] [.] [9] [8] [+/−] [e^x] → 0.00687406 Method A

[4] [.] [9] [8] [+/−] [INV] [ln] → 0.00687406 Method B

Graphics

[2nd] [e^x] [(−)] [4] [.] [9] [8] [ENTER] → 0.0068740626

Answer $e^{-4.98} \approx 0.006\,874\,1$

SELF-CHECK ANSWERS ✓

1 −0.31956 2 3.9849 3 4.7642 4 −10.137

Before calculators were readily available, logarithms and logarithmic tables were used extensively for computations involving multiplication, division, and exponentiation. Logarithmic tables can be used two ways—to find logs of given values and to find numbers whose logs are given. The term **antilogarithm,** or **antilog,** is sometimes used (particularly in conjunction with log tables) to denote a number whose logarithm is known. This terminology is illustrated below.

logarithm	logarithm	logarithm
$\log 100 = 2$	$\text{antilog } 2 = 100$	$10^2 = 100$
antilogarithm	antilogarithm	antilogarithm

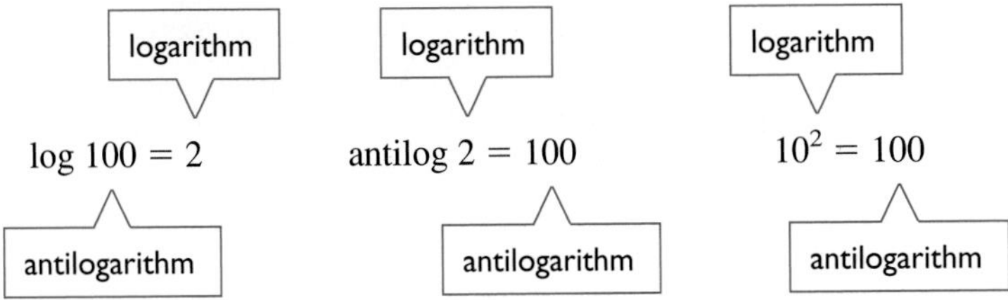

EXAMPLE 4 Using a Calculator to Evaluate Antilogarithms

Use a calculator to approximate each of these expressions to five significant digits.

SOLUTIONS

(a) antilog 0.69897

$\text{antilog } 0.69897 = 10^{0.69897}$ — Rewrite in exponential form.

Keystrokes

Scientific

[0] [.] [6] [9] [8] [9] [7] [10^x] → 4.999999 — Method A

[0] [.] [6] [9] [8] [9] [7] [INV] [log] → 4.999999 — Method B

Graphics

[2nd] [10^x] [0] [.] [6] [9] [8] [9] [7] [ENTER] → 4.99999995

Answer $\text{antilog } 0.69897 \approx 5.0000$

(b) antiln(−1.045)

$\text{antiln}(-1.045) = e^{-1.045}$ — Rewrite in exponential form.

Keystrokes

Scientific

[1] [.] [0] [4] [5] [+/−] [e^x] → 0.35169182 — Method A

[1] [.] [0] [4] [5] [+/−] [INV] [ln] → 0.35169182 — Method B

Graphics

[2nd] [e^x] [(−)] [1] [.] [0] [4] [5] [ENTER] → 0.3516918194

Answer $\text{antiln}(-1.045) \approx 0.35169$ ▼

Most scientific calculators have a [π] key, which will provide an approximation of 3.141 592 654 (give or take a few significant digits) for the irrational constant π. Such calculators usually do not have a key labeled e for the irrational constant e. This constant can easily be obtained, however, by noting that $e = e^1$.

SELF-CHECK 

Use a calculator to approximate each of these expressions to four significant digits.

1 $10^{-4.07}$ **2** $e^{0.0983}$

3 antilog 0.0983

4 antiln 3

EXAMPLE 5 Using a Calculator to Approximate Expressions Containing e

Use a calculator to approximate $\frac{e+7}{11}$ to five significant digits.

SOLUTION $\frac{e+7}{11} = \frac{e^1+7}{11}$

Keystrokes

Scientific

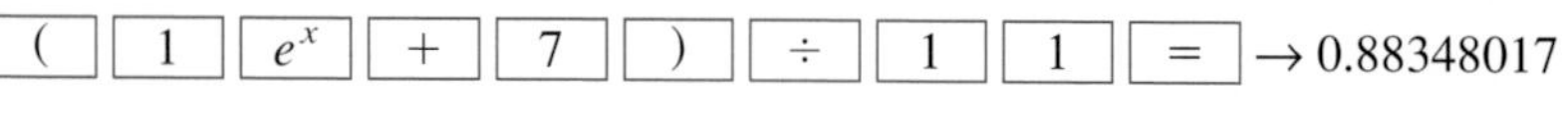

[( ] [1] [e^x] [+] [7] [ )] [÷] [1] [1] [=] → 0.88348017 Method A

[( ] [1] [INV] [ln] [+] [7] [ )] [÷] [1] [1] [=] → 0.88348017 Method B

Graphics

[( ] [2nd] [e^x] [1] [+] [7] [ )] [÷] [1] [1] [ENTER] → 0.8834801662

Answer $\frac{e+7}{11} \approx 0.88348$

To avoid order-of-operations errors, use caution when writing logarithmic expressions. Remember that expressions such as $\log \frac{12}{5}$ and $\frac{\log 12}{\log 5}$ have distinct meanings and values.

EXAMPLE 6 Using a Calculator to Evaluate Logarithmic Expressions

Use a calculator to approximate each of these expressions to five significant digits, and then compare the two values.

SOLUTIONS

(a) $\log \frac{12}{5}$

Keystrokes

Scientific

[( ] [1] [2] [÷] [5] [ )] [log] → 0.38021124

SELF-CHECK ANSWERS 

1 0.000 085 11 **2** 1.103 **3** 1.254 **4** 20.09

Graphics

LOG (1 2 ÷ 5) ENTER → 0.3802112417

Answer $\log \frac{12}{5} \approx 0.38021$

(b) $\frac{\log 12}{\log 5}$

Keystrokes

Scientific

1 2 log ÷ 5 log = → 1.5439593

Graphics

LOG 1 2 ÷ LOG 5 ENTER → 1.543959311

Answer $\frac{\log 12}{\log 5} \approx 1.5440$

Comparison: $\log \frac{12}{5} \neq \frac{\log 12}{\log 5}$. ▼

SELF-CHECK ✓

Use a calculator to approximate each of these expressions to four significant digits.

1 $\frac{12}{\log 5}$ **2** $\frac{\log 12}{5}$

3 $\frac{2e - 5}{13}$

EXAMPLE 7 An Application of Exponential Growth

The number of bacteria in a laboratory culture after t hours is given by $A = 1000e^{0.185t}$. Determine how many bacteria will be present after 8 hours.

SOLUTION

$A = 1000e^{0.185t}$

$= 1000e^{0.185(8)}$ Substitute 8 for t.

≈ 4392.9457 Approximate with a calculator, and then

$A \approx 4400$ bacteria round to two significant digits. ▼

Expressions involving logarithms to bases other than base e or base 10 can be approximated with a calculator. First, however, these logarithmic expressions must be converted to equivalent expressions involving either common logs or natural logs. The procedure for changing bases will be illustrated in Section 10-4.

EXAMPLE 8 Estimation Skills

Mentally estimate the value of each of these expressions.

SOLUTIONS

(a) $2^{3.98765}$

$2^{3.98765} \approx 2^4$

$2^{3.98765} \approx 16$

Since 3.98765 is only slightly less than 4, $2^{3.98765}$ is slightly less than 2^4, which is 16.

(b) $\log_7 50.0017$

$\log_7 50.0017 \approx \log_7 49$

$\log_7 50.0017 \approx 2$

Since 50.0017 is slightly larger than 49, $\log_7 50.0017$ is slightly larger than $\log_7 49$, which is 2. ▼

SELF-CHECK ANSWERS 

1 17.17 **2** 0.2158 **3** 0.03358

Exercises 10-3

A

In Exercises 1–4 determine the value of each logarithm by inspection. Indicate which expressions are undefined.

1 a. $\log 100$ **b.** $\log 0.0001$ **c.** $\log 10^9$ **d.** $\log(-10)$

2 a. $\log 1{,}000{,}000$ **b.** $\log 0.01$ **c.** $\log 10^{-6}$ **d.** $\log 0$

3 a. $\ln e^5$ **b.** $\ln e^{-5}$ **c.** $\ln 0$ **d.** $\ln \sqrt{e}$

4 a. $\ln e^7$ **b.** $\ln \frac{1}{e^4}$ **c.** $\ln(-5e)$ **d.** $\ln \sqrt[3]{e}$

CALCULATOR USAGE (5–42)

In Exercises 5–18 use a calculator to approximate the value of each expression to five significant digits.

5 a. $\log 47$ **b.** $10^{1.6721}$ **c.** $\ln 47$ **d.** $e^{3.8501}$

6 a. $\log 2.47$ **b.** $10^{0.39270}$ **c.** $\ln 2.47$ **d.** $e^{0.90422}$

7 a. $\log 113$ **b.** $10^{2.0531}$ **c.** $\ln 113$ **d.** $e^{4.7274}$

8 a. $\ln 0.0567$ **b.** $e^{-2.8700}$ **c.** $\log 0.0567$ **d.** $10^{-1.2464}$

9 a. $\ln 0.00621$ **b.** $\log 0.00621$ **c.** $\ln 10$ **d.** $\log e$

10 a. $\log 0.00009$ **b.** $\ln 0.00009$ **c.** $\ln \pi$ **d.** $\log \pi$

11 a. $\log(3.87 \times 10^{-7})$ **b.** $\ln(3.87 \times 10^{-7})$ **c.** $\log 45{,}000{,}000{,}000$ **d.** $\ln 45{,}000{,}000{,}000$

12 a. $\log(4.07 \times 10^5)$ **b.** $\ln(4.07 \times 10^5)$ **c.** $\ln 0.000\,000\,000\,06$ **d.** $\log 0.000\,000\,000\,06$

13 a. $\log(-8)$ **b.** $-\log 8$ **c.** $\log(\ln 13)$ **d.** $\ln(\log 13)$

14 a. $\ln(-8)$ **b.** $-\ln 8$ **c.** $\log(\ln 9)$ **d.** $\ln(\log 9)$

15 a. $\ln \frac{11}{4}$ **b.** $\frac{\ln 11}{\ln 4}$

16 a. $\ln \frac{17}{5}$ **b.** $\frac{\ln 17}{\ln 5}$

17 a. $(\ln 11)(\ln 4)$ **b.** $\ln(11 \cdot 4)$

18 a. $(\ln 5)(\ln 17)$ **b.** $\ln(5 \cdot 17)$

In Exercises 19–42 use a calculator to approximate the value of each expression to four significant digits.

19 $10^{3.19}$ **20** $10^{4.87}$ **21** $10^{-0.08}$

22 $10^{-0.035}$ **23** $e^{4.67}$ **24** $e^{2.09}$

25 $e^{-1.78}$ **26** $e^{-2.63}$ **27** antilog 0.35

28 antilog(-3.7) **29** antiln 3.35 **30** antiln(-0.72)

31 $\frac{e-3}{3}$ **32** $\frac{e+12}{8}$ **33** $\frac{\ln 7 + \log 7}{7}$

34 $\frac{\ln 9 - \log 9}{9}$ **35** $(\ln 17)^2 - \ln(17^2)$

36 $(\log 2.83)^2 - \log(2.83^2)$ **37** $\frac{\log 9}{\log 5} - \log \frac{9}{5}$

38 $\frac{\ln 7}{\ln 2} - \ln \frac{7}{2}$

39 $\ln(13 + 15) - (\ln 13 + \ln 15)$

40 $\log(41 + 17) - (\log 41 + \log 17)$

41 $2 \log 13 + 3 \log 4$

42 $5 \ln 9 - \ln 11$

B

ESTIMATION SKILLS (43–54)

In Exercises 43–46 mentally estimate the value of each of these expressions to the nearest integer.

43 $\log_6 35.783$ **44** $\log_5 127.41$

45 $3^{2.0013}$ **46** $4^{1.9957}$

In Exercises 47–54 mentally estimate the value of each expression, and select the answer that is closest to your estimate.

47 $\ln 7.389\,056\,1$

a. 0 **b.** 1 **c.** 2 **d.** 3

48 $\ln \frac{1}{3}$

a. -1.1 **b.** -2.1 **c.** -3.1 **d.** 3.1

49 $\log 105$

a. 1.98 **b.** 2.02 **c.** 2.98 **d.** 3.02

50 log 0.001 034

a. −3.99 **b.** −2.99 **c.** 2.99 **d.** 3.99

51 antilog 1.00111

a. −10.03 **b.** −9.93 **c.** 9.93 **d.** 10.03

52 antilog −2.04

a. 0.0009 **b.** 0.0091 **c.** 0.0912 **d.** 0.9120

53 e^2

a. 3.87 **b.** 17.38 **c.** 83.17 **d.** 7.39

54 $e^{-1.05}$

a. 0.15 **b.** 0.25 **c.** 0.35 **d.** 1.5

C

CALCULATOR USAGE (55–70)

In Exercises 55–66 use a calculator to approximate the value of x to four significant digits.

55 $x = \ln 3$ **56** $x = \log 3$ **57** $\ln x = 3$

58 $\log x = 3$ **59** $x = \text{antilog } 3$ **60** $x = \text{antiln } 3$

61 $\text{antilog } x = 3$ **62** $\text{antiln } x = 3$ **63** $e^x = 5$

64 $10^x = 7$ **65** $\log x = -3.45$ **66** $\ln x = -1.7$

The total amount of money A that accumulates when original principal P is invested at rate r and left to compound continuously for time t is given by $A = Pe^{rt}$. In Exercises 67 and 68 use this formula to calculate each value to the nearest penny.

Compound Interest (67 and 68)

67 Find the value of a $500 investment if interest is compounded continuously for 4 years at 7.5%.

68 Find the value of a $980 investment if interest is compounded continuously for 3 years at 7%.

The number of monthly payments of amount P required to completely pay for a loan of amount A borrowed at interest rate R is given by the formula

$$n = -\frac{\log\left(1 - \frac{AR}{12P}\right)}{\log\left(1 + \frac{R}{12}\right)}$$

In Exercises 69 and 70 use this formula to calculate the number of monthly payments.

Loan Payments (69 and 70)

69 Determine the number of monthly car payments of $342.11 that it will take to pay off a $15,206.23 car loan when the interest rate is 12.5%.

70 Determine the number of monthly car payments of $404.10 that it will take to pay off a $15,206.23 car loan when the interest rate is 12.5%.

DISCUSSION QUESTION

71 A computer programmer wanted to use a logarithmic function called LOG. To determine the base of this function, she evaluated the function using an argument of 10. Explain the logic of her method. What is the base of this logarithmic function if the result of her evaluation was approximately 2.3?

CHALLENGE QUESTION

A new hybrid of water lily starts to grow in one segment of a canal in Florida. A single plant doubles to produce two plants at the beginning of the second day. These plants continue to double each day and as a result fill this segment of the canal in 30 days. How many days would it take to fill this segment of the canal if there were two plants on the first day?

SECTION SKILL REINFORCERS

Graph each of these functions.

1 $y = |x| - 4$ **2** $y = |x| + 4$ **3** $y = |x - 4|$ **4** $y = |x + 4|$ **5** $y = 4|x|$

10-4 Properties of Logarithms

SECTION OBJECTIVES

4 Use the product rule, the quotient rule, and the power rule for logarithms.

5 Use the change-of-base formula for logarithms.

The properties of logarithms follow directly from the definition of a logarithmic function as the inverse of an exponential function. Thus for every exponential property there is a corresponding logarithmic property. The logarithmic properties provide a means of simplifying some algebraic expressions and make it easier to solve exponential equations.

The logarithmic and exponential properties are stated side by side below so that you can compare them. (The capital letters X and Y are used in the logarithmic form to show that these variables are not the same as x and y in the exponential form.)

	Exponential Form	*Logarithmic Form*	
Product rule	$b^x b^y = b^{x+y}$	$\log_b XY = \log_b X + \log_b Y$	In exponential form, we add exponents; in logarithmic form, we add logarithms.
Quotient rule	$\dfrac{b^x}{b^y} = b^{x-y}$	$\log_b \dfrac{X}{Y} = \log_b X - \log_b Y$	In exponential form, we subtract exponents; in logarithmic form, we subtract logarithms.
Power rule	$(b^x)^p = b^{xp}$	$\log_b X^p = p \log_b X$	In exponential form, we multiply exponents; in logarithmic form, we multiply the pth power of X times the log of X.

The proof of the product rule appears below. The quotient rule and the power rule can be proven in a similar fashion; their proofs are left to you in Exercises 68 and 69 at the end of this section.

Proof of the Product Rule for Logarithms Let $x = b^m$ and $y = b^n$; thus $\log_b x = m$ and $\log_b y = n$. Then

$$xy = b^m b^n$$

$$xy = b^{m+n} \qquad \text{Use the product rule for exponents.}$$

$$\log_b xy = \log_b b^{m+n} \qquad \text{Take logs of both sides.}$$

$$\log_b xy = m + n \qquad \text{Simplify the right member of this equation.}$$

$$\log_b xy = \log_b x + \log_b y \qquad \text{Substitute for } m \text{ and } n.$$

PROPERTIES OF LOGARITHMS

For $x, y > 0$, $b > 0$, and $b \neq 1$,

Product Rule: $\log_b xy = \log_b x + \log_b y$ The log of a product is the sum of the logs.

Quotient Rule: $\log_b \dfrac{x}{y} = \log_b x - \log_b y$ The log of a quotient is the difference of the logs.

Power Rule: $\log_b x^p = p \log_b x$ The log of the pth power of x is p times the log of x.

Example 1 illustrates the properties of logarithms with some logarithms that can be determined by inspection.

EXAMPLE 1 Logarithmic Properties

Verify these equalities by computing each member of each equation separately.

SOLUTIONS

(a) $\log_2 4 \cdot 8 = \log_2 4 + \log_2 8$

$$\log_2 32 = \log_2 4 \cdot 8 = \log_2 4 + \log_2 8$$
$$\downarrow \qquad \downarrow \quad \downarrow$$
$$5 = 2 + 3$$

Product rule: Add the logarithms.

(b) $\log_2 \frac{32}{8} = \log_2 32 - \log_2 8$

$$\log_2 4 = \log_2 \frac{32}{8} = \log_2 32 - \log_2 8$$
$$\downarrow \qquad \downarrow \quad \downarrow$$
$$2 = 5 - 3$$

Quotient rule: Subtract the logarithms.

(c) $\log_2 4^3 = 3 \log_2 4$

$$\log_2 64 = \log_2 4^3 = 3 \log_2 4$$
$$\downarrow \qquad \downarrow \quad \downarrow$$
$$6 = 3 \cdot 2$$

Power rule: Multiply the power 3 times the logarithm. ▼

The properties of logarithms are used both to rewrite logarithmic expressions in a simpler form and to combine logarithms. These properties are also used to solve logarithmic equations and to simplify expressions in calculus.

EXAMPLE 2 Using the Properties of Logarithms to Form Simpler Expressions

Use the properties of logarithms to write these expressions in terms of logarithms of simpler expressions. Assume that all arguments are positive real numbers.

SOLUTIONS

(a) $\log xyz$ $\quad \log xyz = \log x + \log y + \log z$ $\quad$ Product rule: Add the logarithms.

(b) $\ln x^7$ $\quad \ln x^7 = 7 \ln x$ $\quad$ Power rule: Multiply the power 7 times the logarithm.

(c) $\log_5 \frac{x+3}{x-2}$ $\quad \log_5 \frac{x+3}{x-2} = \log_5(x+3) - \log_5(x-2)$ $\quad$ Quotient rule: Subtract the logarithms.

(d) $\ln \sqrt{\frac{x}{y}}$

$$\ln \sqrt{\frac{x}{y}} = \ln\left(\frac{x}{y}\right)^{1/2}$$

$$= \frac{1}{2} \ln \frac{x}{y}$$

Power rule: Multiply the power $\frac{1}{2}$ times the logarithm.

$$= \frac{1}{2}(\ln x - \ln y)$$

Quotient rule: Subtract the logarithms. ▼

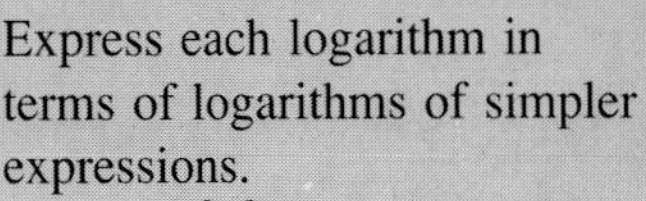

SELF-CHECK

Express each logarithm in terms of logarithms of simpler expressions.

1 $\log_{11} x^3 y^2$

2 $\log_6 \frac{(x+1)^5}{(y-2)^3}$

SELF-CHECK ANSWERS 

1 $3 \log_{11} x + 2 \log_{11} y$ $\quad$ 2 $5 \log_6(x+1) - 3 \log_6(y-2)$

EXAMPLE 3 Using the Properties of Logarithms to Combine Expressions

Combine these logarithms into a single logarithmic expression with a coefficient of 1.

SOLUTIONS

(a) $2 \ln x + 3 \ln y - 4 \ln z$

$$\begin{aligned} 2 \ln x + 3 \ln y - 4 \ln z &= \ln x^2 + \ln y^3 - \ln z^4 && \text{Power rule} \\ &= \ln x^2y^3 - \ln z^4 && \text{Product rule} \\ &= \ln \frac{x^2y^3}{z^4} && \text{Quotient rule} \end{aligned}$$

(b) $5 \log x + \frac{1}{3} \log y$

$$\begin{aligned} 5 \log x + \frac{1}{3} \log y &= \log x^5 + \log y^{1/3} && \text{Power rule} \\ &= \log x^5 + \log \sqrt[3]{y} \\ &= \log x^5\sqrt[3]{y} && \text{Product rule} \end{aligned}$$

▼

SELF-CHECK

Combine these logarithms into a single logarithmic expression with a coefficient of 1.

1 $\log_{12}(x + 3) + \log_{12}(2x - 9)$

2 $3 \log_b x + 2 \log_b y - \frac{1}{2} \log_b z$

EXAMPLE 4 Logarithmic Properties

Use $\log_5 2 \approx 0.43068$, $\log_5 3 \approx 0.68261$, and the properties of logarithms to determine the value of each of these logarithmic expressions.

SOLUTIONS

(a) $\log_5 6$

$$\begin{aligned} \log_5 6 &= \log_5(2 \cdot 3) \\ &\approx \log_5 2 + \log_5 3 && \text{Product rule} \\ &\approx 0.43068 + 0.68261 && \text{Substitute the given values.} \\ &\approx 1.11329 \end{aligned}$$

(b) $\log_5 9$

$$\begin{aligned} \log_5 9 &= \log_5 3^2 \\ &= 2 \log_5 3 && \text{Power rule} \\ &\approx 2(0.68261) && \text{Substitute the given value of } \log_5 3. \\ &\approx 1.36522 \end{aligned}$$

(c) $\log_5 0.4$

$$\begin{aligned} \log_5 0.4 &= \log_5 \frac{2}{5} \\ &= \log_5 2 - \log_5 5 && \text{Quotient rule} \\ &\approx 0.43068 - 1 && \text{The value of } \log_5 2 \text{ is given, and } \log_5 5 \text{ can be determined by inspection.} \\ &\approx -0.56932 \end{aligned}$$

▼

Example 5 proves an identity involving logarithms and exponents.

SELF-CHECK ANSWERS

1 $\log_{12}[(x + 3)(2x - 9)]$ **2** $\log_b \frac{x^3y^2}{\sqrt{z}}$

EXAMPLE 5 Verifying an Identity Involving Logarithms

Verify that $b^{\log_b x} = x$ for any positive real number x.

SOLUTION Let $y = \log_b x$. Then

$b^y = x$ Rewrite the logarithmic expression in exponential form.

and

$b^{\log_b x} = x$ Substitute $\log_b x$ for y.

ANOTHER PROPERTY OF LOGARITHMS

For any positive real number x,

$$b^{\log_b x} = x$$

EXAMPLE 6 Using the Properties of Logarithms

Determine the value of each expression.

	SOLUTIONS	
(a) $7^{\log_7 8}$	$7^{\log_7 8} = 8$	Each of these expressions can be determined by inspection, using the identity $b^{\log_b x} = x$.
(b) $3^{\log_3 0.179}$	$3^{\log_3 0.179} = 0.179$	
(c) $a^{\log_a 231}$	$a^{\log_a 231} = 231$ for $a > 0$, $a \neq 1$	

Some problems give rise to exponential or logarithmic expressions that are neither to base 10 nor to base e. Such expressions can be evaluated with a calculator if they are first converted to either base 10 or base e. The change-of-base formulas given in the following box allow us to convert from one base to another.

CHANGE-OF-BASE FORMULAS

For $a, b > 0$ and $a, b \neq 1$,

1 $\log_a x = \dfrac{\log_b x}{\log_b a}$ for $x > 0$

2 $a^x = b^{x \log_b a}$

Proof of the Change-of-Base Formula for Logarithms Let $\log_a x = y$. Then

$x = a^y$ Rewrite this equation in exponential form.

$\log_b x = \log_b a^y$ Take the log to base b of both members.

$\log_b x = y \log_b a$ Power rule for logarithms

$y = \dfrac{\log_b x}{\log_b a}$ Solve for y.

$\log_a x = \dfrac{\log_b x}{\log_b a}$ Substitute for y.

The most useful forms of this identity are

$$\log_a x = \frac{\log x}{\log a} \qquad \text{and} \qquad \log_a x = \frac{\ln x}{\ln a}$$

The proof of the change-of-base formula for exponents is similar to the proof given for logarithms.

EXAMPLE 7 Using the Change-of-Base Formula for Logarithms

Evaluate $\log_{2.7} 11.45$ to six significant digits.

SOLUTION

$\log_{2.7} 11.45 = \dfrac{\ln 11.45}{\ln 2.7}$ Use the change-of-base formula to convert to natural logs. (Converting to common logs would produce the same answer.)

Keystrokes

Scientific

[1] [1] [.] [4] [5] [ln] [÷] [2] [.] [7] [ln] [=] → 2.4545536

Graphics

[LN] [1] [1] [.] [4] [5] [÷] [LN] [2] [.] [7] [ENTER] → 2.454553615

Answer $\log_{2.7} 11.45 \approx 2.45455$

SELF-CHECK

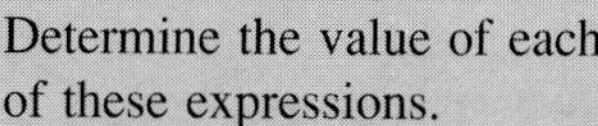

Determine the value of each of these expressions.

1 $11^{\log_{11} 52.3832}$

2 $\log_{14} 52.3832$

EXAMPLE 8 Checking the Value of a Logarithmic Expression

Use the power key on a calculator to verify that $\log_{2.7} 11.45 \approx 2.45455$.

SOLUTION

$\log_{2.7} 11.45 \approx 2.45455$

$2.7^{2.45455} \approx 11.45$ First rewrite this expression in exponential form.

Keystrokes

Scientific

[2] [.] [7] [y^x] [2] [.] [4] [5] [4] [5] [5] [=] → 11.449959 The relatively small difference is due to round-off error.

Graphics

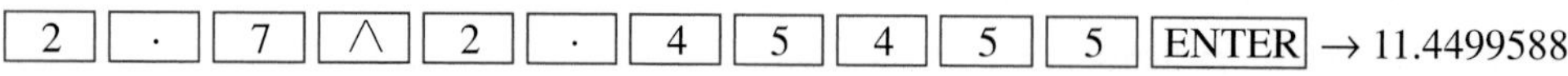

[2] [.] [7] [∧] [2] [.] [4] [5] [4] [5] [5] [ENTER] → 11.44995889

SELF-CHECK ANSWERS

1 52.3832 **2** 1.5

EXAMPLE 9 Using the Change-of-Base Formula for Exponents

Convert the exponential growth function $y = 10^x$ to an exponential function with base e.

SOLUTION $y = 10^x$ Substitute $a = 10$ and $b = e$ into the change-of-base formula $a^x = b^{x \log_b a}$.

$y = e^{x \ln 10}$

$y \approx e^{2.303x}$ Use a calculator approximation for ln 10. ▼

Exercises 10-4

A

In Exercises 1–16 express each logarithm in terms of logarithms of simpler expressions. Assume that the arguments of the logarithms are all positive real numbers.

1 $\log xy^5$ **2** $\log x^3y^4$

3 $\ln x^2y^3z^4$ **4** $\ln(2x + 3)(x + 7)$

5 $\ln \dfrac{2x + 3}{x + 7}$ **6** $\ln \dfrac{5x + 8}{3x + 4}$

7 $\log \sqrt{4x + 7}$ **8** $\log \sqrt[3]{6x + 1}$

9 $\ln \dfrac{\sqrt{x + 4}}{(y + 5)^2}$ **10** $\ln \dfrac{(x + 9)^3}{\sqrt{y + 1}}$

11 $\log \sqrt{\dfrac{xy}{z - 8}}$ **12** $\log \sqrt[4]{\dfrac{x^2y}{z^3}}$

13 $\log \dfrac{x^2(2y + 3)^3}{z^4}$ **14** $\log \dfrac{(x + 1)^3(y - 2)^2}{(z + 4)^5}$

15 $\ln\left(\dfrac{xy^2}{z^3}\right)$ **16** $\ln\left(\dfrac{x^2y^3}{z^5}\right)^3$

In Exercises 17–26 combine these logarithms into a single logarithmic expression with a coefficient of 1.

17 $2 \log x + 5 \log y$ **18** $7 \ln x + 3 \ln y$

19 $3 \ln x + 7 \ln y - \ln z$ **20** $4 \log x + 9 \log y - 5 \log z$

21 $\dfrac{1}{2} \log(x + 1) - \log(2x + 3)$

22 $\ln(3x + 8) - \dfrac{1}{2} \ln(5x + 1)$

23 $\dfrac{1}{3}[\ln(2x + 7) + \ln(7x + 1)]$

24 $\dfrac{1}{4}[\log(x + 3) - \log(2x + 9)]$

25 $2 \log_5 x + \dfrac{2}{3} \log_5 y$

26 $\dfrac{3}{5} \log_7 x - 2 \log_7 y$

In Exercises 27–34 use a calculator and the change-of-base formula to approximate each of these logarithms to five significant digits.

27 $\log_5 37.1$ **28** $\log_7 81.8$ **29** $\log_{13} 7.08$

30 $\log_{11} 4.31$ **31** $\log_{6.8} 0.856$ **32** $\log_{0.49} 3.86$

33 $\log_{0.61} 18.4$ **34** $\log_{7.3} 0.921$

B

In Exercises 35–38 determine the value of each expression.

35 $17^{\log_{17} 34}$ **36** $83^{\log_{83} 51}$

37 $1.93^{\log_{1.93} 0.53}$ **38** $4.6^{\log_{4.6} 0.068}$

In Exercises 39–50 use $\log_b 2 \approx 0.3562$, $\log_b 5 \approx 0.8271$, and the properties of logarithms to determine the value of each logarithmic expression.

39 $\log_b 10$ **40** $\log_b 4$ **41** $\log_b 25$

42 $\log_b \sqrt{5}$ **43** $\log_b \sqrt[3]{2}$ **44** $\log_b 5b$

45 $\log_b 2b$ **46** $\log_b 2.5$ **47** $\log_b 0.4$

48 $\log_b \dfrac{b}{5}$ **49** $\log_b \dfrac{b}{2}$ **50** $\log_b \sqrt{10b}$

In Exercises 51–54 convert each exponential function to an exponential function with base *e*.

51 $y = 2^x$ **52** $y = 3^x$

53 $y = \left(\dfrac{1}{2}\right)^x$ **54** $y = \left(\dfrac{1}{3}\right)^x$

C

55 Which, if any, of the following expressions are equal?

a. $\log_b \dfrac{7}{2}$ **b.** $\dfrac{\log_b 7}{\log_b 2}$

c. $\log_b(7 - 2)$ **d.** $\log_b 7 - \log_b 2$

56 Which, if any, of the following expressions are equal?

a. $(\log_b 9)(\log_b 8)$ **b.** $\log_b 9 + \log_b 8$

c. $\log_b(9 \cdot 8)$ **d.** $\log_b(9 + 8)$

57 If $b > 1$ and $x > 1$, determine whether $\log_b x$ is positive or negative.

58 If $b > 1$ and $0 < x < 1$, determine whether $\log_b x$ is positive or negative.

59 If $0 < b < 1$ and $x > 1$, determine whether $\log_b x$ is positive or negative.

60 If $0 < b < 1$ and $0 < x < 1$, determine whether $\log_b x$ is positive or negative.

ESTIMATION SKILLS (61–64)

In Exercises 61 and 62 mentally estimate the value you would calculate if you used a calculator and the change-of-base formula to approximate each expression, and then select the answer below that is closest to your estimate.

61 $\log_8 63$ **a.** 7.8 **b.** 8.1 **c.** 1.99 **d.** 2.01

62 $\log_2 17$ **a.** 8.5 **b.** 5.8 **c.** 3.99 **d.** 4.09

63 A typical calculus problem involves finding the area under a portion of a curve. Estimate the area of the shaded region of the figure shown, and then select the answer below that is closest to your estimate.

a. 1 **b.** 2 **c.** 3 **d.** 4

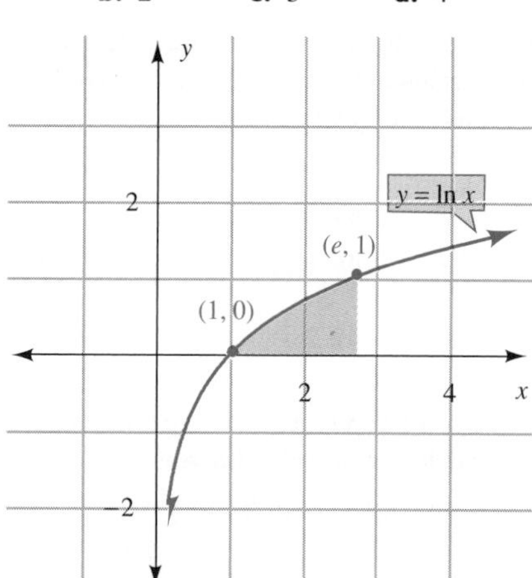

Figure for Exercise 63

64 Estimate the area of the shaded region of the figure shown, and then select the answer below that is closest to your estimate.

a. ln 2 **b.** ln 3 **c.** ln 6 **d.** ln 13

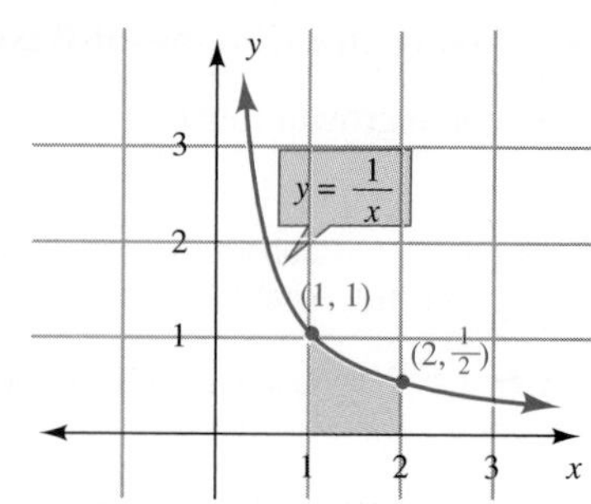

Figure for Exercise 64

CALCULATOR USAGE (65–66)

In Exercises 65 and 66 use the change-of-base formula to rewrite each function in terms of base *e*. Then graph this function on a graphics calculator.

65 $f(x) = \log_{1.5} x$ **66** $f(x) = \log_{2.5} x$

67 Write $\ln x + t = 0$ as an exponential equation that does not contain logarithms.

68 Prove the quotient rule for logarithms.

69 Prove the power rule for logarithms.

DISCUSSION QUESTION

70 Every exponential growth or decay function can be expressed in terms of base *e*. Discuss why this is possible and how one can tell by inspection whether such an exponential function represents growth or decay.

CHALLENGE QUESTION

In a major city 7 patients in various hospitals exhibited a puzzling combination of symptoms. A two-day investigation revealed that the patients had contracted a new strain of virus. Based on its similarity to other viruses, officials determined that this new virus has an incubation period of about a week (this is how long it takes for an infected person to develop symptoms). By the time the investigation was over, the number of patients hospitalized with the new disease had risen to 28. Assuming simple exponential growth in the number of persons hospitalized, estimate how long it will be before the disease has infected 1000 people in the population.

SECTION SKILL REINFORCERS

Graph each of these functions.

1 $y = 2^x$ **2** $y = 2^x + 1$ **3** $y = 2^x - 1$ **4** $y = 2^{x+1}$ **5** $y = 2^{x-1}$

10-5 Exponential and Logarithmic Equations

SECTION OBJECTIVE

7 Solve exponential and logarithmic equations.

An equation in which the variable is in an exponent is called an **exponential equation.** An equation in which the variable is in the argument of a logarithm is called a **logarithmic equation.** We will solve both types of equations in this section.

We will also examine some identities involving exponents and logarithms. Only relatively simple exponential equations, such as the one in Example 1, can be solved by inspection.

EXAMPLE 1 Solving an Exponential Equation by Inspection

Solve $2^{3x+2} = 32$ by inspection.

SOLUTION $2^{3x+2} = 32$

$2^{3x+2} = 2^5$ Express both members of the equation in terms of base 2.

$3x + 2 = 5$ The exponents are equal since the bases are the same.

$3x = 3$

$x = 1$

Answer $x = 1$ ▼

If both members of an exponential equation cannot be easily expressed in terms of a common base, then we will solve the equation by first taking logarithms of both members.

EXAMPLE 2 Solving an Exponential Equation

Solve $5^y = 30$.

SOLUTION

$5^y = 30$

$\log 5^y = \log 30$ Take the common log of both members.

$y \log 5 = \log 30$ Power rule for logarithms

$y = \dfrac{\log 30}{\log 5}$ Use a calculator to determine a numerical approximation of the exact answer obtained in the previous step.

$\left(\textit{Caution: } \dfrac{\log 30}{\log 5} \text{ is not the same as } \log \dfrac{30}{5}.\right)$

Keystrokes

Scientific

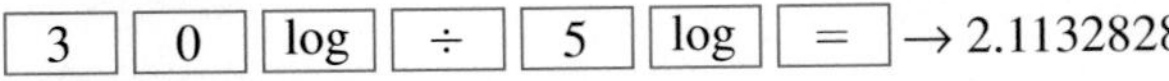
3 0 log ÷ 5 log = → 2.1132828

Graphics

LOG 3 0 ÷ LOG 5 ENTER → 2.113282753

Answer $y \approx 2.1133$

Check $5^{2.1133} \approx 30.000833 \approx 30$

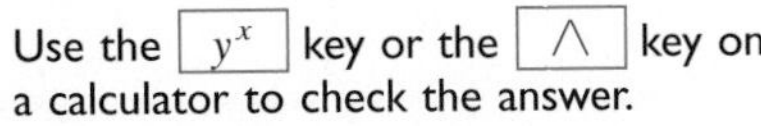
Use the y^x key or the ∧ key on a calculator to check the answer. ▼

SOLVING EXPONENTIAL EQUATIONS

Step 1 **a.** If it is obvious that both members of the equation are powers of the same base, express each member as a power of this base and then equate the exponents.

b. Otherwise, take the logarithm of both members and use the power rule to form an equation that does not contain variable exponents.

Step 2 Solve the equation formed in Step 1.

EXAMPLE 3 Solving an Exponential Equation

Solve $6^{z+3} = 8^{2z-1}$.

SOLUTION

$6^{z+3} = 8^{2z-1}$	
$\ln 6^{z+3} = \ln 8^{2z-1}$	Take the natural log of both members.
$(z + 3) \ln 6 = (2z - 1) \ln 8$	Power rule for logarithms
$z \ln 6 + 3 \ln 6 = 2z \ln 8 - \ln 8$	Multiply using the distributive property, and then combine like terms.
$z \ln 6 - 2z \ln 8 = -\ln 8 - 3 \ln 6$	
$z(\ln 6 - 2 \ln 8) = -(\ln 8 + 3 \ln 6)$	Factor out z.
$z = -\dfrac{\ln 8 + 3 \ln 6}{\ln 6 - 2 \ln 8}$	
$z \approx 3.1492736$	Use a calculator to determine a numerical approximation of the exact answer obtained in the previous step.

Answer $z \approx 3.1493$ Does this value check? 

As Examples 2 and 3 illustrate, either common logs or natural logs can be used to solve exponential equations. If the exponential equation involves base e, then we usually use natural logarithms.

EXAMPLE 4 Solving an Exponential Equation

Solve $3e^{x^2+2} = 49.287$ to five significant digits.

SOLUTION

$3e^{x^2+2} = 49.287$	
$e^{x^2+2} = 16.429$	Divide both members by 3.
$x^2 + 2 = \ln 16.429$	Take the natural log of both members.
$x^2 = \ln 16.429 - 2$	Subtract 2 from both members.
$x^2 \approx 0.79904807$	Calculator approximation
$x \approx \pm 0.89389489$	Solve this quadratic equation by extraction of roots.

Answer $x \approx -0.89389$ or $x \approx 0.89389$ Do these values check? ▼

SELF-CHECK ✓

Solve $3^{2z+1} = 5^z$ to five significant digits.

SELF-CHECK ANSWER 

$z \approx -1.8691$

Since exponential and logarithmic functions are inverses of each other, it is not surprising that we can use logarithms to solve some exponential equations. Likewise, we can solve some logarithmic equations by first rewriting them in exponential form.

EXAMPLE 5 Solving a Logarithmic Equation with One Logarithmic Term

Solve $\log(3w + 7) = 2$.

SOLUTION

$$\log(3w + 7) = 2$$

$$3w + 7 = 10^2$$ Rewrite this logarithmic equation in exponential form with base 10.

$$3w + 7 = 100$$

$$3w = 93$$ Then solve for w.

$$w = 31$$

Check

$$\log[3(31) + 7] \stackrel{?}{=} 2$$

$$\log 100 \stackrel{?}{=} 2$$

$$2 = 2 \text{ checks.}$$

Answer $w = 31$ ▼

Logarithmic functions are one-to-one functions. (The horizontal line test can be applied to the graph of $y = \log_b x$ to show that this function is one-to-one.) Thus if $x, y > 0$ and $\log_b x = \log_b y$, then $x = y$. The fact that arguments are equal when logarithms are equal allows us to replace a logarithmic equation with one that does not involve logarithms. Since logarithms are defined only for positive arguments, however, each possible solution must be checked to make sure that it is not an extraneous value.

EXAMPLE 6 Solving a Logarithmic Equation with Two Logarithmic Terms

Solve $\ln y = \ln(4y + 6)$.

SOLUTION

$$\ln y = \ln(4y + 6)$$ Since the logarithms are equal, the arguments are equal.

$$y = 4y + 6$$

$$-3y = 6$$ Now solve for y.

$$y = -2$$

Check

$$\ln y = \ln(4y + 6)$$

$$\ln(-2) \stackrel{?}{=} \ln[4(-2) + 6]$$

$\ln(-2)$ is undefined. Logarithms of negative arguments are undefined.

Answer There is no solution. -2 is an extraneous value. ▼

SELF-CHECK 

Solve these logarithmic equations.

1 $\log(t - 1) = 1$

2 $\log(v - 5) = \log(1 - v)$

SELF-CHECK ANSWERS 

1 $t = 11$ 2 No solution; $v = 3$ is extraneous.

If a logarithmic equation has more than one logarithmic term on one side of the equation, then we may need to use the properties of logarithms to rewrite these terms as a single logarithm.

EXAMPLE 7 Solving a Logarithmic Equation with Three Logarithmic Terms

Solve $\ln(3 - w) + \ln(1 - w) = \ln(11 - 6w)$.

SOLUTION

$$\ln(3 - w) + \ln(1 - w) = \ln(11 - 6w)$$

$$\ln(3 - w)(1 - w) = \ln(11 - 6w)$$ Express the left member as a single logarithm, using the product rule for logarithms.

$$(3 - w)(1 - w) = 11 - 6w$$ Since the natural logarithms are equal, the arguments are equal.

$$3 - 4w + w^2 = 11 - 6w$$ Simplify this quadratic equation.

$$w^2 + 2w - 8 = 0$$

$$(w + 4)(w - 2) = 0$$ Then factor, and solve for w.

$w + 4 = 0$ or $w - 2 = 0$

$w = -4$ $\qquad$ $w = 2$

Check For $w = -4$,

$$\ln[3 - (-4)] + \ln[1 - (-4)] \stackrel{?}{=} \ln[11 - 6(-4)]$$

$$\ln 7 + \ln 5 \stackrel{?}{=} \ln(11 + 24)$$

$\ln 35 = \ln 35$ checks.

For $w = 2$,

$$\ln(3 - 2) + \ln(1 - 2) \stackrel{?}{=} \ln[11 - 6(2)]$$

$$\ln 1 + \ln(-1) \stackrel{?}{=} \ln(-1)$$

$\ln(-1)$ is undefined. Thus $w = 2$ is an extraneous value.

Answer $w = -4$

SELF-CHECK

Solve $\ln(2z + 5) + \ln z = \ln 3$.

SOLVING LOGARITHMIC EQUATIONS

Step 1 **a.** If possible, rewrite the logarithmic equation in exponential form.

b. Otherwise, use the properties of logarithms to write each side of the equation as a single logarithmic term with the same base. Then form a new equation by equating the arguments of these logarithms.

Step 2 Solve the equation formed in Step 1.

Step 3 Check all possible solutions for extraneous values; logarithms of negative arguments are undefined.

SELF-CHECK ANSWER 

$z = \frac{1}{2}$

EXAMPLE 8 Solving a Logarithmic Equation with Two Logarithmic Terms

Solve $\log(2v + 1) - \log(v - 4) = 1$.

SOLUTION $\log(2v + 1) - \log(v - 4) = 1$

$$\log\left(\frac{2v + 1}{v - 4}\right) = 1$$ Express the left member as a single logarithm, using the quotient rule for logarithms.

$$\frac{2v + 1}{v - 4} = 10^1$$ Rewrite this equation in exponential form, using a base of 10.

$$2v + 1 = 10(v - 4)$$ Then solve for v.

$$2v + 1 = 10v - 40$$

$$-8v = -41$$

$$v = \frac{41}{8}$$

Check $$\log\left[2\left(\frac{41}{8}\right) + 1\right] - \log\left(\frac{41}{8} - 4\right) \stackrel{?}{=} 1$$

$$\log\left(\frac{45}{4}\right) - \log\left(\frac{9}{8}\right) \stackrel{?}{=} 1$$

$$\log\left(\frac{45}{4} \div \frac{9}{8}\right) \stackrel{?}{=} 1$$

$$\log 10 \stackrel{?}{=} 1$$

$$1 = 1 \text{ checks.}$$

Answer $v = \frac{41}{8}$ ▼

The exponential and logarithmic equations that we have examined in this section have been either contradictions with no solution or conditional equations with one or two solutions. We will now examine some exponential and logarithmic equations that are identities.

EXAMPLE 9 Verifying Logarithmic and Exponential Identities

Verify the following identities.

SOLUTIONS

(a) $\log_7 14^x - \log_7 2^x = x$

$$\log_7 14^x - \log_7 2^x = x \log_7 14 - x \log_7 2$$ Rewrite the left member, using the power rule for logarithms.

$$= x(\log_7 14 - \log_7 2)$$ Then factor out x.

$$= x \log_7 \frac{14}{2}$$ Simplify this expression, using the quotient rule for logarithms.

$$= x \log_7 7$$

$$= x$$ Replace $\log_7 7$ by 1 to obtain the right member of this identity.

(b) $e^{-\ln x} = \dfrac{1}{x}$ for $x > 0$

$e^{-\ln x} = e^{\ln x^{-1}}$ — Rewrite the left member, using the power rule for logarithms.

$= x^{-1}$ — Simplify, using the identity $b^{\log_b y} = y$, with $b = e$ and $y = x^{-1}$.

$= \dfrac{1}{x}$ — Replace x^{-1} with $\dfrac{1}{x}$ to obtain the right member of this identity.

The next example illustrates how we can use a graphics calculator to approximate the solution to a logarithmic equation. This procedure can be used to solve a wide variety of algebraic equations including those with exponential expressions or rational expressions.

EXAMPLE 10 Using a Calculator to Solve a Logarithmic Equation

Use a graphics calculator to approximate to the nearest tenth the solution of $\log(27x + 5.5) = x - 1.5$.

SOLUTION

Let $y_1 = \log(27x + 5.5)$
and $y_2 = x - 1.5$

The solution of the given equation will be the point(s) where the graphs of $y_1 = \log(27x + 5.5)$ and $y_2 = x - 1.5$ intersect.

Keystrokes

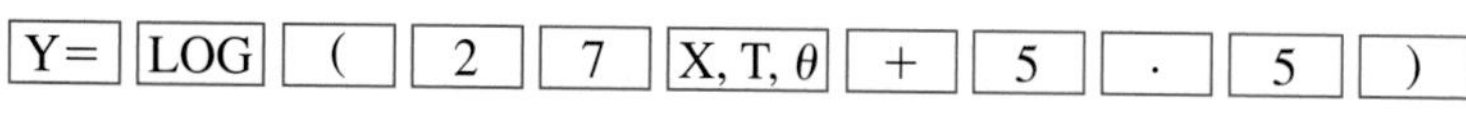

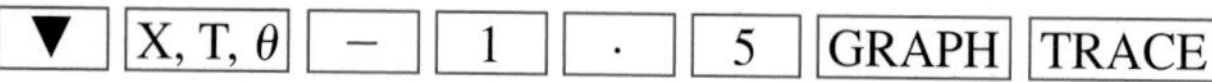

First enter y_1, and then enter y_2. Then graph, assuming the display window of values has already been set to $[-1, 5]$ for x and $[-2, 6]$ for y. The TRACE feature can be used to approximate the point of intersection.

$[-1, 5]$ for x, $[-2, 6]$ for y

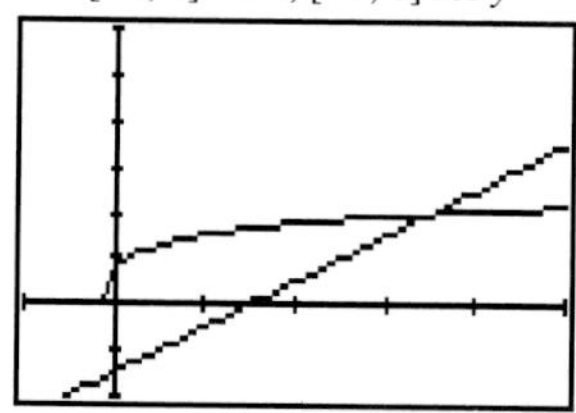

Answer $x = 3.5$ Does this solution check?

Exercises 10-5

In Exercises 1–24 solve each equation without using a calculator.

1 $3^{w-5} = 27$

2 $7^{2w-1} = \sqrt{7}$

3 $\left(\dfrac{2}{3}\right)^{x^2} = \dfrac{16}{81}$

4 $5^{x^2-6} = \dfrac{1}{25}$

5 $\log(y - 5) = 1$

6 $\log(2y + 1) = 1$

7 $\log(3n - 5) = 2$

8 $\log(5n - 4) = 0$

9 $\ln(3m - 7) = \ln(2m + 9)$

10 $\log(5y + 6) = \log(2y - 9)$

11 $\log(4w + 3) = \log(8w + 5)$

12 $\ln(5y - 7) = \ln(2y + 1)$

13 $\ln(3 - x) = \ln(1 - 2x)$

14 $\log(7 - 5x) = \log(4 - 8x)$

15 $\log(t + 3) + \log(t - 1) = \log 5$

16 $\ln(7t + 3) - \ln(t + 1) = \ln(6t + 2)$

17 $\ln(v^2 - 9) - \ln(v + 3) = \ln 7$

18 $\ln(2v + 6) - \ln(v + 1) = \ln(v + 3)$

19 $\ln(5x - 7) - \ln(2x + 3) = \ln 3$

20 $\log(7x + 13) - \log(4x + 13) = \log 5$

21 $\log(1 - y) + \log(4 - y) = \log(18 - 10y)$

22 $\ln(2 - y) + \ln(1 - y) = \ln(32 - 4y)$

23 $\log(w - 3) - \log(w^2 + 9w - 32) = -1$

24 $\log(w^2 + 1) - \log(w - 2) = 1$

In Exercises 25–40 solve each equation, and then use a calculator to approximate the solution to five significant digits.

25 $4^v = 15$

26 $5^v = 18$

27 $3^{-w+7} = 22$

28 $6^{-w+3} = 81$

29 $9.2^{2t+1} = 11.3^t$

30 $8.7^{3t-1} = 10.8^{2t}$

31 $7.6^{-2z} = 5.3^{2z-1}$

32 $8.1^{-3z} = 6.5^{1-2z}$

33 $e^{3x} = 78.9$

34 $e^{3x+5} = 15.9$

35 $10^{2y+1} = 51.3$

36 $10^{3y-4} = 73.8$

37 $0.83^{v^2} = 0.68$

38 $0.045^{v^2} = 0.0039$

39 $3.7e^{x^2+1} = 689.7$

40 $2.5e^{x^2+4} = 193.2$

B

In Exercises 41–50 solve each equation, and then use a calculator to approximate each solution to five significant digits.

41 $\log(5x - 17) = 0.83452$

42 $\ln(11x - 3) = 1.44567$

43 $\ln(\ln y) = 1$

44 $\log(\log y) = 0.48913$

45 $\ln(v - 4) + \ln(v - 3) = \ln(5 - v)$

46 $\log(3x + 1) + \log(x + 2) = \log x$

47 $\ln(11 - 5x) - \ln(x - 2) = \ln(x - 6)$

48 $\ln(3 - x) - \ln(x - 2) = \ln(x + 1)$

49 $(\ln x)^2 = \ln x^2$

50 $(\log x)^2 = \log x^2$

C

Calculator Usage (51–54)

In Exercises 51–54 use a graphics calculator to approximate to the nearest tenth the solution(s) of each equation.

51 $\log(398x + 5) = x + 0.5$

52 $\ln(18x + 3) = 3x - 5$

53 $0.1e^x = 8 - 2x$

54 $e^{0.5x} = 4 - x^2$

In Exercises 55–62 verify each identity, assuming that x is a positive real number.

55 $10^{-\log x} = \dfrac{1}{x}$

56 $100^{\log x} = x^2$

57 $e^{-x \ln 3} = \left(\dfrac{1}{3}\right)^x$

58 $e^{(\ln x)/2} = \sqrt{x}$

59 $\log 60^x - \log 6^x = x$

60 $\log 5^x + \log 2^x = x$

61 $\ln\left(\dfrac{4}{5}\right)^x + \ln\left(\dfrac{5}{3}\right)^x + \ln\left(\dfrac{3}{4}\right)^x = 0$

62 $\ln\left(\dfrac{2}{3}\right)^x + \ln\left(\dfrac{5}{2}\right)^x - \ln\left(\dfrac{5}{3}\right)^x = 0$

Depreciation (63 and 64)

63 The value V of an industrial lathe after t years of depreciation is given by the formula $V = 35{,}000e^{-0.2t} + 1000$. Approximately how many years will it take for the value to depreciate to \$10,000?

64 The value V of an irrigation system after t years of depreciation is given by the formula $V = 59{,}000e^{-0.2t} + 3000$. Approximately how many years will it take for the value to depreciate to \$7000?

Discussion Question

65 The steps used to solve a certain logarithmic equation produced one possible solution that was a negative number. One student claimed this number could not check in the equation since it was negative. Discuss the logic of this student's claim.

Challenge Question

Which is larger: 1994^{1995} or 1995^{1994}?

Section Skill Reinforcers

Graph each of these functions.

1 $y = \log_2 x$ **2** $y = (\log_2 x) - 1$ **3** $y = (\log_2 x) + 1$ **4** $y = \log_2(x - 1)$ **5** $y = \log_2(x + 1)$

10-6 Applications of Exponents and Logarithms

SECTION OBJECTIVE

7 Use exponential and logarithmic equations to solve applied problems.

Exponential and logarithmic equations can be used to describe an increase or decrease in an amount of money, the growth or decline of populations, and the decay of radioactive elements. Logarithmic scales are often used to measure natural phenomena; for example, decibels measure the intensity of sounds, and the Richter scale measures the intensity of earthquakes. In this section we will examine all of these applications.

Two important formulas that we will apply in the following examples are the formulas for periodic growth and continuous growth. Growth that occurs at discrete intervals (such as yearly, monthly, or weekly) is called **periodic growth;** growth that occurs continually at each instant is called **continuous growth.** The formula for continuous growth is a good model for periodic growth when the number of growth periods is relatively large. For example, the population of rabbits over a period of years can be accurately estimated using the continuous growth formula, even though female rabbits do not have offspring continuously.

GROWTH AND DECAY FUNCTIONS

Periodic growth formula: For an original amount P growing at an annual rate r with periodic compounding n times a year for t years,

$$A = P\left(1 + \frac{r}{n}\right)^{nt}$$

Continuous growth and decay formula: For an original amount P growing (decaying) continuously at a rate r for a time t,

$$A = Pe^{rt}$$

If $r > 0$, there is growth; if $r < 0$, there is decay.

EXAMPLE 1 Periodic Growth of an Investment

How many years will it take an investment of \$1000 to double in value if interest on the investment is compounded semiannually at a rate of 11%?

SOLUTION

$$A = P\left(1 + \frac{r}{n}\right)^{nt}$$

$$2000 = 1000\left(1 + \frac{0.11}{2}\right)^{2t}$$

Substitute $A = 2000$, $P = 1000$, $r = 0.11$, and $n = 2$ into the periodic growth formula, and then simplify.

$$2 = (1.055)^{2t}$$

$\ln 2 = \ln(1.055)^{2t}$ Take the natural log of both members.

$\ln 2 = 2t \ln 1.055$ Power rule for logarithms

$t = \dfrac{\ln 2}{2 \ln 1.055}$ Solve for t.

$t \approx 6.4730785$ Calculator approximation

Answer The investment will double in value in approximately 6.5 years. ▼

EXAMPLE 2 Continuous Growth of Bacteria

A new culture of bacteria grows continuously at the rate of 20% per day. If a culture of 10,000 bacteria isolated in a laboratory is allowed to multiply, how many bacteria will there be at the end of one week?

SOLUTION $A = Pe^{rt}$

$= 10{,}000\,e^{0.20(7)}$ Substitute $P = 10{,}000$, $r = 0.20$, and $t = 7$ (1 week = 7 days) into the continuous growth formula, and then simplify.

$= 10{,}000e^{1.4}$

$A \approx 40{,}552$ Calculator approximation

Answer At the end of the week there will be approximately 41,000 bacteria. ▼

EXAMPLE 3 Continuous Decay of Carbon 14

Carbon 14 decays continuously at the rate of 0.01245% per year. When living tissue dies, it no longer absorbs carbon 14; any carbon 14 present decays and is not replaced. An archaeologist has determined that only 20% of the carbon 14 originally in a plant specimen remains. Estimate the age of this specimen.

SOLUTION $A = Pe^{rt}$

$0.20P = Pe^{-0.0001245t}$ Substitute the given values into the continuous decay formula. A 0.01245% decay rate means that $r = -0.0001245$, and 20% of the original amount P left means that $A = 0.20P$.

$0.20 = e^{-0.0001245t}$ Divide both members by P.

$\ln 0.20 = \ln e^{-0.0001245t}$ Take the natural logarithm of both members.

$\ln 0.20 = -0.0001245t$ Simplify, using $\ln e^x = x$.

$t = \dfrac{\ln 0.20}{-0.0001245}$

$t \approx 12{,}927.212$ Calculator approximation

Answer This specimen is approximately 13,000 years old. ▼

SELF-CHECK 

1 Compute the number of years it will take an investment of \$1 to triple in value if interest is compounded annually at a rate of 8%.

2 The population of an island is growing continuously at the rate of 5% per year. Estimate to the nearest thousand the population of the island in 10 years, given that the population is now 100,000.

3 The population of a species of whales is estimated to be decreasing continuously at a rate of 5% per year ($r = -0.05$). If this rate of decrease continues, in how many years will the population have declined from its current level of 20,000 to 8000?

SELF-CHECK ANSWERS 

1 Approximately 14.3 years **2** 165,000
3 18 years (rounded to the nearest year)

Seismologists use the Richter scale to measure the magnitude of earthquakes. The equation $R = \log \frac{A}{a}$ compares the amplitude A of the shock wave of an earthquake to the amplitude a of a reference shock wave of minimal intensity.

EXAMPLE 4 Using the Richter Scale

In the Denver, Colorado area in the 1960s an earthquake with an amplitude 40,000 times the reference amplitude occurred after a liquid was injected under pressure into a well more than 2 miles deep. Calculate the magnitude of this earthquake on the Richter scale.

SOLUTION $R = \log \frac{A}{a}$

$= \log \frac{40{,}000a}{a}$ Substitute the given amplitude into the Richter scale equation.

$= \log 40{,}000$ Divide by a.

$R \approx 4.6020600$ Calculator approximation

Answer This earthquake measured approximately 4.6 on the Richter scale.

The human ear can hear a vast range of sound intensities, so it is more practical to use a logarithmic scale than to use an absolute scale to measure the intensity of sound. The unit of measurement on this scale is the **decibel.** The number of decibels D of a sound is given by the formula $D = 10 \log \frac{I}{I_0}$, which compares the intensity I of the sound to the reference intensity I_0, which is at the threshold of hearing ($I_0 \approx 10^{-16}$ watt per square centimeter).

EXAMPLE 5 Determining the Decibel Level of Music

If a store in a mall plays background music at an intensity of 10^{-14} watt per square centimeter, determine the number of decibels of the music you hear.

SOLUTION $D = 10 \log \frac{I}{I_0}$

$= 10 \log \frac{10^{-14}}{10^{-16}}$ Substitute 10^{-14} for I and 10^{-16} for I_0 into the decibel equation.

$= 10 \log 10^2$ Simplify, and then solve for D by inspection.

$= 10(2)$

$D = 20$

Answer 20 decibels

SELF-CHECK

Compute the number of decibels of a child's cry if its intensity is 10^{-7} watt per square centimeter.

SELF-CHECK ANSWER 

90 decibels

Chemists use pH to measure the hydrogen ion concentration in a solution. Distilled water has a pH of 7, acids have a pH of less than 7, and bases have a pH of more than 7. The formula for the pH of a solution is $\text{pH} = -\log \text{H}^+$, where H^+ measures the concentration of hydrogen ions in moles per liter.

EXAMPLE 6 **Determining the pH of a Beer**

Determine the pH of a light beer if its H^+ is measured at 6.3×10^{-5} mole per liter.

SOLUTION $\text{pH} = -\log \text{H}^+$

$= -\log(6.3 \times 10^{-5})$ Substitute $\text{H}^+ = 6.3 \times 10^{-5}$ into the pH equation.

$\text{pH} \approx 4.2007$ Approximate this value with a calculator.

Answer $\text{pH} \approx 4.2$

Exercises 10-6

A

In Exercises 1–8 use the formula for periodic growth to solve each problem.

Periodic Compound Interest (1–8)

1 Find the value of \$150 invested at 9% with interest compounded quarterly for 5 years.

2 Find the value of \$210 invested at 7% with interest compounded monthly for 9 years.

3 How many years will it take an investment to double in value if interest is compounded annually at 8%?

4 How many years will it take an investment to double in value if interest is compounded semiannually at 10%?

5 How many years will it take a savings account to triple in value if interest is compounded monthly at 6%?

6 How many years will it take a zero-coupon bond to triple in value if interest is compounded monthly at 7.5%?

7 If an investment on which interest is compounded monthly doubles in value in 8 years, what is the annual rate of interest?

8 If an investment on which interest is compounded quarterly doubles in value in 9 years, what is the annual rate of interest?

In Exercises 9–20 use the formula for continuous growth and decay to solve each problem.

Rate of Inflation (9 and 10)

9 If prices will double in 10 years at the current rate of inflation, what is the current rate of inflation? Assume that the effect of inflation is continuous.

10 If prices will double in 9 years at the current rate of inflation, what is the current rate of inflation? Assume that the effect of inflation is continuous.

Continuous Compound Interest (11 and 12)

11 How many years will it take an investment to double in value if interest is compounded continuously at 7%?

12 How many years will it take an investment to double in value if interest is compounded continuously at 9%?

Carbon-14 Dating (13 and 14)

13 Carbon 14 decays continuously at the rate of 0.01245% per year. An archaeologist determines that only 5% of the original carbon 14 from a plant specimen remains. Estimate the age of this specimen.

14 Carbon 14 decays continuously at the rate of 0.01245% per year. An archaeologist has determined that only 10% of the carbon 14 originally in a plant specimen remains. Estimate the age of this specimen.

Radioactive Decay (15 and 16)

15 The radioactive material used to power a satellite decays at a rate that decreases the available power by 0.05% per day. When the power supply reaches $\frac{1}{100}$ of its original level, the satellite is no longer functional. How many days should the power supply last?

16 The radioactive material used to power a satellite decays at a rate that decreases the available power by 0.03% per day. When the power supply reaches $\frac{1}{100}$ of its original level, the satellite is no longer functional. How many days should the power supply last?

Population Decline (17 and 18)

17 The population of a species of whales is estimated to be decreasing at a rate of 4% per year ($r = -0.04$). The current population is approximately 15,000. First estimate the population 10 years from now, and then determine how many years from now the population will have declined to 5000.

18 If the number of white owls in Illinois has decreased from 750 to 500 in 10 years, what is the annual rate of decrease?

Population Growth (19 and 20)

19 If the population of a town has grown from 1200 to 1800 in 3 years, what is the annual rate of increase?

20 The human population in a remote area has doubled in the last 18 years. What is the annual rate of increase?

In Exercises 21–24 use the formula $R = \log \frac{A}{a}$ to solve each problem.

Magnitude of an Earthquake (21–24)

21 The amplitude of the September 19, 1985 earthquake in Mexico City was 63,100,000 times the reference amplitude. Calculate the magnitude of this earthquake on the Richter scale.

22 The amplitude of the September 20, 1965 earthquake in Mexico City was 20,000,000 times the reference amplitude. Calculate the magnitude of this earthquake on the Richter scale.

23 One of the highest Richter readings ever recorded was generated by an earthquake that occurred in an ocean trench. An earthquake with an amplitude 790,000,000 times the reference amplitude lifted the ocean floor several feet. Calculate the magnitude of this earthquake on the Richter scale.

Figure for Exercise 23

24 Calculate how many times more intense an earthquake with a Richter scale reading of 8.6 is than an earthquake with a Richter scale reading of 8.3.

In Exercises 25–28 use the formula $D = 10 \log \frac{I}{I_0}$ to solve each problem. ($I_0 = 10^{-16}$ watt/cm^2.)

Noise Levels (25–28)

25 Find the number of decibels of a whisper if its intensity is 3×10^{-14} watt per square centimeter.

26 Find the number of decibels of city traffic if its intensity is 8.9×10^{-7} watt per square centimeter.

27 The noise level in a bar measures 85 decibels. Calculate the intensity of this noise.

28 The decibel reading near a jet aircraft is 105 decibels. Calculate the intensity of this noise.

In Exercises 29–32 use the formula pH = $-\log H^+$ to solve each problem.

pH Measurements (29–32)

29 Determine the pH of grape juice that has an H^+ concentration of 0.000 109 mole per liter.

30 Determine the pH of saccharin, a sugar substitute that has an H^+ concentration of 4.58×10^{-7} mole per liter.

31 A leading shampoo has a pH of 9.13. What is the H^+ concentration in moles per liter?

32 Blood is buffered (kept constant) at a pH of 7.35. What is the H^+ concentration in moles per liter?

The monthly payment P required to pay off a loan of amount A at an annual interest rate r in n years is given by the formula

$$P = \frac{A\left(\frac{R}{12}\right)}{1 - \left(1 + \frac{R}{12}\right)^{-12n}}$$

In Exercises 33 and 34 use this formula to calculate the monthly payment.

Loan Payments (33 and 34)

33 Determine the monthly payment necessary to pay off a \$47,400 home loan that is financed for 30 years at 9.875%.

34 Determine the monthly payment necessary to pay off a $47,400 home loan that is financed for 30 years at 10.5%.

The number of monthly payments of amount P required to completely pay for a loan of amount A borrowed at interest rate R is given by the formula

$$n = -\frac{\log\left(1 - \frac{AR}{12P}\right)}{\log\left(1 + \frac{R}{12}\right)}$$

In Exercises 35 and 36 use this formula to calculate the number of monthly payments.

Car Payments (35 and 36)

35 Determine the number of monthly car payments of $253.59 required to pay off a $7668.00 car loan when the interest rate is 11.7%.

36 Determine the number of monthly car payments of $200.80 required to pay off a $7668.00 car loan when the interest rate is 11.7%.

Discussion Question

37 Which type of function—exponential or logarithmic—would provide a better model for the amount of learning that an individual acquires during the first 50 years of life? Explain your answer.

Challenge Question

Certain bacteria divide in two every $\frac{1}{2}$ hour. A jar into which some of these bacteria are placed is full after exactly 20 hours. How full is the jar after 19 hours, 40 minutes? First express your answer using a fractional exponent, then express your answer as a percent.

Section Skill Reinforcers

Given that the number in each exercise below is the discriminant of a quadratic equation with rational coefficients, determine whether the solutions of the quadratic equation are distinct rational solutions, distinct irrational solutions, a real solution of multiplicity two, or complex conjugates.

1 -4 **2** 0 **3** 16 **4** 17 **5** -17

Key Concepts for Chapter 10

1 A function is said to be one-to-one if distinct ordered pairs have distinct second components.

2 The vertical line test can be used to determine whether a graph represents a function. The horizontal line test can be used to determine whether the graph of a function represents a one-to-one function.

3 Inverse relations:

a. If f is the set of ordered pairs (x, y), then f^{-1} is the set of ordered pairs (y, x) formed by reversing the ordered pairs of f.

b. The inverse of a one-to-one function is also a function.

c. The graphs of f and f^{-1} are symmetric about the line $x = y$.

4 Exponential function:

a. If $b > 0$ and $b \neq 1$, then $f(x) = b^x$ is an exponential function with base b.

b. If $b > 1$, then $f(x) = b^x$ is called a growth function.

c. If $0 < b < 1$, then $f(x) = b^x$ is called a decay function.

5 Properties of exponential functions:
For real exponents x and y and bases $a > 0$, $b > 0$,

a. for $b \neq 1$, $b^x = b^y$ if and only if $x = y$.
(An exponential function is a one-to-one function.)

b. for $x \neq 0$, $a^x = b^x$ if and only if $a = b$.

6 Logarithmic functions:

a. For $x > 0$, $b > 0$, and $b \neq 1$, $y = \log_b x$ if and only if $b^y = x$.

b. Common logarithms: $\log x$ means $\log_{10} x$.

c. Natural logarithms: $\ln x$ means $\log_e x$.

d. $y = \log x$ is equivalent to $x = 10^y$.

e. $y = \ln x$ is equivalent to $x = e^y$.

7 Properties of logarithmic functions:
For $b > 0$ and $b \neq 1$,

a. $\log_b 1 = 0$

b. $\log_b b = 1$

c. $\log_b \frac{1}{b} = -1$

d. $\log_b b^x = x$

e. $b^{\log_b x} = x$

8 Logarithmic functions and exponential functions are inverses of each other.

9 Properties of logarithms:
For $x, y > 0$, $b > 0$, and $b \neq 1$,

a. Product rule: $\log_b xy = \log_b x + \log_b y$

b. Quotient rule: $\log_b \frac{x}{y} = \log_b x - \log_b y$

c. Power rule: $\log_b x^p = p \log_b x$

10 Change-of-base formulas:
For $a, b > 0$ and $a, b \neq 1$,

a. $\log_a x = \frac{\log_b x}{\log_b a}$ for $x > 0$

b. $a^x = b^{x \log_b a}$

11 Solving exponential equations:

a. If it is obvious that both members of an equation are powers of the same base, express each member as a power of this base and then equate the exponents. Otherwise, take the logarithm of both members, and use the power rule to form an equation that does not contain variable exponents.

b. Then solve the equation formed.

12 Solving logarithmic equations:

a. If possible, rewrite the logarithmic equation in exponential form. Otherwise, use the properties of logarithms to write the two members of the equation as two single logarithms with the same base. Form a new equation by equating the arguments of these logarithms.

b. Then solve the equation formed.

c. Check all possible solutions for extraneous values since logarithms of negative arguments are undefined.

13 a. Periodic growth formula:
$$A = P\left(1 + \frac{r}{n}\right)^{nt}$$

b. Continuous growth formula:
$$A = Pe^{rt}$$

Review Exercises for Chapter 10

In Exercises 1–3 write the inverse of each relation, and state the domain and the range of this inverse relation.

1 $f = \{(-1, 1), \left(-\frac{1}{2}, 2\right), \left(\frac{1}{2}, -2\right), (1, -1)\}$

2 $f(x) = 2x + 7$

3 $f(x) = \dfrac{2x - 1}{x + 2}$

4 Graph the function $y = -\frac{5}{2}x + 2$ and its inverse on the same coordinate system.

In Exercises 5 and 6 graph the inverse of the given function.

5 **6**

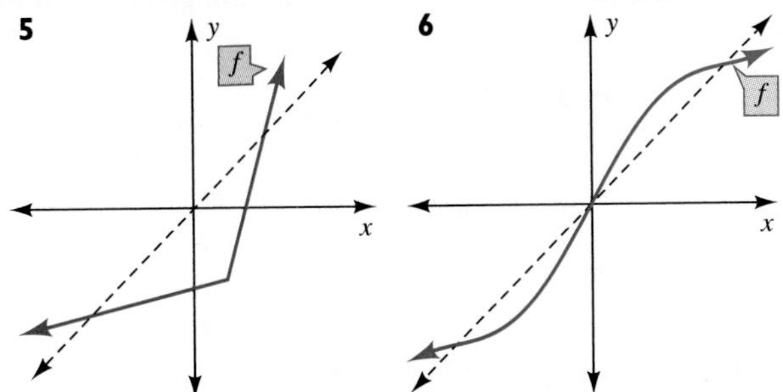

In Exercises 7–10 determine whether each function is a one-to-one function.

7 $\{(3, 7), (4, 5), (5, 6), (6, 7), (7, 10)\}$

8 **9**

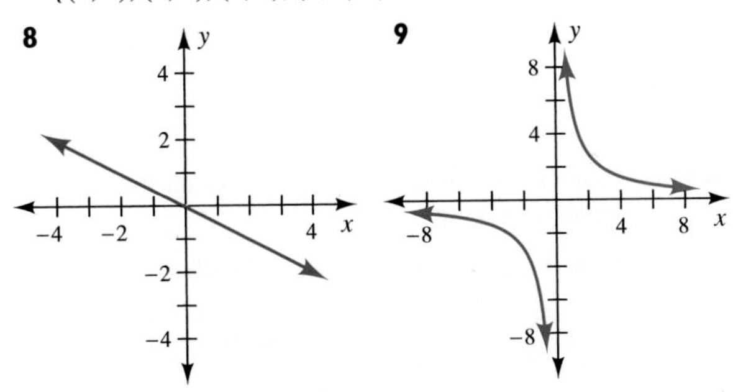

10

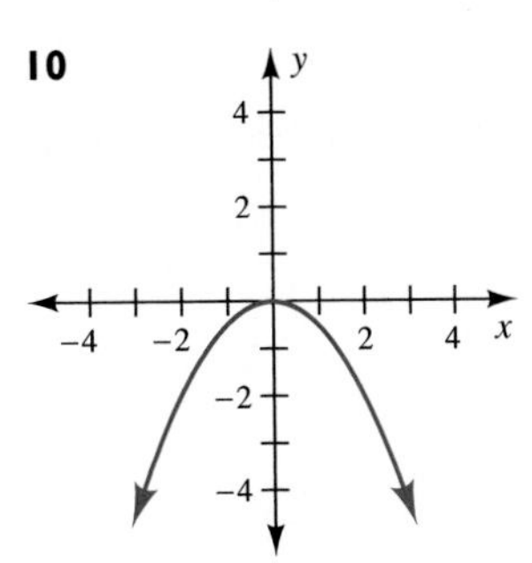

In Exercises 11–16 write each logarithmic equation in exponential form.

11 $\log_6\sqrt{6} = \frac{1}{2}$ **12** $\log_{17}1 = 0$ **13** $\log_8\frac{1}{64} = -2$

14 $\log_b a = c$ **15** $\ln a = c$ **16** $\log c = d$

In Exercises 17–22 write each exponential equation in logarithmic form.

17 $7^3 = 343$ **18** $19^{1/3} = \sqrt[3]{19}$ **19** $\left(\frac{4}{7}\right)^{-2} = \frac{49}{16}$

20 $e^{-1} = \frac{1}{e}$ **21** $10^{-4} = 0.0001$ **22** $8^x = y$

In Exercises 23–54 solve each equation without using a calculator. Indicate which expressions are undefined.

23 $11^x = \frac{1}{121}$ **24** $125^x = 25$ **25** $2^y = \sqrt[3]{4}$

26 $\left(\frac{4}{9}\right)^y = \frac{3}{2}$ **27** $2^{4w-1} = 8$ **28** $9^{2v+1} = 27$

29 $2^{x^2-1} = 8$ **30** $9^{x^2} = 3^{x+1}$ **31** $\log_8 64 = z$

32 $\log_{16}64 = x$ **33** $\log_2 x = 2$ **34** $\log_{13}x = \frac{1}{2}$

35 $\log_3 w = -2$ **36** $\log_3(-2) = w$ **37** $\log_{-2}3 = y$

38 $\log_t 169 = 2$ **39** $\log_t 8 = -3$ **40** $5^{\log_5 11} = x$

41 $\log_5 5^{17} = x$ **42** $\log_7 x = 1$ **43** $\ln 0 = x$

44 $\log(3n - 4) = \log(2n - 1)$

45 $\log_3 20 + \log_3 7 = \log_3 y$

46 $\log(3x + 1) = 2$

47 $\log_3(x^2 - 19) = 4$

48 $\ln(w + 2) + \ln w = \ln 3$

49 $\ln(1 - w) + \ln(1 - 2w) = \ln(7 - 4w)$

50 $\log(5 - 2v) - \log(1 - v) = \log(3 - 2v)$

51 $\ln(5v + 3) = \ln(3v + 9)$

52 $\log_5 27 - \log_5 2 = \log_5 y$

53 $\log(2 - 6v) - \log(2 - v) = \log(1 - v)$

54 $\ln(5 - x) + \ln(x + 1) = \ln(3x - 1)$

In Exercises 55–68 use a calculator to approximate the value of each expression to five significant digits.

55 $(\sqrt{7})^\pi$ **56** $e^{0.7836}$ **57** $e^{-2.4897}$

58 $10^{-0.8107}$ **59** $\log 113.58$ **60** $\ln 113.58$

61 $\log(8.1 \times 10^{-4})$ **62** $\ln(7.3 \times 10^8)$ **63** $\log_5 7$

64 $\frac{e-9}{13}$ **65** $\ln\frac{17.3}{18.3}$ **66** $\frac{\ln 17.3}{\ln 18.3}$

67 $\frac{\ln 17.3}{18.3}$ **68** $\frac{\ln 8}{\log 8}$

In Exercises 69–72 express each logarithm in terms of logarithms of simpler expressions. Assume that the arguments of the logarithms are all positive real numbers.

69 $\log x^3y^4$ **70** $\ln\frac{7x-9}{2x+3}$

71 $\ln\frac{\sqrt{2x+1}}{5x+9}$ **72** $\log\sqrt{\frac{x^2y^3}{z}}$

In Exercises 73–76 combine the logarithms into a single logarithmic expression with a coefficient of 1. Assume that the arguments of the logarithms are all positive real numbers.

73 $2\ln x + 3\ln y$ **74** $5\ln x - 4\ln y$

75 $\ln(x^2-3x-4) - \ln(x-4)$ **76** $\frac{1}{2}(\ln x - \ln y)$

Compound Interest (77–79)

77 How many years will it take a savings bond to double in value if interest is compounded annually at 7.5%?

78 If an investment on which interest is compounded monthly doubles in value in 10 years, what is the annual rate of interest?

79 If an investment on which interest is compounded continuously doubles in value in 8 years, what is the rate of interest?

80 Radioactive Decay The radioactive material used to power a satellite decays at a rate that decreases the available power by 0.045% per day. When the power supply reaches $\frac{1}{100}$ of its original level, the satellite is no longer functional. How many days should the power supply last?

In Exercises 81 and 82 approximate the solution to each equation to four significant digits.

81 $\log(5x-2) + \log(x-1) = \log 10$

82 $\ln(2w+3) + \ln(w+1) = \ln(w+2)$

83 Convert $y = 5^x$ to an exponential function with base e.

84 Verify that $\log 50^x + \log 6^x - \log 3^x = 2x$ is an identity.

85 Verify that $1000^{\log x} = x^3$ is an identity.

In Exercises 86–90, given that $f(x) = 2x + 5$, evaluate each of the following expressions for the given value of x:

a. $f(x)$ **b.** $f^{-1}(x)$ **c.** $\frac{1}{f(x)}$

86 $x = 4$ **87** $x = -4$ **88** $x = -3$

89 $x = 13$ **90** $x = a$

CALCULATOR USAGE (91 AND 92)

91 Use a graphics calculator to graph $y = \left(\frac{5}{4}\right)^x$ and its inverse on the same display window.

92 Use a graphics calculator to approximate to the nearest tenth the solution(s) of $\ln(10x+20) = x^2 + x - 12$.

Mastery Test for Chapter 10

Exercise numbers correspond to Section Objective numbers.

[10-1] 1 Write the inverse of each of these relations.

a. $\{(-1, 4), (8, 9), (-7, 11)\}$ **b.** $f(x) = 2x - \frac{1}{7}$

c.

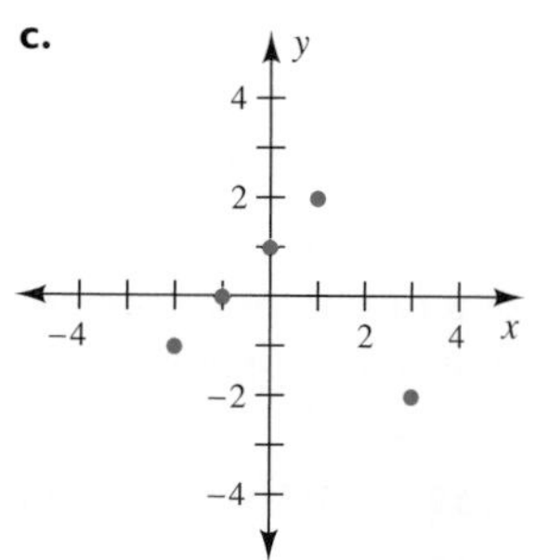

d.

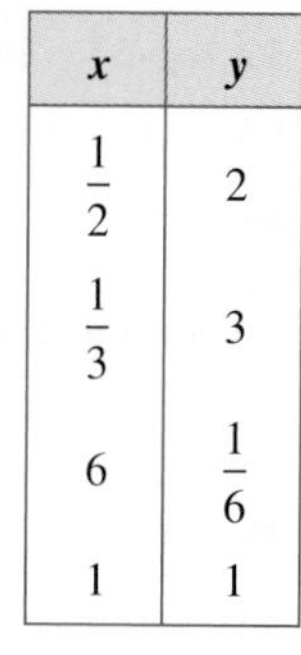

x	y
$\frac{1}{2}$	2
$\frac{1}{3}$	3
6	$\frac{1}{6}$
1	1

[10-1] 2 Graph each function and its inverse on the same coordinate system.

a. $f = \{(-3, 2), (-2, 1), (-1, 3), (2, 4)\}$

b. $f(x) = 3x + 2$ **c.** $f(x) = \left(\frac{1}{2}\right)^x$

d. $f(x) = x^2$ for $x \geq 0$

[10-2] 3 Graph each pair of functions on the same rectangular coordinate system.

a. $y = 4^x$
$y = \log_4 x$

b. $y = \left(\frac{3}{2}\right)^x$
$y = \log_{3/2} x$

[10-2] **4** Translate the following logarithmic equations to exponential form.

a. $\log_5 \sqrt[3]{25} = \frac{2}{3}$ **b.** $\log_5 \frac{1}{125} = -3$

c. $\log_b(y + 1) = x$ **d.** $\log_b x = y + 1$

[10-3] **5** Use a calculator to approximate each of these expressions to five significant digits.

a. $\log 19.1$ **b.** $\ln 0.0043$

c. $e^{3.45}$ **d.** e^1

e. $\log(4.831 \times 10^{-5})$ **f.** $\ln(7.2931 \times 10^4)$

g. $10^{0.9217}$ **h.** $\ln(-9)$

[10-4] **6** Use the properties of logarithms to express each logarithm in terms of logarithms of simpler expressions. Assume that x and y are positive real numbers.

a. $\log x^4y^5$ **b.** $\ln \frac{x^3}{y^6}$

c. $\log \sqrt[3]{x}$ **d.** $\ln \sqrt[5]{xy^2}$

Use the properties of logarithms to combine these logarithms into a single logarithmic expression with a coefficient of 1.

e. $\log(x - 2) + \log(x - 3)$

f. $\ln(5x - 1) - \ln(x - 3)$

g. $2 \ln(7x + 9) - \ln x$

h. $\frac{2}{3} \ln(x^2 + 7) + \ln x$

[10-4] **7** Use the change-of-base formula and a calculator to approximate each of these expressions accurate to four significant digits.

a. $\log_4 11$ **b.** $\log_\pi \sqrt{7}$

c. $\log_2 100$ **d.** $\log_\pi 5\pi$

[10-5] **8** Solve each of these equations without using a calculator.

a. $3^x = \frac{1}{81}$ **b.** $16^w = 64$

c. $\log_2 x = 4$

d. $\log_8 2 = y$

e. $\log(1 - 4t) - \log(5 + t) = \log 3$

f. $\ln(1 - z) + \ln(2 - z) = \ln(17 - z)$

Using a calculator, approximate the solution to each of these equations accurate to four significant digits.

g. $3^{4y+1} = 17.83$

h. $\ln(x + 1) + \ln(3x - 1) = \ln(6x)$

[10-6] **9** **a.** Assume that the population of a new space colony is growing continuously at a rate of 5% per year. How many years will it take the population to grow from 500 to 3000?

Figure for Exercise 9a

b. How many years will it take an investment to double in value if interest is compounded monthly at 8.25%?

Chapter 11

ADDITIONAL TOPICS IN MATHEMATICS

AN APPLICATION

Making Math of Light

As rainbows remind us after every rain, sunlight contains a spectrum of colors ranging from red at one end, through orange, yellow, and green, to blue and violet at the other. However, to the naked eye the appearance of the rainbow is misleading. We seem to see a continuous spread of color with no gaps, but what we are in fact seeing is an array of thin colored bands—more than 700—so closely packed together that they look continuous.

The colored bands represent the radiation spectra of the various chemical elements in the sun. Each element contributes a distinctive set of bands. Interestingly, an element contributes exactly the same set of bands to a light source, no matter what other elements are present. This fact is very important to astronomers, who compare the spectra of distant stars with the spectra of known elements to determine what the stars are made of.

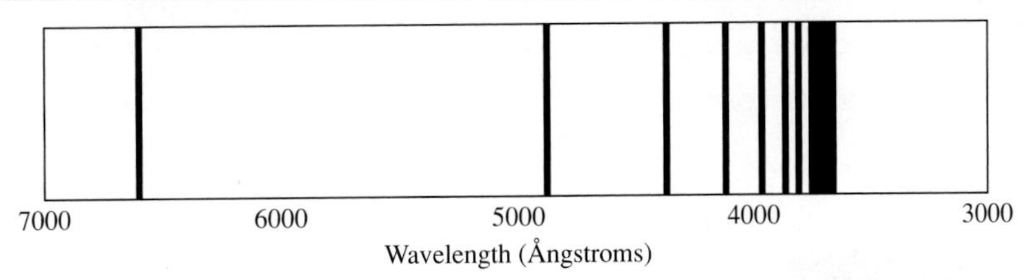

The bands illustrated in the figure are part of the spectrum for hydrogen, the most common element in the universe. In 1913 the physicist Niels Bohr was trying to understand why elements (particularly hydrogen) radiate in discrete bands. He discovered that the wavelengths of the colored bands shown could be generated by the formula $\dfrac{911.5}{\frac{1}{4} - \frac{1}{n^2}}$, where the number n is a variable that takes on the integer values 3, 4, 5, etc. Other, similar formulas generate the rest of the hydrogen spectrum. This discovery was a major step in the development of quantum physics, an important science which describes our physical universe.

The progression generated by Bohr's formula is an example of what mathematicians call a *sequence;* we will examine sequences in Section 11-3. What is the exact value of the first term in the sequence generated by Bohr's formula? What spectrum line does this term correspond to in the figure?

CHAPTER 11 OBJECTIVES

1 **Solve a 2×2 or a 3×3 system of linear equations using Cramer's Rule (Section 11-1).**

2 **Expand a binomial expression (Section 11-2).**

3 **Write the terms of a sequence, given the general term (Section 11-3).**

4 **Use summation notation, and evaluate the series associated with a finite sequence (Section 11-3).**

5 **Calculate the *n*th term of an arithmetic progression (Section 11-4).**

6 **Find the sum of a finite arithmetic progression (Section 11-4).**

7 **Calculate the *n*th term of a geometric progression (Section 11-5).**

8 **Find the sum of a finite geometric progression (Section 11-6).**

9 **Evaluate an infinite geometric series (Section 11-6).**

The various topics included in this chapter are used in many diverse fields, including calculus and statistics. Most of these topics are generally included in texts on discrete mathematics, a branch of mathematics used extensively by those in computer science. In discrete mathematics courses mathematical methods are used to organize, compute, and manipulate data associated with individual or discrete objects.

For those of you who are unfamiliar with these topics, this chapter can serve as an introduction. For those of you who are already taking more advanced courses, it can serve as a source of examples and exercises. The discussions assume that you have no prior knowledge of the various subjects, and the level of the development is consistent with that of the first ten chapters of this text.

11-1 Cramer's Rule

SECTION OBJECTIVE

1 **Solve a 2×2 or a 3×3 system of linear equations using Cramer's Rule.**

The substitution method and the addition-subtraction method are used for solving systems of two or three linear equations in which the coefficients are small integers. However, it is helpful to have a calculator or a computer when solving equations with more complicated coefficients or systems of linear equations with more than three variables. Cramer's Rule, a formula that is appropriate for programming, will be presented in this section. If you apply this method on a programmable calculator or a computer, you can solve a system merely by entering the given coefficients and constants. In this section we will use Cramer's Rule to solve 2×2 linear systems (systems of two linear equations in two variables) and 3×3 linear systems.

For convenience of notation we will state Cramer's Rule in determinant form. A **determinant** is a single value calculated from a square matrix of numbers. (A *square matrix* is a tabular arrangement of numbers in which the number of rows is the same as the number of columns.) A pair of vertical bars is used to denote a determinant. The determinant defined in the following box is called a **second-order determinant,** or a **determinant of order two,** because the matrix of entries has two rows and two columns.

DETERMINANT OF ORDER TWO

$$\begin{vmatrix} a & b \\ c & d \end{vmatrix} = ad - bc$$

An easy way to remember how to evaluate a determinant of order two is to remember that the value of a determinant is equal to the product down the diagonal to the right minus the product down the diagonal to the left.

A Mathematical Note

The formula known as Cramer's Rule was published by Gabriel Cramer in 1750. However, this formula had been known to the Chinese for decades. The Chinese took the concept to Japan, where the Japanese mathematician Seki Kōwa wrote of determinants in 1683.

The term *determinant* was first used in its current sense by Louis Cauchy, who modified the original meaning introduced by Gauss.

EXAMPLE 1 Evaluating Second-Order Determinants

Evaluate each determinant.

SOLUTIONS

(a) $\begin{vmatrix} 2 & 5 \\ 7 & 9 \end{vmatrix}$ $\quad \begin{vmatrix} 2 & 5 \\ 7 & 9 \end{vmatrix} = (2)(9) - (5)(7)$

$= 18 - 35$

$= -17$

Evaluate each determinant by taking the product down the diagonal to the right minus the product down the diagonal to the left.

(b) $\begin{vmatrix} -7 & -3 \\ 8 & 2 \end{vmatrix}$ $\quad \begin{vmatrix} -7 & -3 \\ 8 & 2 \end{vmatrix} = (-7)(2) - (-3)(8)$

$= -14 - (-24)$

$= 10$

SELF-CHECK

Evaluate each determinant.

1 $\begin{vmatrix} 1 & 2 \\ 3 & 4 \end{vmatrix}$ **2** $\begin{vmatrix} 5 & -2 \\ 7 & 3 \end{vmatrix}$

The TI-82 graphics calculator is capable of calculating the determinant of a square matrix. First enter the matrix, and then evaluate the determinant of the matrix as illustrated in Example 2.

EXAMPLE 2 Calculator Usage

Use a graphics calculator to evaluate $\begin{vmatrix} 7.3 & 4.7 \\ 9.2 & 8.6 \end{vmatrix}$.

SOLUTION Enter the matrix $A = \begin{bmatrix} 7.3 & 4.7 \\ 9.2 & 8.6 \end{bmatrix}$.

[MATRX] [▶ (EDIT option)] [1 ([A])] [2] [ENTER] [2] [ENTER] Matrix A has 2 rows and 2 columns.

[7] [.] [3] [ENTER] [4] [.] [7] [ENTER] Then input the entries in rows one and two.

[9] [.] [2] [ENTER] [8] [.] [6] [ENTER]

SELF-CHECK ANSWERS

1 −2 **2** 29

```
MATRIX [A]   2×2
[7.3     4.7     ]
[9.2     8.6     ]
```

Then calculate the determinant of this matrix.

[2nd] [QUIT]

Return to the home screen, and then calculate the determinant of matrix A.

[MATRX] [▶ (MATH option)] [1 (det)] [MATRX] [1 ([A])]

[ENTER]

```
det [A]
        19.54
```

Answer $\begin{vmatrix} 7.3 & 4.7 \\ 9.2 & 8.6 \end{vmatrix} = 19.54$

The derivation of Cramer's Rule follows. We will solve a 2×2 linear system of the form $\begin{Bmatrix} a_1x + b_1y = c_1 \\ a_2x + b_2y = c_2 \end{Bmatrix}$ by means of the addition-subtraction method, and then we will express this solution in determinant notation. For now we will assume that the system is consistent, the equations are independent, and the determinant of the coefficients, $\begin{vmatrix} a_1 & b_1 \\ a_2 & b_2 \end{vmatrix}$, is not zero.

Derivation of Cramer's Rule

$$\begin{Bmatrix} a_1x + b_1y = c_1 \\ a_2x + b_2y = c_2 \end{Bmatrix} \xrightarrow[r_2' = -b_1r_2]{r_1' = b_2r_1} \begin{array}{r} a_1b_2x + b_1b_2y = b_2c_1 \\ -a_2b_1x - b_1b_2y = -b_1c_2 \\ \hline a_1b_2x - a_2b_1x = b_2c_1 - b_1c_2 \\ x(a_1b_2 - a_2b_1) = b_2c_1 - b_1c_2 \end{array}$$

Add these new equations to eliminate the variable y.

$$x = \frac{b_2c_1 - b_1c_2}{a_1b_2 - a_2b_1} \quad \text{for } a_1b_2 - a_2b_1 \neq 0$$

Then solve for x.

$$x = \frac{\begin{vmatrix} c_1 & b_1 \\ c_2 & b_2 \end{vmatrix}}{\begin{vmatrix} a_1 & b_1 \\ a_2 & b_2 \end{vmatrix}}$$

Replace $b_2c_1 - b_1c_2$ with $\begin{vmatrix} c_1 & b_1 \\ c_2 & b_2 \end{vmatrix}$, and replace $a_1b_2 - a_2b_1$ with $\begin{vmatrix} a_1 & b_1 \\ a_2 & b_2 \end{vmatrix}$.

Similarly,

$$y = \frac{a_1c_2 - a_2c_1}{a_1b_2 - a_2b_1}$$

$$y = \frac{\begin{vmatrix} a_1 & c_1 \\ a_2 & c_2 \end{vmatrix}}{\begin{vmatrix} a_1 & b_1 \\ a_2 & b_2 \end{vmatrix}}$$

Replace $a_1c_2 - a_2c_1$ with $\begin{vmatrix} a_1 & c_1 \\ a_2 & c_2 \end{vmatrix}$.

Remembering the result in the concise form made possible by determinants is easier than deriving this formula each time it is needed.

CRAMER'S RULE FOR 2 × 2 LINEAR SYSTEMS

If $D \neq 0$, the solution of $\begin{cases} a_1x + b_1y = c_1 \\ a_2x + b_2y = c_2 \end{cases}$ is $x = \dfrac{D_x}{D}$ and $y = \dfrac{D_y}{D}$,

where

$D = \begin{vmatrix} a_1 & b_1 \\ a_2 & b_2 \end{vmatrix}$ Form the determinant of the coefficients of x and y.

$D_x = \begin{vmatrix} c_1 & b_1 \\ c_2 & b_2 \end{vmatrix}$ Replace the x-coefficients with the constants c_1 and c_2.

$D_y = \begin{vmatrix} a_1 & c_1 \\ a_2 & c_2 \end{vmatrix}$ Replace the y-coefficients with the constants c_1 and c_2.

If $D = D_x = D_y = 0$, the system contains dependent equations. If $D = 0$ but D_x and D_y are not both zero, the system is inconsistent.

EXAMPLE 3 Using Cramer's Rule to Solve a 2 × 2 Linear System

Solve $\begin{cases} 2x + 3y = 1 \\ 5x - 2y = 12 \end{cases}$ using Cramer's Rule.

SOLUTION First compute D, D_x, and D_y.

$$D = \begin{vmatrix} a_1 & b_1 \\ a_2 & b_2 \end{vmatrix} = \begin{vmatrix} 2 & 3 \\ 5 & -2 \end{vmatrix} = (2)(-2) - (3)(5) = -4 - 15 = -19$$

Substitute $a_1 = 2$, $b_1 = 3$, $c_1 = 1$, $a_2 = 5$, $b_2 = -2$, and $c_2 = 12$ into D, D_x, and D_y.

$$D_x = \begin{vmatrix} c_1 & b_1 \\ c_2 & b_2 \end{vmatrix} = \begin{vmatrix} 1 & 3 \\ 12 & -2 \end{vmatrix} = (1)(-2) - (3)(12) = -2 - 36 = -38$$

$$D_y = \begin{vmatrix} a_1 & c_1 \\ a_2 & c_2 \end{vmatrix} = \begin{vmatrix} 2 & 1 \\ 5 & 12 \end{vmatrix} = (2)(12) - (1)(5) = 24 - 5 = 19$$

Then

$$x = \frac{D_x}{D} = \frac{-38}{-19} = 2 \quad \text{and} \quad y = \frac{D_y}{D} = \frac{19}{-19} = -1$$

Substitute the values of D, D_x, and D_y into Cramer's Rule.

Answer $(2, -1)$

Do these values check?

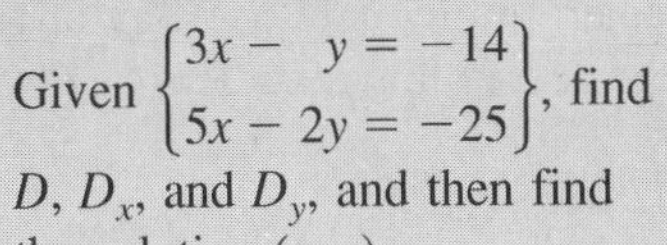

SELF-CHECK

Given $\begin{cases} 3x - y = -14 \\ 5x - 2y = -25 \end{cases}$, find D, D_x, and D_y, and then find the solution (x, y).

If D, the determinant of the coefficients, is zero, then the system will either be inconsistent or contain dependent equations. In such cases it may be easier to use the slope-intercept form (see Section 2-2) than to compute the other determinants, D_x and D_y. Thus if you are doing pencil-and-paper calculations, you should always compute D before you decide whether to compute D_x and D_y.

SELF-CHECK ANSWER 

$D = -1$, $D_x = 3$, $D_y = -5$; the solution is $(-3, 5)$.

EXAMPLE 4 Examining a System of Dependent Equations Using Cramer's Rule

Solve $\begin{cases} 2x - 6y = 10 \\ 3x - 9y = 15 \end{cases}$ using Cramer's Rule.

SOLUTION

$$D = \begin{vmatrix} a_1 & b_1 \\ a_2 & b_2 \end{vmatrix} = \begin{vmatrix} 2 & -6 \\ 3 & -9 \end{vmatrix} = (2)(-9) - (-6)(3)$$
$$= -18 - (-18)$$
$$= 0$$

Since $D = 0$, $\frac{D_x}{D}$ and $\frac{D_y}{D}$ are undefined. Thus we will not compute D_x or D_y.

Since $D = 0$, this system either is inconsistent or contains dependent equations. In this case both equations will simplify to $x - 3y = 5$, so these equations are dependent.

In slope-intercept form both equations are written as $y = \frac{1}{3}x - \frac{5}{3}$. Since they have the same slope and y-intercept, the equations must be dependent.

Answer This system of dependent equations has an infinite number of solutions. Every point (x, y) satisfying $x - 3y = 5$ is a solution of the system. ▼

EXAMPLE 5 Using Cramer's Rule to Solve a 2 × 2 Linear System

Solve $\begin{cases} 5.83x - 7.96y = -54.26 \\ 3.48x - 5.17y = -40.76 \end{cases}$ using Cramer's Rule.

SOLUTION First compute D, D_x, and D_y.

A calculator can be used to evaluate D, D_x, and D_y following the steps used in Example 2.

$$D = \begin{vmatrix} 5.83 & -7.96 \\ 3.48 & -5.17 \end{vmatrix} = -2.4403$$

$$D_x = \begin{vmatrix} -54.26 & -7.96 \\ -40.76 & -5.17 \end{vmatrix} = -43.9254$$

$$D_y = \begin{vmatrix} 5.83 & -54.26 \\ 3.48 & -40.76 \end{vmatrix} = -48.806$$

Then

$$x = \frac{D_x}{D} = \frac{-43.9254}{-2.4403} = 18 \quad \text{and} \quad y = \frac{D_y}{D} = \frac{-48.806}{-2.4403} = 20$$

Substitute the values of D, D_x, and D_y into Cramer's Rule.

Answer (18, 20)

Do these values check?

▼

We will now extend Cramer's Rule to 3 × 3 linear systems. For systems of equations greater than 3 × 3, it is generally advisable to use matrix methods, which are covered in more advanced texts.*

Since our goal is simply to solve 3 × 3 linear systems of equations by Cramer's Rule, only one way of calculating determinants of order three will be presented in this text. There are other methods for evaluating determinants of

*See the augmented matrix method in James W. Hall's *College Algebra with Applications*, 3rd ed. (Boston: PWS-KENT Publishing Company, 1992).

order three, but the method shown here has the advantage that it can be generalized to determinants of higher order. By noting patterns in this calculation, you can eliminate the need for an unreasonable amount of memory work. This formula can also be used by simply entering the coefficients into a computer.

DETERMINANT OF ORDER THREE

$$\begin{vmatrix} a_1 & b_1 & c_1 \\ a_2 & b_2 & c_2 \\ a_3 & b_3 & c_3 \end{vmatrix} = a_1\begin{vmatrix} b_2 & c_2 \\ b_3 & c_3 \end{vmatrix} - a_2\begin{vmatrix} b_1 & c_1 \\ b_3 & c_3 \end{vmatrix} + a_3\begin{vmatrix} b_1 & c_1 \\ b_2 & c_2 \end{vmatrix}$$

You are not expected to memorize this formula symbol by symbol. Rather, you should remember the way the entries are used to calculate a determinant of order three. Each entry is multiplied by the second-order determinant that remains after the row and the column of that entry are deleted. This process is illustrated below.

For a_1 in $\begin{vmatrix} a_1 & b_1 & c_1 \\ a_2 & b_2 & c_2 \\ a_3 & b_3 & c_3 \end{vmatrix}$, we obtain $a_1\begin{vmatrix} b_2 & c_2 \\ b_3 & c_3 \end{vmatrix}$.

For a_2 in $\begin{vmatrix} a_1 & b_1 & c_1 \\ a_2 & b_2 & c_2 \\ a_3 & b_3 & c_3 \end{vmatrix}$, we obtain $-a_2\begin{vmatrix} b_1 & c_1 \\ b_3 & c_3 \end{vmatrix}$.

For a_3 in $\begin{vmatrix} a_1 & b_1 & c_1 \\ a_2 & b_2 & c_2 \\ a_3 & b_3 & c_3 \end{vmatrix}$, we obtain $a_3\begin{vmatrix} b_1 & c_1 \\ b_2 & c_2 \end{vmatrix}$.

Note that in the second expression the entry is preceded by a minus sign. The values of these three expressions are then added to obtain the value of the determinant of order three.

EXAMPLE 6 Evaluating a Third-Order Determinant

Evaluate $\begin{vmatrix} 1 & -2 & -1 \\ 2 & 3 & 1 \\ 1 & -1 & -2 \end{vmatrix}$.

SOLUTION

$$\begin{vmatrix} 1 & -2 & -1 \\ 2 & 3 & 1 \\ 1 & -1 & -2 \end{vmatrix} = 1\begin{vmatrix} 3 & 1 \\ -1 & -2 \end{vmatrix} - 2\begin{vmatrix} -2 & -1 \\ -1 & -2 \end{vmatrix} + 1\begin{vmatrix} -2 & -1 \\ 3 & 1 \end{vmatrix}$$

$$= 1[3(-2) - (-1)(1)] - 2[-2(-2) - (-1)(-1)] + 1[-2(1) - (-1)(3)]$$

$$= (-6 + 1) - 2(4 - 1) + (-2 + 3)$$

$$= -5 - 2(3) + 1$$

$$= -5 - 6 + 1$$

$$= -10$$

Use the pattern illustrated above to form the terms used to calculate this determinant.

▼

EXAMPLE 7 Evaluating a Third-Order Determinant

Evaluate $\begin{vmatrix} 1 & 2 & 0 \\ -4 & 3 & 7 \\ -6 & 5 & -2 \end{vmatrix}$.

SOLUTION

$$\begin{vmatrix} 1 & 2 & 0 \\ -4 & 3 & 7 \\ -6 & 5 & -2 \end{vmatrix} = 1\begin{vmatrix} 3 & 7 \\ 5 & -2 \end{vmatrix} - (-4)\begin{vmatrix} 2 & 0 \\ 5 & -2 \end{vmatrix} + (-6)\begin{vmatrix} 2 & 0 \\ 3 & 7 \end{vmatrix}$$

$$= 1[3(-2)-7(5)] + 4[2(-2) - 0(5)] - 6[2(7) - 0(3)]$$

$$= (-6 - 35) + 4(-4 - 0) - 6(14 - 0)$$

$$= -41 - 16 - 84$$

$$= -141$$

Use the pattern illustrated on the preceding page to form the terms used to calculate this determinant.

Example 8 illustrates how to evaluate a third-order determinant using a TI-82 calculator. Note that the keystrokes follow the same pattern used in Example 2 to evaluate a second-order determinant.

SELF-CHECK

Evaluate $\begin{vmatrix} -3 & 4 & 9 \\ 5 & 1 & -2 \\ 2 & -6 & -5 \end{vmatrix}$.

EXAMPLE 8 Using a Calculator to Evaluate a Determinant

Evaluate $\begin{vmatrix} 5.7 & 4.1 & 6.8 \\ 6.2 & 8.2 & 3.1 \\ 9.3 & 6.7 & 2.6 \end{vmatrix}$.

SOLUTION Enter the matrix $A = \begin{vmatrix} 5.7 & 4.1 & 6.8 \\ 6.2 & 8.2 & 3.1 \\ 9.3 & 6.7 & 2.6 \end{vmatrix}$.

MATRX | ▶ (EDIT option) | 1 ([A]) | 3 | ENTER | 3 | ENTER

Matrix A has 3 rows and 3 columns.

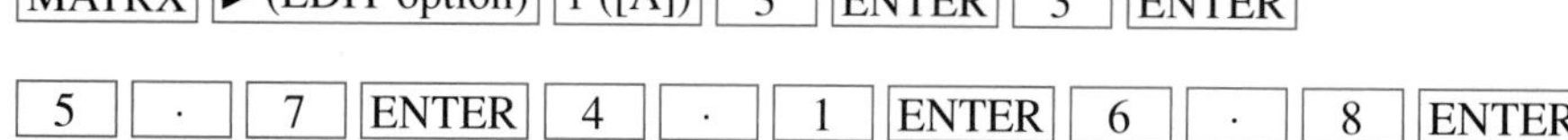

5 | · | 7 | ENTER | 4 | · | 1 | ENTER | 6 | · | 8 | ENTER

6 | · | 2 | ENTER | 8 | · | 2 | ENTER | 3 | · | 1 | ENTER

9 | · | 3 | ENTER | 6 | · | 7 | ENTER | 2 | · | 6 | ENTER

Input the entries in rows one, two, and three.

Then calculate the determinant of this matrix.

2nd | QUIT

MATRX | ▶ (MATH option) | 1 (det) | MATRX | 1 ([A])

ENTER

Return to the home screen, and calculate the determinant of matrix A.

SELF-CHECK ANSWER 

-153

det [A]
−180.85

Answer $\begin{vmatrix} 5.7 & 4.1 & 6.8 \\ 6.2 & 8.2 & 3.1 \\ 9.3 & 6.7 & 2.6 \end{vmatrix} = -180.85$

CRAMER'S RULE FOR 3 × 3 LINEAR SYSTEMS

If $D \neq 0$, the solution of $\begin{cases} a_1x + b_1y + c_1z = d_1 \\ a_2x + b_2y + c_2z = d_2 \\ a_3x + b_3y + c_3z = d_3 \end{cases}$ is $x = \frac{D_x}{D}$, $y = \frac{D_y}{D}$, and $z = \frac{D_z}{D}$, where

$$D = \begin{vmatrix} a_1 & b_1 & c_1 \\ a_2 & b_2 & c_2 \\ a_3 & b_3 & c_3 \end{vmatrix}$$

Form the determinant of the coefficients of x, y, and z.

$$D_x = \begin{vmatrix} d_1 & b_1 & c_1 \\ d_2 & b_2 & c_2 \\ d_3 & b_3 & c_3 \end{vmatrix}$$

Replace the x-coefficients by the constants d_1, d_2, and d_3.

$$D_y = \begin{vmatrix} a_1 & d_1 & c_1 \\ a_2 & d_2 & c_2 \\ a_3 & d_3 & c_3 \end{vmatrix}$$

Replace the y-coefficients by the constants d_1, d_2, and d_3.

$$D_z = \begin{vmatrix} a_1 & b_1 & d_1 \\ a_2 & b_2 & d_2 \\ a_3 & b_3 & d_3 \end{vmatrix}$$

Replace the z-coefficients by the constants d_1, d_2, and d_3.

If $D = 0$, the system either is inconsistent or contains dependent equations. These systems can be solved by other methods, such as the addition-subtraction method.

EXAMPLE 9 Using Cramer's Rule to Solve a 3 × 3 Linear System

Solve $\begin{cases} x + y - z = 2 \\ 2x - y + z = -1 \\ x - y - z = 0 \end{cases}$ using Cramer's Rule.

SOLUTION Evaluate the determinants D, D_x, D_y, and D_z.

These determinants can be evaluated on a calculator if you wish.

$$D = \begin{vmatrix} a_1 & b_1 & c_1 \\ a_2 & b_2 & c_2 \\ a_3 & b_3 & c_3 \end{vmatrix} = \begin{vmatrix} 1 & 1 & -1 \\ 2 & -1 & 1 \\ 1 & -1 & -1 \end{vmatrix} = 6$$

$$D_x = \begin{vmatrix} d_1 & b_1 & c_1 \\ d_2 & b_2 & c_2 \\ d_3 & b_3 & c_3 \end{vmatrix} = \begin{vmatrix} 2 & 1 & -1 \\ -1 & -1 & 1 \\ 0 & -1 & -1 \end{vmatrix} = 2$$

$$D_y = \begin{vmatrix} a_1 & d_1 & c_1 \\ a_2 & d_2 & c_2 \\ a_3 & d_3 & c_3 \end{vmatrix} = \begin{vmatrix} 1 & 2 & -1 \\ 2 & -1 & 1 \\ 1 & 0 & -1 \end{vmatrix} = 6$$

$$D_z = \begin{vmatrix} a_1 & b_1 & d_1 \\ a_2 & b_2 & d_2 \\ a_3 & b_3 & d_3 \end{vmatrix} = \begin{vmatrix} 1 & 1 & 2 \\ 2 & -1 & -1 \\ 1 & -1 & 0 \end{vmatrix} = -4$$

Then

$$x = \frac{D_x}{D} = \frac{2}{6} = \frac{1}{3},\ y = \frac{D_y}{D} = \frac{6}{6} = 1, \text{ and}$$

Now compute x, y, and z using Cramer's Rule.

$$z = \frac{D_z}{D} = \frac{-4}{6} = -\frac{2}{3}$$

Answer $\left(\frac{1}{3}, 1, -\frac{2}{3}\right)$

SELF-CHECK

Given $\begin{Bmatrix} x + y + 2z = 1 \\ 2x + y - z = 0 \\ x - 2y - 4z = 4 \end{Bmatrix}$, find D, D_x, D_y, and D_z; then determine the solution (x, y, z).

Exercises 11-1

A

In Exercises 1–6 evaluate each determinant.

1 $\begin{vmatrix} 4 & 11 \\ 5 & 6 \end{vmatrix}$ **2** $\begin{vmatrix} 7 & 9 \\ 4 & 3 \end{vmatrix}$ **3** $\begin{vmatrix} 6 & -2 \\ 3 & 4 \end{vmatrix}$

4 $\begin{vmatrix} -5 & 9 \\ -6 & 7 \end{vmatrix}$ **5** $\begin{vmatrix} -9 & -7 \\ -4 & -5 \end{vmatrix}$ **6** $\begin{vmatrix} \frac{1}{2} & \frac{-7}{10} \\ \frac{3}{7} & \frac{-2}{5} \end{vmatrix}$

In Exercises 7–20 solve each system of equations using Cramer's Rule.

7 $x + y = 6$, $x - y = 6$

8 $x + 3y = 6$, $2x + 4y = 4$

9 $2x + y = 10$, $3x - y = 5$

10 $3x + y = 7$, $x + 2y = 4$

11 $x = 6$, $2x + y = 4$

12 $4x + 2y = 14$, $x - 3y = -14$

13 $-x - 2y = -3$, $5x + 10y = 10$

14 $2x + y = 6$, $6x + 3y = 18$

15 $\frac{x}{10} - \frac{y}{5} = \frac{1}{2}$, $\frac{x}{2} + \frac{y}{4} = 0$

16 $-\frac{x}{8} + \frac{y}{4} = 1$, $\frac{x}{3} + \frac{y}{5} = -\frac{1}{15}$

17 $0.7x - 0.6y = -1.5$, $0.4x - 0.9y = -0.3$

18 $0.5x + 0.3y = 1.8$, $0.2x - 0.7y = -0.1$

19 $21x - 6y = 5$, $-14x + 3y = -3$

20 $3x + 6y = 8$, $5x + 10y = 11$

SELF-CHECK ANSWER

$D = -9$, $D_x = -18$, $D_y = 27$, $D_z = -9$; the solution is $(2, -3, 1)$.

B

In Exercises 21–30 evaluate each determinant.

21 $\begin{vmatrix} 1 & 0 & -1 \\ 0 & 3 & 4 \\ 5 & -6 & 0 \end{vmatrix}$ **22** $\begin{vmatrix} 3 & 0 & 0 \\ -4 & 1 & 5 \\ 2 & -4 & 0 \end{vmatrix}$

23 $\begin{vmatrix} 2 & 1 & -1 \\ 4 & 2 & 1 \\ 6 & -3 & 2 \end{vmatrix}$ **24** $\begin{vmatrix} 1 & 3 & -2 \\ 2 & -4 & 5 \\ 4 & 5 & -3 \end{vmatrix}$

25 $\begin{vmatrix} 1 & 1 & -1 \\ 3 & 6 & -4 \\ 3 & -3 & 5 \end{vmatrix}$ **26** $\begin{vmatrix} 1 & 1 & 1 \\ 1 & -2 & -1 \\ 2 & 1 & 3 \end{vmatrix}$

27 $\begin{vmatrix} 1 & 0 & 0 \\ 2 & 4 & 0 \\ 3 & 5 & 6 \end{vmatrix}$ **28** $\begin{vmatrix} 6 & 8 & 3 \\ 1 & 0 & 2 \\ 0 & -1 & 0 \end{vmatrix}$

29 $\begin{vmatrix} -1 & 0 & -5 \\ 4 & 0 & 7 \\ 1 & 6 & -3 \end{vmatrix}$ **30** $\begin{vmatrix} 17 & 11 & 0 \\ 18 & -3 & 0 \\ 19 & -5 & 0 \end{vmatrix}$

In Exercises 31–36 solve each system of equations by using Cramer's Rule.

31 $\begin{aligned} x - y + 2z &= 5 \\ 3x + y + z &= 9 \\ x - 2y - 3z &= -6 \end{aligned}$

32 $\begin{aligned} 2x - 3y + z &= -7 \\ x + 4y + 3z &= 0 \\ 3x - y + z &= -2 \end{aligned}$

33 $\begin{aligned} x + y + z &= 3 \\ x - 3y + 2z &= -1 \\ 2x + 2y - z &= 6 \end{aligned}$

34 $\begin{aligned} 3x + 3y + z &= -1 \\ x - 2y + z &= -3 \\ -4x + y + 2z &= 8 \end{aligned}$

35 $\begin{aligned} x + 2y - 3z &= 3 \\ 3x + 2y - 4z &= -2 \\ x - y + 5z &= -1 \end{aligned}$

36 $\begin{aligned} 2x - 3y + 5z &= 5 \\ 4x + 3y - 5z &= -2 \\ 2x + 6y - 5z &= -4 \end{aligned}$

C

CALCULATOR USAGE (37–46)

In Exercises 37–46 use Cramer's Rule and a calculator or computer to solve each system of equations.

37 $\begin{aligned} x - y \quad &= 0 \\ y + z &= -2 \\ x \quad - z &= 12 \end{aligned}$

38 $\begin{aligned} 2x - y + z &= 2 \\ 3x - y - z &= -22 \\ 5x + y - 3z &= 34 \end{aligned}$

39 $\begin{aligned} 0.5x + 0.4y - 0.3z &= 10 \\ 0.4x - 0.5y + 0.3z &= 5 \\ 0.3x + 0.6y - 0.7z &= -4 \end{aligned}$

40 $\begin{aligned} 1.3x + 2.9y + 5.7z &= 57.7 \\ 2.8x - 1.7y - 1.3z &= -27.7 \\ 5.4x + 7.9y - 8.2z &= 86.0 \end{aligned}$

41 $\begin{aligned} 811x + 923y &= 9833.6 \\ 474x + 765y &= 7258.5 \end{aligned}$

42 $\begin{aligned} 703x - 471y &= -9181.8 \\ 654x + 127y &= -3850.9 \end{aligned}$

43 $\begin{aligned} 2.96x - 7.43y &= -57.755 \\ 3.56x + 5.81y &= 20.489 \end{aligned}$

44 $\begin{aligned} 0.076x + 0.109y &= 0.4925 \\ 0.853x + 0.007y &= 9.7850 \end{aligned}$

45 $\begin{aligned} 5.38x + 8.21y - 7.09z &= -77.027 \\ 4.21x - 3.44y + 6.81z &= 75.110 \\ 7.07x - 6.15y + 5.93z &= 78.687 \end{aligned}$

46 $\begin{aligned} 41.5x - 73.2y + 27.3z &= 123.1 \\ 60.4x + 31.8y - 41.1z &= -36.59 \\ 59.9x - 19.3y - 11.8z &= 41.33 \end{aligned}$

In Exercises 47–50 solve each equation for *x*.

47 $\begin{vmatrix} x & 4 \\ x & 7 \end{vmatrix} = -6$ **48** $\begin{vmatrix} 2x & 2x-1 \\ 3x & 3x+1 \end{vmatrix} = -10$

49 $\begin{vmatrix} 2x & x-2 \\ 3x & 3-x \end{vmatrix} = 4$ **50** $\begin{vmatrix} x-1 & 3x-2 \\ x+1 & 2x+3 \end{vmatrix} = -3$

DISCUSSION QUESTION

51 Quadratic equations can sometimes be solved by easier methods than the quadratic formula. Likewise, a linear system of equations can often be solved by easier methods than Cramer's Rule. Write a paragraph discussing the roles of the quadratic formula and Cramer's Rule and when it is advantageous to use these methods.

CHALLENGE QUESTION

A wheel of radius r is rolled around the outside of another wheel of radius $3r$. How many rotations will the smaller wheel make as it travels around the larger circle? (*Hint:* The answer is *not* 3.)

SECTION SKILL REINFORCERS

Determine the slope of each line.

1 $x = 7$ **2** $y = 7$ **3** $x + y = 7$ **4** $x - y - 7 = 0$ **5** $y = 7x + 8$

11-2 Binomial Expansions

SECTION OBJECTIVE

2 Expand a binomial expression.

In Section 4-5 we examined the special product $(a + b)^2$, whose expansion is $a^2 + 2ab + b^2$. This and other binomial expansions of the form $(a + b)^n$ occur so frequently in mathematics that it is useful to be able to expand these expressions without actually performing the repeated multiplications. We can do so by taking advantage of the many patterns that can be observed in these expansions. Before you read further, examine the expansions given below and try to discover the pattern for the terms of $(a + b)^n$. Specifically, try to answer the list of questions that follows the table.

$(a + b)^n$	Expansion of $(a + b)^n$
$(a + b)^0 =$	1
$(a + b)^1 =$	$a + b$
$(a + b)^2 =$	$a^2 + 2ab + b^2$
$(a + b)^3 =$	$a^3 + 3a^2b + 3ab^2 + b^3$
$(a + b)^4 =$	$a^4 + 4a^3b + 6a^2b^2 + 4ab^3 + b^4$
$(a + b)^5 =$	$a^5 + 5a^4b + 10a^3b^2 + 10a^2b^3 + 5ab^4 + b^5$
$(a + b)^6 =$	$a^6 + 6a^5b + 15a^4b^2 + 20a^3b^3 + 15a^2b^4 + 6ab^5 + b^6$

1 How many terms are in this expansion?

2 What is the degree of each term in the expansion?

3 Is there any pattern to the exponents of a and b in the terms?

4 What is the coefficient of the first term and the last term?

5 What is the coefficient of the second term and the next-to-last term?

6 What is the relationship between the coefficients and the exponents?

The answers to the above questions are as follows:

1 There are $n + 1$ terms in the expansion.

2 Each term is of degree n.

3 As the exponents on a decrease by 1, from a^n in the first term to a^0 in the last term, the exponents on b increase by 1, from b^0 to b^n.

4 The coefficient of the first term and the last term is 1.

5 The coefficient of the second term and the next-to-last term is n. These coefficients form a symmetric pattern that starts with the first term and ends with the last term.

6 After the first term, the coefficient of each term is obtained from the preceding term by multiplying its coefficient by the exponent on a and dividing this product by 1 more than the exponent on b.

By taking advantage of these observations, we can write the expansion of $(a + b)^n$ directly.

EXPANDING $(a + b)^n$

Step 1 Write the exponents on all $(n + 1)$ terms. Start with a^n, decreasing the exponents on a by 1 and increasing the exponents on b by 1 until the last term is b^n.

Step 2 Write the coefficients of each term. The first coefficient is 1, the second is n, and the other coefficients are calculated from the preceding term by multiplying its coefficient by the exponent on a and dividing this product by 1 more than the exponent on b. These coefficients form a symmetric pattern that starts with the first term and ends with the last term.

EXAMPLE 1 Expanding a Binomial

Write the expansion of $(a + b)^7$.

SOLUTION

1 $(a + b)^7 = a^7 + (\ )a^6b + (\ )a^5b^2 + (\ )a^4b^3 + (\ )a^3b^4 + (\ )a^2b^5 + (\ )ab^6 + (\ )b^7$

Write the exponents on all 8 terms.

2 $(a + b)^7 = a^7 + 7a^6b + 21a^5b^2 + \cdots + 7ab^6 + b^7$

Coefficient of the second term → $\dfrac{7 \cdot 6}{2}$ ← Exponent on a in the second term

1 more than the exponent on b in the second term (the 2)

Write the first two coefficients, 1 (which is understood) and 7. Then calculate the third coefficient, 21.

$(a + b)^7 = a^7 + 7a^6b + 21a^5b^2 + 35a^4b^3 + 35a^3b^4 + 21a^2b^5 + 7ab^6 + b^7$

Coefficient of the third term → $\dfrac{21 \cdot 5}{3}$ ← Exponent on a in the third term

1 more than the exponent on b in the third term (the 3)

Calculate the fourth coefficient, 35, and then fill in the rest of the coefficients based on the symmetry of these coefficients.

▼

EXAMPLE 2 Expanding a Binomial

Write the first four terms of $(a + b)^9$.

SOLUTION $(a + b)^9 = a^9 + 9a^8b + 36a^7b^2 + 84a^6b^3 + \cdots + b^9$

$36 = \dfrac{9 \cdot 8}{1 + 1}$, $84 = \dfrac{36 \cdot 7}{2 + 1}$

The coefficient of the third term is calculated from the coefficient and the exponents of the second term. The coefficient of the fourth term is calculated from the coefficient and the exponents of the third term.

▼

EXAMPLE 3 Calculating the Seventh Term of a Binomial Expansion

The sixth term of $(a + b)^{15}$ is $3003a^{10}b^5$. Use this information to write the seventh term.

SOLUTION $(\quad)a^9b^6$

Decrease the exponent on a by 1, and increase the exponent on b by 1 from those in the sixth term.

$$\frac{3003 \cdot 10}{5 + 1}a^9b^6 = 5005a^9b^6$$

Calculate the seventh coefficient from the coefficient and the exponents of the sixth term.

▼

EXAMPLE 4 Expanding a Binomial

Expand $(2x - 3y)^4$.

SOLUTION

$$(a + b)^4 = a^4 + 4a^3b + 6a^2b^2 + 4ab^3 + b^4$$

First set up this form for $(a + b)^4$.

$$(2x - 3y)^4 = (\ \ )^4 + 4(\ \ )^3(\ \ ) + 6(\ \ )^2(\ \ )^2 + 4(\ \ )(\ \ )^3 + (\ \ )^4$$
$$= (2x)^4 + 4(2x)^3(-3y) + 6(2x)^2(-3y)^2 + 4(2x)(-3y)^3 + (-3y)^4$$

Then substitute $2x$ for a and $-3y$ for b to expand $[2x + (-3y)]^4$.

$$= 16x^4 - 96x^3y + 216x^2y^2 - 216xy^3 + 81y^4$$

Now simplify each term.

EXAMPLE 5 Expanding a Binomial

Expand $(4x^2 + 3y^3)^4$.

SOLUTION

$$(4x^2 + 3y^3)^4 = (\ \ )^4 + 4(\ \ )^3(\ \ ) + 6(\ \ )^2(\ \ )^2 + 4(\ \ )(\ \ )^3 + (\ \ )^4$$

Set up this form for $(a + b)^4$, leaving space within the parentheses for a and b.

$$= (4x^2)^4 + 4(4x^2)^3(3y^3) + 6(4x^2)^2(3y^3)^2 + 4(4x^2)(3y^3)^3 + (3y^3)^4$$

Then substitute $4x^2$ in the space for a and $3y^3$ in the space for b.

$$= 256x^8 + 768x^6y^3 + 864x^4y^6 + 432x^2y^9 + 81y^{12}$$

Now simplify each term.

Another format for examining the symmetric pattern exhibited by the coefficients of the expansion of $(a + b)^n$ is Pascal's triangle, shown in the table below.

$(a + b)^n$	Expansion of $(a + b)^n$	Pascal's Triangle
$(a + b)^0 =$	1	1
$(a + b)^1 =$	$a + b$	1 1
$(a + b)^2 =$	$a^2 + 2ab + b^2$	1 2 1
$(a + b)^3 =$	$a^3 + 3a^2b + 3ab^2 + b^3$	1 3 3 1
$(a + b)^4 =$	$a^4 + 4a^3b + 6a^2b^2 + 4ab^3 + b^4$	1 4 6 4 1
$(a + b)^5 =$	$a^5 + 5a^4b + 10a^3b^2 + 10a^2b^3 + 5ab^4 + b^5$	1 5 10 10 5 1

Each entry between the 1s is the sum of the two adjacent entries in the preceding row. For example, the first 10 in the bottom row is the sum of the adjacent entries 4 and 6 in the row above it. There are so many other patterns in Pascal's triangle that this triangle of coefficients is a subject of study in its own right. We shall restrict our attention, however, to using Pascal's triangle as an alternative method for determining the coefficients of a binomial expansion.

SELF-CHECK

1 Complete the expansion of $(a + b)^9$ from Example 2.
2 Expand $(2x + 5y^2)^3$.

SELF-CHECK ANSWERS

1 $(a + b)^9 = a^9 + 9a^8b + 36a^7b^2 + 84a^6b^3 + 126a^5b^4 + 126a^4b^5 + 84a^3b^6 + 36a^2b^7 + 9ab^8 + b^9$
2 $(2x + 5y^2)^3 = 8x^3 + 60x^2y^2 + 150xy^4 + 125y^6$

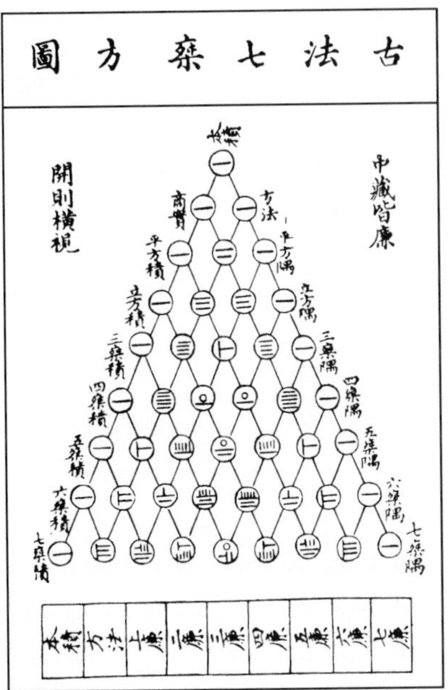

Figure 11-1 *This triangle of numbers, taken from a Chinese manuscript dated 1303, was later named after Pascal.*

A Mathematical Note

Blaise Pascal (1623–1662) was a noted French mathematician and physicist. At the age of 19 he invented an adding machine, copies of which can still be seen in some museums. He also wrote so extensively on the triangular arrangement of the coefficients of the terms of a binomial expansion that this triangular arrangement is now named in his honor. However, this triangular pattern was well known centuries before Pascal. A 1303 manuscript by the Chinese algebraist Chu Shi-kié contains this triangle in the form shown in Figure 11-1.

EXAMPLE 6 The Seventh Row of Pascal's Triangle

Use the sixth row of Pascal's triangle,

$$1 \quad 5 \quad 10 \quad 10 \quad 5 \quad 1$$

to determine the entries of the seventh row.

SOLUTION

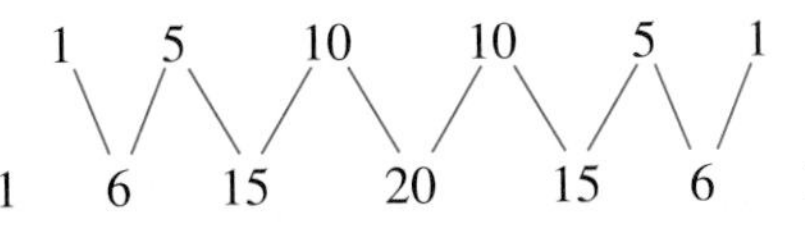

Repeat row 6.

The first and last entries of the next row are 1s. All other entries are the sum of the two adjacent entries in the preceding row. ▼

SELF-CHECK 

Use the seventh row of Pascal's triangle,

$$1 \quad 6 \quad 15 \quad 20 \quad 15 \quad 6 \quad 1$$

to determine the entries in the eighth row.

EXAMPLE 7 Expanding a Binomial Using Pascal's Triangle

Use Pascal's triangle to expand $(x + 3y)^5$.

SOLUTION

$$(x + 3y)^5 = (\ )x^5 + (\ )x^4(3y) + (\ )x^3(3y)^2 + (\ )x^2(3y)^3 + (\ )x(3y)^4 + (\ )(3y)^5$$

First set up the form, including the proper exponents on each term but leaving blanks for the coefficients.

SELF-CHECK ANSWER 

1 7 21 35 35 21 7 1

$= 1x^5 + 5x^4(3y) + 10x^3(3y)^2 + 10x^2(3y)^3 + 5x(3y)^4 + 1(3y)^5$

Next fill in the coefficients, using the entries from the appropriate row of Pascal's triangle.

1
1 1
1 2 1
1 3 3 1
1 4 6 4 1
1 5 10 10 5 1

$= x^5 + 15x^4y + 90x^3y^2 + 270x^2y^3 + 405xy^4 + 243y^5$

Then simplify each term.

Exercises 11-2

A

1 How many terms are there in the expansion of $(3v - 5w)^{11}$?

2 How many terms are there in the expansion of $(7w + 8z)^{14}$?

3 What is the degree of each term of $(3v - 5w)^{11}$?

4 What is the degree of each term of $(7w + 8z)^{14}$?

5 Write the exponents on all nine terms of $(a + b)^8$, leaving blanks for the coefficients of these terms.

6 Write the exponents on all ten terms of $(a + b)^9$, leaving blanks for the coefficients of these terms.

7 Write the exponents on all eleven terms of $(a + b)^{10}$, leaving blanks for the coefficients of these terms.

8 Write the exponents on all twelve terms of $(a + b)^{11}$, leaving blanks for the coefficients of these terms.

9 The coefficients of $(a + b)^7$ from the appropriate row of Pascal's triangle are 1, 7, 21, 35, 35, 21, 7, and 1. Use this row of coefficients to determine the next row of Pascal's triangle. Then complete the expansion of $(a + b)^8$ from Exercise 5.

10 The coefficients of $(a + b)^{10}$ from the appropriate row of Pascal's triangle are 1, 10, 45, 120, 210, 252, 210, 120, 45, 10, and 1. Use this row of coefficients to determine the next row of Pascal's triangle. Then complete the expansion of $(a + b)^{11}$ from Exercise 8.

11 The fifth term of $(a + b)^{16}$ is $1820a^{12}b^4$. Use this information to write the sixth term.

12 The seventh term of $(a + b)^{15}$ is $5005a^9b^6$. Use this information to write the eighth term.

13 The fourth term of $(a + b)^{11}$ is $165a^8b^3$. Use this information to write the fifth term.

14 The eighth term of $(a + b)^{11}$ is $330a^4b^7$. Use this information to write the ninth term.

15 Write the first eight rows of Pascal's triangle.

16 Write the ninth row of Pascal's triangle.

17 Write the tenth row of Pascal's triangle.

18 Write the eleventh row of Pascal's triangle.

In Exercises 19–34 write the binomial expansion of each expression.

19 $(s + t)^4$ **20** $(x + y)^5$ **21** $(x + 2)^5$

22 $(x + 3)^4$ **23** $(y - 3)^3$ **24** $(z - 5)^3$

25 $(2m - 6)^6$ **26** $(3m - 2)^4$ **27** $(2x - 3y)^5$

28 $(3x - 4y)^4$ **29** $(x + y^2)^6$ **30** $(x^2 - y)^5$

31 $(2m^2 + 5n^3)^4$ **32** $(5m^2 + 3n^4)^4$ **33** $\left(\frac{x}{2} - \frac{y}{3}\right)^5$

34 $\left(\frac{a}{5} + \frac{b}{5}\right)^4$

B

In Exercises 35–42 write the first four terms of the binomial expansion of each expression.

35 $(x + y)^{14}$ **36** $(m + n)^{16}$ **37** $(x + 3)^{10}$

38 $(x - 2)^{11}$ **39** $(m - 3n)^{13}$ **40** $(x + 2y)^{12}$

41 $(2x^2 + y)^9$ **42** $(m^3 - n)^{17}$

C

In Exercises 43–50 write the first three terms and the last three terms of the expansion of each expression.

43 $(a + b)^{10}$ **44** $(m + n)^{11}$ **45** $(v + w)^{13}$

46 $(x + y)^{12}$ **47** $(x - y)^{11}$ **48** $(a - b)^{13}$

49 $(2x + y)^7$ **50** $(x - 2y)^8$

51 Using the binomial expansion of $(1 + 0.1)^6$, evaluate $(1.1)^6$.

52 Using the binomial expansion of $(1 - 0.1)^7$, evaluate $(0.9)^7$.

53 Simplify $(1 + i)^3$ by using a binomial expansion.

54 Simplify $(2 - i)^4$ by using a binomial expansion.

55 Simplify $(1 - i)^5$ by using a binomial expansion.

Discussion Question

56 A classmate can correctly expand $(a + b)^4$ but is unable to apply the same procedure to expand $(a - b)^4$. Write an explanation that this student can use to expand $(a - b)^4$.

Challenge Question

Simplify
$(x - 1)^4 + 4(x - 1)^3 + 6(x - 1)^2 + 4(x - 1) + 1.$

Section Skill Reinforcers

Evaluate each of these expressions without using a calculator.

1 log 1000 **2** log 0.001 **3** log 1 **4** log 0 **5** $\log 10^8$

11-3 Sequences, Series, and Summation Notation

Section Objectives

3 Write the terms of a sequence, given the general term.

4 Use summation notation, and evaluate the series associated with a finite sequence.

Objects and natural phenomena often form regular and interesting patterns. The study of these objects and phenomena generally involves data collected sequentially or in a systematic manner. In many cases, the data satisfy the definition of a mathematical sequence.

A **sequence** is a function whose domain is a set of consecutive natural numbers. For example, the sequence 3, 7, 11, 15, 19 is another representation of the function $\{(1, 3), (2, 7), (3, 11), (4, 15), (5, 19)\}$, whose domain $\{1, 2, 3, 4, 5\}$ is understood from the order in which the sequence 3, 7, 11, 15, 19 is written. The range elements are called the **terms of the sequence.** A sequence that has a last term is called a **finite sequence,** whereas a sequence that continues without end is called an **infinite sequence.**

A finite sequence with n terms can be denoted by $a_1, a_2, \ldots, a_n$; an infinite sequence can be denoted by $a_1, a_2, a_3, \ldots, a_n, \ldots$, where a_1 is the first term, a_2 is the second term, and a_n is the nth term. Since a_n can represent any term of the sequence, it is called the **general term.** It is often useful to have a formula for the general term a_n, if one can be determined. Some sample sequences follow:

- 1, 4, 9, 16, 25, 36, 49 is a finite sequence with seven terms. In this sequence, $a_1 = 1, a_2 = 4, \ldots, a_7 = 49$. The general term a_n equals n^2.
- $2, 4, 6, 8, 10, \ldots, 2n, \ldots$ is the infinite sequence of even natural numbers. The general term a_n equals $2n$.

EXAMPLE 1 Calculating the Terms of a Sequence

Write the finite sequence with five terms whose general term a_n equals $3n + 2$.

SOLUTION $a_n = 3n + 2$

$a_1 = 3(1) + 2 = 5$ Substitute the natural numbers 1 through 5 for n.

$a_2 = 3(2) + 2 = 8$

$a_3 = 3(3) + 2 = 11$

$a_4 = 3(4) + 2 = 14$

$a_5 = 3(5) + 2 = 17$

Answer 5, 8, 11, 14, 17

EXAMPLE 2 Calculating the Twentieth Term of a Sequence

Determine a_{20} in an infinite sequence whose general term a_n equals $(-1)^n n^2 - 3n$.

SOLUTION $a_n = (-1)^n n^2 - 3n$

$a_{20} = (-1)^{20}(20)^2 - 3(20)$ Substitute 20 for n in the general term.

$= (+1)(400) - 60$ Note that $(-1)^n$ is $+1$ for any even power n.

$a_{20} = 340$

The general term of a sequence is sometimes defined in terms of one or more of the preceding terms. A sequence defined in this manner is said to be **defined recursively.**

SELF-CHECK

Write the first five terms of the infinite sequence whose general term a_n equals $\dfrac{(-1)^n}{3n}$.

EXAMPLE 3 Calculating the Terms of a Sequence

Write the first five terms of the sequence defined recursively by the formulas $a_1 = 1$, $a_2 = 1$, and $a_n = a_{n-2} + a_{n-1}$.

SOLUTION $a_1 = 1$ The first two terms are given; for $n = 3$, a_n is a_3, a_{n-2} is a_1, and a_{n-1} is a_2.

$a_2 = 1$

$a_3 = a_1 + a_2$

$= 1 + 1$ Substitute 1 for a_1 and 1 for a_2.

$a_3 = 2$

$a_4 = a_2 + a_3$ For $n = 4$, a_n is a_4, a_{n-2} is a_2, and a_{n-1} is a_3.

$= 1 + 2$ Substitute 1 for a_2 and 2 for a_3.

$a_4 = 3$

$a_5 = a_3 + a_4$ For $n = 5$, a_n is a_5, a_{n-2} is a_3, and a_{n-1} is a_4.

$= 2 + 3$ Substitute 2 for a_3 and 3 for a_4.

$a_5 = 5$

Answer 1, 1, 2, 3, 5

A Mathematical Note

The sequence 1, 1, 2, 3, 5, 8, 13, . . . is known as a *Fibonacci sequence,* in honor of the Italian mathematician Leonard Fibonacci (1170–1250). Fibonacci is known as the greatest mathematician of the thirteenth century. His sequence appears in many surprising ways in nature, including the arrangement of seeds in some flowers, the layout of leaves on the stems of some plants, and the spirals on some shells. There are so many applications that in 1963 the Fibonacci Association was founded and began to publish *The Fibonacci Quarterly.* In its first three years the association published nearly 1000 pages of research.

SELF-CHECK ANSWER

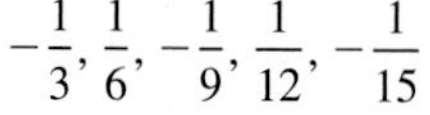

$-\frac{1}{3}, \frac{1}{6}, -\frac{1}{9}, \frac{1}{12}, -\frac{1}{15}$

EXAMPLE 4 Graphing a Sequence

Graph the six-term finite sequence whose general term a_n equals $2n - 5$.

SOLUTION

n	a_n
1	−3
2	−1
3	1
4	3
5	5
6	7

Since the domain consists only of the natural numbers 1, 2, 3, 4, 5, and 6, the graph consists only of six discrete points and is not a solid line. The dashed line is shown solely to emphasize the overall pattern exhibited by this function. ▼

The sum of the terms of a sequence is called a **series.** If $a_1, a_2, \ldots, a_n$ is a finite sequence, then the indicated sum $a_1 + a_2 + \cdots + a_n$ is the series associated with this sequence.

SELF-CHECK

Write the first six terms of the sequence defined recursively by the formulas $a_1 = 3$ and $a_n = -2a_{n-1}$ for $n > 1$.

EXAMPLE 5 Evaluating a Series

Find the value of the six-term series associated with the sequence whose general term a_n equals $3n$.

SOLUTION The series is the sum of the following terms.

$$\begin{aligned} &a_1 + a_2 + a_3 + a_4 + a_5 + a_6 \\ &= 3(1) + 3(2) + 3(3) + 3(4) + 3(5) + 3(6) \\ &= 3 + 6 + 9 + 12 + 15 + 18 \\ &= 63 \end{aligned}$$

Substitute the first six natural numbers into the formula $a_n = 3n$ to determine the first six terms, and then add these terms. ▼

A convenient way of denoting a series is to use **summation notation,** in which the Greek letter Σ (*sigma,* which corresponds to *S* for "sum") indicates the summation.

SUMMATION NOTATION

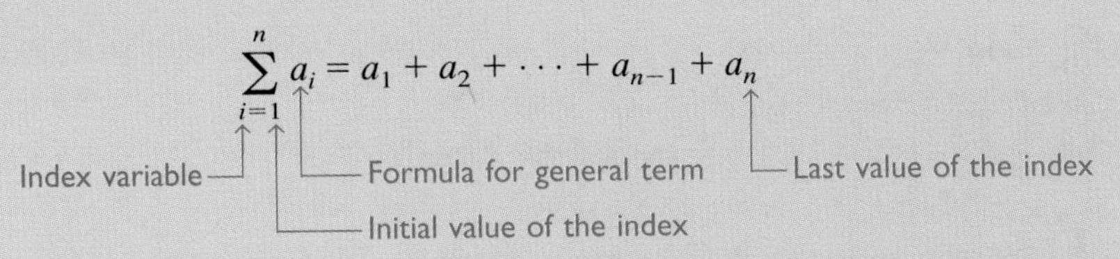

The left-hand member of the above equation is read as "the sum of a sub i from i equals 1 to i equals n."

SELF-CHECK ANSWER

3, −6, 12, −24, 48, −96

Generally the index variable is denoted by i, j, or k. The index variable is always replaced with successive integers from the initial value through the terminal value. For example, in $\sum_{i=1}^{4} a_i$, i is replaced with 1, 2, 3, and then 4, to yield

$$\sum_{i=1}^{4} a_i = a_1 + a_2 + a_3 + a_4$$

EXAMPLE 6 Evaluating a Series Using Summation Notation

Evaluate the series $\sum_{i=1}^{4} 5i$.

SOLUTION

$$\sum_{i=1}^{4} 5i = 5(1) + 5(2) + 5(3) + 5(4)$$

Replace i with 1, 2, 3, and then 4, and indicate the sum of these terms.

$$= 5 + 10 + 15 + 20$$

Evaluate each term, and then add.

$$= 50$$

Although the initial value of the index variable is often 1, it can be any integer. In the next example the initial value of the index variable j is 3.

EXAMPLE 7 Evaluating a Series Using Summation Notation

Evaluate the series $\sum_{j=3}^{5} (2j - 1)$.

SOLUTION

$$\sum_{j=3}^{5} (2j - 1) = [2(3) - 1] + [2(4) - 1] + [2(5) - 1]$$

Replace j with 3, 4, and then 5, and indicate the sum of these terms.

$$= (6 - 1) + (8 - 1) + (10 - 1)$$

Evaluate each term, and then add.

$$= 5 + 7 + 9$$

$$= 21$$

The notation $\sum_{i=1}^{6} a_i$ indicates the six-term sum $a_1 + \cdots + a_6$. The formula for the general term a_i usually involves the index variable, but it does not have to. If each term is 4, then the general term a_i is the constant 4, and the index variable does not need to be written in the formula for a_i.

SELF-CHECK

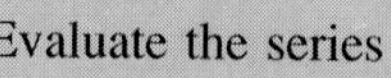

Evaluate the series $\sum_{k=2}^{4} (k^2 - 3k)$.

EXAMPLE 8 Evaluating a Series Whose Terms Are Constant

Evaluate $\sum_{i=1}^{6} 4$.

SOLUTION

$$\sum_{i=1}^{6} 4 = 4 + 4 + 4 + 4 + 4 + 4$$

Each of the six terms is the constant 4.

$$= 24$$

SELF-CHECK ANSWER

2

Graphics calculators have a variety of features that make it easy to explore sequences. The next example uses the TABLE feature on a TI-82 to reexamine the sequence graphed in Example 4.

EXAMPLE 9 Calculator Usage

Use a graphics calculator to determine the first seven terms of a sequence whose general term is $a_n = 2n - 5$.

SOLUTION Use the function $y_1 = 2x - 5$ to enter the equation for $a_n = 2n - 5$.

[2nd] [TblSet] [1 (for TblMin)] [ENTER] [1 (for ▲Tbl)]

[ENTER]

[2nd] [TABLE]

Enter the function $Y_1 = 2X - 5$. Set the table of values to start with X = 1 and then increment by 1. These keys produce a table of values corresponding to $a_n = 2n - 5$ from $n = 1$ to $n = 7$.

X	Y_1	
1 2 3 4 5 6 7	1 1.4142 1.7321 2 2.2361 2.4495 2.6458	
X = 1		

The sequence of values is given in the second column of the table.

Answer $-3, -1, 1, 3, 5, 7, 9$ ▼

The TI-82 graphics calculator has a LIST feature with options that can generate a sequence and sum the terms of a sequence. This capability is illustrated in the next example, which lists the third, fourth, and fifth terms of $a_j = 2j - 1$ and then evaluates the series $\sum_{j=3}^{5} (2j - 1)$. You may wish to compare this example to Example 7.

EXAMPLE 10 Calculator Usage

Use a graphics calculator with a LIST feature to list the terms of $\sum_{j=3}^{5} (2j - 1)$, and then evaluate this series.

SOLUTION The format of the sequence option on a TI-82 calculator is

seq(expression, variable, begin, end, increment)

In our example,

expression = $2j - 1$
variable = j
begin = 3
end = 5
increment = 1

ALPHA J , 3 , 5 , 1) ENTER

2nd LIST ▶ (MATH option) 5: sum 2nd ANS

ENTER

```
seq(2J − 1, J, 3, 5, 1
)
              {5  7  9}
sum Ans
                     21
```

Enter the expression for the general term using the variable J. Start with J = 3 and increment by 1 until J = 5. Pressing the ENTER key the first time will list the three terms of this series. Pressing ENTER the second time will sum the terms of this series.

By modifying the keystrokes to produce sum seq(2J − 1, J, 3, 5, 1), we could find the sum directly without listing the terms.

Answer $\sum_{j=3}^{5} (2j - 1) = 5 + 7 + 9 = 21$

Exercises 11-3

A

In Exercises 1–14 write the first five terms of the sequence whose general term is given.

1 $a_n = 5n$ **2** $a_n = 7n$ **3** $a_n = 3n - 8$

4 $a_n = 2n + 4$ **5** $a_n = \frac{1}{n}$ **6** $a_n = \frac{1}{n+1}$

7 $a_n = 2^n$ **8** $a_n = \left(\frac{1}{3}\right)^n$

9 $a_n = n^2 - 3n + 4$ **10** $a_n = n^2 + 4n - 5$

11 $a_n = (-1)^n\left(\frac{n}{n+1}\right)$ **12** $a_n = (-1)^{n+1}(3n + 2)$

13 $a_1 = 1$, $a_2 = 2$, and $a_n = a_{n-2} + a_{n-1}$ for $n > 2$

14 $a_1 = 1$, $a_2 = 2$, and $a_n = a_{n-2}a_{n-1}$ for $n > 2$

In Exercises 15–22 use the formula for the general term to evaluate the term requested.

15 $a_n = (-1)^n(15n)$, $a_{10} = ?$

16 $a_n = (-1)^n(8n + 3)$, $a_{15} = ?$

17 $a_n = \frac{5n-2}{10n+3}$, $a_{20} = ?$ **18** $a_n = \frac{n^2-1}{n^2+1}$, $a_{30} = ?$

19 $a_1 = 12$, $a_n = a_1 + 7(n - 1)$, $a_{81} = ?$

20 $a_1 = -31$, $a_n = a_1 + 8(n - 1)$, $a_{46} = ?$

21 $a_1 = 81$, $a_n = a_1\left(\frac{1}{3}\right)^{n-1}$, $a_7 = ?$

22 $a_1 = \frac{1}{64}$, $a_n = a_1\left(-\frac{1}{2}\right)^{n-1}$, $a_{10} = ?$

In Exercises 23–44 evaluate each series.

23 $3 + 6 + 9 + 12 + 15$

24 $2 + 4 + 8 + 16 + 32$

25 $5 + 10 + 15 + \cdots + 50$

26 $1 + 3 + 5 + \cdots + 13$

27 $a_1 + a_2 + \cdots + a_6$, where $a_n = 2n + 3$

28 $a_1 + a_2 + \cdots + a_6$, where $a_n = n - 7$

29 $a_1 + a_2 + a_3 + a_4$, where $a_n = \frac{1}{n}$

30 $a_1 + a_2 + a_3 + a_4$, where $a_n = \frac{n}{n+1}$

31 $\sum_{i=1}^{5} 3i$ **32** $\sum_{k=1}^{4} (k^2 - k)$

33 $\sum_{j=2}^{5} (j^2 - j + 1)$ **34** $\sum_{i=3}^{6} (i^2 + i - 4)$

35 $\sum_{k=3}^{5}(k^3-k^2)$

36 $\sum_{j=4}^{6}(2^j-j^2)$

37 $\sum_{i=1}^{5}\frac{1}{2^i}$

38 $\sum_{k=1}^{4}\frac{1}{3^k}$

39 $\sum_{k=1}^{100}5$

40 $\sum_{i=1}^{350}6$

41 $\sum_{i=3}^{40}20$

42 $\sum_{j=6}^{70}9$

43 $\sum_{i=1}^{5}a_i$, where $a_1=24$ and $a_n=\frac{a_{n-1}}{2}$ for $n\geq 2$

44 $\sum_{i=1}^{5}a_i$, where $a_1=2$ and $a_n=3a_{n-1}$ for $n\geq 2$

B

In Exercises 45–50 graph the first five terms of the sequence whose general term is given.

45 $a_n=2n-5$

46 $a_n=-3n+10$

47 $a_n=n^2-10$

48 $a_n=n^2-9$

49 $a_n=(-1)^n3$

50 $a_n=-3$

In Exercises 51–54 write out the terms of both series, and determine whether the two series are the same.

51 $\sum_{i=1}^{5}(2i-1)$ and $\sum_{k=0}^{4}(2k+1)$

52 $\sum_{j=0}^{3}(-1)^j2^j$ and $\sum_{k=1}^{4}(-1)^{k-1}2^{k-1}$

53 $\sum_{k=3}^{7}2k$ and $\sum_{i=2}^{6}(2i+1)$

54 $\sum_{j=3}^{7}j^2$ and $\sum_{j=2}^{6}(j^2+1)$

C

Calculator Usage (55–60)

55 The following form is used on a TI-82 calculator to list the terms of a sequence:

seq(expression, variable, begin, end, increment)

Use this form to determine the result obtained by pressing [ENTER] for the display window.

seq(N² − 5N, N, 1, 8, 1)

In Exercises 56 and 57 use a graphics calculator to write the first eight terms of each sequence.

56 $a_n=n^2-6n+2$

57 $a_n=2^n-n^2$

58 The following form is used on a TI-82 calculator to sum the terms of a sequence:

sum seq(expression, variable, begin, end, increment)

Use this form to determine the result obtained by pressing [ENTER] for the display window.

sum seq(N² − 5N, N, 1, 8, 1)

In Exercises 59 and 60 use a graphics calculator to evaluate each series.

59 $\sum_{n=1}^{8}(n^2-6n+5)$

60 $\sum_{n=1}^{8}(n^3-n^2-4n+1)$

C

Discussion Question

61 Write a paragraph explaining why the ability to see mathematical patterns is such an important skill that many state and national tests have questions designed to test this ability in students.

Challenge Question

The following sequences are similar to those found on IQ and job-placement tests. Determine by inspection or by trial and error the next term in each sequence. Also determine a formula for the general term a_n.

a. 48, 24, 12, 6, 3, ___

b. 1, 5, 9, 13, 17, ___

c. 1, 4, 9, 16, 25, 36, ___

d. $\frac{1}{2}, \frac{2}{3}, \frac{3}{4}, \frac{4}{5}, \frac{5}{6}$, ___

e. −1, 3, −5, 7, −9, ___

f. 0, 7, 26, 63, 124, ___

g. $\frac{1}{2}, \frac{3}{4}, \frac{9}{6}, \frac{27}{8}, \frac{81}{10}$, ___

h. 0, 1, 1, 2, 3, 5, 8, ___

Section Skill Reinforcers

Factor each of these polynomials.

1 x^2-64 **2** x^3-64 **3** x^3+64 **4** x^2+16 **5** x^4-16

11-4 *Arithmetic Progressions*

SECTION OBJECTIVES

5 Calculate the *n*th term of an arithmetic progression.

6 Find the sum of a finite arithmetic progression.

An **arithmetic progression** is a sequence in which consecutive terms differ by a constant. This constant difference is called the **common difference** and is denoted by d.

EXAMPLE 1 Identifying Arithmetic Progressions

Determine which of these sequences are arithmetic progressions. For those that are arithmetic, determine the common difference d.

SOLUTIONS

(a) 5, 8, 11, 14, 17, . . .

This sequence is an arithmetic progression because the difference between any two consecutive terms is 3.

$8 - 5 = 3$, $11 - 8 = 3$, $14 - 11 = 3$, and $17 - 14 = 3$.

(b) 13, 8, 3, −2, −7, . . .

This sequence is an arithmetic progression with $d = 8 - 13 = -5$.

$8 - 13 = -5$, $3 - 8 = -5$, $-2 - 3 = -5$, and $-7 - (-2) = -5$.

(c) 5, 7, 9, 14, 16, . . .

This sequence is not an arithmetic progression since $7 - 5 = 2$ but $14 - 9 = 5$.

The difference between consecutive terms in this sequence is not constant.

EXAMPLE 2 Calculating the Terms of an Arithmetic Progression

Write the first six terms of the arithmetic progression with $a_1 = 5$ and $d = 2$.

SOLUTION

$$a_1 = 5$$
$$a_2 = 5 + 2 = 7$$
$$a_3 = 7 + 2 = 9$$
$$a_4 = 9 + 2 = 11$$
$$a_5 = 11 + 2 = 13$$
$$a_6 = 13 + 2 = 15$$

Add the common difference, 2, to each term to obtain the next term.

Answer 5, 7, 9, 11, 13, 15

SELF-CHECK

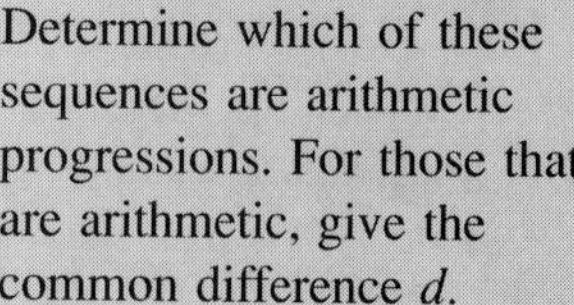

Determine which of these sequences are arithmetic progressions. For those that are arithmetic, give the common difference d.

1 8, 6, 4, 2, . . .

2 2, 4, 8, 16, . . .

3 $\frac{1}{2}, \frac{5}{6}, \frac{7}{6}, \frac{3}{2}, \ldots$

SELF-CHECK ANSWERS 

1 Arithmetic, $d = -2$ **2** Not arithmetic **3** Arithmetic, $d = \frac{1}{3}$

Since the common difference is given by $a_n - a_{n-1} = d$, the general term can be found by using the recursive formula $a_n = a_{n-1} + d$ for $n > 1$. Thus we can rewrite the terms $a_1, a_2, a_3, a_4, \ldots, a_n$ as follows:

$$a_1$$
$$a_2 = a_1 + d$$
$$a_3 = a_2 + d = (a_1 + d) + d = a_1 + 2d$$
$$a_4 = a_3 + d = (a_1 + 2d) + d = a_1 + 3d$$

This pattern continues, with the common difference added once for each term with the exception of the first term.

$$\vdots$$
$$a_n = a_{n-1} + d = a_1 + (n-1)d$$

The last of these equations, $a_n = a_1 + (n-1)d$, is a formula for calculating the nth term directly from the first term a_1 and the common difference d. We will use this formula in the next three examples.

ARITHMETIC PROGRESSION

If $a_n - a_{n-1} = d$ for $n > 1$, then the sequence is called an arithmetic progression and the constant d is called the common difference. The formula for the nth term is $a_n = a_1 + (n-1)d$.

EXAMPLE 3 Calculating a Term of an Arithmetic Progression

Find a_{21} in an arithmetic progression if $a_1 = 8$ and $d = 7$.

SOLUTION

$$a_n = a_1 + (n-1)d$$
$$a_{21} = 8 + (21 - 1)7$$

Substitute the given values into the formula for a_n.

$$a_{21} = 148$$

▼

SELF-CHECK 

Find a_1 in an arithmetic progression if $a_{19} = 46$ and $d = 3\frac{1}{2}$.

EXAMPLE 4 Determining the Common Difference

Find d in an arithmetic progression if $a_1 = -87$ and $a_{57} = 529$.

SOLUTION

$$a_n = a_1 + (n-1)d$$
$$529 = -87 + (57 - 1)d$$

Substitute the given values into the formula for a_n.

$$616 = 56d$$

Then solve for d.

$$d = 11$$

▼

SELF-CHECK ANSWER ✓

$a_1 = -17$

EXAMPLE 5 Determining the Number of Terms

Find the number of terms n in the arithmetic progression

$$-20, -13, -6, \ldots, 281$$

SOLUTION

$d = -13 - (-20) = 7$ First calculate the common difference d.

$a_n = a_1 + (n - 1)d$

$281 = -20 + (n - 1)7$ Substitute $a_1 = -20$, $a_n = 281$, and d into the formula for a_n, and then solve for n.

$301 = 7(n - 1)$

$n - 1 = 43$

$n = 44$

EXAMPLE 6 Calculating a Term of an Arithmetic Progression

Find the 15th term in an arithmetic progression if $a_{12} = 111$ and $a_{14} = 119$.

SOLUTION

$a_{14} = a_{12} + 2d$ $a_{13} = a_{12} + d$, $a_{14} = a_{12} + 2d$

$119 = 111 + 2d$ Substitute the given values, and calculate d.

$8 = 2d$

$d = 4$

$a_{15} = a_{14} + d$

$= 119 + 4$ Substitute for a_{14} and d, and then calculate a_{15}.

$a_{15} = 123$

A Mathematical Note

At the age of 10 Carl Friedrich Gauss (1777–1855) mentally computed the sum of an arithmetic progression like the one shown on this page—a problem that none of his fellow students were able to answer correctly by the end of the hour.

The arithmetic series denoted by $S_n = \sum_{i=1}^{n} a_i$ is the sum of the n terms $a_1 + a_2 + \cdots + a_{n-1} + a_n$ of an arithmetic progression. For a sequence with only a few terms it is easy to obtain the sum by merely adding the terms. For example,

$$S_5 = \sum_{i=1}^{5} 3i = 3 + 6 + 9 + 12 + 15 = 45$$

However, for a sequence with many terms, such as $\sum_{i=1}^{100} i$, it is useful to have a formula that allows us to calculate the sum without actually doing all of the adding.

To develop the logic needed to derive such a formula, let us first look at a simplified way of adding the terms of a long sequence such as $\sum_{i=1}^{100} i$. We first list the terms twice, once in increasing order and once in decreasing order, and then we add the two lines.

$$\sum_{i=1}^{100} i = 1 + 2 + 3 + \cdots + 98 + 99 + 100$$ Terms listed in increasing order

$$\sum_{i=1}^{100} i = 100 + 99 + 98 + \cdots + 3 + 2 + 1$$ Terms listed in decreasing order

$$2\sum_{i=1}^{100} i = 101 + 101 + 101 + \cdots + 101 + 101 + 101$$ Add corresponding terms.

$$2\sum_{i=1}^{100} i = 100(101)$$ Since each of the 100 terms is 101, their sum is 100(101).

$$\sum_{i=1}^{100} i = \frac{100(101)}{2}$$ Solve for the sum by dividing both sides by 2.

$$\sum_{i=1}^{100} i = 5050$$

To develop a formula for $S_n = \sum_{i=1}^{n} a_i$, we will use the same logic.

$$S_n = a_1 + (a_1 + d) + (a_1 + 2d) + \cdots + [a_1 + (n-3)d] + [a_1 + (n-2)d] + [a_1 + (n-1)d]$$

$$S_n = [a_1 + (n-1)d] + [a_1 + (n-2)d] + [a_1 + (n-3)d] + \cdots + (a_1 + 2d) + (a_1 + d) + a_1$$

$$2S_n = [a_1 + a_1 + (n-1)d] + [a_1 + a_1 + (n-1)d] + [a_1 + a_1 + (n-1)d] + \cdots + [a_1 + a_1 + (n-1)d]$$

Add corresponding terms.

$$2S_n = (a_1 + a_n) + (a_1 + a_n) + (a_1 + a_n) + \cdots + (a_1 + a_n)$$ Substitute a_n for $a_1 + (n-1)d$.

$$2S_n = n(a_1 + a_n)$$ Note that $(a_1 + a_n)$ is added n times.

$$S_n = \frac{n(a_1 + a_n)}{2}$$ Solve for S_n by dividing both sides by 2.

$$S_n = \frac{n[2a_1 + (n-1)d]}{2}$$ Substitute $a_1 + (n-1)d$ for a_n for an alternative form of this formula.

ARITHMETIC SERIES

The arithmetic series $S_n = \sum_{i=1}^{n} a_i$ is given by

$$S_n = \frac{n}{2}(a_1 + a_n)$$

or

$$S_n = \frac{n}{2}[2a_1 + (n-1)d]$$

EXAMPLE 7 The Sum of the First 100 Natural Numbers

Find the sum of the first 100 natural numbers using the formula for S_n.

SOLUTION

$$S_n = \frac{n}{2}(a_1 + a_n)$$

The terms of this series form an arithmetic progression, with $a_1 = 1$, $a_{100} = 100$, $d = 1$, and $n = 100$.

$$S_{100} = \frac{100}{2}(1 + 100)$$

Substitute these values into the formula for S_n.

$$= 50(101)$$

$$S_{100} = 5050$$

Note that this sum is the same one we obtained above.

EXAMPLE 8 Evaluating an Arithmetic Series

Find the sum of the first 63 terms of the arithmetic sequence with $a_1 = 42$ and $d = 6$.

SOLUTION

$$S_n = \frac{n}{2}[2a_1 + (n - 1)d]$$

$$S_{63} = \frac{63}{2}[2(42) + (63 - 1)(6)]$$

Substitute the given values into the alternative form for an arithmetic series.

$$= \frac{63}{2}(84 + 372)$$

$$= \frac{63}{2}(456)$$

$$= 63(228)$$

$$S_{63} = 14{,}364$$

SELF-CHECK

Find the sum of the first 84 terms of the arithmetic sequence with $a_1 = -700$ and $a_{84} = 47$.

The next example uses the capabilities of a TI-82 graphics calculator to rework Example 8.

EXAMPLE 9 Calculator Usage

Use a graphics calculator with a LIST feature to evaluate $\sum_{n=1}^{63} (6n + 36)$.

SOLUTION The format of the sum of a sequence option on a TI-82 calculator is

sum seq(expression, variable, begin, end, increment)

In our example,

expression = $6n + 36$

variable = n

begin = 1

end = 63

increment = 1

For $n = 1$, the first term is 42. Can you verify that $d = 6$, as in Example 8?

SELF-CHECK ANSWER 

$S_{84} = -27{,}426$

[2nd] [LIST] [▶(MATH option)] [5: sum]

[2nd] [LIST] [5: seq(] [6] [ALPHA] [N] [+] [3] [6]

[,] [ALPHA] [N] [,] [1] [,] [6] [3] [,] [1] [)] [ENTER]

Select the sum option from the LIST menu. Then enter a description of the sequence to be added.

```
sum seq(6N + 36, N,
1, 63, 1)
                14364
```

Answer $\sum_{n=1}^{63} (6n + 36) = 14{,}364$

▼

EXAMPLE 10 An Application of Arithmetic Series

Rolls of carpet are stacked in a warehouse, with 20 rolls on the first level, 19 on the second level, and so on. The top level has only 1 roll. How many rolls are in this stack? (See the figure.)

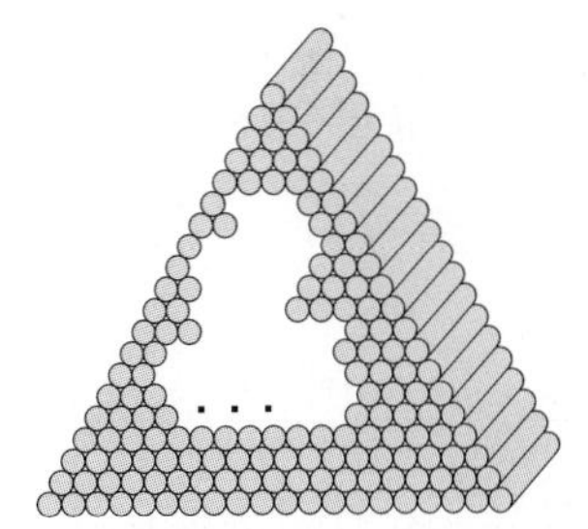

SOLUTION The numbers of rolls on the various levels form an arithmetic progression, with $a_1 = 20$, $n = 20$, and $a_{20} = 1$.

$$S_n = \frac{n(a_1 + a_n)}{2}$$

$$S_{20} = \frac{20(20 + 1)}{2}$$

$$S_{20} = 210$$

There is a constant difference of 1 roll between consecutive levels in this stack. Thus the numbers of rolls form an arithmetic progression, and the total number of rolls is determined by adding the terms of this progression.

Answer There are 210 rolls in this stack.

▼

Exercises 11-4

A

In Exercises 1–12 determine whether the sequences are arithmetic progressions. For those that are arithmetic progressions, find the common difference *d*.

1 $-3, 1, 5, 9, \ldots$

2 $40, 30, 20, 10, 0, \ldots$

3 $20, 17, 21, 18, 22, \ldots$

4 $-11, -6, -1, 4, 9, \ldots$

5 $5.6, 5.8, 6.0, 6.2, \ldots$

6 $3, 9, 27, 81, 243, \ldots$

7 $\frac{1}{2}, \frac{9}{10}, \frac{13}{10}, \frac{17}{10}, \frac{21}{10}, \ldots$

8 $5\frac{1}{2}, 5, 4\frac{1}{2}, 4, \ldots$

9 $a_n = 2^n$

10 $a_n = 2n$

11 $a_n = \frac{n}{3}$

12 $a_n = \frac{3}{n}$

In Exercises 13–24 write the first six terms of the arithmetic progression that satisfies the given conditions.

13 $a_1 = 7, d = -2$

14 $a_1 = -9, d = 3$

15 $a_2 = 4, d = \frac{2}{3}$

16 $a_2 = 2, d = \frac{3}{4}$

17 $a_1 = 6, a_2 = 10$

18 $a_3 = 5, a_4 = 8$

19 $a_1 = -8, a_6 = 12$

20 $a_1 = 13, a_6 = 3$

21 $a_n = 4n - 1$

22 $a_n = 11 - 2n$

23 $a_1 = 1.2, a_n = a_{n-1} + 0.4$ for $n > 1$

24 $a_1 = -4.5, a_n = a_{n-1} + 1.5$ for $n > 1$

In Exercises 25–30 use the information given to find the indicated term of the arithmetic progression.

25 $a_1 = 6, d = 5, a_{83} = ?$

26 $a_1 = -9, d = 3, a_{91} = ?$

27 $a_1 = 17, a_2 = 15, a_{31} = ?$

28 $a_2 = 8, a_3 = 11, a_{36} = ?$

29 $a_{80} = 25, a_{81} = 33, a_{82} = ?$

30 $a_{90} = 4, a_{92} = 8, a_{93} = ?$

In Exercises 31–40 use the information given to evaluate each arithmetic series.

31 $a_1 = 2, a_{40} = 80, S_{40} = ?$

32 $a_1 = 3, a_{51} = 153, S_{51} = ?$

33 $a_1 = \frac{1}{2}, a_{12} = \frac{1}{3}, S_{12} = ?$

34 $a_1 = 0.36, a_{18} = 0.64, S_{18} = ?$

35 $a_1 = 10, d = 4, S_{66} = ?$

36 $a_1 = 11, d = -3, S_{11} = ?$

37 $\sum_{i=1}^{61} (2i + 3)$

38 $\sum_{k=1}^{47} (3k - 2)$

39 $\sum_{k=1}^{24} \frac{k + 3}{5}$

40 $\sum_{j=1}^{40} \frac{j - 5}{3}$

B

In Exercises 41–56 use the information given for the arithmetic progressions to find the quantities indicated.

41 18, 14, 10, . . . , −62; $n = ?$

42 48, 55, . . . , 496; $n = ?$

43 $a_{44} = 216, d = 12, a_1 = ?$

44 $a_{113} = -109, d = -2, a_1 = ?$

45 $a_{11} = 4, a_{31} = 14, d = ?$

46 $a_{47} = 23, a_{62} = 28, d = ?$

47 $a_1 = 5, a_n = 14, d = \frac{1}{5}, n = ?$

48 $a_1 = -12, a_n = 6, d = \frac{1}{2}, n = ?$

49 $a_{77} = 19, d = -11, S_{77} = ?$

50 $a_{85} = 111, d = 3, S_{85} = ?$

51 $S_n = 240, a_1 = 4, a_n = 16, n = ?$

52 $S_n = 1118, a_1 = 15, a_n = 37, n = ?$

53 $S_{30} = 1560, a_{30} = 93, a_1 = ?$

54 $S_{17} = 527, a_1 = 15, a_{17} = ?$

55 $S_{40} = 680, a_1 = 11, d = ?$

56 $S_{25} = 60, a_1 = 17, d = ?$

C

CALCULATOR USAGE (57–62)

57 The following form is used on a TI-82 calculator to list the terms of a sequence:

seq(expression, variable, begin, end, increment)

Use this form to determine the geometric sequence indicated by the display window.

```
seq(4N + 9, N, 11, 19, 1)
```

In Exercises 58 and 59 use a graphics calculator to write the indicated terms of each arithmetic sequence.

58 Given $a_n = 5n - 7$, list a_{31} through a_{40}.

59 Given $a_n = 11n + 9$, list a_{93} through a_{100}.

60 The following form is used on a TI-82 calculator to sum the terms of a sequence:

sum seq(expression, variable, begin, end, increment)

Use this form to determine the sum of the arithmetic sequence indicated by the display window.

```
sum seq(7N − 1, N, 3, 28, 1)
```

In Exercises 61 and 62 use a graphics calculator to evaluate each arithmetic series.

61 $\sum_{n=15}^{38} (6n + 5)$

62 $\sum_{n=11}^{94} (4n + 1)$

63 **Stack of Logs** Logs are stacked so that each layer after the first has 1 less log than the previous layer. If the bottom layer has 24 logs and the top layer has 8 logs, how many logs are in the stack? (See the figure.)

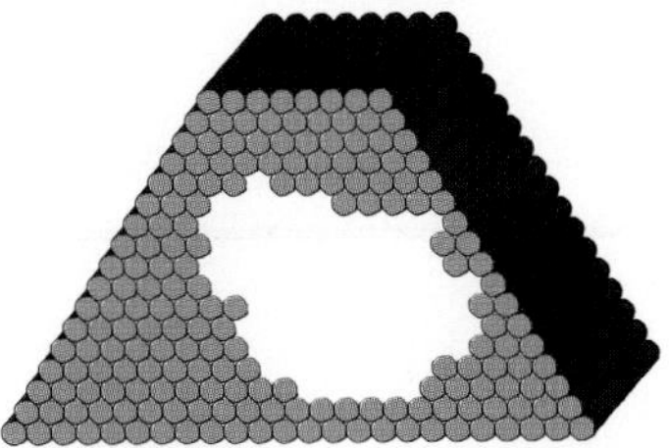

Figure for Exercise 63

64 Rolls of Insulation Rolls of insulation are stacked so that each layer after the first has 8 fewer rolls than the previous layer. How many layers will a lumber yard need to use in order to stack 120 rolls if 40 rolls are placed on the bottom layer? (See the figure.)

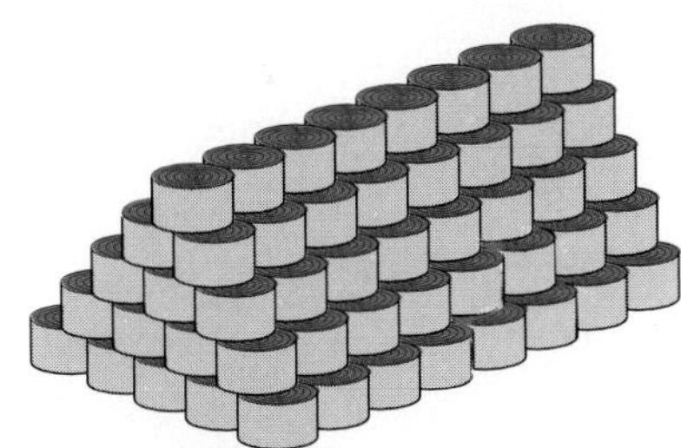

Figure for Exercise 64

65 Increased Productivity The productivity gain from installing a new robot welder on a machinery assembly line is estimated to be \$4000 the first month of operation, \$4500 the second month, and \$5000 the third month. If this trend continues, what will be the total productivity gain for the first 12 months of operation?

66 Integers Divisible by 3 How many integers between 17 and 502 are divisible by 3?

67 Integers Divisible by 4 How many integers between 17 and 502 are divisible by 4?

68 Seats in a Theater A theater has 20 rows of seats, with 100 seats in the back row. Each row has 2 fewer seats than the row immediately behind it. How many seats are in the theater?

69 Magic Square A magic square is a square array of consecutive natural numbers such that the sum of each row and column is the same. A magic square with three rows is shown in the figure. Suppose we constructed an eight-row magic square, with 1 as the smallest entry. What would the sum of all the entries of the square be?

8	1	6
3	5	7
4	9	2

Figure for Exercise 69

70 Area Under a Curve Approximate the area of the shaded region in the figure by adding the areas of the 100 rectangles indicated.

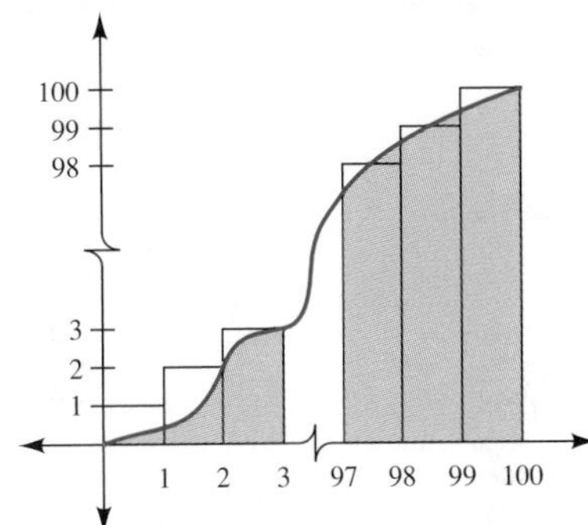

Figure for Exercise 70

DISCUSSION QUESTION

71 Write your own word problem that is algebraically modeled by an arithmetic series with $a_1 = 15$, $d = 2$, and $n = 20$.

CHALLENGE QUESTION

If 10 people in a room shake hands with each other exactly once, how many handshakes will take place?

SECTION SKILL REINFORCERS

Simplify each of these expressions.

1 i^{18} **2** i^{-5} **3** $(6 + i)(6 - i)$ **4** $\dfrac{12}{2 - i}$ **5** $(2 + 3i)^2$

11-5 Geometric Progressions

SECTION OBJECTIVE

7 Calculate the nth term of a geometric progression.

A **geometric progression** is a sequence in which consecutive terms form a constant ratio. This constant, called the **common ratio,** is denoted by r.

EXAMPLE 1 Identifying Geometric Progressions

Determine which of these sequences are geometric progressions. For those that are geometric, determine the common ratio r.

SOLUTIONS

(a) $3, 6, 12, 24, \ldots$

This sequence is a geometric progression because the ratio of any two consecutive terms is 2.

$\frac{6}{3} = 2$

$\frac{12}{6} = 2$

$\frac{24}{12} = 2$

(b) $18, -6, 2, -\frac{2}{3}, \ldots$

This sequence is a geometric progression with $r = \frac{-6}{18} = -\frac{1}{3}$.

$\frac{-6}{18} = -\frac{1}{3}$

$\frac{2}{-6} = -\frac{1}{3}$

$\frac{-2/3}{2} = -\frac{1}{3}$

(c) $2, 6, 12, 36, 108, \ldots$

This sequence is not a geometric progression because $\frac{6}{2} = 3$ but $\frac{12}{6} = 2$.

The ratio of consecutive terms is not constant in this sequence.

SELF-CHECK

Determine which of these sequences are geometric progressions. For those that are geometric, give the value of r.

1 $81, 54, 36, 24, 16, \ldots$
2 $2, 4, 6, 8, 10, \ldots$
3 $3, -6, 12, -24, 48, \ldots$

EXAMPLE 2 Calculating the Terms of a Geometric Progression

Write the first five terms of the geometric progression with $a_1 = 2$ and $r = 3$.

SOLUTION

$a_1 = 2$

$a_2 = 2 \cdot 3 = 6$

$a_3 = 6 \cdot 3 = 18$

$a_4 = 18 \cdot 3 = 54$

$a_5 = 54 \cdot 3 = 162$

Multiply each term by the common ratio, 3, to obtain the next term.

Answer 2, 6, 18, 54, 162

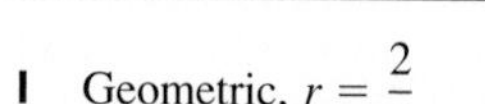

SELF-CHECK ANSWERS

1 Geometric, $r = \frac{2}{3}$ **2** Not geometric **3** Geometric, $r = -2$

Since the common ratio is given by $\frac{a_n}{a_{n-1}} = r$, the general term can be found by using the recursive formula $a_n = ra_{n-1}$ for $n > 1$. Thus we can rewrite the terms $a_1, a_2, a_3\ a_4, \ldots, a_n$ as follows.

a_1

$a_2 = a_1 r$

$a_3 = a_2 r = (a_1 r)r = a_1 r^2$

$a_4 = a_3 r = (a_1 r^2)r = a_1 r^3$ This pattern continues, with the power of the common ratio increased by 1 for each term except the first term.

$\vdots$

$a_n = a_{n-1} r = a_1 r^{n-1}$

The last of these equations, $a_n = a_1 r^{n-1}$, is a formula for calculating the *n*th term directly from the first term a_1 and the common ratio *r*. We will use this formula in the following examples.

GEOMETRIC PROGRESSION

If $\frac{a_n}{a_{n-1}} = r$ for $n > 1$, then the sequence is called a geometric progression and the constant *r* is called the common ratio. The formula for the *n*th term is $a_n = a_1 r^{n-1}$.

EXAMPLE 3 Calculating a Term of a Geometric Progression

Find a_{59} in a geometric progression if $a_1 = 3$ and $r = -1$.

SOLUTION $a_n = a_1 r^{n-1}$

$a_{59} = (3)(-1)^{59-1}$ Substitute the given values into the formula for a_n.

$= (3)(-1)^{58}$

$= (3)(1)$ Note that $(-1)^n = +1$ for any even power n.

$a_{59} = 3$ ▼

EXAMPLE 4 Determining the Common Ratio

Find *r* in a geometric progression if $a_1 = 6$ and $a_3 = 24$.

SOLUTION $a_n = a_1 r^{n-1}$

$24 = (6)(r)^{3-1}$ Substitute the given values into the formula for a_n.

$4 = r^2$ Then solve this quadratic equation for both possible values of *r*.

$r = 2$ or $r = -2$

Check The geometric progressions 6, 12, 24, 48, . . . and 6, −12, 24, −48, . . . both satisfy the stated conditions. ▼

EXAMPLE 5 Determining the Number of Terms

Find the number of terms n in the geometric progression 3125, 1250, . . . , 32.

SOLUTION

$$r = \frac{1250}{3125} = 0.4$$ First calculate the common ratio r.

$$a_n = a_1 r^{n-1}$$

$$32 = (3125)(0.4)^{n-1}$$ Substitute $a_1 = 3125$, $a_n = 32$, and $r = 0.4$ into the formula for a_n.

$$(0.4)^{n-1} = 0.01024$$ Divide both members by 3125.

$$\log 0.4^{n-1} = \log 0.01024$$ Take the common log of both members.

$$(n-1)\log 0.4 = \log 0.01024$$ Simplify by using the power rule for logarithms.

$$n - 1 = \frac{\log 0.01024}{\log 0.4}$$ Divide both members by log 0.4.

$$n - 1 \approx \frac{-1.9897}{-0.3979}$$ Approximate the right member with a calculator.

$$n - 1 \approx 5$$

$$n \approx 6$$

Answer This progression has 6 terms.

You can check this answer by writing the first six terms of this geometric progression.

EXAMPLE 6 Calculating a Term of a Geometric Progression

Find the twenty-first term of a geometric progression if $a_{16} = \frac{729}{64}$ and $a_{19} = -\frac{27}{8}$.

SOLUTION

$$a_{19} = a_{16}r^3$$ $a_{17} = a_{16}r$, $a_{18} = a_{16}r^2$, $a_{19} = a_{16}r^3$

$$-\frac{27}{8} = \frac{729}{64}r^3$$ Substitute the given values for a_{19} and a_{16}.

$$r^3 = -\frac{8}{27}$$ Multiply both members by $\frac{64}{729}$.

$$r = -\frac{2}{3}$$ Solve for r by taking the cube root of both members.

$$a_{21} = a_{19}r^2$$ $a_{20} = a_{19}r$, $a_{21} = a_{19}r^2$

$$= \left(-\frac{27}{8}\right)\left(-\frac{2}{3}\right)^2$$ Substitute the values for a_{19} and r, and then calculate a_{21}.

$$= \left(-\frac{27}{8}\right)\left(\frac{4}{9}\right)$$

$$a_{21} = -\frac{3}{2}$$

The next example uses the capabilities of a TI-82 graphics calculator to list the terms of a geometric sequence.

EXAMPLE 7 Calculator Usage

Use a graphics calculator with a LIST feature to determine the first six terms of a sequence whose general term is $a_n = 2(5^n)$.

SOLUTION The format of the sequence option on a TI-82 calculator is

seq(expression, variable, begin, end, increment)

In our example,

$$\begin{aligned} \text{expression} &= 2(5^n) \\ \text{variable} &= n \\ \text{begin} &= 1 \\ \text{end} &= 6 \\ \text{increment} &= 1 \end{aligned}$$

[2nd] [LIST] [5: seq(] [2] [(] [5] [∧] [ALPHA] [N] [)]

[,] [ALPHA] [N] [,] [1] [,] [6] [,] [1] [)] [ENTER]

Enter the expression for the general term using the variable N. Start with N = 1, and increment by 1 until N = 6.

```
seq(2(5^N), N, 1, 6
, 1)
   {10  50  250  1250...
```

Using the arrow key [▶], you can scroll across the display to determine the other two entries on the line:

6250 31250}

Answer 10, 50, 250, 1250, 6250, 31,250

SELF-CHECK

1 Find a_1 in a geometric progression if $a_7 = \frac{1}{3}$ and $r = -\frac{1}{3}$.

2 Find the number of terms in the geometric progression $\frac{1}{8}, -\frac{1}{4}, \ldots, -16$.

EXAMPLE 8 An Application of Geometric Progressions

A flywheel on a shearing machine is rotating at 250 revolutions per minute (rpm). When the motor driving the flywheel is turned off, the flywheel gradually reduces speed. One minute after the motor is turned off, the flywheel is rotating at 150 revolutions per minute. At the end of each additional minute its speed is three-fifths that of the previous minute. Approximate the number of revolutions per minute made by the flywheel 7 minutes after the motor is turned off.

SOLUTION The numbers of revolutions per minute recorded at the end of each minute form a geometric progression with $r = \frac{3}{5}$.

$a_n = a_1 r^{n-1}$ — The value sought in this problem is a_7.

$a_7 = 150\left(\frac{3}{5}\right)^{7-1}$ — Substitute $a_1 = 150$ and $r = \frac{3}{5}$ into the formula for the nth term of a geometric progression.

$= 150\left(\frac{3}{5}\right)^6$

≈ 6.9984 — Approximate this value with a calculator.

$a_7 \approx 7$

Answer After 7 minutes the flywheel will be rotating at approximately 7 rpm.

SELF-CHECK ANSWERS

1 $a_1 = 243$ 2 $n = 8$

Exercises 11-5

A

In Exercises 1–16 determine whether the sequences are geometric progressions. For those that are, write the common ratio *r*.

1 12, 6, 3, 1.5, . . .

2 24, 36, 54, 81, . . .

3 −125, 25, −5, 1, . . .

4 1, 4, 9, 16, 25, . . .

5 8, 4, 0, −4, . . .

6 4, −4, 4, −4, 4, . . .

7 $\frac{1}{16}, -\frac{1}{4}, 1, -4, 16, \ldots$

8 $\frac{2}{3}, \frac{4}{9}, \frac{8}{27}, \frac{16}{81}, \ldots$

9 0.3, 0.03, 0.003, . . .

10 0.21, 0.0021, 0.000021, . . .

11 $a_n = 3^n$

12 $a_n = 3n$

13 $a_n = -0.1n$

14 $a_n = (-0.1)^n$

15 $a_1 = 24, a_n = 0.5a_{n-1}$ for $n > 1$

16 $a_1 = -24, a_n = -2.0a_{n-1}$ for $n > 1$

In Exercises 17–30 write the first five terms of the geometric progression that satisfies the given conditions.

17 $a_1 = 12, r = 5$

18 $a_1 = 3, r = 4$

19 $a_2 = 12, r = -\frac{2}{3}$

20 $a_2 = 36, r = -\frac{3}{2}$

21 $a_1 = 12, r = 0.01$

22 $a_1 = 0.6, r = 0.1$

23 $a_1 = 3, a_2 = -6$

24 $a_1 = 2, a_2 = 20$

25 $a_1 = 2, a_3 = 18$

26 $a_1 = -3, a_3 = -75$

27 $a_n = \left(\frac{3}{5}\right)^n$

28 $a_n = \left(-\frac{4}{3}\right)^n$

29 $a_1 = 36, a_n = -\frac{1}{2}a_{n-1}$ for $n > 1$

30 $a_1 = -48, a_n = -\frac{3}{2}a_{n-1}$ for $n > 1$

In Exercises 31–40 use the information given to find the indicated term of the geometric progression.

31 $a_1 = \frac{1}{32}, r = 2, a_9 = ?$

32 $a_1 = \frac{1}{81}, r = 3, a_9 = ?$

33 $a_1 = 7, r = 0.1, a_6 = ?$

34 $a_1 = 9, r = -0.1, a_7 = ?$

35 $a_1 = 64, a_2 = -32, a_8 = ?$

36 $a_1 = -4, a_2 = -8, a_9 = ?$

37 $a_{17} = 2, a_{19} = 50, a_{20} = ?$

38 $a_{25} = 18, a_{27} = 2, a_{24} = ?$

39 $a_n = \frac{3}{4}a_{n-1}, a_{11} = \frac{8}{27}, a_{13} = ?$

40 $a_n = 0.3a_{n-1}, a_{20} = 11, a_{22} = ?$

B

In Exercises 41–48 use the information given for the geometric progressions to find the quantities indicated.

41 $243, 81, \ldots, \frac{1}{3}; n = ?$

42 $1024, 512, \ldots, 1; n = ?$

43 $a_5 = 24, r = 2, a_1 = ?$

44 $a_6 = 64, r = 4, a_1 = ?$

45 $a_9 = 32, a_{11} = 288, r = ?$

46 $a_{45} = 17, a_{47} = 425, r = ?$

47 $a_n = 5a_{n-1}, a_1 = \frac{1}{3125}, a_9 = ?$

48 $a_n = 0.1a_{n-1}, a_1 = 7000, a_8 = ?$

C

CALCULATOR USAGE (49–56)

49 The following form is used on a TI-82 calculator to list the terms of a sequence:

seq(expression, variable, begin, end, increment)

Use this form to determine the geometric sequence indicated by the display window.

seq(4^N, N, 3, 8, 1)

In Exercises 50–52 use a graphics calculator to list the indicated terms of each geometric sequence.

50 Given $a_n = 96\left(\frac{1}{2}\right)^n$, list a_5 through a_{10}.

51 Given $a_n = 50{,}000(0.1)^n$, list a_3 through a_7.

52 Given $a_n = 1200(0.8)^n$, list a_4 through a_8.

In Exercises 53–56 use a calculator to determine the quantities indicated.

53 $a_1 = 7.2, r = 2.7, a_5 = ?$

54 $a_1 = 10.5, r = 3.4, a_5 = ?$

55 $a_1 = 44, r = 1.5, a_n = 501.1875, n = ?$

56 $a_1 = 100, r = 0.8, a_n = 26.2144, n = ?$

57 Radioactive Decay A nuclear chemist starts an experiment with 100 grams of a radioactive material. At the end of each time period only one-half of the amount present at the start of the period is left. How much material is present at the end of the fifth time period?

58 Bouncing Ball A ball dropped from a height of 36 meters rebounds to six-tenths of its previous height on each bounce (see the figure). How high does it bounce on its eighth bounce? (Give your answer to the nearest tenth of a meter.)

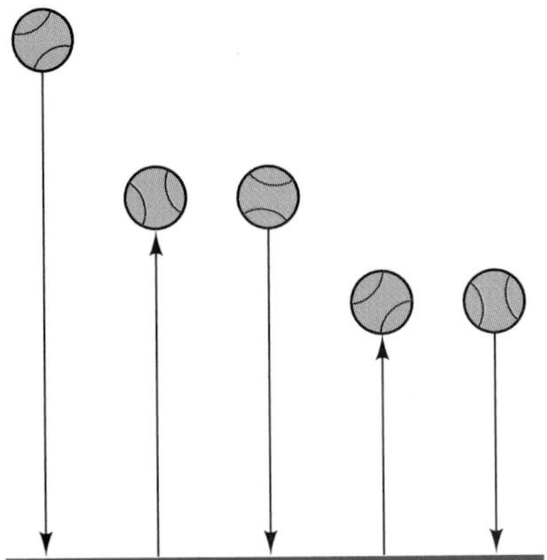

Figure for Exercise 58

59 Shoeing Horses A blacksmith attaches each horseshoe with eight nails. The blacksmith offers to charge by the nail for shoeing all four hooves. The cost for the first nail would be 1 cent, the second 2 cents, the third 4 cents, etc. At this rate, how much would the thirty-second nail cost?

60 Bacterial Growth If a culture of 10,000 bacteria increases by 5% each hour, how many bacteria will be in the culture at the end of the first day?

Discussion Question

61 Describe how to distinguish between an arithmetic progression and a geometric progression.

Challenge Question

An employer offers you a job starting at $1 per day, with a pay increase after each day. Option I is to give you an increase of $50 per day. Option II is to double your salary each day. When will Option II produce a total salary greater than that produced by Option I?

Section Skill Reinforcers

Graph a line satisfying these conditions.

1 Through $(-4, -1)$ with slope $\frac{1}{2}$ **2** Through $(-4, -1)$ with slope 0 **3** Through $(-4, -1)$ with slope undefined **4** Through x-intercept $(-4, 0)$ and y-intercept $(0, -1)$ **5** Through $(-4, -1)$ and the origin

11-6 Geometric Series

Section Objectives

8 Find the sum of a finite geometric progression.

9 Evaluate an infinite geometric series.

The **geometric series** denoted by $S_n = \sum_{i=1}^{n} a_i$ is the sum of the n terms $a_1 + a_2 + \cdots + a_{n-1} + a_n$ of a geometric sequence. For example,

$$S_5 = \sum_{i=1}^{5} 2^i = 2 + 4 + 8 + 16 + 32 = 62$$

Although S_n is used to denote the sum of both an arithmetic and a geometric progression, the context of each problem should make clear whether the series is arithmetic or geometric.

The simplest way to evaluate a geometric series with only a few terms may be to actually add all the terms. For series with more terms, however, it is useful to have a condensed formula for S_n. To develop this formula for S_n, let us consider the expanded forms for S_n and rS_n illustrated below.

$$\begin{array}{rlllllll} S_n = & a_1 + a_1r + a_1r^2 & & + \cdots + a_1r^{n-2} & + a_1r^{n-1} & \\ rS_n = & \quad\; a_1r + a_1r^2 & + a_1r^3 & + \cdots + a_1r^{n-2} & + a_1r^{n-1} & + a_1r^n \\ \hline S_n - rS_n = & a_1 + 0 \quad + 0 & & + \cdots + 0 & + 0 & - a_1r^n \end{array}$$

To obtain the second equation, multiply both sides of the first equation by r and shift terms to the right to align similar terms. Subtract the second equation from the first equation.

$$S_n(1 - r) = a_1(1 - r^n)$$

Factor both sides.

$$S_n = \frac{a_1(1 - r^n)}{1 - r} \quad \text{for } r \neq 1$$

Divide both members by $1 - r$. If $r \neq 1$, $1 - r \neq 0$.

$$S_n = \frac{a_1 - a_1r^n}{1 - r}$$

$$S_n = \frac{a_1 - ra_n}{1 - r}$$

Substitute a_n for a_1r^{n-1} to obtain an alternative form of this formula.

GEOMETRIC SERIES

For $r \neq 1$, the geometric series $S_n = \sum_{i=1}^{n} a_i$ is given by

$$S_n = \frac{a_1(1 - r^n)}{1 - r} \quad \text{or} \quad S_n = \frac{a_1 - ra_n}{1 - r}$$

EXAMPLE 1 Evaluating a Geometric Series

Find the sum of the geometric progression with $a_1 = 1$, $r = \frac{1}{2}$, and $n = 7$.

SOLUTION

$$S_n = \frac{a_1(1 - r^n)}{1 - r}$$

$$S_7 = \frac{1\left[1 - \left(\frac{1}{2}\right)^7\right]}{1 - \frac{1}{2}}$$

Substitute the given values into the formula for S_n, the sum of the terms of a geometric progression.

$$= \frac{1 - \frac{1}{128}}{\frac{1}{2}}$$

Simplify the numerator and the denominator.

$$= \left(\frac{127}{128}\right)\left(\frac{2}{1}\right)$$

Invert the denominator, and multiply.

$$S_7 = \frac{127}{64}, \text{ or } 1\frac{63}{64}$$

SELF-CHECK

Find the sum of the geometric progression with $a_1 = 81$, $r = -\frac{2}{3}$, and $n = 5$.

SELF-CHECK ANSWER

$S_5 = 55$

EXAMPLE 2 Evaluating a Geometric Series Using Summation Notation

Evaluate $S_{10} = \sum_{k=1}^{10} 128\left(\frac{1}{2}\right)^k$.

SOLUTION

$$S_n = \frac{a_1(1 - r^n)}{1 - r}$$

The terms of this series form a geometric progression with $a_1 = 64$, $r = \frac{1}{2}$, and $n = 10$.

$$S_{10} = \frac{64\left[1 - \left(\frac{1}{2}\right)^{10}\right]}{1 - \frac{1}{2}}$$

Substitute these values into the formula for S_n.

$$= \frac{64\left(1 - \frac{1}{1024}\right)}{\frac{1}{2}}$$

$$= 64\left(\frac{1023}{1024}\right)\left(\frac{2}{1}\right)$$

Simplify the numerator, invert the denominator, and multiply.

$$S_{10} = \frac{1023}{8}, \text{ or } 127\frac{7}{8}$$

The next example uses the capabilities of a TI-82 graphics calculator to evaluate a geometric series.

EXAMPLE 3 Calculator Usage

Use a graphics calculator with a LIST feature to evaluate $\sum_{n=1}^{8} [0.005(2^n)]$.

SOLUTION The format of the sum of a sequence option on a TI-82 calculator is

sum seq(expression, variable, begin, end, increment)

In our example

expression = $0.005(2^n)$
variable = n
begin = 1
end = 8
increment = 1

Enter the formula for the nth term. Start the summation at $n = 1$, and terminate with $n = 8$.

| 2nd | | LIST | | ► (MATH option) | | 5: sum |

| 2nd | | LIST | | 5: seq(| | 0 | | . | | 0 | | 0 | | 5 | | (|

| 2 | | ∧ | | ALPHA | | N | |) | | , | | ALPHA | | N | | , |

| 1 | | , | | 8 | | , | | 1 | |) | | ENTER |

Select the sum option from the LIST menu. Then enter the description of the sequence to be added.

```
sum seq(0.005(2^N
), N, 1, 8, 1)
                2.55
```

Answer $\sum_{n=1}^{8} [0.005(2^n)] = 2.55$

EXAMPLE 4 Determining the Common Ratio

Find the value of r in a geometric sequence if $a_1 = 4$, $a_n = 45.5625$, and $S_n = 128.6875$.

SOLUTION

$$S_n = \frac{a_1 - ra_n}{1 - r}$$

$$128.6875 = \frac{4 - r(45.5625)}{1 - r}$$ Substitute the given values into the alternative formula for S_n.

$$128.6875 - 128.6875r = 4 - 45.5625r$$ Multiply both members by $1 - r$.

$$-83.125r = -124.6875$$ Combine like terms.

$$r = 1.5$$ Divide both members by −83.125.

EXAMPLE 5 An Application of Geometric Series

If you could arrange to be paid \$1000 at the end of January, \$2000 at the end of February, \$4000 at the end of March, and so on, what is the total amount you would be paid for the year?

SOLUTION The payments at the end of each month form a geometric progression with $a_1 = 1000$, $r = 2$, and $n = 12$.

$$S_n = \frac{a_1(1 - r^n)}{1 - r}$$

$$S_{12} = \frac{1000(1 - 2^{12})}{1 - 2}$$ The total amount for the year is determined by adding the 12 terms of this geometric progression.

$$= \frac{1000(1 - 4096)}{-1} = \frac{-4095000}{-1}$$ Simplify, and calculate S_{12}.

$$S_{12} = 4{,}095{,}000$$

Answer The total amount for the year would be \$4,095,000. Although this answer may seem unreasonable, it is the doubling pay scheme that makes the total unreasonable—not the arithmetic.

Infinity, symbolized by ∞, is not a specific number; rather, it signifies that numbers continue without ever stopping. For example, the natural numbers 1, 2, 3, . . . tend to infinity. Likewise, the sum

$$S = \frac{1}{2} + \frac{1}{4} + \cdots + \frac{1}{2^n} + \cdots$$

represents a sum of terms that go on forever. Can such a sum be meaningful if the terms never end? Yes, in some cases these sums can be meaningful, although a completely new interpretation of the word *sum* is needed. Observe the pattern of the finite sums in the table below as n increases in the infinite geometric progression whose nth term is $a_n = \left(\frac{1}{2}\right)^n$.

n	1	2	3	4	5	6	7	8	9	10	. . .
a_n	$\frac{1}{2}$	$\frac{1}{4}$	$\frac{1}{8}$	$\frac{1}{16}$	$\frac{1}{32}$	$\frac{1}{64}$	$\frac{1}{128}$	$\frac{1}{256}$	$\frac{1}{512}$	$\frac{1}{1024}$	. . .
S_n	$\frac{1}{2}$	$\frac{3}{4}$	$\frac{7}{8}$	$\frac{15}{16}$	$\frac{31}{32}$	$\frac{63}{64}$	$\frac{127}{128}$	$\frac{255}{256}$	$\frac{511}{512}$	$\frac{1023}{1024}$	. . .

As n increases, a_n decreases and, in fact, tends toward 0. Thus each successive term contributes less and less to the sum S_n, which approaches 1. With each new term the difference between S_n and 1 continues to decrease. Although we could never finish summing these nonending terms, we can see that the limit approached by a_n is 0 and that the limit approached by S_n is 1. An **infinite sum** $S = \sum_{i=1}^{\infty} a_i$ is therefore interpreted to be the limit that S_n approaches as n increases.

EXAMPLE 6 Evaluating an Infinite Geometric Series

Prepare a table of a_n and S_n for the first seven terms of the sequence with $a_n = \left(\frac{1}{3}\right)^n$. Then determine the infinite sum by observing the limit of S_n.

SOLUTION

n	a_n	S_n
1	$\frac{1}{3}$	$\frac{1}{3} \approx 0.3333$
2	$\frac{1}{9}$	$\frac{4}{9} \approx 0.4444$
3	$\frac{1}{27}$	$\frac{13}{27} \approx 0.4815$
4	$\frac{1}{81}$	$\frac{40}{81} \approx 0.4938$
5	$\frac{1}{243}$	$\frac{121}{243} \approx 0.4979$
6	$\frac{1}{729}$	$\frac{364}{729} \approx 0.4993$
7	$\frac{1}{2187}$	$\frac{1093}{2187} \approx 0.4998$

The decimal value of S_n appears to be approaching 0.5.

Answer $S = 0.5$

Since an infinite sum S is defined to be the limit that S_n approaches, this sum is not meaningful if there is no limiting value. If $|r| \geq 1$, the terms do not approach 0 and S_n does not approach any limit; thus S is not meaningful.

On the other hand, if $|r| < 1$ in a geometric series, then each successive multiplication by r produces a term of lesser magnitude and a_n tends toward 0; thus S_n approaches a limit, the infinite sum S.

A general formula for the infinite sum can be obtained by examining the formula for S_n.

$$S_n = \frac{a_1(1 - r^n)}{1 - r}$$

If $|r| < 1$, then $|r|^n$ approaches 0. Thus $S_n = \frac{a_1(1 - r^n)}{1 - r}$ approaches $\frac{a_1(1 - 0)}{1 - r}$; that is, S_n approaches $\frac{a_1}{1 - r}$. The infinite sum S is the limiting value, so $S = \frac{a_1}{1 - r}$.

INFINITE GEOMETRIC SERIES

If $|r| < 1$, then the sum of an infinite geometric progression is

$$S = \frac{a_1}{1 - r}$$

If $|r| \geq 1$, this sum does not exist.

EXAMPLE 7 Evaluating an Infinite Geometric Series

Evaluate $S = \sum_{i=1}^{\infty} \left(\frac{1}{3}\right)^i$.

SOLUTION $S = \frac{a_1}{1 - r}$

$$= \frac{\frac{1}{3}}{1 - \frac{1}{3}}$$ Substitute $a_1 = \frac{1}{3}$ and $r = \frac{1}{3}$ into the formula for the sum of an infinite geometric progression.

$$= \frac{\frac{1}{3}}{\frac{2}{3}}$$ Simplify the denominator.

$= \frac{1}{3} \cdot \frac{3}{2}$ Invert the denominator, and multiply.

$S = \frac{1}{2}$ Note that this sum is the same one that we observed S_n to be approaching in Example 6.

EXAMPLE 8 Evaluating an Infinite Geometric Series

Evaluate $0.6 + 0.06 + 0.006 + \cdots$.

SOLUTION $S = \frac{a_1}{1 - r}$

$= \frac{0.6}{1 - 0.1}$ Substitute $a_1 = 0.6$ and $r = 0.1$ into the formula for the infinite sum S.

$= \frac{0.6}{0.9}$ Simplify, and express S in fractional form.

$= \frac{6}{9}$

$S = \frac{2}{3}$ $0.6 + 0.06 + 0.006 + \cdots = \frac{2}{3}$.

SELF-CHECK

Evaluate $S = \sum_{k=1}^{\infty} (0.1)^{k-1}$.

EXAMPLE 9 Writing a Repeating Decimal in Fractional Form

Write 0.272727 . . . as a fraction.

SOLUTION

$0.272727 \cdots = 0.27 + 0.0027 + 0.000027 + \cdots$ This series is an infinite geometric series with $a_1 = 0.27$ and $r = 0.01$.

$= 27(0.01) + 27(0.01)^2 + 27(0.01)^3 + \cdots$

$= \sum_{i=1}^{\infty} 27(0.01)^i$

$S = \frac{a_1}{1 - r}$

$= \frac{0.27}{1 - 0.01}$ Substitute into the formula for S.

$= \frac{0.27}{0.99}$ Simplify, and express S in fractional form.

$S = \frac{3}{11}$

Answer $0.272727\ldots = \frac{3}{11}$ You can check this answer by dividing 3 by 11.

SELF-CHECK ANSWER

$S = 1\frac{1}{9}$

EXAMPLE 10 An Application of Infinite Geometric Series

A golfball is dropped onto concrete from a height of 6 meters. Each time it bounces on the concrete, it rebounds to two-thirds the height from which it fell. Find the total distance that this bouncing ball travels.

SOLUTION Draw a sketch as shown in the figure, and let

d = Total distance in meters that the ball travels

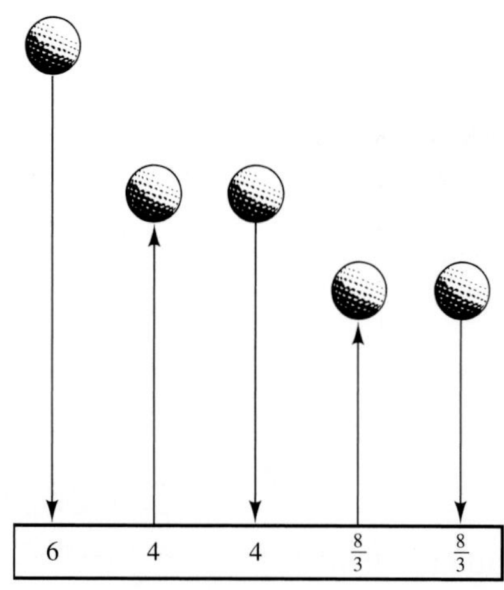

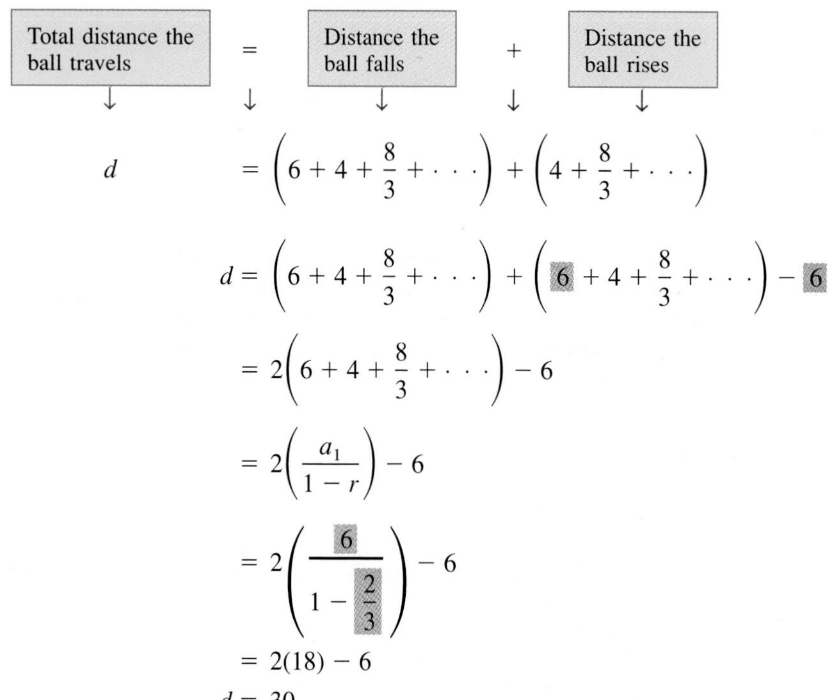

Total distance the ball travels	=	Distance the ball falls	+	Distance the ball rises
↓	↓	↓	↓	↓

Word equation

$$d = \left(6 + 4 + \frac{8}{3} + \cdots\right) + \left(4 + \frac{8}{3} + \cdots\right)$$

The distances that the ball falls and then rises form a geometric progression.

$$d = \left(6 + 4 + \frac{8}{3} + \cdots\right) + \left(6 + 4 + \frac{8}{3} + \cdots\right) - 6$$

Add, and then subtract 6.

$$= 2\left(6 + 4 + \frac{8}{3} + \cdots\right) - 6$$

$$= 2\left(\frac{a_1}{1 - r}\right) - 6$$

Evaluate the sum within parentheses using the formula $S = \frac{a_1}{1 - r}$, with $a_1 = 6$ and $r = \frac{2}{3}$.

$$= 2\left(\frac{6}{1 - \frac{2}{3}}\right) - 6$$

$$= 2(18) - 6$$

$$d = 30$$

Answer The golfball travels 30 m.

Exercises 11-6

A

In Exercises 1–14 use the information given to evaluate each geometric series.

1 $a_1 = 3,\ r = 2,\ S_7 = ?$

2 $a_1 = 2,\ r = 3,\ S_6 = ?$

3 $a_1 = 0.2,\ r = 0.1,\ S_5 = ?$

4 $a_1 = 0.5,\ r = 0.1,\ S_6 = ?$

5 $a_1 = 48,\ r = -\dfrac{1}{2},\ S_8 = ?$

6 $a_1 = 729,\ r = -\dfrac{1}{3},\ S_7 = ?$

7 $a_1 = 1,\ a_n = 3.71293,\ r = 1.3,\ S_n = ?$

8 $a_1 = 3,\ a_n = 4.83153,\ r = 1.1,\ S_n = ?$

9 $64 + 32 + \cdots + \dfrac{1}{8}$

10 $16 + 12 + \cdots + 3.796875$

11 $\displaystyle\sum_{i=1}^{7} 8(0.1)^i$

12 $\displaystyle\sum_{k=1}^{6} 7(0.1)^k$

13 $a_1 = 40,\ a_n = \dfrac{1}{5}a_{n-1}$ for $n > 1,\ S_6 = ?$

14 $a_1 = 81,\ a_n = \dfrac{1}{3}a_{n-1}$ for $n > 1,\ S_6 = ?$

In Exercises 15–28 use the information given to evaluate the infinite geometric series.

15 $a_1 = 5,\ r = \dfrac{2}{3}$

16 $a_1 = 16,\ r = \dfrac{1}{5}$

17 $a_1 = 14,\ r = -\dfrac{3}{4}$

18 $a_1 = 7,\ r = -\dfrac{2}{5}$

19 $a_1 = 4,\ r = \dfrac{3}{5}$

20 $a_1 = 6,\ r = \dfrac{4}{7}$

21 $a_1 = 0.12,\ r = \dfrac{1}{100}$

22 $a_1 = 0.9,\ r = \dfrac{1}{10}$

23 $\displaystyle\sum_{k=1}^{\infty} \left(\frac{4}{9}\right)^k$

24 $\displaystyle\sum_{j=1}^{\infty} \left(\frac{3}{7}\right)^j$

25 $6 - 4 + \dfrac{8}{3} - \dfrac{16}{9} + \cdots$

26 $16 - 12 + 9 - 6.75 + \cdots$

27 $a_1 = 24,\ a_n = \dfrac{3}{8}a_{n-1}$ for $n > 1$

28 $a_1 = -36,\ a_n = \left(-\dfrac{5}{9}\right)a_{n-1}$ for $n > 1$

In Exercises 29–36 write each repeating decimal as a fraction.

29 $0.444\ldots$

30 $0.555\ldots$

31 $0.212121\ldots$

32 $0.656565\ldots$

33 $0.\overline{405}$

34 $0.0\overline{75}$

35 $7.\overline{9}$

36 $8.\overline{3}$

B

In Exercises 37–46 use the information given for the geometric progressions to find the quantities indicated.

37 $r = 2,\ S_{10} = 6138,\ a_1 = ?$

38 $r = 3,\ S_8 = 45920,\ a_1 = ?$

39 $a_1 = 12,\ r = 5,\ S_n = 46872,\ n = ?$

40 $a_1 = 12,\ r = 7,\ S_n = 33612,\ n = ?$

41 $a_1 = 1000,\ a_n = 0.0128,\ S_n = 1249.9968,\ r = ?$

42 $a_1 = 6,\ a_n = 24576,\ S_n = 32766,\ r = ?$

43 $\displaystyle\sum_{i=1}^{\infty} a_i = 27,\ a_1 = 12,\ r = ?$

44 $\displaystyle\sum_{j=1}^{\infty} a_j = 22,\ a_1 = 14,\ r = ?$

45 $\displaystyle\sum_{k=1}^{\infty} a_k = 21,\ r = \frac{2}{9},\ a_1 = ?$

46 $\displaystyle\sum_{i=1}^{\infty} a_i = 2,\ r = \frac{6}{7},\ a_1 = ?$

The trichotomy property of real numbers states that exactly one of the three statements in parts a–c of Exercises 47–54 must be true. Determine which statement is true.

47 a. $0.333 < \dfrac{1}{3}$ **b.** $0.333 = \dfrac{1}{3}$ **c.** $0.333 > \dfrac{1}{3}$

48 a. $0.3334 < \dfrac{1}{3}$ **b.** $0.3334 = \dfrac{1}{3}$ **c.** $0.3334 > \dfrac{1}{3}$

49 a. $0.\overline{3} < 0.333$ **b.** $0.\overline{3} = 0.333$ **c.** $0.\overline{3} > 0.333$

50 a. $0.\overline{3} < 0.3334$ **b.** $0.\overline{3} = 0.3334$ **c.** $0.\overline{3} > 0.3334$

51 a. $0.\overline{3} < \dfrac{1}{3}$ **b.** $0.\overline{3} = \dfrac{1}{3}$ **c.** $0.\overline{3} > \dfrac{1}{3}$

52 a. $0.9999 < 1$ **b.** $0.9999 = 1$ **c.** $0.9999 > 1$

53 a. $0.\overline{9} < 1$ **b.** $0.\overline{9} = 1$ **c.** $0.\overline{9} > 1$

54 a. $0.\overline{9} < 0.9999$ **b.** $0.\overline{9} = 0.9999$ **c.** $0.\overline{9} > 0.9999$

C

CALCULATOR USAGE (55–58)

55 The following form is used on a TI-82 calculator to sum the terms of a sequence:

sum seq(expression, variable, begin, end, increment)

Use this form to determine the sum of the geometric sequence indicated by the display window.

```
sum seq(2.5^N, N, 1, 6, 1)
```

In Exercises 56–58 use a graphics calculator to evaluate each geometric series.

56 $\sum_{n=1}^{5} (6^n)$ **57** $\sum_{n=1}^{9} (4^n)$ **58** $\sum_{n=1}^{7} 10{,}000(0.4^n)$

59 Chain Letter A chain-letter scam requires that each participant persuade four other people to participate. If one person starts this venture as a first-generation participant, determine how many people will have been involved by the time the eighth generation has signed on but not yet contacted anyone.

60 Bouncing Ball A ball dropped from a height of 36 meters rebounds to six-tenths its previous height on each bounce. How far has it traveled when it reaches the apex of its eighth bounce? (Give your answer to the nearest tenth of a meter.)

61 Area Under a Curve Approximate the shaded area in the figure by adding the areas of the four rectangles indicated.

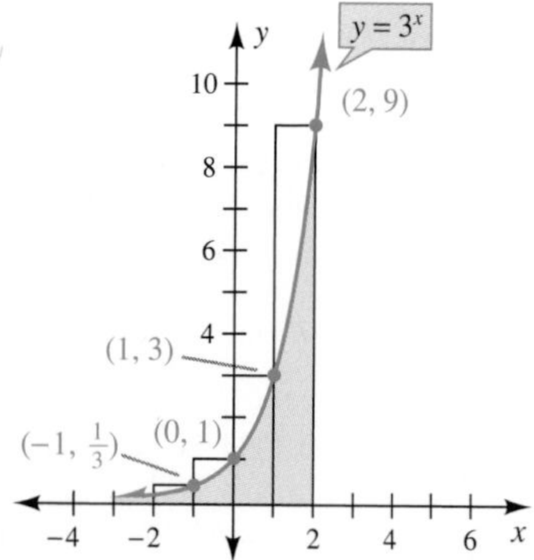

Figure for Exercise 61

62 Golfing Prizes Seven prizes with a total value of \$19,843.75 will be awarded in a golf tournament. The value of each prize after the first is equal to one-half the value of the previous prize. What is the value of the first prize?

63 Vacuum Pump With each cycle a vacuum pump removes one-third of the air in a glass vessel. What percent of the air has been removed after eight cycles?

Bouncing Ball (64–65)

64 A ball dropped from 5 meters rebounds to four-fifths its previous height on each bounce. How far does it travel?

65 A ball dropped from 3 meters rebounds to nine-tenths its previous height on each bounce. How far does it travel?

66 Arc of a Swing A child's swing moves through a 3-meter arc. On each swing it travels only two-thirds the distance it traveled on the previous arc. How far does the swing travel?

Multiplier Effect (67–68)

67 City planners estimate that a new manufacturing plant located in their area will contribute \$600,000 in salaries to the local economy. They estimate that those who earn the salaries will spend three-fourths of this money within the community. The merchants, service providers, and others who receive this money from the salary earners will in turn spend three-fourths of it in the community, and so on. Taking into account the multiplier effect, find the total amount of spending within the local economy that will be generated by this \$600,000.

68 Rework Exercise 67, assuming that two-thirds is the factor for the multiplier effect rather than three-fourths.

DISCUSSION QUESTIONS (69 AND 70)

69 Write your own word problem that is algebraically modeled by an infinite geometric series with $a_1 = 150{,}000$ and $r = 0.6$.

70 Describe in your own words how we can add an infinite number of terms in a finite amount of time.

CHALLENGE QUESTION

Construct you own challenge problem or discussion problem. If you would like to submit this problem for possible inclusion in the next edition of *Intermediate Algebra*, with credit listed to you, please send your typed copy to the author at Parkland College, 2400 West Bradley, Champaign, Illinois, 61821-1899.

SECTION SKILL REINFORCERS

Determine the vertex of each of these parabolas.

1 $y = x^2 - 3$ **2** $y = (x - 3)^2$ **3** $y = (x - 6)^2 + 2$ **4** $y = x^2 - 6x + 10$ **5** $y = 3x^2 + 1$

Key Concepts for Chapter 11

1 Determinants:

a. For second-order determinants,

$$\begin{vmatrix} a & b \\ c & d \end{vmatrix} = ad - bc$$

b. For third-order determinants,

$$\begin{vmatrix} a_1 & b_1 & c_1 \\ a_2 & b_2 & c_2 \\ a_3 & b_3 & c_3 \end{vmatrix} = a_1\begin{vmatrix} b_2 & c_2 \\ b_3 & c_3 \end{vmatrix} - a_2\begin{vmatrix} b_1 & c_1 \\ b_3 & c_3 \end{vmatrix} + a_3\begin{vmatrix} b_1 & c_1 \\ b_2 & c_2 \end{vmatrix}$$

2 Cramer's Rule for 2×2 linear systems:

a. If $D \neq 0$, the solution of $\begin{Bmatrix} a_1x + b_1y = c_1 \\ a_2x + b_2y = c_2 \end{Bmatrix}$ is

$x = \dfrac{D_x}{D}$ and $y = \dfrac{D_y}{D}$, where

$$D = \begin{vmatrix} a_1 & b_1 \\ a_2 & b_2 \end{vmatrix}, \qquad D_x = \begin{vmatrix} c_1 & b_1 \\ c_2 & b_2 \end{vmatrix},$$

and

$$D_y = \begin{vmatrix} a_1 & c_1 \\ a_2 & c_2 \end{vmatrix}$$

b. If $D = D_x = D_y = 0$, the system contains dependent equations.

c. If $D = 0$, but D_x and D_y are not both zero, the system is inconsistent.

3 Cramer's Rule for 3×3 linear systems:

a. If $D \neq 0$, the solution of $\begin{Bmatrix} a_1x + b_1y + c_1z = d_1 \\ a_2x + b_2y + c_2z = d_2 \\ a_3x + b_3y + c_3z = d_3 \end{Bmatrix}$ is

$x = \dfrac{D_x}{D}$, $y = \dfrac{D_y}{D}$, and $z = \dfrac{D_z}{D}$, where

$$D = \begin{vmatrix} a_1 & b_1 & c_1 \\ a_2 & b_2 & c_2 \\ a_3 & b_3 & c_3 \end{vmatrix}, \qquad D_x = \begin{vmatrix} d_1 & b_1 & c_1 \\ d_2 & b_2 & c_2 \\ d_3 & b_3 & c_3 \end{vmatrix},$$

$$D_y = \begin{vmatrix} a_1 & d_1 & c_1 \\ a_2 & d_2 & c_2 \\ a_3 & d_3 & c_3 \end{vmatrix}, \quad \text{and} \quad D_z = \begin{vmatrix} a_1 & b_1 & d_1 \\ a_2 & b_2 & d_2 \\ a_3 & b_3 & d_3 \end{vmatrix}$$

b. If $D = 0$, the system either is inconsistent or contains dependent equations.

4 Methods of solving systems of linear equations:

a. Graphical method **b.** Substitution method

c. Addition-subtraction method **d.** Cramer's Rule

5 Expanding $(a + b)^n$:

a. Write the exponents on all $(n + 1)$ terms. Start with a^n, decreasing the exponents on a by 1 and increasing the exponents on b by 1 until the last term is b^n.

b. Write the coefficients of each term. The first coefficient is 1, the second is n, and the other coefficients are calculated from the preceding term by multiplying its coefficient by the exponent on a and dividing this product by 1 more than the exponent on b.

6 Sequences:

a. A sequence is a function whose domain is a set of consecutive natural numbers.

b. A finite sequence has a last term.

c. The terms of an infinite sequence continue without end.

d. A sequence is defined recursively when the general term is defined in terms of the preceding terms.

e. An arithmetic progression is a sequence in which consecutive terms differ by a constant.

f. A geometric progression is a sequence in which consecutive terms form a constant ratio.

7 Series:
A series is the sum of the terms of a sequence.

8 Summation notation: $\sum_{i=1}^{n} a_i = a_1 + a_2 + \cdots + a_{n-1} + a_n$

9 Arithmetic progressions:

a. Common difference: $d = a_n - a_{n-1}$, $n > 1$

b. nth term: $a_n = a_1 + (n - 1)d$

c. Series of n terms: $S_n = \dfrac{n}{2}(a_1 + a_n)$, or

$$S_n = \frac{n}{2}[2a_1 + (n - 1)d]$$

10 Geometric progressions:

a. Common ratio: $r = \dfrac{a_n}{a_{n-1}}$, $n > 1$

b. nth term: $a_n = a_1r^{n-1}$

c. Series of n terms: $S_n = \dfrac{a_1(1 - r^n)}{1 - r}$, or $S_n = \dfrac{a_1 - ra_n}{1 - r}$

11 Infinite geometric series:

a. If $|r| < 1$, then the sum of an infinite geometric progression is

$$S = \frac{a_1}{1 - r}$$

b. If $|r| \geq 1$, this sum does not exist.

Review Exercises for Chapter 11

In Exercises 1 and 2 evaluate each determinant.

1 $\begin{vmatrix} -5 & 8 \\ 13 & 4 \end{vmatrix}$

2 $\begin{vmatrix} 1 & 2 & 3 \\ 0 & 1 & 7 \\ 2 & 5 & 5 \end{vmatrix}$

In Exercises 3–5 solve each system of equations using Cramer's Rule.

3 $\dfrac{x}{6} + \dfrac{y}{10} = 1$
$\dfrac{x}{3} - \dfrac{3y}{5} = -2$

4 $0.2x - 0.3y = -0.27$
$0.7x + 0.8y = 0.35$

5 $x + y + z = -3$
$2x + 3y - 5z = 22$
$3x - 5y + 2z = -5$

6 Find constants a, b, and c such that $(-2, 1, 7)$ is a solution of the system of equations

$$ax + by + cz = 35$$
$$ax + by - cz = -21$$
$$2ax - by + 3cz = 89$$

In Exercises 7–9 write out the binomial expansion of each expression.

7 $(x + y)^5$ **8** $(5x - 2y)^4$ **9** $(x - 3y)^5$

In Exercises 10–24 write the first five terms of the sequence described.

10 $a_n = 5n$ **11** $a_n = 5^n$

12 $a_n = 5 + n$ **13** $a_n = 5$

14 $a_1 = 5$, $a_n = a_{n-1} + 3$ for $n > 1$

15 $a_1 = 7$, $a_n = 3a_{n-1}$ for $n > 1$

16 $a_n = (-1)^n \dfrac{n-1}{n+1}$

17 $a_n = n^2 - 5n + 8$

18 $a_1 = 1$, $a_2 = 3$, and $a_n = a_{n-1} + a_{n-2}$ for $n > 2$

19 An arithmetic progression with $a_1 = 7$ and $d = -3$

20 An arithmetic progression with $a_1 = -9$ and $a_3 = -1$

21 A geometric progression with $a_1 = 10$ and $r = -2$

22 A geometric progression with $a_1 = 486$ and $r = \dfrac{1}{3}$

23 The sequence of digits of the number π

24 The sequence of prime numbers arranged in order of magnitude

In Exercises 25–32 determine whether each of the sequences is arithmetic, geometric, both, or neither. If the sequence is arithmetic, write the common difference *d*. If the sequence is geometric, write the common ratio *r*.

25 2, 4, 6, 8, 10, . . . **26** 2, 4, 8, 16, 32, . . .

27 2, 4, 6, 10, 16, 26, . . . **28** 3, 6, 11, 18, 27, . . .

29 5, 5, 5, 5, 5, . . . **30** 7, 3, −1, −5, −9, . . .

31 5, −5, 5, −5, 5, . . . **32** 1, 1, 1, 1, 1, . . .

In Exercises 33–46 evaluate each series described.

33 $2 + 4 + 6 + \cdots + 202$

34 $2 + 4 + 8 + \cdots + 128$

35 $a_1 + a_2 + \cdots + a_6$, with $a_n = \left(\dfrac{2}{3}\right)^n$

36 $a_1 + a_2 + \cdots + a_{73}$, with $a_n = 9n - 5$

37 $\displaystyle\sum_{k=1}^{41} (3k - 2)$ **38** $\displaystyle\sum_{j=1}^{8} \left(\frac{2}{3}\right)^j$

39 $\displaystyle\sum_{i=9}^{12} (i^2 - 5i + 2)$ **40** $\displaystyle\sum_{k=5}^{9} (i^2 - 7)$

41 $r = 1.01$, $a_1 = 900$, $S_5 = ?$ (geometric)

42 $d = 1.01$, $a_1 = 9$, $S_{50} = ?$ (arithmetic)

43 $a_1 = 6$, $a_n = a_{n-1} + 5$ for $n > 1$, $S_{17} = ?$

44 $a_1 = 6$, $a_n = 5a_{n-1}$ for $n > 1$, $S_7 = ?$

45 $0.7 + 0.07 + 0.007 + \cdots$
(Write your answer as a fraction.)

46 $0.12 + 0.0012 + 0.000012 + \cdots$
(Write your answer as a fraction.)

In Exercises 47–52 determine the number of terms in each of the sequences.

47 $a_1 = 147$, $d = -6$, $a_n = -153$, $n = ?$ (arithmetic)

48 $a_1 = 18$, $d = \dfrac{2}{3}$, $a_n = 58$, $n = ?$ (arithmetic)

49 $a_1 = 700{,}000$, $r = 0.1$, $a_n = 0.07$, $n = ?$ (geometric)

50 $a_1 = 3200$, $r = \dfrac{1}{5}$, $a_n = 1.024$, $n = ?$ (geometric)

51 $a_1 = 8$, $a_n = 50$, $S_n = 580$, $n = ?$ (arithmetic)

52 $a_1 = 6$, $r = 0.4$, $S_n = 9.744$, $n = ?$ (geometric)

In Exercises 53–60 find the value requested in each sequence.

53 $a_1 = 6$, $r = -1$, $a_{700} = ?$ (geometric)

54 $a_1 = 6, d = -1, a_{700} = ?$ (arithmetic)

55 $a_{11} = 34.6, d = 1.73, a_1 = ?$ (arithmetic)

56 $a_8 = 64, r = \frac{4}{5}, a_1 = ?$ (geometric)

57 $a_1 = 0.5, a_{13} = 8.5, d = ?$ (arithmetic)

58 $a_1 = 11, S_{24} = 126, d = ?$ (arithmetic)

59 $a_1 = 12, a_5 = \frac{243}{4}, r = ?$ (geometric)

60 $a_1 = 200, S_{50} = 0, r = ?$ (geometric)

In Exercises 61–64 write each repeating decimal in fractional form.

61 $8.\overline{3}$ **62** $0.\overline{06}$ **63** $6.0\overline{6}$ **64** 8.999. . .

65 Rolls of Insulation Rolls of insulation are stacked so that each layer after the first has 10 fewer rolls than the previous layer. How many layers will a lumber yard need to use in order to stack 500 rolls if 120 rolls are placed on the bottom layer? (See the figure.)

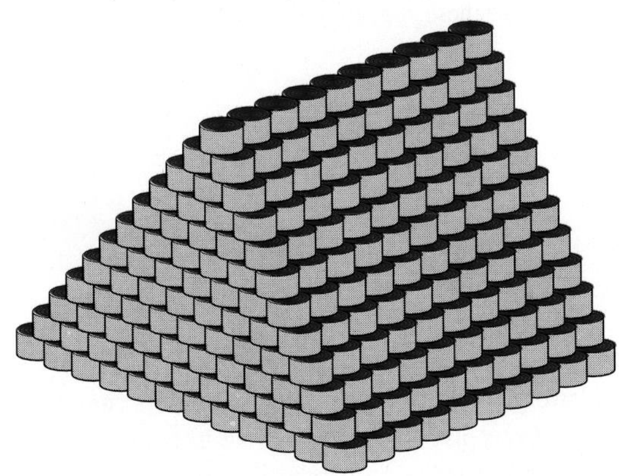

Figure for Exercise 65

66 Multiplier Effect City planners estimate that a new manufacturing plant located in their area will contribute $4,000,000 in salaries to the local economy. They estimate that those who earn the salaries will spend three-fifths of this money within the community. Those who receive the money will in turn spend another three-fifths, and so on. Taking into account the multiplier effect, find the total amount of spending within the local economy that will be generated by this $4,000,000.

CALCULATOR USAGE (67–69)

67 The following form is used on a TI-82 calculator to sum the terms of a sequence:

sum seq(expression, variable, begin, end, increment)

Use this form to determine the sum of the geometric sequence indicated by the display window.

```
sum seq(1.2^N, N, 1, 6, 1)
```

68 Use a graphics calculator to list the terms a_{32} through a_{40} for the arithmetic series whose general term is $a_n = 17.5n - 47$.

69 Use a graphics calculator to evaluate $\begin{vmatrix} 4.2 & 3.9 & 8.3 \\ 5.1 & 9.2 & 7.4 \\ 2.2 & 5.8 & 3.5 \end{vmatrix}$.

Mastery Test for Chapter 11

Exercise numbers correspond to Section Objective numbers.

[11-1] 1 Solve each system of linear equations using Cramer's Rule.

a. $908x - 743y = -1073.5$
$817x - 578y = -558.5$

b. $23x - 17y + 19z = 25.6$
$14x + 15y + 16z = 84.4$
$31x + 13y - 18z = 42.2$

[11-2] 2 Expand each of these binomial expressions.

a. $(x - y)^4$ **b.** $(2x + y)^4$

c. $(x + w)^7$ **d.** $(2x - 5y)^5$

[11-3] 3 Write the first five terms of the sequences defined below.

a. $a_n = 7n + 2$

b. $a_1 = 1, a_n = 3a_{n-1}$ for $n > 1$

c. $a_n = n^2 - 5n + 1$ **d.** $a_n = \frac{3n - 1}{4n + 1}$

[11-3] 4 Evaluate the series given below.

a. $10 + 13 + 16 + 19 + 22$

b. $a_1 + a_2 + \cdots + a_7$, where $a_n = 2n$

c. $\sum_{k=1}^{5} (5k + 3)$ **d.** $\sum_{j=3}^{7} (j^2 - 5j + 2)$

[11-4] 5 Find the indicated quantities from the information given about the arithmetic progressions.

a. $a_1 = 8,\ d = 11,\ a_{47} = ?$

b. $a_{80} = 193,\ d = 3,\ a_1 = ?$

c. $a_1 = 23,\ a_{17} = -9,\ d = ?$

d. $-14, -11, \ldots, 79;\ n = ?$

[11-4] 6 Find the sums of the arithmetic progressions described below.

a. $a_1 = 9,\ a_{73} = 45,\ S_{73} = ?$

b. $a_{79} = 22,\ d = -2,\ S_{79} = ?$

c. $a_1 = -43,\ d = 7,\ S_{60} = ?$

d. $a_1 = 1,\ a_n = 4n - 3,\ S_{111} = ?$

[11-5] 7 Find the indicated quantities from the information given about the geometric progressions.

a. $a_1 = \frac{1}{64},\ r = 4,\ a_7 = ?$

b. $a_6 = 81,\ r = -3,\ a_1 = ?$

c. $a_7 = 5,\ a_9 = 125,\ r = ?$

d. $-\frac{1}{5}, 1, \ldots, -3125;\ n = ?$

[11-6] 8 Find the sums of the geometric progressions described below.

a. $a_1 = 18,\ r = \frac{1}{2},\ S_7 = ?$

b. $a_1 = 11,\ r = -1,\ S_{500} = ?$

c. $a_n = \left(\frac{1}{2}\right)^n,\ S_6 = ?$ **d.** $a_n = \left(\frac{3}{4}\right)^n,\ S_5 = ?$

[11-6] 9 Find the sums of the following infinite geometric progressions.

a. $3 + \frac{3}{2} + \frac{3}{4} + \cdots$ **b.** $a_1 = \frac{2}{7},\ r = \frac{2}{7}$

Write these repeating decimals as fractions.

c. $0.\overline{30}$ **d.** $1.\overline{06}$

ACTIVE APPLICATION

As an example of the significance in applied math of the idea of the sum of a sequence, consider the action of a simple spring. By stretching a spring and measuring the tension, a physicist compiles the following data:

Stretch (cm)	Tension (dynes)
1	.75
2	1.5
3	2.25
4	3
5	3.75

1 If the length of the stretch is n centimeters and the successive measurements of tension are members of a series, what is the general term for the sequence?

2 If we plot the data in bar graph fashion, one rectangle can be thought of as representing a *quantity of work*—the work required to pull against a tension (given by the rectangle's height) through a distance (given by the rectangle's width).

The quantity of work is proportional to the rectangle's area. For example, a rectangle that is 6 dynes "high" and 1 centimeter "wide" (in units of stretch) represents 6 dynes · 1 centimeter = 6 dyne-centimeters worth of work. Using what you have learned about series, calculate the total quantity of work represented by the bar graph below. Why is it important that the rectangles have no space between them?

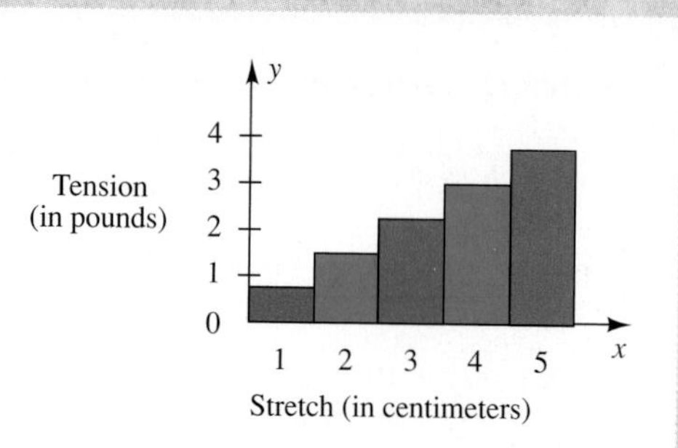

Cumulative Review of Chapters 10 and 11

The limited purpose of this review is to help you gauge your mastery of Chapters 10 and 11. It is not meant to examine each detail from these chapters, nor is it meant to focus on specific sections that may be emphasized at any one particular school.

In Exercises 1–12 solve each equation for x.

1 $5^x = 125$ **2** $5^x = \sqrt[3]{x}$ **3** $2^x = \frac{1}{8}$

4 $16^{3x} = 2$ **5** $\log_8 64 = x$ **6** $\log_4 64 = x$

7 $\log_x 64 = 1$ **8** $\log_{64} x = 0$ **9** $\log_6 x = \frac{1}{2}$

10 $\log_8 x = \frac{2}{3}$ **11** $\log_9 \frac{1}{81} = x$ **12** $\log_b b^7 = x - 2$

In Exercises 13–16 use a calculator to approximate to four significant digits the value of each expression.

13 $\log 137$ **14** $\ln 137$

15 $\log_{29} 137$ **16** $\log 0.000\,000\,000\,007$

In Exercises 17 and 18 use the properties of logarithms to write each of these expressions as a single logarithmic expression.

17 $\log 2 + \log 7 - \log x$ **18** $2 \log x + \frac{1}{3} \log y$

19 Express $\ln \frac{xy^3}{\sqrt{z}}$ in terms of logarithms of simpler expressions.

In Exercises 20–24 solve each equation, and approximate each solution to five significant digits.

20 $6^{3v-2} = 47$ **21** $19e^{4x+1} = 78$

22 $\ln(3x + 2) = 4.23$ **23** $\log(x^2 - 3x) = \log 4$

24 $\ln(x - 5) + \ln(x + 2) = \ln 8$

25 Determine the inverse of $f(x) = \frac{3x}{2} - 5$.

26 Graph $y = \sqrt{5x}$ and its inverse on the same coordinate system.

In Exercises 27 and 28 graph each function.

27 $y = 2^x - 3$ **28** $y = \log_2(x - 4)$

29 Compound Interest If an investment on which interest is compounded continuously doubles in value in 6 years, what is the annual rate of interest?

30 Evaluate $\begin{vmatrix} 4 & 3 & 8 \\ 5 & 9 & 7 \\ 2 & 5 & 3 \end{vmatrix}$.

31 Use Cramer's Rule to solve $\begin{Bmatrix} 2.5x + 5.2y = 31 \\ 4.5x + 6.4y = 41 \end{Bmatrix}$.

32 Expand $(x + 2y)^5$. **33** Expand $(2x - 5)^4$.

In Exercises 34–36 write the first five terms of each sequence.

34 $a_n = 4n + 1$ **35** $a_n = \left(\frac{2}{3}\right)^{n-1}$

36 $a_1 = 72,\ a_n = \frac{a_{n-1}}{n}$

In Exercises 37–40 evaluate each series.

37 $\sum_{j=1}^{4} (2j + 5)$ **38** $\sum_{k=5}^{9} (k^2 - 4k)$

39 $2 + 5 + 8 + 11 + \cdots + 299$

40 $64 + 32 + 16 + \cdots + \frac{1}{4}$

41 Write $0.\overline{34}$ as a fraction in reduced form.

42 In an arithmetic progression, $S_n = 4185$, $a_1 = 5$, and $a_n = 181$. Determine n.

43 In a geometric progression, $a_{11} = 192$ and $a_{13} = 12$. Determine a_{12}.

44 Find the sum of a geometric series with $a_1 = 5$, $r = 1$, and $n = 100$.

45 Evaluate $\sum_{n=1}^{\infty} (0.45^n)$.

Calculator Usage (46–50)

46 The folowing form is used on a TI-82 calculator to sum the terms of a sequence:

sum seq(expression, variable, begin, end, increment)

Use this form to determine the sum of the geometric sequence indicated by the display window.

```
sum seq(1.5^N, N, 1, 5, 1)
```

47 Use a graphics calculator to list the terms a_{43} through a_{50} for the arithmetic sequence whose general term is $a_n = 21.5n - 54$.

48 Use a graphics calculator to evaluate $\sum_{n=1}^{30} (2.5n - 6.5)$.

49 Use a graphics calculator to evaluate $\begin{vmatrix} 5.3 & 4.0 & 9.4 \\ 4.9 & 8.1 & 6.3 \\ 2.4 & 6.0 & 3.7 \end{vmatrix}$.

50 Use a graphics calculator to solve $\ln(3x - 5) = 4 - x$.

Appendix

CALCULATORS

The Role of Calculators

The author fully supports the position of the American Mathematical Association of Two-Year Colleges, the Mathematical Association of America, and the National Council of Mathematics Teachers that the use of calculators and computers should be an integral part of the education process, rather than an artificial adjunct to it. Students who integrate calculators into their coursework will better understand the concepts and be better prepared for their careers. Graphics calculators facilitate teaching through use of the Rule of Three: the examination of functions numerically, graphically, and symbolically.

Types of Calculators

The two types of calculators that are appropriate for this course are scientific calculators and graphics calculators. Keystrokes for both types are illustrated in this textbook. A four-function calculator without a square or square-root key is too limited to perform all the computations necessary for this book.

Scientific Calculators: A scientific calculator will have all the main features needed for this course except for the graphing features. It will have keys for squares, square roots, reciprocals, powers, and logarithms.

Graphics Calculators: A graphics calculator can produce graphs at the press of a key. It will also have all the main features of a scientific calculator. Most graphics calculators can also perform some statistical computations and matrix operations.

In the opinion of the author, the added expense of a graphics calculator is justified for students who will be taking college algebra and, especially, for students who will be taking subsequent mathematics courses. Three popular models of graphics calculators at the time of publication of this text are the TI-81, TI-82, and TI-85, as well as competing brands. The TI-82 is illustrated in this text because it demonstrates the capabilities of this whole family of calculators. It is easy to operate, and it is relatively inexpensive. However, the brand and model of calculator that you choose should depend primarily on how well the calculator will satisfy the needs that you will have. Although cost is certainly a factor, the prices of these calculators are not dramatically different, and the most important factor in your decision should be your intended uses of the calculator over the next few years. Following are some suggestions for choosing a calculator to suit your purposes:

The TI-81 is an excellent calculator for algebra and trigonometry classes. It is easy to operate and relatively inexpensive, but it does not have some of the features that you may want for more advanced courses. (See Figure A-1.)

The TI-82 is almost as easy to operate as the TI-81, but it has several improved features that make it an excellent choice for students who will be taking a statistics course. It also has an I/O port for sharing data with other calculators or a desktop computer. The graphics capabilities can be beneficial in a business calculus course. (See Figure A-2.)

The TI-85 is significantly different from the TI-81 and TI-82. It is considerably more powerful, and thus it is somewhat more difficult to learn to use. This calculator is a good choice for students who plan to take a calculus sequence. (See Figure A-3.)

Figure A-1 *The TI-81*

Figure A-2 *The TI-82*

Figure A-3 *The TI-85*

Advantages of Graphics Calculators

1 The ability to easily graph a function is such a powerful one that it is hard to overstate this advantage. Many algebraic questions can be answered directly by examining the graph of a function.

2 Graphics calculators display the numbers and operations entered, as well as the result of these keystrokes. This means you can visually proof your keystrokes—something you cannot do with an ordinary scientific calculator.

3 Keystroke errors, even those that prevent execution on the calculator, can often be corrected by using the editing feature, which includes an insert key [INS] and a delete key [DEL].

4 Graphics calculators have other advantages besides graphing capability. They can also perform matrix computations, carry out statistical computations, and solve systems of equations.

General Suggestions on Calculator Usage

Some general suggestions on calculator usage follow. For specific instructions consult the manual with your calculator. If you run into problems, ask your instructor for help.

1 Check the operating system of whatever model you have to be sure that it follows **AOS** (Algebraic Operating System) or **EOS**™ (Equation Operating System). For example, $2 + 3 \cdot 4$ should be evaluated as $2 + 12 = 14$. Some four-function calculators will not calculate the correct value for this expression. Calculators using **RPN** (Reverse Polish Notation) are quite powerful, but harder for a beginner to learn to use. Thus RPN calculators are not covered in this book.

2 Go through a problem with simple values whose result is known before you undertake the calculation of similar problems whose results are unknown.

3 For calculations use as many digits of accuracy as your calculator will allow. Since most calculators store more digits than they display, you can increase accuracy by observing the following guidelines:

a. Leave intermediate values in the calculator rather than copying down the display digits and then reentering these values. Learn to use the memory and parentheses keys.

b. To enter fractional values that result in repeating decimals, use the [$1/x$] key or divide the numerator by the denominator. (With calculators that work directly with fractions, this is not an issue.)

c. Enter π and e using the special keys if they are provided on your calculator. Such entries will typically be accurate to two or three more digits than are shown on the display.

Note: EOS is a trademark of Texas Instruments and is used on the TI-81, TI-82, and TI-85 graphics calculators.

TI-82 Program to Solve Quadratic Equations

```
Prgm1 : QUADEQ
      : Disp "ENTER A"
      : Input A
      : Disp "ENTER B"
      : Input B
      : Disp "ENTER C"
      : Input C
      : B² − 4AC → D
      : IF D ≥ 0
      : GO TO R
      : Disp "IMAGINARY ROOTS"
      : −B/(2A) → X
      : √(−D)/(2A) → Y
*     : Disp X ▶ Frac
      : Disp "PLUS OR MINUS"
*     : Disp Y ▶ Frac
      : Disp "          I"
      : GO TO E
      : Lbl R
      : Disp "ROOTS ARE"
      : (−B − √D)/(2A) → X
*     : Disp X ▶ Frac
      : Disp "AND"
      : (−B + √D)/(2A) → X
*     : Disp X ▶ Frac
      : Lbl E
      : END
```

*When using the TI-81 to solve quadratic equations, delete the ▶ Frac instruction from the starred commands listed above. This instruction displays answers in fractional form.

Answers

Chapter 1

EXERCISES 1-1

1 a. 7 **b.** 0, 7 **c.** −7, 0, 7 **d.** −7, 0, $\frac{3}{7}$, 4.$\overline{7}$, 7
e. $-\sqrt{7}, \sqrt{7}$ **f.** −7, $-\sqrt{7}$, 0, $\frac{3}{7}$, $\sqrt{7}$, 4.$\overline{7}$, 7
3 a. −4, −2, 0, 2, 4, 6, 8
b. −7, −5, −3, −1, 1, 3, 5, 7, 9 **c.** 31, 37
d. 1.545
5 a. 2.3 **b.** −2, −1, 0, 1, 2 **c.** 0, 1, 2, 3, 4, 5, 6
d. 12, 14, 15, 16, 18
7 a. $-9, \frac{1}{9}$ **b.** $9, -\frac{1}{9}$ **c.** $\frac{3}{7}, -\frac{7}{3}$
d. $-x - 4, \dfrac{1}{x+4}$ for $x \neq -4$
9 a. 8 **b.** 8 **c.** 0 **d.** 16
11 a. −2 **b.** 2 **c.** −2 **d.** 4
13 a. 3 **b.** x **c.** $3(x + y)$
15 a. $[r + (s + t)]$ **b.** 1 **c.** 0
17 a. vt **b.** vw **c.** 1 **19** nr **21** $7a$
23 Commutative property of addition
25 Multiplicative inverse **27** Additive identity
29 Commutative property of addition
31 Distributive property
33 Associative property of addition
35 Associative property of multiplication
37 Commutative property of multiplication
39 Commutative property of multiplication
41 Distributive property **43** Zero as a factor theorem
45 Addition theorem of equality **47** 0 **49** −7
51 $\sqrt{9}$ **53** 0 **55** 1 **57** 39 **59** 3.$\overline{4}$
61 Division, division by zero **63** 0 **65** Yes
67 Reflexive property
69 Commutative property of multiplication
71 Closure of multiplication of real numbers
73 Associative property of addition
75 $v \leq w$ **77** $m \neq 33$ **79** $|17| = |-17|$ **81** 0
83 The absolute value of 0 is not positive. **85** 2; 1
87 a. 4 **b.** −4 **89** 50(101) = 5050
91 Irrational numbers
93 Reflexive property; additive identity; additive inverses; associative property of addition

EXERCISES 1-2

1 a. 13 **b.** −3 **c.** 3 **d.** −13
3 a. −251 **b.** 95 **c.** 251 **d.** 95
5 a. −6 **b.** −6 **c.** 6 **d.** 9
7 a. −3 **b.** −3 **c.** 3 **d.** $-\frac{1}{3}$
9 a. 2 **b.** 4 **c.** −87 **d.** 0
11 a. 30 **b.** −30 **c.** 0 **d.** 0
13 a. −36 **b.** 36 **c.** 25 **d.** 49
15 a. 33 **b.** 9 **c.** 16 **d.** 64
17 a. 1 **b.** 1 **c.** −1 **d.** −1
19 a. −1 **b.** 1 **c.** −64 **d.** 64
21 a. −2 **b.** 16 **c.** 22 **d.** 0
23 a. 100 **b.** 400 **c.** 40
25 a. 2 **b.** 8 **c.** 8 **27 a.** −17 **b.** −11
29 a. 49 **b.** −95 **31** −89 **33** −16 **35** −8
37 −7 **39** −12 **41** 25 **43** −66 **45** 7
47 −19 **49** 0 **51** $a + b$ **53** $a - b$ **55** $b - a$
57 $a + b$ **59** $b - a$ **61** −3 **63** $3 - x$
65 $-4 - (-y)$ **67** $-x^2$ **69** x^3
71 a. Negative eight **b.** The opposite of x
c. x minus eight
d. The opposite of the quantity x minus eight
73 d **75** b **77** d **79** c **81** b
83 230, 860, −860, −230 **85** $\frac{1}{12}$ c
87 They owe $30. **89** $\frac{1}{20}$
91 a. $-\frac{17}{30}$ **b.** $\frac{2}{7}$ **c.** $\frac{11}{15}$
93 a. $-\frac{7}{5}$ **b.** $-\frac{64}{125}$ **c.** $\frac{16}{81}$
95 a. 10 **b.** $\frac{64}{847}$ **c.** $\frac{95}{36}$

SECTION SKILL REINFORCERS 1-2

1 Commutative property of addition
2 Commutative property of multiplication
3 Distributive property
4 Associative property of addition
5 Commutative property of addition

EXERCISES 1-3

1 a. −14 **b.** 18 **c.** 1
3 a. 21 **b.** 91 **c.** −1
5 a. 4 **b.** 34 **c.** −18
7 a. $\frac{1}{8}$ **b.** $-\frac{1}{2}$ **c.** 1
9 a. 29 **b.** 69 **c.** −11
11 a. 7 **b.** −5 **c.** 1 **13** $-7m$ **15** $7a$
17 $7y$ **19** $4x$ **21** $14x - 15y$ **23** $2m + n$
25 $6x - 11$ **27** $-6x + 10y$ **29** $-2m - 6n$
31 −6 **33** −4 **35** −136 **37** Solution

39 Not a solution **41** Solution **43** 36.92 cm^2
45 \$85 **47** 60 cm^3 **49** d **51** b **53** b
55 $-6x + 20y$ **57** $-\frac{7x}{15} + \frac{9y}{14}$ **59** $-8a + 13b - 10c$
61 $-16v + 122$ **63** 9 **65** $-\frac{1}{3}$ **67** $\frac{7}{12}$ **69** 3
71 57 **73** -182 **75** 5 **77** 160.3125
79 31.36 **81** 590 **83** 495 **85** 1760

SECTION SKILL REINFORCERS 1-3

1 Sum **2** Difference **3** Product **4** Quotient
5 Associative property of addition

EXERCISES 1-4

1 Equivalent **3** Contradiction **5** Linear
7 a. $v = 17$ **b.** $m = 4$ **c.** $x = 4$ **d.** $n = 14$
e. $m = -4$
9 a. $n = -\frac{1}{2}$ **b.** $z = \frac{49}{3}$ **c.** $x = 0$ **d.** $t = 5$
11 $a = -19$ **13** $t = -4$ **15** $n = -1$ **17** $y = 0$
19 $z = 2$ **21** No solution **23** All real numbers
25 $w = -2$ **27** $m = -\frac{3}{2}$ **29** $v = \frac{5}{4}$ **31** $v = -72$
33 $x = \frac{1}{3}$ **35** $y = -4x + 5$ **37** $y = \frac{3x - 7}{2}$
39 $h = \frac{V}{lw}$ **41** $r = \frac{A - p}{pt}$ **43** $h = \frac{3V}{\pi r^2}$
45 $x = \frac{y - b}{m}$ **47** $a = \frac{2A - bh}{h}$ **49** $d = \frac{l - a}{n - 1}$
51 $r = \frac{S - a}{S}$ **53** $h = \frac{S - 2\pi r^2}{2\pi r}$
55 a. $m = -3$ **b.** $2m + 6$
57 a. $x = -22$ **b.** $-x - 22$
59 a. $v = 3$ **b.** $\frac{v - 3}{12}$ **61** d **63** a **65** c
67 $x = 6$ **69** No solution **71** $x = \frac{b}{a + c}$
73 $x = -\frac{7w + 4}{2}$ **75** $y' = \frac{3x - y}{2}$ **77** $y' = -\frac{2x + y}{x + 2y}$
79 $v \approx -50.91$ **81** $x \approx -1.47$
83 $a = 3, b = 2, c = 0$ (Answers may vary.)
85 $a = 2, b = 3, c = 1, d = 4$ (Answers may vary.)
87 Identity

SECTION SKILL REINFORCERS 1-4

1 -100 **2** 100 **3** 4 **4** 16 **5** 48

EXERCISES 1-5

1 $2n + 7 = 13$; 3 **3** $5n + 2 = 2 + 3n$; 0
5 $\frac{n + (-3)}{10} = \frac{n - 2}{3}$; $\frac{11}{7}$ **7** $n + (n + 1) = 85$; 42, 43
9 $n + (n + 2) = 112$; 55, 57
11 $5n = (n + 2) + 26$; 7, 9 **13** $n + (n + 4) = 82$; 39
15 $\frac{n}{n + 20} = \frac{3}{7}$; 15, 35 **17** $\frac{n}{2n - 18} = \frac{5}{6}$; 22.5, 27
19 $3n = n + 6$; 3 **21** 39°, 51° **23** \$70,000
25 $(10.3\pi + 50.6)$ cm $\approx$ 83.0 cm **27** 18 m^2
29 7π $\text{cm}^3 \approx 22.0$ cm^3 **31** 13 km **33** \$1250
35 200 tens, 600 fives **37** \$3000 @ 8%, \$1500 @ 10%
39 62.5 lb @ \$2.05/lb, 437.5 lb @ \$1.25/lb
41 3.5 oz of 10%, 0.5 oz of 50%
43 a. $\frac{1}{5}$ roof/day **b.** $\frac{1}{4}$ yard/h **c.** 18 cars/h
d. $\frac{1}{t}$ report/h
45 1.2 days **47** 17 **49** 24, 20, 12, 26 **51** $1\frac{7}{8}$ h
53 160 quarters, 640 dimes
55 80 grains of 50%, 220 grains of 80% **57** 9.3%
59 4.5 h **61** Loss of \$60.60 **63** 4 cm^2 **65** *ab*
67 d

SECTION SKILL REINFORCERS 1-5

1 $n = 21$ **2** $t = 1$ **3** $x = 4$ **4** No solution
5 All real numbers

EXERCISES 1-6

1 a. 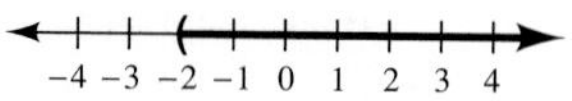 –4 –3 –2 –1 0 1 2 3 4
b. 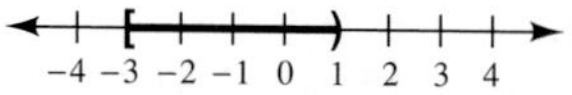 –4 –3 –2 –1 0 1 2 3 4
c. 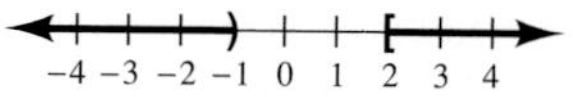 –4 –3 –2 –1 0 1 2 3 4
3 a. $x < -4$ **b.** $-2 < x \leq 3$ **c.** $x \leq -3$ or $x \geq 1$
5 a. $(1, 5)$ **b.** $[-3, 7]$ **c.** $(-\infty, 3)$ **d.** $[-2, +\infty)$
7 a. $-2 \leq x < 3$ **b.** $-4 < x < 9$ **c.** $x \geq 0$
d. $x < 6$
9 a. $\{-2, -1, \frac{1}{2}, 1, 2, 3, 4\}$ **b.** $\{-1, \frac{1}{2}, 1, 2\}$
c. $\{-4, -3, \frac{1}{2}, 1, 2, 3, 4\}$
11 $(-\infty, 5)$ 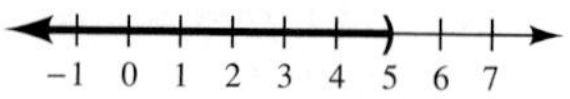 –1 0 1 2 3 4 5 6 7
13 $[1, +\infty)$ –4 –3 –2 –1 0 1 2 3 4
15 $(-2, +\infty)$ –4 –3 –2 –1 0 1 2 3 4
17 $[-3, +\infty)$ –4 –3 –2 –1 0 1 2 3 4
19 $(-\infty, 25)$ **21** $[\frac{8}{5}, +\infty)$ **23** $(-\infty, -\frac{1}{2})$
25 $(-\infty, 2)$ **27** $(4, +\infty)$
29 $(12, 14]$ 9 10 11 12 13 14 15 16 17
31 $[-2, 11)$ 11 –4 –2 0 2 4 6 8 10 12
33 $(-3, 2)$ 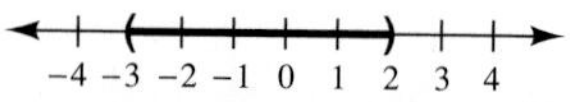 –4 –3 –2 –1 0 1 2 3 4
35 $[-24, -6]$ **37** $[3, 5]$ **39** $[1, \frac{13}{10})$

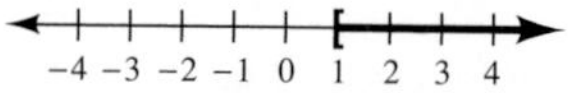

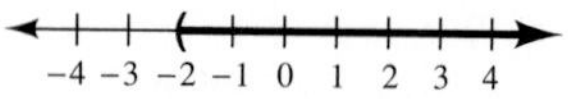

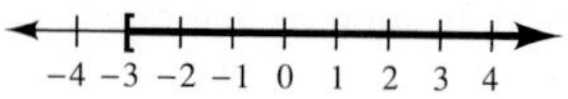

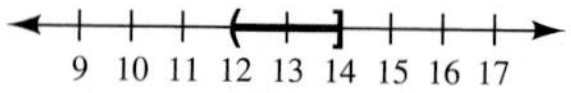

41 $(-5, 3)$

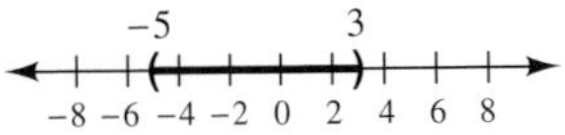

43 $(-\infty, -2] \cup (2, +\infty)$

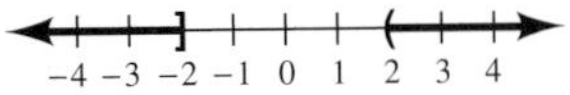

45 $(0, 5)$

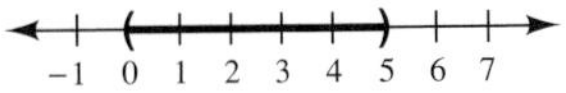

47 $(4, +\infty)$ **49** $[-3, +\infty)$

51 **a.** $(3, 4)$ **b.** $(2, 5]$ **c.** $(2, 5]$

53 **a.** 1 (or any positive value of x)
b. -1 (or any negative value of x) **c.** 0

55 **a.** $x \geq 4$ **b.** $x \leq 9$ **c.** $3 < x < 8$ **57** $[2, +\infty)$

59 $(-2, 3]$ **61** All real numbers **63** No solution

65 $(\frac{1}{3}, \frac{8}{5})$ **67** $(-\infty, -3) \cup (-2, +\infty)$ **69** b **71** a

73 $2(a + 3) \leq 10$; $(-\infty, 2]$

75 $2d - 9 \leq 2(d - 12)$; no solution

77 $2 - 3m > 5$ or $2 - 3m < -1$; $(-\infty, -1) \cup (1, +\infty)$

79 $59° \leq F \leq 86°$ **81** $(38.10, +\infty)$

Section Skill Reinforcers 1-6

1 $\frac{23}{3}$ **2** 8 **3** 0 **4** -14 **5** 35, 37, 39, 41

Exercises 1-7

1 **a.** 7 **b.** 12 **c.** $|3a + 4b|$

3 **a.** $|x| < 5$ **b.** $|x| \geq 2$ **c.** $|x - 1| = 3$
d. $|x| < 4$

5 **a.** $|x - 4| \leq 2$ **b.** $|x - 1| > 2$
c. $|x + 7.5| < 2.5$ **d.** $|x - 2| \leq 2$

7 **a.** $|x| \leq 3$ **b.** $|x| > 3$ **c.** $|x| < 7$

9 **a.** $\{-6, 6\}$ **b.** No solution **c.** $\{-5, -1\}$
d. $\{0\}$

11 **a.** $\{-4, 4\}$ **b.** $\{-25, 25\}$ **c.** $\{3\}$

13 $(-2, 3)$

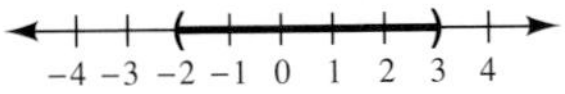

15 $(-\infty, \frac{4}{3}] \cup [2, +\infty)$

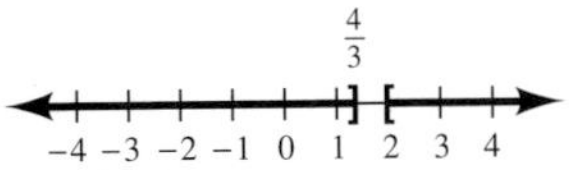

17 $[-4, 4]$

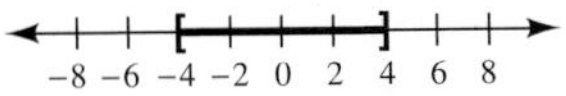

19 $(-\infty, -4) \cup (1, +\infty)$

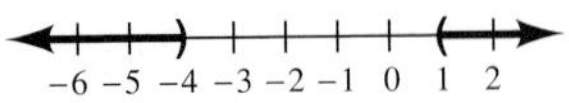

21 $(-\frac{11}{2}, -\frac{5}{2})$

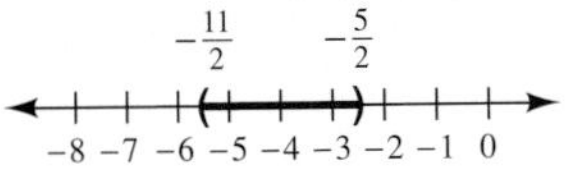

23 No solution **25** All real numbers

27 $(-\infty, -\frac{3}{2}] \cup [\frac{5}{2}, +\infty)$ **29** $[-97, 99]$ **31** $\{-4, 3\}$

33 $\{-\frac{3}{2}, 3\}$ **35** $(-\infty, -2] \cup [6, +\infty)$ **37** $(-4, \frac{10}{3})$

39 $\{-1, \frac{1}{3}\}$ **41** $(-\infty, -\frac{5}{9}) \cup (\frac{35}{9}, +\infty)$

43 $|x| \leq 7$

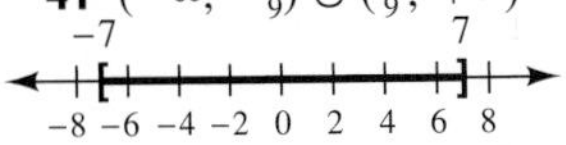

45 $|x - 3| > 2$

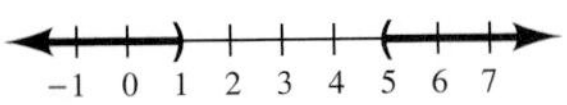

47 $|x + \frac{1}{2}| < \frac{1}{2}$

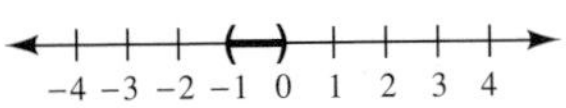

49 $|x - y| > 5$ **51** d **53** $\{-50, 50\}$

55 $\{-\frac{51}{2}, \frac{61}{2}\}$ **57** $[-\frac{5}{3}, \frac{10}{3}]$ **59** $\{-5, -\frac{1}{5}\}$

61 $|L - 15| \leq 0.12$; lower: 14.88 m, upper: 15.12 m

63 $|V - 26.9| \leq 0.9$; lower: 26.0 L, upper: 27.8 L **65** 3

67 89.37

Section Skill Reinforcers 1-7

1 -49 **2** 49 **3** 36 **4** 48 **5** 1

Review Exercises 1

1 **a.** 12 **b.** 18 **c.** -45 **d.** -5

2 **a.** -9 **b.** 9 **c.** -8 **d.** -1

3 **a.** 68 **b.** -100 **c.** 218 **d.** 64

4 **a.** 120 **b.** -120 **c.** 120 **d.** 0

5 **a.** 100 **b.** 160 **c.** 20 **d.** -14

6 **a.** 7 **b.** -24 **c.** 27 **d.** 3

7 **a.** 7 **b.** 4 **c.** 11 **d.** 3 **8** -53 **9** 7

10 $\frac{5}{2}$ **11** $-\frac{7}{5}$ **12** $-\frac{9}{10}$ **13** -1 **14** 7 **15** 90

16 -29 **17** -55 **18** $\frac{189}{100}$ **19** $\frac{3}{4}$ **20** 2

21 -11 **22** 120 **23** -9 **24** -34 **25** $-\frac{1}{12}$

26 -5 **27** $9m$ **28** $4a - 6b$ **29** $-10x - 1$

30 $-4x - 3y + 14$ **31** $-x$ **32** $6x - 5y$

33 $2(-x)^2$ **34** $-7 - (-y)$

35 **a.** Associative property of addition
b. Commutative property of addition
c. Distributive property

36 **a.** Commutative property of multiplication
b. Distributive property
c. Commutative property of addition

37 **a.** $3 - x$ **b.** $\dfrac{1}{x - 3}, x \neq 3$ **c.** 0

38 **a.** Addends or terms **b.** Factors **c.** 3

39 **a.** $\frac{4}{5}$ **b.** $-\frac{5}{4}$ **c.** 0

40 **a.** 2 **b.** 2, 3 **c.** $\sqrt{7}$

41 **a.** 47.4 cm **b.** 135.3 cm^2 **42** 26.2 cm

43 **a.** $4m + 2$ **b.** $m = -\frac{1}{2}$

44 **a.** $-6n - 6$ **b.** $n = -1$

45 **a.** $-\dfrac{w + 60}{10}$ **b.** $w = -60$ **46** No solution

47 $m = 15$ **48** $x = 6$ **49** No solution

50 $(4, +\infty)$

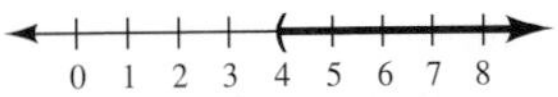

51 $(-\infty, 21.5]$

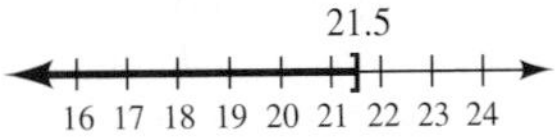

52 $(2, 6]$

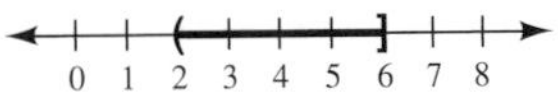

53 $(-4, -3]$

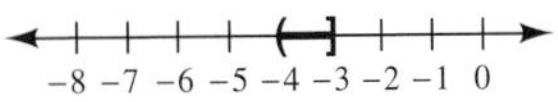

54 $(-29, 3)$

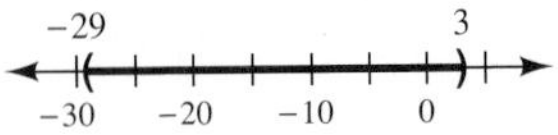

55 $(-\infty, -4] \cup [-2, +\infty)$

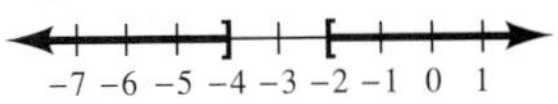

56 $\{-6, 11\}$ **57** $[1, 4]$

58 $(-\infty, -\frac{6}{5}) \cup (2, +\infty)$

59 No solution **60** All real numbers

61 $(5, +\infty)$

62 $z = 7$ **63** $t = 2$ **64** $z = 10$ **65** $[5, +\infty)$

66 $\{-7, 3\}$ **67** $D = \dfrac{W}{F}$ **68** $h = \dfrac{3V}{\pi r^2}$

69 $a = \dfrac{2S - n(n-1)d}{2n}$ **70** $R = \dfrac{A - P}{PT}$ **71** 20, 62

72 173, 175 **73** $(2, 3]$ **74** $-\frac{7}{3}, 5$

75 75 words/min **76** 15 mi/h **77** 11 cm

78 7 km/h, 8 km/h **79** 18 h **80** 69.23 days

81 \$3000 in first investment, \$12,000 in second investment

82 35 of the \$0.23 stamps, 65 of the \$0.29 stamps

83 240 lb **84** 30 units

85 **a.** $|x| > 2$ **b.** $|x| < 2$ **c.** $|x + 4| \leq 3$ **d.** $|x - 1| > 6$

86 **a.** $|x| < 4$ **b.** $|x| \geq 1$

87 **a.** $(-1, 1)$ **b.** $(-2, 3]$ **88** -131.1875

89 64.488 **90** $x/y - 2$ denotes $\dfrac{x}{y} - 2$, not $\dfrac{x}{y-2}$.

91 Parentheses, as in $x(-y)$, must be used to separate the operational symbols.

92 The acceptable format uses properties of equality to form equivalent equations. The unacceptable format uses the equal sign to separate steps between equivalent equations. Using too many equal signs would result in untrue statements, such as $-8 = 2$, in the unacceptable format.

Mastery Test 1

1 **a.** $\sqrt{9}$ **b.** $0, \sqrt{9}$ **c.** $-5, 0, \sqrt{9}$
d. $-5, -1.5, 0, 2.\overline{5}, \sqrt{9}, 3.444\ldots, 5\frac{2}{3}$
e. $-\sqrt{5}, \pi, 4.131131113\ldots$
f. $-5, -\sqrt{5}, -1.5, 0, 2.\overline{5}, \sqrt{9}, \pi, 3.444\ldots, 4.131131113\ldots, 5\frac{2}{3}$
g. 37 **h.** 80, 81, 82, 84, 85, 86

2 **a.** 8 **b.** 8 **c.** 5 **d.** -5

3 **a.** Distributive property
b. Commutative property of multiplication
c. Commutative property of addition
d. Associative property of addition

4 **a.** 2 **b.** -32 **c.** -1.28 **d.** $\frac{1}{6}$ **e.** 150
f. -6 **g.** 100 **h.** $-\frac{6}{5}$

5 **a.** -36 **b.** 36 **c.** 64 **d.** 34

6 **a.** 11 **b.** -50 **c.** 23 **d.** 70

7 **a.** 14 **b.** 26 **c.** 23 **d.** 49

8 **a.** $2x$ **b.** $x + 5y$ **c.** $-x - 5$ **d.** $4x + 7y$

9 **a.** $v = \frac{19}{13}$ **b.** All real numbers **c.** No solution

10 **a.** $x = \dfrac{z}{y}$ **b.** $x = z - y$ **c.** $x = \dfrac{bz - a}{y}$
d. $x = \dfrac{vz - wy}{w}$

11 **a.** 14 **b.** 127, 129 **c.** 75.2 cm^2
d. 300 child tickets, 400 adult tickets **e.** 4 h 15 min

12 **a.** $[-3, 11)$ **b.** $(\pi, +\infty)$ **c.** $(-1, 5]$

13 **a.** $(4, +\infty)$ **b.** $(-\infty, 1) \cup [5, +\infty)$ **c.** $(-5, 7]$
d. $(-29, 6)$

14 **a.** $\{-27, 22\}$ **b.** $(-8, 11)$ **c.** $(-4, 14)$

Chapter 2

Exercises 2-1

1 **a.** $A = (-3, 2)$, quadrant II; $B = (-2, -3)$, quadrant III; $C = (1, 4)$, quadrant I; $D = (4, 1)$, quadrant I

3

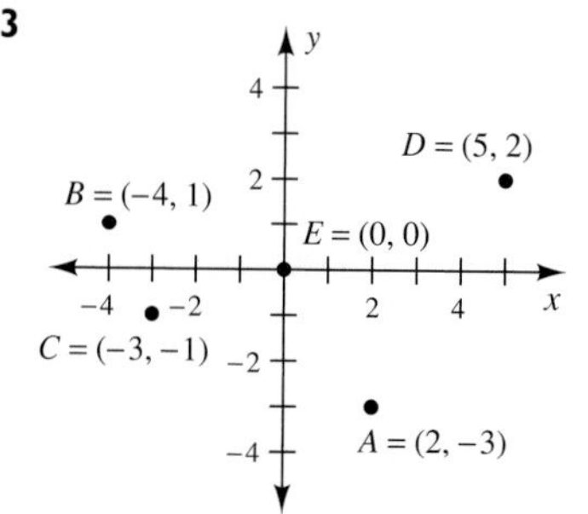

5 A: quadrant IV; B: quadrant II; C: quadrant III; D: quadrant I; E: origin

7 $(1, 3)$, yes; $(-9, -3)$, yes; $(-2, 5)$, no

9 **11** **13** **15**

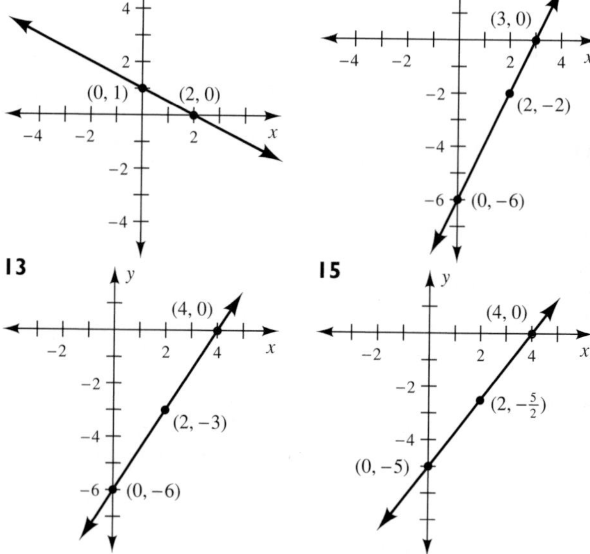

17

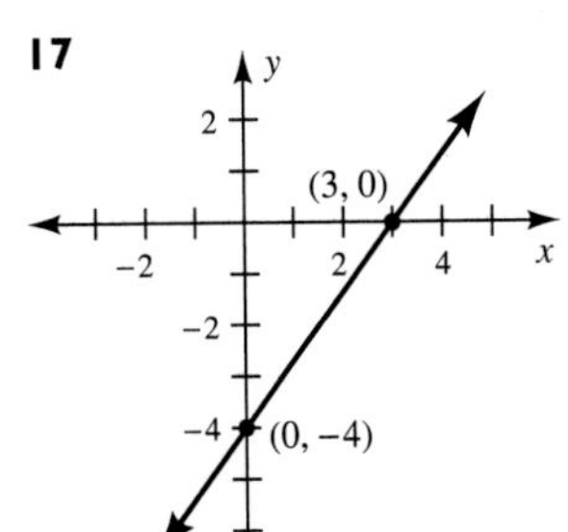

19

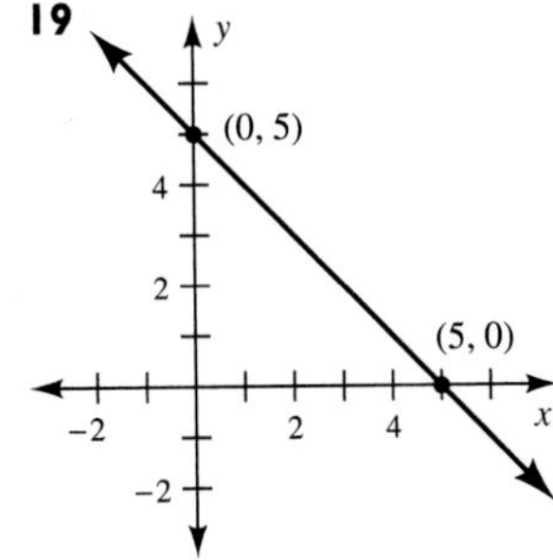

21

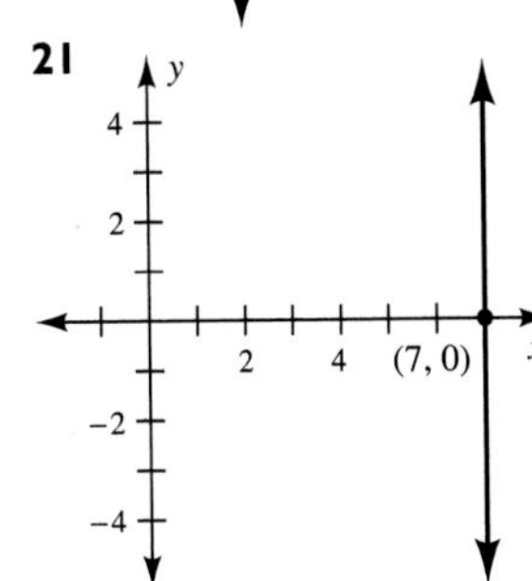

23

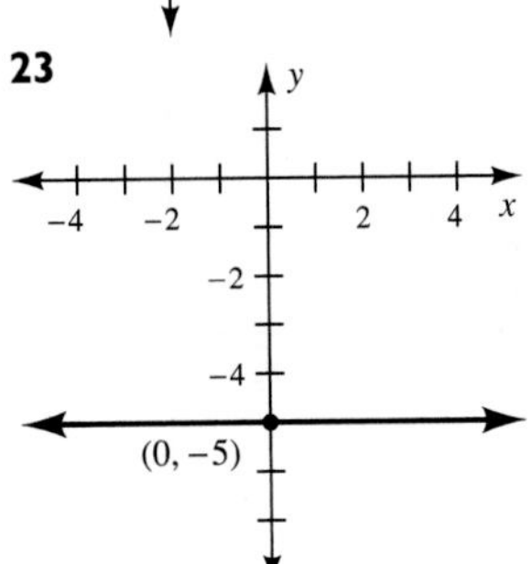

25

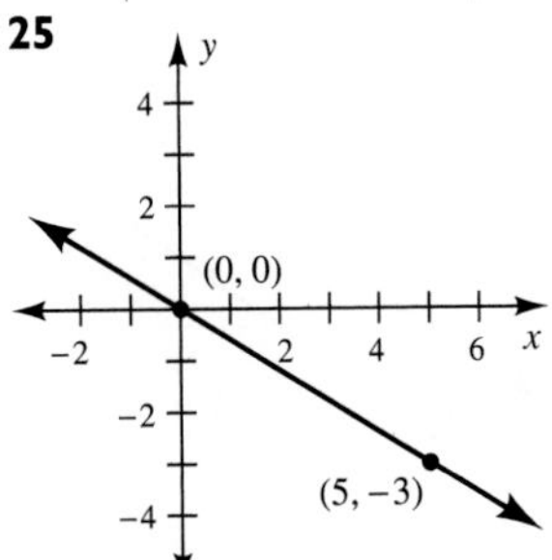

27

x	−2	0	2
y	0	3	6

29

x	$\frac{1}{2}$	0	2
y	0	1	−3

31

x	1	2	3	4	5
y	1	3	4	2	3

33 **35**

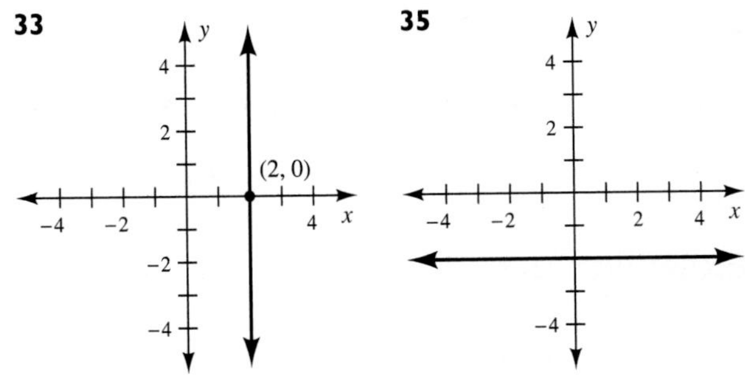

37

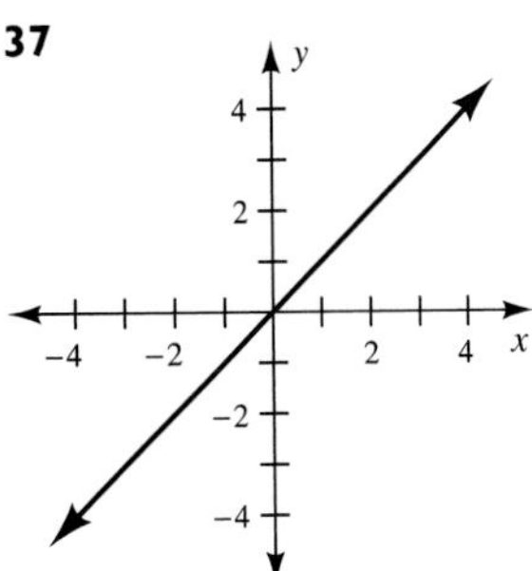

39

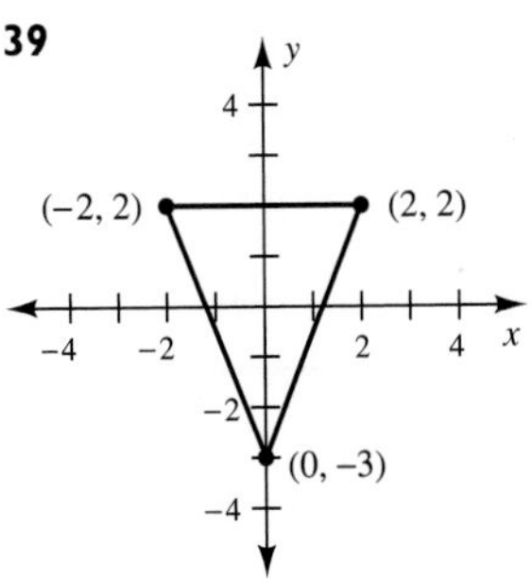

41 **a.** \$2000 **b.** 1987

43 **a.** 10,000 keyboards **b.** 12,000 keyboards **c.** 10,500 keyboards **d.** 9000 keyboards

45

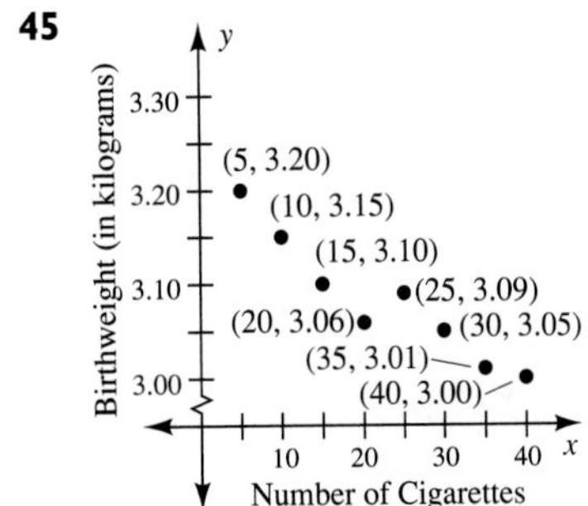

47 Quadrant I **49** Quadrant III **51** On the x-axis

53 $7y = 5x - 9$ **55** $y - 8 = 4(x - 7)$

57

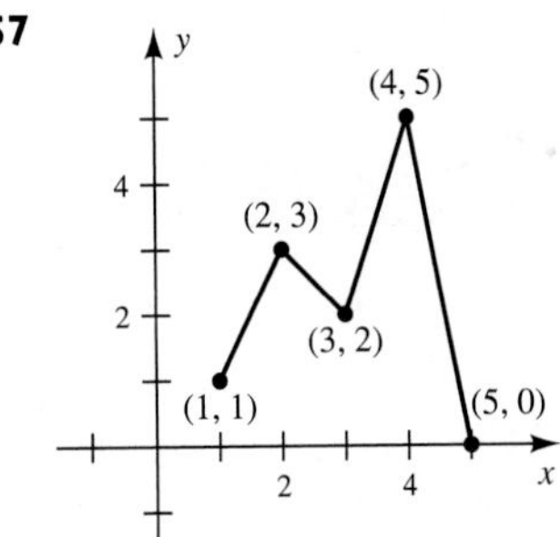

59 10 **61** 3

63

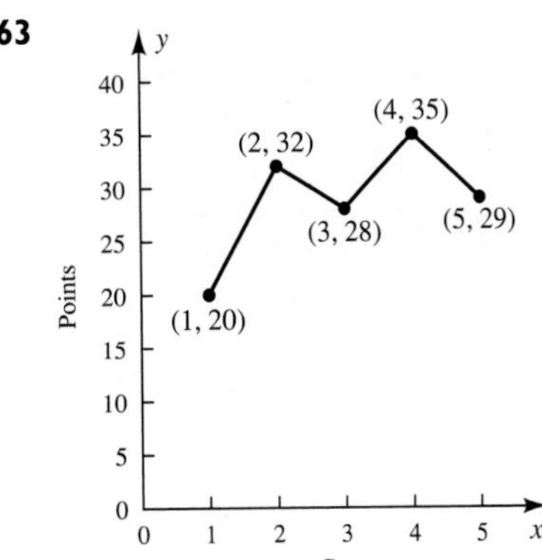

65

x	0	1	2	3	4	5	6
y	8	3	0	−1	0	3	8

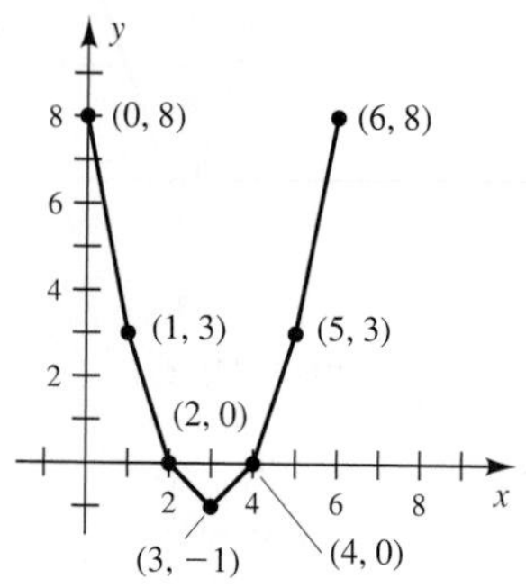

67 a.

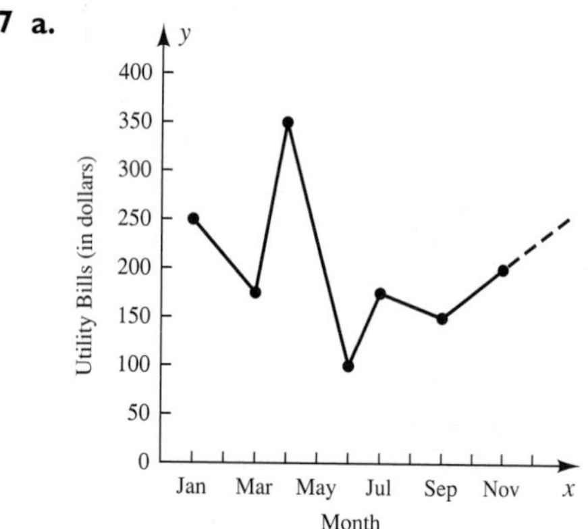

b. \$160 **c.** \$225

Section Skill Reinforcers 2-1

1 $-\frac{1}{2}$ **2** $-\frac{13}{3}$ **3** $\frac{7}{3}$ **4** $\frac{10}{3}$ **5** $\frac{2}{5}$

Exercises 2-2

1 −2 **3** $-\frac{8}{7}$ **5** −1 **7** Undefined **9** 0
11 2 **13** $-\frac{5}{3}$ **15** 0 **17** Undefined **19** $\frac{1}{3}$
21 $-\frac{2}{3}$ **23** 0 **25** Parallel
27 Neither parallel nor perpendicular **29** Perpendicular
31 Neither parallel nor perpendicular **33** Perpendicular
35 Parallel **37** Parallel **39 a.** 4 **b.** $-\frac{1}{4}$

41

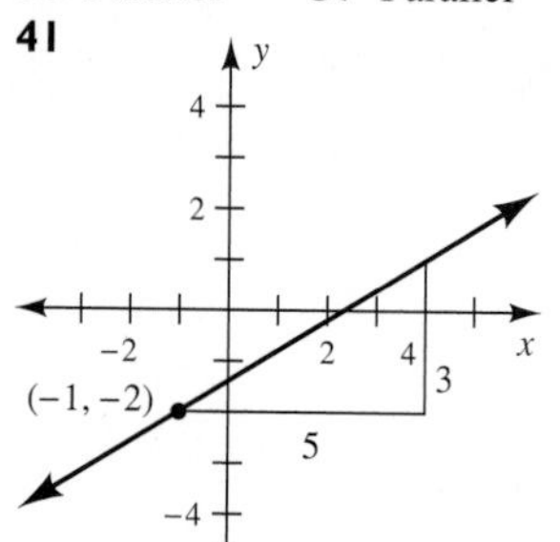

43

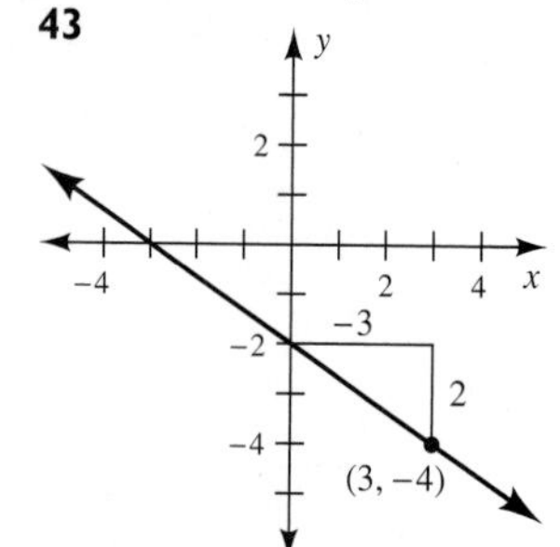

45 **47** **49** **51** **53** **55**

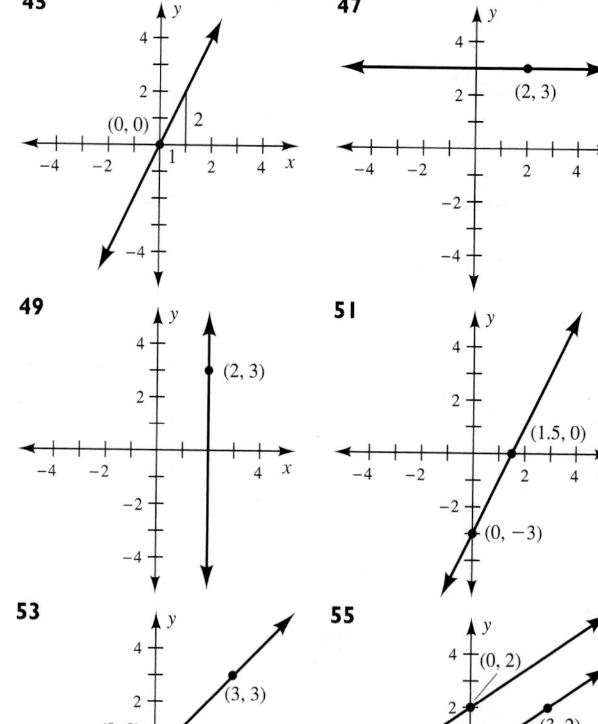

57 The slopes are different for each line; the y-intercepts are the same.
59 The lines are parallel; the slopes are the same, but the y-intercepts are different.
61 The graphed line is not visible in the standard viewing window. To obtain a better viewing window, increase the range of x and y.
63 Not a right triangle **65** A right triangle
67 i. e **ii.** f **iii.** b **iv.** d **v.** a **vi.** c
69 e **71** 96 m **73** 17.5 m **75** Negative
77 Positive **79** Negative **81** Undefined
83 $y = -2x + 3$ **85** $x = -3$
87 a. Decreasing **b.** Increasing

Section Skill Reinforcers 2-2

1 $[-3, 3]$ **2** $(-\infty, -4) \cup (4, +\infty)$ **3** $[-2, 4]$
4 $(-5, 7)$ **5** $[1, 3)$

Exercises 2-3

1 a. $y = \frac{4}{5}x - \frac{7}{5}$; $m = \frac{4}{5}$; $(0, -\frac{7}{5})$
b. $y = 2x$; $m = 2$; $(0, 0)$

3 a. $y - 7 = 0$ **b.** $x + 4 = 0$
5 a. $2x - y - 5 = 0$ **b.** $3x + 4y - 2 = 0$
c. $y - 7 = 0$
7 a. $x + 5 = 0$ **b.** $y - 9 = 0$ **c.** $x - 2 = 0$
9 $3x + y - 1 = 0$ **11** $4x - 7y - 39 = 0$
13 $y - 6 = 0$ **15** $3x + 7y = 0$ **17** $5x - 8y + 20 = 0$
19 $x + y - 5 = 0$ **21** $x - 4 = 0$
23 $2x + 3y - 6 = 0$ **25** $2x - 5y + 34 = 0$
27 $x - 5 = 0$ **29** $x - 3y + 8 = 0$ **31** $m = 0$
33 $m = 8$ **35** Undefined **37** $m = -\frac{1}{8}$
39 $m = -2$ **41** $4x - 5y + 23 = 0$
43 $2x + 9y + 89 = 0$ **45** $x - 8 = 0$ **47** $y - 7 = 0$
49 $5x - 9y - 29 = 0$ **51** $7x - 2y + 24 = 0$
53 $y + 8 = 0$ **55** $x = 0$ **57** b **59** d

61

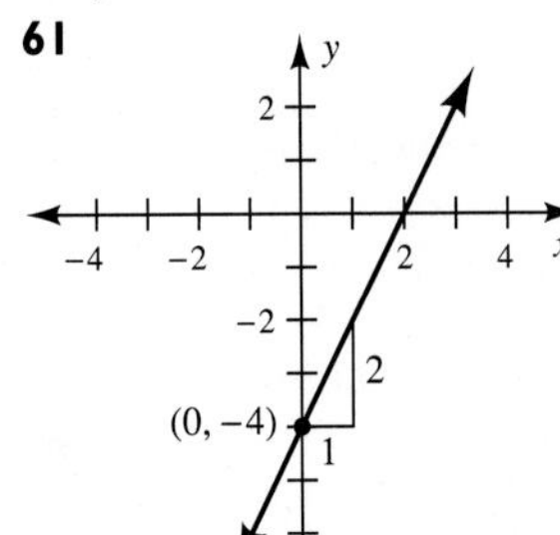

63

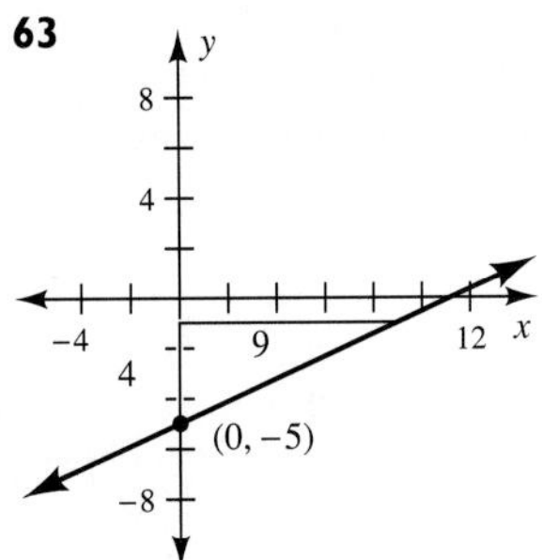

65

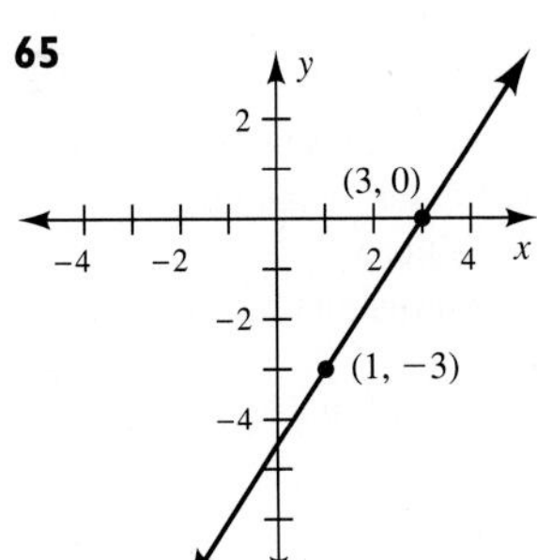

67 $9C - 5F + 160 = 0$

69 a. and **b.**

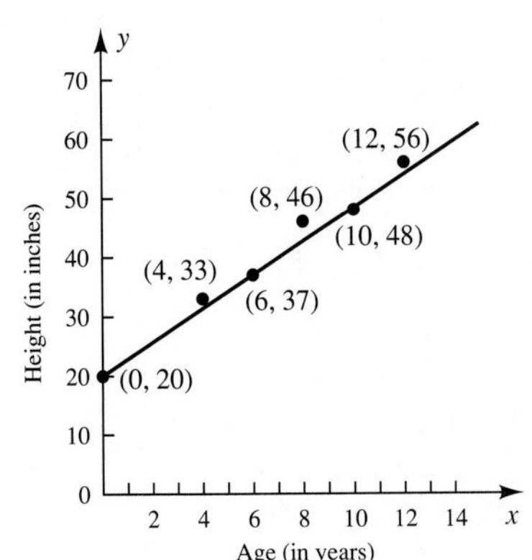

c. $y = \frac{17}{6}x + 20$ (Answers may vary.)
d. 60 in. (Answers may vary.)

71
[−3, 3] for x, [−5, 5] for y

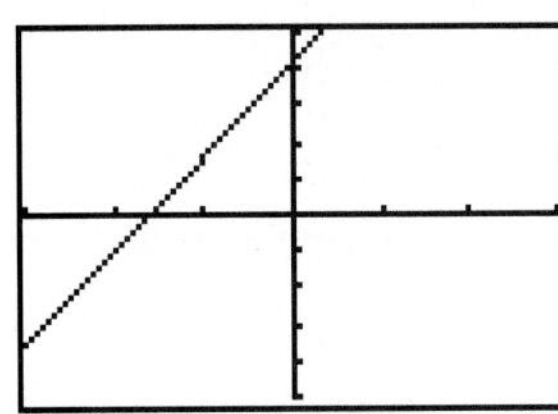

73
[−3, 3] for x, [−5, 5] for y

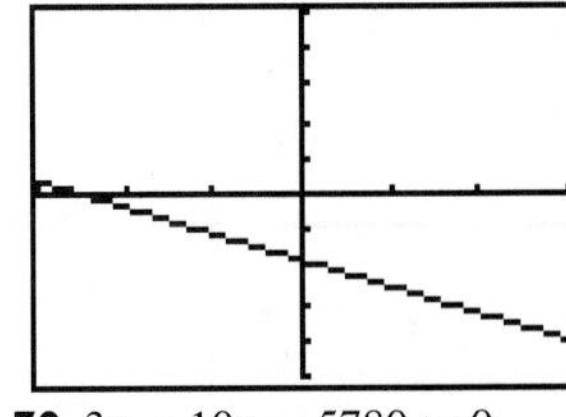

75 $m = -\frac{1}{10}$ **77** $m = \frac{3}{10}$ **79** $3x - 10y - 5780 = 0$

SECTION SKILL REINFORCERS 2-3

1 −2 **2** −157 **3** −67 **4** −20 **5** 5

EXERCISES 2-4

1 Function: $D = \{7, 8, 9\}$, $R = \{3, 4, 6\}$
3 Relation that is not a function: $D = \{7, 8\}$, $R = \{3, 4, 6\}$
5 Function: $D = \{-3, 0, 1, 9\}$, $R = \{\pi\}$ **7** Not a relation
9 Relation that is not a function: $D = \{7\}$, $R = \{11, 2, -4\}$
11 Function: $D = \{1, 2, 3, \pi\}$, $R = \{4\}$
13 Relation that is not a function: $D = \{0, 1, 2, 3, 4\}$,
$R = \{-4, -3, -2, -1, 0, 1, 2, 3, 4\}$
15 Function: $D = \{-3, -2, -1, 0, 1, 2, 3\}$, $R = \{2\}$
17 a.

D		R
−5	→	4
−3	→	2
−2	→	0
0	→	−2
1	→	−3
4	→	−4

b. $\{(-5, 4), (-3, 2), (-2, 0), (0, -2), (1, -3), (4, -4)\}$
c.

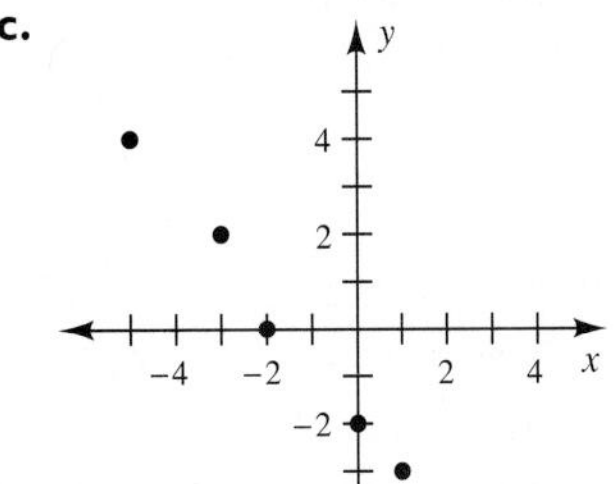

19 a.

D		R
−3	→	1
−2	→	−1
−1	→	3
1	→	−2
3	→	2

b. $\{(-3, 1), (-2, -1), (-1, 3), (1, -2), (3, 2)\}$
c.

x	−3	−2	−1	1	3
y	1	−1	3	−2	2

21 a.

x	-1	1	2	4
y	5	3	-1	-2

b. $\{(-1, 5), (1, 3), (2, -1), (4, -2)\}$

c.

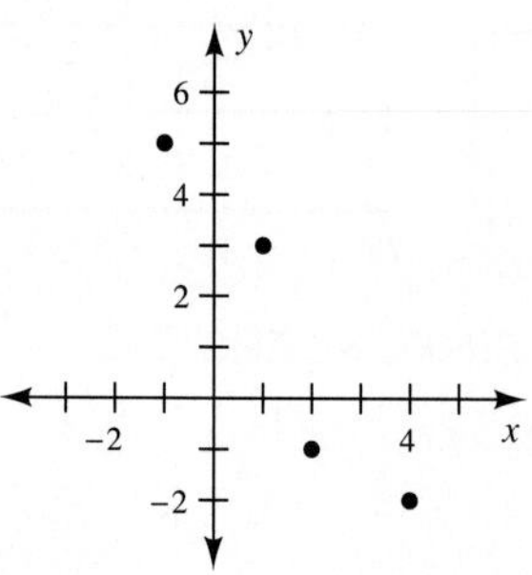

23 a.

x	-5	-3	1	2	3
y	4	4	4	4	4

b. $D \quad R$
$-5 \to 4$
$-3 \to 4$
$1 \to 4$
$2 \to 4$
$3 \to 4$

c.

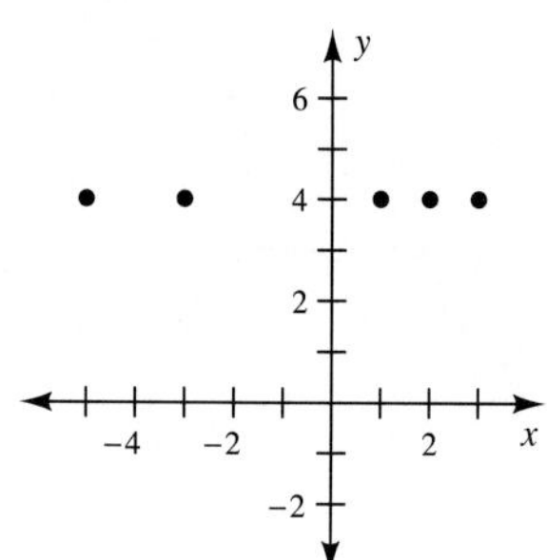

25 a. $D \quad R$
$1 \to 5$
$2 \to 4$
$3 \to 6$
$4 \to 2$
$5 \to 3$

b. $\{(1, 5), (2, 4), (3, 6), (4, 2), (5, 3)\}$

c.

x	1	2	3	4	5
y	5	4	6	2	3

27 a. $D \quad R$
$1960 \to 107$
$1965 \to 104$
$1970 \to 116$
$1975 \to 117$
$1980 \to 122$
$1985 \to 120$
$1990 \to 123$

b. $\{(1960, 107), (1965, 104), (1970, 116), (1975, 117), (1980, 122), (1985, 120), (1990, 123)\}$

c.

x	1960	1965	1970	1975	1980	1985	1990
y	107	104	116	117	122	120	123

29 Function **31** Not a function **33** Not a function
35 Function **37** Function **39** Not a function
41 Not a function **43** Function **45** Not a function
47 Function
49 $D = \{-3, -2, -1, 0, 1, 2, 3\}$, $R = \{-2, -1, 1, 2\}$
51 $D = \mathbb{R}$, $R = \{2\}$ **53** $D = \mathbb{R}$, $R = [-2, +\infty)$
55 $D = [-4, 3)$, $R = (-2, 3]$
57 a. Decreasing **b.** Increasing **c.** Decreasing **d.** (4, 5)
59 (1, D), (2, A), (3, C), (4, B)
61 (Answers may vary.) **63** (Answers may vary.)

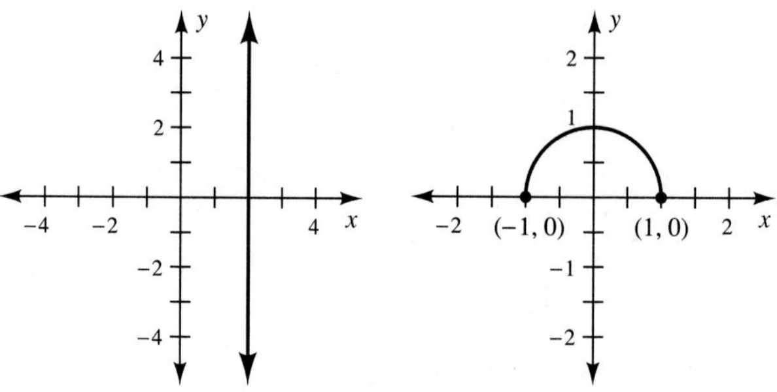

65 A graph that is an ellipse can never represent a function since a vertical line through the center of the ellipse would contain two points on the ellipse.
67 Each person is associated with two distinct elements, a father and a mother.
69 $\{(-4, 4), (-2, 2), (0, 0), (1, 1), (3, 3)\}$; $R = \{0, 1, 2, 3, 4\}$
71 $D \quad R \qquad R = \{0, 1, 2, 3\}$
$0 \to 0$
$1 \to 1$
$4 \to 2$
$9 \to 3$

73

x	-3	-2	-1	$-\frac{1}{2}$	0	$\frac{1}{2}$	1	2	3
y	9	4	1	$\frac{1}{4}$	0	$\frac{1}{4}$	1	4	9

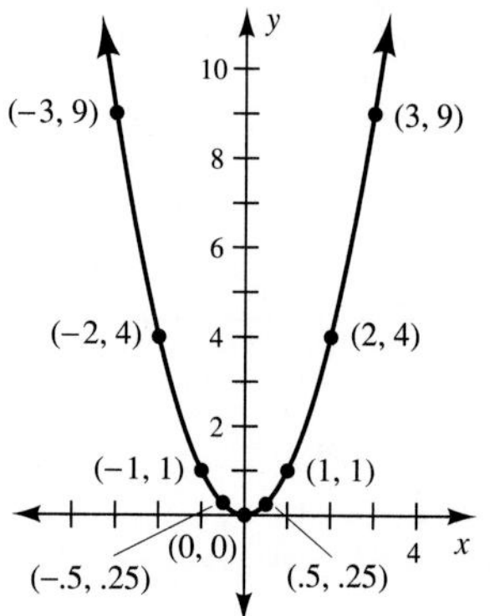

75 $R = [-7, +\infty)$ **77** $R = (-\infty, 2.8]$

Section Skill Reinforcers 2-4

1 100 **2** -100 **3** 100 **4** -75 **5** -384

Exercises 2-5

1 a. -4 **b.** -6 **c.** 16 **d.** -24
3 a. 3 **b.** 0 **c.** 1 **d.** 21
5 a. $\frac{7}{2}$ **b.** 0 **c.** Undefined **d.** $\frac{1}{4}$
7 a. 10 **b.** 3 **c.** 13 **d.** 36
9 a. $3h + 2$ **b.** $3h - 3$ **c.** $3h - 1$
11 $\mathbb{R}$ **13** $\mathbb{R}$ **15** $\mathbb{R}$ **17** $\mathbb{R}$ **19** $\mathbb{R} \sim \{-2\}$
21 $\mathbb{R} \sim \{-2, 2\}$ **23** $[1, +\infty)$ **25** $a = kb$
27 $w = \dfrac{k}{z}$ **29** $a = kbc$ **31** $P = \dfrac{k}{V}$ **33** $V = \dfrac{kT}{P}$
35 y varies inversely with v and w.
37 a varies jointly with b and c and inversely with m and n.
39 $k = \frac{3}{2}$ **41** $k = 6$ **43** $k = 36$ **45** $y = 5$
47 $a = 10$ **49 a.** 2.646 **b.** 0.05556
51 a. $\sqrt{-5}$ is not a real number. **b.** -0.1250
53 $[-4, 4]$ for x, $[-3, 6]$ for y

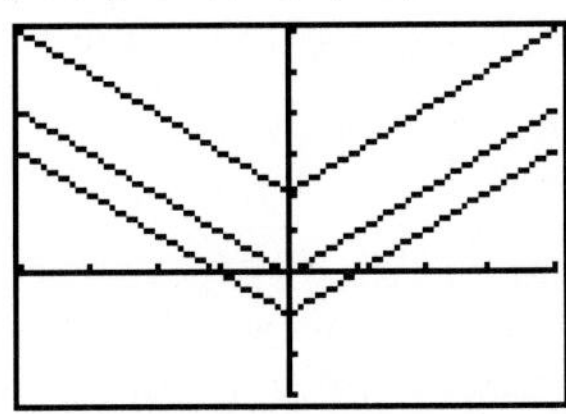

b. The graph is translated up 2 units.
c. The graph is translated down 1 unit.

55 $[-6, 6]$ for x, $[-2, 6]$ for y

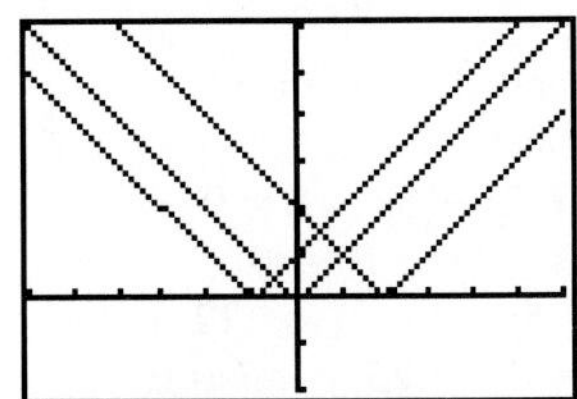

b. The graph is translated to the right 2 units.
c. The graph is translated to the left 1 unit.

57 a.

x	y
-3	-9
-2	-4
-1	-1
0	0
1	-1
2	-4
3	-9

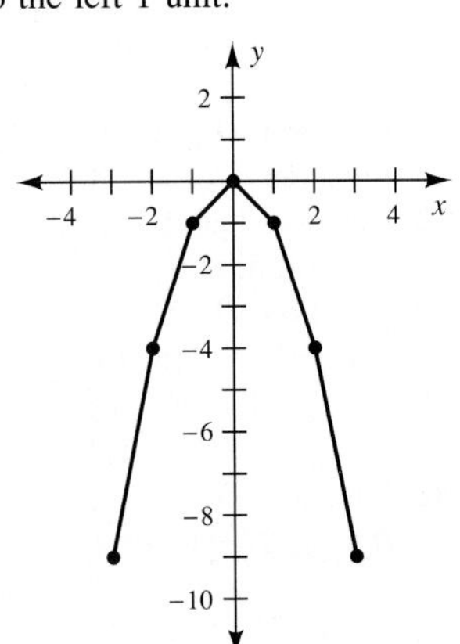

b.

x	y
-3	-9
-2.5	-6.25
-2	-4
-1.5	-2.25
-1	-1
-0.5	-0.25
0	0
0.5	-0.25
1	-1
1.5	-2.25
2	-4
2.5	-6.25
3	-9

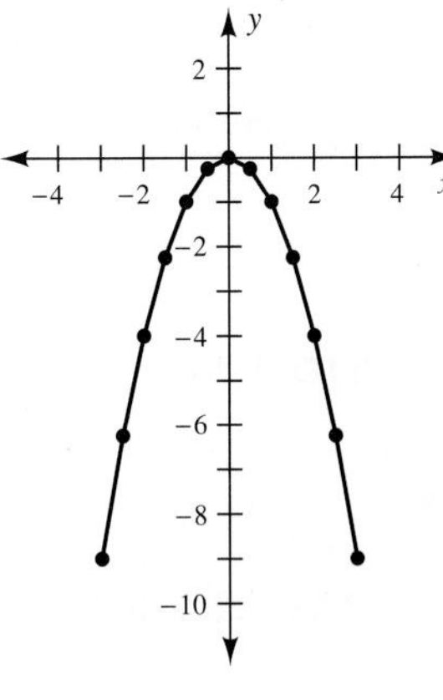

59 a.

x	y
-3	7
-2	2
-1	-1
0	-2
1	-1
2	2
3	7

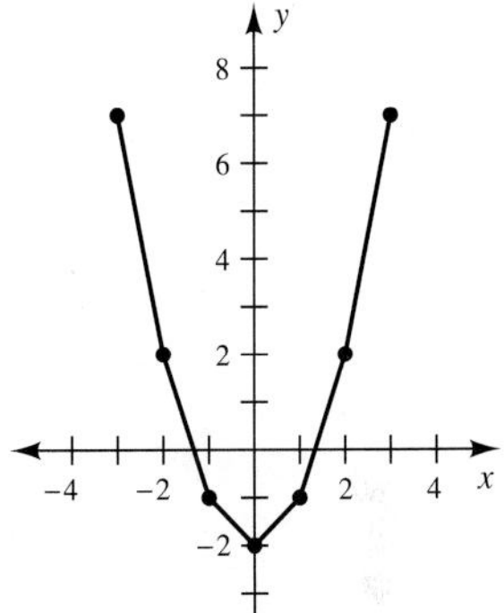

b.

x	y
-3	7
-2.5	4.25
-2	2
-1.5	0.25
-1	-1
-0.5	-1.75
0	-2
0.5	-1.75
1	-1
1.5	0.25
2	2
2.5	4.25
3	7

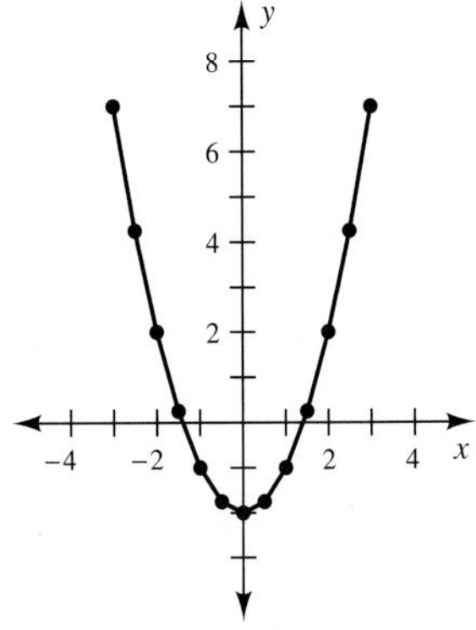

61 6.67 kg **63** 16 amps **65** Satisfy a linear function
67 Do not satisfy a linear function
69 a. ii **b.** iv **c.** i **d.** iii
71 Direct variation; $k = -\frac{1}{2}$
73 Inverse variation; $k = -2$ **75** $k = 2$
77 Direct variation; $k = 3$ **79** Inverse variation; $k = 120$
81 $P = \$700$ **83** $P = \$975$ **85** c

Section Skill Reinforcers 2-5

1 $m = 2$, $(0, -3)$ **2** $m = 4$, $(0, 2)$ **3** $m = \frac{2}{3}$, $(0, 6)$
4 $m = -\frac{4}{3}$, $(0, 12)$ **5** $m = 0$, $(0, 8)$

Review Exercises 2

1 $m = \frac{3}{4}$

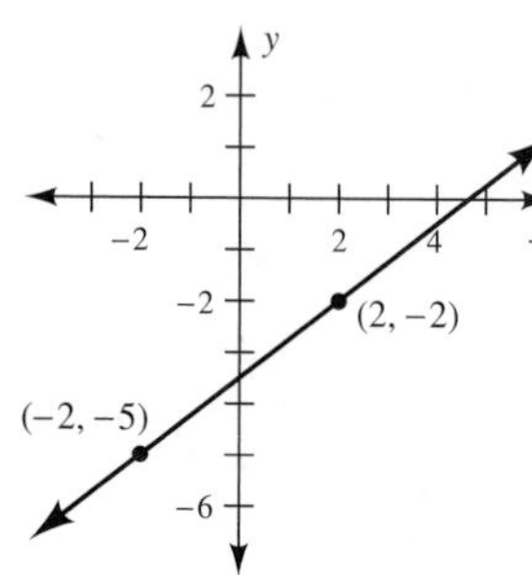

2 $m = 0$

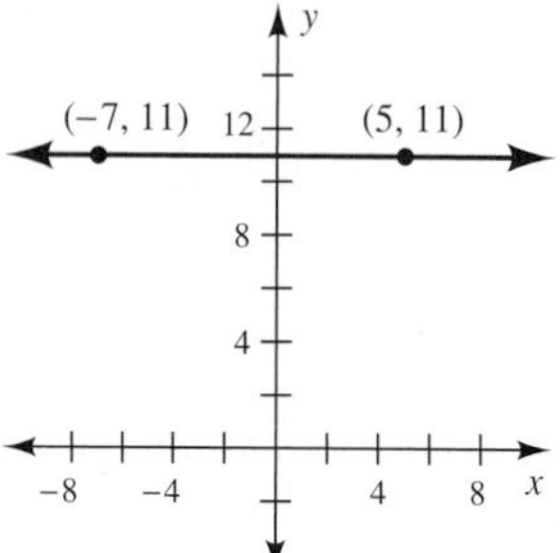

3 $m = 0$

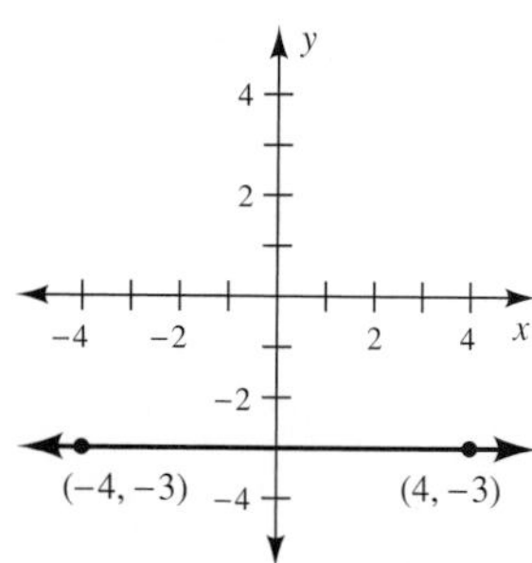

4 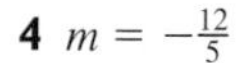 $m = -\frac{12}{5}$

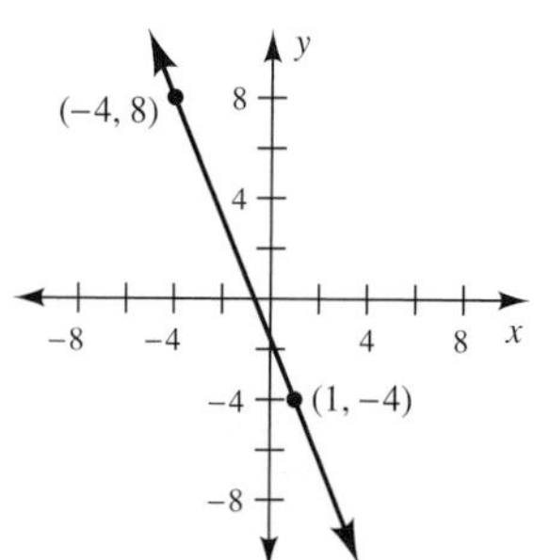

5

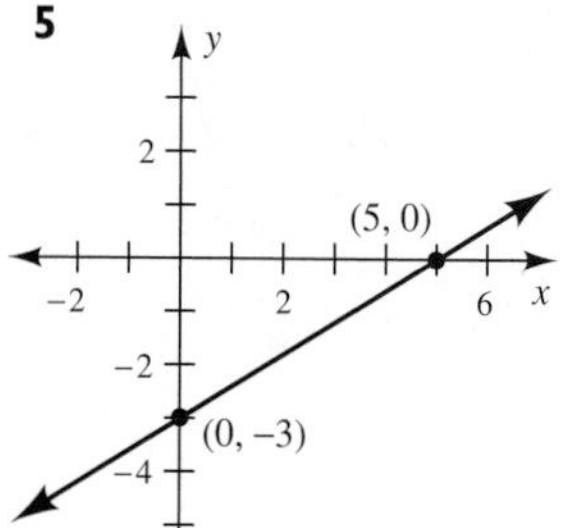

6

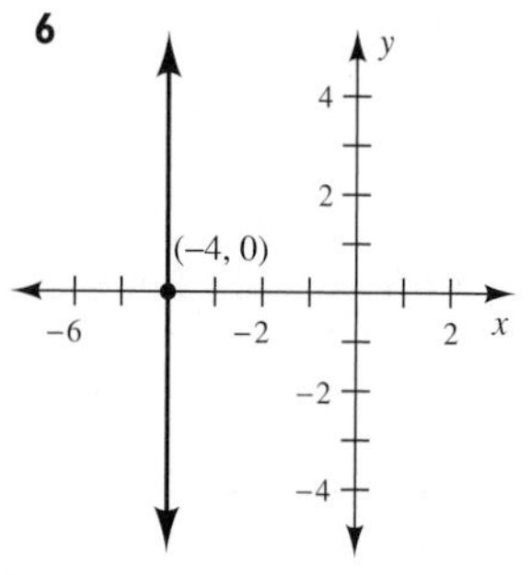

7

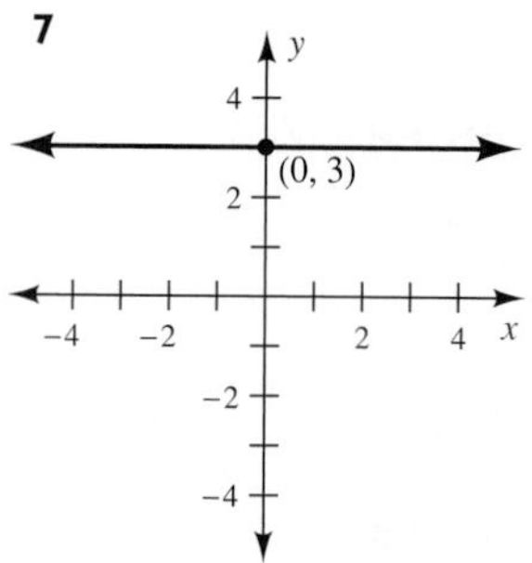

8

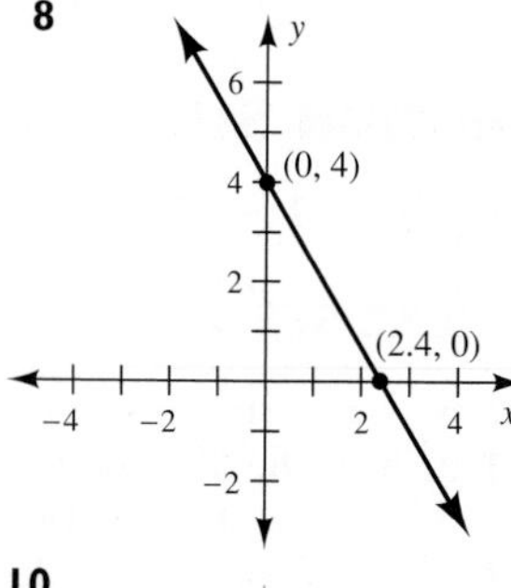

9

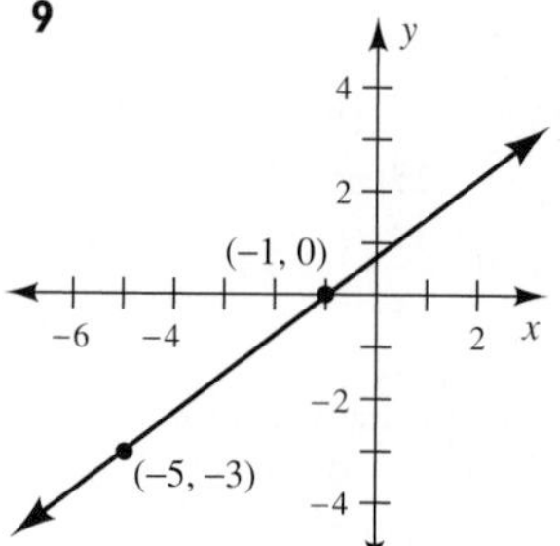

10

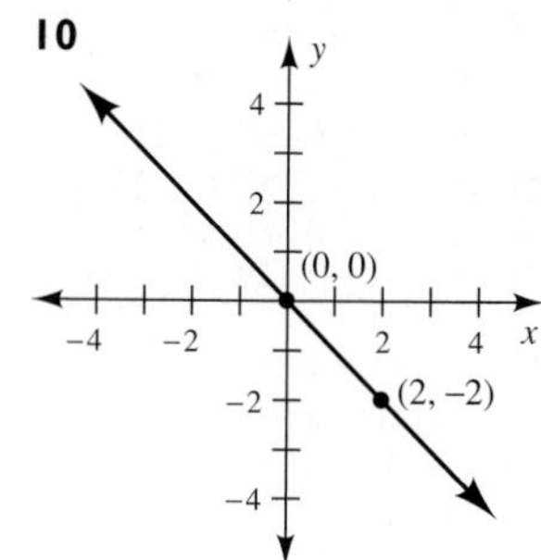

11

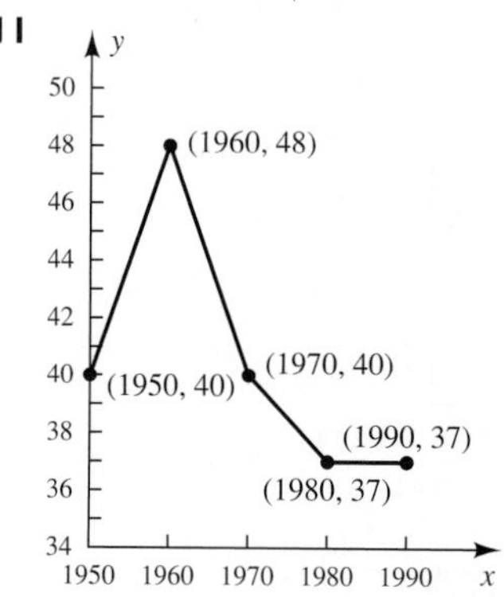

12 Parallel **13** Neither parallel nor perpendicular
14 Perpendicular **15** Parallel
16 Not a right triangle **17** $2x + 3y - 5 = 0$
18 $3x + y - 7 = 0$ **19** $4x + 5y - 10 = 0$
20 $5x + 2y - 10 = 0$ **21** $5x - y = 0$ **22** $x + 6y = 0$
23 $x + 5 = 0$ **24** $y - 4 = 0$ **25** $y - 7 = 0$
26 $x - 1 = 0$ **27** $x + y - 5 = 0$
28 $2x - 3y + 4 = 0$ **29** $4x + y - 4 = 0$
30 $\overline{AB}$: $m = 2$; $\overline{BC}$: $m = -1$; $\overline{CD}$: $m = 0$; $\overline{DE}$: $m = -3$; A to B
31 Function: $D = \{3, 6, 8, 9\}$, $R = \{1, 2, 4, 5\}$
32 Function: $D = \{3, 8, 9\}$, $R = \{1\}$
33 Relation that is not a function: $D = \{1, 6, 7\}$, $R = \{4, 5, 8, 9\}$
34 Function: $D = \{-5, -\pi, 0, \pi, 11\}$, $R = \{13, 16, 18, 19, 25\}$
35 Relation that is not a function: $D = \{-3, -2, -1, 1\}$, $R = \{-3, -2, -1, 0, 1, 2, 3\}$
36 Function: $D = \{-3, -2, -1, 0, 1, 2, 3\}$, $R = \{-2, -1, 0, 1\}$

37 Function: $D = \{1, 2, 3, 4\}$, $R = \{1, 2, 7\}$

38 Relation that is not a function: $D = \{4, 5, 7\}$, $R = \{-7, 2, 3, 8\}$

39 Function: $D = (-3, 3]$, $R = [-1, 2]$

40 Relation that is not a function: $D = [-1, 1]$, $R = [-2, 2]$

41 Relation that is not a function: $D = (-\infty, -2] \cup [2, +\infty)$, $R = \mathbb{R}$

42 Function: $D = (-3, 3]$, $R = [-2, 2]$

43 $\{(-3, 3), (-2, 2), (-1, 1), (0, 0), (1, 1), (2, 2), (3, 3)\}$

44 $\{(-4, -6), (-2, 6), (0, 10), (1, 9), (3, 1), (5, -15)\}$

45 $\{(1984, 3), (1985, 5), (1986, 3), (1987, 4.5), (1988, 6), (1989, 2)\}$

46

x	1940	1950	1960	1970	1980	1990
y	4	5	6	8	14	17

47 -2 **48** 13 **49** -23 **50** -19 **51** 3
52 13 **53** -48 **54** $2h^2 - h + 3$ **55** $2h^2 - h - 6$
56 $2h^2 - 9h + 13$ **57** 4.5 c **58** $R = \dfrac{kL}{r}$ **59** 7.5 L

60 a. \$600 **b.** 8 credit hours

61 a.

x	-3	-2	-1	0	1	2	3
y	12	6	2	0	0	2	6

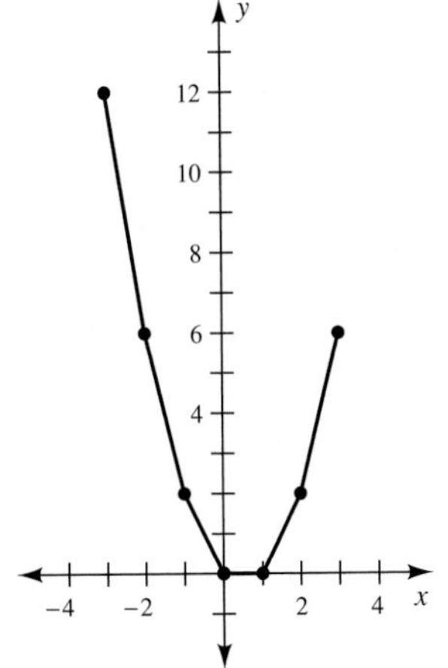

b.

x	-3	-2.5	-2	-1.5	-1	-0.5	0
y	12	8.75	6	3.75	2	0.75	0

x	0.5	1	1.5	2	2.5	3
y	-0.25	0	0.75	2	3.75	6

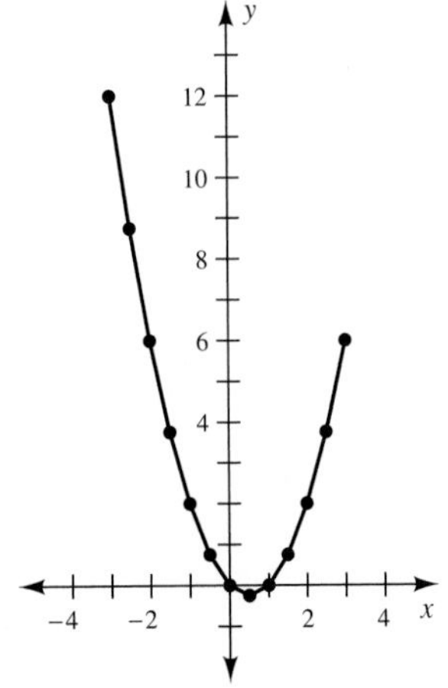

62 y_2 is the graph of y_1, translated down 2 units. y_3 is the graph of y_1, translated down 4 units.

$[-4, 4]$ for x, $[-5, 16]$ for y

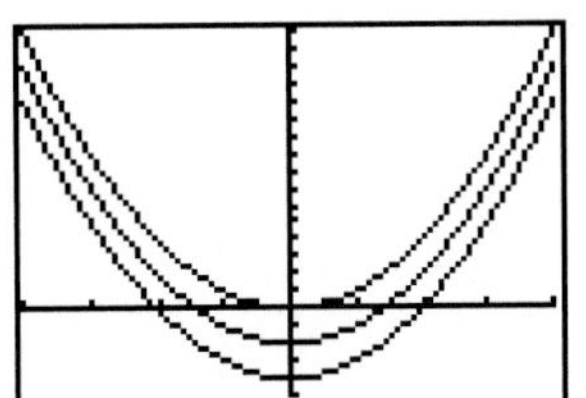

63 (1, B), (2, C), (3, A), (4, D) **64** $\mathbb{R}$ **65** $\mathbb{R}$
66 $\mathbb{R} \sim \{-2\}$ **67** 211.2 ft

MASTERY TEST 2

1

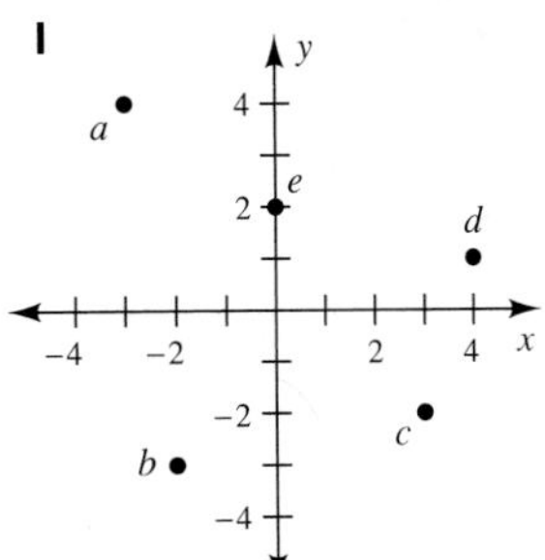

2 a.

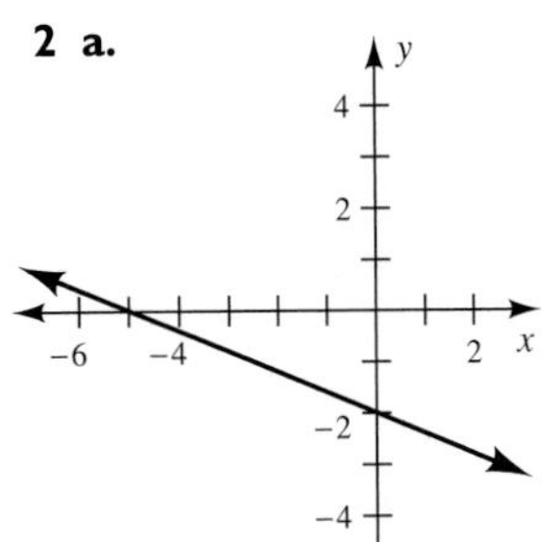

b.

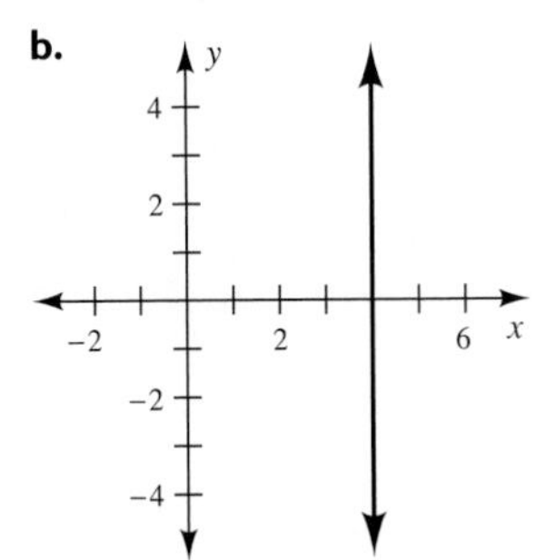

c.

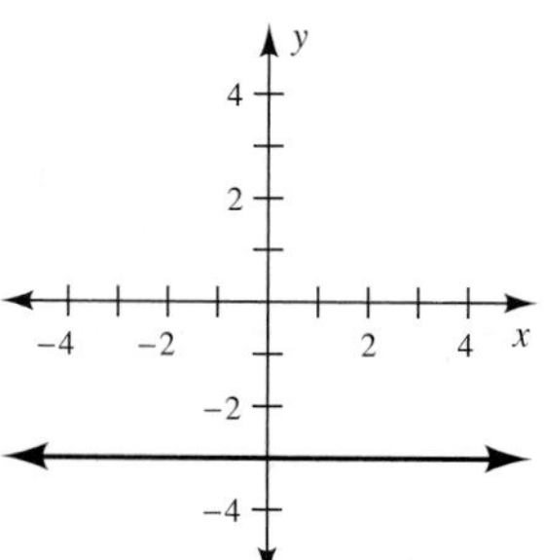

3 a.

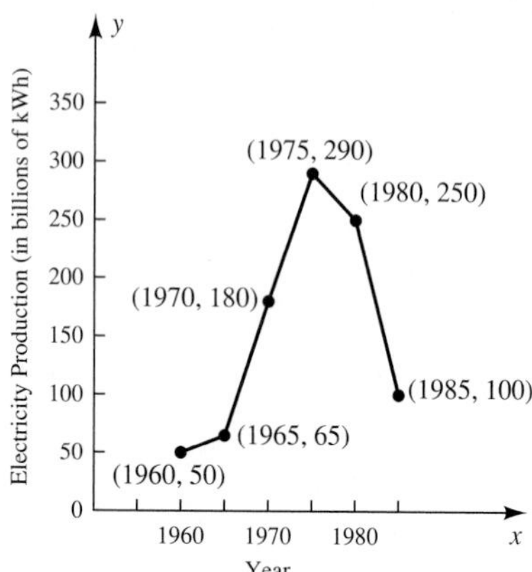

b.

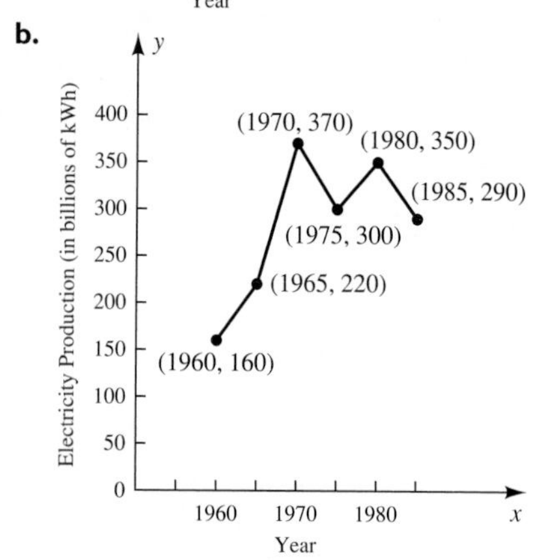

4 a. $m = 0$ **b.** Undefined **c.** $m = 2$ **d.** $m = -2$
e. $m = \frac{3}{2}$

5 a.

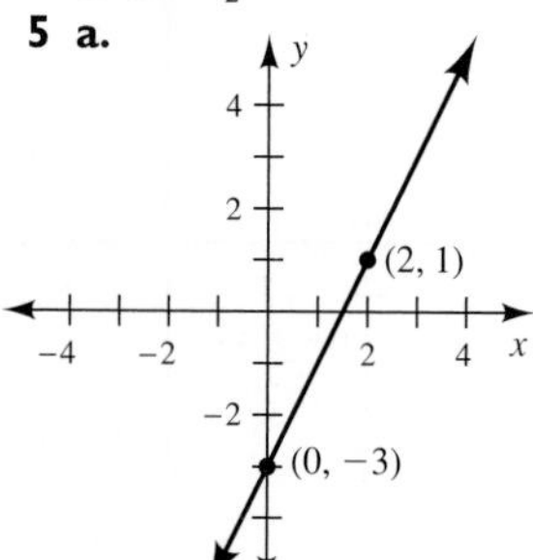

b.

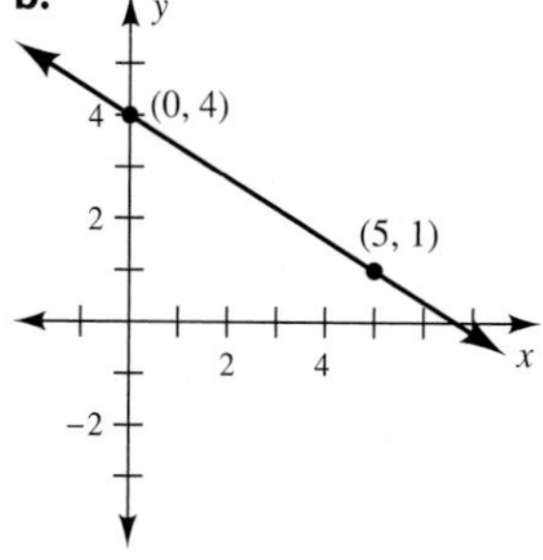

c.

6 a. $y + 2 = 0$ **b.** $x - 7 = 0$ **c.** $x - 5 = 0$
d. $y - 8 = 0$

7 a. $3x - 7y - 26 = 0$ **b.** $x - 2y + 12 = 0$
c. $2x + 3y + 15 = 0$ **d.** $2x + 3y - 8 = 0$

8 a. Function **b.** Not a function **c.** Function
d. Function **e.** Not a function **f.** Function

9 a. $D = \{1, 2, 7, 8\}, R = \{4, 8, 11, 13\}$
b. $D = \mathbb{R}, R = \{7\}$ **c.** $D = [3, +\infty), R = [0, +\infty)$
d. $D = \{-3, -2, -1, 0\}, R = \{0, 1, 2, 7\}$
e. $D = \{-3, -2, -1, 0, 1\}, R = \{-2, -1, 0, 1, 2\}$
f. $D = [-2, +\infty), R = (-\infty, 2]$

10 a. -12 **b.** 2 **c.** -2 **d.** $3h^2 + h - 12$
e. $3h^2 + 13h + 2$ **f.** $3h^2 + h - 10$

11 a. $y = 36$ **b.** $z = \dfrac{kx^2}{y^3}$
c. V varies jointly as r^2 and h.

Chapter 3

EXERCISES 3-1

1 a. Not a solution **b.** Not a solution **c.** Solution

3

5

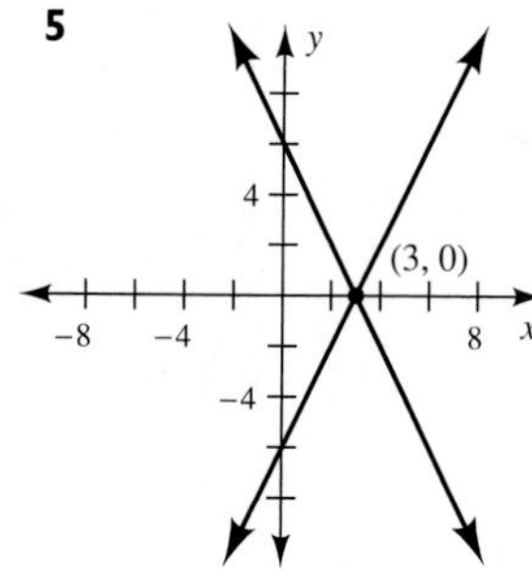

7 No solution

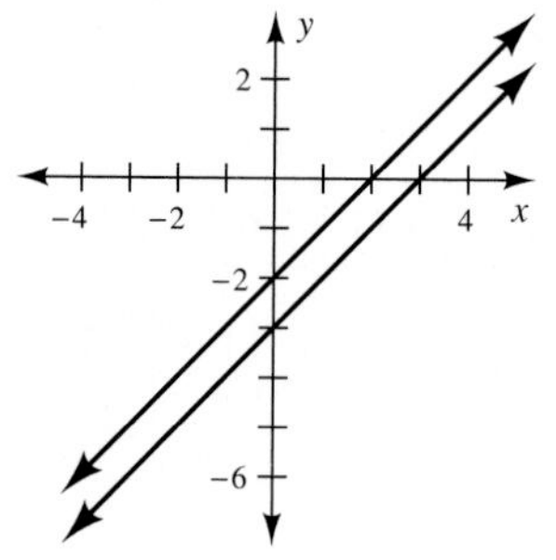

9 Every point of $x - 2y = 3$ is a solution.

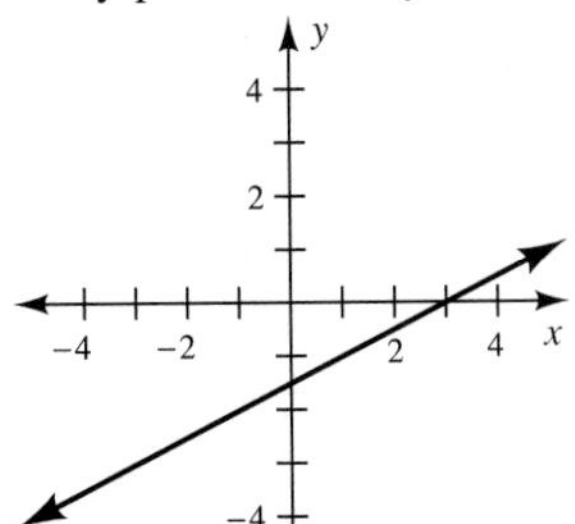

11

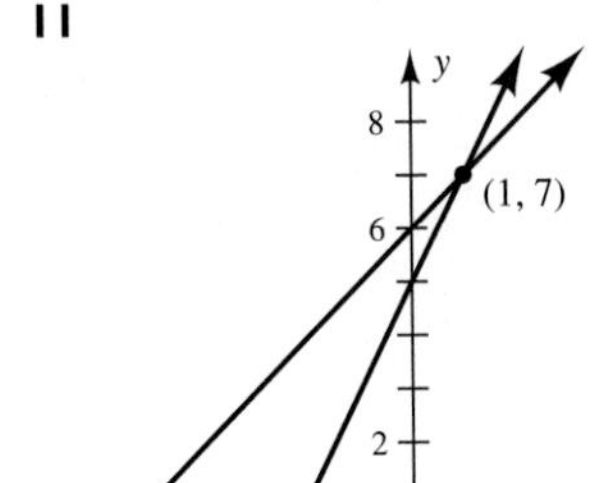

13

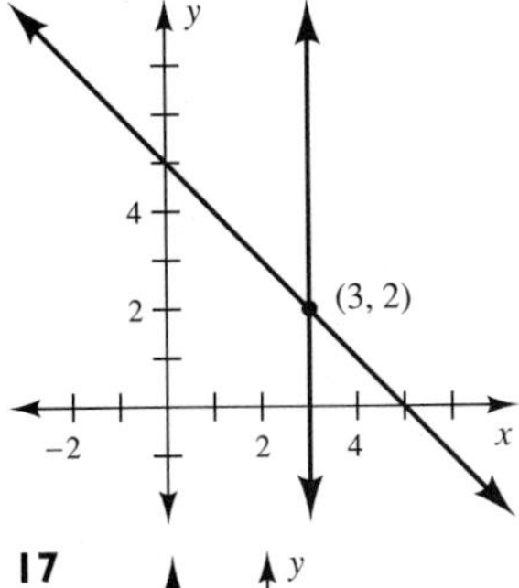

15

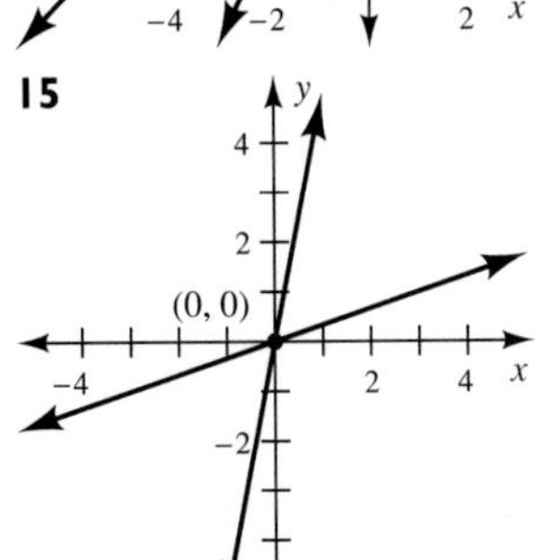

17

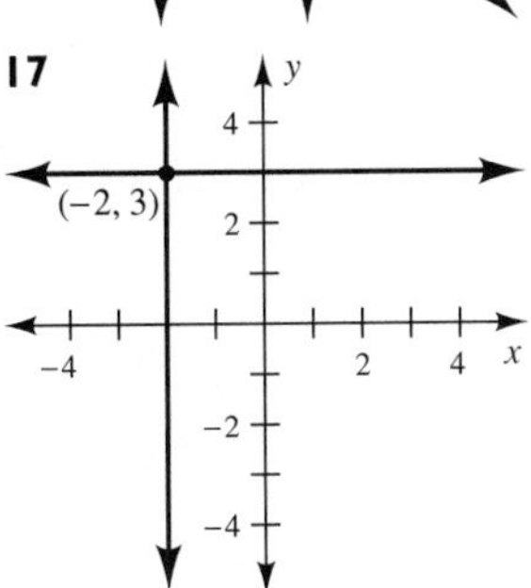

19

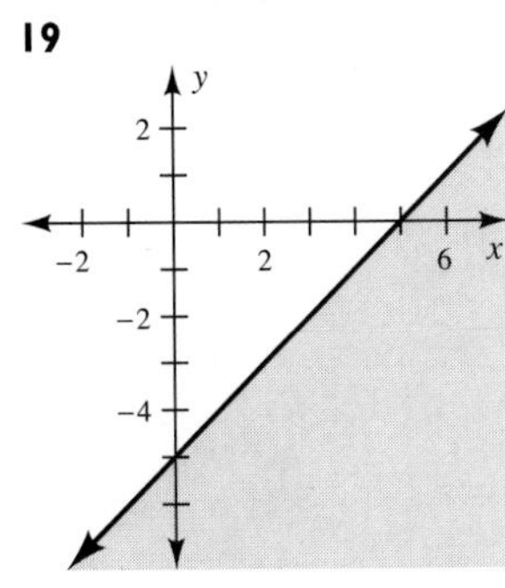

21

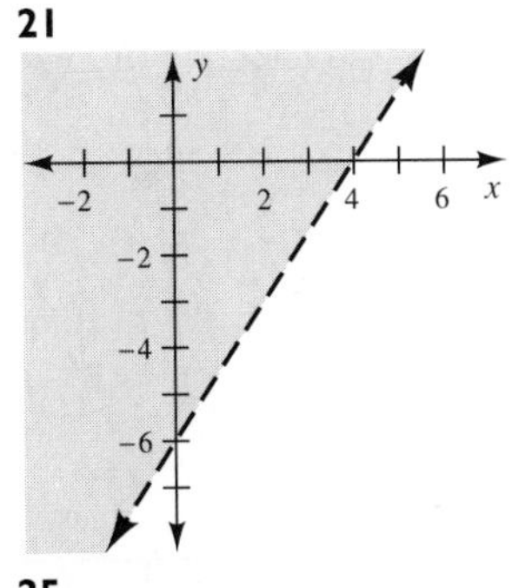

23

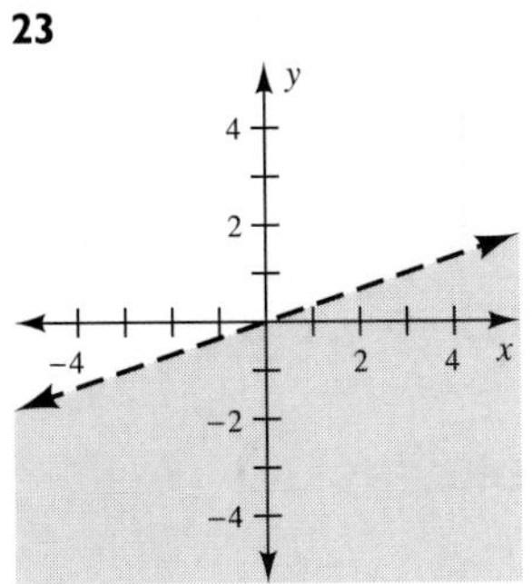

25

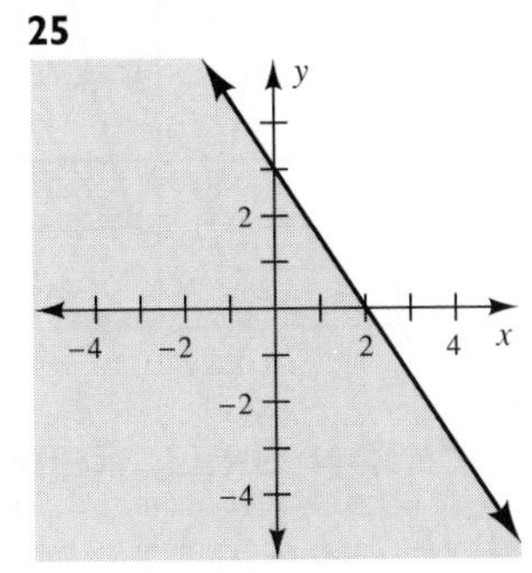

27

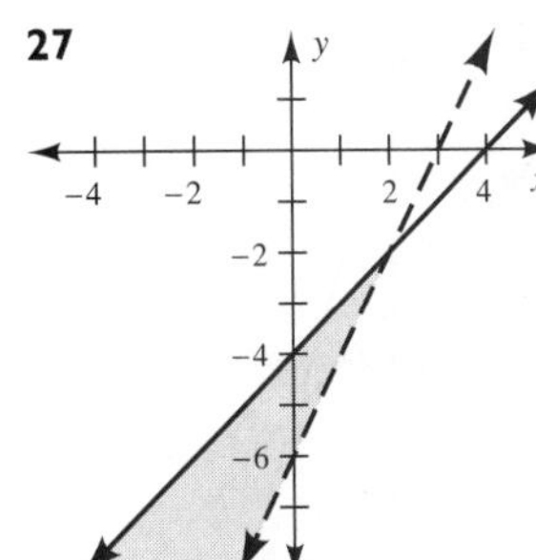

29

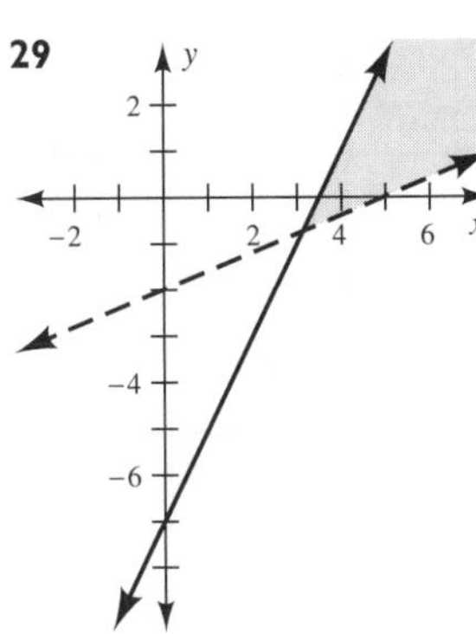

31

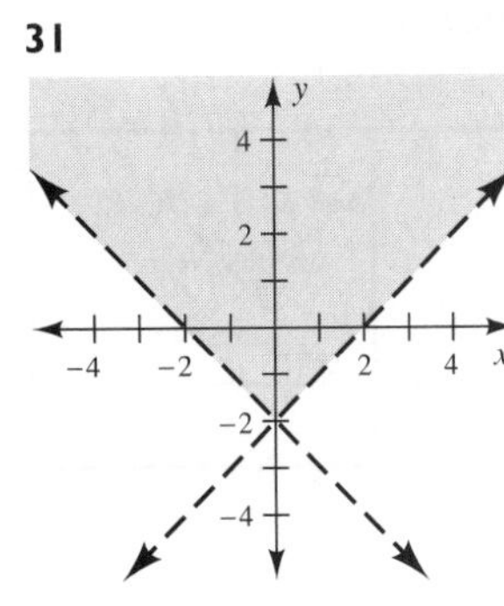

33

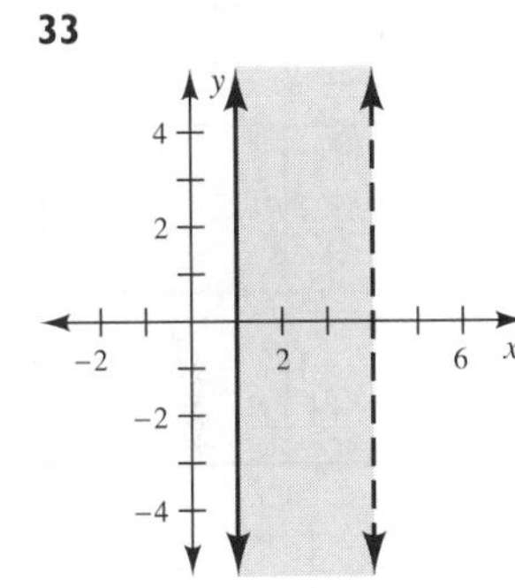

35

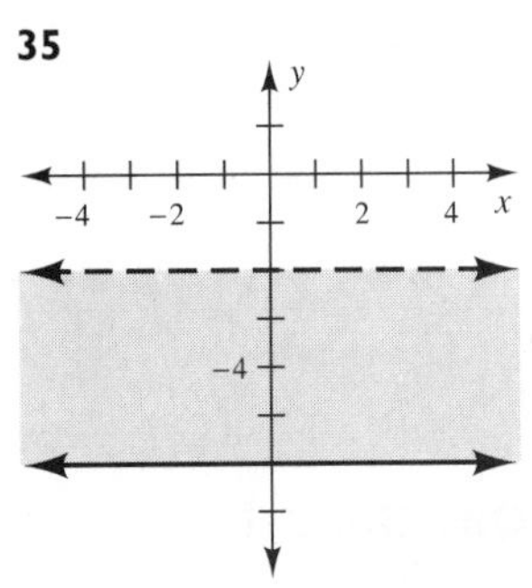

37

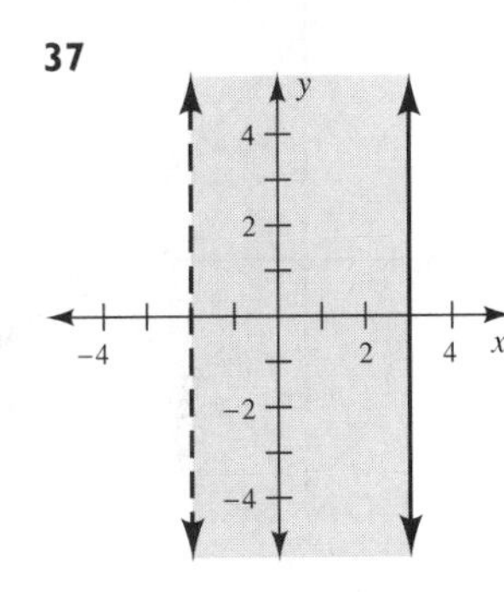

39 No solution **41** 1986; \$400,000 **43 a.** 1 **b.** 1

45 $y = \frac{2}{7}x - 2$, $y = -\frac{3}{5}x - \frac{11}{5}$; one solution

47 $y = 3x - 4$, $y = 3x + \frac{10}{3}$; no solution

49 $y = \frac{2}{3}x - 2$, $y = \frac{2}{3}x - 2$; an infinite number of solutions

51 $y = 1.25x - 7$, $y = -1.5x + 0.3$; one solution

53 $x = \frac{26}{27}$ (a vertical line cannot be put in slope-intercept form), $y = 2x - 5$; one solution

55

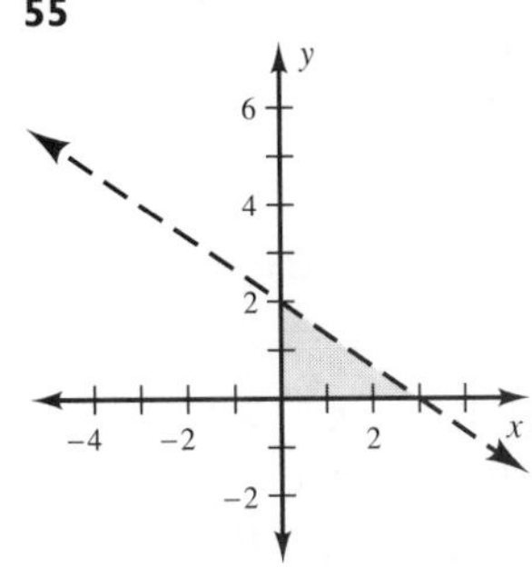

57

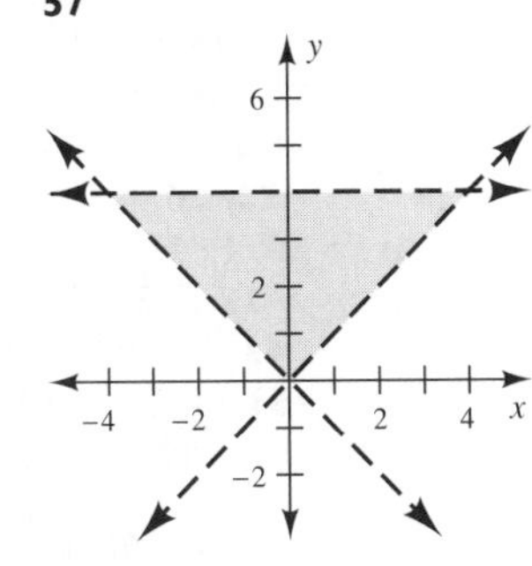

59

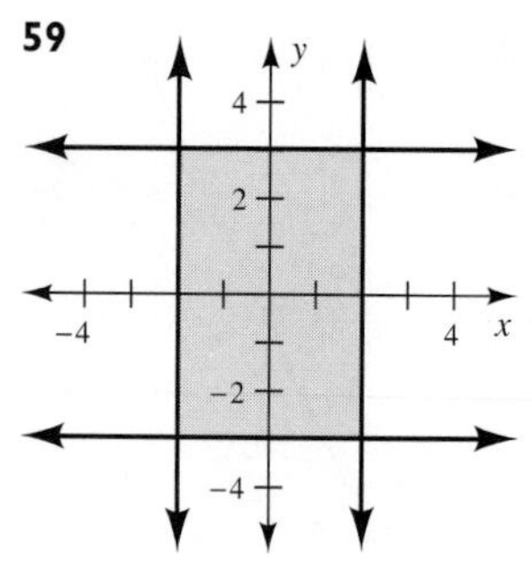

61

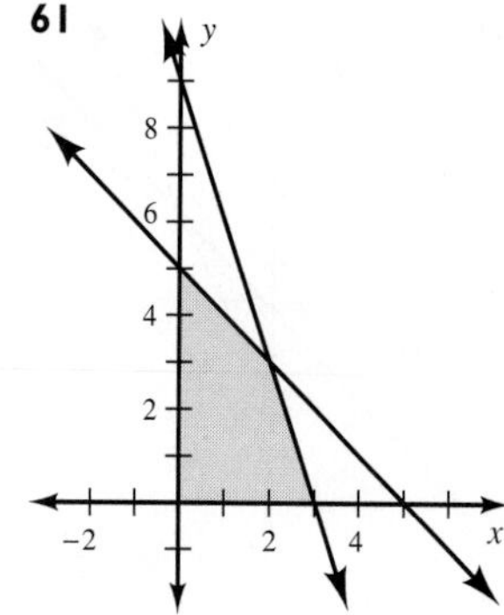

63 (0.1, 0.8)

[−2, 2] for x, [−2, 2] for y

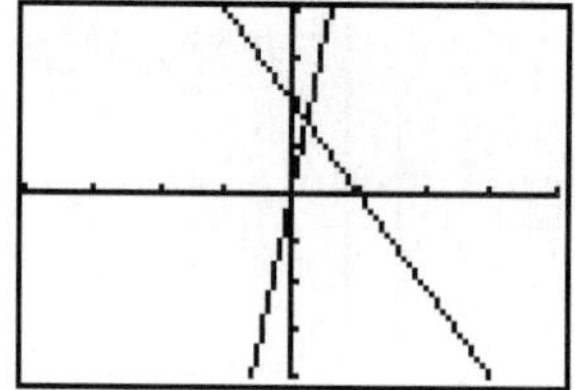

65 (1.9, 2.8)

[−2, 5] for x, [−2, 7] for y

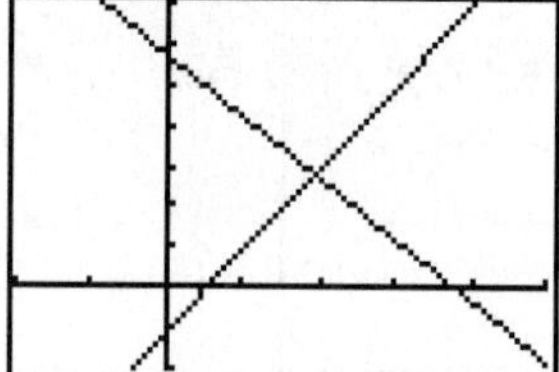

67

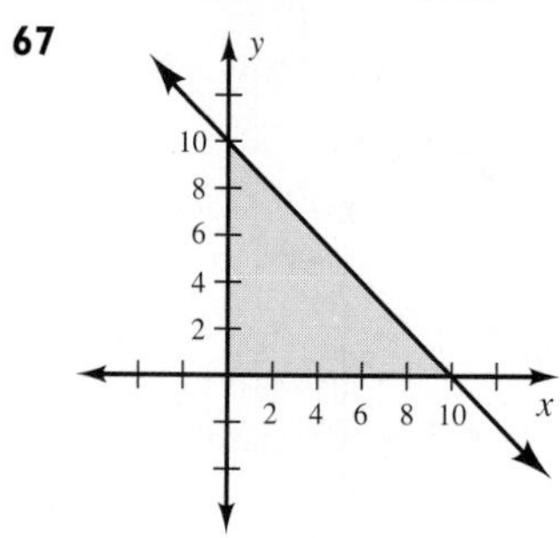

Section Skill Reinforcers 3-1

1 $x + y + 3 = 7$ **2** $x + y - 5 = 2$ **3** $2(x + y) = 8$
4 $2x - y = 11$ **5** $x + 7 = y$

Exercises 3-2

1 (5, 4) **3** $(-1, \frac{1}{2})$ **5** $(\frac{1}{2}, \frac{3}{2})$
7 A system of dependent equations with an infinite number of solutions on the line $x + 2y = 1$
9 (0, 0) **11** An inconsistent system; no solution
13 (−1, −1) **15** (10, 4) **17** (−5, 6)
19 A system of dependent equations with an infinite number of solutions on the line $4x - 5y = 2$
21 $(-\frac{1}{2}, \frac{1}{3})$ **23** An inconsistent system; no solution
25 (0, 4) **27** $(-1, \frac{1}{2})$ **29** (2, −3) **31** (−1, −1)
33 (−4, 3) **35** (−3, −7) **37** (−4, −2) **39** (3, 0)
41 85, 57 **43** 18, 59 **45** 11, 5 **47** 14, 3
49 $2400 **51** (4, 2) **53** $(\frac{1}{2}, \frac{1}{2})$ **55** $(-\frac{2}{3}, \frac{5}{2})$
57 (1.11, 0.45) **59** (−8.14, 9.03) **61** (1, −2)
63 $a = \frac{11}{5}, b = -3$ **65** $\left(\frac{3b}{2}, \frac{b}{2a}\right)$ **67** $\left(\frac{10}{7a}, \frac{1}{7b}\right)$

Section Skill Reinforcers 3-2

1 $-\frac{5}{2}$ **2** $5x + 2y - 29 = 0$ **3** $x = 3$ **4** $x = 7$
5 $y = -3$

Exercises 3-3

1 (2, 3, 3) **3** (1, 1, 1) **5** (1, 0, −2)
7 $(1, -\frac{3}{5}, -4)$ **9** $(-\frac{1}{2}, -\frac{3}{2}, \frac{1}{2})$ **11** (2, 1, 3)
13 An inconsistent system; no solution **15** (1, 1, 1)
17 (2, 0, 2) **19** $(-\frac{1}{11}, -2, \frac{1}{2})$ **21** 100, 83, 15
23 15, 19, 31 **25** 72 cm, 60 cm, 36 cm
27 57°, 28.5°, 94.5° **29** (1, −2, 5) **31** $(1, -\frac{1}{3}, \frac{1}{2})$
33 $a = 2, b = -1, c = 0$ **35** $\left(\frac{1}{a}, \frac{1}{b}, 0\right)$

37

x	y_1	y_2
−2.5	7	−75.94
−2.0	3	−32.00
−1.5	0	−9.19
0.0	−3	0.00
0.5	−2	−1.69
1.5	3	−0.94

a.

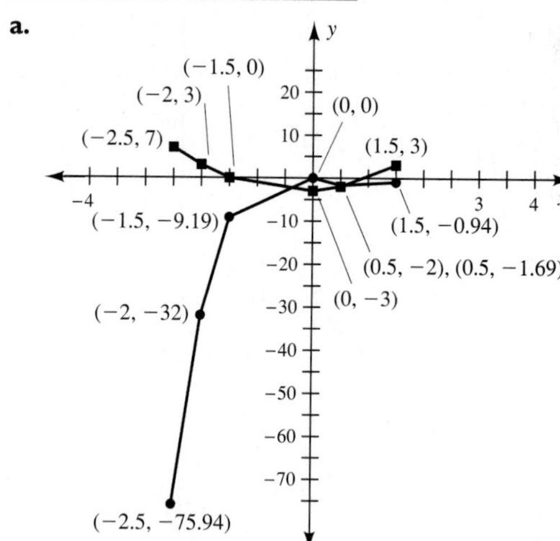

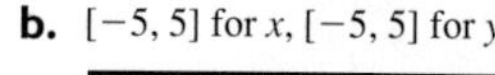
b. [−5, 5] for x, [−5, 5] for y

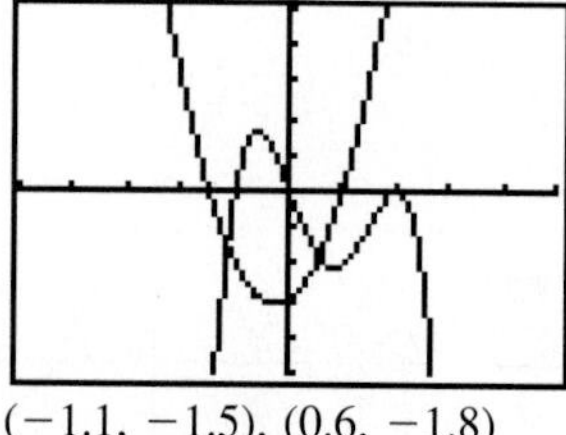

(−1.1, −1.5), (0.6, −1.8)

Section Skill Reinforcers 3-3

1 $y = 7$ **2** $2x + y + 1 = 0$ **3** $7x + 4y = 0$
4 $3x + 5y - 23 = 0$ **5** $x = -4$

EXERCISES 3-4

1 2, −3 **3** 23, 7 **5** Width = 90 ft, length = 160 ft
7 42°, 138° **9** 700 g of A, 1200 g of B
11 40 lb of 10%, 60 lb of 20%
13 \$2500 at 8%, \$1500 at 11%
15 800 adults, 475 youths
17 40 subcompacts, 25 full-sized cars
19 air speed of the plane is 525 mi/h; speed of the wind is 125 mi/h.
21 120 units **23** −30, 85, 100 **25** 20°, 40°, 120°
27 $y = 3x^2 - 5x + 4$ **29** $D = -2, E = 4, F = 1$
31 $A = 2, B = 3, C = -5$
33 600 g of A, 800 g of B, 1300 g of C
35 80 acres of A, 150 acres of B, 50 acres of C

SECTION SKILL REINFORCERS 3-4

1 12 **2** 20 **3** −12

4

x	−3	−2	−1	0	1	2	3
y	−6.75	−2	−0.25	0	0.25	2	6.75

5

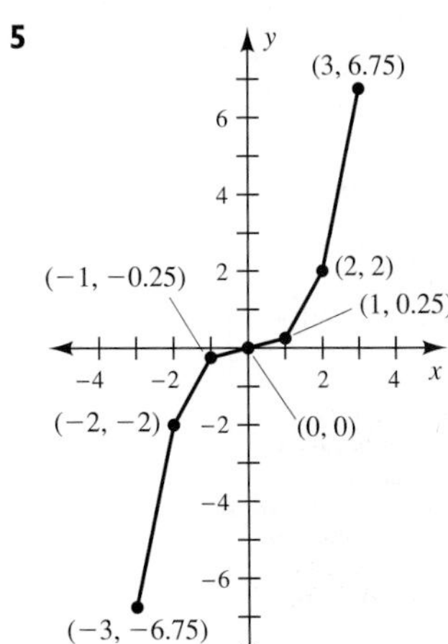

REVIEW EXERCISES 3

1

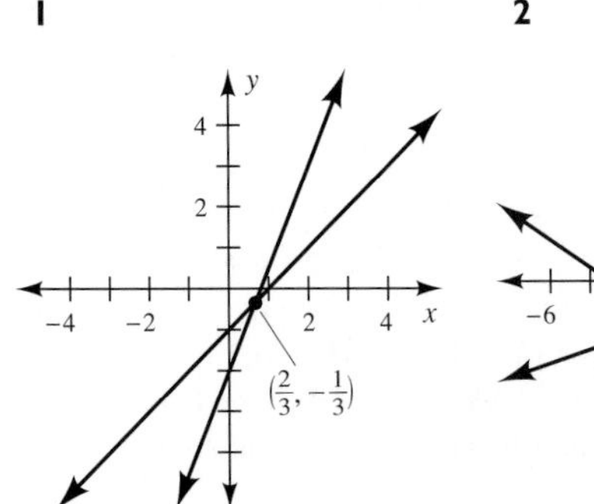

2

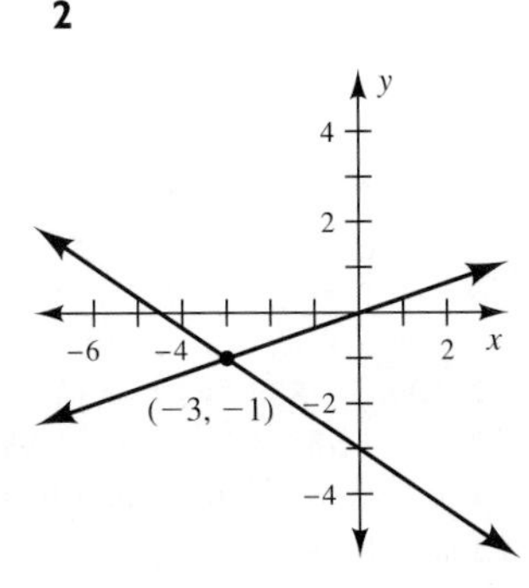

3 No solution

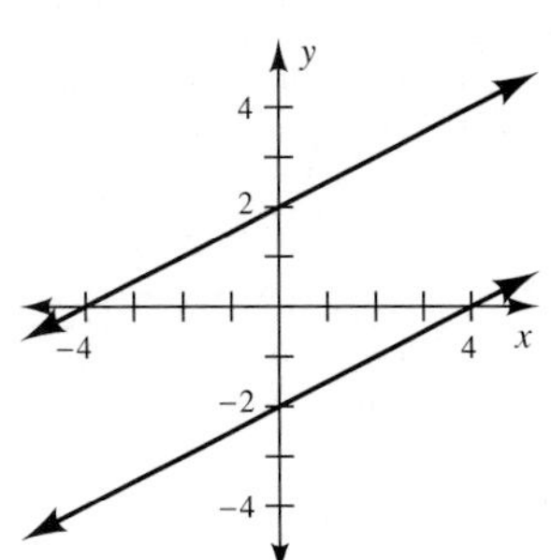

4

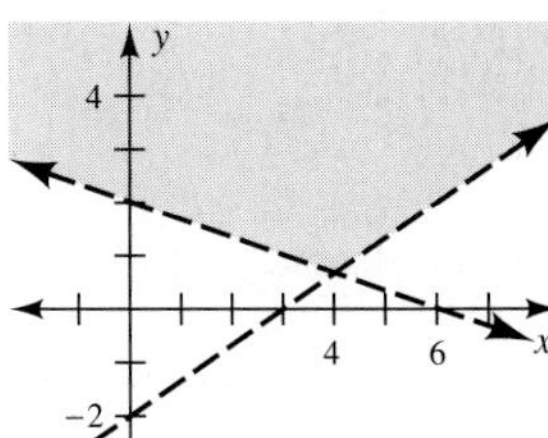

5

6

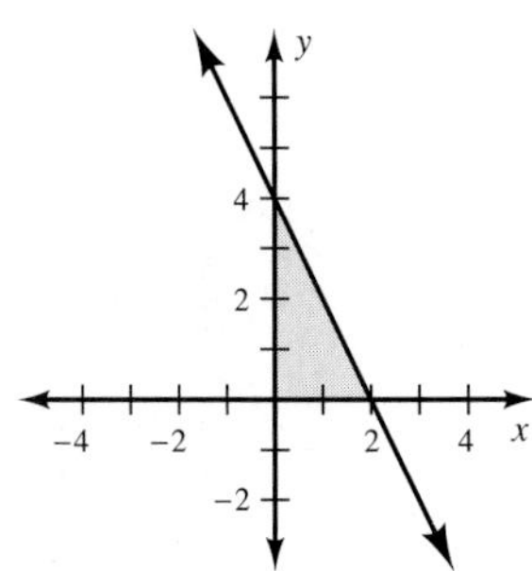

7 (−5, 6) **8** $(-\frac{1}{3}, \frac{3}{2})$
9 An inconsistent system; no solution **10** (8, −5)
11 (7, −2, −3) **12** $(\frac{1}{2}, \frac{1}{3}, -\frac{1}{5})$ **13** (3, 5)
14 (−0.3, 0.7) **15** (1, 0, −4) **16** (2, −3)
17 An inconsistent system; no solution
18 A system of dependent equations with an infinite number of solutions on the line $3x - 5y = -4$
19 An independent system with one solution
20 (1.0, −0.2) **21** (3.0, 5.0) **22** (−0.3, 0.7)
23 $a = -2, b = 3, c = 4$ **24** 11, 4 **25** 24°, 156°
26 200 g of A, 300 g of B **27** 40 cm, 50 cm, 80 cm
28 −5, 21, 9 **29** $y = 6x^2 - 13x + 6$

30

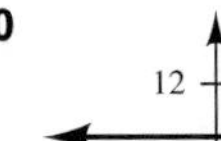

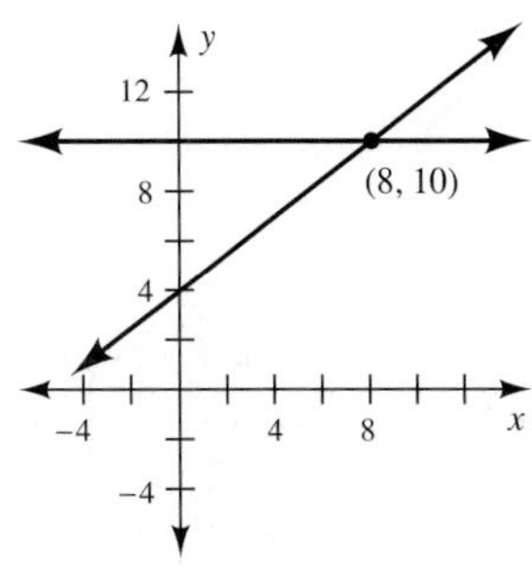

8 mi

31 ≈ 1987

MASTERY TEST 3

1 a.

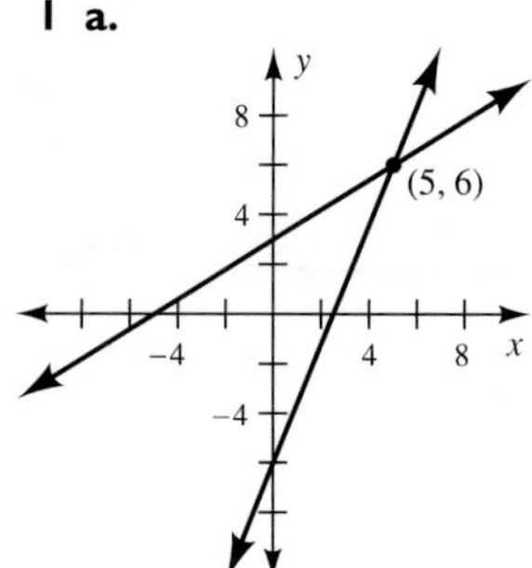

b.

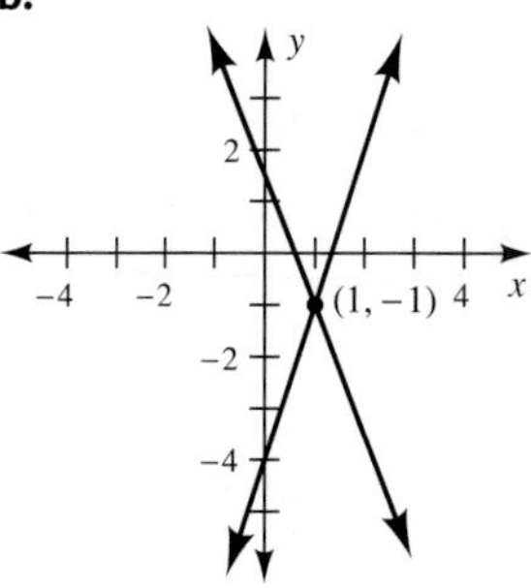

c. No solution

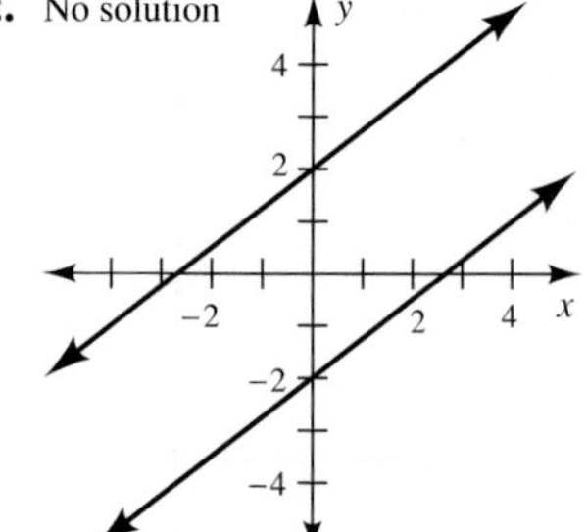

2 a.

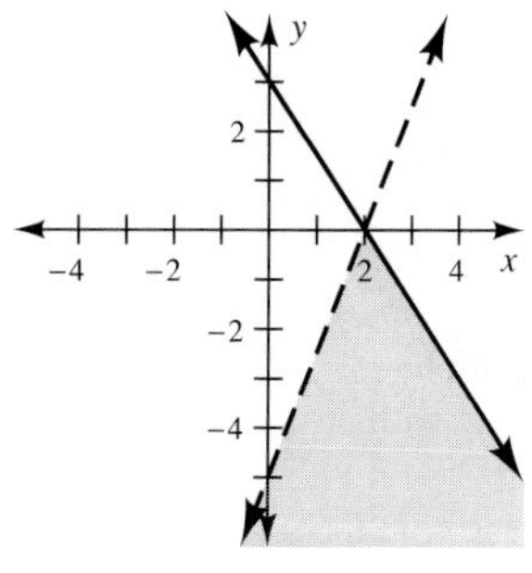

b.

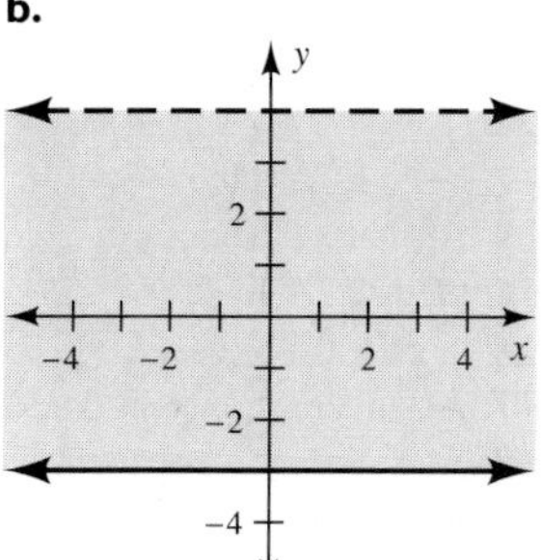

c.

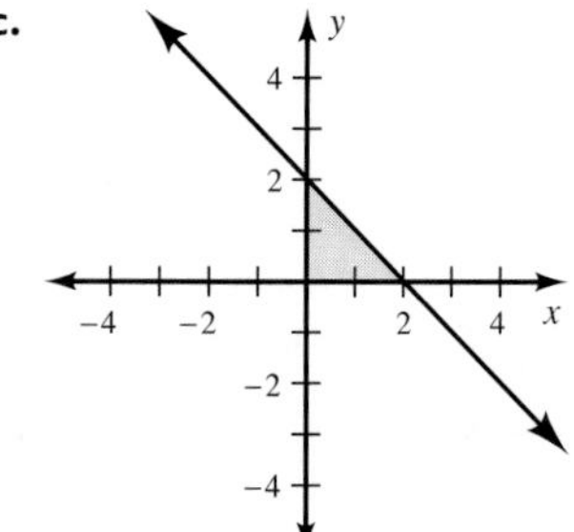

3 a. $(3, -5)$ **b.** $(2, -3)$ **c.** No solution

4 a. $(\frac{1}{2}, -\frac{1}{3})$

b. A system of dependent equations with an infinite number of solutions on the line $x - 4y = 5$

c. $(\frac{1}{7}, -\frac{3}{8})$

5 a. $(5, 2, 1)$ **b.** $(-2, -1, 7)$

6 a. 5 cm, 7 cm, 10 cm

b. \$7000 at 7.5%, \$3000 at 9%

c. Rate of the barge is 14 km/h; rate of the current is 9 km/h.

Cumulative Review Exercises 1–3

1 -14 **2** -20 **3** -120 **4** -2 **5** 1 **6** 9
7 13 **8** 25 **9** 1 **10** 37 **11** 18 **12** $\frac{2}{5}$
13 a. $4m + 2$ **b.** $m = -\frac{1}{2}$ **14 a.** $6x - 6$ **b.** $x = 1$
15 a. -2 **b.** No solution
16 a. 0 **b.** All real numbers **17** $x - y - 13z$
18 $-5a - b + 15c$ **19** $6m - 12$ **20** $-30r + 36$
21 -3 **22** 5 **23** $x = \dfrac{a}{b}$ **24** $x = \dfrac{bc + 3}{2}$
25 $-9, 14$ **26** $-11, 17$ **27** $(3, +\infty)$
28 $(-4, 3)$ **29** $(-6, 4)$ **30** $(-\infty, 3) \cup (7, +\infty)$
31 a. $[2, 4)$ **b.** $(-2, 5]$ **c.** $[-1, 0]$ **d.** $[3, +\infty)$
32 Distributive property
33 Commutative property of addition
34 Commutative property of multiplication
35 Associative property of addition
36 a. $-7, -4.73, -\pi, -\sqrt{4}, -\frac{3}{7}, 0, \sqrt{7}, 17$ **b.** $-\pi, \sqrt{7}$
c. $-7, -4.73, -\sqrt{4}, 0, 17$ **d.** $-7, -\sqrt{4}, 0, 17$
e. 0, 17 **f.** 17
37 2 **38** 3 **39** 0 **40** Undefined **41** $-\frac{3}{4}$

42

43

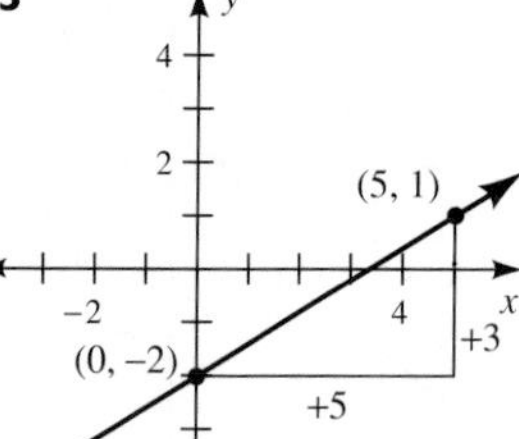

44

45

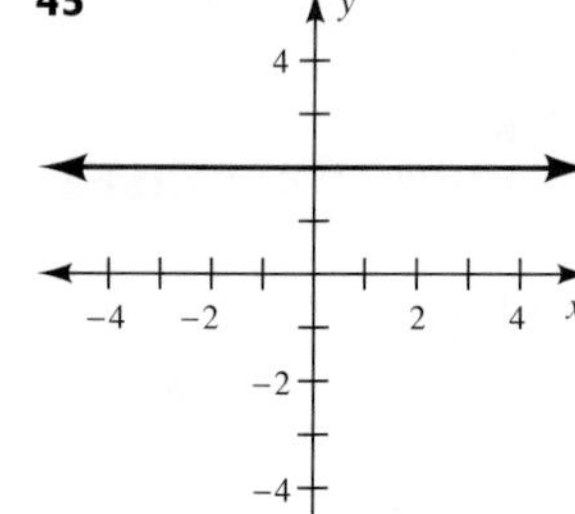

46 $y - 7 = 0$ **47** $x - 3 = 0$ **48** $2x + y - 13 = 0$
49 $x + y - 10 = 0$ **50** $2x + 3y - 27 = 0$ **51** 28
52 109 **53** -81 **54** -11 **55** -0.5 **56** $\frac{4}{3}$
57 0 **58** Undefined **59** $\mathbb{R}$ **60** $\mathbb{R} \sim \{7\}$
61 Function **62** Not a function **63** Function
64 Not a function

65

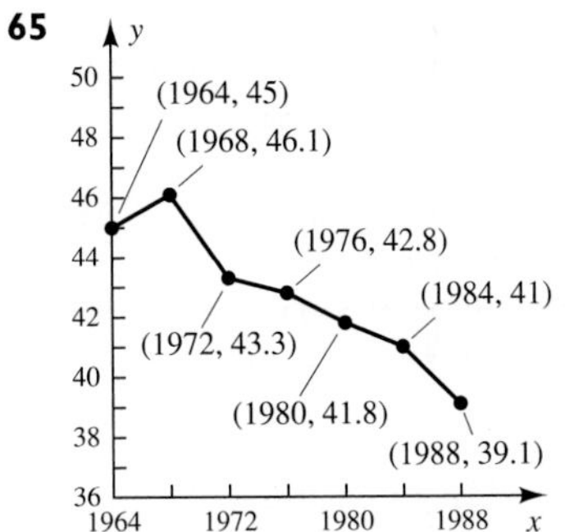

66 a. 1925 **b.** 5 billion **c.** 125 yr **d.** 1960
67 Period 1, event B; Period 2, event D; Period 3, event A; Period 4, event C

68

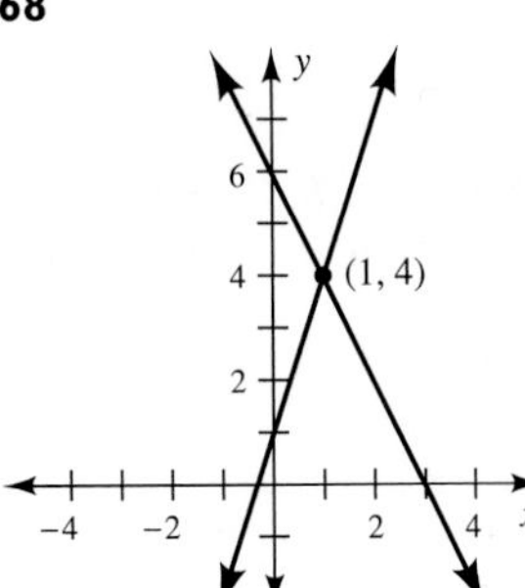

69

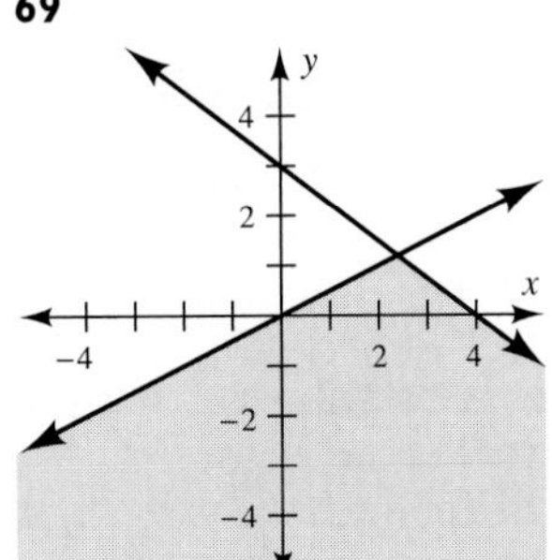

70

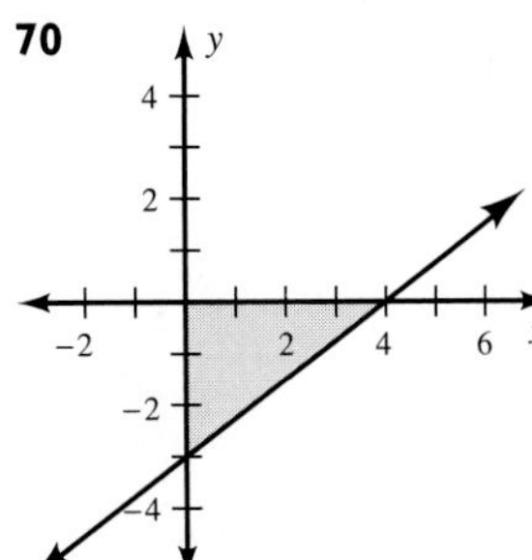

71 $(-6, -2)$ **72** $(\frac{1}{2}, -\frac{1}{3})$ **73** $(2, -1, 1)$
74 $v = 10$ **75** 20 units, 48 units **76** \$600
77 600 g of A, 800 g of B

78 a.

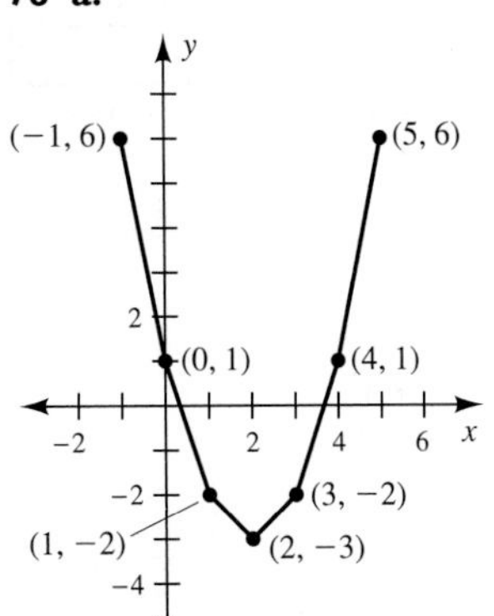

b. $[-4, 8]$ for x, $[-4, 8]$ for y

Chapter 4

Exercises 4-1

1 a. 25 **b.** -25 **c.** 25 **d.** -25
3 a. -1 **b.** 1 **c.** 0 **d.** 1
5 a. 1 **b.** -1 **c.** 1 **d.** -1
7 a. $\frac{1}{2}$ **b.** -2 **c.** $-\frac{1}{2}$ **d.** $-\frac{1}{2}$
9 a. 1 **b.** 0 **c.** -1 **d.** 4
11 a. 100 **b.** 100,000 **c.** .01 **d.** .00001
13 a. $\frac{9}{5}$ **b.** $\frac{9}{4}$ **c.** $\frac{1}{5}$ **d.** $\frac{5}{6}$
15 a. 141 **b.** 44 **17** -2 **19** x^{18} **21** x^4
23 $\frac{1}{x^4}$ **25** a^9 **27** $\frac{1}{y^8}$ **29** $-35x^{12}$ **31** $12a^7b^5$
33 $3v^8$ **35** $-\frac{6x}{7}$ **37** $\frac{1}{x-y}$ **39** $\frac{r+s}{rs}$
41 $\frac{1}{r+s}$ **43** $\frac{2}{x^8}$ **45** $\frac{3b^{14}}{4a^{12}}$ **47** $-\frac{6x}{y^2}$
49 $-\frac{8x^2y^{13}}{3}$ **51** $\frac{2b^3}{a^3}$ **53** 1
55 a. $\frac{1}{64}$ **b.** $\frac{1}{81}$ **c.** $-\frac{1}{64}$ **d.** $-\frac{1}{81}$
57 a. $\frac{1}{x^2}$ **b.** $\frac{2}{x}$ **c.** $2x$ **d.** x^2
59 a. 5 **b.** 9 **c.** -4 **d.** 4
61 a. $-\frac{1}{3}$ **b.** $-\frac{3}{2}$ **c.** $\frac{1}{2}$ **d.** $\frac{1}{2}$
63 x^{m+3} **65** y^{m+5} **67** x^2 **69** d **71** a
73 a. 0 **b.** 1 **c.** Undefined **d.** 0
75 a. -1 **b.** 1 **c.** Undefined **d.** Undefined
77 0.125 **79 a.** 1.6 **b.** -0.5

Section Skill Reinforcers 4-1

1 a. $x + 2$ **b.** $x = -2$
2 a. $5x + 8$ **b.** $x = -1.6$
3 a. $8x - 24$ **b.** $x = 3$ **4 a.** $\frac{x-6}{6}$ **b.** $x = 6$
5 a. $\frac{x-2}{10}$ **b.** $x = 2$

Exercises 4-2

1 a. x^6 **b.** $\frac{1}{x^6}$ **c.** x^2y^2 **d.** x^3y^6
3 a. v^{20} **b.** v^8w^{12} **c.** $\frac{w^{12}}{v^8}$ **d.** 1
5 a. $\frac{v^8}{w^{12}}$ **b.** $\frac{w^{12}}{v^8}$ **c.** $\frac{1}{v^8w^{12}}$ **d.** $w^{12}v^8$
7 a. $8x^6$ **b.** $\frac{9}{x^6}$ **c.** $\frac{x^6}{9}$ **d.** 1
9 a. $\frac{8a^3}{b^6}$ **b.** $\frac{b^6}{8a^3}$ **c.** $\frac{8b^6}{a^3}$ **d.** $\frac{1}{8a^3b^6}$
11 64 **13** 16 **15** 10^8 **17** $32x^{15}y^{20}$ **19** $\frac{81x^8}{y^{12}}$

21 $\frac{9y^6}{4z^2}$ **23** $\frac{9}{4a^2b^{18}}$ **25** $\frac{1}{a^8b^{48}c^{44}}$ **27** $-\frac{v^2}{12w^5}$
29 1 **31** $\frac{a^6c^{24}}{4b^{12}}$ **33** 8 **35** $\frac{25}{4}$
37 a. -30 **b.** 100 **39 a.** 28 **b.** 46
41 29,980,000,000 cm/s **43** .000000003 s
45 6.023×10^{23} molecules/mole
47 6.673×10^{-11} Nm²/kg² **49** 9.87×10^6
51 1.8×10^{-6} **53** a **55** c **57** a **59** x^{2m+6}
61 $\frac{1}{x^{m+2}}$ **63** $x^{3m}y^9$ **65** $\frac{y^{2m}}{x^{2m+2}}$ **67** .0000789
69 $-4{,}710{,}000$ **71** .00000000123 **73** 123,450,000
75 a. 328.509 **b.** 184.145 **77** 209.903

Section Skill Reinforcers 4-2

1 1 **2** 0 **3** 0.01 **4** -1 **5** 1

Exercises 4-3

1 a. Monomial, -7 **b.** Monomial, $\frac{1}{7}$
c. Not a monomial **d.** Not a monomial
3 a. Binomial, 4 **b.** Monomial, 7 **c.** Trinomial, 3
d. Monomial, 0
5 a. $-7a^5b^2c^2$ **b.** $-x^2 + x + 7$ **c.** $-3x^2y + 5xy^2$
d. $x^3y + 3x^2y^2 - 13xy^3$
7 -52 **9** -7 **11** 4 **13** 3 **15** 3 **17** 9
19 $k^2 - k + 3$ **21** 10 **23** 4 **25** -8
27 $-k^3 - 6k^2 + k + 10$ **29** 400 **31** $5x + 7$
33 $-2x^5$ **35** -40; loss of \$40 **37** 0; break even
39 0; break even **41** $3x + y + 13$ **43** $x + y + 13$
45 $\frac{1}{2}yx$ **47** $2x^3 - 5x + 11$ **49** $3w^6 + 5w^4$
51 $x^2 - y^2$ **53** $2w + 7$ **55** e **57** b
59 (I, B), (II, C), (III, D), (IV, A) **61** -12.30
63 85.44
65

x	-4	-3	-2	-1	0	1	2	3	4
$P(x)$	-32	-20	-10	-2	4	8	10	10	8

Section Skill Reinforcers 4-3

1 $[3, +\infty)$ **2** $(-\infty, 4)$ **3** $[1, 4]$ **4** $(-1, 7)$
5 $[-8, -2]$

Exercises 4-4

1 a. Unlike **b.** Like; $4x$ **c.** Unlike
d. Like: $9x^4y^2$
3 $8y + 6$ **5** $5v - 11$ **7** $3x^2 + x + 4$
9 $-x^2 + 5x - 14$ **11** $6ab + 7bc - 3cd$
13 $3m^2 - 12m + 5$ **15** $6a - 4b$ **17** $-2a + 14b$
19 $5x - 6y + 4z$ **21** $-x^2 + 2xy + 12y^2$
23 $a - 3b + 7c$ **25** $-9x - 6y - 23z$
27 $3v - 10w + 5x$ **29** $-a^2 - 3ab + b^2$ **31** $5a$
33 $10x^4 - 3x^3y + 3x^2y^2 - 10xy^3 + 12y^4$
35 $5x^2 - 15x - 7$ **37** $a^2 + 4a - 30$
39 $3.3x - 3.3y + 9.9$ **41** $12a - 5b$
43 $12w^4 + 6w^3 + 3w^2 + 8w - 30$ **45** $x^2 - 4x + 2$
47 $x^2 - y^2$ **49** $x^3 + 7, -x^3$ (Answers may vary.)
51 $x^2 + x, x + 1$ (Answers may vary.)
53 $x^2 + x + 1, -x^2 - x + 1$ (Answers may vary.) **55** a
57 $2t^3 + 5t^2 + 7t + 23$ **59** $2t^4 + 4t^2 + 32$
61 $8t^2 - 8t - 18$ **63** $t^3 - 2t^2 - 8t - 35$
65 a. The entire rectangle **b.** IV **c.** II **d.** I, III
e. Yes
67 9682.958
69 Subtract the quantity $4x^2 - 5x - 9$ to obtain $-x^2 + 14x + 22$. Use parentheses to prevent sign errors.

Section Skill Reinforcers 4-4

1 -12 **2** 180 **3** -120 **4** 54 **5** 144

Exercises 4-5

1 $6x^4$ **3** $-8x^3y^6$ **5** $3x^2 - 15x$ **7** $3x^2 + 3xy$
9 $-8x^2 + 20x$ **11** $-4a^3 + 8a^2 + 12a$
13 $4x^3 - 10x^2 - 6x$ **15** $6x^3y - 8x^2y^2$
17 $-6x^2y^2 + 15xy^3$ **19** $x^2 + xy - 6y^2$
21 $x^2 + 5xy + 6y^2$ **23** $x^2 - 5xy + 6y^2$
25 $x^3 - 2x^2 - 16x - 3$ **27** $10a^2 + 3ab - b^2$
29 $6a^3 + a^2 - a - 21$ **31** $x^2 - 1$ **33** $4x^2 - 25$
35 $x^2 - 64y^2$ **37** $a^2b^2 - 9$ **39** $m^2 + 10m + 25$
41 $4x^2 + 4x + 1$ **43** $49x^2 + 70xy + 25y^2$
45 $x^2 - 10x + 25$ **47** $w^2 - 2wz + z^2$
49 $9m^2 - 6mn + n^2$ **51** $16x^2 - 24xy + 9y^2$
53 $x^2y^2 - 25$ **55** $4x^4 - 81y^2$ **57** $x^4 - y^4$
59 $x^2 + 2xy + y^2 - z^2$ **61** $25x^2 + 10xy + y^2 - 9z^2$
63 $4a^2 - 4ab + b^2 - c^2$ **65** $-17x^3 - 36x^2 + 87x$
67 $-2rs + 2s^2$ **69** $x^3 - 6x^2 + 12x - 8$
71 $24w^3 - 26w^2 - 13w + 10$

73
$[-4, 4]$ for x, $[-10, 10]$ for y

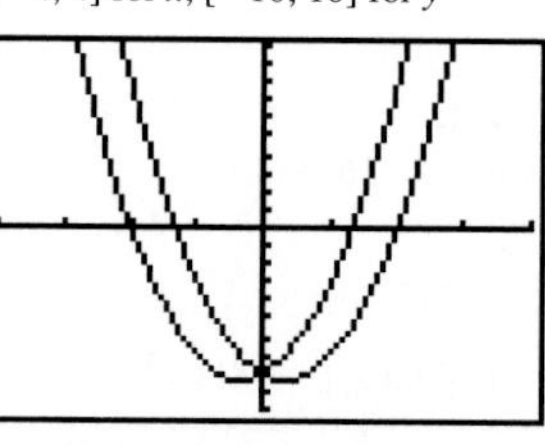

$P_1(x) \neq P_2(x)$

75
$[-4, 2]$ for x, $[-1, 10]$ for y

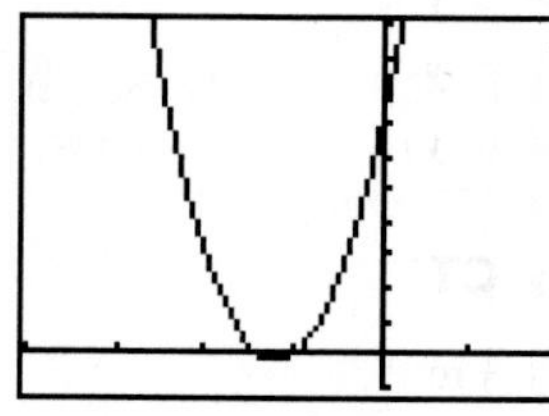

$P_1(x) = P_2(x)$

77
$[-4, 4]$ for x, $[-10, 10]$ for y

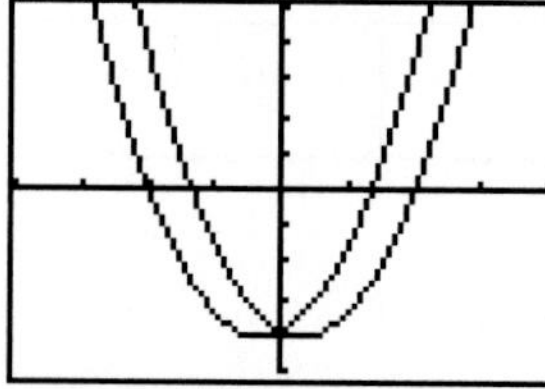

$P_1(x) \neq P_2(x)$

79

x	-2	-1	0	1	2
y_1	-3	-2	-1	0	1
y_2	-1	0	1	2	3
y_3	3	0	-1	0	3

81 $16a^{-4} - 8a^{-2}y^{-5} + y^{-10}$
83 $-150r^3 + 150r^2s + 75rs^2 - 25s^3$ **85** $x^{2m} - x^m - 12$
87 $x^{2m} - y^{2n}$ **89** $x^{3m} - y^{3n}$
91 $-12t^3 + t^2 + 644t + 1015$ **93** $2x^2 + 3x + 1$

Section Skill Reinforcers 4-5

1 x^{16} **2** x^{20} **3** x^{36} **4** x^{20} **5** $\frac{1}{x^{20}}$

Exercises 4-6

1 $3x^2$ **3** $-7a^2b$ **5** $-12x^6y^3$ **7** $-\frac{8x^6y^{12}}{z^6}$
9 $\frac{49b^6}{121a^2c^8}$ **11** $a^{10}b^4c^{10}$ **13** $2x - 3$ **15** $x + 4y$
17 $y^2 - 2y + 3$ **19** $a - 2$ **21** $-5a + 2b - \frac{3b^3}{a}$
23 $-6m^3 + 3m^2 - 2m$ **25** $x - 7$ **27** $b - 2$
29 $5n - 3$ **31** $5v - 2 + \frac{2}{3v + 4}$ **33** $2x + 5y$
35 $2v + 3$ **37** $7x^2 + 9$ **39** $a^4 + a^3 + a^2 + a + 1$
41 $16x^4 + 8x^3y + 4x^2y^2 + 2xy^3 + y^4$ **43** $3v^2 - 2v$
45 $-2x^{2m} - 3x^m + 5$ **47** abc **49** $2c + 3$
51 $b + 23$ **53** $x^2 - 5x - 6$ **55** $9x^{2n} + 3x^n + 1$
57 $\frac{3}{2}x + \frac{1}{2}$ **59** $\frac{4}{3}x - \frac{5}{3}$ **61** $4m + 3$ **63** $k + 3$
65 2 **67** $10x^2 - 3x - 27$ **69** $t - 2$
71 $t^2 - 4t + 16$ **73** a **75** $4.32x - 11y$

77

x	-2	-1	0	1	2
y_1	20	12	6	2	0
y_2	-5	-4	-3	-2	-1
y_3	-4	-3	-2	-1	0

Section Skill Reinforcers 4-6

1 $\frac{1}{2}$ roof/day **2** 2 roofs/day **3** $\frac{1}{5}$ tank/h
4 8 pages/min **5** $\frac{1}{t}$ part/min

Exercises 4-7

1 $x - 2$ **3** $2y + 5$ **5** $5w^2 - 7$ **7** $4a^3 - a + 3$
9 $c + 4 - \frac{2}{c + 3}$ **11** $p^5 + 2p^4 + p^3 + 2p^2 + 4p + 1$
13 $m^3 + m$ **15** $3s^3 + 9s + 3$
17 $b^4 + b^3 + 2b^2 + 2b + 2$ **19** $z^3 + 4z^2 + 16z + 64$
21 243 **23** -32 **25** 4425 **27** 17
29 -609.794 **31** $x^5 + kx^4 - x^3 + 3x + 1$
33 -197.025 **35** 4.579408 **37** -4.659782
39 39.723

Section Skill Reinforcers 4-7

1 $9x^2 + 30xy + 25y^2$ **2** $25x^2 - 30xy + 9y^2$
3 $16a^2 - b^2$ **4** $100a^2 - b^2$ **5** $6a^2 - 5ab - 6b^2$

Review Exercises 4

1 -64 **2** 168 **3** 3 **4** 2 **5** 5 **6** 270
7 121 **8** 169 **9** $6a^5b$ **10** $3v^4w^2$ **11** $\frac{y^8}{4x^6}$
12 $\frac{81}{625x^4y^8}$ **13** $\frac{v^3}{6}$ **14** $\frac{1}{x^{72}z^{12}}$ **15** $2x + 14y$
16 $9x + 49y$ **17** $-8x^2 - 15xy + 10y^2$
18 $6x^3 - 5x^2 - 39x - 27$ **19** $49x^2 - 81$
20 $16x^2 + 40xy + 25y^2$ **21** $9v^2 - 42vw + 49w^2$
22 $6x^2 + 11xy - 35y^2$ **23** $2ab$
24 $4y^2 + y - 10 - \frac{1}{3y + 2}$ **25** $-4m + 5n^2$
26 $9z^2 + 3z + 1$ **27** $2x^2 - 2xy$ **28** $2x + 1$
29 $-2x^2 - 6x + 29$ **30** 1
31 $8v^3 - 36v^2w + 54vw^2 - 27w^3$ **32** -1 **33** -2
34 27 **35** 2 **36** -6 **37** -5 **38** -11
39 61 **40** $-7a^3 + 2a^2 - a - 5$
41 $7a^3 + 2a^2 + a - 5$ **42** $42a^3 - 4a^2 - 5$
43 $x^4 - 3x^3y + 4x^2y^2 + 2xy^3 - 5y^4$; 4 **44** $7x^3 - 3$
45 $4w + 8x$ **46** $w^2 + 4wx$
47 **a.** $10x + 10$ **b.** $x = -1$
48 **a.** $-39x + 78$ **b.** $x = 2$ **49** x^6 **50** x^{2m+6}
51 x^{2m+4} **52** x^{m^2+3m+2}

53

x	-4	-3	-2	-1	0	1	2	3	4
y	18	10	4	0	-2	-2	0	4	10

$[-5, 5]$ for x, $[-4, 20]$ for y

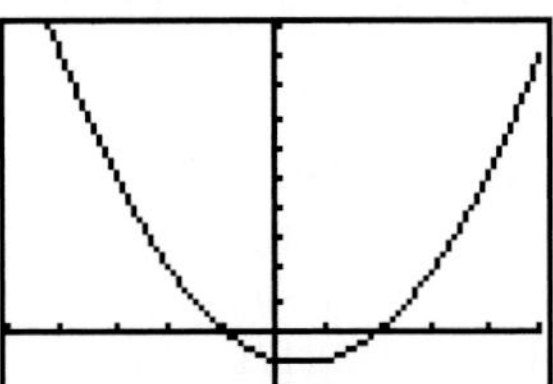

54
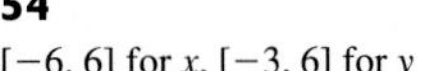
$[-6, 6]$ for x, $[-3, 6]$ for y

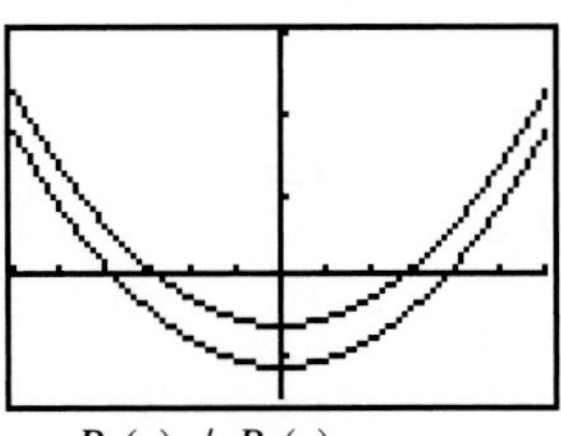
$P_1(x) \neq P_2(x)$

55
$[-2, 4]$ for x, $[-1, 12]$ for y

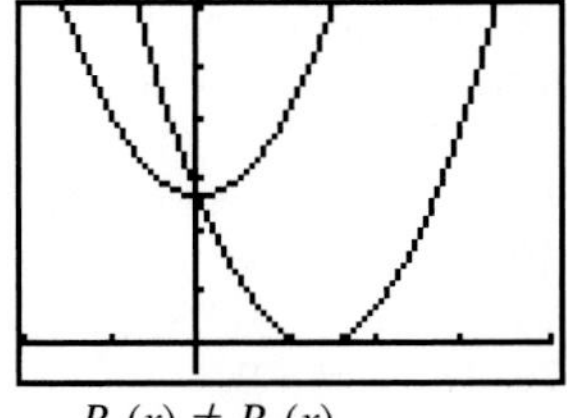
$P_1(x) \neq P_2(x)$

56 $(2x + 5)^2 = 4x^2 + 20x + 25$, not $4x^2 + 25$.

57 $2x^4 + 3x^2 - 9x$ **58** $x^3 - 3x^2 + 15x - 37 + \frac{114}{x+3}$
59 187 **60** -1813 **61** $2x + 7$ **62** $4x + 1$

Mastery Test 4

1 a. -17 **b.** $\frac{100}{49}$ **c.** 29 **d.** 13
2 a. $-\frac{21x^7}{y}$ **b.** $-\frac{3y^2}{x^2}$ **c.** $\frac{49x^{16}}{4y^{16}}$
3 a. 1.7293×10^{-5} **b.** 5,980,000 **c.** 0.243
4 a. $4x^2 - 7xy + 3y^2$; trinomial; degree 2
b. $8x - 9$; binomial; degree 1
c. $-11a^2b^4c^3$; monomial; degree 9
5 a. -9 **b.** 1 **c.** -11
6 a. $12v - 4w$ **b.** $-2v + 22w$ **c.** $-2x^2 + x - 7$
7 a. $-63x^3 - 38x^2 + 44x - 8$
b. $10x^4 + 3x^3 - 8x^2 + 7x - 12$
8 a. $9x^2 - 30xy + 25y^2$ **b.** $9x^2 + 30xy + 25y^2$
c. $121v^2 - 49$ **d.** $26x^2 - 25xy - 21y^2$
9 a. $3ab^2 - 7b^3$ **b.** $3x + 2$ **c.** $z^3 + 3z^2 + 9z + 27$
d. $2v^2 - 3v - 1 + \frac{3v + 10}{2v^2 + 1}$
10 a. $y^2 + 4$ **b.** $w^2 + 4w + 16$
c. $4y^2 - 7y + 5 - \frac{4}{y+1}$
d. $3z^3 + 6z^2 + 10z + 15 + \frac{30}{z-2}$
11 a. 80 **b.** 224 **c.** 62,096 **d.** 905

Chapter 5

Exercises 5-1

1 2 **3** $3a^2$ **5** 3 **7** $4ab^2$ **9** $x - 2y$
11 $5x - 7$ **13** $3m - 5$ **15** $5(y - 7)$ **17** $7(x - y)$
19 $x^2(x - 1)$ **21** $5m^4(3m^2 - 4)$
23 $-11x^3y^3(2x^2 - 3y)$ **25** $2x(2x^2 - 6x - 3)$
27 $-7x^5(2x^5 - 3x^3 - 5)$ **29** $5ab(3a^2b - 5ab^2 - 1)$
31 $(4a - b)(x - 2)$ **33** $-7xy^3(5x^2 - 7xy + 11y^2)$
35 $(x + 3y)(a - 4c)$ **37** $(14x - 3y + z)(a - 2b)$
39 $3(2a - 7c)(5x - 7)$ **41** $(a + 11b)(17x - 9y + 4z)$
43 $(2x - 3y)(a + b)$ **45** $(7v - 4w)(11z - 1)$
47 $18a^2b^2c(117x + 31y)(3a - 5bc)$ **49** $(x - 2)(x + 5)$
51 $(2x - 3)(a - 7)$ **53** $(4x - 3y)(a + b)$
55 $(4a + 3)(a - 2b + 3c)$
57 a. $3vw^2(-2v + 5w)$ **b.** $-3vw^2(2v - 5w)$
59 $x^3(x^n + 1)$ **61** $x^{2m}(x + 7)$ **63** $a^{3n}(2a^n - 3)$
65 $5w^{4n}(3w^{3n} - 5w^n + 4)$ **67** $x^2 + xy + y^2$
69 $2x + 3$ **71** $x^{-2}(3x + 2)$ **73** $a^{-m}(3a^{m+2} - 5)$
75 $x^{-2n}(x^{3n} - 3x^{2n} - 2x^n + 7)$ **77** $2x^2 + 3x + 5$
79 $3x^2 - 2x - 9$
81 $x(x + 1) + 3(x + 1) = (x + 3)(x + 1)$

Section Skill Reinforcers 5-1

1 $25v^2 - 36w^2$ **2** $25v^2 + 60vw + 36w^2$
3 $25v^2 - 60vw + 36w^2$ **4** $125a^3 - b^3$
5 $8m^3 + n^3$

Exercises 5-2

1 $(3a + 1)(3a - 1)$ **3** $(v + 9w)(v - 9w)$ **5** $(a + 3)^2$
7 $(x - 8y)^2$ **9** $(w + z)(w^2 - wz + z^2)$
11 $(2x - 1)(4x^2 + 2x + 1)$ **13** $(4s - 11t)(4s + 11t)$
15 $(5a^2 - 6b^3)(5a^2 + 6b^3)$ **17** $(7s + 2)^2$
19 $(x - 5y)(x^2 + 5xy + 25y^2)$
21 $(3m + 2n)(9m^2 - 6mn + 4n^2)$ **23** $(3x - 2y)^2$
25 Prime **27** $(y^2 - 2)^2$ **29** $(x + 1)(x - 1)(x^2 + 1)$
31 $(3x^3y + 1)(3x^3y - 1)$ **33** $(x - 4)(x^2 + 4x + 16)$
35 Prime **37** Prime **39** $(a + 2b - 2)(a + 2b + 2)$
41 $(5v - 7w + 9)(-5v + 7w + 9)$
43 $(x + y - a + b)(x + y + a - b)$ **45** $(x + y + 1)^2$
47 $(2a - b - 3)^2$
49 $(4v - 3w - 3)(16v^2 - 24vw + 9w^2 + 12v - 9w + 9)$
51 $(2x + 5)(x^2 + 5x + 25)$ **53** $(a + 3b)^2$
55 $(w^3 - 11)^2$ **57** $(13x^3 + 12z)(13x^3 - 12z)$
59 $(5m^5 + 2p)^2$ **61** $3x(x - 11)(x + 11)$
63 $5a(x - 2)(x^2 + 2x + 4)$ **65** $6a(x + 3)^2$
67 $(x^2 + y^2)(x^4 - x^2y^2 + y^4)$ **69** $y(3x^2 + 3xy + y^2)$
71 $(v^{3n} + 3)(v^{3n} - 3)$ **73** $(w^m - 12)^2$
75 $(x^{3m} + y^{4n})(x^{3m} - y^{4n})$
77 $2.86(x^2 - 9) = 2.86(x + 3)(x - 3)$
79 $-5.45(4x^2 + 4x + 1) = -5.45(2x + 1)^2$
81 $[-4, 4]$ for x, $[-10, 10]$ for y

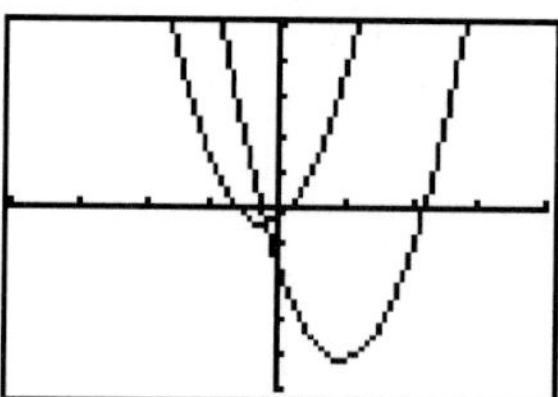

$P_1(x) \neq P_2(x)$

83 $\pi r_2^2 - \pi r_1^2 = \pi(r_2 - r_1)(r_2 + r_1)$
85 $[(2x - 3y)(2x + 3y)]^2$
87 $x^2 - 25 = x(x - 5) + 5(x - 5) = (x + 5)(x - 5)$

Section Skill Reinforcers 5-2

1 $(2a - 3b)(x + y)$ **2** $(3x - 1)(a + b)$
3 $(3a + 2)(x - 5)$ **4** $(4b - 7)(3x - y)$
5 $(3x + 5)(4x - y)$

Exercises 5-3

1 a. $(y - 1)(y - 6)$ **b.** $(y + 1)(y - 6)$
c. $(y + 2)(y - 3)$ **d.** $(y - 2)(y + 3)$
3 a. $(3x + 1)(2x - 5)$ **b.** $(3x - 1)(2x - 5)$
c. $(3x - 1)(2x + 5)$ **d.** $(6x - 5)(x + 1)$

5 $(t + 7)(t + 11)$ **7** $(b - 3)(b + 12)$
9 $(x + 4)(x - 9)$ **11** $(5v + 7)(v + 1)$
13 $(x + 3)(x - 12)$ **15** $(2x + 3)(x + 1)$
17 $-(11x - 5)(x + 1)$ **19** $(3c + d)(2c + 11d)$
21 $-(a + 2b)(a - 3b)$ **23** $(2v + 7w)(v - 3w)$
25 $(2v - 3w)(v + 7w)$ **27** $(x^2 - 7)(x^2 + 2)$
29 $(x^2 + 9y^2)(x^2 - 11y^2)$ **31** $-(7a + 5b)(6a - 5b)$
33 $(a - 2b + 3)(a - 2b - 5)$
35 $(3x - 3y + 1)(2x - 2y + 1)$ **37** $(25x - 21y)(x + y)$
39 $(11a + 11b + 1)(a + b + 9)$ **41** $(x^3 - 4)(x^3 + 2)$
43 $(x^3 + 5y^2)(x^3 - 4y^2)$ **45** $5v(v - 4)(v + 1)$
47 $-6ab(a - 6b)(a + 4b)$ **49** $4xyz(x - 3y)(6x + 13y)$
51 $(a - b)(4c - 3)(2c + 3)$
53 **a.** $(m - 9)(m + 9)$ **b.** Prime **c.** $(m + 9)^2$
d. Prime
55 **a.** $(x + 6)^2$ **b.** $(x - 6)^2$ **c.** $-(x - 6)^2$
d. Prime
57 $(x^n - 4)(x^n + 5)$ **59** $(x^m + 3)(5x^m - 1)$
61 $(x^2 - xy - 2y)(x^2 - xy + 3y)$
63 $(x + 5)(x - 3)$
65 $(x - 24)(x + 6)$ $[-10, 28]$ for x, $[-250, 175]$ for y

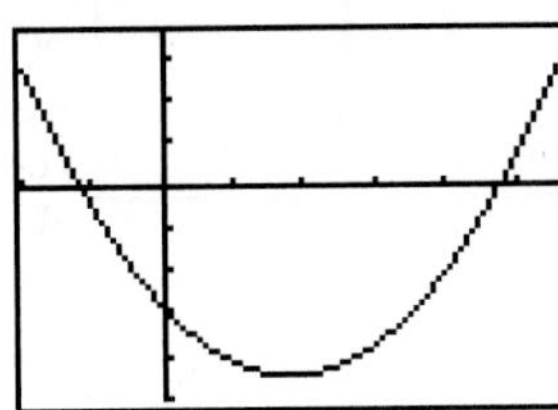

67 Factorable **69** Prime **71** $3a + 4$
73 $P(1 + rt)$
75 $m^3 + 6m^2 + 8m = m(m + 2)(m + 4)$; if m is an even integer, m, $m + 2$, and $m + 4$ are three consecutive even integers.
77 $2t + 1$ dollars/item

SECTION SKILL REINFORCERS 5-3

1 1.23×10^4 **2** 7.89×10^5 **3** 7.8×10^{-2}
4 3.2×10^{-3} **5** 7.1×10^{-5}

EXERCISES 5-4

1 $(a + b)(c + d)$ **3** $(a - 2b)(3 + 5c)$
5 $(x - y)(x + 5)$ **7** $(a + c)(b - d)$ **9** $(v - w)(v - 7)$
11 $(2a + 3 - 4b)(2a + 3 + 4b)$ **13** $(3m - k)(n + 5)$
15 $(x - y + 1)(x + y - 1)$ **17** $(x - y)(x^2 + xy + y^2 - 1)$
19 $(a + 1)(a + 1 + b)$ **21** $(a + b)(z^3 + w^2)$
23 $(3b - 4 + a)(3b - 4 - a)$ **25** $(y + 1)(ay + a - 1)$
27 $(8y - 3z)(8y + 3z)$ **29** Prime
31 $3(4x - 5)(x - 1)$ **33** $(7a - 2)^2$ **35** $(a - b)(x + y)$
37 Prime **39** Prime **41** $(2x^5 + 3y^3)^2$
43 $12xy(x - y)(x + y)$ **45** $(x + y)(c + d)$
47 $(x + 1)(ax + by)$ **49** $(x^3 + 2y)^2$
51 $4b(x - 2)(x^2 + 2x + 4)$ **53** $5(x^2 - 11)$
55 $(5x - y + 6z)(5x - y - 4z)$ **57** $(7b + 9c)^2$
59 $-71a(x - 1)(x + 1)(x^2 + 1)$
61 $(2x^2 - y)(4x^4 + 2x^2y + y^2)$
63 $(x + y + 4z)(x + y - 4z)$ **65** $2(x + y)(x - y + 1)$
67 $8a(x - 9y^2)(x + 9y^2)$ **69** $-6ax(x - 2)(x + 2)$
71 $x(x + 2y)^2$ **73** $(a - 2b)(x + 4)(x - 6)$
75 $(9y^n - 4)(9y^n + 4)$ **77** $(2x^m + 5y^n)^2$
79 $2x^{-3}(x - 7)(x + 2)$
81 $(x + 20)(x - 6)$ $[-24, 10]$ for x, $[-200, 175]$ for y

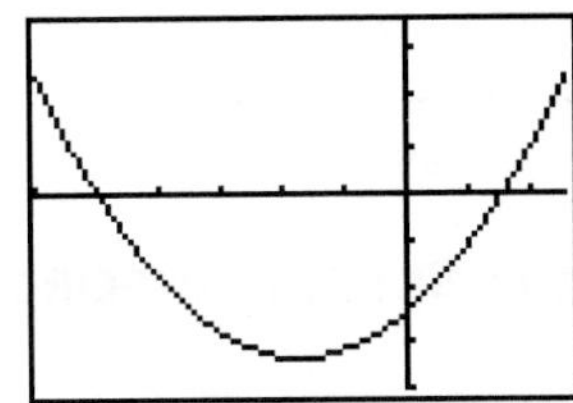

SECTION SKILL REINFORCERS 5-4

1 $x = \frac{5}{3}$ **2** $x = -\frac{3}{5}$ **3** $x = -3$ **4** $x = 7$
5 $x = \dfrac{4y + 5}{y + 3}$

EXERCISES 5-5

1 **a.** Linear equation **b.** Quadratic equation
c. Not an equation
d. Not a linear or quadratic equation
3 **a.** $2x^2 - 7x + 3 = 0$
$a = 2, b = -7, c = 3$
b. $8x^2 - 3x = 0$
$a = 8, b = -3, c = 0$
c. $2x^2 - x - 4 = 0$
$a = 2, b = -1, c = -4$
5 $8, -17$ **7** $\frac{5}{2}, -\frac{1}{3}$ **9** $1, -2, \frac{7}{2}$ **11** $11, -11$
13 $-1, -2$ **15** $6, -3$ **17** $0, -\frac{1}{3}$ **19** $-\frac{3}{2}, 5$
21 $-\frac{2}{3}, -\frac{5}{2}$ **23** $-6, -7$ **25** $\frac{7}{3}$ **27** $-\frac{3}{7}, \frac{1}{2}$
29 $-\frac{5}{3}, \frac{5}{3}$ **31** $-5, 2$ **33** $1, 4$ **35** $4, 7$
37 $4, -1$ **39** $-1, \frac{5}{2}$ **41** $5, -3, -\frac{3}{2}$ **43** $0, 8, -3$
45 $0, -\frac{3}{2}, \frac{2}{3}$ **47** $0, \frac{1}{7}, -\frac{3}{2}$
49 **a.** $x^2 - 9 = 0$ **b.** $7x^2 + 11x - 6 = 0$
c. $x^2 + 4x = 0$ **d.** $x^2 - 6x + 9 = 0$
51 **a.** $\frac{3}{5}, 2$ **b.** $5m^2 - 13m + 6$
53 **a.** $3, -\frac{2}{5}$ **b.** $20x^2 - 52x - 24$
55 $-1, 2$ **57** $-2, 0, 1$ **59** $x^3 - 3x^2 + 2x = 0$
61 $x^3 - 3x^2 - 10x + 24 = 0$
63 $x = \frac{7}{2}, x = 4$; $2x^2 - 15x + 28 = (2x - 7)(x - 4)$
$[-1, 7]$ for x, $[-1, 10]$ for y

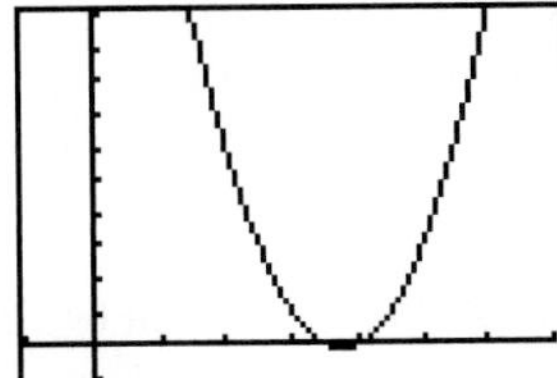

65 $x = \frac{9}{2}, x = 8$; $2x^2 - 25x + 72 = (2x - 9)(x - 8)$

$[-1, 11]$ for x, $[-7, 20]$ for y

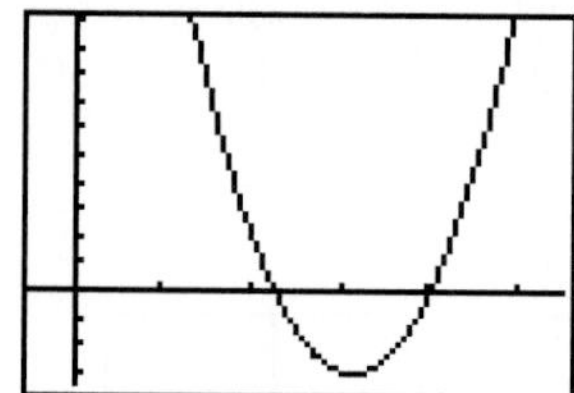

67 $6y, -y$ **69** $-3x, x$

Section Skill Reinforcers 5-5

1 $n + 1$ **2** $n + 2$ **3** $n - 2$ **4** $n - 1$
5 $3n - 5$

Exercises 5-6

1 $-13, -12$ or $12, 13$ **3** $-18, -16$ or $16, 18$
5 $-5, 15$ **7** $\frac{5}{2}, \frac{1}{2}$ or $-\frac{1}{2}, -\frac{5}{2}$ **9** $-8, -7$ or $7, 8$
11 $7, 25$ or $-\frac{25}{3}, -21$ **13** $\frac{7}{3}$ cm by 9 cm
15 6 m by 16 m **17** 3 cm **19** 5, 12, 13 cm
21 4 cm **23** 8 m **25** 20 mi **27** 28 ft
29 200, 210 mi/h **31** 10 cm **33** 200 mi/h, 500 mi/h
35 5 cm

Section Skill Reinforcers 5-6

1 $(x + 8)(x - 8)$ **2** Prime **3** $(x - 4)(x^2 + 4x + 16)$
4 $(x + 4)(x^2 - 4x + 16)$ **5** $(x + 8)^2$

Review Exercises 5

1 $4(x - 9)$ **2** $2x(x - 5)$ **3** $12ax(x - 2)$
4 $(x + 2)(x - 2)$ **5** $(2x + 1)(2x - 1)$ **6** $(x + 2)^2$
7 $7(x + 2)(x - 2)$ **8** $(m + 8)(m - 8)$
9 $(x - 2)(x - 9)$ **10** $(m + 3)(m + 7)$
11 $(a + b)(x + y)$ **12** $(x + 2y)(a - 3)$
13 $(x + 6y)(x - 7y)$ **14** $(v + w + 1)(v - w + 1)$
15 $(3y + 5)^2$ **16** $(x + 4)(x - 4)$
17 $(2x - y)(4x^2 + 2xy + y^2)$ **18** $2x(x - 2)(x^2 + 2x + 4)$
19 $11(x + 2y)(x - 2y)$ **20** $11(x + y)(x - y + 3)$
21 $a(2x + 3y)(2x - 3y)$ **22** $(2x - 3y)(5a - 3b)$
23 $xy(x - 1)(x - 4)$ **24** $(6x + y)(x + 10y)$
25 $(2x + 3y)^2$ **26** $(a + b)(x + 5)(x - 5)$
27 $(x + 5y)(a + b)^2$ **28** $v^2 + 9w^2$ is prime.
29 $10(2x - 1)(x + 3)$ **30** $(v - 1)^2(v^2 + v + 1)$
31 $(x^m + 1)(x^m - 1)$ **32** $(x^m - 1)(x^{2m} + x^m + 1)$
33 $-1, 5$ **34** $-\frac{2}{3}, \frac{3}{2}$ **35** $0, \frac{2}{7}, 7$ **36** $-3, 7$
37 $-\frac{3}{2}, \frac{1}{5}$ **38** $-\frac{7}{6}, 3$ **39** $-3, 5$ **40** $-7, 3$
41 $-6, 0, 6$ **42** $0, 5$ **43** $-4, -2$ **44** $-2, 6$
45 $x^2 - 49 = 0$ **46** $x^2 + 9x - 22 = 0$
47 $10x^2 - x - 3 = 0$ **48** $x^2 - 8x = 0$
49 $x^2 + 4x + 4 = 0$ **50** $x^3 - 12x^2 + 35x = 0$
51 $-7, -6$ or $8, 9$ **52** 6 m by 9 m **53** 6 m by 10 m
54 9 km

55 **a.** $R(t) = -t^3 + 400t$ **b.** 0 units or 20 units
56 $2x^{-3}(3x + 4)(2x - 5)$ **57** $x(x + 2)(x + 5)$
58 $x(x - 2)(x + 3)$
59 $2x^3 + 3x^2 - 23x - 12 = (x + 4)(2x + 1)(x - 3)$ **60** $x^4 + 2x^3 - 9x^2 - 2x + 8 = (x + 4)(x + 1)(x - 1)(x - 2)$

$[-6, 6]$ for x, $[-35, 35]$ for y

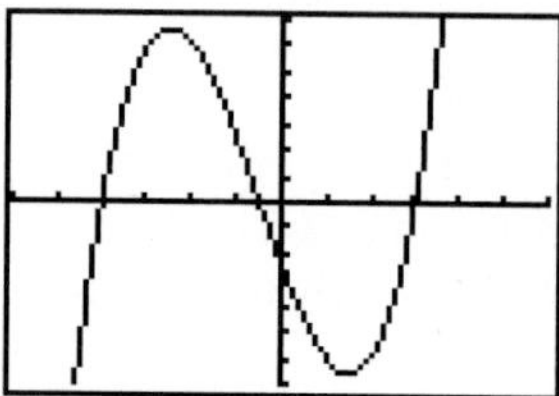

$[-5, 4]$ for x, $[-40, 35]$ for y

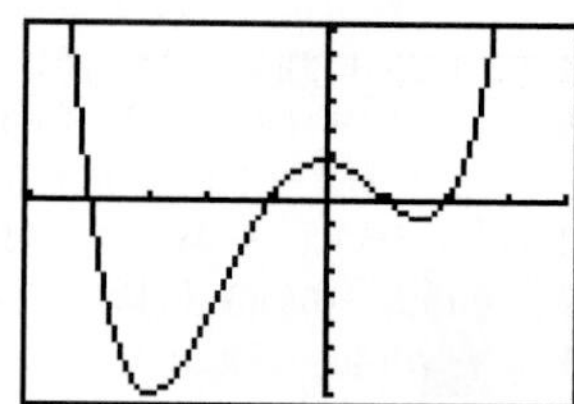

61 $4x^4 + 4x^3 - 49x^2 - x + 12 = (x + 4)(2x + 1)(2x - 1)(x - 3)$

$[-5, 4]$ for x, $[-225, 25]$ for y

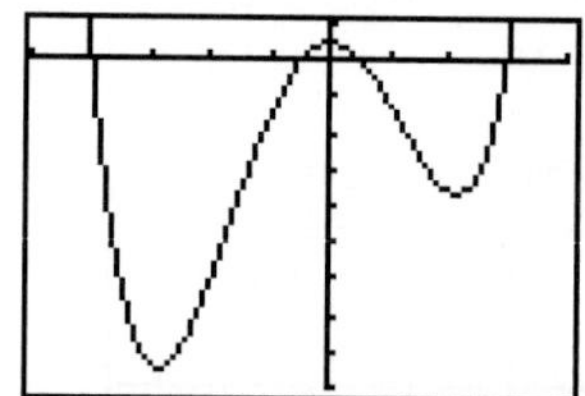

62 $2x^4 - x^3 - 8x^2 + x + 6 = (2x + 3)(x + 1)(x - 1)(x - 2)$

$[-3, 3]$ for x, $[-5, 6.5]$ for y

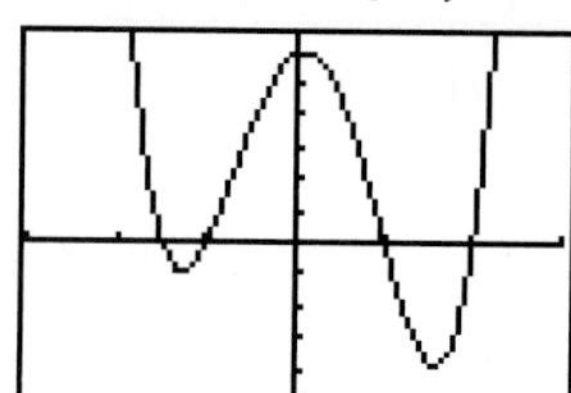

63 $-\frac{1}{2}, \frac{1}{2}$ **64** $-1, 1, 4$
65 $P_1(x) = P_2(x)$ **66** $P_1(x) \neq P_2(x)$

$[-2, 3]$ for x, $[-1, 8]$ for y

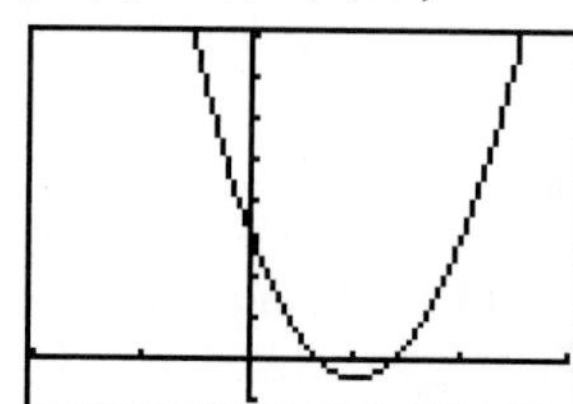

$[-3, 3]$ for x, $[-6, 8]$ for y

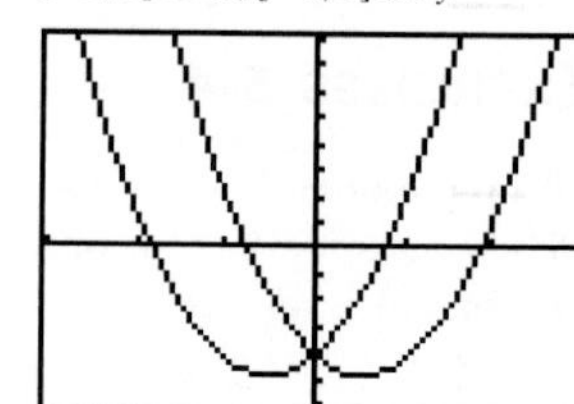

Mastery Test 5

1 **a.** $7(a - 6)$ **b.** $5y(x^2 - 3)$ **c.** $(x - 2y)(x + 3)$
d. $6ab(c - 2)$
2 **a.** $(x - 2y)^2$ **b.** $(x + 2y)(x - 2y)$
c. $(3x - 1)(9x^2 + 3x + 1)$ **d.** $(5x + 1)^2$
3 **a.** $(w - 9)(w + 5)$ **b.** $(w + 9)(w + 5)$
c. $(x - 4y)(x + 3y)$ **d.** $(x - y)(x - 12y)$

4 a. $(2x + 3)(a + b)$ **b.** $(2x - 5y)(7a - 3b)$
c. $(a - 2b)(a + 2b + 1)$ **d.** $(x + y + 2)(x + y - 2)$
5 a. $5(x + 2)(x - 2)$ **b.** $5(x - 1)(x^2 + x + 1)$
c. $2a(x + 5)^2$ **d.** $3b(x - 5y)(x + 4y)$
6 a. $-\frac{1}{2}, 3$ **b.** $-9, 11$ **c.** $-7, 12$ **d.** $-3, 2, 5$
7 a. $-14, -12$ or $12, 14$ **b.** $-7, -5$ or $5, 7$
c. 9 cm
8 a. 10 cm, 24 cm, 26 cm **b.** 18 cm

Chapter 6

EXERCISES 6-1

1 $\frac{2}{3x}$ **3** $\frac{2b^2}{3a}$ **5** $6x - 9$ **7** $\frac{1}{2x - 3}$ **9** $-\frac{a}{b}$
11 -1 **13** $-\frac{a}{b}$ **15** $\frac{x - 2y}{x + 2y}$ **17** $\frac{x - y}{3}$
19 $-\frac{5x + 2}{7}$ **21** $\frac{x}{x - y}$ **23** $\frac{3x - y}{3x + y}$ **25** $\frac{2x + 3y}{7}$
27 $\frac{2x + 3}{5y + 6}$ **29** -1 **31** $\frac{2a + b}{a + b}$ **33** $\frac{x + y}{v + w}$
35 $\frac{x - 5}{3x - 1}$ **37** $\frac{a + b}{a - b}$ **39** $-\frac{a - b + 1}{5a}$
41 $-\frac{x^2 + xy + y^2}{x - y}$ **43** $\frac{3(x + y)}{4(x - y)}$ **45** $\frac{21}{432}$
47 $\frac{-14x + 16y}{10b - 6a}$ **49** $\frac{a^2 - b^2}{10a - 10b}$ **51** $\frac{5x - 5}{x^2 - 7x + 6}$
53 $\frac{2x^2 + xy - y^2}{x^2 + 4xy + 3y^2}$ **55** $\frac{y - 13}{y - 1}$ **57** $-\frac{2x - y}{y}$
59 $\frac{x - y + 1}{2(x^2 - xy + y^2)}$ **61** $\frac{a + 1}{9}$ **63** $\frac{(m - n)^2}{m^2 + mn + n^2}$
65 $R_1(x) = R_2(x)$ for $x \neq \frac{53}{27}$

[−6, 6] for x, [−6, 6] for y

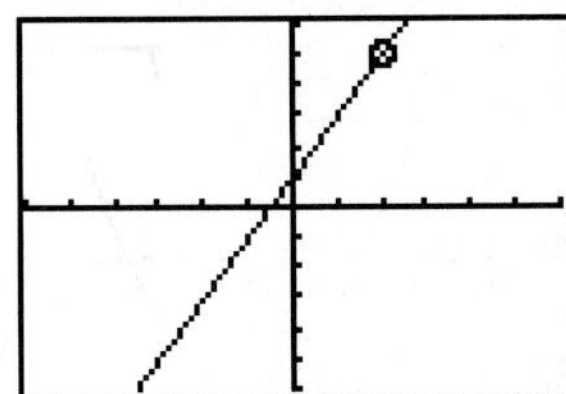

[−6, 6] for x, [−6, 6] for y

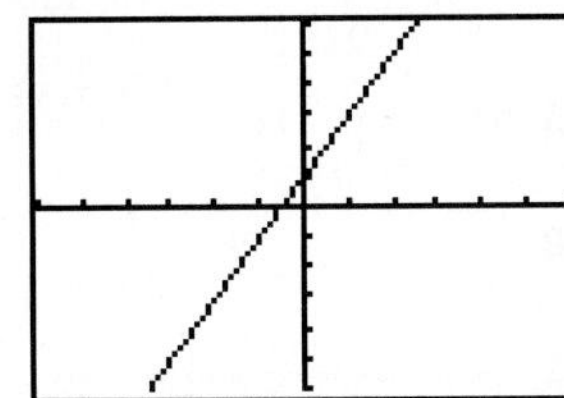

67 $R_1(x) \neq R_2(x)$

[−4, 4] for x, [−30, 10] for y

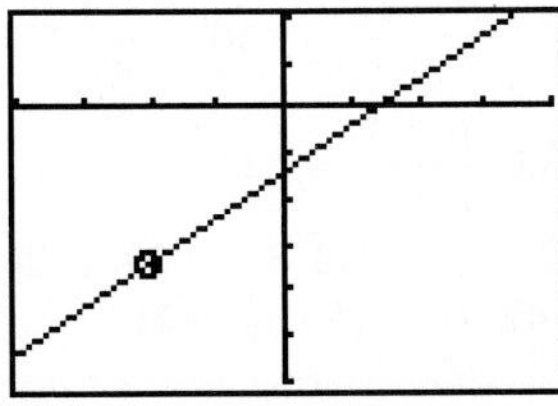

[−4, 4] for x, [−30, 10] for y

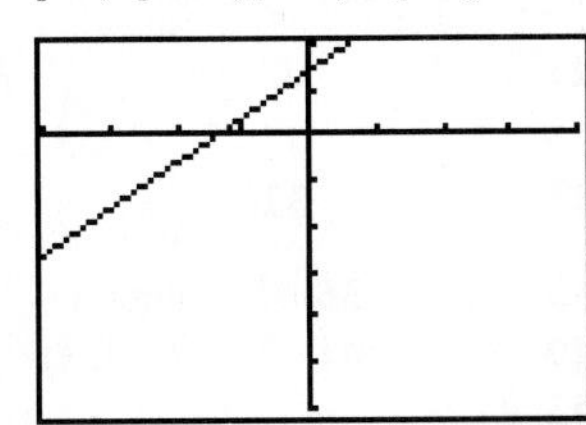

69 $y = \frac{1}{x}$

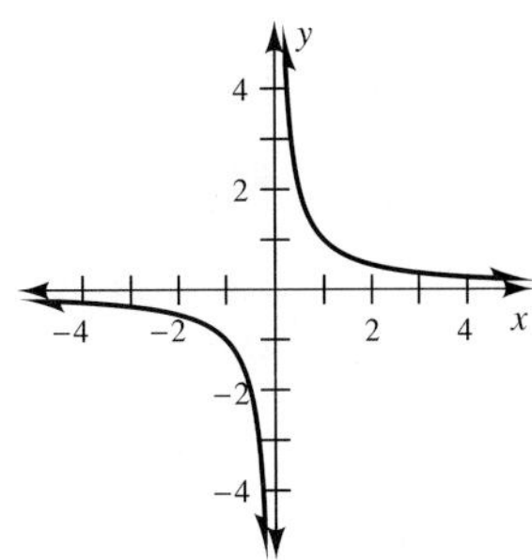

71 -7 **73** 2 **75** -3 **77** $\frac{x^m - 5}{7}$
79 $\frac{2x - 5y}{x + 3y}$ **81** $\frac{2t}{5}$

SECTION SKILL REINFORCERS 6-1

1 1 **2** $\frac{1}{4}$ **3** $\frac{49}{25}$ **4** $\frac{4}{9}$ **5** $\frac{3y}{2z}$

EXERCISES 6-2

1 $-\frac{1}{8}$ **3** 3 **5** $\frac{28w}{3x}$ **7** $\frac{5x}{2y}$ **9** $\frac{-7}{5(a + b)}$
11 $\frac{x - 3}{2(x + 3)}$ **13** $\frac{7(x - 3)}{3(x - y)}$ **15** $\frac{3(x + 2)}{x - 1}$
17 $\frac{7 - a}{c - y}$ **19** $-\frac{2(x - 1)}{3}$ **21** $\frac{6}{5(x + y)^2}$
23 $(x + 5)(x - 5)$ **25** $-\frac{1}{7x(3x - 5)}$ **27** 2
29 $\frac{11x}{9a}$ **31** $\frac{3xy}{4(x + y)}$ **33** $\frac{4}{3}$ **35** $-\frac{3y(x - y)}{14}$
37 $-\frac{2}{ab^2}$ **39** $\frac{3}{11}$ **41** $\frac{7y^2}{2a}$ **43** d **45** c
47 -0.108 **49** -26.600 **51** $\frac{2a - b^2}{a^2 + b^2}$
53 $\frac{4y(x - 2y)}{7x^2(x - y)}$ **55** $3x^{2m}$ **57** 3

SECTION SKILL REINFORCERS 6-2

1 $\frac{29}{35}$ **2** $\frac{1}{35}$ **3** $\frac{3}{5}$ **4** $-\frac{7}{30}$ **5** $\frac{61}{180}$

EXERCISES 6-3

1 $\frac{2b + 3}{b^2}$ **3** -1 **5** $\frac{4}{s - 3}$ **7** $\frac{1}{x + y}$ **9** 90
11 30 **13** $135x^3y^3$ **15** $60(x + y)$ **17** $18x(x + y)$
19 $(x + 1)(x + 6)(x - 6)$ **21** $-\frac{13}{18w}$ **23** $\frac{5}{14}$
25 $\frac{b^2 + 4b + 16}{b(b + 4)}$ **27** $\frac{5x - 1}{x}$ **29** $\frac{x^2 - 2x + 3}{x^3}$
31 $\frac{x + 13}{(x + 3)(x - 2)}$ **33** $\frac{2x}{(x + 1)(x - 1)}$

35 $\dfrac{2(4x+11)}{(x-4)(x+2)}$ **37** $\dfrac{2x^2+2x+5}{(x+2)(x-1)}$ **39** $\frac{1}{77}$

41 $\dfrac{5m+9}{(m+1)(m-2)(m+3)}$ **43** $\dfrac{4m-13}{(m-3)(m-4)}$

45 $\dfrac{2}{a-4b}$ **47** $-\dfrac{x^2-6x+3}{x^2-3x+4}$ **49** $-\dfrac{3}{(x+1)(x-1)}$

51 $\dfrac{2}{x+5}$ **53** $\dfrac{10}{s-2t}$ **55** $\dfrac{1}{z+3}$ **57** $\dfrac{1}{v-w}$

59 $\dfrac{3w-4}{(w-2)(w-1)}$ **61** $\dfrac{1}{w-3}$ **63** $-\dfrac{2m}{m-1}$

65 a **67** $\dfrac{1}{x^m+2}$ **69** $\dfrac{1}{x^{2m}+3x^m+9}$

71 $\dfrac{8(2t+5)}{t(t+5)}$ **73** $\dfrac{2x+5}{(x+5)(2x-5)}$

Section Skill Reinforcers 6-3

1 $\frac{3}{4}$ **2** $\frac{10}{7}$ **3** 5 **4** -31.99 **5** $\frac{7}{2}$

Exercises 6-4

1 x **3** $\dfrac{x+10}{3}$ **5** $2m$ **7** $-\dfrac{29y}{10}$ **9** 1

11 -2 **13** v^2-1 **15** $x-3$ **17** $-\dfrac{4}{x^2}$ **19** $\dfrac{12}{5x}$

21 $\frac{6}{7}$ **23** $\frac{3}{2}$ **25** -2 **27** $\dfrac{9y}{4}$ **29** $\dfrac{2}{xy(w+z)}$

31 $\dfrac{x(x-7)}{10(x+2)}$ **33** $\dfrac{x}{2x+1}$ **35** $\dfrac{v^2w^2}{v+w}$

37 $\dfrac{x}{6(x+1)}$ **39** $\dfrac{3x^2}{x+5}$ **41** $-\dfrac{4aw}{w^2+a^2}$ **43** $\dfrac{v+3}{v+6}$

45 d **47** a **49** $\dfrac{y^2}{x^2+y^2}$ **51** $\dfrac{w+v}{w-v}$ or $-\dfrac{v+w}{v-w}$

53 $-mn(m^2+mn+n^2)$ **55** $\dfrac{a+b}{ab}$ **57** xy

59 $\dfrac{a+b}{a^2b^2}$ **61** 1.601 **63** -0.351 **65** $\dfrac{x^2-x+1}{x(x+2)}$

67 $\dfrac{5a-3b}{4ab}$ **69** x^m+y^n **71** $\dfrac{x^m-1}{x^m+1}$

73 $2x^2-3x$

Section Skill Reinforcers 6-4

1 $27x-20$ **2** $x=1$ **3** $\dfrac{x}{60}$ **4** 0 **5** 120

Exercises 6-5

1 a. 2, 3 **b.** $-\frac{3}{2}, \frac{2}{3}$ **c.** $-3, -\frac{1}{3}, \frac{1}{2}$ **3** 2
5 No solution **7** -5 **9** -16 **11** 1
13 No solution **15** 5 **17** 3, 23 **19** $\frac{2}{3}$, 2
21 0, $\frac{2}{3}$ **23** $\frac{13}{3}$ **25** 2 **27** -2, 3 **29** $-\frac{2}{3}$

31 $d=\dfrac{k}{I}$ **33** $x=\dfrac{yz}{y+z}$ **35** $B=\dfrac{2A-bh}{h}$

37 $r_1=\dfrac{Rr_2}{r_2-R}$ **39 a.** $p=2$ **b.** $-\dfrac{2(p-2)}{(p+1)(p-1)}$

41 a. $x=-2$ or $x=5$ **b.** $-\dfrac{(x-5)(x+2)}{(x+1)(x-2)}$

43 a. $\mathbb{R}\sim\{-\frac{2}{3},\frac{1}{2}\}$ **b.** 0 **45** $\frac{11}{4}$ **47** $\frac{2}{3}$
49 -2, 0, 1 **51** $\mathbb{R}\sim\{-3,\frac{5}{2}\}$ **53** -3, 0, 2 **55** 2
57 $\mathbb{R}\sim\{\frac{5}{3}, 2\}$ **59** 2

Section Skill Reinforcers 6-5

1 $-m$ **2** $\dfrac{1}{m}$ **3** $\dfrac{1}{m}$ **4** $\dfrac{1}{m+3}$ **5** $-(m+3)$

Exercises 6-6

1 3 **3** 5, 6 **5** 7, 9 **7** 8 **9** -3, -2 **11** 6
13 $\frac{2}{3}, \frac{3}{2}$ **15** 5 **17** 4 m, 12 m **19** \$8400
21 60 ohms, 120 ohms **23** 27 **25** 36
27 20 vehicles **29** 5000 units **31** 48 h
33 45 mi/h, 55 mi/h **35** 4 cm
37 All consecutive integers except -1, 0 and 0, 1
39 A: 27 h, B: 13.5 h
41 First pipe: 9 h, second pipe: 12 h **43** 4, 5, 6 cm

Section Skill Reinforcers 6-6

1 $\frac{3}{4}$ **2** $\frac{4}{3}$ **3** $\dfrac{x}{x+1}$ **4** $\dfrac{x+1}{x-2}$ **5** $\dfrac{x-1}{x}$

Review Exercises 6

1 $\dfrac{3x}{y^2}$ **2** $-\frac{3}{2}$ **3** $\dfrac{x+y}{x^2+xy+y^2}$ **4** $\dfrac{x-5}{2x-1}$

5 $\dfrac{3x-4y}{4x-3y}$ **6** $\dfrac{c}{a+b}$ **7** $\dfrac{25v^2-35vw+49w^2}{2a-b}$

8 $\frac{1}{2}$ **9** -6, 6 **10** $-\frac{1}{4}, \frac{7}{5}$ **11** $\dfrac{2x^2-5x-3}{x^2-6x+9}$

12 $\dfrac{2y^2-5y-3}{2y^2+9y+4}$ **13** $126a^3b^2$ **14** $2(x+3y)(x-3y)$

15 $6x-4$ **16** $\dfrac{2y(2x+y)}{x}$ **17** $\dfrac{(3t-2)(2t+1)}{2}$

18 $\dfrac{v+6w}{v-w}$ **19** $\dfrac{1}{2x+1}$ **20** $\dfrac{1}{9y^2+3y+1}$ **21** 0

22 $\dfrac{7v}{(v-1)(2v+1)(3v-2)}$ **23** 3

24 $\dfrac{3w^2+17w+66}{(w-2)(w-3)(w+6)}$ **25** $\dfrac{6v-1}{3v-2}$ **26** 2

27 $y-4$ **28** $\dfrac{2}{3z+2}$ **29** $\dfrac{a(2a+1)}{(a+1)^2}$ **30** $\dfrac{x+1}{x+2}$

31 $\dfrac{v+3}{v-3}$ **32** $-\dfrac{w-4}{3w-4}$ **33** $\dfrac{m+5}{m+4}$ **34** 2

35 -1 **36** No solution **37** $-\frac{8}{3}$ **38** 5 **39** 1, 2
40 -1 **41** $\mathbb{R}\sim\{-4,\frac{1}{2}\}$ **42** $\mathbb{R}\sim\{5,-\frac{1}{2},-2\}$
43 a

44 $\dfrac{2.3x^2 - 12.0x + 11.2}{2.3x - 2.8} = x - 4$ for $x \neq \frac{28}{23}$

$[-2, 6]$ for x, $[-6, 2]$ for y $[-2, 6]$ for x, $[-6, 2]$ for y

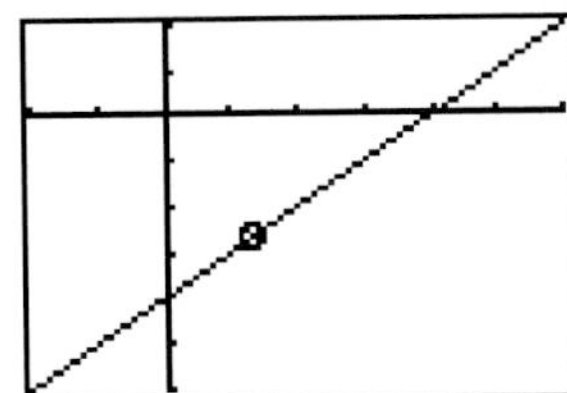

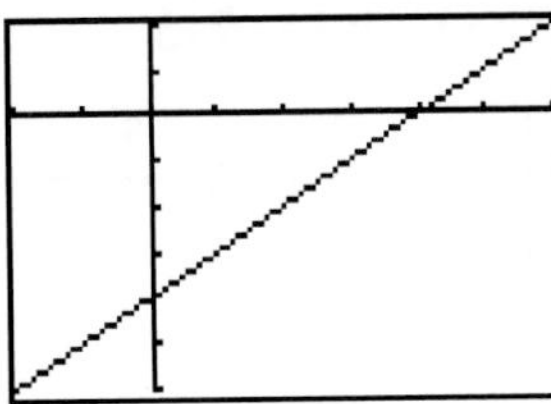

45 a. $m = -1$ **b.** $-\dfrac{2(m+1)}{(m+2)(m+4)}$

46 a. $x = \frac{16}{5}$ **b.** $-\dfrac{5x - 16}{(x-6)(x+1)}$ **47** -3

48 25, 10 **49** 5, 7 **50** 21 min **51** 1000 units

52 8 cm by 14 cm **53** 57 **54** 21 mi/h

55 $\dfrac{x^m - 5}{x^m + 4}$ **56** 2 **57** $\dfrac{1}{x^m + 5}$

Mastery Test 6

1 a. $\dfrac{2}{x+3}$ **b.** $\dfrac{2x+3}{x+5}$ **c.** $-\dfrac{2x - 3y}{a - 2b}$

2 a. $\frac{5}{4}$ **b.** 1 **c.** $\frac{2}{3}$

3 a. 2 **b.** $\dfrac{1}{x-3}$ **c.** $\dfrac{9}{(w+4)(w-5)}$

4 a. x **b.** $\dfrac{x+12}{5}$ **c.** 1

5 a. $\frac{24}{5}$ **b.** $\dfrac{x-3}{3}$ **c.** $-\dfrac{3(2x+3)}{4x-3}$

6 a. 5 **b.** 2, 4 **c.** No solution

7 a. 10 **b.** 8 and 10 **c.** 10 h

Cumulative Review Exercises 4–6

1 $10x^2$ **2** $20x^2$ **3** $75x^4$ **4** 3 **5** -9 **6** 9

7 $\frac{1}{9}$ **8** $\frac{1}{8}$ **9** 1 **10** 2 **11** 5 **12** $\frac{400}{81}$

13 $x + 9y$ **14** $-3a - b$ **15** $6x^3 - 8x^2 + 2xy$

16 $3v^2 + 5vw - 2w^2$ **17** $25x^2 - 1$

18 $4a^2 + 12ab + 9b^2$ **19** $10a^5b^9$ **20** $\dfrac{4x^2}{y}$

21 $2ab - 3b^2$ **22** $z^3 + z^2 + z + 1$ **23** $\dfrac{9a^2}{25b^2}$

24 $3x + 4$ **25** 3

26 a. 0 **b.** Undefined **c.** 1 **d.** 0
e. Undefined **f.** -1 **g.** Undefined

27 7.8×10^{-4} **28** Distributive property

29 $8(x - 3y)(x + 3y)$ **30** $2a(3x - 5y)^2$

31 $(x + 7y)(3x - y)$ **32** $5(a - 2b)(a^2 + 2ab + 4b^2)$

33 $(x + 5)(2x + a)$ **34** $(x + 2y - 3)(x + 2y + 3)$

35 $\dfrac{x+2y}{x+3y}$ **36** $\dfrac{9x^2+3x+1}{2x+5}$ **37** $\dfrac{3}{5mn}$

38 $\dfrac{a^2 - 2ab + 2b^2}{(a-b)(a-2b)}$ **39** $\dfrac{1}{z+3}$ **40** $\dfrac{4}{x+3}$

41 $-\dfrac{7x+1}{x}$ **42** $\dfrac{x-3}{x+2}$ **43** $-4, \frac{3}{2}$ **44** $-\frac{5}{2}, \frac{2}{3}$

45 $-\frac{5}{2}, 0$ **46** $-2, 0, 2$ **47** 4 **48** $-\frac{11}{4}$

49 $x = -5y, x = 2y$ **50** $5x^2 + 11x - 12 = 0$

51 a. $-\dfrac{3x+2}{x-1}$ **b.** $x = -\dfrac{2}{3}$

52 a. $\dfrac{3x+5}{x+2}$ **b.** $x = -\frac{5}{3}$ **53** b

54 $\dfrac{7.36x^2 - 32.87x + 36.4}{3.2x - 6.5} = 2.3x - 5.6$ since graphs coincide for $x \neq \frac{56}{23}$

55 17 ft **56** 70 km/h

Chapter 7

Exercises 7-1

1 a. $x^{1/5}$ **b.** $(2 + z)^{1/3}$ **c.** $2 + z^{1/3}$

3 a. 6 **b.** -6 **c.** Not a real number **d.** $\frac{1}{6}$

5 a. $\frac{2}{5}$ **b.** $-\frac{2}{5}$ **c.** $-\frac{2}{5}$ **d.** $\frac{5}{2}$

7 a. 4 **b.** 2 **c.** -4 **d.** $\frac{1}{2}$

9 a. $\frac{4}{25}$ **b.** $\frac{4}{25}$ **c.** $\frac{25}{4}$ **d.** $\frac{25}{4}$

11 a. 13 **b.** 17 **c.** 5 **d.** 7

13 25 **15** 4 **17** 11 **19** 9 **21** $x^{5/6}$ **23** $x^{1/6}$

25 $z^{3/14}$ **27** w **29** v^2w^3 **31** $\dfrac{64}{v^{3/5}}$ **33** $\dfrac{27}{8n}$

35 $\dfrac{9x}{5y}$ **37** $x - 1$ **39** $2y - 3$

41 $6w^2 - 15w - 27w^{5/11}$ **43** $a - 9$ **45** $b^{6/5} - c^{10/3}$

47 $b^{6/5} - 2b^{3/5}c^{5/3} + c^{10/3}$ **49** $x + 2 + \dfrac{1}{x}$ **51** $y + 8$

53 32 **55** e **57** e **59** $x^{5m/6}$ **61** $x^{2m}y^{3m}$

63 $x^{3/2} - 25x^{-1/2} = x^{-1/2}(x^2 - 25) = x^{-1/2}(x + 5)(x - 5)$

65

x	0	0.027	0.125	1	8
y	0	0.3	0.5	1	2

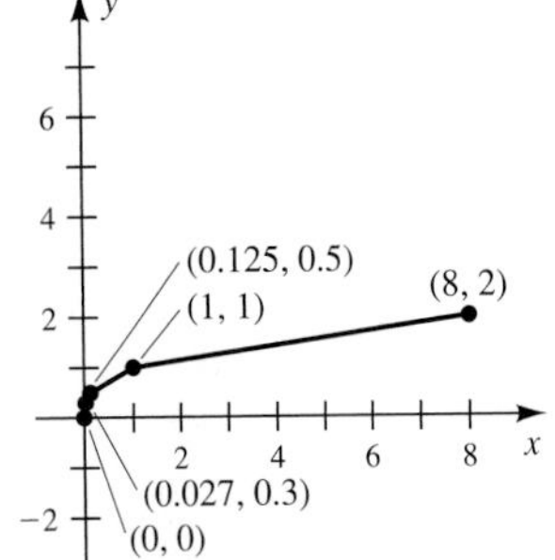

67 2.9230 **69** 4.0000 **71** 0

SECTION SKILL REINFORCERS 7-1

1 $\sqrt{9}$ **2** $\sqrt{63}$ **3** 2.3 (Answers may vary.) **4** 14
5 0

EXERCISES 7-2

1 a. $16^{1/2}$; the principal square root of 16; 4
b. $(-1)^{1/3}$; the principal cube root of -1; -1
c. $16^{1/4}$; the principal fourth root of 16; 2
d. $0^{1/5}$; the principal fifth root of 0; 0
3 a. $\sqrt{5}$ **b.** $\sqrt[3]{5}$ **c.** $(\sqrt[3]{7})^2$ **d.** $(\sqrt[5]{11})^3$
5 a. 10 **b.** -10 **c.** Not a real number **d.** $\frac{1}{10}$
7 a. 3 **b.** -3 **c.** -3 **d.** 3
9 a. 2 **b.** 4 **c.** -4 **d.** 8
11 a. $\frac{4}{7}$ **b.** $-\frac{2}{5}$ **c.** $\frac{1}{2}$ **d.** $\frac{4}{9}$
13 a. 0.5 **b.** 0.01 **c.** -0.2 **d.** 0.1
15 a. 5 **b.** 7 **c.** 15 **d.** 9
17 $5\sqrt{3}$ **19** $2\sqrt{7}$ **21** $6\sqrt{2}$ **23** $2\sqrt[3]{3}$
25 $3\sqrt[3]{2}$ **27** $3\sqrt[4]{5}$ **29** $\frac{\sqrt{3}}{8}$ **31** $\frac{\sqrt{7}}{10}$ **33** 3
35 $5|x|$ **37** $12|x|^3$ **39** $2a$ **41** $x^2|y|^3$ **43** a
45 c **47 a.** $\sqrt[4]{8v^3}$ **b.** $2\sqrt[7]{a^2b^3}$ **49** $3\sqrt[4]{2}$
51 $5y\sqrt{2xy}$ **53** $4v^5w^6\sqrt{6vw}$ **55** $2ab^7\sqrt[3]{3a^2b}$
57 $\frac{\sqrt{a}}{b}$ **59** $\frac{mn\sqrt[3]{m}}{s^2}$ **61** $\frac{ab\sqrt[3]{3a^2b}}{2}$
63 $10vw\sqrt[3]{w^2}$ **65** $2v+3$ **67** $2+\sqrt{11}$
69 $\frac{1+\sqrt{3}}{2}$ **71** 3.0715 **73** 1.8612 **75** 46.3 cm
77 [0, 30] for x, [0, 4] for y

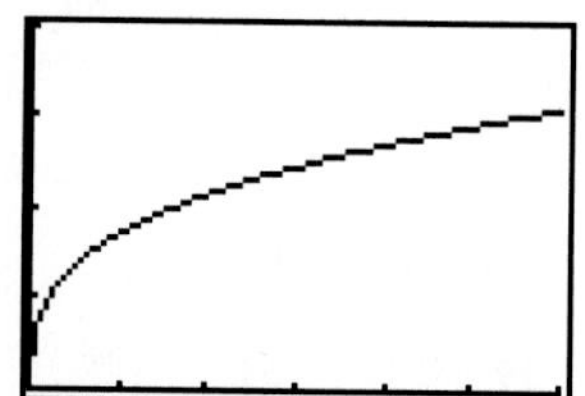

79

d	0	1	4	9	16	25
t	0	0.25	0.50	0.75	1	1.25

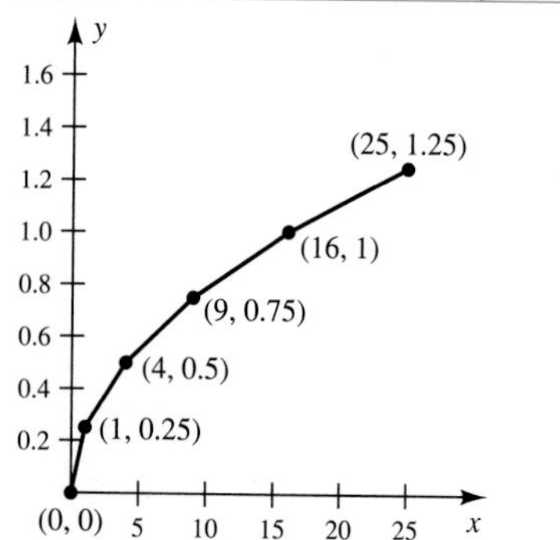

SECTION SKILL REINFORCERS 7-2

1 4 **2** 8 **3** 1 **4** $\frac{1}{8}$ **5** $\frac{1}{64}$

EXERCISES 7-3

1 12 **3** -6 **5** $30\sqrt{2}$ **7** $-3\sqrt{7x}$ **9** $8\sqrt[3]{5}$
11 0 **13** $6\sqrt{5}-4\sqrt{7}$ **15** $5\sqrt{7}$ **17** $\sqrt{3}$
19 $12\sqrt{10}$ **21** $-13\sqrt{2v}$ **23** $-2\sqrt{7w}$ **25** $-3\sqrt[3]{3}$
27 $9\sqrt{6}$ **29** $-74t\sqrt[3]{3}$ **31** $-2\sqrt[3]{5z^2}$ **33** $\frac{5\sqrt{5}}{6}$
35 $\frac{9\sqrt{11x}}{35}$ **37** $7.9\sqrt{2}+0.8\sqrt{3}$ **39** 0 **41** $\sqrt{x}$
43 $\sqrt[3]{3yz^2}$ **45** $-x^3\sqrt[3]{3}$ **47** b **49** c **51** 37
53 14 **55** 109.630 **57** -15.740
59 Translate $y=\sqrt{x}$ upward 2 units to obtain $y=\sqrt{x}+2$. Translate $y=\sqrt{x}$ upward 4 units to obtain $y=\sqrt{x}+4$.
[0, 25] for x, [0, 10] for y

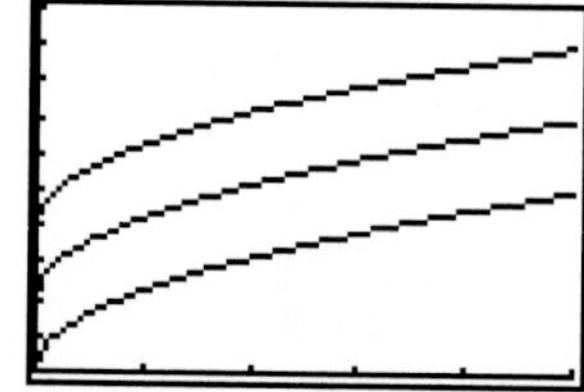

61 $-19a\sqrt{3b}$ **63** $4\sqrt[3]{2}$ **65** $14\sqrt[3]{xy^2}$ **67** $8x^my^{2n}$
69 $\sqrt{64+225}=\sqrt{289}=17$

SECTION SKILL REINFORCERS 7-3

1 a^7 **2** a^3 **3** $2x$ **4** $-4x^9$ **5** $-36x^3$

EXERCISES 7-4

1 $2\sqrt{3}$ **3** $4w$ **5** $8\sqrt{15}$ **7** $-3v$ **9** $24z\sqrt{2}$
11 240 **13** $y\sqrt[3]{x^2y}$ **15** $-132vw$ **17** $2abc$
19 $10-\sqrt{2}$ **21** $-30+35\sqrt{3}$ **23** $\sqrt{6}+\sqrt{15}$
25 $30\sqrt{3}-105\sqrt{7}$ **27** $-4-\sqrt[3]{20}$ **29** $3x\sqrt{6}-15x$
31 -6 **33** $3x-y$ **35** $14-4\sqrt{6}$
37 $a+10\sqrt{3ab}+75b$ **39** $-18+9\sqrt{10}$ **41** $v-11$
43 $2\sqrt{7}-2\sqrt{5}$ **45** 3 **47** -14 **49** $x-3y$
51 $-v+2$ **53** $8\sqrt{30}$ **55** $-\sqrt[3]{70x^2y}$ **57** -123
59 a. $x-y$ **b.** $\sqrt[3]{x^2}-\sqrt[3]{y^2}$ **c.** $x+y$
61 $1+\sqrt{2}$ is a solution.
63 a. $\sqrt[6]{x^3}$ **b.** $\sqrt[6]{y^2}$ **c.** $\sqrt[6]{x^3y^2}$ **65** $\sqrt[12]{432}$
67 -2.28
69

$$\frac{22}{7}\approx 3.143$$
$$\frac{355}{113}\approx 3.1415929$$
$$\sqrt{\sqrt{\frac{2143}{22}}}\approx 3.141592653$$
$$\pi\approx 3.141592654$$

The approximation $\sqrt{\sqrt{\frac{2143}{22}}}$ is closest to π.

Section Skill Reinforcers 7-4

1 4 **2** $4\sqrt{2}$ **3** 8 **4** 1 **5** 0

Exercises 7-5

1 $\frac{\sqrt{6}}{3}$ **3** $\frac{\sqrt{10}}{4}$ **5** $\frac{5\sqrt{3x}}{x}$ **7** $2\sqrt[3]{4}$ **9** $\frac{\sqrt[3]{10}}{5}$
11 $\frac{\sqrt{15ab}}{5b}$ **13** $3\sqrt{6}$ **15** $\frac{\sqrt[3]{vw}}{w}$ **17** $\frac{\sqrt[3]{6xy^2}}{3y}$
19 $\frac{\sqrt[5]{x^2}}{x}$ **21** $-1-\sqrt{3}$ **23** $-3\sqrt{5}-9$
25 $-\sqrt{5}-\sqrt{3}$ **27** $\frac{5-\sqrt{21}}{2}$ **29** $\frac{x+2\sqrt{xy}+y}{x-y}$
31 $\frac{a+\sqrt{ab}}{a-b}$ **33** $4\sqrt{14}+6\sqrt{10}$ **35** $-7+2\sqrt{15}$
37 $\frac{5\sqrt{3x}-5\sqrt{2y}}{3x-2y}$ **39** $\frac{15\sqrt{x}+5\sqrt{2y}}{9x-2y}$
41 $\frac{6x+10\sqrt{6xy}+25y}{6x-25y}$ **43** $\frac{x+2-\sqrt{x^2+4x}}{2}$
45 $-5+2\sqrt{6}$ **47** c **49** e **51** -1.68
53 -0.34
55 **a.** 71 beats/min **b.** 62 in.
c. [50, 80] for x, [64, 86] for y

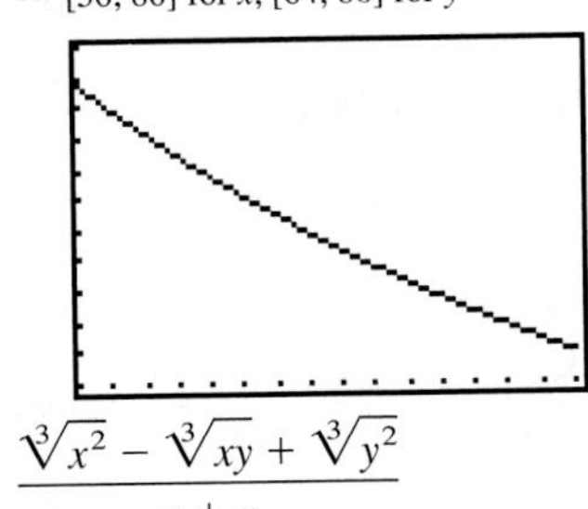

57 $\frac{\sqrt[3]{x^2}-\sqrt[3]{xy}+\sqrt[3]{y^2}}{x+y}$

Section Skill Reinforcers 7-5

1 $x=-2$ **2** $x=\frac{2}{y}$ for $y\neq 0$
3 $x=\frac{7}{2y+3}$ for $y\neq -\frac{3}{2}$ **4** $x=-\frac{1}{2}, x=3$
5 $x=-4, x=5$

Exercises 7-6

1 13 **3** 393 **5** No solution **7** $-\frac{9}{2}$ **9** 3
11 $-4, 8$ **13** 4 **15** 2 **17** $\frac{1}{3}$ **19** 4 **21** 0
23 $-1, \frac{2}{3}$ **25** No solution **27** -10 **29** 5
31 $b=\frac{25c-a}{26}$ **33** $n=\frac{m}{m^2-5}$ **35** 20 ft
37 15 ft **39** $x=6$ **41** $x=2$ **43** 0, 2
45 No solution **47** $-1, 1$ **49** $C(0)=\$450$
51 64 solar cells

Section Skill Reinforcers 7-6

1 $-2x+2y$ **2** $-2x+23y$ **3** $8x^2+2xy-15y^2$
4 5 **5** x^2-4y^2

Exercises 7-7

1 **a.** -6 **b.** $6i$ **c.** $-6i$ **d.** 6
3 **a.** $4+3i$ **b.** $3+4i$ **c.** $-3-4i$ **d.** $-4+3i$
5 **a.** $5i$ **b.** $-5i$ **c.** -6 **d.** 6
7 **a.** -10 **b.** 10 **c.** $-10i$ **d.** 10
9 **a.** $\frac{5}{3}i$ **b.** $\frac{5}{3}i$ **c.** $-\frac{5}{3}i$ **d.** $\frac{5}{3}$
11 **a.** $a=18, b=-5$ **b.** $a=-6, b=-2$
c. $a=4, b=-11$ **d.** $a=24, b=0$
13 $9-i$ **15** $3+i$ **17** $60+24i$ **19** 7
21 -6 **23** $10+6i$ **25** $8-3i$ **27** 53
29 $34+27i$ **31** $24+10i$ **33** $7-24i$
35 $56+33i$ **37** $-3+i\sqrt{6}$ **39** $13i\sqrt{3}$
41 $-1-2i$ **43** 29 **45** 169 **47** $2-2i$
49 $\frac{15}{17}-\frac{8}{17}i$ **51** $7+6i$ **53** i **55** -1
57 $-1-2i, -1+2i$ **59** **a.** Yes **b.** Yes **61** $-i$
63 $2+i$ and $3-i$ (Answers may vary.) **65** 0
67 **a.** $3-2i$ is a solution. **b.** $3+2i$ is a solution.
69 b **71** c **73** $-11+2i$ **75** 1 **77** $11+5i$
79 i **81** $-4-8i$
83 $i^4-2i^3-2i^2-2i-3=1+2i+2-2i-3=0$

Section Skill Reinforcers 7-7

1 $(x+10)^2$ **2** $(x+4y)^2$ **3** $(2x+5y)^2$
4 $(x^2+3y)^2$ **5** $(x^2+1)(x+1)(x-1)$

Review Exercises 7

1 $8|x|$ **2** $4x$ **3** 7 **4** $-2xy^2$ **5** 7 **6** -7
7 $\frac{1}{7}$ **8** $\frac{9}{25}$ **9** $\frac{9}{25}$ **10** $\frac{25}{9}$ **11** 243 **12** 7
13 1 **14** $9x^2$ **15** $3x$ **16** 0 **17** 1 **18** -1
19 $\frac{4}{9}$ **20** $-10x$ **21** $\frac{3y}{5z}$ **22** $\frac{4x}{7}$ **23** $3x\sqrt{10}$
24 $6x$ **25** $2x$ **26** $6x$ **27** $7x\sqrt{2x}$ **28** $2x\sqrt[3]{5x}$
29 $3x^2\sqrt{10}$ **30** $\frac{5x}{3}$ **31** $2\sqrt{7}$ **32** $7\sqrt{7}+7\sqrt{5}$
33 $4\sqrt{2}-3\sqrt{3}$ **34** $-29\sqrt{2}$ **35** -369 **36** -67
37 $83-20\sqrt{6}$ **38** $3\sqrt{5}+3\sqrt{3}$ **39** $-6\sqrt{5z}$
40 $84x-10\sqrt{21xy}-36y$ **41** $6\sqrt[6]{500x^3y^2}$
42 $3y+2$ **43** $\frac{6x}{y}$ **44** $\frac{4y^6}{x^4}$ **45** $4x^{2/3}y^{3/5}$
46 $25v-4w$ **47** $x^2y^2z^2$ **48** $\frac{1}{16}$
49 $42\sqrt{2}-14\sqrt{5}$ **50** $6vw-10w$ **51** -5
52 -3 **53** 7 **54** No solution **55** 25 ft
56 4 m **57** $-5+5i$ **58** $4i$ **59** $-2-36i$
60 $51-27i$ **61** $40-42i$ **62** $4-10i$
63 $\frac{8}{17}+\frac{15}{17}i$ **64** $-1+2i$ **65** 2.352 **66** $x=3$
67 $v=0, v=4$

68 $\sqrt{3x-1} = x - 1$
$$3x - 1 = x^2 - 2x + 1$$
$$0 = x^2 - 5x + 2$$
$$x = \frac{5+\sqrt{17}}{2} \quad \left(\frac{5-\sqrt{17}}{2} \text{ is extraneous.}\right)$$

Mastery Test 7

1 a. 6 **b.** 5 **c.** $2xy^3$ **d.** $4x^6$
2 a. 12 **b.** -5 **c.** $8|x|y^2$ **d.** $4x^8$
3 a. $3x\sqrt{2}$ **b.** $2x\sqrt[3]{5x}$ **c.** $\frac{x}{2}$ **d.** xy
4 a. $5\sqrt{7}$ **b.** $3\sqrt{2} + 4\sqrt{5}$ **c.** $-9\sqrt{7}$ **d.** $-4x\sqrt[3]{2x}$
5 a. $\sqrt{x}$ **b.** $\sqrt{2}$ **c.** $\sqrt[3]{2x}$ **d.** $\sqrt[3]{2xy^2}$
6 a. 30 **b.** $120x\sqrt{2}$ **c.** $30 - 20\sqrt{2}$ **d.** $12x - 5$
7 a. 3 **b.** $3\sqrt{6}$ **c.** $8\sqrt{7} + 8\sqrt{2}$ **d.** $15 - 5\sqrt{7}$
8 a. 3 **b.** No solution **c.** 4 **d.** 7
9 a. $9i$ **b.** $3i$ **c.** $-5 + 4i$ **d.** $-1 - i$
10 a. $2 + 2i$ **b.** $-12 - 26i$ **c.** $\frac{7}{5} + \frac{3}{10}i$ **d.** $8 - 6i$

Chapter 8

Exercises 8-1

1 $-9, 9$ **3** $-2, 2$ **5** $-\frac{7}{6}, \frac{7}{6}$ **7** 4, 10 **9** $-\frac{7}{2}, \frac{5}{2}$
11 $-3\sqrt{2}, 3\sqrt{2}$ **13** $\frac{3-\sqrt{2}}{5}, \frac{3+\sqrt{2}}{5}$ **15** $-4i, 4i$
17 $-2i\sqrt{3}, 2i\sqrt{3}$ **19** $\frac{1}{2} - \frac{3}{2}i, \frac{1}{2} + \frac{3}{2}i$ **21** 25
23 81 **25** 1 **27** $\frac{1}{25}$ **29** 0, 4 **31** $-5, 1$
33 $-1 - i, -1 + i$ **35** $-\frac{1}{4}, \frac{7}{4}$ **37** 2, 3
39 $-2 - \sqrt{3}, -2 + \sqrt{3}$ **41** $\frac{2-\sqrt{5}}{2}, \frac{2+\sqrt{5}}{2}$
43 $2x^2 - 9x - 5 = 0$ **45** $x^2 - 7 = 0$
47 $x^2 + 16 = 0$ **49** d **51** b **53** $-\frac{5}{3}, \frac{3}{2}$
55 $-\frac{11}{3}, 1$ **57** $\frac{-2-\sqrt{6}}{2}, \frac{-2+\sqrt{6}}{2}$
59 $-2i\sqrt{3}, 2i\sqrt{3}$ **61** $\frac{5}{7}$
63 $-3 - \sqrt{17}$ and $3 - \sqrt{17}$ or $-3 + \sqrt{17}$ and $3 + \sqrt{17}$
65 $-2 + 2\sqrt{3}$ cm by $2 + 2\sqrt{3}$ cm **67** 0, 1, 4
69 $\frac{3}{2}, 3 + 2i, 3 - 2i$ **71** $-\frac{1}{2}, \frac{5}{2}$
73 $-0.7, -0.3$

[−2, 1] for x, [−1, 7] for y

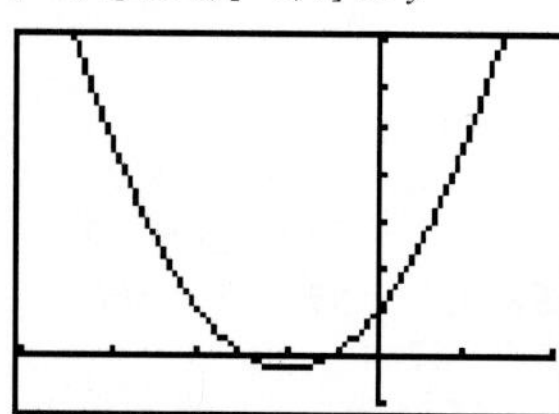

75

x	−5	−4	−3	−2	−1	0	1
y	1	5	7	7	5	1	−5

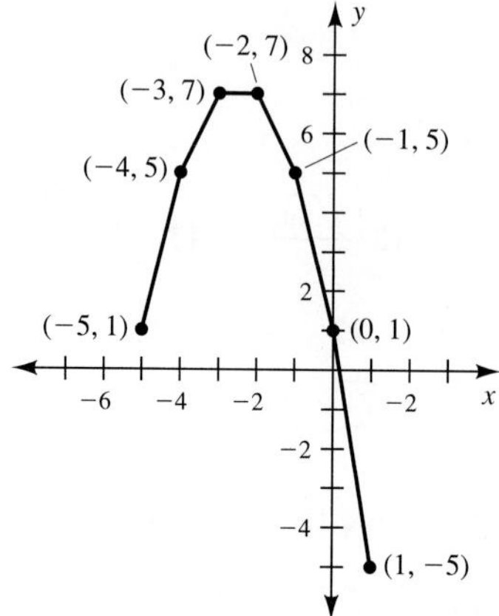

77 $x^2 - 6x + 25 = 0$

Section Skill Reinforcers 8-1

1 $2\sqrt{3}$ **2** $5\sqrt{2}$ **3** $2 + \sqrt{2}$ **4** $2 - \sqrt{2}$
5 $-6 - 5\sqrt{3}$

Exercises 8-2

1 $-\frac{1}{4}, \frac{1}{2}$ **3** $-3, \frac{1}{2}$ **5** $-2\sqrt{2}, 2\sqrt{2}$
7 $-\frac{\sqrt{21}}{3}i, \frac{\sqrt{21}}{3}i$ **9** $0, \frac{6}{5}$ **11** $\frac{3}{2} - \frac{1}{2}i, \frac{3}{2} + \frac{1}{2}i$
13 $3 - 2i, 3 + 2i$ **15** $\frac{3-\sqrt{2}}{2}, \frac{3+\sqrt{2}}{2}$ **17** $-\frac{2}{3}$
19 $\frac{1-\sqrt{6}}{5}, \frac{1+\sqrt{6}}{5}$ **21** Imaginary solutions
23 Distinct irrational solutions
25 A rational solution of multiplicity two
27 Distinct rational solutions
29 Imaginary solutions **31** Distinct rational solutions
33 $-\frac{7}{3}, \frac{5}{2}$ **35** $-15, 15$ **37** $\frac{3-i\sqrt{7}}{4}, \frac{3+i\sqrt{7}}{4}$
39 $\frac{5}{7}$ **41** $-i\sqrt{5}, i\sqrt{5}$ **43** 0, 2 **45** $-\frac{2}{5}, \frac{3}{2}$
47 $1 - \frac{1}{2}i, 1 + \frac{1}{2}i$ **49** $-1 - 2i, -1 + 2i$ **51** 4
53 $-\frac{9}{2}$ **55** $-30, 30$ **57** 0.181, 0.983
59 $-3.993, 0.779$ **61** A real solution of multiplicity two
63 Imaginary solutions **65** A real solution of multiplicity two
67 $\frac{-7-\sqrt{29}}{2}$ and $\frac{7-\sqrt{29}}{2}$ or $\frac{-7+\sqrt{29}}{2}$ and $\frac{7+\sqrt{29}}{2}$
69 $2 - \sqrt{5}, 2 + \sqrt{5}$ **71** 14.75 ft
73 $\left(\frac{-b-\sqrt{b^2-4ac}}{2a}\right)\left(\frac{-b+\sqrt{b^2-4ac}}{2a}\right) = \frac{b^2-(b^2-4ac)}{4a^2} = \frac{4ac}{4a^2} = \frac{c}{a}$

SECTION SKILL REINFORCERS 8-2

1 -2 **2** $\frac{4}{3}$ **3** $-2, 2$ **4** $0, 1$ **5** 0

EXERCISES 8-3

1 $z^2 - 5z + 4 = 0$; $z = x^2$ **3** $z^2 - 2z - 8 = 0$; $z = \sqrt{y}$
5 $z^2 - 2z - 15 = 0$; $z = \dfrac{v-2}{v}$
7 $z^2 + z - 2 = 0$; $z = \dfrac{1}{w}$
9 $z^2 - 2z - 35 = 0$; $z = r^{1/3}$ **11** $-3, -1, 1, 3$
13 $-\sqrt{3}, -\frac{1}{2}, \frac{1}{2}, \sqrt{3}$ **15** $-1-\sqrt{2}, -1, -1+\sqrt{2}$
17 $\frac{1}{4}, 16$ **19** $-\dfrac{\sqrt{13}}{3}, \dfrac{\sqrt{13}}{3}$ **21** $-\frac{3}{2}, 2$ **23** $-\frac{1}{2}, \frac{1}{2}$
25 $0, 2$ **27** $-2, 2, -2i, 2i$ **29** $-32, -\frac{1}{32}$
31 $-3-\sqrt{7}, -3-\sqrt{5}, -3+\sqrt{5}, -3+\sqrt{7}$ **33** -2
35 1 **37** 1 **39** $1, 2$ **41** 2 **43** $-3, -1, 1, 2$
45 $-1, 1$ (both solutions of multiplicity two)

$[-2, 2]$ for x, $[-1, 6]$ for y

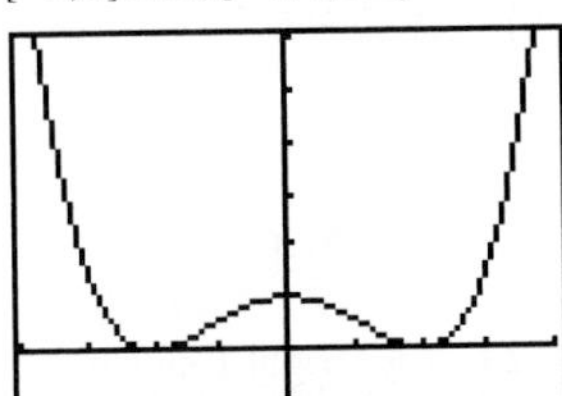

47 $-2.65, -1.00, 1.00, 2.65$ **49** 133.47
51 $\dfrac{1-\sqrt{21}}{10}, \dfrac{1+\sqrt{21}}{10}$ **53** 42.36 cm

SECTION SKILL REINFORCERS 8-3

1 50 mi/h **2** $\dfrac{D}{T}$ **3** $\frac{1}{2}$ job/h **4** 2.5 units/h
5 $\dfrac{1}{T}$ job/h

EXERCISES 8-4

1 $-12, -11$ or $11, 12$ **3** $-11, -9$ or $9, 11$
5 $\dfrac{-5-\sqrt{53}}{2}, \dfrac{9-\sqrt{53}}{2}$ or $\dfrac{-5+\sqrt{53}}{2}, \dfrac{9+\sqrt{53}}{2}$
7 $6-\sqrt{2}, 6+\sqrt{2}$ **9** 5 yd by 8 yd
11 11 m by 13 m **13** 5 cm **15** 12 ft **17** 7%, 8%
19 $\dfrac{7+\sqrt{109}}{2} \approx 8.7$ h, $\dfrac{13+\sqrt{109}}{2} \approx 11.7$ h
21 320 mi/h **23** $\dfrac{5+\sqrt{57}}{2} \approx 6.27$ ohms **25** 3 cm
27 80 mi/h, 150 mi/h **29** 130 h **31** 18 cm by 24 cm
33 $90(\sqrt{2}-1) \approx 37.3$ ft **35** 47.9 m **37** 180 mi
39 \$9/h for husband, \$10.50/h for wife

SECTION SKILL REINFORCERS 8-4

1 (number line) -1 0 1 2 3 4 5 6 7
2 (number line) -3 -2 -1 0 1 2 3 4 5
3 (number line) -2 -1 0 1 2 3 4 5 6
4 (number line) -4 -2 0 2 4 6 8 10 12
5 (number line) -3 -2 -1 0 1 2 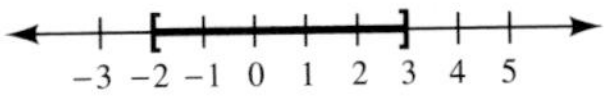 3 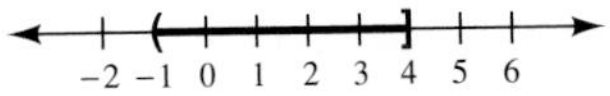 4 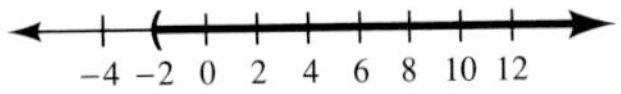 5

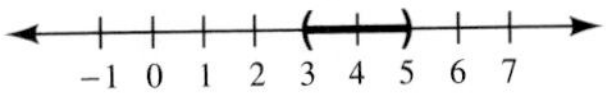

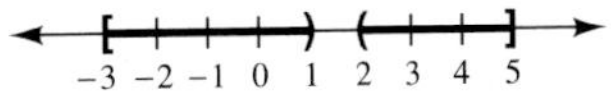

EXERCISES 8-5

1 $(-1, 2)$ **3** $(-\infty, 0] \cup [2, +\infty)$
5 $(-\infty, -4] \cup [6, +\infty)$ **7** $(-\infty, -\sqrt{2}) \cup (\sqrt{2}, +\infty)$
9 $(-6, \frac{1}{3})$ **11** $(-\infty, -\frac{1}{2}] \cup [2, +\infty)$ **13** $(-\frac{5}{2}, \frac{4}{3})$
15 $\left(-\infty, \dfrac{-1-\sqrt{5}}{2}\right] \cup \left[\dfrac{-1+\sqrt{5}}{2}, +\infty\right)$
17 $(-1, 1) \cup (3, +\infty)$ **19** $(-\infty, -3] \cup [0, 3]$
21 $(-\infty, 2) \cup (2, +\infty)$ or $\mathbb{R} \sim \{2\}$
23 $(-\infty, 3) \cup (6, +\infty)$ **25** $(0, 3)$
27 $(-\infty, -2) \cup [1, +\infty)$ **29** $(-\infty, 2) \cup (3, +\infty)$
31 $(-\infty, -2] \cup (-1, +\infty)$ **33** $(1, 2) \cup (3, +\infty)$
35 $(-\infty, -1] \cup (5, 6]$ **37** b **39** c
41 $(-\infty, -6] \cup [6, +\infty)$ **43** $(-\infty, -3] \cup [0, +\infty)$
45 $(-\infty, -1) \cup (0, 4)$ **47** $(-2, 0) \cup (1, +\infty)$
49 $(-7, -4] \cup (-1, 1]$ **51** $(-\infty, -3] \cup [0, 3]$
53 $[-5, 0] \cup [3, +\infty)$ **55** $(-2, 3)$
57 $(0, 2)$ **59** $(-\infty, 0) \cup (1, +\infty)$

$[-2, 4]$ for x, $[-1, 5]$ for y

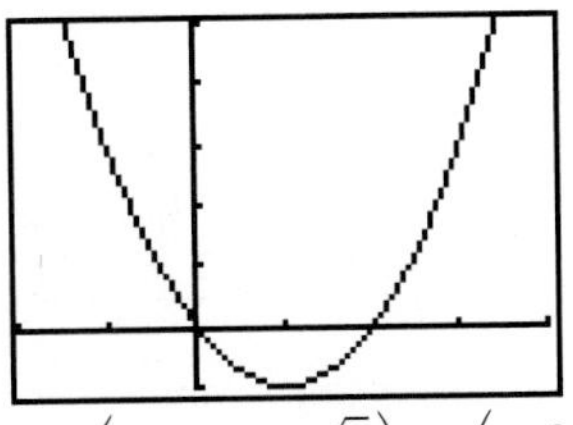

$[-2, 3]$ for x, $[-1, 5]$ for y

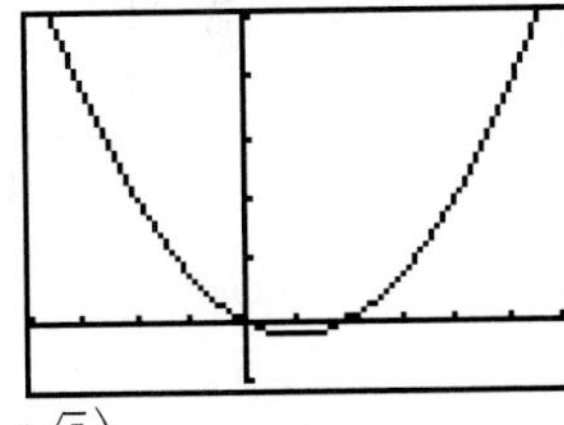

61 $\left(-\infty, \dfrac{2-\sqrt{5}}{2}\right) \cup \left(0, \dfrac{2+\sqrt{5}}{2}\right)$ **63** $(10, 110)$
65 $1\text{ s} < t < 4\text{ s}$

SECTION SKILL REINFORCERS 8-5

1 $y = 6$ **2** $y = -10$ **3** $y = -3$ **4** $y = 6$
5 $y = 3$

REVIEW EXERCISES 8

1 $-12, 12$ **2** $4, 5$ **3** $2-\sqrt{2}, 2+\sqrt{2}$ **4** $-3, 9$
5 $-\frac{2}{5}, \frac{5}{2}$ **6** $\frac{3}{2}-\frac{3}{2}i, \frac{3}{2}+\frac{3}{2}i$ **7** $\frac{4}{3}$ **8** $-\frac{5}{3}, \frac{3}{2}$
9 $-3, 3, -3i, 3i$ **10** $-4, 4, -i\sqrt{2}, i\sqrt{2}$
11 $-5, -1, 4, 8$ **12** $-2\sqrt{22}, 0, 2\sqrt{22}$ **13** $-1, \frac{7}{5}$

14 $-5, \frac{3}{4}$ **15** $-5, \frac{3}{2}, \frac{10}{3}$ **16** $2, -1 - i\sqrt{3}, -1 + i\sqrt{3}$
17 4 **18** 5 **19** $-1, 4$ **20** $-3, 6$
21 $-1, 19683$ **22** $-10 - 3\sqrt{10}, -10 + 3\sqrt{10}$
23 Rational solution of multiplicity two
24 Two imaginary solutions
25 Two distinct irrational solutions **26** $-10, 10$
27 4 **28** 81 **29** 36 **30** $\frac{1 - \sqrt{13}}{2}, \frac{1 + \sqrt{13}}{2}$
31 $\frac{1}{3} - \frac{2\sqrt{5}}{3}i, \frac{1}{3} + \frac{2\sqrt{5}}{3}i$ **32** $x = -2y, x = 5y$
33 $x = -3y, x = 3y$
34 $x = (-2 - \sqrt{3})y, x = (-2 + \sqrt{3})y$ **35** b **36** a
37 b **38** $(-\infty, -5] \cup [7, +\infty)$ **39** $(-\sqrt{7}, \sqrt{7})$
40 $(2 - \sqrt{5}, 2 + \sqrt{5})$ **41** $(0, 3) \cup (3, +\infty)$
42 $(-\infty, -6] \cup (-5, 1]$ **43** $(-4, -3] \cup [2, +\infty)$
44 $(-5, 2)$

45 $(-3, \frac{1}{2})$
$[-4, 2]$ for x, $[-8, 8]$ for y

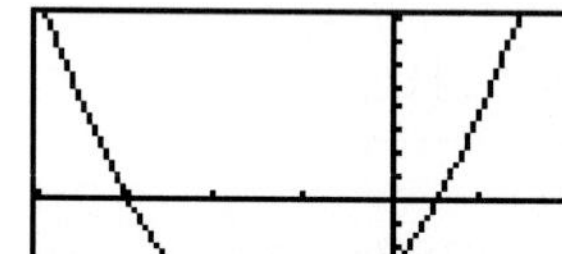

46 $(-\infty, \frac{1}{2}) \cup (4, +\infty)$
$[-2, 5]$ for x, $[-8, 11]$ for y

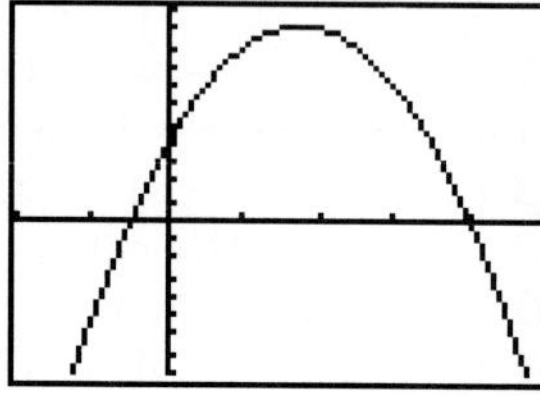

47 $(-3, -1) \cup (1, +\infty)$
$[-4, 2]$ for x, $[-8, 8]$ for y

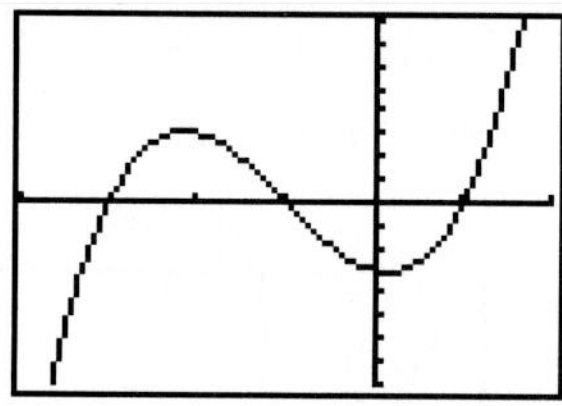

48 $(-\infty, -\frac{1}{2}] \cup [\frac{3}{5}, +\infty)$ **49** $(-\infty, -2] \cup [2, +\infty)$
50 $\frac{-7 - \sqrt{17}}{2}$ and $\frac{7 - \sqrt{17}}{2}$ or $\frac{-7 + \sqrt{17}}{2}$ and $\frac{7 + \sqrt{17}}{2}$
51 $-1 - \sqrt{5}, -1 + \sqrt{5}$ **52** 8 cm, 15 cm, 17 cm
53 $\frac{11 + \sqrt{281}}{2} \approx 13.9$ h **54** 8%, 9% **55** 19%
56 (5, 40) **57** (1, 4)
58 a. $4x^2 - 17x - 15 = 0$ **b.** $2x^2 - 2x - 1 = 0$
c. $x^2 - 4x + 29 = 0$
59 $x^3 - 11x^2 + 28x = 0$
60
$$\begin{aligned}(x - 2)(2x + 3) &= 2x - 6\\ 2x^2 - x - 6 &= 2x - 6\\ 2x^2 - 3x &= 0\\ x(2x - 3) &= 0\\ x = 0 \quad \text{or} \quad 2x - 3 &= 0\\ x = 0 \quad \text{or} \quad x &= \tfrac{3}{2}\end{aligned}$$

MASTERY TEST 8

1 a. $-14, 14$ **b.** $-5i, 5i$ **c.** $-2, 5$
2 a. $-5, 1$ **b.** $1 - \sqrt{2}, 1 + \sqrt{2}$ **c.** $-\frac{3}{2}, -\frac{1}{2}$
3 a. $\frac{2}{3}, \frac{5}{2}$ **b.** $-1 - \sqrt{5}, -1 + \sqrt{5}$
c. $\frac{1}{3} - \frac{\sqrt{2}}{3}i, \frac{1}{3} + \frac{\sqrt{2}}{3}i$
4 a. Two distinct irrational solutions
b. Rational solution of multiplicity two
c. Two imaginary solutions
5 a. $-5, -3, 3, 5$ **b.** 81 **c.** $\frac{4}{3}$
6 a. -6 **b.** $\frac{-1 + \sqrt{13}}{2}$ **c.** $-2, 5$
7 a. $-\sqrt{2}, \sqrt{2}$ **b.** 3.7 cm, 9.7 cm, 10.3 cm
c. 48 h
8 a. $(-\infty, -2) \cup (5, +\infty)$
b. $\left[\frac{-5 - \sqrt{13}}{6}, \frac{-5 + \sqrt{13}}{6}\right]$ **c.** $[-2, 5] \cup (8, +\infty)$
d. $(-\infty, 2) \cup (4, 7)$

Chapter 9

EXERCISES 9-1

1 13 **3** 17 **5** $\sqrt{2}$ **7** 3 **9** (6, 17)
11 $(-10, -2)$ **13** $(\frac{5}{12}, -\frac{3}{20})$ **15** 12 **17** 20
19 39 **21** $\overline{DF} = 2\sqrt{2}$ **23** Not a right triangle
25 A right triangle **27** 4.9 cm **29** $2x + y + 3 = 0$
31 $5x - 9y + 7 = 0$ **33** $x = 0$ **35** 1.09 **37** 17.71
39 (1.83, 2.09) **41** Collinear **43** Not collinear
45 $d = \sqrt{(x - h)^2 + (y - k)^2}$ **47** $4\sqrt{10} \approx 12.65$
49

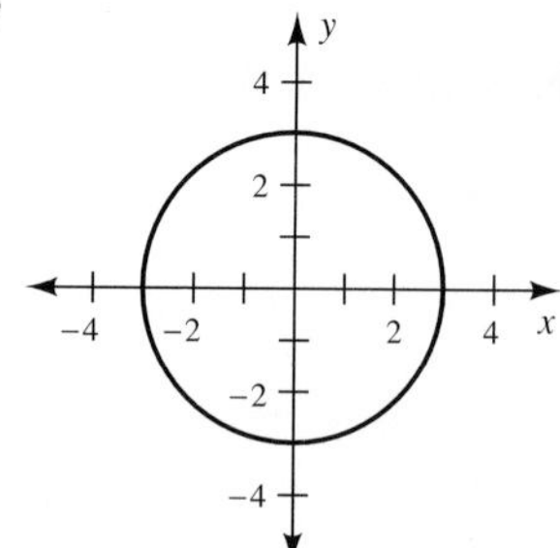

SECTION SKILL REINFORCERS 9-1

1 -2 **2** 2 **3** 10 **4** $h^2 + 4h - 2$ **5** $h + 4$

EXERCISES 9-2

1 a. (0, 7); upward **b.** (0, −7); upward
c. (−7, 0); upward **d.** (7, 0); upward
3 a. (6, −1); upward **b.** (−6, 1); upward
c. (0, −1); upward **d.** (0, 1); downward
5 a. (3, 4); upward **b.** (−3, 4); downward

7 a.

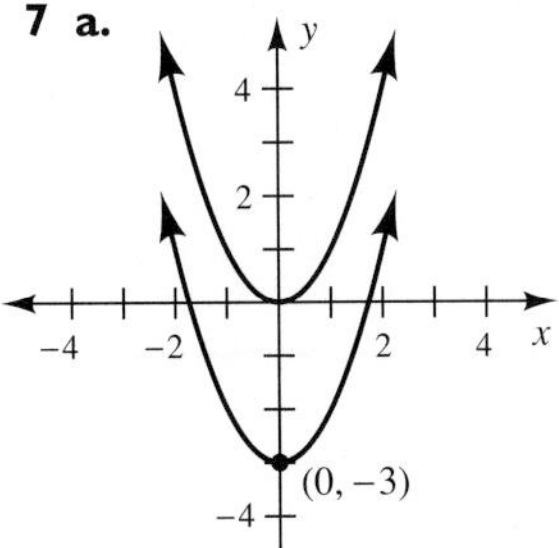

b.

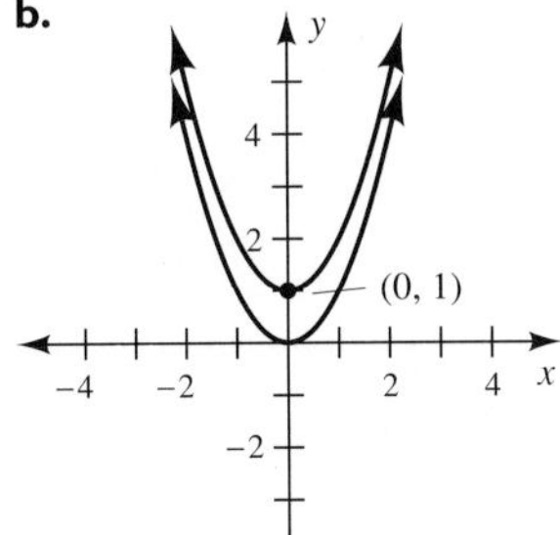

9 a.

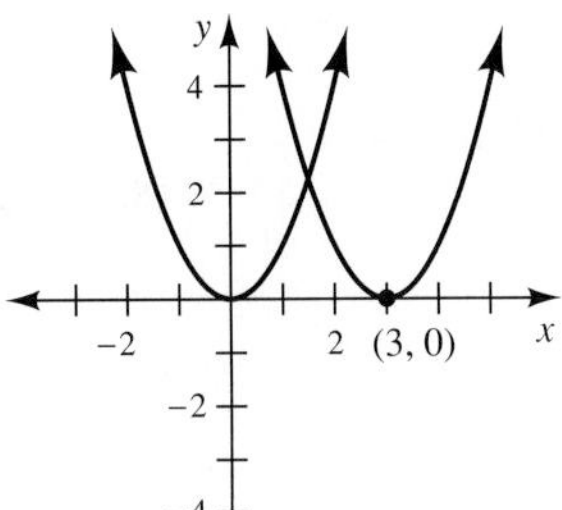

b.

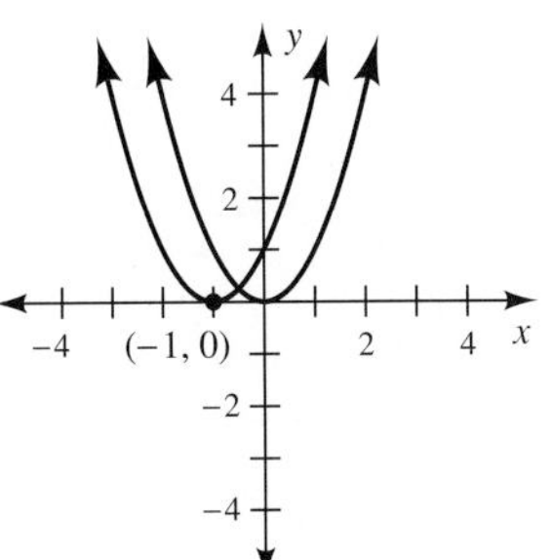

11

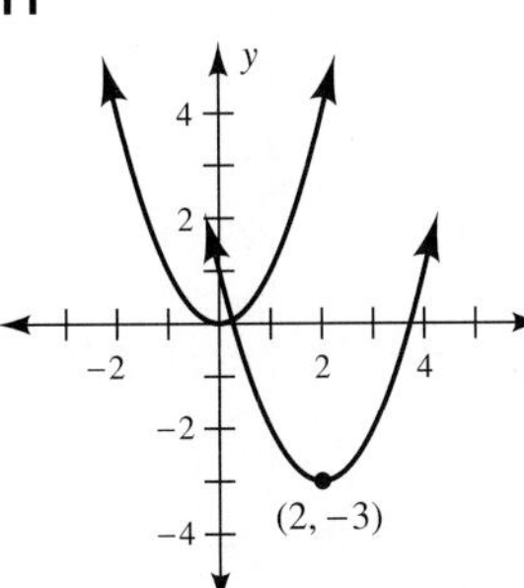

13

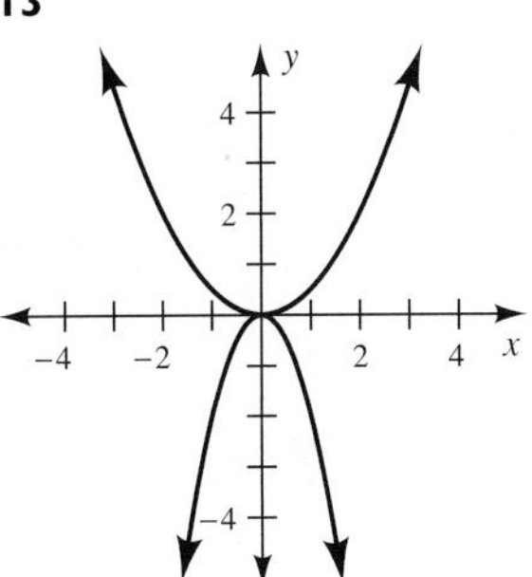

15 Maximum of 11 for $x = -7$

17 Minimum of 5 for $x = 8$

19

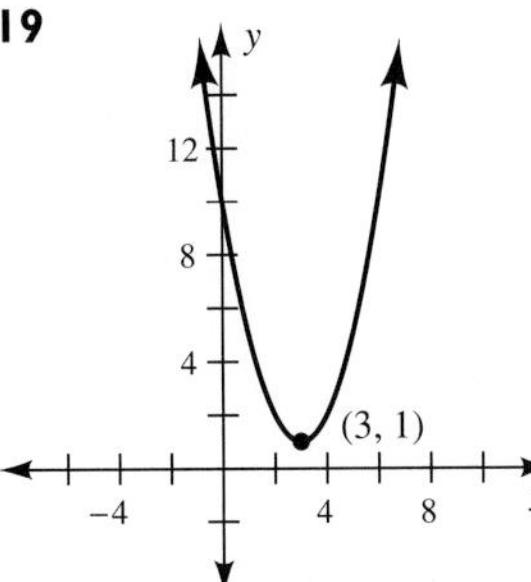

21

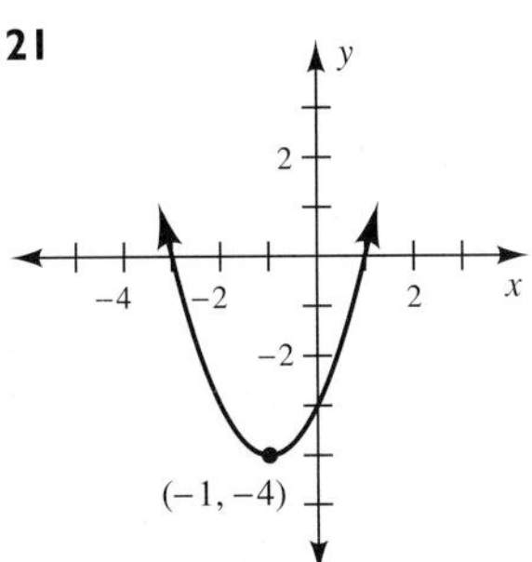

23

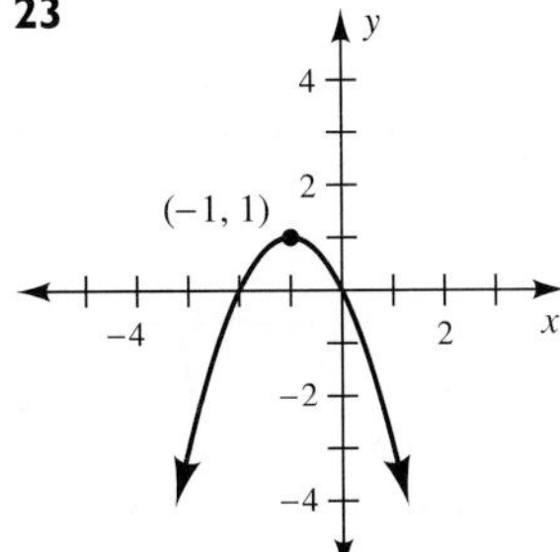

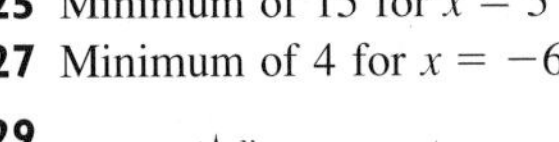

25 Minimum of 15 for $x = 5$

27 Minimum of 4 for $x = -6$

29

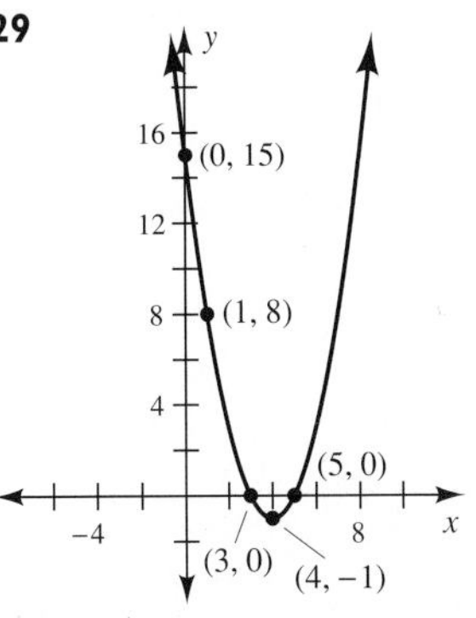

31

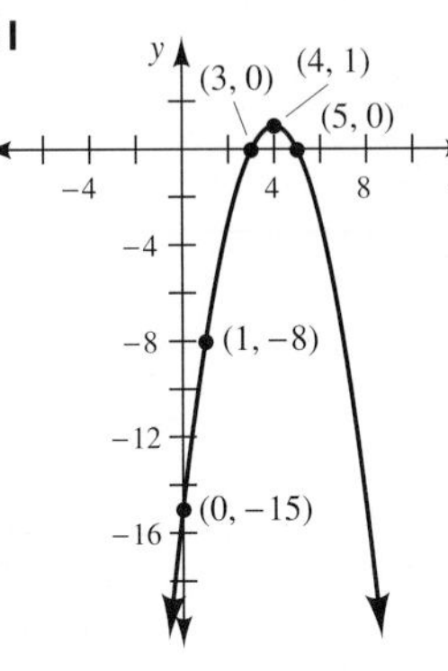

33 (0, 6), (2, 0), (3, 0) **35** (0, −8), (−1, 0), (8, 0)

37 144 ft **39** 100 m^2 **41 a.** −3, 3 **b.** −9

43 a. 25 **b.** \$1000 **45** $D = \mathbb{R}$; $R = [-6, +\infty\}$

47 4

49 $y = (2x - 1)(x + 4)$

$[-6, 2]$ for x, $[-11, 8]$ for y

51 $y = (x + 6)(x - 7)$

$[-8, 10]$ for x, $[-45, 20]$ for y

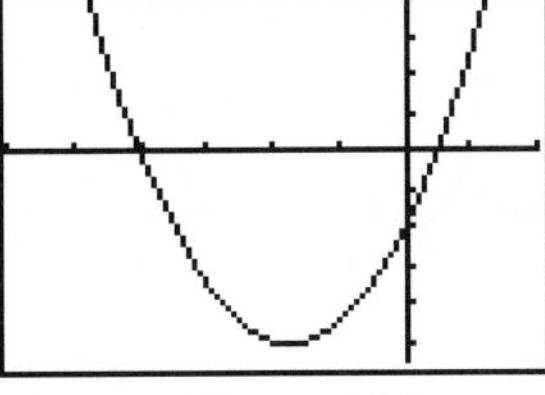

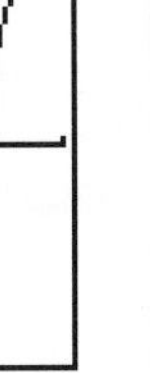

53 At 2100 rpm, 85 hp

$[-500, 4500]$ for x, $[-25, 100]$ for y

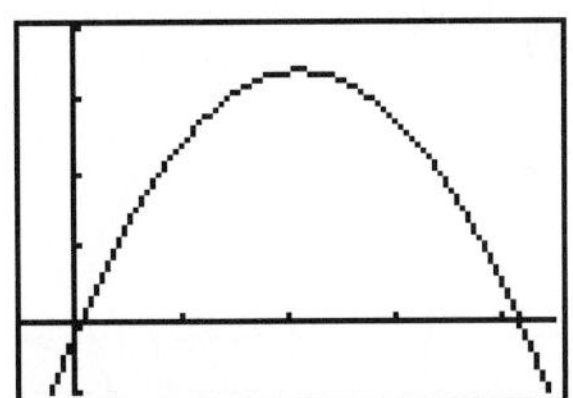

55

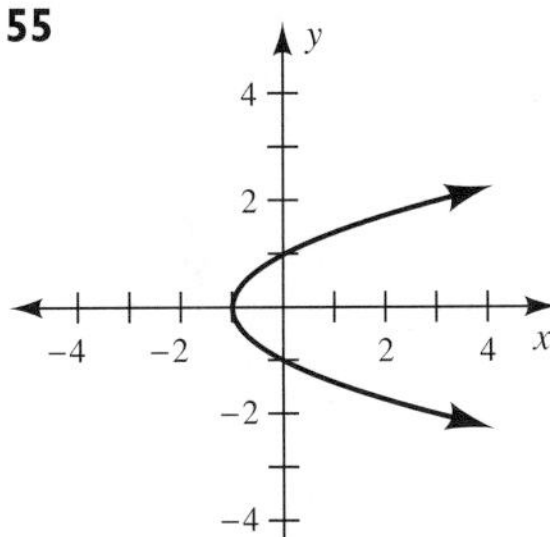

57

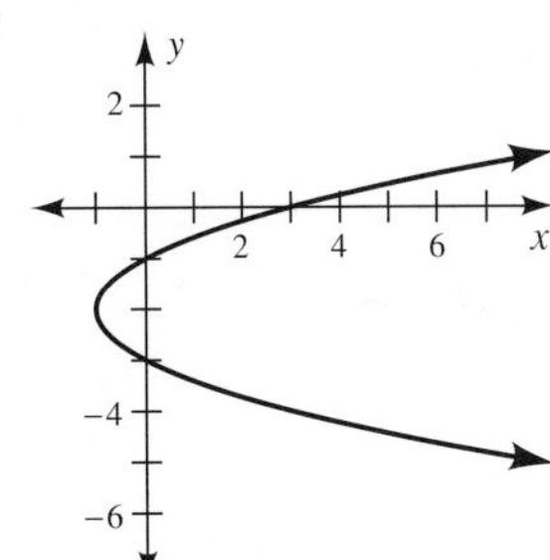

Section Skill Reinforcers 9-2

1 c **2** d **3** e **4** b **5** a

Exercises 9-3

1 $(x - 3)^2 + (y - 7)^2 = 4$ **3** $(x + 2)^2 + (y - 1)^2 = \frac{1}{4}$

5 $(x + 5)^2 + (y + 2)^2 = 3$ **7** $x^2 + y^2 = 16$

9 $x^2 + y^2 = 25$ **11** $\dfrac{x^2}{49} + \dfrac{y^2}{100} = 1$

13 $\dfrac{x^2}{\frac{1}{4}} + \dfrac{y^2}{4} = 1$ **15** $\dfrac{x^2}{2} + \dfrac{y^2}{5} = 1$

17 $x^2 + y^2 = 25$

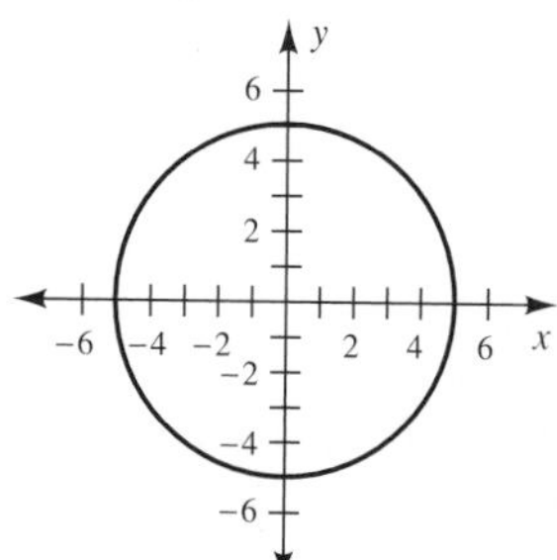

19 $(x + 1)^2 + (y - 3)^2 = 9$

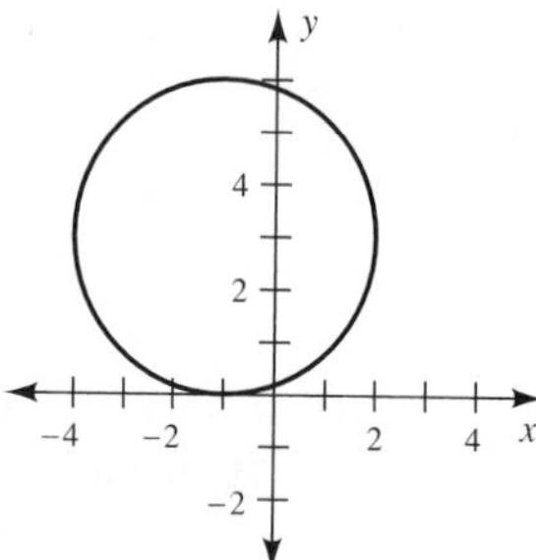

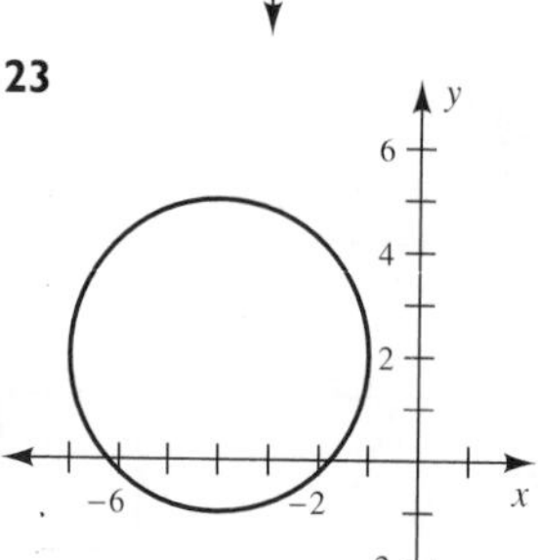

21 $\dfrac{x^2}{25} + \dfrac{x^2}{36} = 1$

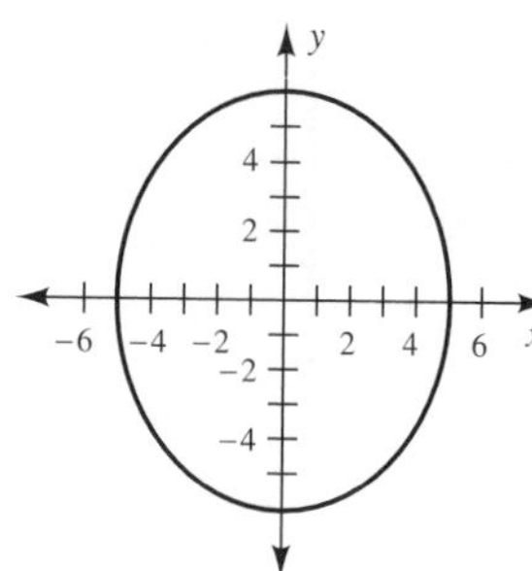

23

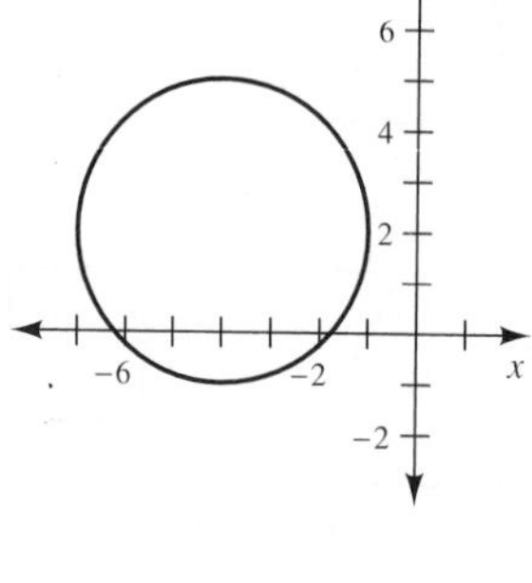

25

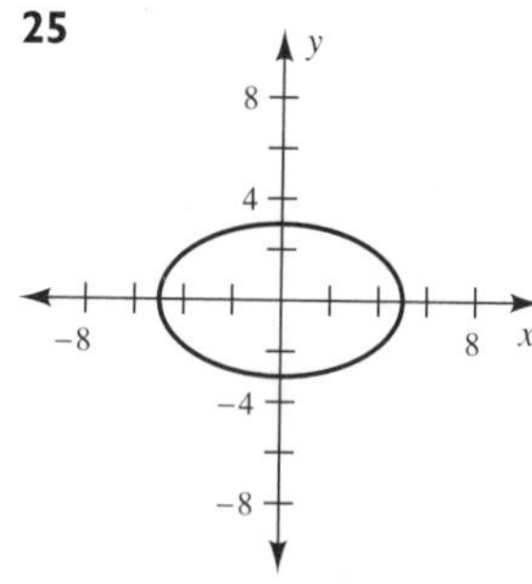

27

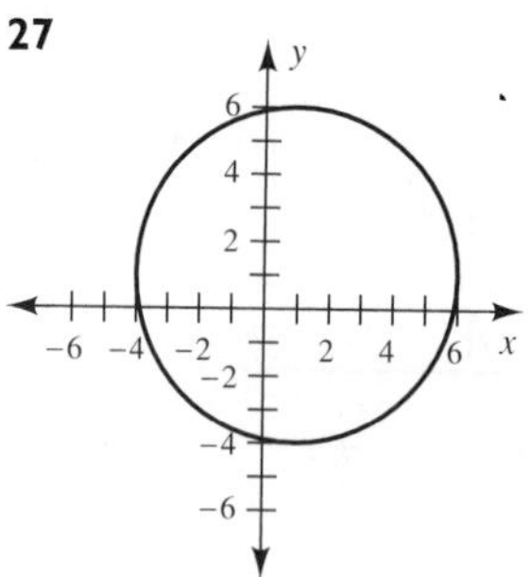

29 c **31** a **33** f

35 Circle

37 Ellipse

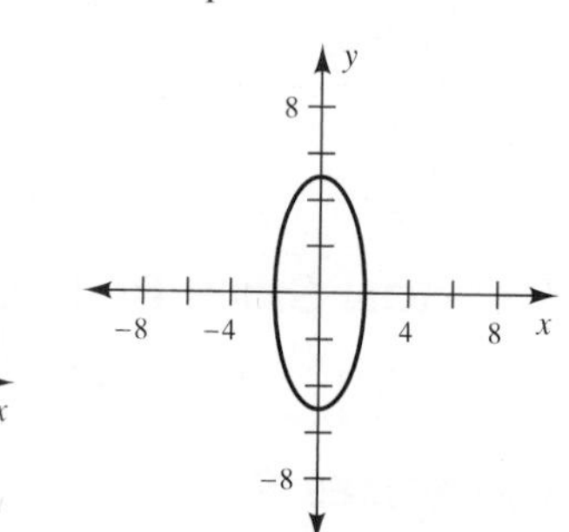

39 Ellipse

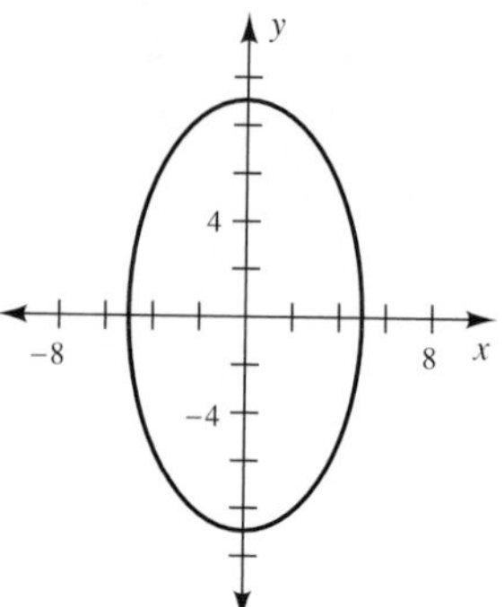

41 Parabola

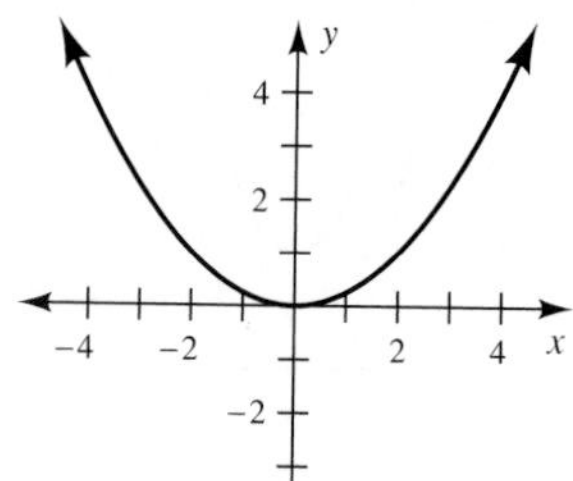

43 Circle

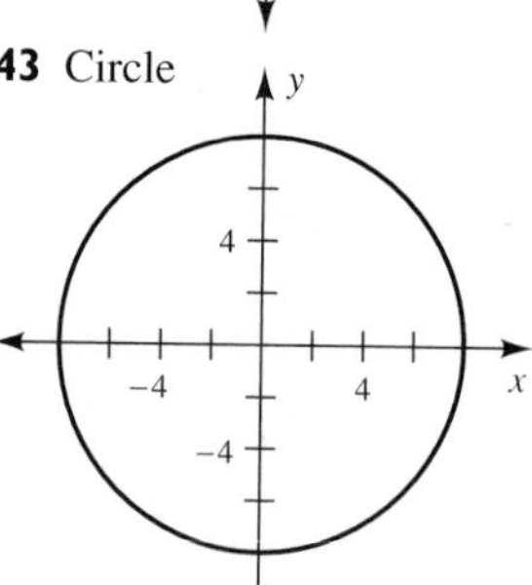

45 Ellipse

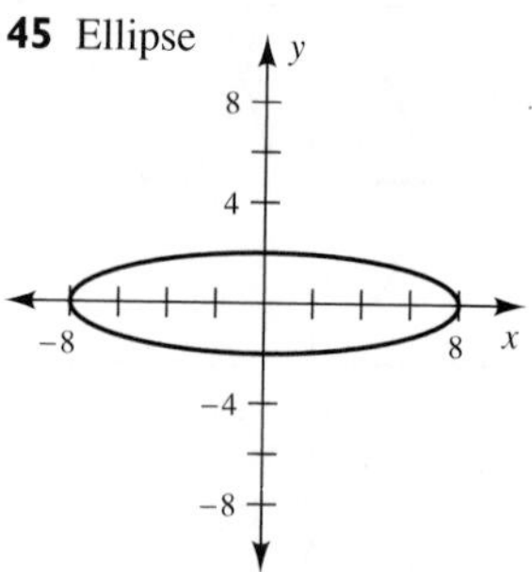

47 Circle

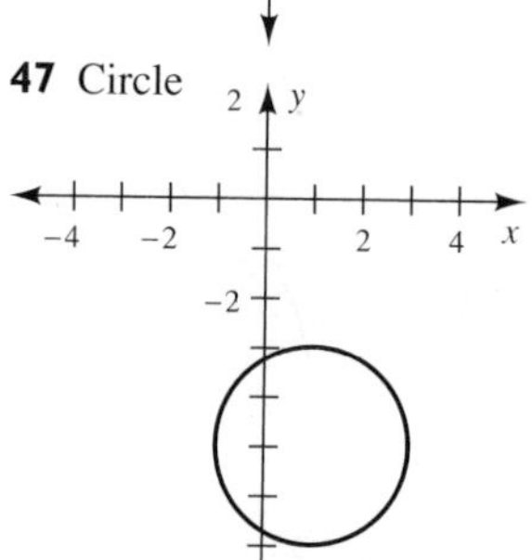

49 Circle

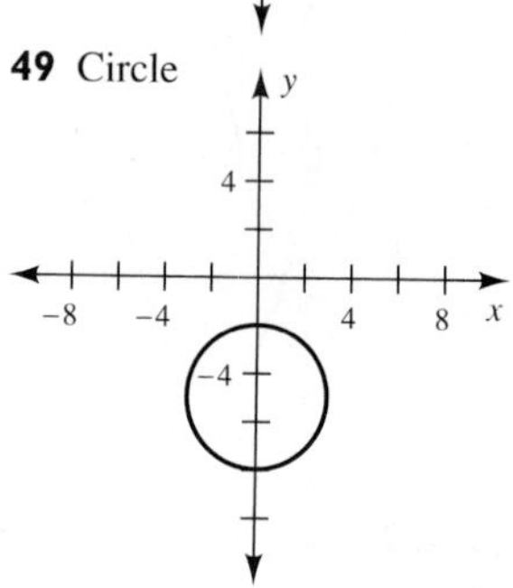

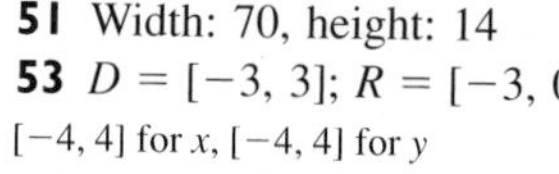

51 Width: 70, height: 14

53 $D = [-3, 3]$; $R = [-3, 0]$

$[-4, 4]$ for x, $[-4, 4]$ for y

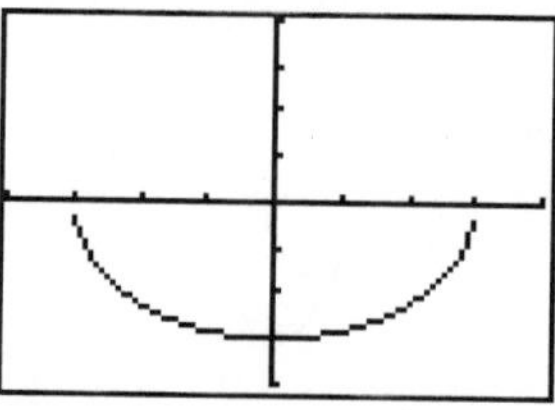

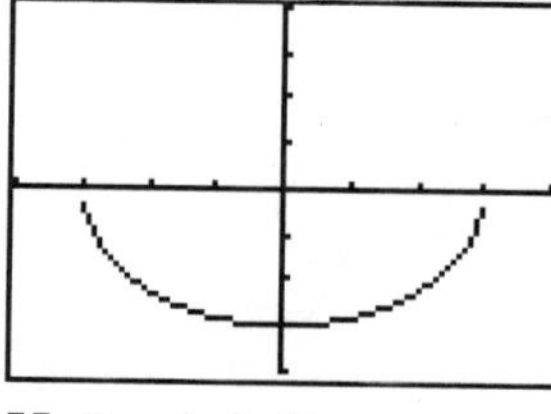

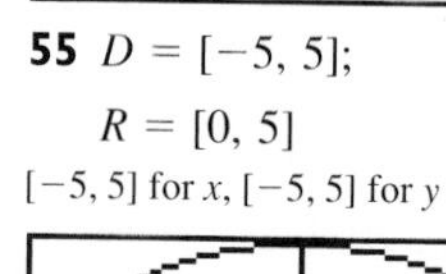

55 $D = [-5, 5]$;

$R = [0, 5]$

$[-5, 5]$ for x, $[-5, 5]$ for y

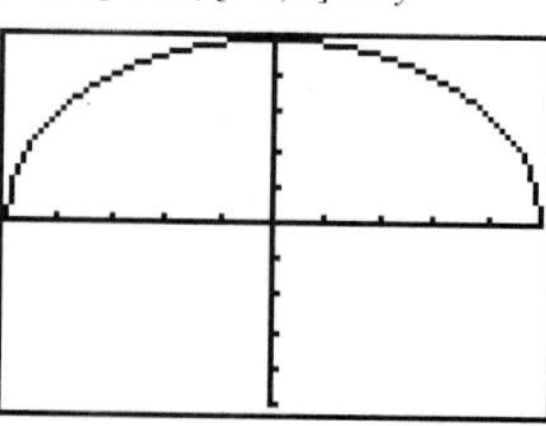

57 $D = [-1.5, 1.5]$;

$R = [-3, 0]$

$[-3.5, 3.5]$ for x, $[-3.5, 3.5]$ for y

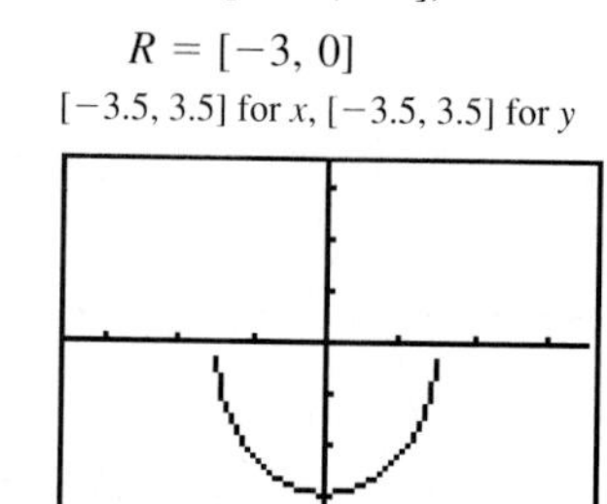

59 $D = (-\infty, 0]$; $R = [-2, +\infty)$

$[-5, 5]$ for x, $[-4, 20]$ for y

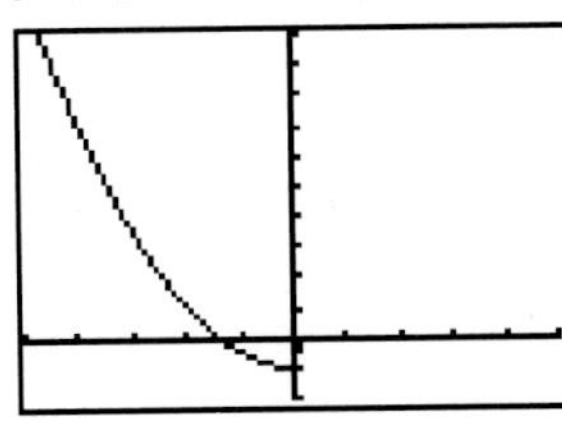

61

$[-9, 9]$ for x, $[-9, 9]$ for y

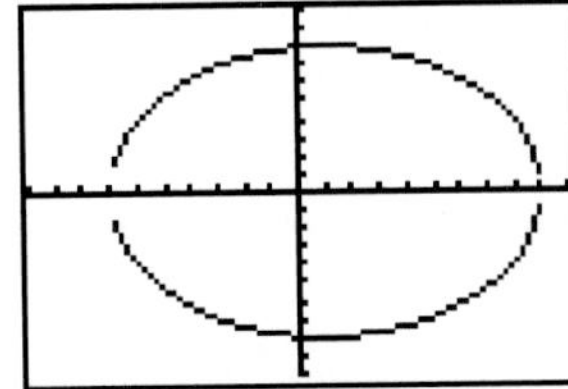

63

$[-8, 8]$ for x, $[-8, 8]$ for y

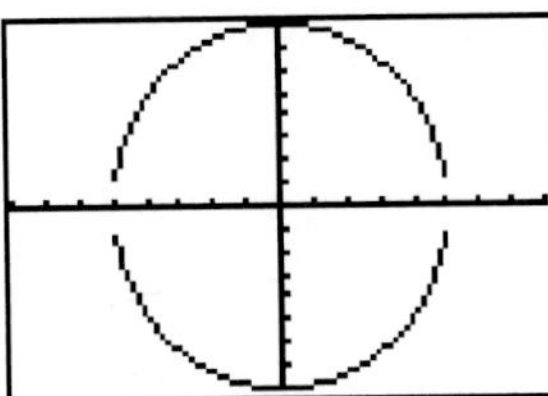

65

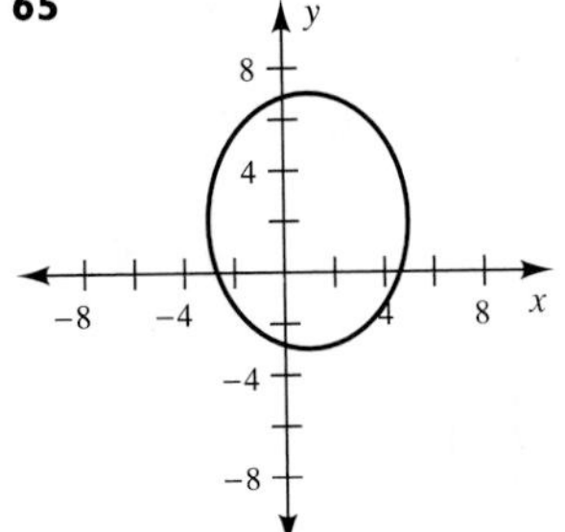

67

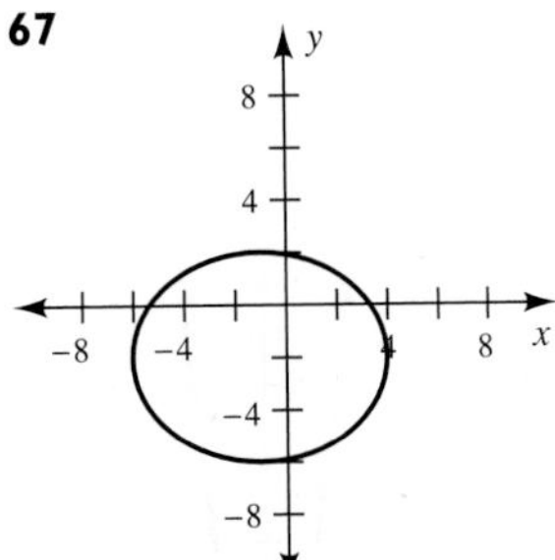

69 $(x + 1)^2 + (y - 4)^2 = 25$

SECTION SKILL REINFORCERS 9-3

1 $D = \mathbb{R}$ **2** $D = \mathbb{R}$ **3** $D = \mathbb{R}$ **4** $D = [3, +\infty)$
5 $D = [-3, 3]$

EXERCISES 9-4

1 Horizontally; $(-7, 0)$, $(7, 0)$
3 Vertically; $(0, -2)$, $(0, 2)$
5 Horizontally; $(-6, 0)$, $(6, 0)$ **7** $\dfrac{x^2}{16} - \dfrac{y^2}{25} = 1$

9 $\dfrac{y^2}{16} - \dfrac{x^2}{25} = 1$ **11** $\dfrac{y^2}{4} - \dfrac{x^2}{9} = 1$

13

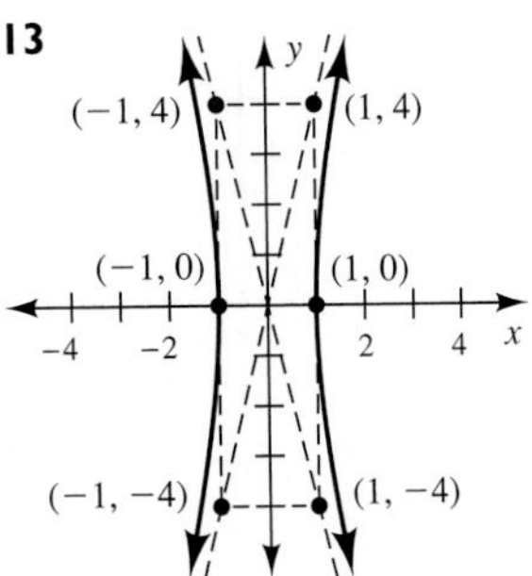

15

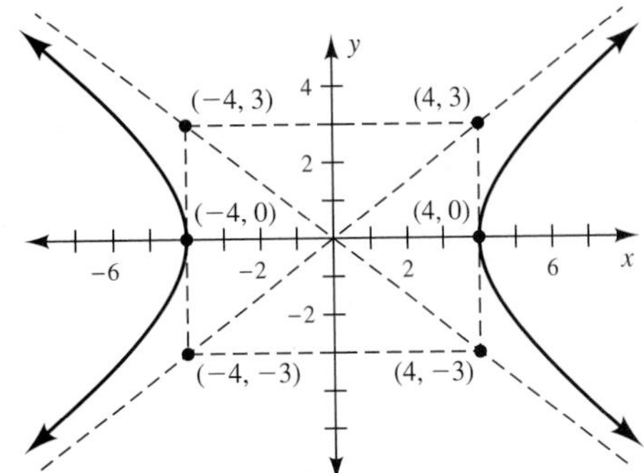

17

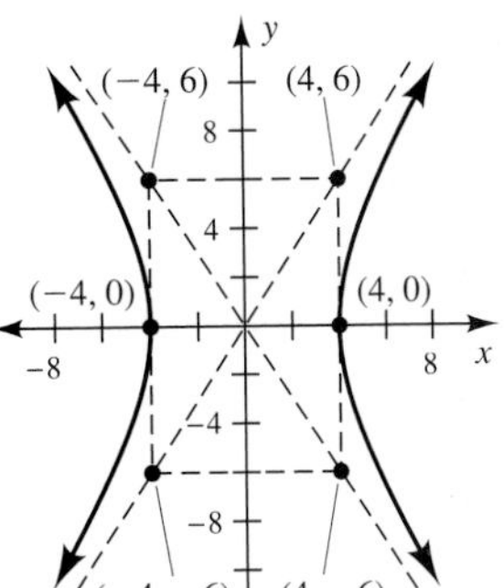

19

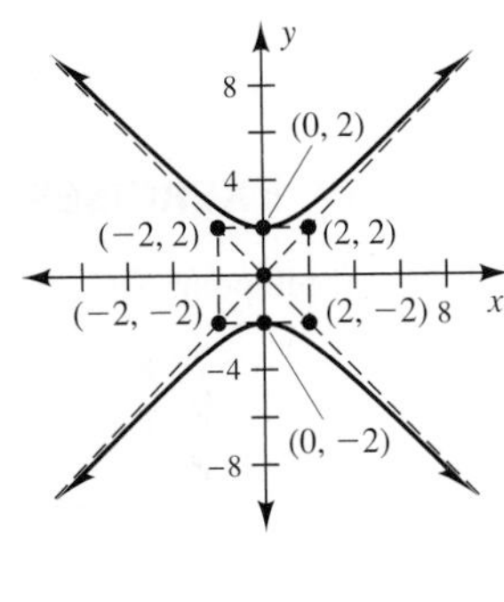

21 Parabola **23** Circle **25** Ellipse **27** Parabola
29 Hyperbola **31** Ellipse **33** d **35** g **37** b
39 e **41** $[-5, 9]$ for x, $[-4, 7]$ for y

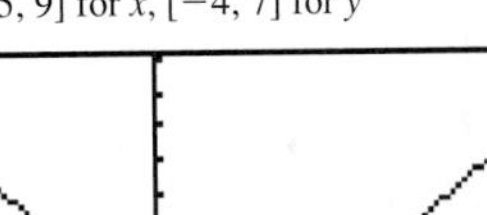

43

$[-1, 10]$ for x, $[-1, 10]$ for y

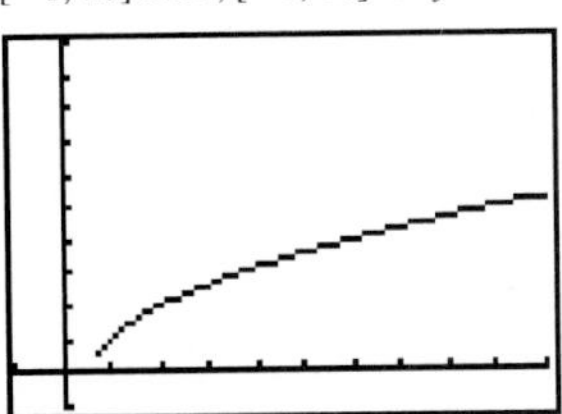

45

$[-5, 5]$ for x, $[-5, 5]$ for y

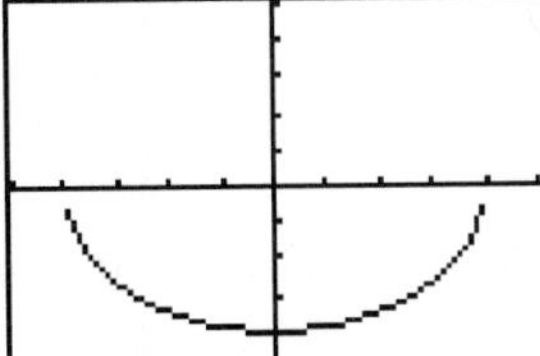

47

$[-9, 9]$ for x, $[-9, 9]$ for y

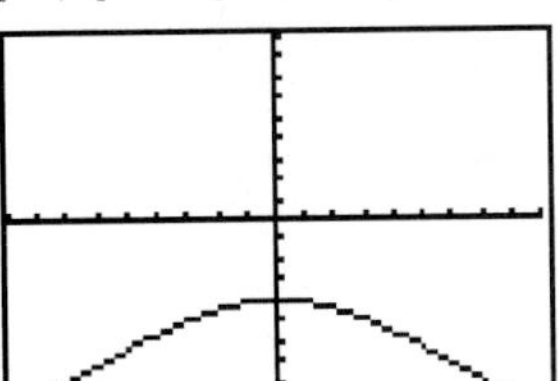

49

$[-8, 8]$ for x, $[-8, 8]$ for y

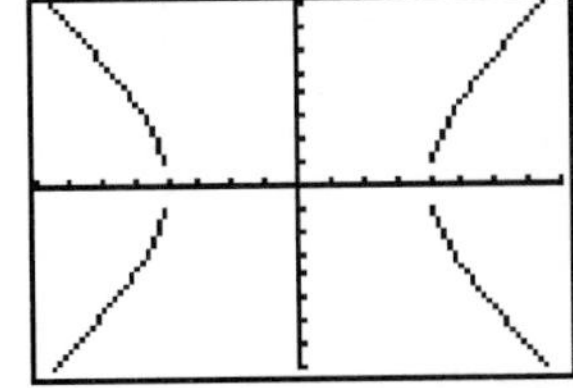

51

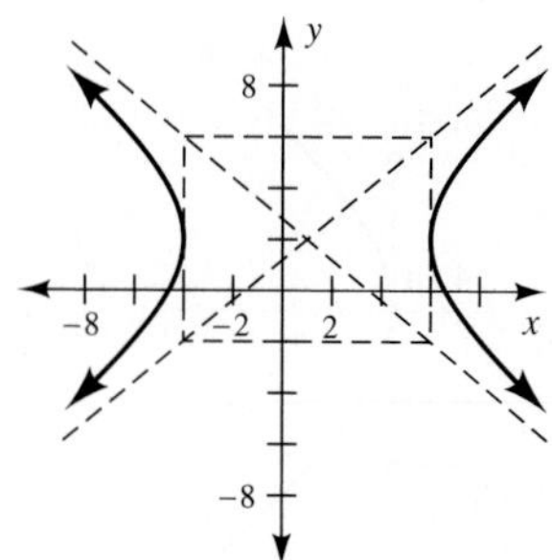

Section Skill Reinforcers 9-4

1 $x - 2 = 0$ **2** $y - 3 = 0$ **3** $7x + 2y - 20 = 0$
4 $4x - y - 5 = 0$ **5** $3x + 2y - 25 = 0$

Review Exercises 9

1 $m = \frac{3}{4}$; midpoint: $(0, -\frac{7}{2})$; $d = 5$

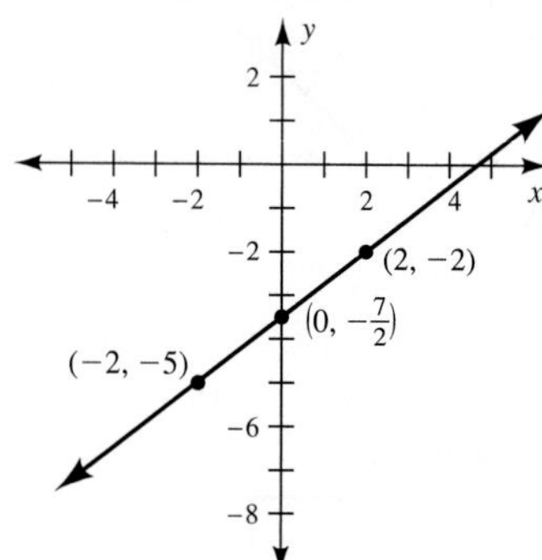

2 $m = 0$; midpoint; $(-1, 11)$; $d = 12$

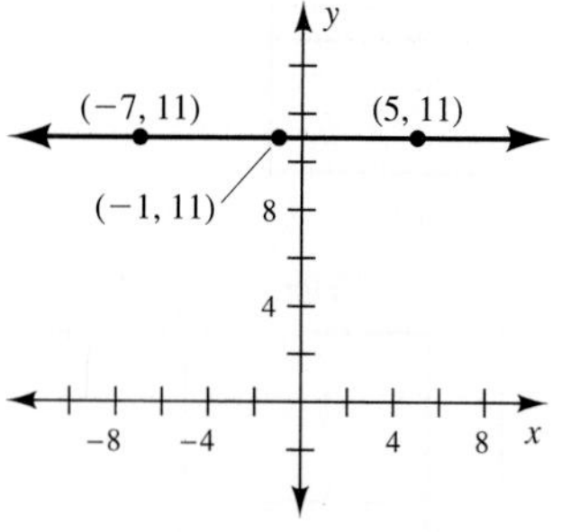

3 $m = 0$; midpoint: $(0, -3)$; $d = 8$

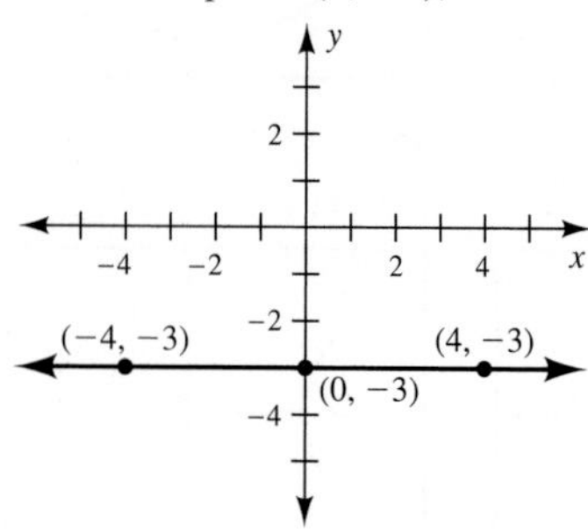

4 $m = -\frac{12}{5}$; midpoint: $(-\frac{3}{2}, 2)$; $d = 13$

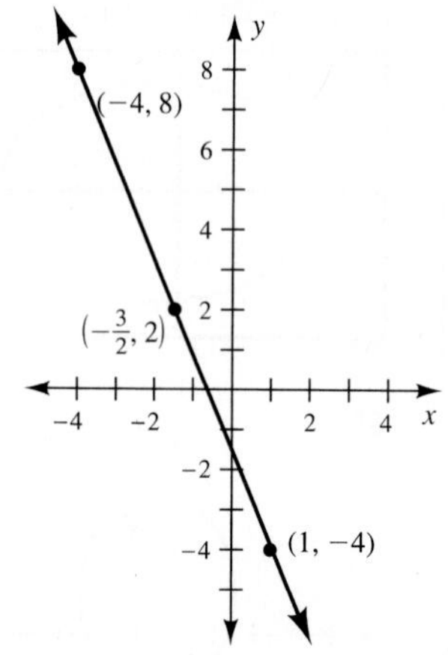

5 A right triangle **6** $x + 2y - 11 = 0$
7 $(0, 7)$; concave downward
8 $(-11, 0)$; concave upward
9 $(-10, -8)$; concave upward
10 $(3, 14)$; concave downward

11

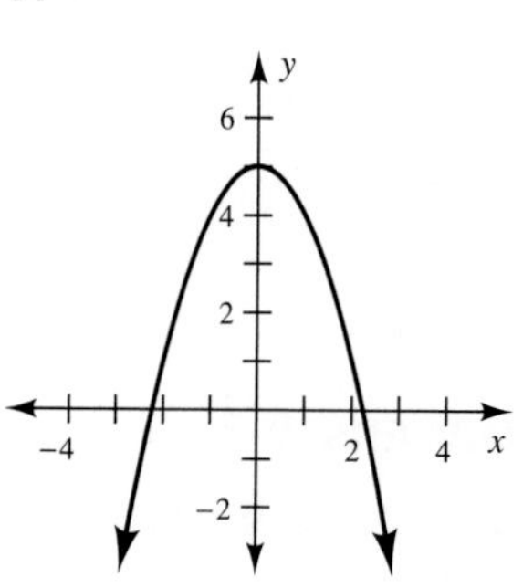

12

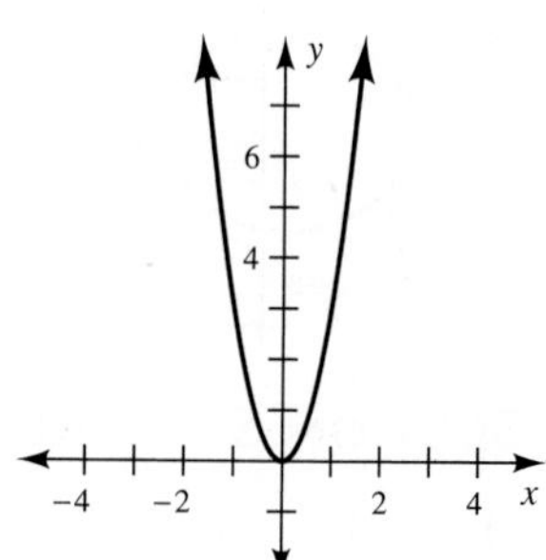

13

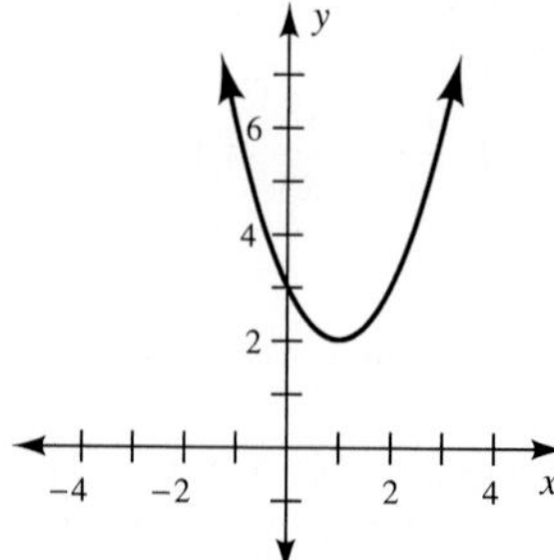

14

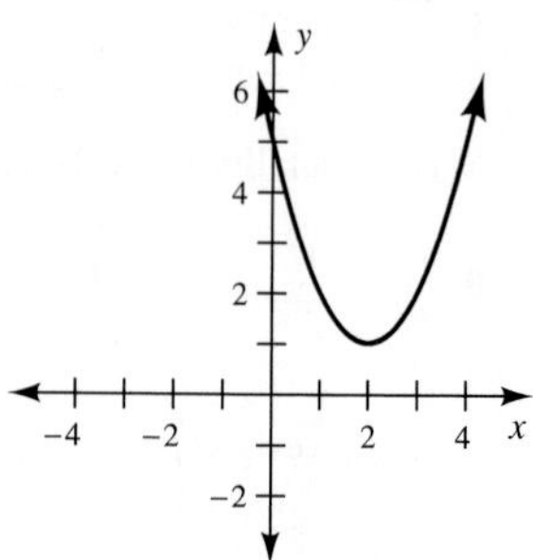

15 2048 m^2 **16** $-5, 5$; product $= -25$ **17** d
18 a **19** g **20** j **21** b **22** c **23** l
24 h **25** e **26** i **27** k **28** f

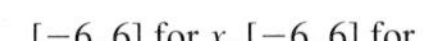

29

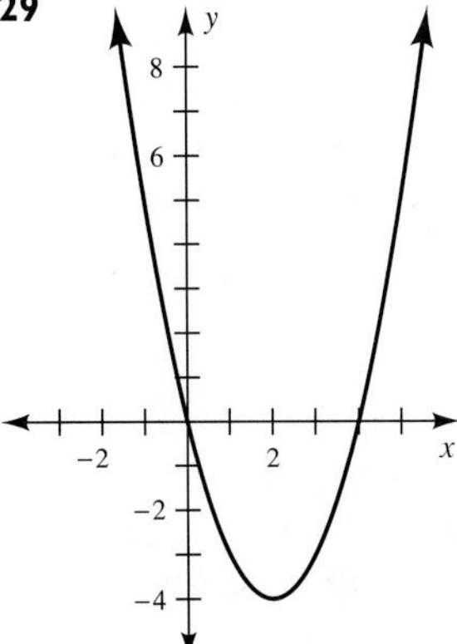

30

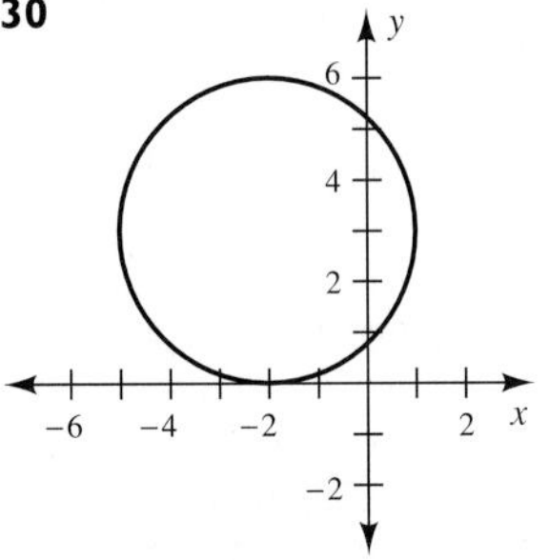

31

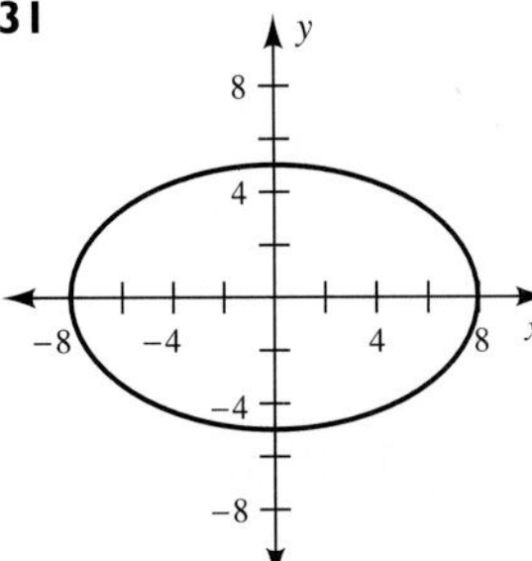

32

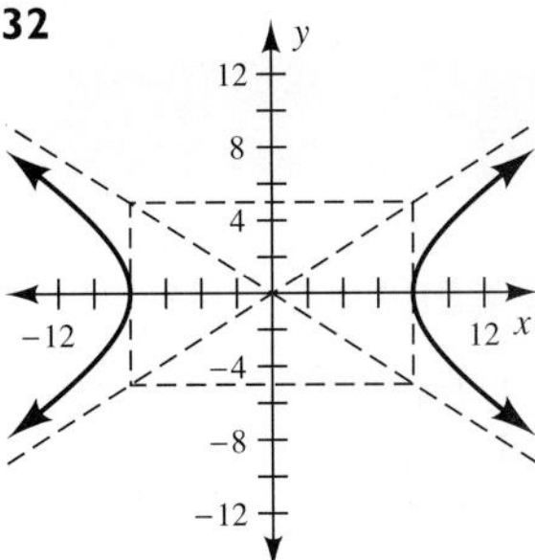

33

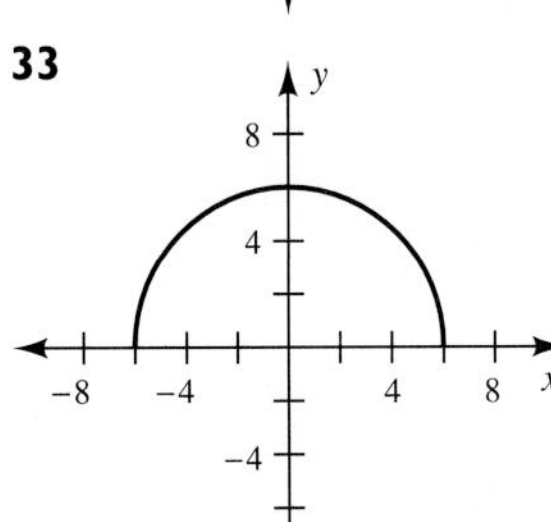

34

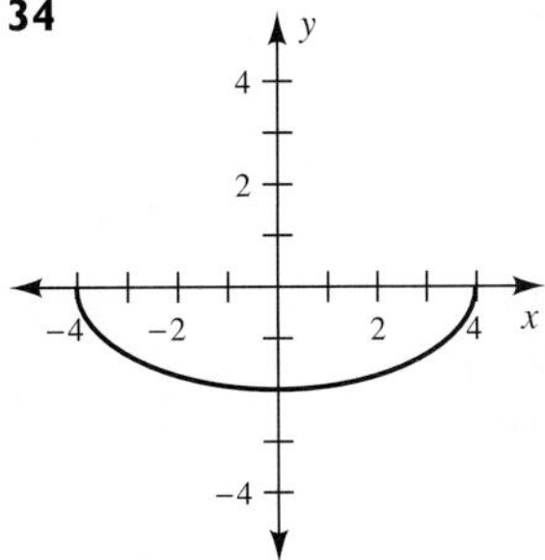

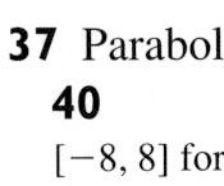

35 Circle **36** Ellipse **37** Parabola **38** Hyperbola

39

$[-7, 1]$ for x, $[-4, 4]$ for y

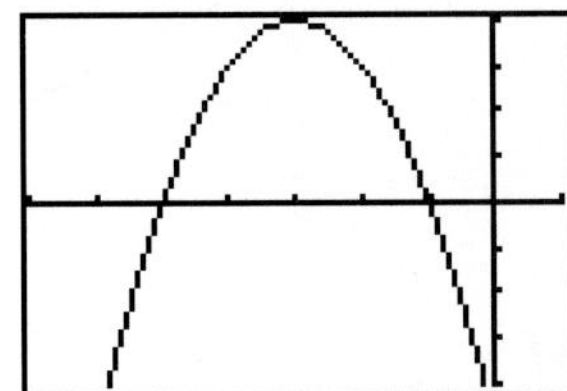

40

$[-8, 8]$ for x, $[-8, 8]$ for y

41

$[-3, 3]$ for x, $[-3, 3]$ for y

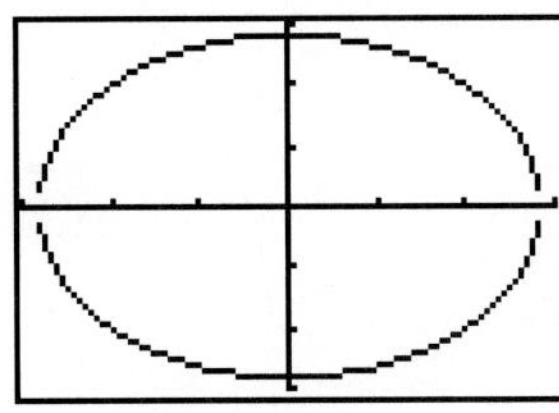

42

$[-5.25, 5.25]$ for x, $[-5.25, 5.25]$ for y

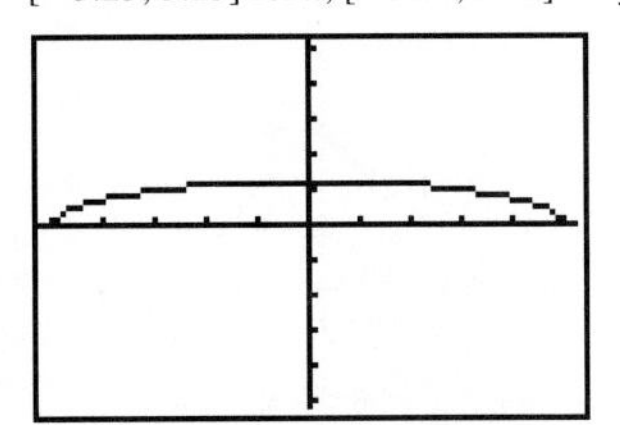

43

$[-2.5, 2.5]$ for x, $[-2.5, 2.5]$ for y

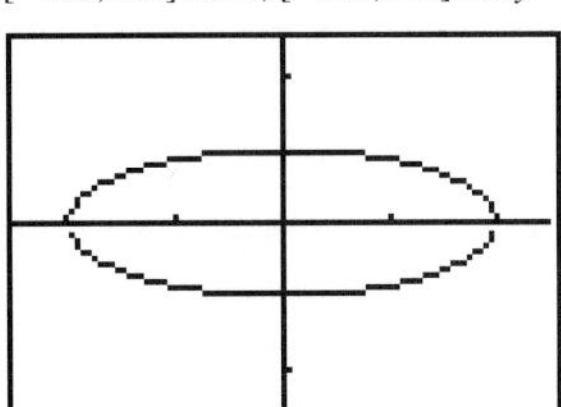

44

$[-6, 6]$ for x, $[-6, 6]$ for y

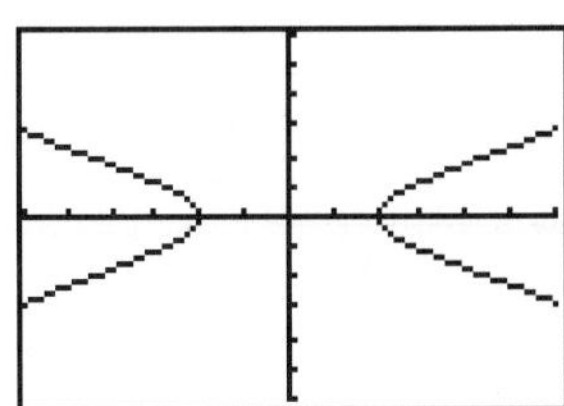

MASTERY TEST 9

1 a. 8 **b.** 10 **c.** 13

2 a. (6, 4) **b.** (8, 2) **c.** (−6, 1)

3 a.

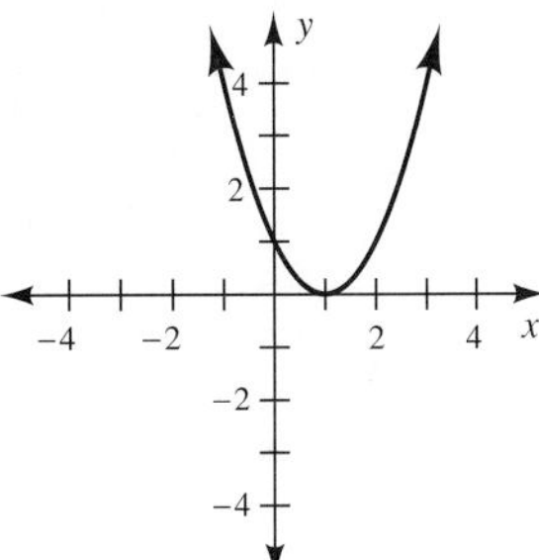

b.

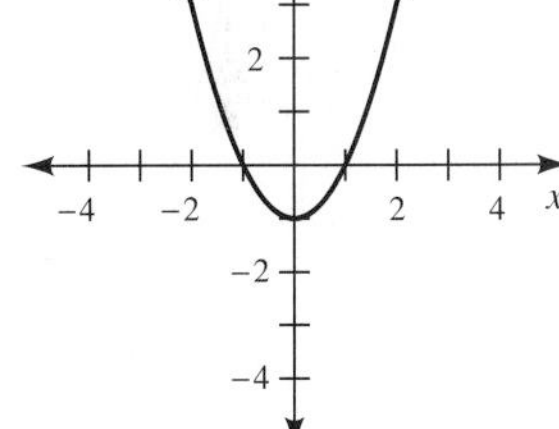

c.

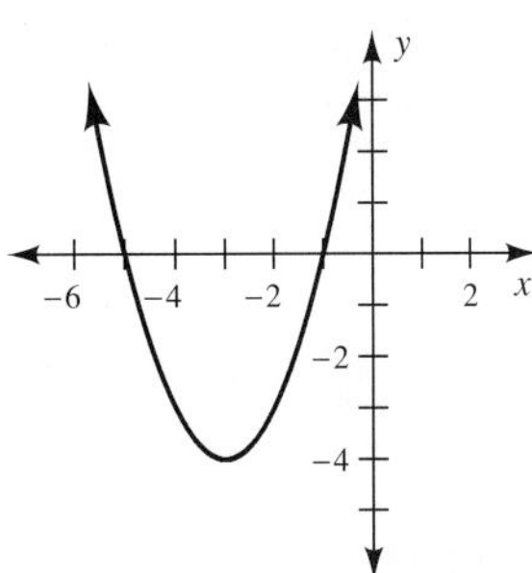

d.

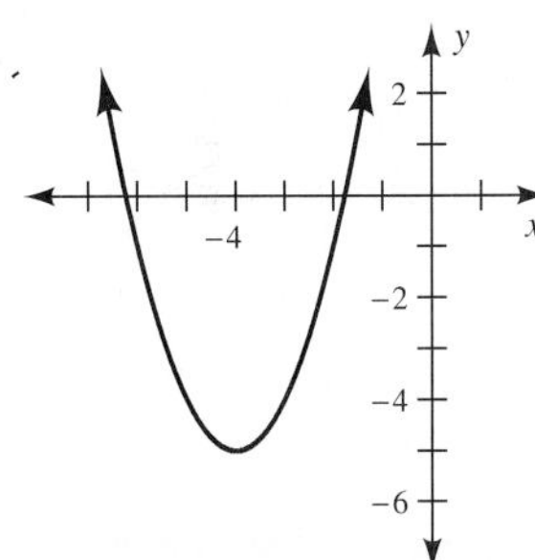

4 $200

5 a.

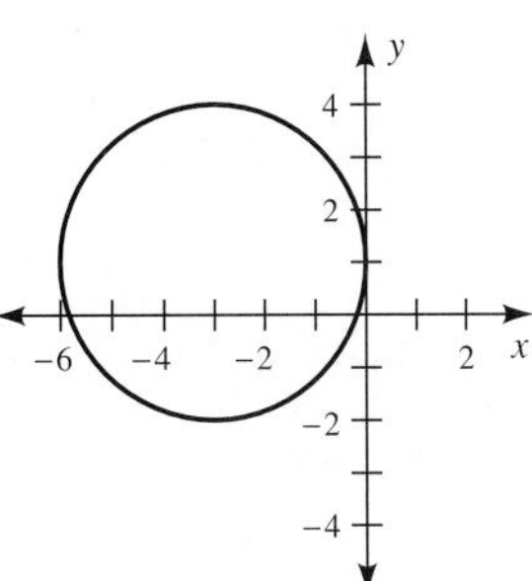

b.

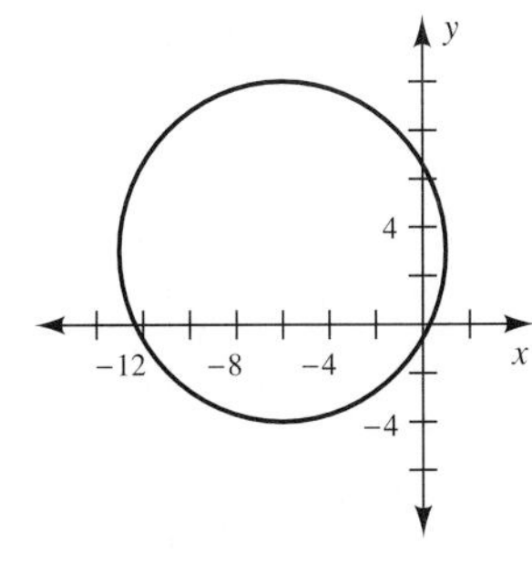

c.

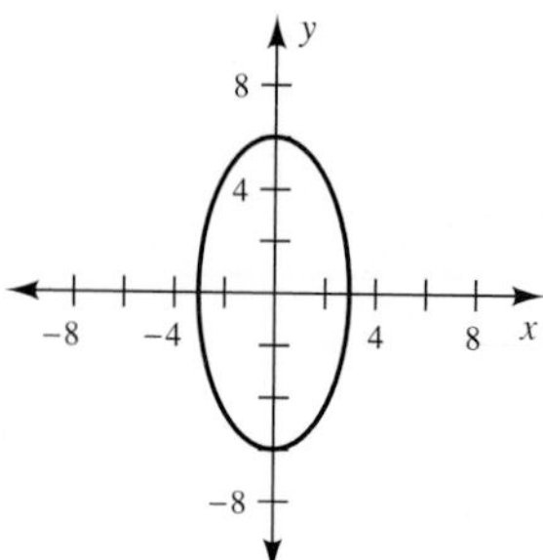

d.

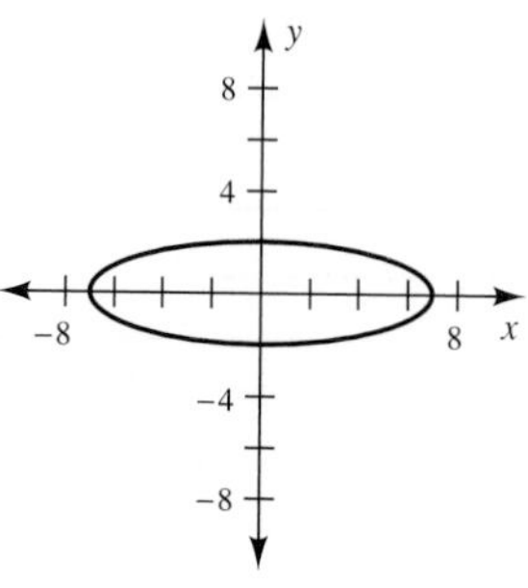

6 a.

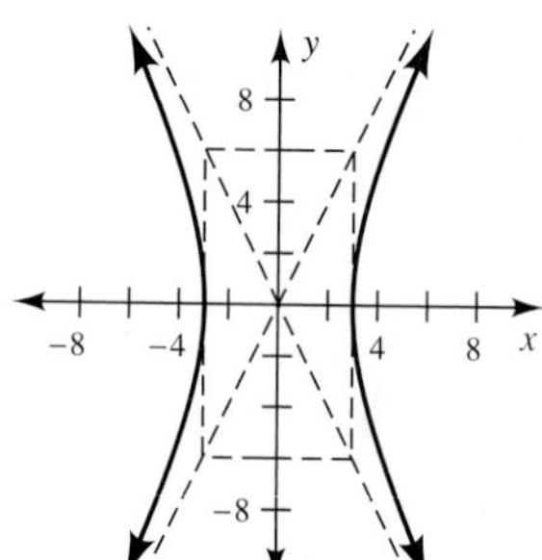

b.

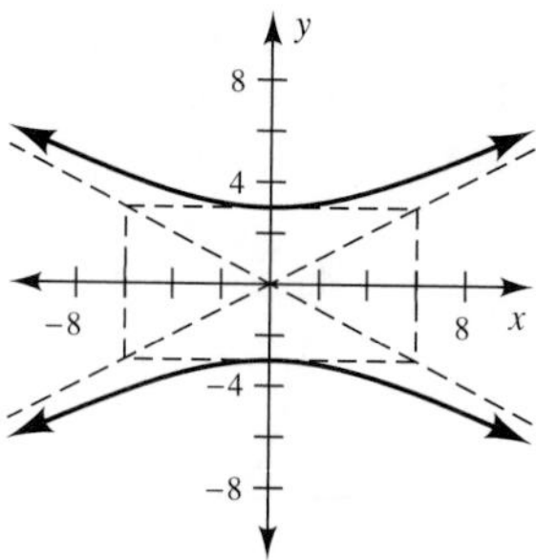

c.

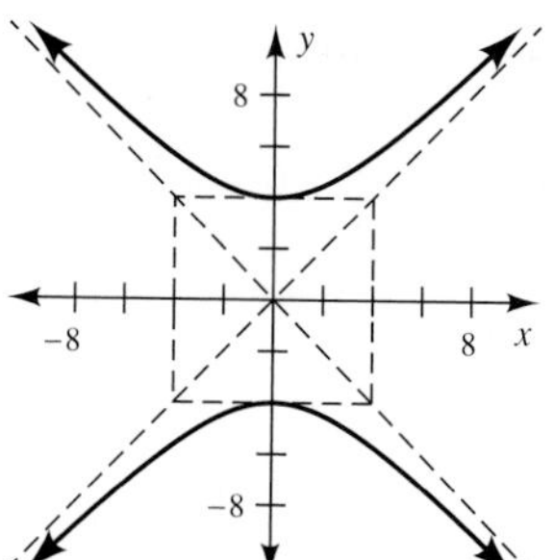

7 a. Parabola **b.** Ellipse **c.** Circle **d.** Hyperbola

Cumulative Review Exercises 7–9

1 $\frac{4}{25}$ **2** $\frac{1}{216}$ **3** 900 **4** $7 + 2\sqrt{10}$
5 $7\sqrt{7} - 7\sqrt{5}$ **6** $-9\sqrt{2}$ **7** $\frac{15x}{y}$ **8** $2xy$
9 $8x^2y$ **10** $x^{1/4}y^{1/4}z^{1/4}$ **11** $xy\sqrt[3]{y^2}$ **12** $|x|y^2\sqrt{xy}$
13 $23 - 31i$ **14** $26 - 22i$ **15** $\frac{3x + 5y + 2\sqrt{15xy}}{3x - 5y}$
16 $-\frac{8}{17} + \frac{15}{17}i$ **17** $-4 + 6i$ **18** $8 - 14i$
19 $\frac{3xy}{4(x + y)}$ **20** $\frac{1}{z + 3}$ **21** $-4, \frac{3}{2}$ **22** $-\frac{5}{2}, \frac{2}{3}$
23 $-\frac{5}{2}, 0$ **24** $-2, 0, 2$ **25** 17.5 **26** -1 **27** 6
28 4 **29** $-\frac{11}{4}$ **30** $-5, -2, 2, 5$
31 $x = -5y, x = 2y$ **32** $5x^2 + 11x - 12 = 0$
33 17 ft **34** 70 km/h **35** $(-\infty, 2] \cup [4, +\infty)$
36 $(-\infty, -4) \cup (2, 5)$ **37** 13 **38** $(2, -1)$ **39** 2
40 $5x - y - 8 = 0$ **41** $2x + 3y - 9 = 0$
42 -8; imaginary solutions **43** \$43.75
44 600 g of source A, 800 g of source B **45** 550 mi/h
46 (3, 4) **47** $(-4, -5)$
48 $(-10, 0), (10, 0), (0, -7), (0, 7)$
49 Horizontally; $(-10, 0), (10, 0)$
50 $(x - 2)^2 + (y + 2)^2 = 169$ **51** 2
52 $(-\infty, -3] \cup [4, +\infty)$ **53** 3 **54** 1, 5 **55** g
56 f **57** b **58** e **59** c **60** h **61** a
62 d

Chapter 10

Exercises 10-1

1 {(4, 1), (11, 3), (2, 8)}
3 {(2, −3), (2, −1), (2, 0), (2, 2)} **5** $\{(b, a), (d, c)\}$
7 {(−5, 5), (0, 5), (5, 5)}
9 {(−1, −11), (−1, −3), (1, 3), (1, 11)}
11 {(1, −3), (2, −2), (0, −1), (−1, 1), (−2, 3)}
13 One-to-one **15** Not one-to-one
17 Not one-to-one **19** One-to-one
21 Not one-to-one **23** $f^{-1}(x) = \frac{x - 2}{5}$
25 $g^{-1}(x) = x + 3$ **27** $h^{-1}(x) = 3(x + 1)$
29 $f^{-1}(x) = -x$
31

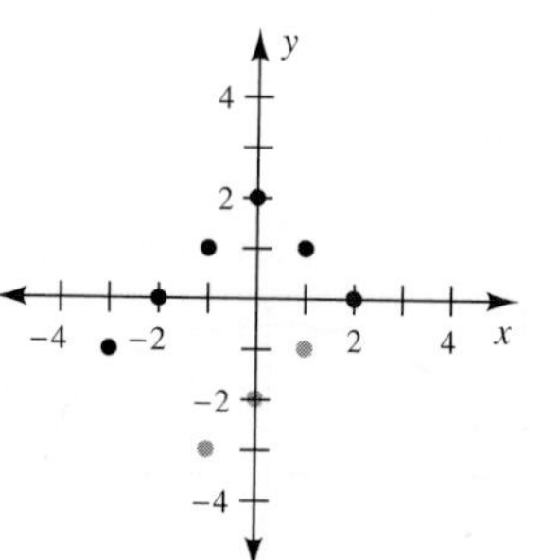

33

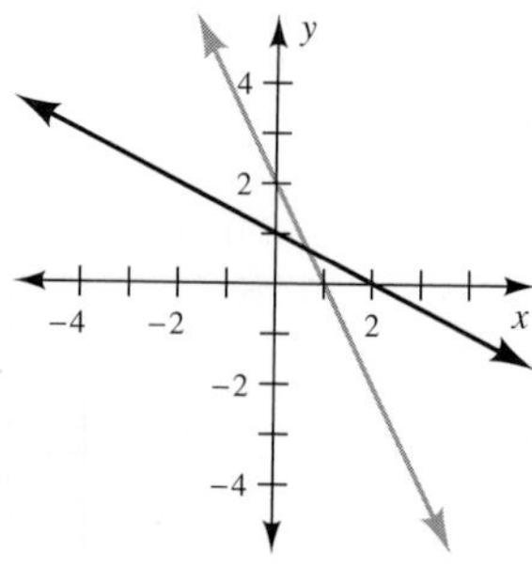

35

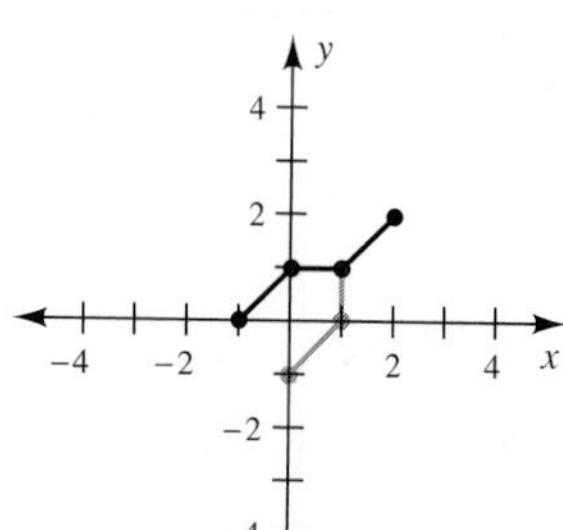

37

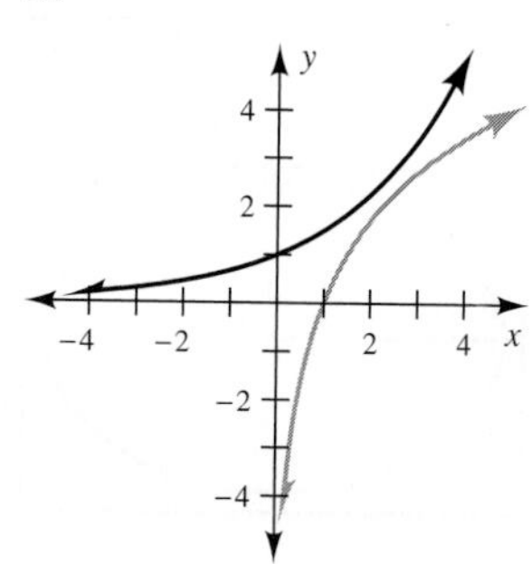

39

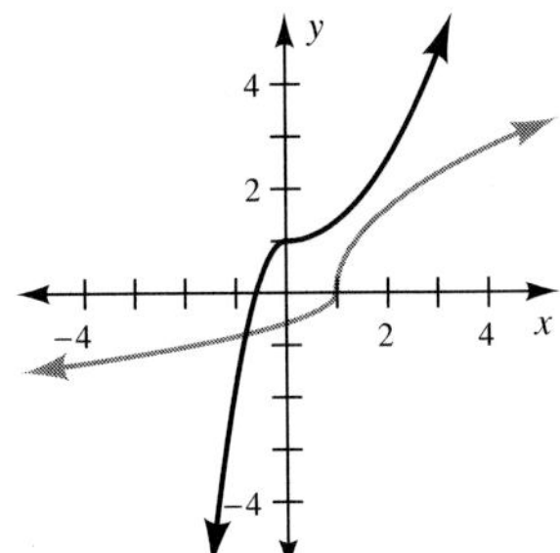

41 28 **43** 5 **45** $5h + 3$ **47** h

49

[−6, 6] for x, [−6, 6] for y

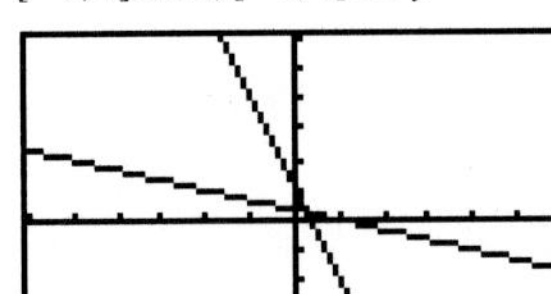

51

[−6, 6] for x, [−6, 6] for y

53

[−6, 6] for x, [−6, 6] for y

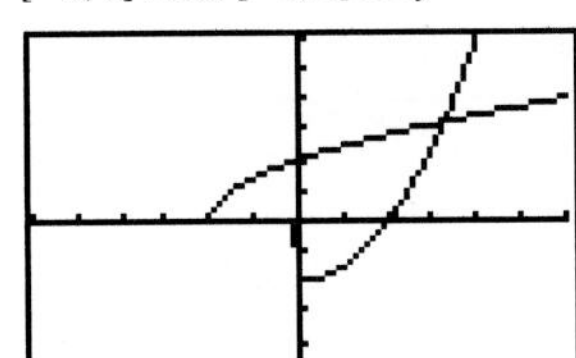

55

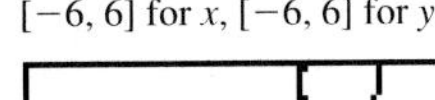

[−6, 6] for x, [−6, 6] for y

57 **a.** 13 **b.** 1 **c.** $\frac{1}{13}$
59 **a.** 88 **b.** 4 **c.** $\frac{1}{88}$
61 **a.** $5a - 2$ **b.** $\dfrac{a + 2}{5}$ **c.** $\dfrac{1}{5a - 2}$

63 $f^{-1}(x) = \dfrac{x + 1}{x}$; domain of f: $\mathbb{R} \sim \{1\}$, range of f: $\mathbb{R} \sim \{0\}$; domain of f^{-1}: $\mathbb{R} \sim \{0\}$, range of f^{-1}: $\mathbb{R} \sim \{1\}$

65 $g^{-1}(x) = \dfrac{4x + 3}{x - 1}$; domain of g: $\mathbb{R} \sim \{4\}$, range of g: $\mathbb{R} \sim \{1\}$; domain of g^{-1}: $\mathbb{R} \sim \{1\}$, range of g^{-1}: $\mathbb{R} \sim \{4\}$

67 $h^{-1}(x) = -\dfrac{x + 2}{2x - 1}$; domain of h: $\mathbb{R} \sim \{-\frac{1}{2}\}$, range of h: $\mathbb{R} \sim \{\frac{1}{2}\}$; domain of h^{-1}: $\mathbb{R} \sim \{\frac{1}{2}\}$, range of h^{-1}: $\mathbb{R} \sim \{-\frac{1}{2}\}$

69 **a.**

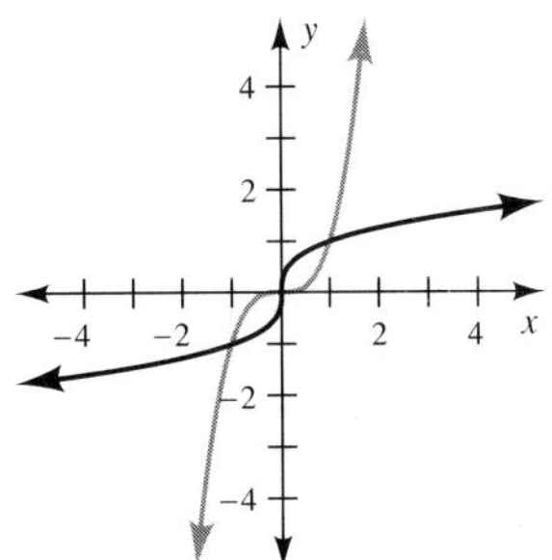

b. $f^{-1}(x) = \sqrt[3]{x}$
c. The domain and range of f and f^{-1} are $\mathbb{R}$.

Section Skill Reinforcers 10-1

1 0 **2** 1 **3** $\frac{1}{10}$ **4** −1 **5** 1

Exercises 10-2

1 **a.** 2 **b.** 5 **c.** 2 **d.** 3
3 **a.** −1 **b.** 0 **c.** $\frac{1}{2}$ **d.** $\frac{1}{6}$
5 **a.** $\frac{26}{15}$ **b.** −2 **c.** −3 **d.** $\frac{3}{2}$
7 **a.** $\frac{1}{2}$ **b.** $\frac{1}{3}$ **c.** 4 **d.** −2
9 **a.** The log of 36 base 6 is 2; $6^2 = 36$.
b. $\log_5 625 = 4$; $5^4 = 625$.
c. $\log_7 1 = 0$; the log of 1 base 7 is 0.
11 **a.** The log of $\sqrt{15}$ base 15 is $\frac{1}{2}$; $15^{1/2} = \sqrt{15}$.
b. $\log_{13} 17 = x$; $13^x = 17$.
c. $\log_{19} 23 = x$; the log of 23 base 19 is x.
13 $10^{-2} = \frac{1}{100}$ **15** $b^k = m$ **17** $\log_{16}(\frac{1}{4}) = -\frac{1}{2}$ **19** 2
21 −1 **23** 5 **25** −1 **27** $\frac{1}{2}$ **29** $-\frac{1}{3}$ **31** 0
33 Undefined
35

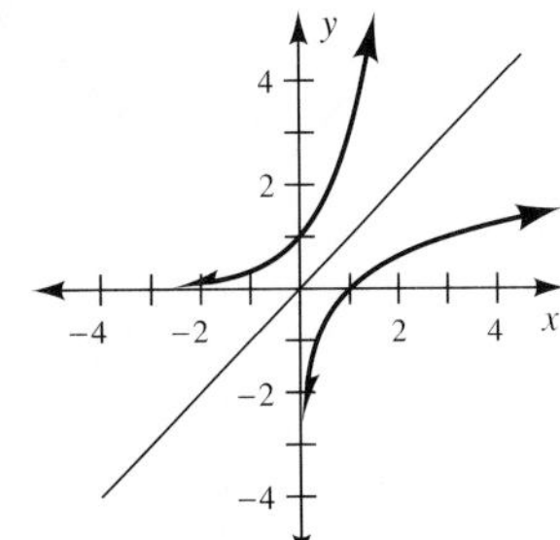

37 −5 **39** $\frac{1}{6}$ **41** −2 **43** −8, 8 **45** −2, 2
47 $-\frac{1}{2}$ **49** 15 **51** $\frac{1}{5}$ **53** 25 **55** $\frac{1}{25}$ **57** $\sqrt{5}$
59 49 **61** $\sqrt{5}$ **63** 5 **65** $\frac{1}{3}$ **67** 6 **69** $\frac{1}{3}$
71 31.544 **73** 2456.7

75

[−2, 6] for x, [−2, 6] for y

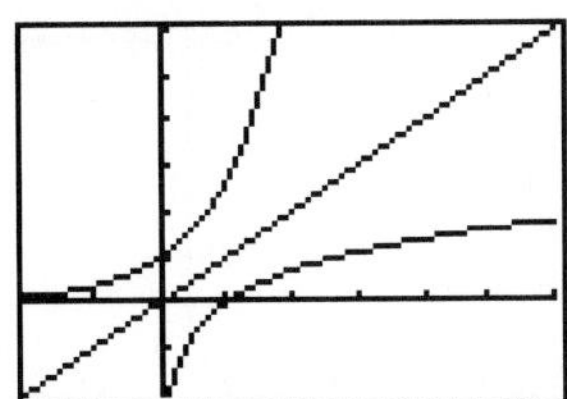

77

[−1, 10] for x, [−1, 10] for y

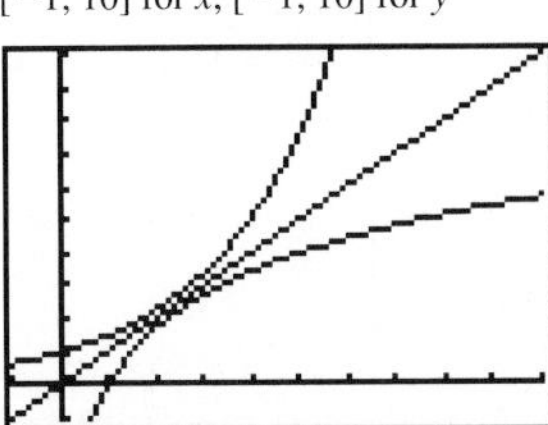

79 \$1734.76

SECTION SKILL REINFORCERS 10-2

1 x^{11} **2** x^5 **3** x^{24} **4** $\dfrac{x^{25}}{y^{20}}$ **5** x^{11}

EXERCISES 10-3

1 **a.** 2 **b.** −4 **c.** 9 **d.** Undefined
3 **a.** 5 **b.** −5 **c.** Undefined **d.** $\frac{1}{2}$
5 **a.** 1.6721 **b.** 47.000 **c.** 3.8501 **d.** 46.998
7 **a.** 2.0531 **b.** 113.01 **c.** 4.7274 **d.** 113.00
9 **a.** −5.0816 **b.** −2.2069 **c.** 2.3026 **d.** 0.43424
11 **a.** −6.4123 **b.** −14.765 **c.** 10.653 **d.** 24.530
13 **a.** Undefined **b.** −0.90309 **c.** 0.40908 **d.** 0.10791
15 **a.** 1.0116 **b.** 1.7297 **17** **a.** 3.3242 **b.** 3.7842
19 1549 **21** 0.8318 **23** 106.7 **25** 0.1686
27 2.239 **29** 28.50 **31** −0.09391 **33** 0.3987
35 2.361 **37** 1.110 **39** −1.941 **41** 4.034
43 2 **45** 9 **47** c **49** b **51** d **53** d
55 1.099 **57** 20.09 **59** 1000. **61** 0.4771
63 1.609 **65** 0.0003548 **67** \$674.93 **69** 60

SECTION SKILL REINFORCERS 10-3

1 **2**

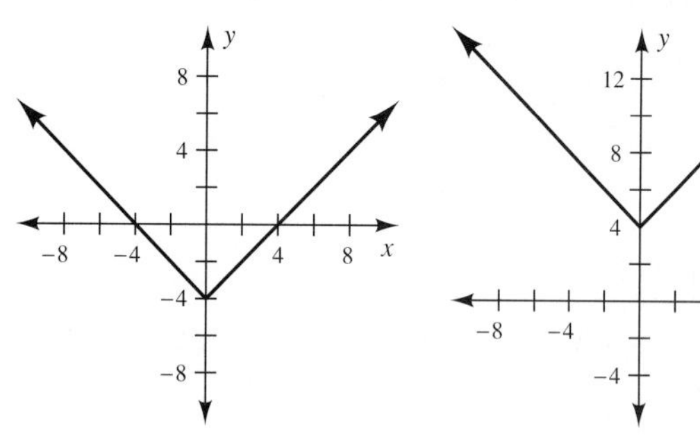

3 **4**

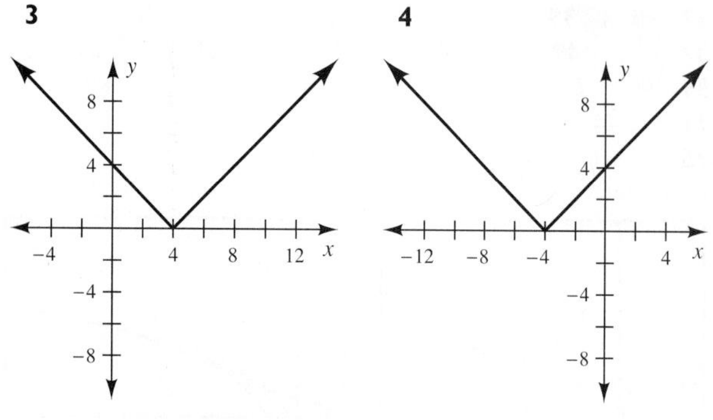

5

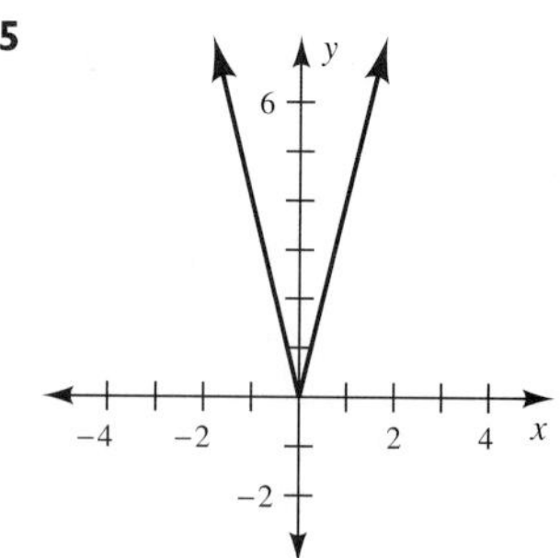

EXERCISES 10-4

1 $\log x + 5 \log y$ **3** $2 \ln x + 3 \ln y + 4 \ln z$
5 $\ln(2x + 3) - \ln(x + 7)$ **7** $\frac{1}{2} \log(4x + 7)$
9 $\frac{1}{2} \ln(x + 4) - 2 \ln(y + 5)$
11 $\frac{1}{2} \log x + \frac{1}{2} \log y - \frac{1}{2} \log(z - 8)$
13 $2 \log x + 3 \log(2y + 3) - 4 \log z$
15 $\ln x + 2 \ln y - 3 \ln z$ **17** $\log x^2y^5$ **19** $\ln \dfrac{x^3y^7}{z}$
21 $\log \dfrac{\sqrt{x+1}}{2x+3}$ **23** $\ln \sqrt[3]{(2x + 7)(7x + 1)}$
25 $\log_5 x^2 \sqrt[3]{y^2}$ **27** 2.2453 **29** 0.76308
31 −0.081112 **33** −5.8919 **35** 34 **37** 0.53
39 1.1833 **41** 1.6542 **43** 0.11873 **45** 1.3562
47 −0.4709 **49** 0.6438 **51** $y = e^{x \ln 2} \approx e^{0.6931x}$
53 $y = e^{-x \ln 2} \approx e^{-0.6931x}$ **55** a and d **57** Positive
59 Negative **61** c **63** a
65 $f(x) = \dfrac{\ln x}{\ln 1.5}$ [−1, 9] for x, [−5, 5] for y

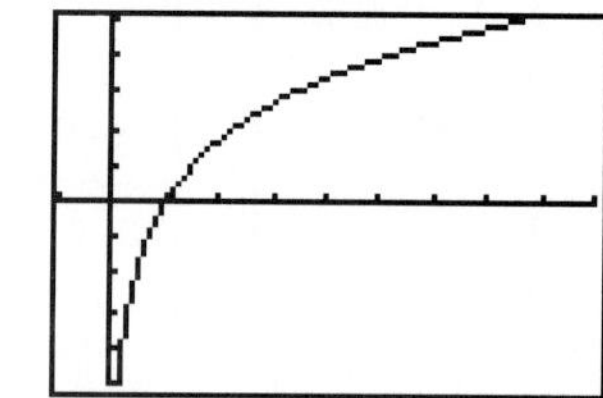

67 $x = e^{-t}$
69 Let $x = b^m$. Then $\log_b x = m$ and $x^p = (b^m)^p \Rightarrow$ $x^p = b^{mp} \Rightarrow \log_b x^p = \log_b b^{mp} \Rightarrow \log_b x^p = mp \Rightarrow$ $\log_b x^p = p \log_b x$.

SECTION SKILL REINFORCERS 10-4

1 **2**

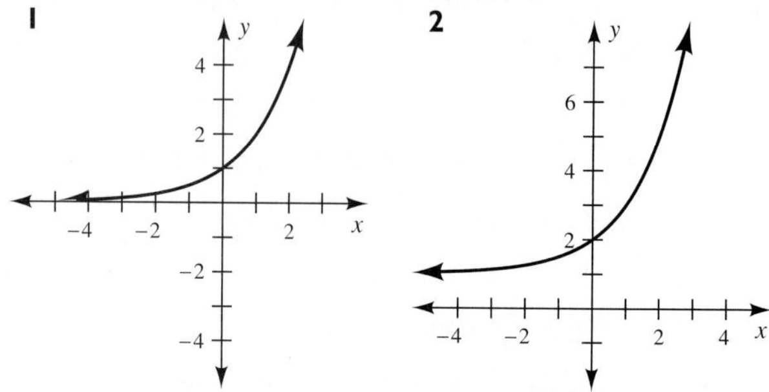

3

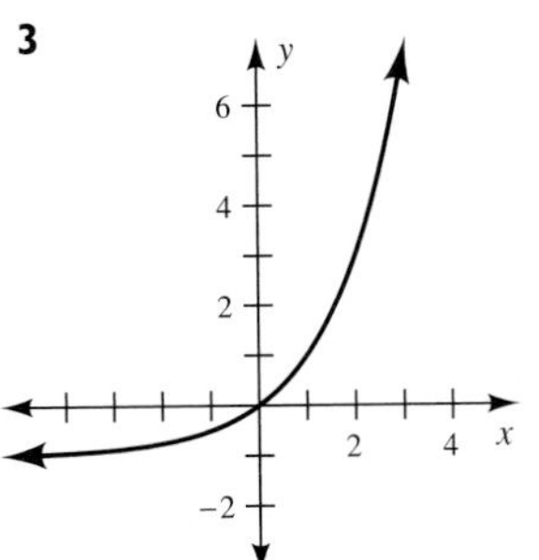

4

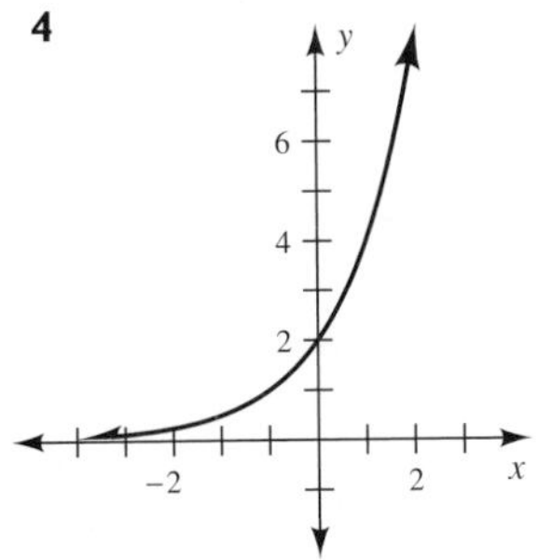

5

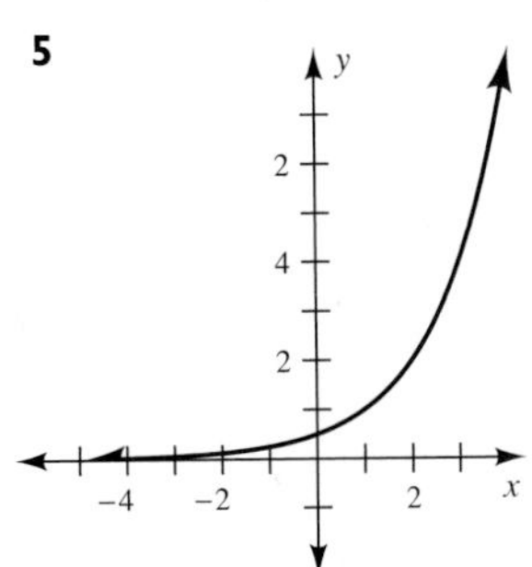

EXERCISES 10-5

1 8 **3** $-2, 2$ **5** 15 **7** 35 **9** 16 **11** $-\frac{1}{2}$
13 -2 **15** 2 **17** 10 **19** No solution **21** -7
23 No solution **25** 1.9534 **27** 4.1864
29 -1.1021 **31** 0.22562 **33** 1.4561 **35** 0.35506
37 $-1.4387, 1.4387$ **39** $-2.0562, 2.0562$
41 $\dfrac{10^{0.83452} + 17}{5} \approx 4.7663$ **43** $e^e \approx 15.154$
45 $3 + \sqrt{2} \approx 4.4142$ **47** No solution
49 $1 = 1.0000$, $e^2 \approx 7.3891$ **51** 0.0, 2.5 **53** 3.0
55 $10^{-\log x} = 10^{\log 1/x} = \dfrac{1}{x}$
57 $e^{-x \ln 3} = e^{x \ln 1/3} = (e^{\ln 1/3})^x = (\frac{1}{3})^x$
59 $\log 60^x - \log 6^x = \log \dfrac{60^x}{6^x} = \log 10^x = x$
61 $\ln(\frac{4}{5})^x + \ln(\frac{5}{3})^x + \ln(\frac{3}{4})^x = \ln[(\frac{4}{5})(\frac{5}{3})(\frac{3}{4})]^x = \ln 1^x =$
$x \ln 1 = x(0) = 0$
63 6.8 yr

SECTION SKILL REINFORCERS 10-5

1

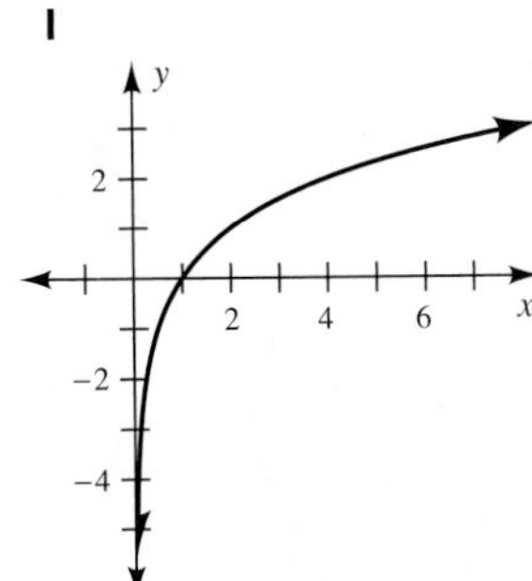

2

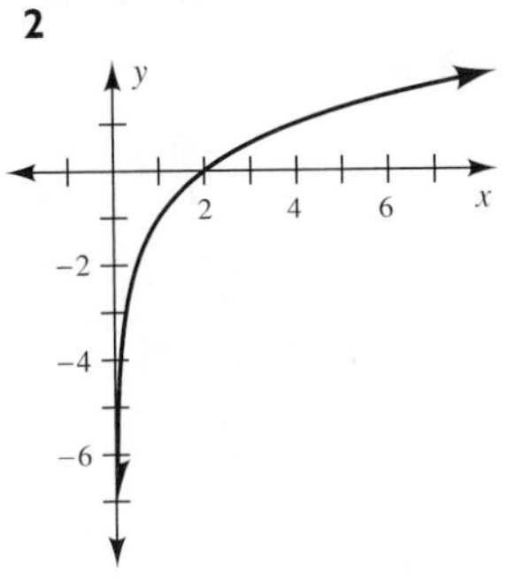

3

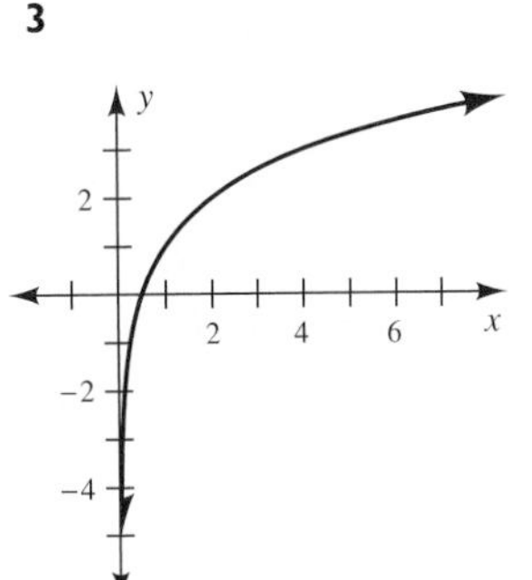

4

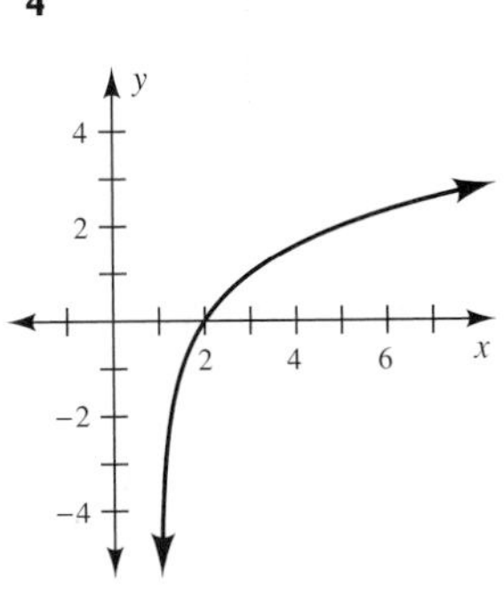

5

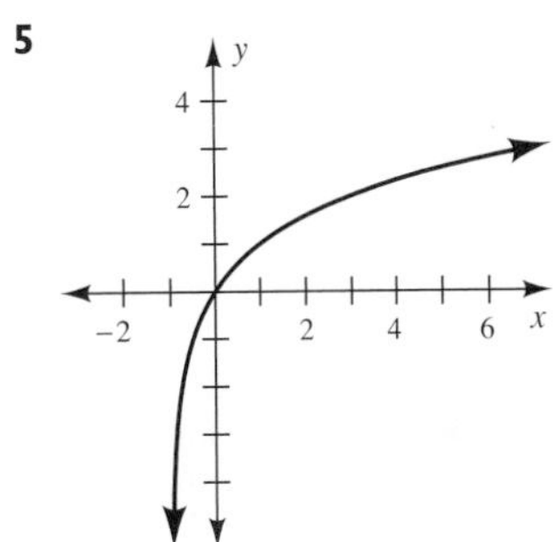

EXERCISES 10-6

1 \$234.08 **3** 9.0 yr **5** 18.4 yr **7** 8.7%
9 6.9% **11** 9.9 yr **13** 24,000 yr
15 9200 days $\approx$ 25 yr **17** 10,000 whales; 27 yr
19 13.5% **21** 7.8 on the Richter scale
23 8.9 on the Richter scale **25** 25 decibels
27 3.2×10^{-8} watt/cm^2 **29** pH of 4.0
31 7.4×10^{-10} mole/L **33** \$411.60
35 36 monthly payments

SECTION SKILL REINFORCERS 10-6

1 Complex conjugates
2 Real solution of multiplicity two
3 Distinct rational solutions
4 Distinct irrational solutions **5** Complex conjugates

REVIEW EXERCISES 10

1 $f^{-1} = \{(1, -1), (2, -\frac{1}{2}), (-2, \frac{1}{2}), (-1, 1)\}$;
$D = \{-2, -1, 1, 2\}$, $R = \{-1, -\frac{1}{2}, \frac{1}{2}, 1\}$
2 $f^{-1}(x) = \dfrac{x - 7}{2}$; $D = \mathbb{R}, R = \mathbb{R}$
3 $f^{-1}(x) = -\dfrac{2x + 1}{x - 2}$; $D = \mathbb{R} \sim \{2\}, R = \mathbb{R} \sim \{-2\}$

4

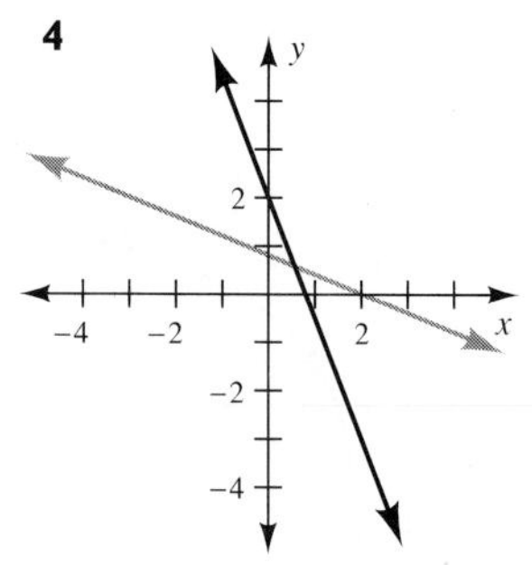

5

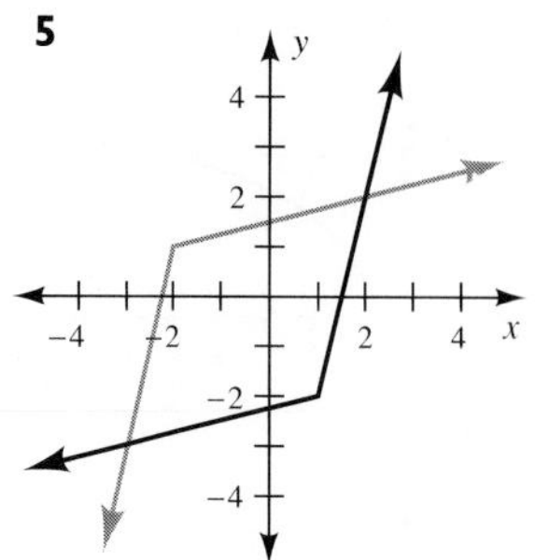

6

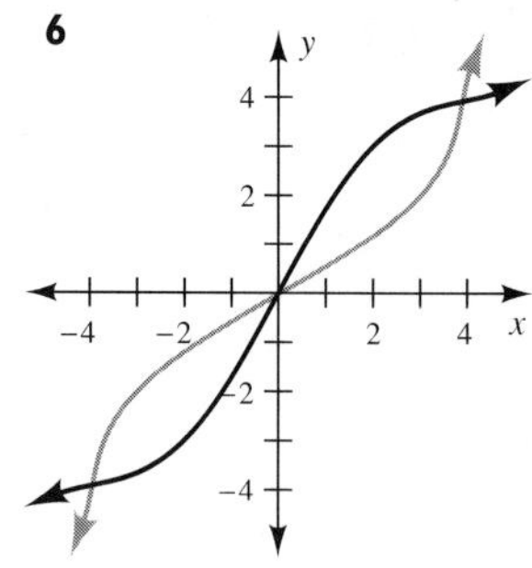

7 Not a one-to-one function **8** One-to-one function
9 Not a one-to-one function **10** One-to-one function
11 $6^{1/2} = \sqrt{6}$ **12** $17^0 = 1$ **13** $8^{-2} = \frac{1}{64}$
14 $b^c = a$ **15** $e^c = a$ **16** $10^d = c$
17 $\log_7 343 = 3$ **18** $\log_{19} \sqrt[3]{19} = \frac{1}{3}$
19 $\log_{4/7} \frac{49}{16} = -2$ **20** $\ln \dfrac{1}{e} = -1$
21 $\log 0.0001 = -4$ **22** $\log_8 y = x$ **23** -2 **24** $\frac{2}{3}$
25 $\frac{2}{3}$ **26** $-\frac{1}{2}$ **27** 1 **28** $\frac{1}{4}$ **29** $-2, 2$
30 $-\frac{1}{2}, 1$ **31** 2 **32** $\frac{3}{2}$ **33** 4 **34** $\sqrt{13}$ **35** $\frac{1}{9}$
36 No solution; $\log_3(-2)$ is undefined.
37 No solution; $\log_{-2} 3$ is undefined. **38** 13 **39** $\frac{1}{2}$
40 11 **41** 17 **42** 7
43 No solution; ln 0 is undefined. **44** 3 **45** 140
46 33 **47** $-10, 10$ **48** 1 **49** -2 **50** -0.5
51 3 **52** 13.5 **53** $-3, 0$ **54** 3 **55** 21.256
56 2.1893 **57** 0.082935 **58** 0.15463 **59** 2.0553
60 4.7325 **61** -3.0915 **62** 20.409 **63** 1.2091
64 -0.48321 **65** -0.056195 **66** 0.98067
67 0.15578 **68** 2.3026 **69** $3 \log x + 4 \log y$
70 $\ln(7x - 9) - \ln(2x + 3)$ **71** $\frac{1}{2} \ln(2x + 1) - \ln(5x + 9)$
72 $\log x + \frac{3}{2} \log y - \frac{1}{2} \log z$ **73** $\ln x^2y^3$ **74** $\ln \dfrac{x^5}{y^4}$
75 $\ln(x + 1)$ **76** $\ln \sqrt{\dfrac{x}{y}}$ **77** 9.6 yr **78** 6.95%
79 8.66% **80** 10,200 days ≈ 28 yr **81** 2.146
82 -0.2929 **83** $y = e^{x \ln 5} \approx e^{1.609x}$
84 $\log 50^x + \log 6^x - \log 3^x = \log\left[\dfrac{50(6)}{3}\right]^x = \log(100)^x = x \log 100 = 2x$
85 $1000^{\log x} = (10^3)^{\log x} = 10^{3 \log x} = 10^{\log x^3} = x^3$
86 **a.** 13 **b.** -0.5 **c.** $\frac{1}{13}$
87 **a.** -3 **b.** -4.5 **c.** $-\frac{1}{3}$
88 **a.** -1 **b.** -4 **c.** -1
89 **a.** 31 **b.** 4 **c.** $\frac{1}{31}$
90 **a.** $2a + 5$ **b.** $\dfrac{a - 5}{2}$ **c.** $\dfrac{1}{2a + 5}$
91 $[-4, 14]$ for x, $[-4, 14]$ for y

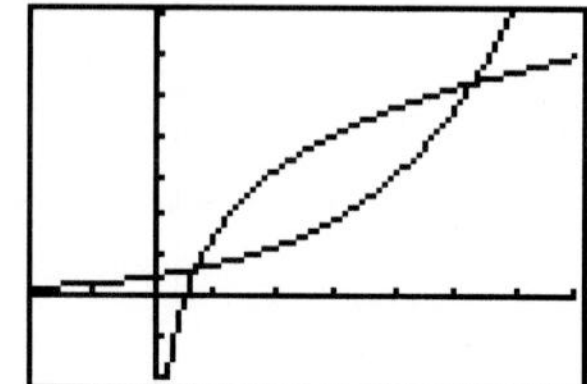

92 3.5

MASTERY TEST 10

1 **a.** $\{(4, -1), (9, 8), (11, -7)\}$ **b.** $f^{-1}(x) = \dfrac{7x + 1}{14}$
c. $\{(-1, -2), (0, -1), (1, 0), (2, 1), (-2, 3)\}$
d. $\{(2, \frac{1}{2}), (3, \frac{1}{3}), (\frac{1}{6}, 6), (1, 1)\}$

2 **a.** **b.**

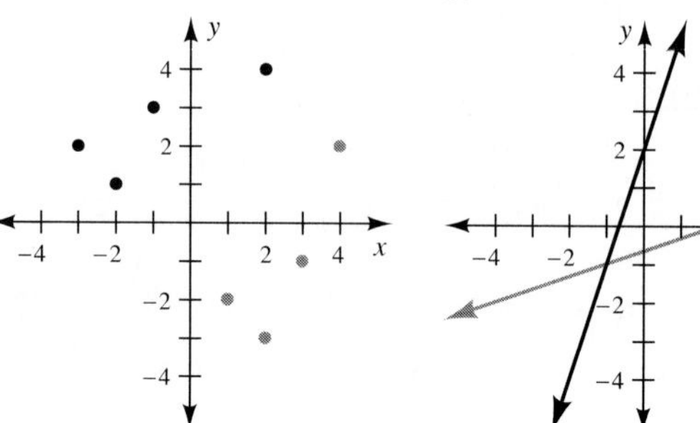

c. **d.**

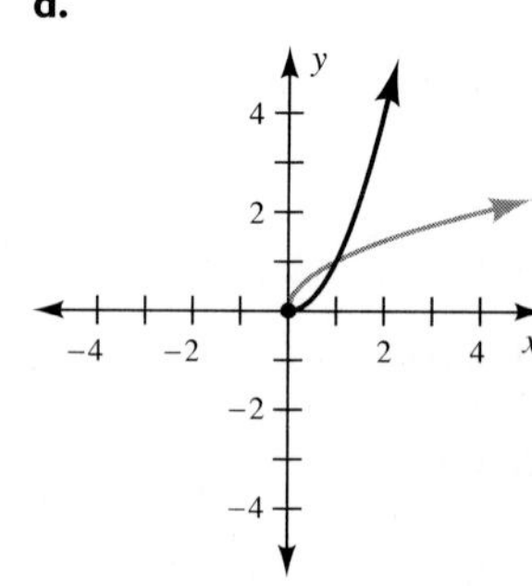

3 **a.**

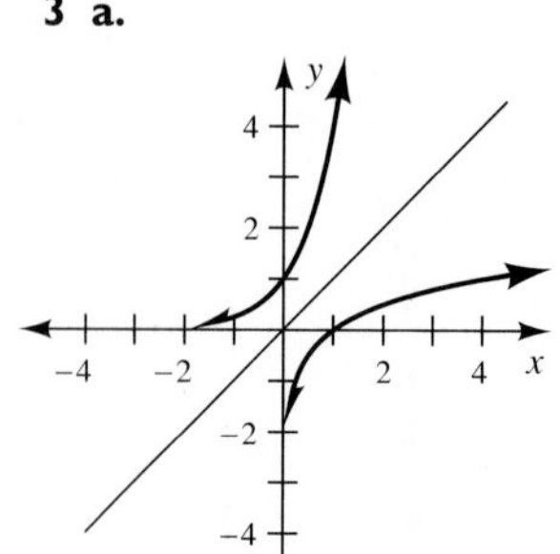

b.

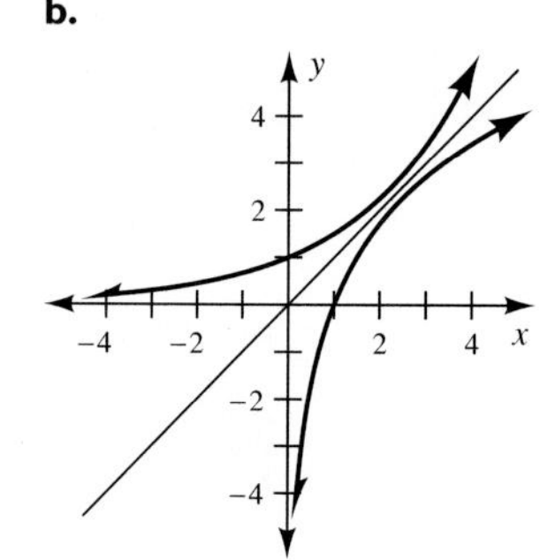

4 a. $5^{2/3} = \sqrt[3]{25}$ **b.** $5^{-3} = \frac{1}{125}$ **c.** $b^x = y + 1$
d. $b^{y+1} = x$

5 a. 1.2810 **b.** −5.4491 **c.** 31.500 **d.** 2.7183
e. −4.3160 **f.** 11.197 **g.** 8.3503
h. Undefined

6 a. $4 \log x + 5 \log y$ **b.** $3 \ln x - 6 \ln y$ **c.** $\frac{1}{3} \log x$
d. $\frac{1}{5} \ln x + \frac{2}{5} \ln y$ **e.** $\log(x - 2)(x - 3)$
f. $\ln \dfrac{5x - 1}{x - 3}$ **g.** $\ln \dfrac{(7x + 9)^2}{x}$ **h.** $\ln x \sqrt[3]{(x^2 + 7)^2}$

7 a. 1.730 **b.** 0.8499 **c.** 6.644 **d.** 2.406

8 a. −4 **b.** $\frac{3}{2}$ **c.** 16 **d.** $\frac{1}{3}$ **e.** −2
f. −3 **g.** 0.4056 **h.** 1.549

9 a. 36 yr **b.** 8.4 yr

Chapter 11

EXERCISES 11-1

1 −31 **3** 30 **5** 17 **7** (6, 0) **9** (3, 4)
11 (6, −8) **13** An inconsistent system; no solution
15 (1, −2) **17** (−3, −1) **19** $(\frac{1}{7}, -\frac{1}{3})$ **21** 39
23 36 **25** 18 **27** 24 **29** −78 **31** (2, 1, 2)
33 (2, 1, 0) **35** (−2, 4, 1) **37** (5, 5, −7)
39 (20, 30, 40) **41** (4.5, 6.7) **43** (−4.2, 6.1)
45 (1.4, −2.7, 8.8) **47** −2 **49** $\frac{2}{5}$, 2

SECTION SKILL REINFORCERS 11-1

1 Undefined **2** 0 **3** −1 **4** 1 **5** 7

EXERCISES 11-2

1 12 terms **3** 11

5 $a^8 + (\ \ )a^7b + (\ \ )a^6b^2 + (\ \ )a^5b^3 + (\ \ )a^4b^4 + (\ \ )a^3b^5 + (\ \ )a^2b^6 + (\ \ )ab^7 + b^8$

7 $a^{10} + (\ \ )a^9b + (\ \ )a^8b^2 + (\ \ )a^7b^3 + (\ \ )a^6b^4 + (\ \ )a^5b^5 + (\ \ )a^4b^6 + (\ \ )a^3b^7 + (\ \ )a^2b^8 + (\ \ )ab^9 + b^{10}$

9 $a^8 + 8a^7b + 28a^6b^2 + 56a^5b^3 + 70a^4b^4 + 56a^3b^5 + 28a^2b^6 + 8ab^7 + b^8$

11 $4368a^{11}b^5$ **13** $330a^7b^4$

15

```
              1
            1   1
          1   2   1
        1   3   3   1
      1   4   6   4   1
    1   5  10  10   5   1
  1   6  15  20  15   6   1
1   7  21  35  35  21   7   1
```

17 1 9 36 84 126 126 84 36 9 1

19 $s^4 + 4s^3t + 6s^2t^2 + 4st^3 + t^4$

21 $x^5 + 10x^4 + 40x^3 + 80x^2 + 80x + 32$

23 $y^3 - 9y^2 + 27y - 27$

25 $64m^6 - 1152m^5 + 8640m^4 - 34{,}560m^3 + 77{,}760m^2 - 93{,}312m + 46{,}656$

27 $32x^5 - 240x^4y + 720x^3y^2 - 1080x^2y^3 + 810xy^4 - 243y^5$

29 $x^6 + 6x^5y^2 + 15x^4y^4 + 20x^3y^6 + 15x^2y^8 + 6xy^{10} + y^{12}$

31 $16m^8 + 160m^6n^3 + 600m^4n^6 + 1000m^2n^9 + 625n^{12}$

33 $\dfrac{x^5}{32} - \dfrac{5x^4y}{48} + \dfrac{5x^3y^2}{36} - \dfrac{5x^2y^3}{54} + \dfrac{5xy^4}{162} - \dfrac{y^5}{243}$

35 $x^{14} + 14x^{13}y + 91x^{12}y^2 + 364x^{11}y^3 + \cdots + y^{14}$

37 $x^{10} + 30x^9 + 405x^8 + 3240x^7 + \cdots + 59{,}049$

39 $m^{13} - 39m^{12}n + 702m^{11}n^2 - 7722m^{10}n^3 + \cdots - 1{,}594{,}323n^{13}$

41 $512x^{18} + 2304x^{16}y + 4608x^{14}y^2 + 5376x^{12}y^3 + \cdots + y^9$

43 $a^{10} + 10a^9b + 45a^8b^2 + \cdots + 45a^2b^8 + 10ab^9 + b^{10}$

45 $v^{13} + 13v^{12}w + 78v^{11}w^2 + \cdots + 78v^2w^{11} + 13vw^{12} + w^{13}$

47 $x^{11} - 11x^{10}y + 55x^9y^2 + \cdots - 55x^2y^9 + 11xy^{10} - y^{11}$

49 $128x^7 + 448x^6y + 672x^5y^2 + \cdots + 84x^2y^5 + 14xy^6 + y^7$

51 $1 + 6(0.1) + 15(0.01) + 20(0.001) + 15(0.0001) + 6(0.00001) + (0.000001) = 1.771561$

53 $-2 + 2i$ **55** $-4 + 4i$

SECTION SKILL REINFORCERS 11-2

1 3 **2** −3 **3** 0 **4** Undefined **5** 8

EXERCISES 11-3

1 5, 10, 15, 20, 25 **3** −5, −2, 1, 4, 7 **5** 1, $\frac{1}{2}$, $\frac{1}{3}$, $\frac{1}{4}$, $\frac{1}{5}$
7 2, 4, 8, 16, 32 **9** 2, 2, 4, 8, 14
11 $-\frac{1}{2}$, $\frac{2}{3}$, $-\frac{3}{4}$, $\frac{4}{5}$, $-\frac{5}{6}$ **13** 1, 2, 3, 5, 8 **15** 150
17 $\frac{14}{29}$ **19** 572 **21** $\frac{1}{9}$ **23** 45 **25** 275 **27** 60
29 $\frac{25}{12}$ **31** 45 **33** 44 **35** 166 **37** $\frac{31}{32}$
39 500 **41** 760 **43** 46.5

45 **47**

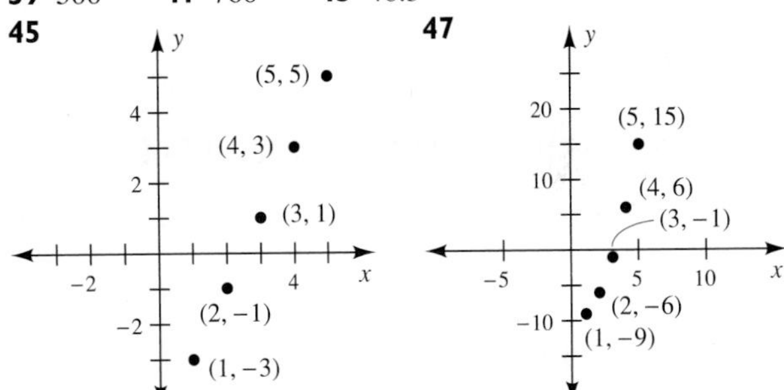

49

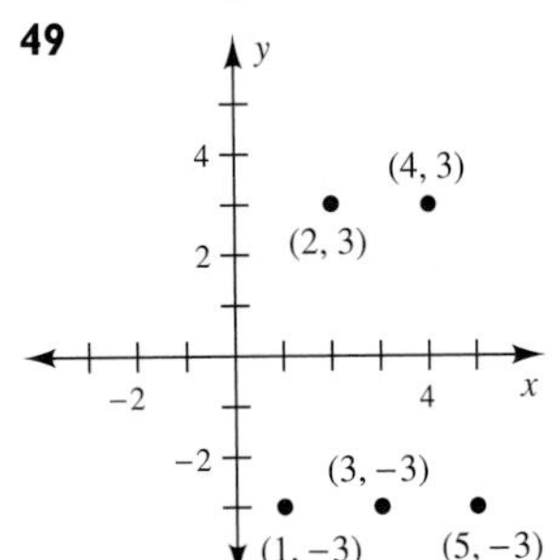

51 $\displaystyle\sum_{i=1}^{5} (2i - 1) = 1 + 3 + 5 + 7 + 9 = \sum_{k=0}^{4} (2k + 1)$

53 $\displaystyle\sum_{k=3}^{7} 2k = 6 + 8 + 10 + 12 + 14 = 50 \neq \sum_{i=2}^{6} (2i + 1) = 5 + 7 + 9 + 11 + 13 = 45$

55 $-4, -6, -6, -4, 0, 6, 14, 24$
57 $1, 0, -1, 0, 7, 28, 79, 192$ **59** 28

Section Skill Reinforcers 11-3

1 $(x+8)(x-8)$ **2** $(x-4)(x^2+4x+16)$
3 $(x+4)(x^2-4x+16)$ **4** Prime
5 $(x^2+4)(x+2)(x-2)$

Exercises 11-4

1 Arithmetic; $d = 4$ **3** Not arithmetic
5 Arithmetic; $d = 0.2$ **7** Arithmetic: $d = \frac{2}{5}$
9 Not arithmetic **11** Arithmetic; $d = \frac{1}{3}$
13 $7, 5, 3, 1, -1, -3$ **15** $3\frac{1}{3}, 4, 4\frac{2}{3}, 5\frac{1}{3}, 6, 6\frac{2}{3}$
17 6, 10, 14, 18, 22, 26 **19** $-8, -4, 0, 4, 8, 12$
21 3, 7, 11, 15, 19, 23 **23** 1.2, 1.6, 2.0, 2.4, 2.8, 3.2
25 416 **27** -43 **29** 41 **31** 1640 **33** 5
35 9240 **37** 3965 **39** 74.4 **41** 21 **43** -300
45 0.5 **47** 46 **49** 33,649 **51** 24 **53** 11
55 $\frac{4}{13}$
57 53, 57, 61, 65, 69, 73, 77, 81, 85
59 1032, 1043, 1054, 1065, 1076, 1087, 1098, 1109
61 3936
63 272 logs **65** \$81,000 **67** 121
69 2080

Section Skill Reinforcers 11-4

1 -1 **2** $-i$ **3** 37 **4** $4.8 + 2.4i$ **5** $-5 + 12i$

Exercises 11-5

1 Geometric; $r = 0.5$ **3** Geometric; $r = -\frac{1}{5}$
5 Not geometric **7** Geometric; $r = -4$
9 Geometric; $r = 0.1$ **11** Geometric; $r = 3$
13 Not geometric **15** Geometric; $r = 0.5$
17 12, 60, 300, 1500, 7500 **19** $-18, 12, -8, \frac{16}{3}, -\frac{32}{9}$
21 12, 0.12, 0.0012, 0.000012, 0.00000012
23 $3, -6, 12, -24, 48$
25 2, 6, 18, 54, 162 or $2, -6, 18, -54, 162$
27 $\frac{3}{5}, \frac{9}{25}, \frac{27}{125}, \frac{81}{625}, \frac{243}{3125}$ **29** $36, -18, 9, -\frac{9}{2}, \frac{9}{4}$ **31** 8
33 0.00007 **35** $-\frac{1}{2}$ **37** -250 or 250 **39** $\frac{1}{6}$
41 7 **43** 1.5 **45** $-3, 3$ **47** 125
49 64, 256, 1024, 4096, 16,384, 65,536
51 50, 5, 0.5, 0.05, 0.005 **53** 382.63752 **55** 7
57 3.125 g **59** \$21,474,836.48

Section Skill Reinforcers 11-5

1

2

3

4

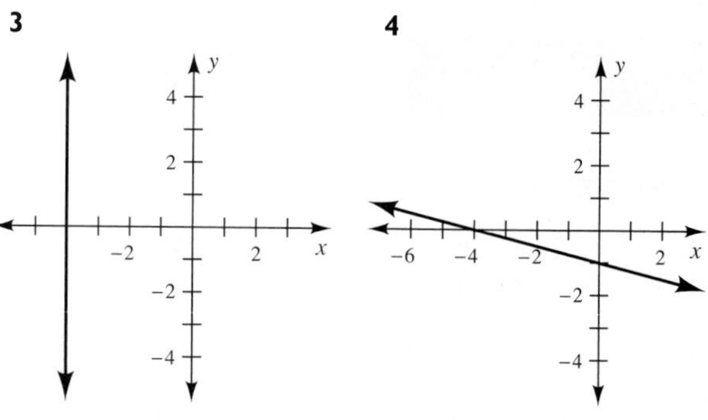

5

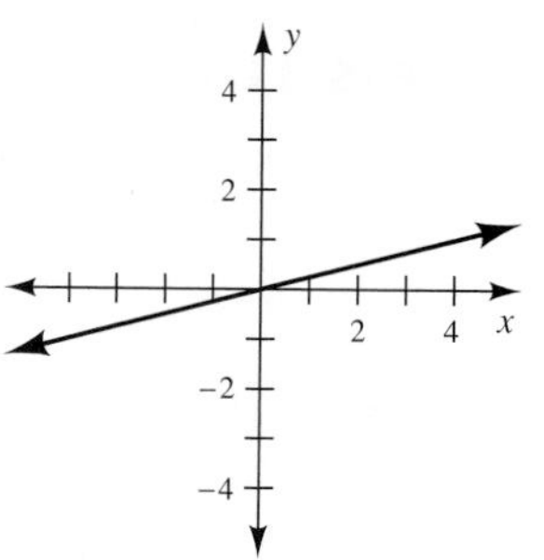

Exercises 11-6

1 381 **3** 0.22222 **5** 31.875 **7** 12.75603
9 127.875 **11** 0.8888888 **13** 49.9968 **15** 15
17 8 **19** 10 **21** $\frac{4}{33}$ **23** 0.8 **25** 3.6
27 38.4 **29** $\frac{4}{9}$ **31** $\frac{7}{33}$ **33** $\frac{15}{37}$ **35** $\frac{8}{1}$ **37** 6
39 6 **41** 0.2 **43** $\frac{5}{9}$ **45** $\frac{49}{3}$ **47** a **49** c
51 b **53** b **55** 405.234375 **57** 349,524
59 21,845 people **61** $13\frac{1}{3}$ square units **63** 49.99%
65 57 m **67** \$2,400,000

Section Skill Reinforcers 11-6

1 $(0, -3)$ **2** $(3, 0)$ **3** $(6, 2)$ **4** $(3, 1)$
5 $(0, 1)$

Review Exercises 11

1 -124 **2** -8 **3** $(3, 5)$ **4** $(-0.3, 0.7)$
5 $(1, 0, -4)$ **6** $a = -2, b = 3, c = 4$
7 $x^5 + 5x^4y + 10x^3y^2 + 10x^2y^3 + 5xy^4 + y^5$
8 $625x^4 - 1000x^3y + 600x^2y^2 - 160xy^3 + 16y^4$
9 $x^5 - 15x^4y + 90x^3y^2 - 270x^2y^3 + 405xy^4 - 243y^5$
10 5, 10, 15, 20, 25 **11** 5, 25, 125, 625, 3125
12 6, 7, 8, 9, 10 **13** 5, 5, 5, 5, 5 **14** 5, 8, 11, 14, 17
15 7, 21, 63, 189, 567 **16** $0, \frac{1}{3}, -\frac{1}{2}, \frac{3}{5}, -\frac{2}{3}$
17 4, 2, 2, 4, 8 **18** 1, 3, 4, 7, 11 **19** $7, 4, 1, -2, -5$
20 $-9, -5, -1, 3, 7$ **21** $10, -20, 40, -80, 160$
22 486, 162, 54, 18, 6 **23** 3, 1, 4, 1, 5
24 2, 3, 5, 7, 11 **25** Arithmetic; $d = 2$
26 Geometric; $r = 2$ **27** Neither arithmetic nor geometric
28 Neither arithmetic nor geometric
29 Arithmetic, $d = 0$; also geometric, $r = 1$
30 Arithmetic; $d = -4$ **31** Geometric; $r = -1$
32 Arithmetic, $d = 0$; also geometric, $r = 1$ **33** 10,302
34 254 **35** $\frac{1330}{729} = 1\frac{601}{729}$ **36** 23,944 **37** 2501
38 $\frac{12{,}610}{6561} = 1\frac{6049}{6561}$ **39** 244 **40** 220 **41** 4590.904509
42 1687.25 **43** 782 **44** 117,186 **45** $\frac{7}{9}$ **46** $\frac{4}{33}$
47 51 **48** 61 **49** 8 **50** 6 **51** 20 **52** 4
53 -6 **54** -693 **55** 17.3 **56** $\frac{78{,}125}{256} \approx 305.176$
57 $\frac{2}{3}$ **58** $-\frac{1}{2}$ **59** $-\frac{3}{2}, \frac{3}{2}$ **60** -1 **61** $\frac{25}{3}$
62 $\frac{2}{33}$ **63** $\frac{91}{15}$ **64** $\frac{9}{1}$ **65** 5 layers
66 \$10,000,000 **67** 11.915904
68 513, 530.5, 548, 565.5, 583, 600.5, 618, 635.5, 653
69 26.375

Mastery Test 11

1 a. $(2.5, 4.5)$ **b.** $(1.4, 2.4, 1.8)$
2 a. $x^4 - 4x^3y + 6x^2y^2 - 4xy^3 + y^4$
b. $16x^4 + 32x^3y + 24x^2y^2 + 8xy^3 + y^4$
c. $x^7 + 7x^6w + 21x^5w^2 + 35x^4w^3 + 35x^3w^4 + 21x^2w^5 + 7xw^6 + w^7$
d. $32x^5 - 400x^4y + 2000x^3y^2 - 5000x^2y^3 + 6250xy^4 - 3125y^5$
3 a. 9, 16, 23, 30, 37 **b.** 1, 3, 9, 27, 81
c. $-3, -5, -5, -3, 1$ **d.** $\frac{2}{5}, \frac{5}{9}, \frac{8}{13}, \frac{11}{17}, \frac{2}{3}$
4 a. 80 **b.** 56 **c.** 90 **d.** 20
5 a. 514 **b.** -44 **c.** -2 **d.** 32
6 a. 1971 **b.** 7900 **c.** 9810 **d.** 24,531
7 a. 64 **b.** $-\frac{1}{3}$ **c.** $-5, 5$ **d.** 7
8 a. $\frac{1143}{32} = 35\frac{23}{32}$ **b.** 0 **c.** $\frac{63}{64}$ **d.** $\frac{2343}{1024} = 2\frac{295}{1024}$
9 a. 6 **b.** $\frac{2}{5}$ **c.** $\frac{10}{33}$ **d.** $\frac{35}{33}$

Cumulative Review Exercises 10 and 11

1 3 **2** $\frac{1}{3}$ **3** -3 **4** $\frac{1}{12}$ **5** 2 **6** 3 **7** 64
8 1 **9** $\sqrt{6}$ **10** 4 **11** -2 **12** 9
13 2.137 **14** 4.920 **15** 1.461 **16** -11.15
17 $\log \frac{14}{x}$ **18** $\log x^2\sqrt[3]{y}$ **19** $\ln x + 3 \ln y - 0.5 \ln z$
20 1.3829 **21** 0.10307 **22** 22.239
23 $-1.0000, 4.0000$ **24** 6.0000
25 $f^{-1}(x) = \dfrac{2x + 10}{3}$
26 **27**

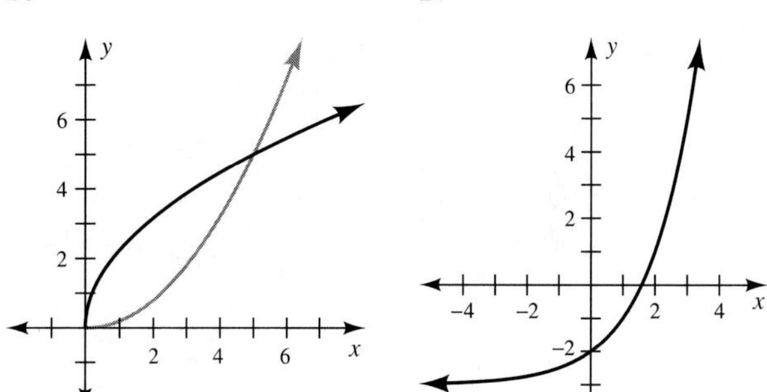

28

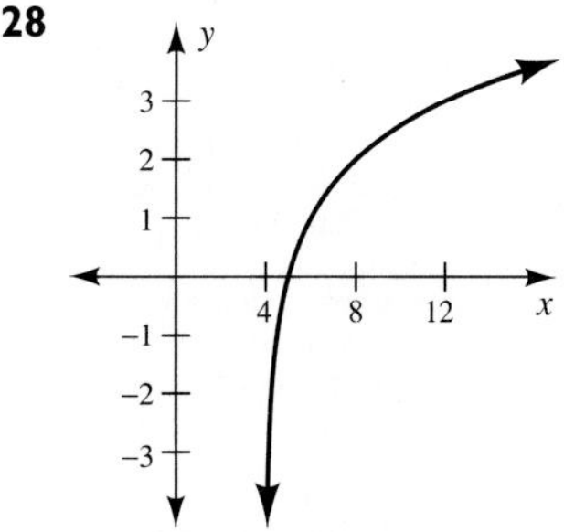

29 11.55% **30** 21 **31** $(2, 5)$
32 $x^5 + 10x^4y + 40x^3y^2 + 80x^2y^3 + 80xy^4 + 32y^5$
33 $16x^4 - 160x^3 + 600x^2 - 1000x + 625$
34 5, 9, 13, 17, 21 **35** $1, \frac{2}{3}, \frac{4}{9}, \frac{8}{27}, \frac{16}{81}$
36 $72, 36, 12, 3, \frac{3}{5}$ **37** 40 **38** 115 **39** 15,050
40 127.75 **41** $\frac{34}{99}$ **42** 45 **43** -48 or 48
44 500 **45** $\frac{9}{11}$ **46** 19.78125
47 870.5, 892, 913.5, 935, 956.5, 978, 999.5, 1021
48 967.5 **49** 40.085 **50** $x \approx 2.8$

Index

Strategies for Forming Equations from Word Problems

Read the problem carefully to determine what you are asked to find.

Apply well-known formulas to the problem.

Translate key phrases into algebraic statements.

Use tables and sketches to clarify relationships among variables and to form equations.

Use the rate principle and the mixture principle to form equations.

General Principles Used to Form Equations

Mixture principle for two ingredients	Amount in first + Amount in second = Amount in mixture	
Rate principle	$A = RB$	Amount = Rate · Base
Work	$W = RT$	Work = Rate · Time
Distance	$D = RT$	Distance = Rate · Time
Interest	$I = PRT$	Interest = Principal · Rate · Time

Statements of Variation

Direct variation	$y = kx$	y varies directly as x
Inverse variation	$y = \dfrac{k}{x}$	y varies inversely as x
Joint variation	$z = kxy$	z varies jointly as x and y

Properties of Exponents and Logarithms

For $y \neq 0$, $b > 0$, $b \neq 1$, $M > 0$, and $N > 0$:

	Exponential Form	Logarithmic Form
	$b^l = M$	$\log_b M = l$
Product Rule	$x^m x^n = x^{m+n}$	$\log_b MN = \log_b M + \log_b N$
Quotient Rule	$\dfrac{x^m}{x^n} = x^{m-n}$	$\log_b \dfrac{M}{N} = \log_b M - \log_b N$
Power Rule	$(x^m)^n = x^{mn}$	$\log_b N^p = p \log_b N$
Product to a Power	$(xy)^m = x^m y^m$	$\log_b (MN)^p = p(\log_b M + \log_b N)$
Quotient to a Power	$\left(\dfrac{x}{y}\right)^m = \dfrac{x^m}{y^m}$	$\log_b \left(\dfrac{M}{N}\right)^p = p(\log_b M - \log_b N)$
	$\left(\dfrac{x}{y}\right)^{-n} = \dfrac{y^n}{x^n}$ for $x \neq 0$	
Change of Base Formula	$M^x = b^{x \log_b M}$	$\log_a M = \dfrac{\log_b M}{\log_b a}$
Special Identities	$b^0 = 1$	$\log_b 1 = 0$
	$b^1 = b$	$\log_b b = 1$
	$b^{-1} = \dfrac{1}{b}$	$\log_b \dfrac{1}{b} = -1$
	$b^{\log_b M} = M$	$\log_b b^x = x$